Geomorphology for Engineers

Geomorphology for Engineers

Edited by
P. G. Fookes,
E. M. Lee and
G. Milligan

Whittles Publishing

CRC PRESS

Typeset by
Mizpah Publishing Services Private Limited, Chennai, India

Published by
Whittles Publishing,
Dunbeath Mains Cottages,
Dunbeath,
Caithness KW6 6EY,
Scotland, UK
www.whittlespublishing.com

Distributed in North America by
CRC Press LLC,
Taylor and Francis Group,
6000 Broken Sound Parkway NW,
Suite 300,
Boca Raton, FL 33487, USA

ISBN 1-870325-03-6
USA ISBN 0-8493-9641-7

Printed and bound in Poland, EU

Contents

16. Hot Drylands
Mark Lee and Peter Fookes

17. Savanna
Michael F. Thomas

18. Hot Wetlands
Ian Douglas

19. Mountain Environments
John Charman and Mark Lee

20. Estuaries and Deltas
Warren E. Grabau, H. Jesse Walker and Mark Lee

21. Coastal Environments
Julian Orford

APPENDICES

Foreword

Denys Brunsden OBE, DSc (Hons), FKC
Emeritus Professor,
University of London, King's College

Ever since the Stone Age, when man hit his thumb with a rock, human beings have of necessity been interested in the earth sciences. Our earliest activities involved an understanding of the environments in which we lived, sought our food, sheltered and survived. Inevitably, we have observed, contemplated and even worshipped natural phenomena. As technology has advanced we have manipulated, altered, exploited and sometimes irrevocably destroyed the systems upon which we depend. Humans are rational beings, however, who are capable of learning from their mistakes and one of our greatest intellectual achievements has been our growing knowledge of the natural world and our increasing willingness to apply that knowledge to the 'proper' management of the earth – the sustainable human use of the land, air, living things and water. Central to this work is the cooperation between earth scientists, planners and engineers, a key element being the part that applied geomorphology can make to the well being of the earth and its inhabitants.

Geomorphology is the study of the forms of the earth's surface: their origin; the processes involved in their development; the properties of the materials from which they are made; predictions about their future form, behaviour and status. Engineering geomorphology is the application of these geomorphological techniques and analysis to an engineering or environmental management problem.

The practice of the subject has provided real opportunities for fundamental research and many of the most important discoveries have emerged from the pragmatic arena. Applied studies have strengthened theoretical developments and improved the quality of inventory, mapping, calibration, evaluation of hazard calculations and stability studies. The need for technical competence has involved geomorphologists in the use of advanced techniques such as scanning electron microscopy, interpretation of satellite imagery for structural geology and hydrocarbon exploration, side scan sonar for landslide identification on the sea bed, hydraulic geometry and discharge of river systems for irrigation, sediment yield monitoring on high mountain rivers and other process modelling.

The recognition that human beings were powerful agents of geomorphological change has a long history and the public concern over the economic costs for remedial works, declining yields and obvious damage led to the establishment of conservation services, erosion control boards, official resource surveys and government research establishments all over the world such as the US Soil Conservation Service, the New Zealand Soils and River Control Council (now Landcare Ltd) and the Land Research Division of the CSIRO, Australia. The work was technical and professional especially where process monitoring, air photograph interpretation, remote sensing, laboratory analysis, geomorphological mapping and land surveying were required. This work was so successful with resource development, engineering and Second World War reconstruction programmes that during the 1960s and 70s there was a rapid expansion of activity sponsored by government agencies such as ODA or UNESCO.

These fundamental changes to the science of geomorphology were accompanied by a rapid growth in international aid that adopted a knowledge-based approach to the solution of environmental problems. Geomorphology was used in the reconnaissance, site investigation and construction, operational and decommissioning stages of engineering developments. The need of newly independent countries to map their natural resources and plan their development programmes, the availability of aid money, the occurrence of natural disasters, the oil-boom development of desert coasts and the spread of development into hostile terrains such as Alaskan permafrost and Middle East sabkhas – all increased public awareness that geomorphology had an important role to play in human development.

The work that evolved was intensely interdisciplinary. Geomorphology suddenly became an essential component of geotechnical education. Engineers, geologists and geomorphologists found themselves involved in the same project and all went through a rapid, sometimes chastening, learning curve. Geomorphologists brought to the geoteam the ability to think in spatial terms, to detect spatial correlations and patterns and to understand and determine the significance of the time dimension in the operation of processes and the evolution of the landscape.

Perhaps the most important aspect has been the growth of professionalism both in terms of the standards of the profession, the quality of the consultants and the employment of geomorphologists within the engineering, planning, environmental management, insurance and related professions. Attempts have been made to establish a code of ethics, standards of education and training, representation in national and local government, legislative, employment and career structures and to provide a good information dissemination service.

A major influence on the development of a professional engineering geomorphology in the UK was recognition by several of Britain's oldest and most respected engineering consultancies who established the first real sponsorships of geomorphology in the UK by retaining the services of geomorphologists and by awarding, in 1971, the first engineering geomorphology studentships at British universities. During the 1970s major projects were carried out by geomorphologists from many British universities and by students who gained professional positions. They worked for numerous national governments and most of the major engineering companies in Britain. This British experience was matched elsewhere, in the USA, Canada, Australia, India and China and geomorphologists found their way into agencies such as the Department of the Environment, FAO, UNESCO, World Bank, the Environment Agency and the British Insurance Association.

The growth of the subject in the UK was largely due to the pioneering vision of Professor P. G. Fookes who is the real champion of engineering geomorphology. Due to his untiring promotion of the subject it is now standard good practice to specify and carry out geomorphological studies within a planning, engineering, resource or hazard assessment programme and to cooperate with planners, developers, architects, engineers, politicians, lawyers and administrators. He demonstrated that an essential task is to cooperate in interdisciplinary teams. To achieve this, a good, clear and well-explained guidebook is an essential. Most engineering companies employ geomorphologists on their fulltime staff and all require a text that directs them to the core of the subject. This book fulfils that need and is a fitting testament to Peter Fookes and his many collaborators.

Preface

This book has been written by geomorphologists, together with some engineers and geologists, and is mainly for engineers. The aim was to present a basic but authoritative handbook of geomorphology for engineers concerned with the implications of earth surface processes (e.g. landslides, erosion and floods) and ground conditions. It should also appeal to others with an interest in understanding the landscape. It does not attempt to be comprehensive and is not particularly systematic. It is not therefore, a text book of geomorphology. The main purpose is to act as an *aide memoire* on a worldwide scale of the different landscapes in which engineers might find themselves working. It is assembled in three parts: firstly the *controls* that have helped shape the landscape; secondly, the *processes*, geomorphological and geological, which, sometimes with the influence of man, have helped shape landforms and can lead to the occurence of natural hazards that are of increasing concern as people rapidly spread over the planet; finally the various *environments and landscapes* that are the product of the controls and processes described earlier.

A hope for this book is that it will help to alert engineers of the issues they may need to take into account especially when working in unfamiliar environments — forewarned is forearmed.

This book is the successor rather than a second edition of the orginal Fookes and Vaughan 1985 '*Engineering Geomorphology*' published by the now ceased Surrey University Press. Copies of the latter are rather difficult to get hold of now, due to the limited print run before the demise of the university press and, in any case, major advances have been made in geomorphology in the last two decades. It was therefore thought timely for the new book. It was also thought better to call it '*Geomorphology for Engineers*' to reflect its contents.

The book has been several years in gestation, in part due to its multi-author nature, some authors replying to requests immediately, others taking somewhat longer. Even with the advent of electronic office wizardry, producing this book has taken as long as, if not longer than, the previous book before the advent of such technology.

Professor Peter Vaughan has now retired (or nearly so) to East Anglia. His role as the engineer on the original team has been ably filled, for this book, by Dr George Milligan, formerly an Oxford academic, now a pragmatic consulting geotechnical engineer and a director of Geotechnical Consulting Group (GCG), London.

The editorial team has been strenghthened by the addition of a geomorphologist, Dr Mark Lee, one of the leaders of the current generation in the newly developing field of engineering geomorphology which now has a handful of full time practitioners in U.K. Mark has worked for a major civil engineering consultancy, has spent time in academia at Newcastle University, and is now an independent consultant. Both Mark and I owe much to the inspiration of Professors Denys Brunsden OBE, Sir Ron Cooke and David Jones and to Dr John Doornkamp. They formed the excellent pioneering Geomorphological Services Limited (GSL) in the 1970s, undoubtedly the first British firm (perhaps first in the world?) to specialise in geomorphology for engineers, and from a professional viewpoint, a very successful company that inspired much of the current practice of applied geomorphology and was a timely training ground for many of the current generation of practitioners. Sadly GSL was taken over and then closed down by accountants. However, its spirit lives on in the way that engineering geomorphology in Britain is becoming an established adjunct to civil engineering and environmental practice.

In addition to the gurus of Geomorphological Services Limited, Mark Lee would like to thank in particular Professor John Pethick (now pretending to be retired) for his friendship and inspiration for what is now one of Mark's principal areas of interest and consulting — coastal engineering.

George Milligan would like to thank Professors Hugh Sutherland and Peter Wroth, Drs. David Hight and Laurie Richards, along with many other colleagues in academia and consulting 'too numerous to mention'.

The editors and publisher thank the many authors for their efforts and stamina without which this book could not have been written.

Special thanks go to Mr Mike Sweeney of British Petroleum, their senior geotechnical engineer who has long been a supporter of geomorphology for engineers and who kindly arranged for the Project Manager Local Environmental Impact, Green Operations, BP, Mr Jim Clarke to make a small financial sponsorship to help keep such a large handbook to a reasonable price.

The original book was dedicated to five given names. Most of my friends knew that these were my children, but there were many questions asking who they were. This book is dedicated to the wives and children or grandchildren of the editors.

Professor P. G. Fookes, FR Eng., Hon. FRGS
Winchester

The printing of this book has been supported by BP in recognition of the importance of geomorphology for engineering of major energy industry projects, and the key role that the authors have played in developing and applying the thinking described herein, with successful results.

Dedication

Edna Fookes and grandchildren Matthew, Christopher, James, Caitriona, Alexander and Niamh
Claire Lee and children Rebecca and James Lee

Barbara Milligan and grandchildren Fraser, Struan, Isabel, Sandy and Carrie

Biographies

Fred Bell is a fellow of the Royal Society of South Africa, a fellow of the Institution of Civil Engineers and the Institution of Mining and Metallurgy, and a fellow of the Geological Society. He is a Visiting Research Associate at the British Geological Survey. Previously, he was Professor and Head of the Department of Geology and Applied Geology, University of Natal, Durban, South Africa. During this time he also was a Distinguished Visiting Professor, Department of Geological Engineering, University of Missouri-Rolla, USA. Professor Bell's consultancy research subjects have included ground stability, subsidence, ground treatment, engineering behaviour of soils and rocks, cement, lime and PFA stabilisation of clay soils, acid mine drainage, mining impacts, landfills, derelict and contaminated ground, rock durability in relation to tunnelling and dam foundations, slope stability, aggregates, bricks, building stone, and geohazards. Professor Bell is author/editor of twenty books and over 200 papers on geotechnical subjects. He has served on the editorial boards of five international journals and has been a series editor for three publishers.

Colin J. R. Braithwaite has a wide research experience centred on the origins and characteristics of carbonate sediments and rocks in marine, lacustrine and subaerial environments in the Seychelles, Florida, the Bahamas, the Red Sea (Sudan) and Tobago, detailing the facies geometry and architecture of reefs and larger scale carbonate systems. The models developed have been tested on relatively recent (Pleistocene, < 2 million years old) limestones on Aldabra, in Kenya and Mauritius and subsurface in Australia (the Great Barrier Reef), in studies that have underlined the importance of sea-level history, and thus of climatic change, to the development of carbonate systems of the more distant past. Over the years additional observations of limestones throughout the Phanerozoic from Britain, Belgium, France, Germany, Norway, Montenegro and Spain have been added and consultancy has provided experience of hydrocarbon reservoirs in Canada, the UK and Libya, and engineering applications in the UK.

John Charman has over 30 years experience as a practising engineering geologist. Before embarking on a career as an independent consultant he worked for a large civil engineering contractor before becoming technical director of a specialist geotechnical consultancy. He now applies his knowledge of geomorphological and engineering geology mapping, site investigation, landslides and soil erosion, soil and rock slope stability, engineering behaviour of temperate, tropical and arid soils, and the location and utilisation of construction materials to civil engineering projects. He works extensively overseas including technical assistance and training assignments for the funding agencies.

Martin Culshaw is in charge of the British Geological Survey's Urban Geoscience and Geological Hazards Programme and is the Survey's Chief Engineering Geologist. He has been involved in engineering geological research, environmental and engineering geological mapping, geohazard assessment and site investigation for over thirty years. During this time he has spent a number of years overseas in Asia, Africa, Europe and Central America. He is a former Chair of the Engineering Group, of the East Midlands

Regional Group, and of the Geological Society. He is Chair of the Engineering Geological Maps Commission No. 1 of the International Association for Engineering Geology and the Environment and is a former editor of Engineering Geological Special Publications for the Geological Society. He has served on various steering groups for the Geological Society, the Institution of Civil Engineers and the Construction Industry Research and Information Association. He is Visiting Professor in Engineering Geology in the School of the Built Environment at the Nottingham Trent University. He has published over ninety papers and articles and more than a hundred technical reports, edited eight conference proceedings, as well as organising numerous meetings and conferences. He has served on the editorial boards of two international journals.

Edward Derbyshire is Research Professor in Quaternary Science in the Department of Geography at Royal Holloway, University of London, England, Emeritus Professor of the University of Leicester, Honorary Professor of the Academy of Science of Gansu Province, China, and an Honorary Member of both the International Union for Quaternary Research (INQUA) and the Quaternary Research Association (UK). He is a holder of the Antarctic Service Medal of the United States, and has conducted field research in all six continents, in pursuit of his interest in the geomorphology and sedimentology of glacial and wind-blown deposits. He was Chairman of the British Geomorphological Research Group in 1982–83, President of Section E (Geography) of the British Association for the Advancement of Science in 1990, Secretary-General of the International Union for Quaternary Research (INQUA) 1991–95, and Chairman of the Scientific Board of the International Geological Correlation Programme (IUGS-UNESCO) 1996–2001. He has served as Editor of *Quaternary Perspectives* and is on the editorial board of *Glacial Geology and Geomorphology* and *Geografia Fisica e Dynamica Quaternaria*. He is currently Chairman of the Committee for Research Directions of the International Union of Geological Sciences

(IUGS). He has authored or edited more than 240 scientific articles, including 6 books and several conference volumes.

Ian Douglas, Professor of Physical Geography at Manchester University, studied erosion rates in tropical rain forests in North Queensland for his doctorate, and has subsequently worked on applied problems in south-east Asia.

Peter Fookes is an independent consultant in engineering geology. He started his private practice in 1971, after two decades working in industry and lecturing at Imperial College, and has worked in more than 90 countries.

He is a visiting or honorary professor at several British universities and has published over 170 papers and books, and continues to lecture and run university field trips. He has chaired many international technical working parties and committees. His work is field investigation and heavy construction orientated and therefore more pragmatic than experimental, and has been recognised by numerous awards from concrete, engineering, geographical, geological and geotechnical institutions worldwide. In 1991, he was made a Fellow of the Royal Academy of Engineering and in 2001 a distinguished research assoicate of Oxford University.

He claims that his work is his hobby and is currently resisting strong attempts by his wife to get him to retire, even in part.

Andrew Goudie is Professor of Geography in the University of Oxford and has worked extensively as a geomorphologist not only in the UK but also in the world's deserts and savanna lands.

Warren E. Grabau, geologist/geographer, retired from the US Army Corps of Engineers in 1984 after a career devoted primarily to multidisciplinary research involving the enhancement of the performances of military materiel, chiefly weapons systems and vehicles, in harsh environments. His interests extend to the dynamics and geomorphology of large alluvial rivers, with special attention to the floodplain and delta of the lower Mississippi. He is also a student of military history and has published material on the impact of geography on military operations.

Kenneth J. Gregory was Lecturer (1962–72) and Reader in Physical Geography (1972–76) at the University of Exeter, Professor of Geography (1976–92), Dean of the Faculty of Science (1984–87), and Deputy Vice Chancellor (1988–92) at the University of Southampton, and Warden of Goldsmiths College University of London (1992–98). From 1998–2001 he held a Leverhulme Emeritus Fellowship. His research and publications are in the fields of river channel changes, palaeo-hydrology, river channel management, and the nature of physical geography. He is President of the Global Continental Palaeohydrology Commission of INQUA (1999–2005). He has received the Back Award (1980) and the Founder's Medal (1993) of the Royal Geographical Society, the Linton award (1999) of the British Geomorphological Research Group, and the Geographical Medal (2000) of the Royal Scottish Geographical Society.

James S. Griffiths has worked as a natural hazard assessment specialist in the UK, Eire, Spain, Switzerland, Chile, California, Liberia, Ethiopia, Syria, Saudi Arabia, Georgia, Pakistan, Hong Kong, Papua New Guinea and Australia. After completing his PhD in flood hazard assessment for highway design in hot deserts he joined Rendel Palmer & Tritton, Consulting Engineers in 1979. For the next 14 years he worked in industry on a range of projects throughout the world specialising in flood estimation for bridge and culvert design, highway earthworks design, slope stabilisation and engineering geomorphological mapping. For the period 1987–90 he was technical director of Geomorphological Services Ltd. In 1993 he joined the University of Plymouth as a lecturer in engineering geology, was made Head of Geological Sciences in 1995, and in 2002 became Enterprise Co-ordinator in the Faculty of Science. He has continued to work as a consultant and in 1998–99 was part of the team carrying out a global assessment of landslide risk in Hong Kong. He is the author or co-author of over 60 publications including

three edited books, 'Landslides', 'Land Surface Evaluation for Engineering Practice' and 'Mapping in Engineering Geology'.

James V. Hengesh has been a Senior Geologist since 1986 specializing in seismic hazard investigations. He has performed investigations for critical facilities and lifelines such as nuclear power plants, liquefied natural gas (LNG) terminals, offshore platforms, pipelines, dams and military installations. During these projects he has conducted geological and seismotectonic assessments to characterize seismic hazards for input to design of new structures or engineering evaluation of existing facilities.

He has been funded by the U.S. National Earthquake Hazard Reduction Program (NEHRP) to investigate faults in northern California and Puerto Rico. He has participated in a number of USGS Working Groups to evaluate seismic hazards in the San Francisco Bay area and has been an invited speaker on seismic hazard issues at conferences in Europe, the Pacific Islands, and New Zealand.

Colin F. Jago graduated in Petroleum Geology from Imperial College, University of London. His doctoral research, also at Imperial, London, was in beach and estuarine sedimentation. He has maintained research interests in the sedimentary dynamics of macrotidal estuaries but his primary research activity is concerned with suspended sediment dynamics of continental shelves. This research is focused on bio-physical interactions that govern the properties and flux of suspended sediments and the geotechnical and hydrodynamic properties of seabed sediments. Dr Jago is Senior Lecturer in Geological Oceanography in the School of Ocean Sciences, University of Wales Bangor.

Mark Lee is an engineering geomorphologist working in private practice. He was formerly Principal Research Associate at the University of Newcastle, UK and prior to that had worked in industry for 18 years. He has had extensive experience of landslide hazard and risk assessment in both natural and man-made terrain throughout the world. He is actively involved in research into the role of geomorphology in Quantitative Risk Assessment. In recent years he has specialised in the development and application of probabilistic methods of modelling coastal cliff recession and landslide reactivation scenarios.

William R. Lettis is responsible for conducting regional and site investigations to assess seismic hazards in both active plate margin and stable intraplate tectonic environments. These investigations typically involve the deterministic and probabilistic assessment of ground motions, surface fault rupture, liquefaction, dynamic slope stability, and tsunami. He has conducted field and office studies for nuclear facilities, transportation systems and bridges, hydroelectric facilities, pipelines and tunnels, and LNG and oil platform facilities throughout the United States and abroad. Many of these projects were successfully completed in a regulatory environment. He is currently a member of the California Earthquake Prediction Evaluation Council, a scientific advisory council to the Governor and Office of Emergency Services in California.

Xingmin Meng is a Geographic Database Manager and Geographic Information System (GIS) Analyst in the Department of Geography, Royal Holloway, University of London. His PhD thesis was on loess and loess instability in North China. He was involved in a major (eight year) international collaborative research programme on geohazards in the loess terrain of North China, financed by the Commission of the European Communities and the Government of Gansu Province, China. He has also held the positions of Postgraduate Research Assistant and Postdoctoral Research Assistant in several NERC-funded research programmes in both the University of Leicester and the University of London (Royal Holloway), as well as being Postdoctoral Research Assistant under a grant from the Leverhulme Trust. His current research interests include: Geographic Information System (GIS) and related applications; landslides and slope instability; aeolian dust and loess deposits.

George Milligan has an MA in Engineering Science from Oxford University, an MEng in Geotechnical Engineering from Glasgow University, and a PhD from Cambridge University, where his research was on flexible retaining walls. He gained experience as a structural and geotechnical engineer with Ove Arup and Partners, Scott Wilson Kirkpatrick and Partners, and Golder Associates, including working in Iraq for two years. He was then a University Lecturer at Oxford University for 17 years, conducting research in soil reinforcement and soil nailing, applications of geotextiles, and small diameter tunnelling, particularly pipe jacking. Since 1997 he has worked for the Geotechnical Consulting Group, on a wide range of projects in the UK and abroad, and has been a Director since 1998.

Julian Orford graduated in Geography and Sociology from Keele University, and undertook research at both Salford and Reading Universities into aspects of beach gravel, beach sedimentation and numerical methods of analysing facies variability in coastal sedimentary environments. He has pursued coastal research in the UK, Ireland, Canada, and the USA. His recent research areas are in coastal development and beach behaviour over decade–century scales, with a specific interest in the behaviour of transgressive sand and gravel barriers and regressive dune systems. He is also concerned with developing effective means of coastal zone management that combine geomorphology and engineering in a sustainable fashion. He has been lecturing in Geography at Queen's University, Belfast, since 1977, where he is now Professor of Physical Geography and head of the School of Geography.

Lewis A. Owen received his doctorate at the University of Leicester in 1988 for his research on the Quaternary geology and geomorphology of the Karakoram Mountains in the northern areas of Pakistan. Since receiving his doctorate he has continued to work in high mountains and drylands to examine aspects of landscape evolution and paleoenvironmental change. This has included research in the Himalaya, Tibet, Sierra Nevada, Alaska and the deserts of Central Asia and Western USA. He has been on the faculty at the Hong Kong Baptist University, Royal Holloway University of London, the University of California and the University of Cincinnati. He is presently an Associate Professor in the Department of Geology at the University of Cincinnati.

Michael F. Thomas is Professor (Emeritus) in Environmental Science, University of Stirling, Scotland, UK. His research includes the role of saprolites in landscape development, and the impact of Quaternary climate changes on sedimentation in humid tropics. He is the author of *Geomorphology in the Tropics* (Wiley, 1994), and was a member of Geological Society working party on Tropical Residual Soils.

H. Jesse Walker, Boyd Professor Emeritus, has taught in and served as Chairman of the Department of Geography and Anthropology at Louisiana State University, Baton Rouge, USA. He has been a visiting professor at the Universities of Hawaii, California, and Tsukuba (Japan), and a Liaison Scientist with the US Office of Naval Research, London. He is a Fellow of the AAAS, the Arctic Institute, and the International Association of Geomorphologists. His major research topics have been devoted to arctic geomorphology and coastal modification.

Tony Waltham was a Senior Lecturer in Engineering Geology at Nottingham Trent University in the UK. After completing degrees at Imperial College London, he turned his geological interests away from mining and towards engineering. His main research has been into the process of ground subsidence and catastrophic failures, especially those related to cave collapse and sinkhole development in limestone karst. This stemmed from his many years as an active caver when he explored and mapped caves while trying to understand their geomorphology. He has travelled widely through almost every country in Europe, Asia and North America, and finds active volcanoes nearly as exciting as underground rivers. He has written numerous books, including the widely used student textbook *Foundations of Engineering Geology*, and now also runs Geophotos, his own picture library that specialises in geology and geomorphology.

"Marcley Hill in the year 1575 after shaking and roaring for the space of three days, to the great horror, fright, and astonishment of the neighbouring inhabitants, began to move about 6 a clock on a Sunday evening and continued moving or walking till 2 a clock on Monday morning; it then stood still and moved no more. It carried along with it the trees that grew upon it, and the sheep folds and flocks of sheep grazing on it. In the place from whence it remov'd, it left a gap of 400 wide, and 320 foot long. The whole spot whereon the Hill stood contained about 20 acres."

Contemporary description of the 1575 'Wonder Landslide' in Herefordshire, written by Fuller and quoted in Coole (1882).

1. Introduction to Engineering Geomorphology

Peter Fookes and Mark Lee

1.1 Preamble

The earth's surface is dynamic and landforms change through time in response to weathering and surface processes (e.g. erosion, mass movement and deposition). Geomorphology is the study of landforms and landform change. The literal meaning of the word is the study of the form of the earth. Geomorphology, therefore, studies the earth's surface forms and processes. It is by no means a new subject, having its modern origins in the 19th century but its specific application to civil engineering (as engineering geomorphology) is a much more recent development.

Engineering geomorphology complements engineering geology in providing a spatial context for explaining the nature and distribution of particular ground-related problems (e.g. the presence of duricrusts, aggressive soils) and resources (e.g. sand and gravel deposits). Importantly, engineering geomorphology is also concerned with evaluating the implications of landform changes for society and the environment. The focus is primarily on the risks from surface processes (i.e. the impact of so-called 'geohazards') and the effects of development on the environment, notably the operation of surface processes and the resulting changes to landforms or the level of risks.

These views form the substance and thrust of the book. We also make the distinction between applied geomorphology — the use of specialist academic expertise to address particular problems — and engineering geomorphology — the integration of the geomorphological approach into engineering project planning, design and construction.

Engineering and geomorphology

The application of some geomorphological acumen by non geomorphologists (i.e. those not specifically trained or experienced in the subject) dates back thousands of years. Ancient examples would be the siting of London (at the last fording place before the sea); the siting of the Nile/Red Sea waterway, finished in the second century BC; the siting of Chinese canals dating back to about 1000 BC or the Qanat systems of Persia dating back to about 1500 BC. More recent examples include aspects of the siting of canals in the canal era in England (c. AD 1760 to 1830), Brunel's or Stephenson's railway alignments (c. 1820 to 1860); aspects of the reclamation of Fenlands (c. AD 600 to present) or polders in Holland (c. AD 600 to present). It is clear that virtually every form of civil engineering involving the ground benefits from some understanding of surface forms and processes (i.e. geomorphology), whether applied or not.

Although geomorphology and engineering geology are complementary subjects, they have different histories and, hence, have experienced different paths towards acceptance by the engineering profession. Engineering geology, which has some of its roots in the attempts to quantify the behaviour of soils, commenced in the early 18th century almost entirely in France (Skempton, 1979; 1985). The engineering works associated with the industrial revolution provided much empirical experience of the behaviour of geotechnical materials. A paradox then developed. While the exposures of the ground provided by engineering works played a major part in the development of geology as a separate subject, the engineering scientists of the day were unable to provide coherent analyses to formulate the behaviour of soils, particularly clays. While such notable scientists as Navier (1833), Poncelet (1840), Collin (1846), Darcy (1856), Rankine (1862), Boussinesq (1883) and Reynolds (1887) — all reviewed by Skempton (1985) — made contributions, it was not until the 1920s when Terzaghi appreciated the role of pore-water pressure in controlling the behaviour of soil, that such analyses became possible. Thus soil mechanics did not become a full part of engineering science until the 1930s, when courses in the subject were first run at Harvard. Its worldwide dissemination did not occur until the 1950s. The importance of geology to engineers was appreciated and taught to engineers, however, long before geology had gone its separate way as a natural science primarily concerned with the history of the earth. It had relatively little interest in the superficial deposits, landforms and the processes that shape them — of primary concern to engineers — and could not offer any quantitative basis for predicting engineering behaviour. It sat uneasily with other quantitative subjects in an engineering education. Much has changed in the last five decades but current attitudes and practices still reflect earlier difficulties.

Geomorphology has, however, developed somewhat as a separate subject related to, but not a part of, geology. This is especially so in Britain, where it has its roots in physical geography — a long way removed from the worldly business of construction. Geomorphology has also developed separately from geotechnical engineering. Inevitably, there are problems of communication, in both directions, some of which may be apparent in this book.

It has been suggested by Cooke and Doornkamp (1974) that applied geomorphology studies date back to about the turn of the last century. In general such work was done by academic geographers with little background in the world of civil engineering, although some learned quickly. Unfortunately, even today there is a general scarcity of papers and textbooks specifically related to applied geomorphology; the books *Environmental Geomorphology* (Coates, ed., 1971); *Geomorphology in Environmental Management* (Cooke and Doornkamp, 1974; 1990); *Applied Geomorphology* (Hails, ed., 1977); *Urban Geomorphology in Drylands* (Cooke

et al., 1982); *Geomorphological Hazards in Los Angeles* (Cooke, 1984) and *Applied Fluvial Geomorphology for River Engineering and Management* (Thorne *et al.*, 1997) are among some of the relatively recent distinguished exceptions.

Geomorphology is without doubt of primary importance to civil engineers involved in ground engineering. However, the authors know of no case where 'geomorphology for engineers' is being taught to engineers. Such a course should incorporate subjects such as basic geomorphology, with the principal topic areas treated with emphasis on surface form, process and near-surface materials; basic climatology and climate change; Quaternary studies; techniques involving photo-interpretation and remote sensing, plus the appropriate geomorphological mapping training in the field, together with the relevant elements of basic geology such as physical geology, historical geology, sedimentology, petrology and structural geology.

This is a book largely by geomorphologists for engineers. It may occasionally show a parallel need for a book on engineering for geomorphologists, but that is a different question. It is not our intention to try and turn engineers into expert geomorphologists, but to help engineers and others to understand the subject and to appreciate the part that geomorphology and geomorphologists can play in engineering. The importance of first understanding a site in broad context, before engaging in detailed investigations and predictions, cannot be overstressed.

Problems have to be identified before they can be solved. This book should thus be of most value at the project planning and site investigation stages. One cautionary comment may be prudent for the inexperienced reader. Many diagrams illustrating landscapes in this book attempt to include most if not all the features which may be associated with a particular landscape. Nature is not always so obliging, or sometimes not so complicated, and this should be remembered when applying these models to a particular site.

The authors can see great benefit in utilising and incorporating geomorphology and geomorphologists within the framework of geotechnical engineering. This is already happening. Such a move will benefit from more specific training and education of both engineers and geomorphologists. It is hoped that this geomorphology for engineers will be found useful in this process.

The book

Many, if not most, difficulties in geotechnical engineering arise either from an unawareness of ground conditions or a failure to appreciate the influence of known ground conditions on a particular engineering problem. A basic understanding of geomorphology is a considerable aid to the solution of the first problem. The second problem requires a specific engineering input.

The aim of this book is, therefore, to present a basic yet authoritative handbook of geomorphology for geotechnical engineers and others. It does not attempt to be comprehensive, is not particularly systematic, and as such is not a textbook of geomorphology. Its main purpose is to act as an aide-memoire, on a worldwide scale, of different landscapes in which engineers may find themselves working. It is hoped that each chapter will help them understand a little better what may be unfamiliar environments.

The essence of the book is in the diagrams around which each chapter is built. The first part (Chapters 2–7) is concerned with *controls* that have helped shape the landscape — climate and weathering, past and present, acting on the land surface; the sediments and sedimentary processes which help build the landscape; stratigraphy, which is concerned with the history of soils and rocks; and tectonics, geological movements, often from deep in the Earth's crust, which may form the bones as well as some of the wrinkles of the landscaped skin of the Earth. This part includes the very important chapter on the Quaternary, the history of the last 2 or 3 million years of the Earth's surface, a study that should figure large in the geological and geomorphological training of any civil engineer. It ends with a chapter on engineering soils in which soil and rock mechanics and their relation to geomorphology are explored.

Often, landscapes have been formed predominantly by past climatic conditions, rather than by present ones. Britain is a good example. The form of the landscape has for the most part been inherited from the Tertiary, but also bears the imprint of relatively recent glacial and periglacial conditions that are no longer present. In terms of engineering at least, there is a need to separate the past effects that have helped form the landscape from current effects which may only be modifying. Current tectonic activity also affects active geomorphological processes, for example if young mountains are slowly moving upwards then landslides, erosion and so on are typically more active. In old mountains which have ceased to go up and are being worn down, their slopes are, in the long term, becoming progressively more stable. Similarly, ground level changes in relation to sea level control river erosion and deposition, the risk of coastal inundation and so on. These effects are discussed mainly in the first part of the book and their results highlighted in the second and third parts.

The second part of the book (Chapters 8–12) is concerned with processes — geomorphological and geological, with and without the influence of man, which help shape landforms and lead to the occurrence of natural hazards such as landslides and floods. The usually slower continual wear of wind, ice and water and the often more spectacular but irregular effects of neotectonics, subsidence and landslips are brought together in this part.

The third part (Chapters 13–27) of the book is concerned with various environments and landforms, mainly natural but also some influenced by man. It starts in the colder, higher-latitude parts of the world, the glacial and periglacial, and moves generally towards the middle and lower latitudes, through temperate environments, hot dryland ('desert'), savanna and hot wetland ('tropical rain forest'). It goes on to outline specific geomorphological features, mountains, estuaries, coasts and shallow seas around the coasts. Three particular terrains which present special problems to engineering are considered: volcanic, soluble rock (mainly limestones) and loess (windblown silt). This part is completed with a chapter about the influence of man on the urban landscape.

However, the three parts should not be seen as rigorous — each chapter contains bits that would fit into other chapters.

The engineering content of the book has necessarily to be brief, yet as useful and relevant as possible. Following from the above, it was thought that a realistic way of dealing with engineering situations was to produce a checklist of the type of engineering difficulties that can arise, with as much emphasis on the relatively unusual problems that can catch you out, as well as the more well known. It is assumed the reader will either know, or will seek elsewhere, the full answer to the problem. The book tries to show and make the reader aware of problems that might arise. It is clear that a separate checklist for each chapter would invoke excess duplication and length. A tabulated master checklist is therefore given in Chapter 7, on Engineering Soils, while each chapter contains further discussion of the relevant issues.

Each chapter thus becomes a summary of the particular types of landforms arising from the particular environments, which concentrate on geometry of form and on types of materials. Each chapter also summarises the particular types of geomorphological agency, with the influence of these agencies, when current, on engineering design and construction.

It should be noted that the specific determination of material types, geometric boundaries, engineering properties and methods of design and analysis, are part of the geotechnical engineering processes which are outside the intent of this book, though difficulties arising in site investigation are not. However, much of the geotechnical process and virtually all the design and analysis process depends on a presumption of the problems with which the proposed structure will have to cope. The engineering aim of this book is, therefore, to aid the engineer in helping to ensure that any relevant problems are not overlooked, hence the checklist method adopted. The experienced reader will doubtless be able to add further points to each list. Each chapter ends with specific references given in the text and a bibliography of related but fuller texts.

1.2 Geomorphology, landforms and the nature of modern engineering geomorphology

One of the key strengths of geomorphology for engineering is the ability to provide a framework for understanding the current and historical behaviour of physical systems — slopes, terrain units, catchments or coastal sediment transport cells. Morphological adjustments (i.e. erosion and deposition) can occur in response to variations in energy inputs into these systems, including variations in river flows (discharge and sediment load) and wave/tidal energy arriving at the coast, over a range of timescales. A number of distinctive modes of change may be recognised:

Episodic, progressive change, where no recovery or retrogressive change occurs. For example, cliff erosion involves sequences of episodic landslide events that result in loss of cliff top land. Cliffs do not recover by advancing forward. Figure 1.1 presents an example of the recession process from the glacial till cliffs of Holderness, Yorkshire, UK.

Periodic or cyclical change, where the landform responds to an event (e.g. a storm or large flood) by altering the morphology, but then gradually recovers (either partially or completely) its original pre-event form. For example, variations in wave energy can cause major changes to beach form, with sediment being eroded from the upper shore and transported offshore under high wave energy conditions and brought onshore in lower energy conditions, leading to beach accumulation (Figure 1.2). Thus, there may be an envelope of beach positions varying between steeper gradients associated with periods of beach deposition and flatter profiles associated with periods of beach erosion.

The availability of sediment for exchange or transfer between different 'stores' is critical to cyclical change. Considering a beach-sand dune system, sand may be deposited on the upper beach during low wave energy events, providing a source of sediment for aeolian transport and fore-dune growth. During high energy events, the beach will flatten and the fore-dunes will erode. This provides an additional source of sand that is

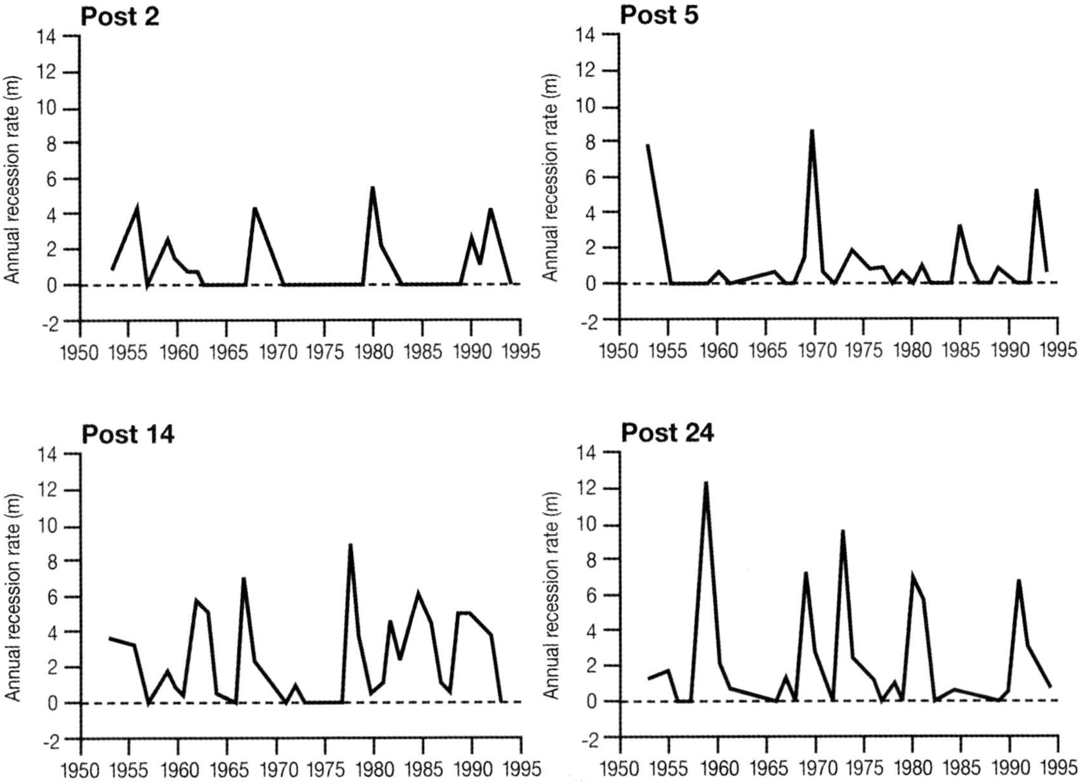

Post 2

Post 5

Post 14

Post 24

Figure 1.1   Representative annual cliff recession measurements on the Holderness coast, UK (redrawn from Pethick, 1996).

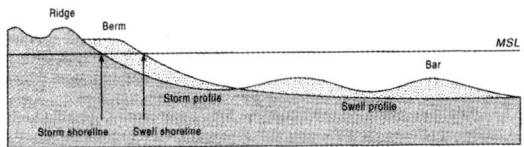

Figure 1.2   A simple beach profile change model (from Hansom, 1998).

transported seaward and deposited on the lower beach.

Secular change, where there is a longer-term trend in the evolution of the systems. For example, Warburton *et al.* (1993) describe how the Ashley River braid-plain, New Zealand, is migrating northwards over a timescale of decades due to bank erosion on the northern margins and bar stabilisation on the southern margin (Figure 1.3). They speculate that this trend may be in response to large-scale sedimentation cycles.

On the coast, the Holocene sea level rise and changes in sediment availability have initiated important long-term trends in landform development. During the early Holocene, for example, rapidly rising relative sea level resulted in onshore movement of large volumes of sand and gravel onto the Scottish coast and is associated with periods of beach and dune formation (Hansom, 1998). However, after around 6500 BP, relative sea level began to fall on parts of the Scottish coast because of the increasing significance of non-uniform isostatic uplift. The resultant sediment deficit had two main effects on coastal evolution. First, erosion of the beach and back-beach dune systems, as sediments were removed from these areas into deeper water in order to maintain a near-shore gradient capable of dissipating wave energy. The widespread trend of beach erosion and eroding dune systems throughout the Western and Northern Isles is believed to be the

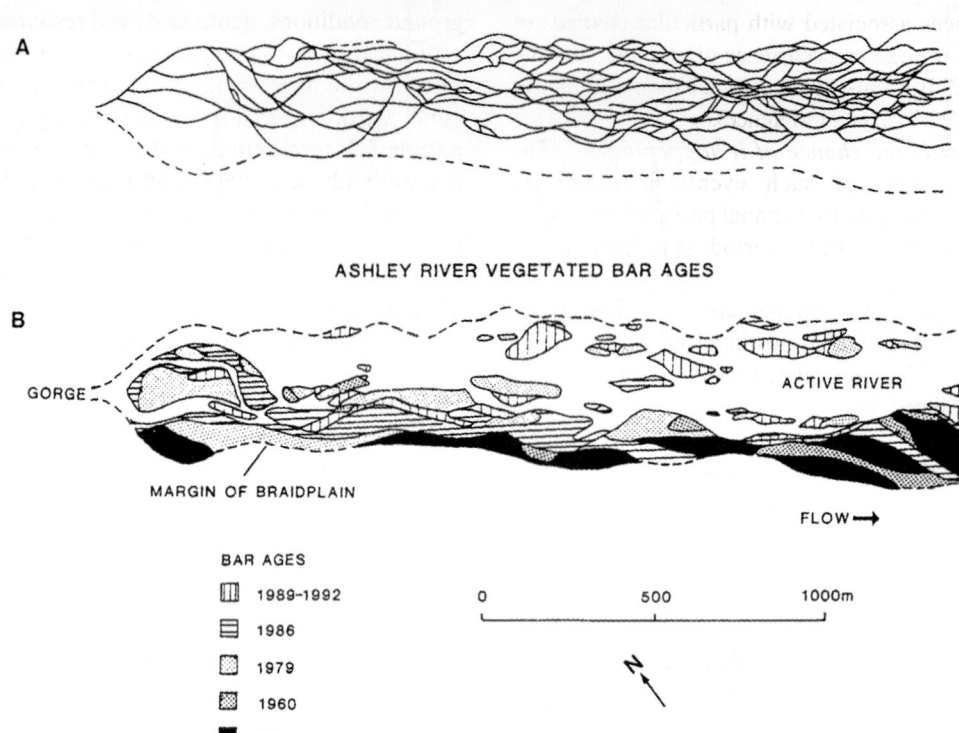

Figure 1.3   Ashley River channel changes: (*A*) composite map of all channel positions 1950–1992 (*B*) age of vegetated bar surfaces showing a migration northwards (from Warburton *et al.*, 1993).

consequence of the long-term sediment deficit. Second, reorganisation of the coastline into progressively smaller sediment transport cells as the emergence of headlands on the retreating shoreline created barriers to long-shore transport. These cells have tended to evolve subsequently by internal reorganisation, erosion and deposition.

The limited duration of engineering time imposes a constraint on the types of landform change that are relevant to engineering geomorphology, viz.:

1. Abrupt and dramatic landscape changes that are likely to be significant over a 1–100+ year timescale. Examples include the establishment of gully systems, migration of sand dunes, river planform changes, coastal cliff recession and the growth and breakdown of shingle barriers.

2. High to relatively high probability events, such as wind blown sand, soil erosion, shallow hillside failures, flooding and river bank erosion, coastal erosion and deposition.

3. Low probability events that would have a major impact on the project or development, such as flash floods, major first-time landslides, reactivation of pre-existing landslides, channel bed scour, fault rupturing and neotectonics, the collapse of solution cavities (especially in areas of gypsum karst) and tsunamis.

Specific questions that frequently have to be addressed include:

1. *What could happen?* — The nature and scale of the changes or events that might occur over the next 50–100 years or so (longer for major projects).

2. *Why might such events happen?* — The circumstances associated with particular changes or events (e.g. the occurrence of extreme intensity rainfall events, wet years, earthquakes, storm surges, human interference).

3. *What is the chance of it happening?* — The probability of such events occurring in any given year (the annual probability) or over a particular time period (the cumulative probability).

4. *What damage could be caused?* — The consequences of such events in terms of losses (e.g. loss of life, property, services, infrastructure damage).

For these reasons, engineering geomorphology is directed towards understanding the way landforms or physical systems respond to relatively short–medium term impulses (e.g. climatic variability, changes in sediment supply, land use change, the effects of man, neotectonics in seismically active regions) rather than longer-term landscape denudation and evolution. However, if the focus is on what in geological terms could be called the 'here and now', there is a need to be aware of the significance of longer term trends (e.g. the Holocene decline in sediment availability experienced on many temperate coastlines) and the presence of potential geohazards or resources inherited from the distant past (e.g. ancient landslides, karst features).

As engineering geomorphology needs to address short- and medium-term changes, notions of gradual landscape denudation (e.g. progressing through the youth, maturity and old age of the so-called Davisian cycle) over millions of years has little relevance. Instead the foundations of engineering geomorphology are awareness of and the distribution of inherited landforms, geomorphological events, interconnected physical systems and their complex behaviour.

Inherited landforms

Landscapes are the product of extremely long periods of weathering and surface processes. It is widely recognised that these processes operate with different intensities in different climatic and tectonic regimes, tending to produce highly regionalised associations between landforms, ground conditions, geohazards and resources — a characteristic geomorphological environment. Some of the major geomorphological environments are discussed later in the book e.g. glacial, periglacial, temperate, hot drylands, savannah, hot wetlands, mountains and highlands. However, over geological timescales the boundaries of these environments are dynamic, drifting across the continents in response to major climatic or tectonic changes. As these environments have shifted, so the intensity and character of geomorphological processes will have changed.

Present day surface processes are acting on a landscape fashioned by ancient processes. In some areas current processes may have behaved in the same or a similar manner for extremely long periods of time, as in the ancient cratonic regions of Australia and South Africa. In other areas, current processes may be actively re-shaping the inherited landscape, destroying it or simply having no effect at all.

It follows that many landscapes are composite forms, containing a range of features of varying antiquity inherited from different environments as major climatic or tectonic changes left their imprint. The landscape rarely reflects any one climate or period of geomorphological change, rather it is usually an assemblage of landforms and systems that have superimposed histories — geomorphologists often describe the landscape as a palimpsest (i.e. like a surface which has been written on many times after previous inscriptions have only been partially erased, e.g. Chorley *et al.*, 1984; Twidale, 1985). Present day conditions do not necessarily give a good guide to the conditions that created particular landforms. The implications for engineering geomorphology include landscape elements of different ages and stabilities; a legacy of pre-existing geohazards; the presence of inherited near surface materials and inherited resources.

Landscape elements under present day climate and environmental controls include:

Actively dynamic landforms: subject to almost continuous adjustment to changes in energy inputs (e.g. rainfall, river discharges, wave attack) or sediment availability. River channels and the coastline are the best examples of this type of landform.

Episodically active landforms: subject to occasional adjustments to extreme events, such as hillsides prone to shallow debris slides in tropical regions; sand dune fields and steep mountain catchments prone to debris flows and flash floods.

Stable, relict landforms: although they are stable under present conditions, they would not have evolved under them. For example, changes in wind velocity and increased rainfall have stabilised many dune systems that would have been active earlier in the Pleistocene (e.g. the stabilised dune fields of the Nebraska Sand Hills, covering $57\,000\,km^2$). Periods of extreme aridity and high wind speeds led to the development of the 100–200+m high primary dunes of the Saharan sand seas. In the current dry phase, sand dune mobility is restricted by relatively low wind speeds and, hence, the primary dunes are effectively stable. However, smaller-scale dune forms (secondary dunes) that are superimposed onto the larger immobile primary dunes are mobile under current conditions. Elsewhere, many stable river channels in temperate regions are relict forms, adjusted to suit climatic and sediment load conditions during and after deglaciation.

Unstable, relict landforms: landforms that developed under past conditions, but are being progressively destabilised by present conditions. Many beaches, for example, were created from material moved onshore as sea levels rose early in the Holocene and, thus, are essentially relict features often with no obvious contemporary source of sediment. Slapton Beach, Devon, comprising chert and flint but surrounded by cliffs of Devonian-aged sedimentary rocks is a good example. The implication for coastal management is clear: unless the erosion of cliffs along adjacent sections of a coastline is able to counter the natural wastage due to attrition, beach mining, loss offshore or along the coast, then these beaches will diminish in size.

Fossil landforms: essentially unchanging elements of the landscape, such as low angled

pediment slopes in deserts and ancient surfaces in cratonic areas.

Although pre-existing geohazards may have been created under very different conditions, these features may continue to present significant risks to engineering projects. Examples include pre-existing landslides prone to reactivation, and karst terrain where there remains a potential for collapse.

Inherited near-surface materials that have accumulated under former environmental conditions include spreads of wind-blown loess derived from glacial outwash plains that bordered the mid-latitude Pleistocene ice sheets, glacial tills and solifluction deposits, and the deeply and intensively weathered regoliths or relict duricrusts that can be found beneath many low relief surfaces, especially in stable cratonic zones.

Inherited resources, such as fluvio-glacial or relict floodplain sands and gravels, are essentially non-renewable under contemporary conditions and processes.

Geomorphological events

A geomorphological event can be described as the occurrence of a process that results in landform change, such as a landslide, flash flood or erosion event. By their nature, those event that are significant for engineers tend to be infrequent episodic events. In the intervening years the landscape might appear completely benign, often instilling a false sense of security before a major damaging event occurs. A good analogy for event history of a landscape is:

'like the life of a soldier, long periods of boredom interrupted by brief moments of terror' (Ager, 1976).

The size of these events is not constant, as illustrated by desert streams where discharge can vary by 2–3 orders of magnitude. However, high-intensity events tend to be rare, whereas low- to medium-intensity events are more frequent. Typical magnitude/frequency distributions tend to be positively skewed in form (Figure 1.4a). However, some geomorphological events such as landslides do not conform to this simple frequency distribution as the events are often constrained within narrow size-bands by internal slope factors (e.g. discontinuity spacing, disposition of strata; Figure 1.4b).

Historical event frequency is often used to estimate the likelihood (probability) of such an event occurring in the future. For example, in flood studies the likelihood of a flood of a particular magnitude is generally expressed by the return period or recurrence interval:

$$\text{Return period} = \frac{N+1}{m}$$

where N is the number of records in the time series and m is the order or rank number within the series of the event in question.

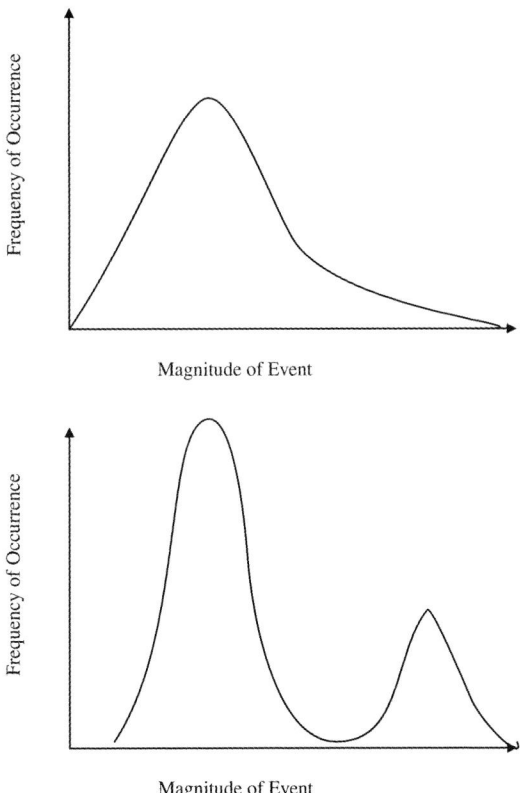

Figure 1.4 Typical event frequency distributions: (a) positively skewed (e.g. wave height) and (b) in narrow, constrained bands (e.g. landslides).

The probability of an event occurring over a particular time period (n years) is:

$$P_n = 1 - (1 - P_o)^n$$

where P_o is the annual probability of the event.

The flood that is expected to be equalled or exceeded, on average, every 100 years, has a return period of 100 years and an annual probability of 0.01. As floods are assumed to be random, rather than regular in occurrence, the 100-year flood event could occur any year (or not), but the chance of its occurrence during a 100-year period is greater than during a single year.

Most estimates of event probability are extremely sensitive to the quality of the historical data set. To achieve 95% reliability on an estimate of the 50-year flood requires 110 years of records. Such lengthy data sets are rare, placing considerable limitations on our ability to predict extreme events.

Where there is no historical record of geomorphological events, the combined body of evidence (site-specific and landform assemblages) does allow certain judgements to be made about the potential for geohazard activity in an area. As the field evidence alone rarely supports the use of precise quantitative results for particular sites, generalised statements applicable to a variety of sites are often used (e.g. high, medium, low potential). In other instances, an absence of peak flow measurements can be partly compensated for by geomorphological or sedimentological information that permits the reconstruction of the maximum heights reached by floodwaters.

The causes of geomorphological events can be very complex, but often relate to a combination of the following factors:

Preparatory factors which make the system susceptible to geomorphological events — for example, antecedent conditions determine the amount of a catchment that is saturated prior to a rainstorm or snowmelt event and, hence, the amount of runoff that will be generated.

Triggering factors which actually initiate the event ('the straw that breaks the camel's back') — for example, gully erosion events are often triggered by intense rainstorms, although land use change is often an important preparatory factor.

Controlling factors that determine the nature, scale and intensity of the event when it occurs — for example, the catchment's network characteristics will influence the speed at which water is transmitted through the channel system. Dendritic networks tend to produce a marked concentration of flow in the lower catchment as floodwaters are delivered down the main tributaries at a similar speed. Trellis networks tend to produce a more muted response.

Events may often be significantly delayed after the initial preparation. For example, although the majority of major rock slope failures in Scotland appear to be associated with glacially steepened slopes, they occurred during the period 10 000–5000 BP (i.e. they are failures delayed many thousands of years after glaciation). These slides are believed to be have been triggered by high magnitude seismic activity associated with differential isostatic uplift following glacial retreat (e.g. Ballantyne, 1991). The concept of a lagged response to impulses of change is illustrated in Figure 1.5, which also highlights that there will be a period of adjustment (the relaxation time), before the system achieves a stable form.

When past records are used to predict the probability of future events, there is an underlying assumption that there will be no change in the factors causing the events (e.g. rainfall, wave climate). This assumption of stasis in the historical record ignores the possibility of climate change or changes to other environmental controls. However, over 'engineering time' there can be significant changes in the probability of triggering events. For example, significant changes in climate and the pattern of geomorphological events are known to be associated with the occurrence of the North Atlantic Oscillation (NAO) and El Niño Southern Oscillation (ENSO) events and volcanic eruptions (Chapters 2 and 23).

Interconnected physical systems

The causes and effects of geomorphological events can be complex and should be viewed as expressions of the operation of large physical systems and not as isolated processes. Indeed, while the effects are readily apparent and well appreciated at a local level, they are not often regarded as the product of broader controls that influence the behaviour of hillsides, river catchments (drainage basins) or coastal cells. For example, to understand the flood character of a river, something must be known of the climatic, geological, topographic and land use controls on the supply of water and sediments from the surrounding hills. On the coast, the development of beaches or shingle banks needs to be seen as the product of sediment transport within dynamic coastal systems.

In addition to providing a framework for understanding the flood behaviour of rivers and streams, the catchment concept can help explain the way water and sediments are transported from supply areas towards the coastal zone. Sediment supply occurs in areas of hill-slope erosion and where river channel migration cuts through areas of stored sediments resulting from past phases of erosion and deposition under different climatic conditions, or from an extreme flood in the recent past (e.g. spreads of glacial deposits or floodplain alluvium). The supply of sediment is generally intermittent, with rare floods among the most effective events in delivering sediment from these stores into the river channel network. Once in the channel, the sediment size is important in determining how far it is carried before being temporarily stored in features such as point bars or as spreads on the riverbed. In short rivers the suspended load may reach the estuary in a single flood, but coarser sediments may become incorporated in the floodplain.

Similar considerations apply on the coast, which can be viewed as a series of interlinked physical systems, comprising both offshore and onshore elements. Sediment (e.g. clay, silt, sand, gravel) is moved around the coast by waves and currents in a series of linked systems. Simple systems comprise an arrangement of sediment source areas (e.g. eroding cliffs, the sea bed), areas where sediment is moved by coastal processes, and sediment sinks (e.g. beaches, estuaries or offshore sinks). Along a particular stretch of coast there may be a series of such systems, often operating at different scales (Figure 1.6). In contrast to catchments, coastal systems have no

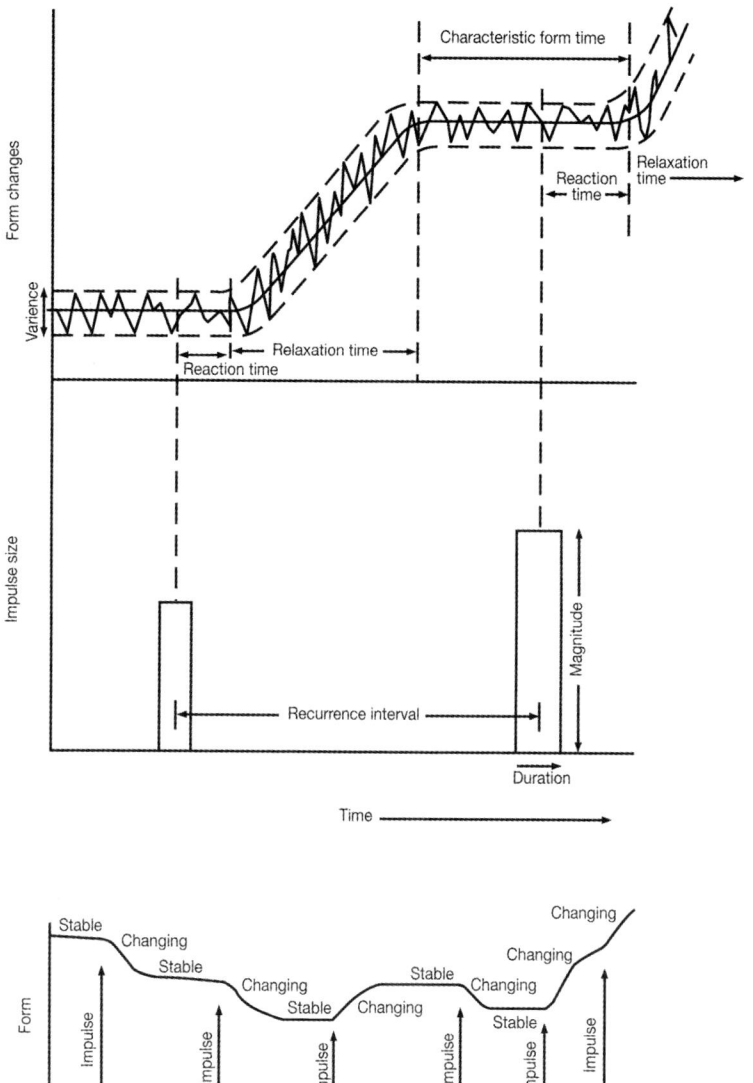

Figure 1.5 Illustration of the lag time following an impulse of change (triggering event), with a subsequent relaxation period over which the landforms adjust by morphological change (redrawn from Brunsden, 1980).

obvious boundaries. Suspended sediments, for example, may be carried thousands of miles around the coast. Although headlands which appear to mark the limits of coarse sediment transport can be identified, they are often not permanent boundaries as material may be moved around these sediment divides in severe storm conditions.

The interconnected nature of physical systems present a number of issues that will need to be addressed by most engineering projects:

Changes at one location can result in a variety of responses elsewhere — for example, an increase in the rate of cliff recession may be of benefit to a nearby low-lying coast. The sand and gravel released by erosion may build up and sustain

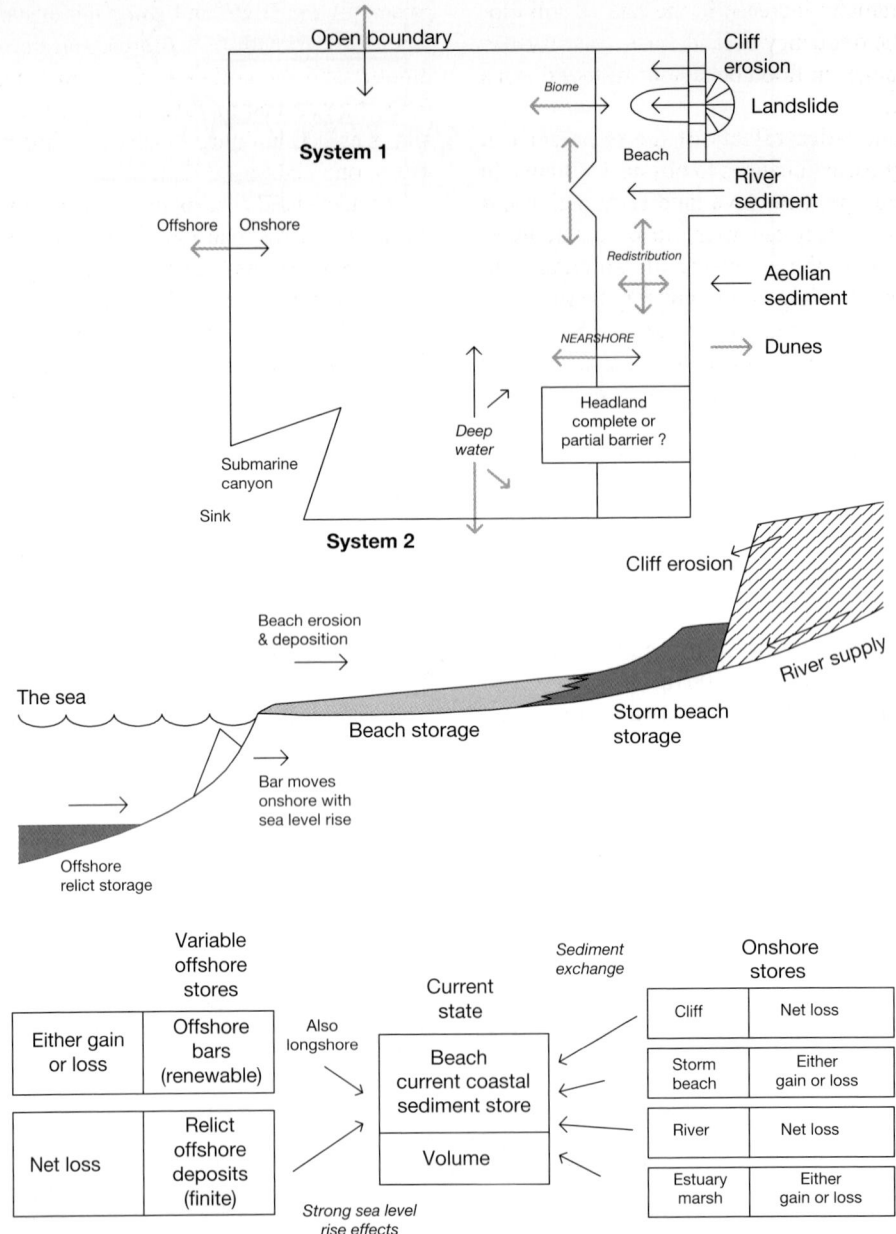

Figure 1.6   A simple coastal cell model (from Brunsden and Goudie, 1997).

beaches or shingle ridges which can form important components of flood defence solutions, either alone or where they front embankments or sea walls. Preventing cliff recession will cause a significant reduction in sediment supply and, hence,

indirectly lead to beach decline and an increase in flood or erosion risk elsewhere.

Man can significantly alter the magnitude and frequency of geomorphological events — upland forestry or land clearance for infrastructure can cause a dramatic increase in the rate of soil erosion and the frequency of landslides, and may also lead to changes in flood behaviour further down a catchment.

The cumulative effects of development can lead to significant changes to physical systems. In Western Europe, extensive land reclamation has taken place in many estuaries, since Roman times, often involving the enclosure of saltmarshes to control tidal flooding and improve grazing conditions. Over time urban and industrial uses have replaced agriculture in many estuaries. The impacts on estuary processes have included the modification of the tidal prism with resultant changes in the pattern of saltmarsh/mudflat erosion and accretion. Tidal delta structures have collapsed in response to the changing tidal prisms, resulting in accelerated erosion on the open coastline. Indeed, it is suspected that collapse of an ebb tide delta structure may have triggered the onset of accelerated erosion at the former port of Dunwich, Suffolk in the UK (Lee, 2001a). Medieval land reclamation probably led to changes in the tidal

prisms of the Blyth and Dunwich estuaries, the southward growth of a shingle spit across their mouths and the collapse of the tidal deltas. By 1328 the port had been rendered virtually useless and over 400 buildings, including churches were lost in one night in 1347.

A widely used concept in geomorphology is that physical systems tend towards equilibrium conditions where the inputs of mass and energy to a specific system are equal to the outputs from the same system. The gross form will remain unchanged throughout these transfers. Unfortunately the concept has tended to be misunderstood, generating a rather simpler view of the behaviour of systems (e.g. 'beaches or rivers are in equilibrium with the forcing conditions'). This is probably because there are a variety of types of equilibrium that are associated with quite different patterns of behaviour (Figure 1.7; Chorley and Kennedy, 1971):

1. *Static equilibrium*: no change over time
2. *Stable equilibrium*: the tendency for the form to return to its original value through internal

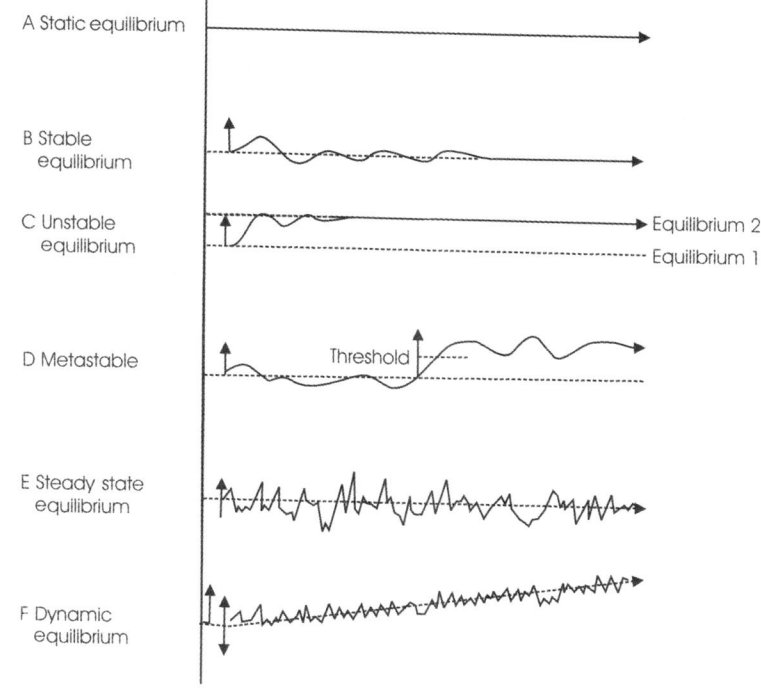

Figure 1.7   Types of equilibrium (after Chorley and Kennedy, 1971).

feedback operations within the system following a disturbance

3. *Unstable equilibrium*: the tendency for the variable to respond to system disturbance by adjustment to a new value

4. *Metastable equilibrium*: a combination of stable and unstable equilibrium except that the variable settles on a new value only after having crossed some threshold value — otherwise it returns to the original value

5. *Steady-state equilibrium*: the variable has short-term fluctuations with a longer-term constant mean value

6. *Dynamic equilibrium*: the variable has short-term fluctuations with a changing longer-term mean value (i.e. an increasing or decreasing trend).

It is worth stressing that steady state is a specific form of dynamic equilibrium when the mean rate is unchanging. Perhaps the most important aspect of dynamic equilibrium is the rate of change of the mean rate, rather than the short-term oscillations around it.

In most systems equilibrium is achieved by feedback mechanisms, involving negative feedback mechanisms that tend to counteract and stabilise system changes by means of internal adjustments (e.g. increased erosion or storage of sediment). If movement along an active fault crossing an alluvial river channel causes an increase in channel gradient, the resulting increase in flow velocity will tend to promote bed lowering and, hence, a reduction in channel gradient. Positive feedback mechanisms tend to propagate or intensify the effects of a change, for a limited time period. Removal of the bed armouring of gravel bed rivers can lead to accelerated erosion of the underlying finer sediments. An intense rainstorm can strip the upper, permeable soil horizons and expose less permeable subsoil. This can lead to the generation of greater runoff depths and velocities and accelerated erosion, exposing another soil horizon with even lower infiltration capacity. In this way the whole soil profile can be stripped off in a single rainstorm.

Complex behaviour

Engineers are familiar with the organised simplicity of machines or engineering works that can be analysed using deterministic mathematical functions. Often, they are comfortable with those systems that behave in a random manner (unorganised complexity) and can be modelled using stochastic methods (e.g. floods, wave climate). Most geomorphological systems lie between these two states (i.e. organised complexity), showing elements of both deterministic and random behaviour but not conforming to either.

Geomorphological events occur when system thresholds are exceeded. These thresholds can be external or internal.

An external threshold is where an event does not occur until an external variable exceeds a critical value. For example, the minimum rainfall intensity required to trigger debris flows and shallow landslides in parts of New Zealand has been found to be (Caine, 1980):

$$i_r = 14.82\ D^{-0.39}$$

where i_r = rainfall intensity (mm/hour) and D = rainfall duration (hours).

Internal thresholds are where gradual changes to the system may result in an event, without a change of external influences. In semi-arid areas, the accumulation of sediment in a valley floor gradually increases the valley slope angle (long profile) until failure (erosion) occurs. Figure 1.8 illustrates how in the oil shale region of western Colorado it has been possible to identify threshold slope angles for drainage basins greater than 4 square miles; above the threshold line, valley slopes tend to be unstable and affected by gullying. Such incipient instability is common in alluvial river channels, where channel sinuosity and meander amplitude increase until a cut-off or channel avulsion occurs. This is due to channel lengthening and gradient reduction and is not necessarily related to external changes.

Often geomorphological events are the result of the variable interaction of a range of factors. Figure 1.9 highlights the complex relationship between preparatory and triggering factors that are active in promoting coastal cliff recession. There are rapid temporal changes in the margin of stability of

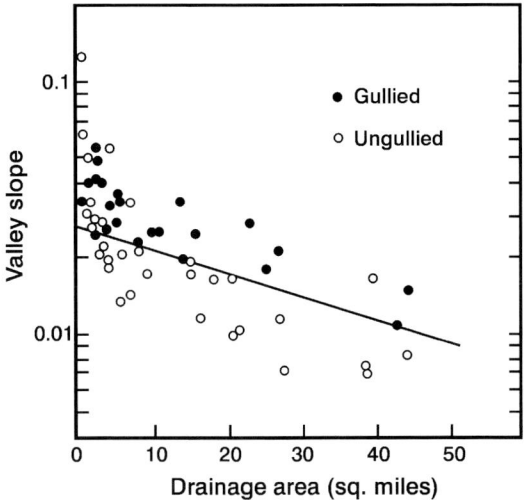

Figure 1.8   Threshold line separating stable and unstable valley floors for drainage basins of different size, Piceance Creek basin, Colorado (redrawn and adapted from Patton and Schumm, 1975).

coastal cliffs due to the superimposition of triggering factors on the trends imposed by relatively steady erosion at the base of the cliff. As the margin of stability is progressively reduced by the operation of preparatory factors, so the minimum size of triggering event required to initiate recession becomes smaller. Thus, triggering events of a particular magnitude may be redundant (i.e. do not initiate cliff recession) until preparatory factors lower the margin of stability to a critical value. As Figure 1.9 indicates, this can mean variable time periods (epochs) between coastal landslides, depending on the sequences of storm or rainfall events. In addition, the same size triggering events may not necessarily lead to landslides.

Another example of this complex behaviour is given in Figure 1.10. As mentioned above, sediment build up causes the gradual steepening of a valley floor, reducing its ability to withstand the effects of potentially erosive flood events. Eventually a flood

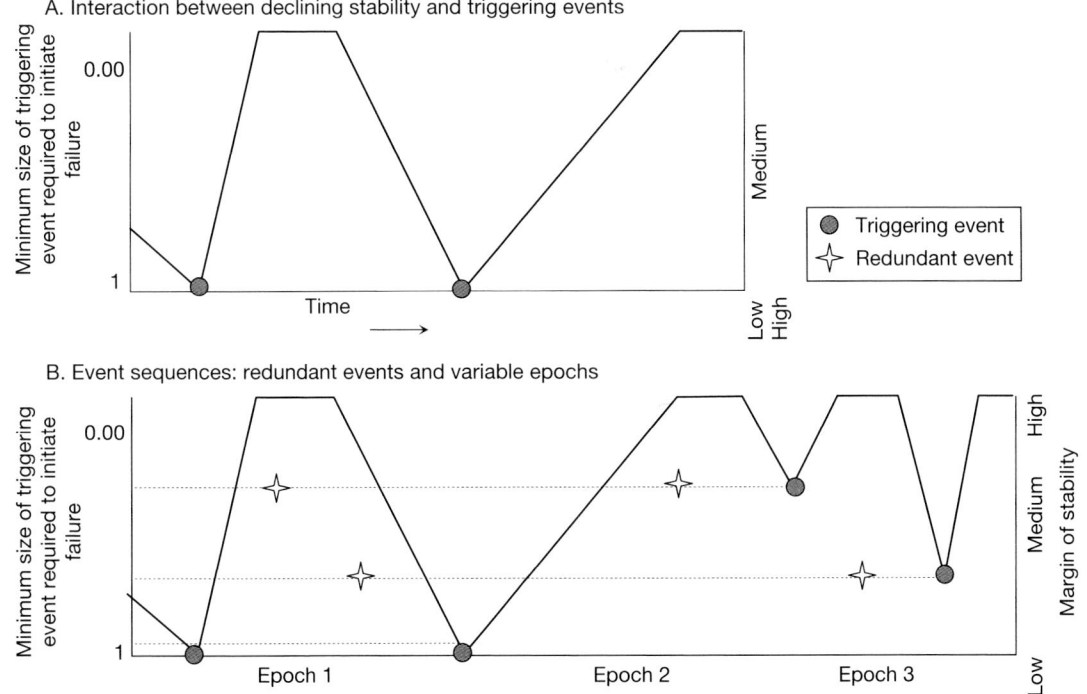

Figure 1.9   A schematic illustration of the variable interaction between potential triggering events and coastal landslide (from Lee *et al.*, 2001).

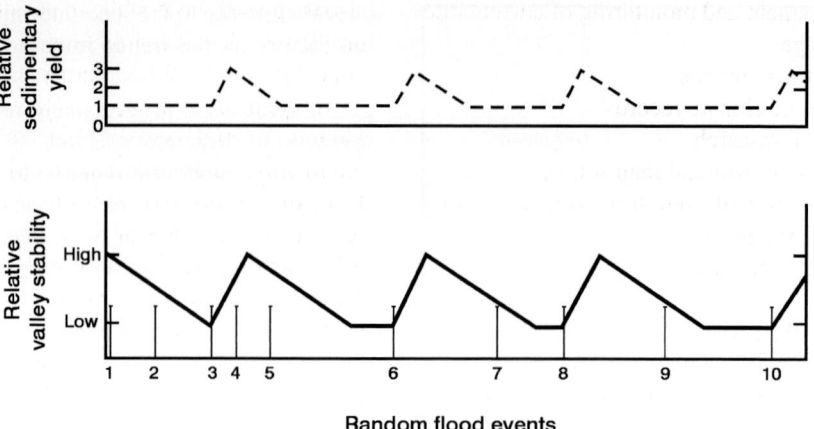

Figure 1.10 The relationship between random flood events of the same magnitude and valley floor erosion. When a flood event coincides with a period of low stability of valley floor alluvium (solid line) it causes incision and removal of sediment (upper dashed line) (redrawn from Schumm, 1977).

of a particular magnitude will exceed the threshold of valley stability and erosion will occur. In Figure 1.10, all the random flood events are of the same size (and return period), however only events 3, 6, 8 and 10 will trigger valley floor erosion, the remainder are redundant.

Many geomorphological events are not independent but are influenced by the size and location of previous events. In other words landslides or gully formation, for example, are processes with a 'memory', insofar as the current and future behaviour is influenced by the effects of past events on the system.

The response of different landscapes to the effects of triggering events or human activity can vary considerably. For example, a storm passing across two catchments may result in very different patterns of erosion, deposition and flooding. This complexity of response is a measure of the sensitivity of a landscape, which can range from fast responding systems, which are very sensitive to disturbing events (this type of system can be morphologically complex because the landforms are subject to rapid change, such as active sand dunes) to slowly responding insensitive systems, such as ancient, cratonic surfaces.

The main controls of sensitivity include material strength, morphological resistance (e.g. relative relief, altitude, slope angle) and the importance of linkages between landforms within a system (i.e. sediment budget). A stable landscape, for example, is one in which the controlling resistances are sufficient to prevent a storm event from having any effect.

1.3 Techniques in engineering geomorphology

Regardless whether an engineering geomorphologist is concerned with the risks from geohazards or the impact of a project on the environment, it will be necessary to develop a 'ground model' or an understanding of the 'total geological and geomorphological history' of the site or region (Fookes, 1997; Fookes, et al., 2000), at an appropriate level of detail. In doing so, an engineering geomorphologist will often seek to establish the geohazard potential in the area, the event history (i.e. frequency and magnitude) and the likely knock-on effects of any modification or disturbance to landforms (e.g. slope re-profiling) or physical systems (e.g. disruption or alteration of sediment transport pathways). The techniques that can be used to achieve these goals are many and varied, but may include:

1. terrain evaluation
2. engineering geomorphological mapping

3. measurement and monitoring of current rates of change
4. field measurements
5. analysis of climate records
6. historical research
7. analysis of historical map sources
8. identification of event triggering conditions
9. the use of tracers
10. the dating of events.

Terrain evaluation

Terrain evaluation is a method of summarising the principal physical characteristics of an area and their implications for engineering projects (e.g. Phipps, 2001). The principles of terrain evaluation involve defining areas of terrain that have similar physical characteristics (i.e. a typical range of topographic, geohazard and constructability factors). Thus, these areas of terrain can be considered to present consistent levels of challenge that will need to be addressed by a project.

Terrain evaluation has its origins in the need to organise and communicate specific earth science information or intelligence in a way that is of direct relevance to the end user (e.g. civil and military engineers, land use planners, agriculturalists and foresters). An interesting summary of the development of the approach can be found in Mitchell (1973). Perhaps the most well-known examples of the use of terrain evaluation have been the Commonwealth Scientific and Industrial Research Organisation (CSIRO) applications in Australia between the 1940s and 1960s to support agricultural development (the Land Research and Regional Survey division) and engineering (the Division of Soil Mechanics), and the Military Engineering Experimental Establishment (MEXE) and Soils Laboratory at Oxford's system for military terrain evaluation, developed in the 1960s. The military need for terrain information includes both strategic (e.g. the gross distribution of terrains and transport corridors) and tactical (e.g. visibility, cross-country mobility, reaction of the terrain to deformation and the availability of water and construction resources).

Terrain evaluation has tended to be associated with rapid assessments of remote, inaccessible regions (e.g. Fookes *et al.*, 2001). However, it has also been used successfully on major infrastructure projects in developed regions (e.g. the Channel Tunnel Rail Link; Waller and Phipps, 1996).

At the broadest scale, landscape types (terrain models or land systems) can be defined (e.g. mountains, desert plateaux, coastal plains); this level of sub-division may be suitable for pre-feasibility overviews of very large areas. Within a landscape type it will be possible to identify a variety of landform assemblages (terrain units), such as river floodplains, escarpment faces, extensive areas of unstable hill-slopes and ridge crests; this level of detail may be sufficient for corridor assessment. Within a terrain unit there will be numerous individual landforms (terrain sub-units) that will each present slightly different levels of challenge to a project; an escarpment face, for example, may contain a variety of sub-units, including bare rock faces and discrete landslide systems separated by stable ridges and spurs (e.g. Figures 1.11*a* and 1.11*b*).

In providing a spatial framework of terrain units the terrain specialist seeks to identify clear associations between surface forms (i.e. the terrain units), near-surface materials and processes; simplify the complexity of ground conditions and surface processes within a particular area, highlighting those of significance to the planning of projects, and provide a tool for predicting terrain characteristics within a particular area or region — this is based on the assumption that the terrain units are sufficiently homogeneous and mutually distinctive to allow valid prediction.

Some basic properties of terrain units and sub-units can be derived from available geological and soil maps (other useful maps include hydrogeology, hydrology, geomorphology, land use and vegetation, but are rarely available). However, it will be necessary to make judgements about the nature and significance of many issues, including geohazards and landscape sensitivity. These judgements should be based on the specialists' experience of the broad association between materials (i.e. geology), landforms (i.e. the terrain units and sub-units) and processes (i.e. geohazards) that are known to occur in different climatic regions (e.g. tundra, temperate, savannah, hot drylands, hot wetlands) and tectonic zones (e.g. plate margins, cratonic areas).

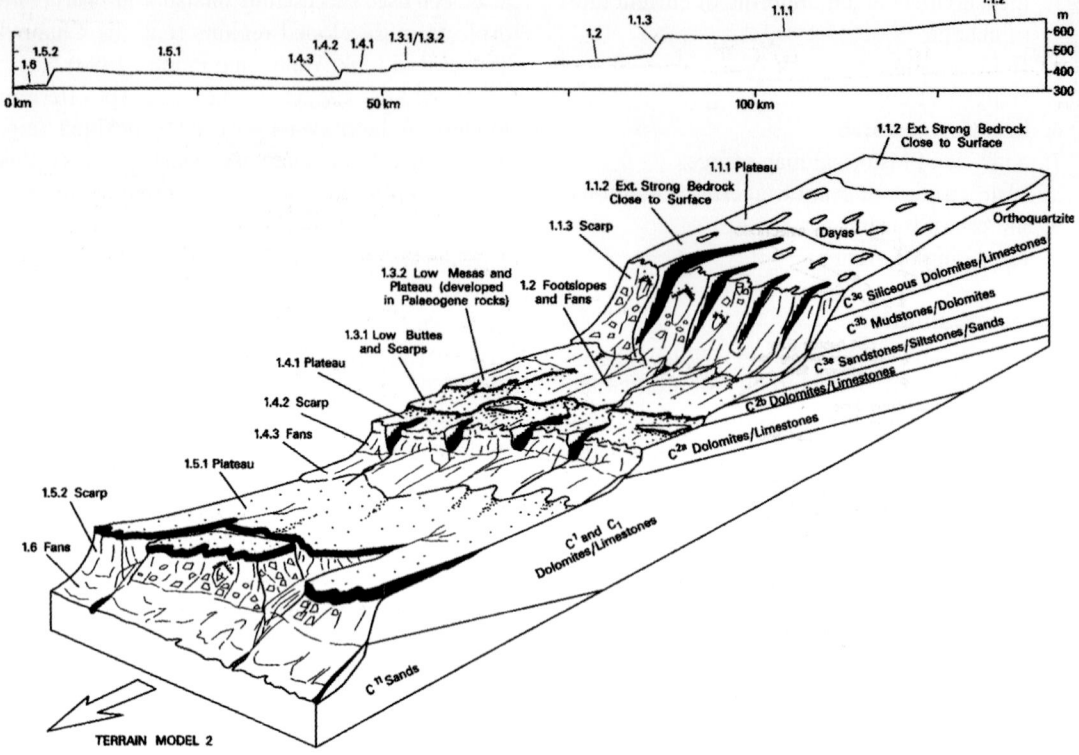

Figure 1.11a   Terrain model 1: Tademait plateau (from Fookes *et al.*, 2001).

Engineering geomorphological mapping

The production of some form of map underpins many engineering geomorphological studies. Training and experience are the keys to producing good quality geomorphological maps. As experienced field geologists produce the best geological maps, so field geomorphologists produce the best geomorphological maps. In addition, the ability to talk the same technical language as the clients will help overcome the all too inevitable misunderstandings over what precisely the landscape means to the end users.

The map might be the product of an intensive fieldwork programme, a sketch map of part of an area based on a walk-over survey, or interpretation of remote imagery (e.g. aerial photographs or satellite images). Most map making follows three stages:

1. The recognition of landforms or landform elements (units) that provide a practical framework for addressing the problem in hand.

2. The characterisation of these units in terms of the significant surface processes (these may be active or relict) and the near-surface materials.

3. The interpretation of the significance of these forms, processes and materials to the problem facing the geomorphologist and the clients. This may involve producing some form of derivative map (e.g. a landslide hazard map, aggregate resource map) or an extended map legend suitable for use by the clients or their technical advisors.

The Geological Society Working Party Report on maps and plans (Anon., 1972) identified examples of geomorphological mapping that could be of use to engineers. However, the value of the technique was best highlighted by its application to road projects in unstable terrain in Nepal and South Wales during the early 1970s (e.g. Brunsden *et al.*, 1975a; 1975b; Doornkamp *et al.*, 1979; Cooke *et al.*, 1982; Jones *et al.*, 1986); soil erosion (e.g.

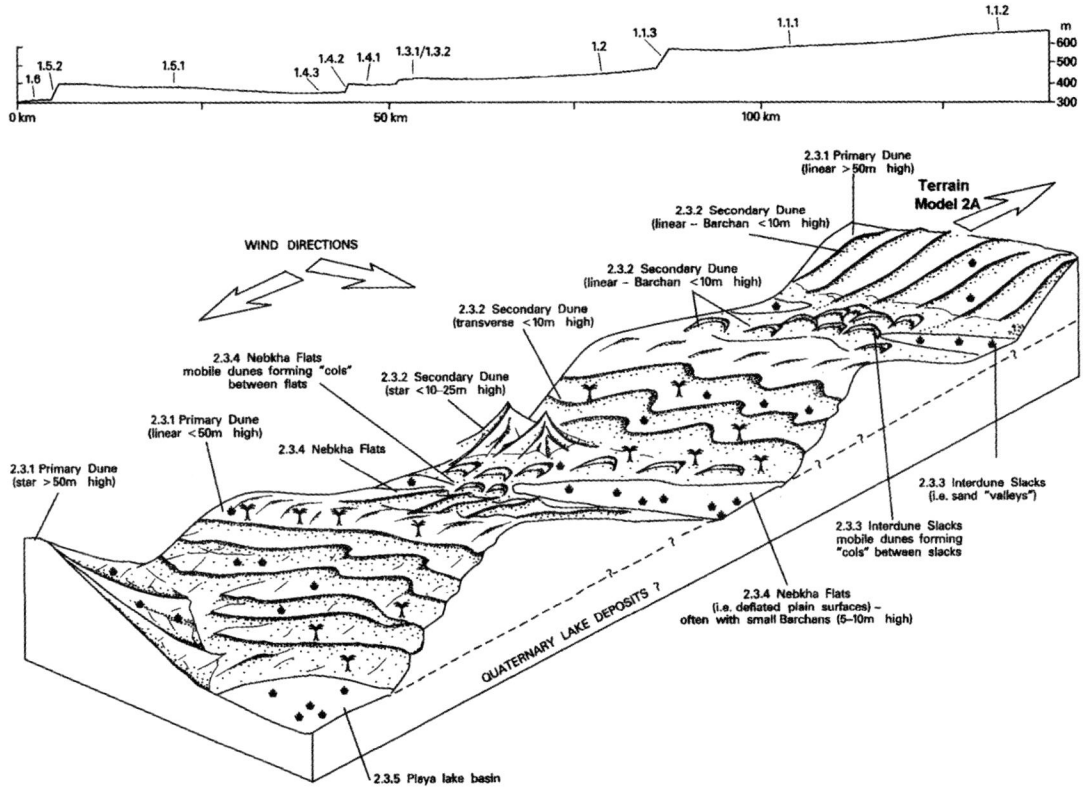

Figure 1.11b   Terrain model 2: Grand Erg Occidental, Algeria (from Fookes *et al.*, 2001).

Morgan, 1986), and river management (e.g. Doornkamp, 1982; Richards *et al.*, 1987). Despite these applications, BS5930 — Code of Practice for Site Investigations, contains little reference to geomorphological maps and the technique remains a marginal skill practised by a few experienced engineering geomorphologists. Lee (2001b) presents an introduction to the preparation of engineering geomorphology maps. Stylised symbols are often used (Demek and Embleton, 1978; Gardiner and Dackcombe, 1983; Cooke and Doornkamp, 1990).

The old adage, 'site and situation', is clearly important in ensuring that potential geohazards are identified at an early stage in the investigation process (e.g. Fookes, 1997). Hillside sites need to be seen in the context of the whole slope, river valleys as part of the whole catchment, coastal sites as part of a sediment transport cell. Geomorphological maps should also provide an indication of the

potential for landscape change, either through major 'natural' events such as neotectonics, gully formation, landslides or river channel migration or as a result of human intervention (e.g. unloading pre-existing landslides or solifluction sheets). Field geomorphologists should seek to understand the inheritance from previous geological periods and the sensitivity of the landscape to change from evidence of surface morphology (e.g. back-tilted ground is often a good indicator of deep-seated landslides), landform assemblages (e.g. the presence of steep, gullied terrain can be interpreted, in some environments, as an indication of the potential for channelled debris flows) and superficial deposits (e.g. clast-supported or matrix-supported deposits can indicate flash flood or debris flow deposits, respectively).

Engineering geomorphological maps, like most other forms of map, incorporate judgements based on the available evidence (i.e. they are based on 'reading the ground', Dearmen and Fookes (1974)). It is important that geomorphologists convey an indication of the reliability of the information and the uncertainties associated with their judgement to the users.

Measurement and monitoring of current rates of change

The direct measurement of river or coastal change is the most obvious method of obtaining information on current erosion or accretion rates. For example, systematic surveys undertaken on a regular basis are a common approach to establishing cliff recession rates. On the Holderness Coast, Yorkshire, for example, the local authority initiated a programme of cliff recession measurement in 1951 which has been continued on an annual basis ever since. A series of seventy-one marker posts, termed 'erosion posts' by the local authority were installed at 500 m intervals along 40 km of the coastline, each post located at a distance of between 50 m and 100 m normal to the coast. These posts are replaced further inland from time to time if they become too close to the cliff top. Annual measurements from each post to the cliff top — defined as the lip of the most recent failure scar — commenced in 1953. The resulting database provides an invaluable source of medium- to long-term measurements of cliff recession (Figure 1.1).

Field measurements

On-site measurement or estimates of surface features and materials can provide the input data for a range of modelling or predictive tools that are used by engineering geomorphologists. For example, in the absence of hydrological records, it is possible to use careful observations of channel cross-section, bed gradient and bed sediment sizes to provide an approximation of the flooding and scour hazard (Chapter 16). Measurements of the wave climate, beach profiles and sediment sizes can be used to predict longshore sediment transport rates using the CERC (Coastal Engineering Research Center) formula (CERC, 1984). Field assessment of soil erodibility, slope length and steepness and crop management are key input data in the Universal Soil Loss Equation for predicting soil erosion rates.

Analysis of climate records

Time series climate records can provide input data to various empirical models, such as the use of rainfall intensity to predict soil erosion loss (e.g. EI_{30}, the energy associated with the maximum 30 minute rainfall intensity) or debris flow activity (e.g. critical rainfall thresholds). Potential sand drift calculation can be calculated using Fryberger's (1979) method, from the average wind velocity at 10 m and the duration of wind.

A common approach to analysing data series is to rank the records (e.g. daily rainfall in order of decreasing magnitude). This can yield the recurrence interval (return period), as described earlier. The data can be plotted on probability paper: data conforming to the log–normal former distribution, plot as a straight line on log–probability graph paper; data having a Gumbel (type 1) distribution, plot as a straight line on extreme probability paper.

Often analysis of the data series can reveal trends (e.g. increased winter rainfall) or cyclic fluctuations (e.g. wet-year sequences) that may be important factors in controlling the frequency and magnitude of geomorphological events. For example, on the West Dorset coast, there is a direct relationship between wet-year sequences and landslide activity. Analysis of rainfall records from Pinhay (3 km west of Lyme Regis) for the period 1868–1998 has revealed that the annual rainfall has increased over time by around 75 mm (i.e. there has been a 10 per cent increase in rainfall over this period; the mean value is 914 mm). A similar trend has occurred for the annual effective rainfall (precipitation minus evapotranspiration). This provides an indication of the amount of rainfall actually available to the groundwater tables within the slopes, i.e. a moisture balance index (Figure 1.12a; the mean annual value is 319 mm). In order to identify wet-year sequences and their frequency, a method of moving averages was used to smooth the effective rainfall (i.e. the moisture balance index) series by replacing each observation with a weighted mean. Figure 1.12b presents the 9-year moving mean of the data series (expressed as values relative to the mean effective rainfall) and indicates several periods of increased rainfall with a possible frequency of 20–25 years.

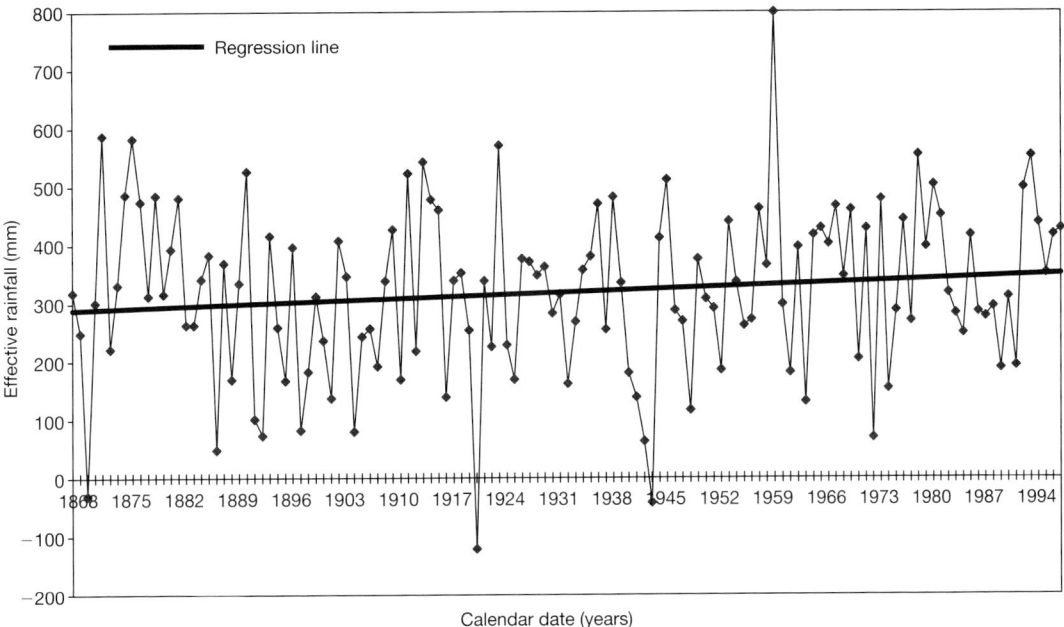

Figure 1.12a Lyme Regis: annual effective rainfall: moisture balance index.

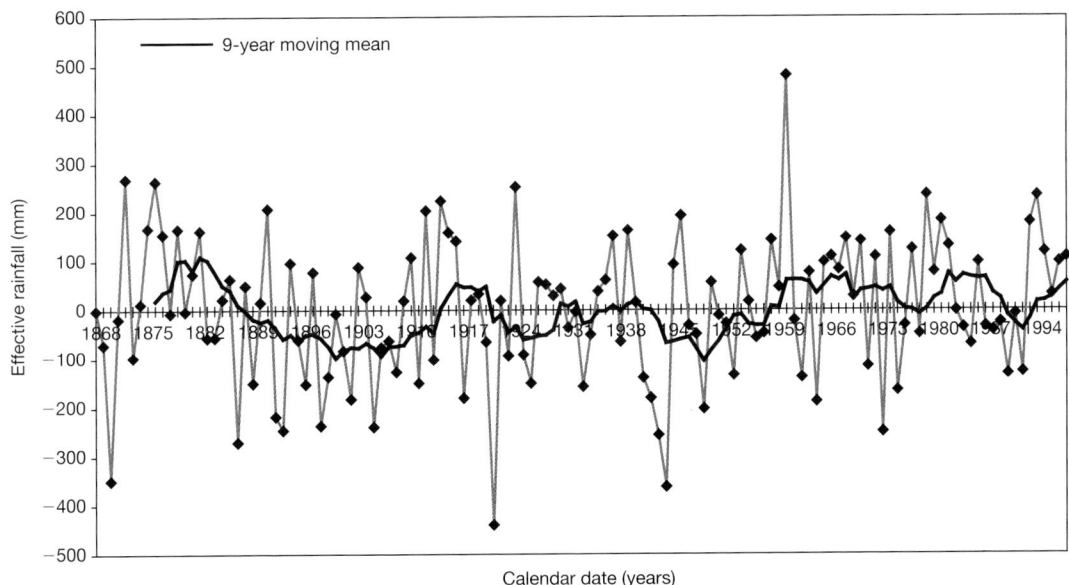

Figure 1.12b Lyme Regis: moisture balance index relative to the mean annual effective rainfall.

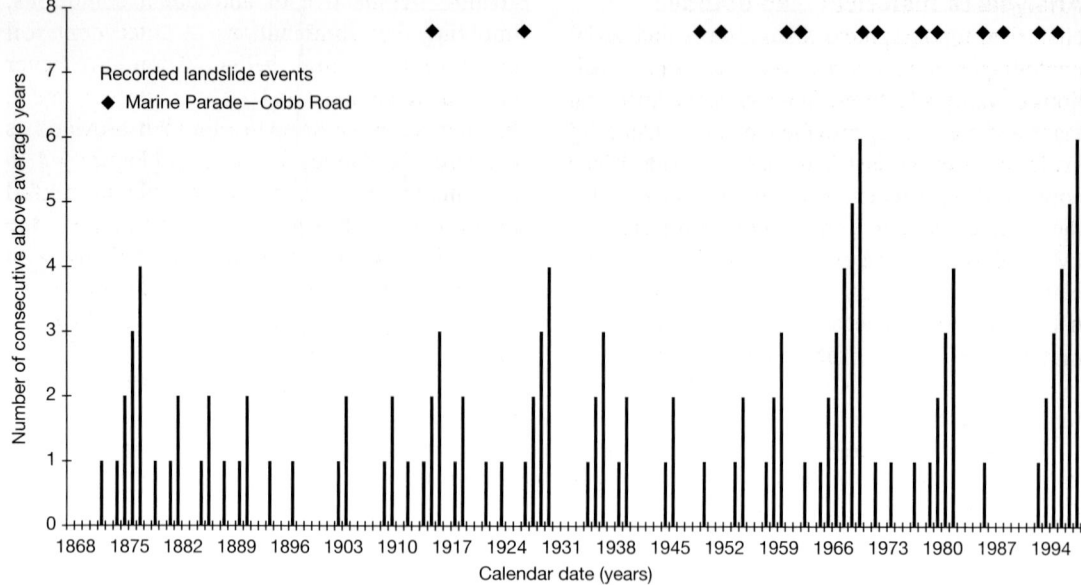

Figure 1.13   Lyme Regis: wet-year sequences and landslide events (after Bray, 1986).

Wet-year sequences were identified by calculating the cumulative number of years with effective rainfall (moisture balance) greater than the mean value (Figure 1.13). This value is set to zero every time it falls below the mean, on the assumption that the groundwater levels only become critical when they rise above the average level.

Historical research

In many countries, there is a wide range of sources that can provide useful information on the past occurrence of events, including: aerial photographs, topographic maps, satellite imagery, public records, local newspapers, consultants, reports, scientific papers, journals and diaries (e.g. Brunsden *et al.*, 1995).

In Ventnor, Isle of Wight, the pattern of contemporary ground movement was established from a systematic search of local newspapers from 1855 to present day (e.g. Lee and Moore, 1991). Over 200 individual incidents of ground movement and coastal erosion were identified, allowing a detailed model of landslip potential to be developed. Close inspection of these records

allowed the identification of distinct phases of landslide activity, often corresponding with periods of prolonged heavy winter rainfall, a range of ground movement problems in different parts of the town and vulnerable areas affected by frequent ground movement.

When researching particular events that have been recorded in local newspapers or documents, it is necessary to make a judgement on the reliability of the data source. Three questions need to be borne in mind:

1. What is the nature of the event being recorded, and with what detail, and is it pertinent to the stated objectives?
2. Who is making the report, in particular what are his or her qualifications to know of the event, i.e. is it a personal observation based on his or her own experience; an editing of reports from other people, who themselves may have edited the information; a plausible rumour; a complete invention, or falsification?
3. In the light of knowledge of this type of event, is the report credible, in whole, in part, or not at all?

Analysis of historical map sources

Historical topographical maps, charts and aerial photographs provide a record of the former positions of various features. In many cases, historical maps and charts may provide the only evidence of evolution over the last 100 years or more. When compared with recent surveys or photographs, these sources can give an estimate of the cumulative land loss and the average annual erosion rate between the survey dates. However, great care is needed in their use because of the potential accuracy and reliability problems (e.g. Carr, 1962; 1980; Hooke and Kain, 1982; Hooke and Redmond, 1989).

As part of the Brahmaputra River Training Study, Thorne *et al.* (1993) describe how Bangladesh survey maps and aerial photographs from the 1950s, together with 1989 'SPOT' satellite imagery were used to assess bank-line migration. The section of the river that was being studied was subdivided into a series of reaches, and point measurements of right and left bank erosion were made at 0.5 km intervals. Mean erosion rates were established by dividing the cumulative losses by the time interval. The survey revealed that bank erosion was more severe on the right bank, where the average annual rate was 90 m (in some reaches the rate was double this figure).

Identification of event triggering conditions

The identification of event triggering conditions often involves establishing initiating thresholds between various parameters (e.g. rainfall, seismic activity) and geomorphological events. The most readily defined threshold is one that identifies the minimum conditions (or envelope) for activity; above this, the conditions are necessary, but not always sufficient to trigger events and, below this, there is insufficient impetus.

For example, Glade (1998) and Crozier and Glade (1999) describe how a rainfall threshold for regions of New Zealand was defined and used to calculate probabilities of landslide occurrence. Figure 1.14a presents the minimum and maximum probability rainfall thresholds for the Wellington region. The minimum threshold required to trigger landslides appears to be 20 mm, irrespective of antecedent conditions, implying that combinations of antecedent soil moisture and rainfall below 20 mm have never been sufficient to attain the critical water content. In contrast, every storm greater than 140 mm has triggered landslides in the past. Figure 1.14b presents the return periods for different rainfall magnitudes — this provides a link between the thresholds and the probability of landsliding, with the return period of the 140 mm rainfall event (the maximum probability threshold) around 20 years. Figure 1.14c highlights the probability of landslides occurring within particular time periods. For example, the maximum threshold rainfall of 140 mm has a probability of occurrence on any day of 0.02%, within 30 days of 0.2% and within a year of 5%.

The use of tracers

A wide variety of methods can be used to model sediment transport and water flow pathways through physical systems. In broad terms, the approach can help assess sediment provenance and transport and water movement.

Fine sediment transport and sediment provenance can be assessed using, for example, mineral magnetic signatures, stable isotopes and radionuclides. For example, Walling and He (1999) used fallout radionuclides to estimate rates of floodplain accretion for a number of British rivers, using ^{137}Cs to date the depositional sequences.

Coarse sediment transport along rivers and the coast can be assessed using artificial sediments (e.g. acoustic, electronic, magnetic or aluminium pebbles), painted or fluorescent coated grains (luminophors), radioactive isotopes or exotic materials. Tracer experiments on Chesil Beach, for example, were used to demonstrate a net eastward drift along the shingle beach (Carr, 1971). Tsoar (1978) used coloured sand to determine sand flow directions in desert dunes.

Water movement through soil, groundwater systems and caves, and in river channels can be traced using, for example, fluorescent dyes, optical brighteners, lycopodium spores and lithium acetate. A useful summary of the geomorphological application of tracing methods can be found in Foster (2000).

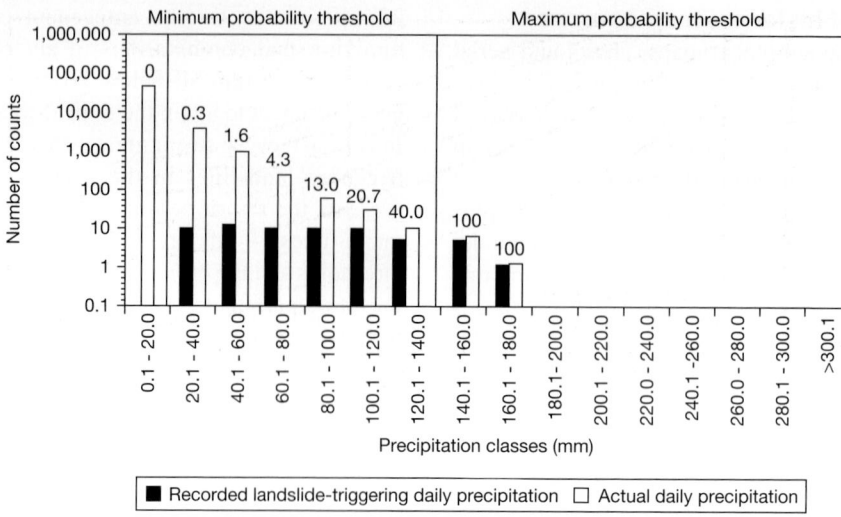

Figure 1.14a   Probabilities (%) of landslide occurrence associated with rainfall of a given magnitude in Wellington, New Zealand, 1862–1995 (note: a value of 50 means that 50% of all measured daily rainfalls in a given category produced landslides in the past) (after Crozier and Glade, 1999).

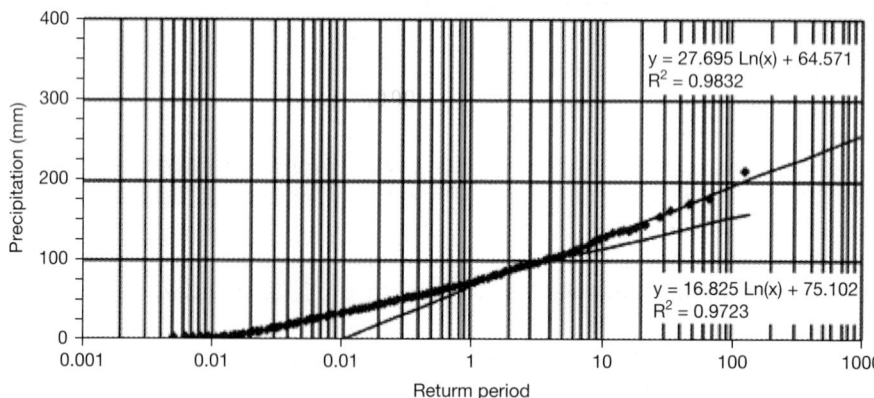

Figure 1.14b   Return periods of daily precipitation and fitted logarithmic trend lines in the Wellington region, New Zealand (after Crozier and Glade, 1999).

Dating of events

A number of methods are available to obtain dates for episodic events and thereby to build up a picture of the event history of the area, including:

Radioactive dating, in which measurement of the ratio of stable (^{12}C) to radioactive (^{14}C) isotopes in an organic sample (e.g. charcoal, wood, peat, roots, macrofossils) allows the determination of the time elapsed since biological or inorganic fixation. Specialist applications include the use of an accelerator mass spectrometer to date the organic material included in varnish layers characteristic of many desert rock surfaces or stone pavements. The age range for radiocarbon dating is from a few centuries to around 40 000 years).

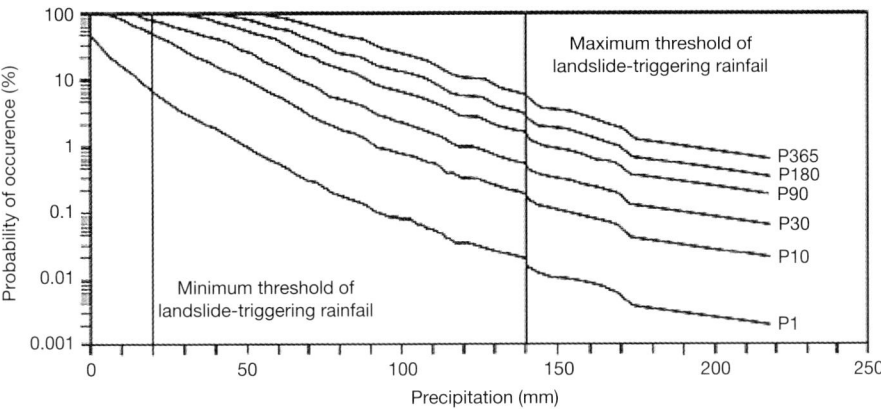

Figure 1.14c   Probability of occurrence of daily precipitation equalling or exceeding given values in Wellington, New Zealand. Note: the different lines refer to the probability of occurrence of specific rainfall magnitude on each single day (P1), within a period of 10 days (P10), within a month (P30), etc. The empirically established minimum and maximum thresholds of landslide triggering rainfall (20 mm and 140 mm) are shown by the thin vertical lines (after Crozier and Glade, 1999).

Lichenometric dating, involving the measurement of the maximum lichen size on exposed surfaces (e.g. boulders) and comparing with lichen-growth curves. This is assumed to indicate the time elapsed since a particular event (e.g. rockfall, debris flow, flash floods).

Dendrochronology, whereby the age of death of a tree (e.g. buried beneath a landslide) may be determined by cross-reference to a master tree-ring chronology for that region. The oldest rings of trees that have colonised a now stabilised slide or debris fan will give a minimum age for the event.

Luminescence dating, based on estimating the time since a sediment was last exposed to daylight, which 'zeroes' the previously accumulated radiation damage to minerals (e.g. quartz or feldspar) in the sample. The age of a sample is derived from:

$$\text{Age} = \frac{\text{Palaeodose}}{\text{Dose rate}}$$

where the palaeodose is the accumulated radiation damage and the dose rate is the rate at which the sample absorbs energy from the immediate proximity. Palaeodoses are calculated by thermoluminescence (energy is supplied by an oven) or optically stimulated luminescence. The maximum age range for these methods depends on the mineral: for quartz it is around 100 000–150 000 years; for feldspars it is around 800 000 years. Loesses have been successfully dated up to 800 000 years old. Other applications include dating alluvial fans and aeolian deposits.

Uranium series dating is applicable for the time range <1000 to over 500 000 years. It involves testing surface sediments to establish the radioactive disequilibrium in the decay chains of Uranium and related isotopes such as Thorium (e.g. the ratio between ^{230}Th and ^{234}U). The method has been used to date calcretes and halite deposits, or crusts beneath landslides.

Cosmogenic nuclide dating, whereby the concentration of cosmogenic nuclides (e.g. ^{3}He, ^{10}Be, ^{14}C, ^{21}Ne, ^{26}Al and ^{36}Cl) produced by *in situ* nuclear reactions (the interaction between cosmic rays and terrestrial atoms) in the upper few metres of the ground surface are detected by accelerator mass spectrometry. The concentrations can be used to determine the length of time that material has spent at or near the ground surface. These methods have been used to date rockfalls, basalt flows and alluvial fan surfaces. The datable range is from 1000 years to several million years.

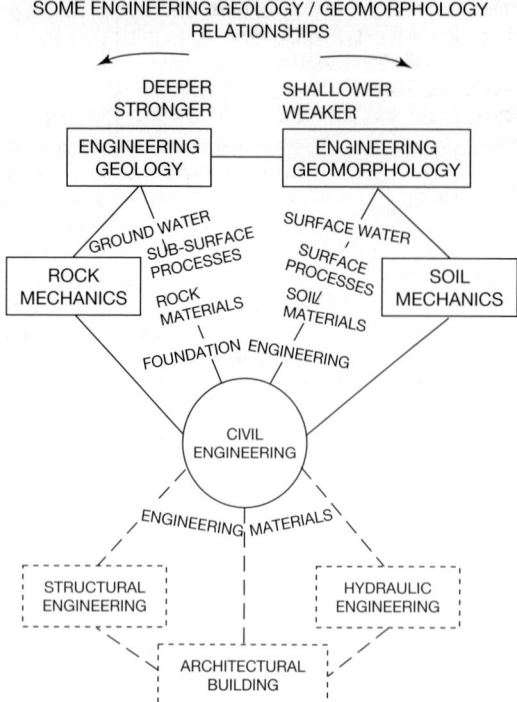

Figure 1.15   Some engineering, geology and geomorphology relationships.

Cation-ratio rock varnish dating method, based on the observation that the ratio of certain cations (K, Ca, Ti) in rock varnish decreases with age as a result of preferential leaching. Cation-ratios can be calculated by photon-induced X-ray emission spectrometry, electron microprobe or inductively coupled plasma spectroscopic (ICP) methods. Comparing ratios from a number of samples can generate relative ages. Calibrated ages can be determined by comparison with curves generated for samples of known age.

Amino acid dating methods provide relative ages by measuring the extent to which certain amino acids within protein (e.g. bones, mollusc shells, eggshells) residues has transformed from one of two chemically identical forms to the other (e.g. the transformation from L-form to D-form amino acids) until equilibrium is reached. The methods can be used for timescales from a few years to hundreds of thousands of years.

Useful discussions of the suitability of the different methods in desert environments and landslide studies are provided by Stokes (1997) and Lang *et al.* (1999) respectively.

1.4 Summary

Geomorphology studies the earth's landforms and landform change. Engineering geomorphology complements engineering geology in providing a spatial context for explaining the nature and distribution of particular ground-related problems and resources.

Importantly, engineering geomorphology is also concerned with evaluating the implications of landform changes for society and the environment.

This book is largely by geomorphologists and its aim is to produce a basic yet authoritative handbook for geotechnical engineers and others. Its essence is in its diagrams around which each chapter is built. The first part is concerned with controls which have shaped the landscape, the second with processes and the third is concerned with the various environments and landforms.

The application of geomorphological knowledge to civil engineering will come about by the education of both engineers and geomorphologists. A relationship between engineering, geomorphology and geology is summarised in Figure 1.15.

References

Ager, D. L. (1976) *The Nature of the Stratigraphic Record.* Methuen, London.

Anon. (1972) The preparation of maps and plans in terms of engineering geology. *Quarterly Journal of Engineering Geology* **5**, 293–381.

Ballantyne, C. K. (1991) Holocene geomorphic activity in the Scottish Highlands. *Scottish Geographical Magazine* **107**, 84–98.

Bray, M. J. (1986) *A Geomorphological Investigation of the South West Dorset Coast, Volume 1: Patterns of Sediment Transport.* Report to Dorset Country Council, 798 pp.

Brunsden, D. (1980) Applicable models of long-term landform evolution. *Zeitschrift für Geomorphologie Supplementband* **36**, 16–26.

Brunsden, D. (2002) Geomorphological roulette for engineers and planners: some insights into an old game. *Quarterly Journal of Engineering Geology and Hydrology* **35**, 101–142.

Part I

Controls

2. Climate and Weathering

Mark Lee and Peter Fookes

2.1 Climate

Climate is the average, long-term summary of the weather patterns in a region, with the short-term extremes and transient weather conditions smoothed out to give average values, for example, of annual or monthly temperature, precipitation and wind regime. It is a major influence on the rate, scale and significance of current geomorphological processes and weathering. The imprint of past climates can be seen in the variety of inherited features that are characteristic of many landscapes, from duricrusts in hot drylands to the relict, periglacial landslides of many temperate regions and solution caverns in limestone terrain. Climate is also an important consideration in many aspects of engineering, from the preparation and performance of concretes, to the timing of earth-moving operations and weathering of building stone.

The importance of climate stems from the fact that solar radiation is a key source of kinetic energy for geomorphological processes, along with tidal energy generated by the gravitational attraction of the Sun and the Moon and rotational energy derived from the momentum of the Earth's rotation (Figure 2.1; note that endogenic processes such as volcanism, seismic activity and tectonics are driven by geothermal energy). Solar energy drives the hydrological cycle which involves the continuous exchange of water from oceans to the continents via the atmosphere and back again as runoff (Figure 2.2). The input of solar radiation varies between the equator and the poles, with the resulting thermal gradients giving rise to the circulation of air in the atmosphere (winds) and water in the ocean (ocean currents).

There are clear links between climate and geomorphological processes. Frost action, for example, can only occur where ground temperatures fall below freezing, while permafrost will only develop where the mean annual temperature is at least below 0 °C. Wind action is most effective in dry regions with less than 200 mm mean annual rainfall. However, in many regions it may be the extremes, such as high intensity rainfall events, that are responsible for triggering processes or the development of landforms and the behaviour of man-made structures.

The variations in surface temperature and available kinetic energy, along with geological structure (tectonics) and lithology (stratigraphy), and local differences in the potential energy arising from the height of material above base level, provide the impetus for global and regional differences in the intensity of geomorphological processes. Three major climate types can be recognised, reflecting global variations in precipitation (Figure 2.3) and temperature (Figure 2.4).

2.2 Tropical climates

Equatorial Trough (Inter-Tropical Convergence Zone): this represents an area of low pressure near the equator towards which persistent winds (trade winds) blow — from the NE in the northern hemisphere and from the SE in the southern hemisphere. This zone is associated with a relatively uniform climate (i.e. little monthly variation) with high rainfall, temperature and humidity, and supporting characteristic tropical rainforest vegetation (Figure 2.5a).

Sub-Tropical High Pressure Zone: this is characterised by subsiding warm, dry air from air circulation cells centred on the equator, known as Hadley cells. These regions are generally cloud free and arid, giving rise to major deserts (e.g. the

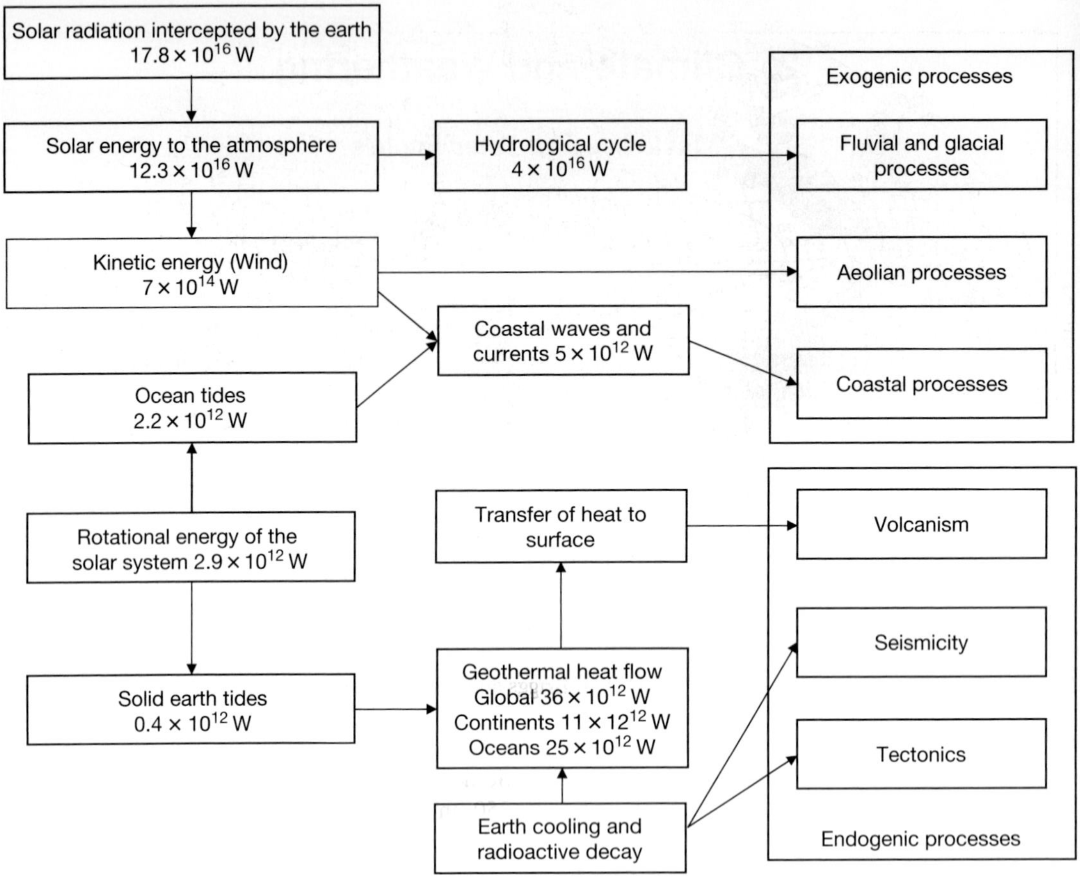

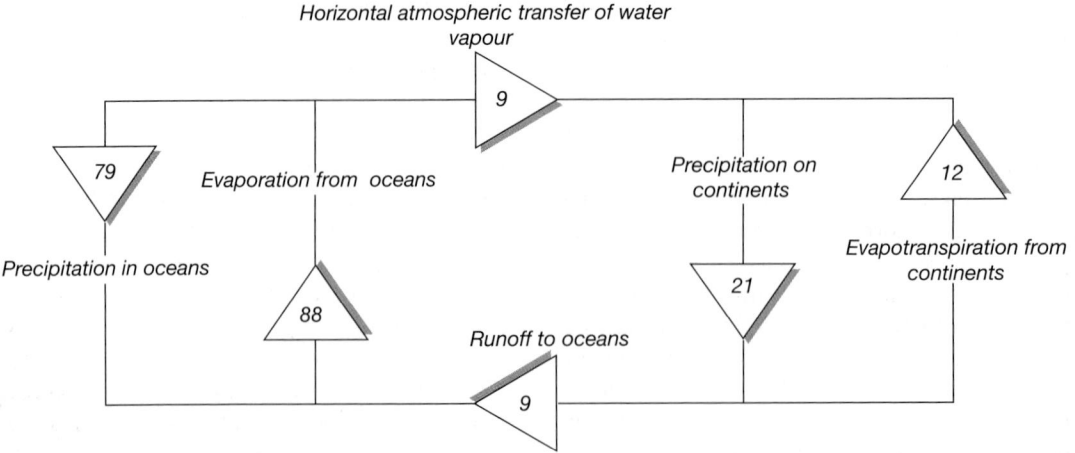

Figure 2.1   Estimated energy flows relevant to different geomorphological processes (from Summerfield, 1991).

Figure 2.2   The hydrological cycle, with figures representing percentages of the mean annual global precipitation of *c.* 1000 mm (from Summerfield, 1991).

Figure 2.3   Global mean annual precipitation (from Briggs and Smithson, 1992).

Sahara, Kalahari and Great Australian Deserts). The climate typically involves extreme temperatures — the desert can be very cold at night, especially in winter — and very rare, intense cloudbursts (Figure 2.5b).

Monsoons: the equatorial climates of West Africa, India and South East Asia are characterised by marked wet and dry seasons. The wet season, the monsoon, is generated by the seasonal migration of the equatorial low-pressure trough as the adjacent continental masses heat up in summer months. This allows moist south and south-west winds to bring intense rain (Figure 2.5c).

Hurricanes, Typhoons and Tropical Cyclones: these intense storms develop over the warmer parts of the oceans during summer months. Once developed they move westwards, within the trade winds, gradually increasing in intensity. The storms decay over land, but not before causing widespread erosion and flooding on exposed coastlines, or landslides in mountainous areas such as Hong Kong and the Philippines.

Equatorial Arid Zone: a few tropical regions lie beyond the influence of tropical storms and experience arid or semi-arid conditions. In Somalia, mean annual precipitation is below 250 mm, although the area is only just north of the equator (Figure 2.5d).

2.3 Temperate climates

Mediterranean: these regions are dominated by variable westerly winds with depressions and anticyclones. In winter, depressions are more frequent because of the migration of the sub-tropical high-pressure zones towards the equator. This leads to a climate of mild wet winters and hot dry summers (Figure 2.5e). Rainfall can be highly variable, with very dry years interspersed with sudden flooding or prolonged heavy rainfall often resulting in widespread hillside erosion and landslide activity.

Maritime: on the western side of continents in higher latitudes, the influence of depressions is felt throughout the year, producing less seasonality to the rainfall pattern (Figure 2.5f). However, extreme storms can occur. In Britain, the heaviest daily falls recorded include 297.4 mm on 18 July 1955 at

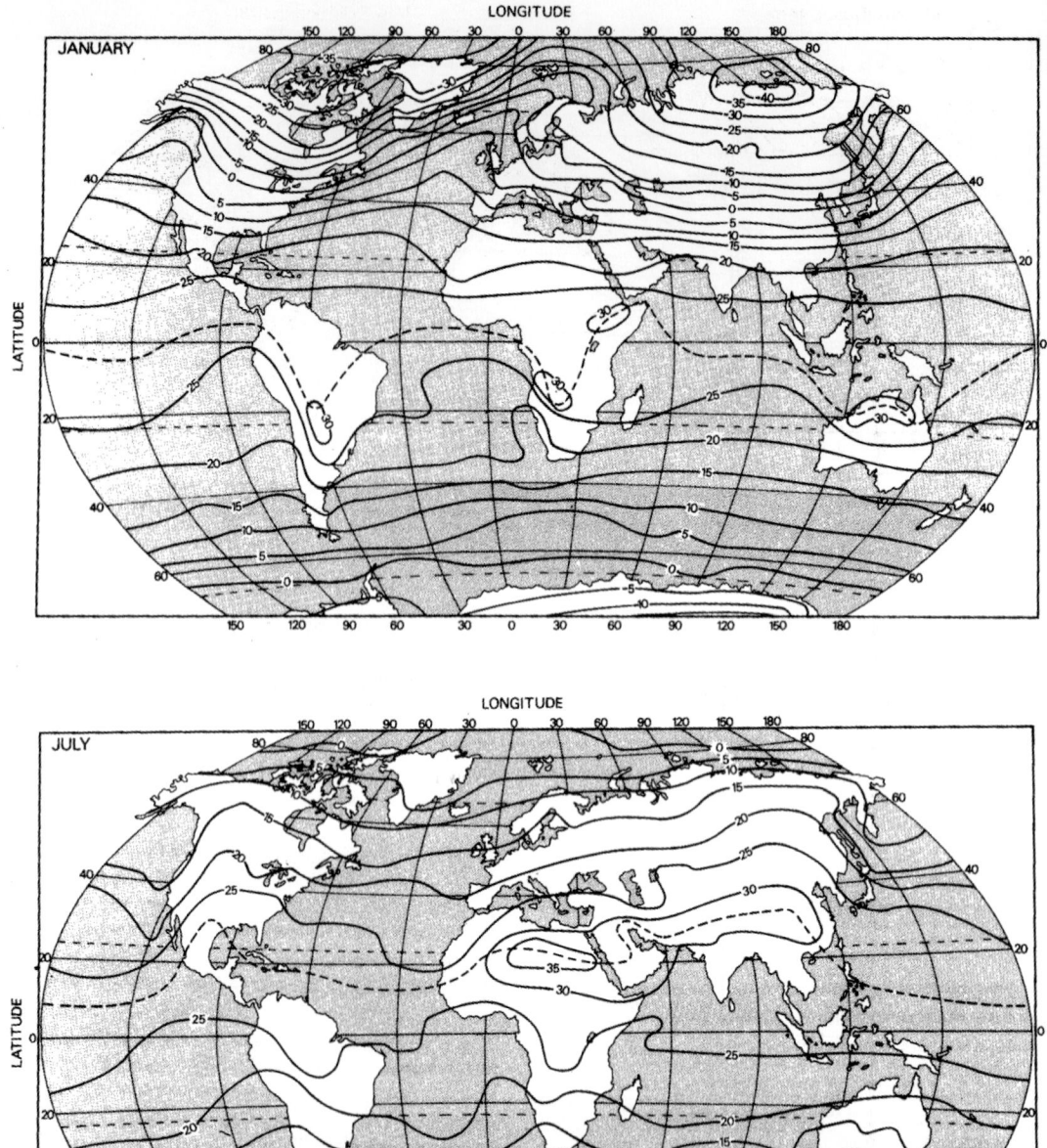

Figure 2.4   Mean sea-level temperatures in January (above) and July (below) in degrees centigrade (from Briggs and Smithson, 1992).

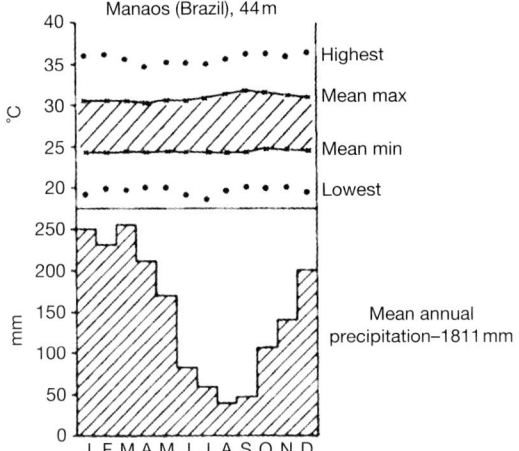

Figure 2.5a   Climate graphs for the equatorial trough zone: Manaos, Brazil (from Briggs and Smithson, 1992).

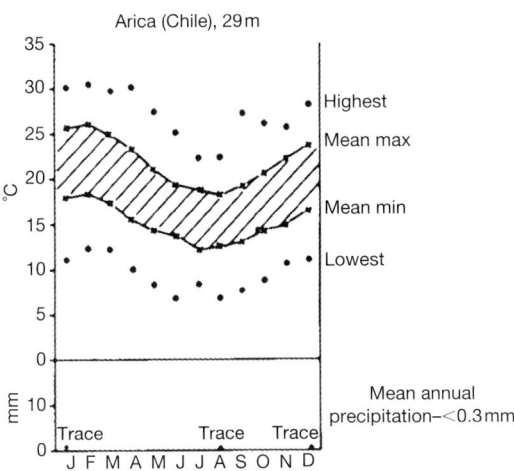

Figure 2.5b   Climate graphs for the sub-tropical high pressure zone: Arica, Chile (from Briggs and Smithson, 1992).

Martinstown, Dorset; 242.8 mm on 28 June 1917 at Bruton, Somerset and 238.8 mm on 18 August 1924 at Cannington, Somerset. The cooling influence of the sea reduces summer temperature but generally minimises the duration and severity of winter freezing.

Continental: on the eastern side of high latitude continents the prevailing airflow is offshore, so the maritime influence is reduced. Near the

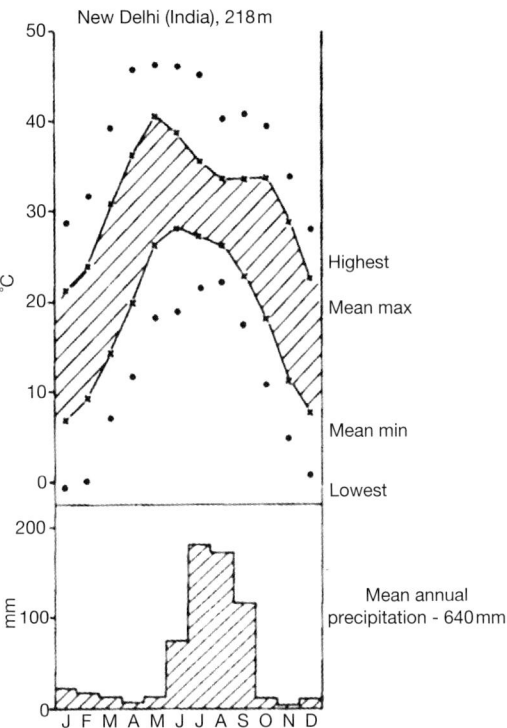

Figure 2.5c   Climate graphs for the monsoon climate: New Delhi, India (from Briggs and Smithson, 1992).

coast, this climate is associated with hot summers, cold winters and precipitation spread evenly through the year. Away from the coast, temperatures become more extreme. Winnipeg, Canada experiences a temperature range from 42 °C in the summer to −48 °C in the winter. Rainfall gradually decreases away from the coast, leading to semi-arid or arid deserts (Figure 2.5g).

Northern Extremes: relatively warm ocean currents and high latitude depression tracks allow the fringes of the maritime climate zone to extend towards the poles. In these areas the climate is cool, moist and cloudy with frequent gales (Figure 2.5h). The coastal waters generally remain ice-free.

2.4 Polar climates

Tundra: this region is associated with permanently frozen ground (permafrost) and a characteristic vegetation of mosses, lichens, sedges, grasses and

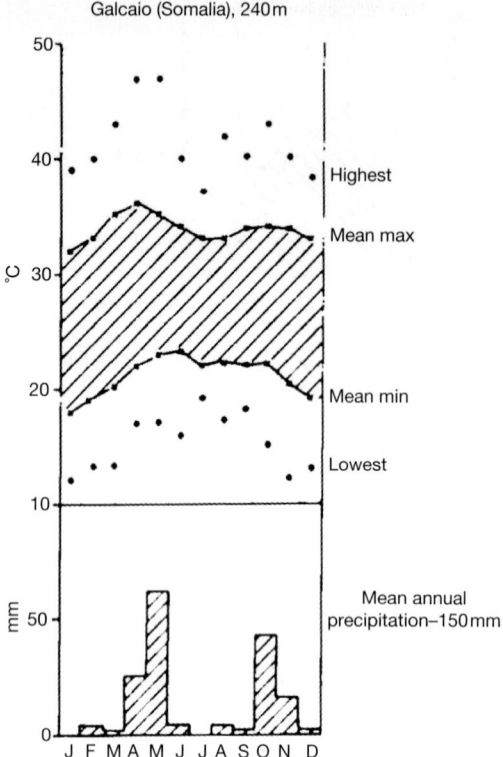

Figure 2.5d   Climate graphs for the equatorial dry zone: Galcaio, Somalia (from Briggs and Smithson, 1992).

a few birch trees. Winter temperatures are very low. Snowfalls may be low because of continental effects. Summer temperatures can rise as high as 30 °C, although precipitation remains low (Figure 2.5i).

Arctic: this region is dominated by sea-ice and experiences low winter temperatures and precipitation. In summer, 24-hour daylight allows increased receipt of solar radiation. However, much of the energy is spent on melting surface ice and evaporation, so that temperatures rarely rise above 5 °C. The combination of moisture from melting ice and relatively low temperatures leads to dense, saturated air and abundant cloud cover. However, precipitation remains low because conditions for rising air are infrequent.

Antarctic: the effects of latitude, altitude (over 50% of the continent is above 2000 m) and isolation from other continents results in much of the region

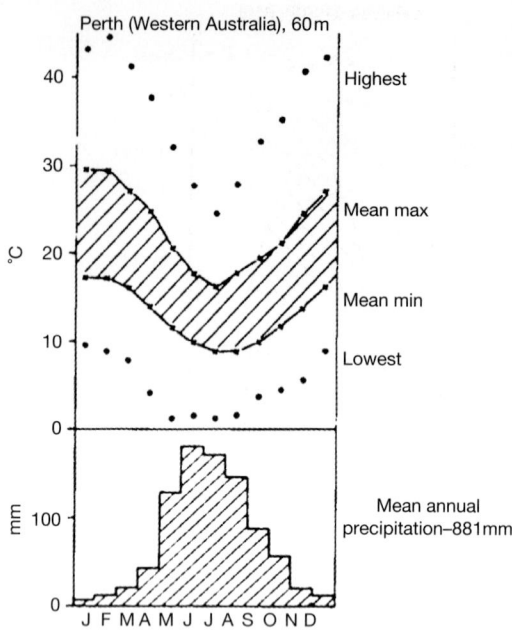

Figure 2.5e   Climate graphs for the mediterranean climate: Perth, Australia (from Briggs and Smithson, 1992).

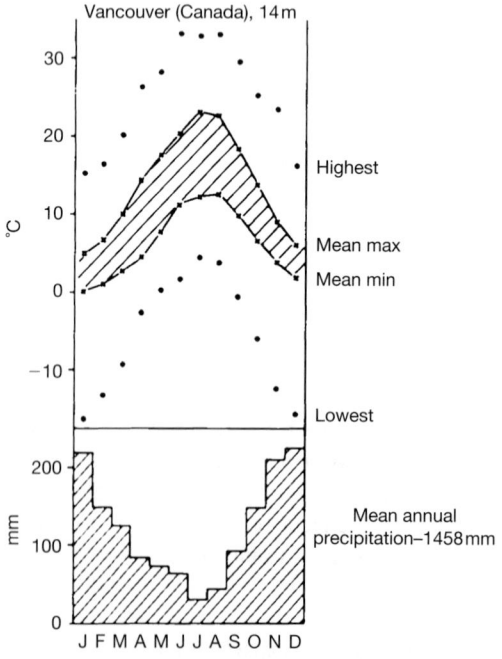

Figure 2.5f   Climate graphs for the maritime climate, Vancouver, Canada (from Briggs and Smithson, 1992).

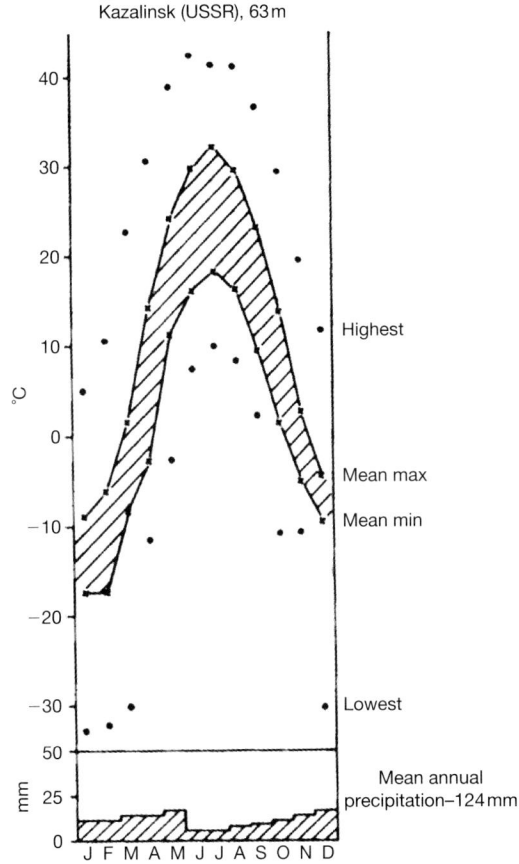

Figure 2.5g   Climate graphs for the continental climate, Kazalinsk (CIS, former Soviet Union) (from Briggs and Smithson, 1992).

being ice covered. Depressions rarely penetrate into the region, although low pressure prevails because of the dense cold air. Precipitation is very low. Cold temperatures prevail throughout the year, seldom rising above 0 °C (Figure 2.5j). Persistent, strong winds blow off the ice caps, reaching average speeds of over 6 m/s. Near the coast the winds strengthen as they are funnelled by steep glacier valleys. At Cape Denison, for example, the annual mean wind speed is around 20 m/s.

2.5 Climate classification

The classifications developed by Köppen (1931; 1936) and Thornthwaite (1933) provide a useful

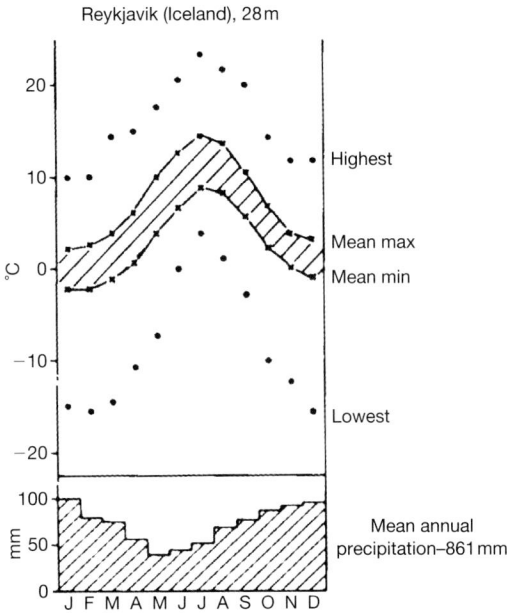

Figure 2.5h   Climate graphs for the maritime northern extreme, Reykjavik, Iceland (from Briggs and Smithson, 1992).

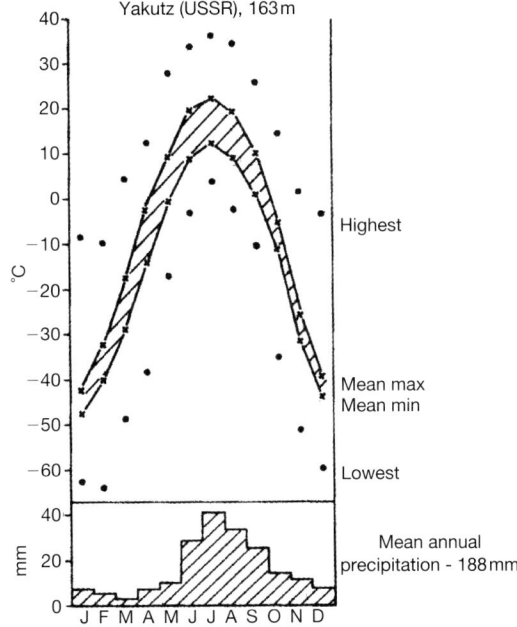

Figure 2.5i   Climate graphs for the tundra climate, Yakutz (Siberia) (from Briggs and Smithson, 1992).

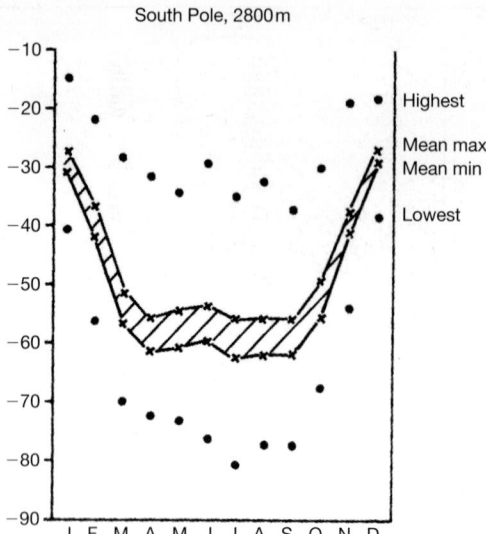

Figure 2.5j   Climate graphs for the Antarctic climate: South Pole (from Briggs and Smithson, 1992).

guide to support the development of a preliminary ground model for a region. The latter is based on a combination of effective precipitation (i.e. precipitation minus evaporation) and thermal efficiency. The former is related to the *P/E* index (i.e. the total monthly rainfall is divided by the total monthly evaporation summed for each of the twelve months of the year). Five humidity provinces were proposed:

Humidity province	Characteristic vegetation	*P/E* index
A Wet	Rainforest	> 128
B Humid	Forest	64–127
C Sub-humid	Grassland	32–63
D Semi-arid	Steppe	16–31
E Arid	Desert	< 16

These five provinces can be further sub-divided on the basis of rainfall seasonality:

r   rainfall abundant in all seasons
s   rainfall deficient in summer
w   rainfall deficient in winter
d   rainfall deficient in all seasons.

The thermal efficiency is expressed as the *T/E* index, defined as the monthly temperature divided by the total monthly evaporation summed for each of the twelve months of the year. Six temperature provinces can be defined:

Temperature province	*T/E* index
A′ Tropical	> 128
B′ Mesothermal	64–127
C′ Microthermal	32–63
D′ Taiga	16–31
E′ Tundra	1–15
F′ Frost	0

These two indexes were combined to produce a world map showing thirty-two climatic types (Figure 2.6). For example, the climate of western Britain is described as BC′r, whereas the Amazon Basin is BA′r.

The important influence of climate on geomorphological processes has been expressed in terms of morphoclimatic zones within which there are distinctive landform assemblages and dominant processes (e.g. Figure 2.7; Table 2.1). Two broad groupings of morphoclimatic regions can be identified:

1. First-order morphoclimatic zones: glacial, arid and humid tropical. These have non-seasonal processes, generally low average erosion rates, highly infrequent and episodic erosive activity (e.g. desert rainstorms).

2. Second-order morphoclimatic zones: humid–arid tropical, semi-arid tropical, dry continental, wet mid-latitude and periglacial. These have distinctly seasonal processes. In regions with warmer climates, the processes differ most significantly in terms of the length of the wet season. In cooler regions, variations in summer temperature and precipitation are important.

While these morphoclimatic zones provide a starting point for developing global or site-scale models of anticipated ground conditions, it can be misleading to seek a link between present-day climate and a particular landscape. All current processes act on a stage fashioned by ancient processes (i.e. many landscapes comprise a

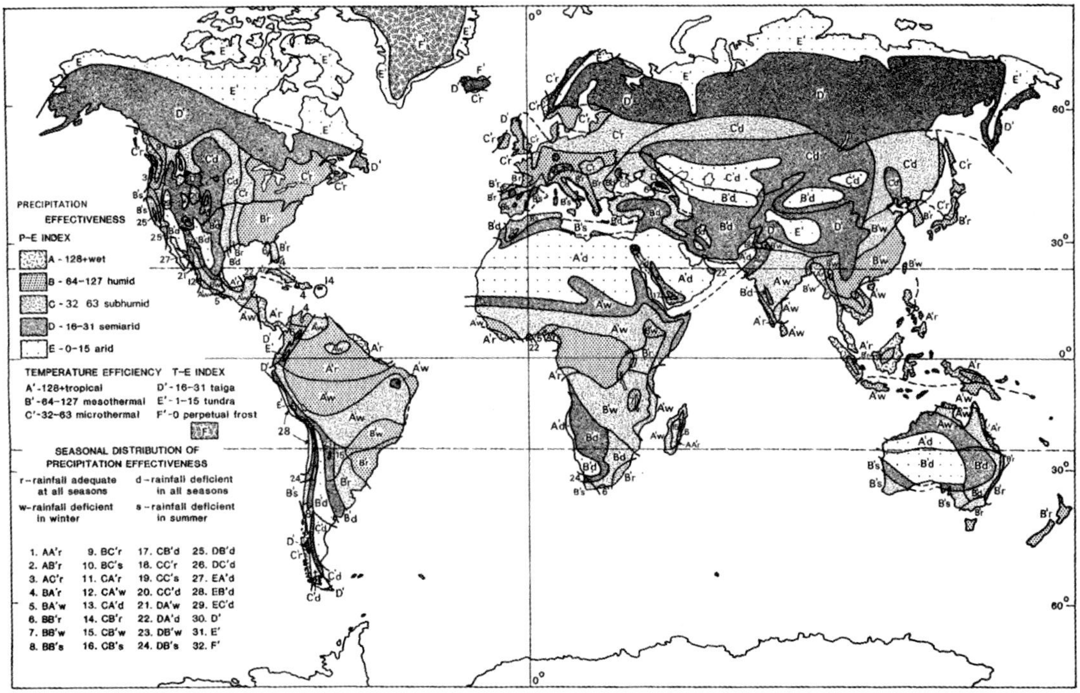

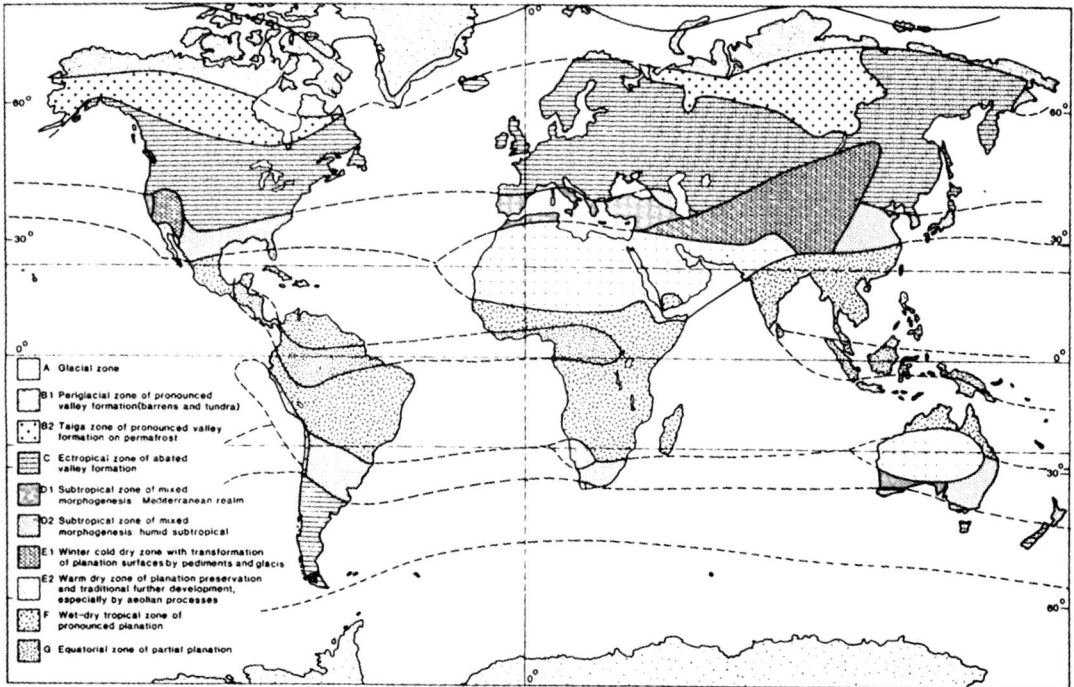

Figure 2.6 World climate based on Thornthwaite's classification (from Doornkamp, 1986).

Figure 2.7 World morphoclimatic zones, excluding highlands (from Kiewiet de Jonge, 1984).

Table 2.1   Major morphoclimatic zones (from Fookes *et al.*, 2000).

Morphoclimatic zone	Mean annual temperature (°C)	Mean annual precipitation (mm)	Relative importance of geomorphological processes
Azonal mountain zone	Highly variable	Highly variable	Rates of all processes vary significantly with altitude; mechanical and glacial action become significant at high elevations.
Glacial	<0	0–1000	Mechanical weathering rates (especially frost action) high; chemical weathering rates low; mass movement rates low except locally; fluvial action confined to seasonal melt; glacial action at a maximum; wind action significant.
Periglacial	−1 to +2	100–1000	Mechanical weathering very active with frost action at a maximum; chemical weathering rates low to moderate; mass movement very active; fluvial processes seasonally active; wind action rates locally high. Effects of repeated formation and decay of permafrost.
Wet mid-latitude	0–20	400–1800	Chemical weathering rates moderate, increasing to high at lower latitudes; mechanical weathering activity moderate with frost action important at higher latitudes; mass movement activity moderate to high; moderate rates of fluvial processes; wind action confined to coasts.
Dry continental	0–10	100–400	Chemical weathering rates low to moderate; mechanical weathering, especially frost action, seasonally active; mass movement moderate and episodic; fluvial processes active in wet season; wind action locally moderate.
Hot dry (arid tropical)	10–30	0–300	Mechanical weathering rates high (especially salt weathering); chemical weathering minimal; mass movement minimal; rates of fluvial activity generally very low but sporadically high; wind action at maximum.
Hot semi-dry (semi-arid tropical)	10–30	300–600	Chemical weathering rates moderate to low; mechanical weathering locally active especially on drier and cooler margins; mass movement locally active but sporadic; fluvial action rates high but episodic; wind action moderate to high.
Hot wet-dry (humid-arid tropical)	20–30	600–1500	Chemical weathering active during wet season; rates of mechanical weathering low to moderate; mass movement fairly active; fluvial action high during wet season with overland and channel flow; wind action generally minimal but locally moderate in dry season.
Hot wet (humid tropical)	20–30	> 1500	High potential rates of chemical weathering; mechanical weathering limited; active, highly episodic mass movement; moderate to low rates of stream corrasion but locally high rates of dissolved and suspended load transport.

mosaic of features of different origins and ages, some of which may be of considerable antiquity). Current processes may be gradually changing this inherited relief, rapidly destroying it or having no effect on it at all. The presence of relict, deep-seated rotational landslides on scarp faces in the Sahara (Figure 16.6, Chapter 16) highlights the importance of looking beyond the present-day climate–process associations in developing a comprehensive ground model.

2.6 Climate variability

Climate has not remained constant throughout the long history of the Earth (Figure 6.1), primarily due to variations in the orbit around the Sun and solar radiation. Other possible mechanisms causing rapid climatic changes include the cold iceberg melting events (7000–13 000 year intervals: so-called 'Heinrich' events) and the warm cycles (1000–3000 year intervals, 1500-year duration: so-called 'Dansgaard-Oeschger' cycles). However, there has been a growing appreciation of the link between plate tectonics and climate changes over tens of millions of years (Figure 2.8). This arises from variations in the rate of sea-floor spreading and the release of CO_2 that, along with other factors, leads to adjustments in global temperatures.

In addition to the climatic changes that occur over geological time (e.g. the repeated glacial and interglacial episodes of the Quaternary), there are important climatic variations that influence geomorphological processes over shorter periods (Figure 2.9; Brunsden, 1996). Over a 10 000-year period average temperatures can change by c. 5–10 °C with recognisable sequences of cool–wet and warm–dry periods. For example, there have been several major fluctuations over the Holocene including the cooling of c. 8000–4800 BP, the cool event at c. 3500 BP and the move from the warm-dry sub-Boreal to the cooler-wetter sub-Atlantic of c. 2500 BP. These events caused significant changes in rates of geomorphological activity.

Over the 1000-year period, major climatic trends may be due to solar activity variation related to sunspot cycles, North Atlantic Oscillation (NAO) and El Niño Southern Oscillation (ENSO) events, volcanic eruptions, and stochastic variations (Barber et al., 1994; Roberts, 1998; Waple, 1999). There are also sustained periods of climate change lasting several hundreds of years or more. Over the last millennium the UK has experienced two important natural perturbations on the 100-year scale, the so-called 'Little Optimum' (or Medieval Warm Period) warming of c. 700–1200 AD (Lamb, 1997) and the 'Little Ice Age' (widest definition 1420–1850 AD with a period of amelioration from 1500–1550 AD). There is quite strong coincidental evidence that these are related to the 'Grand Maximum' and the two great minima of sunspot activity, the Spörer and the Maunder Minima (Adams et al., 1999). The 'Little Ice Age' was associated with an increase in debris flow and avalanche activity (Grove, 1988).

Shorter episodes such as droughts or sequences of wet years can give rise to noticeable changes in the rate of geomorphological processes. In the Caspian Sea a post-1975 rise in sea-level of over 2 m (reversing the trend of falling sea-levels; Figure 2.10) that is believed to be associated with the development of a more humid climate in the region has been accompanied by a 40% increase in the extent of coastal erosion (Bird, 2000).

2.7 Extreme events

It is common for significant geomorphological changes and natural hazards to be associated with extreme climatic events, be they hurricanes, intense rainstorms or periods of heavy rainfall or strong winds. These climatic events can be regarded as a random part of a natural series of events of varying magnitude and frequency, whose distribution can be established from a sequence of records and whose probability of occurrence in a given period can be calculated using standard statistical methods. For example, the likelihood of a rainfall of a particular magnitude is generally expressed by the return period or recurrence interval. Thus, the rainfall that is expected to be equalled or exceeded on average

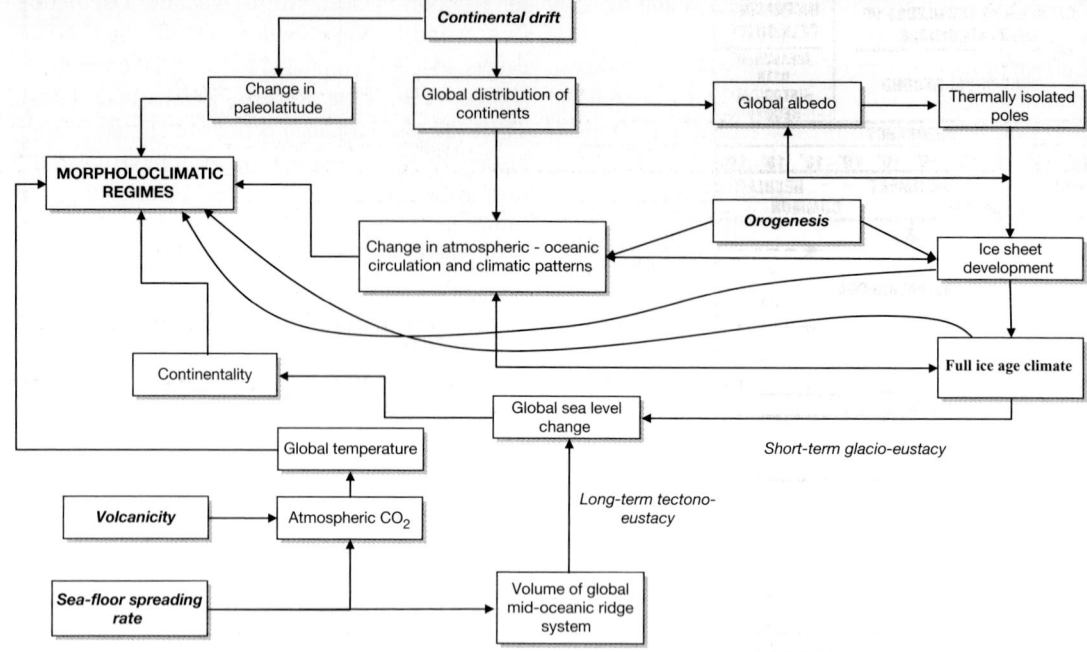

Figure 2.8   A schematic representation of some of the ways in which tectonic processes generate morphoclimatic zone change (from Summerfield, 1991).

every 100 years has a return period of 100 years. This event could occur any year, but the probability of its occurrence during 100 years is much greater than during a one-year period.

The likelihood of a particular rainfall event (e.g. an annual rainfall total) occurring in any single year (i.e. the return period) may be calculated from a series of events (i.e. annual rainfall totals over many years) as follows:

$$\text{Return period} = \frac{\begin{array}{c}\text{Number or years}\\ \text{in the sequence}\end{array} + 1}{\begin{array}{c}\text{Ranking of the event}\\ \text{in the sequence}\end{array}}$$

The relationship between probability (expressed as a percentage), return period and the length of period under consideration is shown in Table 2.2. This indicates that a 1000-year event has a 9% chance of occurring during the 100-year lifetime of an engineering structure. The likelihood of encountering an event equal to or exceeding a

rainfall with return period T during a time period of N years is given by the following equation:

$$P = 1 - (1 - 1/T)^N$$

This provides the 'encounter probability' which is useful for communicating risk as it is less likely to be misunderstood than return period.

2.8 Climate and flooding

Climate is a major control of flooding events, along with other factors such as landslides and dam failures, and flood intensifying factors (Figure 2.11). Figure 2.12 presents a global classification of flood zones. The latitude, position and strength of the circumpolar vortex (or jet stream) that controls depression tracks govern the occurrence of floods in the high mid latitudes. Changes in atmospheric circulation can lead to changes in flood frequency. More frequent floods in the Upper Mississippi valley during the late

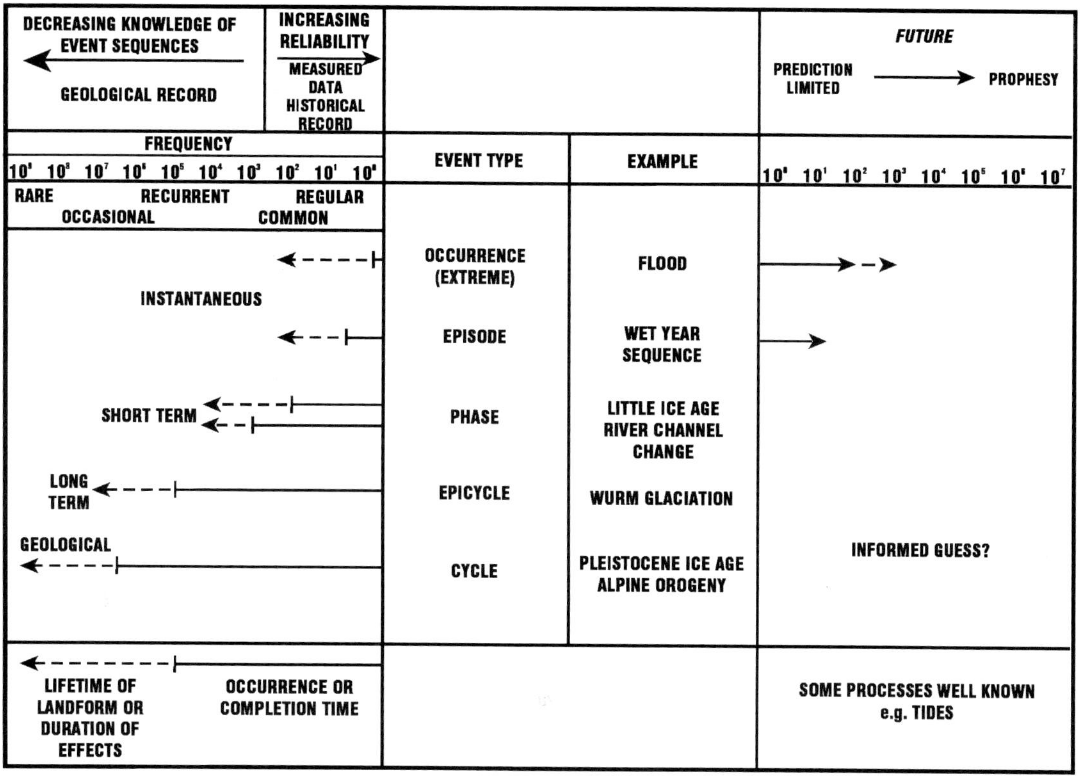

DECREASING KNOWLEDGE OF EVENT SEQUENCES
GEOLOGICAL RECORD
INCREASING RELIABILITY
MEASURED DATA
HISTORICAL RECORD
FUTURE
PREDICTION LIMITED
PROPHESY
FREQUENCY
10⁹ 10⁸ 10⁷ 10⁶ 10⁵ 10⁴ 10³ 10² 10¹ 10⁰
EVENT TYPE
EXAMPLE
10⁰ 10¹ 10² 10³ 10⁴ 10⁵ 10⁶ 10⁷
RARE
RECURRENT
REGULAR
OCCASIONAL
COMMON
INSTANTANEOUS
OCCURRENCE (EXTREME)
FLOOD
EPISODE
WET YEAR SEQUENCE
SHORT TERM
PHASE
LITTLE ICE AGE RIVER CHANNEL CHANGE
LONG TERM
EPICYCLE
WURM GLACIATION
GEOLOGICAL
CYCLE
PLEISTOCENE ICE AGE ALPINE OROGENY
INFORMED GUESS?
LIFETIME OF LANDFORM OR DURATION OF EFFECTS
OCCURRENCE OR COMPLETION TIME
SOME PROCESSES WELL KNOWN e.g. TIDES

Figure 2.9 Timescales and terminology of geomorphological events (redrawn from Brunsden, 1996).

SEA LEVEL CHANGES
Caspian Sea
Metres below Mediterranean sea level
26
27
28
29
1930 1935 1940 1945 1950 1955 1960 1965 1970 1975 1980 1985 1990 1995 2000
Year

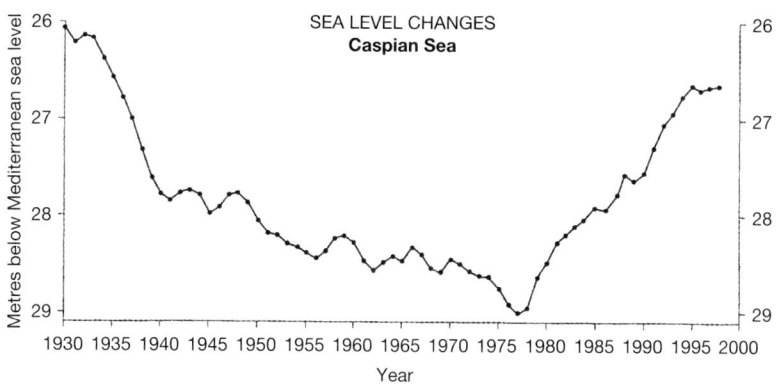

Figure 2.10 Changes in the level of the Caspian Sea since 1930, as indicated by the Baku tide gage, Azerbaijan (from Bird, 2000).

1800s and since 1950 were due to a weak westerly circulation in mid latitudes (Knox, 1984; 1988). In low latitudes floods and/or drought events are associated with tropical storms, the monsoon and the El Niño Southern Oscillation (ENSO). The latter is associated with high sea-surface temperature anomalies that spread eastwards across the equatorial Pacific Ocean. ENSO events occur every 3–10 years and trigger extreme floods and droughts in many regions (Figure 2.13).

Table 2.2   Percentage probability of the N-year rainfall occurring in a particular period.

Number of years in period	\multicolumn{8}{c}{N = average return period in years}							
	5	10	20	50	100	200	500	1000
1	20	10	5	2	1	0.5	0.2	0.1
5	67	41	23	10	4	2	1	0.5
10	89	65	40	18	10	5	2	1
30	99	95	79	45	26	14	6	3
60	–	98	95	70	31	26	11	6
100	–	99.9	99.4	87	65	39	18	9
300	–	–	–	99.8	95	78	45	26
600	–	–	–	–	99.8	95	70	45
1000	–	–	–	–	–	99.3	87	64

Where no figure is inserted the percentage probability >99.9.
In the shaded box: there is, on average, a 9% chance that a 1000-year event (annual probability of 0.001) would occur within a 100 year time period.

An important consequence of short-term climatic variation is that the probability of flood events may not be constant over time. Many of the world's major rivers experienced 'humid' and 'dry' periods throughout the last century (Probst and Tardy, 1987). A degree of caution is needed, therefore, when using return period statistics as they assume that the probability of flooding (or any other geohazard) of a given magnitude remains constant from year to year, over the historical record.

2.9 Global climate change

Global climatic changes are occurring as the result of man-induced accumulation of so-called greenhouse gasses such as CO_2 in the atmosphere. Evidence from ice cores supplemented by direct measurements since the mid-1950s shows a steady rise in greenhouse gas concentrations from the late 1700s changing to a rapid rise post 1950. Atmospheric concentrations of carbon dioxide (CO_2), the primary anthropogenic greenhouse gas, have risen from about 270 ppm in pre-industrial times to over 360 ppm.

It is expected that the mean temperature will rise by 1–3.5 °C by 2100, accompanied by a sea-level rise of 15–95 cm (IPCC, 2001). Warming at the higher end of this range would shift climatic zones poleward by about 550 km. Precipitation has increased by 0.5 to 1% per decade in the twentieth century over the Northern Hemisphere continents and there has been a 2 to 4% increase in the frequency of heavy precipitation events.

The impacts of climate change will vary from region to region. The UK Climate Impacts Programme (Hulme *et al.*, 2002), for example, has predicted that the climate changes as a result of human activity over the next century are expected to increase risks from coastal flooding and erosion in two ways:

1. by the 2050s the rise in sea level is predicted to increase the frequency of extreme high water levels from once a century to, typically, once a decade — this situation would be further exacerbated if storminess were to increase
2. days with heavy rainfall will become typically three or four times more common, increasing the risk of non-tidal floods in estuaries and promoting accelerated cliff instability.

2.10 The hydrological cycle

The hydrological cycle (Figure 2.2) provides an important link between climate, geomorphological processes and weathering, particularly the

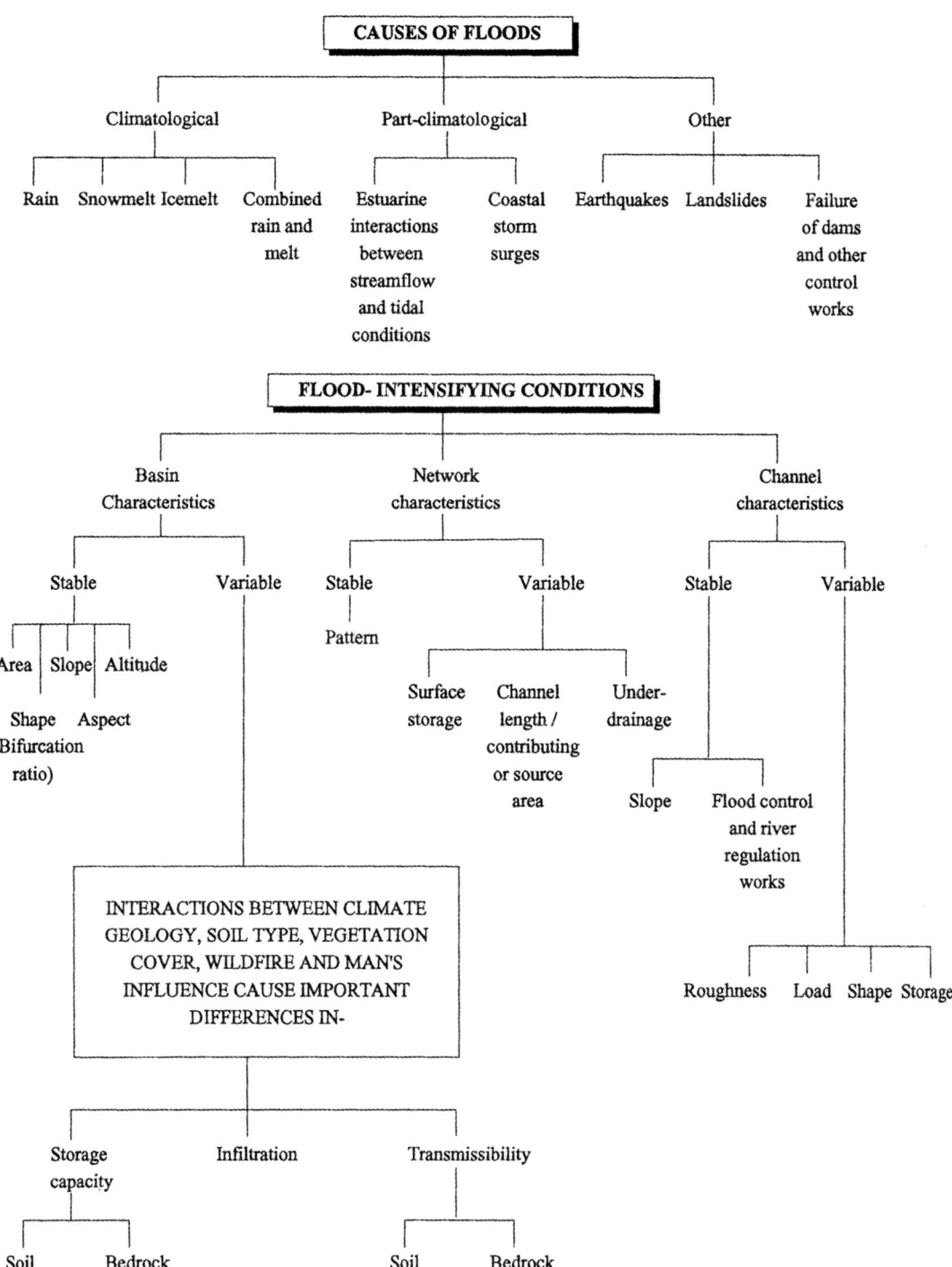

Figure 2.11   Causes of floods and flood intensifying factors (after Ward, 1978).

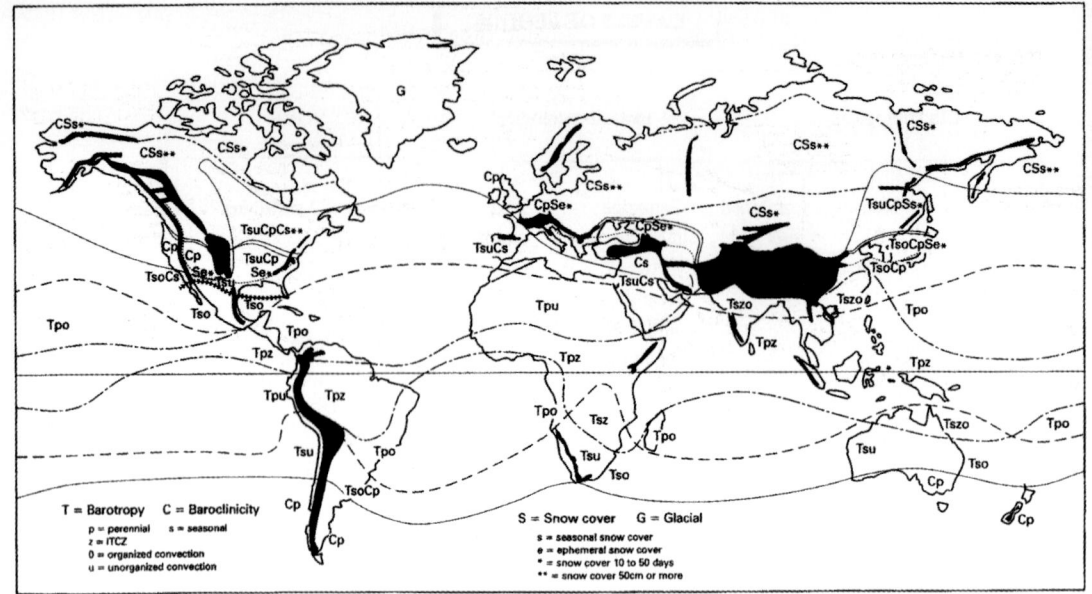

Figure 2.12   Flood climate regions of the world (from Hayden, 1988; Macklin and Lewin, 1997).

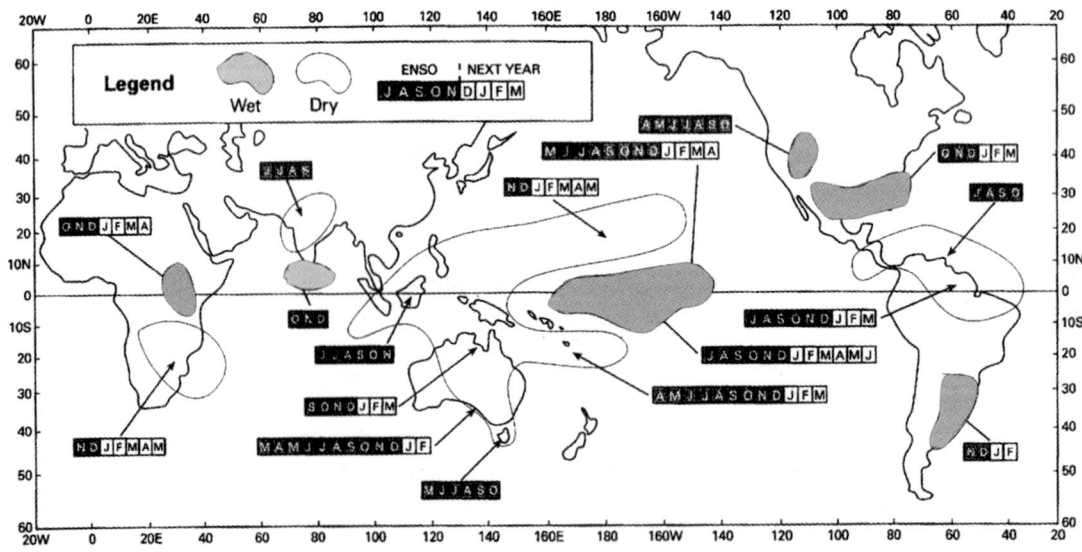

Figure 2.13   Regions affected month by month by the El Niño Southern Oscillation (from Folland, *et al.*, 1990; Macklin and Lewin, 1997).

balance between runoff, infiltration into the soil and subsequent evaporation, through-flow or percolation towards the groundwater table (Figure 2.14). The balance between these processes can control the occurrence of geohazards such as

flooding (e.g. the speed of transmission of rainfall into the drainage channels), gully formation (e.g. the concentration of runoff), soil piping (e.g. concentration of through-flow), landslides (e.g. the rapid build up of pore water pressures in

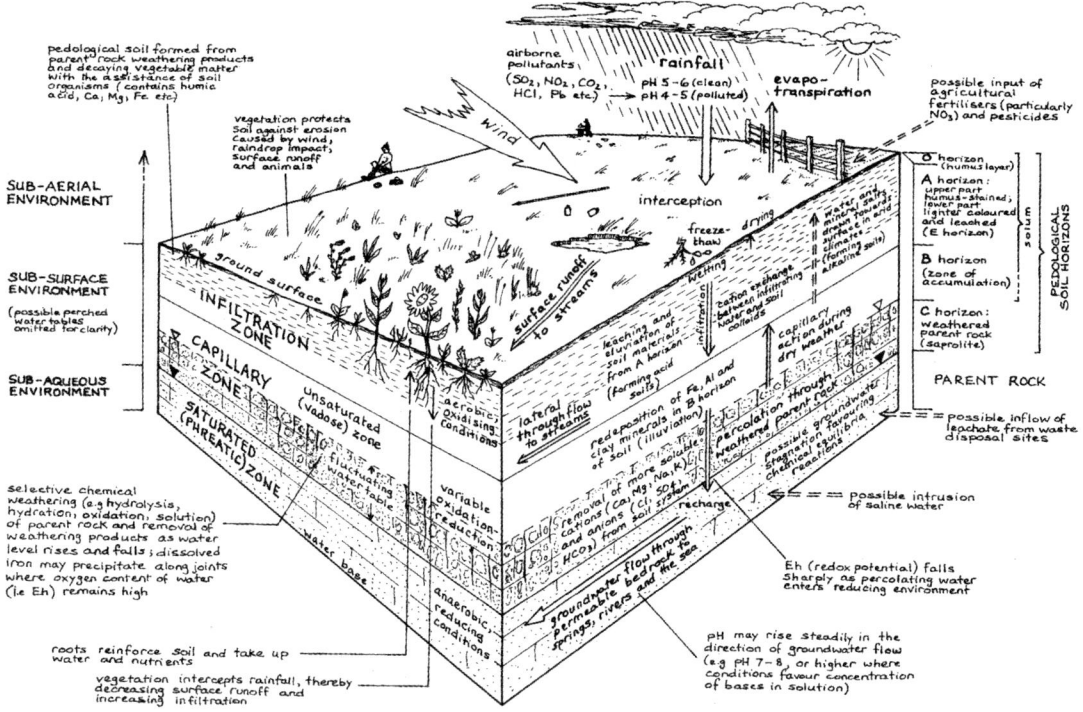

Figure 2.14   Idealised characteristics of near-surface hydrological environments (from Fookes, 1997).

soils prone to shallow slides or the slow rise of the groundwater table in areas prone to deep-seated slides), high groundwater tables, the height of capillary rise and associated aggressive soil problems. Table 2.3 highlights some engineering-related issues associated with climate and groundwater.

2.11 Weathering

Weathering involves the gradual breakdown (i.e. disintegration) and alteration (i.e. decomposition) of materials through a combination of chemical and physical (mechanical) processes (Table 2.4; Figure 2.15), leading to the development of weathering profiles and residual soils. The relative significance of any weathering process depends on the weathering environment (e.g. climate), the nature of the geological materials and biological conditions.

Climate has an important control on both the nature and rate of weathering. Strakhov (1967)

expressed the global variation in weathering in terms of a series of weathering zones related to temperature and precipitation (Figures 2.16 and 2.17). Peltier (1950) used mean annual temperature and mean annual precipitation to define zones characterised by different intensity of chemical and frost weathering (Figure 2.18). While both approaches allow a broad indication of the anticipated weathering regime in a region, at a detailed level weathering will reflect site conditions, notably lithology and structure of the parent material (including the features inherited from past climates), availability of water (e.g. groundwater or through-flow), vegetation and human activity.

2.12 Weathering and landscape

Weathering plays an important role in landscape development, notably etchplain formation, duricrusts and karst landforms.

Table 2.3   Some engineering issues associated with climate and groundwater (modified from Fookes, 1997).

Condition	Comments
Annual evapotranspiration exceeds infiltration	Low water tables controlled by stream levels and presence of more permeable strata, upward flow due to evaporation. Partly saturated soils, sometimes to great depth. Long-term desiccation. Possibility of collapse on wetting (or when evaporation is prevented by sealing the ground surface) in porous residual soils, weathered sands, loess, and heave on wetting of plastic clay. Soil suctions facilitate construction (e.g. steep temporary slopes, shafts) Erosion protection of slopes by vegetation often impractical. Chemically active zone at ground level due to evaporation of ground water, recementation and cap-rock. Tendency for flashy sheet floods and ephemeral channel flows in response to cloudbursts. Tendency for shallow landslides (e.g. rock and debris slides) and debris flows in response to cloudbursts.
Annual infiltration exceeds evapotranspiration	Deep percolation dominates. High groundwater tables usual in fine-grained soils, even in slopes. Saturated soils. Flow into excavations and underground works. Ground movements due to dewatering and internal erosion. Erosion protection of slopes by vegetation usually practical. Flood character varies with catchment type and antecedent conditions, but can range from flash floods in steep catchments to prolonged lowland floods. Landslide character varies with slope conditions (e.g. geology, slope angle, antecedent conditions), but includes deep-seated failures on clay slopes or slopes with mudrock beds.
Seasonal shorter term variations	Evapotranspiration and infiltration rates often vary seasonally (and from year to year) and also over shorter timescales. Surface conditions and water contents (and related engineering properties of superficial fine-grained soils) vary accordingly. Seasonal heave and settlement of superficial clays and foundations. Variation of superficial undrained strength of clays and suitability as embankment foundations. Heave and settlement of superficial structural foundations depend on conditions prior to construction. Undrained strength and suitability of superficial fine-grained soils as fill also varies. Conditions during site investigation may vary from those during construction. Earthmoving of fine-grained soils very difficult when rainfall exceeds evapotranspiration. Seasonal variations control earth moving season and short-term variation controls working time. These variations depend on temperature, wind and rainfall duration, rather than on rainfall amount, problems usually worst in temperate climates and are often strongly influenced by site elevation. Supplies of water for conditioning fills may vary seasonally. Seasonal variation in probability of flood, erosion and landslide events. Risk greatest during intense cyclonic storms and periods of snowmelt or prolonged rainfall, although antecedent conditions will determine whether events actually occur.

Table 2.4   Classification of weathering processes (from Cooke and Doornkamp, 1990).

Processes of disintegration	Processes of decomposition

Crystallisation processes:
    Salt weathering (crystal growth,
    hydration, thermal expansion)
    Frost weathering
	Hydration and hydrolysis

Temperature/pressure change processes:
    Insolation weathering
    Sheeting, unloading
	Oxidation and reduction

Weathering by wetting and drying:
    Moisture swelling
    Alternate wetting and drying
    Water-layer weathering
	Solution, carbonation, sulphation

Organic processes:
    Root wedging
    Colloidal plucking
    Lichen activity
	Chelation

	Bio-chemical changes
	    Micro-organism decay
	    Bacteria
	    Lichens

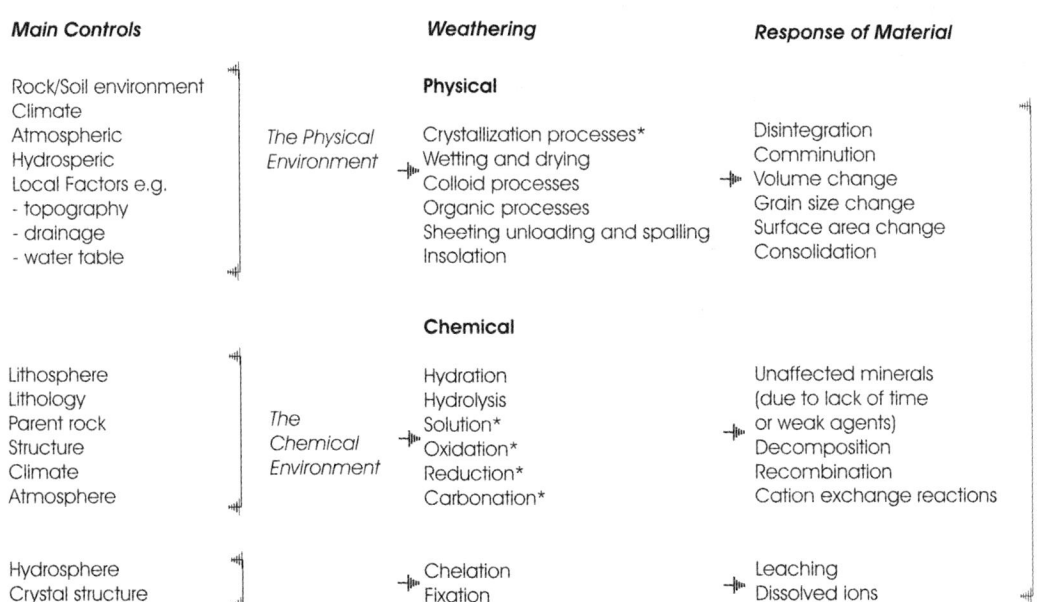

Main Controls	Weathering	Response of Material
Rock/Soil environment Climate Atmospheric Hydrosperic Local Factors e.g. - topography - drainage - water table	**Physical** *The Physical Environment* Crystallization processes* Wetting and drying Colloid processes Organic processes Sheeting unloading and spalling Insolation	Disintegration Comminution Volume change Grain size change Surface area change Consolidation
Lithosphere Lithology Parent rock Structure Climate Atmosphere	**Chemical** *The Chemical Environment* Hydration Hydrolysis Solution* Oxidation* Reduction* Carbonation*	Unaffected minerals (due to lack of time or weak agents) Decomposition Recombination Cation exchange reactions
Hydrosphere Crystal structure	Chelation Fixation	Leaching Dissolved ions

Residual Soil Profile

* Indicates those processes considered to be most applicable over an engineering timescale

Figure 2.15   Factors and processes important in weathering (adapted from Brunsden, 1979).

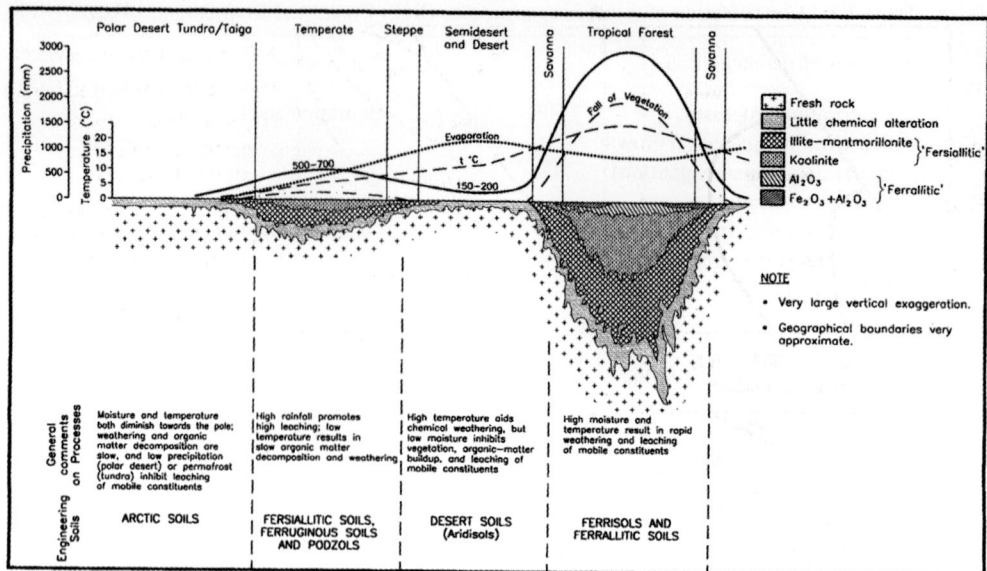

Figure 2.16 Simple cross-section from pole to equator showing climate and weathering characteristics (adapted from Strakhov, 1967).

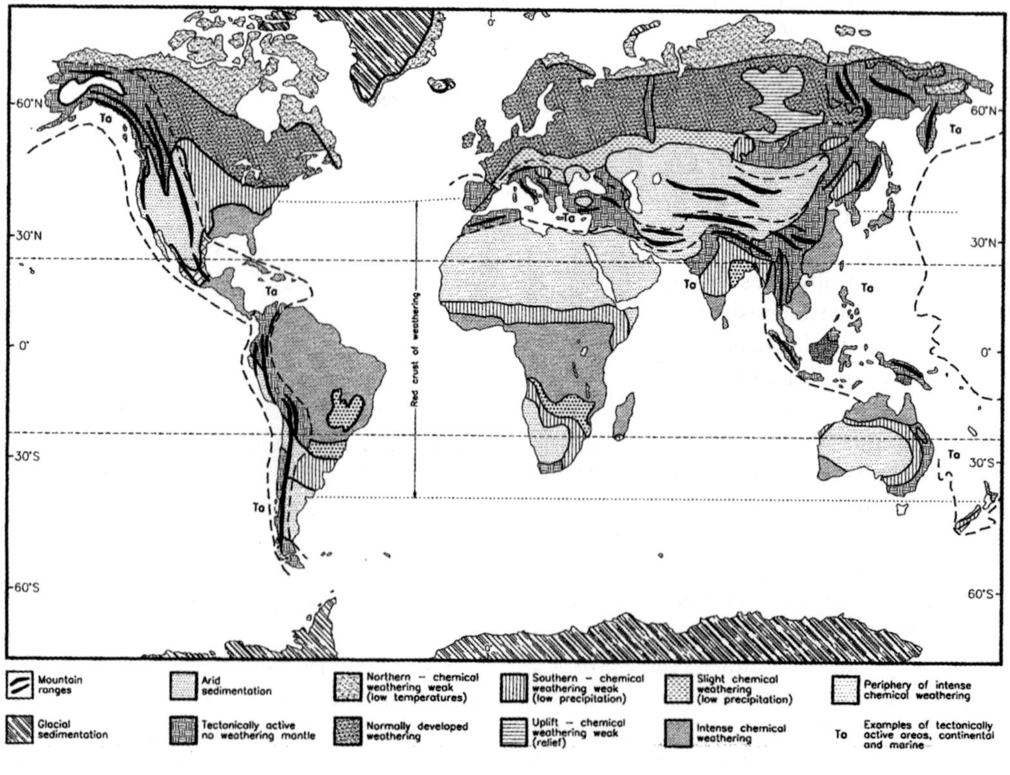

Figure 2.17 World weathering zones (adapted from Strakhov, 1967).

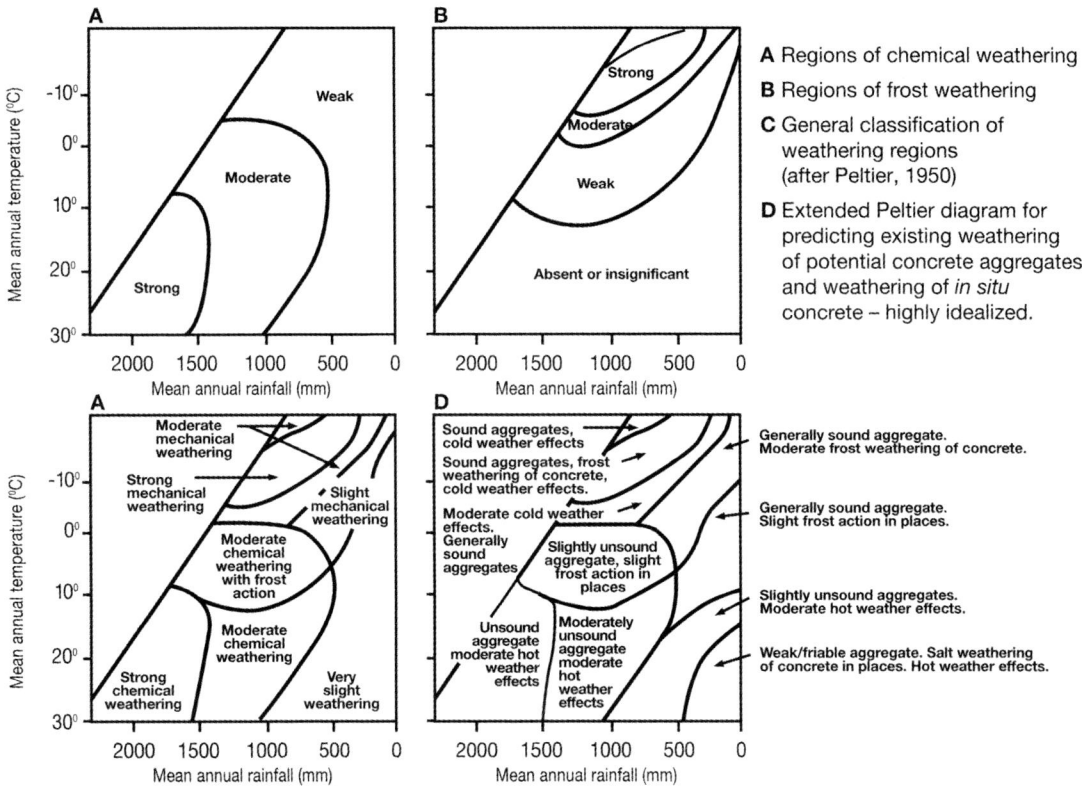

Figure 2.18 Characteristic climate and weathering patterns, *A–C* redrawn from Peltier (1950); *D* redrawn from Fookes (1980).

Etchplain development

In many areas, especially the humid tropics, landscape denudation involves a combination of ongoing chemical decomposition at the weathering front under the influence of groundwater through-flow, and the removal of weathered material from the ground surface by sheet-wash or wind erosion (Figure 2.19). Etch forms can cut across lithologies and can result in deep and intensely weathered residuals soils (Twidale, 1990). Acting over millions of years, etching can result in remarkably flat plains and plateau surfaces, such as those developed in limestones in the Sahara and in granites and gneiss in Western Australia and Namibia. Climate change or tectonic uplift can stimulate incision through parts of the etchplain surface, resulting in a pattern of soils that appear to be unrelated to the current climate or landscape.

Duricrusts

Duricrusts are indurated horizons at or near the ground surface, generally formed through the accumulation of iron or aluminium (ferricrete or alcrete), silica (silcrete), calcium carbonate (calcrete), magnesium carbonate (dolocrete) or gypsum (gypcrete). Ferricretes and alcretes form by the relative accumulation of iron and aluminium oxides in the soil, as more mobile compounds are leached out of the weathering profile. For this reason, they tend to be associated with high rainfall climates (Chapter 18). The other duricrusts form through an absolute accumulation in the weathering profile. Accumulation may occur as a result of capillary rise (*per ascensum* model), downward percolation (*per descensum* model) or through-flow of solute-rich groundwater. These duricrusts are frequently associated with desert environments (Chapter 16). Duricrusts are generally

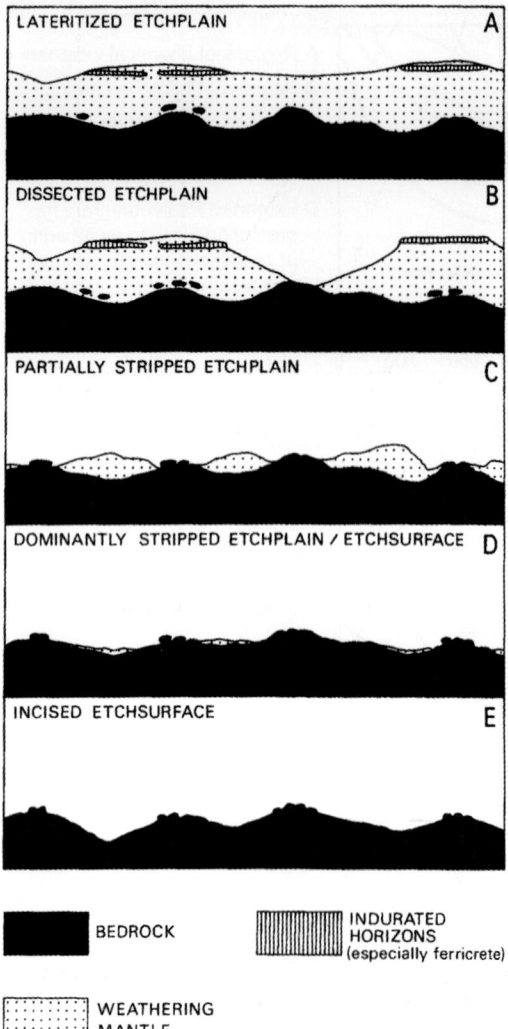

LATERITIZED ETCHPLAIN   A

DISSECTED ETCHPLAIN   B

PARTIALLY STRIPPED ETCHPLAIN   C

DOMINANTLY STRIPPED ETCHPLAIN / ETCHSURFACE   D

INCISED ETCHSURFACE   E

■ BEDROCK

▦ INDURATED HORIZONS (especially ferricrete)

▨ WEATHERING MANTLE

Figure 2.19   The development of different types of etch-plains and etchsurfaces. The diagrams do not necessarily represent an evolutionary sequence as repeated episodes of accelerated erosion may only succeed in partially removing the weathering mantle. The types of etchplains and etchsurfaces illustrated are: (A) lateritized etchplains comprising a surface of low local relief underlain by a thick weathering mantle, including indurated lateritic horizons (ferricretes), which has been subject to only limited stream incision; (B) dissected etchplains in which accelerated stream downcutting promoted by climatic change or uplift leads to the development of well-defined valleys, fringed in places by duricrust breakaways, and the very localized exposure of bedrock and the formation of tors; (C) partially stripped etchplains characterized by widespread stream dissection and the extensive stripping of the weathering mantle (after Thomas, 1994).

stronger than the underlying or surrounding materials and, hence, tend to armour the landscape by forming a hard cap to flat-topped hills and plateaux surfaces. The presence of these indurated horizons may have significant implications for excavation operations and sources of construction aggregates.

Karst landforms

Karst landforms are distinctive landscapes that have developed where chemical weathering of soluble rocks such as limestones, gypsum and halite has created features such as thin soil covers, ground depressions and underground cave systems. These landscapes present a unique range of engineering problems that are described in Chapter 24.

2.13 Weathering products: engineering soils

In situ weathering produces a weathering profile comprising various grades of weathered rock and residual soil (Appendix Tables A3.3a, A3.3b and A3.3c and Chapter 7 Table 7.4). These changes are accompanied by significant changes in material properties, notably a decrease in strength and changes in permeability, porosity and water content. The nature of the weathering profiles tends to vary with rock type and climate, as will be discussed in later Chapters. The thickness of the weathering profile reflects the relative balance between the rate of bedrock weathering and the rate of removal by soil erosion or landslides. Depths of weathering may exceed 100 m, especially in humid tropical environments.

The type of clay mineral generated through weathering depends, in part, on the climate. For example, the relative abundance of kaolinite, gibbsite and iron and aluminium oxides is related to the mean annual precipitation (Figure 2.20), the nature of the original parent material and the timescale over which the material has experienced weathering activity. In addition, the intensity of soil leaching (a function of temperature, precipitation and soil drainage)

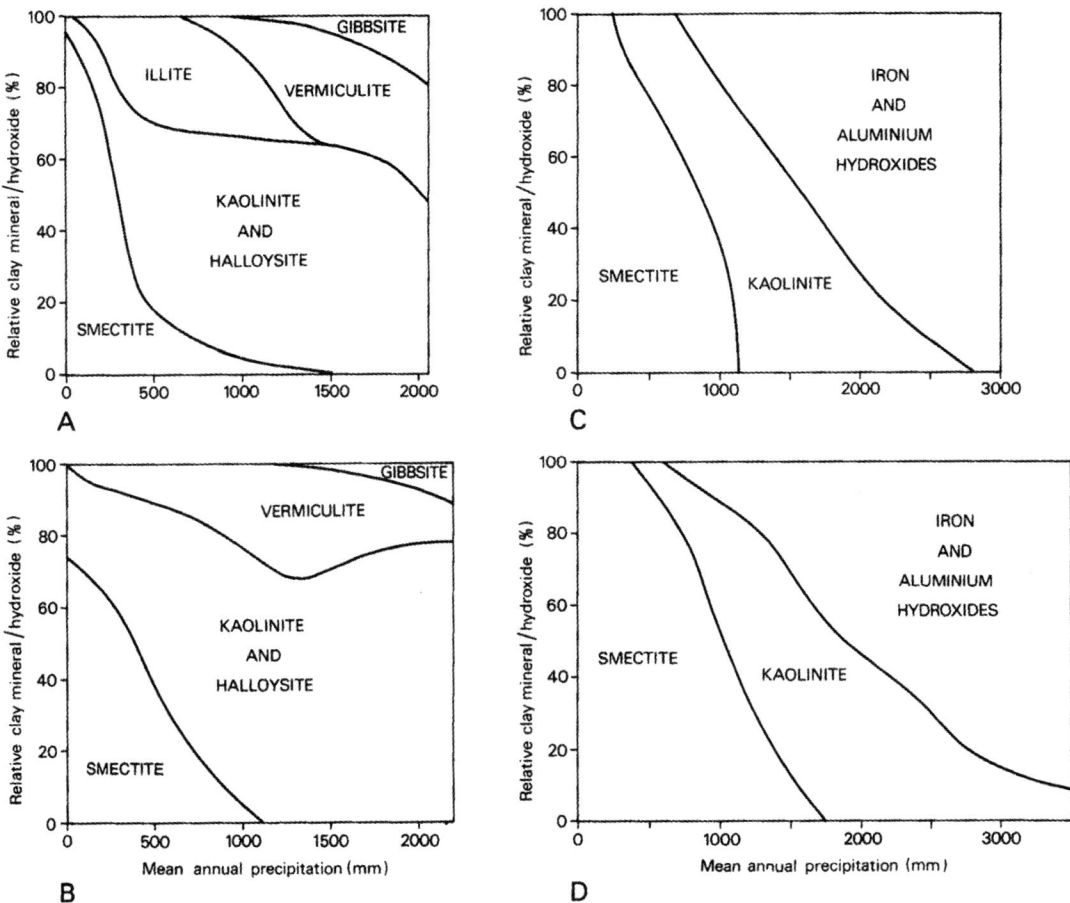

Figure 2.20   Variations in the clay mineral and residual hydroxide composition of soils in relation to mean annual precipitation: (*A*) surface soil samples on feldspar and quartz rich igneous rocks; (*B*) olivine, amphibole and pyroxene-rich igneous rocks; (*C*) soils developed under an alternating wet and dry climate and (*D*) soils developed in a continuously humid climate (from Summerfield, 1991).

by water through-flow is an important factor (Table 2.5).

Intense leaching environments are associated with the removal of metal cations, iron and aluminium oxides and hydroxides from the soil in solution. Kaolinites tend to be the dominant clay mineral.

Moderate leaching environments are associated with the build up of cations released by weathering, leading to the formation of cation-bearing clays such as illite and smectite.

Low intensity leaching environments are associated with very high concentrations of dissolved minerals. Calcium carbonate often accumulates with soils in arid environments, with more soluble salts such as gypsum or halite able to build up in areas of extreme aridity.

2.14 Weathering and geohazards

Weathering and the resultant changes to material properties often act as a preparatory factor in the generation of geohazards, especially debris flows and landslides. For example, in many mountainous areas, landslide activity follows a progressive

Table 2.5 Weathering products and environmental controls (from Loughnan, 1969; Summerfield, 1991).

Environment	pH	Redox potential (Eh)	Behaviour of major elements	Mineralogy of weathering products
Non-leaching. Mean annual precipitation < 300 m. Hot	Alkaline	Oxidising	Some loss of Na^+ and K^+. Iron present in ferric state.	Partly decomposed parent minerals. Illite, chlorite, smectite and mixed layered clay minerals. Hematite, carbonates, secondary silica and salts. Organic matter absent or sparse.
Non-leaching below water table	Alkaline to neutral	Reducing	Some loss of Na^+ and K^+. Iron present in ferrous state.	Partly decomposed parent minerals. Illite, chlorite, smectite and mixed layered clay minerals. Siderite (iron carbonate) and pyrite (iron sulphide). Organic matter present.
Moderate leaching. Mean annual precipitation 600–1300 mm. Temperate	Acid	Oxidising to reducing	Loss of Na^+, K^+, Ca^{2+}, Mg^{2+} and some loss of SiO_2. Concentration of Al_2O_3, Fe_2O_3, TiO_2.	Kaolinite with or without degraded (K-deficient) illite. Some hematite present. Organic matter generally present.
Intense leaching. Mean annual precipitation > 1300 mm. Hot	Acid	Oxidising	Loss of Na^+, K^+, Ca^{2+}, Mg^{2+} and SiO_2. Concentration of Al_2O_3, Fe_2O_3, TiO_2.	Hematite, goethite, gibbsite and boehmite with some kaolinite. Organic matter absent or sparse.
Intense leaching. Mean annual precipitation > 1300 mm. Cold	Very acid	Reducing	Loss of Na^+, K^+, Ca^{2+}, Mg^{2+} and some iron and Al_2O_3. SiO_2. and TiO_2 retained.	Kaolinite, possibly with some gibbsite or degraded Illite. Organic matter abundant.

soil-stripping model (Crozier, 1986), with the pattern and frequency of failures controlled by the availability of unstable material. Soil development proceeds until a critical depth and a landslide triggering event (e.g. intense rainfall) occur in combination, generating shallow hillside failures. After the initial failure, the rate of weathering and soil formation will be the limiting factor on the frequency of failures at a particular site.

In cuttings in stiff clay, the combination of the slow recovery in pore water pressures, progressive failure and strain-softening will result in a gradual decline in effective shear strength and may result in the eventual failure of the slope. On London Clay railway cuttings in the UK, delayed failures have been reported over 100 years after the initial excavation (e.g. Skempton, 1964). In other circumstances, frost weathering may act as a trigger for rockfalls.

2.15 Summary

Climate — both current and past — has a strong influence on engineering through the operation of geomorphological processes and geohazards that might impact on the site, and through the production of a weathering profile and residual soil. All of these have a significant influence on both engineering design and construction. Climate should not be regarded as constant over 'engineering time', with obvious implications for assessments of the probability of climatically triggered geohazards. Global climate change and sea-level rise may have important implications for geohazards and associated risks. The various climatic, morphoclimatic, weathering zone and soils maps presented in the Chapter should provide a starting point for developing a ground model for the area of interest. In the following Chapters more specific accounts are provided of the particular conditions to be found, and engineering problems that can occur, in selected environments.

References

Adams, J., Maslin, M. and Thomas, E. (1999) Sudden climatic transitions during the Quaternary. *Progress in Physical Geography* **23**, 1–36.

British Standards Institution (1999) *BS5930-1999 Code of Practice for Site Investigations.*

Bird, E. (2000) *Coastal Geomorphology: An Introduction.* Wiley, London.

Barber, K. E., Chambers, F. M., Maddy, D., Stoneman, R. and Brew, J. S. (1994) A sensitive high resolution record of late Holocene climatic change from a raised bog in northern England. *The Holocene* **4**, 198–205.

Briggs, D. and Smithson, P. (1992) *Fundamentals of Physical Geography.* Routledge, London.

Brunsden, D. (1979) Weathering. In Embleton, C. and Thornes, J. (eds), *Process in Geomorphology.* Arnold, London, 73–129.

Brunsden, D. (1996) Geomorphological events and landform change. *Zeitschrift für Geomorphologie* **40**, 273–288.

Cooke, R. U. and Doornkamp, J. C. (1990) *Geomorphology in Environmental Management.* Oxford University Press.

Crozier, M. J. (1986) *Landslides: Causes, Consequences and Environment.* Routledge, London.

Doornkamp, J. C. (1986) Climate and weathering. In Fookes, P. G. and Vaughan, P. R. (eds) *A Handbook of Engineering Geomorphology.* Surrey University Press, Blackie and Son, London, 10–24.

Folland, C. K., Karl, T. R. and Vinnikov, V. (1990) Observed climate variations and change. In Houghton, J. T., Jenkins, G. J. and Ephraums, J. (eds) *Climate Change: the IPCC Scientific Assessment.* Cambridge University Press, 200–228.

Fookes, P. G. (1980) An introduction to the influence of natural aggregates on the performance and durability of concrete. *Quarterly Journal of Engineering Geology* **13**, 207–229.

Fookes, P. G. (1997) First Glossop Lecture: Geology for engineers: the geological model, prediction and performance. *Quarterly Journal of Engineering Geology* **30**, 293–424.

Fookes, P. G., Baynes, F. J. and Hutchinson, J. N. (2000) Keynote Lecture: Total geological history: a model approach to the anticipation, observation and understanding of site conditions. *GeoEng 2000, International Conference on Geotechnical and Geological Engineering, Melbourne* **1**, 370–460.

Grove, J. M. (1988) *The Little Ice Age.* Methuen, London.

Hayden, B. P. (1988) Flood climates. In Baker, V. R., Kochel, R. C. and Patton, P. C. (eds) *Flood Geomorphology.* Wiley, New York, 13–26.

Hulme, M., Jenkins, G. J., Lu, X., Turnpenny, J. R., Mitchell, T. D., Jones, R. G., Lowe, J., Murphy, J. M., Hassell, D., Boorman, P., McDonald, R. and Hill, S. (2002) *Climate Change Scenarios for the United Kingdom: The UKCIPO2 Scientific Report.* Tyndall Centre for Climate Change Research, University of East Anglia, UK.

IPCC (Intergovernmental Panel on Climate Change) (2001) The IPCC third assessment report – summary for policy markers. *http://www/ipcc.ch/ index.html.*

Kiewiet de Jonge, C. J. (1984) Budel's geomorphology II. *Progress in Physical Geography* **8**, 365–397.

Knox, J. C. (1984) Flurial responses to small scale climatic changes. In Costa, J. E. and Fliesher, P. J. (eds) *Developments and Application of Geomorphology.* Springer-Verlag, New York, 318–342.

Knox, J. C. (1988) Climatic influence on Upper Mississippi Valley. *Annals of the Association of American Geographers* **77**, 224–244.

Köppen, W. (1931) *Grundriss der Klimakunde.* De Gruyter, Berlin.

Köppen, W. (1936) *Das Geographischem System der Klimate.* Gebr. Bortraeger, Berlin.

Lamb, H. H. (1997) *Climate, History and the Modern World.* Routledge, London.

Loughnan, F. C. (1969) *Chemical Weathering of the Silicate Minerals.* Elsevier, Amsterdam.

Macklin, M. G. and Lewin, J. (1997) Channel, floodplain and drainage basin response to environmental change. In Thorne, C. R., Hey, R. D. and Newson, M. D. (eds) *Applied Fluvial Geomorphology for River Engineering and Management.* Wiley, Chichester, 15–45.

Peltier, L. C. (1950) The geographic cycle in periglacial regions as related to climatic geomorphology. *Annals of the Association of American Geographers* **40**, 214–236.

Probst, J. L. and Tardy, Y. (1987) Long range streamflow and world continental runoff fluctuations since the beginning of this century. *Journal of Hydrology* **94**, 289–311.

Roberts, N. (1998) *The Holocene: an Environmental History.* Blackwell, Oxford.

Skempton, A. W. (1964) Long term stability of clay slopes. *Geotechnique* **14**, 77–101.

Strakhov, N. M. (1967) *Principles of Lithogenesis*, Vol. 1. Oliver and Boyd, Edinburgh.

Summerfield, M. A. (1991) *Global Geomorphology*. Longman, Harlow.

Thomas, M. F. (1974) *Tropical Geomorphology*. Macmillan, London and Halstead Press, New York.

Thornthwaite, C. W. (1933) The climates of the earth. *Geographical Review* **23**, 433–440.

Twidale, C. R. (1990) The origin and implications of some erosional landforms. *Journal of Geology* **98**, 343–364.

Waple, A. M. (1999) The sun–climate relationship in recent centuries: a review. *Progress in Physical Geography* **51**, 309–328.

Ward, R. (1978) *Floods: A Geographical Perspective*. Macmillan, London.

3. Sedimentology

Colin J. R. Braithwaite

3.1 Introduction

Sedimentology first grew as a distinct branch of Earth Science in response to the demands of the petroleum exploration industry for more detailed information on the nature and origin of sedimentary rocks involved in the formation and accumulation of hydrocarbons. It is concerned with the ways in which sedimentary grains may be generated, transported, deposited and buried to become rocks. Observations may be at scales varying from microns (using the scanning electron microscope) to global. The geometry of sedimentary rock sequences can be used to define environmental systems, their assembly in basins and their global distribution. Almost all of these data are also, to some degree, of concern to the engineer involved in site investigation and in the evaluation of sediments or sedimentary rocks as construction materials.

A principal goal of the sedimentologist is to interpret sedimentary sequences in order to reconstruct the environment in which they formed. This might seem an academic exercise but it provides a framework for predicting the specific characteristics and distribution of the sediments or rocks involved. These are important issues in site investigation, mineral resources and hydrology. The approach is based on the concept of uniformitarianism, which assumes that the rules governing the physics and chemistry of environmental systems have remained the same throughout geological time. From this premise, study of the processes and products of present-day environments allows sedimentologists to interpret sequences that formed over thousands of millions of years. The preserved record of these ancient environments (Chapter 5) may be used to reconstruct the geological history of the Earth. For this reason it is important to understand how surface environments have come about in the last few hundred thousand years, and to appreciate that they are dynamic systems that experience changes observable within our lifetimes and affecting all human constructions. This perspective may be gained through Part II of this book. However, there are two important caveats that must be applied. Present environmental conditions do not necessarily represent all of those that may be possible and that may have been present in the past. There is good evidence, for example, that climate and atmospheric and ocean compositions have been different in the past. In addition, there are now clear indications of major events, shaping Earth environments that have never been directly observed. The evidence of major meteorite impacts falls into this category. This chapter reviews concepts surrounding the processes of weathering, transport and deposition that form sediments, together with the processes and products of their conversion to rocks.

From an engineering perspective, the environments described are those in which structures will be built, and to a large extent each develops a characteristic architecture. These models provide a basis on which the distributions of sedimentary materials can be predicted. The petrography of sedimentary rocks addresses the major influences on their strength and behavioural characteristics.

3.2 Weathering

The group of processes referred to as 'weathering' is discussed in relation to climate in Chapter 2. Igneous and metamorphic rocks, and by inference a proportion of sediments, consist of mineral assemblages that formed at elevated temperatures and pressures, for the most part deep within the

Earth's crust. They are commonly unstable when exposed to surface environments and as a consequence slowly disintegrate. Their breakdown may be by physical processes that include the unloading resulting from their exposure, or result from chemical attack by percolating surface waters, decomposing and changing their mineralogical and chemical compositions. Although weathering is normally described as 'physical' or 'chemical', neither operates in isolation. One or the other may dominate at different times and places, but more often they aid each other and are commonly augmented by biological processes. Rock surfaces freshly exposed by physical weathering are typically attacked by chemical processes. It can be argued that the main function of physical weathering is to increase the surface area available for chemical attack. Dividing a cube into eight smaller cubes by only three cuts at right angles doubles the surface area. Generally, chemical weathering is most effective in temperate and tropical regions where temperatures are moderate or high and water is plentiful. In such environments, a thick cover of weathered material may mantle land surfaces. Table 3.1 summarises the weathering products that are produced from various rock-forming minerals. It is important to realise that in all rocks, weathering may begin along grain boundaries and

may dramatically reduce strength without any change being visible to the naked eye. In areas such as Brazil the early effects of weathering can be detected at depths of well over 100 metres.

3.3 Sediment transport and deposition

Introduction

Erosion is the progressive removal from the land surface of material that is typically weathered. If erosion did not take place, a thickening layer of weathered debris would accumulate over rock surfaces. The relative protection afforded by this cover would progressively slow the weathering process until eventually little further change would take place. The rate at which weathering can occur therefore depends in part on the rate at which weathering products are removed by erosion. Conversely, however, the rate of erosion is itself influenced by the rate at which particles are released from the rock surface by weathering. The same factors, principally climate and relief, influence the rates of both weathering and erosion. It is notoriously difficult to measure rates of erosion, and regional patterns of denudation are commonly derived from estimates of sediment yield

Table 3.1   Summary of weathering processes and products.

Process	Materials attacked	Soluble components	Insoluble residue
Physical or mechanical disintegration	Outcrops of all rock types		Successively smaller rock fragments Unaltered silicate grains
Chemical reactions, hydrolysis and dissolution	Olivine	Mg^{2+}, SiO_2	Fe^{3+} oxides and hydroxides
	Pyroxene	Mg^{2+}, SiO_2	Fe^{3+} oxides and hydroxides, plus hydrous aluminium silicates (clay minerals)
	Amphiboles	Mg^{2+}, Ca^{2+}, Na^+, SiO_2	
	Biotite mica	Mg^{2+}, SiO_2	Fe^{3+} oxides and hydroxides, plus hydrous aluminium silicates (clay minerals)
	Muscovite mica	K^+, SiO_2	No change
	Feldspars	Ca^{2+}, Na^+, K^+, SiO_2	Hydrous aluminium silicates (clays)
	Quartz		Quartz (no change)
	Calcite and dolomite	Ca^{2+}, Mg^{2+}, HCO_3^-	None

Table 3.2 Comparison of rates of denudation for high-altitude (mountainous) and lowland regions under contrasting climatic regimes (data from Kukal, 1971, Blatt *et al.*, 1980 and others).

Climate regime	High altitude	Denudation rate (mm/1000 years)	Lowland	Denudation rate (mm/1000 years)
Arid to semi-arid. Very low precipitation, Sporadic floods, no seasonal pattern	Low latitude desert mountains	100–200	Low latitude Sahara Desert	1
	California, Fort Sage Mountains	50	Eyre Peninsula, South Australia	0.5–1
Warm temperate: moderate precipitation throughout the year	Mid latitude, Coastal Range of the USA	3000–8000	Mid-latitude Rhine Basin	30
Warm temperate, moderate to high precipitation throughout the year	Tropical to low latitudes e.g. Himalayas	720	Tropical to low latitudes, Northern India	20
Cool Temperate, moderate to heavy precipitation throughout the year	New Zealand Alps	50–130		
	European Alps	50–910		

(Table 3.2). However, weathering is greatest where rainfall is moderately high all year, and lowest in arid to semi-arid regions, regardless of altitude, whereas rates of denudation are greater in high altitude mountainous regions, regardless of climate. Recent years have seen the development of direct methods of dating exposed surfaces by the use of apatite fission track measurements and cosmogenic nucleides, and these promise more detailed and more accurate measures in the future. Rates of both weathering and erosion are modified by vegetation, the distribution of which generally parallels that of rainfall. The organic acids generated by plant decay enhance weathering, while the formation of a canopy of shrubs or trees may stabilise slopes and slow erosion. In some areas there have been dramatic (and deleterious) changes in rates of erosion in response to deforestation and changes in farming practice.

Mass movement

In the absence of vegetation, weathered debris on a slope is usually stable until the friction between the debris and the slope is overcome. In theory, friction or pore suction can hold weathered debris on slopes of up to about 45°, but in practice movement may occur on slopes as low as 1° and it is

obvious that other factors act to reduce friction. Among the most important of these are porewater pressures and the expansion and contraction of the soil due to temperature changes, including freeze and thaw. Triggering mechanisms that overcome friction most commonly relate to variations in these factors but include events such as exceptional rainfall and earthquakes. To emphasise the importance of these events, between 1950 and 1993 a succession of wet years in southern California resulted in damage from mass movements that on average totalled $500M.

Some of the varied forms of mass movement are illustrated in Figure 3.1 but see also Chapter 8. The principal features that differentiate these from one another are the volume of water included with the rock debris, the speed at which they move, and the degree of internal deformation within the moving mass. In one respect, mass movement can be regarded as gradational between river transport, where the movement of debris is the direct result of the movement of water, and the movement of rock debris by gravity alone.

Creep is a slow and almost continuous downslope movement, typically occurring in a moderately thick soil layer. The displacement that occurs represents the cumulative effect of individual small movements resulting from wetting and drying or

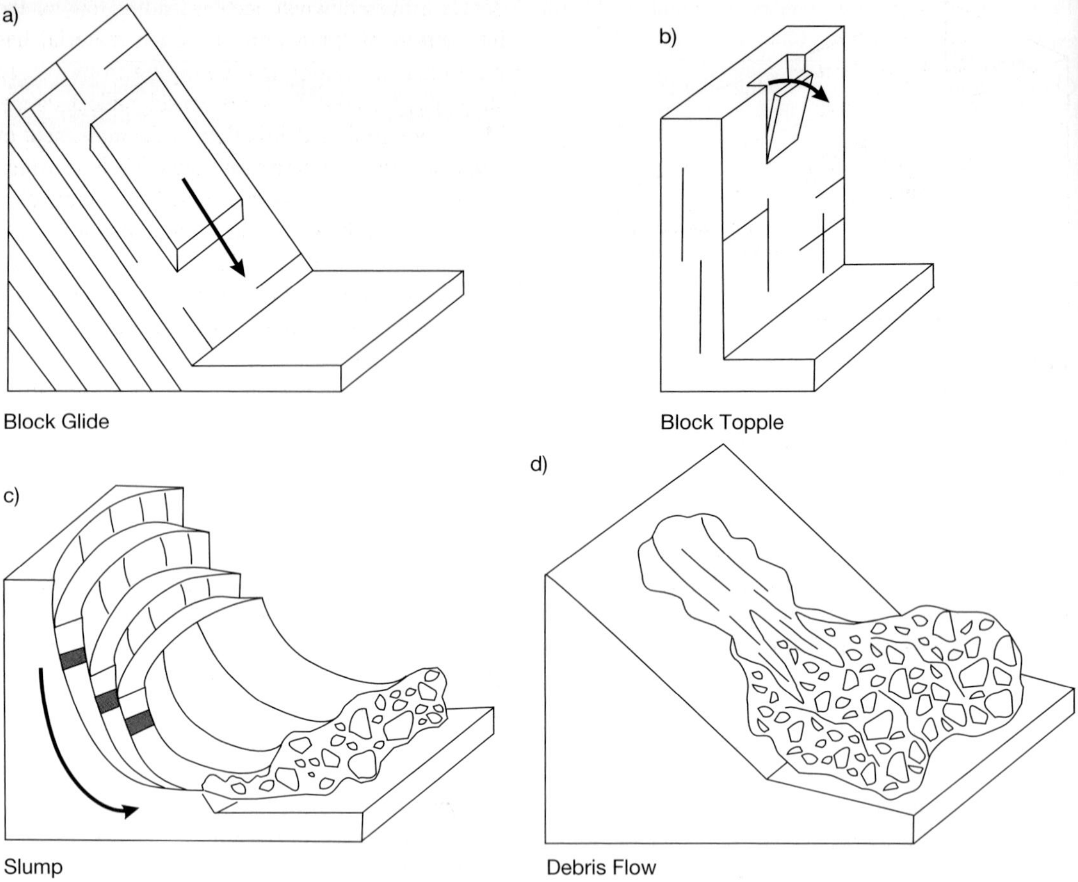

a)

Block Glide

b)

Block Topple

c)

Slump

d)

Debris Flow

Figure 3.1   Summary of the varied mechanisms of slope failure. In block glide (*A*) movement takes place along a bedding plane, cleavage or foliation surface inclined at a lower angle than the slope. Elevated pore pressures on the plane may aid movement. Block topple (*B*) occurs where steeply inclined rectilinear jointing frees blocks that topple from the exposed face. In slumping (*C*), failure occurs along arcuate surfaces that daylight on the exposed face. The detached material rotates to push forwards a raised toe of debris. In debris flows (*D*) high pore fluid pressures mean that the material of the slope loses cohesion and moves as a semi-fluid mass of variable viscosity.

freezing and thawing. The mechanism is illustrated in Figure 3.2. Expansion causes displacements perpendicular to the ground surface while subsequent shrinkage, aided by gravity, is vertical, so that there is a net movement downslope. Although such movements are slow, their presence may be revealed by bulging soil surfaces, by bent trees, where growth is compensating for realignment, or by disturbed fencing. Any of these indicators should be a cause for caution.

The movement of landslides is generally much more rapid and is of two kinds. Block glides are the displacement of coherent and competent masses of rock on relatively planar surfaces such as bedding or foliation planes. Mapping the distribution and orientation of such planes is important in assessing slope stability. Typically there is little or no internal deformation within the moving mass. Slumps are more common in less cohesive materials such as soils or clay-rich rocks, and involve displacements on curved, concave-up shear surfaces (Figure 3.1c). These create tensional stress in the material at the upper end of the shear plane, and compression at the foot. The moving mass rotates, and the toe of the slide moves upwards until its mass counterbalances the

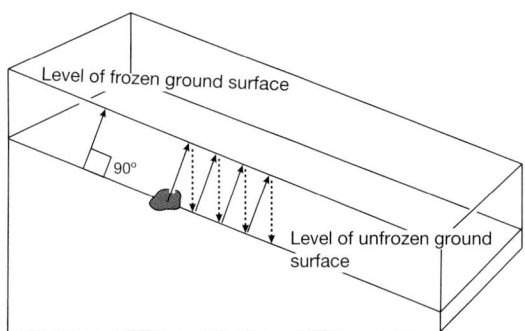

Figure 3.2 The process of soil creep under the influence of freeze-and-thaw. Note the expansion of the active layer at 90° to the slope surface as the contained water freezes and the vertical return under gravity as the ice melts.

downward force exerted by the body of the slide and movement ceases. Arcuate fractures and a concave scar are left on the hillside where material has been removed, and a convex hump forms where it comes to rest. There may be little internal deformation within the rotating mass but slumps are also transitional to other movements.

Both block glides and slumps move rapidly, and their 'instantaneous' failure, the reason why they are considered hazards, is commonly due to rising water seepage pressures. Although small amounts of water in dry materials may actually increase friction, pore pressures that approach overburden pressure cause a dramatic reduction. Pore pressures may be focused in joints or bedding planes, or along the junction between two rock or sediment (soil) types of varying permeability, but in poorly consolidated materials are commonly more dispersed. Once initiated, slides may occur intermittently along the same shear plane over many years, whenever pore pressures rise sufficiently.

As more and more water is added to the system, by direct rainfall, by groundwater seepage or from leaking pipelines, pore pressures rise and the material loses strength, either by the dilation of coarser materials or the loss of suction in clays. This may mean that the slump deforms internally, sometimes generating complex folds of existing sedimentary layers, but with increasing volumes of water the moving mass becomes a debris flow. Examples are known of rock flows in which unconsolidated debris moves downslope buoyed up only by the interaction of grain impacts. Such material has little internal friction and emerging from a steep slope may 'run out' for some considerable distance. More commonly a debris flow is the movement of a viscous mass of water-saturated debris in which pebbles or boulders are suspended in a slurry consisting of sand, silt or clay. Rates of movement range from barely perceptible in relatively dry systems to almost instantaneous where large volumes of water are incorporated and the mass becomes fluid, travelling downslope at speeds in excess of 4 m/s. Debris avalanches at Mount Huscarán in Peru in 1962 are thought to have travelled at speeds in excess of 320 kph, forming a wave 80 m high. As densities may reach 1.5–2.0 they are capable of moving blocks weighing many tonnes and of destroying large structures.

Mass movements may result in significant hazards, either within the area of failure or as a result of subsequent deposition. The distinctive characteristics of the deposits formed allow the contrasting mechanisms of transport to be identified. A mass of rock moved by a block glide (Figure 3.1a) usually retains its internal structure and any original stratification. By contrast, strata contained within a slump are typically tilted by the rotational movement (Figure 3.1c) and, where strength begins to be lost, may become folded. As the mass continues to move downslope internal structures may be completely destroyed. The deposits of debris flows usually contain a wide range of particle sizes, from clay-sized particles through to pebbles or boulders, and are described as 'poorly sorted' (see 'siliciclastic sediments and sedimentary rocks'). Those from rock falls (Figure 3.1b) or avalanches are likely to consist of angular fragments of varying sizes piled into a heap and lacking any form of layering. However, although debris-flow deposits are typically poorly sorted, they may sometimes develop a crude stratification as a result of shear generated by their movement. The more water a flow contains the more likely it is to develop stratification as the conditions approach those of fluid flow.

In areas where environmental factors are closely monitored some element of prediction may be possible. In some areas in California, for example, it is known that where rainfall exceeds about 15 mm/hour the threshold time for the onset of debris flows varies from 8–14 hours depending on local slopes, materials and runoff.

Although they are commonly a significant natural hazard, the environmental importance of mass movements lies in the fact that they are the means by which weathered debris can be moved from hillsides and cliffs into a river, on to a glacier or into the sea to begin a more substantial phase of transport.

Transport and deposition by fluids

The principle differences in the behaviour of fluids and gasses lie in their relative viscosities, essentially how 'stiff' they are. Two general modes of fluid flow are recognised, laminar flow and turbulent flow. This distinction was first recognised and demonstrated by Osborne Reynolds (1842–1912), a British physicist who released dye streams into long straight tubes full of flowing water. Reynolds described flow by reference to a dimensionless coefficient that has since become known as the 'Reynolds number' (Re).

For sedimentary particles:

$$Re = \frac{Ud\rho}{\mu}$$

where U is the velocity of the particle or flow; ρ is the density of the particle; μ is the viscosity of the fluid and d is the diameter of the pipe or grain.

At low Reynolds numbers, the threads of dye remain straight and the flow is therefore laminar. As velocity increases the dye streams become highly disturbed, indicating that the flow is turbulent. For flow in tubes this transition occurs when the value of Re is about 2000. The important feature here is that although the net movement of the fluid is large and is still downstream, there are now random secondary motions that at a fixed point vary erratically in both direction and velocity. These may provide lift for grains in erosion but are primarily responsible for the suspension of smaller grains within the fluid.

In open channels a second important coefficient, the 'Froude number' differentiates the flow conditions, referred to as 'Lower' and 'Upper Flow Regime' conditions, required for two discrete sets of bedforms. Like the Reynolds number this is a dimensionless coefficient that allows the comparison of flow conditions in channels, pipes and around objects at different scales. William Froude (1810–1878) was a naval engineer and used the coefficient to scale the behaviour of model ships.

The Froude number (F) is defined as:

$$F = \frac{U}{\sqrt{gD}}$$

where U is the average velocity of the current; D is the depth of the channel and g is the acceleration due to gravity.

However, even in laminar flow a velocity gradient is set up normal to the direction of flow. Fluid in contact with the bed surface is brought to rest by friction and successive 'layers' above this can be thought of as moving at increasing rates. With a cohesionless granular bed, flow effectively extends into the bed surface. As the velocity of the fluid over the bed increases, a point is reached when the stress applied to the surface is large enough to tear particles away. This threshold of movement is the critical erosion velocity.

The physics of sediment movement

During fluid transport, the particles of sediment (grains) are buoyed up by turbulence in the fluid that moves in response to gravity. The range of size of particles that can be carried, and the characteristics of movement of the particles, depend on the relationship between the size and density of the particles and the density and viscosity of the fluid, together with the rate of flow. In natural systems the fluid may be either water or air. The following discussion relates largely to water, but the principles of transport by air are essentially the same (see the discussion of the Reynolds number).

Particles placed in water will settle to the floor because of the force of gravity acting on them. There are, however, two forces that oppose this process. First there is frictional drag between the surface of the particles and the water molecules. This depends largely on particle shape. Second, there is the natural buoyancy of the particles, expressed as the difference in density between the particles and the water (or other fluid). The velocity at which particles may be expected to settle through a fluid takes both of these forces into account and can be expressed by the 'Stokes' law equation'.

Stokes' law and sediment transport

Stokes related the erosion and transport of sediment particles to their size and density relative to the velocity and viscosity of the fluid around them, as described in the equation:

$$v = \frac{2r^2 g \Delta \rho}{9 \eta} \tag{1}$$

where v is the settling velocity (m/s); g, the acceleration due to gravity; r, the radius of the particle (m); $\Delta \rho$, the difference in density between the particle and fluid (kg/m^3) and η, the dynamic viscosity of the fluid (N s/m^2 or Pa s).

For most mineral grains the size of the particle has the greatest influence on the settling velocity, and as this is proportional to the square of the radius, equation (1) can be simplified to:

$$v \propto r^2 \text{ or } v = kr^2 \tag{2}$$

where k is a factor that embodies all of the other factors in equation (1) except for particle radius.

However, the settling velocities of sediments in natural environments such as lakes and rivers are much lower than those calculated theoretically from equation (1). There are several reasons for this, but the two most important are:

1. The Stokes' law equation was derived for perfectly spherical particles, as these offer the least frictional resistance. Natural sedimentary particles are not perfect spheres. Although quartz grains are sometimes reasonably spherical, grains of other minerals, as a result of fundamental properties such as cleavage, are more likely to have prolate or bladed shapes. Particles of clay minerals present in mudstones and shales are flake-like and so offer the maximum frictional resistance for their volume. Carbonate grains, consisting of fragments of animal or plant

skeletons may have very elaborate shapes but also a relatively low density, reflecting the presence of internal pores rather than crystalline calcite or aragonite. Finally, the fall velocities of single grains are somewhat faster than the hindered settling of large numbers of similar grains in a sediment mixture because they avoid grain-to-grain collisions.

2. Equation (1) assumes that the fluid is stationary, whereas natural waters are rarely completely still.

Stokes' Law can only be applied when the molecular flow of the water around the particles is smooth, or laminar (Figure 3.3a). In practice, at velocities of only a few centimetres per second, water flow becomes turbulent and chaotic, and complex multidirectional eddies are superimposed on the overall flow direction (Figure 3.3b). These oppose particle settling. In completely turbulent flow, the relationship between velocity and particle radius expressed by equation (2), is no longer valid and changes to:

$$v \propto r^{1/2}$$

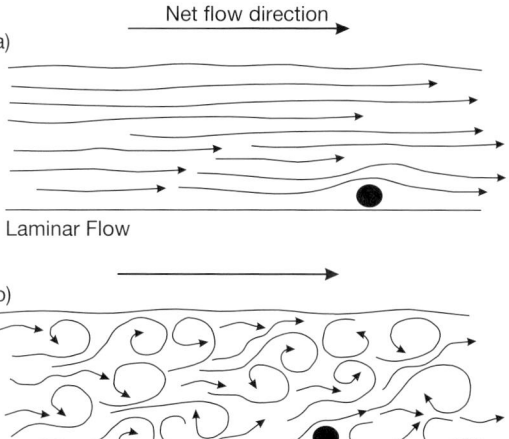

Net flow direction

a)

Laminar Flow

b)

Turbulent Flow

Figure 3.3   Changes in state during fluid flow: (*A*) laminar flow in which streamlines are broadly parallel and movement around a particle is smooth with essentially linear flow lines and (*B*) turbulent flow with chaotic multidirectional eddies superimposed on the overall flow direction.

As a result, the difference in settling velocities between relatively large and small particles is less than would be expected. Equation (1) indicates that fine-grained particles are deposited more slowly than coarse ones. Consequently, when a mixture of grain sizes enters moving water, coarse material is deposited first and finer material is carried down current. Coupled with reworking this leads to a progressive size sorting of the original mixture. Sorting, together with variations in sediment loads and current velocities, leads to the deposition of laminated sediments.

Turbulent flow increases the ability of moving water to sort sediment. This is because only the coarser particles are able to reach the stream bed and smaller grains travel progressively further downstream before they are deposited. The relationship between the velocity of a current and the sizes of particles that may be transported has been determined experimentally and is shown, along with other data in Figure 3.4. The lower curve in the diagram represents the current velocities that are required to keep grains of different sizes in motion, and divides the graph into a zone of transport and a zone of deposition.

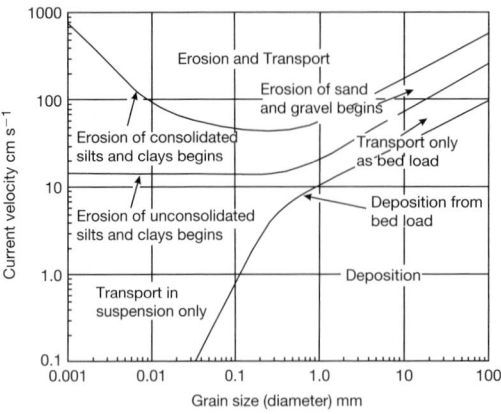

Figure 3.4 Graph illustrating the relationship between current velocity in a fluid and transport and deposition of sedimentary particles. The graph is for fresh water and would be displaced downwards slightly for salt water or upwards for air. Note the lack of erosion where finer-grained sediments have become cohesive and that transport may continue after current velocities have fallen below those required for erosion and entrainment.

Very fine grained sediments carried in suspension by laminar flow (i.e. within the moving water column) form suspension deposits when the current decreases. When the flow is turbulent, coarse sand grains and even pebbles can be transported, but these are initially carried along the river or channel bed, and are *not* in suspension. This coarser sediment forms the bed load of the moving water. A much higher current velocity is required to lift the bed load into suspension. Bed load movement takes place in two ways. Sand grains or pebbles may simply be rolled downstream along the bed as the traction load, or they may bounce downstream due to turbulence in the fluid and impact with larger particles on the river bed, a process known as saltation. Typically, at moderate rates of flow in water, sand grains saltate to a height of 2–3 mm before sinking back to the channel bed. If the current velocity decreases, such that part of the bed load ceases to move, the resulting deposits are referred to as traction deposits (Chapter 10).

A similar relationship between sediment movement and current velocity applies to transport in air. However, as turbulence is directly related to the viscosity of the fluid and air has a much lower viscosity than that of water, flow is always turbulent. In addition, a grain of a given size has a much higher settling velocity in air than in water because, as a result of the greater difference in density, air provides less buoyancy. As a consequence, much greater wind speeds are required to hold or lift material into suspension than are needed in water. Generally, only silt and clay-sized grains ($<150 \, \mu m$) form wind-blown (aeolian) suspension deposits (Chapter 16). The sediment load carried in this way may be very large and it has been calculated that dust storms in the American Middle West may contain in the order of 1000 tonnes per km^3. Such storms may be 2000–3000 km in length, with cross-sectional areas of 10s or 100s of square kilometres. In 1977, a 24 hour storm in California with winds up to 300 kph is estimated to have removed nearly 100 million tonnes of topsoil.

Saltation is a much more important method of sediment movement in wind than in water, especially for sand-sized grains. A typical sand grain of 0.25 mm diameter saltating in air may rise steeply to heights of 12–25 cm because of the low viscosity of air, before striking the ground about 10 cm downwind. On a large scale this creates a cloud of sand grains flowing over the surface of windblown sand dunes. Eye-witness reports of storms refer to sand rising 1–2 m above the surface in this way. Although rolling of sand grains by the wind is not common, saltating grains may bombard those resting on the ground so that they are effectively pushed along. This type of movement is referred to as surface creep, and should not to be confused with soil creep described earlier. However, in this way, saltating grains can move grains with a diameter up to five or six times greater than their own. Finally, as a consequence of the low viscosity, impacts between grains in air are much greater than those in water with the consequence that sand-sized grains more rapidly become well rounded and acquire a 'frosted' surface texture. Paint and windscreens of vehicles, and structures such as wooden poles and supporting cables, may be at considerable risk from this sand-blast abrasion and pebbles exposed on the surface commonly acquire planar facets forming 'dreikanter'.

Erosion of sediments

It might be assumed that, once deposited, sediment would automatically become buried and compacted to form an irreversible addition to the sedimentary record. This rarely happens, most of it is soon eroded and reworked to undergo one or more further phases of transport before 'final' deposition elsewhere. In rivers, transport may not end until the sediment has reached the sea. To begin moving a grain requires a significantly higher velocity than that at which it was deposited (See again Figure 3.4). This reflects the need to overcome the friction between the grain and the channel bed. The velocities at which grains of different sizes begin to move can be determined experimentally, but not as precisely as the settling velocities. A graph such as Figure 3.4 plotting results, therefore, reflects a spread of data rather than a narrowly constrained relationship.

The area between the two upper curves on Figure 3.4 indicates the velocity ranges within which sediment erosion will occur. A sand grain 2 mm in diameter might begin rolling or saltating

along a river bed when the current velocity reaches a value between about 25 and 75 cm/s, depending on conditions at the river bed. Where wind transport is concerned, wind speeds of the order of thirty times greater than those of water are required to set the same grain in motion. This is a reflection of the greater contrast in densities incorporated in the Stokes equation. It seems to imply that wind is less effective as a transport mechanism but this view neglects the size of wind systems that, as indicated, are commonly orders of magnitude larger than those of rivers.

Wind deposits, loess, commonly form light porous soils that may retain low densities (as low as 1.3) long after deposition. These carry particular risks in construction (Chapter 25). They are sensitive to disturbance and particularly wetting and may undergo sudden structural collapse. Such soils occur in about 30% of North America, including 500 000 km^2 in Illinois and Indiana, and are common in arid and semi-arid areas in Wyoming, Colorado, Washington, Utah and California. In China they are locally over 300 m thick. Spectacular collapse has resulted where attempts have been made to provide irrigation through the construction of canals, because on wetting the open fabric undergoes rapid hydrocompaction.

For grains larger than 0.2 mm transported in water there is an essentially linear relationship between the velocity required for erosion and transport, and grain size (Figure 3.4). However, for finer-grained particles, the resistance to erosion not only ceases to decrease but actually increases progressively for consolidated sediments. This is because finer silts and clays show a cohesive strength due to attractive forces between the particles. Clay minerals in particular carry surface charges and become linked together by water dipoles. As a result they behave as plastic solids. Such sediments are referred to as 'cohesive' and cohesion increases as the porosity of the sediment decreases. A substantial increase in water velocity is required to lift the particle back into suspension. Once a particle of silt 0.01 mm in diameter has been deposited, a velocity between 1.4 and 10 cm/s is required to erode it, depending on how dense the sediment has become. However, it requires the *same* current velocity (around 500 cm/sec) to strip a consolidated surface of grains of 0.001 mm diameter as it does to move pebbles of 100 mm or so in diameter!

Cohesive properties are important in the preservation of fine-grained sediments laid down by rivers, in lakes or in coastal and offshore environments on the continental shelf. They are also the reason why material transported into harbours or similar enclosed areas by incoming tidal flow, and deposited at slack water cannot be removed by ebb currents. The resulting deposition can only be remedied by dredging. The resistance to erosion of consolidated clays is exploited in the 'puddled' clays used locally to protect river banks from erosion.

The attractive forces that provide cohesion are dependent on water chemistry, and may be destroyed in certain conditions. They may also be responsible for deposition. In a weak electrolyte such as sea water the particles stick together or flocculate. Flocculation is responsible for much of the deposition in estuaries where, because flocs have a relatively open structure, deposited muds that remain submerged may initially retain porosities as high as 70 to 80%. Slopes consisting of such materials are prone to failure and sub-marine slumps have been reported on slopes of less than one degree. Silt-sized particles are also cohesive because of their small size in comparison to the strength of their surface charges. Silts and mixtures of silts and clays may retain high porosity after deposition and during shallow burial and may liquefy if subjected to vibration or shock. Some show dramatic changes in behaviour when subject to stress. For example, sediments that at first appear solid and capable of bearing a load may, when subject to vibration as when driving piles, suddenly fluidify and lose any bearing capacity. Such behaviour is described as 'thixotropic'.

Sediments and sedimentary bed forms resulting from deposition in flowing water

Suspension and traction deposition lead to characteristic types of layering in sediments that are preserved in sedimentary rocks. When the discrete

Table 3.3　General terminology of beds and laminae.

>1 m	Very thick bedding
$300 \rightarrow 1000$ mm	Thick bedding
$100 \rightarrow 300$ mm	Medium bedding
$30 \rightarrow 100$ mm	Thin bedding
$10 \rightarrow 30$ mm	Very thin bedding
$3 \rightarrow 10$ mm	Thick lamination
<3 mm	Thin lamination

layers are only a few millimetres thick they are referred to as laminae and the sediments or rocks are described as laminated. Layers thicker than about 1 cm are called beds (Table 3.3). Horizontal or planar laminae are generally characteristic of suspension deposits, but coarse horizontal layering, planar bedding, also occurs in some traction deposits. As current speeds increase, the sediment surface is built up into a series of bed forms that include the familiar ripple marks seen on beaches and river bars swept by strong currents. (Bed forms produced by wind action are described in Chapter 16.)

The origins and structures of bed forms have been extensively studied in laboratory flume tank experiments. At low velocities flow is essentially laminar and there is no bed motion, giving a planar surface. As the flow rate increases grains begin to move and there is a rapid transition from laminar to turbulent flow. Current ripples appear almost instantaneously and generate a separation in the flow close to the bed (Figure 3.5; 3.6a). The current adheres closely to the upstream (stoss) slope of the ripple, separating at the crest, with a rolling cylindrical eddy forming beneath, and reattaching on the next ripple downstream. The net result is a decrease in the overall drag exerted on the bed by the flow. Sediment is typically eroded on the stoss slope of the ripple and carried as bed load to the crest where grains are deposited as the flow ejects them into suspension. Grains settle down the lee slope and periodically avalanche, accumulating to form a series of parallel cross-laminae. These are the dominant features preserved. However, where relatively large amounts of coarser sediment are in suspension, erosion may be prevented and sediment is deposited on the stoss side as well, giving structures referred to as

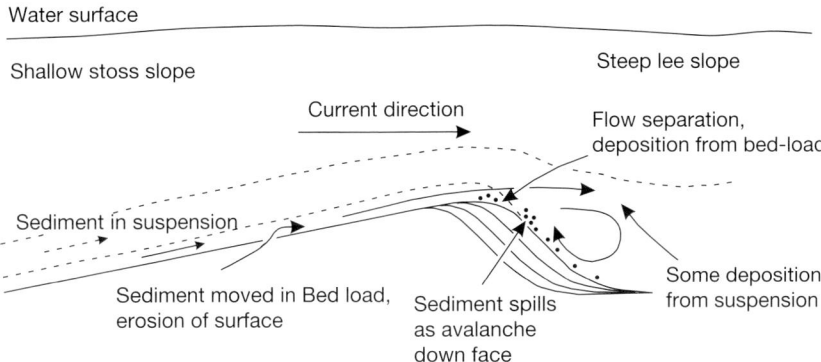

Water surface

Shallow stoss slope
Steep lee slope

Current direction

Flow separation,
deposition from bed-load

Sediment in suspension

Sediment moved in Bed load,
erosion of surface

Sediment spills
as avalanche
down face

Some deposition
from suspension

Figure 3.5 Movement and deposition of sediment in a ripple. Solid arrows represent transport and deposition from the bed load. Sediment is deposited as the flow separates at the ripple crest and may then avalanche down the lee slope. Successive positions of the lee slope are indicated by inclined laminae formed as the ripple migrates downstream. Megaripple and dune forms (including those deposited by wind) have a similar structure and mechanism of movement.

Figure 3.6 (*A*) Current ripples exposed on the foreshore at Wells-next-the-sea, Norfolk (UK) (*B*) Megaripples with a wavelength of 8–10 m in a tidal channel at Wells-next-the-sea, Norfolk (UK) (*C*) Planar bedding an parting lineation formed in the upper flow regime. Preserved in Jurassic sandstones at Cloughton Wyke, Yorkshire (UK) (*D*) Standing waves generated in the upper flow regime. Channel on an intertidal beach, Monifieth, Angus (UK).

climbing ripples (see 'Transport and deposition by density flows').

As velocities increase, both the amplitude and the wavelength of the bed forms change to produce larger features variously described as megaripples (Figure 3.6*b*), sand waves (Chapter 16) and dunes (Figure 3.7). These migrate downstream in the same way as ripples and are characterised by similar (but larger scale) flow separation. They therefore generate large-scale sets of cross-laminae. However, as the current velocity continues to increase, ripples are swept away. Large quantities of sediment are now in suspension and are deposited as planar beds in a carpet across the stream channel. The resulting bed surface is commonly marked by linear sand streaks (a parting lineation: Figure 3.6*c*) aligned parallel to the flow. In hydrodynamic terms the appearance of these planar beds heralds the onset of the 'Upper Flow Regime' conditions described by Froude. However, at higher velocities, the planar bed forms are disturbed, and a new wave form, about the same size as a dune, appears. This is commonly accompanied by the appearance of standing waves (Figure 3.6*d*) on the water surface that periodically break upstream. The steeper face of the bed form also faces upstream and bursts of deposition occur on the upstream face, although some downstream motion has also been observed. For this reason the structures are known as antidunes. Antidunes have been observed in flume-tank experiments but are unlikely to be preserved in the sedimentary record because, as the current decreases, planar beds are re-established. In any event they would be difficult to differentiate from downstream migrating dunes.

The gradual downstream migration of sediment and bed forms under conditions where sedimentation from the bed load exceeds reworking, produces a carpet of material that may form a prograding bar. Bars may also form in channels during periods of low stage flow. The successive accreting layers formed in these generate a cross-stratification or cross-bedding in which individual layers are inclined to the main bedding planes that reflect the depositional surface (Figures 3.5, 3.7, 3.8, 3.9*a*). The assembly of layers in each unit is known as a set and in a large-scale bar or dune successive sets may be separated by erosion surfaces. Laminae within a set may be planar or trough shaped, depending on the morphology of the migrating bedforms (Figure 3.8). It is important to note that features such as bars may accrete laterally, and thus cross-bedding does not necessarily face downstream. Commonly in bar accretion there is a general downstream fining as well as a fining upwards of the sediments involved. Large-scale cross-beds, of the order of tens of metres (Figure 3.9*b*), can be generated by the bar accretion. Delta formation may also generate cross-bedding on a similar scale but many deltas are of kilometre scale and in these it may be difficult to identify relatively low-angle depositional units.

So far, the precise current velocities that generate the various bed forms have not been stated.

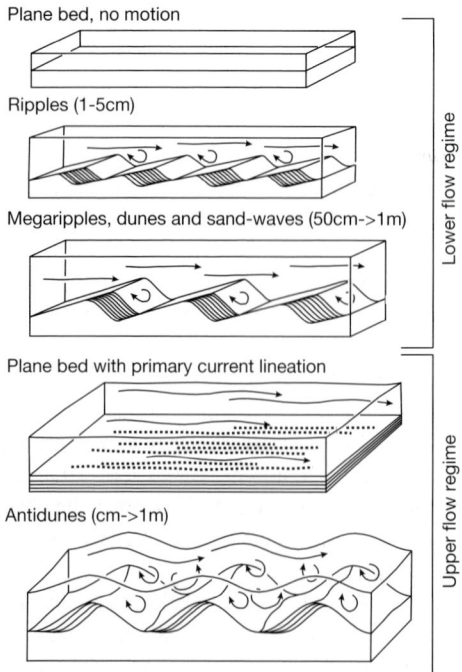

Figure 3.7   The sequence of bed forms produced in response to increasing current velocity and their relationship to flow regimes and water movements. The layering generated is shown schematically. Note the appearance of in-phase water surface waves when antidunes are formed.

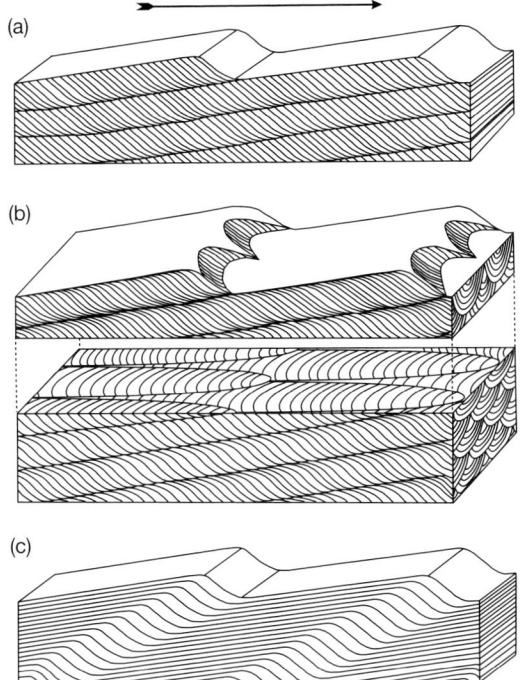

Current direction

(a)

(b)

(c)

Figure 3.8 Planar and trough cross-lamination formed by the lateral migration of ripples. In (A) linear ripples migrate as their stoss faces are eroded. By contrast, in (B) similar structures beneath linguoid ripples form a series of troughs visible in horizontal section and surfaces cut transverse to the flow direction. In (C) climbing ripples are formed where large quantities of sand are carried in suspension and there is therefore no erosion of stoss surfaces.

This is because although there is an obvious and direct relationship between current velocity and bed forms, the depth of the flowing water and the mean grain size of the sediment forming the bed load are also significant. Water flowing in a deep channel may cause no movement of the traction deposits on the channel floor, whereas water flowing at the same speed in a shallow channel may cause current ripples to form. Decreasing the depth of flow while keeping the velocity constant has the same effect as increasing the velocity of the water at a constant depth.

The effect of the mean grain size of the bed load is shown in Figure 3.10. Planar beds arise by deposition from the bed load of material coarser than about 0.6 mm at low flow velocities. Current ripples are restricted to sands finer than about 0.6 mm but do not form on sediments with a grain size of less than about 0.1 mm. Sand waves and dunes do not form in fine-grained sands, silts or clays.

It is clearly impossible to define bed forms and cross-stratification simultaneously in terms of the current velocity and water depth under which they formed. It is therefore usual to describe them in terms of the flow regime under which deposition takes place. Planar beds, ripples, sand waves and sand dunes form in the Lower Flow Regime, and planar beds with sand streaks (parting lineation) and antidunes form in the Upper Flow Regime (Figure 3.10).

Cross-lamination or cross-bedding preserved in sedimentary rocks may be used to determine which way the current was flowing when the sediments were originally deposited and may help to define the orientation of a buried channel. This is referred to as the palaeocurrent direction. It is important to note, however, that depositional systems vary in the nature of the record that they preserve. A meandering stream, for example, with a highly sinuous channel may preserve palaeocurrent directions that point in virtually any direction, whereas in a braided stream or a more linear channel they are more narrowly constrained to the mean transport direction. From the range of bed forms present in a sequence of sediments, estimates can be made about how flow conditions varied with time.

It may be difficult to recognise whether cross-stratification represents a cross-section of a bed form parallel to the flow direction and errors may lead to the determination of incorrect palaeocurrent directions. Figure 3.8 illustrates the three-dimensional geometry produced by straight- and curve-crested ripples. It is clear that steeply dipping tabular cross-stratification may appear to be parallel laminated when viewed in a plane normal to current flow. For this reason, it is essential to identify the true direction of the depositional dip and at least two dimensions are typically required, a single planar face in a quarry is never sufficient. Finally it is also necessary to return bedding to the horizontal (to unfold any folds) before the direction is determined. This can be

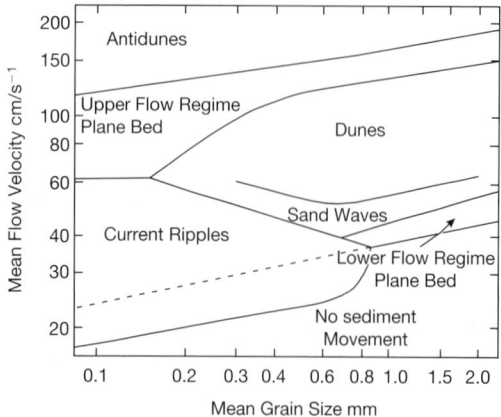

Figure 3.9 (*A*) Large-scale trough cross-bedding in Quaternary aeolian carbonate sands, Sainte François, Rodrigues, Mauritius (*B*) Large-scale cross-bedding in Jurassic carbonate sands, south of Dijon, France (*C*) Tidally formed herringbone cross-lamination, south of Mombasa, Kenya (*D*) Flute casts on sole of turbiditic greywacke, Bude, Cornwall (UK).

Figure 3.10 Graph illustrating the relationship between mean flow velocity, mean grain size of the sediment and the bed forms generated. Note that these variables are not independent.

achieved by simple geometry using stereographic projections.

Sedimentary bed forms resulting from oscillatory water movements

The movement of shallow sea water is controlled by the ebb and flow of tidal currents and by wave action. Both involve movement in more than one direction. If the currents in tidal flows are approximately equal in strength, then the inclination of the cross-stratification generated will reverse in direction on each tide. If sets reflecting opposing current directions are preserved, then a distinctive herringbone pattern is produced (Figures 3.11 and 3.9c). However, opposing tidal currents do not necessarily generate herringbone cross-stratification. Where one tide is stronger, transport may leave only an erosion horizon, or

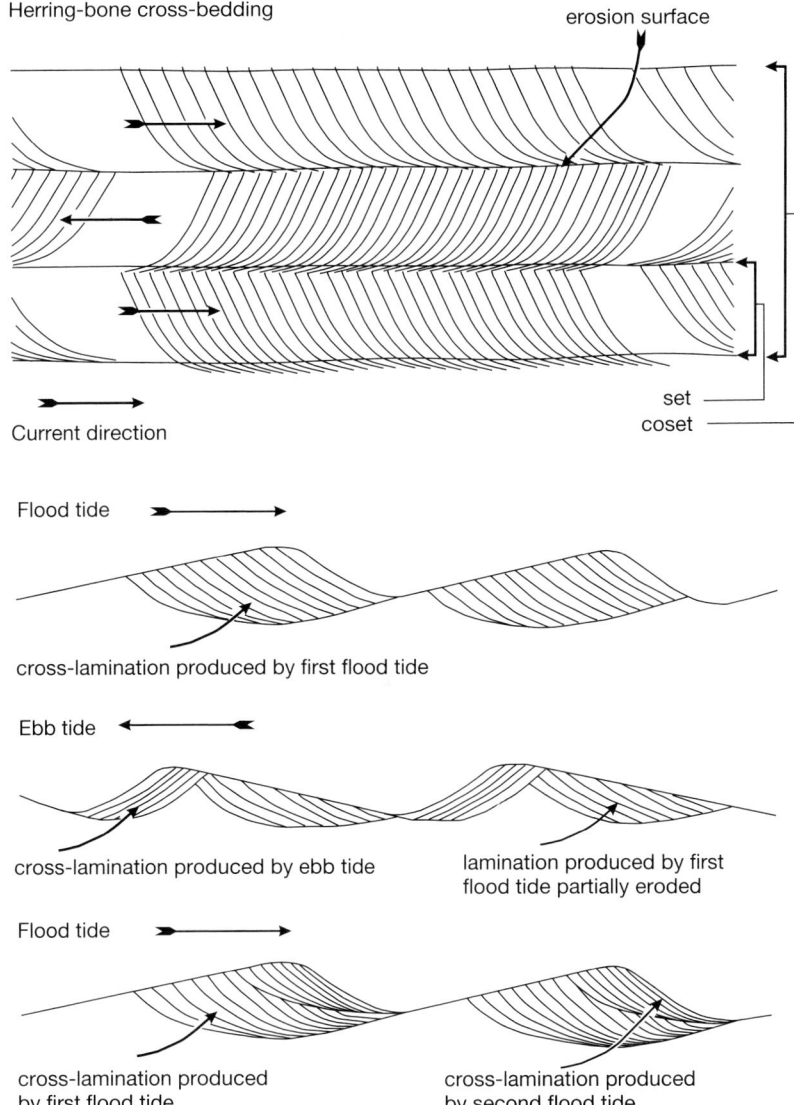

Figure 3.11   Bed forms and sedimentary structures generated by tidal flow. Herringbone cross-bedding (indicating reversing tidal flow), and ripples generated by unidirectional, bi-directional and repeated unidirectional tidal currents. Note the asymmetry resulting where one tide is dominant.

reactivation surface, as a record of the reversal (Figure 3.11). Unfortunately, reactivation surfaces may also form where flow is unidirectional. This occurs if the current velocity increases so that erosion occurs but is insufficient to produce a change in the type of bed form generated.

As a result of the greater variation in water movement the ripples that form on sea or lake beds in response to wave action have more complex internal structures than those formed in response to unidirectional currents. Waves are produced by friction as wind blows over the surface of a body of water. Within a wave, water particles move with a circular (cylindrical) orbit with the direction of motion parallel to the direction of wave motion as the wave-form passes (Figure 3.12*a*).

The radius of the motion decreases with depth, such that when the depth is equal to half the wavelength it can be considered negligible. Thus, in deep water surface waves have no direct effect on the sea bed. When the water shallows, as the wave approaches the shore, the orbit becomes flattened to an ellipse due to friction with the sea bed (Figure 3.12*b*). The orbital paths mean that as the wave crest passes there is an onshore surge of

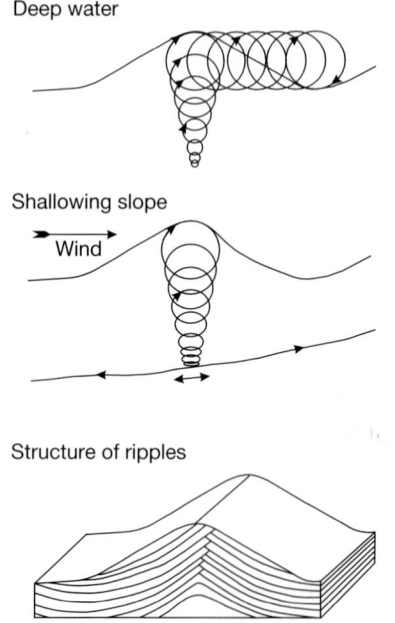

Figure 3.12 captions appear below.

water, but as the trough passes there is an offshore movement that is retarded by bed friction. The onshore movement tends to carry sediment landwards (up the beach as the wave breaks) whereas the offshore movement carries it seawards. This typically leads to the formation of symmetrical ripples with complex cross-laminae facing in alternate directions. However, if the movement is more pronounced in one direction, asymmetrical ripples may develop. When exposed on a bedding plane in sedimentary rocks, wave-formed ripples can usually be distinguished from current-formed ripples by their relative symmetry and by their relatively straight and sharply defined crests.

Transport and deposition by density flows

Density flows form a transition between mass gravity flows and fluid flow. In this transition, the initial movement of the sediment is due to gravity but here the particles form a concentrated dispersion within the fluid. Once movement has begun, it is the force of the moving fluid that keeps the sediment in motion. Density flows may occur in either air or water, but sub-aqueous flows are by far the most important in terms of sediment movement and erosion. In 1929 the Grand Banks earthquake, off Newfoundland, triggered a gigantic slump that devastated the continental slope (Figure 3.13). It also produced a series of breaks in sub-marine telephone cables but it was some twenty years before it was realised that it was the slump rather than the earthquake that was responsible for these. As it moved down the continental slope the slump formed a concentrated dispersion of sediment in sea water referred to as a turbidity current, moving down the slope and spreading across the abyssal plain (Chapter 22), decreasing in velocity.

Concentrated suspensions of sediment in water behave as a fluid with greater density than the water alone, and therefore, driven by gravity, are able to flow downslope beneath relatively sediment-free water of normal density. The study of simulated small-scale turbidity currents in laboratory tanks has provided an understanding of this type of density flow. Each flow comprises a 'head', a bulbous area with an overhanging nose,

Figure 3.12   The orbital motion of water particles in waves in deep and shallow water. The particles remain effectively stationary whereas the wave form moves in the direction of wind movement. In deep water (*A*) the wave form is symmetrical, travelling in the direction of the motion of water particles at the top of circular (cylindrical) orbits. The diameter of the orbits is equal to wave height (trough to crest) at the surface but decreases with depth and where depth is equal to half the wavelength the motion become negligible. As the water shallows (*B*) the counter flow at the base of the orbit is retarded by friction with the sea floor and as the seawards element of the motion slows the crest of the wave advances, becoming asymmetrical and ultimately breaking. At the sea bed orbits become progressively elliptical and are finally reduced to a back-and-forth motion as a result of which the sediment forms symmetrical ripples (*C*) with laminae facing both shoreward and seaward.

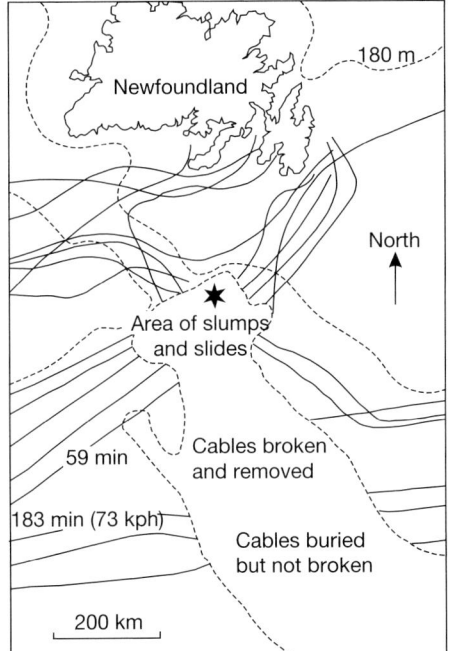

Figure 3.13   The inferred extent of the Grand Banks turbidity current of 1929. The epicentre of the earthquake and the area of the resulting slumps and turbidity currents are shown together with the effects on sub-marine telephone cables. Note the times of breakage of the cables and the implied current speed calculated for the flow at a specific cable given the distance travelled and time since the earthquake.

and a generally thinner body. The head is a turbulent chaotic mass in which water and sediment are mixed together. The nose is present because the bottom water into which it is intruding cannot be deformed rapidly enough to be thrust upwards, and so adheres to the bed, forcing the head upwards. On the upper surface of the head, transverse vortices, billows, are generated by friction with the overlying (still stationary) water, forming a wake. There is little mixing. Behind the head, the flow moves in turbulent eddies and here deposition of the coarsest sediment begins at the same time as the muddy sediment of the sea floor at the head is being eroded by the passage of the flow. It is believed that successive turbidity currents are responsible for gouging sub-marine canyons, locally hundreds of metres deep, into the continental slopes.

Density currents are also known in air. The most extensively studied form *nuées ardentes*, violent surges of hot gases and incandescent ash ejected from certain volcanoes. In the eruption of Mont Pelée in 1902 turbulent clouds with an internal temperature as high as 1075 °C rolled down slopes at speeds in excess of 160 kph. The burial of Pompei was probably the result of such a flow. Large-scale dust-laden gravity currents have also been recorded on the cold-fronts of weather systems (Chapter 23). By contrast, some forms of snow avalanche are also effectively density flows.

The physics of sediment movement in a turbidity current is complex; it is sufficient to say that a mass of sediment and water will begin to flow down a sub-aqueous slope when the friction between the sediment–water mass and both the slope and the overlying water, is overcome. The greater the density contrast between the turbidity current and the surrounding water, and the thicker the flow, the faster it will move. The Grand Banks earthquake broke twelve sub-marine cables in its path, and the precise timing of these breaks indicated that during its early stages the current velocity must have been in the order of 100 kph. The head of the flow is a region of devastating erosion, with eddies generating characteristic scour features (flute casts: Figure 3.9d) on the sediments it passes over. Larger pebbles or other objects carried in the flow may form deep scratches (tool marks) on the sediment surface beneath.

Because many factors are involved in the initiation and maintenance of turbidity currents, it is not yet possible to arrive at a comprehensive theoretical model. The driving force can be expressed (approximately) with reference to a unit of flow of fixed width and length:

$$\text{Driving force} = x h_{B} (\rho_2 - \rho_1) g \mathrm{Sin}\beta \ \text{(N/m)}$$

where: x is the length of the element; h_B its thickness (or height above the base); ρ_1 and ρ_2 the density of the medium and the current respectively; g the acceleration due to gravity and β the angle of slope.

Although the initial velocity of the current is a function of the contrast in density between the sediment-laden water mass and the surrounding sea water, and of the thickness of the flow, the decrease in gradient towards the base of the continental slope leads to a rapid decrease in velocity and results in deposition. The deposit formed is referred to as a turbidite. However, deceleration of the flow is not instantaneous, and deposition therefore takes place in stages according to conditions within the flow and the settling velocities of the various particles present. Typically, turbidites exhibit graded bedding at the base, with the coarsest particles in contact with the eroded sea bed, becoming progressively finer-grained towards the top. However, because the flow is dense and turbulent the sediment initially deposited is commonly a poorly sorted muddy sand or greywacke. Once this period of rapid deposition is over, the concentration of sediment within the flow declines so that it begins to behave more like a conventional fluid. Deposition then takes place from the bed load and changes in bed forms reflect changes in the flow velocity (the sequence of bed forms is shown in Figure 3.14). Flow is still rapid (Upper Flow Regime), and

because of the large volume of sediment still held in suspension the first structures to appear are commonly parallel laminae with a parting lineation and these are rapidly replaced by climbing ripples.

The ideal turbidite sequence (first outlined by Arnold Bouma in 1962 and now referred to as a Bouma sequence) ranges from a few centimetres to several metres in thickness. It is divided into five distinct units, conventionally designated A to E (Figure 3.14), that can be interpreted in terms of the velocity of fluid flow. The thickness and completeness of the Bouma sequence formed where the flow is at its maximum, decreases down current as the current velocity decreases and sediment is deposited. The scour and impact structures at the base of the sequence become smaller, shallower and more localized. Frequent turbidity currents spilling from sub-marine canyons onto the base of the continental slope build a submarine fan.

The main types of transport and the varied sedimentary deposits that result from these are summarised in Table 3.4. Transport by glacial action is treated fully in Chapter 13, and transport by wind is discussed in Chapter 16.

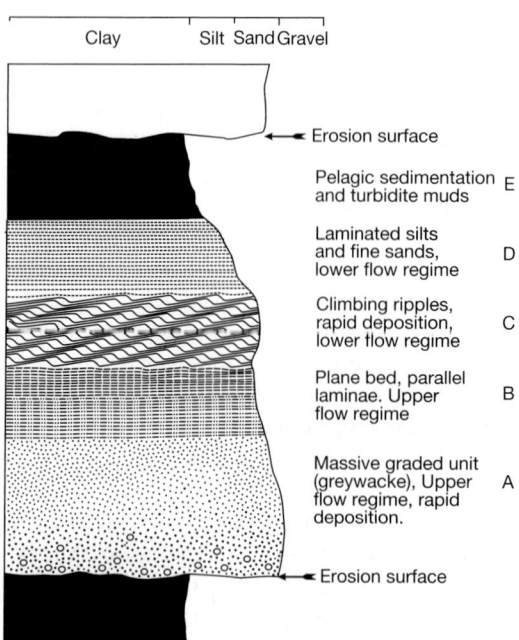

Figure 3.14 Model Bouma sequence for a density (turbidity) current illustrating the sequence of lithologies and bed forms that may be expected from a single event.

3.4 Sedimentary rocks

Lithification

The processes that transform loose unconsolidated sediments into rocks are collectively referred to as lithification. Lithification may be very rapid in geological terms, taking only a few years' or may be much slower. There are for example sands well over 200 Ma old that are still loose sediments. The changes that take place are referred to as diagenesis, and the two most important stages in this are compaction of the sediment as it consolidates under gravitational stress, and the cementation of individual grains.

Compaction. Freshly deposited sediments have high porosities (the ratio of pore volume to total volume). In freshly deposited sand the porosity depends on a number of factors including the way in which grains are packed together, their shape (their fabric: Figure 3.15), and whether the

Table 3.4 Summary of mechanisms of sediment transport and characteristics of the resulting deposits.

Mechanism	Fluid flow	Density flow	Mass movement	
Method of movement	Fluid flows downslope. Movement initiated by fluid current: material carried as bedload or in suspension	Movement of fluid initiated by density contrasts. flow turbulent, large volumes of sediment in suspension	Grain interaction driven by gravity	Material carried on or within flowing ice
Agent of transport	*Water* (Traction saltation and suspension)	*Turbidity currents (in water)*	*Slumps and debris flows*	*Brittle deformation and plastic flow of ice*
	Air (wind) (Saltation and surface creep but very large volumes of sediment in suspension)	*Turbulent flow in air (Nuées ardentes)*		*Sliding, deformation, melting and refreezing of ice*
Nature of deposits	Planar laminated clays, silts and fine grained sands (suspension load)	Planar laminated clays and silts at top of sequence (suspended load)	Massive deposits, poorly bedded	Diamict deposits unsorted, but may be locally layered by shear
	Cross-bedded and planar laminated sands to gravels (traction and saltation)	Massive graded and poorly sorted sands at base of sequence (Turbulent suspension)	Slumps may contain deformed bedding, debris flows unsorted or crudely laminated	
	Planar laminated clays and silts (suspension load)			
	Cross-bedded sands, with steep foresets (traction and saltation)			

a) Packing and porosity

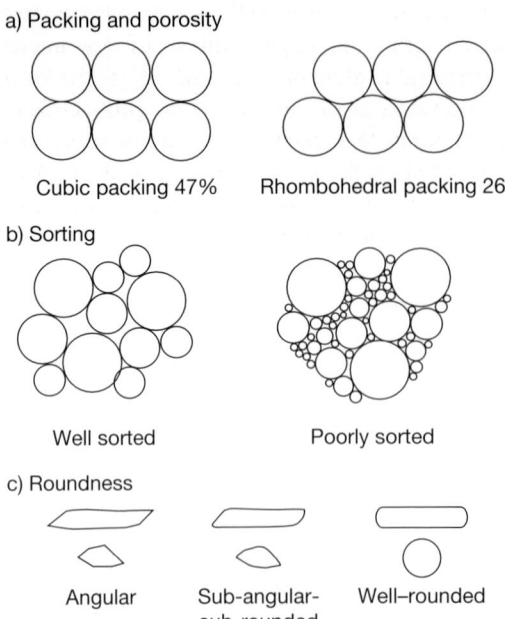

Cubic packing 47%   Rhombohedral packing 26%

b) Sorting

Well sorted   Poorly sorted

c) Roundness

Angular   Sub-angular-   Well–rounded
             sub-rounded

Figure 3.15 The effects of grain packing and grain size on porosity of a granular material: (*A*) spherical grains of uniform size arranged in cubic packing have a porosity of ~47%; in rhombohedral packing the same grains have a porosity of ~26%. In (*B*) a well-sorted sediment with a narrow range of grain sizes has a relatively high porosity whereas a poorly sorted sediment with a wide range of grain sizes has a relatively low porosity. (*C*) illustrates the effects of rounding, the degree to which grain margins are worn. Note that this is distinct from sphericity, which describes the degree of approximation to a sphere — an elongated grain can be as well rounded as a sphere.

sand also contains fine clay or silt particles. For relatively well-sorted quartz sands, porosity on deposition may be as high as 40%. By contrast, bioclastic sands, consisting of the skeletal remains of a variety of organisms and made up of irregularly shaped grains may locally achieve porosities of 80–90%.

As successive layers of sediment are deposited, burial causes compression and initially brings about a reduction in porosity by rearranging grains, the geological process known as compaction. In quartz sand with near spherical grains, porosity may be reduced to around 30%, but values of this order are exceptional. Sands, consisting mostly of

quartz or similar silicate grains generally have a low compressibility and their porosity after burial may still be close to that at deposition even after large stress increases. A few types of grains may be able to deform plastically to reduce the total volume and some may fracture. Large stress increases are concentrated at the points of contact between grains and result in an increase in the solubility of stressed areas, leading to pressure-dissolution. Typically, however, this requires more than a kilometre of burial. Pressure-dissolution results in an increase in both the areas of contact and the number of contacts between grains. Porosity decreases and may ultimately approach zero. This process alone may firmly bind the particles together to form a rock without the need for any added cement. There is a linear relationship between the bulk density of the rock and its strength, reflecting the increasing size of the areas of bonding between grains and the parallel decreasing pore space.

The porosity of freshly deposited clay is much higher than that of sand. Because the clay mineral particles have a plate-like form and carry differing charges on their faces and edges, they are able to aggregate together to form loose box-like structures, retaining porosities as high as 70–90% when first deposited. As a result, clays are much more compressible than sands and porosity decreases rapidly with increasing stress, controlled almost entirely by burial. The large volumes of water that are expelled escape through the overlying sediment where they play an important role in later cementation. The final compaction of the clay to a rock requires a substantial depth of burial and temperatures of 100°C or more, resulting in chemical changes in the clay minerals.

High porosity is an economically important feature of sedimentary rocks that form natural reservoirs for the accumulation of oil, gas or water, and porosity has a reciprocal relationship to strength. Overpressured formations are those in which porewaters are trapped, sealed in by overlying impermeable formations, with the result that normal compaction cannot occur. Pumping of water or oil from these may lead to significant surface subsidence as overburden pressures are transferred from pore fluids to grains, resulting in compaction.

Cementation. This involves the growth of new mineral crystals, referred to as cement, within the pore spaces of sediment, and is common in both sandstones and limestones. The cement is most often precipitated from porewaters expelled from inter-bedded or underlying clay sediments during compaction, but may also result from basin-wide fluid movements. The dissolved salts forming the cement are commonly derived from water–rock interactions that occur during these large-scale movements but may also be generated locally by late stage pressure-dissolution. The material dissolved is reprecipitated as cement (Figure 3.16*a*),

and may form overgrowths on existing coarse detrital grains or large numbers of new nuclei scattered at random on grain surfaces. In the latter case crystals grow outwards from grain surfaces to fill pores, their relative sizes and shapes being controlled by the space available to them. As a result, crystals orientated parallel to grain surfaces are rapidly overgrown and remain small whereas those that extend towards the centres of pores are free to grow and become significantly larger.

Sands that are deposited in highly oxidizing, arid environments are commonly red coloured and this is why the 'red beds' found in the geological record are usually interpreted as indicating

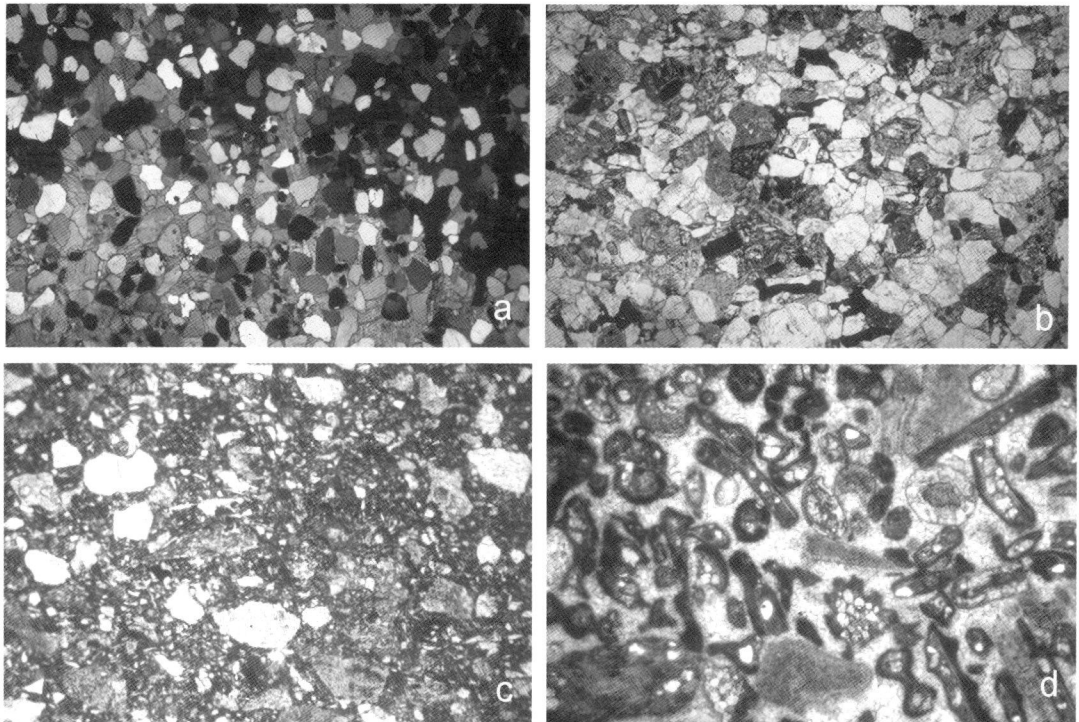

Figure 3.16 (*a*) Thin section of quartz arenite with coarse calcite cement. Note that calcite crystals are so large that they enclose many quartz grains that are approximately 0.25 mm diameter. Crossed polars. (*b*) Thin section of lithic sandstone. A wide variety of grains, including lithic (rock) fragments are squeezed together by compaction and boundaries have been modified by pressure-dissolution, there is no added cement. Grains are approximately 0.5 mm diameter. Plain polarized light. (*c*) Thin section of greywacke, note poor sorting, angular sand-sized grains including quartz, feldspar and lithic fragments, in a dense muddy matrix. Largest grains are approximately 0.5 mm diameter. Plain polarized light. (*d*) Thin section of bioclastic grainstone. Note fragments of a variety of fossils surrounded by a granular calcite cement. Note the lack of compaction, implying early cementation. Grains are approximately 0.5 mm diameter. Plain polarized light.

a highly oxidizing atmosphere, and sometimes even desert conditions. However, detailed investigations have shown that this interpretation must be made with caution, as the sands in modern deserts are not typically red. The colour in these ancient rocks reflects the weathering of grains of ferromagnesian minerals. These may be dissolved long after deposition — tens of thousands or possibly millions of years — and form a red iron oxide stain that spreads over adjacent grain surfaces. As little as 0.1% of iron oxide is enough to redden sediments. The red coloured surfaces may later be overgrown by quartz or other cements. It is important to note that in the area where these processes have been studied in greatest detail, intertidal and even marine sediments bordering the desert have also been stained red.

Although quartz cements are common, carbonates of calcium, magnesium and iron also form cements in sandstones. In all but a few very young marine limestones calcite is the dominant cement. The initial growth of cement within a sediment binds the loose grains together, forming a rock. Subsequent growth is largely responsible for decreasing the porosity and increasing the strength of the material.

Siliciclastic sediments and sedimentary rocks

Siliciclastic sedimentary rocks consist principally of the silicate minerals that remain after weathering; they are dominated by quartz. Sandstones are defined in terms of their grain size (Table 3.5) and may be classified on the basis of the relative proportions of quartz, feldspar, and lithic (rock) fragments (Figure 3.16b), providing information on the provenance, transport and depositional history of the sediment. The common names are illustrated in Figure 3.17. Sandstones with a carbonate cement are described as calcareous (e.g. calcareous quartz arenite). Those in which a large proportion of grains are carbonate fragments such as bioclasts (see below) are described as calcarenaceous. Both may weather in a similar manner to limestones as the carbonate dissolves.

Although (as indicated in 'Transport and deposition by fluids') there is a close correlation between current speed and the size of grains that

Table 3.5 The Udden-Wentworth scheme of grain-size classification of sediments. Compare with the engineering classification given in Chapter 7.

Class terms (sediments)	Grain size	Sedimentary rock types
Boulders		Conglomerate (rounded fragments or clasts), or breccia (angular fragments or clasts)
	256 mm	
Cobbles		
	64 mm	
Pebbles		
	4 mm	
Granules		
	2 mm	
Very coarse sand		
	1 mm	
Coarse sand		Sandstone
	0.5 mm	
Medium sand		
	0.25 mm	
Fine Sand		
	0.125 mm	
Very fine sand		
	0.0625 mm	
Silt		Siltstone Mudstone when non-fissile, shale
	0.039 mm	
Clay		Claystone when fissile

can be transported, small fluctuations in current velocity and in the size and density of individual grains mean that sediments consist of a population of grains rather than a single grain size. The most important descriptor of this grain-size distribution is sorting which characterises the range of particle sizes present in a sedimentary rock (Figure 3.15). Traditionally, grain size has been determined by sieving the sediment, but it can be measured more accurately by allowing grains to settle in a water column, or by direct measurement. It should be noted that the sizes attributed to grains by sieving bear little relationship to their behaviour in fluids if they are of irregular shapes or show much variation in density, both grain shape and density may vary

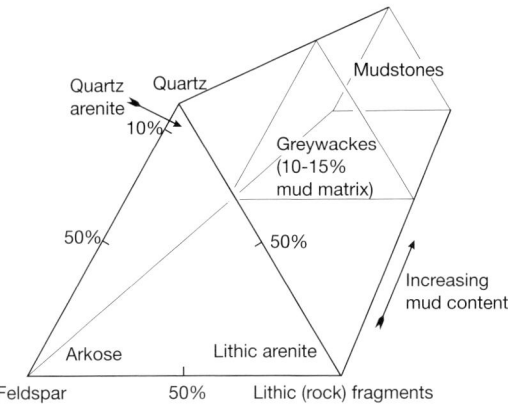

Figure 3.17 Illustration of the classification of clastic (siliciclastic) sandstones based on the relative proportions of quartz, feldspar and rock (lithic) fragments (QFL). A variety of similar schemes are available but there is no agreement on precisely where boundaries should be drawn. Nevertheless, most sandstones can be plotted within the space defined by this geometrical form.

widely in carbonate sediments. Settling has two advantages. First the results reflect the real hydrodynamic behaviour of the grains rather than an extrapolation based on their size alone. Second, it measures the behaviour of a population (hindered settling) rather than that of individual grains. Irrespective of the method, a mean or average grain size present in the sample can be calculated, and sorting represents the spread of grain sizes about this mean size.

So far as depositional processes are concerned, good sorting is achieved either where grains have undergone aqueous transport for long periods, or where sustained currents in the transporting medium are selective with respect to the size of grain that they are able to move as, for example, in wind transport. Poorly sorted sediments are typically formed when a mass of material is transported and deposited rapidly without being reworked. This may occur when there is a sudden reduction in the velocity of a current, or when the transporting medium is non-selective with respect to grain sizes, as in debris flows, turbidity currents and glacial transport. The typical deposits of turbidity current transport are poorly sorted greywackes (Figure 3.16c). However, it is important to appreciate that the degree of sorting of sediment (or sedimentary rock) is also a function of the source rock from which it was derived. Erosion of a poorly cemented, well-sorted sandstone cannot produce a poorly sorted sediment, unless material is added from another source. The more cycles of erosion and deposition a sediment suffers the more 'well sorted' (consisting of a narrower range of grain sizes) it is likely to become. 'Poorly sorted' sediments (consisting of a wide and commonly discontinuous range of grain sizes) may be generated after deposition by mixing of originally separate grain-size populations by the activities of burrowing animals (bioturbation) and by later infiltration of a finer-grained component, a feature of some desert sands that would otherwise be well sorted. It should be noted that there is an important divergence in terminology that may lead to confusion. Many engineers refer to sediments comprising a broad range of sizes of grains (commonly referred to as particles) as 'well graded'.

The fabric or texture of a rock is a reflection of sorting and, with the exceptions noted, is a product of the way in which the original sediment was transported. It reflects the relationship between coarser grains, which form the dominant component, and a finer-grained matrix that occupies inter-grain spaces. In sediments where the dominant grains form a self-supporting framework, the fabric is described as grain supported. Where the larger grains are dispersed in a finer-grained matrix, with little or no inter-grain contact, rather in the way currents are dispersed in a fruit-cake, the fabric is said to show matrix-support. Matrix-support is common in sediments deposited by glacial action, by mass movement and from density flows, where sorting is inefficient. Grain-support is common in suspension and traction deposits from fluid flow (wind and water) that are efficiently sorted.

The attrition between grains during transport causes rounding. Sediments that have been transported far enough for the grains to become well rounded (Figure 3.15) and also efficiently separated so that the sediment is well sorted are described as texturally mature. During aqueous transport, inter-grain impact is reduced due to the cushioning effect of the relatively viscous water. Attrition is therefore reduced although sorting is moderately efficient. The effects of attrition decrease as grain size decreases. By contrast, during wind transport inter-grain impact is great, attrition is high and sorting is also very efficient. Both water transported and aeolian (wind-blown) sands show increasing textural maturity as they are transported further. However, glacial diamicts and deposits from mass wasting are characteristically texturally immature because very little sorting or attrition is possible during transport.

The mineral content of a sediment also indicates 'maturity' because both physical and chemical weathering continue during transport and lead to the progressive disintegration and decomposition of the less stable silicate minerals. These processes continue long after burial. Thus, although feldspars are far more abundant than quartz in igneous and metamorphic rocks, quartz is one of the most abundant minerals in sedimentary rocks. This is because feldspars are chemically weathered more rapidly than quartz and decompose to form clay minerals. The presence of feldspars in a water-lain sediment indicates that the periods of both weathering and transport must have been brief, with physical weathering dominant and deposition occurring close to the source area. Limited weathering, rapid erosion and rapid deposition commonly also result in the formation of sands containing large numbers of rock (lithic) fragments (Figure 3.17). These may dominate all other grain types. It is important to note that rocks resulting from the deposition of sediments containing these less stable components weather more rapidly when used as building stones.

With increasing intensity of chemical weathering and longer aqueous transport, almost all silicate minerals except quartz are decomposed, and so the end products are either quartz sand or clay-rich mud. These sediments are said to be mineralogically mature and, by contrast, those that contain a high proportion of minerals such as feldspar, mica, ferromagnesian minerals or rock fragments are said to be mineralogically immature.

Textural maturity is a sign of prolonged transport during one or many cycles in water or of aeolian transport. Mineralogical maturity may result from a period of prolonged chemical weathering or substantial transport by water that may include multiple sediment cycles. Glacial diamicts, debris flows and sub-aerial density deposits are commonly both texturally and mineralogically immature. Aeolian deposits may be texturally mature but mineralogically immature due to the absence of significant chemical weathering in arid climates.

Limestones

Although limestones and dolomites (Ca–Mg carbonate rocks) form only about 15% of the Earth's exposed sediments, they have great economic importance. Some 40% of the world's oil reservoirs are found in them. They are host rocks to some of the largest lead–zinc deposits and are closely associated with phosphorites and bauxites. Many form important aquifers. They are the key component in the manufacture of cement and are widely used in aggregate production. Because of their solubility and the unusual landforms that develop on their surfaces they provide some of the most intractable foundation problems in civil engineering (Chapters 24 and 26).

Among the soluble products of weathering, that tend to end up in the sea, are Ca^{2+} and HCO_3^- ions (Table 3.1). These are the raw materials from which calcareous sediments, and eventually limestones, are formed. Surface sea water is supersaturated with respect to both calcite and aragonite, the common marine carbonate minerals. The majority of both modern and ancient carbonate sediments are biogenic, that is, they owe their origin to biological activity. A vast number of microscopic planktonic organisms that float in the surface layers of the oceans and macroscopic creatures living on the sea floor extract $CaCO_3$ from sea water to build their skeletons. These include molluscs, echinoderms (sea-urchins and their relatives), corals and calcareous algae (Chapter 22). Skeletons may consist of the minerals aragonite or calcite. After the death of the organisms they accumulate on the sea floor to form calcareous sediments. At the present day, about 95% by area of marine calcareous sediments are accumulating in the ocean basins beyond the continental shelves. The chalk of southern England is an example of a carbonate mud formed in this way. The remaining 5% by area of carbonate sediments are found in the shallow waters on continental shelves or fringing volcanic islands and comprise a high proportion of macro-skeletal debris.

Regardless of where carbonates form, one overriding factor controls their accumulation as relatively pure carbonate sediments. This is the absence, or virtual absence, of any siliciclastic sediment. A high influx of siliciclastic material will dilute the skeletal remains that are deposited so that mixed calcareous–siliciclastic sediments are formed (these are generally referred to using terms such as 'sandy limestone'). Muds in suspension may cloud the water and inhibit the growth of photosynthetic organisms including algae in reef-building corals. The most widely distributed carbonate sediments are found in the tropics although there are occurrences even in polar regions.

Not all carbonates are biogenic. Because sea water is supersaturated with respect to calcite and aragonite a slight increase in temperature and or salinity may be enough to cause precipitation. The presence of Mg^{2+} ions in sea water apparently inhibits the nucleation of calcite and so it is the mineral aragonite that most commonly appears. In the Bahamas at present this may take one of two forms. In shallow lagoon areas, carbonate muds are formed in part as the breakdown products of calcareous algae but also by direct precipitation. In addition, towards the margins of the banks, current-swept shoals are sites for the accumulation of oolites. These sand-sized grains are made up of concentric layers of aragonite crystals that also form directly from sea water. Oolitic limestones are common in the rock record and many are used as building stones or form important aquifers. Direct precipitation of carbonate from sea water is also responsible for local cementation of the sea floor.

Limestones are classified on the basis of the relative proportions of their principle components, grains, and muddy matrix or cement, following Dunham (1962) as shown in Table 3.6. In addition, however, the dominant grain types such as bioclasts (skeletal fragments of organisms) or ooids may be identified (Figure 3.16d).

In the geological past, when sea level was higher, shelf seas were far more extensive than they are now, and so we find that shelf carbonates make a very important contribution to the geological record. At present, extensive shelves occur off the north-east coast of Australia (the Great Barrier Reef, 2000 km long), off Honduras in the Caribbean, in the Arabian Gulf and Red Sea and forming the Bahamas Bank in the western Atlantic and the Seychelles Bank in the western Indian Ocean. In many locations the shallow-water areas generated by reef systems have provided the foundations for land reclaimed for the construction of airports, harbour works and housing.

Table 3.6 Dunham's (1962) classification of limestones based on the relative proportions of the dominant components. Most limestones can be accommodated within this scheme with prefixes to indicate the dominant grain type e.g. oolitic grainstone.

Grain supported (grains in contact)		Mud supported (muddy matrix)		Components bound by organic growth during deposition	Depositional texture obscured by diagenetic change
No mud present (crystalline cement)	Sparse mud between grains	More than 10% grains	Less than 10% grains		
Grainstone	Packstone	Wackestone	Mudstone	Boundstone	Crystalline carbonate

Chemical and organic sediments

Although generally beyond the scope of this book, the reader should be aware of a variety of chemical and organic sediments, many of which are of economic importance. Under the broad heading of 'coals' are two distinct groups. Those that formed as land-based deposits of woody plant material, and accumulated where the trees grew and died, and those composed of algae, spores and fragmented plant debris that were washed into lakes or the sea to accumulate. Like carbonates, these accumulations require the absence of any other sediment. The gradual conversion of decaying vegetation to coal takes place with increasing temperature, typically resulting from an increasing depth of burial. Volatiles are lost and as a result the residue comes to consist of an increasing proportion of carbon. During the formation of peat, there is little alteration of plant material, but there is a gradual progression from this to brown coals and on to bituminous and anthracitic coals, hard, black and shiny materials with few traces of the original plant structure. It is important to remember that the gas methane is generated as a result of the coal-forming process and may accumulate in tunnels through coal-bearing rocks, potentially forming an explosive hazard.

Petroleum hydrocarbons (crude oil and gas) are also formed by the alteration of organic matter buried in sediments. They originate in organic-rich source rocks. Almost all mud rocks contain some organic matter (typically 2–3%) but source rocks such as oil shales may have as much as 10–12%. Like coals, these mature with increasing temperature, with optimum conditions on average 2–3 km beneath the surface but they retain a high hydrogen content. The dispersed droplets of oil and dissolved gas that result migrate upwards to be trapped and accumulate in porous reservoir rocks.

Bedded cherts originate as biogenic siliceous oozes on the sea floor. The oozes comprise the siliceous skeletal remains of microscopic marine plankton, diatoms and radiolaria. These are normally found in deeper waters than calcareous oozes, but some mixed siliceous–calcareous oozes also occur in shallow waters. These skeletal remains consist of silica in the form of opal. This has a poorly defined crystal structure that contains relatively large amounts of water. It is unstable and recrystallises relatively rapidly to microcrystalline quartz; cherts generally consist of this. However, it is important to note that many limestones and dolomites also contain chert or flint nodules that form by a process of chemical replacement of the original carbonate by silica and are not biogenic. They may be decimetres or even metres in diameter and have important implications in construction, and in aggregate and cement production.

In addition to the organic sediments described above, a number of sedimentary rocks have an unequivocally chemical origin. As their name indicates, evaporites form where sea water or inland drainage basins evaporate, and their presence in the sedimentary record is therefore a good indication of a former arid climate. The principal evaporite minerals are gypsum (a hydrated form of calcium sulphate, $CaSO_4 \cdot 2H_2O$), anhydrite (the anhydrous form of calcium sulphate, $CaSO_4$) and halite (NaCl), although potassium and magnesium sulphates and chlorides and a variety of other salts may also be present. In experiments, carbonates form when sea water is evaporated to about 50% of its original volume, gypsum at 20% and halite at 10%. Evaporite minerals are an economic resource but in some areas play a major role as impermeable seals to petroleum reservoirs. Because of their high solubility they represent a significant hazard to construction even at low concentrations. They may also play an important detrimental role in weathering.

Ironstones are sedimentary rocks containing more than 15% iron that may be in the form of iron oxides, carbonates, silicate or sulphides.

3.5 Sedimentary facies and environments

Ancient sedimentary environments are reconstructed on the basis of the interpretation of features of the individual rock units, of their relationships with units above and below, and of the way in they change laterally. Geologists use the term 'facies' to describe the individual units, but there are variations in meaning.

Strictly, 'facies' describes the specific set of features that characterise a body of rock and is therefore wholly objective. Although it is now being used to refer to both igneous and metamorphic rock associations, the term originated as a descriptor of sedimentary sequences. The features that are considered can be observed in the field and laboratory. They include the overall geometry of the sediment body, the thickness, shape and distribution of beds; their colour, mineral composition, texture, including grain size, shape and sorting; sedimentary structures and fossil content. Two derivatives have appeared that conform to this objective approach. Lithofacies refers solely to the lithological features of the unit, geometry, textures, and structures, while biofacies is concerned with the nature and distribution of the fossil flora and fauna present in the unit. However, other more subjective variations are also in use and for some authors the term regrettably implies a specific origin, as in deltaic facies or turbidite facies. Further discussion of the concept of facies is included in Chapter 5 and in Reading (1978).

Facies models result from interpretations of both modern and ancient facies sequences, and for some it is the ancient record that is dominant. The 'essence' of the model is achieved by 'boiling away the local details, but distilling and concentrating the important features they have in common into a general summary' (Walker, 1984, p.6). Such models may be specific to a particular sequence or generalised for an environment (see examples in Chapters 7, 8 or 15).

An implicit element in this distillation is the principle of uniformitarianism, that the geological record can be interpreted in terms of processes acting at the present time. Earth history has, for the most part, unfolded in a gradual evolutionary manner, and has not been dominated by a series of *universal* catastrophes, although there have been significant rare events in geological history. It is nevertheless important to realise that the individual units observed within a rock sequence are records of specific events. While some of these were protracted in geological terms, many were of short duration, measured in weeks, or even hours. Most of geological time is represented by the

omissions, by bedding planes and by erosion surfaces (Chapter 5).

As depositional models have become more refined, there has been a growing realisation that although our observations of the Recent and Quaternary strata are invaluable, it is unlikely that they indicate the relative importance of the varied processes and depositional environments that have existed through geological time (Chapters 5 and 6).

3.6 Summary

Sedimentology is concerned with the manner in which sedimentary grains are produced, transported, deposited and buried to become sedimentary rocks. It includes the study of sequences of sedimentary rocks aimed at interpreting the nature of past environments and much of the remainder of this book outlines the characteristics of such sequences.

References

Blatt, H., Middleton, G. and Murray, R. (1980) *Origin of Sedimentary Rocks* (2nd edn). Prentice-Hall, Englewood Cliffs, New Jersey, 782pp.

Dunham, R. J. (1962) Classification of carbonate rocks according to depositional texture. In Ham, W. E. (ed.) *Classification of Carbonate Rocks*. American Association of Petroleum Geologists, Memoir 1, 108–121.

Kukal, Z. (1971) *Geology of Recent Sediments* (translated by Helena Zarubova). Academic Publishing House of the Czechoslovak Academy of Science and Academic Press, London, 490pp.

Reading, H. G. (ed.) (1978) *Sedimentary Environments and Facies*. Blackwell Scientific, Oxford, 615pp.

Walker, R. G. (1984) General introduction. In Walker R. G. (ed.) *Facies Models* (2nd edn). Geoscience Canada Reprint Series 1. Geological Association of Canada, 1–10.

Further reading

The following texts provide more detailed introductions to sedimentology.

Allen, J. R. L. (1985) *Principles of Physical Sedimentology*. George Allen & Unwin, London, 272pp.

An accessible and well-illustrated introduction to the physics of sediment movement.

Collinson, J. D. and Thompson, B. D. (1982) *Sedimentary Structures*. George Allen & Unwin, London, 194pp.
A well-illustrated introduction to the origin of bedforms and sedimentary structures.

Leeder, M. R. (1982) *Sedimentology: Process and Product*. HarperCollins Academic, London, 344pp.
A wider-ranging account of sedimentary rocks and environments.

Pettijohn, F. J., Potter, P. E. and Siever, R. (1972) *Sand and Sandstone*. Springer, Berlin and New York, 618pp.

Tucker, M. E. (2001) *Sedimentary Petrology: An Introduction to the Origins of Sedimentary Rocks* (3rd edn) Blackwell Science, Oxford, 262pp.

Scoffin, T. P. (1987) *An Introduction to Carbonate Sediments and Rocks*. Blackie, Glasgow and London, 274pp.
A clear and well-written introduction to carbonate sediments and rocks (limestones and dolomites) addressed only superficially in the texts above.
This book provides both general descriptions of sedimentary environments and an introduction to sedimentary petrography.

4. Tectonics

Tony Waltham

4.1 Geological structures

All rocks have been subjected to some degree of structural deformation, the results of which influence the physical properties of the rock mass. The only materials that have not been deformed are some of the younger sediments, whose low strengths mean that engineers classify them as soils, rather than rocks. The small-scale effect of the deformation is to create fractures and structural weaknesses that are the prime factors in determining the engineering properties of *in situ* rock masses (Figure 4.1). After their deformation, the upper parts of those rocks that constitute the landmasses are partially removed by erosion. The large-scale effect of the rock deformation then becomes apparent by determining outcrop patterns within the landscape, the shape and distribution of some larger landforms and also the larger structural elements that remain beneath the surface.

Tectonic structures are by definition those that originate by large-scale geological processes. They are ultimately the consequence of plate movements, and the vast proportion of rock deformation takes place within or adjacent to the orogenic belts that develop along convergent boundaries. Non-tectonic structures are smaller features that develop by surface processes, including unloading fractures, ice deformation (see Chapters 13 and 14), cambering and sliding (see Chapter 8).

Plate tectonics

All geological processes require energy, which has its ultimate origin in the residual and radioactive heat energy of the Earth's core. This creates convection cells within the mantle by heating its base. The rising currents roll over the top of the cells, creating horizontal movements at the base of the crust. The relatively brittle crust is therefore broken into a number of plates (Figure 4.2), which move relative to each other over the mantle cells. Plate boundaries are the zones of movement and deformation. There are three types.

Plates move sideways past each other along conservative boundaries. These have little outward significance, except for the development of large earthquakes. The San Andreas Fault is part of the Pacific/American plate boundary roughly along the Californian coast.

Plates move apart at divergent boundaries, which are also constructive, as new crust is created to add to the plates and fill the gap. Basaltic magma is fractionated from the mantle and rises to form new oceanic floor of sheeted dykes, with the excess creating submarine volcanoes. Only oceanic crust is created, and the Mid-Atlantic Ridge is over the boundary where the American plate diverges from the African and European plates.

Plates move together at convergent boundaries, which are destructive, as crust must be destroyed to accommodate the convergence. Only oceanic plate can be totally destroyed when it is subducted and melted into the mantle. Continental plate can only be crumpled, shortened and therefore thickened, to create an orogenic belt along a mountain chain over the convergence. The Andes have formed where the American plates overrides the Pacific floor. The Himalayas are an even greater mountain chain formed at the collision of the Indian and Eurasian plates where both are formed of continental crust.

The processes of orogenesis that occur within an orogenic belt encompass nearly all the major geological processes (Figure 4.3). Most igneous rocks (except oceanic basalt), sedimentary rocks, metamorphic rocks and rock deformation owe

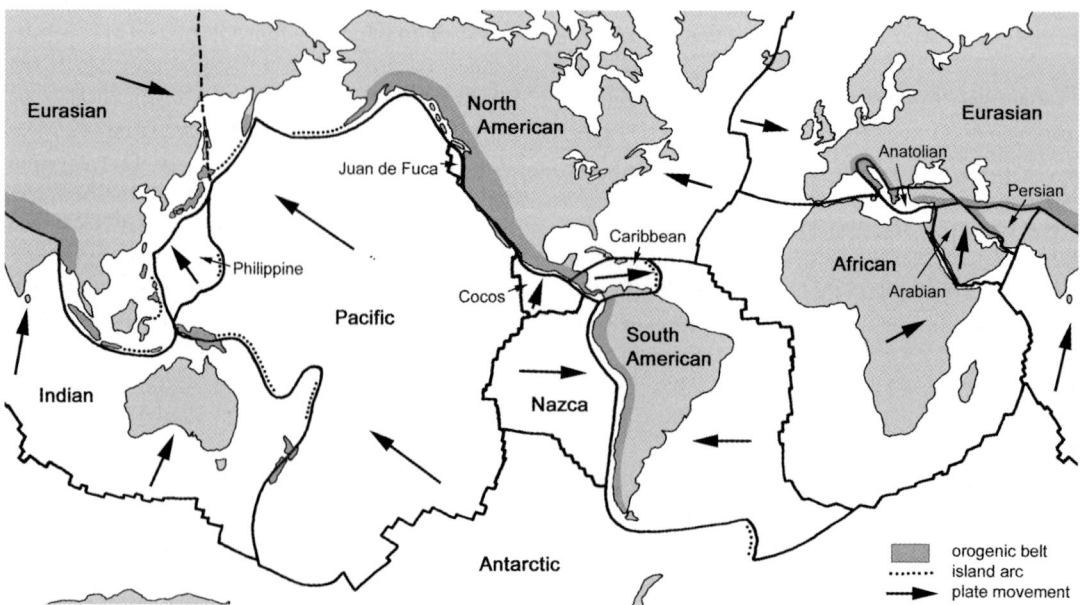

Figure 4.1 Tectonic deformation of limestones exposed in a road cut about 5 m high in Greece. Both the sub-horizontal reverse faults in the thick bed of limestone and the tight folding in some of the more thinly bedded rock were formed by the same compressive earth movements.

Figure 4.2 Outline map of the major crustal plates and the active orogenic belts along the convergent boundaries.

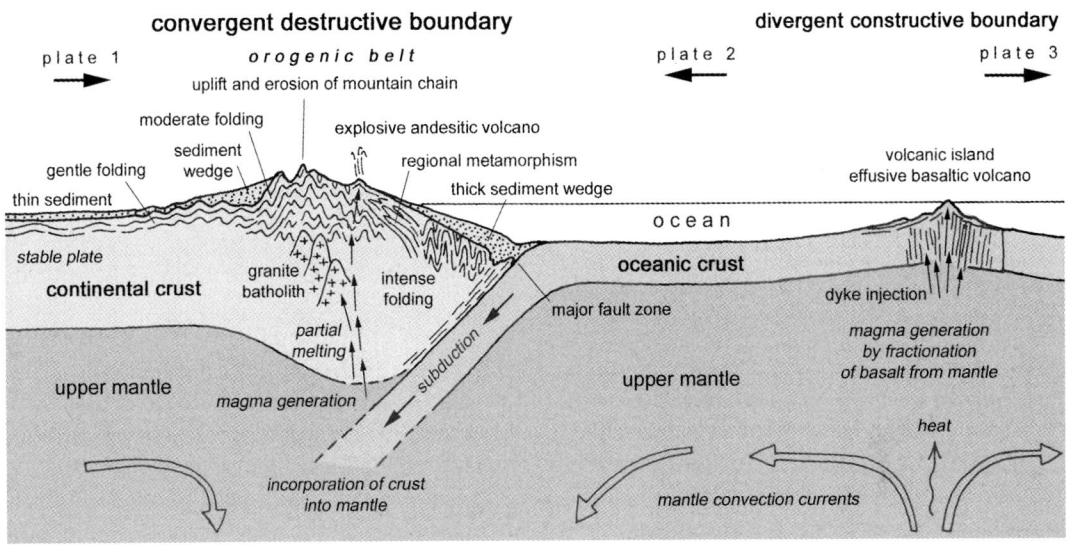

Figure 4.3 Diagrammatic cross-section through convergent and divergent plate boundaries, indicating the relative positions where major tectonic processes take place.

their origins to processes that are part of, or consequent on, orogenesis at convergent plate boundaries. In addition, mountain uplift provides the energy for rapid erosion and sediment transport. It is therefore ultimately responsible for landscape evolution; geomorphological processes evolve much more slowly on stable plates away from active convergent boundaries.

Rock deformation

The behaviour of a rock or rock mass during tectonic deformation depends on the nature of the rock, the ambient environment of temperature and water pressure, and the degree of stress imposed upon it. These factors ultimately determine whether a rock is folded or fractured, though rock strength and the rate of deformation are also significant. The stresses imposed on rock are a function of tectonic processes; they increase steadily when it sinks due to tectonic subsidence, and dramatically when it is caught within zones of plate convergence. Within the ground, vertical stress increases by 25 MPa (and temperature increases by about 20 °C) with each kilometre of burial. At depths of more than a few kilometres, horizontal stresses are generally less than vertical stresses, but they vary by an order of magnitude between

maximum and minimum values in boundary zones of plate convergence and divergence respectively.

The state of stress within the ground is defined in terms of the directions of maximum, intermediate and minimum stress, which are mutually perpendicular. The scale of rock deformation depends on the differential stress, which is the difference between the maximum and minimum stresses. The style of rock deformation depends on the confining stress, which is the value in the minimum direction. Under lower confining stresses and at lower temperatures, brittle behaviour leads to the development of fractures, both joints and faults. At higher confining stresses, at higher temperatures and aided by higher pore water pressures, ductile behaviour allows plastic deformation and the creation of fold structures.

Tectonic subsidence may carry sedimentary rocks to depths of 5–8 km, where burial stress rises to 125–200 MPa. Clays are nearly always ductile, and therefore deform into fold structures, though most clays also contain networks of micro-fractures. Sandstones and limestones may also be ductile at these low confining stresses where pore water pressures are high and strain rates are low. Typical rates of strain within tectonic deformation are around 1% per 30 000 years. Far more rapid strain can be seen in the sagging of unsupported roof slabs in built structures. At depths of less than a few kilometres, horizontal tectonic compression deforms rocks upwards, towards the minimum stress. Open symmetrical folds are therefore the dominant structures in sequences of sedimentary rocks.

Granite becomes ductile at confining stresses of about 500 MPa and temperatures of around 400 °C, conditions which pertain at depths of around 20 km within the cores of orogenic belts over convergent plate boundaries. The same conditions account for the main part of regional metamorphic processes. Metamorphic rocks are therefore characterised by plastic deformation and complex folding.

Faults are brittle failures that can develop at depths where locally high differential stress exceeds rock strength. Failure is aided by high strain rates, most conspicuously imposed along convergent boundaries. High confining stress means that frictional resistance on the fault plane is high, so that strain energy accumulates in the rock adjacent to the locked fault. Ultimately, the fault does slip with a sudden movement when the strain energy is released as ground vibrations — expressed on the ground surface as earthquakes.

High burial stresses mean that rocks are rarely subjected to absolute tensile stress, but effective tension is created in the direction of minimum stress. Normal faults and small tension gashes are therefore created normal to the minimum stress in deeply buried rocks. Rifts and grabens subside where gravitational stress exceeds horizontal stresses that are reduced by crustal divergence, and the consequently low confining stress means that brittle failure occurs.

At shallower depths, brittle fracture increasingly dominates and joint systems become the normal mode of rock failure. Within about 100 m of the ground surface, vertical stress is nearly always less than horizontal stress, except in the faces of high cliffs. Denudation and surface lowering progressively reduce burial stress, and stress relief fractures develop parallel to the exposed surface as the ground relaxes upwards towards the minimum stress. In roughly level ground, this creates the sub-horizontal unloading fractures that

are a normal component of rock weakening within the weathering profile. In steep slopes and cliffs, the unloading fractures are steeply inclined and become potential slip surfaces for many of the landslides and large rock falls that are major components of slope degradation in mountainous terrains. Numerous massive rockfalls have occurred on the precipitous slopes of Norway's fjord country, and the most catastrophic have been where tectonic joints, faults or foliation sub-parallel to the cliffs have opened up due to this stress relief. A large-scale extension of this process is seen in the Holocene 'deglaciation landslides' of Iceland, which developed when ice support was lost from the walls of U-shaped valleys. On the small scale, rock bursts are failures of the walls in deep mines, where the loss of confining stress was due to excavation of the mine tunnel. These generally occur at depths of > 600 m where overburden stress concentrated in tunnel walls exceeds the tensile strength of the rock: dangerously explosive rock bursts occur in strong rocks (unconfined compressive strength, $UCS > 140$ MPa) where stress can accumulate before failure; in weaker rocks, the unloading effect is largely an increase in rock spalling.

All rocks have a stress history of multiple phases, created by deep burial and denudation exposure and by stages of their tectonic plate evolution. They therefore have multiple suites of deformation structures. Whether these are brittle or ductile may depend on the smallest of variations in the ground environment; rock structures are not mutually exclusive. A ground engineer must anticipate finding both folds and fractures in any rock mass under assessment.

Fold structures and fold mountains

An infinite variety of three-dimensional fold structures can be created by the intense plastic deformation that is common in metamorphic rocks, though folds are generally more simple in sedimentary rock sequences. Fold terminology can be comparably complex, but may often be reduced to a few varieties of up-folded anticlines and down-folded synclines (Figure 4.4).

The intensity and style of folding are also variable and are part of any full description of rock

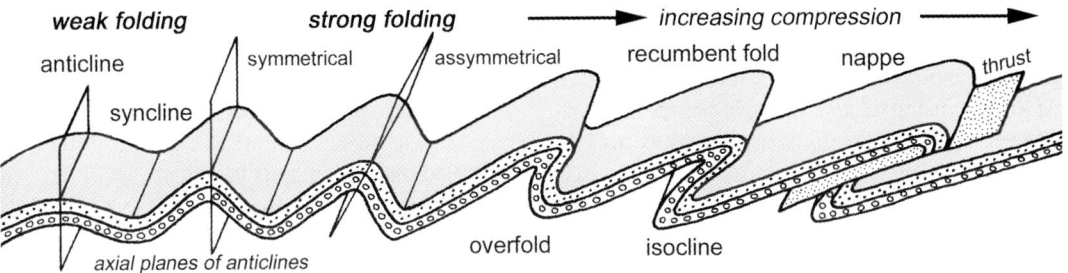

Figure 4.4   Shapes and types of folds in a sedimentary rock sequence.

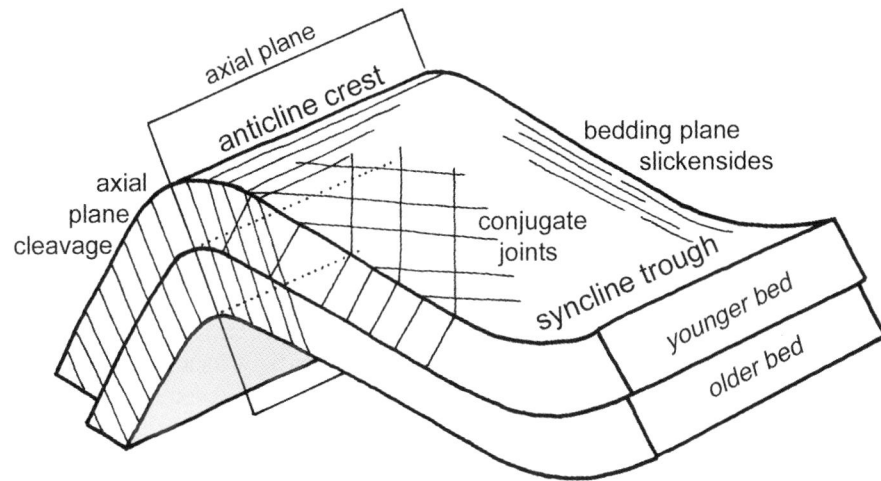

Figure 4.5   Structural relationships within a fold.

structure. Gentle folding, with dips < 20°, generally forms rounded structures, but more intense folding may be very angular. Both rounded and angular folds may occur with wavelengths down to millimetre scales, but the largest folds, with wavelengths on the kilometre scale tend to have more rounded profiles.

A single phase of rock deformation tends to produce fold structures that are of roughly uniform intensity across many kilometres of terrain due to the uniform stress distribution within rocks of comparable strength. A major exception is the development of a monocline where near-surface rocks are draped over a fault step in the underlying basement, while rocks on top of an intact basement block remain relatively undeformed. Displacements within an unseen basement

account for many of the structural contrasts in cover rocks at outcrops.

The axial plane of a fold is the plane that contains the hinge-line axes on each bed. It defines the geometry of the fold and is created normal to the maximum stress that caused the folding (Figure 4.5). Therefore in low-grade metamorphism new micas grow parallel to the axial plane and create the cleavage in slates; these important planes of structural weakness are also known as 'slaty cleavage' or 'axial-plane cleavage'. Strong folding and squeezing may induce shear along the axial-plane cleavage and so create parallel folds in the more deformable rocks; these contrast with the concentric folds produced by bending of less deformable sequences (Figure 4.4). Slaty cleavage is not the only rock fabric induced by metamorphism.

Schistosity is defined by sub-parallel micas with grain size larger than in slate, but its pattern is generally more complex and variable than that of cleavage, typically due to multiple phases of development. Foliation banding in gneiss may also be complex, but is less important as a structural weakness within the rock.

The folding that is produced by crustal convergence accommodates lateral shortening at the cost of vertical thickening. Most of the world's mountain ranges and chains are essentially fold mountains (notable exceptions are the volcanic chains, and the granite mountains, which are largely remnants exposed by erosion of folded cover rocks). The Himalayas are the prime example, typified by Dhaulagiri and Annapurna which are formed by a bed of carbonate over 2000 m thick stacked up to triple height by almost isoclinal recumbent folding (Figure 4.6). Plate convergence creates overfolds and recumbent folds where rocks override each other, and these develop into nappes in the cores of the orogenic belts. Nappes may have moved more than 50 km over their basal thrust planes, and stacked nappes create the famously complex layered structure of the Alps where the

rules of stratigraphy cease to apply (Figure 4.7). Both the Alps and the Himalayas are heavily eroded fold mountains, structurally very distinct from the stratimorphic fold mountains (see below).

Fractures, joints and faults

Natural fractures occur in all rocks. The most widespread are those induced by tectonic stress, when the rock fails, normally along shear planes. Joints are fractures with no measurable displacement. Faults are fractures along which there has been significant movement, which often, but not always, leaves telltale slickenside scratches on parts of the fault plane. Both may be formed by the same tectonic stress, where the strain is accommodated either by large movements on a few faults or by small movements on many joints.

Joints may be irregular in shape, but most are close to planar and they generally occur in sets or systems of sub-parallel fractures. Variable imposed strains may be accommodated by three sets of joints that are roughly perpendicular to each other. This creates the simplest pattern of joints, commonly found in sedimentary rocks, where one set is aligned on the bedding planes,

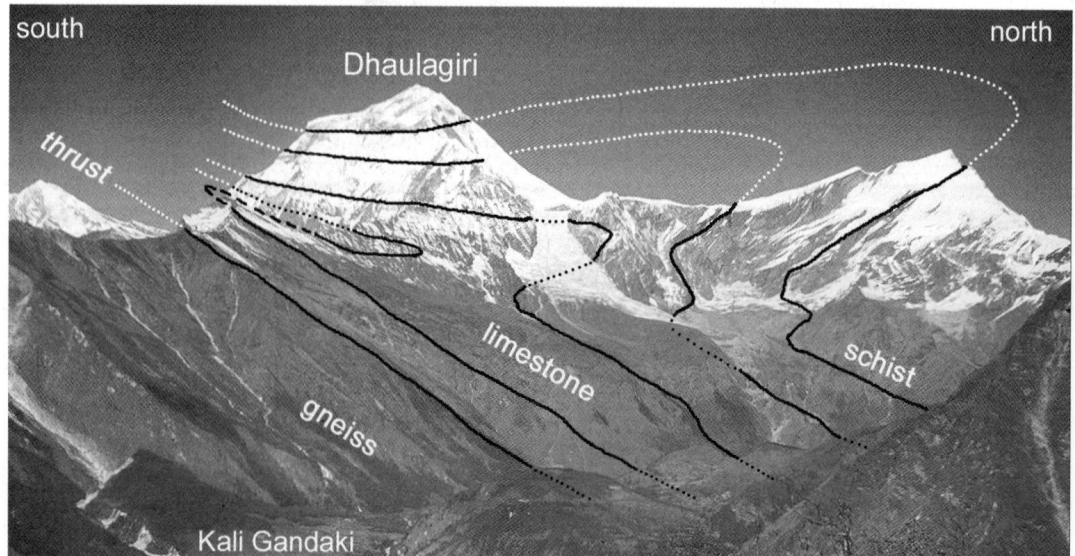

Figure 4.6   The recumbent folds that stack up a thick carbonate unit to form Dhaulagiri in the plate convergence zone of the Nepal Himalayas. The thinning of the beds towards the south is only apparent in this view, as the South Face of Dhaulagiri recedes far beyond the gneiss ridge in the foreground.

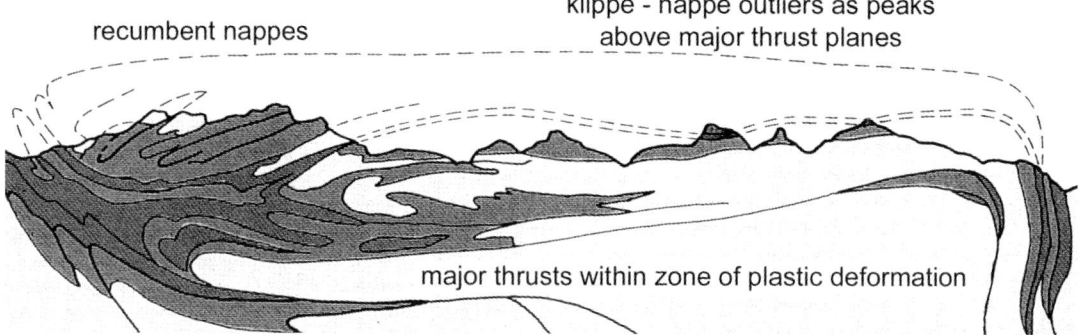

recumbent nappes

klippe - nappe outliers as peaks
above major thrust planes

major thrusts within zone of plastic deformation

Figure 4.7   Structural complexity in the over-thrust nappes of the central Alps.

Figure 4.8   Slate exposed in the Nantlle Quarry, in North Wales, has seven systems of planar weaknesses, but only four are recognisable in this view, where the ubiquitous cleavage is vertical and straight into the face.

with the other two sets across the beds. As these intersecting conjugate joints are produced in shear, at about 45° to the maximum stress, the angle between them is bisected by the associated fold axes (Figure 4.5). In reality, most rock masses have been subjected to multiple phases of deformation and their joint patterns are more complex. Slate in the Nantlle quarries of North Wales (Figure 4.8) contains five sets of joints in addition to the cleavage and the bedding. Each system of planar weakness is recognised (and individually named) by the quarrymen — whose livelihoods depend on extracting blocks of slate devoid of fractures other than the ubiquitous cleavage.

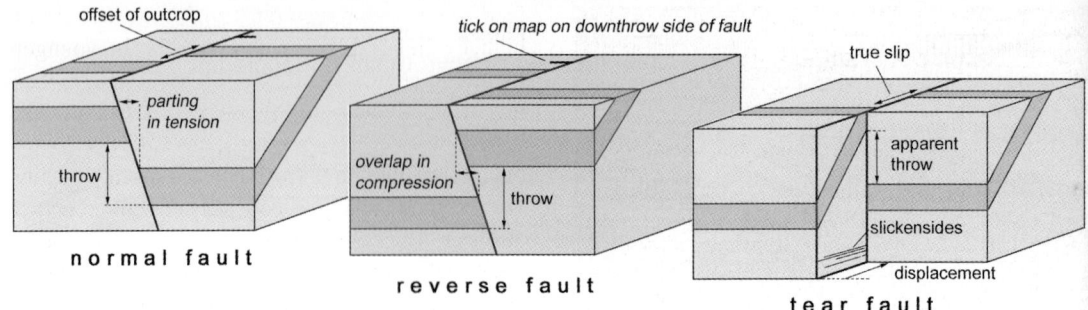

Figure 4.9   The three main types of fault. The tear fault is also known as a strike slip fault.

Joint density, or conversely joint spacing, may vary considerably, and is always a key factor in the engineering properties of a rock mass. A mean fracture spacing of < 200 mm defines a rock mass of poor quality, with low bearing capacity and high permeability. Some granites may be truly massive, with fracture spacing of > 10 m, and hence constitute very solid ground. Many clays have few or no visible fractures, but their properties at low water contents may be defined by networks of irregular and curved joints that are spaced at < 10 mm in all directions.

Faults are significant structural breaks through rock structures. In many cases they are spaced apart on a scale of kilometres, but structurally complex rock units can have many smaller faults that lie anywhere across a spectrum of rock fractures from rare large faults to abundant small joints. The scale of a fault's movement is generally expressed by its 'throw', its vertical component of displacement. Faults are described as normal when formed under tension, reverse when under compression, or tear when in shear (Figure 4.9), though there are many other terms that can be used to describe fault movement. Normal and reverse faults may be recognised by the relationship between their dip direction and downthrow side (Figure 4.9) but many vertical faults are not easily classified. In a simple stress field, reverse faults lie parallel to fold axes, and both are perpendicular to normal faults, but many structures are complicated by multiple stress phases.

A fault zone is commonly a notable element of weakness in a rock structure, regardless of any potential for future displacement. Broken rock, shattered between the moving fault blocks, is known as fault breccia (Figure 4.10), while fault gouge is the same material ground to a fine paste. Fault breccia can occupy zones many metres across, and a single fault zone may have dozens of sub-parallel or braided faults, each with breccia, or passing laterally into a single brecciated zone. All this faulted, broken rock represents locally weak material that may constitute difficult ground for an engineer; also most fault breccia has enhanced permeability. There is no direct link between the scale of displacement and the extent of brecciation on a fault, as the latter depends on the confining stress at the time of movement. Fault gouge is rarely more than 300 mm thick but its presence makes subsequent sliding on the fault plane more easy and more likely.

Fractures that constitute structural weakness in rocks are not limited to tectonic faults and joints. Bedding planes are the dominant planar weaknesses in sedimentary rocks. Most bedding planes were formed as breaks or changes in the original sedimentation and many are actually very thin layers or partings of clay or shale. With or without the clay parting, bedding planes are structural features and are therefore commonly developed by tectonic stress, effectively into joints. Strong concentric folding causes slip along the bedding planes, which therefore become faults with slickensides normal to the fold axes. Cleavage, schistosity, columnar joints (formed by cooling contraction) and unloading joints (formed by stress relief close to the surface) are all planar

Figure 4.10   A fault breccia of broken rock constituting a zone of weak ground a metre wide.

structures that the engineer should regard as types of fracture and therefore potential weaknesses in a rock mass.

Active earth movements

Absolute rates of long-term tectonic movement are so low that they are rarely of engineering significance. Mean relative lateral movements of crustal plates are only 10–100 mm/year, and vertical movements are generally an order of magnitude less. The uplift of mountain chains, currently by about 15 mm/year in the Himalayas and Karakoram, powers more active erosion and river incision, and consequently creates steep slopes that are unstable on engineering timescales (Chapter 8). Subsidence of sedimentary basins is essential for thick sediment accumulation (Chapter 3) and can be significant to man-made structures where deltaic regions are subsiding partly due to crustal sag and partly due to ongoing sediment compaction (Chapter 12).

The balance between tectonic uplift and surface denudation determines the long-term evolution of landforms. Denudation rates are higher in areas of higher relief and higher elevation, in the younger mountain chains and in environments with higher precipitation and runoff (Table 4.1). Rates of mean denudation are less than those for valley incision, but steep slopes have erosive losses that are locally very high. Rock type accounts for the contrast between slow denudation the ice-scoured basement rocks of the St Lawrence basin and the rapid erosion of the loess lands of the Huang He basin (Table 4.1).

Localised crustal uplift is still occurring due to isostatic rebound after the melting of thick Pleistocene ice caps. The fastest current uplift of this style is around 9 mm/year in the Gulf of Bothnia, where coastlines are receding as a consequence. Surface uplift and decline can also occur on and around volcanoes as they inflate or deflate (Chapter 23). Though the main significance of these movements relates to eruption hazards, Pozzuoli's harbour in Italy was rendered partly unusable due to 2 m of volcanic inflation in the late 1980s.

Tectonic movements do become significant when they are localised on active faults, most of which are in well-defined zones on or related to plate boundaries. Frictional resistance, under the confining stress of burial depth, ensures that most faults move in a series of jerks, each of which creates the radiating shock waves that are earthquakes. Seismology (Bolt, 1999) and earthquake engineering (Ambraseys, 1988) are major subjects with extensive literature. The prediction of earthquakes has proved almost impossible, to the extent that America and Japan have now virtually cut research funding. The more productive earthquake research is currently into ground behaviour and the means of engineering structures to withstand vibration damage. The minimal earthquake death tolls of recent events in Japan and USA contrast the massive tolls in Asian regions of mainly adobe buildings, which indicates the value of properly designed and built modern structures of steel and reinforced concrete.

Infrastructure can be built successfully across active faults if an allowance is made for movement. The Alaskan oil pipeline crosses the Denali Fault

Table 4.1 Typical rates of surface lowering by erosion. Environment rates are from various sources (largely after Burbank and Anderson, 2000). River basin rates are estimated from sediment yields (after Milliman and Syvitski, 1992). Valley incision rates are identified by drainage of cave systems.

Mean denudation rates in environments	*mm/year*
Temperate hill country, mean relief 500 m	0.04
Temperate mountains, mean relief 2500 m	0.40
Old mountain chains at altitudes around 1500 m	0.08
Young mountain chains at altitudes around 1500 m	0.40
Himalayan region	0.4–0.5
Salt Range, Pakistan	1.5–2.0
Arabian desert	0.05
New Mexico semi-arid lands, USA	0.017
Active glacier basins	0.1–50.0
Landslide-prone slopes in New Zealand Alps	5.0–12.0
Mean rates in complete river basins	
St Lawrence, Canada	0.001
Niger, Africa	0.012
Mississippi, USA	0.044
Amazon, Brazil	0.07
Huang He, China	0.52
Valley incision rates	
Cumberland Plateau, USA, fluvial late Pleistocene mean	0.06
Yorkshire Dales, UK, fluvial and glacial late Pleistocene mean	0.12
Gunung Mulu, Sarawak, fluvial Pleistocene mean	0.19

Figure 4.11 The Alaskan oil pipeline crossing the active Denali Fault. The pipeline trestles rest on concrete beams with steel caps and a teflon surface so that they can slide freely within the flexible limits of the pipe, while the ground is displaced laterally beneath them.

ground, as opposed to the strength of unfractured, intact rock. The mean fracture spacing is difficult to estimate in poorly exposed rock without being subjective. A convenient objective measure is Rock Quality Designation (RQD), calculated from the lengths of pieces of unbroken core recovered from a borehole:

$$RQD = (\Sigma \text{ lengths of cores each}$$
$$> 100 \text{ mm long}) \times 100$$
$$/ \text{ borehole length}$$

A value of RQD > 70 indicates a generally sound rock mass; RQD can be 0 for well-fractured or thinly bedded materials.

The strength of a well-fractured rock mass may be no more than 10% of the strength of the same material almost devoid of fractures. This relationship may be expressed as the Rock Mass Factor (RMF), which can be correlated with mean fracture spacing or with RQD (Figure 4.13). All classifications of rock mass (Barton et al., 1974; Bienawski, 1973) are based largely on fracture spacing and other properties of the fractures. A simplified guide to the safe bearing pressures applied by foundation loading of engineered structures is also based on fracture spacing, besides the intact rock strength (Table 4.2).

Besides joints and true fractures, all other rock structures and rock fabrics influence the engineering properties of rock masses. Cleavage, due to mica parallelism, may be barely visible in a compact slate, but its orientation is critical to potential shear failure and can influence unconfined compressive strength (UCS) by a factor of five (Figure 4.13). Schistosity has a similar influence, except strengths of coarse-grained schists never match those of compact slate. Foliation in gneiss is due to mineral orientation and banding, but is less significant in this generally strong rock. Most sedimentary rocks are anisotropic to some extent, even where they have no visible structure in between their major bedding planes, which are always weak. The compressive strength of apparently homogeneous Triassic sandstone from the English Midlands declines by about 20% under loading that is oblique to the bedding. The degree of parallelism within the

Figure 4.12 Sixth Street in Hollister, California, built across a branch of the San Andreas Fault. Slow, smooth, creeping displacement along the fault deforms structures on its outcrop but shock waves typical of significant earthquakes are not created.

on sliding foundations designed to tolerate 5 m of lateral displacement (Figure 4.11). Some faults with smooth, sliding displacement do not create destructive earthquakes but do cause structural damage directly over them (Figure 4.12). Vertical movements on faults can be very significant in coastal regions. Large areas along Alaska's southern coast were transformed to wetlands by as much as 2.5 m of subsidence that accompanied the 1964 earthquake. Some offshore islands were uplifted by 10 m and moved seaward by 14 m in the same earthquake, as the fault movement allowed the stress relief and the unfolding of rocks that had been crumpled against the fault until it slipped.

Engineering implications

The density or spacing of fractures is the key parameter in defining rock mass strength — the strength of natural, disturbed rock within the

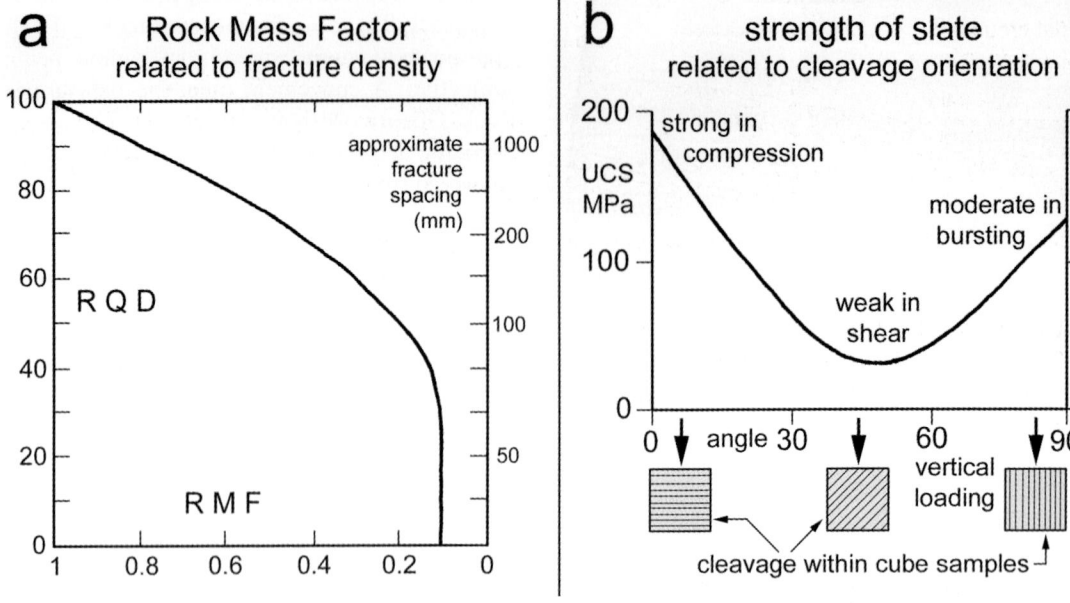

Figure 4.13 The influence of fractures on rock properties: (*a*) the decline of the Rock Mass Factor (RMF) with decrease of mean fracture spacing (and decline of RQD) and (*b*) the variation in unconfined compressive strength (UCS) of slate with respect to orientation of its cleavage.

Table 4.2 Guideline values (expressed in MPa) of safe bearing pressure (SBP) for the engineering design of foundations, with respect to the fracturing of the rock mass (indicated by RQD) and the unconfined compressive strength of the intact rock (UCS).

Fracture spacing (mm)	60	200	600
RQD	25	75	90
UCS 100 MPa	4	8	12
UCS 25 MPa	1	3	5
UCS 10 MPa	0.2	1	2

orientation of clay mineral particles accounts for some of the variation in the shear strengths of mudstones and clays, and is reflected in the more conspicuous contrast between massive mudstone and laminated shale.

Patterns and orientation of systems of fractures and other structural elements greatly determine the ease of excavation back to a clean face, and stable faces are most easily engineered parallel to the dominant fracture system. Similarly, fracture patterns define the extent of over-break in underground excavations. A well-defined fracture system, or even just one individual fracture, can determine the stability of a cut face, which may have to cut back to the planar weakness, regardless of design concepts that were not related to rock structure.

The main impact of rock folding on engineering is the creation of dipping bedding planes, which control most large landslides in rock (Chapter 8). Nearly all slope failures in strong rock are planar, or partially planar, and are therefore related to inherent rock structure. Bedding planes in sedimentary rocks are critical to stability where the dip is greater than the angle of friction — generally around 17°. Stability is achieved in steeper dips by cohesion, which is difficult to determine *in situ* and is best estimated from local empirical data. Bedding plane irregularities are a component of fracture roughness, which is critical to sliding resistance, but is difficult to quantify in the field. Fractures and bedding planes that are slickensided or contain significant fault gouge or clay parting are best regarded as having nil cohesion; this matches the concept of the drop to residual strength in

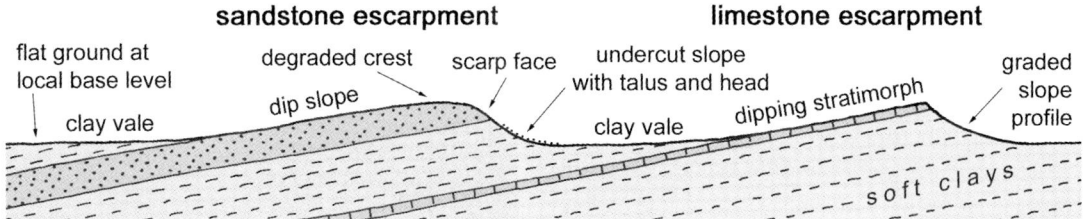

Figure 4.14   Profiles through escarpments formed in sedimentary rock sequences.

sheared soils. Rock structures that dip at angles of 20–30° towards an exposed face are therefore the least stable; steeper structures are likely to have failed already, and gentler structural dips are generally stable.

Faults that can be seen to displace Holocene sediments are classed as active and then define construction practice in accordance with local building codes for seismic regions. Unbroken Holocene sediments over a fault outcrop define the fault as inactive. In such cases, and in all cases in seismically quiet regions such as Britain, a fault may have little impact on engineering works. The exceptions are faults that may be reactivated — notably by mining. All the deformation within a migrating subsidence wave over an active longwall mine panel (Chapter 12) may be accommodated on any available fault plane, causing a ground step, or more commonly a narrow zone of abnormally high subsidence damage. Now that mining has ceased under many parts of Britain, a major cause of fault reactivation is water table rebound when mine pumping systems are finally switched off; rising groundwater pressures reduce effective stress across the fault planes, and thereby permit renewed displacements in shear.

4.2 Tectonic landforms

Stratimorphs and escarpments

A stratimorph is defined as a component of the landscape morphology that is an element of the geological structure — generally a bedding plane on top of a strong rock unit exhumed by erosive stripping of weaker overlying rocks.

Escarpments are the most widespread landforms in areas of dipping sedimentary rocks,

formed where strong limestones or sandstones are undercut and stripped bare by more rapid differential erosion of weaker clays or shales (Figure 4.14). Most of England south and east of the Pennine crest consists of scarplands, with the limestone Cotswolds and the chalk Downs as just the largest of a succession of escarpments. A gently inclined dipping stratimorph is one element of an escarpment, except where the dip is steeper than the naturally stable slope so that a more symmetrical hogsback is formed. A scarp face may form dramatic cliffs where it is efficiently undercut in a mountain environment, but it is generally degraded into a rounded step in a more mature landscape. The well-known gritstone edges of the Derbyshire Peak District (Figure 4.15) are all scarp faces above gentler shale foot slopes, but their rugged scars are all

Figure 4.15   A gritstone edge in the Derbyshire Pennines (UK), Curbar Edge, a natural scarp face that has been modified by quarrying.

the remains of old quarry faces that were cut into much less rocky natural hill crests.

The most dramatic stratimorph is an anticlinal mountain whose entire surface is the rounded arch of a strong limestone or sandstone before it is breached and reduced by erosion to leave a pair of facing escarpments. The classic examples are in the Zagros Mountains of southern Iran (Figure 4.16), where soft clays and gypsum have been stripped off the strong limestones to reveal the anticlines that are the first pointers to deeply buried petroleum reservoirs.

Plateaus, mesas and buttes

A large stratimorph in horizontal rocks forms a plateau, of which South Africa's Karoo Plateau and Table Mountain are the perfect examples on different scales. At many other sites strong beds form only wide stratimorphic steps across hilly terrains. Where these are horizontal they can be misinterpreted as remnants of old erosion surfaces. The wide limestone benches around Ingleborough, in the Yorkshire Pennines (UK), are both erosion surfaces and stratimorphs (Figure 4.17); erosion surfaces cutting across dipping rock sequences do also exist, though they are less well defined, and there are also some sloping benches that are dipping stratimorphic surfaces.

Arizona's Monument Valley is the finest demonstration of erosive dissection at a plateau margin to leave residual hills with profiles that create one of the world's most spectacular landscapes. Widely spaced vertical joints in the strong sandstone, 150 m thick, define vertical cliffs above the slopes of the underlying shale–sandstone sequence. The larger remnants are known as mesas, but they are known as buttes when their width is reduced to less than their height (Figure 4.18); the thinnest remnants are variously described as towers, spires or pillars.

Fault scarps and fault-line scarps

As faults are major breaks in bedrock structure, they commonly define linear features within the landscape. The simplest forms are fault scarps — steps in the landscape created by the fault movements. Some are formed during a single displacement with its associated earthquake, but

Figure 4.16 The classic stratimorph of Kuh-e-Kailan formed by an anticline of Asmari Limestone in the Zagros Mountains of Iran (photo: Hunting).

these can only be a few metres high (Figure 4.19), and rarely survive in long-term landscape evolution. Larger fault scarps are the product of repeated movements, with some of the clearest examples occurring along the margins of rift valleys (see below).

Far more common are fault-line scarps. These are steps in the landscape along the line of a fault, where the step is not the direct consequence of the fault throw (though they are often described incorrectly as fault scarps). In most cases the landscape step has been formed by differential surface lowering of rocks of contrasting resistance on either side of the fault; the lower side is on the weaker rock and is not necessarily the down-throw side of the fault. In many cases, erosion has left the scarp as a degraded slope that has retreated from the line of the fault. The Craven Faults have defined some major landscape features in the Yorkshire Pennines (Figure 4.17). They mark the edge of the Craven Uplands with 30 km of eroded hillsides that step down 200 m to the south, though the fault throw is well over 1000 m. Giggleswick Scar is a more cleanly

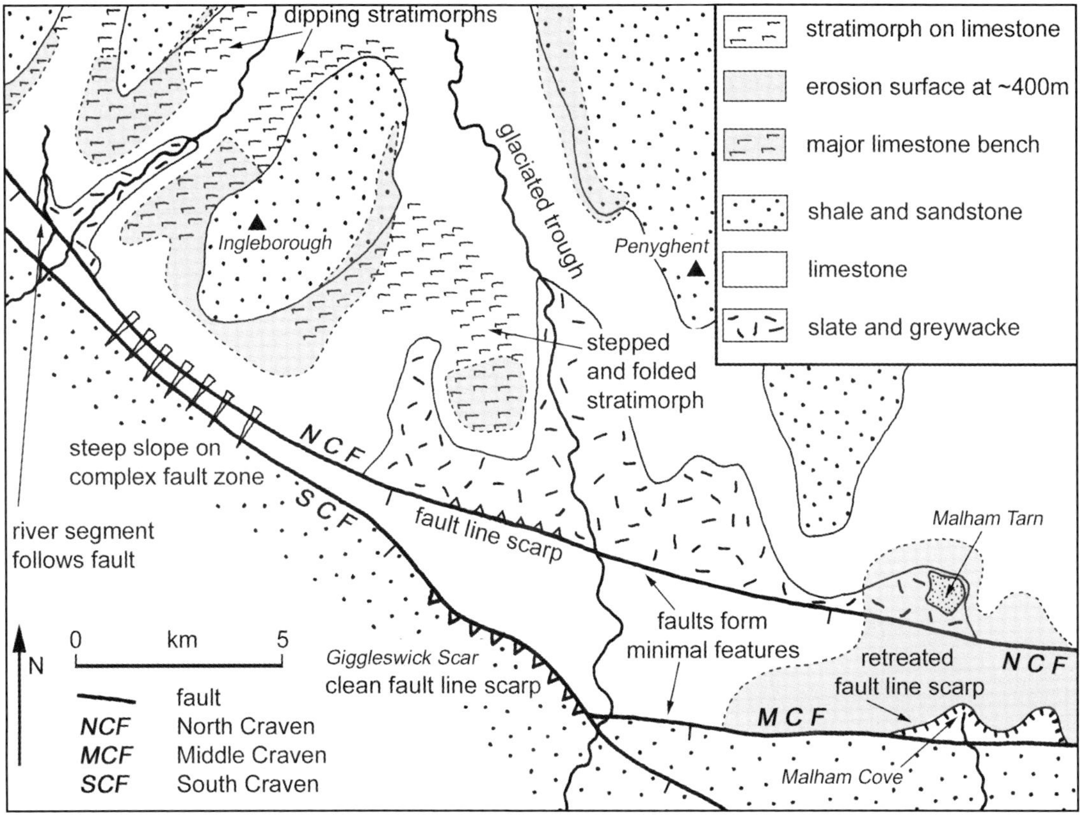

Figure 4.17 Outline map of the major landforms of Ingleborough and the Craven fault zone in the Yorkshire Pennines (UK). The largest limestone benches lie where the 400 m erosion surface coincides with the stratimorph on the top of the strong limestone. All three faults down-throw to the south.

Figure 4.18 The spectacular residual hills of Monument Valley, Arizona, with the small East Mitten Butte in the near right, and the wide Sentinel Mesa in the left distance. Between them are the slender Stagecoach Buttes, with a narrow unnamed pillar on the mesa's right flank.

Figure 4.19 A fault scarp created by surface displacement on the Hebgen Fault, in Montana; the associated earthquake was of Richter magnitude 7.1.

defined fault-line scarp of strong limestone that stands 100 m above a shale lowland; it is defined by one clean fault within the wider fault zone. The 70 m high limestone cliff of Malham Cove is a fluvial feature that has retreated a kilometre from the fault; though its origins lie at the fault-line scarp, it should not now be described as a fault feature.

The broken rock along many faults constitute zones of weakness that are commonly picked out by erosion. Straight segments of river courses are common indicators of fault lines, as are linear lakes in shield terrains. Glaciers are little influenced by small faults, but may excavate major troughs where fault-guided fluvial valleys are invaded by ice. The Great Glen of Scotland cuts right through the Highlands along the line of a massive tear fault that has been excavated over millions of years; there has been no recent movement on it. Wave action is the most selective of erosion processes, and faults are commonly marked by bays, inlets, fissures, caves and geos within marine cliff lines.

Rift valleys, horst blocks, fins and arches

A graben is an elongated crustal block displaced downwards between two normal faults. A rift valley, bounded by inward-facing fault-line scarps, is created by recent graben sinking where the faults effects have not yet been obliterated by surface degradation. Long histories of graben sinking leave faults with kilometres of throw, but rift valley scarps are normally much smaller (Figure 4.20). The Rhine Graben in western Germany has subsided 4400 m in 45 million years, and has filled with sediment so that the modern valley floor is only about 500 m below the marginal hills. The normal faults are the product of regional tension over a zone of deep crust that has been heated and expanded laterally; commensurate vertical expansion accounts for the highlands that flank most rift valleys.

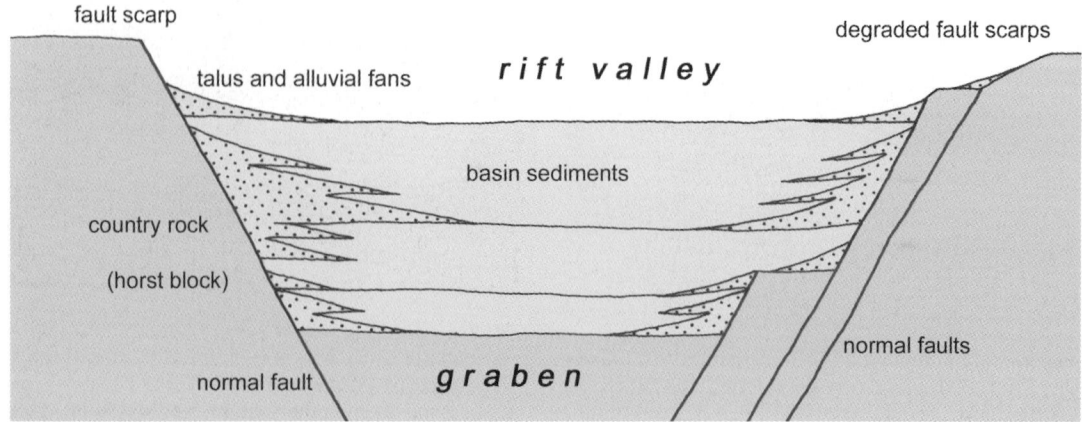

Figure 4.20 Profile through the marginal faults of a graben beneath a rift valley.

More complex crustal tension creates fault blocks that rise or subside relevant to each other along bounding normal faults. The Basin-and-Range country of Nevada, USA, is the classic example, with uplifted horsts of Precambrian rocks forming mountains that rise above intervening basins floored with thick young sediments. The tension structures are continued southwards to Death Valley and Panamint Valley, both deep rift valleys in the California desert.

Small grabens are common in the head zones of landslides, where wedges of ground subside behind the moving landslide block. The Grabens, in Utah's Canyonlands, are elongated blocks bounded by normal faults that originally opened in tension over a rising salt dome. Subsequently, they have subsided behind sandstone slabs that are sliding very slowly over the salt towards the canyon of the Colorado River. They are so fresh that they now form rift valleys. Maintained tension in the ground causes joints and faults to open. The famous Joint Trail leads right through a sandstone ridge in a single straight fissure 400 m long, 40 m deep and only 1 m wide (Figure 4.21); elsewhere, joints open beneath the soil cover and cause localised linear collapses.

The same tension joints on the flanks of anticlinal salt domes in Utah have created the narrow ribs of rock known as fins in the strong Entrada Sandstone. The faces of the fins have then retreated by frost action, spalling and undercutting as a part of long-term weathering, and the numerous breaches of the fins have enlarged into natural arches (Figure 4.22). Patchy selective weathering is not uncommon, but arches are rarely formed except in structurally defined fins.

Engineering implications

The rock strength and denudational resistance of a stratimorph makes it inherently stable, until it is

Figure 4.21 The open tension fracture that carries the Joint Trail right through a sandstone ridge in Canyonlands, Utah; there has been minimal additional opening of the fissure by weathering and face retreat.

Figure 4.22 Double O Arch, weathered through a narrow fin between open joints in the Entrada Sandstone of Arches National Park, Utah. Visible beyond and through the arch are fins within the same joint system, formed in tension over an elongated salt dome.

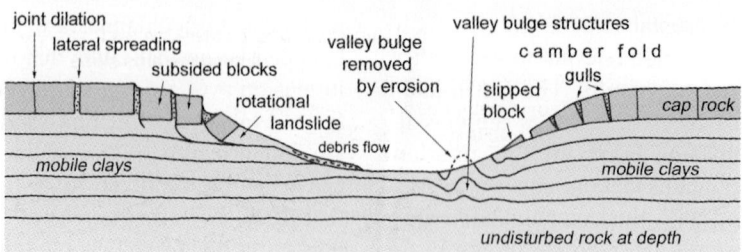

Figure 4.23   Valley side deformation by camber, gulls, joint dilation and plateau spreading.

undermined by erosive excavation of an adjacent weaker rock. The world's largest recognisable landslide was caused when a slab of dipping limestone failed where support was partially removed by excavation of a river valley. In the Saidmarreh slide, dated to 10 400 BP, 22 km³ of rock slipped off an anticlinal stratimorph in the Zagros Mountains of Iran (Watson and Wright, 1969). Smaller landslide events are a normal component of landscape evolution in mountainous areas of folded rocks.

Plateau edges, scarp faces and fault scarps are characterised less by planar slides than by rockfalls and toppling failures, which become a significant engineering hazard where strong caprocks overlie weak clays or shales. Squeezing of soft underlying clays causes ground flowage towards open slopes, and thereby deforms and weakens them. Camber folding, gull opening, lateral spreading and fracture dilation (Figure 4.23) are all precursors to landslide development along plateau edges and scarp faces of low-dipping escarpments.

4.3 Lithology and landforms

Differential erosion of sedimentary rocks

The lithology and structure of bedrock are paramount to geomorphological evolution and landscape development. Different rock types have most expression within the landforms of mature terrains (Figure 4.24), where they provide the data for engineering geomorphological analysis and ground model construction (Fookes, 1997; Hutchinson, 2001). Rapid uplift of youthful mountain ranges causes rapid and unselective erosion, with steep and unstable slopes that only reflect

geological structure in their smaller features and in their landslides. Similarly, old-age terrains have minimal expression of lithology. The great shields of Precambrian metamorphic rocks have largely been reduced to lowlands with minimal relief, though major faults can control much of the drainage pattern. On Canada's Laurentian Shield, faults can be traced for hundreds of kilometres by their lines of elongated lakes. In contrast, mature landscapes commonly reveal their bedrock geology. Canada's Front Ranges and the limestone mountains that fringe the Alps in south-eastern France both have all their major landforms tightly controlled by their geological structure.

Sedimentary rock sequences in mature lowlands are especially sensitive to selective erosion as the weaker clays, shales and mudstones are stripped away from the more resistant sandstones and limestones. The results are scarplands or terraced flatlands depending on the regional dips. Clays form broad valleys with the rich soils most suited to arable farming. Sandstones provide the dry sandy soils that support low-value heathland and also in England the more extensive uncleared woodlands; Sherwood Forest housed Robin Hood because it was so extensive along the outcrop of the Triassic sandstones.

Carbonate rocks almost invariably form positive elements of a landscape. The older and stronger limestones and dolomites are generally strong in their own right forming white scars and cliffs in often very spectacular landscapes (Chapter 24). Chalk is generally a weak rock, but still forms positive landforms, notably the Downs and Wolds of England (Chapter 26). This is partly because its outcrops are everywhere surrounded by clays, but also because its underground drainage reduces

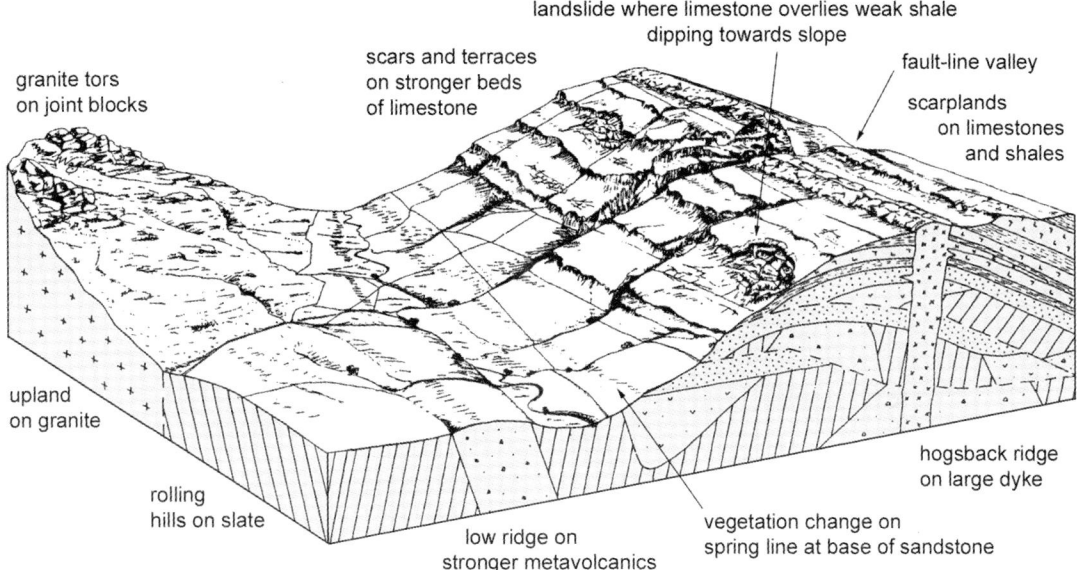

granite tors
on joint blocks

scars and terraces
on stronger beds
of limestone

landslide where limestone overlies weak shale
dipping towards slope

fault-line valley

scarplands
on limestones
and shales

upland
on granite

rolling
hills on slate

low ridge on
stronger metavolcanics

vegetation change on
spring line at base of sandstone

hogsback ridge
on large dyke

Figure 4.24   Some influences of rock type and structure on landforms (after Fookes, 1997).

fluvial erosion of its surface. The distinctive short grass of the chalk downlands is not a true function of the bedrock, it is classic 'sheepwalk' where all but grass is kept down by grazing sheep since their introduction by man (Figure 4.25). The natural vegetation of English chalk is oak woodland.

Salt and gypsum are both weak rocks. They are also highly soluble, and consequently can develop some elements of karst terrains (Chapter 24). Salt can only survive at the surface in a desert, as in Israel's Negev, and is otherwise geomorphologically more significant with respect to its ground

Figure 4.25   Downland on chalk, with the typical short turf of 'sheepwalk', on the sides and floor of a deep dry valley in the Yorkshire Wolds.

Figure 4.26 Curved exfoliation joints in the granite of Half Dome above Yosemite Valley, California. The scale is given by people on the cable ladder on the left.

subsidence (Chapter 12). Gypsum can form karstic uplands in the moderately dry regimes of central Turkey, western Ukraine and northern Russia, but it is rapidly eroded and rarely reaches to outcrop in more temperate or wet environments.

Landforms of igneous rocks

Granite is one of the most distinctive rocks in terms of its landforms. Its high intact strength (typically with UCS of 200 MPa), massive structure (giving it a high rock mass strength) and the large size of typical batholiths, combine to make granite a major positive feature in most landscapes. Large domes are granite's primary landform, best known in California's Yosemite Park. They are clearly developed by progressive exfoliation and stripping of curved sheets separated by relaxation (unloading) joints parallel to the surface (Figure 4.26). The massive vertical cliffs of granite country are largely formed on faults or major joints exposed by 'ice plucking' in the plane of weakness. Half Dome, at the head of the Yosemite Valley, has a clean vertical face through the bisecting of what was once an almost perfect dome (Figure 4.27). In periglacial regimes, selective weathering of granite with varying joint densities leaves residual tors (Chapter 14).

All igneous rocks are strong, and the smaller intrusions tend to create their own distinctive landforms when exhumed during denudation. Selective removal of softer and weaker country

rock can leave dykes as upstanding ridges or walls (Figure 4.28), while plugs and volcanic vent fills can form isolated high crags (Chapter 23). Sills can form stratimorphs and escarpments similar to those in a strong sedimentary rock, though dolerites are distinguished by their dark colour and commonly by massive columnar jointing. Slow underground cooling creates columns larger than those typical of lava flows. The Whin Sill is a conspicuous feature of the northern Pennines

Figure 4.27 Half Dome, Yosemite, California, with the vertical face formed on a single major joint cut into the rounded dome defined by exfoliation sheet jointing.

Figure 4.28   An exhumed dyke forms a wall of rock 10–30 m high across the plains of New Mexico. It is a radial feature from Ship Rock, itself a product of differential erosion around a sub-volcano plug that now stands 500 m high (and is 5 km away in this view).

Table 4.3   Typical values for safe bearing pressures and safe slope angles on different rock types.

Lithology, rock type	Safe bearing pressure (SBP) (MPa)	Safe cut slope (degrees)
Strong igneous rocks	4–12	80–90
Strong metamorphic rocks	3–10	60–90
Stronger limestones	2–4	80–90
Weaker limestones	0.8–3	70–90
Chalk	0.5–1	45–80
Stronger sandstones	2–4	70–90
Weaker sandstones	0.7–2	50–70
Mudstone	0.7–2	40–70
Shale	0.4–1.5	25–50
Heavily fractured rock	0.3–1	20–40
Weathered rock and soil	0.01–0.4	15–25

(UK), forming inland cliffs (High Cup Nick), coastal headlands (Bamburgh) and waterfalls (High Force), and it was utilised by Roman engineers to enhance the impregnability of Hadrian's Wall, which was built largely on its escarpment.

Youthful volcanic rocks create their own positive landforms and environments (Chapter 23). Older volcanic rocks are generally strong and also form mountain terrains. England's and Wales's highest mountains, in the Lake District and Snowdonia, are both formed of Ordovician rhyo-lites, but folding and erosion have obliterated the original volcanic landforms.

Engineering implications

Rock type is the prime factor in determining ground conditions for engineering. Though fractures are critical to rock mass strengths (see above), most rocks have characteristic styles of fracturing that allow broad concepts of engineering conditions to relate to the rock type. Table 4.3 shows the ranges of bearing capacities and also stable angles for cut slopes for a variety of rocks. These figures are entirely based on empirical data, and any more accurate assessment of ground conditions must be based on a specific investigation of the local rock structure.

Sedimentary rocks provide the greatest range of difficult ground conditions. Clay is the only widespread rock whose strength is so low that settlement under imposed load can be the limiting factor in engineering design. The potential compaction of clay is a function of its age, consolidation history, clay mineral content, silt or sand content, thickness and drainage state. Pre-Miocene, over-consolidated, illite-rich, silty clays (that geologists would define as a rock) cause minimal settlement, while Holocene, superficial, smectite-rich, silt-free clays (that anyone would define as soil) can cause severe ground movements. Most of the science of soil mechanics (Craig, 1997) is devoted to the engineering behaviour of clays. Peat is the only natural material with greater potential compaction than clay (Chapters 7 and 12). Naturally cavernous ground is almost restricted to the soluble limestones, gypsum and salt (Chapters 12 and 19). The sandstones provide the better ground with fewer problems for the engineer.

The intrusive igneous rocks are generally strong, though instabilities within them may derive from their jointing, notably exfoliation in granites and columnar in dolerites. Perhaps the major influence on their fractures is with respect to their extraction and utilisation. Granite is typically favoured as a dimension stone by its widely spaced fractures, while the higher joint densities of most dolerites reduce quarrying costs in the supply of hard aggregate.

4.4 Summary

The internal structures of rock masses are critical to the evolution of terrain morphology and to all aspects of engineering geomorphology. On the small scale, fractures determine the strength of rock masses, with implications on their erosion resistance, slope stability and bearing capacity for structures. On the larger scale, the tectonic history and the larger inherited structures of a terrain determine the overall nature of its ground conditions, the evolution of its landforms and the potential for the larger geohazards including earthquakes, major landslides and regional subsidence.

References

Ambraseys, N. N. (1988) Engineering seismology. *International Journal of Earthquake Engineering and Structural Dynamics* **17**, 1–106.

Barton, N., Lien, R. and Lunde, J. (1974) Engineering classification of rock masses for tunnel design. *Rock Mechanics* **6**, 189–236.

Bieniawski, Z. T. (1973) Engineering classification of jointed rock masses. *Transactions of the South African Institute of Civil Engineers* **15**, 335–343.

Bolt, B. A. (1999) *Earthquakes*. W. H. Freeman, New York.

Burbank, D. and Anderson, R. (2000) *Tectonic Geomorphology*. Blackwell Science: Oxford.

Craig, R. F. (1997) *Soil Mechanics*. Spon, London.

Fookes, P. G. (1997) Geology for engineers: the geological model, prediction and performance. *Quart. Journ. Eng. Geol.* **30**, 293–431.

Hutchinson, J. N. (2001) Reading the ground: morphology and geology in site appraisal. *Quart. Journ. Eng. Geol. Hydrogeol.* **34**, 7–50.

Milliman, J. D. and Syvitski, J. P. M. (1992) Geomorphic/tectonic control of sediment discharge to the ocean: the importance of small mountainous rivers. *Journal of Geology* **100**, 525–544.

Watson, R. A. and Wright, H. E. (1969) The Saidmarreh landslide, Iran. *Geological Society of America Special Paper* **123**, 115–139.

Further reading

Burbank, D. and Anderson, R. (2000) *Tectonic Geomorphology*. Blackwell Science, Oxford.

Hills, E. S. (1963) *Elements of Structural Geology*. Wiley, New York.

Kearney, P. and Vine, F. J. (1996) *Global Tectonics*. Blackwell Science, Oxford.

Ollier, C. and Pain, C. (2000) *The Origins of Mountains*. Routledge, London.

Prentice, J. E. (1990) *Geology of Construction Materials*. Chapman & Hall, New York.

Ramsay, J. G. and Huber, M. I. (1967) *The Techniques of Modern Structural Geology, Volume 2: Folds and Fractures*. Academic Press, New York.

Rikitake, T. (ed.) (1981) *Current Research in Earthquake Prediction*. Reidel, Dortrecht.

Twiss, R. J. and Moores, E. M. (1992) *Structural Geology*. W. H. Freeman, New York.

5. Stratigraphy

Colin J. R. Braithwaite

5.1 Introduction and basic principles

Although the term 'stratigraphy' suggests simply logging (recording) sequences of rock strata, stratigraphy is more commonly seen as the study of Earth History as interpreted from these sequences. The methods used to determine the conditions of formation of a particular sedimentary unit are outlined in Chapter 3, but the three-dimensional ordering of these to create a conceptual (and predictive) model of the environmental system depends on the exercise of stratigraphic methods. It is the predictive capacity of such models that is of particular value to the engineer.

Nicolaus Steno, in 1667, was the first to outline the basic principles relating to the ordering of sequences of rocks, which he stated in three general laws pertaining to:

1. original horizontality, which assumes that regardless of their present attitude layered sedimentary rocks were originally deposited as horizontal beds or strata
2. lateral continuity of layers
3. the law of superposition, which determines that layers can only be formed in sequence. Inherent in the idea of wide distribution of similar layers of rocks is the realisation that in normally layered sequences, older beds are overlain by younger, on the assumption that new layers cannot be deposited underneath existing deposits.

These were important observations, predating by almost a hundred years acceptance of the fact that most of the rocks concerned were sedimentary sequences laid down in the oceans. The 'law' of superposition, and its application was particularly important as it led to the development of a relative time scale of events in the evolution of the Earth.

James Hutton (1726–1797) in Edinburgh, realised that the sequences of layers were not continuous. At Siccar Point on the Berwickshire coast he saw steeply inclined (Silurian) turbidite layers overlain by gently dipping (Devonian) fluvial sandstones and conglomerates. He recognised that the contact between the two sequences represented a break in the stratigraphic record. He referred to such breaks as unconformities and noted that they seemed to demand long periods of time for their formation, time in which folding, uplift and erosion could take place before deposition was resumed. Subsequently others, principally in the Geological Survey in Britain, used unconformities to subdivide the stratigraphical sequence into units that were identified as geological time periods: Cambrian, Ordovician, Silurian, Devonian and so forth. (Table 5.1). These were generally named after the areas in which the particular sequences were first recognised (Table 5.2). The names defined relative ages and, for any area, provided and provide a guide to the relative induration of the rock. Rocks of greater age are more likely to have suffered greater burial and greater diagenetic or metamorphic change and are therefore likely to be stronger. However, there are variations between different rock types of the same age and, on a global scale, the history of particular areas may result in relatively young sequences being significantly altered or relatively old sequences remaining quite fresh relative to the experience of rocks of the same age elsewhere.

Table 5.1   The stratigraphical column, showing the divisions of geological time and the relative ages of major events.

EON	ERA	Period	Age (Ma)	Duration (Ma)	Orogenic Phases	Major Events
Phanerozoic (Evident life)	Cenozoic (Recent life)	Quaternary (Pleistocene)		1.8	Pyrenean	Major glaciations of northern hemisphere
			1.8			First Hominids
		Tertiary		63.1		Age of mammals, birds and flowering plants
			65		Alpine-Laramide	Major meteorite impact, extinction of the Dinosaurs
	Mesozoic (Middle life)	Cretaceous		64.7		Indian and southern oceans open
			142			First birds, modern bony fishes, rudist bivalves and flowering plants
		Jurassic		63.8	Nevadan	Opening of North Atlantic Ocean
			205.7			First Dinosaus
		Triassic		42.5	Hercynian-Appalachian	First Mammals
			248.2			Break-up of Gondwanaland
	Palaeozoic (Ancient life)	Permian		41.8		Mass extinction of rugose corals, trilobites and many others
			290			Glaciation in the southern hemisphere
		Carboniferous		64	Bretonian-Acadian	First reptiles
			354			Last graptolites
		Devonian		63		First insect and amphibians
			417			First land-living animals and plants
		Silurian		26		First fish with jaws
			443			
		Ordovician		52		First vertebrates (Jawless fish
			495			
		Cambrian		50		First graptolites
			545		Cadomian	First skeletal organisms
	Proterozoic	Precambrian (All rocks older than Palaeozoic)			Huronian	First soft-bodied animals ~590 Ma forming tracks and trails
			2500			
	Archaean				Laurentian	Increasing atmosphic O_2 at 1.7 Ga
			4000			Earliest bacteria 3.5 Ga
	Priscoan		4600			Major cratering on the Moon 4.2 Ga

Table 5.2   Derivations of names of the principle units of geological time.

Cenozoic = recent life
   Pleistocene = most newly formed
   Tertiary = the third Era
      Neogene = later formed
      Palaeogene = earlier formed
Mesozoic = middle life
   Cretaceous = Greek (Creta) for Chalk
   Jurassic = from the French Jura Mountains
   Triassic = from the three divisions in Germany
Palaeozoic = ancient life
   Permian = Permia, ancient kingdom of the Urals
   Carboniferous = coal-bearing
      Pennsylvanian
      Mississippian (In North America)
   Devonian = Devon
   Silurian = area of the Silures, an ancient Welsh tribe
   Ordovician = area of the Ordovices, an ancient
      Welsh tribe
   Cambrian = Cambria, ancient name for Wales
Proterozoic = before life
Archaean = first or beginning
Priscoan = accretion of the Earth and Moon

5.2 Correlation

William Smith (1769–1839) a civil engineer and surveyor, who worked on the construction of roads and canals in England, is regarded as the father of stratigraphic mapping. Using Steno's principles he recognised that the same sequences of sedimentary rocks, containing the same fossils and with the same physical characteristics, could found in different parts of Britain, and that their distribution could be mapped to provide a predictive framework for excavation, saving him time and his clients money. His maps are recognised as among the first of their kind in the world and stand comparison with maps of the same areas produced by present-day geologists. At the core of Smith's observations was the realisation that he could use fossils to differentiate rocks that were otherwise similar on the basis of their relative age. Such correlation was an objective exercise made simply by recognising the fossils concerned, but it formed the basis for what became the 'Law of Faunal Succession'. Georges Cuvier (1769–1832)

in France was at this time arguing that the changes that could be seen in fossil faunas in the Paris basin were a reflection of repeated catastrophes, each followed by a new creation. It was nearly a hundred years before Charles Darwin (1809–1882) was able to demonstrate that they changed in response to evolution. It is to William Smith's particular credit that he made careful observations and used them to construct a conceptual framework that allowed him to make predictions that were of significance to his work of construction long before the existence of a supporting framework of evolutionary theory.

5.3 Bedding and time

The layering (bedding) in a rock sequence is defined by bedding planes. These form as a result of pauses in deposition or intervals of erosion interrupting the original succession (Chapter 3), and rock sequences rarely provide a complete record of the time that has elapsed. The early stratigraphers, following Steno's principles, measured individual rock units and sequences and used the values obtained to calculate their relative ages on the basis of their maximum thickness and assumed rates of deposition. As no reliable estimates of rates of deposition were available there were wide variations in the results, some of which suggested total ages for the Earth of only a few million years. Even today there are considerable disparities in estimates of rates of both denudation and accumulation — commonly used to assess denudation (Chapter 3, Table 3.2) and in recent years our view of rates of accumulation has changed. We now believe that most of the time represented by a rock or sediment sequence is reflected in the bedding planes (the divisions between the beds) whereas the beds themselves represent events that took place over relatively short periods of time that in some cases may be measured in hours or days. Derek Ager (1923–1993) succinctly described this realisation in terms of his experience of National (military) Service, 'long periods of boredom punctuated by moments of blind panic'. In sedimentological terms (Chapter 3) all depositional environments, apart perhaps from the pelagic realm of the deep oceans, are places where the effects of commonly catastrophic events (storms or tidal surges, flooding of rivers, volcanic eruptions, or extreme desiccation) have a greater potential for effect and for preservation than those resulting from 'normal' day-to-day processes. For this reason, although the time represented by an entire rock succession may total millions of years, that recorded by individual beds may account for only a tiny fraction of this.

It is important to note that although the general concept of accumulating sedimentary layers can be applied to the entire geological record (accepting the occasional intrusion or extrusion of igneous rocks) it is not easily applied to Quaternary deposits on which most engineering works are founded. In these, as will be shown, sedimentary units are seldom laterally continuous and the sequences of layers are most commonly characterised by discontinuities.

Estimates of the passage of time represented by rock sequences and of the age of the Earth have become reasonably accurate and reliable, but have not always been so. Much traditional stratigraphy, predicated on the concept of evolution has made little impact on this area, although Charles Lyell (1797–1875) attempted to calculate geological times based on the supposed rate of change of molluscan faunas. His figure of 240 Ma (million years) elapsed since the beginning of the Ordovician was plausible but incorrect, and even now estimates of absolute age resting solely on supposed evolutionary rates would not be considered reliable.

For reasons indicated above, estimates based on accumulation rates fared no better. Some produced figures in excess of 1500 Ma others were as low as 3 Ma. At the turn of the 19th century Lord Kelvin (William Thomson, 1824–1907) calculated the supposed heat loss from a cooling Earth and concluded that only 20–40 Ma were necessary for it to have reached its present temperature. Kelvin was an influential figure and because his estimates were based on objective mathematical calculations his conclusions dominated much geological thought. However, following the discovery of radioactivity by Henri Becquerel in

1896 and Ernest Rutherford's (1906) attempts to use this to measure the ages of minerals, in 1911 Arthur Holmes, then working with R. J. Strutt (who was the first to show that the Earth is not actually cooling as a result of heat loss), set out the principles of radiometric dating. As methods of analysis have improved, these have become correspondingly sophisticated. Although correlation using fossils remains widespread, and is cheap and practical, the age ranges of the various organisms used can now be established with some certainty, using radiometric dating of interbedded lavas or ashes, or minerals such as glauconite, phosphorite or carbonates that formed authigenically within the sediments. Table 5.1 also includes the main events recorded in the rock record during the history of the Earth and the radiometric ages currently assigned to them.

The rock units defined may be mapped on the basis of their gross lithology or facies, representing the sum of all the observable characteristics, of rock type, geometry, sedimentary structures and fossil content. They may be correlated on the basis of containing similar fossil assemblages.

Radiometric dating

Where age dates are required it is important to select material that contains an isotope or isotopes with an appropriate half-life. The half-lives determine the useful range of the isotopes commonly used in dating Quaternary deposits.

Potassium– $^{40}K \rightarrow ^{40}Ar$: 110 Ma   (igneous and
argon   metamorphic
                                              rocks)

Uranium $^{234}U \rightarrow ^{230}Th$: 248 ka   (corals,
series   speleothems
                                              and bone)

Radio- $^{14}C \rightarrow ^{14}N$:   5730 a   (wood,
carbon   charcoal,
                                              shells and
                                              coral)

Lead $^{210}Pb \rightarrow ^{206}Pb$: 22.2 a   (siliciclastic
                                              sediments).

However, this is only part of the story. There has to be some degree of certainty that the proportions of the original element and its isotope are indeed a reflection of the time elapsed. If the rock is heated, for example, elements may become mobile and the 'clock' reset, with the decay process restarting as the mineral cools below a critical temperature, the blocking temperature, at which diffusion ceases and the mineral grain becomes a closed system. Different minerals have different blocking temperatures that are independent of the isotopes they contain.

5.4 Three kinds of stratigraphy

Lithostratigraphy

Lithostratigraphy involves the recognition of mappable rock bodies. Whereas lithology defines an individual rock type such as sandstone or limestone, lithostratigraphy refers to a facies. This represents a body of rock that may include several lithologies but is identifiable on the basis of the association of lithology, texture, fossils and sedimentary structures, together with the overall geometry of the sediment body, the shape, distribution and thickness of beds.

Defined in this way the facies reflects the sum of all the processes operating when the sediment was deposited in a particular environment at a particular place and time. It might represent, for example, the products of part of a delta (Chapter 20), a barrier beach, or a turbidite sequence. Facies form lithostratigraphic units that may vary in both time and space. A beach, for example, will respond to follow a rising or falling sea level and change its position on a sloping shoreline. The direction in which it moves will depend on the supply of sediment relative to the rate of sea level change. Thus, the beach deposits will retain an internal continuity but are to be found in different places at different times. A unit that behaves in this way is described as diachronous (literally through time), because, when viewed in three dimensions, it crosses time-defined surfaces. Most facies cross time planes to some extent.

It might appear from this that facies could vary at random but in fact they maintain a precise spatial relationship. This was first appreciated by Johannes Walther (1860–1937), who in 1893–4 set out the 'Law of Correlation', now more commonly known as Walther's Law. This is the key idea on which the

concepts of facies relationships were founded and underpins all palaeoenvironmental interpretations of facies descriptions. Translated this states:

> The various deposits of the same facies area and, similarly, the sum of the rocks of different facies areas, were formed beside each other in space, but in crustal profile we see them lying on top of each other . . . It is a basic statement of far reaching significance that only those facies and facies areas can be superimposed, without a break, that can be observed beside each other at the present time. Translation from Blatt *et al.* (1980).

This observation, illustrated in Figure 5.1, is at once very simple but profound and provides the geologist and the engineer with a valuable tool, predicting rock relationships in three dimensions. It can be demonstrated by considering a common coastal environment — a tidal flat. Tidal flats typically show lateral zones of sediment types (each with characteristic faunas and floras). A borehole sunk at locality D would encounter the vertically changing sequence of sediments shown in Figure 5.1 that can be related to the horizontal changes in sediment type. The vertical sequence of superimposed sediment types occurs because the tidal flats are prograding, that is, building seawards. Moving back through time, there must have been a period when the zone now below the saltmarsh above mean high tide level was below low mean tide level and covered by the sea. As the sediments prograded, the saltmarsh grew over the muds of the former high tidal flats and these in turn extended over mid-flat sediments. However, not all rock sequences record a history of continuous deposition. Where erosion has occurred, some sediment must be missing. If the missing deposits represent a geologically long period of time so that the deposits of an entire environment have disappeared, then Walther's law no longer applies.

Detailed description provides the starting point for any facies study, irrespective of the objective or the time available. Harms *et al.* (1982) suggest that once lithologies have been identified the next stage is the recognition of a series of 'building blocks' that simplify and summarise descriptive data. Such

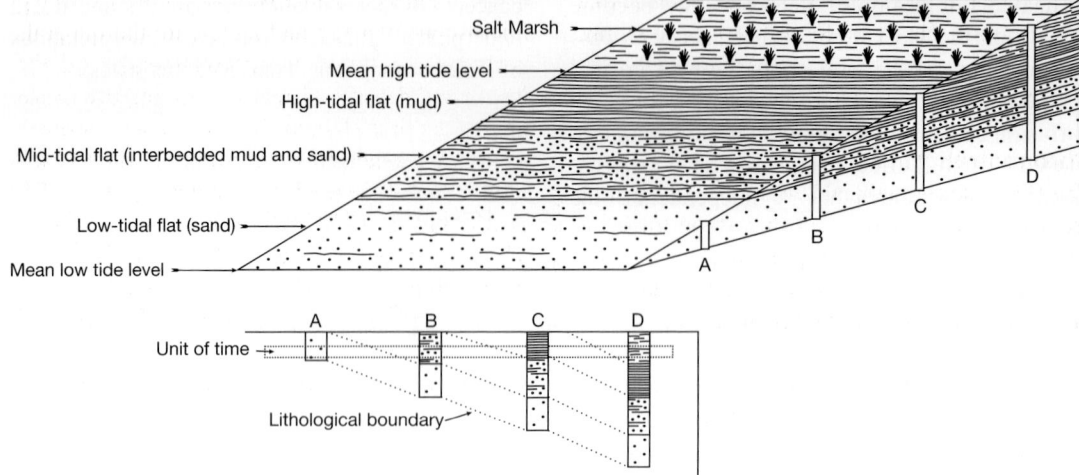

Figure 5.1   Illustration of Walther's Law. Cross-section through the sediments of a tidal flat and boreholes A–D. Note the vertical stacking (seen in cross-section) of the various lithologies as a result of progradation. The same sequence is seen in the vertical succession at D (representing an extended period of time) as may be seen on the present surface moving from A to D. Note that the horizontal lithological boundaries cross-cut the inclined time planes as the slope progrades seawards. Thus, the same lithology occurs at different time intervals in successive boreholes and is therefore diachronous.

blocks are defined as 'homogeneous bodies of rock that differ from vertically and laterally adjacent bodies of rock by their physical, biological or chemical characteristics' that is, they represent a facies. Although many facies sequences are subjectively defined, there are methods of producing statistical summaries of relationships within a succession. These methods are described in reviews by Harms *et al.* (1982), Reading (1986) and Walker (1984).

Regrettably, whereas the term facies originated with a holistic view of its components there are now three extensions in its use. Lithofacies describe the physical characteristics of a unit, lithology, bedding thickness, sedimentary structures and the like. These are the characters of most value to the engineer and are relatively easily mapped. Biofacies describe the characteristic fossil associations and are useful in refining the age and nature of the environment but of less relevance to ground conditions. Neither of these is likely to cause any confusion so long as their point of reference is stated. However, the term facies is also used subjectively in the sense of deltaic facies or turbidite facies. Such usage is only acceptable in interpretation and should be avoided as a substitute for description because the

interpretation may not be correct! An incorrect interpretation can lead to false assumptions concerning the characteristics and distribution of the deposit and the engineer should be wary of alleged descriptions offered in these terms.

Biostratigraphy

Correlation by fossils offers a more reliable way of obtaining the age equivalence of rocks than their lithological appearance, and fossils may be used to define palaeontological 'zones' (chrons or units of time). The most useful 'zonal fossils' are those that are abundant, easily recognisable, have the widest geographical distribution and, most importantly, have evolved rapidly so that a particular form lived for only a short geological time. However, organisms are environmentally selective and there are considerable difficulties in correlating between contrasting environments, the most dramatic contrast being between marine and non-marine systems. Few organisms are able to leave their remains in both. Nevertheless, relatively good global correlations are available for most Phanerozoic sequences (Table 5.1) and define a biostratigraphy. A zone may sometimes be defined by the appearance and disappearance of a single species, an 'index fossil'. More commonly, however, the distribution of fossil assemblages provides a more sensitive subdivision of time.

Chronostratigraphy

The chronostratigraphical subdivision of the rock record depends on the fact that some minerals contain naturally occurring elements that have radioactive isotopes that decay to known daughter products over specified time periods.

Ages can be obtained from minerals formed within rocks at the time of deposition, for example crystals grown within cooling magmas, or clay minerals such as glauconite formed in marine sediment. Individual volcanic lavas inter-bedded with fossiliferous sedimentary sequences provide reference time points and have allowed the construction of an absolute global timescale (Table 5.1).

In younger rocks and sediments, which are most commonly those of concern to the engineer, other methods are available. In the Pleistocene limestones common on many tropical shores, ^{234}U decays to ^{230}Th with a half-life of 248 ka (thousands of years), while in more recent deposits, the proportion of ^{14}C provides an accurate measure up to about 40 ka. Radiocarbon is particularly valuable in carbon-based materials such as wood that may be found in young deposits formed on or close to land, but is also present in marine carbonates and in cave deposits such as stalactites (speleothems). Some success has also been achieved using the decay rate of amino acids in skeletal carbonate grains. Indirect methods such as tree ring dating (dendrochronology) locally provide a very accurate recent time scale extending over some 2000 years, while methods such as optically stimulated luminescence (OSL), fission track analysis and cosmogenic dating provide means of estimating the age of eroded surfaces and rates of erosion.

5.5 Non-radiometric chronology

Fission-track dating is based on the principle that spontaneous fission occurs in minerals bearing ^{238}U and that the energy released produces visible damage (fission tracks) in surrounding minerals that can be seen with a reasonably high-powered microscope. Apatite and zircon are the minerals commonly affected in igneous rocks and the length and density of tracks ($\sim$10 μm) are measures of the time elapsed since the mineral formed. Tracks are annealed by heating, the time measured is that since the last such heating event. For apatite this is $\sim$100°C and for zircon $\sim$250°C. The time since these relatively low temperatures were achieved provides a useful record of the cooling and erosion (unroofing) of igneous or metamorphic bodies.

Luminescence dating depends on the fact that any material that contains Uranium, Thorium or Potassium is continuously bombarded with α, β and γ particles. The effect of this radiation is to produce local ionisation and this leads to trapping of metastable electrons within the atomic structure. These can be freed by heating. This causes a characteristic emission of light (Thermoluminescence or TL) from the mineral. It is quite separate from the light emitted when any material is heated and can only be measured once.

Optically stimulated luminescence (OSL) measures the light emitted from the most light-sensitive electron traps in minerals such a quartz and feldspars. In this case trapping is a result of movements of electrons in response to solar radiation. It is particularly valuable for determining how long sediment or rock surfaces have been exposed on the surface but is most effective from about 1–150 ka.

Electron spin resonance (ESR) again measures damage induced by α, β and γ radiation from naturally occurring isotopes. Free electrons accumulate within the atomic structure of the mineral and high frequency electromagnetic radiation excites these to the point where their resonance can be detected. The numbers of electrons is a direct reflection of the age of the material. The method can be applied to teeth, mollusc shells and corals as well as speleothem deposits and deep-sea cores. Ages obtained range from 0–2000 ka.

The increased use of mass-spectrometers has made cosmogenic dating more practical. ^{36}Cl, ^{26}Al and ^{10}Be are produced on exposed rock surfaces as a result of damage caused by cosmic rays and can be used to estimate erosion rates and exposure histories.

Magnetostratigraphy depends on the response of rocks and sediments to the Earth's magnetic field at the time when they were formed. Both small-scale secular variations and long-term field reversals can be used to provide a time scale. However, this is not absolute and the pattern of change must be matched to sequences of events in order to arrive at a time correlation.

5.6 Seismic stratigraphy

Seismic reflection profiling, used universally in the oil industry to support exploratory drilling programmes, has given birth to the technique of seismic stratigraphy, formalising the division of sedimentary successions on the basis of discontinuities. Seismic reflections are produced by shock waves generated at the surface of the earth by explosions or mechanical vibrations. These travel downwards and are reflected by boundaries such as bedding planes and unconformities within the rock sequence where there are relatively abrupt changes in rock density. Seismic reflections are typically only capable of resolving units of a few metres thickness, but higher frequency investigations of shallow depths (for example the seabed) may resolve layering of a few centimetres and are increasingly used in engineering investigations. Discontinuities commonly correspond with time-stratigraphic boundaries, that is, interruptions in the sedimentary sequence that represent pauses in deposition that occurred during some specific interval of time.

Major conformable depositional packages (sequences) are often bounded above and below by unconformities. The lower surfaces of these packages may show relationships that are described as onlap, downlap or concordant, whereas upper surfaces may be truncational, or show toplap or concordance (Figure 5.2). By correlating sequences on a chronostratigraphic chart (Figure 5.3), where unconformities show up as major time gaps, relative changes of ancient sea levels in a region may be inferred. In the simplest case a relative rise in sea level is indicated by coastal onlap, the magnitude of which can be used to infer the amount of sea level change. The realisation that the timing of major offlap and onlap events has commonly been the same worldwide, has led to the construction of a global sea level curve (Figure 5.4) and the recognition of long-term cycles of sea level change.

5.7 The geological time scale

The earth formed about 4.6×10^9 years (Ga) ago and meteorites found here, and the oldest rocks recovered from the moon, have a similar age. The oldest terrestrial rocks that have been discovered so far are continental crustal gneisses in West Greenland that are between 3.7 and 3.8 Ga old. Areas of very ancient rocks (>2.5 Ga) form the Archean Shields or cratons within continents (Figure 5.5). Of particular importance in these is evidence of the first emergence of life on our planet. The oldest fossils known are microscopic spheres and filaments found in cherts from Swaziland that are around 3.5 Ga old. Over many millions of years the photosynthetic activities of primitive cyanobacteria and algae eventually led to the evolution of our oxygen-rich atmosphere. The build-up of oxygen over 2.5 Ga years induced a massive precipitation of iron oxide as banded iron formations, exploited as iron ores in Canada, Australia and South America.

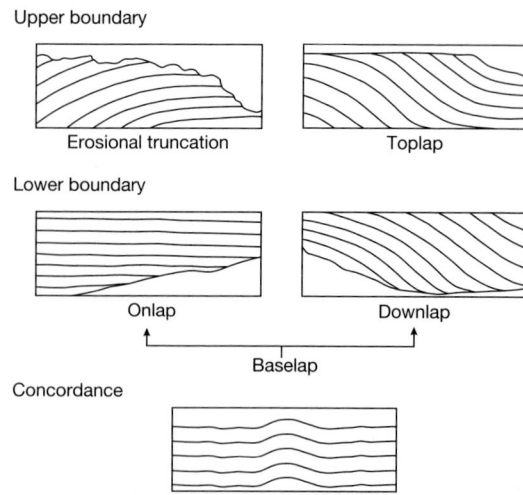

Figure 5.2   Stratal relationships recognised in seismic stratigraphy (after Vail *et al.*, 1977).

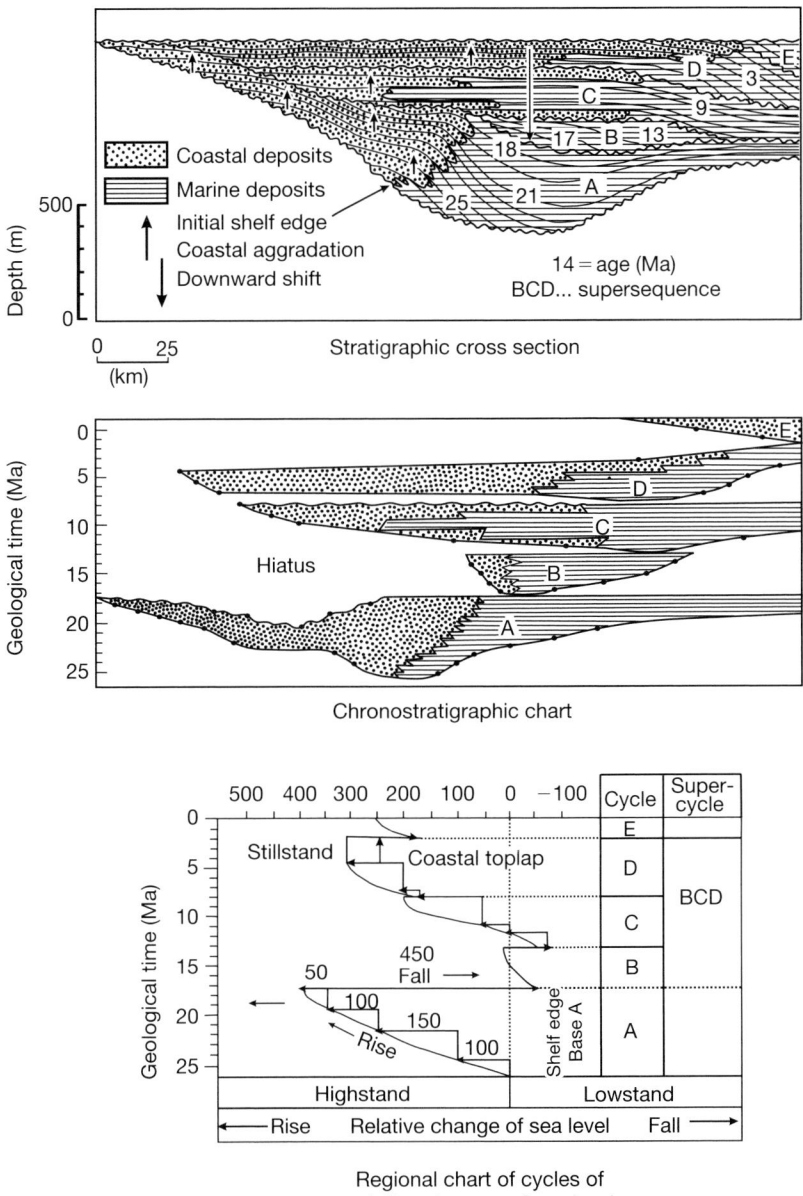

Figure 5.3 Correlation of stratigraphic sequence relationships, chronostratigraphic charts and charts of cycles of relative sea level change (after Vail *et al.*, 1977).

Although traces of soft-bodied animals are known from a few rocks around 1 Ga years old, the first widespread and well-preserved shelled fossils that can be used in dating and correlation only occur globally at about 600×10^6 years

(Ma) ago (Table 5.1). The three eras of the Phanerozoic epoch (Table 5.1) are characterised by the dominance of particular groups of fossils and the boundaries between them correspond with major phases of worldwide extinction. They

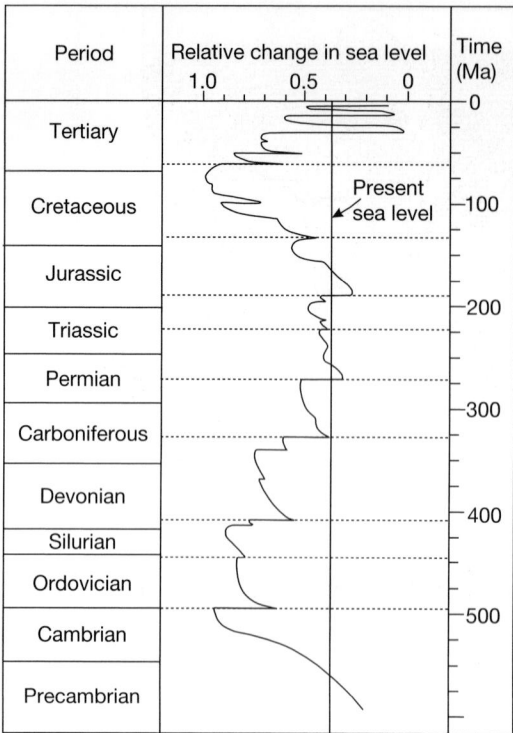

Period	Relative change in sea level	Time (Ma)
Tertiary		
Cretaceous	Present sea level	100
Jurassic		
Triassic		200
Permian		
Carboniferous		300
Devonian		
Silurian		400
Ordovician		
Cambrian		500
Precambrian		

Figure 5.4 Global time scale and changes in sea level (identified by Vail *et al.*, 1977).

also broadly coincide with widespread mountain-building and folding episodes (Table 5.1).

5.8 Correlating and interpreting sequences

It is impossible here to review the stratigraphic evolution of the world but the principles of correlation and interpretion of rock sequences can be demonstrated. The succession of mid-Jurassic rocks in southern England consists of limestones, claystones and rarely, sandstones. The relative ages of these rocks are indicated by their fossil contents and world correlation. Their interrelationships have been unravelled by methods that include mapping, logging outcrops and boreholes, and seismic stratigraphy. Correlation allows a thickness (isopach) map to be constructed (Figure 5.6) that defines the form and limits of accumulation of particular lithologies and time-stratigraphic units. Interpretative cross-

sections derived from this, (Figure 5.7) may be used to generate a palaegeographical map (Figure 5.8), illustrating the regional depositional environment during a specified time interval and the form of the original sedimentary basin. In the hydrocarbon industry such models are used extensively to provide a predictive framework for exploration, combining facies modelling with a structural geological appraisal of a region, but the same techniques are also of value in large-scale hydrological modelling of aquifers.

On a global scale, regional stratigraphical reconstructions provide the basis for plate-tectonic models. For example, subduction (Chapter 4) adjacent to a trench margin results in under-thrusting of successive units of sea-floor sediments as oceanic crust slips beneath the adjacent continental margin. As a result, a prism or wedge of sediment is accreted in which individual thrust slices become progressively younger in the direction of the original oceanic abyssal plain. Thus, in Figure 5.9, slice A is younger than slice B. Within individual slices, however, the beds form a normally stacked sequence and are younger from bottom to top. Detailed mapping and stratigraphic appraisal of the terrain forming the Southern Uplands of Scotland has revealed an assemblage of rocks of Silurian age that apparently shows these characteristics (Anderton *et al.*, 1979). However, no matter how elegant they may appear to be, the reliability of such models ultimately depends on the accuracy of stratigraphic correlation.

Some rock sequences appear to be monotonous and are consequently difficult to subdivide. However, a detailed examination can sometimes provide subdivisions in a variety of ways:

1. macro-palaeontological
2. micro-palaeontological
3. mineralogical
4. lithological.

Macro-palaeontological zonation involves the detailed collection and identification of macrofossils and their comparison with known standard specimens. Where no macrofossils are available, either because the depositional environment was

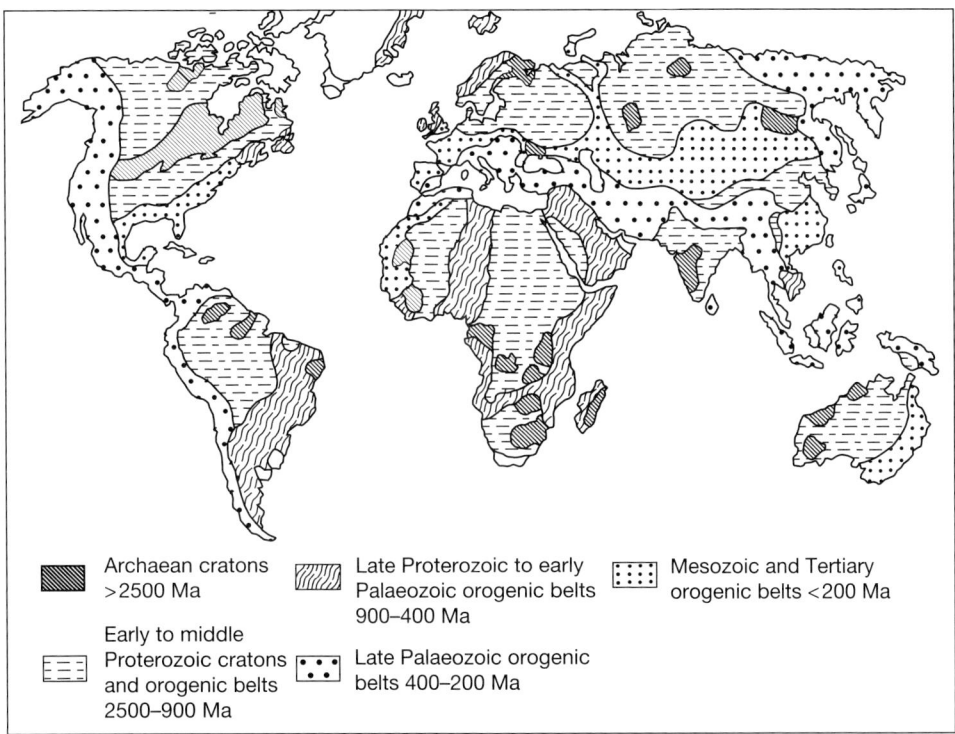

Figure 5.5   Generalised geological map of the world showing major tectonic provinces (after Anderton *et al.*, 1979; see also Chapters 4 and 9).

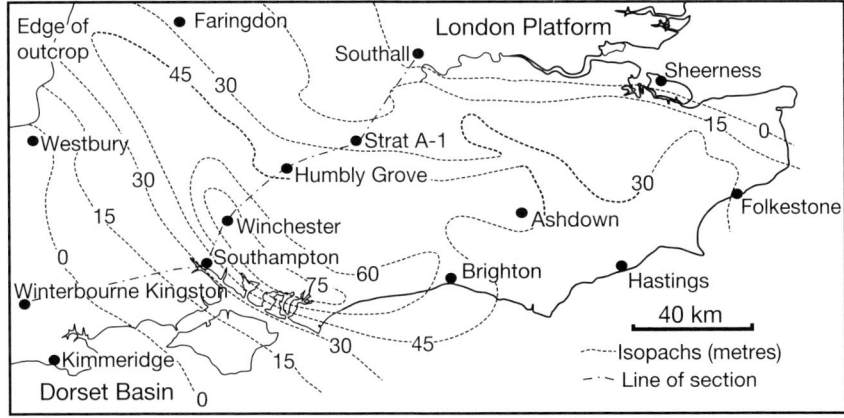

Figure 5.6   Isopach (equal thickness) map of the Great Oolite Limestone Formation (Middle Jurassic) in south-east England (after Sellwood *et al.*, 1985).

unfavourable for their life or because burial conditions were unsuitable for their preservation, or simply because an uncored borehole sequence provides only fragments in drill cuttings, then microfossils may provide an alternative means of zonation. The methods of identification, comparison and correlation are essentially the same as those used for macrofossils but as the fossils

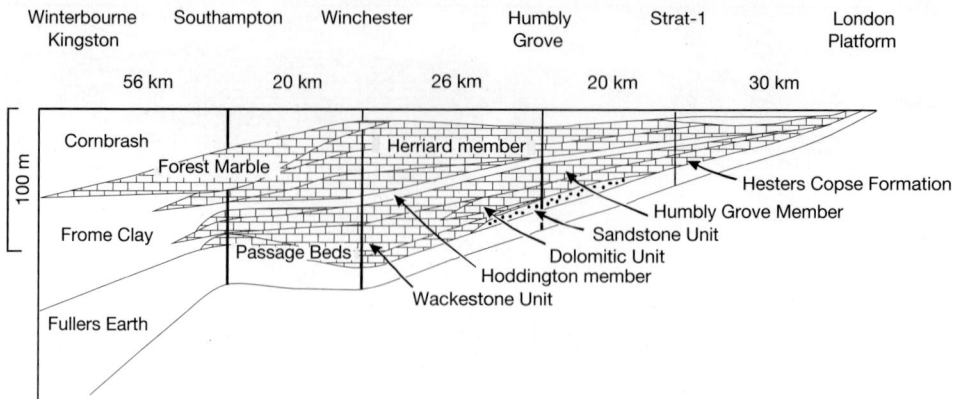

Figure 5.7   Lithostratigraphic correlation in the middle Jurassic Great Oolite Formation of south-east England showing facies variation. The line of section is shown in Figure 5.5 (after Sellwood *et al.*, 1985).

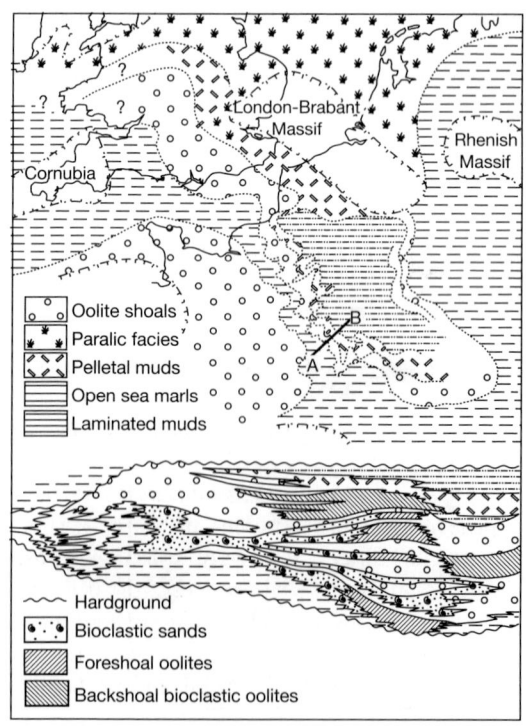

Figure 5.8   Generalised palaeogeography and facies distribution of the Middle Jurassic (Bathonian) of southern Britain and north-east France, Belgium and the Netherlands (after Sellwood and Sladen, 1981; Purser, 1969, and Dubois and Yapaudjian, 1979). The section below illustrates the facies variation along the line of section A–B (after Sellwood, 1985).

themselves are so small, large numbers may be obtained from small samples. Foraminifera, ostracodes, pollen, spores and dinoflagellates are all commonly employed.

Mineralogical variation in a sequence can be cryptic and may be of two kinds. The characteristic detrital minerals of the sediment, or indeed the composition of these minerals, may change abruptly as a result of a change in the geometry of the basin and thus the origin, provenance, of the grains. Such changes may be visible over a wide area. However, mineralogical variation may also reflect post-depositional processes. The diagenetic degradation of feldspars in sandstone may generate a rock rich in the clay mineral kaolinite. During burial, and with rising temperature, the kaolinite is progressively replaced by the clay illite. Thick hydrocarbon-bearing reservoir sandstones are sometimes subdivided on the basis of this mineral transition that, in the North Sea, occurs at a temperature of 120–150° C. It is an irreversible transformation and may be recognised even when rocks formerly buried into the illite zone have later been uplifted to shallower depths. However, as it is a reflection of both depth of burial and local geothermal gradient it does not relate to conventional stratigraphical correlatives. The presence of fibrous illite has serious implications for the reservoir quality of a sandstone because the fibres of illite

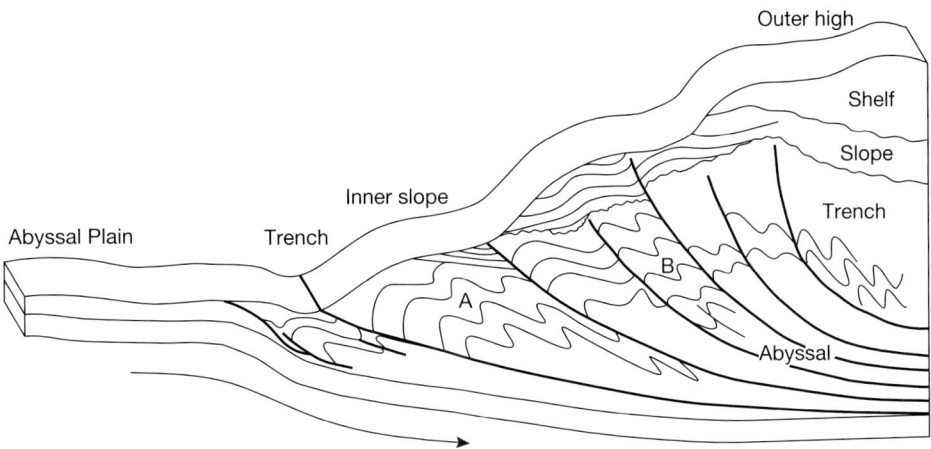

Figure 5.9   Model of a subducting margin. Note the successive under-thrusting of slices and the parallel rotation of folds (after Anderton *et al.*, 1979).

restrict pore-throats in the rock and thus reduce permeability.

Lithological subdivision relies on the development of easily recognisable 'marker beds'. These may be distinctive deposits such as volcanogenic tuff bands and bentonites, glauconitic layers, caliche (calcrete) profiles in alluvial sequences, or diagenetic nodular and concretionary beds. Provided that it is laterally extensive, any form of marker unit may provide a datum upon which the geometry of stratal units can be modelled. In the North Sea area, for example, the base of the Eocene is marked by a regionally extensive volcanic tuff, which is easily recognised in otherwise featureless mudstones.

Problems in the Quaternary

Quaternary sediments and rocks (Chapter 6), deposited in the last 2 million years, dominate the problems encountered in engineering excavation. However, their stratigraphy and to some extent their correlation differs in several important respects from those of older rocks. There are two reasons for this. The first is one of distribution. Typical marine deposits have remained beneath the sea and are laterally extensive, but the successions that concern us were mostly deposited either sub-aerially or in the shallow marine margins where they have been subject to recent changes in sea level. There have been four major

glacial–interglacial cycles, reflected in changes in sea level of the order of 130 m, but in addition, superimposed on these there have been around twenty minor cycles. Thus, all Quaternary successions are punctuated by erosion surfaces (Figure 5.10). The amount of incision that these represent is locally of the same order as the sea-level changes and may be greater than 100 m. As a result, wide variations in thickness are common and particular units are missing locally. The record is at best fragmented.

In ancient rocks the idea of a stratotype, a type succession showing a typical facies succession is widely applied. However, the variability of the Quaternary means that such a system is unworkable. Although units of different ages must appear in chronological order, their local characteristics may differ dramatically, reflecting deposition in contrasting environments. Depositional units may acquire names but these normally have only local validity. The contrast can best be demonstrated by reference to high and low latitude successions.

In northern Europe, and in high latitudes, the Quaternary is generally unlithified. A thick over-consolidated diamict deposit, representing glacial debris overridden by a later ice surge, may form the base of a succession resting on a glacially generated erosion surface. Laminated silts and clays, that reflect accumulation in a glacially dammed

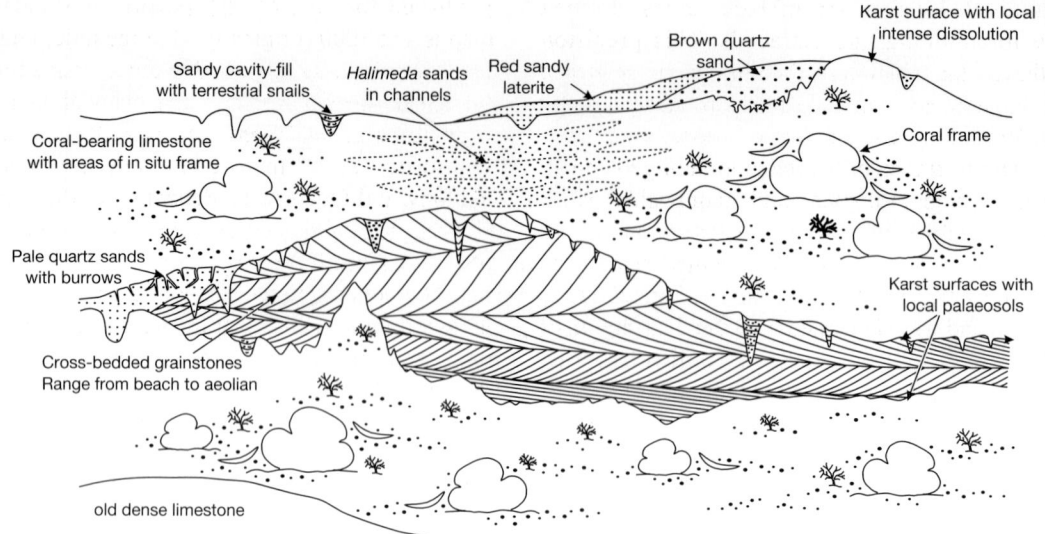

Figure 5.10   Schematic section of the late Pleistocene succession of the Kenya coast illustrating the wide lateral variation and punctuation of the succession by karst erosion surfaces (after Braithwaite, 1984).

lake may overlie it, and these may be followed in turn by cross-bedded sands and gravels recording the advance of outwash streams. These may be capped by unconsolidated diamicts containing sand and silt lenses that reflect meltout from a later glacial advance. However, there will be considerable lateral variability and it is unlikely that precisely the same succession will be identified over distances of more than a few tens of metres. In addition, successions elsewhere that appear similar may be of different ages and not correlatable. The time between depositional intervals is generally too short to generate a satisfactory palaeontological timescale, although there have been some advances using teeth of rodents, beetle carapaces and the jaws of chironomids (non-biting midges: Lowe and Walker, 1997) to define sequences of climatic change. The methods of dating such materials include radiocarbon (for younger deposits) and pollen analysis.

In the tropics the most common deposits are Quaternary limestones. In these, depositional environments include coral reef related systems, tidal sands, wind-blown deposits (aeolianites) and soils (Figure 5.10). Changes in sea level produce a rapid lithification and deep erosion, principally by dissolution. Erosion surfaces and palaeosols

offer a convenient means of dividing successions. Hearty (1998) has identified as many as eight palaeosol-bounded sequences in succession in the Bahamas but as few as two may be present in any one place and they do not provide a means of correlation. There are, however, many more methods of absolute dating available. Uranium series dates, magneto-stratigraphy, amino-stratigraphy, electron spin resonance (ESR) and stable isotope stages ($^{16}O/^{18}O$ ratios) defined both by reference to deep ocean cores and ice cores from the Arctic and Antarctic provide alternative means of determining age, although diagenetic changes may render them unreliable.

Applications

Horswill and Horton (1976) provide an illustration of the engineering application of micro-palaeontological zonation in a relatively homogenous sequence. They investigated the effects of cambering at the Empingham Dam site in the Gwash valley (English Midlands). Although the Upper Lias clays at the site may be crudely divided using markers such as a silty micaceous mudstone, an ammonite nodule bed, a phosphatic nodule bed, and a pisolitic bed, the thick intervening claystones are more closely subdivided using assemblage zones defined by foraminifera and ostracods. This precision allowed the valleyward thinning of the sequence to be more accurately defined, providing a better understanding of the nature of the deformation.

An illustration of the integration of geological, macro-palaeontological and geophysical work in an engineering appraisal, is provided by Bruckshaw *et al.*, 1961. This formed part of the site investigation for the Channel Tunnel Study Group and was concerned with the distribution of the Lower Chalk beneath the English Channel. It also provides a lesson in nomenclature. The Lower Chalk is easily characterised as a rock formation in engineering terms and the geological and engineering terminology closely correspond. By contrast, Dearman and Coffey (1981) attempted to assess the engineering characteristics of the late Permian Magnesian Limestone of north-east England. They found a considerable diversity in terminology, if not of rock type, identified by a variety of geologists that they properly described as 'bewildering'. Some of the reasons for this have been discussed and centre on the use of subjective interpretative terms as descriptors. For engineering purposes the plethora of terms was reduced to only four main lithological types: bedded limestone, reef limestone, concretionary limestone and silty clay. These simple divisions proved locally effective and were easily adapted to include variously brecciated versions. However, the term 'reef limestones' is clearly a subjective interpretation and bodies of concretions are not bounded by time equivalent surfaces. It would therefore be unwise to attempt to apply the same scheme elsewhere.

5.9 Geological maps

Geological mapping forms part of the training of all geological earth scientists and geological maps are a fundamental tool in any preliminary site investigation for engineering purposes. In Britain the Geological Survey has traditionally mapped at a scale of around 1 : 10 000 (originally 6″ to the mile). Such maps are available for consultation at offices of the Survey, but relatively few are published for sale. More typically the standard map is 1 : 50 000 (originally 1″ to the mile) and a few special sheets at 1 : 25 000 are published for 'classical' areas of geology. For many areas two forms are available, referred to as solid and solid-with-drift. 'Solid' maps show only the 'solid' geology and Quaternary deposits are therefore omitted. The solid geology shows the distribution of bedrock and is organised on the basis of age but differentiates major lithostratigraphic units. It is important to realise that the names applied are often traditional ones and do not necessarily describe the characteristics of the formations concerned. For example, the 'Devonian Old Red Sandstone' may not be red and locally may include considerable thicknesses of conglomerate, shale and even volcanic rocks. The terminology applied: sandstone, limestone, shale etc. is a basic one and conveys nothing of the physical attributes of the material or of its likely behaviour in engineering. The local or regional Guides published by the British Survey go some way to explaining variations and providing interpretations of the major features. However, not all of these maps or guides are in print and it may be necessary to go to libraries in order to obtain them. The modern British Geological Survey offers consultancy on issues that require further clarification and in some areas will be able to offer advice on engineering characteristics, but most large companies now have their own engineering geologists or earth scientists who can interpret the information and draw attention to critical issues.

In Britain so-called 'drift' maps, also on a scale of 1 : 50 000, include the distribution of Quaternary alluvial, glacial or marine deposits. These may be differentiated on the basis of age, the distribution of marine or alluvial terraces, or different ages of diamict deposits. The areas covered now include the continental shelf around Britain. Maps are also available for some areas on a scale of 1 : 25 000 that provide more detailed distributions of such deposits, their thicknesses and engineering characteristics. In a very few areas specialised maps describe the distribution of hazards relating to flooding, subsidence, hydrology or other issues and indicate areas suitable for development or excluded for these and other sound geological reasons.

In the rest of Europe, high quality geological maps are available but are seldom at scales larger than 1 : 50 000. Exceptions are found in Germany where excellent 1 : 25 000 maps are available for many areas and Switzerland where 30% of the country is available at this scale. Many former European colonies and dependencies have limited numbers of maps on scales from 1 : 50 000 or, more commonly, 1 : 200 000 or 1 : 500 000. Regrettably, large areas of the world have not been geologically surveyed in detail or have air photograph interpretations that may not provide adequate 'ground truth', and small-scale maps offer only the most general guidance for the engineer.

The United States Geological Survey not only publishes excellent standard geological maps, with and without superficial deposits, but also specialist maps dealing with depth to bedrock, engineering characteristics and hydrology, flood impacts and the distribution of landslides, together with marine geology and compilations illustrating the distributions of earthquakes, volcanoes and even diagenetic characteristics. Useful scales vary from 1 : 24 000 to 1 : 48 000 but compilations commonly cover State or larger areas. The Survey's web catalogue also includes publications by the US Bureau of Mines and State Surveys and independent organisations such as the influential American Association of Petroleum Geologists and the Society for Sedimentary Geology (SEPM).

Information on the availability of maps of particular areas of interest can be obtained from the Geological Surveys of the countries concerned or from texts such as *International Maps and Atlases in Print* (edited by K. L. Winch). However, access to many regional catalogues is now also available using the Internet and for the US, for example, exhaustive lists can be available on the desk of the enquirer within seconds.

5.10 Summary

Stratigraphy is the description and interpretation of layered rock sequences and provides a basis for the study of earth history through 3.8 Ga. The principal applications are in the exploration for hydrocarbons and minerals, but an accurate stratigraphical interpretation of both bedrock and superficial deposits is essential to the engineer in site investigations for dams, tunnels and all major construction works.

Regional interpretations of strata depend on correlation. This is best achieved regionally by fossils, and locally by the mapping of distinctive marker beds such as volcanic tuffs. Age equivalence in sequences dated by fossils may be correlated with an absolute timescale based on the half-lives of radioactive isotopes in inter-bedded igneous or sedimentary rocks. Recent developments have involved regional correlation by geophysical means, employing the technique of seismic stratigraphy. This has helped to refine models of global sea level change over the past 600 Ma.

References

Ager, D. V. (1981) *The Nature of the Stratigraphical Record* (2nd edn). Macmillan, London.

Anderton, R., Bridges, P. H., Leeder, M. R. and Sellwood, B. (1979) *A Dynamic Stratigraphy of the British Isles; a Study in Crustal Evolution*. George Allen & Unwin, London.

Blatt, H., Midelleton, G. and Murray, R. (1980) *Origin of Sedimentary Rocks* (2nd edn). Prentice-Hall, Englewood Cliffs, New Jersey, 782pp.

Braithwaite, C. J. R. (1984) Depositional history of the late Pleistocene limestones of the Kenya coast. *Journal of the Geological Society* **141**, 685–699.

Bruckshaw, J. M., Goguel, J., Harding, H. J. B. and Malcor, R. (1961) The work of the Channel Tunnel Study Group, 1958–1960. *Proceedings of the Institution of Civil Engineers* **18**, 149–178.

Dearman, W. R. and Coffey, J. R. (1981). An engineering zonal Map of the Permian Limestones of NE England. *Quarterly Journal of Engineering Geology* **14**, 41–57.

Dubois, P. and Yapandjian, L. (1979) Jurassique moyen. In Mégnien, C. (ed.) *Synthèse Géologique du Bassin de Paris*. JM1–JM4, Mem. BRGM.

Harms, J. C., Southard, J. B., Spearing, D. R. and Walker, R. G. (1982) *Structure and Sequence in Clastic Rocks*. Society of Economic Paleontologists and Mineralogists (SEPM) Short Course Notes 9 Tulsa, Oklahoma, 249pp.

Hearty, P. J. (1998) The geology of Eleuthera Island, Bahamas: a Rosetta Stone of Quaternary stratigraphy and sea-level history. *Quaternary Science Reviews* **17**, 333–355.

Horswill, P. and Horton, A. (1976) Cambering and valley bulging in the Gwash Valley at Empingham, Rutland. *Philosophical Transactions of the Royal Society, London* A **283**, 427–462.

Lowe, J. J. and Walker, M. J. C. (1997) *Reconstructing Quaternary Environments* (2nd edn). Prentice-Hall, London.

Purser, B. H. (1969) Syn-sedimentary marine lithification of middle Jurassic limestones in the Paris Basin. *Sedimentology* **12**, 205–230.

Reading, H. G. (1986) Facies. In Reading, H. G. (ed.) *Sedimentary Environments and Facies* (2nd edn). Blackwell, Oxford, 4–19, 615pp.

Sellwood, B. W. (1985) Shallow-marine carbonate environments. Reading, H. G. (ed.) *Sedimentary Environments and Facies*. Blackwell, Oxford, Chapter 10, 283–342.

Sellwood, B. W., Scott, J., Mikkelsen, O. and Akroyd, P. (1985) Stratigraphy and sedimentology of the Great Ooolite Group in the Humbly Grove oilfield, Hampshire. *Marine and Petroleum Geology* **2**, 44–55.

Sellwood, B. W. and Sladen, C. P. (1981) Mesozoic and Tertiary argillaceous units: distribution and compositions. *Quarterly Journal of Engineering Geology* **14**, 263–275.

Vail, P. R., Mitchum, R. M. and Thompson, S. III. (1977) Seismic Stratigraphy and global changes of sea level. In *Seismic Stratigraphy — Applications to Hydrocarbon Exploration. American Association of Petroleum Geologists*, Memoir **26**, 49–212.

Walker, R. G. (1984) General introduction. In Walker, R. G. (ed.) *Facies Models* (2nd edn). Geoscience Canada Reprint Series 1. Geological Association of Canada, 1–9.

Winch, K. L. (ed.) (1976) *International Maps and Atlases in Print* (2nd edn). Bowker, London and New York. GeoPubs, 4 Glebe Crescent, Minehead. TA24 5SN UK.

6. The Quaternary

Andrew Goudie

6.1 Introduction

Environmental changes can have a huge significance for engineering projects. The degree, frequency and abruptness of change, particularly over the last few millions of years (i.e. the Quaternary), has been remarkable (Figure 6.1). Landscapes often display the legacy of forms (e.g. relict landslides that may be susceptible to reactivation) and materials (e.g. sands and gravels that are possibly not being replenished today) inherited from significantly different climatic conditions to those experienced at present. There are some parts of the world, termed paraglacial environments, where processes and forms are still dominated by the effects of the glacial advances that happened over that period. The form of river systems and the nature of their sediment stores may relate to those times. Moreover, as higher resolution environmental histories are reconstructed from ocean, lake and ice cores, it becomes clear that in addition to the major glacials and interglacials of the Quaternary, there have been a whole series of changes at timescales of the decade, century and millennium. In terms of such considerations as the return periods (i.e. probability) of hazardous events, the present may be highly atypical of the past and of the future.

6.2 The late Cainozoic

During the Tertiary, which started after the Cretaceous about 65 million years (Ma) ago, the world's climate suffered the so-called Cainozoic climate decline (Goudie, 1992). Temperatures showed a general tendency, though not steady or uninterrupted, to fall (Figure 6.2). Thus in the North Atlantic region in the early Tertiary, conditions favoured a widespread, tropical moist forest.

By the Pliocene, around 5 Ma ago, the degree of cooling was such that a more temperate flora was present in the North Atlantic region, and at 2.4 Ma ago glaciers started to develop in mid-latitude areas and many of the world's deserts came into being (Williams *et al.*, 1998).

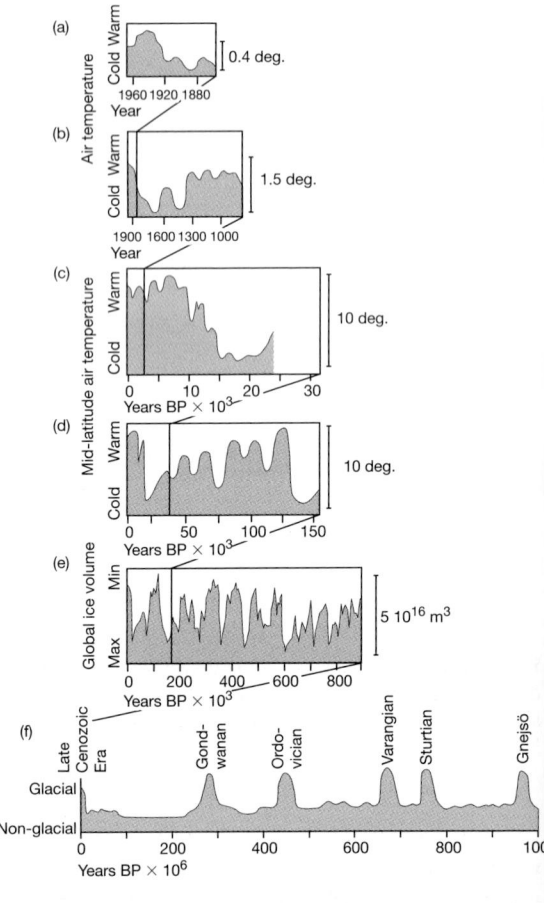

Figure 6.1 The variation in world climate over the past billion years: (*A*) by decade; (*B*) by century; (*C*) by millennium; (*D*) by tens of millennium; (*E*) by hundreds of millennium and (*F*) by era.

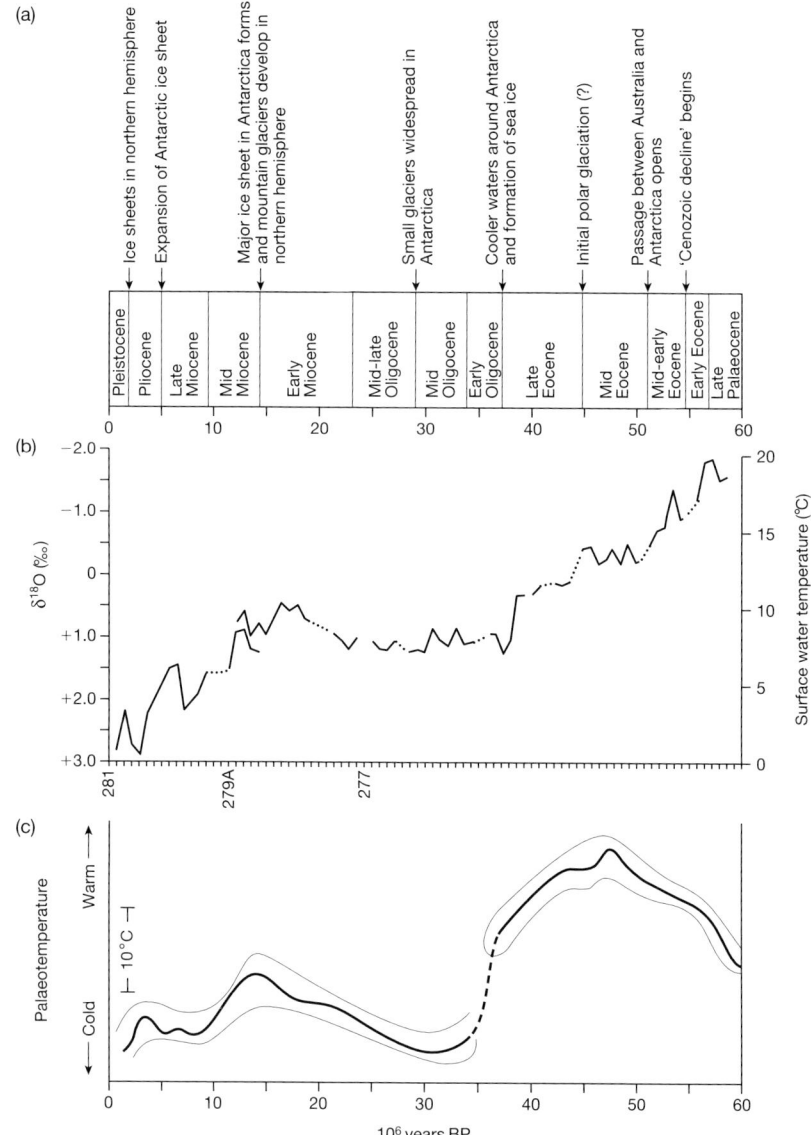

Figure 6.2   The Cenozoic climate decline: (*A*) a generalised outline of significant events in the Cenozoic climate decline; (*B*) Oxygen isotopic data and palaeotemperature indicated for planktonic foraminifera at three sub-antarctic sites (277, 279A, and 281) and (*C*) temperature changes calculated from Oxygen isotope values of shells in the North Sea (after Goudie, 1992).

In the Quaternary (which comprises the Pleistocene and the Holocene) the gradual and uneven progression towards cooler conditions, which had characterised the earth during the Tertiary, gave way to extraordinary climatic instability (Lowe and Walker, 1997). Temperatures oscillated wildly from values similar to, or slightly higher than, today in interglacials to levels that were sufficiently cold to treble the volume of ice sheets on land during the glacials (Figure 6.3). Not only was the degree of change remarkable but so also, according to evidence from the

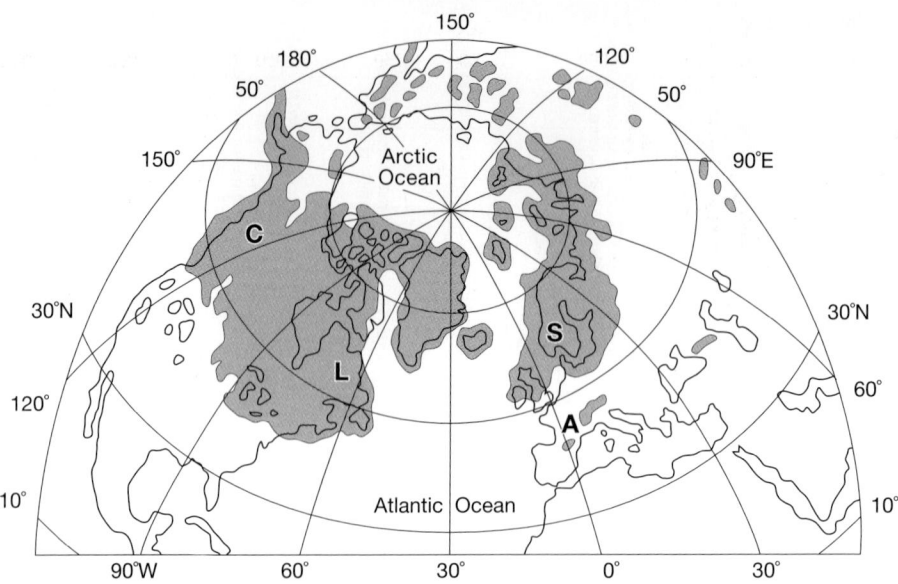

Figure 6.3   The possible maximum extent of glaciation in the Pleistocene in the northern hemisphere: C = Cordilleran ice; L = Laurentide ice; S = Alpine ice (after Goudie, 1992).

sedimentary record retrieved from deep-sea cores, was its frequency. In all there have been about seventeen glacial–interglacial cycles in the last 1.6 million years. The cycles tend to be characterised by a gradual build up of ice volume (over a period of c. 90 000 years), followed by a dramatic glacial 'termination' in only about 8000 years. Furthermore, over the three or so millions of years during which humans have inhabited the earth, conditions such as those we experience today have been relatively short lived and atypical of the Quaternary as a whole (Wilson et al., 2000).

6.3 Abrupt change

One of the features of palaeoclimatic research in the past decade has been the realisation of just how abruptly climatic change can occur (Adams et al., 1999). An indication of this came from a deep peat core at Grande Pile in France (Woillard, 1979) where it was argued that the temperate forest of the Last Interglacial (Eemian), c. 120 000 years ago (BP) was replaced by taiga within approximately 150 years. This was not a finding that received universal approval (see Frenzel and Blundau, 1987).

Ice cores, such as that from Vostok in Antarctica (Petit et al., 1999), provide a particularly fine temporal resolution and contain large numbers of valuable environmental indicators (Figure 6.4). High frequency swings in isotopic and dust content and other indicators suggest that dramatic oscillations have taken place in environmental conditions over quite short periods of time. The rapid temperature oscillations that have been identified from ice cores are known as Dansgaard-Oeschger events. Dansgaard et al. (1993), for example, documented no less than twenty-four interstades (warmer phases) in the last glacial period from an ice core from Greenland. However, high frequency shifts are also known from the last interglacial (the Eemian of Europe and the Ipswichian in Britain), but some workers fear that the ice core records could have been corrupted by deformation within the ice. That said, there is also evidence of a sharp Eemian cold phase in ocean cores (Maslin and Tzedakis, 1996).

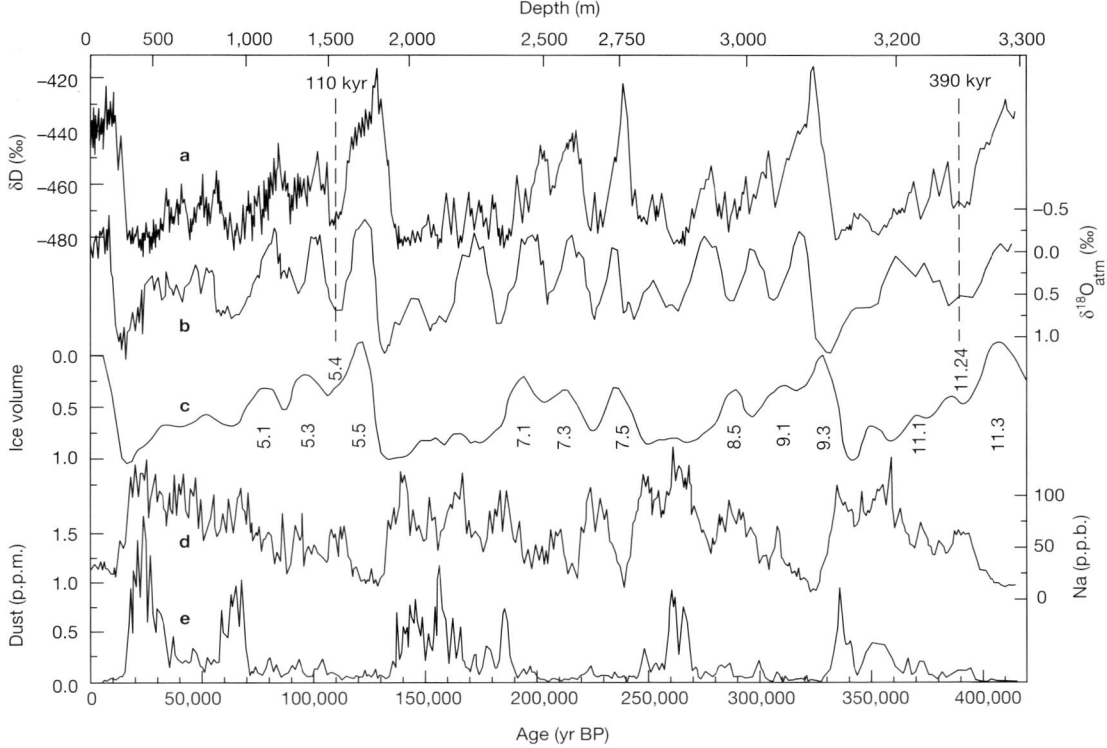

Figure 6.4 Environmental change as revealed in the Vostok ice core. The values along line 'c' are marine isotope stage (after Petit *et al.*, 1999).

High frequency abrupt changes have also been identified from ocean cores (Oppo *et al.*, 1998), where the observed saw-tooth patterns of climatic variation have been named Bond cycles. Also within the ocean core sediment record there are layers of sediment that are rich in dolomite and limestone detritus but poor in foraminifera. Each layer is interpreted as being the result of deposition by massive armadas of icebergs released from ice caps around the North Atlantic (Bond *et al.*, 1992). The records of these iceberg flotillas are termed Heinrich events (Andrews, 1998) and it is evident that they represent cold stadials of short duration — less than 1000 years (ka). The Younger Dryas event towards the end of the Last Glacial (*c.* 12 000 years BP) was an example of a very short-lived stadial (no more than a thousand years) that came and went within decades (Anderson, 1997), perhaps in response to changes in the circulation of the ocean currents.

Recent years have seen an increasing appreciation of the importance of rapid changes over shorter timescales (Viles and Goudie, 2003), including decadal and inter-decadal fluctuations such as El Niño Southern Oscillation (ENSO) and the North Atlantic Oscillation (NAO) (Figure 6.5). Such fluctuations can have a significant effect on such geomorphological phenomena as debris flows (Grosjean *et al.*, 1997), floods (Magilligan and Goldstein, 2001), wave climate (Wang and Surail, 2001), lake levels (Birkett *et al.*, 1999), glacial damming of drainage (Depetris and Pasquini, 2000), hurricane landfalls (Elsner and Kera, 1999), valley-bottom incision (Bull, 1997) and dune reactivation (Forman *et al.*, 2001). Changes in the frequency and magnitude of such phenomena are of great engineering concern, particularly in sensitive

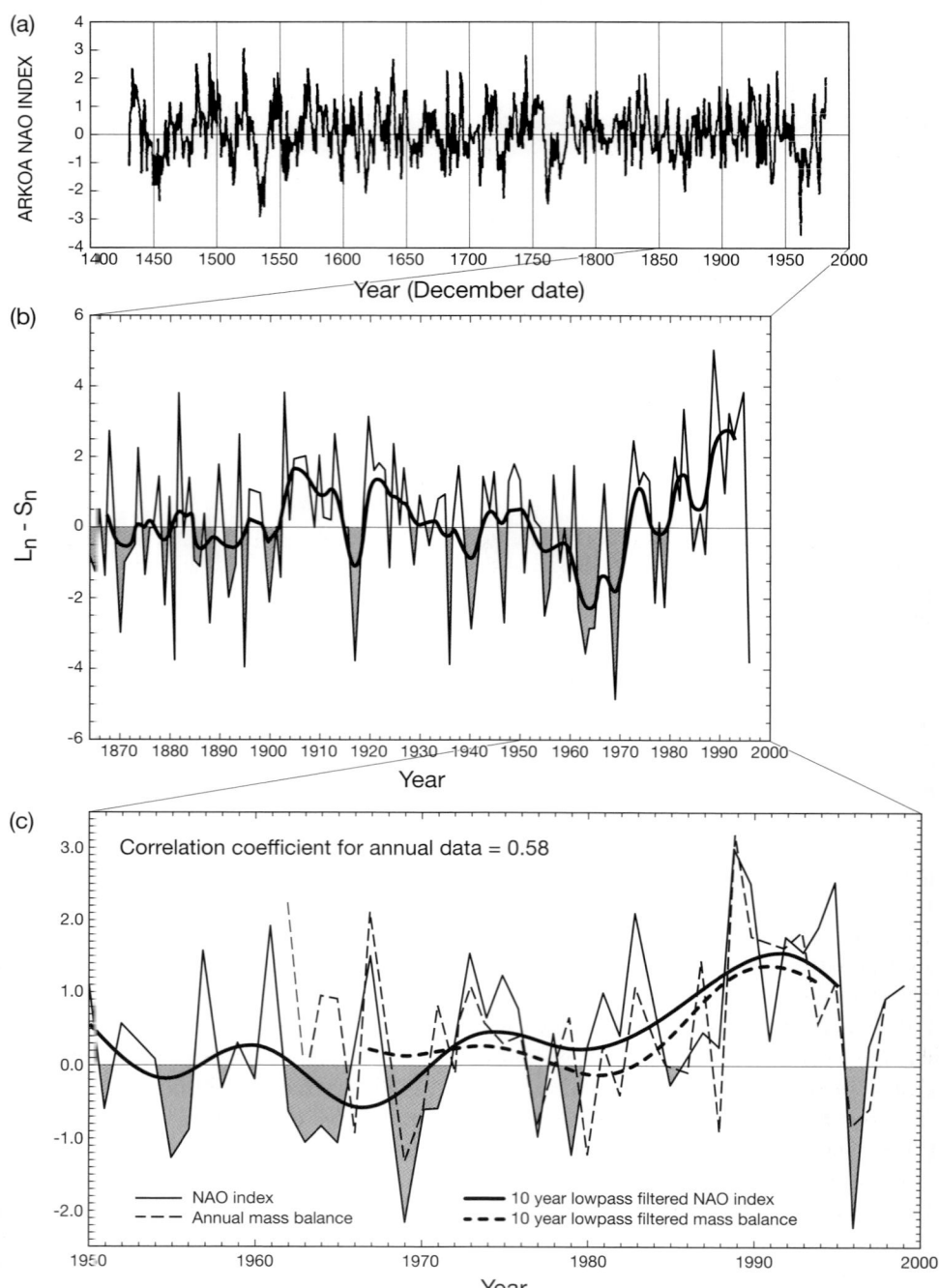

Figure 6.5 The North Atlantic Oscillation (NAO): (*a*) A 555-year reconstruction of the NAO based on tree-ring analyses; (*b*) NAO index for December–March from 1864 to 1996 based on the difference in normalized mean sea-level pressure between Lisbon and Iceland; (*c*) NAO index (1950–1999) and mass balance of Nigardsbreen Glacier, Norway (1962–1998). (Modified from various sources in Viles and Goudie, 2003.)

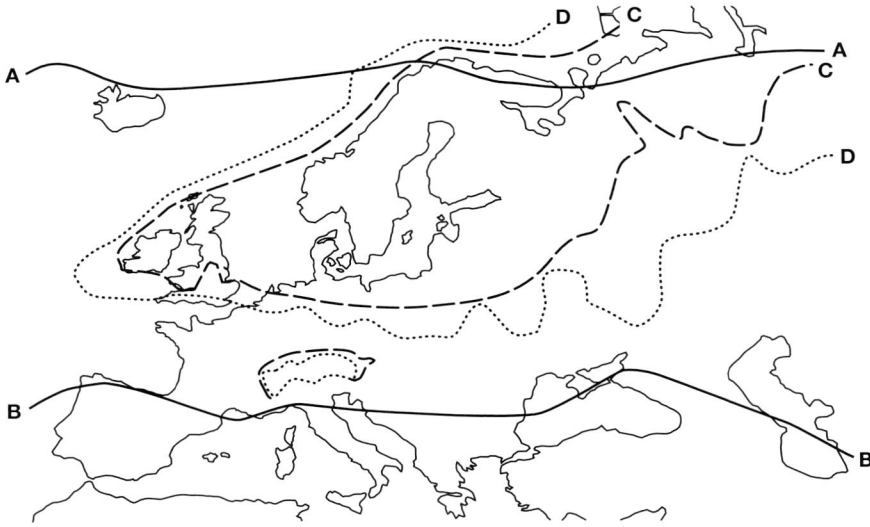

Figure 6.6 Glacial conditions in Europe: (*A*) the position of the present polar timberline in Europe; (*B*) the position of the timberline during the maximum of the last (Würm) glacial; (*C*) the extent of north European drift deposits that can be attributed to the Last Glacial; (*D*) the drift borders of the Riss-Saale and Mindel-Elster in North Europe.

environments where critical thresholds may be crossed (Chapter 1). These are dealt with in detail elsewhere in this book.

6.4 The last glacial cycle

The last glacial cycle (the Devensian of Britain, the Würm and Weichselian of Europe and the Wisconsinian of the USA) reached its peak about 18 000 years BP, with ice sheets extending over Scandinavia to the north German plain, over all but the south of Britain, and over North America to 39 °N. To the south of the Scandinavian ice sheet (Figure 6.6) was a tundra steppe underlain by permafrost, and forest was relatively sparse to the north of the Mediterranean. The Last Glacial had a major impact on ground conditions over extensive areas and accounts for the present day inheritance of frost-shattered rock, sand and gravel resources, loess and unstable slopes.

Ice covered nearly one-third of the Earth's land area, but the additional ice-covered area was almost all in the Northern Hemisphere, with no more than about 3 per cent in the Southern. Nonetheless, substantial ice sheets developed over Patagonia and New Zealand. The thickness of the ice sheets may have exceeded 4 km, with typical depths of 2 to 3 km.

Highly important changes also took place in the oceans. During the present interglacial conditions of the Holocene, the Atlantic is at least seasonally ice free as far north as 78 °N in the Norwegian Sea. This condition reflects the bringing of warm water into this region by the Gulf Stream (North Atlantic Current). During the Last Glacial Maximum, however, the oceanic polar front probably lay at about 45 °N, and north of this latitude the ocean was mainly covered by sea ice during the winter.

The degree of temperature change that occurred over land was particularly great in the vicinity of the great ice sheets. The presence of permafrost (permanently frozen subsoil) in southern Britain (Ballantyne and Harris, 1994) suggests a temperature depression of the order of 15 °C. Mid-latitude areas probably underwent a lesser decline (perhaps 5–8 °C) though in areas subject to maritime air masses temperatures were more likely to have been depressed by 4–5 °C.

Periglacial conditions (i.e. conditions characterised by snow and frost though not by glaciation) caused many slopes in areas like southern England to be unstable, especially as a result of solifluction and landslides. Cambering and valley bulging were widespread phenomena. Relict instability is widespread and is susceptible to reactivation (Chapter 14).

Another feature of areas to the south of the great northern-hemisphere ice sheets was the deposition of silt-sized aeolian material (loess) derived from deflation of outwash plains and moraines (Pye, 1987). Over vast areas (at least $1.6 \times 10^6 \, km^2$ in Europe) it blankets pre-existing relief, and reaches thicknesses in excess of 200 m in Tajikistan, Uzbekistan and in China. Other important areas include the Missouri–Mississippi plains, the North German Plain and the Pampas of Argentina and New Zealand. Loess presents its own particular problems for engineering structures (Derbyshire *et al.*, 2000; Chapter 25).

6.5 Interglacials

In general terms the Quaternary interglacials were short lived but appear to have been essentially similar in their climate, fauna, flora and landforms to the present Holocene interglacial. One of their most important characteristics was that they witnessed the rapid retreat and decay of the great ice sheets and saw the replacement of tundra conditions by forest over the now temperate lands of the Northern Hemisphere. At their peak they may have been a degree or two warmer than now.

The general sequence of vegetational and soil development during an interglacial has been described for north-west Europe by Birks (1986). The first, or cryocratic phase, represents cold glacial conditions, with sparse assemblages of pioneer plants growing on base-rich skeletal mineral soils under dry, continental conditions. In the second, or protocratic phase, there is the onset of interglacial conditions. Rising temperatures allow the base-loving, shade-intolerant herbs, shrubs and trees to migrate and expand quickly to form widespread species-rich grasslands, scrub and open woodlands, which grow on unleached, fertile soils with a still low humus content. In the third, or mesocratic phase, temperate deciduous forest and fertile, brown-earth soils develop under warm conditions, allowing the expansion of shade-giving forest genera such as *Quercus, Ulmus, Fraxinus* and *Corylus,* followed by slower immigrants such as *Fagus* and *Carpinus.* In the fourth and last retrogressive phases, the telocratic phase, soil deterioration and climatic decline leads to the development of open conifer-dominated woods, ericaceous heaths and bogs growing on less fertile humus-rich podzols and peats.

6.6 Holocene

The maximum of the Last Glacial occurred at around 18 000 years BP. Studies of the Oxygen isotope composition of deep-sea cores suggest that deglaciation started at around 15 000 to 14 500 years BP in the North Atlantic and at 16 500 to 13 000 years BP in the Southern Ocean (Bard *et al.*, 1990). The years between the glacial maximum and the beginning of the Holocene are usually termed the Late Glacial and they were marked by various minor stadials and interstadials, but the character, identification and correlation of the Late Glacial interstadials is a matter which is still in need of clarification (Anderson, 1997).

The ending of the Last Glacial period, was, however, not the end of substantial environmental change. Indeed, as the Holocene progressed the impact of climatic change was augmented as a cause of environmental fluctuation by the increasing role of humans (Roberts, 1998).

Some portions of the Holocene may have been slightly warmer than now and terms like 'climatic optimum' have been used to denote the existence of a possible phase of mid-Holocene warmth, when conditions may have been 1–2 °C warmer than present. There may also have been a Medieval Climatic Optimum between AD 750 and AD 1300. However, there have also been times which have been rather colder than today, as is made evident by phases of glacial re-advance (neoglaciations) in alpine valleys. The latest of these was 'the Little Ice Age' which peaked around AD 1700 and ended towards the end of the nineteenth century (Grove, 2004).

Fluctuations of climate also occurred in lower latitudes, and of especial importance for vegetation and human activities was the early to mid-Holocene pluvial, which transformed the Sahara from a hyper-arid region into a savanna (Ritchie *et al.*, 1985). The hyper-arid belt more or less disappeared for one or two millennia before 7000 BP. The northern limit of the Sahel shifted about 1000 km to the south around 18 000 BP and about 600 km to the north at 6000 BP compared to the present.

6.7 The Quaternary in low latitudes

The events which led to the expansions and contractions of the great ice sheets in middle and high latitudes led to major changes in lower latitudes. Periods of greater moisture (pluvials) were interspersed with periods of less moisture (interpluvials). The evidence for such changes is particularly evident on the margins of great deserts, where dry phases saw the development and advance of great sand seas, whereas in moister phases the dunes were stabilised by vegetation and large lakes filled with water in areas that had previously been salty wastes.

Some of the lakes that developed in pluvial phases were enormous. One of the greatest concentrations of pluvial lakes developed in the Basin and Range Province of the American Southwest. Between 100 and 120 depressions, formed by high-angle faulting, were occupied wholly or in part by pluvial lakes during various phases of the Pleistocene.

By contrast, in interpluvials large dune fields expanded. Relict forms occur in areas where there is now a well-developed vegetation cover and annual precipitation totals of around 800 mm (Figure 6.7). The dunes probably formed when vegetation cover was much less capable of inhibiting sand movement under annual precipitation totals that were less than 100 to 300 mm (Goudie, 1992; 1999).

Temperature depression and reduced glacial atmospheric CO_2 levels also played a role in changing environmental conditions in low latitudes. The substantial degree of change in the altitude of vegetation zones on tropical mountains during cold phases can be demonstrated by detailed pollen analysis from lakes and swamps from tropical highlands. Vegetation zones and geomorphological processes may have moved through as much as 1700 m of altitude.

A combination of temperature and precipitation change had a dramatic impact on the nature and extent of rainforests in Africa and South America (Goudie, 1999). This is brought out by a consideration of pollen analyses undertaken in Lake Bosumtwi in southern Ghana (Maley, 1989). At the time of the Last Glacial Maximum, between 20 000 and 15 000 years BP, the lake had a very low level and arboreal (tree) pollen percentages reached minimum values of between 4 and 5 per cent. Trees were in effect replaced at that time by herbaceous plants. This compares with the situation since *c.* 8500 BP, when arboreal pollen percentages have oscillated between 75 and 85 per cent. The shifts in the position of the monsoon in the Last Glacial Maximum and the Holocene are shown in Figure 6.8. This clearly had a significant impact on processes, especially erosion and instability (Chapters 8 and 11).

6.8 Sea-level changes

In addition to the climatic and vegetational changes which have been discussed so far, it is important to remember that the Quaternary has also been a time of major changes in sea level (Cronin, 1999). Many factors influence sea level in any particular location (Table 6.1), but among the most important are glacial eustasy and glacio-isostasy. The former is the worldwide change in sea level that results from the waxing and waning of the great ice sheets, which took up or released fresh water into the oceans. Thus since the last glaciation, the melting of the ice caps has caused world average sea levels to rise by something between 100 and 150 m. The Flandrian transgression, as this event is called

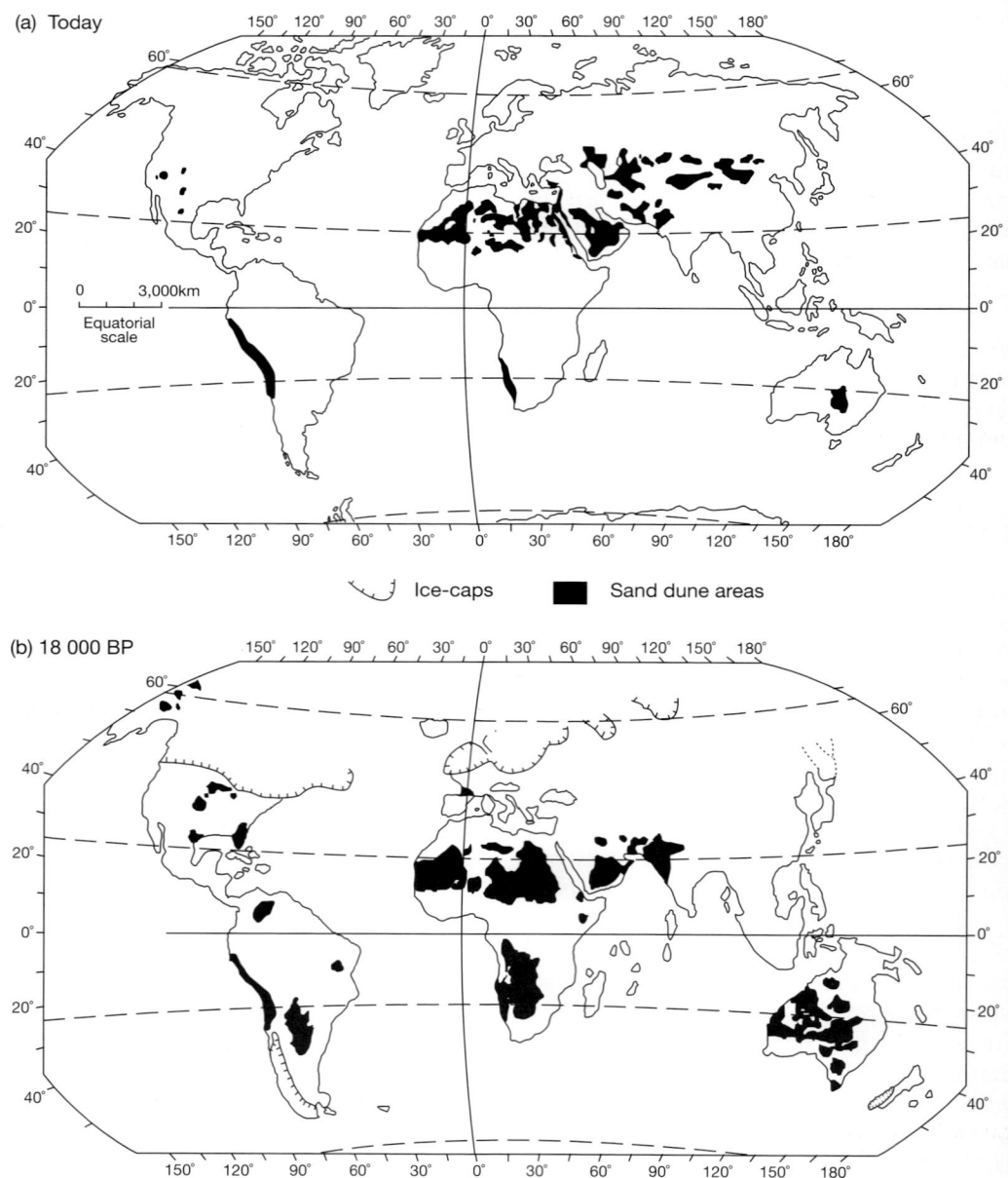

Figure 6.7   The extent of sand seas at *c*. 18 000 years ago (lower) in comparison with the situation today.

(Figure 6.9), flooded the continental shelves and created many of the flooded river valleys (rias, fjords etc.) and embayments of present-day coastlines. Some of these inlets have since been filled in, partially or totally, by complex alternations of marine and terrestrial sediments including potentially unstable clays and peats.

Sea level became stabilised at around its present height (give or take a few metres) at 6000 years BP. The significance of this cessation of sea-level rise

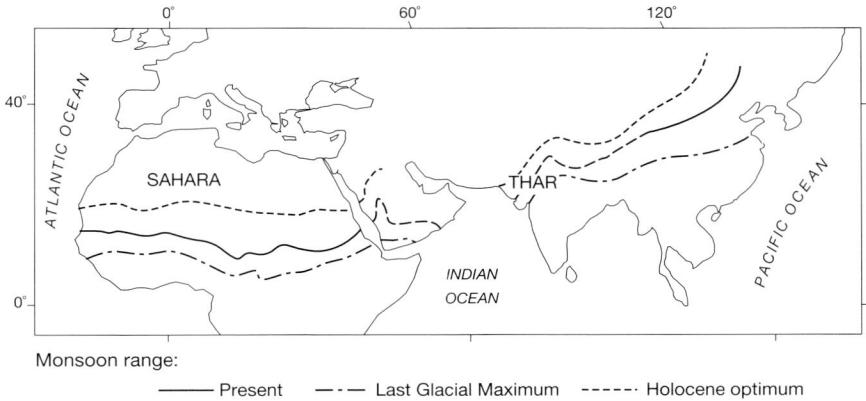

Figure 6.8   The range of the monsoon between the Last Glacial Maximum and the Holocene (from Goudie, 1999).

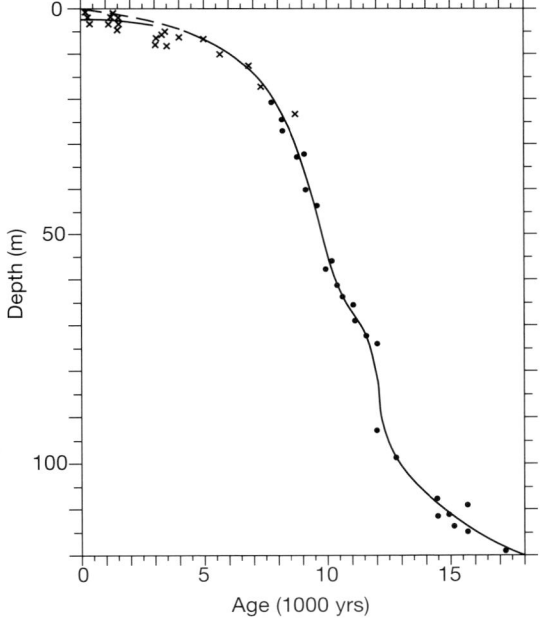

Figure 6.9   The sea-level curve for Barbados, based on radio-carbon dated corals (*Acropara palmata*). The circles represent data from Barbados corrected for estimated uplift. The crosses represent data from four other Caribbean islands. The curve reflects the depth range of live corals while the broken line has been adjusted to sea-level (after Fairbanks, 1989).

Table 6.1   Factors in sea-level change.

Eustatic (worldwide)	Local
Glacio-eustasy	Glacio-isostasy
Infilling of basins	Hydro-isostasy
Orogenic-eustasy	Erosional and depositional isostasy
Decantation	Compaction of sediments (autocompaction)
Transfer from lakes to oceans	Orogeny
Expansion or contraction of water volume because of temperature change	Epeirogeny
	Ice–water gravitational attraction
Juvenile water	
Geoidal changes	

is that many of the shingle structures of the British coast (e.g. Chesil Beach, Dorset) resulted from the upward combing of sediments by a rising sea. They are now largely relict features receiving relatively modest amounts of replenishment, and thus may be prone to serious decay if they are used as sources of aggregates (Chapters 21 and 22).

Glacio-isostasy results from the application and release of pressure by glaciers on the earth's crust. As already noted, many of the great ice caps were very thick and caused crustal depression of the order of 200–300 m in certain parts of the high latitudes (Figure 6.10). Since the ice burden has been released, the land has risen (and continues to do so), but in areas more distant from the former ice sheets some compensatory depression has occurred, leading to accelerated flooding of low-lying coastal areas.

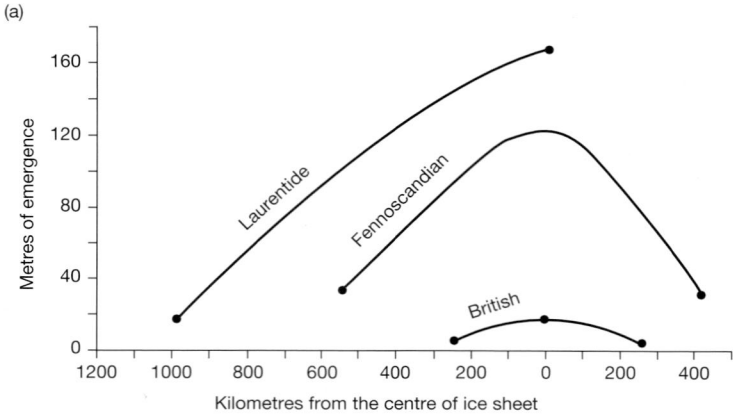

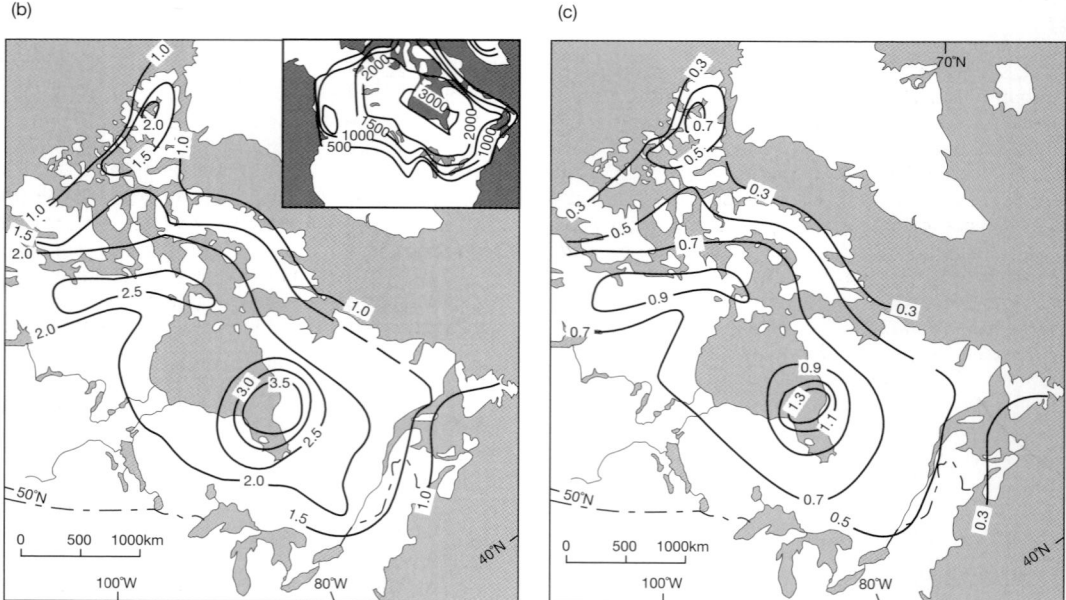

Figure 6.10   Glacio-isostatic adjustment: (*A*) cross-section through three Northern Hemisphere ice sheets, showing the amounts of isostatic recovery over the past 7000 years; (*B*) Average rate of Holocene uplift in metres per 1000 years for northern and eastern Canada — the inset shows the thickness of the Laurentide ice sheet at 18 000 years ago (contours in metres), and (*C*) present rate of uplift in metres per 100 years of northern and eastern Canada.

6.9 Summary

Over the last two million years the world has seen many major environmental changes, and has been in a state of constant flux. Many of the landforms and surface materials upon which human activities are conducted date from this time and many of them are not necessarily the result of currently active processes. Even within recent decades substantial changes have taken place in river discharges, lake dimensions, glacial snout positions etc. and these need to be considered when engineering schemes are being planned.

References

Adams, J., Maslin, M. and Thomas, E. (1999) Sudden climate transitions during the Quaternary. *Progress in Physical Geography* **23**, 1–36.

Anderson, D. (1997) Younger Dryas research and its implications for understanding abrupt climatic change. *Progress in Physical Geography* **21**, 30–249.

Andrews, J. T. (1998) Abrupt changes (Heinrich events) in late Quaternary North Atlantic marine environments. *Journal of Quaternary Science* **13**, 3–16.

Ballantyne, C. K. and Harris, C. (1994) *The periglaciation of Great Britain*. Cambridge University Press.

Bard, E., Labergrue, L. D., Pichon, J. J., Labracherie, M., Arnold, M., Duprat, J., Moyes, J. and Duplessy, J. C. (1990) The last deglaciation in the southern and northern hemispheres. In Bleil, V. and Thiede, J. (eds) *Geological History of the Polar Oceans: Arctic versus Antarctic*. Kluwer, Dordrecht, 405–415.

Birkett, C., Murtugudde, R. and Allan J. A. (1999) Indian ocean climate event brings floods to East Africa's lakes and the Sudd Marsh. *Geophysical Research Letters* **26**, 1031–1034.

Birks, H. J. B. (1986) Quaternary biotic changes in terrestrial and lacustrine environments, with particular reference to north-west Europe. In Bergland, B. E. (ed.) *Handbook of Holocene Palaeoecology and Palaeohydrology*. Wiley, Chichester, 3–65.

Bond, G. and 13 collaborators (1992) Evidence for massive discharges into the North Atlantic Ocean during the last glacial period. *Nature* **360**, 245–249.

Bull, W. B. (1997) Discontinuous ephemeral streams. *Geomorphology* **19**, 227–276.

Cronin, T. M. (1999) *Principles of Palaeoclimatology*. Columbia University Press, New York.

Dansgaard, W. and 8 collaborators (1993) Evidence for general instability of past climate from a 250 k yr ice-core record. *Nature* **364**, 218–220.

Depetris, P. J. and Pasquini, A. I. (2000) The hydrological signal of the Perito Moreno Glacier damming of Lake Argentino (Southern Andean Patagonic): the connection to climate anomalies. *Global and Planetary Change* **26**, 367–374.

Derbyshire, E., Meng, X. M. and Dijkstra, T. A. (eds) (2000) *Landslides in the Thick Loess Terrain of Northwest China*. Wiley, Chichester.

Dickson, R. R., Osborn, T. J., Hurrell, J. W., Meincke, J., Blindheim, J., Adlandsvik, B., Vinje, T., Alekseev, G. and Maslowski, W. (2000) The Arctic Ocean response to the North Atlantic Oscillation. *Journal of Climate* **13**, 2671–2696.

Elsner, J. B. and Kera, A. B. (1999) *Hurricanes of the North Atlantic*. Oxford University Press, New York.

Fairbanks, R. G. (1989) A 17 000-year glacio-eustatic sea level record: influence of glacial melting rates on the Younger Dryas event and deep ocean circulation. *Nature* **342**, 637–42

Forman, S. L., Oglesby, R. and Webb, R. S. (2001) Temporal and spatial patterns of Holocene dune activity on the Great Plains of North America: megadroughts and climate links. *Global and Planetary Change* **29**, 1–29.

Frenzel, B. and Blundau, W. (1987) On the duration of the interglacial to glacial transition at the end of the Eemian Interglacial (Deep Sea Stage 5E): botanical and sedimentological evidence. In Berger, W. H. and Labeyrie, L. D. (eds) *Abrupt Climatic Change*. D. Reidel, Dordrecht and Boston. 151–162.

Goudie, A. S. (1992) *Environmental Change* (3rd edition). Oxford University Press.

Goudie, A. S. (1999) The Ice Age in the Tropics. In Slack, P. (ed.) *Environments and Historical Change*. Oxford University Press, 10–32.

Grove, J. M. (2004) *Little Ice Age: Ancient and Modern*. Routledge, London.

Grosjean, M., Nunez, L., Castajena, I. and Messerli, B. (1997) Mid-Holocene climate and culture changes in the Atacama Desert, Northern Chile. *Quaternary Research* **48**, 239–246.

Lowe, J. J. and Walker, M. J. C. (1997) *Reconstructing Quaternary Environments* (2nd edition). Longman, Harlow.

Magilligan, F. J. and Goldstein, P. S. (2001) El Niño floods and culture change: a late Holocene flood history of the Rio Moquegua, Southern Peru. *Geology* **29**, 431–434.

Maley, J. (1989) Late Quaternary climatic changes in the African rainforest: forest refugia and the major role of sea surface temperature variations. In Leinen, M. and Sarnthein, M. (eds) *Palaeoclimatology and Palaeometeorology*. Reidel, Dordrecht, 585–616.

Maslin, M. A. and Tzedakis, C. (1996) Sultry last interglacial gets sudden chill. *Eos* **77**, 353–354.

Oppo, D. W., McManus, J. F. and Cullen, J. L. (1998) Abrupt climate events 500 000 to 340 000 years ago: evidence from sub-polar North Atlantic sediments. *Science* **279**, 1335–1338.

Petit, R. J. and 18 collaborators (1999) Climate and atmospheric history of the past 420 000 years from the Vostok ice core, Antarctica. *Nature* **399**, 429–436.

Pye, K. (1987) *Aeolian dust and dust deposits*. Academic Press, London.

Ritchie, J. C., Gyles, C. H. and Haynes, C. B. (1985) Sediment and pollen evidence for an early to mid-Holocene humid period in the eastern Sahara. *Nature* **314**, 252–255.

Roberts, N. (1998) *The Holocene*. (2nd edition). Blackwell, Oxford.

Viles, H. A. and Goudie, A. S. (2003) Internannual, decadal and multidecadal scale climatic variability and geomorphology. *Earth-Science Reviews* **61**, 105–131.

Wang, X. L. and Surail, V. R. (2001) Changes of extreme wave heights in Northern Hemisphere oceans and related atmospheric circulation regimes. *Journal of Climate* **14**, 2204–2221.

Williams, M. A. J., Dunkerly, D., De Dekker, P., Kershaw, P. and Chappell, J. (1998) *Quaternary Environments* (2nd edition). Edward Arnold, London.

Wilson, R. C. L., Drury, S. A. and Chapman, J. L. (2000) *The Great Ice Age. Climate Change and Life*. Routledge and the Open University, London and New York.

Woillard, G. (1979) Abrupt end of the last interglacial s.s. in north-east France. *Nature* **281**, 558–562.

Further reading

Goudie, A. S. (2002) *Great Warm Deserts of the World: Landscapes and Evolution*. Oxford University Press.

Ruddiman, W. F. (2001) *Earth's Climate: Past and Future*. W. H. Freeman, New York.

Wilson, R., Drury, S. A. and Chapman, J. L. (2000) *The Great Ice Age*. Routledge, London.

7. Engineering Behaviour of Soils and Rocks

George Milligan, Peter Fookes and Mark Lee

7.1 Introduction

A brief introduction to the engineering properties of soils and rocks is appropriate in this book for two reasons: the physical properties of soils and rocks, in particular strength and permeability, affect many geomorphological processes, and the ground conditions encountered in many engineering projects are the result of such processes. Ground material properties, therefore, constitute both input to and output from these processes. For example, the ground conditions on a valley side may be the result of past glacial or periglacial actions; the current nature and state of those soils will then determine how and if the slope will fail due to erosion of its toe by a present-day river, or as a result of a cut into the hillside for a new road.

No attempt will or should be made here to give extensive coverage of the sciences of soil and rock mechanics or their application in geotechnical engineering. What can be attempted is to provide a general understanding of the range of material behaviour and some of the key factors affecting that behaviour. Soils and rocks are mechanically complex, even in their simplest states, with strength and stiffness related to their geological origin, past stress and strain history, current stress state and loading path. Natural ground conditions commonly comprise layers of different soil and rock types within each of which there will be variations in properties both spatially and in direction, i.e. the materials are neither homogeneous nor isotropic.

Concepts in soil mechanics, and applications of geotechnical engineering, have both developed very rapidly in recent decades, and this has been the most active research area in civil engineering, so that the need for specialist practitioners is now well recognised. Nevertheless, a fairly simple basic framework for the behaviour of soil can be established, with natural variations being accounted for by modifications to or differences from this model. For further reading, there are a number of excellent modern references dealing with soil mechanics and the classical problems of geotechnical engineering, such as Bolton (1991), Atkinson (1981; 1993) and Powrie (1997).

The development of rock mechanics as the science of rock engineering has been even more recent. Useful introductions to rock mechanics and rock engineering are provided by Hudson (1989) and Waltham (2001).

The main ideas to be covered in this chapter include the important differences between coarse- and fine-grained soils, the importance of plasticity in fine-grained soils, the concept of effective stress in soil, the interrelation of density, volume change and strength, and some of the ways of assessing the properties of soil. Because geomorphology is mainly concerned with relatively superficial deposits, the main emphasis is on the behaviour of soils. However, a brief discussion is also included on the engineering behaviour of rock masses, in particular the influence of discontinuities on near-surface processes.

Many geomorphological processes and engineering constructions involve issues of slope stability (Chapter 8), so a brief discussion of the methods and some of the results of slope stability analysis are included (7.3 Slope instability); for further information, reference should be made to Simons *et al.* (2001) or other specialist text on slope stability. The particular problems or risks associated with different soil types resulting from geomorphological processes are summarised (7.4 Geotechnical problems associated with different soil types). Typical values of engineering properties of a range of soil types are also provided (7.5 Typical values of soil properties).

7.2 Elements of behaviour

Coarse and fine soils

Soil particles exist in a range of sizes from less than 1 μm to more than 500 mm. The words clay, silt, sand, gravel, cobbles and boulders are used in British engineering practice in a standard way to describe particular ranges of particle sizes, as shown in Figure 7.1. They are also used, along with suitable qualifying terms, to describe natural soils in which the properties of a particular size range predominate.

Coarse soils are those mainly in the sand size and upward; they consist predominantly of particles of strong crystalline minerals such as silica, which have been broken down from parent rock by physical processes of weathering. The particles are of the same material as the parent rock and approximately equi-dimensional, but may be very angular if recently broken, or smooth and rounded if they have suffered abrasion and polishing by water or wind action.

Fine soils are those consisting primarily of particles in the clay and silt size ranges. Clay particles are produced mainly by chemical weathering of parent rocks *in situ*, often followed by erosion, transportation and deposition elsewhere, or by direct precipitation from water. The resulting clay minerals are complex, but the most important are built up from basic crystalline units of silica and alumina to form rod or sheet structures, which may combine in various ways to form different clay minerals. The three commonest clay minerals are kaolinite, illite and montmorillonite (a member of the smectite family of clay minerals); some basic properties are given in Table 7.1. Montmorillonite is particularly notable in that it is able to store water within the particles between the crystal layers, which are much less strongly bonded together than in other clay minerals. It is able to change volume by a much greater amount than other clay minerals, so that soil containing a high proportion of montmorillonite tends to swell on wetting and contract on drying much more than other soils. At the other extreme, the layers in kaolinite are fairly strongly bonded and particles of kaolinite are relatively bulky for a clay.

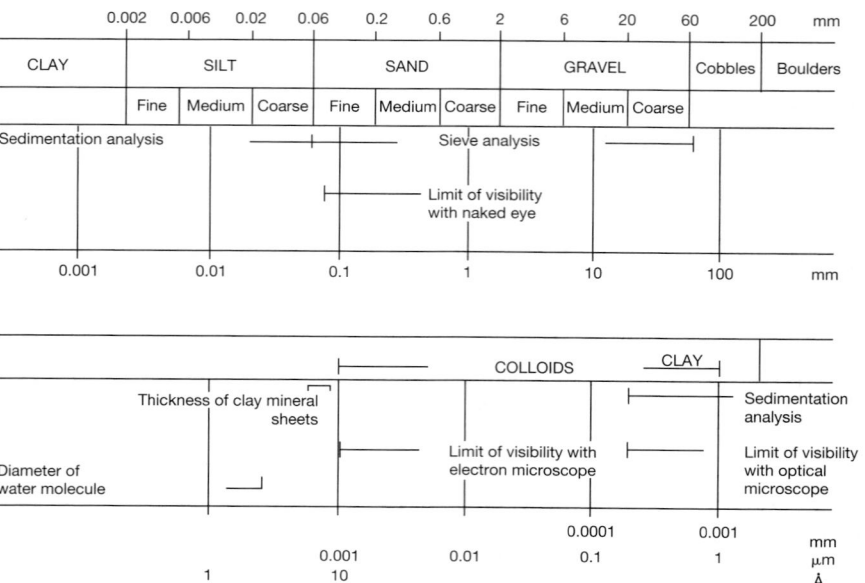

Figure 7.1   Range of particle sizes of soils.

Table 7.1   Some basic properties of soil minerals.

	Quartz Sand	Clays		
		Kaolinites	Illites	Montmorillonites
Particle diameter (μm)	∼ 1000	0.1∼ 4	0.1–∼2	0.01–∼2
Particle thickness (Å)	∼ 1000	500	50–300	10–100
Specific surface (m²/g)	0.002	5–20	80–120	700–800

Natural soils described as clays may contain as little as 10% clay-sized particles, and seldom more than 50%. This is because the clay-sized particles have a profound effect on the soil behaviour, in two main ways, both related to the pore size and packing of the soil particles. Soil is a particulate material, and as such must have spaces or pores between the solid grains; these are filled with fluid (usually water), gas (usually air), or a combination of both. The presence of this pore fluid, and the ease with which it may move through the soil, are of fundamental importance to the behaviour of soils. The size of the pores is obviously related by geometry to the size of the particles, and in clay soils will be very small. In mixed soils the pore sizes may also be very small provided there is sufficient clay-sized material to fit into the spaces between the larger particles. Such soils will be nearly impermeable. Also, very large capillary forces may be developed in small pores, and the resulting suction will hold the material together even when no external stresses are applied. This ability to hold together in a lump and undergo substantial distortion without breaking up makes clay appear cohesive and plastic. Suction forces also exist in damp sand (hence sand castles), but the effects are very limited and easily lost on the wetting or drying out of the sand.

Silts often cause particular problems in identification and mechanical properties. Silt particles are similar to sand particles, and silt may exhibit many of the properties of fine sands, but the small particle size results in silt having significant plasticity and relatively low permeability. Most natural clay soils contain a substantial proportion of silt, but their properties are dominated by the clay fraction.

The different natures of fine and coarse particles greatly affect the possible particle packing arrangements. There is only a limited range of density between the loosest and densest arrangements

of approximately equi-dimensional particles. Thin clay particles can pack so that the same volume of solid material occupies very different total volumes; limits can be thought of as akin to a pack of cards (extremely densely packed), or the same cards built into a card house (extremely loose packing). The properties of the soils will clearly be very different. The packing of particles is described numerically by the porosity (usual symbol n), which is the volume of voids divided by the total volume; the void ratio (symbol e), which is the volume of voids divided by the volume of soil particles; or the specific volume (symbol v), which is the volume containing unit volume of solid material. They are related by:

$$n = e/(1 + e), \quad v = (1 + e)$$

For coarse soils the value of e is generally in the range 0.3 to 1.0, but for clays it may range from below 0.2 to above 9 as a result of variations in the geological origin and stress history of the deposit (see Compression of soils).

If the voids are completely filled with air, the soil is dry; if completely filled with water it is fully saturated. For intermediate states the soil is partially saturated, with the degree of saturation S_r (volume of water over volume of voids) ranging from zero to unity. Fine soils are often fully saturated in their natural state, except in arid conditions; the water content of the soil (w), which in soil mechanics is defined as the mass of water divided by the mass of soil particles, is then directly related to the void ratio by:

$$w = e/G_s$$

where G_s is the specific gravity of the soil particles.

The susceptibility of soil to erosion by water is an important factor in many geomorphological processes and also in the failure of many types of construction. Erosion may be due to impact of rain drops, overland flow, underground seepage, or water flow in a river, sea or ocean. It is largely controlled by the particle size of the soil. Large particles of gravel size or greater need large amounts of energy to lift and transport them. They can be displaced by rapid rivers in flood or storm waves, but are quickly redeposited when water velocities reduce. At the other extreme, clay may be resistant to erosion due to its innate cohesion, except in the case of dispersive clays. Sands and (especially) silts are the most susceptible to erosion, and silt in suspension may be carried great distances before being deposited. Wind can also act as the agent of erosion; the particle sizes affected are more restricted to silt and fine sand but the size of a wind system may be such that very large amounts of material may be carried for very large distances. The importance of erosion in geomorphology is addressed in Chapter 11, and to a lesser extent in Chapters 3 and 10.

In engineering calculations soil is usually treated as a continuum material, but its properties are closely related to its particle arrangement or microstructure. Its bulk properties may also be significantly affected by its 'fabric', such as fine laminations of sand in a clay soil, cracks or fissures, or inclusions of organic material (Rowe, 1972). An extreme example of an organic soil is peat, which consists almost entirely of decayed vegetable matter.

Many real soils have complex structures and their mechanical behaviour may depart significantly from that of the 'ideal' soils discussed in this section. This should not be taken as a reason to avoid attempting to apply quantitative methods of analysis, but these should always be undertaken on the basis of a good understanding of the geological, hydrogeological and geomorphological origins of a site, along with a clear picture of the resulting likely variability in ground and groundwater conditions.

Identification and classification

Investigation of a site will usually start with a survey of existing information from geological maps and memoirs, historical maps, aerial and satellite photographs, records from adjacent construction, and other local knowledge. Perry and West (1996) provide useful guidance on sources

of material for the UK. Much information is available from databases and statutory registers relating to possible geological and environmental risks such as unstable ground, old mines or land contaminated by past industrial use.

Detailed examination of the succession of soil strata will then be carried out using trial pits and boreholes; these will also be used for obtaining samples of the soil for laboratory testing, and for undertaking *in situ* tests such as the standard penetration test (SPT) (Stroud, 1989), vane shear test or pressuremeter test. Methods of investigation are covered by British Standard 5930: *Code of Practice for Site Investigations* (BS, 1999), and other national standards. There are also various methods of probing and *in situ* testing which may be done without boreholes, such as cone or self-boring pressuremeter tests (Meigh, 1987; Mair and Wood, 1987). In some cases geophysical methods involving seismic, electrical, gamma radiation, magnetic or gravimetric surveys undertaken from the surface or down boreholes can provide much additional information, but usually require interpretation by specialists. A good general introduction to the purpose and methods of site investigation is provided by Waltham (2001), while Fookes (1997) provides an extensive discussion of site investigation as a means of establishing the basic geological model for a site. The organisation of a typical simple site investigation is summarised in Figure 7.2; for small projects a single stage of field work is normal, but for large projects there may be several stages, each building on the information from earlier work. For further coverage see Clayton *et al.* (1995) and Fookes (1997). Fookes *et al.* (2001) provide the basis for a total geological history approach to site investigation.

The soils recovered are first described and classified on the basis of visual and tactile properties; methods are given in British Standard 5930 (BS, 1999). Classification is then assisted by simple laboratory tests, details of which are given in British Standard 1377 (BS, 1991), *Methods of Test for Soils for Civil Engineering Purposes*. Particle size analyses, using sieving for coarse soils and sedimentation methods for fine soils, produce curves showing the distribution of particle sizes within the soil, typical examples being shown in

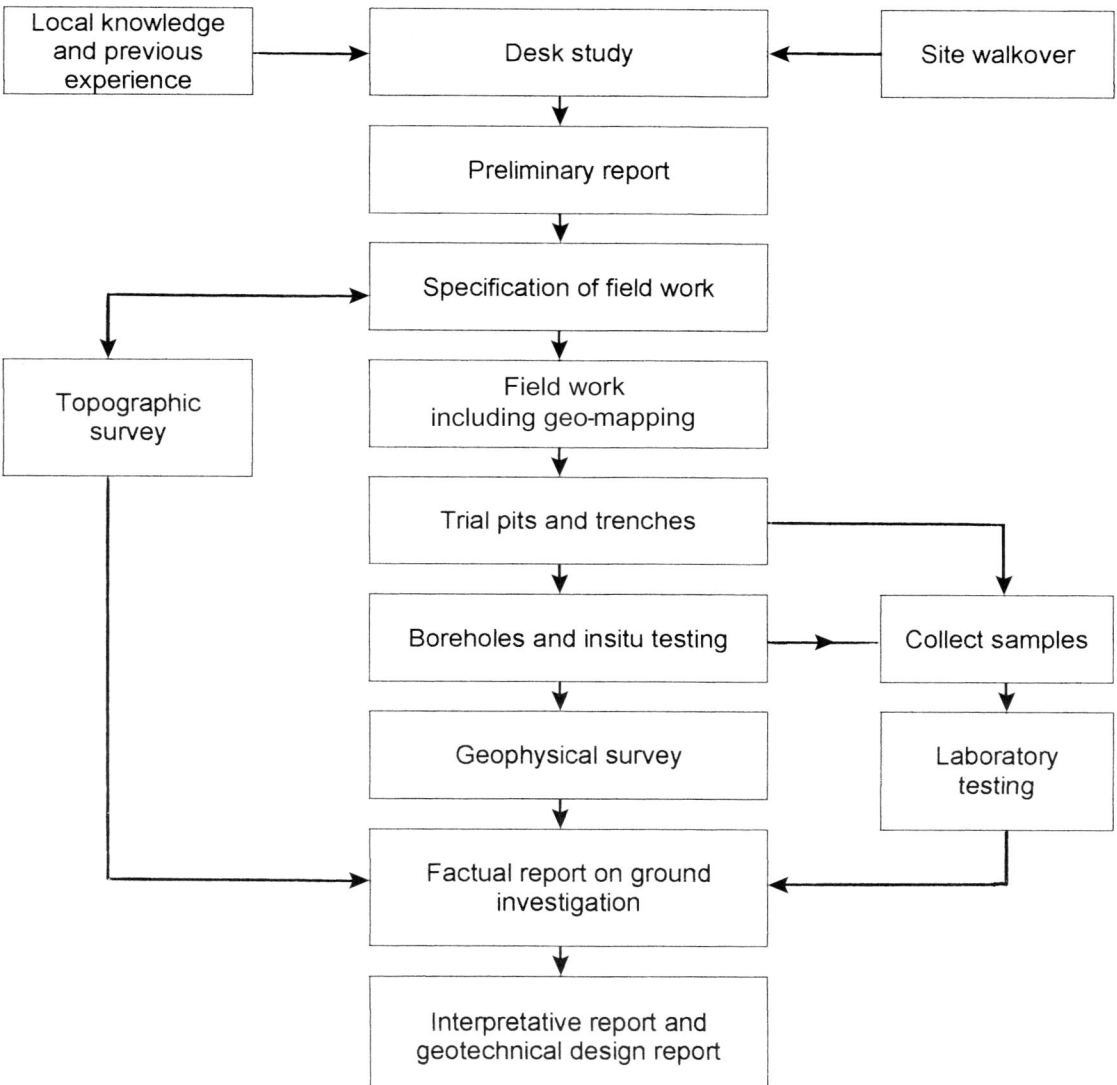

Figure 7.2   Flow chart for a simple one-stage site investigation; large or complex projects may have several stages of investigation (see Fookes, 1997).

Figure 7.3. For historical reasons these are plotted as cumulative curves rather than frequency distribution curves.

For fine soils, much useful information about the soil may be obtained by determining the natural water content as it exists in the ground and the water contents corresponding to two arbitrarily defined limits of consistency known as the Atterberg limits. The liquid limit (LL) defines the change in behaviour from that of a plastic solid to a viscous liquid as the soil becomes wetter, and the plastic limit (PL) the change from a plastic to a brittle solid as the soil becomes drier. These water contents are expressed as percentages, the LL and PL to the nearest whole number; the difference between the LL and PL is called the plasticity index (PI). LL, PL and PI all tend to increase with the clay content (the percentage of clay-sized

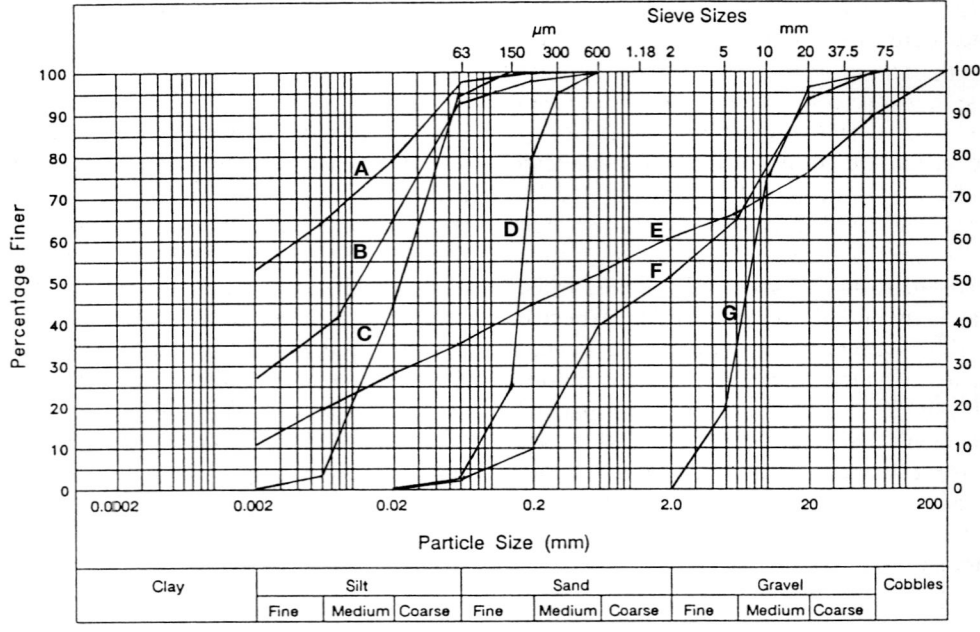

Figure 7.3   Typical particle size distribution plots for various soils: (*A*) marine clay; (*B*) alluvial silty clay (estuarine); (*C*) uniform silt (loess); (*D*) uniform fine sand (dune); (*E*) glacial till (boulder clay); (*F*) sandy gravel (coarse alluvium); and (*G*) uniform (clean) gravel.

Table 7.2   Typical plasticity indexes and activities of clays.

	Plasticity index	Activity
Kaolinites	30	0.4–0.5
Illites	70	0.5–1.0
Montmorillonites	200	1–7
Silty clay	45–60	0.6
Glacial clay	10–30	0.5–0.75
Loess	4–9	0.5–0.75
Weathered mudstone	17–35	0.56–0.70
Marine clay	60–100	1.6–1.8
Organic estuarine clay	50–90	~1.3
Lateritic clay	20–45	0.3–1.15
London clay	35–60	0.75–1.25

particles in the soil), and also depend on the nature of the clay minerals. The activity of the clay is indicated by the ratio of plasticity index to clay content; typical values for common clay minerals and some natural clays are given in Table 7.2.

Classification of clay soils is assisted by the use of the Casagrande plasticity chart which grades soils from low to high plasticity and divides soils that are predominantly silty or organic from those which have significant clay content. Detailed classification systems have been developed based on the particle size distributions and plasticity tests, see Appendix A2.

These simple tests give an initial indication of the likely engineering properties of fine soils. Natural water contents in the ground generally range between somewhat below the PL to just above the LL. The former would indicate a stiff, strong soil, the latter a very weak and compressible soil. The relative consistency is described by the liquidity index (LI) given by:

$$LI = (w - PL)/PI$$

In some countries a consistency index (CI) is used, which is given by:

$$CI = (LL - w)/PI$$

and hence:

$$CI = (1 - LI)$$

However, it must be remembered that these index tests are carried out on 'remoulded' specimens of the soil, in which all the natural soil fabric has been destroyed. Thus while they can provide a very useful baseline to likely engineering properties, the behaviour of the natural soils in the field may be significantly affected by structural features. For example, stiff clays will generally have a smaller mass strength than expected due to the presence of fissures, while soft clays may initially be stronger than expected due to slight cementation between particle contacts; soils of volcanic origin may have porous particles which hold water and give natural water contents and liquid limit values which are not consistent with 'normal behaviour', while fine soils in a partially saturated state will appear much stronger than expected due to the effects of capillary suction.

Effective stress and soil strength

A key to the understanding of soil behaviour is the concept of effective stress. The pore spaces in soils are interconnected and fluid pressure may therefore be transmitted through the pore fluid; in an unsaturated soil there will be both air and water pressures, which may be different. In saturated soils, fluid pressure will vary hydrostatically with depth below the groundwater level (water table); above the water table the soil may still be saturated but pore pressures will be negative relative to atmospheric pressure. Water may flow through soil via the pore spaces, but additional pressure differentials must exist to drive the flow.

The pore pressure has the effect of reducing the contact forces between the soil particles; it is the summation of these forces over a unit area that is known as effective stress. Thus the total normal stress (σ) across a plane in a body of soil will be carried partly by pore pressure (u) and partly by effective stress (σ') so that:

$$\sigma = \sigma' + u$$

and

$$\sigma' = \sigma - u$$

Since stationary fluid cannot sustain shear stress, total and effective shear stresses are the same, thus:

$$\tau = \tau'$$

Note that effective stresses cannot be measured, but only determined from total stresses and pore pressures. However it is change in effective stress that results in deformation and failure of soil and hence controls the behaviour. Similar concepts apply in unsaturated soils, but inevitably become more complex and are still the subject of active research.

Soils generally fail in shear, in an essentially frictional manner, failure occurring when the shear stress on some plane through the soil reaches a certain proportion of the effective normal stress across the plane:

$$\tau' = \tau = \mu\sigma' = \sigma' \tan \phi'$$

The constant of proportionality (μ) is commonly expressed as the tangent of the angle of internal friction of the soil (ϕ'). Some heavily compressed or partially cemented soils exhibit shear resistance even under zero normal stress; they have an effective cohesion (c'), and the failure condition becomes:

$$\tau' = c' + \sigma' \tan \phi'$$

This cohesion should not be confused with the undrained strength or cohesion (c_u), which is discussed below.

Shear failure and critical states

The shear failure of soil may be considered in terms of a direct shear test carried out in a simple piece of laboratory equipment known as a shear box (Figure 7.4). The test is normally carried out with a constant vertical stress from a load P and the soil specimen forced to undergo shear displacement Δx by a shear force Q. The soil is confined horizontally, but free to expand or dilate vertically, and the vertical expansion Δy is measured. Pore pressures are usually zero — either the soil is dry or testing is slow so that pore pressures are allowed to dissipate. Typical plots of results are shown in Figure 7.4. If the soil is initially dense, the shear force is seen to rise to a maximum, then drop back to a constant value. The peak in strength arises because the densely packed soil

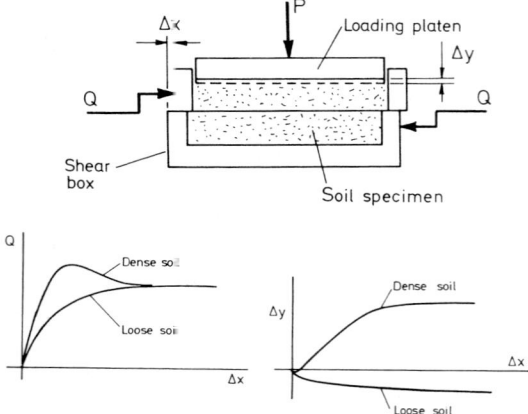

Figure 7.4 Schematic of shear box apparatus for measuring the shear strength of soils, with typical results plotted as: shear resistance (Q) against displacement (Δx), and dilation (Δy) against Δx.

grains have to be forced apart before the soil can shear; this is seen as initial dilation of the specimen followed by further shearing at constant volume. The peak of strength corresponds to the maximum rate of dilation. If on the other hand the soil is initially loose, it will at first compress slightly on shearing, with the shear resistance increasing monotonically, then shear further at constant volume. For an important discussion of the relation between strength and volume change see Bolton (1986).

The final state of density and shear strength under the same vertical stress will be the same for both initially dense and initially loose states, and is known as a critical state. The angle of shearing resistance in such states is referred to as the critical state or constant-volume value, and given the symbol ϕ_{cs}, ϕ_{crit} or ϕ_{cv}. A series of such states at different vertical stresses defines the critical state line in a plot of shear stress against vertical stress, or, under more general stress conditions, of deviator stress against mean normal stress. Note that each such state also has a void ratio or specific volume associated with it, so that the critical state line may also be presented on a plot of void ratio against mean normal stress (see Compression of soils).

An alternative form of the test would be to prevent any change in height (and therefore volume) of the specimen, but allow the vertical stress to

vary. On shearing, the dense soil will try to expand and the vertical stress must increase to prevent it from doing so; eventually a critical state will be reached with the same initial density but a higher stress. Conversely, the loose soil will tend to contract and the vertical stress decrease.

A third type of test could have a constant vertical applied stress, but the soil fully saturated, with no water allowed to move in or out of the specimen. Since both soil grains and water are effectively incompressible, this specimen also cannot change volume on shearing. To prevent volume change in the dense soil, the effective stress must increase, and this occurs by negative pore pressure developing; in the loose soil, reduction in volume will be prevented by positive pore pressure that reduces the effective stress.

In the first type of test the soil is sheared under 'fully drained' conditions, in the third under 'undrained' conditions; note that either drained or undrained strength may be the higher, depending on the initial conditions of the soil and also on the changes in stress imposed on the soil. In field conditions, changes in pore pressures in coarse soils usually dissipate quickly due to the high permeability of the soil, but fine soils will respond in an undrained manner initially and then revert to drained behaviour over a long period of time. The period involved may be many years; for example failures started to occur in cuttings for railways in Weald Clay (a dense engineering soil) more than fifty years after their construction, as pore pressures gradually re-established equilibrium conditions and drained failure states were reached in the soil.

The differences in behaviour of small specimens of loose and dense soils have implications for failure mechanisms in large masses of soil. In dense soil, small local variations in stress or initial density will allow parts of the soil to reach peak strengths first. These parts will then start to get weaker (post-peak) with further shear distortion, while the soil around is still pre-peak and therefore getting stronger. Further deformation will tend to concentrate into these weakening zones, which develop into relatively narrow failure 'planes' separating and allowing relative movement between non-deforming blocks of soil.

In loose soils the strain-softening behaviour in shear is absent and there is less tendency for deformation to concentrate into such narrow zones.

Critical states are reached as the final states in tests on coarse soils and also with clays of low plasticity. In clays of high plasticity, with a relatively high proportion of thin plate-shaped clay particles, the strength continues to drop, with continuing shear in a thin failure zone causing particles to be broken or rotated to align along the failure plane. The friction angle may then fall to very low values, known as residual values. Figure 7.5 gives the variation of critical state friction angle with the angularity of the grains for coarse soils, while Figure 7.6 shows how the residual angle of friction varies with clay content for fine soils. (For further information on soil strength see 7.4.)

Compression of soils

Just as the shear strength was discussed above in terms of a simple laboratory shear test, so the compression of soil may be considered in terms of a simple compression test in laboratory apparatus known as an oedometer (Figure 7.7). In this, the soil specimen is confined so that it cannot deform laterally and is loaded in a vertical direction. The specimen is usually kept saturated, but water is allowed to move in or out of the specimen as it changes volume. The test is normally only used for clays, as coarse soils are extremely stiff in one-dimensional compression. The test gives results that may be applied directly to certain engineering problems, for instance the settlement of foundations on clay strata. It also elucidates what has happened to natural sedimentary soils as they have formed by deposition, been compressed by burial under further sediment or ice loading, and then unloaded due to melting of the ice or erosion of overlying ground.

As the stress is increased the specimen gradually decreases in volume as water is squeezed out, a process known as consolidation. For most soils, the volume is found experimentally to be related approximately linearly to the logarithm of

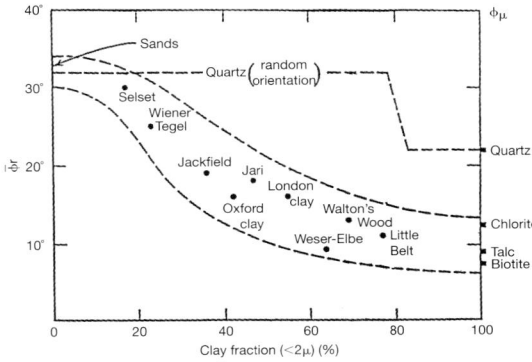

Figure 7.6 Variation of the residual angle of friction for fine soils with clay fraction, compared with characteristic friction coefficients for some soil minerals.

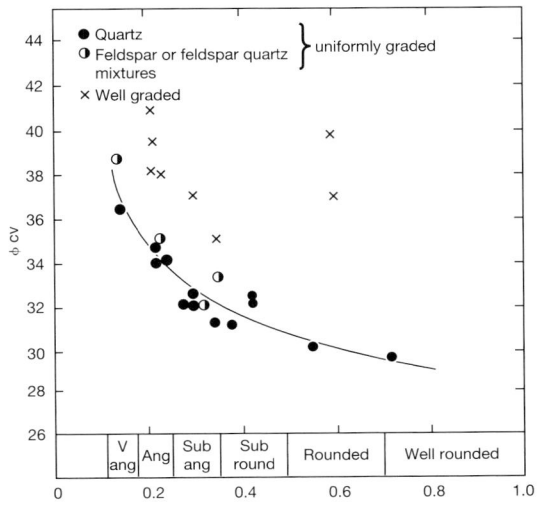

Figure 7.5 Variation of the critical state angle of friction for coarse soils with grading and particle shape.

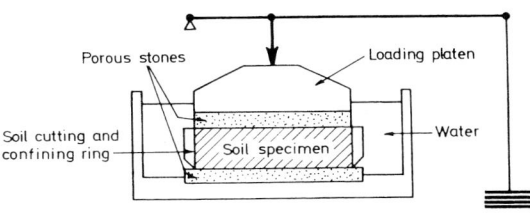

Figure 7.7 Schematic of oedometer apparatus for measuring the compressibility and swelling behaviour of soils in one-dimensional compression (no lateral strain) under incremental loading.

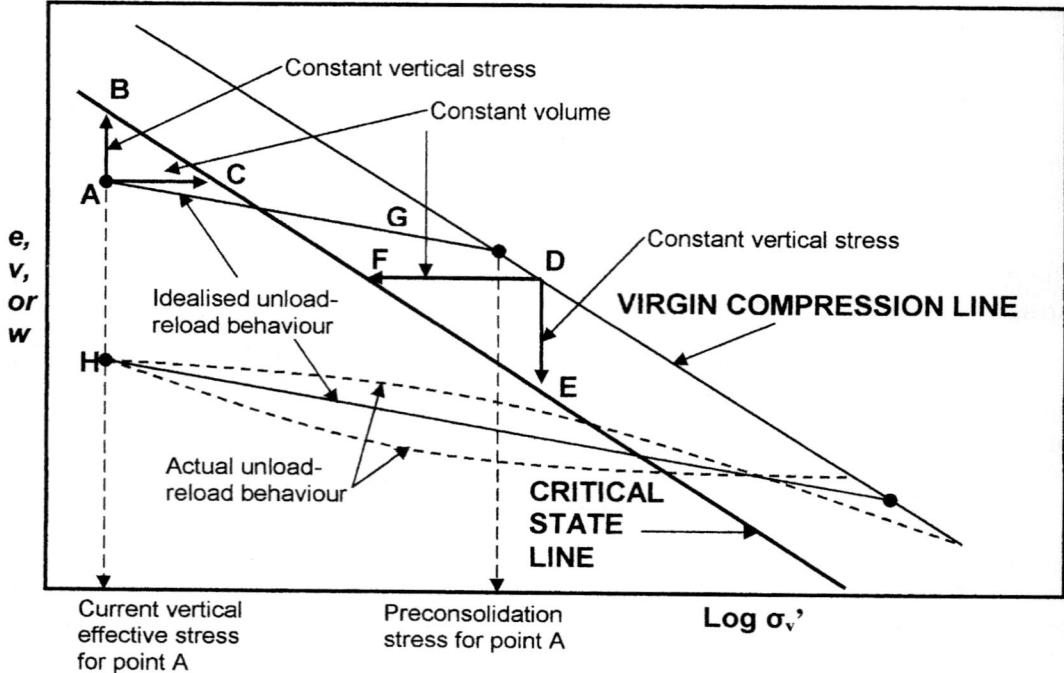

Figure 7.8 Compression and swelling behaviour of soils plotted as specific volume, void ratio or water content against the effective vertical stress (logarithmic scale). Also shows typical stress paths for specimens subjected to shearing under constant vertical stress or constant volume conditions.

the applied effective stress (Figure 7.8). If the specimen has not previously experienced a higher stress, the plot of volume against pressure is known as the virgin compression line (VCL). The slope of this line (the Compression Index C_c) depends on the plasticity of the clay, an empirical relation being:

$$C_c = 0.007 \,(LL - 10)$$

If the specimen is unloaded, it increases in volume but at a very much lower rate than during virgin compression. On reloading, it returns approximately along this 'swelling line' until the VCL is rejoined and the rate of compression increases again. Soil whose state lies on the VCL (such as point D) is described as normally consolidated; soil whose state lies below the VCL (point A) is said to be over-consolidated. The highest pressure to which it has been subjected is its pre-consolidation pressure, and the ratio of

this pressure to its current pressure is its over-consolidation ratio (OCR).

Many natural soils are heavily over-consolidated, having been compressed by considerable depths of later sediment that has subsequently been eroded, or by great depths of ice which has subsequently melted. Over-consolidation can also be caused by large suctions developed due to evaporation from the surface or lowering of the groundwater table after the soil has been formed. On the other hand, recently deposited alluvial soils may still be normally consolidated, except perhaps for a crust of over-consolidated material at the surface. In some situations, such as in the Gulf of Mexico, where clays are being deposited under water at a high rate, the dissipation of pore pressure may not keep up with the rate of new deposition and the soil may be under-consolidated.

The critical states (discussed in Shear failure and critical states, above) in relation to soil strength

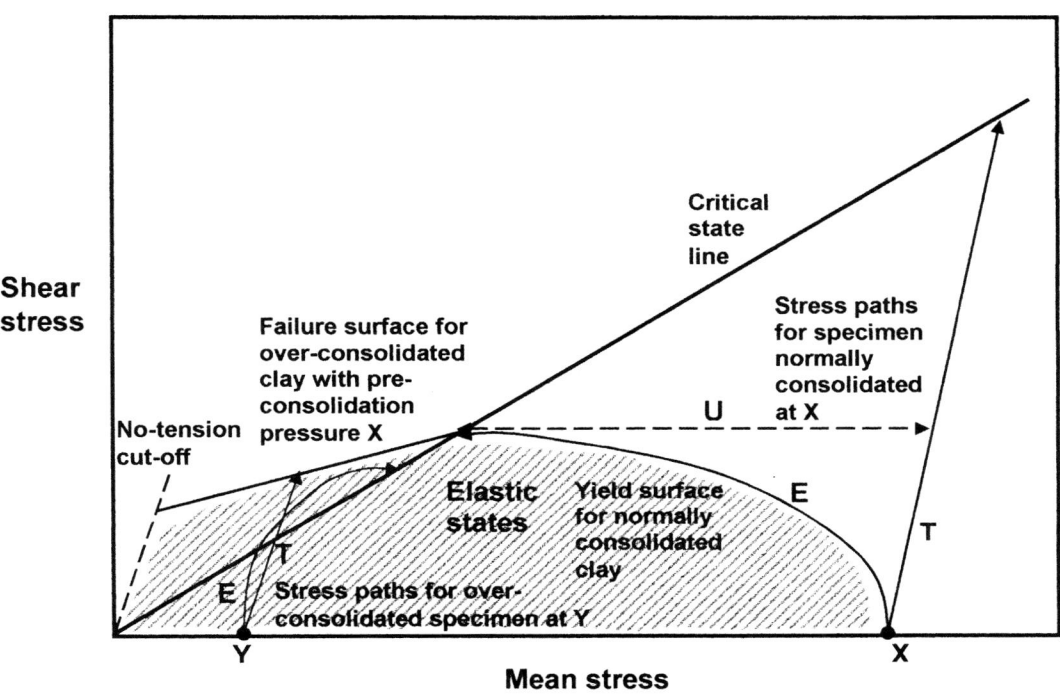

Figure 7.9   Simple unified model for soil behaviour, showing stress paths for tests on normally consolidated and over-consolidated soil specimens. Total stress paths are labelled T, effective stress paths E. The pore pressure U is the difference in mean stress, positive or negative (suction), between total and effective stress.

tests may also be shown on this plot and are found to give a line parallel to, but somewhat below, the VCL. Note that a soil element starting at a state below this critical state line (CSL), if tested under drained conditions at constant vertical stress, must increase in volume as it moves to the CSL during failure (path AB); while if tested at constant volume (AC) it must develop increased vertical effective stress by a reduction in pore pressure. This is consistent with the behaviour observed for 'initially dense' specimens; thus 'dense' soils are essentially those starting at states below the CSL. Conversely, 'loose' soils are those starting at states above the CSL, so that failure under constant stress involves reduction in volume (DE), and failure at constant volume requires a reduction in effective stress and increase in pore pressure (DF). The former are relatively heavily over-consolidated, the latter either normally consolidated or lightly over-consolidated (such as point G).

The concept of OCR is less clear when applied to residual soils that have formed from the *in situ* chemical weathering of rock. However the concept of a critical state line, which can be defined by testing, can still apply, and the initial state of the soil in relation to it will define its general engineering behaviour as discussed above.

Unified models of soil behaviour have been developed based on these concepts. Essentially they invoke the existence of a yield surface that controls the failure of loose soils, a failure surface that defines the peak strength of dense soils, and elastic soil behaviour for soil states within these bounds (Figure 7.9). Elastic analyses may usefully be used in many geotechnical problems when imposed stresses are significantly below those needed to cause yield or failure; Poulos and Davis (1974) provide a compilation of relevant solutions. The basic models are subject to ongoing research and development to incorporate more complex features of soil behaviour, including the response to cyclic loading, unsaturated soils, anisotropy, and non-linear and stress-path-dependent elastic behaviour. The primary purpose of these models is for incorporation into numerical analyses by finite element or finite difference methods, which allow calculation of ground deformations as well as failure conditions in complex engineering problems.

In situ stress conditions

Initial stress conditions in the ground may often be of importance, particularly in the design of retaining structures or tunnels. The total vertical stress on an element in the ground may usually be taken as being equal to the weight of the ground above the element. The vertical effective stress is then the total stress less the pore water pressure within the element.

The horizontal stress is more difficult to establish, and may be smaller or greater than the vertical stress. The ratio between the vertical and horizontal effective stresses is the coefficient of earth pressure at rest and given the symbol K_0. For normally consolidated soils the value of K_0 is given approximately by $(1 - \sin \phi')$. During unloading, horizontal stress reduces more slowly than vertical stress and the value of K_0 therefore increases with over-consolidation ratio; an approximate value of K_0 for over-consolidated soils is given by:

$$K_0 = (1 - \sin \phi')\sqrt{OCR}.$$

The state of initial stress in residual soils is difficult to define, since the soil is produced by weathering processes from rock in which a wide range of stresses may have been left by complex geological processes. Weakening and softening in the rock by weathering is likely to reduce horizontal stresses, but conversion of inert material to swelling clays may lead to an increase in lateral stress. In addition, the inevitable heterogeneous nature of residual soils is likely to mean that initial stresses are also variable over short distances.

Undrained shear strength

When saturated soils are tested or loaded in the ground under conditions in which no drainage of

water can occur, the strength at failure depends only on the initial water content or void ratio and is not affected by increases in confining stress, which are carried by increased pore pressure. The failure criterion then becomes simply:

$$\tau = c_u$$

where c_u is the undrained strength or cohesion; the alternative symbol s_u is sometimes preferred. The undrained strength is measured in the laboratory by simple, quick and cheap tests, or may be deduced from in situ tests. The failure condition applies in terms of total stresses and allows simple calculations to be performed without knowledge of pore pressure conditions, provided the problems involve loading or unloading over a time period that allows a negligible amount of drainage to occur. Such analyses are, therefore, usually only applicable to fine-grained soils of low permeability. In situations where the stresses are increasing, such as beneath a foundation on clay, the short-term undrained analysis is usually the most critical, since the soil will consolidate with time and become stronger. However when the stresses are reducing, such as adjacent to an excavation, the long-term state will be more critical and the more stable undrained conditions will only apply in the short term. This basic fact explains the large number of (often fatal) accidents involving workers being buried in trenches due to collapse of the sides.

For normally consolidated soils the undrained strength increases linearly with the vertical effective stress; the rate of increase is related to the plasticity of the soil, an empirical relation due to Skempton (1957) being:

$$c_u/\sigma'_v = 0.11 + 0.0037\,PI$$

The undrained strength of over-consolidated clay depends on the stress history of the soil element, since it can be in equilibrium at different values of water content under a particular vertical stress (Figure 7.9: points A and H). A useful empirical relation (Ladd et al., 1977) gives:

$$(c_u/p')_{\text{over-consolidated}} = (c_u/p')_{\substack{\text{normally} \\ \text{consolidated}}} \times OCR^{0.8}$$

where p' is the mean effective stress.

Properties of rocks

Rocks essentially differ from soils in that they have significant bond strength, so that they can sustain tensile stress and shear stress under zero normal stress. The unconfined compression strength of intact material can be very high, over 200 MPa in hard igneous rocks such as basalt or granite; while at the other end, the strengths of weak shale rocks merge with those of hard clays at a strength around 1.0 MPa (Tables A3.2$a-c$ in Appendix A3). Young's modulus may similarly vary from about 1 to 100 GPa. The initial stresses in the ground can be complex, reflecting the geological processes that have formed the rock, from cooling of magma to severe compression and folding. Horizontal stresses may be very substantially higher than vertical, and vary in different horizontal orientations.

In many engineering situations the properties of the rock mass are controlled less by the intact rock strength than by the properties of the discontinuities in the rock. These discontinuities (known as 'defects' in the USA and Australia) can range from major faults to closely spaced fractures. The spacing, orientation and continuity of these features must be considered in relation to the nature and scale of the problem. For a large excavation in heavily fractured rock, the ground may essentially be treated as dense coarse soil and calculations undertaken using a continuum model of the ground. On the other hand, rock falls into a shallow tunnel will be controlled by a few sets of joints that allow discrete blocks to fall or slide into the open space. The effect of discontinuities on the stability of rock slopes is considered briefly below (7.3 Slope instability).

Where failures are controlled by sliding along discontinuities, it is the properties of the discontinuities rather than the intact rock that are important. Discontinuities may be tightly closed, with clean rough surfaces, or may be open, or filled with weaker material. Joints that are open or filled with compressible material will obviously reduce the stiffness of the rock in compression as well as its strength in shear. Beneath the water table, water pressures in connected fissures will increase hydrostatically with depth, as in a soil, and the concept of effective stress will apply to contact

stresses between rock surfaces. Discontinuities also reduce the tensile strength to zero. The various factors affecting the behaviour of discontinuities in rock masses and quantitative methods of describing them have been drawn up by the International Society for Rock Mechanics (ISRM, 1981); these are summarised in Appendix A3.

Various systems of classification have been devised to relate rock mass properties to intact strength; core recovery during drilling; fracture spacing, orientation and condition, and groundwater conditions. These classifications have been related empirically to rock behaviour in various situations, such as for slope stability and tunnel support requirements. A classification for geomorphic purposes primarily related to slope stability is given in Table A3.13 of Appendix A3.

7.3 Slope instability

Stability of soil slopes

Many geomorphological processes involve failure of slopes, and slope stability is also a common issue in engineering construction. Some applications of the soil models introduced above to slope stability problems are considered in this section, and a brief discussion of the stability of rock slopes. For further coverage of both topics, and of their applications to slope engineering, see Simons *et al.* (2001).

The stability of slopes may be introduced by consideration of 'infinite' slopes, slopes that are sufficiently long that any element of the slope may be considered as typical and end effects are negligible. By considering such a typical element (Figure 7.10), it can be shown that:

$$F = \frac{\tan \phi'}{\tan \beta}\left[1 - \frac{\gamma_w}{\gamma}\frac{\left(1 + \tan^2 \beta\right)}{\left(1 + \tan \alpha \tan \beta\right)}\right]$$

where F is the ratio of available to mobilised soil strength on an assumed failure plane referred to as the factor of safety, γ is the saturated unit weight of the soil and γ_w the unit weight of water. The slope angle is β, and α defines the direction of seepage flow within the slope, with the water table at the ground surface. There are a number of interesting

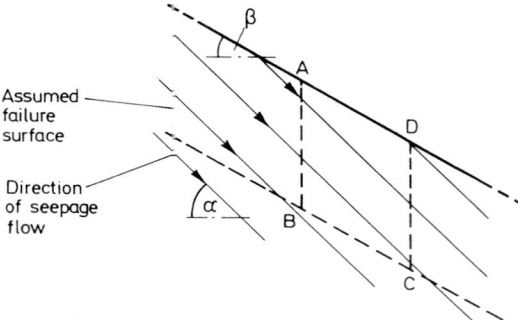

Figure 7.10 Typical element of an 'infinite' slope; the stability of the slope is determined by considering the equilibrium of the element along the line of the slope under gravitational forces, water pressures and friction forces.

special cases. For instance, if the slope is dry rather than saturated the expression simplifies to:

$$F = \frac{\tan \phi'}{\tan \beta}$$

The same expression is reached for vertical percolation of water into the slope ($\alpha = 90°$). The slope is then just stable ($F = 1.0$) when the slope angle is equal to the angle of friction of the soil. This condition exists in many natural slopes, such as in talus material, which are freely drained.

If the seepage flow is parallel to the slope ($\alpha = \beta$), the expression reduces to:

$$F = \frac{\tan \phi'}{\tan \beta} \left[1 - \frac{\gamma_w}{\gamma} \right]$$

The maximum stable slope angle is approximately halved. If seepage is out of the slope ($\alpha < \beta$), the stable slope angle is further reduced. Such conditions may well exist in thin soil cover on the sides of hills. Thus it is not uncommon for natural slopes with gradients as low as 10–15° to have a factor of safety close to unity; even minor excavations in such slopes may trigger further movement of the slope.

In all these situations the factor of safety (F) is independent of the depth assumed for the failure surface. Failure will tend to occur either on a plane of weaker soil, perhaps on a shallow rock surface or close to the surface where the soil is

loosest and the changes needed to trigger a failure are most likely to happen.

However, in the more realistic case where the water table is below the surface, the calculated factor of safety is found to get smaller the deeper into the slope the failure surface is assumed to occur. A similar result is obtained for short-term (undrained) stability of an infinite slope in fine soil. It then becomes unlikely that the assumption of an infinite slope will be reasonable, and real slips are more likely to occur along approximately circular failure surfaces. This assumption is made in standard methods of analysing slope stability, in which a number of potential slip surfaces are considered and a search made (usually nowadays by computer) to find the most critical with the lowest factor of safety.

The short-term stability of slopes in clay of uniform undrained strength is conveniently expressed in terms of a stability number N_s, where:

$$N_s = c_u/F \, \gamma \, H$$

or:

$$F = c_u/N_s \, \gamma \, H$$

where H is the height of the slope. Charts were produced by Taylor relating values of N_s and slope angle β for various values of the depth from the base of the slope to the strong layer (Y) (Figure 7.11). If Y is not restricted and β is less than 53°, then N_s has a constant value of 0.18. For other conditions reference should be made to the charts, which are reproduced in many textbooks (Smith and Smith, 1998; Whitlow, 1995; Powrie, 1997). N_s increases for $\beta > 53°$ and is 0.25 for a vertical cut, and decreases for $\beta < 53°$ when the depth of slip is limited by a firm stratum (Figure 7.11).

General methods for analysing soil slopes, which may be used for undrained total stress analysis or effective stress analysis including pore pressures, are based on dividing the slope into a number of (usually vertical) slices (Figure 7.12). The equilibrium conditions between these slices are considered, making simplifying assumptions that vary between different methods. Pore pressures within the slope are commonly expressed in terms of the pore pressure ratio r_u, which is the ratio between the pore pressure and the overburden

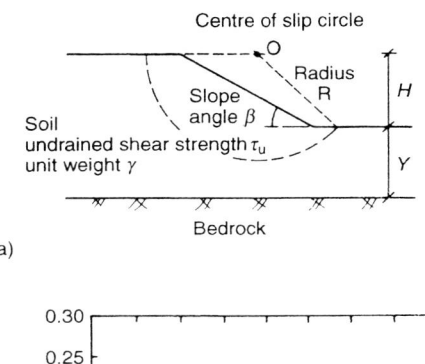

(a)

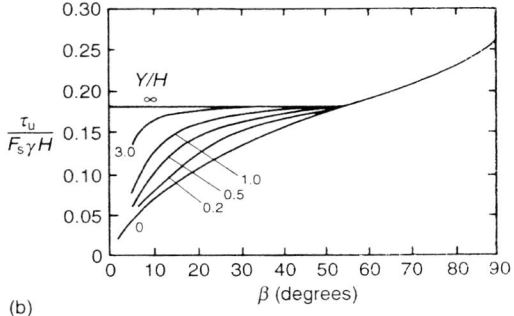

(b)

Figure 7.11   Stability of a slope in fine soil in terms of total stresses (undrained behaviour): (*A*) definition of terms and (*B*) Taylor's chart showing variation of stability number with slope geometry.

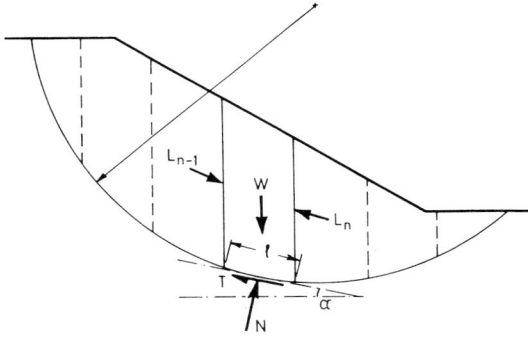

Figure 7.12   General analysis for slope stability using the method of slices; the equilibrium of a series of slices bounded by a circular failure surface is considered, making varying assumptions about the forces acting between the slices. W is the weight of the slice; L are the inter-slice forces; l is the length along the failure surface at the base of the slice; N is the normal reaction force on the base of the slice; T is the tangential force on the base of the slice and α is the inclination to the horizontal of the base of the slice.

(a) Parallel flow, no slope seepage

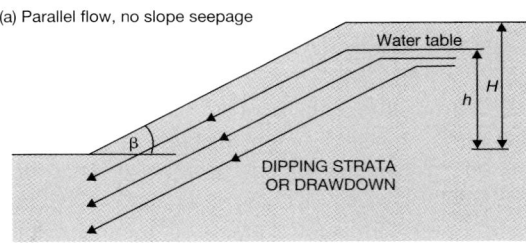

$$r_u = \frac{\gamma_w}{\gamma}\cos^2\beta \text{ for } \frac{h}{H} > 0.8 \text{ or } (H-h) < 3 \text{ m}$$

(b) Horizontal flow, full slope seepage

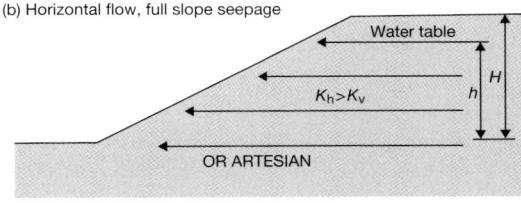

$$r_u = \frac{\gamma_w}{\gamma} \text{ for } \frac{h}{H} > 0.8 \text{ or } (H-h) < 3 \text{ m}$$

(c) Parabolic top full line

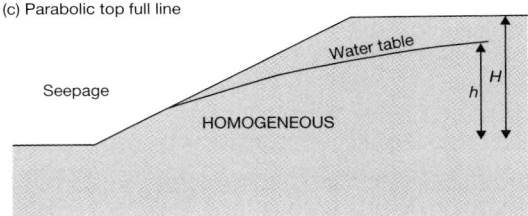

$$r_u = \frac{\gamma_w}{\gamma}\cos\beta \text{ for } \frac{h}{H} > 0.8 \text{ or } (H-h) < 3 \text{ m}$$

Figure 7.13   Determination of approximate average values for the pore pressure parameter r_u in a slope for different seepage conditions (after Mitchell, 1993).

pressure at any point. It will normally vary throughout a slope but in some cases may be considered as approximately a constant for the slope (Figure 7.13). In a homogeneous soil, the factor of safety then varies linearly with r_u, and charts have been produced by Bishop and Morgenstern (1960) giving stability coefficients m and n as functions of $c'/\gamma H$, β, ϕ' and D, such that

$$F = m - n\,r_u$$

where D is the depth factor R/H (Figure 7.11).

These charts are also reproduced in many standard text books (e.g. Smith and Smith, 1998; Whitlow, 1995). In many cases they allow preliminary estimates to be made of the stability of a slope and of its sensitivity to variations in the main parameters. For example, consider a 15 m high slope in London Clay, with assessed values of ϕ' and c' of $25°$ and $7\,kPa$ respectively, unit weight $19\,kN/m^3$, and a pore pressure regime approximating to $r_u = 0.3$. It is required to determine what is the steepest angle at which this slope will be stable, the factor of safety when the slope angle is reduced to $1:3$, and the factor of safety for the latter slope if the effective cohesion is reduced to zero by stress relief and weathering. In the first two cases $c'/\gamma H = 0.025$ approximately. On Bishop and Morgenstern's chart for parameter n, a series of dotted lines show that, provided the slope is steeper than about 1 in 4 (for $\phi' = 25°$), the failure mechanism does not extend below the base of the slope $(D = H)$ and values of m and n for different slope angles are read off as in the following table:

Slope angle	1:4	1:3.5	1:3	1:2.5	1:2
Parameter m	2.39	2.13	1.87	1.61	1.36
Parameter n	2.15	1.91	1.69	1.38	1.29
Factor of safety	1.745	1.557	1.363	1.196	0.973

From these values it can be seen that a safety factor of 1.0 (just stable) is achieved with a slope of about $1:2.1$ or $25.5°$; the factor of safety is increased to 1.36 for a slope of $1:3$ or $18°$. If $c' = 0$, the chart for $c'/\gamma H = 0$ gives $m = 1.40$, $n = 1.55$ and hence the factor of safety is 0.935 and the slope becomes unstable.

In many cases geological or other restraints may result in slope failures that approximate to neither the infinite slope nor a circular slip (Figure 7.14). These may often be analysed with sufficient accuracy by assuming a failure mechanism involving three sliding bocks and considering the equilibrium of the central block when acted on by gravity, normal and shear forces on the base of the block, a driving force from the upper wedge and a resisting force from the lower wedge.

Stability of rock slopes

Except in the case of very heavily fractured rock, failure of rock slopes will be dominated by discontinuities, with the failing mass sliding or falling as a block or collection of blocks: there are four basic mechanisms, as shown in Figure 7.15. Planar and wedge failures depend on the shear strength along the discontinuities, toppling depends purely on block geometry, while flexural toppling involves tensile failure in bending followed by toppling. The interrelation of sliding and toppling failures is shown in Figure 7.16 for the simple cases with joint planes parallel to the slope.

More complex cases, including wedge failures, involve three-dimensional interactions between discontinuity and slope planes. Stereographic projections provide a useful method of representing three-dimensional information about discontinuities in two dimensions and may also be used to determine likely slope instabilities (see Hudson, 1989 for an introduction and Hoek and Bray, 1981 for further information). Limit equilibrium methods

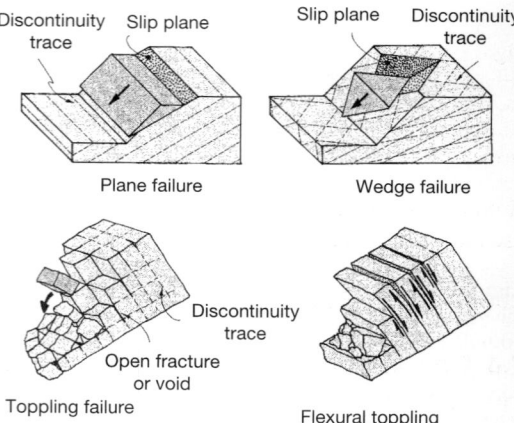

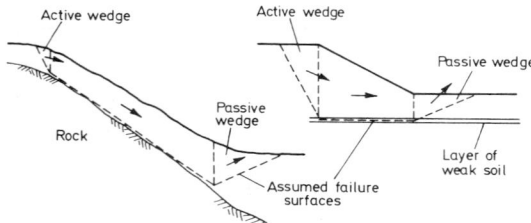

Figure 7.14 Examples of two cases where stability of a slope may be analysed by considering the equilibrium of three sliding blocks, where neither an infinite slope nor circular failure surface analysis is appropriate.

Figure 7.15 The four basic failure mechanisms of rock slopes, involving either sliding on one or two slip planes (discontinuities) or toppling (from Hudson, 1989).

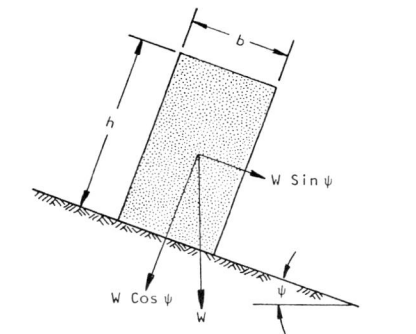

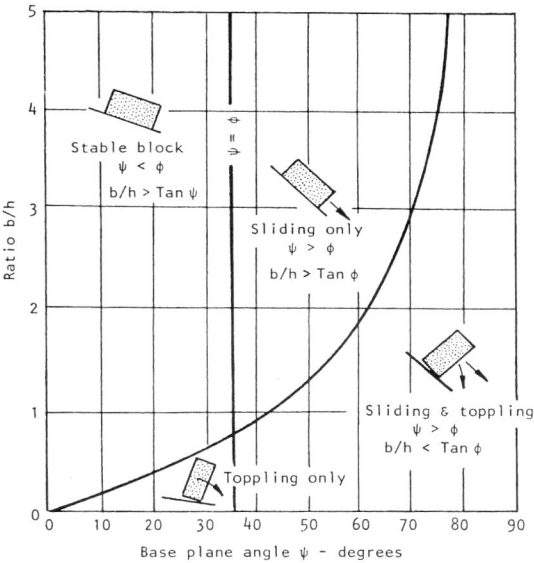

Figure 7.16   The relationship between the stability or failure of a rock slope in sliding or toppling mode and the geometry of a typical block (as defined by sets of discontinuities) and the coefficient of friction at the base of the block (ϕ) (from Hoek and Bray 1981).

similar to those used for soil slopes may then be used to analyse rock slopes and determine the reinforcement needed to stabilise the slope (Hoek and Bray, 1981; Simons *et al.*, 2001).

7.4 Geotechnical problems associated with different soil types

Soil types according to origin

Soil types and their related problems are presented below in relation to their origin, as set out

in Table 7.3. Detailed discussion of the processes involved in the formation of the soils and the resulting nature of the deposits is provided in subsequent chapters.

Effects of climate and groundwater conditions

The different climatic and groundwater regimes affecting geomorphological processes have been summarised in Chapter 2, and are discussed further in relation to different specific environments in later chapters. This section provides a brief summary of the main influences of climate and groundwater conditions on the formation and nature of soils in relation to their likely engineering properties and the geotechnical problems that may arise from them.

Where annual evapotranspiration exceeds infiltration, water tables are generally low, controlled by local stream levels and the presence of more permeable strata. In areas without tree cover, upward moisture movement and evaporation from the ground surface may produce a chemically active zone at ground level, leading to cementation and the formation of duricrust. Surface soils will be partially saturated, sometimes to great depth. Soils may have been subjected to desiccation over a long period of time. Plastic clays will then tend to heave on wetting or when evaporation is prevented by sealing of the ground surface, for example by construction of new buildings; on the other hand, porous residual soils, weathered sands or loess may collapse. Soil suctions can facilitate construction of temporary steep slopes, excavations and shafts. However long-term protection of slopes from erosion by vegetation may not be practical.

Where annual infiltration exceeds evapotranspiration, groundwater flow will be downward from the surface, soils will be mainly saturated and the water table is likely to be high in fine-grained soils, even in slopes. Excavations and underground works will generally suffer from water inflows, while dewatering may lead to settlements due to the consolidation of fine soils or internal erosion and the loss of fines from coarse soils. Erosion protection of slopes by vegetation is usually practical.

Table 7.3   Soil types according to origin.

Classification	Formation	Nature of deposit examples	Chapters
Residual			
Residual	Formed in place by chemical weathering of parent rock or soil	Type depends mainly on weathering process, less on parent material. Pronounced structure, usually bonded, sometimes porous. Heterogeneous and of variable depth. Transitional with parent material	15, 17, 18, 23
Evaporites	Salts from saline water by precipitation or evaporation	Carbonate oolites precipitated from sea water. Gypsum commonly precipitated from sulphate-rich desert playa lakes, often forms cemented soil and hard sub-surface crust	16
Organic	Formed in place by growth and decay of vegetation	Wholly organic fibrous and amorphous peat. Organic silts and clays	
Transported and deposited			
Aeolian	Transportation and deposition by wind	Usually silt and fine sand with uniform grading. Loess often contains vertical cracks, joints and root holes	16, 25
Alluvial	Transportation and deposition in water	Fine clay to coarse gravel. Coarse particles usually rounded. Soils usually sorted and often show pronounced stratification	10, 11, 15, 16, 19
Colluvial	Transport by gravity, freeze–thaw and water	Local in origin. Hillside creep, downwash and solifluction deposits. Variable grading from clay to fine gravel. Often termed 'head'	14, 15, 17, 26
Estuarine and coastal	Transport and deposition by water and wind	Fine clay to cobbles. Coarse particles usually rounded. Soft clays and silts in estuaries; sand in coastal dunes; sand, gravel and cobbles on beaches. Carbonate deposits	20
Glacial	Transportation and deposition from ice or meltwater from ice	Tills and moraines, usually heterogeneous and can have a wide range of grading. Outwash material becomes finer with distance from meltwater source. Finer material usually laminated and varved (glacial lake deposits)	13, 19
Marine	Transportation and deposition by water; also aeolian material deposited in water	Mainly clays and silts, some sand. Clays may be very deep, and initially very soft, possibly under-consolidated or sensitive and liable to liquefaction. Outwash material from rivers becomes finer with distance from coast. Biogenic materials may predominate in deep water	22
Periglacial	Heavily influenced by freeze–thaw processes	Deep colluvial deposits. Ground contorted by freeze–thaw. Parent soils and rocks fractured, brecciated and cementing destroyed. Cambering and valley bulging	14
Taluvial	Transportation by gravity	Local in origin. Landslide debris, screes and coarse variety of colluvium. Heterogeneous. Widely variable grading	8, 13, 19

Table 7.3   (*Continued*).

Classification	Formation	Nature of deposit examples	Chapters
Volcanic	Ash and pumice deposited during volcanic eruptions	Silt size particles with larger volcanic debris. Highly angular particles, often vesicular. Weathering often produces highly plastic clay	23
Man-made			
Fill	Deposited by man	Loose dumped, hydraulically placed or compacted. Old fills often heterogeneous and may contain organic and toxic material, and voids associated with human artefacts	27

Evapotranspiration and infiltration rates often vary seasonally and from year to year; they may also vary on a much shorter time scale, owing for example to individual storms. Surface conditions, water contents of fine-grained soils and their related engineering properties will vary accordingly. Conditions during construction may differ significantly from those found during site investigation.

Seasonal variations will lead to heave and settlement of superficial clays and shallow foundations; the effects on new structures will depend on the conditions prior to construction. The undrained strength of superficial clays will vary and affect the behaviour of embankment foundations, and also the suitability of the soils as fill materials. Earthmoving of fine-grained soils becomes very difficult when rainfall exceeds evapotranspiration; seasonal variations will, therefore, control the season for earthmoving and short-term variations will control working time. These variations depend on temperature, wind and rainfall duration, rather than simply rainfall amount. Problems are usually worst in temperate climates and are often strongly influenced by elevation of the site. At the other extreme, supplies of water for treating fills may vary seasonally and be in short supply during hot dry spells. Severe cyclonic rainfall is a major cause of erosion and failure of natural and man-made slopes.

The presence of vegetation increases transpiration from soil and causes an increase in soil suction and a decrease in water content; root systems may provide structural reinforcement of the soil. Development of the root system as a tree grows will tend to cause shrinkage of clays and settlement of shallow foundations. Conversely, clearance of vegetation leads to the swelling of clays, heave of shallow foundations and a loss of soil strength in slopes, while decay of the root system removes the reinforcing effect; surface erosion and slope instability may follow.

Seasonal variations may lead to complex groundwater conditions, with different pressures in permeable strata separated by less permeable strata. The permeability of superficial strata may be strongly influenced by for example structure and stress relief. Faults may form impermeable features, or provide zones of high permeability. Adverse ground water flows may affect tunnels, shafts and pile bores, while dewatered excavations may be affected by high inflows, base uplift or 'quicksand' conditions. Temporary or permanent dewatering may lead to ground settlement.

Artesian groundwater pressure is common below the base of valleys in layered strata or below glacial till on the lower slopes of valleys, and may occur within alluvium near to slopes. It exacerbates all problems involving uplift below structures and excavations, and causes severe problems with boreholes, pile bores and anchors if the flow is significant.

Common geotechnical problems associated with different soil types

A checklist of the principal geotechnical problems associated with different soil types is provided in Table 7.4. In discussing the properties

Table 7.4   Geotechnical problems associated with different soil types (adapted and extended from Fookes, 1997).

Soil type	Investigation	Properties/characteristics	Applications
Alluvial and other sedimentary soils (These soils are conveniently listed as below according to dominant particle size.)		Typically strength decreases and compressibility increases with increasing fines content.	
Gravels	Difficult to investigate by boring. Difficult to obtain representative samples from below water table as fines are washed out during drilling and sampling. Thin layers of other soils may not be detected.	Densities vary widely according to deposition. Strength high and compressibility low. Permeability variable, depending on grading and packing; often very high and can only be determined by pumping tests. Can be gap-graded with voids only partially filled with fines, which may then migrate if hydraulic gradient increased.	Generally good foundation, full consolidation occurs during construction. Large flows into excavations. Generally good fill material; single size gravels are self-compacting when deposited in water. However fills which are 'choked' with silt and clay matrix material can be difficult to compact when wet, and may not lose water readily under gravity drainage.
Sands and silts	Often weakly bonded or with interlocking grains, but effects lost on disturbance and difficult to detect. Very difficult to obtain undisturbed samples, ground below boreholes often disturbed by water inflow during drilling below the water table. SPT results may be low due to soil disturbance or, in silt, high due to pore pressure effects.	Properties usually improve with geological age. Densities vary widely according to deposition. Loose sands and silts very susceptible to liquefaction during earthquakes, can develop flow slides, large settlements when subject to vibration. Permeability moderate to high. Very erodible, piping risks when subject to internal water flow. Surface erosion by water flow and wind. Possibility of collapse of dry sands on wetting, particularly if weathered. Silts are subject to frost heave.	Generally good foundation, consolidation during construction. In excavations, beware of base failures by piping, loss of soil through sheet pile clutches etc. Danger of soil and water inflows into tunnels, high abrasion of tunnelling machines. Generally good fill material, but dry single size sands and silts have poor trafficking characteristics. Silty soils prone to rapid deterioration and poor trafficking during wet weather, and may not lose water under drainage by gravity. Silty soils may 'bounce' when trafficked, creating difficulties in forming graded surfaces prior to laying road bases etc., and increasing fuel consumption of plant. If loose dumped, moist

Alluvial clays (General properties: significant differences between over-, normally- and under-consolidated clays listed below.)	Thin layers of other soil soil may not be detected. Measured undrained strength different from different types of test.	Properties vary widely with mineralogy and grading (proportion of silt and sand). Drained strength decreases and compressibility increases with increasing clay content and plasticity. In clays with more than 30–40% platy clay minerals present, shearing produces discontinuities of low residual strength. Permeability low unless sand or silt layers are present. Strength likely to be anisotropic. Density and *in situ* strength depend on stress history due to burial, desiccation etc.	sands and silts may have low density, and collapse may occur on inundation. For saturated clays, short-term undrained strength governs stability in loading cases (foundations). Long-term drained strength governs unloading cases (excavations). Little consolidation or swelling during construction except near drainage boundaries; post-construction consolidation and swelling occurs.
Normally-consolidated 'marine' clays (Current effective stress is maximum to which soil has been subjected.)	Sensitive clays difficult to sample without undue disturbance, need to use thin-walled piston sampling. *In situ* vane tests useful; allow measurement of loss of strength on remoulding (sensitivity). Cone testing good for locating thin layers of silt and sand.	Low undrained strength, high compressibility and secondary compression (creep), which increase with plasticity. Exposed surfaces usually over-consolidated by desiccation.	Low allowable loading pressures under structures and embankments. Large post-construction settlements. Base heave and failure in strutted excavations, and high strut loads. Down-drag on piles. Low strength and difficult working conditions for plant during excavation.
Over-consolidated 'marine' clays (Current effective stress is less than previous maximum — the pre-consolidation pressure)	Test results may be affected by sampling disturbance, presence of fissures etc. In heavily over-consolidated soils rotary coring better than push-sampling.	If effective stresses due to engineering work exceed pre-consolidation pressure, behaviour reverts to that of normally consolidated clay; otherwise compressibility much less. Undrained strengths higher but difficult to predict. Permeability may be controlled by flow through fissures.	Mass strength of foundations often affected by fissuring. Pre-existing shear surfaces, particularly in highly plastic clays, may control stability of slopes. Probable high *in situ* horizontal stress in heavily over-consolidated clays, large horizontal movements during and after excavation, high lateral stresses on buried structures.

Table 7.4 (*Continued*).

Soil type	Investigation	Properties/characteristics	Applications
	SPT tests often useful back-up to laboratory testing.		
Under-consolidated 'marine' clays (Not yet fully compressed under current stresses, due to rapid deposition or recent additional loading)	Very difficult due to very low strength.	Excess pore pressures still present. Very low undrained strength and high compressibility relative to depth.	Ground surface still settling. Stability of excavated slopes may be controlled by undrained strength.
Clay fill		Bonding of parent clay usually destroyed by excavation and fill placing. Laminated and varved clays usually mixed, and produce a clay of low permeability. Low-plasticity clays lose significant undrained strength with small increases in water content in wet weather. Rutting of wet plastic clay fills under trafficking causes shear surfaces that reduce bulk strength. Smooth drum rollers can cause horizontal surfaces of low strength.	Trafficability and height of undrained slope construction controlled by remoulded undrained strength. Degree of saturation important. Strength and stability of fill slope usually governed by long-term conditions (drained strength plus pore pressures due to rainfall infiltration etc.). Deep-seated failure may be governed by short-term undrained strength. Clay haul road surfaces become too slippery for trafficking in rain. Early compaction and profiling of clay important to seal surface and reduce infiltration and softening by rain. Drying of clay prior to placement requires reliable warm dry weather.
Colluvial soils	Soils usually thin and subject to seasonal water content change. Time of investigation and of engineering construction can be important.	Formation often involves shearing. Low strength shears may be present, often continuous and at base of soil, on slopes currently too flat for slope movement. Properties may differ from those measured at time of site investigation due to seasonal variations.	Problems for embankment foundations on sidelong ground and in excavations.

Talluvial materials		Landslides are often at limiting equilibrium, on major shear surfaces. Strata disturbed in massive landslides; open fissures and fracture zones formed. Soil more porous and wetter than parent soil. Perched water table effects.	Development of equilibrium slope in clays may take many decades. Instability due to toe erosion may continue long after erosion prevented. Excavation and filling likely to initiate new movements. Screes may develop avalanche/flow slide behaviour when disturbed.
Hot desert soils		Generally have little or no fines. Usually granular, uniformly graded and of low density when wind blown or coastal. Often coarse, well-graded and with angular particle shape when deposited by ephemeral spate flow in wadis or fans. Water table near to the surface leads to precipitation of evaporite salts.	Soil may be highly erodible once thin protective pavement removed or disturbed. Duricrusts or densely packed boulders in wadis may cause excavation difficulty. Aggregate may be in short supply or contaminated by salts. Problems of sediment movement by water in sudden storms and by wind. Deposits of wind-blown sand liable to collapse on wetting. Salty ground highly aggressive to structures and road pavements.
Glacial soils	Often variable and heterogeneous, horizontally and vertically. Original land forms obscured — buried valleys etc. High and low permeability strata give complex groundwater conditions. Severe artesian pressure conditions often exist below tills which mantle valley slopes. In tills, boulders cause problems with drilling. Interface with rock may be difficult to determine.	Often variable and heterogeneous, horizontally and vertically, with complex groundwater conditions, including artesian. Density and strength of tills depend mainly on method of deposition, not on stress history as for alluvial soils. Grading curve may be almost straight line over wide range of particle size. Nature of fine matrix material and presence of clay minerals control properties. Dense materials may be very strong and stiff. Often contain local inclusions of water-deposited laminated silts and clays. Often have discontinuities as for alluvial clays.	Problems with water-deposited soils as for alluvial soils. Tills generally good foundation material. Boulders cause problems in piling, tunnelling, excavation and filling. Drag structures at base of tills on weak rocks etc. cause errors in rock level estimation, problems with piles etc. Water flows into excavations and stability of slopes in short term may be heavily influenced by layers or lenses of high permeability. Problems with fills as for alluvial soils.

Table 7.4   (*Continued*).

Soil type	Investigation	Properties/characteristics	Applications
Periglacial conditions	Valley slopes will be affected by strata disturbance and shearing due to valley bulging and cambering. Near-vertical fissures and gulls difficult to locate by conventional drilling.	Permafrost in active conditions. In relic conditions, past ground freezing is likely to have produced extensive colluvium deposits on slopes. Ground contorted by freeze–thaw features e.g. cryoturbation, frost wedges, pipes. Parent soils and rocks will have been fractured, brecciated and un-cemented by ground freezing. Such effects may be present generally or locally.	Effects of cambering and valley bulging may cause problems with foundations, excavations and tunnels.
Organic soils	Most methods of sampling and testing not suitable for highly organic materials, particularly when fibrous. Compressibility usually more important than strength, best measured in large Rowe cells or, for shallow deposits, by large scale loading tests — plate bearing, skip test, trial embankment etc.	Highly compressible and subject to severe long-term creep. Often of very low unit weight. Methane gas may be present.	Very large settlements of foundations and embankments underlain by organic soils, requiring light weight fills or pre-compression by surcharging. Problems in slope stability due to low passive resistance. Non-saturated peat deposits may float when flooded. Wastage of peats when exposed and subject to drying. Usually difficult to run plant on and to handle as spoil.
Volcanic soils	Layered and complex deposition. Old weathering surfaces and residual soil often covered with fresh deposits. Thin ash layers, weathered to clay, within other	Properties significantly different from those of sedimentary soils due to porosity and crushability of silt and sand sized particles: *in situ* moisture content higher than usual; greater reduction in strength, but smaller reduction in compressibility, with increase in stress; no clear peak in compaction curve. Fine soils often of high plasticity (with	Low particle density makes earthwork fills very susceptible to erosion. May soften with compaction, are easily damaged by earth-moving machinery, leading to loss of trafficability and sometimes flow. Drying produces non-reversible improvement, addition of quicklime effective.

	deposits. Investigation by cone testing shows behaviour for both loose and dense soils to be similar to that for loose quartz sand, due to crushability of soil grains.		smectite, allophane), but strength higher and less dependent on plasticity than with sedimentary soils.
Residual soils (see Fookes, 1997, Appendix 5, for further details)	Difficult to obtain 'undisturbed' samples without destroying structure. Core stones (unweathered rock) within weathered profile cause problems in drilling. May be difficult to establish rockhead, which varies rapidly. Relic discontinuities from parent rock (often containing iron and manganese salts) give planes of low-drained strength, very difficult to locate during site investigation.	Wide ranging of grading, plasticity, mineralogy, etc., depending primarily on weathering processes. Grading often depends on unweathered quartz particles present. Recementation of soils may occur. 'Black' soils, usually formed with high drainage have high plasticity. Expansive soils with large volume changes on wetting and drying. 'Red' soils, usually formed with good drainage, but with pronounced structure due to weathering process. Engineering properties depend more on structure than on grading, mineralogy etc. Usually behave as if bonded, with structure yielding at a certain stress level. Strength and compressibility depend on this yield as much as on density. Mineralogy and properties can be changed by drying. Structure usually gives high *in situ* permeability. Porous soils with a high degree of saturation may be sensitive, giving low undrained strength on remoulding and destruction of structure; most likely in soils from volcanic rocks.	Core stones within weathered profile cause problems in piling, can influence excavation methods in open cut. Relic low-strength discontinuities can cause slope instability if of critical extent and angle. Soils may be readily eroded by water, causing severe gullying during heavy rain. Rapid consolidation during construction. Porous soils existing in dry conditions may collapse on wetting. Porous soils with high degree of saturation become weak with reworking and cause problems with operating plant during excavation, forming roads etc. Strength due to structure lost in fills, then strength and compressibility depend on density achieved by compaction, which depends on water content at source. Compaction and loss of structure usually gives substantial reduction in permeability. General properties of fill are similar to those of alluvial clays of similar grading and mineralogy, but mineralogy often differs from that of sedimentary clays.
Man-made fills	Old, non-engineered fills likely to be heterogeneous and difficult to investigate. Site history is of great importance. Usually	Fills will generally be loose and compressible. Ground and groundwater may be toxic. Methane gas may be generated. Hydraulic fills likely to be of low density, subject to liquefaction if of appropriate particle size. Difficult to characterise in engineering terms.	Major costs may be involved in rendering the site environmentally acceptable for development. Likely to be very poor foundation conditions, requiring piles or ground improvement. Special precautions with boreholes and piling to prevent connecting polluted and clean aquifers.

Table 7.4 (Continued).

Soil type	Investigation	Properties/characteristics	Applications
	investigated by trial pits, with simple *in situ* tests or as for organic soils. May contain toxic and organic materials, materials subject to decay, voids associated with human artefacts, remains of old constructions, sludge lagoons from sewage, agricultural and industrial processes, obscured by subsequent filling. The original topography is obscured by tipping. Compressible and weak alluvial and organic soils associated with streams and ponds may be buried locally. Back filling of old mineral workings may involve overburden and waste material similar to original geological strata and may be difficult to detect from boring.		Infilled quarries likely to have had steep sides, giving rapid transition between natural ground and deep fill material. Waste materials such as mine tailings and pulverised fuel ash may be useful as engineered fills, particularly where of relatively low weight.

and associated problems of clay soils in particular, consideration must always be given to the influence of soil structure or fabric, which may have a profound influence on the behaviour of the soil and make it significantly different from that of a remoulded soil of the same mineralogy and water content. Some aspects of the structure are fissures, joints and sheared surfaces; bonding; sensitivity; dispersion, and laminations or varves of sand and silt with clay.

Fissures, joints and sheared surfaces are frequently present, particularly in over-consolidated clays. Joints and fissures in low plasticity clays may have coatings of higher-plasticity clay. Bulk clay strength is reduced, particularly by pre-sheared surfaces in plastic clays of low residual strength. Conversely, permeability may be increased, particularly if the joints and fissures are opened by stress relief during excavation.

Some bonding, mainly due to cementation between clay particles, is often present, varying from very slight to significant (forming weak mudstone). Bonding is usually destroyed by weathering and also by mechanical disturbance. Bonding improves intact strength and reduces compressibility, but may lead to sensitivity, or loss of undrained strength on remoulding and large settlements following yield. It may allow fissures to remain open, giving higher bulk permeability and accelerated swelling and consolidation.

Sensitivity is usual in normally consolidated clays and can be severe in 'quick' clays, usually due to bonding or to leaching of low-plasticity marine clay by fresh seepage water. It can also occur in residual soil.

Inter-particle attractive forces usually give clays resistance to erosion by water. In dispersive clays such forces do not exist, making them highly erodible both *in situ* and if used as fills.

Laminated and varved clays are highly anisotropic in strength and permeability, with drained strength along the layers controlled by the most plastic layer present. High horizontal permeability allows rapid consolidation or swelling, giving rapid strength gain under loading. However the short-term undrained strength cannot be relied on for temporary works. Water from sand layers

in exposed excavation faces can cause internal and surface erosion, swelling of clay layers and rapid deterioration; similar effects can occur in borings for piles and piers. Laminated clays are easily sampled in borings, but clay layers tend to swell immediately after sampling, taking in water from sand and silt layers. Incorrect water contents and undrained strengths will then be measured. Smear of borehole walls by clay during drilling influences the results of *in situ* permeability tests and reduces the efficiency of vertical drains.

7.5 Typical values of soil properties

This section provides a selection of typical values of some basic and derived properties of soils.

Specific gravity and density

The full range of specific gravity for soil minerals is from about 2.3 for gypsum to around 5 for magnetite. However for most of the commonest components of soil it falls within a narrow range. The specific gravity of quartz, and hence of the particles of most coarse-grained soils, is 2.65 to 2.66. The specific gravity of clay minerals is more variable, but typically in the range 2.55 to 2.75, while for calcareous sand it is between 2.64 and 2.73. As a first approximation, a value of 2.65 is often assumed.

Some typical values of the bulk density of natural soils and compacted fills are given in Table 7.5. Densities of around 2000 to 2100 kg/m^3 are typical of dense soils, and 1700 to 1800 kg/m^3 for loose soils. Much lower densities are sometimes found in highly organic soils and in soils with hollow or porous particles such as volcanic ash or pulverised fuel ash (PFA).

Permeability

Typical ranges of the permeability of soils are given in Table 7.6; note that the value can vary over ten orders of magnitude. Permeability may be very significantly affected by the structure of the soil. For example, it may be greatly increased by fissuring in otherwise very impermeable clay, particularly around excavations or tunnels where

Table 7.5   Typical values of density and moisture content (after Seed *et al.*, 1962).

Material	Natural densities		BS/AASHTO compaction	
	Bulk density* (kg/m^3)	Dry density (kg/m^3)	Dry density (kg/m^3)	Optimum m/c (%)
Sands and gravels: very loose	1700–1800	1300–1400	–	–
loose	1800–1900	1400–1500	–	–
medium dense	1900–2100	1500–1800	–	–
dense	2000–2200	1700–2000	–	–
very dense	2200–2300	2000–2200	–	–
Poorly-graded sands	1700–1900	1300–1500	1500–1800	0–7
Well-graded sands	1800–2300	1400–2200	1700–2100	0–8
Well-graded sand/gravel mixtures	1900–2300	1500–2200	1800–2200	5–10
Clays: unconsolidated muds	1600–1700	900–1100	–	–
soft, open-structured	1700–1900	1100–1400	–	–
typical, normally consolidated	1800–2200	1300–1900	–	–
tills (boulder clays)	2000–2400	1700–2200	–	–
Compared sandy clays	–	–	1800–2200	15–30
Tropical red clays	1700–2100	1300–1800	1400–2100	20–40

* Assumes saturated or nearly saturated soil.

Table 7.6   Typical permeability values (after Casagrande and Fadum, 1940).

Soil types	Homogeneous clays below the zone of weathering	Silts, fine sands, silty sands, glacial till, stratified clays	Clean sands, sand and gravel mixtures	Clean gravels
		Fissured and weathered clays and clays modified by the effects of vegetation		

Coefficient of permeability (log scale)	10^{-11} 10^{-10} 10^{-9} 10^{-8} 10^{-7} 10^{-6} 10^{-5} 10^{-4} 10^{-3} 10^{-2} 10^{-1} 1
	m/sec
	10^{-9} 10^{-8} 10^{-7} 10^{-6} 10^{-5} 10^{-4} 10^{-3} 10^{-2} 10^{-1} 1 10 100
	cm/sec
	10^{-10} 10^{-9} 10^{-8} 10^{-7} 10^{-6} 10^{-5} 10^{-4} 10^{-3} 10^{-2} 10^{-1} 1
	ft/sec
	Practically Impermeable | Very low Low | Medium | High

Drainage conditions	practically impermeable | Poor | Good

Estimation of coefficient of permeability: for granular soils, the coefficient of permeability can be estimated using Hazen's formula:

$$k = c_1 D^2{}_{10}$$

where k is the coefficient of permeability in m/s, D^{10} is the effective particle size in mm, and c_1 is a factor varying between 100 and 150.

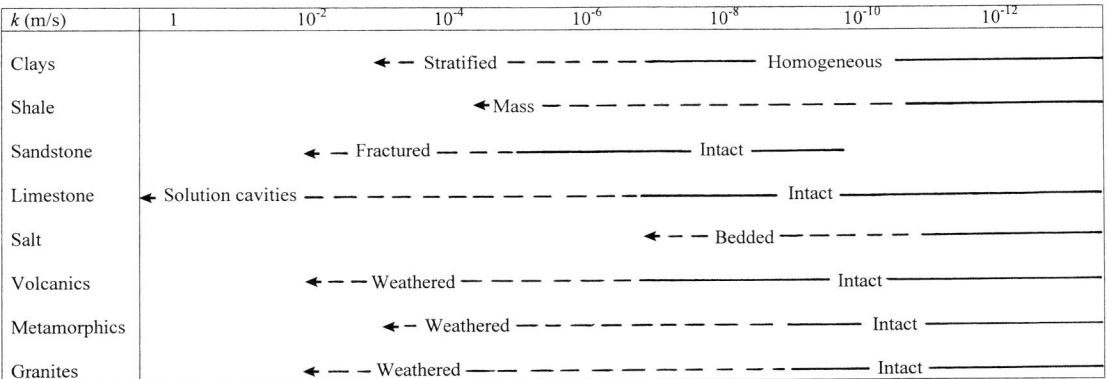

Figure 7.17   Typical ranges of coefficient of permeability (k) for different types and conditions of rock.

the reduction in stress allows the fissures to open. It may also be highly anisotropic: old root holes can provide vertical drainage paths in clay, so that the bulk vertical permeability is greater than the horizontal, while sand and silt laminations may make the horizontal permeability much greater than the vertical. In heterogeneous materials like glacial tills, lenses of sand and gravel provide zones of high permeability, the consequences of which depend critically on the extent of the lenses and whether they are connected to a water source such as a river.

Except in porous rocks, such as sandstone and chalk, the permeability of rock is largely controlled by the flow of water through joints and fissures. Typical values for intact and fractured or weathered materials are shown in Figure 7.17.

Strength
Ranges of values for unconfined compressive strength (UCS) of rocks and clays are given in Tables A3.2a–c in Appendix A3. Note that for clays the UCS is double the undrained strength in compression.

Typical values of the angle of shearing resistance (ϕ') for coarse soils are given in Table 7.7 and Figure 7.18, and the variation of ϕ' with plasticity index for clays in Figure 7.19. Relations between the value of ϕ', relative density and the corrected blow count (N) from standard penetration tests (SPTs) in sand are given Figure 7.20.

Table 7.7   Typical values of shearing resistance of cohesionless soils.

Material	ϕ (degrees)	
	Loose	Dense
Uniform sand, round grains	27	34
Well-graded sand, angular grains	33	45
Sandy gravels	35	50
Silty sand	27–33	30–34
Inorganic silt	27–30	30–35

SPTs are also useful for providing a measure of the undrained strength of over-consolidated clays, using the relation:

$$c_u = f_1 N$$

with f_1 as given by the plot in Figure 7.21.

Compressibility and swelling potential
The compressibility of clays under load and their tendency to shrink on drying or swell on wetting or unloading are both related to the plasticity of the clay and its natural water content. Thus clays of high plasticity are more compressible than those of similar strength but of low plasticity, while any particular clay becomes less compressible as its void ratio decreases. Typical values for

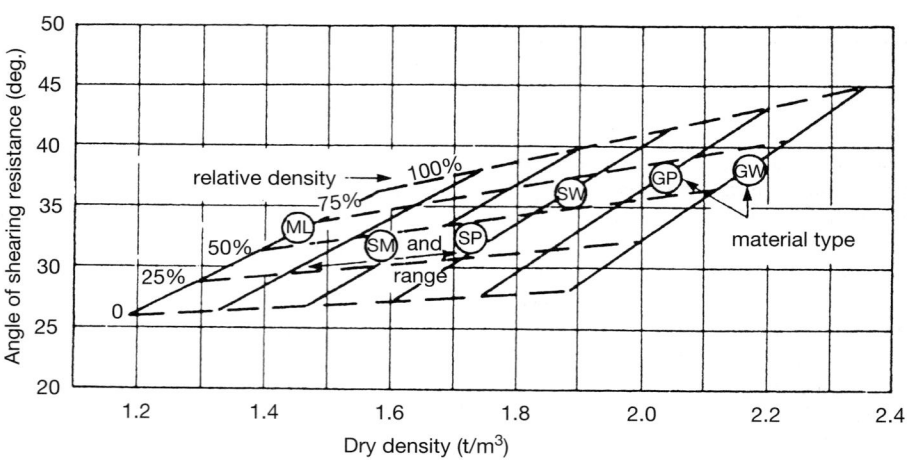

Figure 7.18   Variation of angle of shearing resistance with density and soil type: ML: silt of low plasticity; SM: silty sand; SP: poorly graded (uniform) sand; SW: well-graded sand; GP: poorly graded gravel and GW: well-graded gravel.

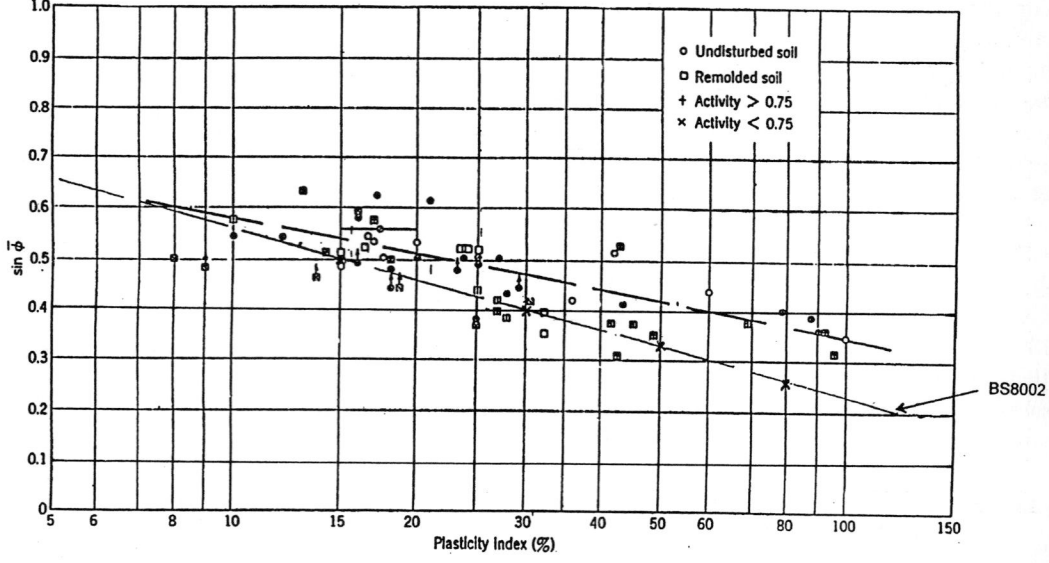

Figure 7.19   Variation of angle of shearing resistance (expressed as sin ϕ) with the plasticity index of a soil (after Kenney, 1959).

natural soils are given in Table 7.8. The coefficient of compressibility is the ratio of compressive strain to applied stress for one-dimensional compression (i.e. with no lateral strain). On drying, wet plastic clays tend to shrink and crack; conversely dry clay which is wetted tends to swell, the potential for swelling being related to plasticity index as shown in Table 7.9.

Bearing capacity
Typical values of allowable bearing pressure for a range of soils and rocks are given in Table 7.10.

Table 7.8 Typical compressibility values and descriptive terms used.

Type of clay	Descriptive term	Coefficient of compressibility, m_v	
		(m²/MN)	(ft²/ton)
Heavy over-consolidated tills, stiff weathered rocks (e.g. weathered mudstone) and hard clays	Very low compressibility	< 0.05	< 0.005
Tills, marls, very stiff clays and stiff tropical red clays	Low compressibility	0.05–0.1	0.005–0.01
Firm clays, glacial outwash clays, lake deposits, weathered marls, firm tills, normally consolidated clays at depth and firm tropical red clays	Medium compressibility	0.1–0.3	0.01–0.03
Normally consolidated alluvial clays such as estuarine and delta deposits, and sensitive clays	High compressibility	0.3–1.5	0.03–0.15
Highly organic alluvial clays and peats	Very high	>1.5	> 0.15

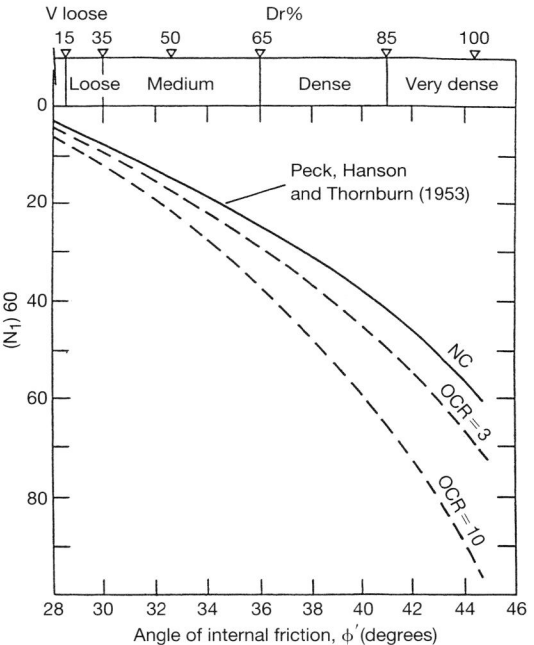

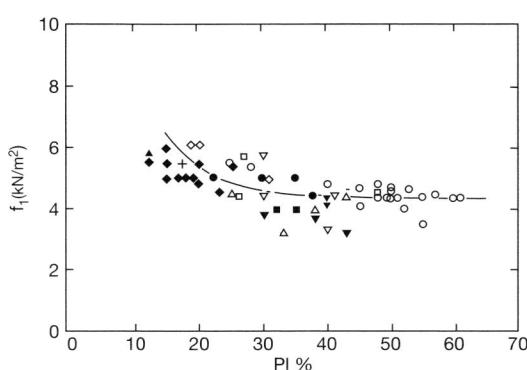

Table 7.9 Typical values of plasticity index and inherent swelling capacity.

Plasticity index (%)	Inherent swelling capacity
0–15	Low
10–35	Medium
20–55	High
35+	Very high

Figure 7.20 Determination of the density and angle of internal friction of coarse soils from SPT blow count (N); NC refers to normally consolidated soils, OCR is over-consolidation ratio (see Peck *et al.*, 1953).

Figure 7.21 Plot showing the relation between the factor f_1 and the plasticity index (PI) of a fine soil; f_1 relates the undrained strength of the soil to the SPT blow count (N) (from Stroud and Butler 1975).

Note that for clays the allowable bearing pressure is about twice the undrained strength of the clay. A good general reference for the design and construction of foundations is Tomlinson (2001). For more extensive information on foundations on rock, see Wyllie (1999).

7.6 Closing remarks

It should always be remembered that the influence of water on the behaviour of soils is profound. Damaging effects may be due to erosion or scour either at the surface or underground, inflow of water

Table 7.10 Typical values of allowable bearing pressures (after Carter, 1983).

Type of bearing material	Allowable bearing pressure (kN/m²)
Rocks	
Massive hard crystalline igneous and metamorphic rocks	6000–10 000
Massive hard crystalline limestones; thoroughly cemented sandstones and conglomerates	4000–6000
Unweathered schists and slates	3000–4000
Hard shales and mudstones, moderate and weakly cemented sandstones; hard unweathered marl or chalk	1500–2500
Weathered and broken bedrock; clayey shales and soft mudstones	800–1200
Clays	
Hard clays; cohesive strength >300 kN/m²	600
Very stiff clays; cohesive strength 150–300 kN/m²	300–600
Stiff clays; cohesive strength 75–150 kN/m²	150–300
Firm clays; cohesive strength 35–75 kN/m	70–150
Soft and very soft clays: cohesive strength <35 kN/m²	Negligible
Sands	
Very dense sands and gravels; SPT N-value > 50	400
Dense sands and gravels; SPT N-value 30–50	300–400
Medium dense sands and gravels; SPT N-value 10–30	100–300
Loose sands and gravels; SPT N-value 5–10	50–100

The table above gives typical values of allowable bearing pressures for shallow spread foundations subjected to vertical static loading. It is presumed that the founding levels is at about 1 m depth in soils and that the ground surface is fairly level. Allowable bearing pressure is the pressure applied by the foundations under design loads; the pressure is limited to a value such that the risk of collapse is minimal and settlements are restricted to acceptable values.

into excavations and tunnels, softening of fill materials, initiation of collapse settlements due to inundation of sensitive soils, or increase in pore pressures leading to slope failures. On the other hand, suctions in fine soils may provide useful short-term stability of excavations and slopes, while lack of water may also lead to problems, for example of drying shrinkage, upward migration of harmful salts, or difficulty in compaction. Proper understanding of the water regime, past present and future, affecting a project is often of prime importance, yet often not given the attention it deserves.

References

Atkinson, J. E. (1981) *Foundations and Slopes*. McGraw-Hill, London.

Atkinson, J. H. (1993) *AN Introduction to the Mechanics of Soils and Foundations*. McGraw-Hill, London.

Bishop, A. W. and Morgenstern, N. (1960) Stability coefficients for earth slopes. *Geotechnique* **10**, (4 December 1960).

Bolton, M. D. (1986) The strength and dilatancy of sands. *Geotechnique* **36**(1), 65–78.

Bolton, M. D. (1991) *A Guide to Soil Mechanics*. M. D. and K. Bolton, Cambridge.

BS (British Standards Institution) (1991) BS1377: *Methods of Test for Soils for Civil Engineering Purposes*. BSI, London.

BS (British Standards Institution) (1999) BS5930: *Code of Practice for Site Investigation*. BSI, London.

Carter, M. (1983) *Geotechnical Engineering Handbook*. Pentech Press, London.

Casagrande, A. and Fadum, R. E. (1940) *Notes on Soil Testing for Engineering Purposes*. Harvard University Graduate School of Engineering Publication 268.

Clayton, C. R. I., Simons, N. E. and Matthews, M. C. (1995) *Site Investigation*. Blackwell, Oxford.

Fookes, P. G. (1997) Geology for engineers. *Quarterly Journal of Engng. Geol.* **30**, 293–424.

Fookes, P. G., Baynes, F., and Hutchison, J. (2001) Total geological history: a model approach to understanding site conditions. *Ground Engineering*. **34**(3), 42–47.

Hoek, E. and Bray, J. W. (1981) *Rock Slope Engineering* (Revised 3rd edn). Instn. of Mining and Metallurgy/Spon, London.

Hudson, J. A. (1989) *Rock Mechanics Principles in Engineerig Practice*. CIRIA, London.

ISRM (International Society for Rock Mechanics) Brown, E. T. (ed.) (1981) *Suggested Methods for the Quantitative Description of Discontinuities in Rock Masses*. Pergamon Press, Oxford.

Kenney, K. C. (1959) Discussion. *J. Soil Mech. and Foundn. Div. ASCE* **85**, No.SM3, 67-69.

Ladd, C. C., Foote, R., Isihara, K., Schlosser, F. and Poulos, H. G. (1977) Stress-deformation and strength characteristics. IX Int. Conf. Soil Mech. and Foundn. Engng, Vol. 2, 421–494.

Mair, R. J. and Wood, D. M. (1987) *Pressuremeter Testing, Methods and Interpretation*. CIRIA, London.

Meigh, A. C. (1987) *Cone Penetration Testing, Methods and Interpretation*. CIRIA, London.

Mitchell, J. K. (1993) *Fundamental of Soil Behaviour* (2nd edn). Wiley, New York.

Peck, R. B., Harson, W. E. and Thorburn, T. H. (1953) *Foundation Engineering* (2nd edn 1974).

Perry, J. and West, G. (1996) Sources of information for site investigations in Britain. Transport Research Laboratory (TRL) Report LR 192.

Poulos, H. G. and Davis, E. H. (1974) *Elastic Solutions for Soil and Rock Mechanics*. Wiley, New York.

Powrie, W. (1997) *Soil Mechanics: Concepts and Applications*. Spon, London.

Rowe, P. W. (1972) The relevance of soil fabric to site investigation practice. *Geotechnique* **22** (2), 195–300.

Seed, H. B., Woodward Jr, R. J. and Lundgren, R. (1962) Prediction of swelling potential for compacted clays. *J. ASCE* **88**, (SM3).

Simons, N., Menzies, B. and Matthews, M. (2001) *A Short Course in Soil and Rock Slope Engineering*. Thomas Telford, London.

Skempton, A. W. (1957) Discussion: the planning and design of new Hong Kong airport. *Proc. Instn. Civ. Engrs.* **7**, 305–307.

Smith, G. N. and Smith, I. G. N. (1998) *Elements of Soil Mechanics* (7th edn). Blackwell, Oxford.

Stroud, M. A. (1989) The standard penetration test – its application and interpretation. *Penetration Testing in the UK*. Thomas Telford, London, 29–49.

Stroud, M. A. and Butler, F. G. (1975) The standard penetration test and the engineering properties of glacial materials. *Proc. Symp. Engng. Behaviour of Glacial Materials*. University of Birmingham, 124–125.

Tomlinson, M. J. (2001) *Foundation Design and Construction*. Pearson, Essex, UK.

Waltham, A. C. (2001) *Foundtions of Engineering Geology* (2nd edn). Blackie, London.

Whitlow, R. (1995) *Basic Soil Mechanics* (3rd edn). Longman, Harlow.

Wyllie, D. C. (1999) *Foundations on Rock* (2nd edn). Spon, London.

Further reading

Atkinson, J. H. (1981) *Foundations and Slopes*. McGraw-Hill, London.

Atkinson, J. H. (1993) *An Introduction to the Mechanics of Soils and Foundations*. McGraw-Hill, London.

Bolton, M. D. (1991) *A Guide to Soil Mechanics*. M. D and K. Bolton, Cambridge.

Hudson, J. A. (1989) *Rock Mechanics Principles in Engineering Practice*. CIRIA, London.

Powrie, W. (1997) *Soil Mechanics: Concepts and Applications*. Spon, London.

Waltham, A. C. (2001) *Foundations of Engineering Geology* (2nd edn). Blackie, London.

Part II
Geomorphological Processes

8. Landslides

James S Griffiths

8.1 Introduction

There are many different definitions of landslides in the literature but the simplest and most useful is that provided by Cruden (1991) for the UNESCO Working Party on World Landslide Inventory:

a landslide is the movement of a mass of rock, earth or debris down a slope.

The term 'landslide' does not mean investigations are limited only to failures on land that move with a sliding mechanism. With the increasing use of the seafloor for resource exploitation sub-marine slope failures are an important area of study in engineering geomorphology. In the literature the terms 'landslip' or 'mass movement' are often used rather than 'landslide'. Whilst these terms may be regarded as describing the same geomorphological phenomena, detailed investigations of the movement of landslides have shown that sliding along a basal separation or shear surface is one of the main mechanisms involved, hence the term 'landslide' is preferred. However, the movement mechanisms involved in landsliding are complex and can include sliding, flowing and falling.

The losses and damage resulting from landslides vary considerably around the world. Brabb (1991) estimated that during the early 1970s the average annual number of deaths in the world caused by landslides was c. 600 but twenty years later the actual number of deaths had increased to several thousand. In the USA, it is estimated that landslides annually result in over US$1.5 billion in losses and 25 to 50 deaths whilst in Japan the annual economic losses exceed US$1 billion (Smith, 2001). Apart from these general figures there have been some single very tragic events that have resulted in appalling losses of life and examples of these are listed in Table 8.1. Clearly landslides are an important natural hazard that require investigation and mitigation. However, the scale of losses are considerably less than those associated with other natural hazards such as floods, earthquakes, famine and epidemics. As a result, the development of techniques through research for anticipating landslide occurrence is still in its infancy. Whilst there is a good understanding of the processes and causes of landsliding, our ability to anticipate where and when a landslide will occur is still very limited compared to other forms of discrete ground movement, such as faults. Also, within landslide zones there are complex patterns of compression and tension that make them difficult to manage, and there is considerable uncertainty about the scale and nature of the landslide response to movement triggering events.

The landslide literature is vast, with the definitive general work on landslides presently probably best represented by Turner and Schuster (1996). Almost every year there are landslide symposia and conferences, each generating a wide range of theoretical models and practical case studies (e.g. Bromhead et al., 2000). In addition there are numerous articles in the scientific literature on various aspects of landslide investigation (Hutchinson, 2001). For detailed landslide investigations reference should be made to this impressive wealth of material. However, in engineering geomorphology the requirement is usually to go out to a site where there has been little previous work and geomorphologists will have to rely on their own knowledge and skills to study the problem. The typical questions the geomorphologists are asked in these situations are:

1. What is the distribution of pre-existing landslides?

Table 8.1 Examples of high magnitude/low frequency landslide events in the latter part of the 20th century (after Schuster, 1995; Godt and Savage, 1999, and Guadagno and Perriello Zampelli, 2000).

Year	Location	Name and Type*	Volume (m³)	Trigger	Impact	Comment
1962	Peru (Ancash)	Huascaran Debris avalanche	13×10^6	Unknown	4000–5000 killed; much of Ranrahirca village destroyed	Major debris avalanche from Nevado Huscaran; average velocity 170 km/hr
1963	Italy	Vaiont rock slide	250×10^6	Filling Vaiont reservoir	c. 2000 killed; city of Longarone badly damaged; total cost in 1963 US$ 200 million	High velocity rock slide into reservoir caused 100 m wave to overtop the dam. Estimated maximum velocity 50 m/s
1964	Alaska	The 1964 Alaskan landslide (spreading failures)	Unknown	M9.4 earthquake	Damages in 1964 cost US$180 million	Spreading failures caused major landslide damage in Anchorage, Valdez, Whittler & Seward
1966	Brazil (Rio de Janeiro	Rio de Janeiro avalanches debris & mud flows	Unknown	Heavy rains	~1000 killed	Many landslides around Rio de Janeiro
1966	Wales	Aberfan Tip flow slide (debris flow)	1.1×10^5	Loose tipping on a spring	144 killed including 112 school children	Flowslide of loose tipped colliery waste, estimated maximum velocity 8.8 m/s; site of previous slides that did not reach the village
1970	Peru	Nevados Huascaran rock/debris avalanche	Unknown	M7.7 earthquake	18 000 killed; town of Yungay destroyed; Ranrahirca partially destroyed	Rock/debris avalanche from same peak as 1962; average velocity 280 km/hr
1974	Peru	Mayunmarca rock slide – debris avalanche	1.6×10^9	Rainfall — river erosion	Mayunmarca village destroyed ~450 killed; failure of 150 m high land-slide dam caused major downstream flooding	Debris avalanche with average velocity 140 km/hr; dammed Maataro river
1980	USA (Wash.)	Mount St Helens rotational rock slide followed by debris avalanche	2.8×10^9	Eruption of Mt St Helens	World's largest historic landslide; 5–10 killed — most people evacuated; major destruction to infrastructure	Began as rotational rock slide, degraded to 23 km long debris avalanche with average velocity 125 km/hr; surface remobilised in to 95 km long debris flow
1982	San Francisco Bay Region	18 000 debris flows	Unknown	January 3–5 rainstorm	25 deaths and landslide damage estimated at $65 million	

1983	USA (Utah)	Thistle debris slide	21×10^6	Snowmelt and heavy rain	No deaths; destruction of infrastructure; dammed Spanish Fork flooding town of Thistle. Total losses in 1983 US$400 million	Loess landslide
1983	China (Gansu)	Salasham landslide	35×10^6	Unknown	237 killed; 4 villages buried; 2 reservoirs filled with debris	Death toll unnecessarily large because hazard warnings not passed to residents
1985	Colombia (Tolina)	Nevado del Ruiz debris flows	Unknown	Eruption of Navado del Ruiz	4 town & villages destroyed; flow in valley of Langunillas River killed 20 000 + in city of Armero	
1986	Papua New Guinea (East New Britain)	Bairaman rock slide — debris avalanche	200×10^6	M7.1 earthquake	Village of Bairaman destroyed by debris flow from breached landslide dam; evacuation prevented casualties; major environmental effects	Debris avalanche formed 210 m high dam that impounded 50 million m^3 lake; dam failed causing 100 m high debris flow — flood downstream
1987	Ecuador (Mapo)	Reventador landslides (mainly debris flows)	$75 - 110 \times 10^6$	M6.1 + 6.9 earthquake	~1000 killed; many kms of oil pipeline & highway destroyed; 1987 costs US$1 billion	Landslides mainly in saturated residual soils on steep slopes; thousands of thin debris flows in catchments
1994	Colombia (Cauca)	Paez landslides (mainly debris flows)	Area = 250 km^2	M6.4 earthquake	271 killed; 1700 missing; 32 000 displaced; several villages destroyed	Thousands of thin, residual soil slides on steep slope becoming debris flows
1997/8	San Francisco Bay region	c. 300 landslide events of all types	Largest individual failure: 13 million m^3 Mission Peak Earthflow	High rainfall	Landslide damage estimated at $158 million; only one fatality	Rainfall recorded at more than twice the annual average as a result of a Type 1 El Niño Southern Oscillation
1998	Campania, southern Italy	129 separate initial slides – predominantly debris flows	Area of sliding = 70 km^2	Very high daily and antecedent rainfall	161 lives lost; towns of Quindici, Sarno, Bracigliano and Siano devastated	Rainfall return period calculated as 100 years; minimum flow rates of 10–20 km/hr; failures mainly in colluvium and weathered pyroclastic material from Mt Vesuvius

* For a description of the different types of failure refer to Table 8.2.

2. Are there typical landslide-prone strata or sequences in the area?
3. What is the potential for first-time failures?
4. Where are landslides likely to occur?
5. Can an estimate be made of the probability of landslide events, including the re-activation of any pre-existing failures?
6. What are the likely consequences or impact of any landslide events?
7. If a pre-existing failure exists at the site, what was the cause, can it be stabilised or avoided?

To investigate and understand landslides, it is necessary to classify the landslide type, identify the mechanisms of movement and establish the causes of the failure. Engineering geomorphology has a key role in providing this information. Once this understanding has been acquired it might then be possible to provide comments on the landslide stability, susceptibility, hazard and risk of an area or a particular site. Broader scale susceptibility, hazard and risk assessments provides end-users with a product that has widespread application in development planning and engineering feasibility studies (Cruden and Fell, 1997; Lee and Jones, 2004). Detailed geomorphological surveys of a single landslide can be used in conjunction with conventional ground investigation techniques to provide data that will feed directly in to the engineering design process (Griffiths et al., 1995).

8.2 Description

In order to be able to collate landslide studies from a range of environments and geomorphological situations it is necessary to adopt a standard method of describing the basic types of landslide, their features and dimensions. The recommended approaches, as proposed by the European Commission (Dikau et al., 1996) and UNESCO World Landslide Inventory (1990), are presented in Figures 8.1 and 8.2 and Tables 8.2, 8.3 and 8.4.

Landslide types and mechanisms
The classification of landslide types presented in Figure 8.1 and Table 8.2 was proposed by Dikau et al. (1996) in their report to the European

Commission Environment Programme on 'Landslide Recognition'. Whilst this classification has its detractors (see the discussion in Hutchinson, 2001), it provides a relatively simple and easy to apply system that has gained widespread acceptance in engineering geomorphology. The system identifies only five types of landslide and these are defined by their initial mechanism of movement: 'fall' (Figure 8.3), 'topple' (Figure 8.4), 'slide' (Figure 8.5), 'spread' (Figure 8.6), and 'flow' (Figure 8.7) plus a 'complex' landslide involving two or more of the failure mechanisms. Slides and flows are further subdivided based on the form of their initial failure surface. In addition, the complex category does identify one specific type of failure, the rock or debris avalanche, as this is a widely recognised and particularly devastating form of landslide. As shown in Table 8.2, after the initial movement all landslides deform further as the movement continues, until the failure stops at a maximum runout distance.

The classification proposed by Dikau et al. (1996) is based on that originally suggested by Varnes (1978). However, whilst the Varnes classification used the same main categories of fall, topple, slide, spread and flow, it differentiated between the failures occurring on rock, debris and soil. This is no longer deemed to be a useful additional complication as all the failure types can occur in all materials. Alternative systems of classification, such as that proposed by Hutchinson (1988), are of most value in detailed geotechnical investigations of landslides and landslide research. For general purposes in engineering geomorphology, therefore, the Dikau system is recommended.

Examples of the various landslide types that occurred in the latter part of the 20th Century and their details are presented in Table 8.1. This illustrates that the rapidly moving rock and debris avalanches are the most devastating forms of failure, in terms of deaths and economic losses.

Landslide dimensions
The UNESCO World Landslide Inventory (1993) proposed the standard system for describing the dimensions and features of a landslide which is presented in Figure 8.2 and Tables 8.3

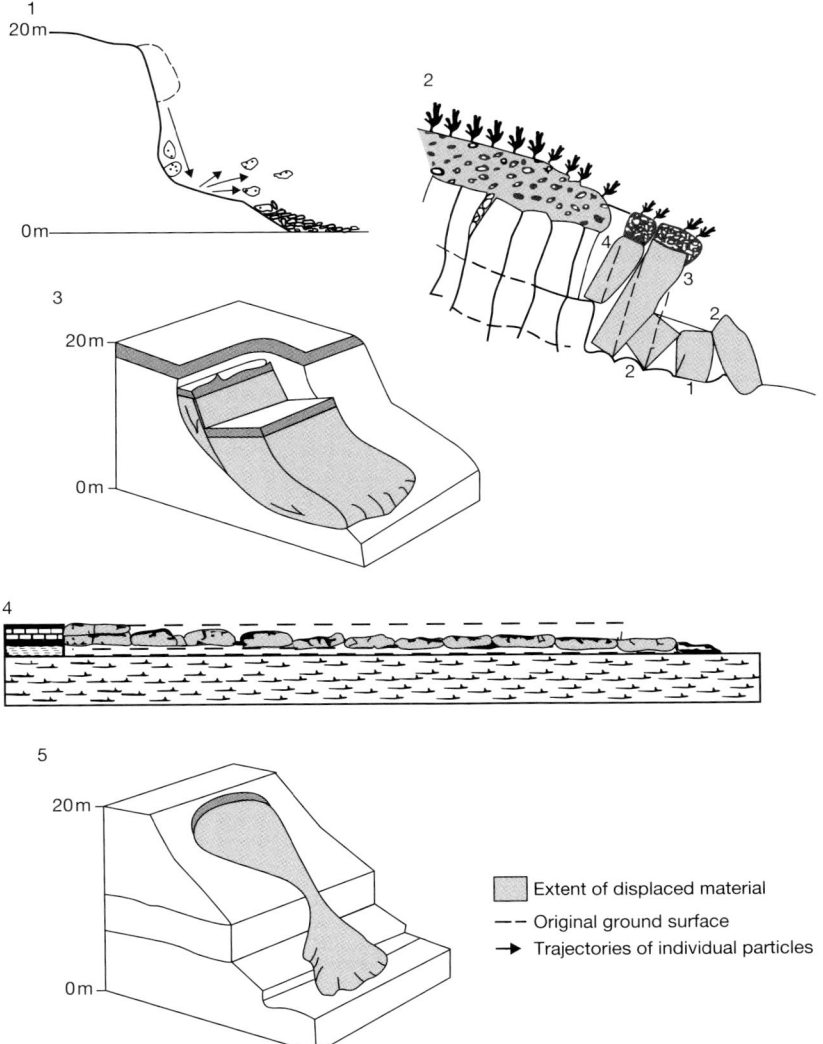

Figure 8.1 Landslide types from Dikau *et al.*, 1996, p. A2.

Types of landslides:
(1) A fall starts with detachment of soil or rock from a steep slope along a surface on which little or no shear displacement takes place. The material then descends largely through the air by falling, saltation or rolling.
(2) A topple is the forward rotation, out of the slope, of a mass of soil or rock about a point or axis below the centre of gravity of the displaced mass.
(3) A slide is the downslope movement of a soil or rock mass occurring dominantly on surfaces of rupture or relatively thin zones of intense shear strain.
(4) A spread is an extension of a cohesive soil or rock mass combined with a general subsidence of the fractured mass of cohesive material into softer underlying material. The rupture surface is not a surface of intense shear. Spreads may result from liquefaction or flow (and extrusion) of the softer material.
(5) A flow is a spatially continous movement in which surfaces of shear are short-lived, closely spaced and usually preserved. The distribution of velocities in the displacing mass resembles that in a viscous fluid.

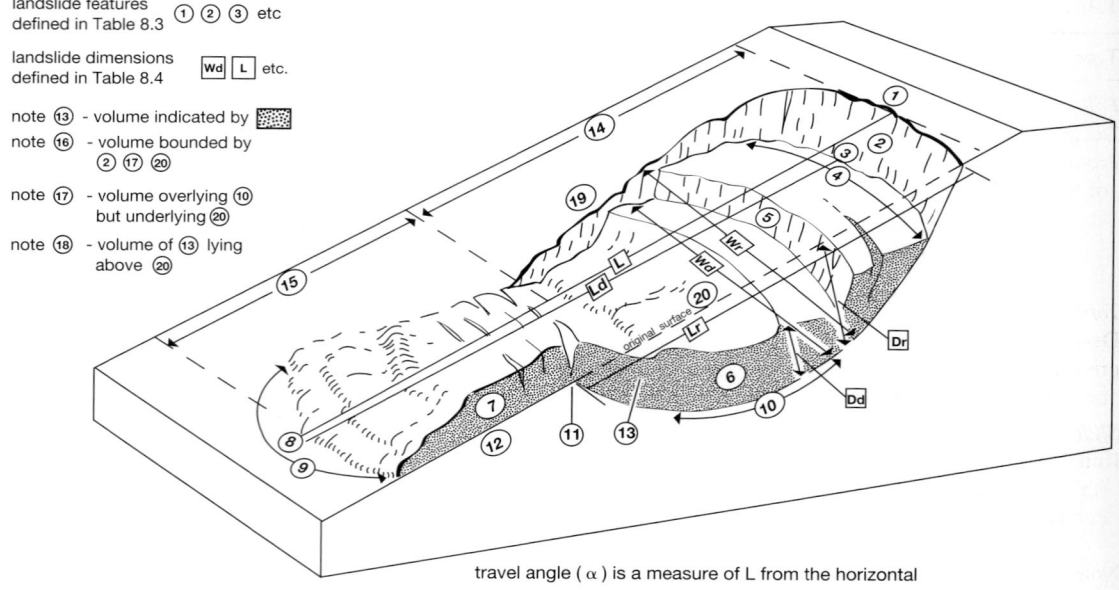

Figure 8.2   Dimensions of a landslide (based on the UNESCO World Landslides Inventory, 1990).

and 8.4. The landslide type shown in Figure 8.2 is a multiple rotational slide but the terminology is appropriate for all types of failure. In Table 8.4 two methods are given for calculating the volume of displaced material in a landslide. The first is more suitable for rotational sides and some flows and is based on a general shape that can be approximated by half an ellipsoid. The second method is suitable for failures controlled by discontinuities such as planar slides, topples and falls.

8.3 Landslide cause

Landslides occur when the strength of the materials in a slope is exceeded by the stresses acting on the slope. The actual causes of the failure are often complex and a slope is likely to have been subject to many 'causes' over a long period of time before a single, possibly quite small, event actually triggers a significant landslide movement (triggering factors in relation to all geomorphological events are discussed in Chapter 1). These changes to the slope may be the result of both natural processes and human interference. In

Table 8.5 a method for classifying the processes that cause a landslide is presented (based on Brunsden, 1993). This identifies nine main 'external processes' that change the situation in or on a slope: weathering, erosion, subsidence, deposition, seismic events, air fall, changes in water regime, complex 'follow-on' processes that occur after the initial failure and human interference. These result in a number of different effects to the slope that either reduce the strength of the materials (e.g. decomposition of the rock by chemical weathering) or increase the stress (e.g. erosion at the toe of a slope changing its geometry).

Human interference as an external process creates the same effects and changes to a slope as a range of natural processes. For example, excavation at the toe of slope is essentially the same as natural fluvial erosion (Griffiths *et al.*, 2004), except that it is likely to take place in quite different locations. Likewise, one effect of flooding an area for a reservoir will be to raise the regional and local water table, possibly causing widespread landsliding around the rim. If the reservoir is subsequently drained, for whatever reason, this rapid drawdown can create transient very high porewater

Table 8.2 Classification of landslides and landslide mechanisms (*cf.* Figure 8.1) (after Dikau *et al.*, 1996).

Type	Form of initial failure surface	Subsequent deformation
Fall		
Detachment form: Pre-existing discontinuities or tension failure surfaces	a) Planar surface b) Wedge (two intersecting joints) c) Stepped surface d) Vertical surface	Free fall, may break up, bounce, slide or flow down slopes. May involve fluidisation, liquefaction, cohesionless grain flow, heat generation or other secondary effects
Topple		
Detachment form: pre-existing discontinuities or tension failure surfaces	a) Single b) Multiple	As above
Slide		
Rotational movement (failure surface essentially circular)	a) Single b) Multiple c) Successive	Toe area may deform in a complex way. The ground can bulge, the slide may creep or even flow, it can override
Non-rotational compound movement (non-circular failure surface; may be listric or bi-planar)	a) Single b) Progressive c) Multistoried	existing failures. Failure might be retrogressive or progressive. Graben often develops at the head of the landslide. It may include a toe failure of a different type
Translational movement (often associated with discontinuity controlled failures in bedded or foliated rocks)	a) Planar b) Stepped c) Wedge d) Non-rotational	May develop complex runout forms after disintegrating (see falls and flows)
Spread		
Lateral spreading of ductile or soft material that deforms	a) Soft layer beneath a hard rock b) Weak interstratified layer c) Collapsing structure	Can develop sudden spreading failures in quick clays when the slope opens up in blocks and fissures followed by liquefaction. Might be a slow movement associated with denudational unloading. Can be represented by cambering and valley bulging
Flow		
Debris movement by flow	a) Unconfined b) Channelled	Flow involves complex runout mechanisms. It may be catastrophic in effect and it may move in sheets or lobes. The form of movement is a function of the rheological properties of the material

Table 8.2   (*Continued*).

Type	Form of initial failure surface	Subsequent deformation
Creep movement	Failure surface rarely clearly defined	Creep may be a superficial gravity movement, seasonal movements or it might represent pre-failure and progressive movements prior to a larger scale failure
Rock flow (sometimes referred to as sagging or Sackung). Usually associated with mountain terrain or areas of rapid and deep incision	a) Single sided b) Double sided c) Stepped (failure surface may be rotational, compound, listric, biplanar or intermittent)	May be slow gravity creep or the early stages of larger scale movements that only show as bulging in the topography without a clearly defined toe deformation. Where controlled by discontinuities it may involve toppling
Complex a) Movements involving two or more of the above mechanisms (referred to as compound when two types of movement occur currently)	Dependent on the form of failure as described above	As described for the various categories above
b) Rock or debris avalanche	Often initiated as fall/slide of rock and/or debris	Complex long-runout mechanisms, including fluidisation and cohesionless grain flow

conditions causing quite extensive landslide activity. Similarly, blasting, or plant and machinery vibrations can replicate natural seismic activity.

The system presented in Table 8.5 provides a useful method for understanding how the range of natural and man-made causal processes can affect a slope. However, during the investigation of a single landslide it will be necessary to identify the various causes and the triggering causal event. In such situations it is more appropriate to have a checklist of the possible causes and a useful example is presented in Table 8.6, from Cruden and Varnes (1996).

that indicate the presence of a landslide in the terrain. The book *Landslide Recognition* by Dikau *et al.* (1996) provides the most comprehensive collection of landslide indicators as well as utilising the landslide type classification system recommended above. A simpler list of features, that can be observed both in the field and while using aerial photographs, is provided in Table 8.7 (after Soeters and Van Westen, 1996). As with all aspects of geomorphology, the ability to recognise landslide features, particularly when they are subtle or degraded, is a skill that develops with experience. This experience can only be obtained through fieldwork.

8.4 Landslide recognition

Armed with the appropriate terminology to describe the features and dimensions of a landslide and an understanding of the possible causes of failure, it is necessary to identify the landscape features

8.5 Landslide activity and behaviour

Whilst the general population tends to envisage a landslide as being a single, often catastrophic, event that occurs rapidly, in geomorphological

Table 8.3   Landslide features in Figure 8.2.

1. *Crown*: the practically *in situ* material adjacent to the highest parts of the main backscar scarp
2. *Main scarp*: a steep surface on the undisturbed ground at the upper edge of the landslide representing the backscar, caused by movement of the slide material (13) away from the undisturbed ground; it is a visible part of the surface of rupture (10)
3. *Top*: the highest point of contact between the displaced material (13) and the main scarp (2)
4. *Head*: the upper parts of the landslide along the contact between the displaced material and the main scarp (2)
5. *Minor scarp*: a steep surface on the displaced material of the landslide, produced by differential movements within the sliding mass
6. *Main body*: the part of the displaced material of the landslide that overlies the surface of rupture (10) between the main scarp (2) and the toe of the surface of rupture (11)
7. *Foot*: the portion of the landslide that has moved beyond the toe of the surface of rupture (11) and overlies the original ground surface (20)
8. *Tip*: the point of the toe (9) furthest from the top of the landslide (3)
9. *Toe*: the lower, usually curved, margin of the displaced material of a landslide, it is the most distant from the main scarp (2)
10. *Surface of rupture*: the surface that forms the lower boundary of displaced material (13) below the original ground surface (20)
11. *Toe of surface of rupture*: the intersection (sometimes buried) between the lower part of the surface of rupture (10) of a landslide and the original ground surface (20)
12. *Surface of separation*: the part of the original ground surface (20) overlain by the foot (7) of the landslide
13. *Displaced material*: material displaced from its original position on the slope by the movement of the landslide
14. *Zone of depletion*: the area of the landslide within which the displaced material lies below the original ground surface (20)
15. *Zone of accumulation*: the area of the landslide within which the displaced material lies above the original ground surface (20)
16. *Depletion*: the volume bounded by the main scarp (2), the depleted mass (17) and the original ground surface (20)
17. *Depleted mass*: part of the displaced material that overlies the rupture surface (10) but underlies the original ground surface (20)
18. *Accumulation*: the volume of the displaced material (13) which overlies the original ground surface (20)
19. *Flank*: undisplaced material adjacent to sides of surface of rupture
20. *Original ground surface*: surface of slope that existed before the landslide took place

Table 8.4   Landslide dimensions shown in Figure 8.2.

Wd	Width of displaced mass: maximum breadth of displaced mass (13) perpendicular to Ld
Wr	Width of rupture surface: maximum width between flanks (19) of the landslide perpendicular to Lr
Ld	Length of displaced mass: minimum distance from tip (8) to top (3)
Lr	Length of surface of rupture: minimum distance from the toe of surface of rupture (11) to crown (1)
Dd	Depth of displaced mass: maximum depth of displaced mass (13) measured perpendicular to a plane containing Wd and Ld
Dr	Depth of surface of rupture: maximum depth of surface of rupture (10) below original ground surface (20) measured perpendicular to a plane containing Wr and Lr
L	Total length: minimum distance from tip of landslide (8) to crown (1)
α	Travel angle: difference in elevation between the crown (1) and the tip (8) of the landslide divided by L

Landslide volume estimation

Approximate volume of the ground displaced by a landslide (Vol$_{ls}$) based on the failure being represented by half an ellipsoid:

$$Vol_{ls} = (1/6).\pi.Dr.Wr.Lr$$

Approximate volume of the ground displaced by a landslide (Vol$_{ls}$) based on the failure being planar in form and lateral limits determined by vertical discontinuities:

$$Vol_{ls} = Lr. Wr . Dr$$

Figure 8.3   Rockfall associated with fluvial undercutting in a river cliff near Sorbas, south-east Spain.

Figure 8.4   Backtilted topple blocks on the outside of a meander bend near Sorbas, south-east Spain.

Figure 8.5   Small translational slide in middle ground, centre of the image, moved from right to left. On the western flanks of Pen-y-Fan in the Brecon Beacons, south Wales.

investigations it is useful to distinguish between first-time failures at sites with no evidence of previous movement, and pre-existing slides that might be prone to re-activation. This is because experience has shown that some landslides can remain dormant for long periods of time and then become active again. Other landslides, whilst active and continually moving, may only creep along at a few millimetres a year, despite being long established features in the contemporary landscape. Landslides tend to have the typical frequency–magnitude distribution associated with natural geomorphological processes where low magnitude events occur more often than high magnitude events. However, it is not easy to apply the concept of recurrence interval to landslides, as, for example, is possible with river flood or rainfall events, because landslides vary both in location and over time. This variation in landslide activity is highly relevant to engineering, and for geomorphological investigations it is useful to adopt a

landslide activity classification system, as suggested in Table 8.8. This system divides landslides into active, suspended, young dormant, mature dormant, old dormant and fossil forms. With decreasing activity the landslides become increasingly degraded and more difficult to identify in the landscape. However, even dormant and some fossil landslides that remain constrained within the bounding shear surface will move when the residual strength along the surface is exceeded. The debris accumulation from these landslides will also be loose and poorly consolidated with a relatively low strength. These landslides, therefore, constitute an increased engineering risk and it is necessary they be identified in any engineering geomorphological investigation.

The other facet of landslide activity that has major engineering implications is the behaviour of the ground during movement, including its velocity. Figure 8.8 illustrates how a landslide typically deforms the ground. In the crown and

Figure 8.6   Tension cracks associated with lateral spreading of a limestone caprock overlying marls east of the village of Los Perales, south-east Spain.

Figure 8.7   Removal of mudflow debris from a track west of Lyme Regis, Dorset (UK).

Table 8.5   Classification of the processes that cause landslides (based on Brunsden, 1993).

External process(es)	Causal effects	Description of typical changes	Examples of specific changes on slope
Weathering: physical, chemical & biological	Changes in physical and chemical properties; horizonation; changes in regolith thickness	Changes in grading; cation exchange; cementation; formation of weak discontinuities or hard bands; increased depth of low strength materials	Changes in: density, strength, permeability; stress, pore and cleft water pressure
Erosion of material from face or base of slope by fluvial, glacial and/or coastal processes	Changes in slope geometry; unloading	Alterations to: relief; slope height, length, angle and aspect	Changes in stress, permeability and strength
Ground subsidence	Undermining	Mechanical eluviation of fines; solution; loss of cement; leaching seepage erosion; backsapping; piping	Loss of support; consolidation; changes in porewater pressure; loss of strength
Deposition of material to face or top of slope by fluvial, glacial or mass movement processes	Loading; long term (drained) or short term (undrained)	Alterations to: relief; slope height, length, angle and aspect	Changes in stress, permeability, strength, loading and porewater pressure
Seismic activity and general shocks and vibrations	Rapid and repeated vertical and horizontal displacements	Disturbance to intergranular bonds; transient high porewater pressures; materials subject to transient and repeated periods of compression and tension	Changes in stress; loss of strength; high porewater pressures; potential for liquefaction
Air fall of loess or tephra	Mantling slopes with fines; adding fines to existing soils	New slope created with well defined discontinuity boundary	Changes in stress; strength; water content and water pressure
Water regime change	Rising or falling groundwater; development of perched water tables; saturation of surface; flooding	Piping, floods, lake bursts; 'wet' years; intense precipitation; snow and ice melt; rapid drawdown	Excess porewater pressures; changes in bulk density; reduction in effective shear strength
Complex 'follow-on' or runout processes after initial failure	Liquefaction; remoulding; fluidisation; 'acoustic grain flow'	Long runout landslides; low values for ratio of initial failure volume to total failure volume; low angles of reach; low breadth to length ratios	Changes in effective shear strength, water distribution, bulk density and rheological characteristics
Human interference	Excavation at toe of slope Top loading of slopes Flooding (e.g. leaking services; reservoir construction)	Same as natural erosion Same as natural deposition Same as natural water regime change	Same as natural erosion Same as natural deposition Same as natural water regime change

Table 8.6   Checklist of landslide causes (after Cruden and Varnes, 1996).

1. *Geological causes*	a) weak materials
	b) sensitive materials
	c) weathered materials
	d) sheared materials
	e) jointed or fissured materials
	f) adversely orientated mass discontinuity (bedding, schistosity etc.)
	g) adversely orientated structural discontinuity (fault, unconformity etc.)
	h) contrast in permeability
	i) contrast in stiffness (stiff, dense material over plastic material)
2. *Morphological causes*	a) tectonic or volcanic uplift
	b) glacial rebound
	c) fluvial erosion of slope toe
	d) wave erosion of slope toe
	e) glacial erosion of slope toe
	f) erosion of lateral margins
	g) subterranean erosion (solution; piping)
	h) deposition loading slope or its crest
	i) vegetation removal (by forest fire, drought)
3. *Physical causes*	a) intense rainfall
	b) rapid snow melt
	c) prolonged exceptional precipitation
	d) rapid drawdown (of floods and tides)
	e) earthquake
	f) volcanic eruption
	g) thawing
	h) freeze-and-thaw weathering
	i) shrink-and-swell weathering
4. *Human causes*	a) excavation of slope or its toe
	b) loading of slope or its crest
	c) drawdown (or reservoirs)
	d) deforestation
	e) irrigation
	f) mining
	g) artificial vibration
	h) water leakage from utilities

backscar areas there will be differential horizontal and vertical movements with tension cracks and scarps. Within the main body of the landslide there will be areas in compression, secondary tension scarps but generally an area of subsidence. Discrete shears or *en échelon* cracks may mark the lateral limits of the slide area. The toe area will be an area of uplift or toe bulge. Material in this area is likely to be under compression at the base of the landslide debris whereas at the surface there is likely to be cracking as the debris comes under tension. Knowing the state of the ground within a landslide is the key to understanding how structures will respond to instability and how to design structures to accommodate landslide movements.

With respect to landslide velocity, Cruden and Varnes (1996) introduced a useful guide to the destructive significance of landslides, reproduced in Table 8.9. This guide follows the style of the Modified Mercalli scale used to describe the local effects of earthquakes. The table illustrates that it is only the Velocity Class 7 landslides that are truly catastrophic and that for all slower moving landslides, loss of life can be either minimised or avoided. It also demonstrates that the economic affects of landslides are far more widespread as it is only in the slowest moving failures in Velocity Classes 1 and 2 that structures can be maintained. The concern with 'creeping' landslides in these slow and extremely slow moving velocity classes, however, is that they may be creeping towards a threshold situation when a larger, more rapid movement is triggered. This emphasises the need to monitor any landslides in the vicinity of both planned and existing developments to ensure that movements are not accelerating.

8.6 Landslide distribution and occurrence

The four main controls on landscape development are climate, geology, relief and time. In broad terms geomorphological studies have shown that certain processes and landforms are often specific to 'morphoclimatic' regions

Table 8.7   Landslide recognition (after Soeters and van Westen, 1996).

Terrain features	Relation to slope instability	Aerial photograph characteristics
Morphology		
Concave/convex slopes	Landslide niche and associated deposits	Concave/convex anomalies in stereo model
Step-like morphology	Retrogressive sliding	Step-like appearance of slope
Semi-circular backscarp and steps	Head of slide with outcrop of failure plane	Light-toned scarp, associated with small, slightly curved lineaments
Back-tilting of slope facets	Rotational movement of slide blocks	Oval or elongate depressions with imperfect drainage
Hummocky and irregular slope	Micro-relief associated with shallow movements or small retrogressive slide blocks	Coarse surface texture contrasting with smooth surroundings
Infilled valleys with slight convex floor where V-shaped valleys are normal	Mass movement deposit of flow-type form	Anomaly in valley morphology often with lobate form and identifiable flow pattern
Vegetation		
Vegetational clearance on steep scarps coinciding with morphological steps	Absence of vegetation on head scarp or on steps in main body of slide	Light-toned elongated areas at crown of slide
Irregular linear clearances along slope	Slip surface of translational slides and track of flows and avalanches	Denuded areas showing light tones often linear pattern in direction of movement
Disrupted, disordered and partly dead vegetation	Slide blocks and differential movements in main body	Irregular sometimes mottled grey tones
Differential vegetation associated with changing drainage conditions	Stagnated drainage on back-tilted blocks, seepage at frontal lobe and varying conditions on main body	Tonal differences displayed in pattern associated with morphological anomalies in stereo model
Stretched and exposed roots	Separation of blocks as a result of downslope movements	Rarely visible, might appear as disrupted vegetation
Drainage		
Areas with stagnant drainage	Landslide niche, back-tilting block and hummocky relief on main body	Tonal differences with darker tones associated with wetter areas (shows up well in the near infra-red)
Excessively drained areas	Overbulging landslide body with differential vegetation and possibly soil erosion	Light-toned zones in association with convex relief forms
Seepage and spring levels	Springs along frontal lobe and at places where failure surface crops out	Dark patches sometimes in slightly curved pattern and enhanced by differential vegetation
Interruption of drainage lines	Drainage anomaly caused by head scarp	Drainage line abruptly broken off on slope by steeper relief
Anomalous drainage pattern	Streams curving around frontal lobe or streams on both sides of main body of landslide	Curved drainage pattern upstream with sedimentation or meandering in asymmetric valley

Table 8.8   Activity classification for landslides (after Mather et al., 2003, adapted from Cruden and Varnes 1996; Keaton and DeGraft 1996).

Activity state	Identification of causes of movement	Condition of main scarp	Condition of lateral margins	Internal morphology	Estimated age (years)
Active — currently moving (includes inactive landslides that have been reactivated)	Causes of movement identifiable and active	Sharp; unvegetated	Sharp; unvegetated; streams at edge	Hummocky; fresh scarps; reverse slopes; undrained depressions; fresh tension cracks	<100 (historic)
Suspended — moved within the last 12 months and likely to become active again	Causes of movement identifiable and likely to re-occur	Sharp; unvegetated	Sharp; unvegetated; streams at edge	Hummocky; fresh scarps; reverse slopes; undrained depressions; identifiable tension cracks	<100 (historic)
Dormant young (inactive)	Cause of movement still identifiable and could re-occur	Relatively sharp; partially vegetated	Relatively sharp; partially vegetated; lateral streams fed by small tributaries flowing off the main body of the slide	Hummocky; relatively sharp and fresh scarps; reverse slopes; undrained depressions; tension cracks closed and vegetated but marked by small depressions	100–5000 (Late Holocene)
Dormant mature (inactive) — described as abandoned by Dikau et al. (1996)	Cause of movement still identifiable but not likely to re-occur	Smooth; vegetated	Smooth vegetated; lateral streams fed by tributaries flowing off the main body of the slide	Smooth, rolling topography; disrupted and disjointed internal drainage network.	5000–10 000 (Early Holocene)
Dormant old (inactive) or relict*	Causes of movement may be inferred but associated with different climatic or geomorphological conditions	Dissected; vegetated	Vague lateral margins; no lateral drainage	Smooth, undulating topography; normal stream pattern	10 000–100 000 (Late Pleistocene)
Fossil (inactive) or ancient*	Causes of movement unknown but associated with different climatic and geomorphological conditions	May not be identifiable; likely to be at least partially if not completely removed by erosion	May not be identifiable; likely to be at least partially if not completely removed by erosion	Fully integrated into the existing topography and very little indication of the former landslide morphology remains	>100 000

* Relict or fossil landslides may also be covered by subsequent deposits. Rib and Liang (1978) described these as 'hidden' or 'buried' landslides, and they always have the potential to be exhumed and reactivated. This is the situation described by Schultz and Harper (1996) for the Late-Carboniferous palaeo-landslides reactivated by building excavations in Pennsylvania, USA.

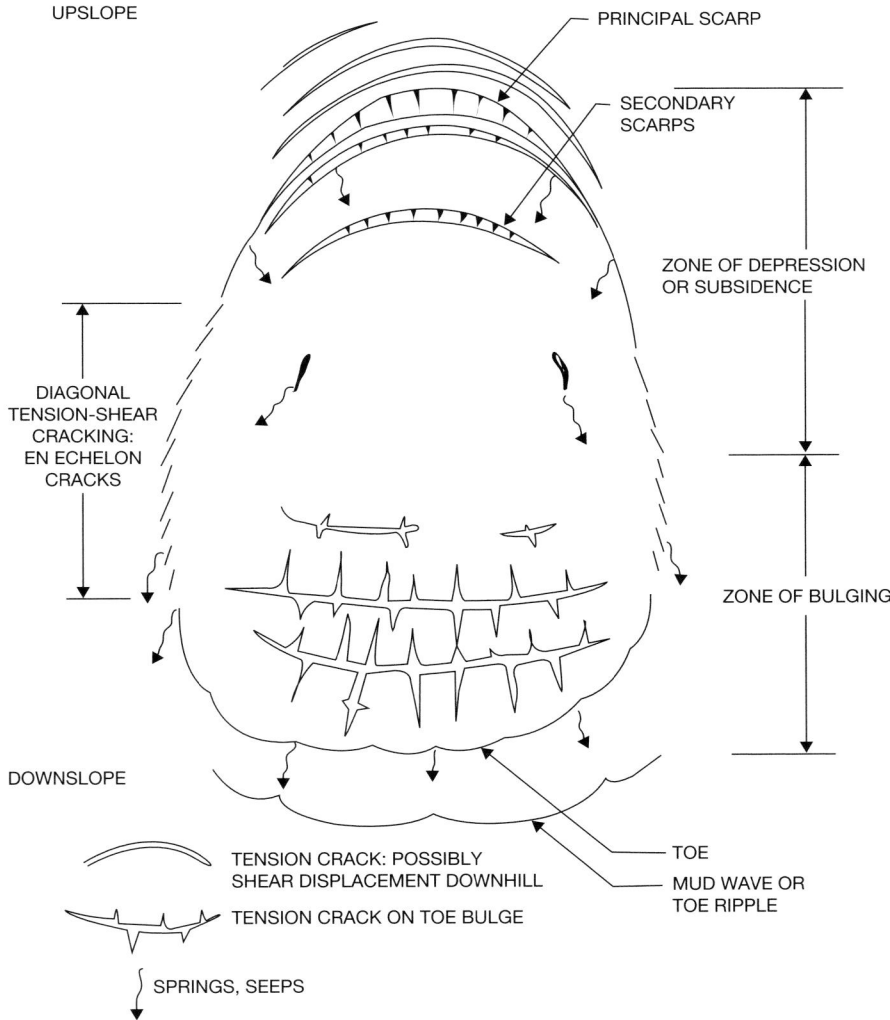

Figure 8.8   Landslide behaviour (after Sowers and Royster, 1978).

(Chapter 1). Within the morphoclimatic region the local geology (structure, lithology, rock and soil mass characteristics) and relief or topography controls the shape and nature of individual landforms. The occurrence and distribution of landslides, which are just one type of landform, fit this general model of landscape development and whilst individual landslides can be a site-specific problem anywhere in the world, in some situations landslides represent a major component of the landscape.

Landslide-prone terrain

The concept of morphoclimatic regions controlled by climate can be used to identify the broad distribution of different landslide types, and these are presented in Table 8.10. Fookes (1997) took this broad scale terrain system concept further by producing a sequence of climate controlled landscape models highlighting facets of geomorphological processes that affect engineering design. In many of these models landsliding represents a significant process that must be taken into account.

Table 8.9   Destructive significance of landslides of different Velocity Classes (after Cruden and Varnes, 1996).

Velocity class	Description	Velocity (mm/s)	Typical velocity	Probable destructive significance
7	Extremely rapid			Catastrophe of major violence; buildings destroyed by impact of displaced material; many deaths; escape unlikely
		5×10^3	5 m/s	
6	Very rapid			Some lives lost; velocity too great to allow all persons to escape
		5×10^1	3 m/minute	
5	Rapid			Escape evacuation possible; structures, possessions and equipment destroyed
		5×10^{-1}	1.8 m/hour	
4	Moderate			Some temporary and insensitive structures can be temporarily maintained
		5×10^{-3}	13 m/month	
3	Slow			Remedial construction can be undertaken during movement; insensitive structures can be maintained with frequent maintenance work if total movement is not too large during a particular acceleration phase
		5×10^{-5}	1.6 m/year	
2	Very slow			Some permanent structures undamaged by movement
		5×10^{-7}	16 mm/year	
1	Extremely slow			Imperceptible without instruments; construction possible with precautions

A number of morphoclimatic regions can be identified where landsliding is a very significant process in landscape development. High mountainous areas (Figure 8.9) subject to seismic activity (Figure 8.10) and where rainfall is high stand out as being particularly landslide prone. An examination of the major landslide events listed in Table 8.1 supports this, the South American Andean Mountains, for example, have been subject to a number of major landslide catastrophes during the latter part of the 20th Century. In contrast, other mountainous areas where slopes are gentle, rainfall is only moderate and which are no longer seismically active are unlikely to be subject to extensive contemporary landsliding, although there may be relict features within the landscape. For high mountainous terrain in Nepal, Fookes (1997) produced the landscape model presented in Figure 19.3 which illustrates the range of landslide processes that can be anticipated in these regions.

Hong Kong provides an example of the tropical wet–dry morphoclimatic region where landslide occurrence is also significant. The average annual rainfall in Hong Kong is 2225 mm with a marked seasonal pattern and periodic typhoons. Hong Kong has been subject to extensive landslide research since a major failure occurred at Po Shan Road in 1972 killing 67 people (Li *et al.*, 1998). The landscape model developed for Hong Kong arose out of a territory-wide terrain evaluation study carried out during the 1980s in order to provide the basis for planning new building and infrastructure construction. The terrain model, presented in Figure 8.11, was based on the creation of a detailed study of land forming processes illustrated in Figure 8.12 (Hansen, 1984). Subsequent detailed studies have shown that the overall

Table 8.10   Landslide occurrence in the global morphoclimatic regions (after Chorley *et al.*, 1984); see Table 2.1.

Morphoclimatic region[1] (typical name)	General geomorphic processes	Morphological features	Typical landslide types
Humid tropical (rainforest)[2]	Maximum chemical weathering; episodic mass wasting; moderate slope wash and fluvial processes; high solute and suspended loads in rivers	Low gradient rivers; wide flat floodplains; steep slope arising abruptly from valleys, stabilised by vegetation; knife-edged ridges	Earthflows and rotational slides in weathered debris
Tropical wet–dry (savannah)[3]	High chemical and moderate mechanical weathering; moderate to high mass wasting; seasonal sheetfloods; moderate wind action	Steep irregular slopes of coarse debris; wide planation surfaces; isolated inselbergs; some badlands	Flows and slides in weathered debris; rock falls and topples from exposed rock faces
Arid (desert)[2]	Maximum thermal and salt weathering; rare episodic fluvial action (often relict); maximum wind action	Dunes, playas, pediments, debris covered slopes, ephemeral stream channels (some fossil), alluvial fans	Minor rockfalls; debris flows on alluvial fans
Semi-arid (Mediterranean)[3]	Moderate chemical and mechanical weathering; episodic moderate mass wasting; maximum fluvial erosion in ephemeral rivers	Pediments backed by cliffs and talus slopes; inselbergs; integrated ephemeral stream networks; badlands; alluvial fans; local dunes	Wide range: rockfalls and larger scale translational slides; debris flows on fans; rotational slides in weathered debris
Dry continental (steppe)[3]	Moderate rates for all forms of weathering; moderate mass wasting but seasonally important; moderate to high fluvial activity; moderate wind action and fossil loess	Very similar to semi-arid and tropical wet–dry but with increased local effects of frost action, e.g. angular talus slopes	Similar to semi-arid landscapes
Humid mid-latitude (temperate marine)[3]	Full range of weathering processes at moderate rates; moderate fluvial processes; evidence of fossil glacial activity	Smooth soil covered slopes with rounded ridges and valleys; wide range of grain sizes in alluvium	High rates of soil creep; wide range of landslides but often relict or only slowly moving
Periglacial (tundra)[3]	Maximum frost shattering and mechanical weathering; maximum talus creep and gelifluction; moderate fluvial action in thaw season; relict glacial scour; moderate to high wind action	Permafrost and seasonally frozen ground; screes; solifluction sheets cryoplanation surfaces; outwash plains; patterned ground; loess	Rockfalls; talus creep; solifluction
Glacial (arctic)[2]	Maximum frost shattering and mechanical weathering; maximum glacial scour; maximum wind action	Alpine topography; abrasion surfaces; glacial and fluvio-glacial features; glaciers and ice caps	Rockfalls; talus creep

Table 8.10   (*Continued*).

Morphoclimatic region[1] (typical name)	General geomorphic processes	Morphological features	Typical landslide types
Mountains[4]	Maximum mechanical and moderate chemical weathering extensive mass high stream gradients and high sediment concentrations; often recent or active glaciation	High mountains; deeply incised valley with over-steepened slopes; braided rivers with terraces	Maximum landslide activity of all types

[1] A morphoclimatic region is a large areal unit within which distinctive associations of geomorphological processes operate and tend towards a morphoclimatic equilibrium wherein regional landforms reflect regional climates.
[2] First order morphoclimatic region with non-seasonal processes, generally low average erosion rates, highly infrequent and episodic erosive activity (including mass slope failure), and a tendency for the location of their cores to persist in a certain latitude during periods of climate change.
[3] Second order morphoclimatic region have processes that are more seasonal in operation, often with high rates of erosion which may be episodic although is often consistent over long periods. Location and extent of region varies considerably as climate changes. Landforms often left as relicts in other morphoclimatic regions when climate changes.
[4] Azonal unit that crosses morphoclimatic regions.

Figure 8.9   Talus cone associated with large-scale rock slide, Rocky Mountains, British Columbia.

frequency of landsliding within the $1000 \, km^2$ area of Hong Kong is 325 per year (Evans, 1998).

In temperate regions such as the UK, whilst the overall density of landsliding is relatively low (Jones and Lee, 1994), there are a range of old dormant periglacial mass movement features in the landscape that were last active during the Late-Glacial period (*c.* 10 000 BP). These features continue to pose a threat to engineering design unless correctly identified and fully investigated. An example of a terrain model highlighting landslide activity in this type of landscape is provided by Croot and Griffiths (2001) for the south-west of England (Figure 8.13).

Another type of 'terrain' found throughout the world that is particularly landslide prone, although it has no clearly defined relationship with the climate, is an eroding coastline (Figures 8.14 and 8.15). Landslides are a major component in the process of coastal cliff retreat. Not all the landslides that occur on the coast are caused solely by sea erosion at the base of cliffs, many are also the result of weathering or changes to the water regime in the slope. Lee and Clark (2002) presents a classification of coastal cliffs that treats them as open sediment systems characterised by inputs, throughputs and outputs of material. The

Figure 8.10 The Carrasco translational landslide near Los Molinos, south-east Spain.

cliff system is coupled with the foreshore system (supply and removal of debris; undercutting etc) and the cliff top (supply of water; inland spread of tension cracks etc). Lee and Clark (2002) recognises a range of types of cliff system on the basis of the throughput and storage of sediment within the system (Figure 8.16):

1. Simple cliff systems: comprising a single sequence of sediment inputs, from falls or slides, and outputs with limited storage. A distinction is made between cliffs prone to falls and topples and those shaped by simple landslides. The former is characterised by limited storage of sediment within the cliff system, with material from the cliff top and face reaching the foreshore in a single event. By contrast simple landslide systems comprise a single sequence of inputs and outputs with variable amounts of storage within the failed mass. Debris from the cliff may only reach the foreshore after a sequence of events involving landslide reactivation.

2. Composite systems: comprising a partly coupled sequence of contrasting simple sub-systems. The output from one system may not necessarily form input for the next, for example where material from the upper unit falls directly onto the foreshore.

3. Complex systems: comprising strongly linked sequences of sub-systems, each with their own inputs and outputs of sediment. The output from one sub-system forms the input for the next. Such systems are often characterised by a high level of adjustment between process and form, with complex feedback mechanisms.

4. Relict systems: comprising sequences of pre-existing landslide units which are being gradually reactivated and exhumed by the progressive retreat of the current sea cliff.

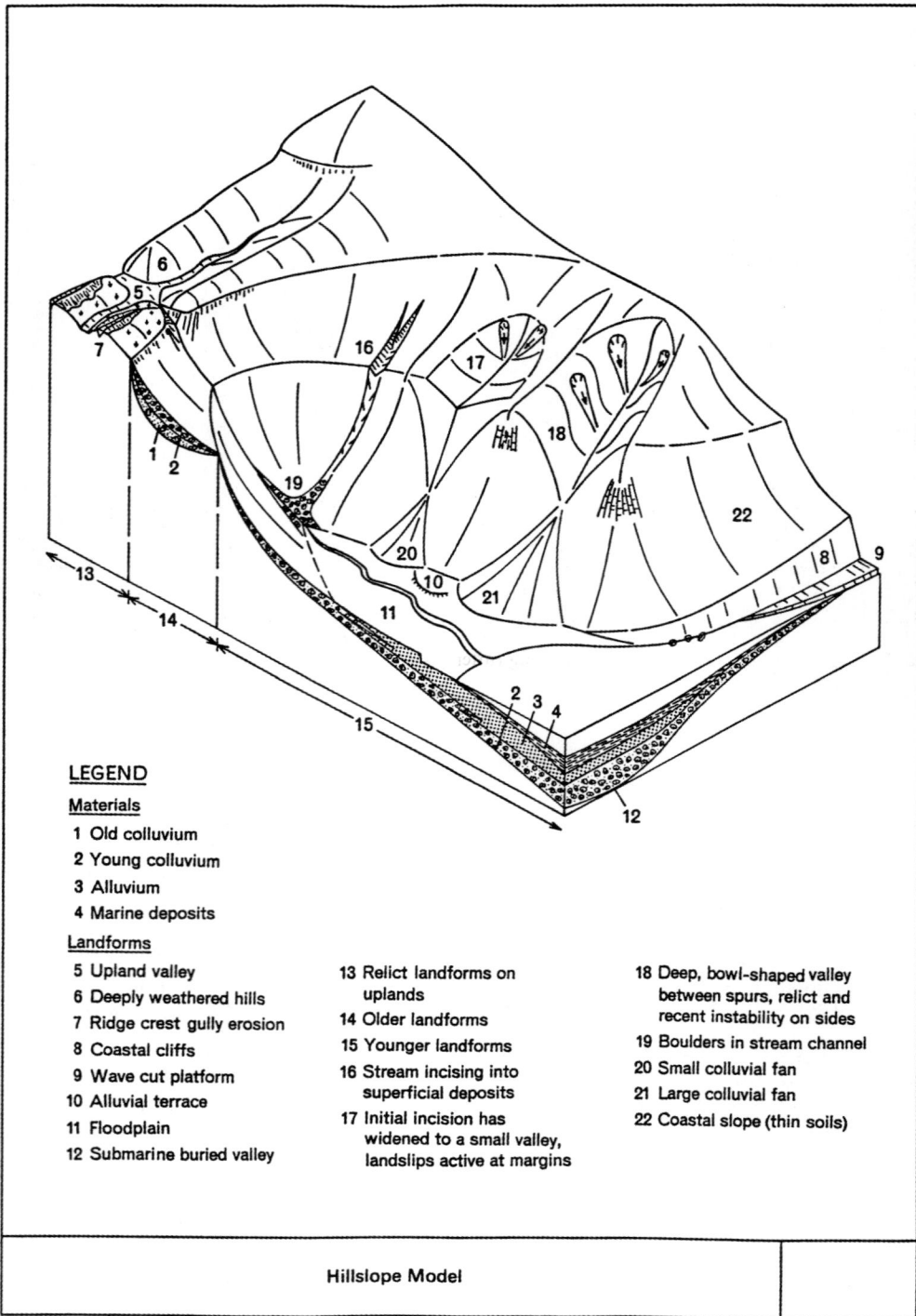

LEGEND

Materials

1 Old colluvium
2 Young colluvium
3 Alluvium
4 Marine deposits

Landforms

5 Upland valley
6 Deeply weathered hills
7 Ridge crest gully erosion
8 Coastal cliffs
9 Wave cut platform
10 Alluvial terrace
11 Floodplain
12 Submarine buried valley

13 Relict landforms on
   uplands
14 Older landforms
15 Younger landforms
16 Stream incising into
   superficial deposits
17 Initial incision has
   widened to a small valley,
   landslips active at margins

18 Deep, bowl-shaped valley
   between spurs, relict and
   recent instability on sides
19 Boulders in stream channel
20 Small colluvial fan
21 Large colluvial fan
22 Coastal slope (thin soils)

Hillslope Model

Figure 8.11   Landform model for Hong Kong (after Hansen, 1984).

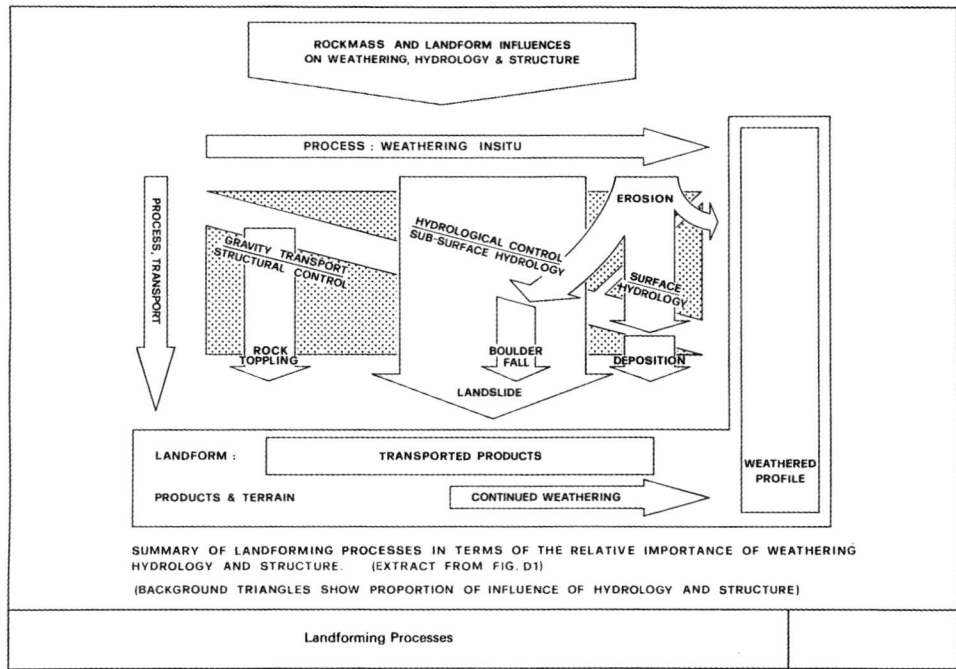

Figure 8.12   Land forming processes in Hong Kong (after Styles and Hansen, 1989).

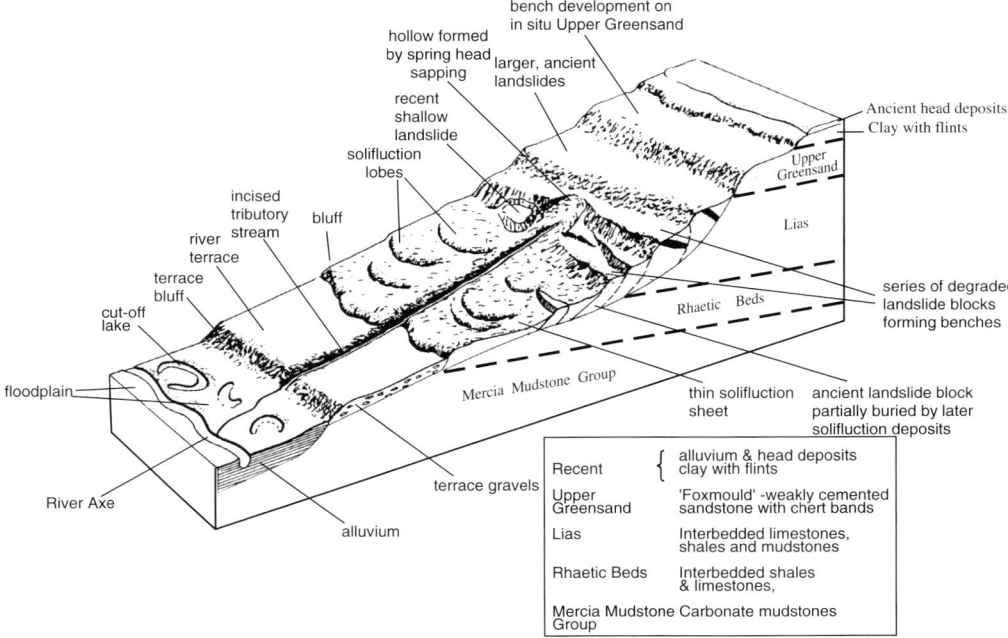

Figure 8.13   Schematic representation of the landscape in the Axminster area, south-west England (after Croot and Griffiths, 2001) reprinted with the permission of the Geological Society of London.

Figure 8.14   Mudslides and rotational failure in the West Dorset coast (UK).

Figure 8.15   Major fall of chalk, Alum Bay, Isle of Wight (UK).

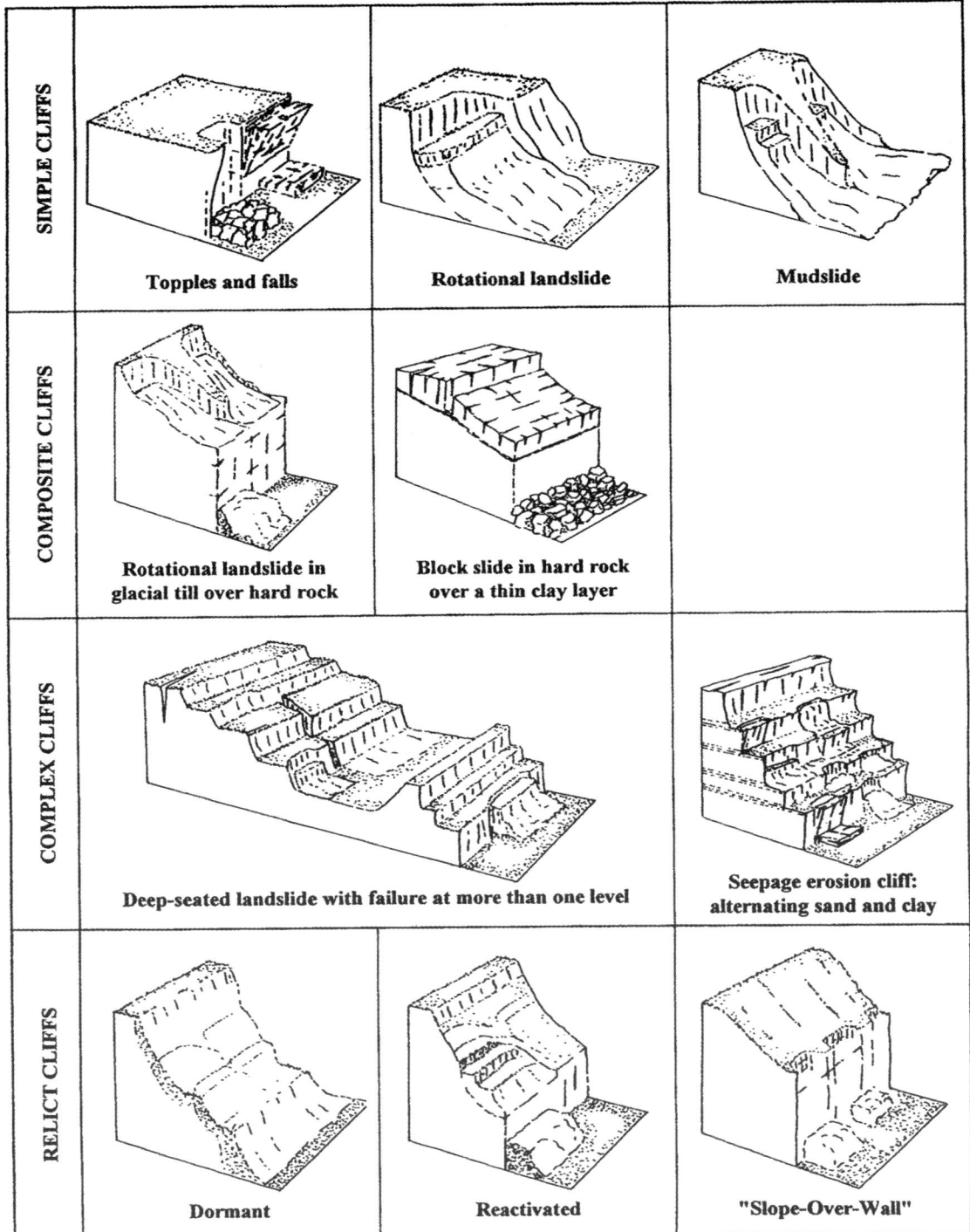

Figure 8.16   Classification of coastal cliffs (after Lee and Clark, 2002).

For all types of morphoclimatic region, or where non-climatically controlled landscapes are identified such as coasts and mountains, it is important for the engineering geomorphologist to develop landscape models (as exemplified by Figures 8.11, 8.13 and 8.16). These allow the engineering geomorphologist to identify potentially hazardous situations in the field or when using remote sensing data. For landslide investigations the model establishes the typical combination of landscape elements that are likely to result in 'mass movement'. Thus when undertaking any form of geomorphological survey where a terrain model is available, the geomorphologist is provided with a graphical 'checklist' of landscape features. Such models are also an effective way of illustrating the scale of the natural hazards in a region, including landslides, to less experienced staff and clients. This whole subject of developing geological and geomorphological models for engineering is discussed in detail in Fookes *et al.* (2000).

Landslide situations

In any general study of the distribution of landslides for engineering purposes within a given morphoclimatic region it is necessary to identify typical landslide situations that are a function of the materials, the groundwater conditions, the particular geological structure or lithology, or the geomorphological circumstances. Whilst the morphoclimatic landscape models provide general indicators of landslide incidence, the occurrence of individual landslides in specific locations or situations is controlled by the local conditions.

When a single specific landslide is examined the geotechnical properties of the materials and the hydrogeological conditions become the critical factors. For example the shape of the rockfall in Figure 8.3 is controlled by the discontinuity patterns in the sandstones that form the river cliff, although the primary cause was the result of undercutting by the river increasing the stress on the slope and exceeding the shear strength of the materials. In an engineering geomorphological study, the identification of these landslide 'situations' will allow both the creation of a distribution map of existing landslides and provide an

indication of landslide susceptibility, hazard and risk. For example, the work in Hong Kong (Evans, 1998) established that the highest density of landsliding was associated with the weathered zone of particular rock types, notably tuffs and lavas, and slope angles between 35° and 40°. Where possible a schematic representation of the landslide situation can be a very useful tool to illustrate the nature of the landslide problem to clients and non-specialists. An excellent example of a landslide summary diagram for the Undercliff landslide on the Isle of Wight, by Geomorphological Services Ltd (Lee and Moore, 1991), is presented in Figure 8.17. This falls into the 'complex system' category within the Lee and Clark (2002) classification of coastal cliffs (Figure 8.16). Similarly, where landslides might result from the injudicious use of the landscape by humans, a synoptic model can be created that highlights some of the potential landslide sites. Figure 8.18 illustrates how human activity affected slopes in South Wales, an area of extensive mining.

8.7 Landslides: susceptibility, hazard and risk

Landslide studies in engineering geomorphology can be very site specific but the track record in the subject has primarily been in the investigation of large areas, often for highways (Brunsden *et al.*, 1975; TRL, 1997). Whilst detailed surveys using large-scale geomorphological mapping techniques typically at scales of 1 : 5000 or larger are most suited to site-specific investigations, when large areas have to be covered at smaller scales for engineering feasibility studies or planning purposes other forms of analysis have to be adopted. Mapping at scales smaller than 1 : 25 000 is ideal for the development of general landscape models (see above) and falls generally within the broad category of 'land surface evaluation' (Griffiths, 2001). However, over the past decade in landslide studies increasing use is being made of the techniques associated with hazard and risk assessment both at a site-specific and regional level. The definitions of the terms

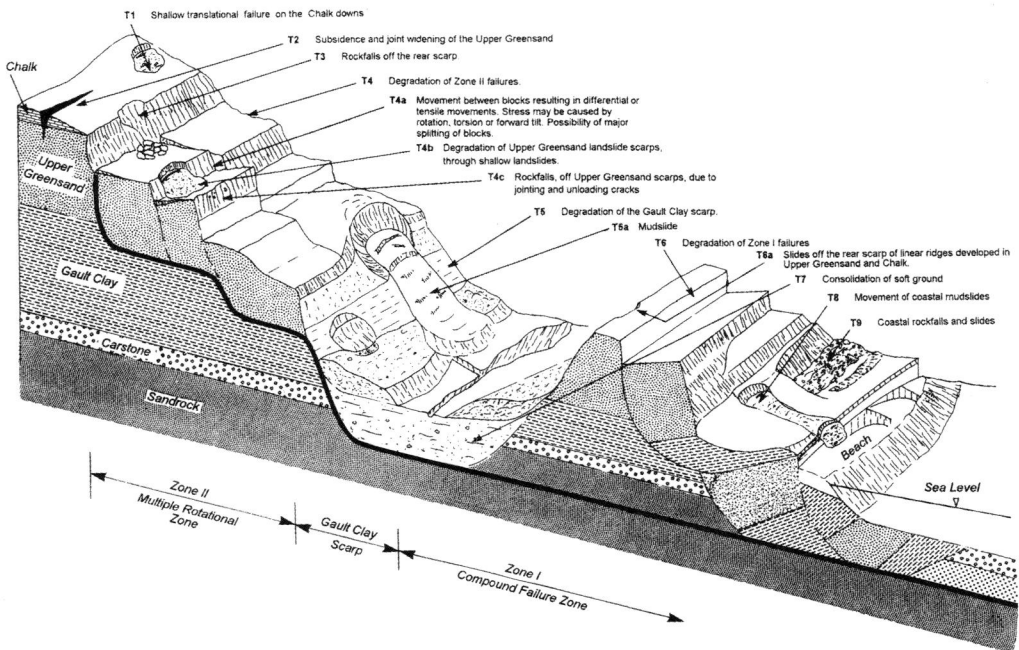

Figure 8.17 Types of contemporary movement in the Undercliff (after Lee and Moore 1991; Moore *et al.*, 1995).

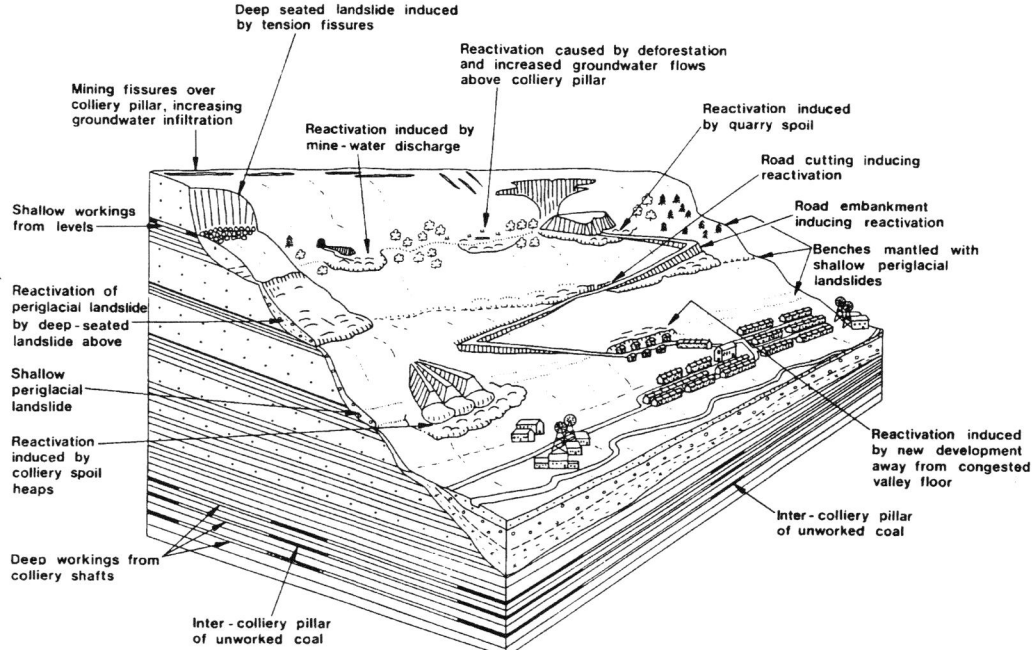

Figure 8.18 Effects of human activity on the stability of South Wales slopes (from Halcrow, 1989, Rhonda Landslip Potential Assessment Summary Report, for Department of the Environmental and Welsh Office).

Table 8.11   Definitions in risk assessment and management adapted for landslide studies (based on the Royal Society, 1992).

Landslide susceptibility:	slopes, materials and geomorpho-logical situations where landslides are likely to occur.
Landslide hazard:	the probability that a landslide event of given magnitude will occur within a given time period.
Landslide risk:	a combination of the probability, or frequency, of occurrence of a defined landslide hazard and the magnitude of the consequences of the occurrence.
Environmental risk:	a measure of the potential threats to the environment which combines the probability that events will cause or lead to degradation of the environment, and the severity of that degradation.
Risk criteria:	a qualitative and quantitative statement of the acceptable standard of risk with which the assessed risk needs to be compared.
Landslide risk assessment:	the integrated analysis of the landslide risks inherent in a region and their significance.
Risk quantification:	the estimation of a given risk by a statistical and/or analytical modelling process.
Risk evaluation:	the appraisal of the significance of a given quantitative (or, when appropriate, qualitative) measure of risk.
Societal risk:	the relation between frequency of occurrence and the number of people in a given population suffering from a specified level of harm from the specified hazards, including landslides.
Risk management:	the process whereby decisions are made to accept a known or assessed risk and/or the implementation of actions to reduce the consequences or probability of occurrence.

used in these types of investigation are presented in Table 8.11.

Landslide susceptibility studies specifically attempt to identify the geological and geomorpho-logical situations where landslides would most likely be encountered (see above). These analyses can be based on the known distribution of landslides and the factors controlling them, or the distribution of the typical controls on landsliding (e.g. soft clays underlying permeable hard cap-rocks). The outcome from these studies will normally be landslide inventory maps and databases. Such analyses are clearly ideal for showing where landslides are likely to occur but they do not address the question of 'when'.

To establish when a landslide might occur (i.e. to define the landslide hazard) it is necessary to know the landslide frequency. To establish frequency requires a record of the historical occurrence of landslides in a defined area. One of the few places in the world where this has been possible is Hong Kong where a landslide database has been developed that contains the records of

nearly 27 000 landslides. Given this quantity and quality of data a genuine attempt to evaluate the hazard is possible (Evans, 1998). Many previous landslide 'hazard' studies only really established susceptibility or used qualitative statements about the level of 'hazard'.

An alternative to hazard analyses having to complete a historical record of landsliding in an area is to establish the relationship between landslide activity and causal factors for which the frequency is known. Typically these might mean investigating the relationship between rainfall (intensity, antecedent conditions, totals etc.) or seismic activity and landsliding. Normally these studies would indicate that landslide conditions prevail once certain thresholds are reached but they would not necessarily be able to identify where specific failures could be expected. One non-quantitative example of this approach was presented by TRL (1997). As shown in Figure 8.19, a combination of maps on slope angle, physiography aspect and land use distribution was compiled in order to produce an ordinal rating of

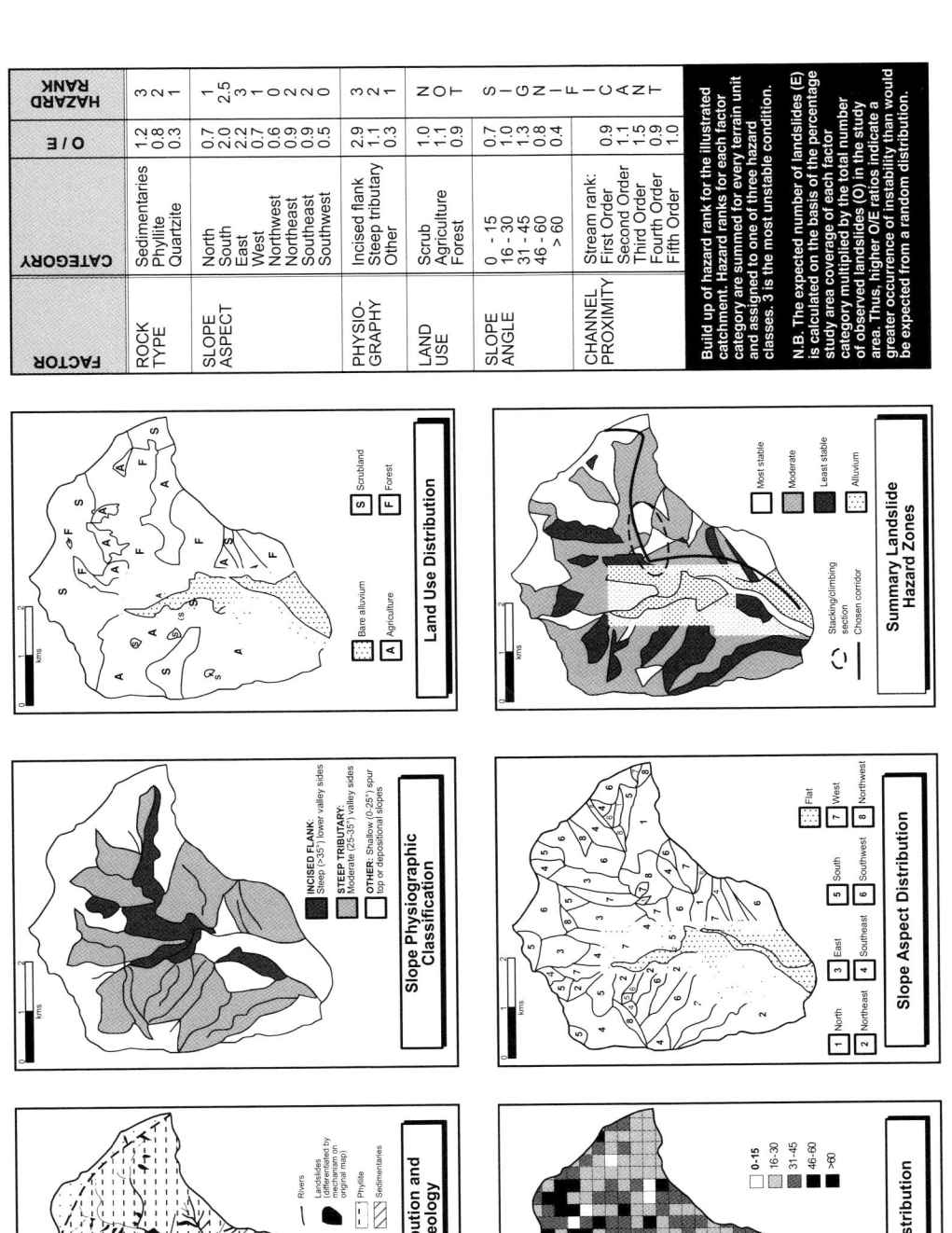

FACTOR	CATEGORY	O/E	HAZARD RANK
ROCK TYPE	Sedimentaries	1.2	3
	Phyllite	0.8	2
	Quartzite	0.3	1
SLOPE ASPECT	North	0.7	1
	South	2.0	2.5
	East	2.2	3
	West	0.7	1
	Northwest	0.6	0
	Northeast	0.9	2
	Southeast	0.9	2
	Southwest	0.5	0
PHYSIOGRAPHY	Incised flank	2.9	3
	Steep tributary	1.1	2
	Other	0.3	1
LAND USE	Scrub	1.0	N
	Agriculture	1.1	O
	Forest	0.9	T
SLOPE ANGLE	0 – 15	0.7	S
	16 – 30	1.0	I
	31 – 45	1.3	G
	46 – 60	0.8	N
	> 60	0.4	I
CHANNEL PROXIMITY	Stream rank: First Order	0.9	F
	Second Order	1.1	I
	Third Order	1.5	C
	Fourth Order	0.9	A
	Fifth Order	1.0	N T

(HAZARD RANK for LAND USE, SLOPE ANGLE and CHANNEL PROXIMITY = "NOT SIGNIFICANT")

Build up of hazard rank for the illustrated catchment. Hazard ranks for each factor category are summed for every terrain unit and assigned to one of three hazard classes. 3 is the most unstable condition.

N.B. The expected number of landslides (E) is calculated on the basis of the percentage study area coverage of each factor category multiplied by the total number of observed landslides (O) in the study area. Thus, higher O/E ratios indicate a greater occurrence of instability than would be expected from a random distribution.

Figure 8.19 Landslide hazard mapping for route alignment through an unstable river basin in east Nepal (with acknowledgement to the Transportation Research Laboratory (TRL), 1997).

landslide hazard for an area in east Nepal. Another example of this type of analysis is provided by the study of the Undercliff landslide at Ventnor on the Isle of Wight, Figures 8.20 and 8.21 (Lee *et al.*, 1991a; 1991b). In Figure 8.20 a geomorphological map of the Undercliff is presented that illustrates the complexity of the landslide system. This was used as the basis for an investigation of the relationship between landslide movements and rainfall to establish the 'sensitivity' of the system to rainfall input of different recurrence intervals (Figure 8.21). This work identified the components of the landslide system that were likely to be active even during moderate rainfall conditions, and this information was used to help plan both new developments and remedial works in the Undercliff.

The Undercliff example raises another issue in landslide hazard and risk studies, which is the potential influence of climatic change on landslide susceptibility and, hence, hazard. As shown in the Undercliff, landslide frequency is closely related to antecedent rainfall, and the trigger for a major failure is often an extreme climatic event, perhaps after a long period of gradually increasing susceptibility due to loss of strength through weathering. However, current projections of climate change may significantly alter the landslide hazard around the globe. For example rising sea levels and increases in winter storms and rainfall may accelerate erosion rates of soft rock cliffs along the south coast of England. Thus even where research has managed to establish the levels of landslide hazard, future projections will need to take into account the potential changes in climatic conditions.

Another component in landslide hazard studies is calculating how far the landslide is likely to travel should it occur. Various models of landslide runout have been developed (e.g. Hungr, 1995) but this is a subject that requires further research. At present a useful guide to the length of runout of landslides in a region is to investigate the travel angle (Figure 8.2). This can provide an indication of the maximum travel distance of landslides of different types and volumes that can be expected (see Corominas, 1986). TRL (1997) provides an illustration of an attempt to establish the runout

hazard for a town in Papua New Guinea based on geomorphological mapping, slope classification, a hazard classification and previous runout routes (Figure 8.22).

The ultimate stage in more regional landslide studies is to establish the landslide risk. This is the relationship between the landslide hazard and the human environment. Within the human environment it is necessary to establish what is at risk (people, infrastructure, economic activity), how vulnerable the components are, and what the consequence would be of any landslide activity (Lee and Jones, 2004). Cruden and Fell (1997) provide a 'state of the art' review in this complex and rapidly developing field. To date it is rare that engineering geomorphological studies have been able to complete a full landslide risk assessment except for very specific sites (e.g Lyme Regis, Dorset (UK), Lee *et al.*, 2000; Hong Kong building development, Hardingham *et al.*, 1998). To assist in the process of risk assessment the problem can be broken down into separate components and the probability of occurrence of different events happening can be calculated. One technique suggested by Wu *et al.* (1996) to achieve this is the use of an 'event tree', and an example from the Lyme Regis site in Dorset by Lee *et al.* (2000) is presented in Figure 8.23.

As part of the risk assessment it is necessary to include an evaluation of the consequences of a landslide occurring, either in terms of building damage or potential for loss of life. Indeed, it is possible to develop consequence models that use a rational framework to take into account the key factors such as runout distance, location and type of structures affected, and the spatial and temporal distribution of the population at risk from a landslide event. This approach was used by Bunce *et al.* (1997) to back-calculate the risk of the fatalities that occurred on a highway in British Columbia, Canada. To date the technique appears to have mainly been used in hindsight, however the approach clearly has merit. In Table 8.12 an 'Infrastructure Damage Classification' system is presented (after Geomorphological Services Ltd., 1991) that could be used in consequence models. The classification can be used both for surveying existing building and road damage or for estimating

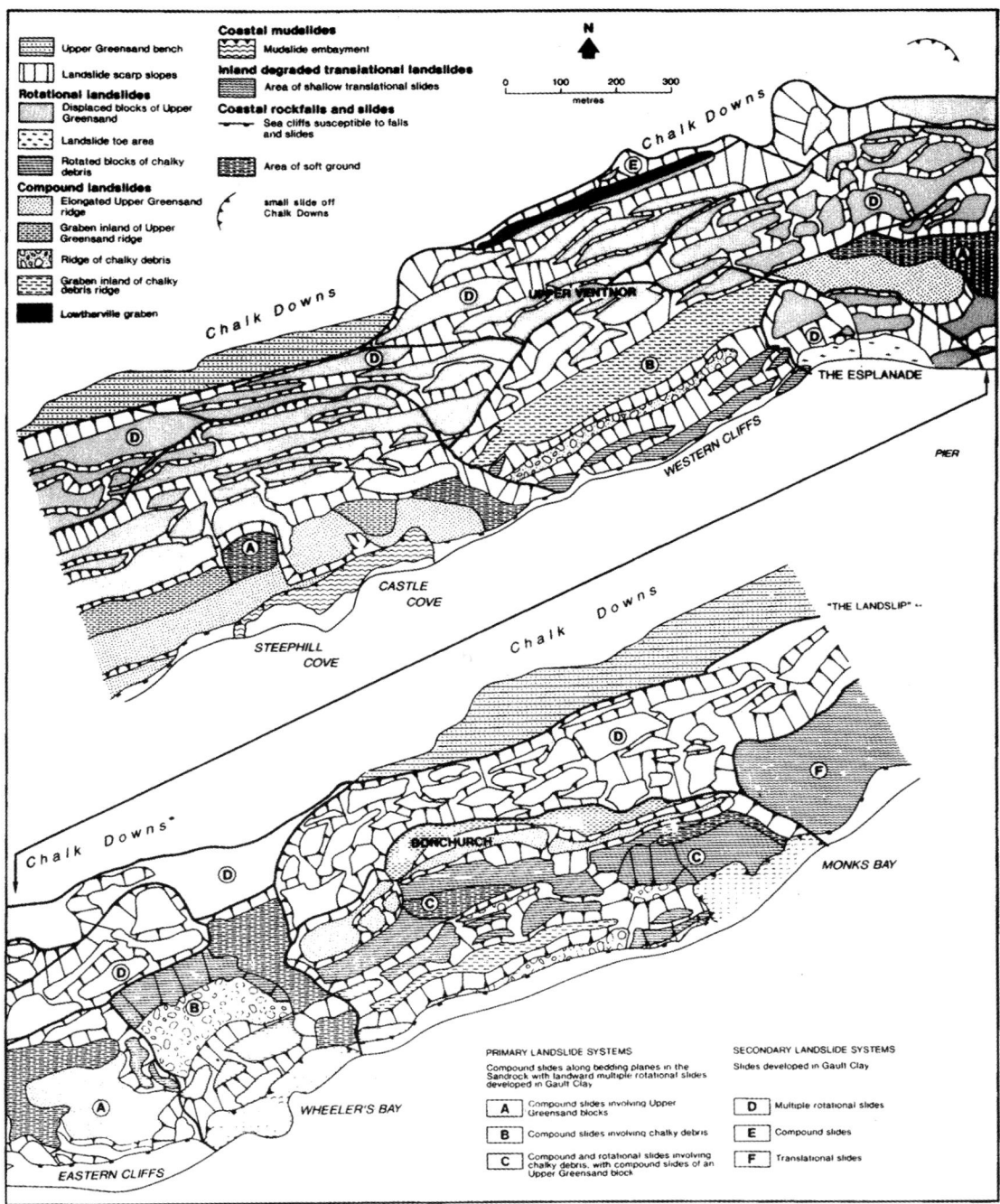

Figure 8.20   Geomorphological maps of the Ventnor Undercliff, Isle of Wight (UK), after Lee and Moore (1991).

SUSCEPTIBILITY TO GROUND MOVEMENT
VERY HIGH
HIGH
MODERATE
LOW
VERY LOW
1000
800
600
400
200
Winter Rainfall (Sept-Jan) Total (mm)
The Landslip
Luccombe
Mirables
Blackgang
Monks Bay
Wheelers Bay
Steephill Cove
Havensbush play area
Luccombe Coombe
Bonchurch
Ventnor
St. Lawrence
Old Park area
Upper Bonchurch
Upper Ventnor
Ventnor Bay
The Orchard, Niton
Undercliff Drive west of St. Lawrence
Binnel Bay
Castle Cove
1.1
1.5
2
3
5
10
25
50
100
200
Winter Rainfall Return Period (years)
(See Box K for explanation)

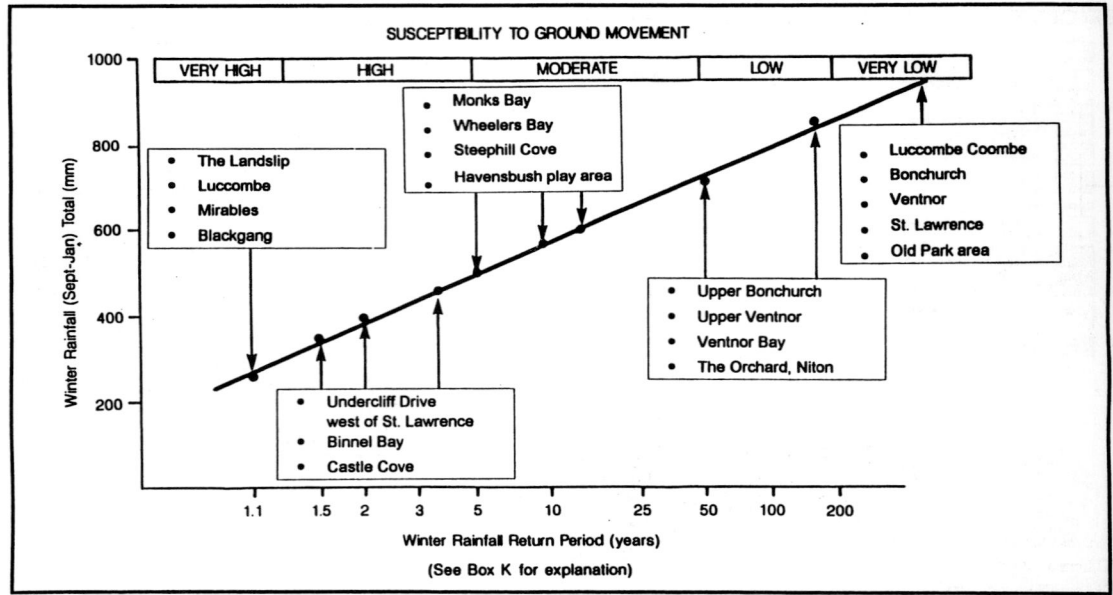

Figure 8.21 Return period rainfall and susceptibility of landslide systems, Ventnor, Isle of Wight (UK) (Moore *et al.*, 1995; Lee *et al.*, 1998).

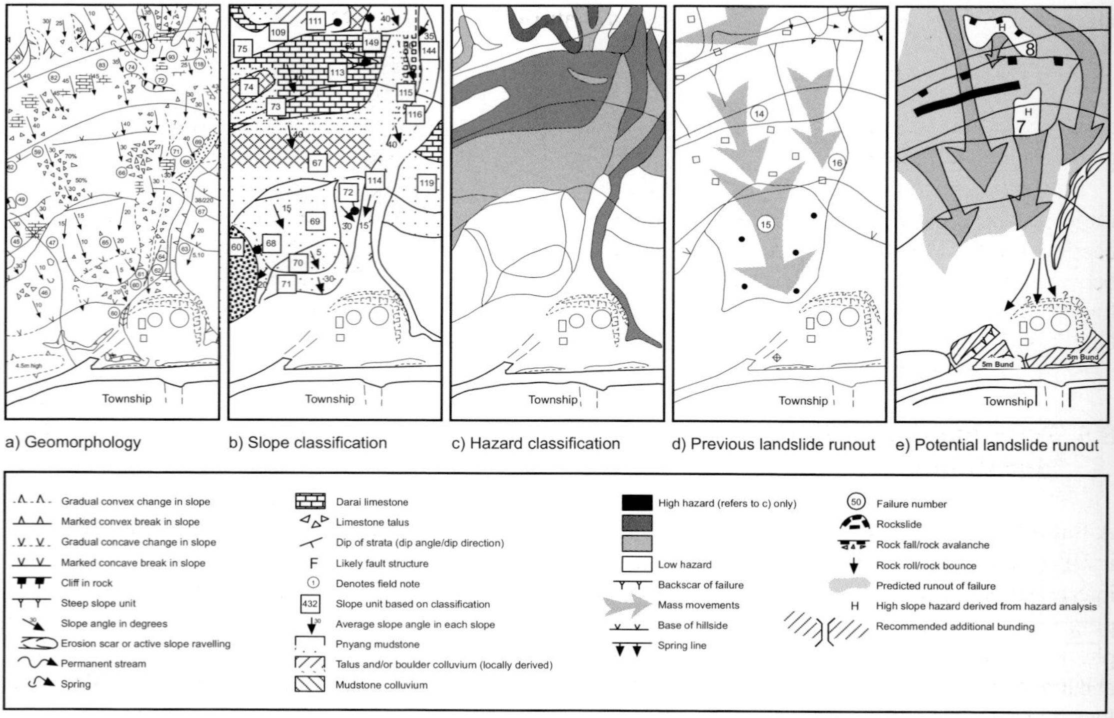

Township
a) Geomorphology
b) Slope classification
c) Hazard classification
d) Previous landslide runout
e) Potential landslide runout
5m Bund
Gradual convex change in slope
Marked convex break in slope
Gradual concave change in slope
Marked concave break in slope
Cliff in rock
Steep slope unit
Slope angle in degrees
Erosion scar or active slope ravelling
Permanent stream
Spring
Darai limestone
Limestone talus
Dip of strata (dip angle/dip direction)
F Likely fault structure
Denotes field note
432 Slope unit based on classification
Average slope angle in each slope
Pnyang mudstone
Talus and/or boulder colluvium (locally derived)
Mudstone colluvium
High hazard (refers to c) only)
Low hazard
Backscar of failure
Mass movements
Base of hillside
Spring line
50 Failure number
Rockslide
Rock fall/rock avalanche
Rock roll/rock bounce
Predicted runout of failure
H High slope hazard derived from hazard analysis
Recommended additional bunding

Figure 8.22 Landslide hazard and runout mapping in Papua New Guinea (with acknowledgement to the Transportation Research Laboratory (TRL), 1997).

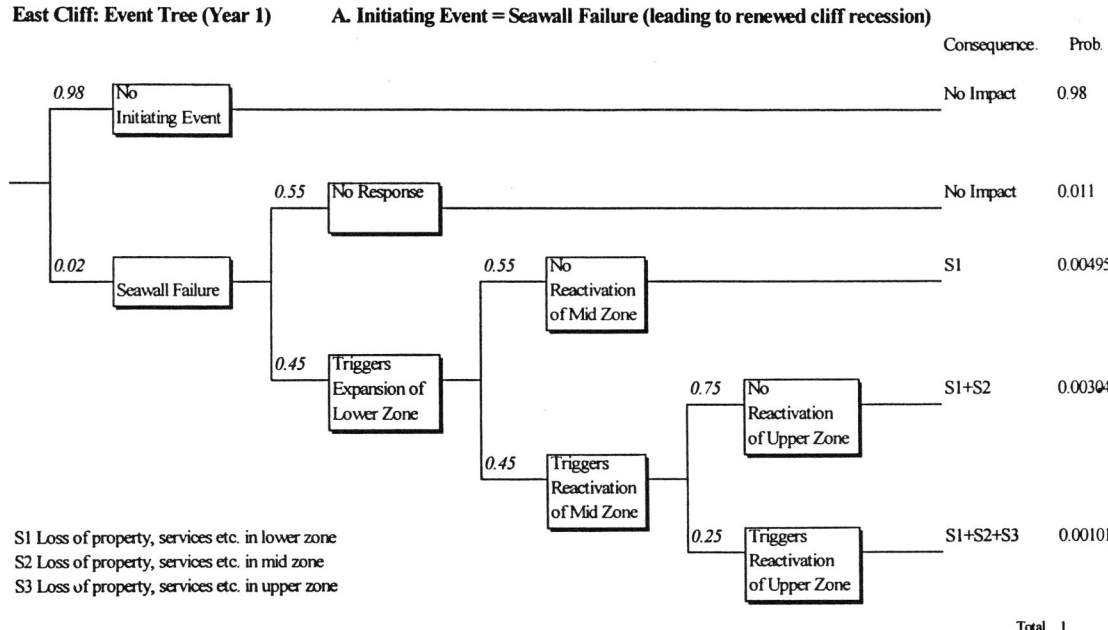

East Cliff: Event Tree (Year 1)   **A. Initiating Event = Seawall Failure (leading to renewed cliff recession)**

	Consequence.	Prob.
No Impact	0.98
No Impact	0.011
S1	0.00495
S1+S2	0.00304
S1+S2+S3	0.00101
Total	1

S1 Loss of property, services etc. in lower zone
S2 Loss of property, services etc. in mid zone
S3 Loss of property, services etc. in upper zone

An example event tree: East Cliff, Lyme Regis

Figure 8.23   Event tree for Lyme Regis, Dorset (UK) (after Lee *et al.*, 2000).

the likely damage resulting for a landslide event of given magnitude and frequency.

It should be recognised that as databases improve and geographical information systems become more sophisticated, risk assessment will become part of a routine engineering geomorphological appraisal for landslide areas (Hearn and Griffiths, 2001).

8.8 Planning and undertaking a landslide investigation

A full-scale site investigation of a landslide would involve desk studies, including aerial photograph interpretation, field reconnaissance, detailed mapping and a ground investigation involving geophysics, drilling, pitting, instrumentation, sampling and a materials' testing programme. This would follow the long established procedures for site investigation presently covered in the UK by BS 5930: 1999 (BSI, 1999). The full scope of a preliminary

site investigation of the type typically required in landslide studies is provided in Figure 8.24 (Fookes, 1997). Whilst engineering geomorphologists should and do become involved with the ground investigation stage, in this Chapter the emphasis is on the earlier stages of a landslide investigation where the geomorphologist is primarily using a combination of desk studies, remote sensing interpretation and field mapping. Table 8.13 (after Sowers and Royster, 1978) provides a checklist of all the data the geomorphologists should be collating in their investigation. Initial interpretation of remote sensing images and photographs, by the identification of terrain features listed in Table 8.7, provides an invaluable database before going out into the field. This is the case whether it is the investigation of a single landslide or a 200 km highway through mountainous terrain. However, no matter how good the remote sensing interpretation, the main part of any engineering geomorphological investigation must be in the field. The techniques for data collection will be dependent on the scale of

Table 8.12 Infrastructure Damage Classification (after Lee and Moore, 1991).

Damage rating	Description
Negligible	Hairline cracks to roads; pavements and structures with no appreciable vertical displacement or separation
Slight	Occasional cracks; distortion, separation or vertical displacement apparent; small fragments of debris may occasionally fall onto roads and structures causing light damage; repairs are not urgent
Moderate	Widespread cracks; settlement may cause slight tilt to walls and fractures to structural members and service pipes
Serious	Extensive cracking; settlement may cause open cracks and considerable distortion to structures; walls out of plumb and the road surface may be affected by subsidence; parts of roads and structures may be covered with landslide debris from above; repairs urgent to safeguard the future use of roads and structures
Severe	Extensive cracking; settlement may cause rotation or slewing of road; gross distortion to roads and structures; repairs will require partial or complete rebuilding and may not be feasible; severe movements likely to lead to the abandonment of the site or area

the task and the requirements of the client. Normally some form of map output would be expected, and the accuracy of these will be totally dependent on the quality of the base maps. If a large area is to be covered, serious consideration should be given to producing a synoptic terrain model (Fookes *et al.*, 2000). The mapping techniques used should draw upon the best practices of geological (Barnes, 2004) engineering geological (Dearman, 1991; Griffiths, 2002) and geomorphological mapping (Gardner and Dackombe, 1983; Lee, 2001). Where data are available and the client requires such output, it might be possible to carry out landslide susceptibility, hazard and risk surveys. At all

times the emphasis must not just be on giving geomorphological detail but providing the information the client requires for engineering or planning purposes.

8.9 Landslide stabilisation

Remediation: design options

Where a landslide could affect any proposed new development it will be necessary to undertake some form of remediation, in the form of stabilisation or preventative measures. This generally involves some or all of the following (Holtz and Schuster, 1996):

1. avoiding the problem
2. reducing the driving forces
3. increasing resisting forces by either applying an external force or increasing the internal shear strength.

These nature and extent of the various measures covered by these three broad categories are summarised in Table 8.14. The selection of the appropriate engineering measures is crucial to effective slope stabilisation. The scope of works may vary in their applicability according to the size and mechanism of failure, the soil and groundwater conditions and financial constraints. The manner in which remedial measures are implemented is also important to the success of the works as, for example, incorrect placement of fill and drainage can lead to a reduction in the factor of safety. It must also be emphasised that the effective stabilisation of most landslides can generally only be achieved following a detailed ground investigation and stability analysis. This will have considerable resource implications and, if part of a larger development programme, it may delay the project. Whilst these aspects of design are quite rightly mainly the concern of the geotechnical engineers, engineering geomorphologists must have an understanding of the scope and nature of the design process if they are going to provide an effective input.

The manner in which geomorphology can be included in this design process can be illustrated with two UK case studies.

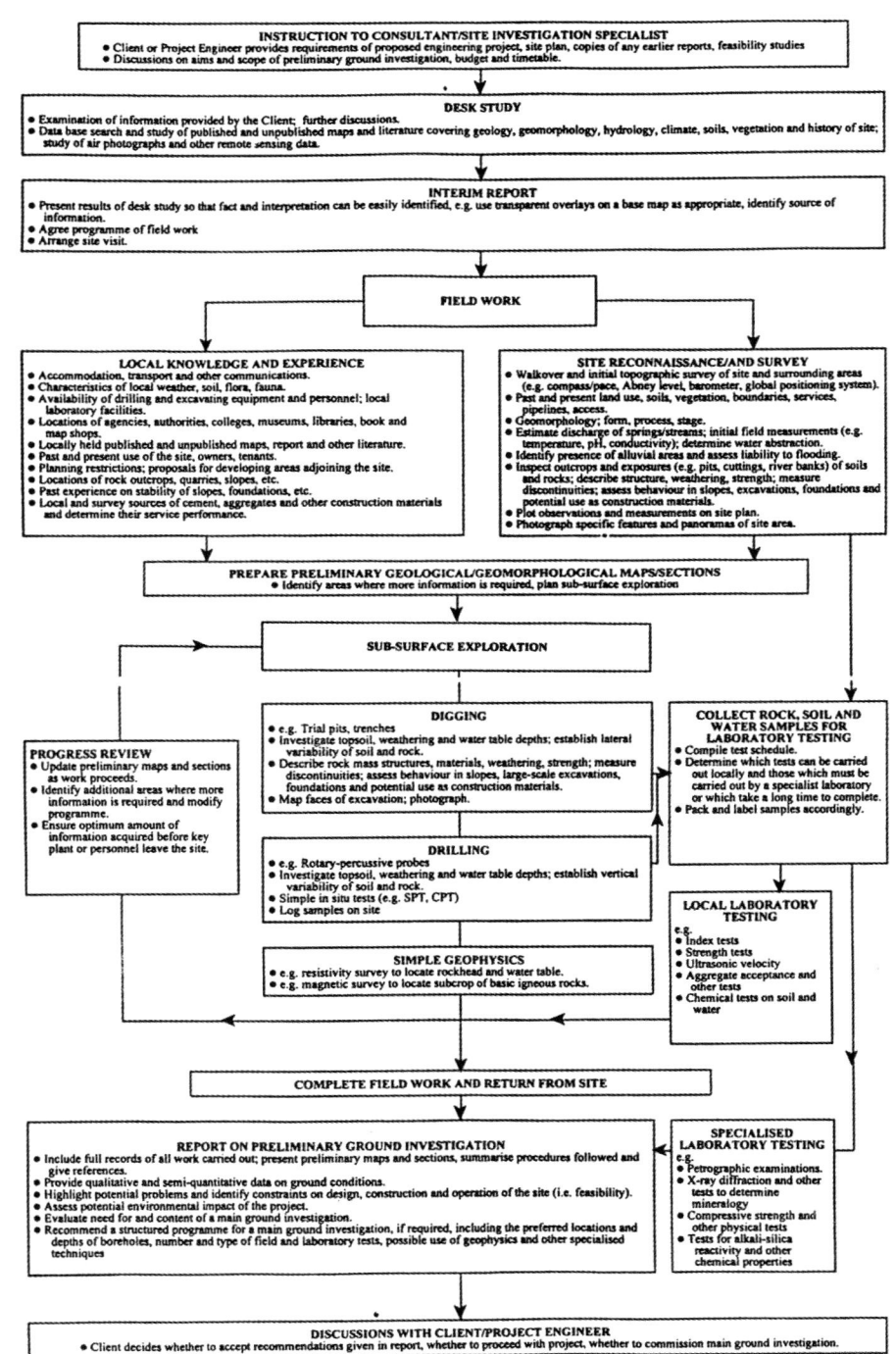

INSTRUCTION TO CONSULTANT/SITE INVESTIGATION SPECIALIST
- Client or Project Engineer provides requirements of proposed engineering project, site plan, copies of any earlier reports, feasibility studies
- Discussions on aims and scope of preliminary ground investigation, budget and timetable.

DESK STUDY
- Examination of information provided by the Client; further discussions.
- Data base search and study of published and unpublished maps and literature covering geology, geomorphology, hydrology, climate, soils, vegetation and history of site; study of air photographs and other remote sensing data.

INTERIM REPORT
- Present results of desk study so that fact and interpretation can be easily identified, e.g. use transparent overlays on a base map as appropriate, identify source of information.
- Agree programme of field work
- Arrange site visit.

FIELD WORK

LOCAL KNOWLEDGE AND EXPERIENCE
- Accommodation, transport and other communications.
- Characteristics of local weather, soil, flora, fauna.
- Availability of drilling and excavating equipment and personnel; local laboratory facilities.
- Locations of agencies, authorities, colleges, museums, libraries, book and map shops.
- Locally held published and unpublished maps, report and other literature.
- Past and present use of the site, owners, tenants.
- Planning restrictions; proposals for developing areas adjoining the site.
- Locations of rock outcrops, quarries, slopes, etc.
- Past experience on stability of slopes, foundations, etc.
- Local and survey sources of cement, aggregates and other construction materials and determine their service performance.

SITE RECONNAISSANCE/AND SURVEY
- Walkover and initial topographic survey of site and surrounding areas (e.g. compass/pace, Abney level, barometer, global positioning system).
- Past and present land use, soils, vegetation, boundaries, services, pipelines, access.
- Geomorphology; form, process, stage.
- Estimate discharge of springs/streams; initial field measurements (e.g. temperature, pH, conductivity); determine water abstraction.
- Identify presence of alluvial areas and assess liability to flooding.
- Inspect outcrops and exposures (e.g. pits, cuttings, river banks) of soils and rocks; describe structure, weathering, strength; measure discontinuities; assess behaviour in slopes, excavations, foundations and potential use as construction materials.
- Plot observations and measurements on site plan.
- Photograph specific features and panoramas of site area.

PREPARE PRELIMINARY GEOLOGICAL/GEOMORPHOLOGICAL MAPS/SECTIONS
- Identify areas where more information is required, plan sub-surface exploration

SUB-SURFACE EXPLORATION

DIGGING
- e.g. Trial pits, trenches
- Investigate topsoil, weathering and water table depths; establish lateral variability of soil and rock.
- Describe rock mass structures, materials, weathering, strength; measure discontinuities; assess behaviour in slopes, large-scale excavations, foundations and potential use as construction materials.
- Map faces of excavation; photograph.

PROGRESS REVIEW
- Update preliminary maps and sections as work proceeds.
- Identify additional areas where more information is required and modify programme.
- Ensure optimum amount of information acquired before key plant or personnel leave the site.

COLLECT ROCK, SOIL AND WATER SAMPLES FOR LABORATORY TESTING
- Compile test schedule.
- Determine which tests can be carried out locally and those which must be carried out by a specialist laboratory or which take a long time to complete.
- Pack and label samples accordingly.

DRILLING
- e.g. Rotary-percussive probes
- Investigate topsoil, weathering and water table depths; establish vertical variability of soil and rock.
- Simple in situ tests (e.g. SPT, CPT)
- Log samples on site

LOCAL LABORATORY TESTING
e.g.
- Index tests
- Strength tests
- Ultrasonic velocity
- Aggregate acceptance and other tests
- Chemical tests on soil and water

SIMPLE GEOPHYSICS
- e.g. resistivity survey to locate rockhead and water table.
- e.g. magnetic survey to locate subcrop of basic igneous rocks.

COMPLETE FIELD WORK AND RETURN FROM SITE

REPORT ON PRELIMINARY GROUND INVESTIGATION
- Include full records of all work carried out; present preliminary maps and sections, summarise procedures followed and give references.
- Provide qualitative and semi-quantitative data on ground conditions.
- Highlight potential problems and identify constraints on design, construction and operation of the site (i.e. feasibility).
- Assess potential environmental impact of the project.
- Evaluate need for and content of a main ground investigation.
- Recommend a structured programme for a main ground investigation, if required, including the preferred locations and depths of boreholes, number and type of field and laboratory tests, possible use of geophysics and other specialised techniques

SPECIALISED LABORATORY TESTING
e.g.
- Petrographic examinations.
- X-ray diffraction and other tests to determine mineralogy
- Compressive strength and other physical tests
- Tests for alkali-silica reactivity and other chemical properties

DISCUSSIONS WITH CLIENT/PROJECT ENGINEER
- Client decides whether to accept recommendations given in report, whether to proceed with project, whether to commission main ground investigation.

Organization of a preliminary site investigation for a large project. Not all projects and a preliminary ground investigation.

Figure 8.24   Organisation of a typical preliminary site investigation (after Fookes, 1997).

Table 8.13   Checklist for planning a landslide investigation (after Sowers and Royster, 1978).

1. *Topography*	a) contour map (land forms; morphology; anomalous patterns)
	b) surface drainage
	c) slope profiles (correlation with geology; correlation with contour maps)
	d) topographic changes (rate of change over time; correlation with weather, groundwater, seismic activity)
2. *Geology*	a) geological formations at the site (stratigraphy; superficial deposits; landslide prone sequences; minerals subject to alteration)
	b) structure, i.e. three dimensional geometry (stratification; folding; dip and strike of all discontinuities including bedding, foliation, joints; faults; shear zones)
	c) weathering (nature and depth)
	d) development of the general geological model for the area
3. *Groundwater and surface water*	a) piezometric levels (normal; perched; artesian)
	b) variations in pieometric level (correlate with weather; seismic activity; slope changes)
	c) groundwater chemistry (dissolved salts and gases; changes in radioactive gases)
	d) nature and extent of surface water (streams; springs; seepage; ponds; lakes; tidal reaches)
	e) surface indications of subsurface water (seepage zones; areas of poor drainage; changes in vegetation)
	f) effects of human activity (groundwater utilisation, impoundment, restriction and recharge; surface water abstraction and recharge; changes to infiltration rate and capacity, e.g. by covering in concrete or tarmac)
4. *History of slope changes*	a) natural processes (long term geological change; erosion; past movements; submergence and emergence)
	b) human activity (earthworks; changes in surface and groundwater; changes in surface cover, including paving and deforestation; flooding; rapid reservoir drawdown)
	c) rate of movement (visual accounts; topographic evidence; remote sensing evidence; instrumentation results; review of historical maps and reports)
	d) correlation of movement with other factors (surface and groundwater; seismic activity; weather; human activity, including earthworks and vibrations)
	e) development of the site-specific landslide ground model incorporating geology, geomorphology, anthropology and the range of external processes (Tables 8.5 and 8.6).
5. *Weather*	a) precipitation (type; continuous through to annual rates)
	b) temperature (diurnal means and ranges; cumulative degree-day deficit (freezing index); thaws)
	c) barometric changes
6. *Vibration*	a) seismicity (events magnitude, frequency and duration; microseismic intensity and changes)
	b) human-induced (blasting; transport; heavy machinery)

Table 8.14 Approaches to landslide remediation (after Holtz and Schuster, 1996).

Method	Procedure	Best application	Limitations	Remarks
Avoid problem	1. Relocate facility	Alternative site off the area of landsliding	None if identified during planning phase; large cost implications if planning and design is complete or if construction has already started	Detailed studies of proposed relocation should ensure alternative site is an improvement
	2. Completely or partially remove unstable materials	Where small volumes of excavation are involved	Can be costly to control excavations; unlikely to be the best option for large landslides; may not be feasible because of human factors such as Rights of Way	Detailed analyses of the stability of the site are required; depth of excavation must be sufficient to ensure firm foundations
	3. Install bridge	On sidelong slopes with relatively shallow slope failures	May be costly and needs to provide adequate support capacity to withstand the lateral pressures resulting from a moving landslide	Analyses must be performed on anticipated loadings over the long term as well as the structural capability
Reduce driving forces	1. Change exact location site, alignment or grade	During preliminary design phase of project	For highways it will affect alignment adjacent to landslide area; for all projects it may increase land-take requirements	
	2. Drain surface	Appropriate for any scheme both as part of design and in remedial measures	Will only correct surface infiltration or seepage to surface infiltration	Slope vegetation should be considered in all cases
	3. Drain subsurface	On any slope where lowering of groundwater table will increase slope stability	Not effective when sliding mass is impervious	Stability analyses should include consideration of seepage forces
	4. Reduce weight	At any existing or potential landslide	May require use of lightweight materials that may be costly or unavailable; excavation waste may create problems; may require access to land beyond site area	Very careful stability analyses required to ensure excavation or placement of lightweight fill will improve stability
Increase resisting forces by applying external force	1. Use buttress and counterweight fills; toe berms	At an existing landslide in combination with other methods	May not be effective on deep-seated landslides; must be placed on firm foundations	Where space is limited reinforced walls and slopes with geotextiles maybe valid alternative

Table 8.14 (*continued*).

Method	Procedure	Best application	Limitations	Remarks
	2. Use structural systems	To prevent movement before excavation; where space is limited	Will not withstand large deformations; must be founded well below the sliding surface	Stability analyses must incorporate soil–structure interaction
	3. Install anchors (appropriate for both soils and rocks)	Where space is limited	Foundation materials must be able to resist shear forces by anchor tension; can inhibit future development on adjacent land	*In situ* strengths of soils and rocks must be known; economics dependent on anchor capacity, depth and frequency
Increase resisting forces by increasing internal strength	1. Drain subsurface	At any landslide where water table is above the shear surface	Requires expertise to install and ensure long-term effective operation	
	2. Use geosynthetic reinforcement in backfill	For embankments and fill slopes on steep sidelong ground; in landslide reconstruction	Geosynthetic reinforcement must be durable in the long term	Need to consider stresses imposed on reinforcement during construction
	3. Install *in situ* reinforcement	As temporary structures in stiff soils; for small scale failures on rock slopes	Requires long-term durability of nails, anchors, micropiles and masonry dentition	Design methods still undergoing development
	4. Use biotechnical stabilisation	On soil slopes of modest heights	Climate: may require irrigation in dry seasons; longevity of selected plants	Design is by trial and error linked to local experience
	5. Treat chemically	Where sliding surface is well-defined and soil reacts positively to treatment	May be reversible; long-term effectiveness has not been proved; environmental stability unknown	Field installation must evaluate possible long-term environmental effects
	6. Use electro-osmosis	To relieve excess porewater pressures and increase shear strength at required construction rate	Only a short-term construction solution as requires constant diect current power supply and maintenance	Used when nothing else works; possible application for emergency stabilisation of landslides
	7. Treat thermally	To reduce sensitivity of clay soils to action of water	Requires expensive and carefully designed systems to artificially dry or freeze subsoils	Methods are experimental and costly

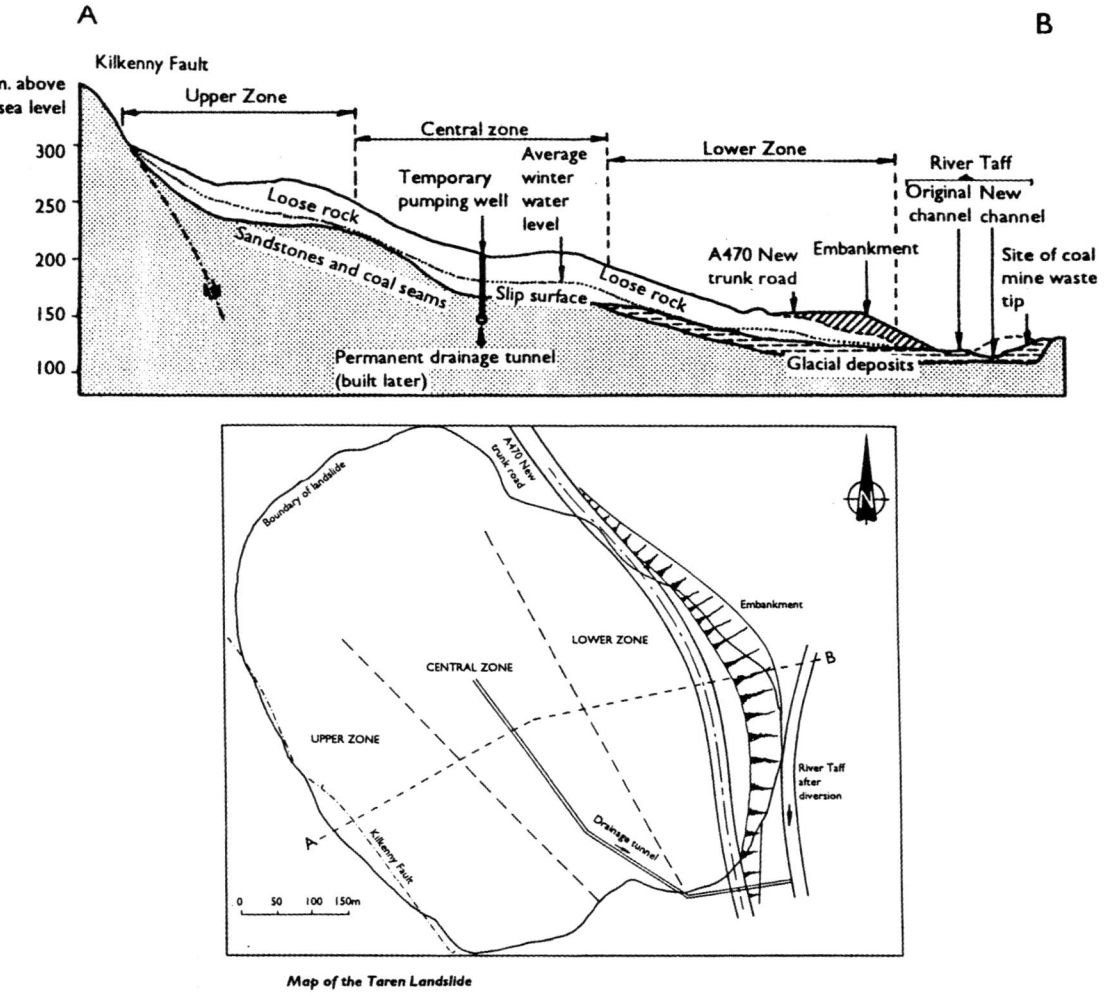

Map of the Taren Landslide

Figure 8.25 Plan and cross-section of the Taren landslide south Wales (after Kelly and Martin, 1985).

A470 crossing of the Taren landslide (after Kelly and Martin, 1985; Cobb, 2000)

Just south of the village of Aberfan in South Wales lies the late-glacial landslide of Taren. During the 1980s, Stage IV of the A470 trunk road was constructed along the upper valley of the River Taff and this had to cross the Taren landslide and bypass the village of Aberfan. Because of the tragedy in Aberfan in 1966 when 144 people, mainly children, were killed by a flowside failure of a mine-waste tip (Bishop et al., 1969), there was sensitivity to any form of construction that might cause landslide movement. The Taren landslide is 640 m long,

rises 220 m between toe and crest, and involves nearly 8 million m³ of material. In the 19th Century a railway and canal had been constructed across the foot of the landslide but the A470 required far more extensive earthworks. The investigations for the road in the area of the landslide involved detailed engineering geomorphological mapping and comprehensive ground investigations over a ten-year period with extensive monitoring of ground movements and porewater conditions. These studies produced a general model (Figure 8.25) that divided the landslide into three zones where the main movements had taken place (upper,

Figure 8.26   Construction works for the Channel Tunnel terminal with the Castle Hill landslide in the foreground, Folkestone, Kent (UK).

middle and lower zone), and two adjacent areas (the NW block and the northern wing) that had been slightly displaced. Inclinometer readings showed that the main landslide was still moving at 3–4 mm/year during the 1970s predominantly along seat-earths in a minor carboniferous coal sequence called the Cefn Glas. In order to achieve a suitable level of safety for the construction of the road, two techniques were employed: weighting through the placement of a major embankment on the toe of the landslide, and drainage of the central zone of the landslide via an adit constructed under the basal shear surface. Construction was completed in 1984 and no movements have been recorded in the landslide since.

UK Channel Tunnel portal at Castle Hill (after Griffiths *et al.*, 1995; Varley *et al.*, 1996)

The UK portal to the Channel Tunnel enters the ground through a late to early post-glacial

landslide at Castle Hill near Folkestone in Kent (Figure 8.26). The failure had formed in Cretacaeous Lower Chalk overlying Gault Clay. Engineering geomorphological mapping of the landslide (Figure 8.27) showed it had a multiple rotational form along a basal shear surface in the Gault Clay. There was inter-digitation of the toe with periglacial coombe-rock deposits, and subsequent secondary shallower degradational failures within the main complex (Figure 8.28). Inclinometer readings and historic map studies indicated the landslide had been moving at 1–2 mm/year for the previous 200 years. The Channel Tunnel alignment required the two separate railway tunnels to be driven through the slide mass. In order to achieve this a combination of stabilisation measures and a strictly controlled construction programme was required. A 100 000 tonne toe-weighting berm was placed on the foot of the landslide and three drainage adits were driven into the slide to lower groundwater levels.

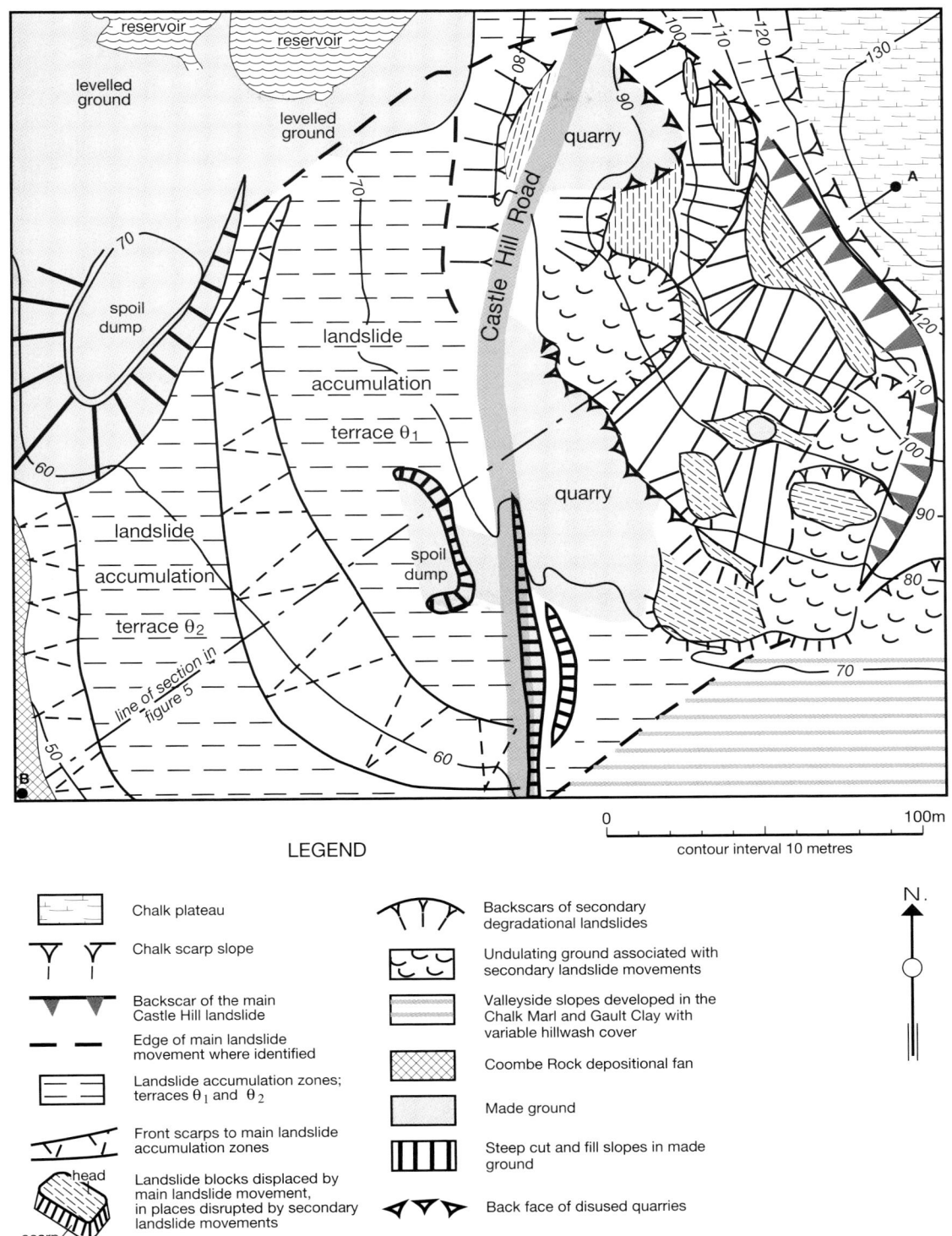

Figure 8.27   Geomorphological map of the Castle Hill landslide (after Griffiths *et al*, 1995).

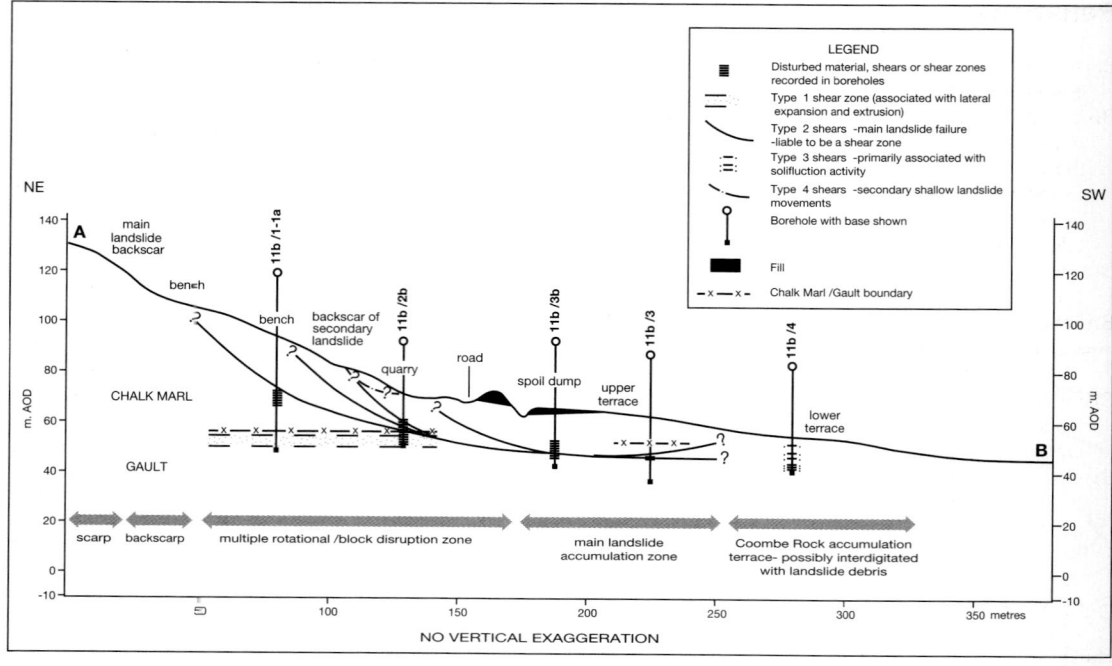

Figure 8.28   Cross-section of the Castle Hill landslide Folkestone, Kent (UK) (after Griffiths *et al.*, 1995) reprinted with the permission of the Royal Geographical Society.

A carefully phased 'top-down' construction approach was adopted to minimise changes in overburden pressure, and ground movements were closely monitored throughout construction. Whilst increased movements were recorded during construction over the period 1989–90, on completion of the works the landslide stabilised and no further activity has been recorded.

8.10 Conclusion

As a landscape feature, landslides in either natural or artificial slopes are a well-recognised engineering problem. Geotechnical engineers have developed sophisticated methods for modelling and analysing the stability of slopes as part of normal earthworks design. These analyses are predominantly based on detailed ground investigations and testing of materials. In contrast, as a consequence of their training, engineering geomorphologists are more able to take a broader-scale view of the landscape and provide data on natural landslides through analyses of the terrain and field observation. Such studies must precede any form of detailed ground investigation if it is to be designed in a cost-effective manner, and by identifying potential landslide problems at an early stage in a project it may be possible to alter sites or change alignments in order to avoid major problem areas. The engineering geomorphologist is also able to provide the spatial context within which the subsequent site investigation results can be interpreted. In addition engineering geomorphology is now developing the tools for landslide hazard and risk assessment that will be able to be incorporated into planning and feasibility studies for all types of development project.

Landslides are a geomorphological process of landscape development. To deal with problems arising from slope failures, geomorphologists have expertise and experience of immense value and importance to the construction industry.

References

Barnes, J. W. (2004) *Basic Geological Mapping* (4th edition). Wiley, Chichester, 184pp.

Bishop, A. W., Hutchinson, J. N., Penman. A. D. M. and Evans, H. E. (1969) Geotechnical investigation into the causes and circumstances of the disaster of 21 October 1966. In *A Selection of Technical Reports Submitted in the Aberfan Tribunal*. Welsh Office, HMSO, London, 1–82.

Brabb, E. E. (1991) The world landslide problem. *Episodes* **14**, 52–61.

Bromhead, E., Dixon, N. and Ibsen, M.-L. (eds) (2000) *Landslides in Research, Theory and Practice*. Thomas Telford, 1684pp.

Brunsden, D. (1993) Mass movement: the research frontier and beyond: a geomorphological approach. *Geomorphology* **7**, 85–128.

Brunsden, D., Doornkamp, J. C., Fookes, P. G., Jones, D. K. C. and Kelly, J. M. H. (1975) Large-scale geomorphological mapping for highway engineering. *Quarterly Journal of Engineering Geology* **8**, 227–253.

BSI (British Standards Institute) (1999) *BS 5930: Code of Practice for Site Investigations*. London.

Bunce, C. M., Cruden, D. M. and Morgensten, N. R. (1997) The assessment of the hazard of rock fall on a highway. *Canadian Geotechnical Journal* **34**, 344–356.

Chorley, R. J., Schumm, S. A. and Sugden, D. E. (1984) Geomorphology. Methuen, London, 605pp.

Cobb, A. E. (2000) Taren landslide: road construction across a landslide. In Siddle, H. J., Bromhead, E. N. and Bassett, M. G. *Landslides and Landslide Management in South Wales*. National Museum & Galleries of Wales, Geological Series No.18, Cardiff, 97–98.

Cooke, W. H. (1882) *Hereford*. John Murray, London.

Corominas, J. (1986) The angle of reach as a mobility index for small and large landslides. *Canadian Geotechnical Journal* **33**, 260–271.

Croot, D. and Griffiths, J. S. (2001) Examples of periglacial activity in south-west England and their engineering geological significance. *Quarterly Journal of Engineering Geology & Hydrogeology* **34**, 269–282.

Cruden, D. M. (1991) A simple definition of a landslide. *Bulletin of the International Association of Engineering Geology* **43**, 27–29.

Cruden, D. and Varnes, D. J. (1996) Landslide types and processes. Chapter 3 in Turner, A. K. and Schuster, R. L. (eds) *Landslides Investigation and Mitigation*. Special report 247 of the Transportation Research Board, National Research Council, National Academy Press, Washington DC.

Cruden, D. and Fell, R. (eds) (1997) *Landslide Risk Assessment*. Balkema, Rotterdam, 371pp.

Dearman, W. R. (1991) *Engineering Geological Mapping*. Butterworth-Heinemann, Oxford, 387pp.

Dikau, R., Brunsden, D., Schrott, L. and Ibsen, M.-L. (1996) *Landslide Recognition*. Wiley, Chichester, 251pp.

Evans, N. C. (1998) The natural terrain landslide study. In Li, K. S. Kay, J. N. and Ho, K. K. S. (eds) *Slope Engineering in Hong Kong*. Balkema, Rotterdam, 137–143.

Fookes, P. G. (1997) Geology for engineers: the geological model, prediction and performance. *Quarterly Journal of Engineering Geology* **30**, 293–424.

Fookes, P. G., Baynes, F. and Hutchinson, J. N. (2000) Total geological history: a model approach to understanding site conditions. Proceedings of GeoEng2000, Melbourne **1**, 370–460.

Gardner, V. and Dackombe, R. (1983) *Geomorphological Field Manual*. Allen & Unwin, London, 254pp.

Geomorphological Services Ltd (1991) Coastal landslip potential, Isle of Wight Undercliff, Ventnor. Technical Report to the Department of the Environment, Research Contract PECD 7/1/272, 68pp.

Godt, J. W. and Savage, W. Z. (1999) El Niño 1997–98: direct costs of damaging landslides in the San Francisco Bay Region. In Griffiths, J. S., Stokes, M. R. and Thomas, R. G. (eds) *Landslides*. Balkema, Rotterdam, 162pp.

Griffiths, J. S. (ed.) (2001) *Land Surface Evaluation in Engineering Practice*. Geological Society Engineering Geology Special Publication No.18, 249pp.

Griffiths, J. S., Brunsden, D., Lee, E. M. and Jones, D. K. C. (1995) Geomorphological investigations for the Channel Tunnel terminal and portal. *The Geographical Journal* **161**, 275–284.

Guadagno, F. M. and Perriello Zampelli, S. (2000) Triggering mechanisms of the landslides that inundated Sarno, Quindici, Siano and Bracigliano (S. Italy) on May 5–6, 1998. In Bromhead, E., Dixon, N. and Ibsen, M.-L. (eds) (2000) *Landslides in Research, Theory and Practice*. Thomas Telford, 671–676.

Hansen, A. (1984) Engineering geomorphology: the application of an evolutionary model of Hong Kong terrain. *Zeitschrift für Geomorphologie*, supplementary volume 51, 39–50.

Hardingham, A. D., Ho, K. K. S., Smallwood, A. R. H. and Ditchfield, C. S. (1998) Quantitative risk assessment of landslides – a case study from Hong Kong. In Li, K. S., Kay, J. N. and Ho, K. K. S. (eds) *Slope Engineering in Hong Kong*. Balkema, Rotterdam, 145–152.

Hearn, G. J. and Griffiths, J. S. (2001) Landslide hazard assessment. In Griffiths, J. S. (ed.) *Land Surface Evaluation in Engineering Practice*. Geological Society Engineering Geology Special Publication No.18, 43–52.

Holtz, R. D. and Schuster, R. L. (1996) Stabilisation of soil slopes. Chapter 17 in Turner, A. K. and Schuster R. L. (eds) *Landslides Investigation and Mitigation*. Special report 247 of the Transportation Research Board, National Research Council, National Academy Press, Washington DC.

Hungr, O. (1995) A model for the runout analysis of rapid flow slides, debris flows and avalanches. *Canadian Geotechnical Journal* **32**, 610–623.

Hutchinson, J. N. (1988) General report: morphological and geotechnical parameters of landslides in relation to geology and geohydrology. *Proceedings of the 5th International Symposium on Landslides*, Lausanne **1**, 3–35.

Hutchinson, J. N. (2001) Reading the ground: morphology and geology in site appraisal. *Quarterly Journal of Engineering Geology and Hydrogeology* **34**, 5–50.

Jones, D. K. C. and Lee, E. M. (1994) *Landsliding in Great Britain*. HMSO, London.

Keaton, J. R. and DeGraff, J. V. (1996) Surface observation and geologic mapping. In Turner A. K. and Schuster R. L. (eds.) *Landslides: Investigation and Mitigation*. Transportation Research Board, Special Report 247, National Research Council, National Academy Press, Washington DC, 178–230.

Kelly, J. M. H. and Martin, P. L. (1985) Construction works on or near landslides. In Morgan, C. (ed.) *Proceedings of the Symposium on Landslides in South Wales*. Polytechnic of Wales, Pontypridd, 85–102.

Li, K. S., Kay, J. N. and Ho, K. K. S. (1998) Slope Engineering in Hong Kong. Balkema: Rotterdam, 342pp.

Lee, E. M. (2001) Geomorphological mapping. InÿGriffiths, J. S. (ed.), *Land Surface Evaluation in Engineering Practice*. Geological Society Engineering Geology Special Publication No.18, 53–56.

Lee, E. M. and Moore, R. (1991) *Coastal landslip potential assessment: Isle of Wight Undercliff, Ventnor*. Department of the Environment.

Lee, E. M. and Clark, A. R. (2002) *The Investigation and Management of Soft Rock Cliffs*. Thomas Telford, London, 392pp.

Lee, E. M. and Jones, D. K. C. (2004) *Landslide Risk Assessment*. Thomas Telford, London, 454pp.

Lee, E. M., Doornkamp, J. C., Brunsden, D. and Noton, N. H. (1991a) Ground movement in Ventnor, Isle of Wight. Report for the Department of the Environment.

Lee, E. M., Moore, R., Brunsden, D. and Siddle, H. J. (1991b) The assessment of ground behaviour at Ventnor, Isle of Wight. In Chandler, R. J. (ed.) *Slope Stability: Engineering Developments and Applications*. Thomas Telford, London, 219–225.

Lee, E.M., Moore, R., and McInnes, R. G. (1998) Assessment of the probability of landslides reactivation: Isle of Wight Undercliff, UK. In Moore D. and Hungr O. (eds.) *Engineering Geology: The View from the Pacific Rim*, 1315–1321.

Lee, E. M., Brunsden, D. and Sellwood, M. (2000) Quantitative risk assessment of coastal landslide problems, Lyme Regis, UK. In Bromhead, E., Dixon, N. and Ibsen, M. L. (eds) *Landslides in Research, Theory and Practice*. Thomas Telford, London, 899–904.

Mather, A. E., Griffiths, J. S. and Stokes, M. R. (2003) Anatomy of a 'fossil' landslide from the Pleistocene of SE Spain. *Geomorphology* **50**, 135–149.

Moore, R., Lee, E. M., and Clark, A. R. (1995) *The Undercliff of the Isle of Wight: a review of ground behaviour*. South Wight Borough Council.

Rib, H. T. and Liang, T. (1978) Recognition and identification. Chapter 3 in Schuster, R. L. and Krizek, R. J. (eds) *Landslides – Analysis and Control*. National Academy of Sciences, Washington DC, 33–79.

Royal Society (1992) *Risk: Analysis, Perception and Management*. Report of the Royal Society Study Group, The Royal Society, London.

Schultz, Ch. H. and Harper, J. A. (1996) Pittsburgh red beds cause renewed landsliding after a *c.* 310 Ma pause. In Chacon, J., Irigaray, C. and Fernandez, T. (eds) *Landslides*. Balkema, Rotterdam, 189–196.

Schuster, R. L. (1995) The 25 most catastrophic landslides in the 20th century. In Chacon, J., Irigaray, C. and Fernandez, T. (eds) *Landslide*. Balkema, Rotterdam, 53–62.

Styles, K. A. and Hansen, A. (1989) Geotechnical area studies programme: Territory of Hong Kong. GASP Report XII, Geotechnical Control Office, Civil Engineering Services Department, Government of Hong Kong, 346pp.

Smith, K. (2001) *Environmental Hazards: Assessing Risk and Reducing Disaster* (3rd edition). Routledge, London, 392pp.

Soeters, R. and van Westen, C. J. (1996) Slope instability recognition. Chapter 8 in Turner, A. K. and Schuster, R. L. (eds) *Landslides Investigation and Mitigation*. Special report 247 of the Transportation Research Board, National Research Council, National Academy Press, Washington DC.

Sowers, G. F. and Royster, D. L. (1978) Field investigation. Chapter 4 in Schuster, R. L. and Krizek, R. J. (eds) *Landslides – Analysis and Control*. National Academy of Sciences, Washington DC, 80–111.

TRL (Transportation Research Laboratory) (1997) Principles of low cost road engineering in mountainous regions. Overseas Road Note 16. TRL, Crowthorne, 149pp.

Turner, A. K. and Schuster, R. L. (eds) (1996) *Landslides Investigation and Mitigation*. Special report 247 of the Transportation Research Board, National Research Council, National Academy Press, Washington DC.

UNESCO World Landslide Inventory (1990) A suggested method for reporting a landslide. *Bulletin of the International Association of Engineering Geology* **41**, 5–16.

UNESCO World Landslide Inventory (1993) A suggested method for describing the activity of a landslide. *Bulletin of the International Association of Engineering Geology* **47**, 53–57.

Varnes, D. J. (1978) Slope movement types and processes. Chapter 2 In Schuster, R. L. and Krizek, R. J. (eds) *Landslides – Analysis and Control*. National Academy of Sciences, Washington DC, 11–33.

Varley, P. M., Warren, C. D. and Avgherinos, P. (1996) Castle Hill west landslip. In Harris, C. S., Hart, M. B., Varley, P. M. and Warren, C. D. *Engineering Geology of the Channel Tunnel*. Thomas Telford, London, 295–309.

Wu, T. H., Tang, W. H. and Einstein, H. H. (1996) Landslide hazard and risk assessment. Chapter 6 in Turner, A. K. and Schuster, R. L. (eds) *Landslides Investigation and Mitigation*. Special report 247 of the Transportation Research Board, National Research Council, National Academy Press, Washington DC.

Acknowledgements

The author wishes to thank Prof. Peter Fookes for permission to use Figure 8.24; Dr Mark Lee for permission to use Figures 8.16, 8.17, 8.20, 8.21 and 8.23; the Geological Society of London for permission to use Figure 8.14; the Transportation Research Laboratory for permission to use Figures 8.19 and 8.22, and the Royal Geographical Society for permission to use Figures 8.27 and 8.28. The photographs have all been supplied by the author.

9. Active Tectonic Environments and Seismic Hazards

James V. Hengesh and William R. Lettis

9.1 Introduction

Geological hazards are greatest along plate boundary regions of the world due to high rates of seismic and volcanic activity and dynamic geological and geomorphological processes (Figure 9.1). Although some major cities such as San Francisco, California or Wellington, New Zealand are built along active plate boundaries, geological hazards in these areas are being mitigated through hazard mapping programs, and implementation and enforcement of geological hazards characterisation and seismic design in building code provisions. However, vast populations in cities and rural areas of developing countries live along highly active plate boundaries and lack hazard information, building codes, or code enforcement to adequately mitigate risk (e.g. Indonesia, the Mediterranean, Asia, Latin and South America).

Geological hazards also are important considerations within stable continental regions (Johnston, 1996; Bakun and McGarr, 2002). However, the processes that produce earthquakes, i.e. strain accumulation and release, are less well understood in stable continental regions than plate boundary regions (Kenner and Segall, 2000), and therefore, there is greater uncertainty in estimates of hazard and risk in these areas. Despite uncertainties in our understanding of the earthquake generation process in stable continental regions, a number of factors indicate that high levels of risk exist for some parts of these areas. Historical examples of large magnitude earthquakes within stable continental regions include the three M $\sim$ 7.3 to 8.1 events that occurred in the central United States within a three month period between December 1811 and February 1812 (Tuttle et al., 2002a), the M $\sim$ 8 1819 and

M$_w$ 7.72001 earthquakes that occurred in north-western India (Hengesh and Lettis, 2002), and the M $\sim$ 8 1755 Lisbon earthquake that occurred in Portugal (Johnston, 1996).

It should be noted that:

1. maximum earthquake magnitudes in stable continental regions are of similar size to maximum earthquake magnitudes in plate boundary regions (excluding subduction zones)
2. larger areas of stable continental regions are affected by earthquake strong ground shaking than areas of plate boundary regions for comparable sized earthquakes
3. recent studies have identified short earthquake recurrence intervals in some stable continental regions (Kelson et al., 1996; Tuttle et al., 2002a).

Earthquake 'Risk' for engineered structures and systems

Recent earthquakes in Turkey, Taiwan and India in 1999 and 2001 dramatically illustrate the damaging effects of strong ground shaking, permanent ground deformation and liquefaction on structures, buildings, and lifelines, as well as the negative impact on the local economy (Figure 9.2; EERI 2000, 2001, 2002). As in many previous earthquakes, these three events produced damage as a result of surface fault rupture, strong ground shaking, amplified ground shaking, liquefaction, and ground settlement.

The term 'risk' is commonly used within earthquake-related studies, but often has no clear definition. In this chapter the term 'risk' has a very specific definition. Risk describes the losses that may occur during an earthquake. Losses can be measured in a variety of ways, such as: (1) direct economic losses resulting from damaged buildings

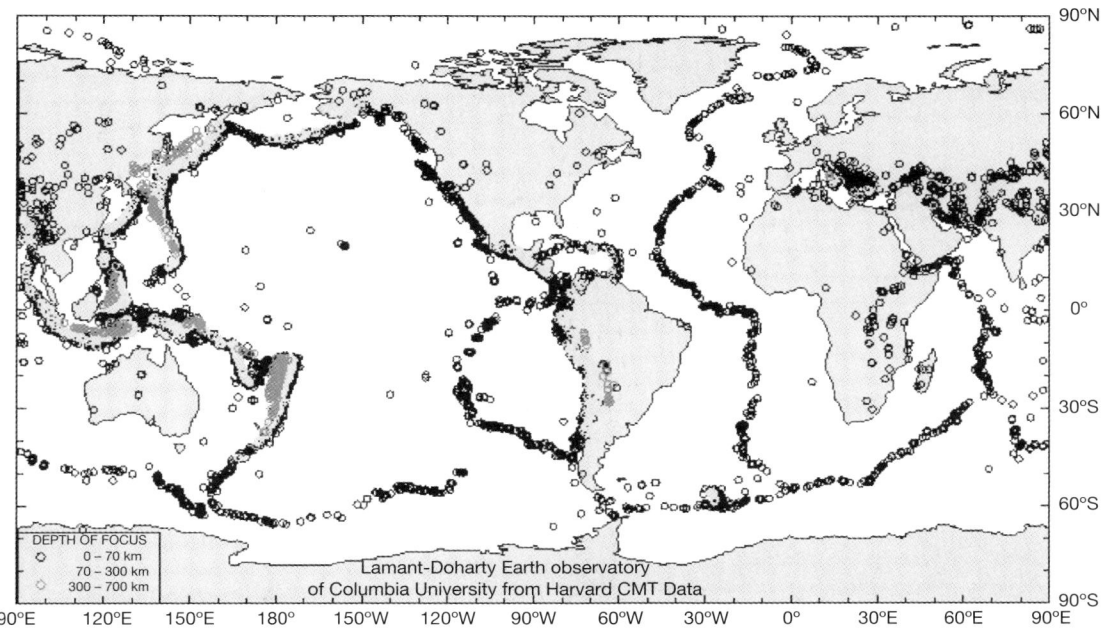

Figure 9.1   Map showing global seismicity that defines major plate boundaries. Note the lack of seismicity within the continental interiors compared to seimicity along plate boundaries. Black circles represent shallow seismicity, light gray clusters represent intermediate depth seismicity. Map shows events >M_w 5.5 for period 1977 to 1992.

or other facilities; (2) indirect losses resulting from business interruption and economic impacts; (3) injuries and fatalities; and (4) impact on the environment.

Risk, as described above, is actually the product of three important parameters including: geological hazard, structural vulnerability, and socio-economic factors. This can be thought of conceptually as:

Risk = Hazard × Vulnerability × Consequence

where the term 'hazard' describes parameters such as level of ground shaking, liquefaction potential, site response, surface fault rupture, and slope instability. The term vulnerability describes the susceptibility of buildings or lifeline systems to damage from the effects of earthquakes, and consequence reflects factors such as damage, death/injury and economic performance. Therefore, assessing 'risk' requires information on geological hazards, vulnerability of structures in a specific

area, and the distribution of people and economic networks (McGuire, 2004).

Types of seismic hazards
Earthquake-related geological hazards include:

1. primary surface fault rupture i.e. the sudden displacement of adjacent crustal blocks along a fault which may cause permanent displacement of the ground surface (Figure 9.2)
2. strong ground shaking
3. liquefaction and related ground deformation (Figure 9.2)
4. seismically induced slope failure
5. tsunami.

Strong ground shaking and tsunami are seismic waves generated from the primary fault rupture and ground displacement, and are measured in terms of their frequency, amplitude, velocity, particle displacement, and acceleration. Liquefaction and seismically induced slope failure, in turn, are

Figure 9.2 Photographs of damage from the 1999 Turkey earthquake. (*A*) Collapse of apartment buildings due to surface fault rupture and near-field strong ground shaking. Note approximate 5-metre offset of concrete wall. (*B*) 500 metres of waterfront property catastrophically submerged due to liquefaction induced lateral spreading by translation/rotation. Ismit Bay, Turkey. Photographs by James Hengesh.

produced as seismic waves propagate through soils (Figure 9.2).

9.2 Earthquake occurrence and magnitude scaling

Identifying and characterising active faults is essential both for the development of earthquake design ground motions and the evaluation of site-specific surface fault rupture hazards. Seismic geologists attempt to link observations such as location and type of tectonic landforms, locations of instrumentally recorded earthquakes, and geodetic measurements of crustal plate motion with processes occurring deep within the earth's crust that generate earthquakes. A number

of techniques often are used to integrate these observations and develop interpretations that can be translated into engineering terms for use in seismic hazard assessments during the feasibility, siting, design, or construction phases of major projects (Gutenberg and Richter, 1954; Cornell, 1968; Molnar, 1979; Schwartz and Coppersmith, 1984; Youngs and Coppersmith, 1985; Weznouski, 1988; Krinitsky and Slemmons, 1990; SSHAC, 1995 and Yeats *et al.*, 1997).

Fault characteristics most commonly required for engineering projects include:

1. location of fault or tectonic province
2. fault type, i.e. strike-slip, reverse, normal or combination
3. fault length and seismogenic width (vertically through the crust)
4. width of a fault zone on ground surface
5. earthquake magnitude
6. coseismic displacement
7. fault slip rate
8. earthquake recurrence interval
9. recency of fault movement.

For seismic hazard analysis projects this information is used to develop a seismic source model, which may include a combination of individual fault segments (referred to as line sources or planar sources if dipping), seismotectonic provinces (areal source zones), and point sources. Each line, plane, point, or areal source zone in a seismic source model must be assigned coordinates, maximum earthquake magnitude, and earthquake recurrence model.

Earthquake mechanics

Large damaging earthquakes occur along plate boundaries and within continental interiors (Figure 9.1). Plate boundary earthquakes occur due to the concentration of stress along the edges of crustal-scale plates, or blocks, that define planes of weakness known as faults. The stress is caused when adjacent crustal-scale plates or blocks have different rates and directions of relative motion (Figure 9.3). Stress is accommodated by elastic strain of large crustal blocks, and

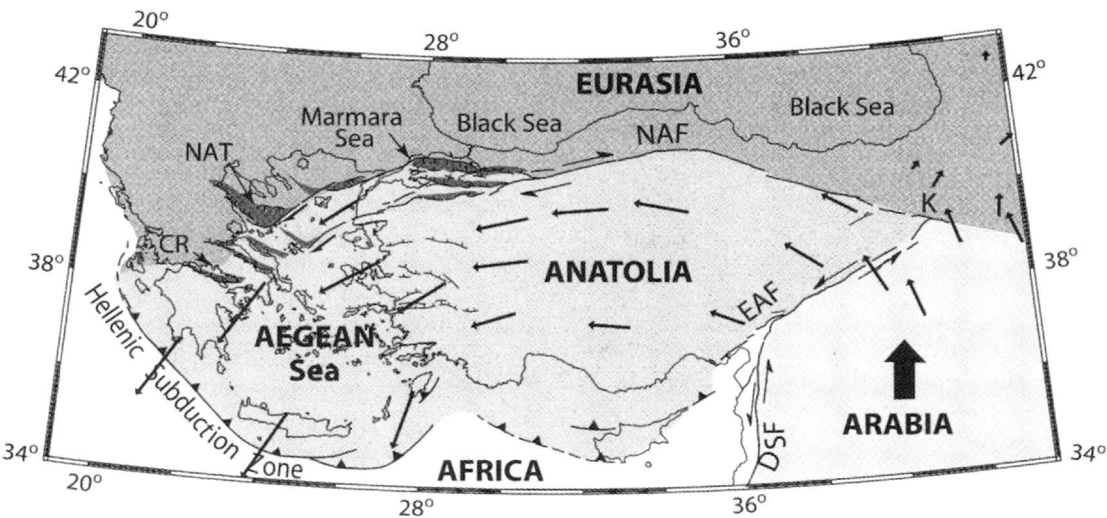

Figure 9.3 Continental extrusion in the eastern Mediterranean. The Aegean-Anatolia block is escaping westward from the Arabia-Eurasia collision zone toward the Hellenic subduction zone. Current velocity vectors relative to Eurasia in mm/yr (arrows), using GPS (Global Positioning System) and SLR (satellite laser ranging) are from Reilinger *et al.* (1977). In the Aegean, the westward propagation of the North Anatolian fault is associated with localised and rapid transtension (Armijo *et al.*, 1996). CR, Corinth Rift; NAT, North Aegean Trough; NAF North Anatolian fault; K, Karliova triple junction; EAF, East Anatolian fault; DSF, Dead Sea fault. Modified from Hubert-Ferrari *et al.* (2002).

distributed brittle faulting along the margins of the blocks.

An earthquake occurs when the dynamic stress exceeds the static frictional stress of the fault plane, and the two adjacent plates slip with a discrete displacement across a fault plane (Figure 9.4). As discussed below, the size of an earthquake is related to the amount of stress released along the displacement, ($\sigma_o - \sigma_f$ on the lower diagram of Figure 9.4). The rate of earthquake occurrence is related to the rate that stress accumulates across the plate boundary, and is proportional to the velocity differential between the two plates. Therefore, because earthquake occurrence rates are related to the plate velocity differential across a fault plane, many faults in plate boundary regions follow a time dependent earthquake occurrence model of strain accumulation and release (Kanamori and Brodsky, 2004). Figure 9.5 illustrates the time dependent earthquake occurrence model for a portion of the Aleutian subduction zone that produced the 1965 M_w 8.7 Rat Island earthquake. The model shows a constant

accumulation of stress (measured as seismic moment) along three alternative fault segments of the Aleutian subduction zone (e.g. single-block, double-block, of triple-block ruptures). Once the static stress (expressed as characteristic seismic moment) of the fault plane is exceeded, coseismic displacement occurs and produces an earthquake, the seismic moment is released as energy, and the process of accumulating stress repeats itself.

Earthquakes within plate interiors also are due to the concentration of stress along fault planes, however, the model of strain accumulation and release differs dramatically. In plate interiors, earthquakes may occur from residual stress following prior tectonic episodes, or due to the mechanical concentration of broad regional stress on localized zones of weakness, or due to non-tectonic processes such as isostatic loading or unloading (EPRI, 1994; Kenner and Segall, 2000). The concentration and release of strain energy within plate interiors may not follow a simple time-dependent model and as such, there is greater uncertainty in the location, magnitude

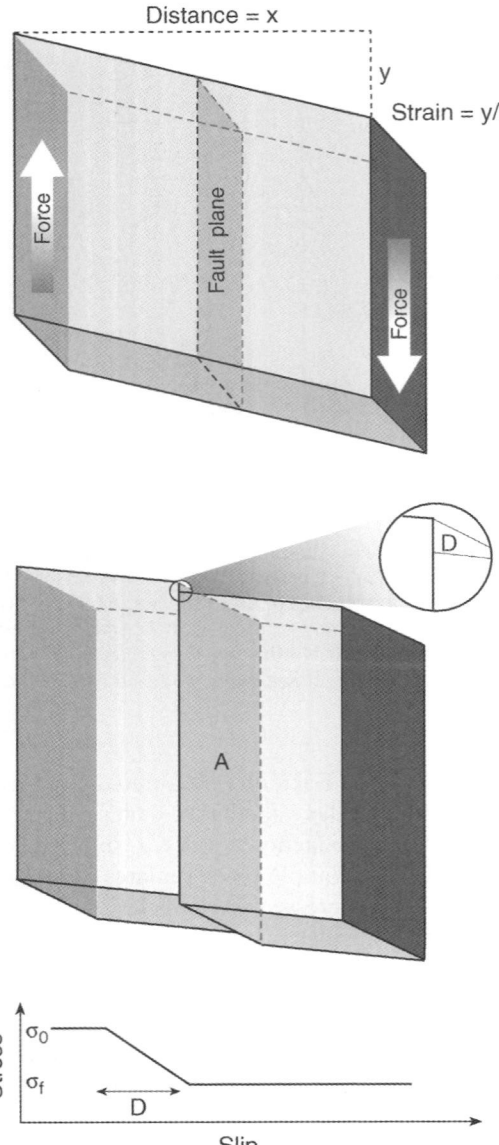

and frequency of these events. Strain rates for plate interiors may be several orders of magnitude lower than for plate boundaries; however, areas such as the New Madrid Seismic Zone in the central United States (e.g. Tuttle *et al.*, 2002a; Obermeier *et al.*, 2001; Kenner and Segall, 2000) have earthquake recurrence intervals that are similar to those of active plate boundary regions during the last 2000 years or more. This points out a major source of uncertainty as geologists grapple to understand the earthquake occurrence process in stable continental regions.

Earthquake magnitude

Because it is difficult to directly measure the stress released during an earthquake, the amount of coseismic displacement, slip area, and rigidity of crustal blocks are used to estimate the energy released (or moment) during an earthquake (Hanks and Kanamori, 1979; Kanamori and Brodsky, 2004). The moment magnitude scale is now the most commonly used measure of earthquake magnitude in earthquake engineering studies. Whereas previous earthquake magnitude scales measure a particular type of ground motion (e.g. surface wave (M_s) or body wave (M_b) magnitudes), the moment magnitude (M_w) scale is a physical measure of the size of an earthquake (Kanamori, 1983). The energy released or moment (M_o) of an earthquake is defined by:

$$M_o = AD\mu$$

where (A) = rupture area, (D) = displacement, and (μ) = crustal rigidity. The moment magnitude of an earthquake (Hanks and Kanamori, 1979; Kanamori and Brodsky, 2004) is related to seismic moment by the equation:

$$M_w = \log M_o - 16.1/1.5$$

Figure 9.4   This figure illustrates the response of the earth's crust to tectonic forces. Equal, but oppositely directed forces act tangent to the light and dark shaded planes on the boundaries of the slab. The forces cause stress, which is defined as the magnitude of the force divided by the area of the planes. As stress occurs within the slab, the dark shaded plane is displaced a distance *y* relative to the light colored plane, and the distance between the two planes is *x*; strain of the slab is defined as y/x (upper diagram). An earthquake occurs when stress on a fault plane exceeds the static frictional stress of the fault plane, σ_o. When an earthquake occurs, plates on opposite sides of the fault experience a relative displacement (D) over an area (A) (middle diagram). As displacement (D) continues, the frictional stress drops from the original static frictional stress σ_o to a lower kinetic stress σ_f, shown on the lower diagram. Figure modified from Kanamori and Brodsky, 2004.

Model of major block rupture followed by a "doublet" of smaller events

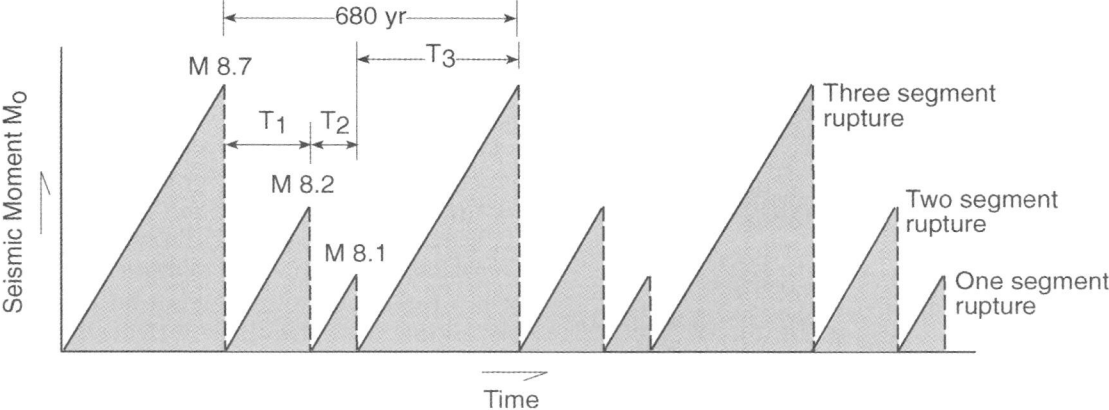

Figure 9.5   Hypothetical rupture model for the Aleutian Subduction Zone near Rat Islands, Alaska. Model illustrates the process of stress accumulation and release along the subduction zone plate interface. The top of each triangle represents the characteristic moment for each characteristic segment rupture. The sloping line represents the constant rate of stress accumulation, or seismic moment rate, across the Aleutian subduction zone. Once the accumulated moment reaches the characteristic moment, the strength of the fault plane is exceeded and a fault displacement and earthquake occur. During a particular earthquake cycle, the characteristic event may involve a single-block, double-block, or triple-block rupture. Because the characteristic moment varies by the size of the rupture segment, the recurrence time (T) varies for each segment rupture.

The moment magnitude scale is preferred for engineering studies because it is stable throughout the entire range of earthquake magnitudes, whereas other magnitude scales are appropriate for a limited range of earthquake magnitudes, or particular geographic regions. Furthermore, the moment magnitude scale is used in empirical relationships among fault rupture dimensions and magnitude (Wells and Coppersmith, 1994), and most modern ground motion attenuation equations (SRL, 1997; EPRI, 2003).

9.3 Types of fault displacement

The coseismic displacement depicted in Figure 9.4 can occur along different types of faults (e.g. strike-slip, normal, reverse, or subduction zone) depending on the orientation of the fault with respect to the regional stress field. Figures 9.6 and 9.7 show the different types of faults and the orientation in which they form with respect to a

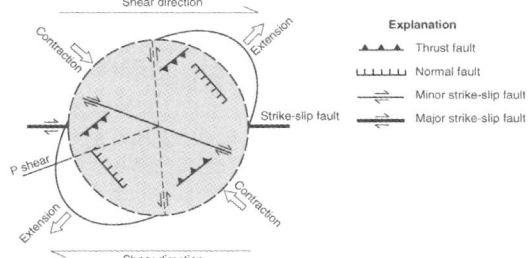

Figure 9.6   Simple shear associated with strike-slip faulting produces preferred orientation of fractures, faults, and folds, as well as extensional and contractional landforms. Figure modified from Burbank and Anderson, 2001. Examples of fault types shown on Figure 9.7.

region undergoing simple shear. The repeated occurrence of these displacements and long term response of the earth's crust to regional stress patterns, climatic conditions and surficial geological processes also forms characteristic landforms, such as linear valleys, faceted mountain fronts, or fold and thrust belts (Figure 9.8). Principals of

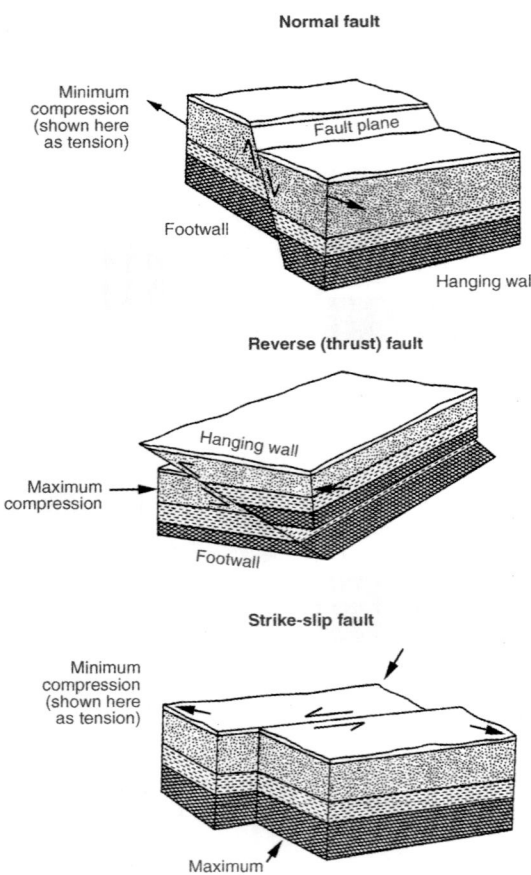

Figure 9.7   Types of faults and illustrations of fault slip. The preferred orientations of each fault type with respect to the regional stress direction are shown on Figure 9.6.

tectonic geomorphology are used to evaluate landforms, locate faults, and characterise potential earthquake magnitude and recurrence intervals (e.g. Yeats *et al.*, 1997; Burbank and Anderson, 2001).

Strike-slip faults

A strike-slip fault is a fault along which most movement is parallel to the fault strike (Figures 9.6 and 9.7; Bates and Jackson, 1987). Strike-slip faults are either right-lateral (dextral) or left lateral (sinistral) depending on the relative motion of the block on the opposite side of the

fault. These faults are important sources of large damaging earthquakes in many transcurrent or transform plate boundary settings, examples of which include the San Andreas, North Anatolian, Great Sumatra, and Alpine fault systems.

Strike-slip faults are very high angle to vertical structures and as such form very straight fault traces along the ground surface. As shown on Figure 9.8, strike-slip faults form a number of characteristic landforms due to the lateral translation of adjacent crustal blocks on opposite sides of the fault. These may include offset or deflected drainages, offset ridges, sag ponds, shutter ridges, and/or fault scarps (Figure 9.8). Due to high slip rates, faults such as the San Andreas often form pronounced linear valleys (Figure 9.9).

Complexities along the trace of a strike-slip fault produce secondary features that vary from metre to kilometre scale. Releasing bends or stepovers form pull-apart basins due to localised extension within the fault zone (Figure 9.10*a*). During the 1999 Turkey earthquake subsidence within pull-apart basins caused dramatic coastal subsidence and extensive damage (Figure 9.11). Restraining bends or stepovers form localised uplift, folding and/or faulting due to contraction within the fault zone (Figure 9.10*b*). Perhaps one of the best examples of a restraining bend is the 'Big Bend' along the San Andreas Fault in southern California (Figure 9.12; Yeats *et al.*, 1994).

Earthquakes along strike-slip faults form a relatively narrow zone of permanent ground deformation (Figure 9.13). Primary surface fault ruptures on strike-slip faults cause a zone of ground cracking, bulging and tearing of near surface materials that is commonly 5- to 15-metres wide.

Normal faults

Normal faults have dip-slip displacement in which the hanging wall has moved downward relative to the footwall (Figure 9.7). These faults are important sources of large damaging earthquakes in continental rift zones and extensional tectonic provinces. Examples of continental rift zones include the Rio Grande New Mexico, East African,

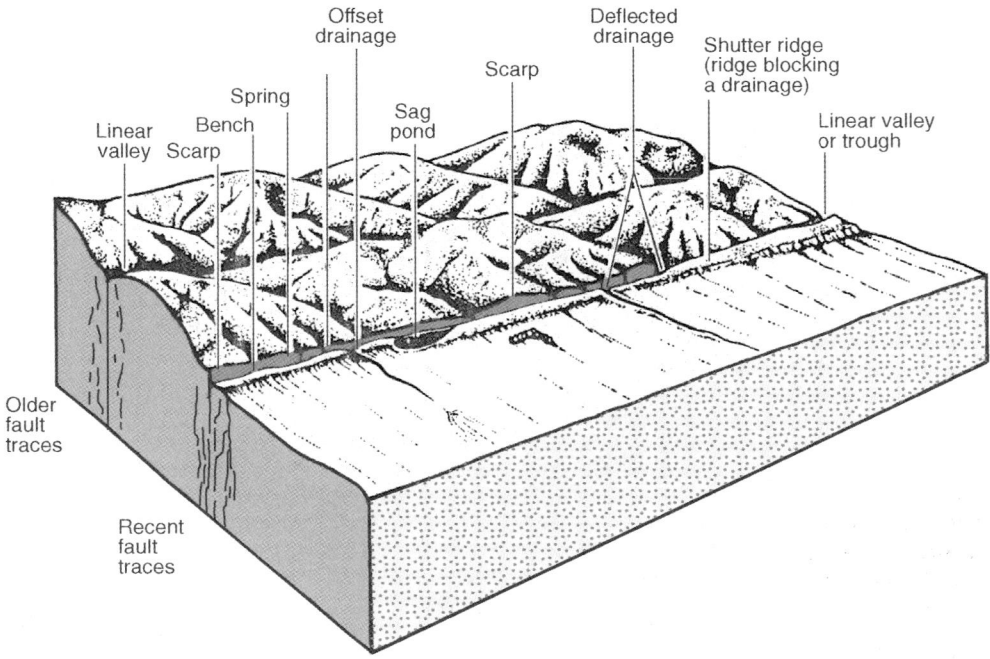

Figure 9.8   Examples of landforms associated with active strike-slip faulting. Modified from Wesson *et al.* (1975).

Figure 9.9a   Sag pond and side-hill bench along North Anatolian fault. Photograph courtesy of Professor Aykut Barka.

and Taupo New Zealand rift zones; examples of extensional tectonic provinces include the Rhine Graben, Basin and Range province, and Tibetan Plateau rift zone.

Normal faults are moderate to steeply dipping structures and, as such, fault traces typically form irregular and discontinuous traces across the ground surface (Figure 9.14*a*). Vertical

Linear valley along
San Andreas fault

Figure 9.9b   Linear valley formed along the Gualala segment of the San Andreas fault. From Wallace, R.E., ed. (1990).

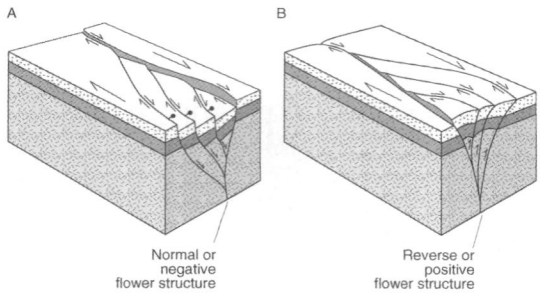

A

B

Normal or
negative
flower structure

Reverse or
positive
flower structure

Figure 9.10   (*A*) Formation of a pull apart basin at an extensional bend on a right lateral strike-slip fault. (*B*) Formation of a contractional duplex at a restraining bend on a right lateral strike-slip fault.

displacement of adjacent crustal blocks on opposite sides of the fault affects erosion of mountain fronts and drainage basin evolution. Landforms associated with normal faults may include offset geomorphic surfaces such as alluvial fans, fluvial terraces or glacial moraines, linear mountain fronts, triangular facets, and/or fault scarps (Figure 9.15).

Complexities along the traces of normal faults form secondary features that vary from metre to kilometre scale, and include relay ramps, transfer faults, synthetic and antithetic faults, and grabens (Figure 9.14). During the 1956 Dixie Valley/ Fairview Peak, 1959 Hebgen Lake, 1915 Pleasant Valley, and 1983 Borah Peak earthquakes, ground rupture occurred along multiple fault splays across zones 5 to 10 kilometres wide (Figure 9.16; Wheeler, 1989). During some of these earthquakes surface fault ruptures occurred on range-bounding faults on opposite sides of a basin and had opposite directions of displacement (e.g. east-side-down vs. west-side-down).

Earthquakes along normal faults form a relatively wide zone of permanent ground deformation (Figures 9.16 and 9.17). Primary surface fault rupture on normal faults typically forms

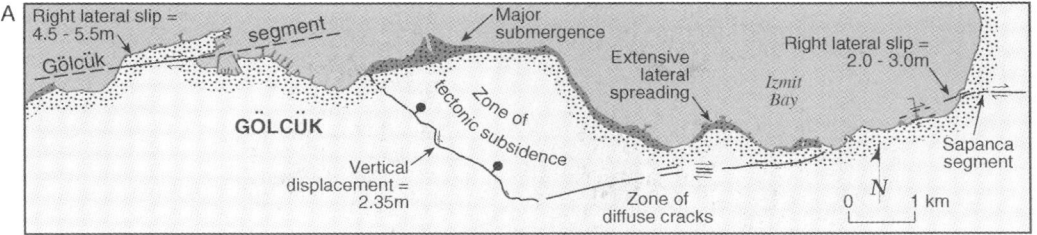

Figure 9.11 (*A*) Map showing extensional step-over and related area of tectonic subsidence and coastal submergence. (*B*) Tectonic subsidence in extensional step-over during the 1999 earthquake, Ismit Bay, Turkey. Note submerged buildings and apartment complexes in foreground. Photograph by William Lettis.

zones of ground cracking, bulging and tearing of near-surface materials that are up to 100-metres wide (Figure 9.17). Individual fault splays may be up to 10-km apart and secondary deformation may occur within approximately one kilometre of these fault splays.

Reverse faults

Reverse faults are dip-slip faults in which the hanging wall block has moved up relative to the footwall block (Figure 9.7). In general, thrust faults place older rocks or sediments over younger rocks or sediments, resulting in repetition of stratigraphy in cross-section above the fault. Reverse faults are moderately to shallowly dipping structures and, as such, fault traces typically form a sinuous, discontinuous trace across the ground surface. Reverse faults form a number of characteristic landforms due to the vertical displacement of the hanging wall above the footwall, including fault scarps, folds, and terrace or alluvial fan sequences (Figure 9.18).

Complexities along the traces of reverse faults form secondary features that vary from 10's of metres to kilometre scale. Secondary features such as lateral ramps or tear faults can produce a strike-slip sense of deformation within the reverse fault system where the strike of the fault approximates the slip direction, or where a reverse fault terminates or steps between fault splays. The 1999 M_w 7.6 Chi-Chi Taiwan reverse earthquake produced four primary types of ground deformation, including primary fault rupture, secondary ground deformation, rotated

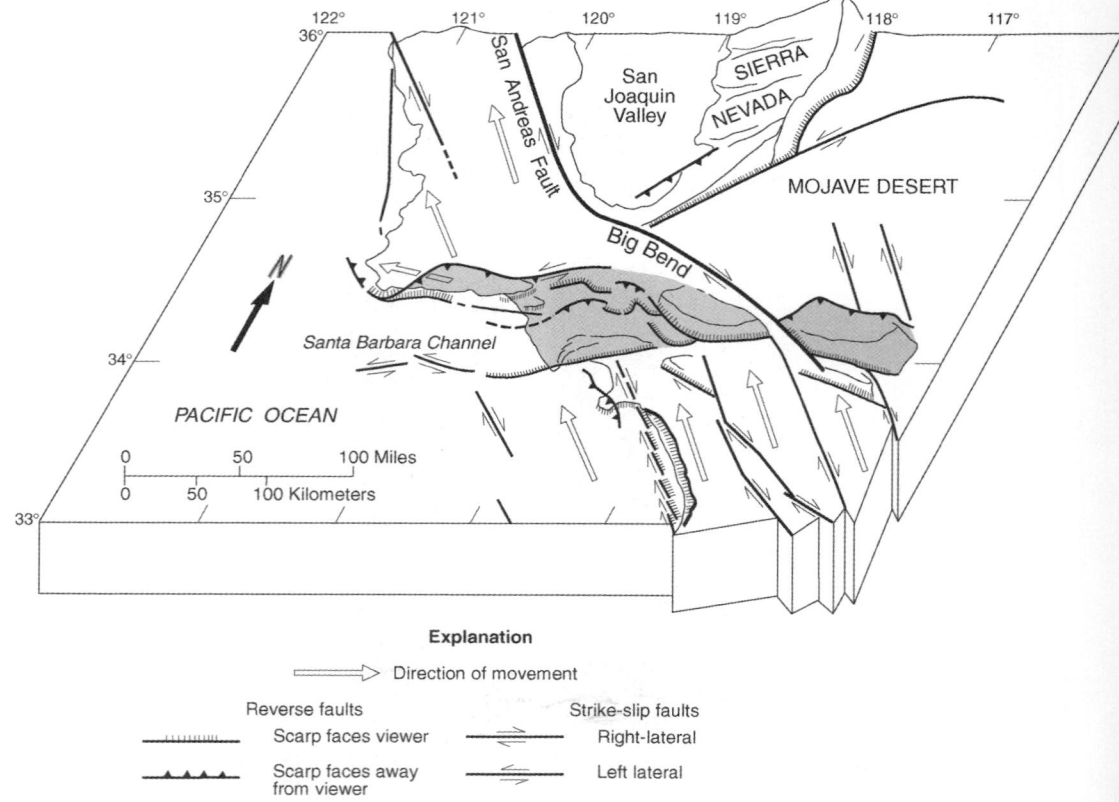

Figure 9.12   The 'Big Bend' in the San Andreas fault in southern California forms a zone of contractional deformation that has caused uplift of the Transverse Ranges. The Transverse Ranges (shaded) are generally bounded by reverse faults and prominent scarps. Modified from Anderson (1971).

fold limb, and hingeline deformation (EERI, 2001; Kelson *et al.*, 2001; Figures 9.19 to 9.23). Up to 13 metres of permanent ground deformation occurred along the fault rupture. Ground deformation resulted from a combination of primary tectonic displacement and secondary 'plowing' of the ground surface by the fault tip, as well as internal deformation of the hanging-wall block. The amount of total ground deformation can be 2 to 3 times greater than the tectonic displacement due to the plowing effects of the fault tip (Figure 9.23).

Earthquakes along reverse faults form a relatively broad zone of permanent ground deformation. Primary surface fault ruptures on reverse faults can form zones of secondary folding, ground

cracking, bulging and tearing of near surface materials that are up to 1 kilometre wide (Figures 9.20, 9.21 and 9.22). Secondary deformation such as back-thrusts resulting from surface rupture on reverse faults may occur several kilometres from the main fault trace.

Subduction-related submergence and emergence

Subduction zones produce the world's largest earthquakes, including the M_w 9.5 1960 Chile earthquake, the M_w 9.3 2004 Sumatra earthquake and the M_w 9.2 1964 Alaska earthquake (Plafker, 1969; Ruff and Kanamori, 1983; Thatcher, 1990). However, these earthquakes do not produce primary surface fault rupture because the slip surface

Figure 9.13   Surface fault rupture produced during the 1999 Turkey earthquake along North Anatolian fault. Lateral displacement in this area was 4.7 metres. The width of ground deformation was typically less than 10 metres. This form of ground deformation is referred to as a mole track. Photographs by James Hengesh.

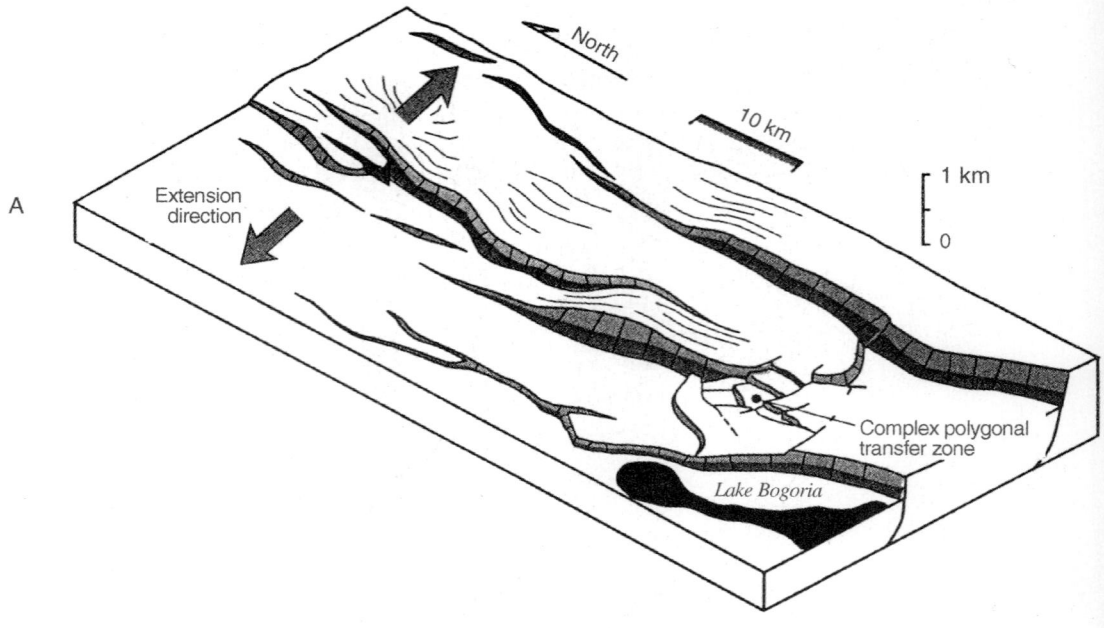

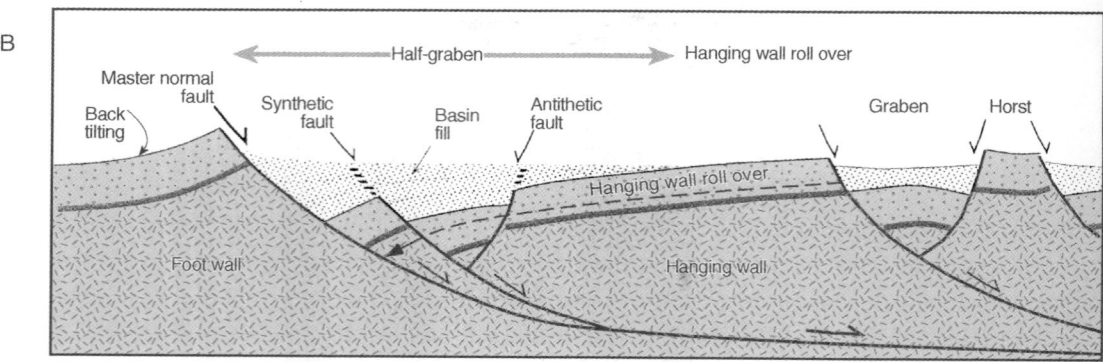

Figure 9.14   (*A*) Schematic fault geometry in the East African Rift System. (*B*) Schematic of cross section of normal faults in an extensional regime. Modified from Burbank and Anderson (2001).

on the plate interface occurs at depths of 10- to 50-km. The process of strain accumulation and release along subduction zones does, however, produce a combination of recognizable coastal uplift and subsidence both coseismically and interseismically (Plafker, 1969; Figures 9.24 and 9.25). Figures 9.25*a* and *b* show examples of reef and beach ridges that emerged coseismically during the 1855 Wellington, New Zealand and 1964 Alaska earthquakes, respectively. Figures 9.25*c*

and *d* depict tectonic subsidence and submergence during the 1964 Alaska Earthquake, and the previous earthquakes along the 1964 rupture zone. Subduction zone earthquakes can produce coseismic land level changes of up to 10 metres, and interseismic land level changes that may be locally greater than the coseismic land level changes, and in the opposite direction (Plafker and Rubin, 1992). Where uplift or subsidence occurs along the intertidal zone, depositional environments can

Figure 9.15   Photographs of normal faults along the Sierra Nevada range front fault zone, near Bishop California. Photograph A shows fault scarp along offset debris flow fan. Photograph B shows offset crests (arrows) of approximate 20 000 year old Tioga glacial moraine (inner) and 114 000 year old Tahoe glacial moraine (outer). Photographs by James Hengesh.

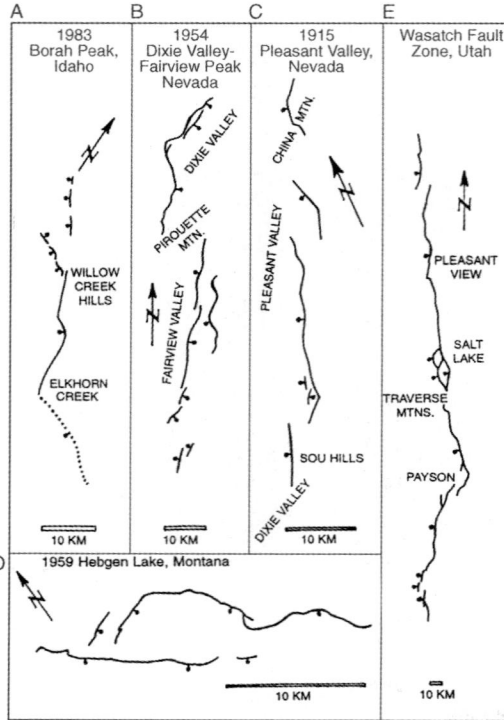

Figure 9.16   Sketch maps of normal fault surface ruptures in the Basin and Range Province, western United States. Maps A through D show historic rupture. Map E shows the late Quaternary traces of the Wasatch fault zone. Modified from Wheeler, 1989.

change from subarial to intertidal conditions. As shown on Figure 9.25 (photograph D), these environmental changes can be used to identify and date paleoseismic events along subduction zones.

Earthquakes along subduction zones produce uplift and subsidence over large areas. The 1964 Alaska earthquake caused land-level changes in an area over 140 000 km² (Figure 9.24; Plafker, 1969). Secondary faults above the plate interface zone also can produce large surface displacements. For example, eight metres of secondary fault rupture occurred along the Patton Bay fault during the 1964 Alaska earthquake (Plafker, 1969), and 12.5-metres of secondary displacement occurred along the Wairarapa fault during the 1855 earthquake near Wellington, New Zealand (Grapes and Wellman, 1988).

9.4 Seismic source characterisation

Three primary geological questions need to be considered in earthquake engineering studies: where are earthquakes likely to occur; how big an event needs to be addressed; how often will these events occur? Each of these questions need to be addressed both for probabilistic and deterministic seismic hazard analyses (PSHA and DSHA, respectively) and for evaluation of surface fault rupture, liquefaction, and seismically induced slope failure. The questions typically are considered by conducting a series of investigations that often include one or more of the following steps:

1.   data compilation and review
2.   analysis of regional tectonic setting
3.   analysis of geological, seismological, geophysical and geodetic data
4.   aerial and field reconnaissance
5.   analysis of local geological setting
6.   geological and geomorphic mapping of specific faults
7.   subsurface investigations such as paleoseismic trenching, boring programs
8.   laboratory and dating studies
9.   development of parameters for specific project requirements
10.   treatment of uncertainty.

Location—Where are earthquakes likely to occur?

Identifying the location and geometry of seismogenic faults (or seismic source zones) is accomplished through interpretation of regional geology, seismic activity, geodetic data, and Quaternary geology and geomorphology. The regional geologic, seismologic and geodetic data provide general information on the location, style and rate of crustal deformation. However, identifying and characterising specific faults and seismic source zones commonly requires detailed assessment of the Quaternary geology, geomorphology and seismology of a study region (Yeats *et al.*, 1997). In order to identify the locations of active faults, it is necessary to have deposits or

Figure 9.17   Graben formed in the fault rupture of the 1983 M_w 7.3 Borah Peak, Idaho earthquake. The zone of ground deformation is approximately 30 metres wide at the intersection of the road. Photograph from Gori and Hayes, 1992.

landforms of Pleistocene or Holocene age (Hanson *et al.*, 1999).

Although different techniques are used to characterise strike-slip, normal, and reverse faults, the overriding principles are the same. Investigations are conducted to identify where repeated moderate to large magnitude earthquakes have produced permanent ground deformation, affected the distribution of surficial deposits, and modified the landscape in Quaternary time. The primary objective of geologic and geomorphologic field investigations is to identify landforms such as fault scarps, drainage systems, glacial or volcanic deposits, alluvial fans, and fluvial or marine terraces, which indicate that normal landscape evolution processes have been perturbated by tectonic activity. These landforms provide a spatial datum (e.g. a surface or lineation) that serves as a

strain gauge, which can be dated to document the location, rate of activity, and whether or not the most recent movement of a fault is older or younger than the activity threshold applicable to the project (e.g. 10 000 or 50 000 years before present).

A stepwise process typically is followed to identify and map faults. Following literature review and analysis of regional geological, seismological, and geodetic data, these steps include:

1. *Aerial photograph interpretation.*   Figure 9.26 provides an example of an interpreted aerial photograph and illustrates the types of features that can be recognised. The photograph is from the eastern side of the Sierra Nevada, near Bishop, California. This area was blanketed by the Bishop Tuff, an ignimbrite flow,

Figure 9.18   Uplifted and dissected alluvial fans along the eastern Andes range-front, Mendoza, Argentina. Note the offset alluvial fan complexes and fault scarps (arrows) on A; and, deep stream incision of alluvial fan complexes shown on B. Stream incision indicates a change in stream base-level due to uplift of the hanging wall (typically the mountain front). Photographs by James Hengesh.

approximately 700 000 years ago. The tuff forms a plateau surface which can be used as a strain gauge. The tuff has been incised by the Owens River and a smaller stream. A series of terraces (Qt_{1-3}) are inset into the tuff surface. On the lower left of the photo, an alluvial fan (Qa_1) has eroded and removed the terraces. The terraces and alluvial fan surfaces are related to climatic changes during late Quaternary glacial and interglacial periods. In addition to these geomorphic surfaces, scarps, offset drainages, and tonal lineaments are mapped.

Based on published information on the geology and seismicity, field reconnaissance and interpretation of the aerial photographs,

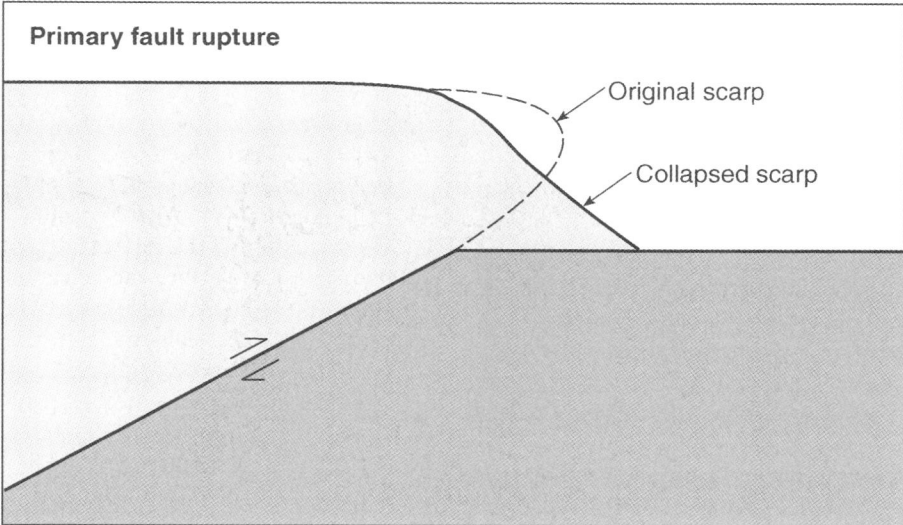

Figure 9.19   Photograph and schematic cross section of the primary fault rupture produced during the 1999 M_w 7.6 Chi-Chi, Taiwan earthquake (Kelson *et al.*, 2001). Collapse of the overriding tip of the hanging wall block forms a wedge of material above the ground surface. Note the difference in the angle of the fault scarp where track material prevented collapse and in the field where collapse of the scarp occurred.

multiple lines of evidence suggest that the Tungsten Hills fault is a capable tectonic feature. The Tungsten Hills fault offsets and forms a scarp on the 700 000 year old Bishop Tuff surface, offsets the Owens River Gorge, and forms lineaments across the alluvial fan surface. Uplift on the southeastern side of the fault may have changed the stream base-level

and formed the Qt_{1-3} terraces. The 1986 Chelfant Valley earthquakes (M5.7, 6.3, 5.4) occurred along the eastern projection of the Tungsten Hills fault. Therefore, this fault is recently active, offsets mid-Pleistocene deposits, offsets the mid- to late Pleistocene Owens River Gorge, and forms a series of late Quaternary terraces. Beause of the fault's geological and

Secondary fault rupture

Figure 9.20 Photograph and schematic cross section of secondary deformation produced during the 1999 M$_w$ 7.6 Chi-Chi, Taiwan earthquake (Kelson *et al.*, 2001). Three types of secondary deformation were observed: tensional faults in the fault propagation fold; back-thrusts; and complex deformation involving collapsed fault scarps, back thrusts, and penetrative fracturing or cracking.

geomorphic expression, long-term history of movement, and location it was considered in developing seismic design criteria for the Pleasant Valley dam. Aerial photograph interpretation of the area provided a basis to locate the Tungsten Hills fault and to identify geomorphic features that could be used to estimate the Quaternary fault slip rate, fault length, and

potential earthquake magnitude and recurrence interval.

2. *Surface Geological and Geomorphological Mapping.*   Geological and geomorphological mapping provides detailed field verification of the distribution of Quaternary features such as marine or fluvial terraces, volcanic or lacaustrine deposits, soil types, landslides, and erosional and/or depositional surfaces, such as stream terraces and alluvial fans. The example of an interpreted aerial photograph, illustrated on Figure 9.26, shows the variety of features that can be mapped in a small area. Maps showing the distribution of Quaternary deposits are particularly useful for defining areas undergoing active deformation or alluviation. The scale and level of detail shown on geologic and geomorphic maps is dependent on the level of documentation and geotechnical requirements of a project.

Quaternary deposits and geomorphic surfaces are mapped on the basis of several stratigraphic, geomorphic, and pedologic criteria, including: (1) topographic position in a sequence of inset deposits or surfaces; (2) relative degree of surface modification (e.g. erosional dissection, mima-mound relief, etc.); (3) relative degree of soil-profile development or other surface-weathering phenomena; (4) superposition of deposits separated by erosional unconformities and/or buried soils; (5) relative or numerical age of individual deposits; (6) physical continuity and lateral correlation with other stratigraphic units; (7) distinctive lithology or mineralogy; and (8) textural or lithologic uniqueness such as inclusion of distinctive volcanic ash, lacustrine sediments, or exotic clast lithologies.

3. *Soil profile development.*   The analysis of soil-profile development is an effective means to assess the relative age of Quaternary deposits and features and is an integral part of Quaternary geologic mapping (Sowers *et al.*, 1998; Birkeland, 1984). For example, the use of soil-profile development can be a time- and cost-effective approach to evaluate relative ages of related surfaces, or to correlate a

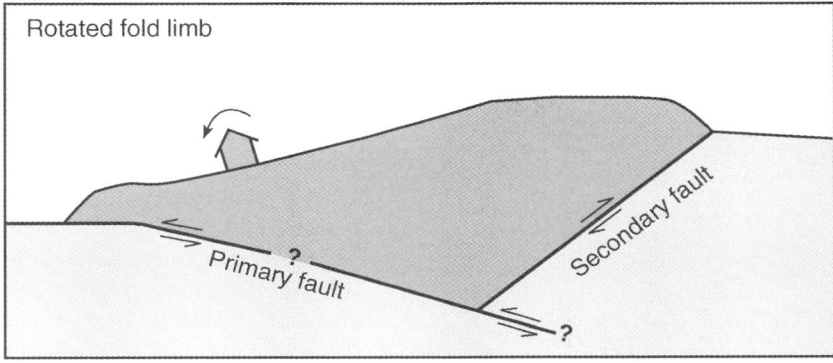

Figure 9.21   Photograph and schematic cross section of a rotated fold limb, produced during the 1999 M_w 7.6 Chi-Chi, Taiwan earthquake (modified from Kelson *et al.*, 2001). The block between the primary fault trace and the secondary fault trace is rotated in a counterclockwise direction.

surface of unknown age to one of known age (Sowers *et al.*, 1998).

Soil profile development typically is analysed in shallow soil pits on Quaternary surfaces, such as glacial moraines, alluvial fans, stream terraces, floodplains, and landslide deposits. Soil characteristics that typically are assessed include: depth of profile development; soil texture; clay and silt accumulation and orientation (flows); salt accumulation (e.g. carbonate); soil structure; soil color; and degree of horizonation (Sowers *et al.*, 1998; Birkeland, 1984). Comparison of the relative degree of soil-profile development between like surfaces with similar parent material leads to the development of a

relative chronology for landforms, sometimes referred to as a soil chronosequence. The estimated ages of landforms across a fault zone can then be used to locate and evaluate the activity of a fault, and to estimate a fault slip rate. These data are used in characterising faults for seismic hazard assessment.

Buried soils represent former periods of landscape stability/instability, and can be indicators of past climatic, tectonic and depositional conditions. Buried soils can be recognised in natural exposures, exploratory excavations, and boreholes. Buried soils in outcrops typically have many diagnostic features similar to those of surface soils (e.g. soil

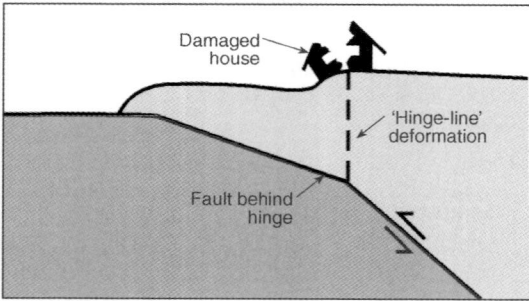

Figure 9.22 Photographs and schematic cross section of 'hinge-line' fault produced during the 1999 M_w 7.6 Chi-Chi, Taiwan earthquake (modified from Kelson *et al.*, 2001). A zone of cracking and ground deformation occurs above the fault-bend, at depth.

horizonation, soil structure, color). In a rapidly aggrading floodplain or alluvial-basin depositional environment, buried organic- rich soils commonly are preserved and useful for establishing the stratigraphic correlation and depositional history of alluvial sequences.

4. *Geomorphic analysis.* Geomorphic techniques are an integral part of Quaternary geologic mapping and provide useful information on the timing, rates, and locations of ground deformation. Geomorphic analyses commonly involve inferring the amount of fault displacement during individual earthquakes, and over specified lengths of time, from measurements of landform deformation. Geomorphic features such as stream channels (thalwegs), stream terraces, marine terraces, and glacial moraines commonly form well preserved datum's from which to assess fault slip rate, recency of activity, and the direction and amount of displacement during an earthquake.

Fluvial terraces, marine terraces, or glacial landforms are examples of geomorphic features that can be used as strain gauges to evaluate location, timing, amount and rates of deformation where these types of features overlie a fault (Figure 9.27). For example, streams form at a relatively constant gradient; however, where crossed by faults, the position and gradient of stream thalwegs and terraces may be affected by fault displacement, tilting or folding. Terraces on the upthown side of a fault may be higher in elevation than terraces of equivalent age on the downthrown side of the fault (Figure 9.27). Increasing scarp height on progressively older terrace surfaces indicates recurrent fault moment as shown by terraces Qt_1 and Qt_2 on Figure 9.27*b*. Measurements of scarp height across these surfaces and estimates of surface ages based on soil development or other properties may yield data on fault slip rate. The degree of post-earthquake scarp modification also can be an indicator of the age of an event (Hanks, 1998). Terraces that cross a fault and are not deformed provide a useful datum to confirm the inactivity of a fault.

Magnitude—How large an event needs to be considered?

Once an active fault is located, the maximum earthquake that the fault is capable of producing in the current tectonic environment needs to be estimated.

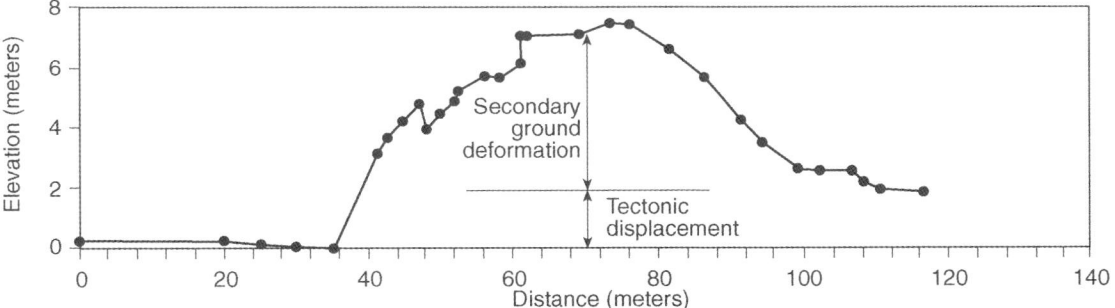

Figure 9.23   Photograph and topographic profile of scarp produced during the 1999 M_w 7.6 Chi-Chi, Taiwan earthquake. Note that the total permanent ground deformation is approximately 8 metres, but only 2 metres of the displacement is due to tectonic offset. The scarp has been amplified due to the 'plowing' effect of the fault tip. The photograph is of an originally horizontal rice paddy. From Kelson *et al.*, 2001.

For engineering purposes the maximum earthquake magnitude (M_{max}) is usually expressed in terms of its moment magnitude (M_w).

The M_{max} is commonly estimated by identifying the length, area or displacement of a particular fault segment, and deriving the characteristic magnitude from empirical relationships (e.g. Figure 9.28; Schwartz and Coppersmith, 1984; Wells and Coppersmith, 1994, Hanks and Bakun, 2002). Defining the length of a fault segment to be used in estimating earthquake magnitude requires evaluation of surficial geology and geomorphology to define the fault geometry and segmentation points (Yeats *et al.*, 1997; McCalpin 1996; Schwartz and Sibson, 1989). Investigations following the 1999 Turkey earthquake provided valuable new information regarding the role of fault geometry (e.g. fault stepovers) in the termination of fault rupture (Lettis *et al.*, 2000). Detailed mapping of the North Anatolian fault showed a correlation between width of a fault stepover and the ability of the stepover to arrest

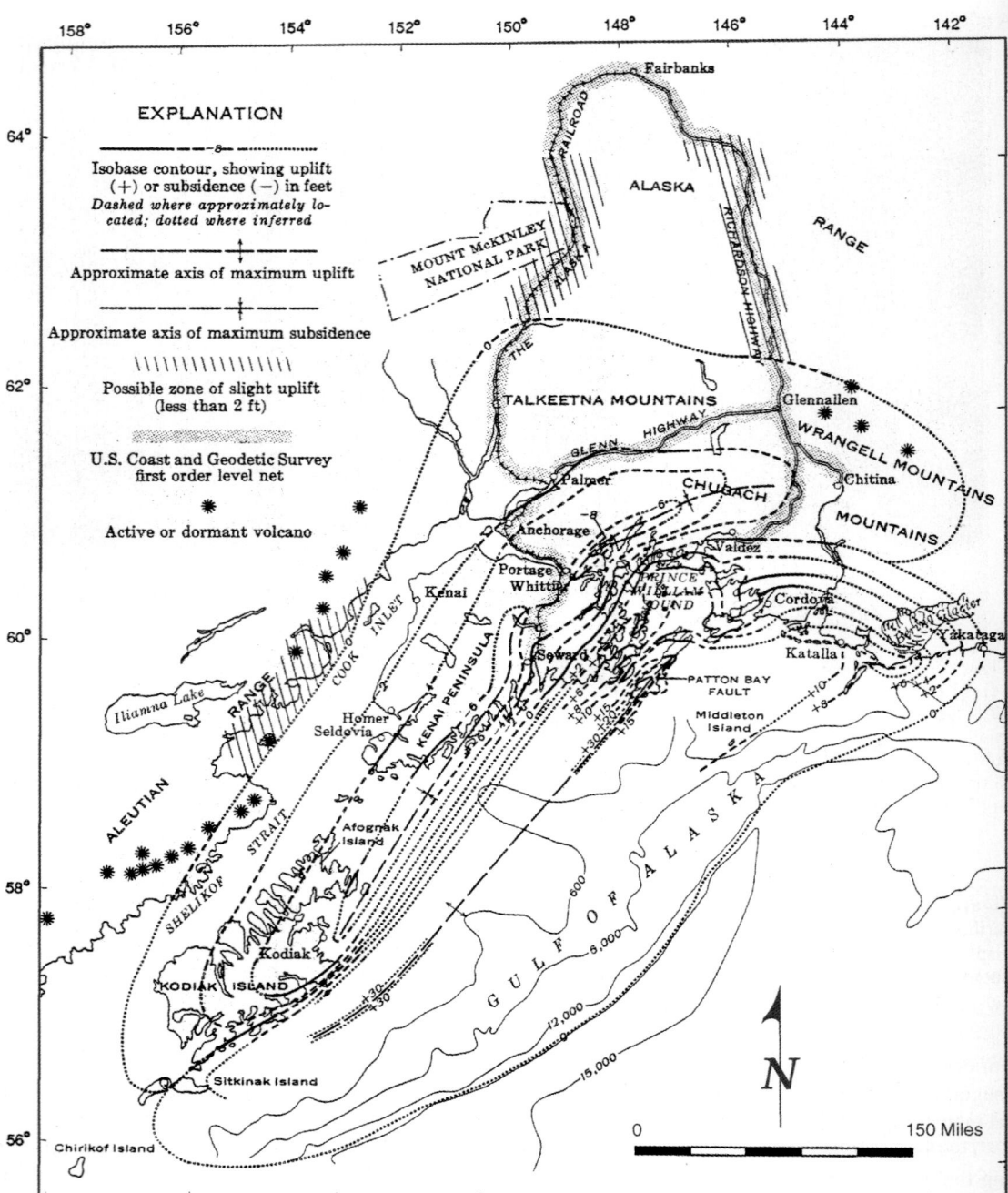

Figure 9.24 Area of surface deformation associated with the 1964 M$_w$ 9.2 Great Alaska earthquake. The earthquake caused up to 30 feet of uplift and 8 feet of subsidence. Modified from Plafker, 1969.

Figure 9.25   Examples of emergent and submergent land forms produced during subduction zone earthquakes: (*A*) Raised wave-cut platform in Wellington Harbor formed during the 1855 Wellington, New Zealand earthquake. (*B*) Storm berm and driftwood lines raised approximately 14 feet during M_w 9.2 Great Alaska Earthquake (Kayak Island). Upper driftwood line represents the pre-1964 earthquake shoreline. (*C*) Spruce forest killed due to tectonic subsidence and salt-water innundation during 1964 Alaska earthquake. (*D*) Spruce stump and subareal peat layer buried by intertidal mud, Hinchinbrook Island, Alaska. Tree killed due to tectonic subsidence and submergence, as in Photo C, approximately 1336 to 1390 years ago. Photographs by James Hengesh.

propagation of the fault rupture (Lettis *et al.*, 2002). For example, the 5-km wide Karamursel stepover was sufficient to arrest 4 to 5.5 metres of fault displacement near the Hersek peninsula, while 3- to 5-metres of fault displacement ruptured through the 2-km wide Gulcuk and Sapanca stepovers (Figure 9.29). Data from the 1999 Turkey earthquake combined with data from other strike slip earthquakes were used to develop a preliminary empirical relationship between surface rupture displacement and fault stepover width. The graph on Figure 9.29 provides an initial constraint on the size of a fault stepover that can be considered a segmentation point in estimating fault rupture length and earthquake magnitude for strike slip faults (Lettis *et al.*, 2002). Observations from the 1999 Turkey earthquake

underscore the importance of conducting careful geological mapping and analysis to identify the geometry of fault segmentation points when estimating maximum earthquake magnitudes.

Maximum earthquake magnitude can be difficult to estimate in areas characterised by low seismic activity or where faults are obscured by youthful surficial geological deposits. For example, a sequence of three M ~ 7.3 to 8.1 earthquakes occurred along the New Madrid Seismic Zone in the Mississippi Embayment province of the central United States in 1811–1812. However, no surface expression of the faults that caused two of these events has been recognised; therefore, empirical relations between fault length, area and magnitude cannot be readily used to estimate the future M_{max}. This

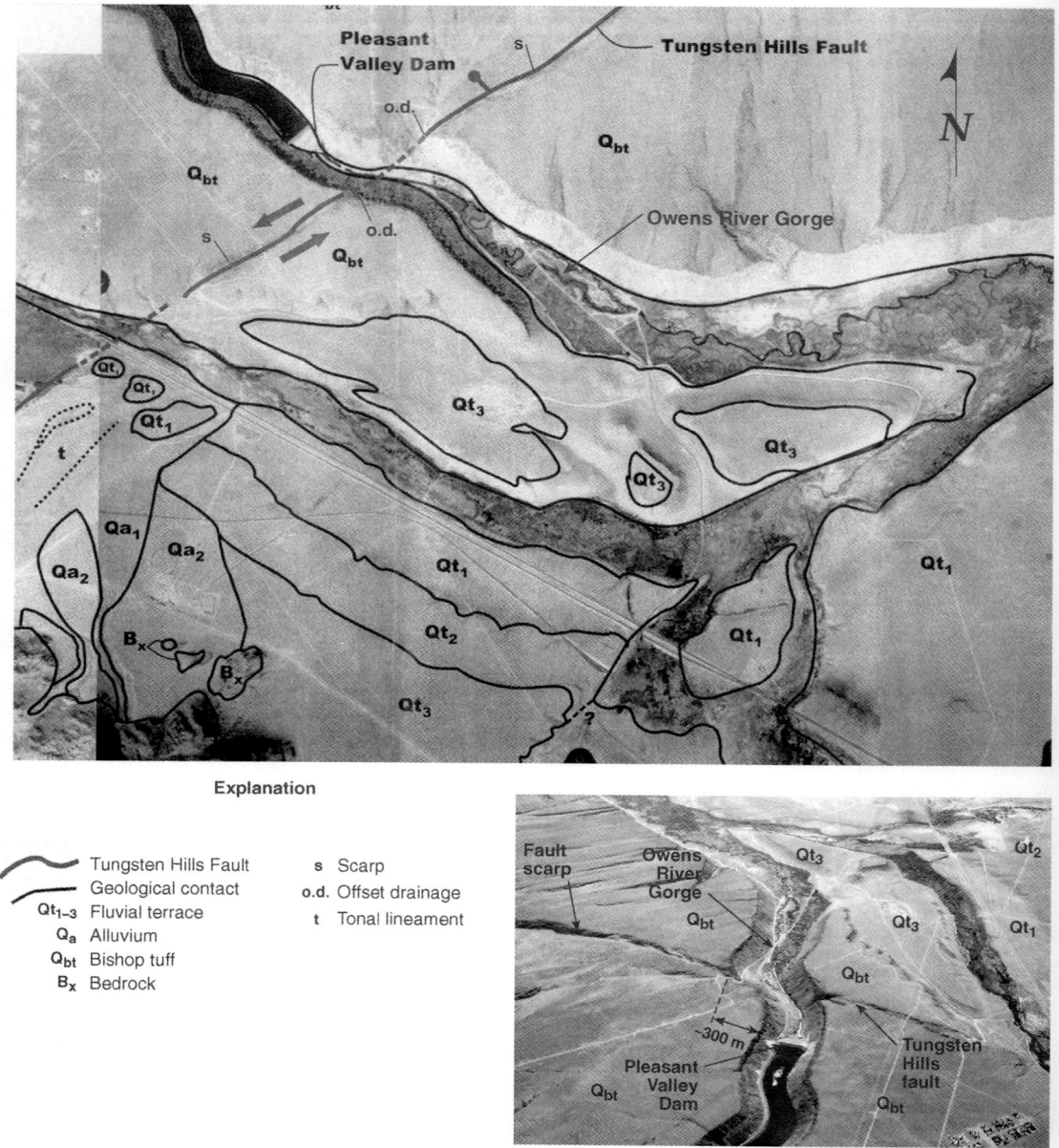

Figure 9.26   Interpreted aerial photograph of Tungsten Hills fault, near Bishop, California (US Bureau of Land Management (date 10.02.77), series CA01-77, no. 4-28-7, scale 1:24 000). The interpreted photograph identifies mid-Quaternary volcanic deposits, late Quaternary fluvial terrace and alluvial fan surfaces, as well as scarps, offset drainages, and tonal lineaments. Lower photograph is an oblique southeastward view of same area in upper photograph. Note ~300 metre offset of post 700 ka Owens River Gorge across Tungsten Hills fault. Lower photograph by James Hengesh.

historical earthquake sequence represents the expected future maximum earthquake magnitude for the New Madrid Seismic Zone. Therefore, historical seismicity, e.g. the magnitude of the

1811–1812 earthquakes, can be used to estimate the maximum magnitude of future earthquakes in this region. The M_{max} of the New Madrid Seismic Zone has been estimated by Hough *et al.* (2000)

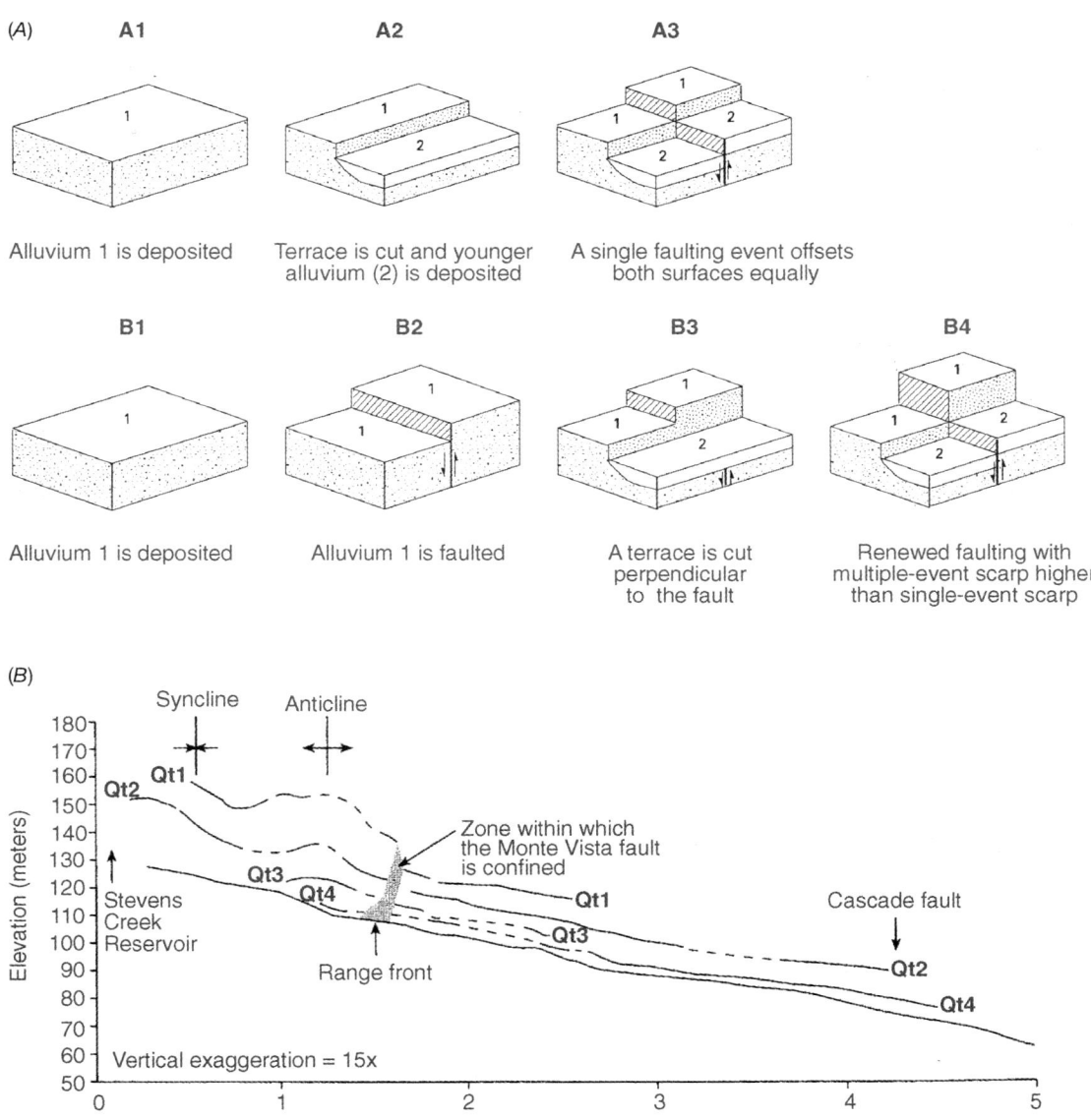

Figure 9.27   (*A*) Idealised block diagrams illustrating the use of stream terraces to evaluate recurrent fault activity. Sequence A shows the development of two river terraces that are subsequently faulted. Sequence B shows the development of two terraces separated by a faulting event and followed by a second faulting event. The fault scarp is progressively larger across the older terrace surface (modified from McCalpin, 1996). (*B*) Longitudinal profile of stream terraces (Qt1–Qt4) along Stevens Creek across Monte Vista and Cascade faults, Santa Cruz Mountains, California (Hitchcock and Kelson, 1999).

and Bakun and Hopper (2004) using historical intensity data from the 1811–1812 earthquake sequence. These approaches use Modified Mercalli Intensity (MMI) data from historical damage reports to estimate the moment magnitude of the earthquakes.

In areas such as the New Madrid Seismic Zone where fault exposures are not available, several

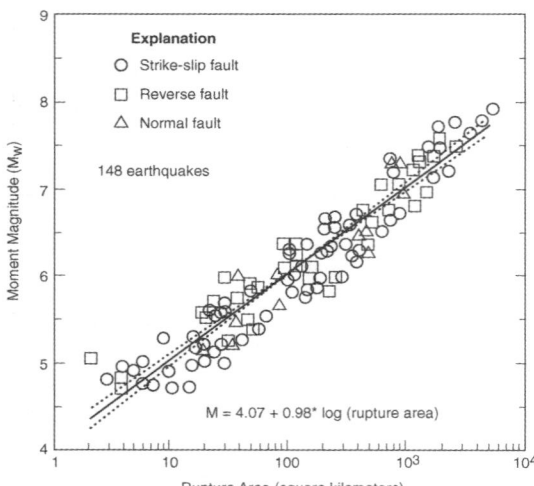

Figure 9.28 Relation between earthquake magnitude and rupture area (after Wells and Coppersmith, 1994).

additional approaches are available to estimate the magnitude of an earthquake. If paleoliquefaction features are present, the areal distribution of liquefaction features, size and characteristics of sand blow deposits, and the geotechnical properties of subsurface soils can all be used to estimate earthquake magnitudes. Table 9.1 summarises the data requirements and outputs for a number of these techniques.

There are considerable uncertainties in the use of paleoliquefaction features to derive earthquake magnitude. Some of these uncertainties may include variation in groundwater elevations, site response effects, and crustal attenuation properties (Tuttle, 2001). However, in areas such as the Mississippi Embayment where large magnitude earthquakes are known to have occurred, but little data are available regarding fault dimensions, use of liquefaction fields can provide valuable constraints on the size of past earthquakes (Tuttle *et al.*, 2002a; 2002b; Obermeier *et al.*, 2001)

Earthquake recurrence — how often will events occur?

Earthquake recurrence interval is a measure of the frequency of occurrence of an earthquake with a given magnitude. Earthquake recurrence intervals

Table 9.1   Summary of methods used to estimate magnitude or ground motion acceleration from paleoliquefaction features.

Method	Input Data Required	Output	Reference
Magnitude-bound	Maximum distance from source to liquefaction field	Magnitude	Ambraseys, 1988; Obermeier *et al.*, 2001; Pond, 1996
Cyclic Stress	Geotechnical borings (blow counts), ground motion attenuation models	Magnitude	Hwang et al., 1995; Seed and Idriss 1971; Seed *et al.*, 1983; Whitman, 1971; Youd *et al.* 2001; Idriss and Boulanger, 2004; Cetin *et al.*, 2004
Energy	Geotechnical data, Hypocenter data	Back calculated PGA for assumed magnitude	Berrill and Davis, 1985
Ishihara	Thickness of cap, thickness of source sand	Back calculated PGA; thickness of fractured cap if PGA is known	Obermeier *et al.*, 2001; Ishihara, 1985; Youd and Garris, 1995
Energy Based Approach	Geotechnical borings (blow counts)	Energy and stress conditions at site	Dobry *et al.*, 1982; Green, 2001; Law *et al.*, 1990; Obermeier *et al.*, 2001; Pond, 1996
Comparison to Modern Earthquakes	Modern earthquake analogs, size and spatial distribution, with known magnitudes	Magnitude	Tuttle *et al.*, 2002a
Arias Intensity	Ground motion, field penetration tests	Magnitude, source distance	Tuttle *et al.*, 2001; Arias, 1970; Kayen and Mitchell, 1997, Obermeier *et al.*, 2001

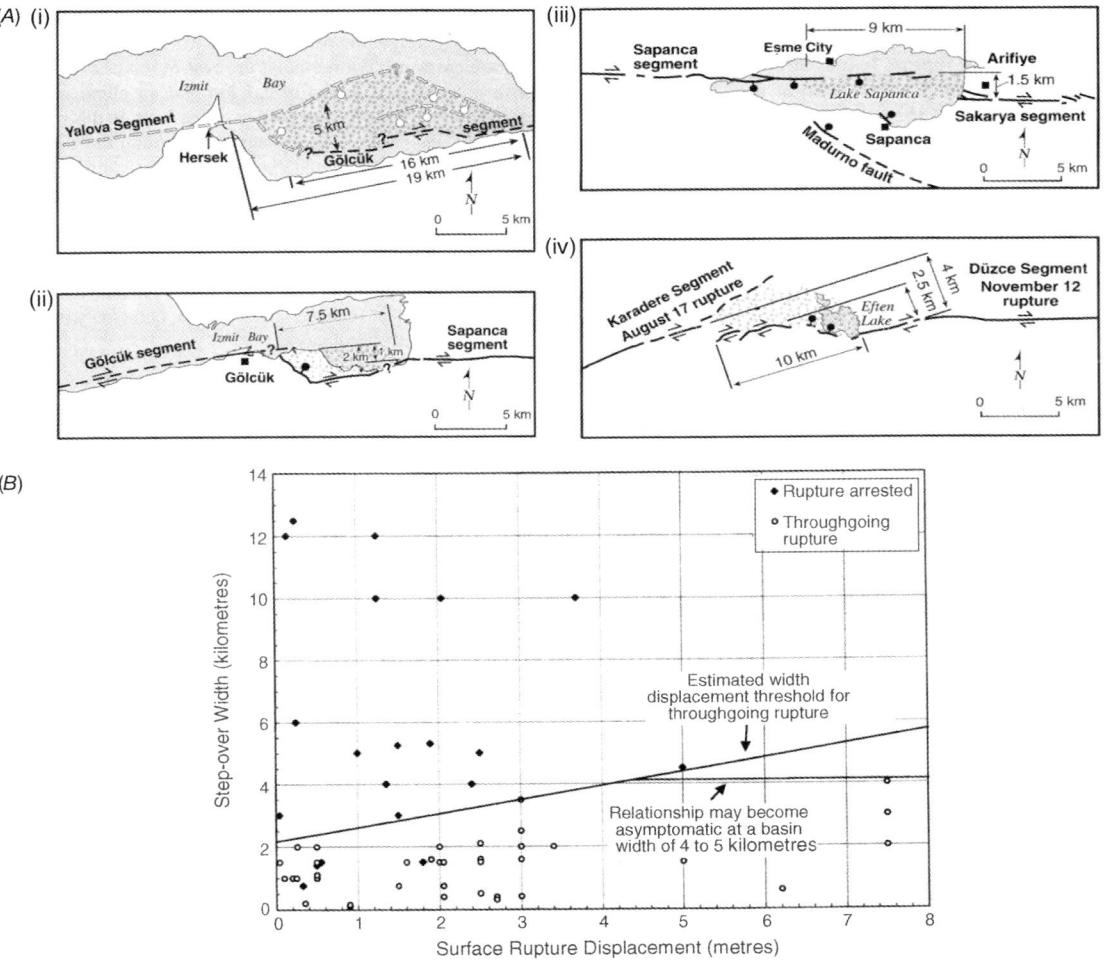

Figure 9.29 (*A*) Comparison of pull-apart basins along the 17 August 1999 Izmit earthquake rupture. (i) Karamursel step-over; (ii) Golcuk step-over; (iii) Lake Sapanca step-over; (iv) Eften Lake step-over. Modified from Lettis *et al.*, 2002. (*B*) Relationship between average surface displacement or maximum displacement entering the step-over (which is largest) and average width of the step-over. Modified from Lettis *et al.*, 2002.

usually are estimated using one of three general approaches:

1. Paleoseismic Approach – direct observation and dating of past earthquakes from offset geological units, paleoliquefaction features, or coseismically produced landforms.
2. Slip Rate Approach – quotient of fault displacement and fault slip rate.

3. Seismological Approach – statistical analysis of an earthquake catalogue to develop an earthquake magnitude-frequency distribution.

Paleoseismic Approach The paleoseismic approach offers the benefit of directly describing and dating evidence of past earthquakes. This typically is accomplished through subsurface investigations, which provide the most definitive information on fault location, fault behaviour, features produced during paleoliquefaction, and the timing of past earthquakes. Subsurface investigations include exploratory trenching, large- and small-diameter boreholes, and geophysical profiling. Exploratory trenching is the most commonly used, and generally most successful, method for assessing paleoseismic activity. Sites for trenching should be carefully chosen following initial geologic observation and mapping. Preferred sites include those with datable late Quaternary deposits across the fault trace or paleoliquefaction feature, minimal erosion, and continuous deposition to provide a complete record of earthquake activity. Periods of erosion and/or non-deposition may destroy of fail to preserve evidence of earthquakes.

Fault rupture associated with past earthquakes produces a range of features that can be recognised in trenches to evaluate the presence and timing of individual paleoseismic events. As shown on Figure 9.30, these include:

1. deposits offset by a fault
2. abrupt upward termination of a fault strand
3. deposits and surfaces deformed by folding, tilting or warping
4. lithologic variations across a fault
5. variation in thickness of stratigraphic units or soils across a fault
6. deposits directly related to faulting, including scarp-derived colluvium, impounded or ponded alluvium (e.g. sag pond deposits), or fissure-fill deposits
7. transformed deposits including liquefied sediments and sediment sheared by faulting
8. systematic and abrupt or stepped increases in displacement downsection
9. intruded material such as fissure fills, fault gouge, or liquefied clastic deposits
10. open fissures or pockets along a fault plane; and fault planes exposed at the surface (i.e. along scarp).

Figures 9.31a through 9.31d illustrate the development of multiple colluvial wedges by recurrent surface fault rupture. Each fault rupture surface produces a scarp with a steep 'free face' (Figure 9.31a). Sediment eroded from the free face is shed

1. Faulted Rock or Sediment

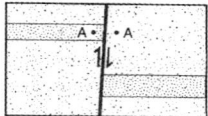

2. Upward Fault Termination (UFT) at Unconformity

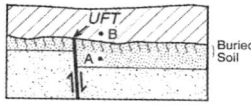

3. Deformed Rock, Sediment or Unconformity

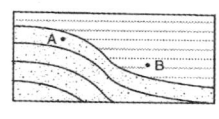

4. Juxtaposition of Unlike Lithologies

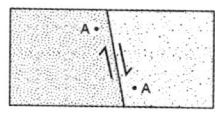

5. Thickness Variation in Stratigraphic Unit and/or Soil

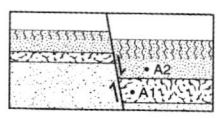

6. Colluvial Wedge

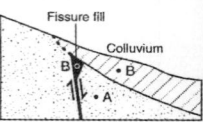

7. Transformed Material

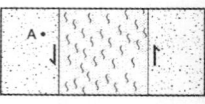

8. Increase in Displacement Downsection

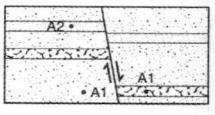

9. Intruded Material

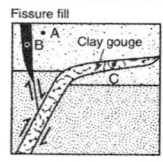

10. Open Fissures along Fault Plane and Exposed Fault Plane

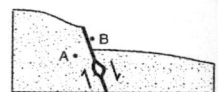

Figure 9.30 Schematic diagram illustrating the principal stratigraphic and structural criteria used to identify the occurrence and timing of past earthquakes. Dated samples at locations A predate the earthquake. Dated samples at locations B post-date the earthquake. Dated samples at C are not helpful for deciphering chronology of past earthquakes because the sample may pre-date or post-date the earthquake depending on geologic context. Dated samples at locations A and B, therefore, constrain the timing of the earthquake. Where evidence of multiple earthquakes is present, locations A1 pre-date the earlier event and locations A2 post-date the earlier event and pre-date the later event. Modified from Lettis and Kelson, 1998.

across the fault, producing a colluvial wedge (Figure 9.31b). Given sufficient time between surface faulting events, weathering and soil development will occur on the upper surface of the

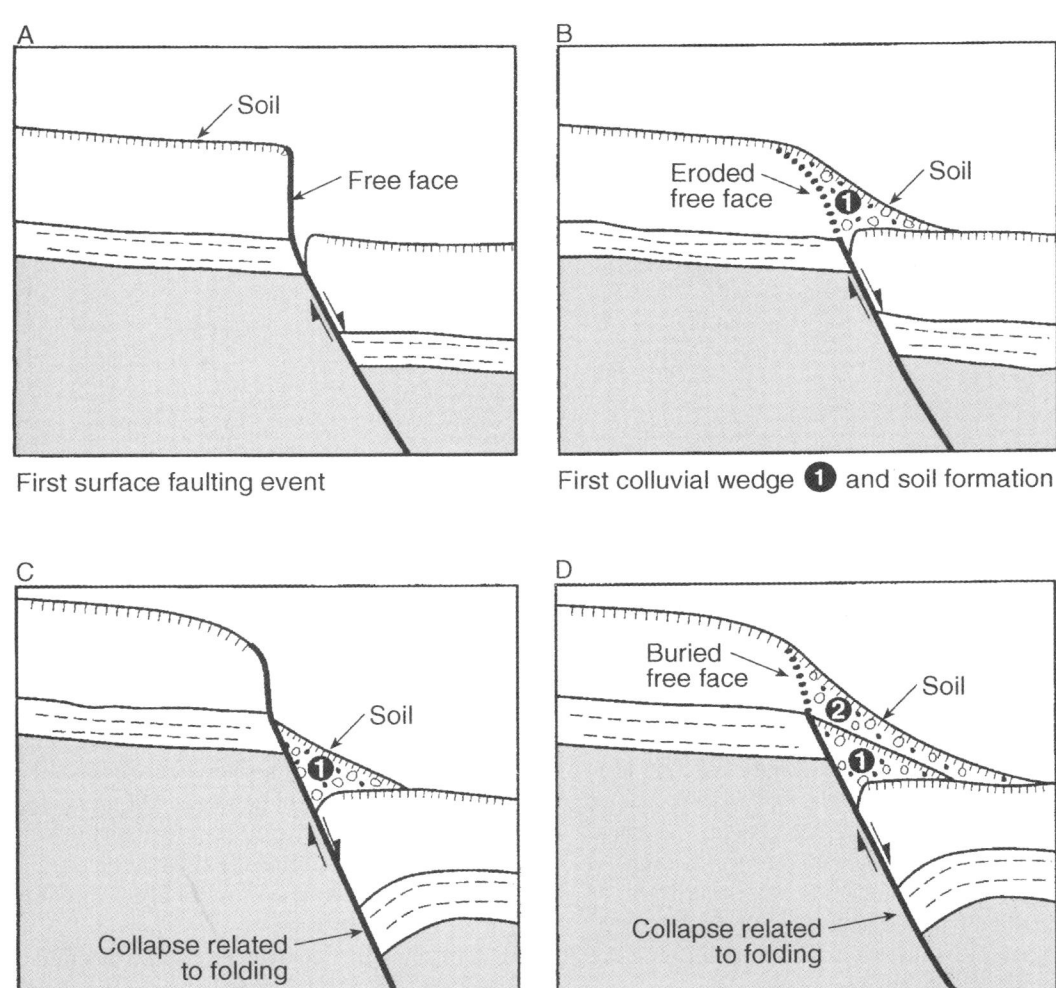

A

Soil

Free face

First surface faulting event

B

Eroded
free face

Soil

First colluvial wedge ❶ and soil formation

C

Soil

Collapse related
to folding

Second surface faulting event

D

Buried
free face

Soil

Collapse related
to folding

Second colluvial wedge ❷ and soil formation

Figure 9.31   Formation of colluvial wedges from recurrent surface faulting on a single normal fault (modified from Schwarz and Coppersmith 1984).

colluvial wedge (Figure 9.31c). The degree of weathering and soil development provides an indication of the length of time between earthquakes. If a second faulting event occurs (Figure 9.31d) the first colluvial wedge will be displaced, and the renewed scarp will give rise to a second colluvial wedge. The use of colluvial wedge stratigraphy, combined with the other features illustrated on Figure 9.30, enables paleoseismologists to decipher an 'event-stratigraphy' of past earthquakes. When logging a trench it is critical to document

and describe the full range of features observed, and special attention should be given to the textures and orientations of clasts in trench walls and floor. Figure 9.32 provides an example of the type of detail that should be included in a trench log. This detail allows recognition of subtle folding, warping, or tilting, or structural fabrics such as oriented clasts, that might not be readily observed if only major stratigraphic units are logged.

An event stratigraphy may be dated by geochronologic methods (Sowers *et al.*, 1998). The

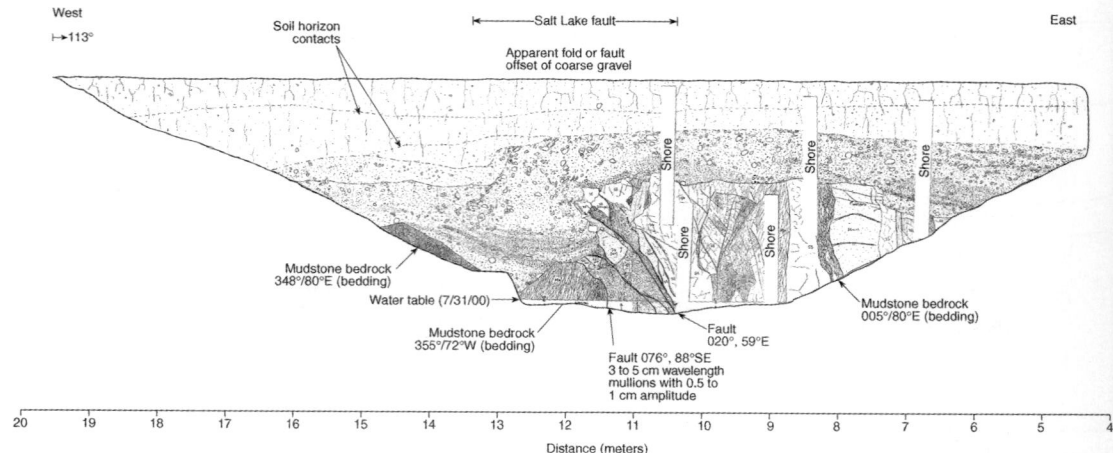

Figure 9.32 Log of trench from Salt Lake fault, California. Note how the detail of the log shows tectonic fabric of sand and gravel deposits. The fabric results from displacement along fault plane and rotation, tilting or alignment of clasts or layers. Exposure shows bedrock, fluvial stratigraphic units, and soil horizonisation.

method used is dependent on the type of feature being dated, the approximate age of the feature (different methods are applicable for different time periods and sample materials, and the presence of suitable material for dating e.g. charcoal, volcanic ash, etc.) in a clear relationship to the fault related feature. Generally, the identification and dating of two or more of these features is required to document a paleoseismic event and to determine the relative timing between past earthquakes.

In areas such as the central and eastern United States where earthquakes have occurred without producing surface fault rupture, paleoliquefaction features can be used to date the occurrence of past earthquakes (Obermeier *et al.*, 2001; Tuttle *et al.*, 2002a). Figure 9.33 illustrates the form of a typical paleo-sand vent and the approach to dating the earthquake history using paleoliquefaction features. Samples should be collected from within the vented sand material or below the vented horizon to provide a maximum limiting age of the event, and samples should be collected from above the vented sand horizon to provide a minimum limiting age of the event. The age of the event horizon is then bracked and the event can be inferred to have

occurred within the time interval represented by the samples. Care should be taken in describing the stratigraphy at a liquefaction site to determine the geological and geotechnical properties of the soils, the geological context of the site, and to provide descriptions that could be used in interpreting whether one or more events are represented in the exposure.

Establishing the recurrence of large magnitude subduction zone earthquakes involves the dating of either submergent or emergent landforms that were produced by coseismic crustal deformation of coastal areas above the plate interface part of the subduction zone complex. Although subduction zone earthquakes do not produce primary surface fault rupture, areas affected by these earthquakes experience both coseismic and interseismic land-level changes (Figure 9.34; Plafker and Rubin, 1992). Where land-level changes occur along the coast, rapid environmental changes (e.g. from subareal to intertidal or visa versa) take place (Figures 9.24, 9.25 and 9.34). The deposits representing the time of change of distinct depositional environments can be dated and serve as a proxy to bracket the time of the earthquake.

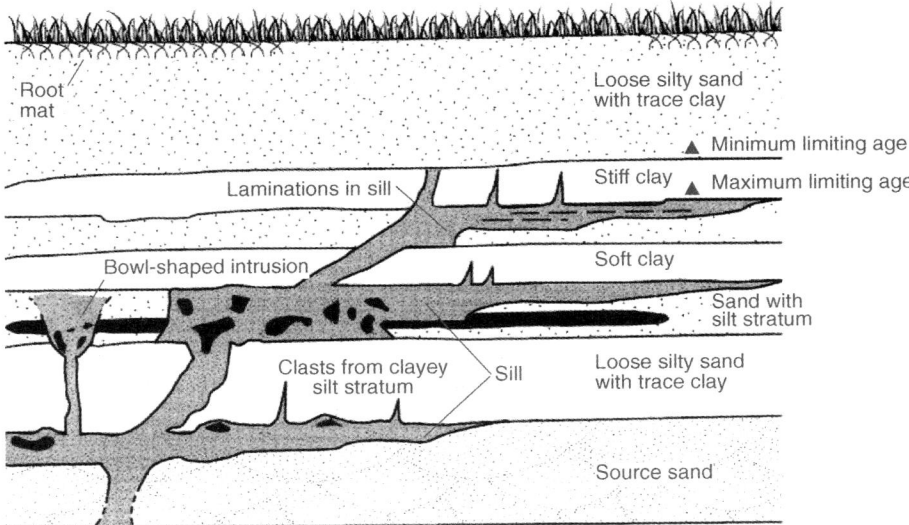

Figure 9.33 Schematic vertical section showing where liquefaction related sand dykes and sills form preferentially. Length of sketch represents 10 to 100 metres. Height represents 3 to 5 metres. Such severe sill development is typically accompanied by venting of sand to the surface. Triangles indicate position of samples that could be collected and dated to bracket the age of the liquefaction event. Modified from Obermeier, *et al.*, 2001.

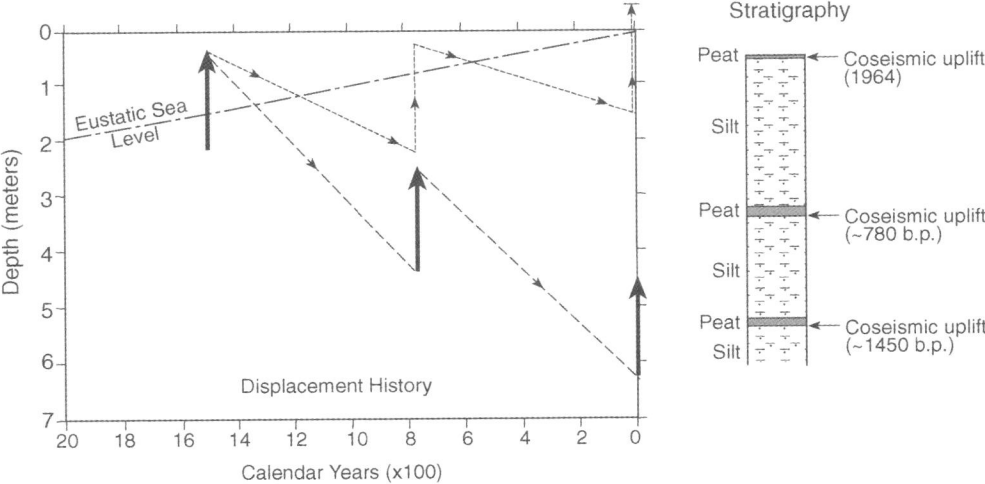

Figure 9.34 Simplified displacement history and stratigraphy for the younger part of the Copper River delta sequence. Column shows stratigraphy and ages of the last three coseismic uplift events. Graph shows an idealised tectonic displacement history for the last two pre-1964 events to illustrate the 'yo-yo' effect of coseismic uplift (heavy arrows) alternating with larger interseismic subsidence (dashed lines). Also shown is the vertical displacement path of the ground surface (dotted line), the sum of the tectonic displacements plus interseismic sediment accumulation. The amount of uplift for the pre-1964 events is arbitrarily assumed to be comparable to the 2 m uplift that occurred during the 1964 Alaska earthquake, the interseismic subsidence rate shown is about 5.5 mm/yr, and eustatic sea level rise is taken as +1mm/yr. Modified from Plafker and Rubin, 1992.

Figure 9.25*d* shows an example of an inter-tidal marsh with evidence of sudden changes from intertidal to subareal environments. The site shown on this photograph has estuarine silt overlain by a subareal peat, which is overlain by a second estuarine silt. The lower silt, dated 1390 to 1720 2σ calibrated years before present (cal. ybp), was coseismically uplifted out of the tidal range and colonised as a spruce forest. The overlying subareal peat is dated 1060–1336 2σ cal ybp. These dates represent the minimum and maximum limiting ages, which bracket the time of the earthquake. The forest subsequently was submerged interseismically and was again uplifted during the 1964 Alaska earthquake. The current vegetated surface represents the modern analog of the environment that formed the buried peat. An intervening event dated approximately 700 ybp was not preserved at this site, which highlights one source of uncertainty that must be considered in estimating earthquake recurrence in marsh deposits (i.e. lack of preservation).

Slip Rate Approach   The slip rate approach for estimating earthquake recurrence is based on the assumption that the earthquake occurrence process for many major faults follows a time-dependent model of strain accumulation and release (Kanamori and Brodsky, 2004; Figure 9.5), and that these faults produce characteristic earthquake magnitudes with characteristic displacements. This is referred to as the characteristic earthquake model (Schwartz and Coppersmith, 1984). Using the slip rate approach to estimate earthquake recurrence requires knowledge of both amount of coseismic displacement and fault slip rate. Given these parameters the earthquake recurrence interval (T_r) is:

$$T_r = \text{Displacement/Fault Slip rate.}$$

The amount of coseismic displacement can be determined directly from an event stratigraphy established from paleoseismic investigations (described above), or estimated using an empirical relationship between earthquake magnitude and fault displacement (Wells and Coppersmith, 1994). The fault slip rate is determined using geological data from paleoseismic or geomorphic

studies (Figures 9.26, 9.27, and 9.31), tectonic observations, or geodetically-based crustal velocity surveys (Figure 9.3).

The geomorphic feature identified on the aerial photographs in Figure 9.26 provide an example of the types of features that can be used to estimated fault slip rate. In this example, the Owens River gorge is offset approximately 300 metres across the Tungsten Hills fault. Assuming a maximum age of 700 000 years for the Owens River gorge, based on the age of the Bishop tuff surface, the amount of offset and age of the gorge yield a minimum fault slip rate of approximately 0.4 mm/yr.

Although the fault slip rate can be estimated based on geomorphic data, no direct information is available regarding the amount of displacement per event. In this case, the displacement can be estimated using empirical relationships between fault length and earthquake magnitude, and earthquake magnitude and displacement (Wells and Coppersmith, 1994). Based on the 16-km length of the Tungsten Hills fault, the maximum earthquake magnitude is estimated to be M_w 6.5 using the empirical relationship between fault length and earthquake magnitude (Wells and Coppersmith, 1994). The displacement for a M_w 6.5 earthquake is estimated to be 0.3 metres using the empirical relationship between displacement and magnitude for all fault types (Wells and Coppersmith, 1994). Therefore, the estimated recurrence interval for a M_w 6.5 earthquake on the Tungsten Hills fault, calculated by dividing 0.3 metres (300 mm) by 0.4 mm/yr, is 750 years.

The assumptions made using this approach can significantly affect the estimated earthquake magnitude, fault slip rate and earthquake recurrence parameters. Therefore, care must be used in estimating the data (e.g. age of Owens River Gorge, amount of displacement across the Tungsten Hills fault, length of fault) used in calculating these parameters. A range of values can be developed to capture the uncertainties in the data. The range of displacement and slip rate estimates would then be used to compute a range of recurrence estimates for input to a probabilistic assessment of seismic hazard for this fault.

Seismological Approach   The seismological approach uses the instrumental and historical (i.e. pre-instrumental) record of earthquakes to estimate earthquake recurrence. An earthquake catalogue for a region is statistically analysed to develop a magnitude-frequency distribution (Gutenberg and Richter, 1954). The occurrence of earthquakes follows the form:

$$\text{Log}_{10}\, N(m) = a - b(m)$$

where $N(m)$ is the cumulative number of earthquakes greater than magnitude (m), and the a- and b-values are constants.

The recurrence curve developed using only seismological data is referred to as the exponential earthquake recurrence model and is represented by the upper part of the graph on Figure 9.35. The exponential recurrence model adequately predicts the rate of earthquake occurrence for large regions, but underestimates the occurrence of large events for individual fault sources.

The standard of practice for developing an earthquake recurrence curve for individual faults involves combining the exponential and characteristic earthquake recurrence models (Figure 9.35). This approach uses the seismicity record to describe the recurrence of small to moderate events, and the geological record to describe the recurrence of large 'characteristic' events (Schwartz and Coppersmith, 1984; Youngs and Coppersmith, 1985). The recurrence of a 'characteristic earthquake' determined from the geological data is combined with the seismological earthquake recurrence curve to develop the characteristic earthquake recurrence relationship (Figure 9.35). The characteristic event is best determined from large magnitude historical earthquakes or paleoseismic data. However, the magnitude and recurrence interval of the characteristic event can also be estimated using the indirect approach, described in the Slip Rate Approach section above. When using this indirect approach, a range of values should be considered to adequately capture the uncertainty in the estimates of maximum earthquake magnitude and recurrence interval.

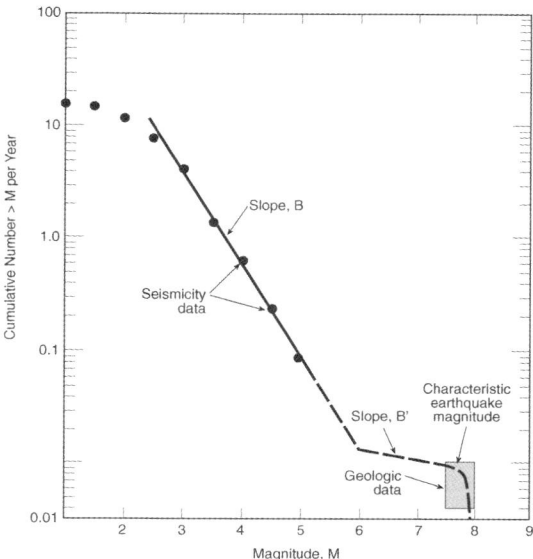

Figure 9.35   Diagrammatic characteristic earthquake recurrence relationship for an individual fault or fault segment. Modified from Schwartz and Coppersmith, 1984.

9.5 Probabilistic seismic hazard assessment and treatment of uncertainty

The location, magnitude, and rates of activity for seismic source zones provide the primary inputs for a probabilistic seismic hazard assessment (PSHA). The probabilistic seismic hazard methodology is now the preferred approach for developing design ground motions for most critical facilities, and is the basis for seismic design provisions in international building codes. The input parameters typically are described in terms of a seismic source model (e.g. Figure 9.36).

The objective of a PSHA is to estimate the probability that different levels of ground motion may occur at a site during some future period of time (e.g. 1, 50, 100, 500, 1000 years). The probabilistic seismic hazard is computed by (1) developing a seismic source model for a region that describes each significant seismic source in terms of maximum earthquake magnitude, and earthquake recurrence, (2) identifying the appropriate ground-motion attenuation equations for the

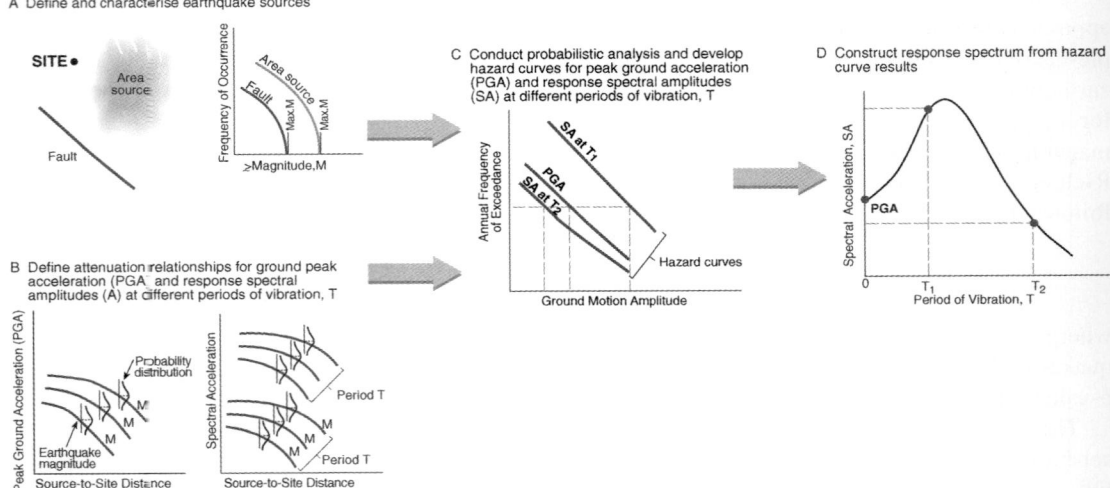

Figure 9.36   Illustration of major steps used in conducting a probabilistic seismic hazard assessment.

region, and (3) calculating the contribution to the hazard from each model source for a given period of time (Figure 9.36). The results are typically displayed in terms of a hazard curve, which gives the annual probability of a ground motion being exceeded (e.g. 10^{-2}, 10^{-3}, 10^{-4} annual probability of exceedence; Figure 9.36). The hazard calculation explicitly considers the randomness of ground motions for a given magnitude and source-to-site distance (aleatory variability). In addition to the aleatory variability there are also scientific uncertainties due to choice of models (e.g. earthquake recurrence models) and limitations of data (epistemic uncertainties). Identifying and quantifying both aleatory and epistemic uncertainties has become an integral part of conducting PSHA (SSHAC, 1995).

When completing a PSHA for critical facilities such as nuclear reactor sites, LNG terminals, dams, and deep-water platforms, it is necessary to consider the uncertainties (e.g. SSHAC, 1995; McGuire, 2004). Logic trees are used to explicitly treat multiple interpretations and to track the decision process used in computing the hazard for a site (Figure 9.37). Use of the logic tree approach allows multiple parameters and multiple models to be considered in the hazard assessment, and thus the uncertainty in the hazard computation to be

estimated. However, the input parameters to the logic tree must still be developed. Two approaches have traditionally been used: (1) expert elicitation to develop input to a logic tree (SSHAC, 1995); and (2) analysis by an individual or team to develop the input parameters to the logic tree. Expert elicitation involves convening a panel of experts on a range of issues related to the characterisation of seismic sources, seismological characteristics of a region, ground motion attenuation characteristics, and calculation of hazard. Each of these experts is asked to develop the inputs to the source models and the results are added to a logic tree for eventual hazard analysis. This process can be a time and cost intensive approach. The second approach is more efficient and more widely employed, where a selected team estimates the range of parameters for input to the logic tree, rather than expert opinions being individually elicited from a group of experts.

The results of a PSHA typically are expressed in terms of rock ground motions accelerations. However, many sites are located on soft rock or soil and a site-specific site-response analysis is conducted to modify the rock ground motions for the specific subsurface conditions at the site. Figure 9.38 is a schematic illustration of the effect of soil response on rock ground motions. The site

Segmentation	Segments	Total length	Maximum rupture length	Maximum magnitude approach	Recurrence approach	Slip/activity rates

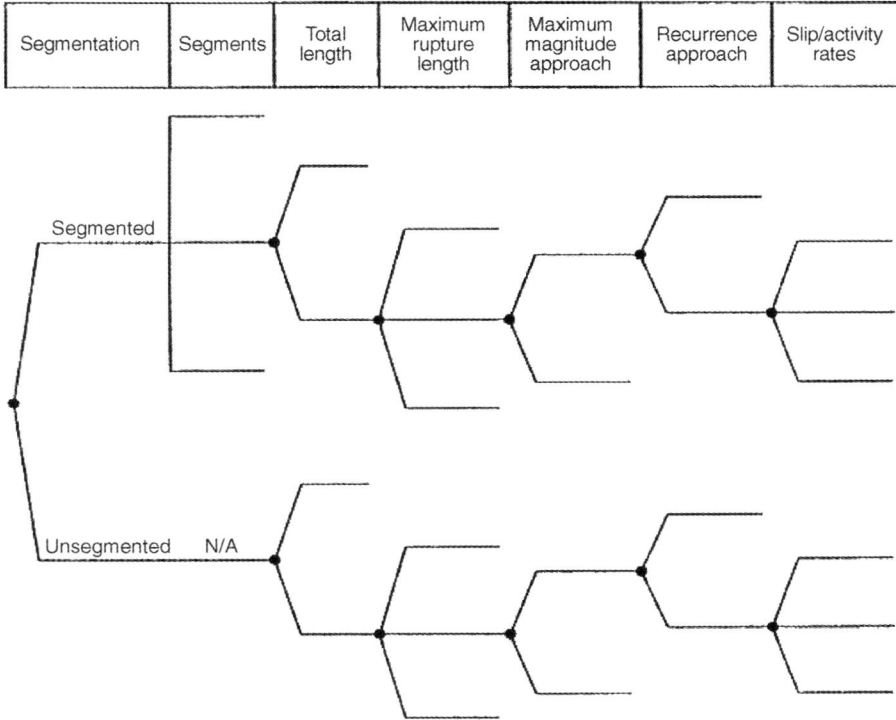

Figure 9.37   Logic tree showing parametres used in fault characterisation.

response due to soil conditions can have the effect of dampening short period ground motions and amplifying long-period ground motions compared to rock conditions.

9.6 Liquefaction and related ground deformation

The 1964 Niigata, Japan earthquake (M 7.5) caused widespread and spectacular building and ground damage as a consequence of soil liquefaction. Loss of soil bearing strength caused buildings to settle and tilt, including several 4-story apartment blocks in the Kawagishicho complex that were otherwise structurally undamaged by shaking effects of the earthquake. Widely publicised photographs of the tilted buildings dramatically focused world attention on the liquefaction process and associated hazards. During the Niigata earthquake, liquefaction-induced lateral-spreading displacements of up to 10 meters tore apart buildings, sheared piles, severed pipelines, compressed or collapsed bridges, and caused general destruction. Overall, liquefaction induced ground-failures caused severe damage to tens of square kilometers of Niigata and its environs (Hamada *et al.*, 1986).

The 1964 M_w 9.2 Great Alaska Earthquake triggered large liquefaction induced flow failures that demolished port facilities in Valdez, Seward and Whittier and carried large parts of those towns into the sea (Plafker, 1969). Earthquake shaking and flow failures generated seiches in surrounding bays, some of which over-ran coastal areas, causing additional damage and many deaths. The earthquake also triggered numerous lateral spread failures that severely damaged the Alaska highway and railway systems, and damaged 266 bridges, many beyond repair. That destruction prevented use of much of the highway and rail systems for months following the earthquake.

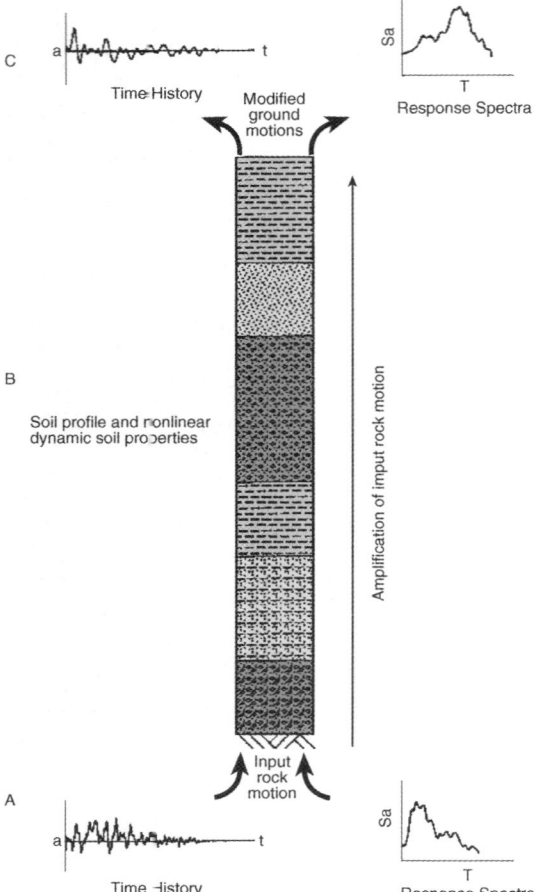

Figure 9.38   Illustration of ground shaking amplification process. (*A*) Ground motions from an earthquake are transmitted through rock and enter a soil column. (*B*) Properties of the soil column modify the rock motion and can cause amplification or dampening of ground motion, depending on the frequency of motion. (*C*) After transmission through soil column, the duration, frequency content, and amplitude of ground shaking is modified from rock input ground motions.

The 1999 Turkey and 2001 India earthquakes also caused widespread damage from liquefaction and related ground damage (Figure 9.39). The Turkey earthquake caused catastrophic submergence of coastal development (Figure 9.2b) and the India earthquake caused liquefaction of approximately 15 000 km² and contributed to the failure of many small and moderate-sized dams (Hengesh and Lettis, 2002; Tuttle *et al.*, 2002b).

Causes and mechanisms

Perhaps the most widely accepted definitions of terms related to seismic liquefaction are those recommended by the American Society of Civil Engineers (ASCE Committee, 1978):

Liquefaction   The act or process of transforming cohesionless soils from a solid state to a liquefied state as a consequence of increased pore pressure and reduced effective stress.

Several general comments can be made regarding the liquefaction process:

1. Liquefaction usually is associated with, and initiated by, strong shaking during earthquakes, which causes certain soils (mainly cohesionless, uniformly-graded fine sands and coarse silts) to densify. Rearrangement of soil particles causes increases in intergranular pore water pressure with resulting decrease in shear strength. The liquefaction process always produces a transient loss of shear resistance, but not always a longer-term loss of shear strength.

2. Liquefaction is most likely to occur: 1) in saturated, relatively uniform, cohesionless, fine sands, silty sands, or coarse silts of low relative density (loose); 2) generally above depths from 15 to 20-metres; and 3) in areas where the water table is within 5 metres of the ground surface. Although liquefaction effects are normally observed only in loose soils, dense sands and silts may show initial liquefaction (pore pressure buildup) effects, but these are rapidly inhibited by the dilatancy characteristics of such soils.

Characteristic forms and ground behaviour

Liquefaction damage during earthquakes usually is related to lateral spreads or settlement in flat, low-lying areas (Seed, 1968; Youd and Perkins, 1978; Seed and Idriss, 1982; and Youd 1991). The four main types of ground failures and typical ground damage effects due to liquefaction are summarised in Table 9.2.

Assessment of liquefaction potential

The assessment of liquefaction *potential* involves three steps:

1. *Evaluating liquefaction susceptibility*. This involves identification of areas or layers that

Figure 9.39   (*A*) Example of lateral spread failure, from 2001 M$_w$ 7.7 Bhuj India earthquake; (*B*) Sand boils associated with extensional cracking in lateral spread failure, Bhuj earthquake; (*C*) Example of bearing capacity failure, 1999 Turkey earthquake. Photographs A and B by James Hengesh, C from EERI, 2000.

Table 9.2   Liquefaction failure modes and their characteristic ground damage effects (after Tinsley *et al.*, 1985).

Liquefaction failure mode	Typical ground damage and effects
1. Lateral Spreads	Small to large lateral displacements of surficial blocks of sediments, on gentle slopes ($< 3°$). Movements, commonly of several metres to tens of metres, usually are toward a free face, particularly in incised stream channels, canals, or open cuts. Lateral spreads are particularly damaging to pipelines, bridges, and structures with shallow foundations, especially on flood plains adjacent to river channels.
2. Flow Failures	Flow failures, commonly the most catastrophic mode of liquefaction failure, usually are developed on slopes greater than $3°$, with movements ranging from tens of metres to several kilometres, at very rapid velocities. Such flows involve great volumes of material, and are highly damaging to structures located on them or in their paths.
3. Ground Oscillation	Oscillation occurs when liquefaction is triggered at depth, or within confined liquefied layers. May produce visible ground oscillation waves, ground settlements, opening and closing of fissures, and ejections of sand and water from cracks and fissures (sand 'boils'). Subsurface structures (pipes, tanks, etc.) may be damaged from this phenomenon, but damage typically is less than from lateral spread or flow failures.
4. Loss of Bearing Strength	Strength loss caused by liquefaction can cause ground collapse and settlements. Structures may settle and topple, and buried structures (pipelines, septic tanks, etc.) may float to the surface. Spreading and collapse of embankment fills may occur from liquefaction of foundation soils.

have the physical characteristics of liquefiable soil.

2. *Assessing the opportunity for strong ground shaking.* This involves identifying seismic sources that are capable of generating moderate to large magnitude earthquakes and estimating the occurrence of ground shaking that is strong enough to generate liquefaction in susceptible materials.

3. *Assessing liquefaction potential.* This is completed by comparing the distribution of susceptible deposits and the opportunity for strong ground shaking to identify the areas in which the level of ground shaking is greater than the liquefaction triggering threshold of susceptible deposits. Liquefaction opportunity maps typically are produced for specified probability levels of ground motion.

Liquefaction potential is site-dependent because certain soils are more susceptible to liquefaction than others. Liquefaction can occur at low levels of ground shaking (e.g. 0.10 percent gravity [g]) in saturated, relatively uniform fine sands or coarse silts in a loose state, at depths less than 20 metres, and where the groundwater level is within about 5 metres of the ground surface. However, liquefaction can occur in other less susceptible deposits at higher levels of ground shaking, or from long duration earthquakes.

Geomorphological characterisation of liquefaction susceptibility and related ground failure

Liquefaction does not occur randomly, but is restricted to areas with a narrow range of geologic and hydrologic characteristics that can be identified and mapped based on established Quaternary mapping techniques. Most liquefaction occurs in areas of poorly engineered hydraulic fills and in fluvial deposits less than 1500 years old. Mapping liquefaction susceptibility (e.g. Figures 9.40 and 9.41) involves:

1. *Quaternary geological mapping.* This provides an important first step in assessing liquefaction susceptibility mapping (Hengesh and Bachhuber, 2005; Pyke, 2001). The distribution

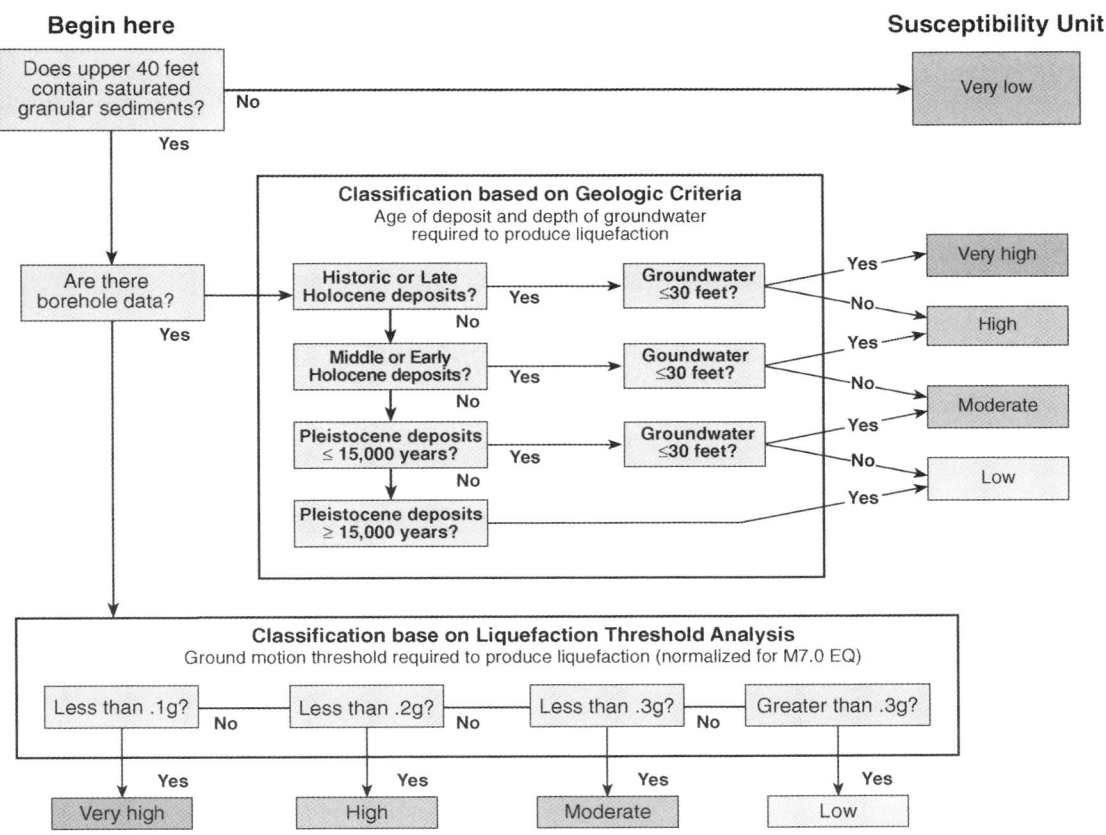

Figure 9.40   Decision tree used to evaluate relative liquefaction hazards.

of Quaternary deposits at the surface and in the shallow subsurface can be mapped to differentiate units based on age, texture, and depositional environment. Quaternary geological mapping for liquefaction susceptibility analyses is based on: (1) interpretation of existing geologic and soil maps; (2) compilation and interpretation of geotechnical borings; (3) interpretation of topographic maps and aerial photography, in particular pre-development 1:35 000 scale or larger photography; (4) construction of local Quaternary stratigraphic columns, including age estimates, and a correlation chart; (5) construction of geologic cross sections; and (6) field reconnaissance.

2. *Interpreting geotechnical and geophysical data.* Geotechnical boring logs and water well logs from previous explorations in an area can

provide lithologic and engineering data that are useful for assessing liquefaction susceptibility. The soil characteristics from SPT, Cone Penetrometer Test (CPT), and laboratory data can be compiled for each Quaternary geological map unit in a study area. These data are used to develop quantitative estimates of liquefaction triggering thresholds or relative susceptibility ratings for each map unit (Seed and Harder, 1990; Pyke, 2001).

3. *Depth to groundwater maps.* Depth to groundwater is a significant factor governing liquefaction hazard. Saturation reduces the normal effective stress acting on loose, sandy sediments. This condition, particularly in the upper 20 metres of the ground surface, increases the likelihood of liquefaction and resulting ground failure (Youd, 1973). Because groundwater levels

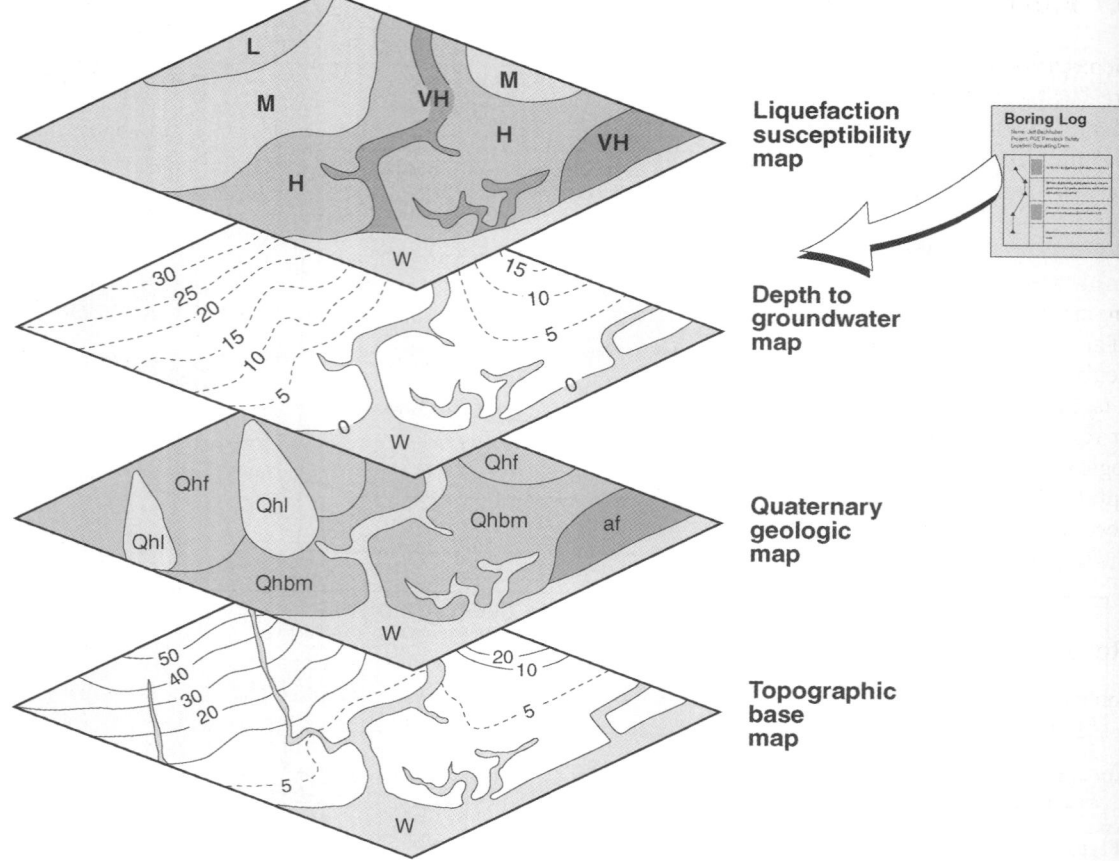

Figure 9.41   Data layers used in creating liquefaction susceptibility map.

may vary due to seasonal variations and historic groundwater use, the highest reasonable water levels should be considered in the liquefaction susceptibility analysis.

4. *Liquefaction susceptibility maps*. Figure 9.40 illustrates an approach used in assigning relative liquefaction susceptibility hazard levels. This flow chart illustrates a process that systematically assesses degrees of liquefaction susceptibility, and was developed in collaboration with the State of California Geological Survey.

Quantification of liquefaction susceptibility typically is performed for the uppermost 20 metres of sediment based on the Seed 'simplified procedure' and subsequent revisions (Seed and Idriss, 1971; 1982; Seed *et al.*,

1983; 1985; Seed and Harder, 1990; Youd, 1997; Idriss and Boulanger, 2004; Cetin *et al.*, 2004). In the absence of detailed subsurface data, liquefaction susceptibility ratings can be assigned by using decision tree analysis (Figure 9.40). Keying liquefaction susceptibility ratings to estimated threshold triggering ground motion values allows evaluation of changes in liquefaction potential for various earthquake scenarios or probabilistic motions that might be used for risk assessment. In addition, the liquefaction susceptibility ratings incorporate parameters that may change with time, including depth to groundwater and, therefore, can be modified in the future to reflect new geotechnical borings or other additional data.

9.7 Implications for engineering

Geological and geomorphological investigations are carried out for engineering projects in order to identify and characterise the location, severity and frequency of seismic hazards including primary surface fault rupture, strong ground shaking, liquefaction, as well as seismically induced slope failure and tsunami. This information is, in turn, used to evaluate site suitability, develop engineering design criteria, assess the probability of an event exceeding the design criteria for a particular facility, and to evaluate the residual risk to a facility from the geological conditions at a site or region. Due to the inherent variability in geological processes, and uncertainties in analytical approaches, the treatment of uncertainty has become a fundamental part of geological hazards investigations.

References

Ambraseys, N. N. (1988) Engineering seismology. *Earthquake Engineering and Structural Dynamics,* **16** 985–1006.

Anderson, D. L. (1971) The San Andreas fault. *Scientific American* **225**(5), 52–68.

Arias, A. (1970) A measure of earthquake intensity. In Hansen, R. J. (ed) *Seismic design for nuclear power plants.* MIT Press, Cambridge, Massachusetts.

Armijo, R., Meyer, B., King, G. C. P., Rigo, A. and Papanastassiou, D. (1996) Quaternary evolution of the Corinth Rift and its implications for the late Cenozoic evolution of the Aegean. *International Journal of Geophysics* **126**, 11–53.

ASCE Committee (1978) Definition of terms related to liquefaction: Report submitted by the Committee on Soil Dynamics of the Geotechnical Engineering Division of the American Society of Civil Engineers (ASCE). *Journal of the Geotechnical Engineering Division,* **104**, No GT9.

Bakun, W. H. and Hopper, M. G. (2004) Magnitudes and locations of the 1811–1812 New Madrid, Missouri, and the 1886 Charleston, South Carolina earthquakes. *Bulletin of the Seismological Society of America* **94**(1), 64–75.

Bakun, W. H. and McGarr, A. (2002) Differences in attenuation among the stable continental regions. *Geophysical Research Letters* **29**(23), 2121.

Bates, R. L. and Jackson, J. A. (1987) *Glossary of Geology* (3rd edition), American Geological Institute, Alexandria, Virginia.

Berrill, J. B. and Davis, R. O. (1985) Energy dissipation and seismic liquefaction of sands; revised model. *Soils and Foundations* **25**(2), 106–118.

Birkeland, P. W. (1984) *Soils and Geomorphology,* Oxford University Press.

Bonilla, M. G. (1970) Surface faulting and related effects, in, Earthquake Engineering, R. L. Wiegel (ed.) Prentice-Hall, p. 47–74.

Burbank, D. W. and Anderson, R. S. (2001) *Tectonic Geomorphology,* Blackwell Science, 274p.

Cetin, K. O., Seed, R. B., Der Kiureghian, A., Tokimatsu, K., Harder, L. F., Keyen, R. E. and Moss, R. E. S. (2004) Standard penetration test-based probabilistic and deterministic assessment of seismic soil liquefaction potential. *Journal of Geotechnical and Geoenvironmental Engineering* **130**(12), 1314–1340.

Cornell, C. A. (1968) Engineering seismic risk analysis. *Bulletin of the Seismological Society of America* **58**, 1583–1606.

Dobry, R., Ladd, R. S., Yokel, F. Y., Chung, R. M. and Powell, D. (1982) Prediction of pore water pressure build-up and liquefaction of sands during earthquakes by the cyclic strain method. NBS Building Science Series 138, U.S. Dept. of Commerce, 152 p.

Earthquake Engineering Research Institute (2000) Reconnaissance Report on Kocaeli, Turkey, Earthquake of August 17, 1999. Youd, T. L., Bardet, J. P. and Bray J. D. (editors) Supplement A to Vol. 16.

Earthquake Engineering Research Institute (2001) Chi-Chi, Taiwan Earthquake of September 21, 1999 Reconnaissance Report, Uzarski, J. and Arnold, C. (eds) Supplement A to Vol. 17, Earthquake Spectra.

Earthquake Engineering Research Institute (2002) The 2001 Bhuj, India Earthquake Reconnaissance Report, Jam, S. K., Lettis, W. R. Murty C. V. R., Bardet, J. P. (eds) Supplement A to Vol. 18, Earthquake Spectra.

Electric Power Research Institute (1994) The earthquakes of stable continental regions; Volume 1: Assessment of large earthquake potential. EPRT Technical Report, TR-102261-V1.

Electric Power Research institute (2003) Ground motion attenuation.

Electric Power Research Institute (2003) (Rev. 1) 1008910 EUS ground motion project-model development and results.

Gori, P. L. and Hayes, W. W. (1992) Assessment of regional earthquake hazards and risk along the Wasatch Front, Utah. U.S. Geological Survey Professional Paper, 1500-A-J.

Grapes, R. and Wellman, H. (1988) The Wairarapa fault. Victoria University of Wellington, New Zealand, p.54.

Green, R. A. (2001) Energy-based evaluation and remediation of liquefiable soils. Ph.D. Thesis, Civil Engineering Dept., Virginia Tech, Blacksburg, Virginia, 397p. (http://scholar.lib.vt.edu/theses/available/etd-08132001-170900/)

Gutenberg, B. and Richter, C. F. (1954) *Seismicity of the earth.* Princeton University Press, 440p.

Hamada, M., Yasuda, S., Isoyama, R. and Emoto, K. (1986) Study of Liquefaction induced permanent ground displacements. *Association for the Development of Earthquake Prediction,* Japan, 87pp.

Hanks, T. C. (1998) The age of scarp-like landforms from diffusion-equation analysis, In Sowers J. M., Noller, J. S., and Lettis, W. R. (eds) Dating and Earthquakes: Review of Quaternary geochronology and its application to Paleoseismology NUREG/CR-5562, U.S. Nuclear Regulatory Commission.

Hanks, T. C. and Bakun, W. H. (2002) A bilinear source-scaling model for M-log A observations of continental earthquakes. *Bulletin of the Seismological Society of America* 92(5).

Hanks, T. C. and Kanamori, H. (1979) A moment magnitude scale. *Journal of Geophysical Research* 84, 2348–2350.

Hanson, K., Kelson, K. I., Angell, M. A. and Lettis, W. R. (1999) Techniques for identifying faults and their origins, NUREG/CR-5503, U.S. Nuclear Regulatory Commission.

Hengesh, J. V. and Bachhuber, J. L. (2005) Liquefaction susceptibility zonation map of San Juan, Puerto Rico. In Mann, P. (ed) *Active tectonics and seismic hazards of Puerto Rico, Virgin Islands and offshore areas.* Geological Society of America Special Paper 385, 249–262.

Hengesh, J. V. and Lettis, W. R. (2002) Tectonic setting and geological effects of the M_w 7.7 2001 Bhuj Earthquake. *Earthquake Spectra* A(18), 19–28.

Hitchcock, C. S. and Kelson, K. I. (1999) Growth of late Quaternary folds in southwest Santa Clara Valley, San Francisco Bay area, California: Implications of 'triggered slip' for seismic hazard and earthquake recurrence. *Geology* 26(5), 391–394.

Hough, S. E., Armbruster, J. G., Seeber, L. and Hough, J. F. (2000) On the Modified Mercalli intensities and magnitudes of the 18 11–12 New Madrid earthquakes. *Journal of Geophysical Research* 105, 23 839–23 864.

Hubert-Ferrari, A., Armijo, R., King, G., Meyer, B. and Barka, A. (2002) Morphology, displacement, and slip rates along the North Anatolian Fault, Turkey, 2002. *Journal of Geophysical Research* 107(B10), 2235.

Hwang, J. H., Chang, C. T. and Chen, C. H. (1995) Study on stress reduction factor rd for liquefaction analysis. Proceedings of 1st International Conference on Earthquake Geotechnical Engineering, Tokyo, Japan 1, 617–622.

Idriss, I. M. and Boulanger, R. W. (2004) Semi-empirical procedures for evaluating liquefaction potential during earthquakes (invited paper). 3rd International Conference on Earthquake Geotechnical Engineering, University of California, Berkeley, California, USA.

Ishihara, K. (1985) Stability of natural deposits during earthquakes. Proceedings of 11th International Conference on Soil Mechanics and Foundation Engineering, San Francisco, California 1, 321–376.

Johnston, A. C. (1996) Seismic moment assessment of earthquakes in stable continental regions — III. New Madrid 1811-1812, Charleston 1886, and Lisbon 1755. *Geophysical Journal International* 126, 314–344.

Kanamori, H. (1983) Magnitude scale and quantification of earthquakes. *Tectonophysics* 93, 185–199.

Kanamori, H. and Brodsky, E. E. (2004), *The physics of earthquakes.* Reports on Progress in Physics, Institute of Physics Publishing 76, 1429–1496. PII: S0034-4885(04)25227–7.

Kayen, R. E. and Mitchell, J. K. (1996) Arias intensity assessment of liquefaction test sites on the east side of San Francisco Bay affected by the Loma Prieta, California, earthquake of 17 October 1989. *Natural Hazards* 16(2–3), 243–265.

Kelson, K. I., Kang, K.-H., Page, W. D., Lee, C.-T. and Cluff, L. S. (2001) Representative styles of deformation along the Chelungpu fault from the 1999 Chi-Chi (Taiwan) earthquake: Geomorphic characteristics and responses of man-made structures. *Bulletin of Seismological Society America* 91(5), 930–952.

Kelson, K. I., Simpson, G. D., Van Arsdale, R. B., Haraden, C. C. and Lettis, W. R. (1996) Multiple Late Holocene Earthquakes along the Reelfoot Fault, Central New Madrid Seismic Zone. *Journal of Geophysical Research* 101(83), 6151–6170.

Kenner, S. J. and Segall, P. (2000) A mechanical model for intraplate earthquakes: Application to the New Madrid Seismic Zone. *Science* 289, 2329–2332.

Krinitsky, E. L. and Slemmons, D. B. (1990) Neotectonics in earthquake evaluation. *Reviews in Engineering Geology,* Vol. VIII.

Law, K. T., Cao, Y. L. and He, G. N. (1990) An energy approach for assessing seismic liquefaction potential. *Canadian Geotechnical Journal* 27(3), 320–329.

Lettis, W. R. and Kelson, K. I. (1998) Applying geochronology in paleoseismology. In Sowers, J. M., Noller, J. S. and Lettis, W. R. (eds) *Dating and Earthquakes: Reviews of Quaternary Geochronology and its application to Paleoseismology.* US Nuclear Regulatory Commission, Washington DC, NUREG/CR 5562: 3–1 to 3–26.

Lettis, W. R., Bachhuber, J., Witter, R., Barka, A., Bray, J., Page, W., Swan, F., Altunel, E., Bardet, J. P., Boulanger, R., Brankman, C., Cakir, Z., Christofferson, S., Cluff, L., Dawson, T., Fumal, T., Guneysu, A. C., Hengesh, J., Kaya, A., Langridge, R., Rathje, E., and Stenner, H. (2000) Surface fault rupture, in Kocaeli, Turkey, Earthquake of August 17, 1999, Reconnaissance Report (Supplement A to vol. 16), in T. L. Youd, J. P. Bardet and J. D. Bray (Editors), Chapter 2, 11–53.

Lettis, W., Bachhuber J., Witter, R., Brankman, C., Randolph, C. E., Barka, A., Page, W. D. and Kaya, A. (2002) Influence of releasing stcpovers on surface fault rupture and fault segmentation: Examples from the 17 August 1999 Izmit earthquake on the north Anatolian fault, Turkey. *Bulletin of the Seismological Society of America* **92**(1), 19–42.

McCalpin, J. P. (1996) *Paleoseismology*. Academic Press, 588p.

McGuire, R. K. (2004) *Seismic hazard and risk analysis*. Earthquake Engineering Research Institute, Monograph No. 010, 240p.

Molnar, P. (1979) Earthquake recurrence intervals and plate tectonics. *Bulletin of the Seismological Society of America* **70**, 1321–1335.

Obermeicr, S. F., Pond, E. C., and Olson, S. M., with contributions by Green, R. A., Stark, T. D., and Mitchell, J. K. (2001) Paleoliquefaction studies in continental settings: Geologic and geotechnical factors in interpretations and back-analysis. U.S. Geological Survey Open-File Report 0 1–29.

Plafker, G. (1969) Tectonics of the March 27, 1964, Alaska earthquake. U.S. Geol. Surv. Professional Paper 543-I, 74p.

Plafker, G. and Rubin, M. (1992) "Yo-yo" tectonics above the eastern Aleutian subduction zone: coseismic uplift alternating with even larger interseismic submergence. Proceedings of the Wadati Conference on great subduction earthquakes, Geophysical Institute, Univ. Alaska, Fairbanks.

Pond, E. C. (1996) Seismic parameters for the central United States based on paleoliquefaction evidence in the Wabash Valley. PhD Thesis, Virginia Tech, Civil Engineering Dept., Blacksburg, Virginia, 583p.

Pyke, R. (2001) Discussion of "liquefaction resistance of soils: Summary report from the 1996 NCEER and 1998 NCEER/NSF Workshops on evaluation of liquefaction resistence of soils", Jour. of Geotech. And Geoenviron. Eng. Robertson, P. K., and Wride, C. E., 1997, Cyclic liquefaction and its evaluation based on SPT and CPT: Seismic Short Course on Evaluation and Mitigation of Earthquake Induced Liquefaction Hazards, NCEER Workshop, San Francisco.

Reilinger, R. E., McClusky, S. C., Oral, M. B., King, R. W., Toksoz, M. N., Barka, A. A., Kinik, I., Lenk, O. and Sanli, I. (1997) Global Positioning System measurements of the present day crustal movements in the Arabia-Africa-Eurasia plate collision. *Journal of Geophysical Research* **102**, 9983–9999.

Ruff L. and Kanamori, H. (1983) The rupture process and asperity distribution of three great earthquakes from long-period diffracted P-waves. *Phys. Earth Planet. Interior* **31**, 203–230.

Schwartz, D. P. and Coppersmith, K. J. (1984) Fault behaviour and characteristic earthquakes: examples from the Wasatch and San Andreas faults. *Journal of Geophysical Research* **89**, 5681–5698.

Schwartz, D. P. and Sibson, R. H. (1989) Fault segmentation and controls of rupture initiation and termination. U.S. Geol. Surv. Open File Report 89–315.

Seed, H. B. (1968) Landslides during earthquakes due to liquefaction. *Journal of Geotechnical Engineering* **94**(5), 1055–1122.

Seed, R. B. and Harder, L. F. (1990) SPT-based analysis of cyclic pore pressure generation and undrained residual strength, in Duncan, M. J., editor, H. Bolton Seed Memorial Symposium Proceedings, May, 1990: BiTech Publishers, Vancouver, B. C., Canada, 351–376.

Seed, H. B. and Idriss, I. M. (1971) Simplified procedure for evaluating soil liquefaction potential. Proceeding of the American Society of Civil Engineers, Journal of the Soil Mechanics and Foundations Division, 93, no. SM9, 1249–1273.

Seed, H. B. and Idriss, I. M. (1982) Ground Motions and Soil Liquefaction During Earthquakes. Earthquake Engineering Research Institute Monograph, Berkeley, California, 134pp.

Seed, H. B., Idriss, I. M. and Arango, I. (1983) Evaluation of liquefaction potential using field performance data. *Journal of Geotechnical Engineering* **109**(3), 458–482.

Seed, H. B., Tokimatsu, K., Harder, L. F. and Chung, R. M. (1985) Influence of SPT procedures in soil liquefaction resistance evaluations. *Journal of Geotechnical Engineering* **111**(12), 1425–1445.

Senior Seismic Hazard Analysis Committee (SSHAC) (1995) Recommendations for Probabilistic Seismic Hazard Analysis: Guidance on Uncertainty and Use of Experts, Lawrence Livermore National Laboratory, UCRL-ld-122l60, NUREG/CR-6372.

Sowers J. M., Noller, J. S. and Lettis, W. R. (1998) Dating and Earthquakes: Review of Quaternary geochronology and its application to Paleoseismology; NUREG/CR5562, U.S. Nuclear Regulatory Commission.

Seismological Research Letters (1997) Special volume on ground motin attenuation equations, Seismological Society of America, **68**(1).

Tinsley, J. C., Youd, T. L., Perkins, D. M. and Chen, A. T. F. (1985) Evaluating Liquefaction Potential. In Ziony, J. I. (ed.) *Evaluating earthquake hazards in the Los Angeles Region -an earth-science perspective*. United States Geological Survey Professional Paper **1360**, 263–316.

Thatcher, W. (1990) Order and diversity in the modes of circum-Pacific earthquake recurrence. *Journal of Geophysical Research* **95B3**, 2609–2623.

Tuttle, M. P (2001) The use of liquefaction features in paleoseismology: Lessons learned in the New Madrid seismic zone, central United States. *Journal of Seismology* **5**, 361–380.

Tuttle, M. P., Schweig, E. S., Sims, J. D., Lafferty, R. H., Wolf, L. W. and Haynes, M. L. (2002a) The earthquake potential of the New Madrid seismic zone. *Bulletin of the Seismological Society of America* **92**(6), 2080–2089.

Tuttle, M. P., Hengesh, J. V., Tucker, K. B., Lettis, W., Deaton, S. L. and Frost, J. D. (2002b), Observations and comparisons of liquefaction features and related effects induced by the Bhuj earthquake. *Earthquake Spectra* **A**(18), 79–100.

Wallace, R. E. (1990) The San Andreas Fault System. US Geological Survey Professional Paper 1515, 283p.

Wells, D. L. and Coppersmith, K. J. (1994) New empirical relationships among magnitude, rupture length, rupture width, rupture area, and surface displacement. *Bulletin of the Seismological Society of America* **84**(4), 974–1002.

Wesson, R. L., Helley, E. J., Lajoie, K. R. and Wentworth, C. M. (1975) Faults and future earthquakes. In Borcherdt, R. D. (ed) *Studies for seismic zonation of the San Francisco Bay region*. US Geological Survey Professional Paper, 941-A, A5–30.

Weznouski, S. G. (1988) Seismological and structural evolution of strike-slip faults. *Nature*, **335**(6188), 340–343.

Wheeler, R. L. (1989) Persistent segment boundaries on basin and range normal faults. In Schwartz D. P. and Sibson R. H. (eds) *Fault Segmentation and Controls of Rupture Initiation and Termination*, U.S. Geol. Surv. Open File Report 89-3 15, 432–444.

Whitman, R.V. (1971) Resistance of soil to liquefaction and settlement. *Soils and Foundations* **11**(4), 59–68.

Yeats, R. S., Huftile, G. J. and Stiff, L. J. (1994) Late Cenozoic tectonics of the cast Ventura basin, Transverse Ranges, California. *AAPG Bull.*, **78**, 1040–1074.

Yeats, R. S., Sieh, K. and Allen, C. R. (1997) *The geology of earthquakes*. Oxford University Press, 568p.

Yeh H., Liu, P. L. -F. and Synolakis, C. E. (1997) *Long Wave Runup Models*. World Scientific, Singapore, 405pp.

Youd, T. L. and Garris, C. T. (1995) Liquefaction-induced ground-surface disruption. *Journal of Geotechnical Engineering* **121**(11), 805–809.

Youd, I. M., Idriss, L. M., Andrus, R. D., Arango, I., Castro, G., Christian, J. T., Dobry, R., Liam, F., Harder, W. D., Hynes, L. F., Ishihara, M. E. , Koester, K., Liao, J. P., Marcuson III, S. S. C. W. F., Martin, G. R., Mitchell, J. K., Moriwaki, Y., Power, M. S., Robertson, P. K., Seed, R. B., and Stokoe III, K. H. (2001) Liquefaction resistance of soils: Summary report from the 1996 and 1998 NCEERINSF workshops on evaluation of liquefaction resistance of soils. *Journal of Geotechnical and Geoenvironmental Engineering* **127**(10), 817–833.

Youd, T. L. (1973) Liquefaction, flow and associated ground failure. US Geological Survey, Circular 688.

Youd, T. L. (1991) Mapping of earthquake-induced liquefaction for seismic zonation. Proceedings, Fourth International Conference on Seismic Zonation, Earthquake Engineering Research Institute, Stanford, **1**, 111–147.

Youd, T. L. and Perkins, D. M. (1978) Mapping liquefaction-induced ground failure potential. Journal of Geotechnical Engineering, American Society of Engineers **104**(4), 433–446.

Youd, T. L. (1997) Updates on the simplified procedure: An overview of NCEER workshop in Salt Lake City on liquefaction resistance of soils. Seismic Short Course on Evaluation and Mitigation of Earthquake Induced Liquefaction Hazards, NCEER Workshop, San Francisco, CA, March, 1997.

Youngs, R. R. and Coppersmith, K. J. (1985) Implications of fault slip rates and earthquake recurrence models to probabilistic seismic hazard estimates, *Bulletin of Seismological Society of America* **75**(4), 939–964.

10. Rivers

Mark Lee

10.1 Introduction

River systems are open channels that carry water and sediment. The overall system comprises three broad zones, each with a characteristic form and function (Figure 10.1):

1. Zone 1: the drainage basin or water and sediment source area (i.e. mountain or hillslope catchments)
2. Zone 2: the sediment transport zone (i.e. the main river channel)
3. Zone 3: the sediment store or deposition zone (i.e. the river floodplain, delta or estuary).

This sub-division should only be viewed as illustrating how the dominant processes in different parts of the system are interlinked into a water and sediment cascade, from source areas to sinks; sediments are, in reality, supplied (i.e. by erosion), transported and deposited in all of the Zones.

The drainage basin (catchment or watershed) is the fundamental unit for studying river channels and hydrology. River channel adjustments

can be seen, in part, as reflecting the need to maintain continuity of water and sediment transport. Among the key features of a basin are the drainage network (Figure 10.2) — which often reflects a combination of climatic (Figure 10.3) and geological (Figure 10.4) controls — and the stream ordering (Figure 10.5).

10.2 River channel changes

Between around 1865 and 1915 many broad valleys and plains in the arid and semi-arid south-west USA underwent significant changes, developing trench-like gulleys with rectangular cross sections and near-vertical side-walls (arroyos). For example, Kaneb Creek in south Utah developed a 25 m deep and 80–120 m wide channel between 1880–1914. At the time the changes were blamed on land use change, especially over-grazing, and the resultant increase in runoff. However, the importance of climate change and high magnitude, low frequency events has been recognised as a factor in triggering periods of entrenchment (Figure 10.6). In south Utah, arroyos such as Kaneb Creek developed during a period of increased flood magnitude that had been preceded by over 350 years without significant floods.

In 1982 an intense rainstorm over the Howgill Fells, UK triggered widespread gulley erosion, delivering large volumes of sediment to the stream channels. This in turn prompted localised switches from meandering to braided channels. Since the event some channels have changed back to meandering forms as the sediment supply has diminished (Harvey, 1986). Similar changes have followed mining activities. In the northern Pennines and mid-Wales, UK the dumping of mine waste in streams during the eighteenth and nineteenth

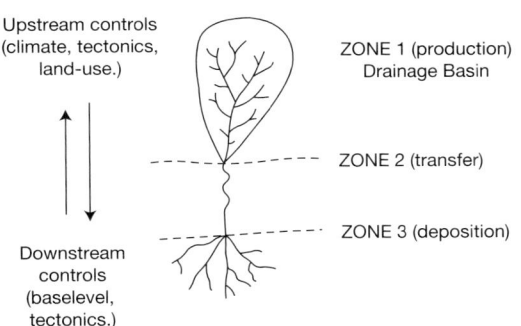

Figure 10.1   An idealised river system (from Schumm, 1977).

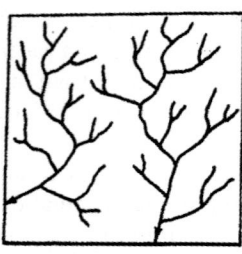

Dendritic: the commonest pattern. Indicates uniform materials.

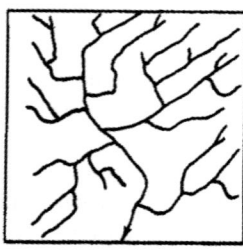

Rectangular: implies strong bedrock jointing and thin soil cover. The stronger the pattern, the thinner the soil.

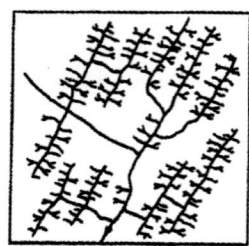

Trellis: implies strike ridge topography.

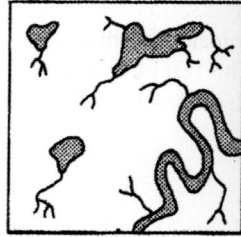

Deranged: with many ponds, bogs, or lakes. Indicates flattish landscape often glaciated.

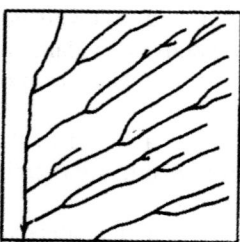

Parallel: characteristic of outwash areas of low topography, where main stream may indicate a fault.

Anastomosing or braiding: in alluvial areas where sediment load exceeds carrying capacity of a stream.

Radial (centrifugal): in isolated circular hill masses.

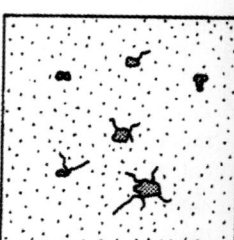

Internal: indicates highly porous level materials or karst conditions.

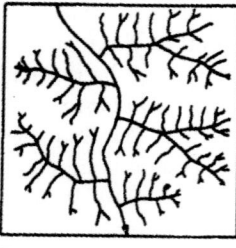

Pinnate: generally indicates high silt content as in loess or on flood plains.

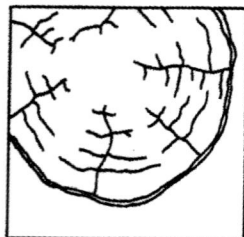

Annular: indicates igneous or sedimentary domes with concentric fractures or escarpments.

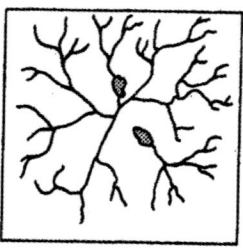

Centripetal: a variation of the radial pattern with drainage towards a central point, usually a sink or the centre of an eroded anticline or syncline.

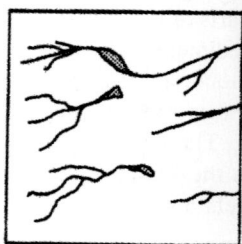

Dislocated: due to interruptions of drainage lines by faults or extrusions.

Figure 10.2 Types of drainage networks (after Mitchell, 1973).

centuries led to the switch from meandering to braided channels with valley floor aggradation. On some streams, incision and reversal to a meandering channel has taken place since the end of waste dumping (Macklin and Lewin, 1997).

Some rivers may undergo major course changes. The Huang He (Yellow River) in China, for example, has had at least twenty changes during the last 4000 years or so. Some of these changes have been very substantial. The decline of the civilisation of Mohenjo Daro is believed to have coincided with course changes of the River Indus. The 'Old Testament' city states such as Ur and Uruk declined when the Euphrates moved and left

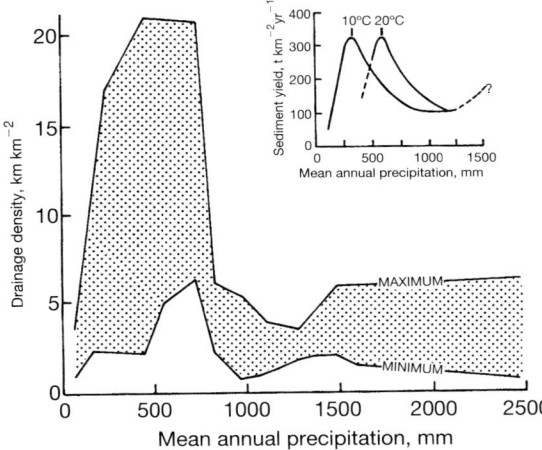

Figure 10.3   Variation of drainage density with climate: mean annual precipitation (after Gregory, 1976).

them high and dry. Doornkamp (1982) describes how the Kosi River, which supplies irrigation water to the Ganges plains shifted course away from the entrance to the feeder canals in the early 1970s, leaving the irrigation system short of water.

Changes in both location and depth of channels can be of particular importance in relation to river crossings. Major river training may be necessary to keep a migrating river flowing under a bridge (e.g. the Jamuna River crossing in Bangladesh), while increase in scour at bridge piers can lead to bridge collapse. Examples include the loss of a section of railway embankment at Dalguise and structural damage to Fortevoit rail bridge during the 1993 Perth floods in Scotland; in October 1987 four people died when channel scour and erosion around the piers of a railway bridge at Glanrhyd, Dyfed in Wales led to the bridge collapsing under the weight of a train (Lee, 1995).

Significant channel changes can take place as a result of water impoundment by reservoirs (e.g. Petts, 1984). Generally dams decrease peak flows downstream and impound practically all sediment.

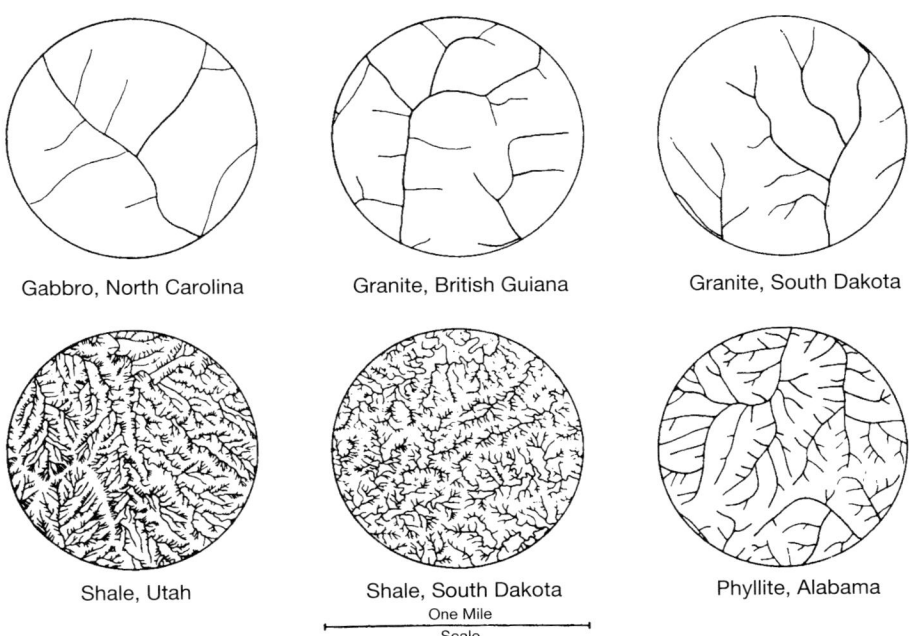

Figure 10.4   Drainage densities on different rock types (from Ray and Fischer, 1960).

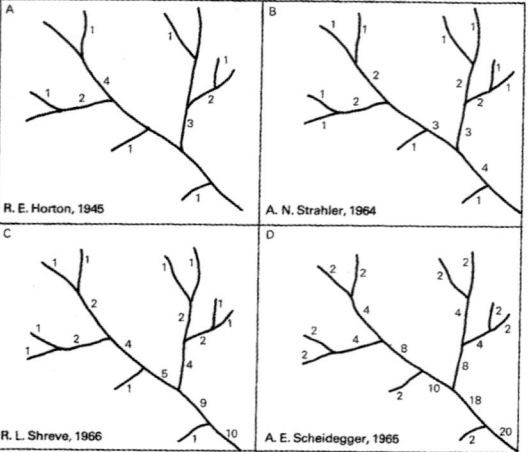

Figure 10.5 Approaches to stream ordering (after Gregory and Walling, 1973).

Channels tend to reduce in size downstream but also tend to incise because of the lack of sediment. Most of the adjustment seems to be by cross-section form and slope but even along one river there may be both increases and decreases in channel capacity. Channel degradation and scour will persist until the reduction of channel slope reduces the flow velocity below the threshold for sediment transport. The timescales of adjustment to river impoundment also vary but five years may be required before any channel response is observed. Continuing channel changes have been reported more than 50 years after dam construction, and stability in terms of sediment transport and channel form may take 200 years.

Urbanisation can lead to channel modifications downstream following an increase in the size of peak flows and a decrease in sediment supply (e.g. Knight, 1979; Hollis and Luckett, 1976). Among the most common changes are bed and bank erosion, loss of riverbank trees and undermining of structures. Following the development of Cumbernauld New Town in Strathclyde, UK some streams had enlarged through the urban area (Roberts, 1989). Locally there were instances of extensive erosion and bank collapse; up to 10 m of vertical incision through glacial till was observed in gullies draining industrial areas, and some

channels developed gravel bars and became more braided. River channel engineering has had notable upstream and downstream effects. Such channelling, which generally involves attempts to control the channel form at a site, can initiate instability not only in the improved channel reach but also upstream and downstream as the channel adjusts to a new state of equilibrium.

Gravel extraction increases the channel gradient through a reach and increases bed roughness, leading to enhanced bed scour and gravel/cobble transport. If the rate of removal is greater than the rate of supply from upstream, channel down-cutting will occur. This incision is self-enhancing because flow becomes more confined to an entrenched channel with higher flow velocity and bed shear stresses. Entrenchment may continue long after extraction ceases, as the river channel continues to adjust to these changes.

These examples serve to illustrate how significant river metamorphosis (i.e. modification of the river channel form) can occur over 'engineering time'. Changes in discharge or sediment supply, together with the calibre of bed or bank materials, vegetation or slope angle, can lead to adjustments of channel form (Table 10.1). This is because of the close links between different parts of the channel and sediment erosion, transport and deposition processes (Figures 10.7a and b). However, as system thresholds need to be exceeded, not all changes will lead to sustained river metamorphosis.

10.3 Channel form: regime theory

The ability to respond to changes in the controlling factors depends on the resistance (i.e. erodibility) of the bed and bank materials. Channels that have been formed in strong bedrock reflect geological control (i.e. lithology and structure) and can be described as confined in that they are not generally prone to significant channel change. In contrast, channels formed in erodible sediments can be described as unconfined (or 'alluvial') channels.

Under stable climatic conditions, alluvial channel geometry can be considered to be in

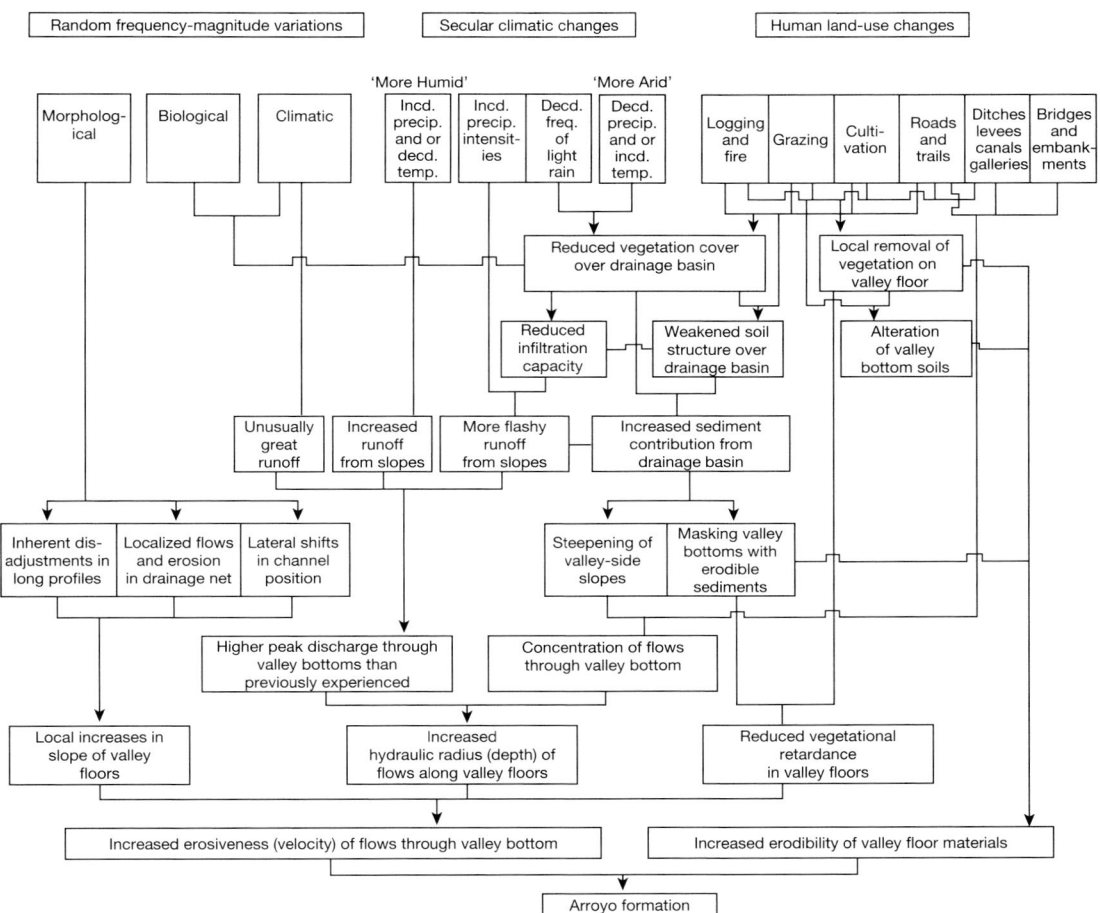

Figure 10.6 A model of arroyo formation (after Cooke and Reeves, 1976).

Table 10.1 Channel changes in response to changes in discharge and sediment load (after Schumm, 1977).

Change	River bed morphology	Change	River bed morphology
$Q_s + Q_w =$	Aggradation, channel instability, wider and shallower channel	$Q_w + Q_s =$	Incision, channel instability, wider and deeper channel
$Q_s - Q_w =$	Incision, channel instability, narrower and deeper channel	$Q_w - Q_s =$	Aggradation, channel instability, narrower and shallower channel
$Q_s + Q_w -$	Aggradation	$Q_w - Q_s -$	Processes decreased in intensity
$Q_s + Q_w +$	Processes increased in intensity	$Q_w + Q_s -$	Incision, channel instability, wider channel

Note: Q_s: sediment discharge; Q_w: water discharge; $+$: increase; $-$: decrease; $=$: remains constant.

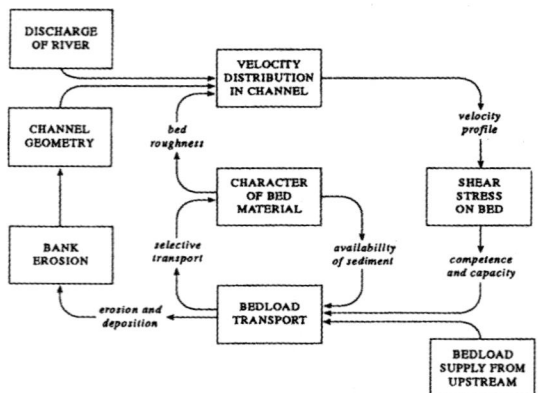

Figure 10.7a A feedback model of channel change (after Richards and Lane, 1997).

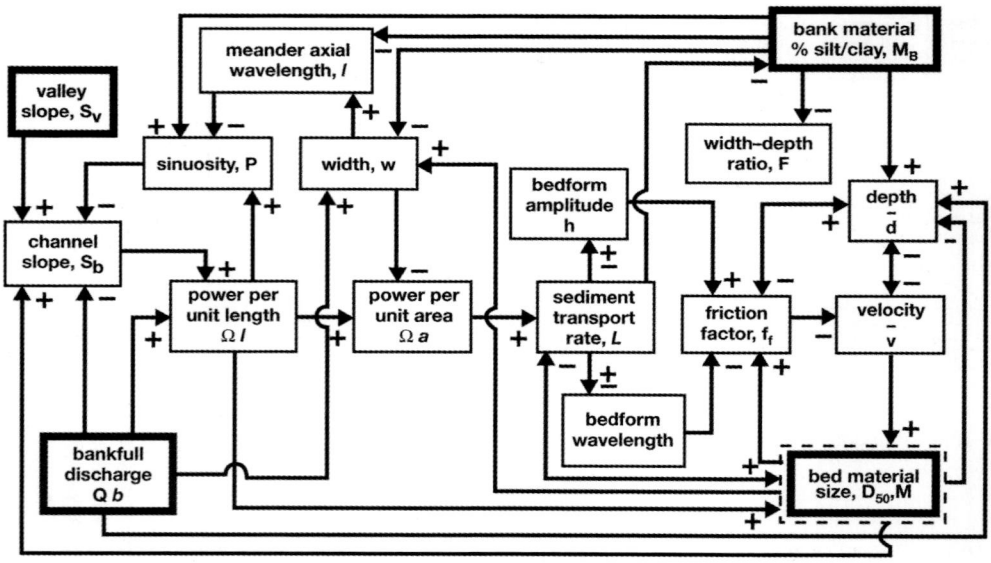

Figure 10.7b A summary of the alluvial channel system, highlighting the interlinkages between variables (redrawn from Richards, 1982).

equilibrium (i.e. in 'regime') with the prevailing flows, along with slope gradient, sediment type, bank vegetation and valley constrictions. Channel cross-sectional form adjusts by erosion and deposition to accommodate the sediment load and all but the highest flows. As discharge (Q) increases so width (w), mean depth (d) and velocity (v) all increase. These adjustments can be described by power functions: $w = a\ Q^b$;

$d = c\ Q^f$; $v = k\ Q^m$, as determined by Leopold and Maddock (1953) for rivers in mid-west USA and elsewhere (Table 10.2; Figure 10.8). These relationships have formed the basis for predicting stable channel forms.

These hydraulic geometry equations highlight the tendency for mean velocity to increase downstream. Although the long-profile gradient normally decreases, velocity may increase because of

Table 10.2   A selection of downstream hydraulic geometry relations (from Knighton, 1984).

Location	Discharge	b	f	m
Midwest USA	Q_m	0.50	0.40	0.10
Pennsylvania, USA	Q_{50}	0.34	0.45	0.32
	Q_{15}	0.38	0.42	0.32
	Q_b	0.42	0.45	0.05
Appalachians, USA	$Q_{2.33}$	0.55	0.36	0.09
Cheshire, UK	Q_{50}	0.46	0.16	0.38
	Q_{15}	0.54	0.23	0.23
Idaho, USA	Q_b	0.54	0.34	0.12
Gravel bed rivers, UK	Q_b	0.45	0.40	0.15

Discharge (Q) defined as mean annual flood ($Q_{2.33}$); bankfull (Q_b); mean annual (Q_m); and the one equalled or exceeded 50% (Q_{50}); 15% (Q_{15}).

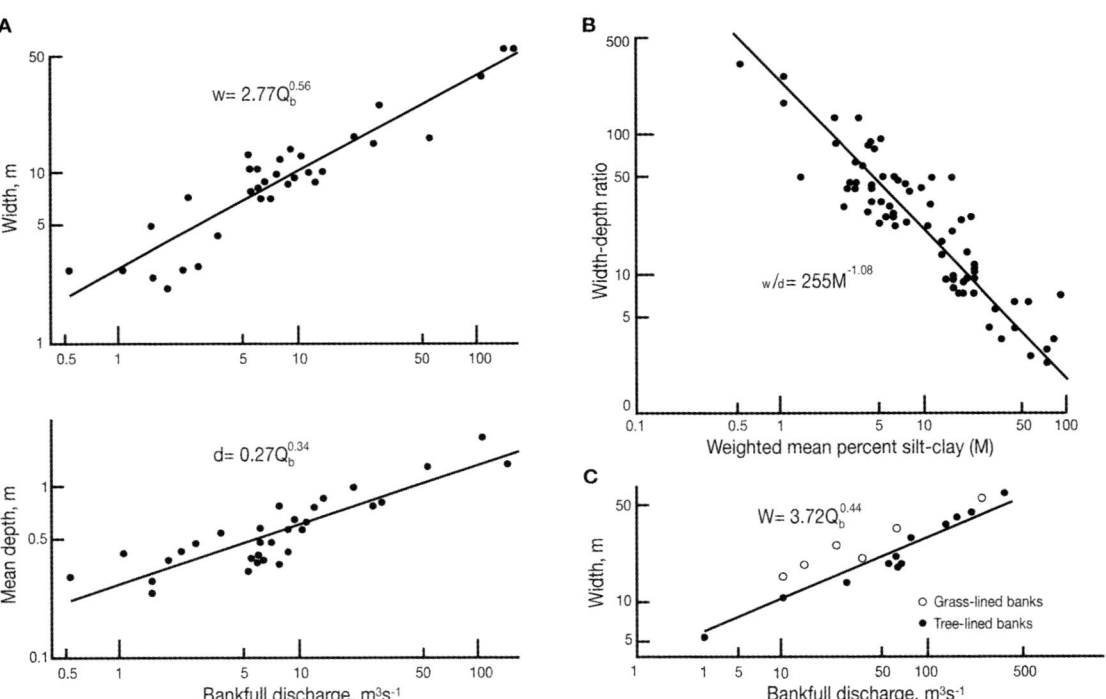

Figure 10.8   Relationships between bankfull discharge and various channel variables (redrawn from Knighton, 1984).

the increased cross-section efficiency (i.e. less friction) and decreased bed roughness if the bed materials become finer downstream.

The dominant discharge that controls the channel geometry is generally assumed to be the 'bankfull discharge' i.e. the flow that, over time, yields the maximum bed and bank erosion, sediment transport and deposition. Relatively high frequency flows are thus more significant than rare floods in terms of their cumulative impact on channel form, hence the assumption that a regime equilibrium form is adjusted to the bankfull discharge.

The frequency of bankfull discharge varies between rivers and can range from 1–32 years. However, the importance of bankfull discharge varies between river environments (Wolman and Miller, 1960). It is the dominant discharge where the entrainment threshold of the bed sediments is low (i.e. in sand/mud bed rivers). The more variable the flow, as in semi-arid and arid environments, the greater the effectiveness of rare flash floods.

Many changes initiated by large rare floods may be subsequently removed by the action of smaller, more frequent floods. The duration of this 'healing' process is termed the recovery time, which varies between river environments. In deserts, the forms may be very persistent and channels will not undergo significant modification until the next major flood passes through. However, in temperate rivers the recovery time may be relatively short.

Stream power is a useful guide to the erosive potential of a river and is defined by:

$$\text{Stream power } \omega \text{ (W/m}^2) = \rho g Q S$$

where ρ is the fluid density; g is the acceleration due to gravity; Q is the discharge; S is the channel slope.

The areas most vulnerable to channel change in a catchment often coincide with the middle reaches, where the river emerges from the upland zone. Here, rivers are sufficiently steep to have the potential for erosion, while entering softer rocks or alluvial areas where the beds and banks are more readily eroded.

The recognition that climate and flood frequency has been, and will be, variable over 'engineering time' puts severe constraints on the assumption of equilibrium channel forms developed under static conditions. Indeed, many rivers appear to adjust their form more frequently and rapidly than had previously been appreciated. It is probably more realistic to assume that an alluvial channel form is transient or unstable unless it can be demonstrated otherwise (Macklin and Lewin, 1997). Where stable channels exist it is

possible that they are relict forms, adjusted to suit previous climatic conditions. Although the forms may be stable under current conditions they might never have evolved under them. This has been described as 'passive disequilibrium' (Ferguson, 1981).

10.4 Planform types

River channel form reflects the interaction between the discharge and sediment load with the materials through which the river flows and the vegetation along its course. These interactions are expressed in terms of a characteristic cross-section, long-profile and planform. The planform is often of greatest interest in erosion hazard studies. Despite the great variety of flow regimes, sediment loads, bed and bank materials and vegetation types, there are a limited number of planform types: straight, meandering, braided and anastomosing channels (Figures 10.9a and b).

Straight alluvial channels are not common. The channel bed tends to have a sequence of regularly spaced riffles (shallows) and pools (deeps), generated by non-uniform flows that spiral downstream in three-dimensional circulatory cells. These cells generate both vertical and lateral variations in flow velocity, with deep scour adjacent to one bank and a shoaling bar by the opposite bank. These features alternate from one side of the channel to the other downstream and lead to a tendency for meandering (Figure 10.10). Straight channels tend to exhibit very slow lateral migration and planform evolution.

Two broad groups of meandering channel can be recognised. Passive meanders occur where a river does not have sufficient power to adjust its channel through bed scour and bank erosion, with the course tending to be diverted across the floodplain by hard-points or interlocking spurs. Active meanders are associated with alluvial channels that undergo almost constant readjustment in response to variations in discharge and sediment load. Although most actively meandering rivers comprise a complex pattern of bends of varying sizes and intervening straights, there appears to be a good relationship between the meander

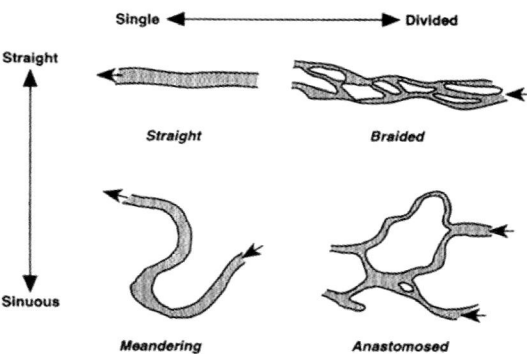

Figure 10.9a Classification of channel pattern, based on sinuosity and degree of division (after Thorne, 1997).

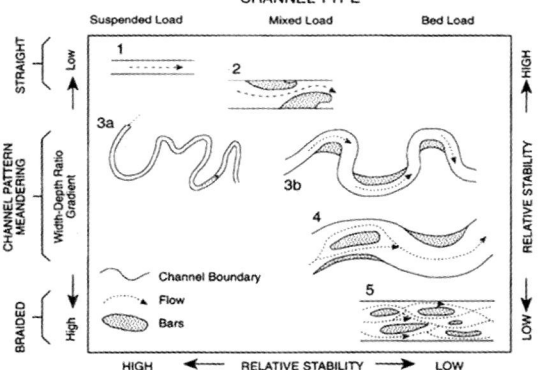

Figure 10.9b Classification of channel pattern, based on sediment load and system stability (from Thorne, 1997).

wavelength (Figure 10.11) and other channel parameters, notably:

$$L = 12.34\ \overline{w} \qquad \text{(Richards, 1982)}$$

$$L = 54.3\ Q_b^{0.5} \qquad \text{(Dury, 1955)}$$

where L is the meander wavelength, $\overline{w}$ is the mean channel width and Q_b is the bankfull discharge. It should be appreciated, however, that outcrops of stronger materials (e.g. clay plugs) within the floodplain sediments can have a major influence on meander form and pattern.

Meanders may migrate downstream, although this process can be impeded when less erodible materials are encountered in the riverbanks. Continued growth and elongation of meanders

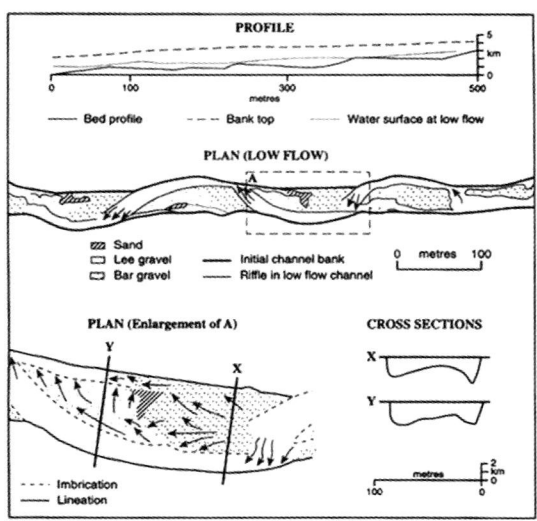

Figure 10.10 Formation of asymmetric pools, bars and riffle crossings (from Thorne, 1997).

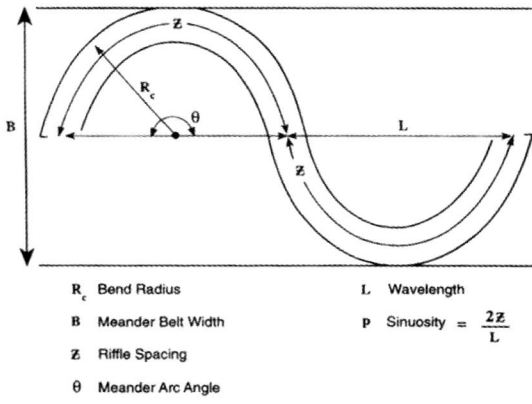

R_c	Bend Radius	L	Wavelength
B	Meander Belt Width	P	Sinuosity $= \dfrac{2z}{L}$
z	Riffle Spacing		
θ	Meander Arc Angle		

Figure 10.11 Meander planform: terminology (from Thorne, 1997), note that the wavelength is the complete width of the schematic planform shown (i.e. $2L$).

is a common feature of many alluvial channels, rather than the maintenance of an equilibrium form. Eventually cut-offs occur (Figure 10.12), creating oxbow lakes that will infill over time. Some meandering rivers appear to experience continuous change and increased instability over time (Figure 10.13) i.e. change is inherent and that ongoing change with little stability might occur.

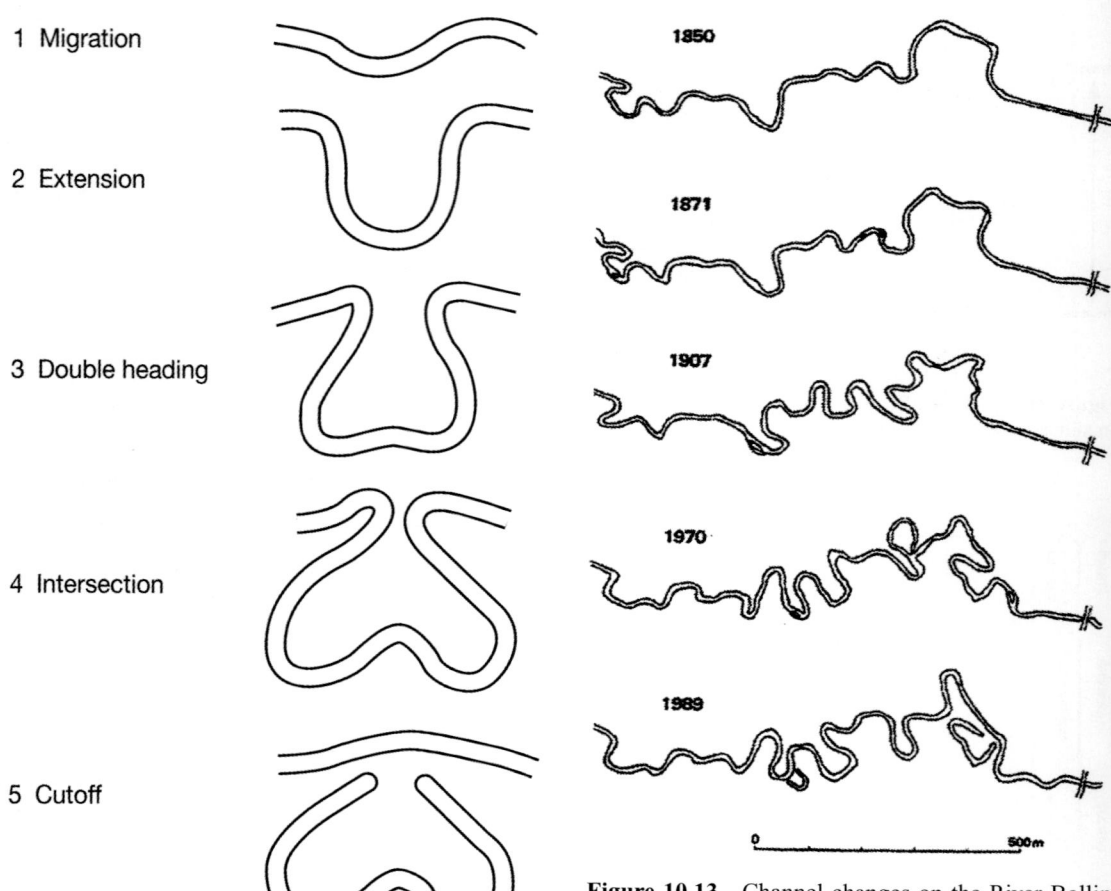

Figure 10.12   Examples of meander change (redrawn from Hooke, 1991).

Figure 10.13   Channel changes on the River Bollin, Cheshire, UK (from Hooke, 1991).

Braided channels tend to be associated with high-powered rivers where there is an abundance of relatively coarse material moving as bed-load, and with weak banks. Typically there is a sequence of sub-channels (anabranches) separated by braid bars that are flooded at bankfull discharge (i.e. at bankfull stage the flow is in a single channel). This type of planform can resemble a string of beads with relatively long, wide multi-thread island reaches separated by short, narrower single-thread reaches (nodes). Braided rivers are susceptible to rapid bank erosion and channel change (avulsion). Figure 10.14 illustrates how the position of bars within an island reach can evolve through time.

Anastomosed channels are associated with low-energy rivers. The typical planform is of two or more highly sinuous channels separated by large, semi-permanent vegetated islands at a similar elevation to the surrounding floodplain (i.e. at bankfull stage the flow remains multi-thread). These tend to be relatively stable channels. Rates of bank erosion, bend migration and planform change are characteristically small.

These four major planform types represent a continuum of forms. Whilst it is tempting to seek thresholds between these forms and to use these as a measure of the potential instability of a channel, in reality a range of transient patterns occurs. Thorne (1997) offers some practical advice to

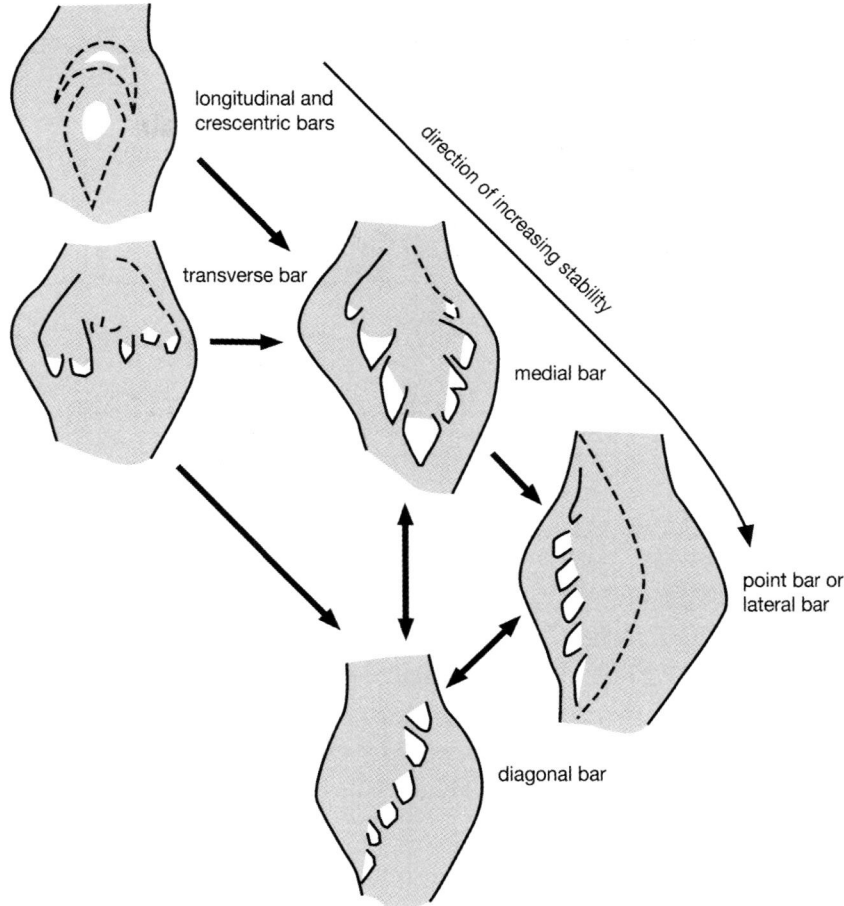

Figure 10.14 Evolution of gravel bars in a braided channel (redrawn from Church and Jones, 1982).

help determine whether a river is likely to be vulnerable to major planform change. If a river is susceptible to switching between meandering and braiding this should be apparent in the current and abandoned channel forms. If a channel displays only typical meandering characteristics, then it is probably safe to assume it is not prone to braiding.

Figure 10.15 and Table 10.3 present a useful system for classifying channel forms.

Figure 10.16 presents an indication of the patterns of dynamic morphological adjustment that may occur in alluvial channels. Analysis of historical maps, photographs and satellite imagery can

help determine past trends. However, great care is needed in their use because of the potential accuracy or reliability problems (e.g. Carr, 1962; Hooke and Kain, 1982). Table 10.4 provides a guide to some field indicators of channel instability. However, caution is needed before simply extrapolating these trends into the future.

10.5 Bank erosion and instability

Bank retreat generally involves a combination of flow erosion and mass movement followed by the removal of the debris from the bank toe by

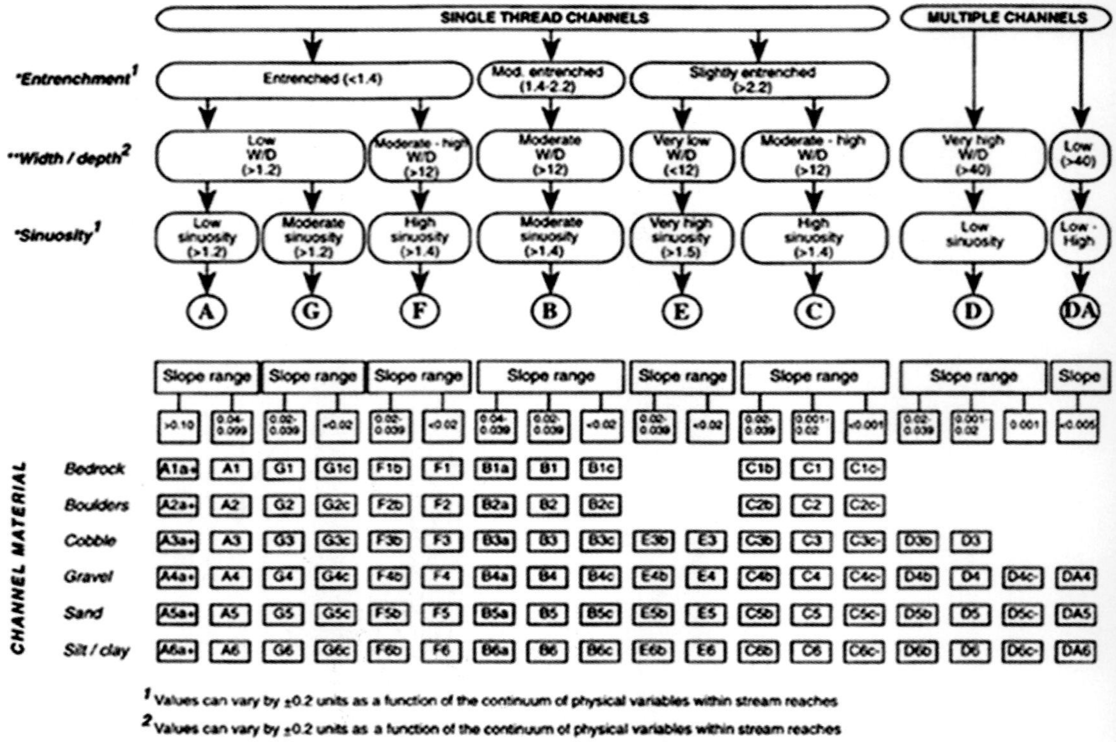

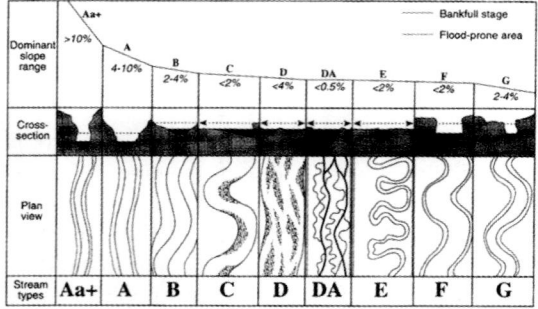

Figure 10.15*a*   Rosgen's classification of rivers (from Rosgen, 1994).

Figure 10.15*b*   Stream types (from Rosgen, 1994).

running water. Bank erosion rates can be determined from a comparison of maps and charts, aerial photographs and satellite images of different dates, repeated surveying along established baselines or through monitoring with erosion posts or automatic systems (e.g. the Photo-Electronic Erosion Pin; Lawler, 1992).

A range of typical bank failure types is presented in Figure 10.17. Bank stability can be analysed using conventional stability analysis techniques (e.g. Bromhead, 1986). Alternatively, a useful tool is the dimensionless stability chart developed by Osman and Thorne (1988) for rotational and slab-type failures in cohesive materials (Figure 10.18).

10.6 Bed scour

Many serious bank erosion problems are usually related to the meander planform. Erosion often occurs at the outer banks in a bend, as a result of both elevated near-bank velocities and deep bed

Table 10.3 A summary of the criteria used in the Rosgen channel classification system (after Rosgen, 1994).

Stream type	Description	Entrenchment ratio	W/D ratio	Sinuosity	Slope	Landform/soils/features
Aa+	Very steep, deeply entrenched, debris transport streams.	<1.4	<12	1.0–1.1	>0.10	Very high relief. Erosion, bedrock or deposition features; debris flow potential. Deeply entrenched streams. Vertical steps with deep scour pools, waterfalls.
A	Steep, entrenched, cascading, step/pool streams. High energy/debris transport associated with depositional soils. Very stable if bedrock or bedrock dominated channel.	<1.4	<12	1.0–1.2	0.04–0.10	High relief. Erosive or depositional and bedrock forms. Entrenched and confined streams with cascading reaches. Frequently spaced deep pools associated step-pool bed morphology.
B	Moderately entrenched, moderate gradient, riffle–dominated channel with infrequently spaced pools. Very stable plan and profile. Stable banks.	1.4–2.2	>12	>1.2	0.02–0.039	Moderate relief, colluvial deposition and/or residual soils. Moderate entrenchment and width/depth ratio. Narrow, gently sloping valleys. Rapids predominate with occasional pools.
C	Low gradient, meandering, point bar, riffle-pool, alluvial channels with broad well defined floodplains.	>2.2	>12	>1.4	<0.02	Broad valleys with terraces, in association with floodplains, alluvial soils. Slightly entrenched with well defined meandering channel. Riffle-pool bed morphology.
D	Braided channel with longitudinal and transverse bars. Very wide channel with eroding banks.	N/A	>40	N/A	<0.04	Broad valleys with alluvial and colluvial fans. Glacial debris and depositional features. Active lateral adjustment, with abundance of sediment supply.

Table 10.3 (*Continued*).

	Description	Entrenchment	W/D	Sinuosity	Slope	Description
DA	Anastomosing narrow and deep with expansive well vegetated floodplain and associated wetlands. Very gentle relief with highly variable sinuosities. Stable streambanks.	>4.0	<40	Variable	<0.005	Broad, low gradient valleys with fine alluvium and/or lacustrine soils. Anastomosed, fine deposition with well vegetated bars that are laterally stable with broad wetland floodplains.
E	Low gradient, meandering, riffle-pool stream with low width/depth ratio and little deposition. Very efficient and stable. High meander width ratio	>2.2	<12	>1.5	<0.02	Broad valley/meadows. Alluvial materials with floodplain. Highly sinuous with stable, well vegetated banks. Riffle-pool morphology with very low width/depth ratio.
F	Entrenched meandering riffle-pool channel on low gradients with high width/depth ratio.	<1.4	>12	>1.4	<0.02	Entrenched in highly weathered material. Gentle gradients, with a high width/depth ratio. Meandering, laterally unstable with high bank erosion rates. Riffle-pool morphology.
G	Entrenched gulley step-pool and low width/depth ratio on moderate gradients.	<1.4	<12	>1.2	0.02–0.039	Gulley, step-pool morphology with moderate slope and low width/depth ratio. Narrow valleys or deeply incised in alluvial or colluvial materials i.e. fans, deltas. Unstable with grade control problems and high bank erosion rates

Note that 'entrenchment' is the ratio of channel width to bank height and 'W/D ratio' is the ratio of the channel width to depth.

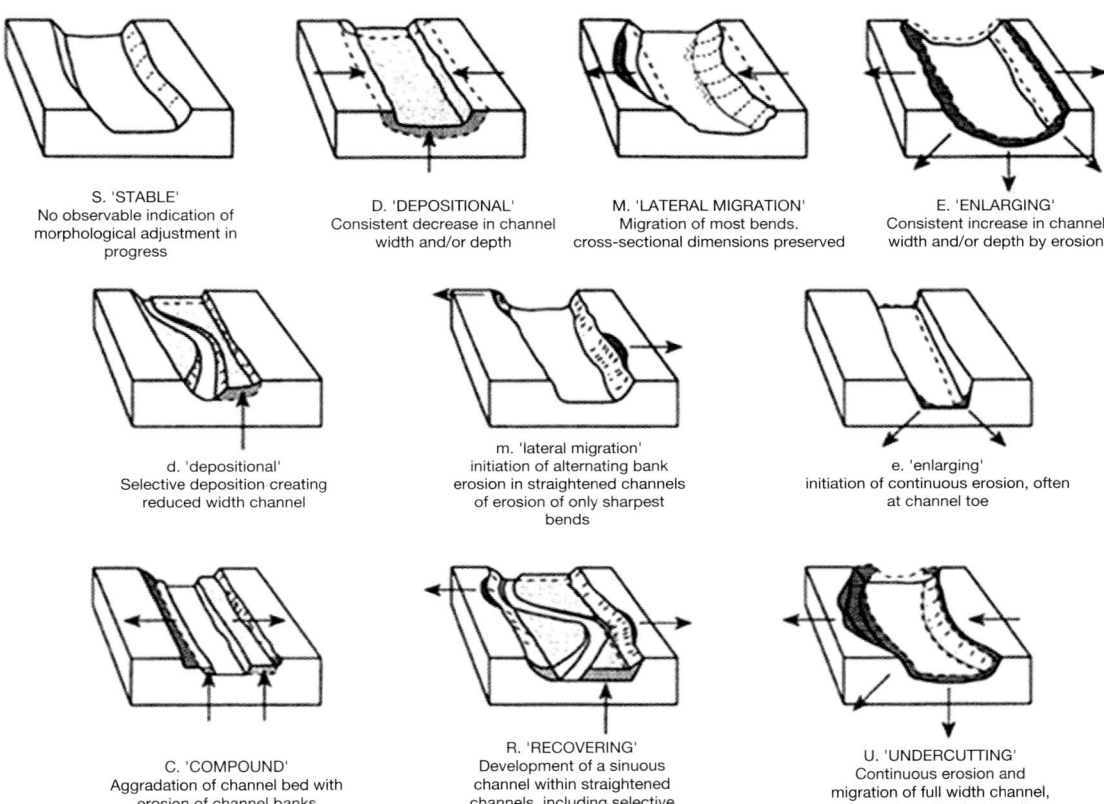

Figure 10.16   Channel classification, based on types of change (from Downs, 1995).

scour adjacent to the bank. The magnitude of the near-bed velocity increases and the additional scour depth due to the bend effects are related to the shape and, particularly, to the curvature of the bend.

Potential scour depth can be estimated from the maximum flow depth during flood flows, using a simple empirical approach (Neill, 1973). A multiplication factor is applied to the maximum flow depth as shown in Table 10.5.

Scour depth = (Mean depth × Multiplication
Factor) − Mean depth

The estimates are *potential* scour depths. In many channels this predicted depth may not be realised as there may be only a thin veneer of loose sediment above bedrock.

10.7 Floodplain accretion

Floodplains accumulate by two processes:

1. Vertical accretion from over-bank flow, producing thin laminae of fine sediments
2. Lateral accretion from the migration of channels across a floodplain. This results in a fining upwards sequence of gravels and sands, capped by finer sediments.

Floodplain accretion rates in temperate rivers generally vary between fractions of mm per year to a few cm per year (Reid and Frostick, 1994). However, the potential for faster rates should be considered along rivers prone to flash floods or debris flows. In Bijou Creek, Kansas, USA, 1 m of sediment was deposited in a flash flood (McKee *et al.*, 1967). At Shishkat in the Hunza

Table 10.4   Field indicators of channel instability (from Sear and Newson, 1992; Newson *et al.*, 1997).

	Uplands	Upland margins/Piedmont	Lowlands
Evidence of erosion	Perched boulder berms Terraces Old channels Old slope features Undermined structures Exposed tree roots Narrow/deep channel Bank failures both banks Armoured/compacted bed Gravel exposed in banks	Terraces Old channels Narrow/deep channels Undermined structures Exposed tree roots Narrow/deep channel Bank failures Armoured/compacted bed Deep gravel exposed	Old channels Undermined structures Exposed tree roots Narrow/deep channel Bank failures Deep gravel exposed
Evidence of aggradation	Buried structures Buried soils Large uncompacted bars Eroded banks at shallows Contracting bridge space Deep fine sediment in bank Many unvegetated bars	Buried structures Buried soils Large uncompacted bars Eroded banks at shallows Contracting bridge space Deep fine sediment in bank Many unvegetated bars	Buried structures Buried soils Large silt/clay banks Eroded banks at shallows Contracting bridge space Deep fine sediment in bank Many unvegetated bars
Evidence of stability	Vegetated bars and banks Compacted weed-covered bed Bank erosion rare Old structures in position	Vegetated bars and banks Compacted weed-covered bed Bank erosion rare Old structures in position	Vegetated bars and banks Weed-covered bed Bank erosion rare Old structures in position

Valley, Pakistan, over 100 m of sediment was deposited by a single debris flow event in 1973 (Brunsden, 1996).

Figure 10.19 illustrates a range of typical floodplain landforms. Levees develop on channel margins as a result of fall-out of the coarser components of the suspended load. There is a rapid decline in grain size, sediment thickness and, hence, elevation away from the channel. Splays can develop around localised breaches at low points along the levees. Away from the levees the over-bank deposits are predominantly fine silts and clays. Ground conditions are likely to be very variable both horizontally and vertically over relatively short distances in floodplain environments, especially where river channels have been continually changing.

Old river channels can present problems for new developments. For example they may have become filled with soft organic silts and clays and

lead to excessive settlements of foundations and embankments. Alternatively they may be filled with coarse material and constitute high-permeability channels incised into low-permeability rocks or soils and cause problems with dewatering of excavations or inflows into tunnels. They may have been buried under later floodplain sediment and not be apparent at the surface. Buried river channels may be bigger, deeper and at lower levels than would be expected from present-day river systems.

10.8 River terraces: phases of aggradation and incision

River terraces are relatively flat surfaces within a river valley, separated from the active floodplain by a scarp slope (terrace riser). These terraces are the product of channel incision into

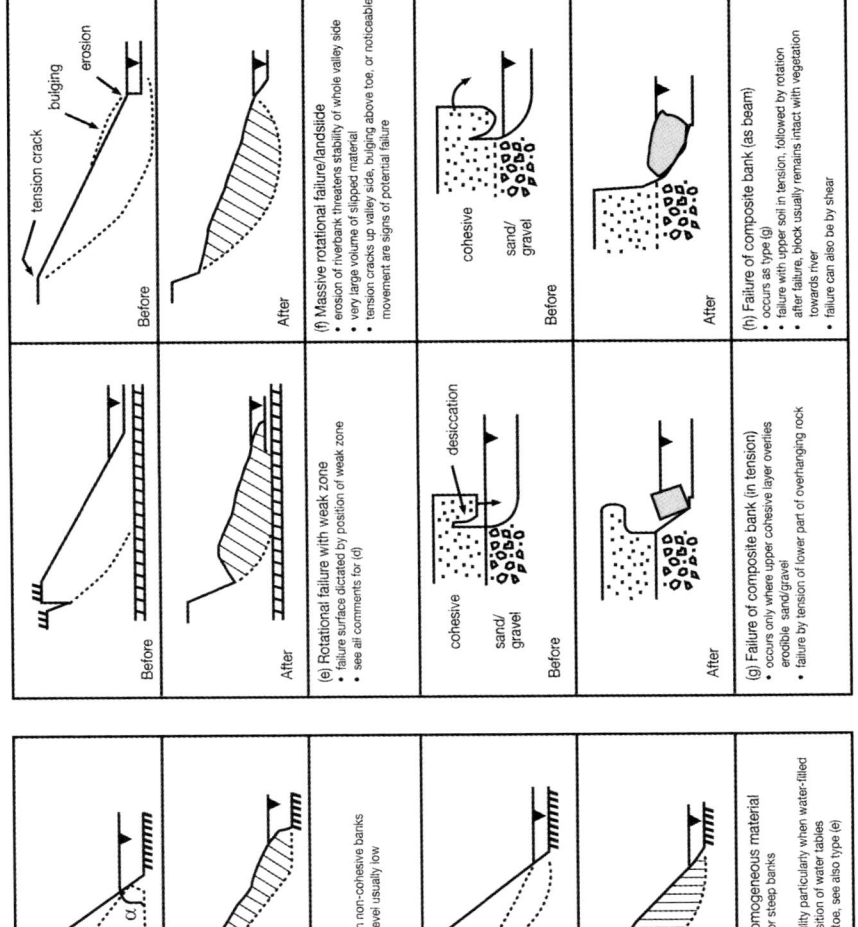

Figure 10.17 Bank failure modes (redrawn from Hey *et al.*, 1991).

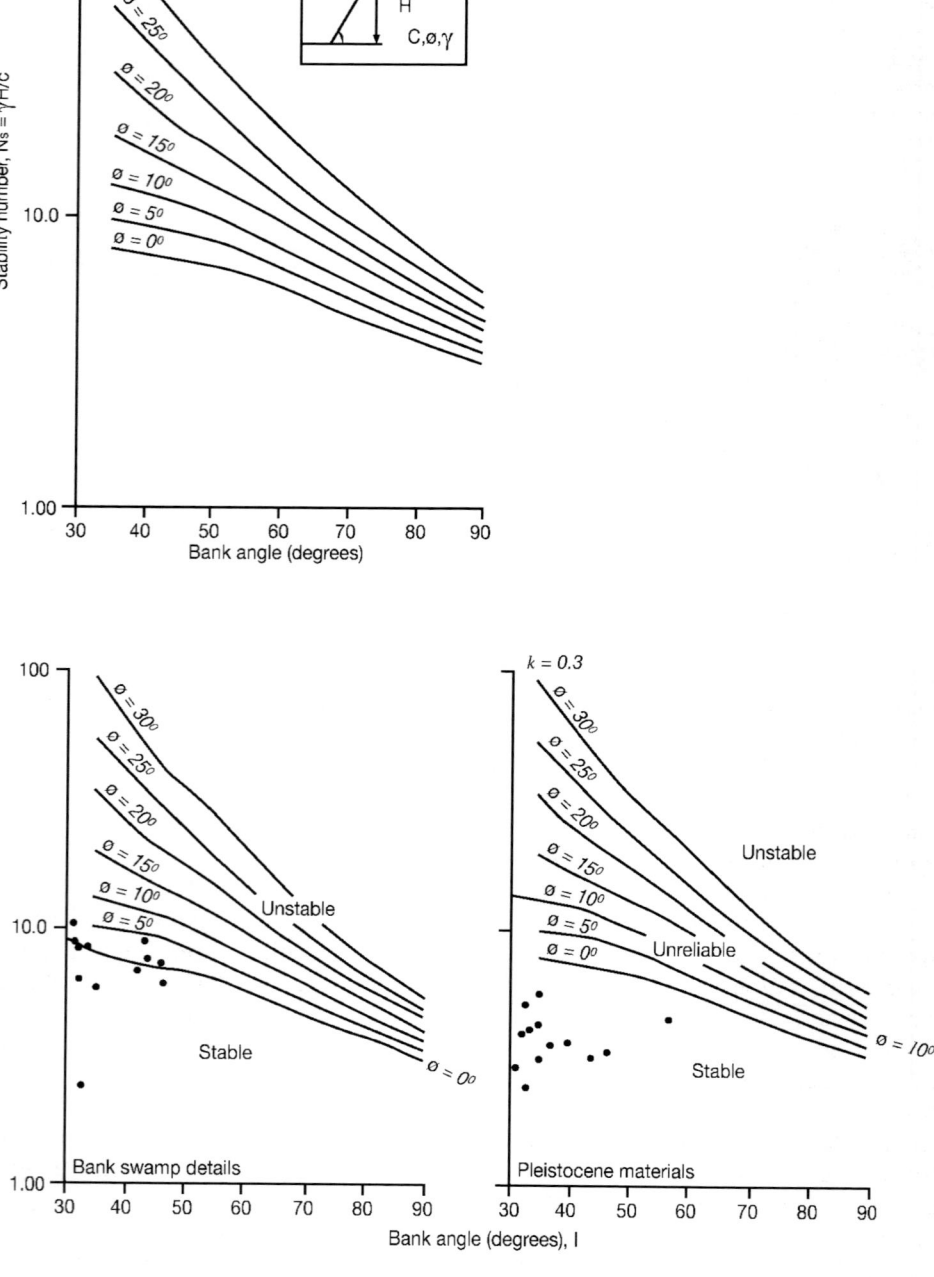

Figure 10.18 A simple method for assessing bank stability: dimensionless stability chart for rotational and slab-type failures in cohesive materials (redrawn from Osman and Thorne, 1988).

Table 10.5 Estimated potential scour depth (after Neill, 1973).

Location	Multiplying Factor	Scour Depth (10m Flow)
Straight channel reach	1.25	2.5
As above with mobile bed dunes	1.50	5.0
Moderate channel bend	1.50	5.0
Severe channel bend	1.75	7.5
Right-angle abrupt channel bend	2.00	10.0

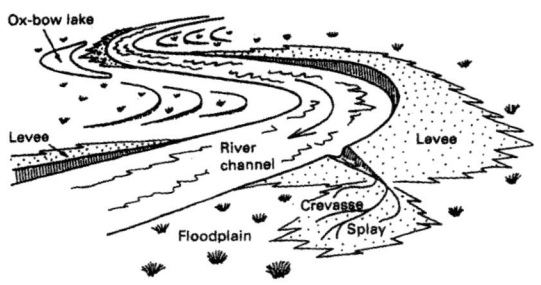

Figure 10.19 Typical floodplain morphology (from Reid and Frostick, 1994).

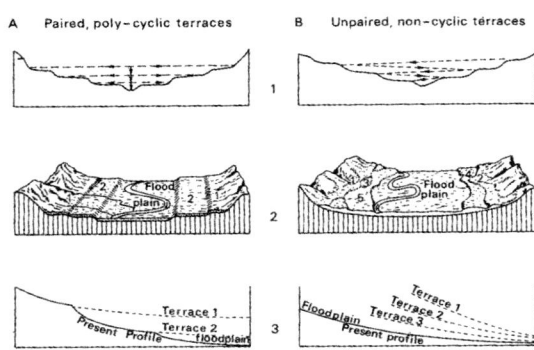

Figure 10.20 Paired and unpaired river terraces (from Sparks, 1972; Thornbury, 1969).

the floodplain. Paired terraces occur where incision is rapid compared with the rate of lateral channel migration; unpaired terraces tend to develop along rapidly migrating channels (Figure 10.20).

River incision and terrace formation are generally the product of changes in channel gradient (e.g. base level change caused by neotectonics or sea-level change), sediment load (e.g. as a result of land use change) or discharge (e.g. as a result of climate change). However, terrace formation may also be an inherent feature of the river system behaviour and not necessarily the product of external factors. For example, a stream channel may experience repeated phases of aggradation and incision under a relatively constant climate and sediment supply; valley floor aggradation occurs until the channel gradient exceeds a critical angle, initiating a phase of rapid bed erosion. Figure 10.21 illustrates how a single impulse (base level fall) can set in motion a pattern of episodic incision and aggradation. In this model, valley infills and terrace development are simply

the product of how the system seeks to adjust to a new equilibrium form (i.e. complex behaviour; Chapter 1).

10.9 Floods

Floods occur when water levels rise to overflow land not normally submerged; this can occur in a variety of settings from 'dry valleys' to river floodplains and coastal lowlands. Although floods can result from a variety of factors (Chapter 2 Figure 2.11), the most common causes are rainfall and snowmelt. Amongst the most important factors include the following regime characteristics:

The catchment area affects the total volume of stream-flow generated by a catchment-wide event (i.e. the larger the catchment the greater the potential rainfall input).

Slope characteristics influence the amount of runoff produced by an event (i.e. slope angle, bedrock geology, soil type, land use and vegetation cover).

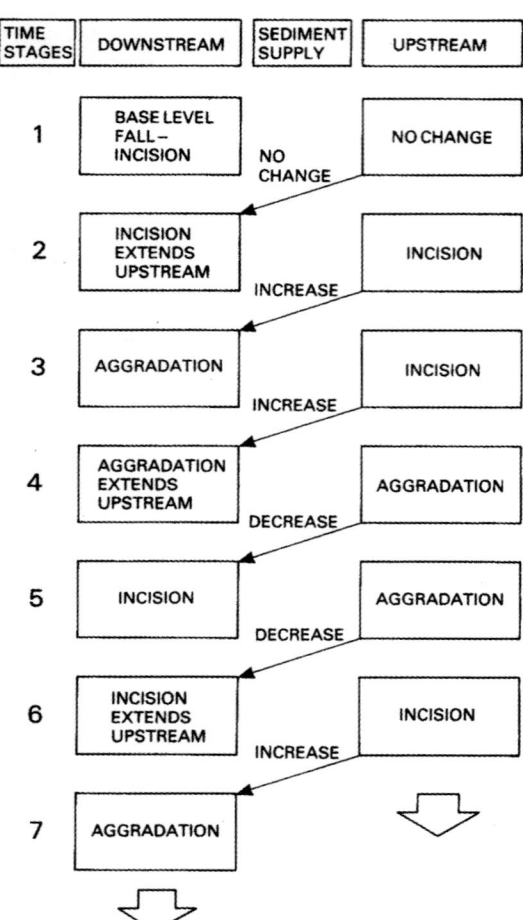

Figure 10.21 Schematic representation of the complex behaviour of a river system to a change in base level (from Summerfield, 1991).

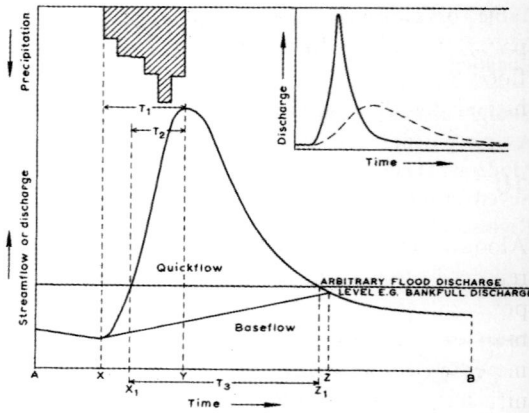

Figure 10.22 A river flood hydrograph (from Ward, 1978).

Network characteristics influence the speed at which water is transmitted through the channel system. In general, the time of rise of floodwaters after a rainstorm or snow melt event will be determined by a range of factors related to the nature of the drainage network itself. Dendritic networks (Figure 10.2) tend to produce a marked concentration of flow in the lower catchment as floodwaters are delivered down the major tributaries at a similar speed; trellised networks tend to produce a more muted response. The number and size of lakes, reservoirs or other storage areas can be a significant factor in reducing the size of a flood.

By contrast, artificial drainage, such as field drains, helps speed up the movement of water towards the channel network.

Channel characteristics influence the ability of the channel to carry a flood flow. Channel capacity is not constant; deposition of eroded sediments can significantly reduce the channel depth and cross-section. Entrapment of debris behind structures may cause 'backing-up' behind these temporary dams, and lead to over-topping of the banks.

Antecedent conditions determine the amount of the catchment that is saturated prior to a rainstorm or snowmelt event and, hence, the amount of runoff that is generated. The river or stream level prior to an event is also critical as this will influence whether the channel system can carry the additional runoff.

The hydrogeological characteristics of a river flood can be described with reference to the flood hydrograph, the continuous trace of discharge over time (Figure 10.22). Flooding does not begin immediately at the onset of rain, as the initial increase in discharge is contained within the channel system. The rate of water level rise, the magnitude of the peak flow and duration of flooding are central in defining the nature of a flood event and its impact; these attributes can vary significantly from 'flashy' streams which can have a high peak and short duration and 'sluggish' streams which may have a relatively low peak and long duration.

In other types of flooding, climate is only partly or indirectly responsible. In many estuaries flooding is often caused by the ponding back of high river discharges by rising tides.

10.10 Sediment transport

Along many rivers, sediment has to be regularly removed to maintain navigable channels or control potential flooding problems (sediment accumulation reduces channel capacity and, hence, increases flood risk). In order to address the significance of sedimentation problems and identify appropriate solutions (e.g. sediment traps or narrowing of the cross-section) it is helpful to develop a sediment budget for the catchment or section of interest. In addition to identifying potential sediment source areas (e.g. eroding hillsides and channel banks) it is useful to estimate the volumes of sediment that can be carried into the reach (i.e. the sediment transport rates).

Rivers transport sediment in a variety of ways:

In solution: soluble materials may be dissolved in water and flushed into rivers, especially during the initial stages of a storm. In terms of the total amount of material removed by rivers, the solute component can be high.

In suspension: suspended sediments are transported in the flowing body of the water where they are supported by turbulent eddies in the flow. The finest materials (clays) may be evenly distributed through the water column, whereas coarser materials may be concentrated near the bed.

Bed load: material is moved along the stream bed.

For river management the suspended and bed loads are the most significant. Suspended sediment concentrations may be estimated from a sediment rating curve (C) of the form:

$$C = a\, Q^b$$

A range of values of the exponents a and b for different types of river are presented in Table 10.6 (Q is the discharge).

Estimates of bed load can be made by using one of a number of empirical or semi-theoretical equations that have been developed. Amongst the most useful are the Meyer-Peter and

Table 10.6 A range of exponents that can be used for estimating suspended sediment yield in different river environments (from Reid and Frostick, 1994).

Country	Environment	Exponent a	Exponent b
Kenya	Arid	2570	0.512
Mexico	Arid	100	0.700
Israel	Arid	4217	0.159
Austria	Temperate, humid	0.004	2.200
USA	Temperate, sub-humid	0.01	1.600
USA	Temperate, humid	40	2.500
Germany	Temperate, humid	31	1.391

Muller (1948) equation and the Bagnold (1980) equation. The Meyer-Peter and Muller (1948) equation is:

$$q_b = \frac{\gamma_s}{\gamma_s - \gamma}$$

$$\times \left[\frac{(K_b/K_g)^{2/3}\, Y\, S - 0.047\{(\gamma_s - \gamma)/\gamma\} D_m}{(0.25/\gamma)(\gamma/g)^{1/3}} \right]^{3/2}$$

where q_b is the specific bed load flux rate in sub-aerial mass terms (kg/s/m); γ is the specific weight of the fluid; γ_s is the specific weight of the sediment; K_b the bed roughness as $u/Y^{2/3}S^{1/2}$ (where u is the mean water velocity); K_g specifies grain roughness as $26/D_{90}^{1/6}$ (D_{90} is the ninetieth percentile of the bed surface grain size distribution); Y is water depth; S is the slope of the water surface; D_m is the effective diameter of the surface bed sediments defined as $\Sigma D_i\, P_i/100$ (where D_i and P_i are the average diameter and percentage fraction by weight of the ith size fraction — often approximates D_{64} in gravel bed rivers); g is the acceleration due to gravity.

The Bagnold (1980) equation is:

$$i_b = \frac{\gamma_s}{\gamma_s - \gamma} \cdot i_{br} \cdot \left[\frac{\omega - \omega_o}{(\omega - \omega_o)r} \right]^{3/2}$$

$$\times (Y/Y_r)^{-2/3} \cdot (D/D_r)^{-1/2}$$

where i_b is the specific bed load flux; ω is the stream power (W/m^2) $(\omega = \rho g Q S$ as per secion 10.3); $\omega_o = 5.75 \times (0.04(\gamma_s - \gamma)\rho)^{3/2}(g/\rho)^{1/2}D^{3/2}$ $\log 10(12Y/D)$ (W/m^2); $D = D_{50}$ for unimodal bed sediment size distributions (m); $i_{br} = 0.01$ kg/s/m and is an arbitrary reference value for bed load flux rate; $(\omega - \omega_o)_r = 0.5$ (W/m^2) and is a reference value for excess specific stream power; $Y_r = 0.1$ m; $D_r = 0.0011$ m.

These equations give an indication of potential sediment transport rather than actual. The availability of sediment often limits the transport rate. Bed armouring can also affect the transport rate, especially in gravel bed rivers where a coarse surface layer often protects underlying finer sediments, preventing its entrainment and removal. In such cases rivers may experience two phases of sediment transport. Phase I involves a 'throughput' load with little or no bed disturbance. Once the armour is disrupted (e.g. by higher velocity flows) the finer material can be rapidly removed during Phase II transport. This provides an indication of the accelerated bed erosion that can follow inadvertent disruption of the bed armour (e.g. during pipeline construction).

10.11 The significance for conservation

River beds, banks and floodplains are all important in supporting particular species or communities of plants and animals throughout or during parts of their life cycle. For example, some fish require shallow water and gravels, as found on riffles, in which to spawn but commonly live in the deeper water at other times.

Natural channel changes can be vital to the maintenance of both channel and floodplain sites. Floods tend to maintain ecological diversity by removing vegetation which may have clogged a channel, by recreating bare vertical banks which have become overgrown, or by forming new riffles and pools (Lewis and Williams, 1984). In addition, maintenance of natural vegetation and ecological diversity can help maintain water quality through natural processes of purification.

The conservation need is for the continuance of the processes not for preservation of forms created. Channelling of mobile river reaches commonly involves direct destruction and removal of plants and habitats, particularly the loss of pools and riffles, vertical eroding banks and sinuous courses. Changes in fish populations can result from the loss of natural rifle-pool sequences that provide a variety of low flow conditions suitable as cover for both fish and the organisms on which they feed (Brookes, 1988). Fish require sheltered water in fast flowing rivers, these conditions may be absent where a meandering stream has been artificially straightened. Changes to the channel width and depth may create unsuitable habitats or present topographical difficulties for fish migration. Clearance of bank-side vegetation may destroy valuable cover for fish.

10.12 Soft engineering and river restoration

Often river engineering works are an integral part of flood defence and channel stabilisation measures. It is now recognised that so-called 'soft engineering' designs (that work with natural processes rather than attempting to impose solutions) are likely to be the most cost-effective solutions, requiring less maintenance and minimising the environmental impact. Soft engineering solutions can also help restore or rehabilitate reaches that have been damaged by previous channelling works.

Regime equations can be used to predict the geometry of stable channels, from discharge, sediment load, bed and bank materials and valley slope (section 10.3). For gravel bed rivers, Hey and Thorne's (1986) regime equations offer a method for predicting stable channels.

10.13 Summary: the catchment approach to river engineering

Although it is tempting to treat river channel change hazards as site-specific problems, this can

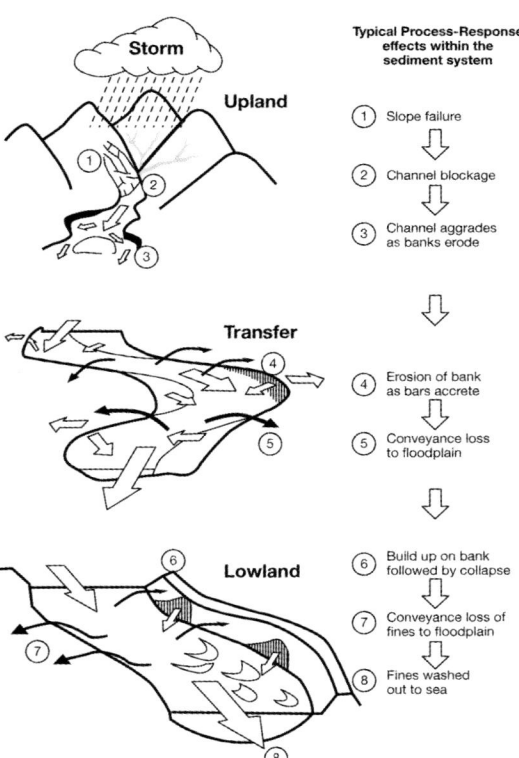

Typical Process-Response effects within the sediment system

1. Slope failure
2. Channel blockage
3. Channel aggrades as banks erode
4. Erosion of bank as bars accrete
5. Conveyance loss to floodplain
6. Build up on bank followed by collapse
7. Conveyance loss of fines to floodplain
8. Fines washed out to sea

Figure 10.23 Typical 'knock-on' effects within a drainage basin (redrawn from Sear and Newson, 1992).

lead to frequent and costly maintenance. A more efficient approach is to view the site problems as a symptom of broader-scale processes that operate throughout a catchment. Many problems can be related to the way water and sediments are transported from supply areas towards the coast and held in, or released from, temporary stores such as bars or spreads on the river bed. Figure 10.23 highlights the linkages between different parts of a river system and illustrates the knock-on effects of a large sediment input in the uplands. Within the catchment framework, engineering geomorphology can be used to identify the causes and significance of site specific problems; identify the potential impacts (upstream and downstream) of proposed works and identify suitable mitigation measures, including channel restoration operations.

References

Bagnold, R. A. (1980) An empirical correlation of bedload transport rates in flumes and natural rivers. *Proceedings of the Royal Society, London* A, **372**, 453–473.

Bromhead, E. N. (1986) *The Stability of Slopes*. Surrey University Press.

Brookes, A. B. (1988) *Channelized Rivers: Perspective for Environmental Management*. Wiley, Chichester.

Brunsden, D. (1996) Geomorphological events and landform change. *Zeitschrift für Geomorphologie* **40**, 273–288.

Carr, A. P. (1962) Cartographic error and historical accuracy. *Geography* **47**, 135–144.

Church, M. and Jones, D. (1982) Channel bars in gravel bed rivers. In Hey, R. D., Bathurst, J. C. and Thorne, C. R. (eds) *Gravel-bed Rivers*. Wiley, Chichester, 291–338.

Cooke, R. U. and Reeves, R. W. (1976) *Arroyos and Environmental Changes in the American Southwest*. Oxford University Press.

Doornkamp, J. C. (1982) The physical basis for planning in the Third World. *Third World Planning Review* **4**, 11–31; 111–128; 213–246.

Downs, J. (1995) Estimating the probability of river channel adjustment. *Earth Surface Processes and Landforms* **20**, 687–705.

Dury, G. H. (1955) Bed width and wave-length in meandering valleys. *Nature* **176**, 31.

Ferguson, R. I. (1981) Channel forms and channel changes. In Lewin, J. (ed.) *British Rivers*. George Allen & Unwin, London, 90–125.

Gregory, K. J. (1976) Drainage networks and climate. In Derbyshire, E. (ed.) *Geomorphology and Climate*. Wiley, Chichester, 289–315.

Gregory, K. J. and Walling, D. (1973) *Drainage Basin Form and Process*. Arnold, London.

Harvey, A. M. (1986) Geomorphic effects of a 100 year storm in the Howgill Fells, Northwest England. *Zeitschrift für Geomorphologie* **30**, 71–91.

Hey, R. D. and Thorne, C. R. (1986) Stable channels with mobile gravel beds. *Journal of Hydraulic Engineering, American Socity of Civil Engineers* **112** (8), 671–689.

Hey, R. D., Heritage, G. L., Tovey, N. K., Boar, R. R., Grant, N. and Turner, R. K. (1991) *Streambank Protection in England and Wales*. NRA, London R & D Note 22.

Hollis, G. E. and Luckett, J. K. (1976) The response of natural streams to urbanisation: two cast studies from south-east England. *Journal of Hydrology* **30**, 351–363.

Hooke, J. M. (1991) *Non-linearity in River Meander Development*. University of Portsmouth Working Paper No. 19.

Hooke, J. M. and Kain, R. J. P. (1982) *Historical Change in the Physical Environment: a Guide to Sources and Techniques*. Butterworths, Sevenoaks.

Knight, C. R. (1979) Urbanisation and natural stream channel morphology: the case of two English new towns. In Hollis, G. E. (ed.) *Man's impact on the hydrological cycle in the United Kingdom*. Geobooks, Norwich, 181–198.

Knighton, D. (1984) *Fluvial Forms and Processes*. Arnold, London.

Lawler, D. M. (1992) Design and installation of a novel automatic erosion monitoring system. *Earth Surface Processes and Landforms* 17, 455–463.

Lee, E. M. (1995) *The Occurrence and Significance of Erosion, Deposition and Flooding in Great Britain*. HMSO, London.

Leopold, L. B. and Maddock, T. (1953) *The Hydraulic Geometry of Stream Channels and some Physiographic Implications*. United States Geological Survey Professional Paper 252.

Lewis, G. and Williams, G. (eds) (1984) *Rivers and Wildlife Handbook*. RSPB, Lincoln.

Macklin, M. G. and Lewin, J. (1997) Channel, floodplain and drainage basin response to environmental change. In Thorne, C. R., Hey, R. D. and Newson, M. D. (eds) *Applied Fluvial Geomorphology for River Engineering and Management*. Wiley, London, 15–45.

McKee, E. D., Crosby, E. J. and Berryhill, H. L. (1967) Flood deposits of Bijou Creek, Colorado, June 1965. *Journal of Sedimentary Petrology* 37, 829–851.

Meyer-Peter, E. and Muller, R. (1948) Formulas of bedload transport. *International Association for Hydraulic Structures Research, Report of Second Meeting, Stockholm*, 39–64.

Mitchell, C. (1973) *Terrain Evaluation*. Longman, London.

Neill, C. R. (1973) *Guide to Bridge Hydraulics*. Roads and Transportation Association of Canada, University of Toronto Press.

Newson, M. D., Hey, R. D., Bathurst, J. C., Brookes, A., Carling, P. A., Petts, G. E. and Sear, D. A. (1997) Case studies in the application of geomorphology to river management. In Thorne, C. R., Hey, R. D. and Newson, M. D. (eds) *Applied Fluvial Geomorphology for River Engineering and Management*. Wiley, London, 311–363.

Osman, A. M. and Thorne, C. R. (1988) Riverbank stability analysis I: Theory. Proceedings of the American Society of Civil Engineers, *Journal of Hydraulic Engineering* 117 (8), 1091–1092.

Petts, G. E. (1984) *Impounded Rivers*. Wiley, London.

Ray, R. G. and Fischer, W. A. (1960) Quantitative photography — a geologic research tool. *Photogrammetric Engineering* 25, 143–150.

Reid, I. and Frostick, L. E. (1994) Fluvial sediment transport and deposition. In Pye, K. (ed.) *Sediment Transport and Depositional Processes*. Blackwell, Oxford, 89–155.

Richards, K. (1982) *Rivers: Form and Process in Alluvial Channels*. Methuen, London.

Richards, K. S. and Lane, S. N. (1997) Prediction of morphological change in unstable channels. In Thorne, C. R., Hey, R. D. and Newson, M. D. (eds) *Applied Fluvial Geomorphology for River Engineering and Management*. Wiley, London, 269–292.

Roberts, C. R. (1989) Flood frequency and urban-induced channel changes: some British examples. In Beven, K. and Carling, P. A. (eds) *Floods: Hydrological, Sedimentological and Geomorphological Implications*. Wiley, London, 57–82.

Rosgen, D. L. (1994) A classification of natural rivers. *Catena* 22, 169–199.

Schumm, S. A. (1977) *The Fluvial System*. Wiley, New York.

Sear, D. A. and Newson, M. D. (1992) *Sediment and Gravel Transportation in Rivers including the use of Gravel Traps*. NRA Project Report 232/1/T.

Sparks, B. W. (1972) *Geomorphology*. Longman, London.

Summerfield, M. A. (1991) *Global Geomorphology*. Longman, Harlow.

Thornbury, W. D. (1969) *Principles of Geomorphology*. Wiley, New York and London.

Thorne, C. R. (1997) Channel types and morphological classification. In Thorne, C. R., Hey, R. D. and Newson, M. D. (eds) *Applied Fluvial Geomorphology for River Engineering and Management*. Wiley, London, 175–222.

Ward, R. (1978) *Floods*. Macmillan, London.

Wolman, M. G. and Miller, J. P. (1960) Magnitude and frequency of forces in geomorphic processes. *Journal of Geology* 68, 54–74.

11. Soil Erosion

Mark Lee and John Charman

11.1 Introduction

Soil erosion is a three-stage process comprising detachment of material, its transport by water or, less frequently, wind and the subsequent deposition when sufficient energy is no longer available to transport the material. The main factors involved in the process are summarised in Figure 11.1. The severity of erosion depends on the quantity of material supplied by detachment and the ability of the running water or wind to carry it. As described in the following sections, erosion can be viewed as a function of the power of water or wind (erosivity) and the resistance of the material (erodibility). Recognition of the relative significance of these factors provides the basis for developing appropriate soil erosion control or soil conservation strategies.

Climate is an important control on the soil erosion process (Chapter 2). Temperature, both seasonal and daily, together with rainfall, influences the rate and type of weathering. Mechanical weathering may cause breakage of rock into more closely fractured components while chemical weathering causes decomposition of the rock and the disaggregation of minerals into a soil comprising a collection of discrete particles. Rainfall quantity, duration and intensity influence the rate of erosion in which disaggregated particles are detached and transported.

The highest rates of erosion are often associated with deforestation or during the construction phase of development. Sediment yields during construction can be up to 2000 times greater than those from undisturbed forested areas (Table 11.1a). For example, in Maryland USA, Wolman and Schick (1967) demonstrated that sediment yields during construction reached 55 000 t/km² per year, whereas neighbouring undisturbed areas yielded less than 400 t/km² per year.

Erosion, by water or wind, presents a variety of problems for engineers and environmental managers. It can severely limit the long-term sustainable use of agricultural land and hinder policies that seek to deliver greater, more reliable crop yields. The Ethiopian Highlands, for example, have lost around 1 Mt of topsoil per year as a result of overuse. In El Salvador, 77% of the land area is severely eroded. The 'dust bowl' years of the 1930s in southern USA are perhaps the most

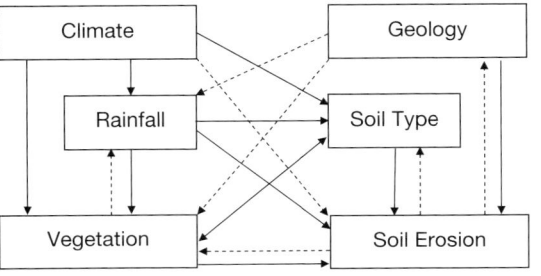

Figure 11.1 The main factors involved in the process of soil erosion by water (from Selby, 1987).

Table 11.1a Representative rates of erosion in the USA (from US Environmental Protection Agency, 1973).

Land Use	Sediment Yield t/km² per year	Relative to Forestry = 1
Forest	8.5	1
Grassland	85	10
Abandoned surface mines	850	100
Cropland	1 700	200
Construction sites	17 000	2000

Table 11.1b   Annual rates of erosion in selected countries (sediment yield t/km² per year) (from Morgan, 1995).

	Natural	Cultivated	Bare Soil
China	10–200	15 000–20 000	28 000–36 000
USA	3–300	500–17 000	400–900
Australia	0–6400	10–15 000	4400–8700
Nigeria	50–100	10–3500	300–15 000
India	50–500	30–4000	1000–18 500
Ethiopia	100–500	800–4200	500–7000

famous example of disastrous wind erosion, damaging 9 Mha and generating large dust storms (Worster, 1979).

The 'on-site' problems of erosion include the encroachment of gully-heads onto an alignment or site, erosion loss of land or exposure of services. 'Off-site' impacts, include blocked roads and ditches and damage to property. In the UK, over thirty separate damaging erosion events (so-called 'mudfloods') were reported in the South Downs between 1976 and 1990 affecting around 200 houses, with the scale of problems ranging from inundation by soil-laden water to damage to outbuildings and gardens (Boardman, 1990). Estimates of costs associated with the events suggest a minimum figure of £836,000 over a 15-year period since 1976.

Erosion can cause the loss of water storage capacity as soil removed from upland areas is deposited in reservoirs. For example, the capacity of the San Gabriel Dam, one of the Los Angeles County Drainage Area reservoirs declined from 65 Mm³ in 1938 to 55 Mm³ by 1980. During this period some 22 Mm³ of sediment had been removed at a cost of $20M (1981 prices, Bruington, 1982).

A decline in water quality, involving the transfer of agricultural chemicals to watercourses, discoloration and increased cost of treatment, due to high sediment yields from eroding catchments may also create problems for fisheries. Of particular concern is the possible impact of increased sediment loads on salmon and trout rivers. For example, Drakeford (1979) correlated the reduction in catches of salmonoid fish in the River Fleet, Scotland, with the expansion of forestry in the catchment, noting a 90% decline in sea trout

catches between 1960 and 1978. Sedimentation in watercourses leads to reduced channel capacity and, hence, increased flood risk. High sediment loads may also trigger significant alluvial river channel changes (Chapter 10).

Construction and post-construction problems associated with accelerated sediment yields off bare ground and subsequent reinstatement difficulties can be major issues for the environmental impact of engineering projects such as roads, railways or pipelines.

11.2 Types of erosion features

Figure 11.2 presents a simple model that highlights the dominant slope processes that generally operate on different units within a slope profile. Soil erosion tends to be concentrated in units 5 and 6. A variety of distinctive erosion types and forms can be recognised, including:

Sheet erosion (inter-rill erosion): the washing of surface soil from hillslopes.

Pedestal erosion: where stones or tree roots protect erodible soil from rain-splash erosion. Although these features generally develop slowly over many years, they may also occur on agricultural land in response to intense storms.

Rills: small, linear channels that have been cut into a slope as a result of the concentration of overland flow. Rills often occur as a dense network of parallel channels, occasionally feeding a master rill. Although they tend to be ephemeral features they can have considerable erosive power. Once rills have formed they may migrate upslope by headcut retreat. Maximum movement occurs when the depth of water flow is about equal to the particle diameter, so that as the water becomes concentrated into rills so its ability to carry larger particles increases. Thus, still at a small scale, the aggregated particles become at risk and the process self perpetuates as the water/sediment mixture scours the bottoms and sides of the rills, erodes the head of the channel and causes mass slumping from the over-steepened head and sides. The amount of soil detached is in proportion to the square of the velocity, whereas the sediment transport potential increases in proportion to the fifth power of the velocity.

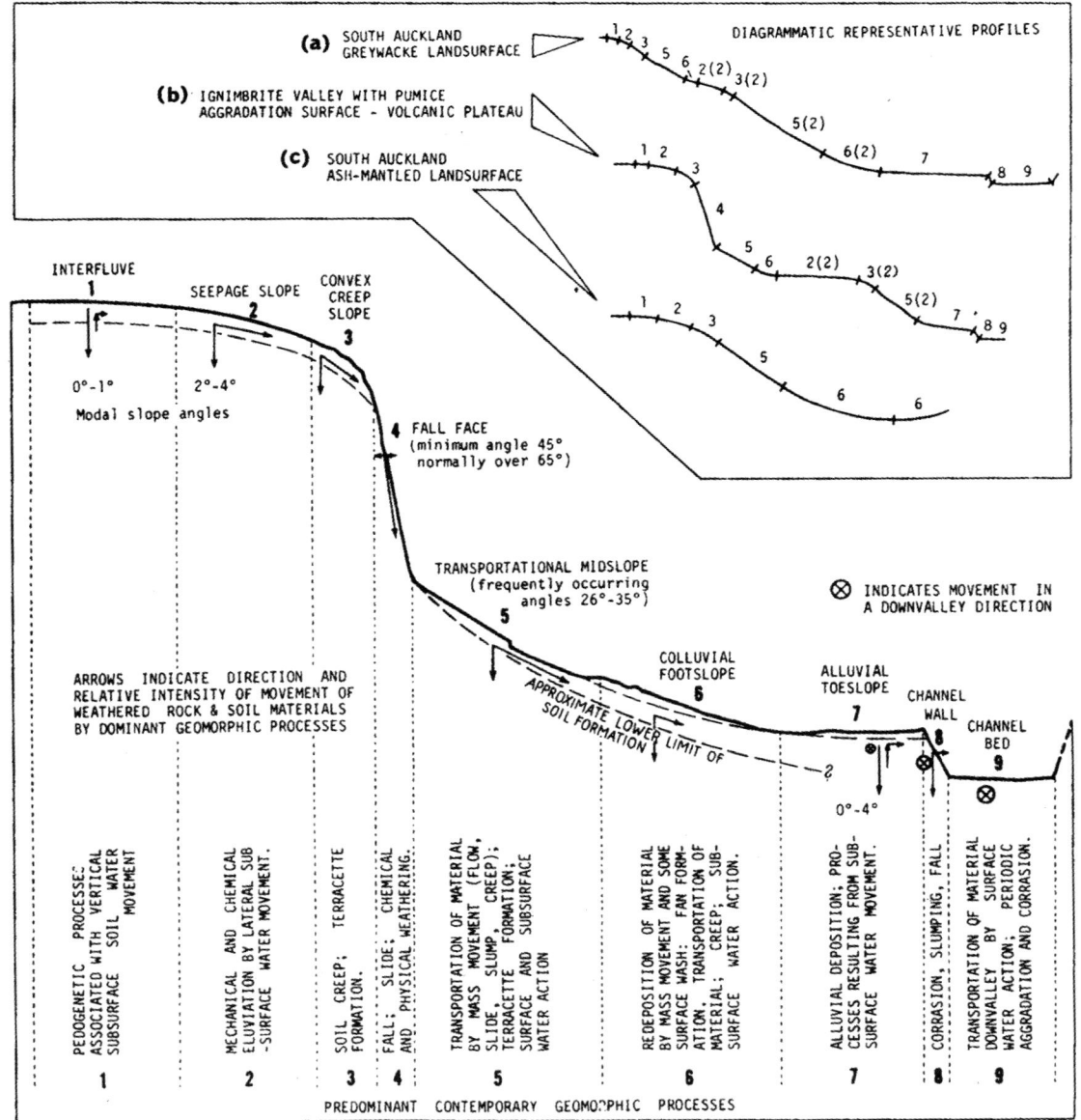

Figure 11.2   A simple slope profile model relating slope units to dominant processes (from Dalrymple *et al.*, 1968).

Piping: subsurface tunnels or pipes can develop where through-flow is concentrated along erodible soil layers (Jones, 1981), often forming a separate subsurface drainage system that can cut across topographic divides. There is a number of common settings that are prone to piping; a permeable soil horizon below an impermeable horizon, the occurrence of a weak clay horizon or gulley heads and landslide backscars (Figure 11.3). Piping can be prevalent in semi-arid regions where the soils contain swelling clays that crack on drying. Pipes often collapse and initiate gullying. Collapses under roads or buildings can cause significant problems.

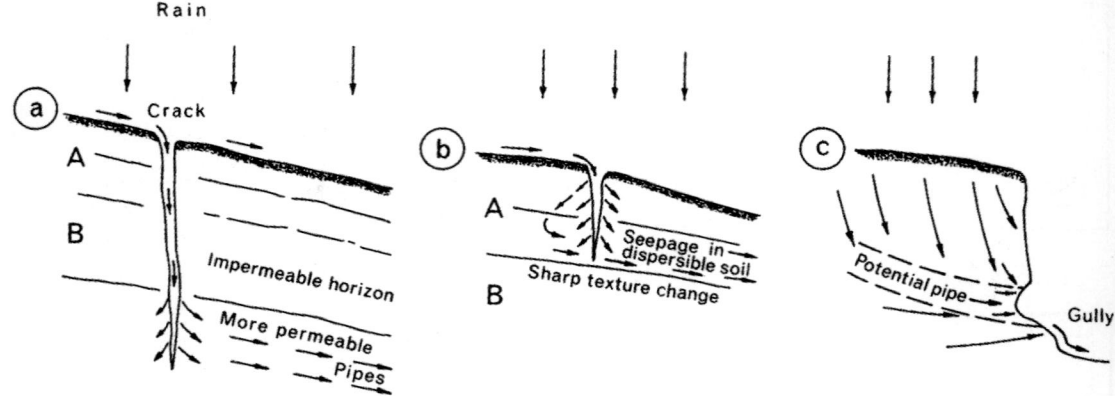

Figure 11.3   Settings prone to soil piping: (*a*) cracking and a permeable horizon below an impermeable horizon; (*b*) an horizon of dispersible clay and (*c*) a gully head (from Selby, 1982).

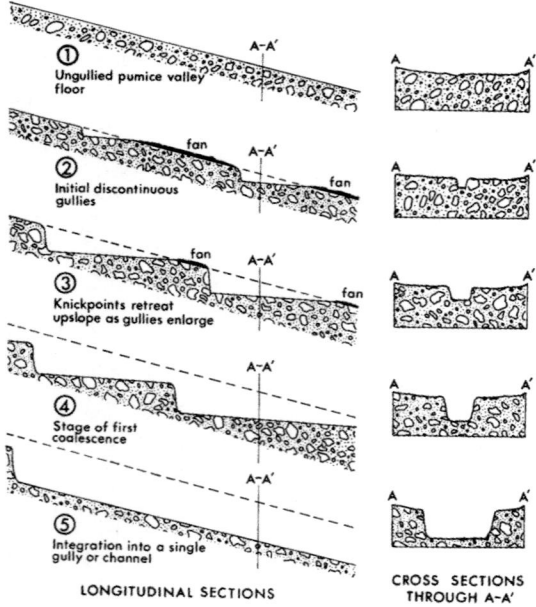

Figure 11.4   Stages in gully development (after Leopold *et al.*, 1964).

Gullying: from a practical viewpoint gullies are rills that are so large and well established that they cannot be crossed by traffic (Hudson, 1995). They are steep-sided, deeply incised features with characteristic steep headcuts and a stepped long-profile (Figure 11.4). These features are generally

associated with an acceleration in the degradation of an area, in response to increased runoff (e.g. as a result of changing patterns of rainfall intensity or land use change), or a change in slope gradient (e.g. undercutting by a stream or pipe collapse). Gullies are most common in weak materials (e.g. loess, alluvium, colluvium, gravels and deeply-weathered soils).

The characteristic stepped profile can take two forms. Knickpoints occur where there is a channel both upstream and downstream, and are related to the upslope limit of the current phase of erosive activity (e.g. in response to a change in base level or climate) or a change in lithology. Gully-heads occur where there is a switch from un-channelled hill-slope to channelled flow.

Gully development may involve the random convergence of flow around vegetation; local disturbance of the ground surface which reduces its resistance to shear (e.g. grazing or burrowing) and the through-flow or groundwater seepage (i.e. spring sapping). This process requires large flows and is most efficient in unconsolidated layered sediments where the limiting factor is the transport capacity of the spring-fed flows (Thornes, 1994). Gully development may also involve mass movement of soil into a gully from the sides and head, which have been over-steepened by the scouring effect of the channel flow.

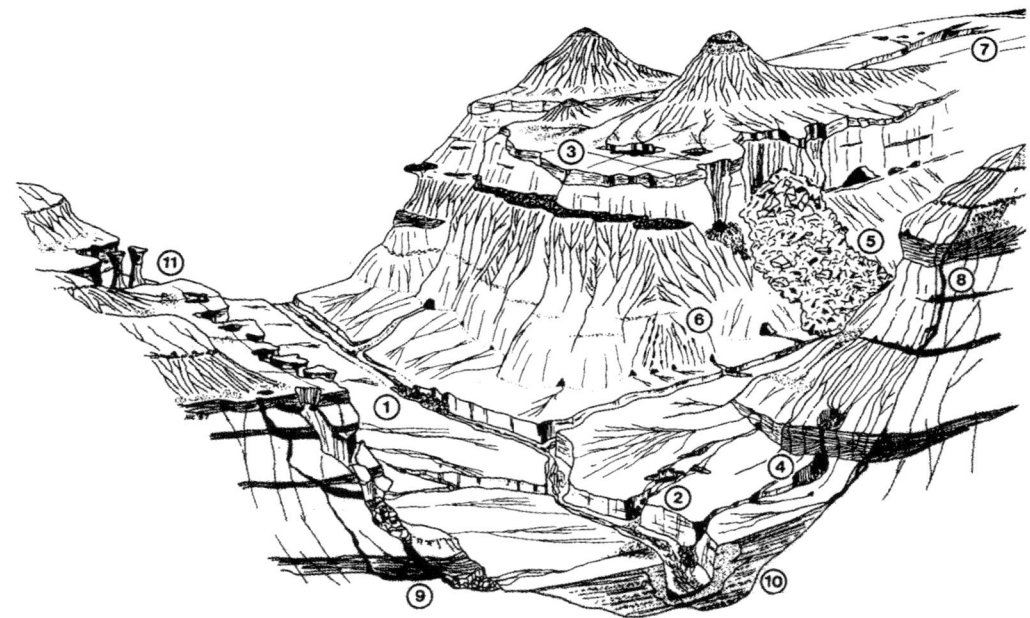

Figure 11.5   Composite of badlands lithogeomorphic characteristics. (1) Headcutting by valley-bottom gully in alluvial fill with accompanying slump and collapse by basal sapping. (2) Pipe-induced collapse initiates discontinuous gully. (3) Surface drainage into sinks and pipe shafts controlled by structural discontinuities. (4) Valley-side gullies fed by convergent rill flow. (5) Percoline-controlled large-scale piping generates slope collapsed rills. (7) Gully heads expanding into undissected surface. (8) Early stage of meso-scale pipe development. (9) Mature meso-scale piping initiating slope collapse, large gulies and tributary valley formation. (10) Multiple cut-and-fill gully deposits in bedrock-floored valley. (11) Hoodoos formed by dissection of resistant caprock (from Campbell 1997).

There may be successive cycles of gully activity. Initiation is often accompanied by rapid incision, channel widening and gully-head retreat. Over time the channel gradient is reduced and the gully may stabilise as the channel infills and the flanks degrade and re-vegetate. The next phase of activity may be triggered by climate or land use change. In many instances, different sections of a gully may experience stability or activity at different times. For example, the head may be actively retreating while the lower section is stabilising through active deposition.

Two main types of gully occur:

1. *Valley-floor gullies*, including ephemeral features to more deeply incised channels (e.g. entrenched arroyos and wadis). These features occur where runoff is concentrated in alluvial valley floors.

2. *Valley-side gullies*, formed where runoff is concentrated on hillslopes, subsurface pipes collapse or landslides create elongated scars. The typical form comprises a broad arcuate head and a narrow outlet channel downslope. They tend to occur in swarms across a hillside or entire landscape.

Badlands: badland terrain has a characteristic assemblage of steep slopes dissected by rills and gullies, narrow elongated ridgelines, high drainage density, thin bare soils, rapid erosion and shallow landsliding (Figure 11.5; Howard, 1994; Campbell, 1997). In many badlands the steep eroding slopes abut almost flat alluvial surfaces. Among the best known examples are the Mancos Shale badlands in the Henry Mountains, Utah (Howard, 1994), the Brule Formation shale badlands of South Dakota (Schumm, 1956) and the

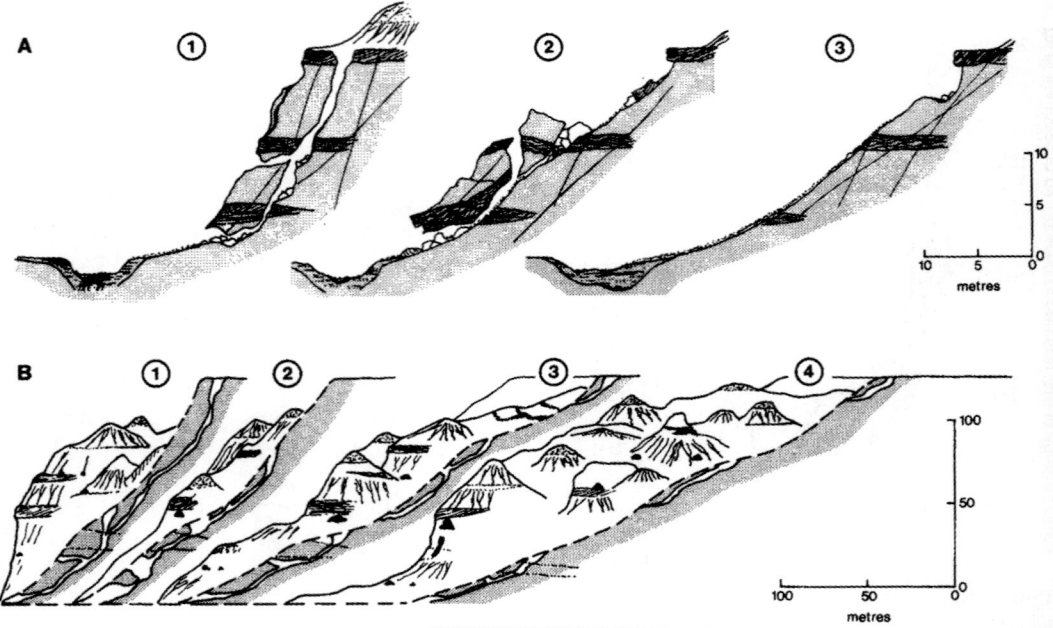

Figure 11.6   Slope and valley development in badland terrain by combination piping, collapse and gullying. (A) Three-stage meso-scale evolution of slope and valley information from initial piping (1) through pipe-induced collapse at (2), to final valley from at (3), where hydraulic gradients are no longer steep enough to generate extensive large pipe systems. (B) Macrovalley development from the early deep incision of trunk streams at (1) and extensive pipe formation with steep topographic and hydraulic gradients, through (2) and (3) gradually diminishing as badlands extend into upper surface with large pipe collapse triggering mass movements, to the stage at (4), where piping ceases to play the major role in badland valley evolution because of reduced hydraulic and topographic gradients (from Campbell, 1997).

Loess Plateau badlands in China (Chen, 1983; Chapter 25).

These erodible terrains can produce extremely high sediment yields. The Loess Plateau of China, for example, generates around $38\,000\,t/km^2$ per year (Chen, 1983). The estimated annual yield from the Cheyenne River Badlands of South Dakota is $13\,500\,t/km^2$ (Hadley and Schumm, 1961). The Alberta Badlands were shown to erode at up to 83 mm/year, with an average annual rate of around 38 mm/year (Campbell, 1981).

Badlands are generally associated with weak, impermeable, smectite-rich mudrocks in high-relief areas with semi-arid climates. Often the surface exhibits narrow polygonal cracking or an irregular, loose 'popcorn' texture with large intervening voids. The side-slopes may be at the angle of repose of dry weathered debris. Consequently these slopes are almost permanently on the verge of failure when wetted.

Variations in form range from smooth rounded ridge crests and convex side-slopes to pinnacle badlands with knife-edge divides, reflecting changes in lithology and climate. Pinnacle badlands develop where mudrocks occur beneath a caprock which erodes very slowly compared with the surrounding slopes. Eventually the caprock is lost and the underlying rocks erode rapidly, developing fluted forms because of rapid rill erosion on the steep slopes.

Figure 11.6 illustrates the development of badland valley slopes through a combination of piping, collapse and gullying.

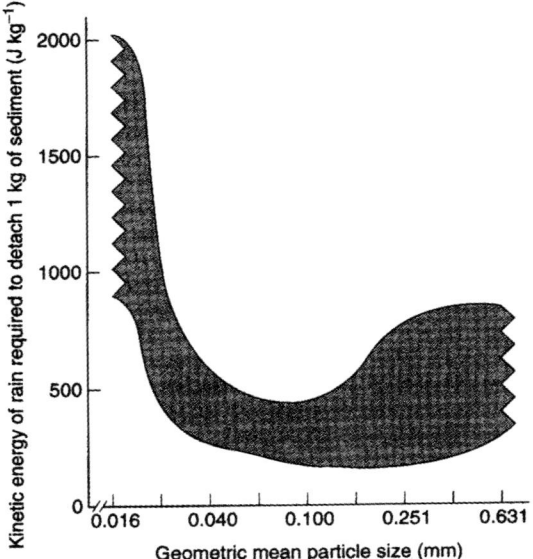

Figure 11.7a The relationship between particle size and raindrop erosion.

11.3 Detachment: erosivity of rainfall

The impact of raindrops is widely regarded to be the most significant detachment agent, although weathering, running water and wind can also loosen the soil so that it may be removed. Silts and fine sand particles tend to be the most prone to detachment — clay particles are more resistant because of adhesive or chemical bonding (Figure 11.7a).

The rate of detachment (D_r) varies with the instantaneous kinetic energy of rainfall:

$$D_r \propto I^a S^c$$

where I is the rainfall intensity (mm/hour) and S is the slope (m/m). The exponents a and c are defined as follows: a = 2.0 − (0.01 × % clay) and c is in the range 0.2–0.3, depending on the grading.

A general relationship between rainfall intensity (I) and kinetic energy of the rain (KE) (i.e. the erosivity of rainfall) is (Wischmeier and Smith, 1958):

$$KE \text{ (J/m}^2\text{per mm rainfall)} = 11.87 + 8.73 \log I$$

For tropical rainfall, Hudson (1965) developed a revised equation:

$$KE = 29.8 - \frac{127.5}{I}$$

These equations show that at rainfall intensities greater than 75 mm/hour, kinetic energy levels off at around 29 J/m² per millimetre of rainfall.

A number of indices have been developed to express the relationship between kinetic energy and soil loss, including EI_{30}, the product of the kinetic energy of a storm and the maximum 30 minute rainfall intensity (Wischmeier and Smith, 1958) and $KE > 25$ (Hudson, 1965). A worked example is presented in Table 11.2. To estimate soil loss from high intensity rainfall, a maximum value of 28.3 J/m²/mm for all rains above 76.2 mm/hour and a maximum value of 63.5 mm/hour for I_{30} (the maximum 30-minute intensity) can be used. For temperate latitudes, Morgan (1980) has suggested that the Hudson $KE > 25$ index can be modified by using a lower threshold value such as $KE > 10$.

Raindrop impact may, however, lead to soil compaction and the development of a surface crust a few millimetres thick. Surface crusts may limit the rate of detachment but, by reducing the infiltration rate, will also promote greater surface runoff and, hence, greater potential for particle transport elsewhere on a slope.

11.4 Transport: running water

Surface water flow (runoff) occurs when the rainfall intensity during a storm exceeds the rate of infitration (infiltration capacity) into the soil or when the soil is saturated. Runoff can occur as shallow flows across a slope (sheet flow or overland flow) or as channelled flows in rills, gullies and streams. Saturated through-flow can also be important in generating concentrated discharge adjacent to streams or gully-heads or where soils are thin or impermeable (Kirkby 1969; Figures 11.8a and b).

The amount of runoff from a catchment can be estimated in a number of ways, depending on the

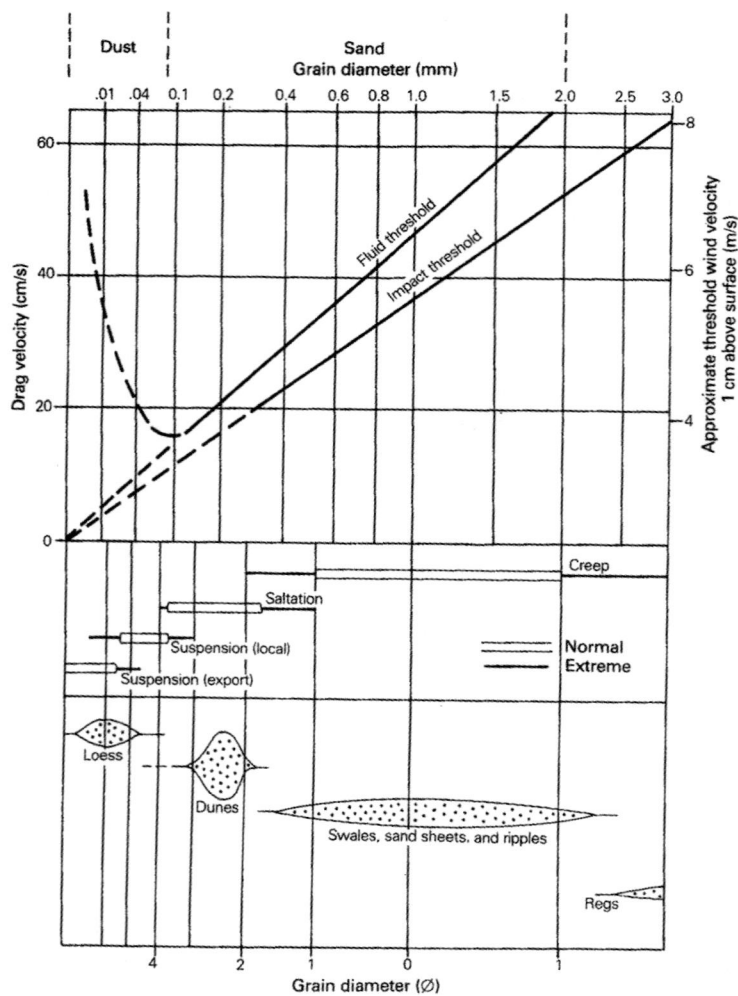

Figure 11.7b   The relationship between particle size, fluid and impact threshold velocities, characteristic modes of aeolian transport, and resulting grading of aeolian sand formations (after Mabbutt, 1977).

data availability and the degree of precision required. For example, the US Soil Conservation Service Method was developed in the USA, but is applicable in other settings:

1 Determine the design storm event: in this example 100 mm/hour.
2 Determine the run-off curve number (Table 11.3): for soil group C and paved roads the run-off curve is 90.
3 Determine the run-off during the design storm (Figure 11.9): for 100 mm rainfall and

run-off curve 90, the depth of run-off is 72.5 mm.

4 Determine the time of concentration (t = $0.02L^{0.77}\ S^{-0.385}$. L is the maximum length of flow (m) S is the average stream gradient (m/m) for a steep gradient (0.1), 1000 m long catchment; the time of concentration is 4.5 hours.

5 Determine the unit peak discharge expected from the design storm (Table 11.4): for a time of concentration of 4.5 hours, the unit discharge is 0.058m³/s.

Table 11.2 Calculation of rainfall erosivity (from Morgan, 1995).

Time from start (min)	Rainfall (mm)	Intensity (I) (mm/hour)	Kinetic energy (J/m^2 per mm rainfall)	Total kinetic energy (KE) (Col. 2 $\times$ Col. 4) (J/m^2)
0–14	1.52	6.08	8.83	13.42
15–29	14.22	56.88	27.56	391.90
30–44	26.16	104.64	28.58	747.65
45–59	31.50	126.00	28.79	906.89
60–74	8.38	33.52	26.00	217.88
75–89	0.25	1.00		

Kinetic energy $(KE) = 29.8 - \dfrac{127.5}{I}$ (Hudson, 1965).

Wischmeier Index (EI_{30})
1. Maximum 30-minute rainfall
        = 26.6 + 31.50 mm
        = 57.66 mm
2. Maximum 30-minute intensity (I_{30}) = 57.66 $\times$ 2
        = 115.32 mm/h
3. Total kinetic energy   = total of col 5
        = 2277.74 J/m^2
4. EI_{30} = 2277.74 $\times$ 115.32
        = 262 668.98 J mm/m^2 hour

Hudson Index ($KE > 25$)
1. Total kinetic energy (rainfall > 25 mm/h)
        = Total Column 5, lines 2, 3, 4 and 5 only
        = 2264.32 J/m^2

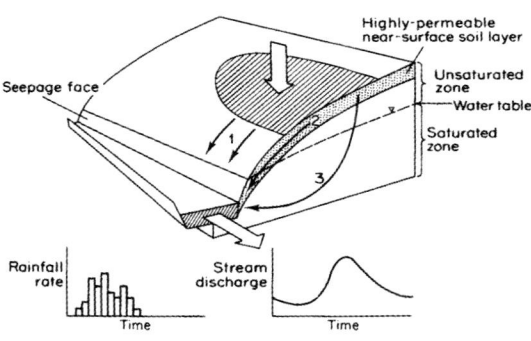

Figure 11.8a Mechanisms of delivery of rainfall to a stream channel from a hillslope: (1) overland flow (2) through-flow (3) groundwater flow (from Freeze, 1978).

6 Convert the unit peak discharge to the actual discharge.
 Peak discharge = Unit discharge $\times$ catchment area (km^2) $\times$ run-off depth (mm)
     = 0.058 $\times$ 1 $\times$ 72.5
     = 4.205 m^3/s

A specific nomograph has been developed for identifying appropriate curve numbers for dryland conditions (Figure 11.10), taking account of different plant covers. The key limitation of the method is that the choice of curve number is critical, but remains largely subjective.

The velocity of flow is critical in initiating soil erosion, as the flow must attain a threshold value before entrainment of particles begins. Once this critical condition is reached (it varies with particle size: Figure 3.4), the entrainment rate is dependent on the shear velocity of the flow and the discharge.

The greater the turbulence, the greater the erosive power generated by the flow. The velocity (v) of fully turbulent flow can be determined using Manning's equation:

$$v = \frac{r^{2/3} S^{1/2}}{n}$$

where r is the hydraulic radius, S is the slope (m/m) and n is the Manning roughness coefficient.

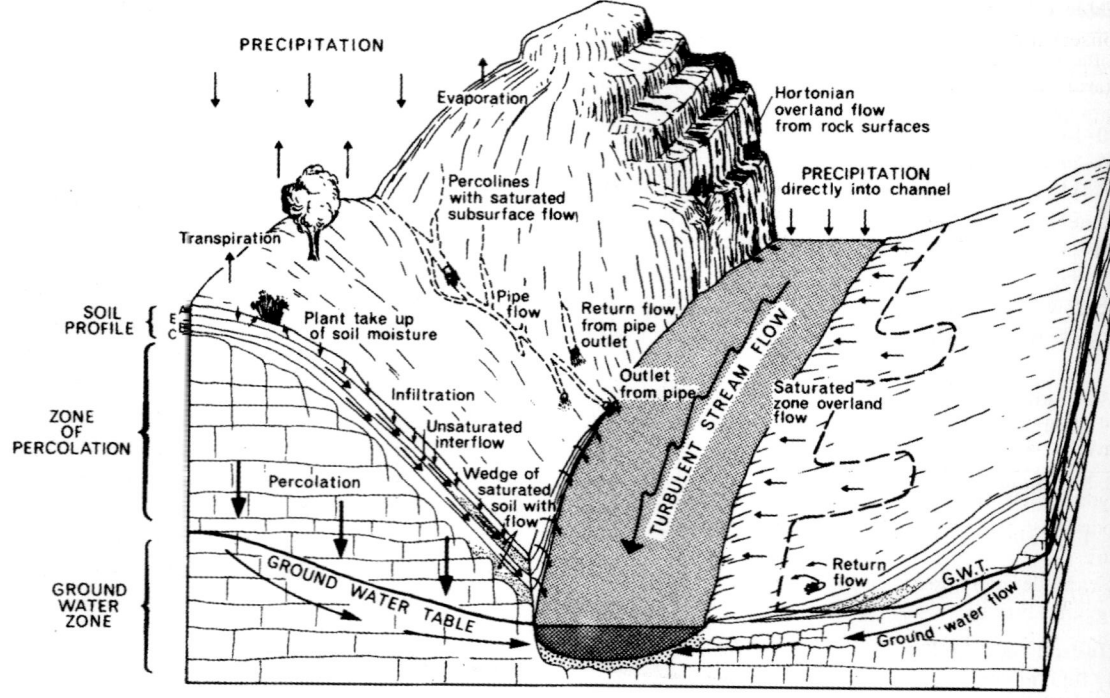

Figure 11.8*b*   A schematic runoff model (from Selby, 1982).

Tables A5.1 to A5.3 (Appendix A5) provide a guide for Manning's *n* for different plant covers and stream bed types. For flows that are not fully turbulent (Reynold's number, *Re*, < 2000; see Chapter 3), the following equations can be used:

$$v \propto r^{1.7} S^{0.95} \quad Re = 250$$

$$v \propto r^{0.95} S^{0.7} \quad Re = 500$$

$$v \propto r^{0.5} S^{0.4} \quad Re = 1000$$

The Reynold's number (*Re*) is an index of the turbulence of flow.

The transport capacity of flow (T_f) can be determined using either empirical or theoretical sediment transport equations:

$$T_f \propto Q^{5/3} S^{5/3} \quad \text{(Meyer and Wischmeier, 1969)}$$

$$T_f = 0.0085 \, Q^{1.75} S^{1.625} D_{84}^{-1.11} \quad \text{(Carson and Kirkby, 1972)}$$

where Q is the discharge or flow rate, S is the slope gradient and D_{84} is the 84th percentile grain size.

11.5 Transport: erosivity of wind

Wind erosion commences when air pressure on a loose, dry soil surface overcomes the force of gravity acting on the particles. Particles are moved through the air by saltation (bouncing), surface creep or suspension within the flow. The relationship between soil particle size and the wind velocity required for entrainment is shown in Figure 11.7*b*; the fluid threshold is the velocity needed to initiate saltation, the impact threshold is that needed to initiate entrainment by the impact of saltating particles. However, most soil surfaces comprise aggregates (collections of soil particles held in a single structure such as a clod) or crusts that limit the potential for wind erosion. As wind erosion can selectively remove silt and fine sand particles there can be a tendency for the remaining material (the lag deposit) to armour the ground surface, protecting it from further erosion unless or until it is disrupted.

Table 11.3 Runoff curve numbers for use in the Soil Conservation Service Method for estimating runoff (for use in Figure 11.9).

Land Use	Hydrologic Condition	Soil Group (see below)			
		A	B	C	D
Range or Pasture	Poor	68	79	86	89
	Fair	49	69	79	84
	Good	39	61	74	80
Meadow	Good	30	58	71	78
Woodland	Poor	45	66	77	83
	Fair	36	60	73	79
	Good	25	55	70	77
Dirt Roads		72	84	87	89
Paved Roads		74	84	90	92

Hydrologic Conditions:
Poor: heavily grazed or plant cover $< 50\%$ or regularly burned
Fair: moderately grazed or plant cover 50–75% or not regularly burned
Good: lightly grazed, plant cover $> 75\%$.

Soil Groups:
A: high infiltration capacity—sands, gravels, deeply weathered, well-drained
B: moderate infiltration capacity, moderately to deeply weathered, moderately to well-drained, moderately fine to moderately coarse texture
C: low infiltration capacity, moderately fine to fine texture, usually with a horizon that impedes drainage
D: very low infiltration capacity, swelling clays, soils with permanent high water tables, soils with clay lenses, shallow soils over impervious materials.

A simple wind erosivity index is based on velocity and duration of the wind, rather than the kinetic energy approach used for assessing rainfall erosivity (Skidmore and Woodruff, 1968):

$$EW_j = \sum_{i=1}^{n} Vt^3_{ij} f_{ij}$$

where EW_j is the wind erosivity value for vector j, Vt is the mean velocity in the ith speed group for vector **j** above a threshold velocity (taken as 19 km/hour) and f_{ij} is the duration of the wind for vector **j** in the ith speed group. The equation can be expanded for the total wind erosivity (EW) over all vectors:

$$EW_j = \sum_{j=0}^{15} \sum_{i=1}^{n} Vt^3_{ij} f_{ij}$$

where vectors **j** = 0 to 15 represent the principal compass directions beginning with East: **j** = 0 and working anticlockwise ENE = 1 etc.

The eroded material is either sand sized or dusts (<0.8 mm diameter). Whereas sand is moved mainly in saltation close to the ground surface, dust can be carried in suspension, often at great heights and for considerable distances. Figure 11.11 illustrates the concept of sediment sources, transport pathways and permanent or temporary stores for both sand and dust movement in a desert environment (Chapter 16).

The rate of soil movement (Q) can be described by:

$$Q \text{ (g/cm per second)} = a (D_e)^{1/2} \rho/g (V)^3$$

where V is the drag velocity above the eroding surface; D_e is the average equivalent diameter of soil particles moved by the wind; ρ/g is the mass density of air; a is a coefficient that varies with soil type (adapted from Chepil and Woodruff, 1963). The quantity of material removed from a given area (X) is:

$$X \text{ (tons/acre)} = a (V)^5$$

11.6 Resistance to erosion: soil erodibility and plant cover

The soil erodibility is a key factor in controlling the erosion process. A range of conditions can be recognised, as illustrated in Figure 11.12. Materials may be completely undetachable and the rate of erosion will be weathering limited. In other circumstances, the soil is readily detachable, but the rate of erosion will be limited by the transport capacity of the wind or water. The erodibility of soil varies with the grading (soil texture), stability of the aggregates and clods, shear strength, infiltration capacity and the organic/chemical content (Chapter 7).

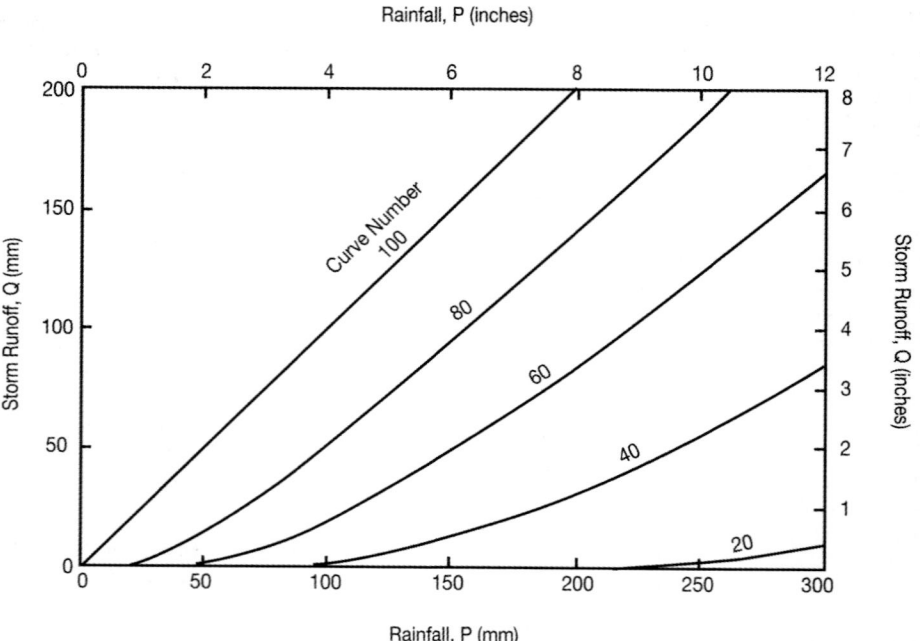

Figure 11.9   Nomograph for estimating runoff from rainfall. Runoff curve numbers represent hydrologic and soil conditions as outlined in Table 11.3. (redrawn from US Soil Conservation Service, 1972).

Table 11.4   Unit Peak Discharges*

Time of Concentration (hours)	Peak Discharge (m³/s)	Time of Concentration (hours)	Peak Discharge (m³/s)	Time of Concentration (hours)	Peak Discharge (m³/s)
0.1	0.337	1.0	0.158	4.0	0.063
0.2	0.300	1.5	0.120	5.0	0.054
0.3	0.271	2.0	0.100	10.0	0.034
0.4	0.246	2.5	0.086	20.0	0.021
0.5	0.226	3.0	0.076	24.0	0.019

*Discharge rate in m³/s from a run-off depth of 1 mm and a discharge area of 1 km².

The size and density of particles above about 0.1 mm in diameter govern the initial resistance to displacement by wind or rain-splash erosion and their susceptibility to transportation in running water. Coarser grained particles also form a soil with high porosity, which encourages infiltration so that in short duration storms runoff may be minimised. However, if particles below this size exhibit plasticity this provides inter-particle cohesion.

Successively smaller sizes below 0.1 mm tend to require higher forces to displace and transport them. For these reasons the soils most susceptible to erosion are silts and fine sands. Other soils that are particularly susceptible to erosion are dry organic soils, some residual soils such as those derived from volcanic ash, and lightweight fill materials such as pulverised fuel ash (PFA). This is mainly because the soil particles are of low density

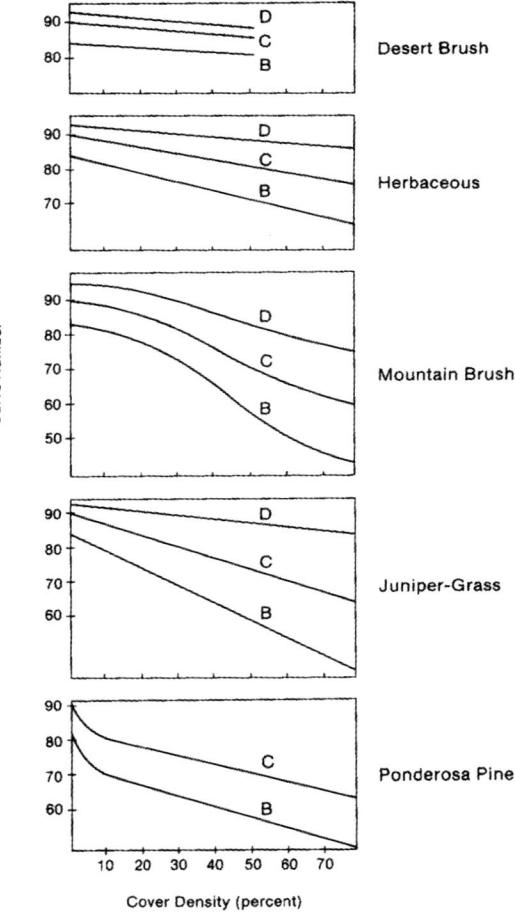

Figure 11.10   Nomograph for estimating runoff curve numbers in drylands for use in the US Soil Conservation Service method (after Jencsok, 1968).

due to their material content or form — fibrous, porous or hollow — as well as having particles falling in the critical size range.

The most widely used erodibility index is the K value that represents the soil loss per unit of EI_{30} (as measured on a standard bare soil plot, 22 m long and on a 5° slope). Nomographs have been developed to assist the estimation of K (Figure 11.13; Wischmeier *et al.*, 1971; Landon, 1984), although there may be problems when it is necessary to extrapolate beyond the nomograph values. The procedure for estimating K is illustrated

by the heavy line in Figure 11.13, for a soil with the following properties:

Proportion of sand and silt	=	65%
Proportion of sand	=	5%
Amount of organic matter	=	2.8%
Structural class	=	2
Permeability class	=	4

From the nomograph, the resulting erodibility index (K) is 0.30.

Vegetation cover can reduce soil erosion by providing a protective layer or buffer against rain-splash (by absorbing raindrop energy), runoff (by dissipating the energy of running water through increased roughness) or wind erosion (by reducing the shear velocity of wind by imparting roughness). It also modifies the moisture content of the soil and thus its shear strength. Mechanically, vegetation increases the strength and competence of the soil in which it is growing and therefore contributes to its stability, for example it decreases pore water pressure and increases soil suction because of its own water requirement. Vegetation also improves soil structure and porosity through enrichment with organic material and protects the soil from trampling by humans and animals.

Undisturbed forest is effective in controlling erosion because the tree canopy intercepts rainfall and reduces its energy. Drops from the canopy are absorbed in the leaf litter and thence into a porous soil surface. Once the forest is disturbed by tree removal or grazing, the gaps in tree cover remove the erosion protection.

Plants vary in their effectiveness, depending on the height and continuity of the canopy, stage of growth and amount of bare earth exposed to erosion. For adequate protection around 70% cover is necessary, but reasonable protection can be achieved with over 40% (Morgan, 1995).

Agricultural land is more susceptible to erosion under some crops than others because of different planting dates in respect of rainfall, different row spacing, growth rates and, therefore, effective ground protection against rain-splash. Certain management practices associated with particular crops may encourage erosion: in the UK the frequency of vehicle traffic producing

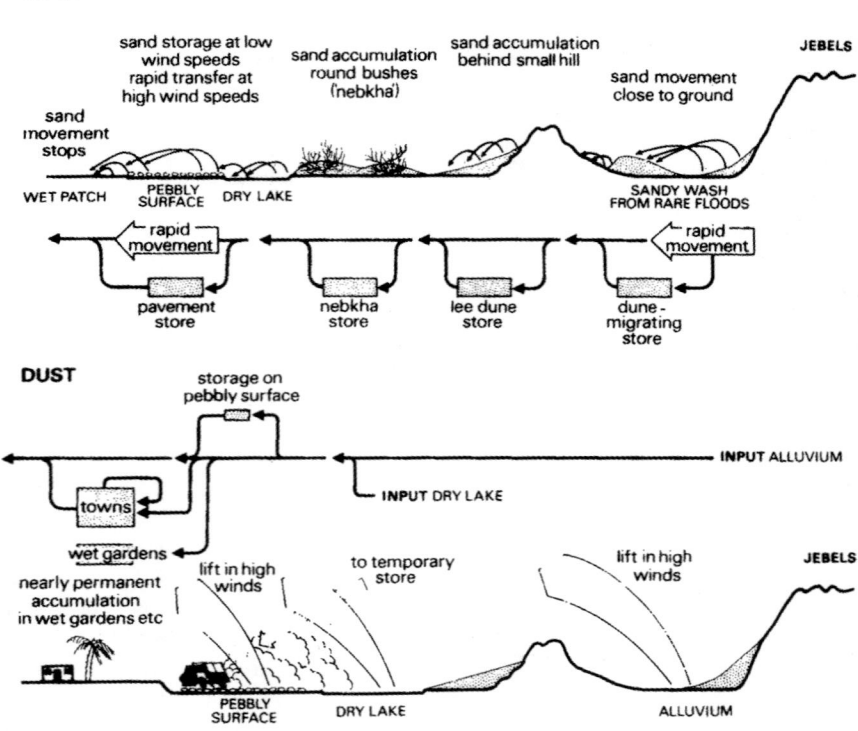

Figure 11.11   Sources and stores for sand and dust (from Jones *et al.*, 1986).

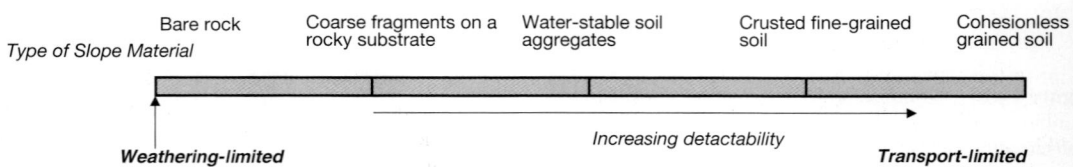

Figure 11.12   The detachability continuum for soils (from Parsons, 1988).

compaction is a problem on sugar beet and veg-etable crops, as is the rolling of seedbeds on win-ter cereals and maize. There are also frequently occurring associations of crops and soils, for example sugar beet with sandy soils, which influ-ence erosion rates. There is general agreement that the reported increase in significant lowland erosion events in temperate areas is due to the adoption of winter cereals and the consequent expansion of the area left bare in autumn and winter (e.g. Evans and Cook, 1986).

The interaction between the various land man-agement and physical factors results in risk peri-ods associated with different crops defined by the bare ground associated with the growing of a par-ticular crop and rainfall. Each of the six periods of erosion risk identified by Boardman (1991) for the UK is associated with a different land use or management practice:

1. in late summer and early autumn in fields sown with oil seed rape or grass ley

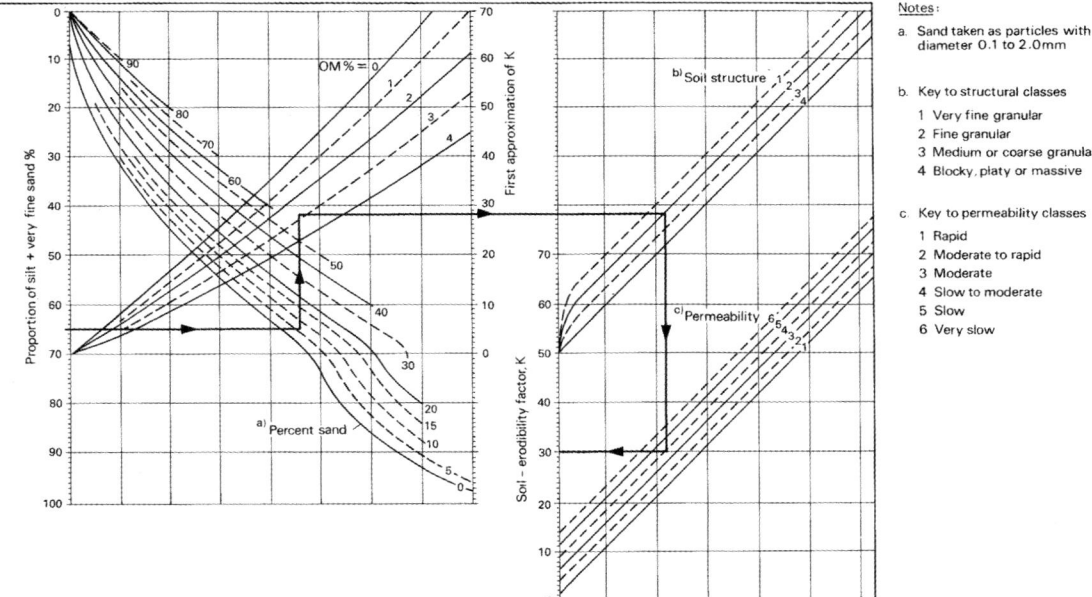

Figure 11.13   Nomograph for estimating the erodibility factor K (from Wischmeier *et al.*, 1971; Landon, 1984).

2. a short period in autumn when land has been cultivated before drilling winter cereals

3. land drilled with winter cereals the period of risk depends on rainfall distribution, drilling date and growth rate of the crop—in some years crop growth may be insufficient to inhibit erosion until April, in others the period of risk may extend only to December

4. land ploughed and cultivated over winter and spring before sowing spring cereals

5. a short period of risk in spring on land drilling with spring cereals, although rapid crop growth can limit the risk period

6. in May and June for land planted with late spring crops such as maize.

A similar situation can be identified in forestry practice, with the risk period corresponding to the 5–10 year interval between ploughing and the establishment of an effective tree cover or at the end of clear-felling and timber extraction.

Soils tend to be at greatest risk of wind erosion when they are left bare with a fine, smooth surface in spring and early summer. Removal of

hedgerows increases the erosivity of the wind by reducing the number of windbreaks.

11.7 Erosion hazard assessment

A range of approaches is available for assessing the erosion hazard within an area, ranging from simple measures derived from readily available climatic data (these consider erosion hazard in terms of the erosivity and not soil erodibility) to more detailed mapping-based methods that consider field evidence of erosion. Widely used methods are detailed below.

Erosivity indices: maps of erosivity can be prepared using a rainfall index $R(J/m^2)$, from the rainfall energy $E(J)$ and maximum 30-minute rainfall intensity $I_{30}(mm/hr)$:

$$R = EI_{30}/1000$$

Note that this produces an R value in metric units (Figure 11.14). Where rainfall intensity data is not available, erosivity can be estimated from the mean

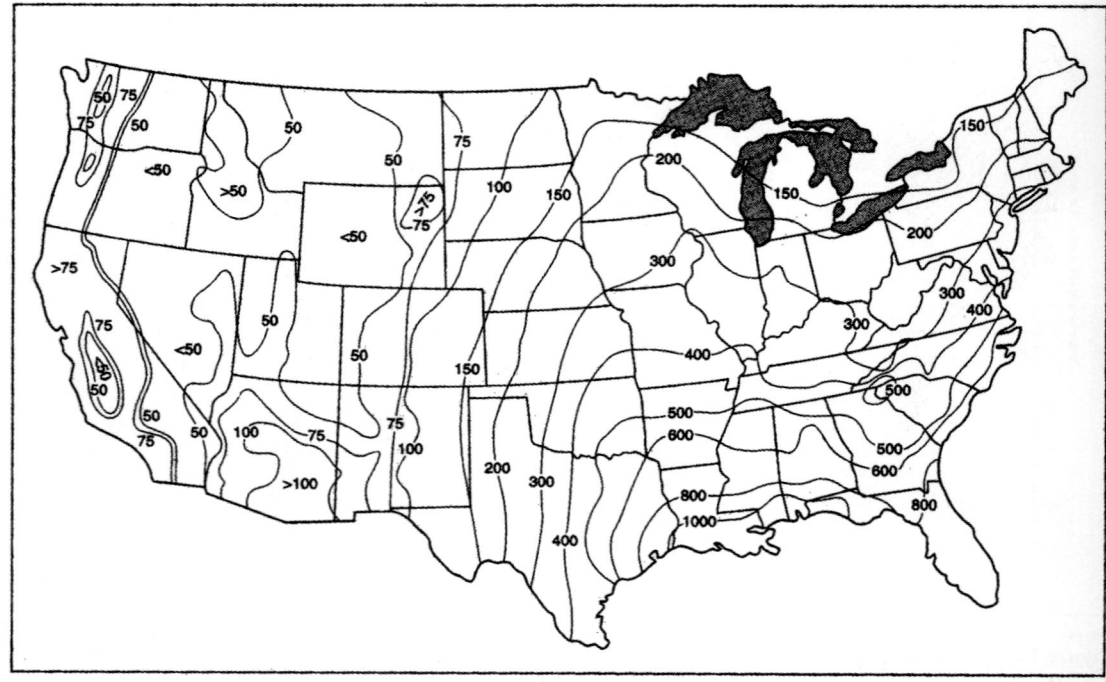

Figure 11.14 Mean annual values of the rainfall erosion index R (10^7 J/ha) (from Wischmeier and Smith, 1978).

annual rainfall. For example, in Malaysia, Morgan (1974) defined the mean annual erosivity (EV_a) as follows:

$$EV_a = 9.28\ P - 8838.15$$

Rainfall aggressiveness: Fournier's (1960) p^2/P index is a measure of the concentration of precipitation (P) into a single month and, hence, is a measure of rainfall intensity (p = the highest mean monthly precipitation, P = mean annual precipitation). The mean annual sediment yield Q_s (g/m²) from a catchment was defined as:

$$\log Q_s = 2.65 \log p^2/P \\ + 0.46\ (\log H)\ (\tan S) - 1.56$$

where H is the mean altitude (m) and S the mean slope (degrees). An alternative approach is to identify which relief and climate type the drainage basin falls into and calculate p^2/P, as follows:

Ia (low relief, $E = 6.14\ (p^2/P) - 49.78$
temperate climate)

Ib (low relief, $E = 27.12\ (p^2/P) - 475.4$
tropical,
subtropical,
semi-arid climate)

II (high relief, $E = 52.49\ (p^2/P) - 513.21$
humid climate)

III (high relief, $E = 91.78\ (p^2/P) - 737.62$
semi-arid climate)

Sediment yield tends to be greatest in tropical areas with seasonally concentrated rainfall (Figure 11.15).

Land classification: a wide variety of approaches have been developed to describe soils or land in terms of the limitations for agriculture or other land uses. Many of these methods use the degree of erosion as a factor in the classification system (e.g. Table 11.5; Figure 11.16).

Soil erosion survey: field mapping or aerial photograph interpretation of evidence of soil erosion activity (e.g. sheet wash, rills, gullies) can provide a measure of erosion hazard across

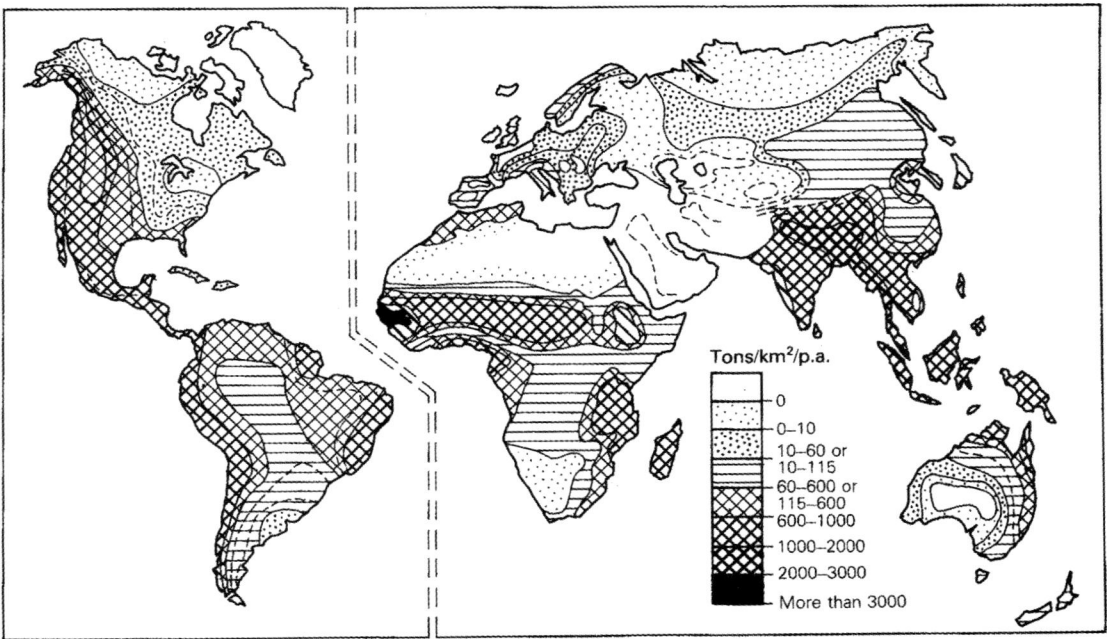

Figure 11.15 World distribution of suspended sediment yield (from Fournier, 1960).

Table 11.5 Land capability classes (US system, modified from Stallings, 1957; Morgan, 1995).

Class	Characteristics
I	Deep, productive soils easily worked, on nearly level ground; not subject to overland flow; no or slight risk of damage when cultivated; use of fertilizers and lime, cover crops, crop rotations required to maintain soil fertility and soil structure.
II	Productive soils on gentle slopes; moderate depth; subject to occasional overland flow; may require drainage; moderate risk of damage when cultivated; use of crop rotations, water-control systems or special tillage to control erosion.
III	Soils of moderate fertility on moderately steep slopes, subject to more severe erosion; subject to severe risk of damage but can be used for crops provided plant cover is maintained; hay or other sod crops should be grown instead of row crops.
IV	Good soils on steep slopes, subject to severe erosion; very severe risk of damage but may be cultivated if handled with great care; keep in hay or pasture but a grain crop may be grown once in 5 or 6 years.
V	Land too wet or stony for cultivation but of nearly level slope; subject to only slight erosion if properly managed; should be used for pasture or forestry but grazing should be regulated to prevent plant cover from being destroyed.
VI	Shallow soils on steep slopes; use for grazing and forestry; grazing should be regulated to preserve plant cover; if plant cover is destroyed, use should be restricted until cover is re-established.
VII	Steep, rough, eroded land with shallow soils; also includes droughty or swampy land; severe risk of damage even when used for pasture or forestry; strict grazing or forest management must be applied.
VIII	Very rough land; not suitable for woodland or grazing; reserve for wildlife, recreation or watershed conservation.

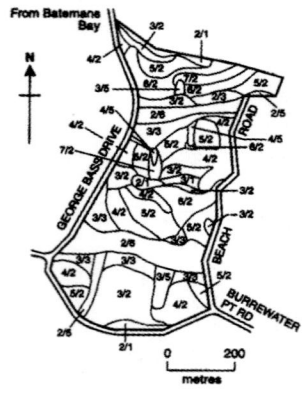

LANDFORM

Slope

0–1%	1
1–5%	2
5–10%	3
10–15%	4
15–20%	5
20–25%	6
25–30%	7

Terrain

Hillcrest	1
Sideslope	2
Footslope	3
Drainage plan	5
Swamp	6

LANDFORM

Physical Criteria and Urban Landuse

Slope class	Terrain component	Potential hazards related to topographic location and slope and which will affect urban landuse	Suitable urban landuse
0–5%	Drainage plain	Flooding, seasonally high water-tables, high shrink-swell soils, high erosion hazard	Drainage reserves/stormwater disposal
	Floodplain	Flooding, seasonally high water-tables, high shrink-swell soils, saline soils, gravelly soils	Open space areas, playing fields
	Hillcrests Sideslopes	Shallow soils, stony/gravelly soils Overland flow, poor surface drainage and profile damage	Residential: all types of recreation; large-scale industrial, commercial and institutional development
	Footslopes	Impedence in lower terrain positions, deep soils Others–swelling soils, erodible soils, dispersible soils	
5–10%	Hillcrests Sideslopes Footslopes	Shallow soils Overland flow Deep soils, poor drainage Others–swelling soils, erodible soils, dispersible soils	Residential subdivisions, detached housing, medium-density housing/ unit complexes, modular industrial, active recreational pursuits
10–15%	Sideslopes	Overland flow Geological constraints–possibility of mass movement Swelling soils Erodible soils	Residential subdivisions, detached housing, medium-density housing/ unit complexes, modular industrial, passive recreational
15–20%	Sideslopes	Overland flow Geological constraints–possibility of mass movement Swelling soils Erodible soils	Residential subdivisions, detached housing, medium-density housing/ unit complexes, modular industrial, passive recreational
20–25%	Sideslopes	Geological constraints Mass movement High to very high erosion hazard	Residential subdivision, passive recreational
25–30%	Sideslopes	Geological constraints Possible mass movement High to very high erosion hazard	Upper limit for selective residential use, low-density housing on lots greater than 1 ha, passive recreation
>30%	Sideslopes	Geological constraints Mass movement Severe erosion hazard	Recommend against any disturbance for urban development

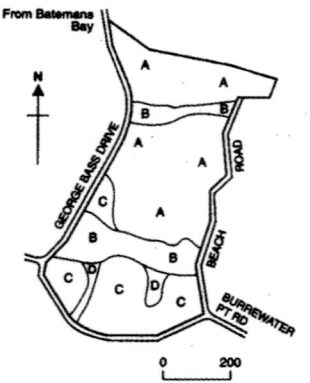

SOILS

Summary of Properties of Soils

Map unit	Dominant soils	Lithology and physiography	Erosion hazard	Limitations
A	Shallow gravelly soils	Metasediments; cherts, phyllites, etc. with quartz veination Ridges, sideslopes and some footslopes	High	Impeded soil drainage, shallow soil depth, high stone and gravel contents
B	Swamp alluvial soils	Metasediments; cherts, phyllites, etc. with quartz veination Alluvial parent materials, swamp and drainage plains	Very high to extreme	Seasonally high water-tables, poor to impeded soil drainage
C	Yellow duplex soils	Metasediments; cherts, phyllites, etc. Crests, sideslopes and some footslopes	High to very high	Low to moderate shrink-swell potential, poor soil drainage
D	Drainage plain alluvial soils	Metasediments; cherts, phyllites, etc. Alluvial/colluvial parent materials and surface materials	Very high	Seasonally high water-tables, poor to impeded soil drainage

Figure 11.16 Urban capability classification for soil erosion control, Australia (from Hannam and Hicks, 1980).

different terrain units. A simple coding system for assessing the severity of erosion is included as Table 11.6.

Potential sand drift calculation: this can be calculated using Fryberger's (1979) method which uses Lettau's formula:

$$Q(V - v_t)V^2 \cdot t$$

where Q is the proportional amount of sand drift, V is the average wind velocity at 10 m, v_t is the

Table 11.6   Coding system for soil erosion appraisal (from Morgan, 1995).

Code	Indicators
0	No exposure of tree roots; no surface crusting; no splash pedestals; over 70% plant cover (ground and canopy).
0.5	Slight exposure of tree roots; slight crusting of the surface; no splash pedestals; soil level slightly higher on upslope or windward sides of plants and boulders; 30–70% plant cover.
1	Exposure of tree roots, formation of splash pedestals, soil mounds protected by vegetation, all to depths of 1–10 mm; slight surface crusting; 30–70% plant cover.
2	Tree root exposure, splash pedestals and soil mounds to depths of 1–5 cm; crusting of the surface; 30–70% plant cover.
3	Tree root exposure, splash pedestals and soils mounds to depths of 5–10 cm; 2–5 mm thickness of surface crust; grass muddied by wash and turned downslope; splays of coarse material due to wash and wind; less than 30% cover.
4	Tree root exposure, splash pedestals and soil mounds to depths of 5–10 cm; splays of coarse material; rills up to 8 cm deep; bare soil.
5	Gullies; rills over 8 cm deep; blow-outs and dunes; bare soil.

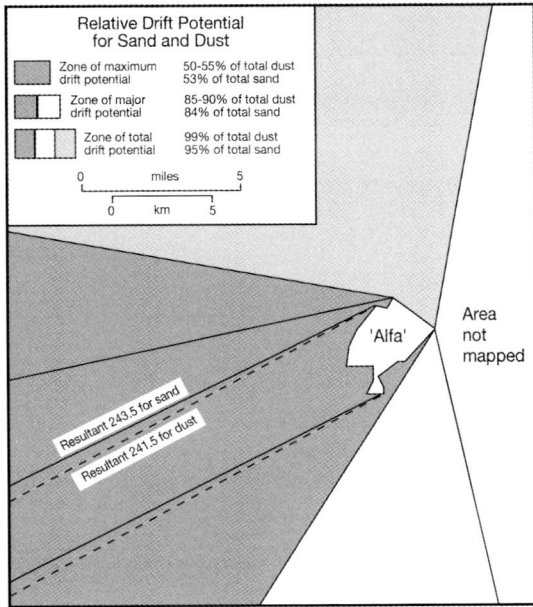

Figure 11.17a   Sand drift potential (redrawn from Jones *et al.*, 1986).

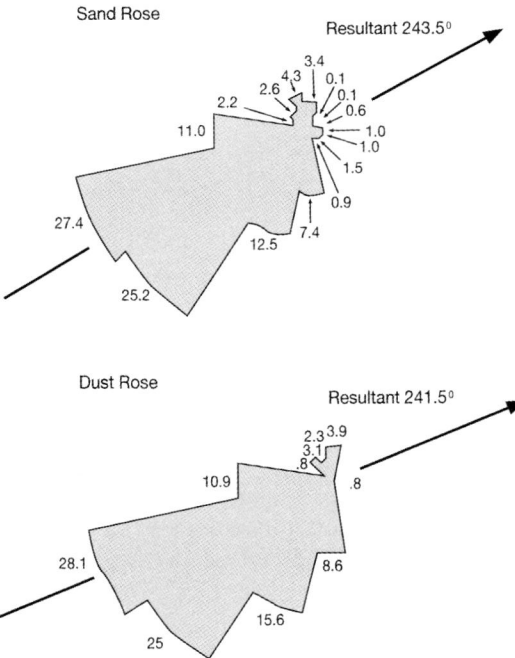

Figure 11.17b   Sand rose and dust rose for airport site in Saudi Arabia, showing percentage of mean annual drift potential of winds from 16 sectors (redrawn from Jones *et al.*, 1986).

impact threshold wind velocity (e.g. 12 knots), and *t* is the duration of wind). Figure 11.17a shows the distribution of sand drift potential by month for a site in Saudi Arabia, developed in this

way by Jones *et al.* (1986), using Fryberger's tables. Sand rose diagrams can be developed to illustrate the dominance of particular wind directions (Figure 11.17b). It should be noted that the

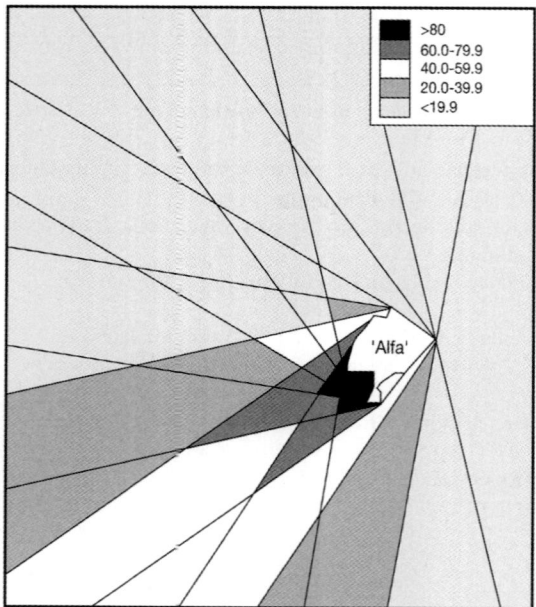

Figure 11.18a Dust hazard assessment, Saudi Arabia: dust source significance map (redrawn from Jones *et al.*, 1986).

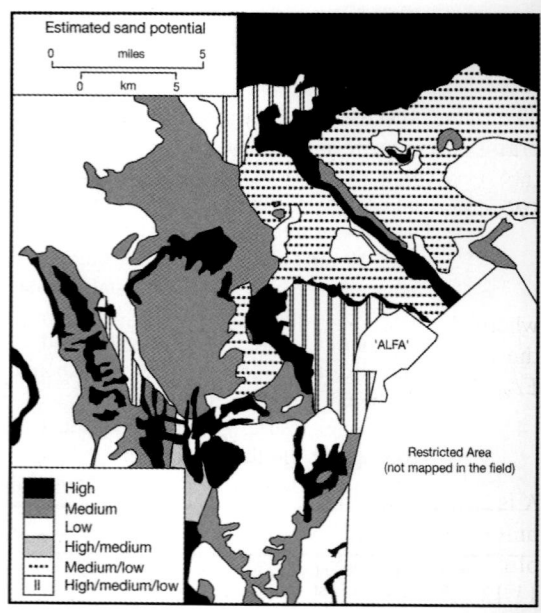

Figure 11.18b Dust hazard assessment, Saudi Arabia: potential dust source areas (redrawn from Jones *et al.*, 1986).

approach estimates potential rather than actual sand drift. The potential drift rates can only be realised if there is an unlimited supply of sand—actual drift rates can be considerably lower. Geomorphological mapping should be used to identify sediment source areas and transport pathways.

Dust hazard assessment: Jones *et al.* (1986) describe a method for assessing dust hazard around a town in Saudi Arabia. The method relates a dust transport rose (Figure 11.17b) to the outline of the urban area, producing overlapping rays of differing width. The values for each segment of the dust rose were then applied to the respective rays and the values summed for those areas where the rays overlap (Figure 11.18a). The higher the value the greater the likelihood that dust will be blown into the town. A geomorphological map of potential dust sources (Figure 11.18b) and the dust drift assessment (Figure 11.18a) were combined to produce a hazard score for the area around the town (Figure 11.18c).

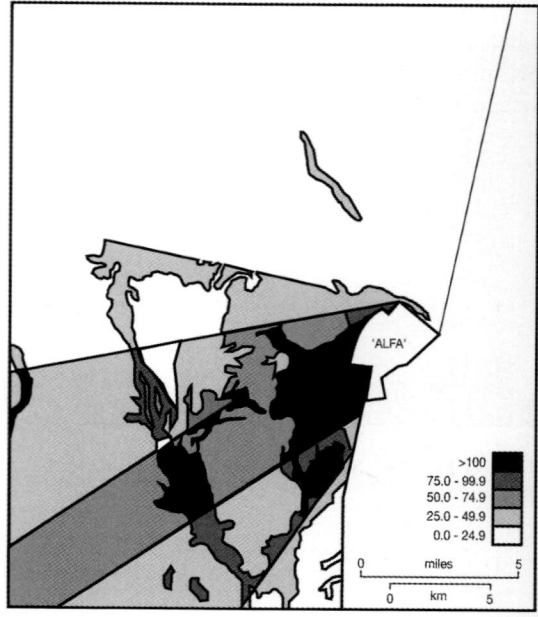

Figure 11.18c Dust hazard assessment, Saudi Arabia: dust hazard map, prepared by multiplying the values on the dust source significance map by 1, 2 or 3 depending on whether areas were estimated to be of 'low', 'medium' or 'high' drift potential (redrawn from Jones *et al.*, 1986).

11.8 The Universal Soil Loss Equation

The most important equation for predicting general soil loss from rain-splash and runoff is the Universal Soil Loss Equation (USLE) (Wischmeier and Smith, 1978):

$$E = R \times K \times LS \times C \times P$$

where E = average annual soil loss (t/ha) and R is the rainfall erosivity factor based on the mean EI_{30} (see above). For R to be in metric units:

$$R = EI_{30}/1000$$

K is the soil erodibility index, i.e. the soil loss per unit of EI_{30} (as measured on a standard bare soil plot, 22 m long and on a 5° slope; see Figure 11.13). LS is the combined slope length (L) and slope steepness (S) factor and can be derived from Figure 11.19 or the following equation:

$$LS = (x/22.13)^n (0.065 + 0.045s + 0.0065s^2)$$

where x is the slope length (m); s is the slope gradient in percent, the exponent n varies with slope steepness: for 3° slopes, $n = 0.4$; for 2° slopes, $n = 0.2$ for; <1° slopes, $n = 0.1$; for slopes, >6° $n = 0.6$.

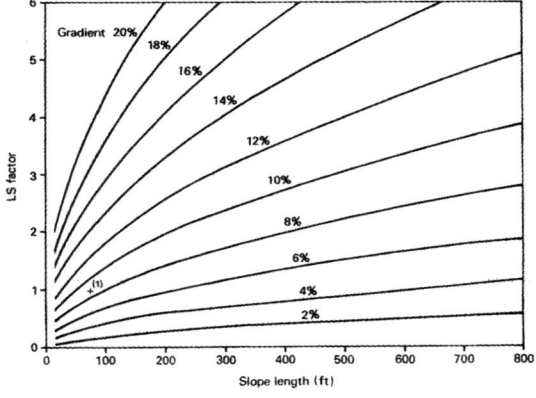

Note: (1) Standard condition: LS = 1 when slope = 9% and length = 72.5 ft.

Figure 11.19 Relationship of the USLE slope factor (*LS*) to gradient and length of slope (from Hudson, 1995).

C is the crop management factor and represents the ratio of soil loss under a given crop to that from bare soil (see Table A5.4, Appendix A5). P is the erosion control factor. With no erosion control in place $P = 1.0$ (see Table A5.5, Appendix A5).

Table A5.6 (Appendix A5) provides a worked example of the use of the USLE for predicting mean annual soil loss (i.e. long-term erosion rates and not loss associated with single storms). It is a widely used tool for soil conservation planning, but caution is needed when applying it outside its research base, i.e. eastern and central USA, on slopes less than 7°.

For a summary of other soil loss prediction models see Morgan (1995).

11.9 Erosion control methods and materials

A variety of methods have been used to control surface erosion in different settings, with varying degrees of success. Although many of the methods originated in soil conservation practice for agriculture and forestry projects, they have become increasingly used in civil engineering, especially as a way of reducing visual impact and enhancing the environment.

The following sections can only draw attention to the more successful methods; more details can be found in a range of manuals and guidebooks, including FAO (1996), Schiechtl and Stern (1996, 1997), Gray and Leiser (1982), Coppin and Richards (1990), Barker (1995) and Hudson (1995). It should be appreciated that each soil erosion problem will tend to be unique, because of the great range of slope forms and processes and the inherent variability of the soils and bedrock materials. Solutions, therefore, need to reflect site conditions and cannot be provided 'off-the-shelf'. An essential starting point is to understand the erosion processes operating at the site. Sympathetic design and construction; an understanding of the relative risks, a mechanism for observation and monitoring of the development, and a plan for future maintenance and mitigation of problems are all necessary.

Wind erosion

Approaches to mitigating wind erosion either aim to reduce the force of the wind or to improve the ground surface characteristics so that particle movement is restricted. There are four general methods of approach (Figure 11.20): (1) establish and maintain vegetation and organic residues (2) produce or bring to the surface non-erodible aggregates or clods (3) reduce field width (exposure) along the prevailing wind-erosion direction or (4) roughen the land surface.

All of these methods are encompassed in good land husbandry practice (FAO, 1996). The more widely used methods include the use of wind-breaks and land management practice.

Windbreaks: placing a barrier across the path of the wind reduces velocity at the ground surface both in front of and behind the barrier, and reduces the field length. Some measured reductions for average tree shelter belts are provided in Table 11.7. Barriers may be relatively permanent live vegetation structures (hedges or lines of trees) or they may be constructed of artificial materials such as geotextiles, stakes or palm fronds.

Windbreaks should be set as closely as possible at right angles to the dominant wind erosion

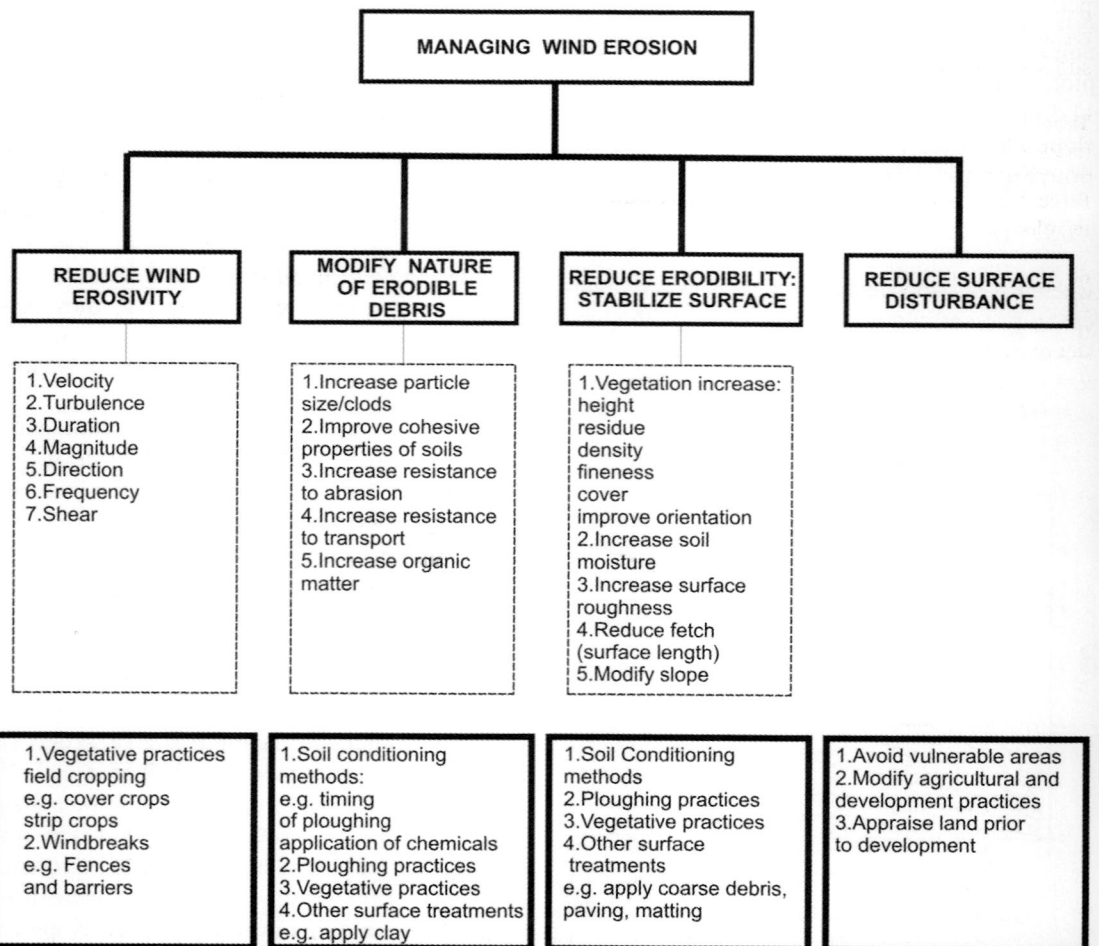

Figure 11.20   Approaches to management of wind erosion.

force. The degree of protection is determined by the spacing between barriers and by their width, height and porosity. In approximate terms wind velocity is reduced to about five to ten times the windbreak height on the windward side and about ten to thirty times the windbreak height on the leeward side. Other factors that should be considered are that the ends of barriers tend to cause funnelling and local increases in velocity and therefore fewer longer barriers are preferable to a greater number of shorter ones. Barriers that are semi-permeable are also preferable to those providing a complete obstacle to the wind, which can cause eddying, turbulence and local increases in velocity. It should also be remembered that in periods when wind speed is particularly high even reductions effected by windbreaks might not be sufficient to prevent particle transport.

Table 11.7 Effect of barriers in reducing wind velocity (after FAO, 1996).

Percentage reduction in velocity	Distance from barrier (multiples of height)
60–80	0
20	20
0	30–40

Land management practice: protecting the surface from attack and trapping moving particles can be achieved by keeping the surface covered throughout the year. Planting 'cover' crops to protect the surface in windy seasons when they occur outside the main crop growing period is an effective and cheap method which may produce another useful crop or provide an effective green manure or mulch. Crops of differing type can be mixed so that the differing heights, or rates of germination and growth, increase surface roughness or provide strips of vegetation that protect intervening strips of still-bare soil. Table 11.8 illustrates typical widths of vegetated strip required for different soil types and wind direction.

The management of crop residue and stubble can also be significant, since these also trap moving particles, provide a rough surface and contribute organic matter to the soil. Again relationships exist between stubble height, width of the stubble strip and the type of stubble. Ploughing creates a rough surface and can contribute to preventing soil erosion particularly if the ridges and furrows are created at right angles to the prevailing winds. Care is needed to ensure that the choice of equipment is suited to the soil type, particularly if erosion prevention is of major concern.

Table 11.8 Strip dimensions for the control of wind erosion (from Chepil and Woodruff, 1963).

Soil class	Width (m) of strips		
	Wind at right angles	Wind deviating 20° from a right angle	Wind deviating 45° from a right angle
Sand	6.1	5.5	4.3
Loamy sand	7.6	6.7	5.5
Granulated clay	24.4	22.9	16.5
Sand loam	30.5	28.0	21.3
Silty clay	45.7	42.7	33.5
Loam	76.2	71.6	51.8
Silt loam	85.4	79.3	57.9
Clay loam	106.7	99.1	76.2

Note: the table shows average width of strips required to control wind erosion equally on different soil classes and for different wind directions, for conditions of negligible surface roughness, average soil cloddiness, no crop residue, 300 mm high erosion resistant stubble to windward, 64.4 km/h wind at 15.24 m height and a tolerable maximum rate of soil flow of 203.2 kg per 5 m width per hour.

Conditioning the soil by increasing its cohesion with the addition of organic matter, mulching to retain its moisture or even irrigating to keep the surface moist all help to resist erosion. Moisture retention may merely involve a change in the timing of ploughing in relation to seeding. A relatively new technique is the conditioning of soil by the spraying of artificial additives.

Rain and sheet erosion

Good land management is also the key to controlling the loss of soil from rainfall and sheet flow. When land is under active production careful consideration needs to be given to land use, crop management, tillage methods and the application of manures and fertilisers. If the land is fallow then the establishment, re-establishment or maintenance of vegetation cover is important. Alternatively, reductions in runoff velocity can be achieved by dividing land into small plots or benching to reduce slope steepness and the introduction drainage ditches and sediment traps can conserve soil cover. In addition specific measures may be necessary to address particular problem areas. Such measures may include contouring, strip cropping, terracing, and the construction of drainage measures or structures. The control of runoff and its effects on nearby watercourses is now an important aspect of major engineering schemes.

In contour farming, rows are orientated across the slope and thus act as a barrier to the down-slope flow of water. It is most effective on shallow slopes. It becomes difficult to operate machinery on steeper slopes, because it needs to work across the slope to create ruts that act as small dams. Contour farming reduces runoff and, therefore soil erosion. Generally, as the slope becomes steeper the contour strips need to be closer together (Table 11.9).

Contour ridging and ridge drains are used to produce specific ridge features, rather than relying on the cross-slope texture produced by contour farming, which significantly improves the ability of the system to reduce flow velocities. The ridges are simple water control structures that act to dam the flow and provide a temporary storage until infiltration can occur. They are less effective as slopes become steeper because flow

Table 11.9 A guide to contour spacing on sloping ground.

Slope angle		Contour spacing (m)
Per cent	Degrees	
<6	<4	100
8	5	60
10	6	30
12	7	25
>12	>7	20

velocities increase rapidly over short distances and the ridges can be easily breached. A solution is to use the ridges as a drainage control by sloping them obliquely down the slope at a *very* shallow angle to encourage water behind them to flow across the slope to a collection and distribution system.

The effectiveness of this method can be extended by using the ridges in conjunction with a drainage channel, or by using a geotextile separator to prevent the soil from being carried into the drain. Thus the soil is preserved while the water is drained away through the system.

Gully erosion

Methods to protect the gully-head from further erosion involve either the reduction of the volume and velocity of flow into the gully or the direct protection of the gully-head from erosion due to excessive flow. If the gully is in a state of active development then the source of the flow and its velocity should be calculated to determine if reductions are achievable. Reducing the volume of flow may be achievable by modifying farming practices on the slopes above the gully using the techniques described earlier. If water flow has recently been diverted, for example during the construction of a new road without attention to accommodating the pre-existing drainage regime, the preferable step is to establish a suitable drainage system to accommodate the additional flow.

Artificial methods may be required to protect the gully-head and the measures adopted depend on the size and slope of the gully and on the typical maximum flows (Figure 11.21). If the duration

PRINCIPLE

The measures are intended to arrest regressive erosion of the gully head thus preventing it from progressing any further uphill.

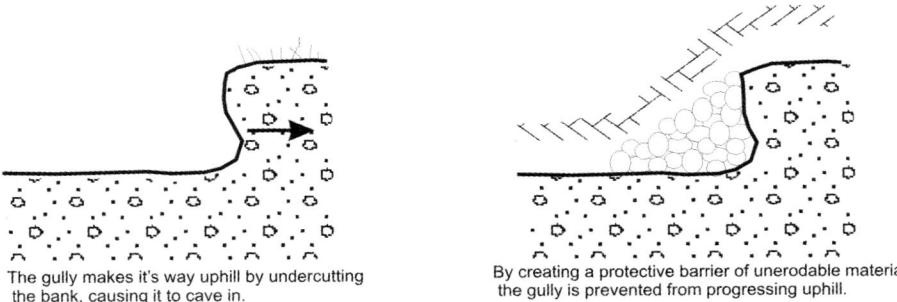

The gully makes it's way uphill by undercutting the bank, causing it to cave in.

By creating a protective barrier of unerodable material the gully is prevented from progressing uphill.

CONDITIONS   This measure only suitable for use in the first stages of erosion

CONSTRUCTION

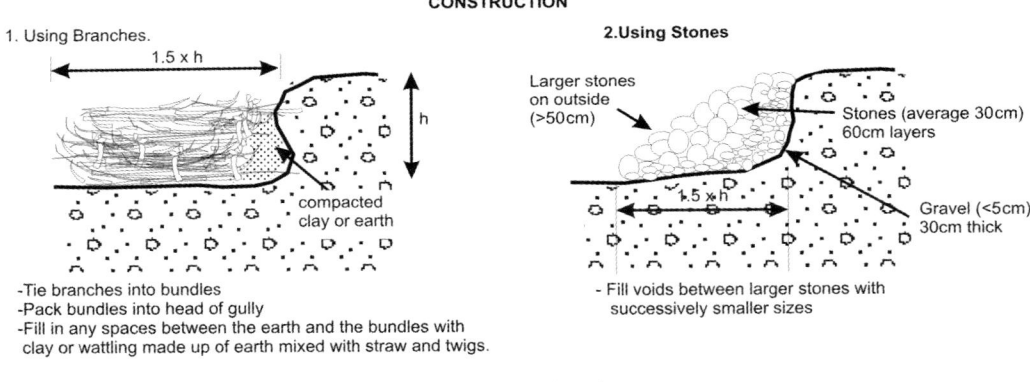

1. Using Branches.

1.5 x h

h

compacted
clay or earth

-Tie branches into bundles
-Pack bundles into head of gully
-Fill in any spaces between the earth and the bundles with
 clay or wattling made up of earth mixed with straw and twigs.

2.Using Stones

Larger stones
on outside
(>50cm)

Stones (average 30cm)
60cm layers

1.5 x h

Gravel (<5cm)
30cm thick

- Fill voids between larger stones with
  successively smaller sizes

3. Using Fascines and Gravel.

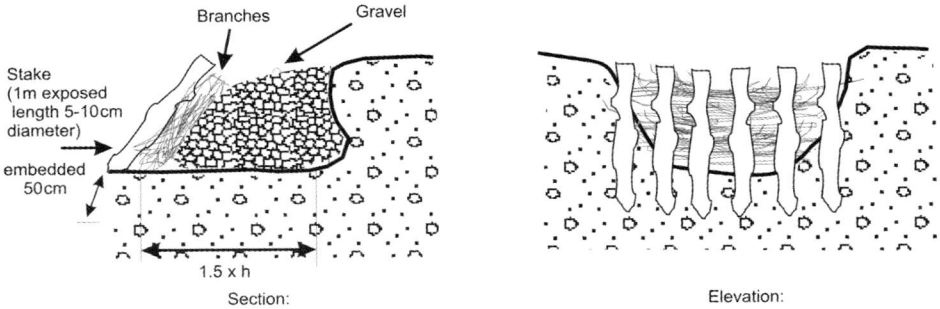

Branches   Gravel

Stake
(1m exposed
length 5-10cm
diameter)

embedded
50cm

1.5 x h

Section:

Elevation:

Figure 11.21   Methods for protection of the gully head.

of a potential event can be estimated and the channel geometry is known then flow velocity can be calculated from standard open-channel hydraulic relationships and an appropriate structure selected.

For low flow regimes it may be possible to check erosion by establishing vegetation. Grasses and legumes are effective in providing soil binding, and bamboo with its hardy stems and foliage is effective in diffusing flow. A distinct gully-head feature points to the fact that at least moderate flow has occurred in the past.

In moderate flow regimes a simple structure may be needed to provide an erosion resistant gully-head. Brushwood bundles can be laid and pegged in the gully-head. Alternatively a rubble bank can be constructed in the gully head, using large stones. These stones must be of sufficient individual size to resist potential detachment and transportation during peak flow. The main principles to follow are that the flow of water should not be impeded by the structure, otherwise flow will be diverted around or behind and under the structure. Ideally, the structure should also help to dissipate the energy of the water.

In areas of very high flows gabion structures may be necessary. Masonry structures are not recommended because they are impermeable and resist flow and the mortar inevitably disintegrates after a few years. They are also rigid and crack as erosion develops in front of the apron. Gabions are highly permeable and tend to break up and dissipate the flow and they are also flexible. They should have a long aprons so that the energy is dispersed along the length of the structure; gabion aprons deform to accommodate the erosion at their toe.

Any structure should mimic the slope profile at the head of the gully so that flow continues unimpeded onto the structure.

Prevention of channel scour can be achieved in low flow regimes by a grass lining to channels (Figure 11.22). The sward reduces the flow velocity at the soil surface by interfering with the flow, and when it is deflected under higher velocity flow it provides a protective cover to the soil. The litter layer also provides protection to the soil surface and the roots provide mechanical stability to the soil particles, anchoring the soil into the underlying subsoil. The use of geomeshes or geomats can further improve the stability of grass-lined channels. Indications of the scale of improvement are illustrated in Figure 11.23, which shows the limiting velocities that can be withstood by various grass or reinforced grass covers.

In moderate flow regimes live branches reduce erosion by initially providing a vegetative cover

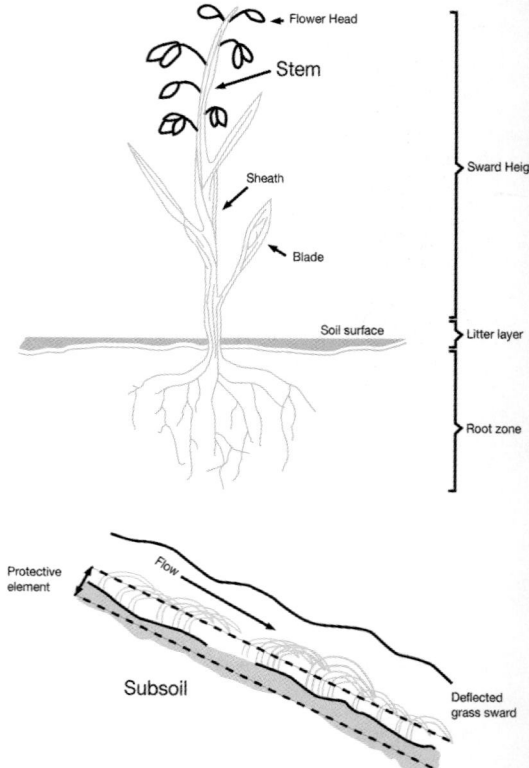

Figure 11.22   Grass components in gully scour protection.

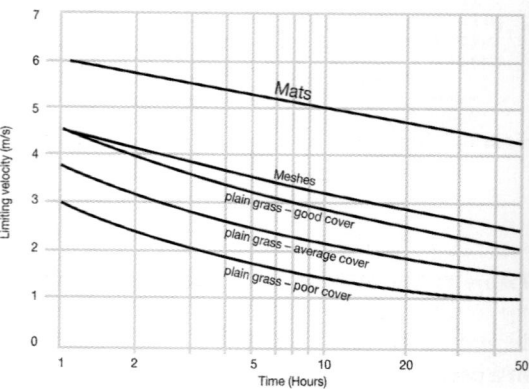

Figure 11.23   Limiting velocities for various grass and reinforced grass covers.

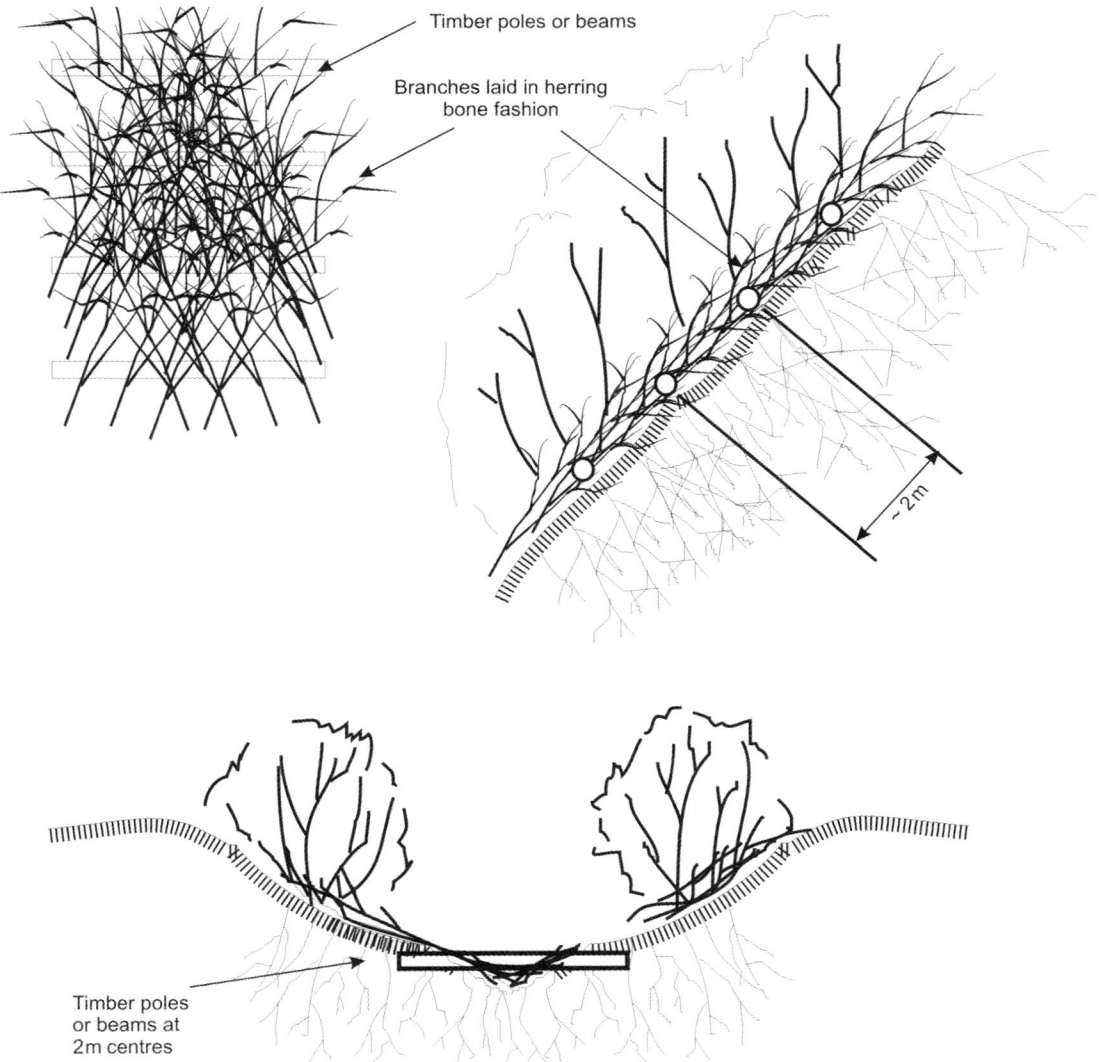

Timber poles or beams

Branches laid in herring
bone fashion

~ 2 m

Timber poles
or beams at
2 m centres

Figure 11.24   Gully protection using live branches.

over the gully floor, which reduces velocity. As root development takes place this provides a binding to the gully floor and sides which continues the protection even during dormancy. A layer of branches is laid in a herringbone pattern over the gully floor and extending to the gully sides (Figure 11.24). The branches are then covered by a soil layer, ensuring that the tips of the branches are left uncovered. A further layer of branches is laid, staggered down the gully and covered in turn by a soil layer. The process is repeated until the required area is covered. Initially, the live branches must be held in place until the roots develop sufficiently to provide resistance to flow. Cross-poles can be used at approximately 2 m intervals. They are placed over the live branches and embedded into the gully sides to at least 0.5 m.

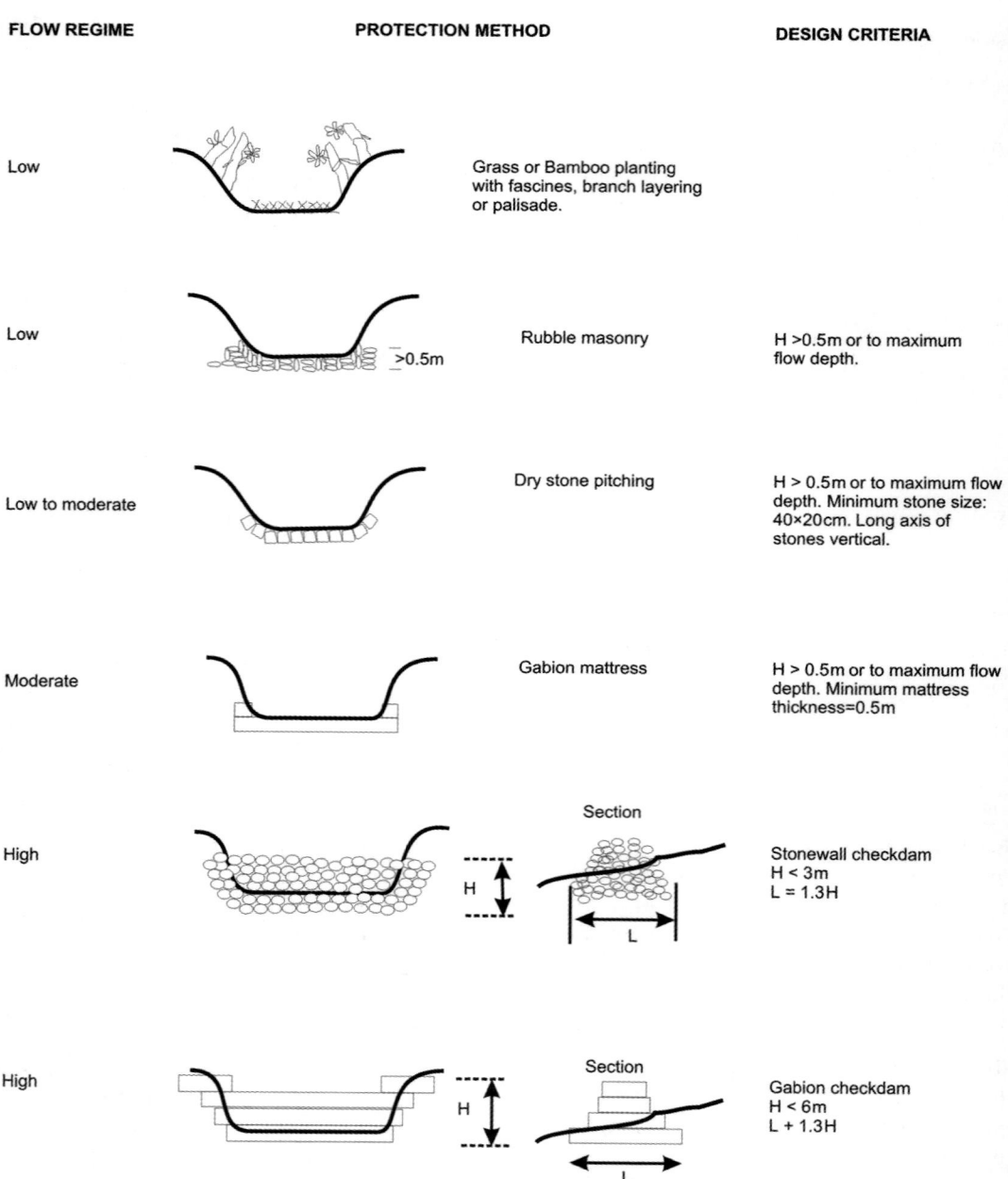

FLOW REGIME

Low

Low

Low to moderate

Moderate

High

High

PROTECTION METHOD

Grass or Bamboo planting
with fascines, branch layering
or palisade.

Rubble masonry

Dry stone pitching

Gabion mattress

Section

Stonewall checkdam

Gabion checkdam

DESIGN CRITERIA

H >0.5m or to maximum
flow depth.

H > 0.5m or to maximum flow
depth. Minimum stone size:
40×20cm. Long axis of
stones vertical.

H > 0.5m or to maximum flow
depth. Minimum mattress
thickness=0.5m

Stonewall checkdam
H < 3m
L = 1.3H

Gabion checkdam
H < 6m
L + 1.3H

Figure 11.25   Methods of gully scour protection.

Check dams are constructed along the length of gullies in order to decrease the gradient of the gully floor (Figure 11.25). Check dams slow the water flow because they create a small reservoir that eventually overtops. The water drops its sediment load and the sediment accumulates until it reaches the top of the check dam. The result is a shallower gradient along the length of the gully over which

Purpose:
  - Decrease gradient of gully
  - Reduce velocity of flow

But:
  Remember checkdams will increase soil erosion
  further downstream as the surface water flow
  strives for equilibruim. Local increases in velocity will
  lead to eddying and turbulence.

If high flows use impact apron

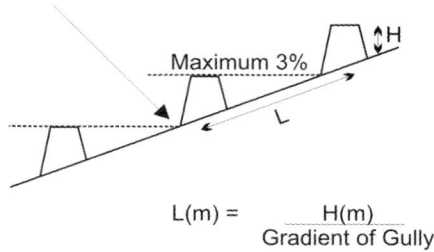

$$L(m) = \frac{H(m)}{\text{Gradient of Gully}}$$

Example Calculation:

  If the gradient = 1/25 and the engineer wishes to use
  1m high dams.

  Spacing (L) = $\dfrac{1}{1/25}$ = 25 metres

Figure 11.26   Dimensioning and spacing of check dams.

the check dams have been constructed. If a greater
separation is employed sediment will not accumulate to the necessary extent and erosion will work
back to undermine the next dam upstream.
Eventually successive dams will be undermined
until the gully-head protective works are destroyed.

   Check dams can be made with vegetation,
rock-fill, timber, dry-stone masonry and gabions.
As they hinder water flow, extreme care is needed
in their design to ensure that they do not cause
such an obstruction as to promote increased
erosion of the side banks, or cause the gully flow
to divert around the check dam.

   There are two main rules for the siting of
check dams:

1. The top of each dam should be at or just
    below the base of the next dam up-gully. The
    maximum gradient between the top of the
    dam and the base of the next dam up-gully
    should be 3% (Figure 11.26).
2. Dams should be positioned so that they are
    perpendicular to the flow (Figure 11.27) — if
    they are not, they divert the flow to one side
    of the gully and cause erosion in the adjacent
    bank.

   It should be remembered that as check dams
effectively decrease the velocity over the length

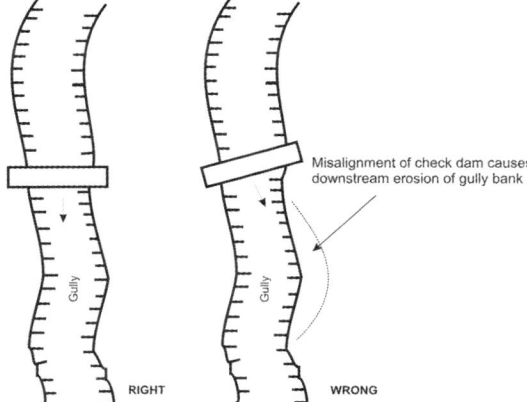

Figure 11.27   Orientation of check dam structures.
Check dams must be perpendicualr to gully banks.

through which they have been constructed, the
velocity will increase further downstream and
may cause enhanced erosion in that area. Ideally,
the natural gully gradient below the lowest check
dam should be equal to or less than the gradient
between the top of the lowest check dam and the
base of the next check dam up-gully. If this is
not the case, erosion will occur immediately
below the lowest check dam and eventually
undermine it.

11.10 Summary

Soil erosion by water or wind can cause a range of problems for engineers and environmental managers, from loss of productive soils to blocked infrastructure routes, loss of reservoir storage capacity and environmental degradation. Gullying and the development of badland terrain can develop over engineering timescales and, hence, can present a significant problem for some projects. Often serious problems arise because of a combination of changes to land use or management practice and the occurrence of low–medium frequency triggering events—heavy rainfall or high wind speeds. Once the nature and cause of the problems have been established, simple but effective measures can be implemented to reduce or control erosion.

References

Barker, D. H. (ed.) (1995) *Vegetation and Slopes: Stabilisation, Protection and Ecology*. Thomas Telford, London.

Boardman, J. (1990) Soil erosion on the South Downs: a review. In Boardman, J., Foster, I. D. L. and Dearing, J.A. (eds) *Soil Erosion on Agricultural Land*. John Wiley and Sons, Chichester, 87–105.

Boardman, J. (1991) Land use, rainfall and erosion risk on the South Downs. *Soil Use and Management* **7**, 34–38.

Bruington, A. E. (1982) Fire-loosened sediment menaces the city. In Proceedings of the Symposium on Dynamics and Management of Mediterranean-type ecosystems, June 22–26, 1981, San Diego, California. *USDAFS General Technical Report PSW-58*, 40–422.

Campbell, I. A. (1981) Spatial and temporal variations in erosion measurements. In Symposium on Erosion and Sediment Transport Measurements. *Proceedings of the Florence Symposium International Association of Hydrological Sciences* **133**, 447–456.

Campbell, I. A. (1997) Badlands and badland gullies. In Thomas, D. S. (ed.) *Arid Zone Geomorphology: Process, Form and Change in Drylands*. Wiley, Chichester, 261–291.

Carson, M. A. and Kirkby, M. J. (1972) *Hillslope Form and Process*. Cambridge University Press

Chen, Y. (1983) A preliminary analysis of the processes of sediment production in a small catchment on the Loess Plateau. *Geographical Research (China)* **2**, 35–47.

Chepil, W. S. and Woodruff, N. P. (1963) The physics of wind erosion and its control. *Advances in Agronomy* **15**, 211–302.

Coppin, N. J. and Richards, I. G. (1990) *Use of Vegetation in Civil Engineering*. CIRIA (Construction Industry Research and Information Association) Report, Butterworth, London.

Dalrymple, J. B., Blong, R. J. and Conacher, A. J. (1968) A hypothetical nine-unit landsurface model. *Zeitschrift für Geomorphologie* **12**, 60–76.

Drakeford, T. (1979) *Report of Survey of the Afforested Spawning Grounds of the Fleet Catchment*. Forestry Commission.

Evans, R. and Cook, S. (1986) Soil erosion in Britain. *SEESOIL* **3**, 28–58.

FAO (Food and Agriculture Organization) (1996) *Methods and Materials in Soil Conservation*. FAO Soils Bulletin 70, FAO, Rome.

Fournier, F. (1960) *Climat et Érosion: La Relation Entre L'érosion du Sol par L'eau et Les Précipitations Atmosphériques*. Presses Universitaires de France, Paris.

Freeze, R. A. (1978) Mathematical modelling of hillslope hydrology. In Kirkby, M. J. (ed.) *Hillslope Hydrology*. Wiley, Chichester, 177–225.

Fryberger, S. G. (1979) Dune forms and wind regimes. In McKee, E. D. (ed.) A Study of Global Sand Seas. *US Geological Survey Professional Paper* **1052**, 137–169.

Gray, D. H. and Leiser, A. J. (1982) *Biotechnical Slope Protection and Erosion Control*. Van Nostrand Reinhold, New York.

Hadley, R. F. and Schumm, S. A. (1961) Sediment sources and drainage basin characteristics in upper Cheyenne River basin. *US Geological Survey Water Supply Paper* **1531B**, 137–198.

Hannam, I. D. and Hicks, R. W. (1980) Soil conservation and urban land use planning. *Journal of the Soil Conservation Service NSW* **36**, 134–145.

Howard, A. (1994) Badlands. In Abrahams, A. A. and Parsons, A. J. (eds), *Geomorphology of Desert Environments*. Chapman and Hall, London, 213–242.

Hudson, N. W. (1965) *The Influence of Rainfall on the Mechanics of Soil Erosion with Particular Reference to Southern Rhodesia*. Unpublished MSc. thesis, University of Cape Town.

Hudson, N. W. (1995) *Soil Conservation*. Batsford, London.

Jencsok, E. I. (1968) *Hydrologic Design for Highway Drainage in Arizona*. Arizona Highway Department, Phoenix.

Jones, D. K. C., Cooke, R. U. and Warren, A. (1986) Geomorphological investigation, for engineering purposes, of blowing sand and dust hazard. *Quarterly Journal of Engineering Geology* **19**, 251–270.

Jones, J. A. A. (1981) *The Nature of Soil Piping — a Review of Research*. British Geomorphological Research Group Monograph, Geo Books, Norwich.

Kirkby, M. J. (1969) Infiltration, throughflow and overland flow: and erosion by water on hillslopes. In Chorley, R. J. (ed.) *Water Earth and Man*, Metheun, London, 215–238.

Landon, J. R. (ed.) (1984) *Booker Tropical Soil Manual.* Longman, Harlow.

Leopold, L. B., Wolman, M. G. and Miller, J. P. (1964). *Fluvial Processes in Geomorphology*. Freeman, San Francisco.

Mabbutt, J. A. (1977) *Desert Landforms*. MIT Press, Cambridge, Mass.

Meyer, L. D. and Wischmeier, W. H. (1969) Mathematical simulation of the process of soil erosion by water. *Trans. Am. Soc. Agr. Engnr.* **12**, 754–758, 762.

Morgan, R. P. C. (1974) Estimating regional variations in soil erosion hazard in Peninsular Malaysia. *Malaysia Nat. J.* **28**, 94–106.

Morgan, R. P. C. (1980) *Soil Erosion*. Longman, London.

Morgan, R. P. C. (1995) *Soil Erosion and Conservation*. Longman, London.

Parsons, A. J. (1988) *Hillslope Form*. Routledge, London and New York.

Schiechtl, H. M. and Stern, R. (1996) (English translation) *Ground Bioengineering Techniques for Slope Protection and Erosion Control.* Blackwell Science, Oxford.

Schiechtl, H. M. and Stern, R. (1997) (English translation) *Water Bioengineering Techniques for Watercourse Bank and Shoreline Protection.* Blackwell Science, Oxford.

Schumm, S. A. (1956) The role of creep and rainwash on the retreat of badland slopes. *American Journal of Science* **254**, 693–706.

Selby, M. J. (1982) *Hillslope Materials and Processes*. Oxford University Press.

Selby, M. J. (1987) Slopes and weathering. In Gregory, K. J. and Walling D. E. (eds), *Human Activity and Environmental Processes*. Wiley, Chichester, 183–205.

Skidmore, E. L. and Woodruff, N. P. (1968) Wind erosion forces in the United States and their use in predicting soil loss. *USDA Agricultural Research Service Handbook* **346**.

Stallings, J. H. (1957) *Soil Conservation*. Prentice Hall Englewood Cliffs NJ.

Thornes, J. B. (1994) Channel processes and forms. In Abrahams, A. A Parsons A. J. (eds), *Geomorphology of Desert Environments*. Chapman and Hall, London, 288–318.

US Environmental Protection Agency (1973) *Methods for Identifying and Evaluating the Nature and Extent of Nonpoint Sources of Pollutants*. US Department of Agriculture, Washington DC.

US Soil Conservation Service (1972) *National Engineering Handbook*, Section 4: Hydrology. US Department of Agriculture, Washington DC.

Wischmeier, W. H. and Smith, D. D. (1958) Rainfall energy and its relationship to soil loss. *Trans. Am. Geophys. Un.* **39**, 285–291.

Wischmeier, W. H., and Smith, D. D. (1978) Predicting rainfall erosion losses. *USDA Agricultural Research Service Handbook* **537**.

Wischmeier, W. H., Johnson, C. B. and Cross, B. V. (1971) A soil erodibility nomograph for farmland and construction sites. *J. Soil and Water Conservation* **26**, 189–193.

Wolman, M. G. and Schick, A. P. (1967) Effects of construction on fluvial sediment, urban and suburban areas of Maryland. *Water Resources Research* **3**, 451–464.

Worster, D. (1979) *The Dust Bowl*. OUP, New York.

12. Subsidence

Tony Waltham

12.1 Subsidence environments

Downward movement of the ground, or ground subsidence, can only occur where the subsurface conditions are such that space exists for the ground to move into. Most of the Earth's surface is on stable rock, where subsidence cannot occur. But there are specific environments where subsidence is a significant geohazard, and these fall into four distinct categories that may be summarised as:

1. Porous and deformable rocks and soils that are therefore compressible. Compaction and consequent surface subsidence occur as the rock or soil restructures with declining pore space, normally accompanied by abstraction or expulsion of interstitial groundwater. The main material involved is clay (12.2 Subsidence on clay), but the process can apply in peat (12.3 Subsidence on peat) and silt (12.4 Hydrocompaction of collapsing soils), as well as in most types of artificial landfill and made ground, in permafrost when ground ice is melted (12.5 Subsidence on permafrost), and in some sands when vibrated by earthquakes (12.11 Tectonic subsidence). Subsidence may be local, due to structural loading, or may be regional, due to changes in groundwater conditions.
2. Rocks that contain large cavities, into which ground can fall, when rock forming the roof collapses or deforms. Natural caves occur mainly in limestone or gypsum where subsidence is therefore a frequent geohazard (Chapter 24) and in basalt lava flows (Chapter 23). Subsidence does occur on salt, where few caves survive to maturity (12.7 Salt subsidence), but natural caves are not significant in other rocks. Mined cavities can be left in almost any rock type (12.10 Subsidence due to mining). All forms of

open cavities create subsidence that extends over only small areas, but it may involve catastrophic collapse within those areas.
3. Tectonic subsidence, where plastic rock moves away at great depth from beneath a site. Crustal deformation can cause regional subsidence (12.11 Tectonic subsidence), and magma movement can cause volcanic deflation of smaller areas.
4. Landslides, where the head zone subsides and the toe moves outwards. This geohazard is a function of slope profiles (Chapter 23) and is not true ground subsidence.

It is possible for a single site to have more than one subsidence hazard. A thick soil over a karst limestone may develop both compaction subsidence and sinkhole events, and piping failures may further complicate an understanding of the processes, while mining collapse and tectonic movements can be superimposed on any geomorphological environment. However, a constructive engineering approach to remediation of any subsiding ground is best achieved by modelling the primary processes — which must fall into one or more of the above four categories. When regional subsidences are separated out from localised ground failures, each of the latter can occur in only a few geological environments. It is also significant that most subsidence is caused, induced or exacerbated by man's own activities; very little ground subsidence is entirely natural.

12.2 Subsidence on clay

The special properties of clay minerals, particularly their water retention or expulsion and their low strength, account for the singular plastic deformation properties of the clay soils and rocks, which are therefore the most widespread cause of destructive ground subsidence. These clay materials include those like the Paleogene London Clay that are viewed as rocks by a geologist (because they are within old lithified sequences) and are treated as soils by an engineer (because they can be excavated without pre-breaking). Their plasticity is derived from the water that is loosely bonded to the chemically complex clay mineral particles (Chapter 3). They may therefore compact (and cause subsidence) when the water is squeezed out under load or in response to water abstraction. In either case, the loss of water causes consolidation, as increased grain-to-grain contact creates increased strength. Subsidence is the precursor of consolidation, and may be either regional or localised.

Clay subsidence due to fluid abstraction

Worldwide, this ranks as the most widespread and most destructive subsidence process. Entire cities have subsided; Venice (Figure 12.1), Shanghai, Mexico City and Bangkok are just examples that are well known (Waltham, 2002). Clays gain a part of their internal support (and therefore their load-bearing capacity) from their pore water pressure. If this pressure is reduced, as water is removed, the clays compact. The natural process is for water to be squeezed out during very slow consolidation and lithification. Artificial removal of the water, by abstraction pumping, induces far more rapid compaction. The subsidence hazard lies in alluvial sequences with alternating beds of poorly consolidated sand and clay beneath large cities. Cheap and convenient water supplies are pumped from the productive sand aquifers, but these are almost incompressible, so negligible subsidence is induced. But the inevitable consequence is equilibrium of pore water pressures in the sands and the clays, when the clays then compact.

The amount of subsidence is directly proportional to the groundwater head decline, but is also a function of the age and mineralogy of the clay. For equal head decline and clay thickness, greater subsidence occurs on the younger clays (which are therefore less consolidated by self-weight) and on those with higher contents of the unstable smectite. The clay minerals kaolinite and illite are both more stable with far less removable water

Figure 12.1   Boardwalks over one of the piazzas of Venice, which have now subsided below the level of most winter high tides.

Table 12.1   Comparison of subsidence parameters for seven sites on clay.

Site	Subsided area (km²)	Maximum subsidence (m)	Clay thickness (m)	Head decline (m)	Specific subsidence (m/m)	Compressibility (×10⁻³m/m/m)	Smectite content (%)	Age
London	300	0.35	60	100	0.0035	0.06	0	Eocene
Savannah	200	0.19	50	48	0.004	0.08	60	Miocene
Venice	150	0.12	130	9	0.013	0.10	10	Recent
Houston	6000	2.3	150	90	0.025	0.21	50	Recent
Bangkok	600	1.2	200	30	0.05*	0.5*	40	Recent
Santa Clara	650	5.3	145	49	0.11	0.74	70	Recent
Mexico City	220	9.0	50	55	0.16	32.0	80	Recent

* Specific subsidence and compressibility values for Bangkok are for the upper part of the clay sequence only.

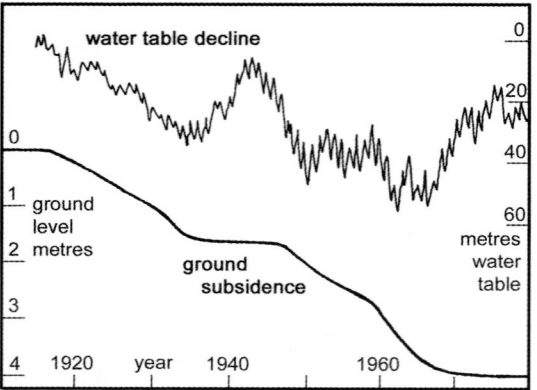

Figure 12.2   Correlation of aquifer head decline and ground subsidence in the Santa Clara Valley, California.

than smectite, which is formed primarily by weathering of volcanic rocks in wet tropical environments, and therefore imposes some geological and climatic control on the scale of clay subsidence. These factors combine to influence the compressibility of a clay (measured as metres of subsidence per metre of head decline per metre of clay thickness). Table 12.1 demonstrates the roles of smectite and age, even though the data are generalised because there is rarely a clearly defined base to the clays that are compressed due to groundwater abstraction that is mainly at shallow depths. The compressibility of the Bangkok clay may appear high because it is based on data from a short period of rapid subsidence, whereas data at the other sites covers a longer term, during which stability is approached due to consolidation.

Subsidence rates on clay may therefore be predicted for any site, except that local geological factors preclude accurate predictions without a database of recorded movements.

The role of pressure decline within overpumped aquifers was first recognised in the Santa Clara Valley of California (Figure 12.2). Subsidence was virtually stopped by water table recovery during 1935–45, largely due to higher rainfalls, but restarted as soon as water levels declined past their previous minimum, when pumping increased in drier years. The only engineering response to prevent or reduce subsidence is to facilitate groundwater recovery. Reduction of abstractions is critical, and aquifer recharge can be employed where seasonal water excesses can be injected via existing abstraction wells. Both were applied at Santa Clara after 1965, and subsidence has stopped. Similarly, the subsidence of many other cities (notably Tokyo, Osaka, Shanghai and Houston) has been controlled, but Bangkok continues to subside while a new water supply system is developed to replace the thousands of shallow wells in the city.

Injection of surface water into an aquifer may cause substantial head recovery, but only stops the subsidence. Reversal of subsidence, known as rebound, due to re-inflation of the aquifer is minimal. Reversible elastic deformation of sand aquifers are recorded through seasonal fluctuations of water table levels, but the amounts are only a tiny fraction of the largely inelastic compression of the interbedded clays (Poland, 1984). Ground rebound is therefore normally only 1–6% of the

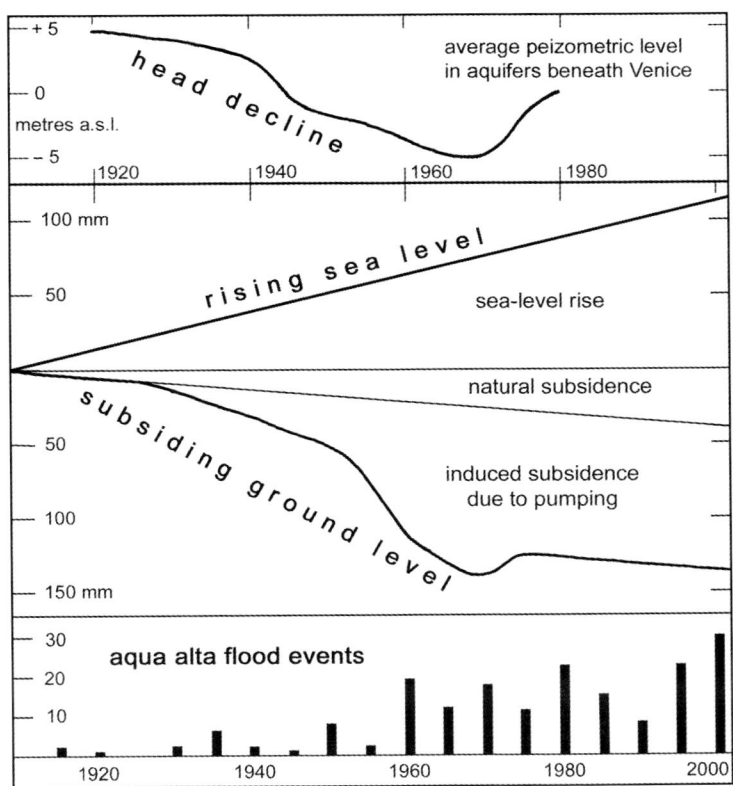

Figure 12.3   Causes and effects of the subsidence of Venice over the last 100 years. The bar graph shows, for each 5–year period, the numbers of flood events when high tides reach more than 600 mm above the level that initiates flooding in the lowest part of Piazza San Marco.

amount of previous subsidence. This did give rise to claims that Venice had started to rise instead of subside soon after groundwater pumping was severely reduced by legislation. The claims were however short-lived, as rebound soon stopped, while world sea levels continue to rise so that the winter flood events continue to be ever more frequent (Figure 12.3).

Groundwater withdrawals can be matched by extraction of petroleum, natural gas and steam as causes of major ground subsidence, where the loss of hydraulic support is the common factor, though it may also affect rocks other than clay. Oilfield subsidence includes the classic case at Wilmington, which caused the Long Beach harbour area of Los Angeles to subside by nearly 9 m in the period 1935–1960. Though part of this was due to compression of inter-bedded clays, about

two-thirds of the movement was due to crushing of the feldspar grains at contact points once hydrostatic support was lost in the reservoir sandstones at depths of 600–1200 m. Secondary oil production by water injection then stopped the subsidence, and caused about 6% rebound. Chalk reservoir rock in the North Sea's Ekofisk oilfield was similarly crushed, necessitating raising of the production platform during its abstraction lifetime. Natural gas exploitation has caused comparable subsidence over reservoirs in Japan, Holland and Italy. Depletion of geothermal steam has caused over 4 m of ground subsidence above parts of the Wairakei field in New Zealand. This movement developed after pressure decline in reservoirs of porous rhyolitic pyroclastics; geothermal fields elsewhere have suffered no ground subsidence over reservoirs of stronger rocks.

Table 12.2 Safe distances for houses to stand away from various species of trees to avoid subsidence damage by root water abstraction (extracted from ISE, 2000).

Species	Mature height (m)	Recorded influence (m)	Safe distance (m)
Willow	24	18	40
Poplar	28	20	35
Oak	24	18	3
Elm	25	19	30
Cherry	17	6	11

At shallow depths, where stronger rocks are stable, shrinkage due to water loss only causes subsidence in clays. This is well known when trees extract excessive amounts of soil water during times of drought. The amount of subsidence is related to potential shrinkage of the soil, but all clays may be affected, including those in the weathering profile of mudstones that are stable at depth. Oaks, poplar and willow are the most powerful tree species, and flowering cherries have a bad reputation because they are often planted close to houses. All trees should be placed away from houses by a safe distance, which is usually rather more than the tree's height (Table 12.2),

and removal of any closer is often the only way to eliminate building subsidence. Movement may also be the consequence of site drainage to allow deep excavations, and in such cases installation of a cut-off wall may be the only way to protect adjacent structures. Climatic effects impact on larger areas, and the famously dry summers of 1976 and 1989 in the UK started a wave of subsidence damage to older houses on foundations so shallow that clay beneath them suffered first-time shrinkage in the new regime of drier climates.

Loading compaction of clay

All soils and rocks compact under load, but the movement is only more than negligible on weak soils, and notably on clay. The normal engineering response is to distribute loads vertically or horizontally by foundation designs that include rafts and friction piles respectively. Acceptable stresses on clay soils have been established by decades of soils engineering practice (Table 12.3). These values are determined by the clay mineralogy and consolidation history, but are always determined by soil properties measured by routine testing for each site, as geological classifications cannot adequately define the variations in soil properties. Precise loading limits are calculated with respect to water content and

Table 12.3 Acceptable Bearing Pressures (ABPs) (or Presumed Bearing Values) for foundations with static loading on cohesive clay soils, where long-term settlement is $<50\,mm$; acceptable values are also modified, upwards or downwards, with respect to foundation width and shape.

Description	Example in Britain	Shear strength (kN/m^2)	CPT (MN/m^2)	Liquidity Index	Acceptable Bearing Pressure (kN/m^2)
Soft soil	Alluvium	20–40	0.3–0.5	>0.5	<50
Firm soil	Drift at depth	40–75	0.5–1.0	0.2–0.5	75–100
Stiff soil	Weathered London Clay	75–150	1–4	−0.1–0.2	150–250
Very stiff soil	Fresh London Clay	150–300	2–8	−0.4−−0.1	300–500
Hard soil	Gault Clay	>300	>4	<−0.4	600
Weak rock	Carboniferous shale	500–1000			400–750
Strong rock	Silurian mudstone	>2000			1000–2000

Note: the comparable values for rock are only approximate and their ABPs are lower because fracture spacing influences bearing parameters in rocks.

CPT = end resistance in a Cone Penetration Test (which cannot be applied to rock).

Liquidity Index = (water content − plastic limit)/plasticity index, and is a measure of soil strength at a given water content.

The mudstone is only strong in the context of the clay rocks.

the Mohr-Coulomb parameters of internal friction and cohesion (Chapter 3; Craig, 1997). Structural loading may easily cause shear failure of a soil, and safe limits are defined by a suitable factor of safety (normally 3) in relation to ultimate failure stress. In practice, the acceptable stress limit is normally determined by the acceptability of the induced settlement.

Exceeding the safe and acceptable loading may lead to gross settlement, which is even more serious when it is differential (Figure 12.4). The ultimate example is the Leaning Tower of Pisa, in Italy (Figure 12.5). Its foundations were hopelessly small, and a detail of lateral variation within the ground instigated differential subsidence; this then became self-enhancing by differential loading, as the centre of gravity of the tall narrow tower moved with the tilt. The remedial engineering was a

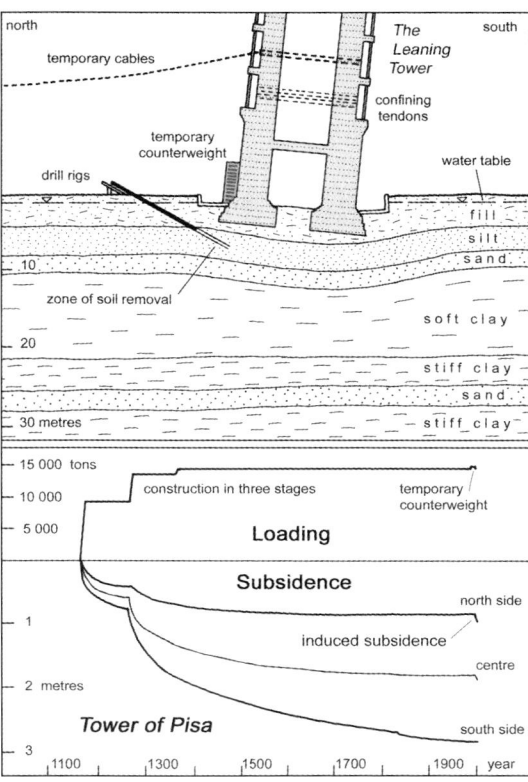

Figure 12.5   Differential subsidence and its remediation at the Leaning Tower of Pisa. The diagram above shows the ground profile and remedial methods to stabilise the tower, and the lower graph shows the rates of subsidence since the tower was built.

Figure 12.4   Differential subsidence of an old building that stands on short timber piles into the soft young clay that underlies Amsterdam.

classic, with a temporary surcharge followed by the drilled removal of soil beneath the foundations, both on the north side, to induce a tilt towards the north, counter to the long-term southerly tilt (Burland, 1997).

Extreme subsidence problems are created where a soft clay is both loaded and de-watered, and the prime example is provided by Mexico City (Figueroa Vega, 1984). The downtown area is built on an old lake bed of thick, young and very soft, volcanic-derived, smectite-rich clays, and these are inter-bedded with sands that have been over pumped for water supply for many decades. Subsidence is therefore twofold:

1. The clays are so soft that all buildings subside under self-load. The Palace of Fine Arts, built on

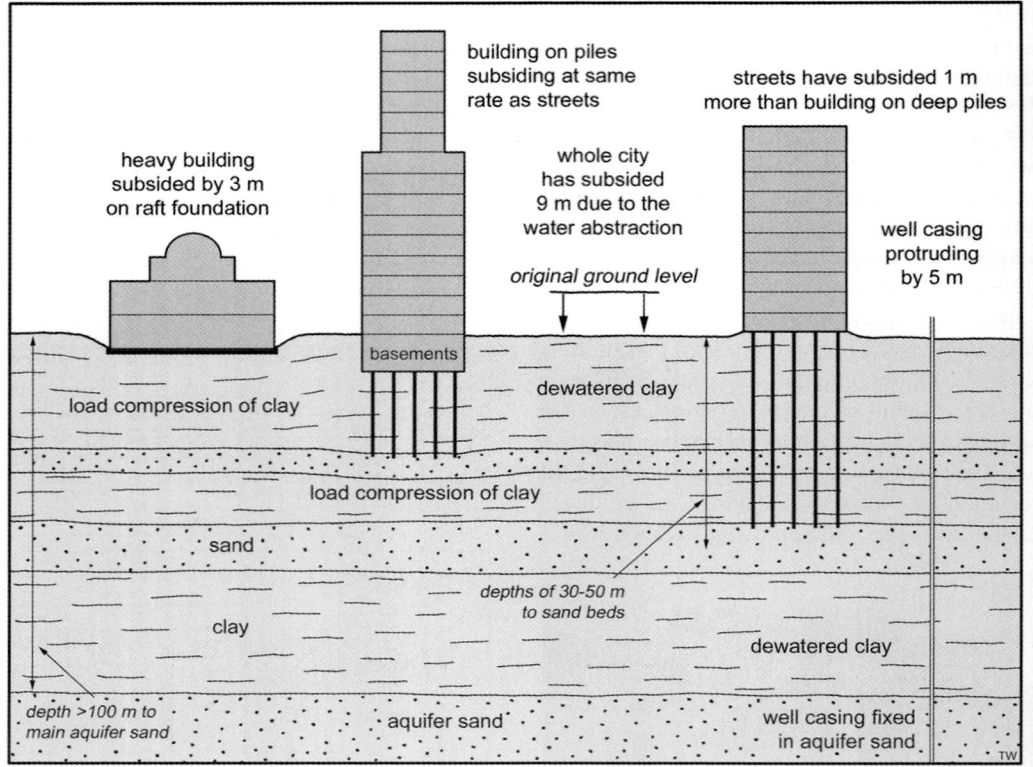

Figure 12.6   Diagrammatic profile through Mexico City with, from the left, the subsided Palace of Fine Arts, the successful Latin American Tower, a 'perched' building on deep foundations, and an old well casing that now protrudes above street level (because it is founded below the clays that it has helped to de-water by abstraction from the sands).

a massive raft, has subsided by 3 m. Movements are exacerbated by earthquakes that cause partial liquefaction of some of the soils.

2. The entire city centre has subsided due to the groundwater abstraction. This has reached a maximum of about 9 m. Subsidence rates are now reduced by pumping controls.

The engineering response has been the evolution of survivable building designs. All new structures are placed on raft foundations that can tolerate settlements reaching up to 1 m. Modern high-rise buildings have deep piles to sand horizons, and many also have deep basements that act as compensated foundations by reducing effective loads. However, deep-piled foundations leave a stable building standing higher than surrounding ground that has subsided due to de-watering. The Latin American Tower is successfully founded on piles to a shallow sand, so that loading compaction of the underlying clay matches de-watering compaction of the higher clay — and the tower's entrance remains at street level (Figure 12.6).

12.3 Subsidence on peat

Formed entirely of partially decayed plant material, peat is the weakest natural soil that exists. Blanket peats are spread over wet upland, but the main peat areas are the extensive fenlands, distinguished by their totally flat landscapes and black soils. Peat has a bearing capacity that is effectively nil, until it has been drained or consolidated by surcharge, and even then it is stable only under very small loads. Excessive subsidence is almost inevitable on peat.

The organic structure of peat varies from fibrous to granular and amorphous, but its engineering properties relate largely to the local burial and drainage history. Water contents are typically 500–2000%, but may be as low as 100% from above the water table. Undrained, peat acts as a liquid with negligible strength, but after drainage and structural loading a safe bearing pressure of 70 kPa may be applied to the partially consolidated peat.

Subsidence on peat is induced by either loading or drainage, and is therefore similar to clay except that the compressibility of peat is far greater. Subsidence under load means that few structures can be placed on peat without special treatment. Most hill or moor peat is only a few metres thick, while lake-fills, muskeg and fenland peat are rarely more than about 10 m thick. On many construction sites the response to peat is therefore its entire removal. An alternative is placing end-bearing piles through to the base of the peat, as is required by state law for all buildings in the peatlands of Florida. Options on thick peat include floating structures on footings of polystyrene, sawdust, brushwood, timber corduroy or baled peat (which are stable when depressed below the water table). Raft foundations (maybe with compensating basements) can support buildings on drained peat, and highway construction is successful after surcharging of peat.

Drainage causes immediate subsidence of peat. This is widely recorded, but nowhere better than by the Holme Post in the English Fens (Hutchinson,

Figure 12.7 A sequence of images that show the emergence of the Holme Post from its drained and subsiding peat. The 1850 image is a reconstruction, and the 1932 image is from Fowler (1933).

1980). As the Holme Fen was the last to be drained (to permit useful agriculture) subsidence was anticipated and the post was placed in the peat but was founded on the stable clay beneath (Waltham, 2000). Through multiple phases of land drainage, the post emerged from the subsiding peat (Figure 12.7). Each pumped fall of the water table causes almost immediate subsidence on the peat (Figure 12.8). Across sites worldwide, this subsidence on peat (that was originally 5–10 m thick) is about 50–65% of the head decline on first drainage and

20–30% of head decline on subsequent phases of renewed drainage.

After the initial drainage and compaction, subsidence on peat continues due to wastage — loss to the atmosphere of the peat left above the water table and thereby exposed to microbial oxidation. The subsidence rate depends on the type of peat, and increases with greater depth to the water table and with higher soil temperatures (Figure 12.9). Annual surface lowering is in the range of 5–100 mm and is typically 1–7% of the depth to the water table. The effect of wastage subsidence is that agricultural land inevitably becomes too low and too wet, thus requiring renewed drainage, causing renewed subsidence. Rivers have to be maintained between high banks across the subsidence bowl, land drainage water has to be pumped up into the elevated rivers, and buildings on piles progressively rise from the peat requiring increasing flights of steps up to their entrances. All these features can be seen in the English Fens.

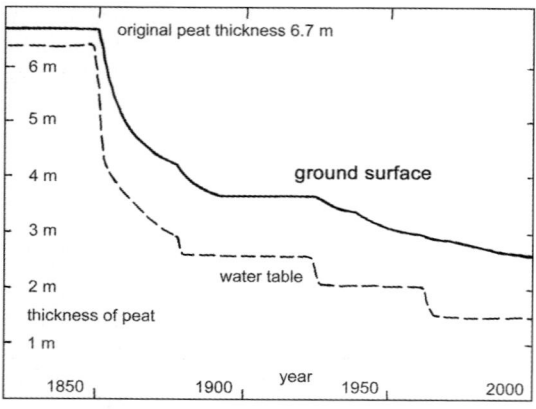

Figure 12.8   The record of 150 years of fenland drainage and ground subsidence on the peat, that is provided by the Holme Post.

12.4 Hydrocompaction of collapsing soils

Sediments prone to internal structural collapse when water is added to them are generally known as collapsing soils. Weak clay bonds between the grains of loosely packed silt sediments are broken by the introduction of water; the soil then densifies by repacking under self-load or imposed load, in the process known as hydrocompaction. Potentially collapsible soils are wind-deposited loess and some alluvial silts that were rapidly deposited and then desiccated in large basin fans.

Hydrocompaction can promote ground subsidence of up to 5 m over wide areas, and this represents up to 10% of the original thickness of the collapsible soil. This only occurs in semi-arid regions where the soils have not been previously wetted, and collapse potential is normally limited at depth by the water table or by prior consolidation. Subsidence occurs most commonly where water is added to the soils artificially, notably by irrigation schemes in semi-desert terrains. The rapid progress of hydrocompaction can be seen both in laboratory consolidation tests where water

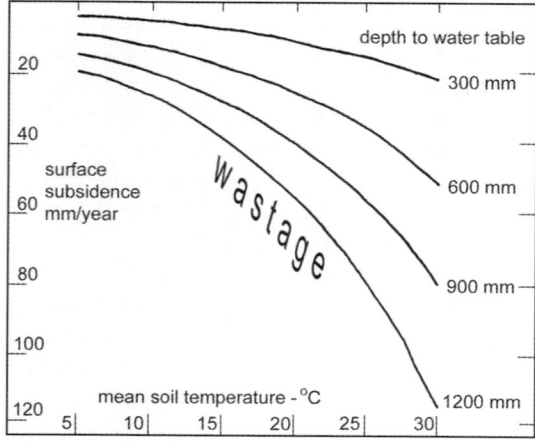

Figure 12.9   Peat wastage rates related to mean temperatures and depths to the water table (after Stephens and Stewart, 1977).

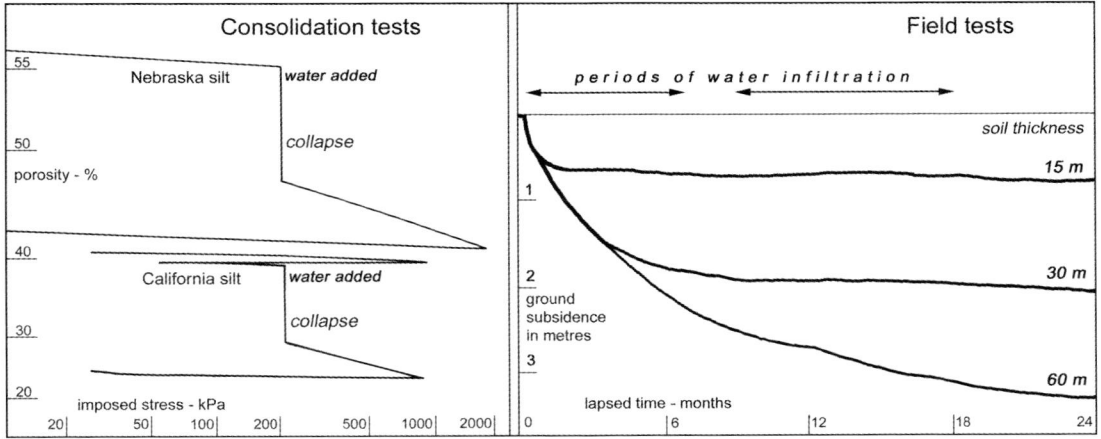

Figure 12.10   Hydrocompaction of collapsible silt soils when water is added to them, recorded in laboratory consolidation tests (from Waltham, 1989) and in field tests in the Central Valley of California, which reveal the time delay of subsidence on thicker soils (after Lofgren, 1969).

is added during loading, and also in field tests where surface water is allowed to infiltrate the soils (Figure 12.10). Collapse only occurs in soils with initial porosities that are > 40% (Dudley, 1970), and hydrocompaction is at a maximum in soils with 12% clay (that provides the metastable bonds between silt particles). The amounts of potential subsidence can be approximately estimated from the laboratory tests, which can also identify those soils that may compact by more than 10% when also subject to engineering loads.

The San Joaquin Valley, the southern part of the Central Valley of California, is a prime site of massive ground subsidence by hydrocompaction due to the introduction of irrigation water (as well as including even larger areas that have subsided due to water abstraction from underlying clays). Rolling profiles are acceptable in the newly irrigated fields, but the subsidence has a serious impact on the canals that carry the irrigation water into and through the area (Prokopovich, 1986). Pre-construction flooding over a period of two years has successfully instigated complete hydrocompaction to leave zones of subsided but stable ground. Selective soil wetting by sprinkle irrigation, and mechanical compaction, can improve shallow or thin collapsible soils, but may leave dormant the potential for a subsequent phase of hydrocompaction.

12.5 Subsidence on permafrost

Continuous, permanently frozen ground exists where mean air temperatures are lower than −8 °C, and discontinuous permafrost occurs where mean air temperatures are between −1 °C and −8 °C. Permafrost may be only 20 m thick where it is discontinuous, but can reach to more than 500 m deep in the coldest regions. Below the permafrost, the talik zone is kept unfrozen by geothermal heat (Chapter 14).

The top 1–3 m of the ground is known as the active layer as it thaws and re-freezes annually, creating the huge unstable bogs that characterise the Arctic in summer. Gravels that retain grain-to-grain contact are thaw-stable, as are all rocks. Clays and silts are solid when frozen, but are very weak when wet and saturated. Sands are thaw-stable with low ice contents, but some permafrost contains over 60% ice, and this must collapse when it is thawed.

Structures can be founded over permafrost, but subsidence is inevitable if they cause thawing of any soils that obtain their strength largely or wholly from their interstitial ice. Dawson City grew out of the Klondike goldfields in the Canadian Yukon, and its original houses were built directly on thaw-stable gravels that overlie frozen clay soils. Downward heat loss from the warm houses thawed the unstable clays, and every

Figure 12.11   Houses subsided on permafrost thawed by their own heat in Dawson, Canada.

Figure 12.12   Utilidors and buildings on piles in Longyearban, Svalbard. The tunnels in the foreground take the elevated utilidors through a gravel bank supporting a road, and the pile-supported buildings have free-air underspace closed only by slatted boarding.

one of the original timber buildings subsided and tilted (Figure 12.11). The key to subsidence prevention is conservation of the permafrost.

Stable roads are built on thick gravel pads that provide the insulation beneath the bare road surface. The roadbed for Alaska's Dalton Highway was built by end-tipping gravel directly onto the natural surface. The vegetation and organic muck were never stripped out, as they provide the natural insulation and removing them would have let construction traffic sink into an unfrozen quagmire (as happened on the original military Alaska Highway). A layer of expanded polystyrene 70–150 mm thick can preserve more fragile ground and reduce the required thickness of gravel.

Low-rise houses can rest on wooden blocks, preferably 1 m above the ground, with open access beneath for cold winter air. Larger buildings rest on thick gravel pads, like those for roads. Heated buildings often require extra cooling of the gravel by air ducts aligned with prevailing winter winds; some have ammonia-filled refrigeration tubes. Alternatively, structures are built on adfreeze piles which are sunk into stable ground ice below the active layer: oversize pile holes are drilled before a pile is inserted and the hole is then filled with slurry that freezes after a few days or weeks to create a stable bearing pile. Water and sewer lines are installed above ground in utilidors (Figure 12.12); the Trans-Alaska Oil Pipeline works on the same principle of keeping warm pipe contents away from fragile permafrost, though with the addition of refrigerated piles for added protection.

12.6 Subsidence on limestone and gypsum

The vast majority of the world's natural cavities are in limestone and their existence provides the potential for ground subsidence wherever limestone is at outcrop, just below outcrop or lying beneath a soil cover. Collapses of rock over caves are very rare events, and voids within the limestone present only a very small but unpredictable risk to engineering works. A much more widespread hazard is the development of subsidence sinkholes in soil covers that are washed into caves or fissures in underlying limestone. Karst terrains have a very distinctive geomorphology of closed depressions, isolated hills and bare rock outcrops, that are fully considered in Chapter 24 — this covers the processes, hazards and remediation of both cave collapses and subsidence sinkholes (Waltham *et al.*, 2005).

Gypsum is the world's second most cavernous rock, and can develop extensive karst landscapes that have their own subsidence hazards (Chapter 24). However, gypsum is much more rapidly dissolved in natural environments than is limestone and can therefore cause gentle subsidence of large areas of ground by extensive dissolution at its rockhead (or at a buried contact with an aquifer). Quarries in gypsum in England and Germany have provided cut sections through subsidence bowls that are hundreds of metres across and only a few metres deep. A comparable subsidence bowl has deformed a terrace of houses in Ripon, UK, within their lifetime of less than 200 years. It is unclear how much of this subsidence is due to active dissolution of the gypsum rockhead, and how much is due to compression of peat soils that have accumulated in an older subsidence bowl that had formed more slowly. The rapid dissolution of gypsum at its

rockhead can cause its almost total removal within the weathering zone, leaving a complex breccia of collapsed insoluble inter-beds, remnant gypsum blocks, slumped soils and isolated open voids. These collapse breccias provide very unstable ground, that is unmatched in limestone karst, and sites in England have required extensive grouting and ground improvement for safe construction on them (Cooper and Saunders, 2002).

12.7 Salt subsidence

Salt, as halite or rock salt, is totally and rapidly soluble in natural waters. Significant voids and total removal of salt beds can develop within a few years, orders of magnitude faster than mineral losses in limestone and gypsum, and salt is so soluble that it cannot survive at outcrop except in desert climates. Most salt bodies occur within clay sequences where they remain dry because groundwater cannot reach them. The major subsidence process is dissolution at a rockhead beneath thick soil and drift covers that contain flowing groundwater.

In lowland sites like the English Cheshire Plain, rockhead dissolution of salt is limited by the layering of the drift groundwater where heavier brine (saturated with salt) lies beneath the fresh water that is capable of dissolving the salt. Where salt beds are dissolved at a rockhead, insoluble residues and the typically abundant inter-bedded clays are left as a residual dissolution breccia (Figure 12.13). This further restricts groundwater flow, but dissolution does continue on the top of the salt beneath (known in Cheshire as the 'wet rockhead'). Dissolution rates and subsidence rates are a function of groundwater flow patterns, and overall subsidence is generally > 0.1 mm/year. Continuing through a large part of the Pleistocene, this has created subsidence hollows that became lakes as they sank below the water table; these are now known as 'meres', many of which are over 1000 m across, and are diagnostic of the Cheshire-style of lowland salt karst. These natural movements are generally too slow to record, except where tilting of new houses show that they do continue today. The most severe movements occur in linear subsidences

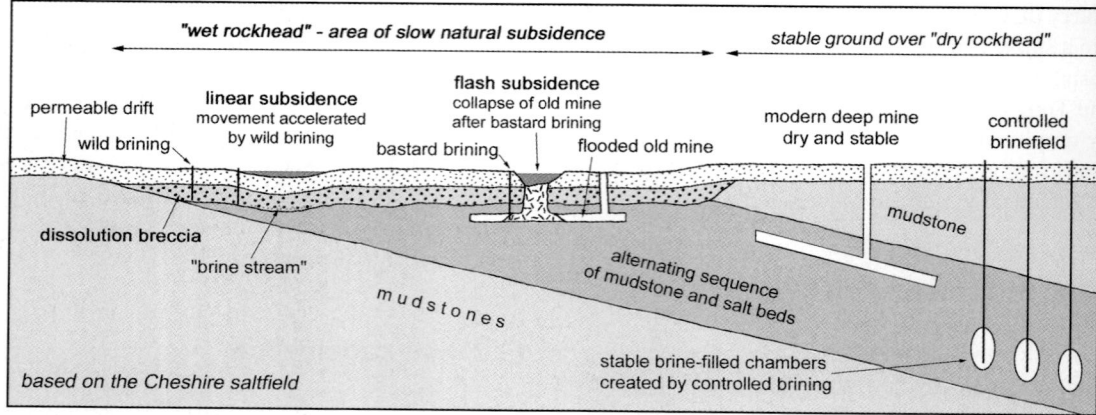

Figure 12.13 Diagrammatic profile through zones of active subsidence and stable ground over salt beds in the Cheshire Plain (UK). Subsidence can occur anywhere where the salt beds are in contact with fresh groundwater in the drift sediments — along the zone known colloquially as 'wet rockhead'. Under a protective cap of dry mudstone, the salt is stable.

Figure 12.14 The linear subsidence occupied by Moston Flash in the Cheshire salt field (UK). The road required repeated raising when subsidence was accelerated by wild brining, but has become almost stable since the local brine pumping ceased.

(typically 1–5 km long, 100–400 m wide and 1–10 m deep) that appear as valleys except that they continue over interfluves. These can form over rockhead outcrops of thicker beds of pure salt and 'brine streams' that are zones of enhanced groundwater flow just above rockhead (Waltham, 1989).

All salt subsidence is vastly accelerated by any brine-pumping operations that draw in new supplies of chemically aggressive freshwater to replace the abstracted brine. Traditional wild brining targets the brine streams, and thereby causes the linear subsidences above them to deepen by 100 mm/year or more, ultimately creating new ribbon lakes, known as 'flashes', within them (Figure 12.14). Catastrophic subsidences were caused by small mines beneath wet rockhead, which were abandoned when unsafe, allowed to flood, then brine-pumped (in the 'bastard-brining' operations)

Figure 12.15 A house in Northwich, UK, seriously subsided over an old mine that collapsed during bastard brining. It survived intact because the brickwork formed only panels within a strong timber frame (from an old postcard).

until the remnant pillars were dissolved by new inputs of freshwater. Major collapses caused subsidence lakes to form overnight — the true, rapidly formed flashes.

All the severe salt subsidence was induced by the mining and brining. Bastard brining was banned in Cheshire in the 1930s (but continued in Thailand until the 1980s). Since then, wild brining has steadily been replaced by stable controlled brining from caverns in deep dry rock, where Cheshire's one pillar-and-stall salt mine is also dry and stable. The engineering response in the salt fields has been to create solid structures that can be jacked up after they have subsided. Timber frames, steel frames or concrete rafts have been used in the Cheshire salt fields for 100 years (Figure 12.15), but simple reinforced strip footings are only regarded as an optional precaution for today's minimal movements.

Jacking and repair costs for any houses were covered by the Cheshire Brine Compensation Board, which was funded from a levy on all wild-brining operations, but the almost complete end of wild brining and its associated major subsidence made the Board obsolete. The risk of any one house now subsiding on the salt is statistically very small, and is sensibly carried by appropriate insurance surcharges.

12.8 Volcanic subsidence

Active volcanoes offer a unique suite of subsidence processes whereby liquid rock can flow away from beneath particular sites. On the small scale, liquid lava can drain out from the cores of lava flows. Drainage from a lava pond normally causes immediate subsidence of its unsupported solid crust, but this is a process of eruption activity, with no influence on engineering. In contrast, drainage from a single tube leaves a lava cave with a thin rock roof that is liable to be stable in the short term but offers the long-term threat of its potential collapse (Chapter 23).

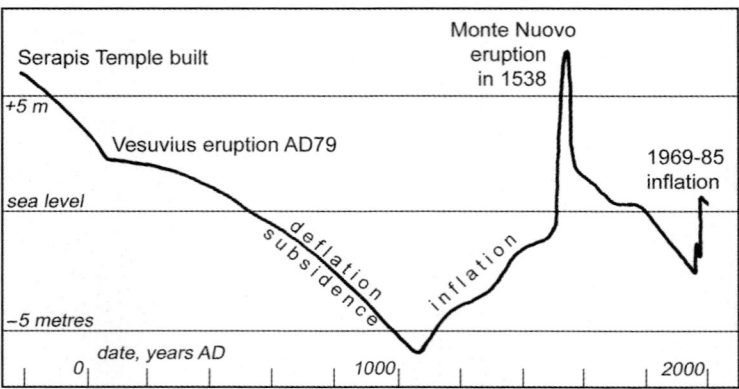

Figure 12.16 Subsidence and uplift of the Temple of Serapis, over the active magma chamber of Campi Flegri, near Naples, Italy. Monte Nuovo erupted from a new vent just 3 km away in 1538, after a period of rapid uplift.

On the large scale, entire volcanoes and their immediate surrounds, can deflate due to migration or retreat of magma from beneath them. Deflation of Italy's Campi Flegri volcanic centre, just west of Naples, caused inundation of the coastal Temple of Serapis, which subsided by over 10 m in 1200 years (Figure 12.16). This is, however, one case where ground subsidence is welcome, as the reverse uplift is due to volcanic inflation that normally precedes an even more destructive eruption.

12.9 Implications for engineering on naturally cavernous ground

The major engineering problem posed by natural ground cavities is that their locations are generally unpredictable and those with no open entrance can be very difficult or very expensive to find prior to construction activity (Culshaw and Waltham, 1987). Fortunately, natural caves only occur in a limited number of rock types and their main characterstics may be summarised as follows:

1. Caves in limestone are the most abundant. They may be many tens of metres across, but most are at depths that make them irrelevant to surface subsidence. The scale of cave development in limestone, and therefore of potential subsidence, increases in the hot and wet climatic regions (Chapter 24).
2. Caves in dolomite are essentially the same as those in limestone but are less common.

3. Caves in chalk are rarely more than a metre across and the main subsidence hazard is the development of sinkholes when soil is washed down narrow fissures (Chapter 24).
4. Caves in gypsum are widespread. They are mostly smaller than caves in limestone but they form and expand by dissolution more rapidly and collapse more readily in the weaker rock.
5. Caves in salt occur only in arid environments and are mostly small.
6. Caves in quartzite are formed by very slow dissolution of silica and only grow to significant size in very ancient and long-exposed craton terrains. Some collapse cavities in quartzite in Venezuela are over 100 m across, but they are stable, incredibly old and in high, remote mountains.
7. Caves in granite are mainly complex systems of narrow fissures and breakdown rooms, and are not a significant feature of most granite terrains.
8. Caves in soil are known as 'pipes'. They develop headwards along seepage lines as the fine and then the coarse fractions of the soil are progressively washed out (Figure 12.17). Piping failures are common in soils that are washed into broken drains and sewers, and thereby exacerbate subsidence of houses on shrinkable clays (where the clay movement caused an initial drain break).

The only other natural caves are sea caves and river cliff notches which all have conspicuous

Figure 12.17   Subsidence due to collapse of piping cavities in a terrace of fine-grained soils in southern Tunisia, where uncontrolled drainage developed pipes to point outlets on a lower terrace.

entrances and present no hidden geohazard; tension fissures (also known as 'tectonic caves', 'gulls' and 'windypits') that develop in the head zones of landslides (Chapter 8); caves in clastic rocks that have soluble calcite cements and/or clasts; isolated fissure caves that have opened in tension zones or due to glacial drag, and caves in glacier ice (that are irrelevant to foundation engineering). All these are rare but, along with mined cavities, can occur in any rock.

In any of these geological conditions where caves could occur, practising engineers must consider the subsidence hazard that they present. The engineering risk must be evaluated for each site with respect to the impact of a ground collapse and the chance of an unknown cave underlying the structure. Even in mature karst terrains, the statistical chance of encountering a cave at a critical location under a single structure is very small, but its potential impact may be disastrous. A highway across karst is more likely to encounter a cave, but it probably creates minimal engineering inconvenience, and many long road excavations reveal no caves at all.

Each site in potentially cavernous ground must be treated on its own merits. Engineers have the option of two responses. They may find and remediate every significant cavity. This may demand

Figure 12.18   A heavily reinforced raft for structural foundations in cavernous pinnacle karst on limestone at Shilin, China.

Figure 12.19   Collapse sinkholes in the gypsum karst of central Turkey: (*A*) a very large old sinkhole with no sign of recent collapse and (*B*) the small-scale progressive collapse active within a corner of a nearby sinkhole whose total area equals that of the large old feature.

very expensive, close-spaced grids of boreholes to eliminate every possibility, as geophysical searches cannot be totally reliable (Chapter 24). A rock arch over a natural cave can develop a stable span that is at least twice the width of its thickness in sound rock (thereby excluding soils prone to piping and most chalks). Under-drilling to prove between 3 and 5 m of sound rock is therefore sufficient to ensure the integrity of foundations in almost any area of cavernous ground (Waltham *et al.*, 2005). However, every site must be assessed for its own conditions.

Alternatively, engineers may design structures safely to span any unseen ground cavity that could possibly collapse. Reinforced strip footings,

extended ground beams, reinforced rafts (Figure 12.18) and sacrificial piers may be employed as appropriate. Though collapse sinkholes and collapse-modified caverns up to hundreds of metres across are known in terrains of limestone, gypsum and quartzite, no surface collapse more than about 20 m across has ever been seen or can be recognised as a single past event (Figure 12.19). The larger features have developed by progressive, multiple collapses over geological timescales. This implies a worst-case scenario for sensible design of bridging structures. In most terrains, the largest potential collapse is smaller. The exception is provided by large subsidence sinkholes that can develop almost instantaneously by adequately thick soils slumping into cavernous limestone below.

Excavating a tunnel through cavernous ground involves an unavoidable element of risk, which can only be evaluated by reference to local records on known caves and voids revealed in boreholes. Exploration ahead of the face is best by long advance probes, and large voids can often be grouted or treated before the tunnel reaches them, though tunnel diversion may be required round large unstable caverns (Milanovic, 2000). The benefits of flexibility make drill-and-blast preferable to tunnel boring machines (TBMs) on many projects in cavernous ground.

12.10 Subsidence due to mining

The extraction of minerals for economic benefit becomes a geomorphological process when man is accepted as part of the natural world, and the resultant subsidence is certainly of major concern to engineers. Also, it cannot be ignored as it is one of the most widespread and destructive surface processes, especially in the developed world. There is immense variation in underground mining techniques that have been applied in different shapes of orebodies and have evolved around the world and over the centuries, but their subsidence hazards fall into four groups — longwall mining, block caving (and stope mining), old pillar-and-stall workings and old shafts.

Longwall mining is a total extraction method. It removes all of a thin roughly horizontal ore bed, on a usually mechanised face, that retreats and lets the ground collapse behind its travelling zone of temporary support (Figure 12.20). Ground subsidence is inevitable, immediate and broadly predictable (Waltham, 1989), but is normally completed within about two years of its undermining, when complete closure of the mine workings removes most of the potential for further surface decline. Longwall mining has been widespread in coal workings since about 1930, and its ground distortion in active coalfields is generally referred to as 'mining subsidence' (Figure 12.21).

Block caving is also a total extraction method. It is applied in massive orebodies where the ore is blasted and then tapped off from beneath into underlying transport tunnels while the walls above are allowed to collapse into a mass of broken rock that subsides as more ore is removed from below.

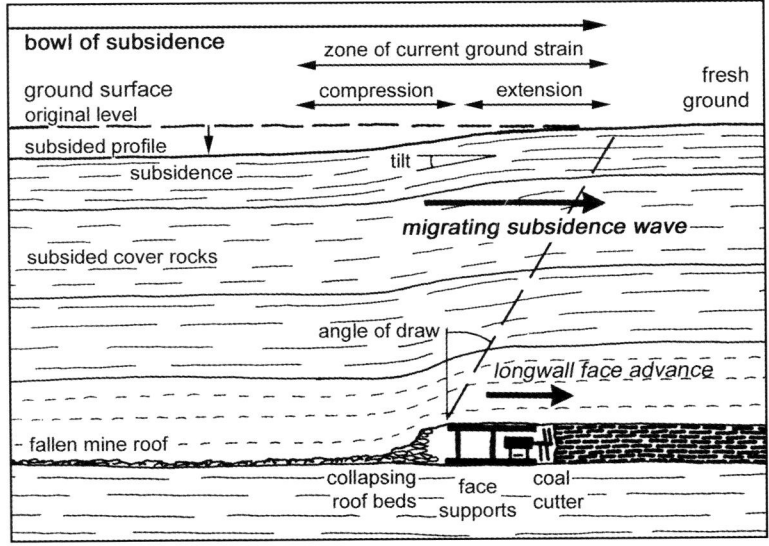

Figure 12.20   Diagrammatic profile through an advancing longwall mine face, showing the components of movement within the migrating wave of ground subsidence. Note the scale distortion: a coal seam is typically 1–2 m thick, and is generally worked at depths of 200–400 m.

Figure 12.21 'Very severe' mining subsidence damage to a house on the Nottinghamshire coalfield (UK), where ground strain has damage the brickwork.

Total removal of the ore leaves no support for the ground above, and surface subsidence may be massive, commonly lowering the floors of the abandoned open-cut pits that worked steeply dipping orebodies before extraction was forced to continue underground at greater depths. A smaller version of block caving is the stope working of narrow and steeply inclined vein deposits, which is very widespread in mining fields all across the world.

Pillar-and-stall mining is the main method of partial extraction. Ore is excavated from grids or networks of stalls (tunnels) while intervening blocks of ore are left in place as pillars that provide roof support. Dimensions and patterns of working vary enormously, and there is no ground subsidence over pillar-and-stall mines that are designed to be stable with respect to the rock mass strengths of the pillars and roofs. However, many old mines were left seriously unstable, notably through 'robbing the pillars', until collapse was imminent (Figure 12.22). Pillar failures cause major 'areal subsidence', while

roof failures are followed by upward void migration and crown hole failures at the surface; both types of subsidence can occur over mines long abandoned. Most thin, sub-horizontal ore was worked by this method, including most coal before about 1930.

Mine shafts were essential features above and adjacent to all underground mines. Even where the actual mining creates no subsidence threat, abandoned shafts create serious hazards especially where their walls are unstable or they lie unseen beneath inadequate covers. A shaft failure creates only a small sinkhole-type subsidence, but this can totally destroy any house or structure inadvertently placed on a ground cover that hides it.

Engineering implications of old and modern mining

The subsidence hazard from modern mining should be predictable and controllable as long as the mine is properly designed and managed. Subsidence over longwall coal mines involves ground compression, extension and tilt besides ultimate subsidence, and all parameters are predictable on the basis of both theoretical deformation and long-term empirical records (Whittaker and Reddish, 1989) except where geological factors localise and distort movements along faults and joints. The scale of movement increases over thicker seams, and also over wider workings closer to the surface. Ground strain is the most destructive component and it controls the classification of subsidence damage used in coalfields but is also applicable more widely (Table 12.4). Strain is typically 1–10 mm/m; values at the lower end of the range are normal for modern deep mines, so damage to houses is minimal, though bridges and larger structures may still be significantly affected. Vertical subsidence is typically 200–1500 mm over a seam 1–2 m thick, and is often most significant with respect to drainage in lowland areas where levees and bridges may require raising, while ground tilt develops in the marginal zones of the subsidence bowls (Figure 12.23). With the decline of deep mining in Britain, most workings are now kept under agricultural land where subsidence damage is minimised.

Block caving and vein stoping produce such large zones of unstable and cavernous ground that land above them is generally regarded as sterile

Figure 12.22 Narrow pillars of limestone left in an old pillar-and-stall stone mine in Dorset (UK). Though extraction was about 90%, this mine is stable because it has only a shallow cover resting on strong pillars, but pillars of coal of similar proportions are liable to collapse causing 'areal subsidence'.

Table 12.4 Classifications of subsidence damage (after NCB, 1975 and ISE, 2000). The NCB system is related primarily to ground strains over longwall mines, while the ISE system relates to the crack damage in houses regardless of cause.

Classification terms		Class descriptions	Crack width (mm)	Length change (mm)	Causative ground strain (mm/m)	
NCB (1975)	ISE (2000)				House 20 m across	Bridge 100 m long
0	Hairline	Negligible cracks that can be covered	< 0.1			
1 Very slight	Fine	Barely noticeable, small plaster cracks	0.1–1.0	< 30	1.0	0.2
2 Slight	Moderate	Slight internal cracks, doors/windows may stick	1–5	30–60	2.5	0.5
3 Appreciable	Serious	Slight external fractures, service pipes may fracture	5–15	60–120	5.0	1.0
4 Severe	Severe	Floors slope and walls lean, door frames are distorted, extensive repairs needed	15–25	120–180	7.5	1.5
5 Very severe	Very severe	Floors slope badly, walls bulge, floors and beams lose bearing, partial or complete rebuilding	> 25	> 180	10.0	2.0

Figure 12.23   Tilt of a building in the West Midlands coalfield (UK), which is unusually severe as it stands over the edge of very thick workings and also on the edge of a zone of soft valley sediment fill.

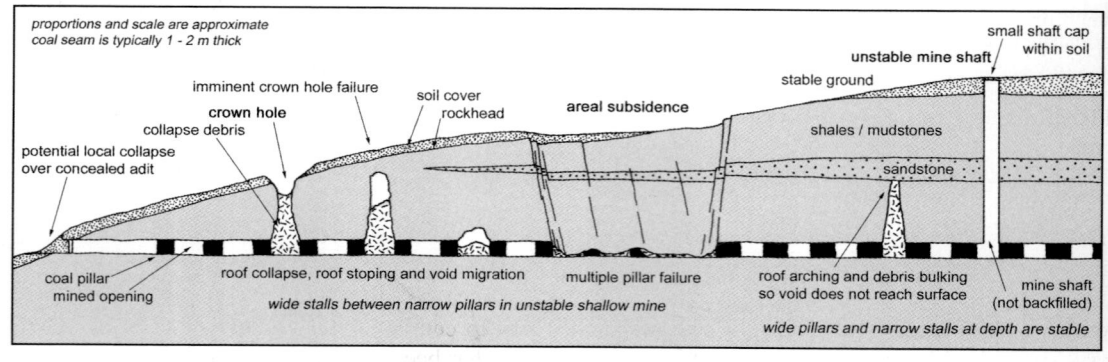

Figure 12.24   Diagrammatic profile through a shallow old coal mine, with subsidence hazards created by an unstable hillside *adit*, crown holes and an 'areal subsidence' over multiple pillar failure, while deeper workings with wide pillars are stable except where an old shaft is inadequately capped.

both during and after active mining. On dipping orebodies, the footwall is normally solid, but the hanging wall may be unstable for considerable distances from the outcrop. Roads can bridge narrow-vein workings on reinforced ground slabs, but large-scale remediation of old workings is rarely cost-effective.

Old pillar-and-stall mines can constitute a major subsidence hazard. Many failures occur soon after the mining is abandoned but failures can occur 100 years or more later. The main danger lies in shallow mines (Figure 12.24). Roof failures are unlikely to migrate more than about 30 m, so crown holes are rare over deeper mines. Thin pillars in shallow mines may be critically overloaded by new structures placed on the surface, whereas the wider pillars left to support thicker cover in deep mines are less influenced by engineering works. Guideline values for the cover depth required to render a mine of no general concern vary from 3 m in intact strong rock to 30 m in thinly bedded weak rock (Waltham and Swift, 2004); local conditions and experience are always critical to engineering design and practice. Treatment of shallow old mines may demand expensive grout injection through close grids of bored holes, though more specifically tailored methods can be applied at many sites (Littlejohn, 1979; Waltham, 1989).

Old mine shafts are only rendered safe for construction over them by an appropriate combination of filling, grouting and/or capping by over-sized concrete slabs (NCB, 1982). Most old shafts were inadequately treated before they were covered and forgotten, and buildings subsequently placed on them can suffer, many years later, subsidence that ranges from distressing to catastrophic (Figure 12.25). The main difficulty with concealed old shafts is locating them when records of them are imprecise or non-existent. Geophysical surveys can be useful (McCann *et al*., 1987) and ever-improving techniques mean that any engineer faced with the problem is well advised to contact a specialist subcontractor to investigate optimum approaches, which are usually site-specific.

12.11 Tectonic subsidence

True tectonic subsidence occurs in two-plate environments. The first is where the marginal zones of plates are being subducted close to convergent boundaries — the Nagoya region of eastern Japan is subsiding by 10 mm/year and is among the few places where this crustal warping is not restricted to ocean floors. The second is where the stretching of a plate causes its necking and thinning and therefore its surface lowering, where mantle flow beneath maintains isostatic balance — the North Sea is a prime example of this. While these processes affect very large areas, subsidence rates are mostly very low: the London area is subsiding by more than 2 mm/year due to thinning and sinking of the North Sea Basin. Such rates of subsidence have no impact inland, but may be critical at coastal sites, especially when combined with the current rise in world sea-levels (at 1–2 mm/year),

Figure 12.25 Catastrophic subsidence of a house over a forgotten mine shaft in the West Midlands (UK), where unconsolidated fill had settled within the shaft and also run into the workings below (the shaft was already being backfilled when the photograph was taken).

largely due to global warming that has been continuous for about the last 500 years. The combination has made the Thames Barrier necessary to combat tidal flooding of London where there has been no risk in the past.

Major deltaic basins are also sites of very slow but relentless subsidence that is due to a combination of factors. Soft clay sediments, with typical porosities of about 55% at depths down to tens of metres, are consolidated into mudstones with porosities of about 20% at depths around 3000 m (Skempton, 1970), and this compaction causes subsidence. Crustal sag of the basin floor matches sediment accumulation in continuously active deltas to provide the second component of subsidence. Under the Po Valley of northern Italy, about 2400 m of Quaternary sediment has partially consolidated into a sequence about 2000 m thick. A mean

Figure 12.26   The drowned forest of Portage after the rapid ground subsidence associated with the 1964 earthquake in Alaska.

deposition rate of 1.2 mm/year has been accommodated by 0.2 mm/year of compaction and 1.0 mm/year of crustal sag. Venice lies towards the margin of this deltaic basin, where the long-term mean subsidence is 0.4 mm/year. This is due to both compaction and sag, and is the natural, uncontrollable component of Venice's continued subsidence (Figure 12.3). In the centre of the Mississippi delta, surface subsidence is about 8 mm/year, due to the combination of sediment compaction and crustal sag. The latter occurs under the sediment load and there is no scope for subduction down-warping of the crust, as may be a contributory factor at orogenic sites including the Ganges and Po valleys.

Crustal subsidence also occurs under the loads imposed by ice caps and reservoirs. Scandinavia temporarily subsided by up to 200 m under its Pleistocene icecaps. Sea levels were lower at the same time, but isostatic rebound by the landmass was slower than sea level recovery when ice melted at the end of each glacial stage, leaving successions of raised beaches, marine cliff lines and the famous Norwegian strand-flat in the coastal region profiles. Very large masses of water impounded by high dams provide the only cases of crustal subsidence that are due to man. Lake Mead, behind the Hoover Dam on the Colorado River, caused a maximum down-warping of about 200 mm; as at other sites, this appeared to be crustal sag as it occurred on strong rocks with very little potential compaction.

Accelerated subsidence may be accompanied by earthquakes when crustal deformation is transferred to fault displacement. Large areas in Alaska, between Prince William Sound and the Chugach Mountains, subsided by up to 2.5 m during the 1964 earthquake, causing permanent drowning of coastal forests (Figure 12.26). Both the earthquake and the subsidence were results of relaxation of the American plate that had been buckled and overstressed over the subducting Pacific plate (Figure 12.27). Tectonic subsidence is a common result of earthquakes, though usually on smaller scales than that in Alaska. It is uncontrollable, and is no more predictable than are the earthquakes themselves, except that subsidence may be anticipated if any pre-event uplift has been monitored.

Subsidence is more widespread as a secondary effect of earthquakes, due to liquefaction of unconsolidated sand soils. During the period of earthquake vibration, sands may temporarily behave as a liquid and offer greatly reduced support to structures, which therefore subside into them, as happened famously to the Kawagishi-cho buildings in Niigata, Japan, in 1964. The vibrations also cause densification of loose sands by their improved grain packing, while inter-granular water is expelled through sand eruptions, and this causes the ground surface to subside permanently. Soil liquefaction only occurs in sands of fine and uniform grain size that are poorly consolidated

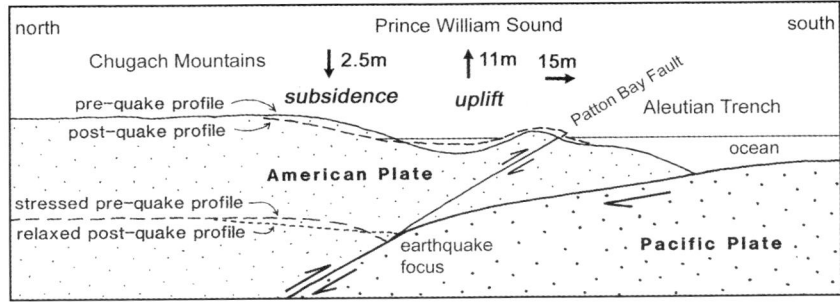

Figure 12.27   Diagrammatic profile through the subsidence zone that was created during the 1964 earthquake over the plate boundary subduction zone of Alaska.

(recognised by low SPT N-values on a standard penetration test) and lie below the water table. Dry, dense or poorly sorted sands do not liquefy, and mapping the soil parameters allows the definition of subsidence hazard zones for future earthquakes (Tinsley *et al.*, 1985). In high-value urban areas, improved land drainage may then be an effective protective measure against subsidence.

12.12 Summary

Ground subsidence must be recognised as a significant geomorphic process and as a potential geohazard in a range of specific geological environments dictated mainly by rock type. The feature common to nearly all subsidence processes, with the notable exception of tectonic subsidence, is that they are largely initiated and exacerbated by man's own activities. The two most widespread 'natural' processes of clay compaction and subsidence sinkhole development over limestone occur when the truly natural processes of groundwater movement and drainage are disturbed, unbalanced or accelerated by man. Subsidence on mined ground is entirely due to man's actions, and most other subsidence events, on salt, peat, silt and permafrost, follow a history of local engineering activity. The implication is that subsidence is a geohazard created by man, and is therefore avoidable or controllable by man. It is a classic example of 'man as a geomorphological agent', and should be in the mind of every engineer embarking on a ground project.

References

Burland, J. (1997) *Propping up Pisa*. Royal Academy of Engineering, London, 20pp.

Cooper, A. H. and Saunders, J. M. (2002) Road and bridge construction across gypsum karst in England. *Engineering Geology* **65**, 217–223.

Craig, R. F. (1997) *Soil Mechanics*. Spon, London, 486pp.

Culshaw, M. G. and Waltham, A. C. (1987) Natural and artificial cavities as ground engineering hazards. *Quart. Journ. Eng. Geol.* **20**, 139–150.

Dudley, J. H. (1970) Review of collapsing soils. *Proc. Amer. Soc. Civil. Engrs.* **96** (SM3) 925–947.

Figueroa Vega, G. E. (1984) Land subsidence case history: Mexico. *Studies and Reports in Hydrology* **40**, UNESCO, Paris, 217–232.

Fowler, G. (1933) Shrinkage of the peat-covered fenlands. *Geographical Journal* **81**, 149–150.

Hutchinson, J. N. (1980) The record of peat wastage in the East Anglia fenlands at Holme Post, 1846–1978 AD. *Journal of Ecology* **68**, 229–249.

ISE (2000) *Subsidence of Low-rise Buildings*. Institution of Structural Engineers, London, 126pp.

Littlejohn, G. S. (1979) Consolidation of old coal workings. *Ground Engineering* **12** (4), 15–21.

Lofgren, B. E. (1969) Land subsidence due to the application of water. *Geol. Soc. Amer. Reviews in Engineering Geology* **2**, 271–303.

McCann, D. M., Jackson, P. D. and Culshaw, M. G. (1987) The use of geophysical surveying methods in the detection of natural caves and mineshafts. *Quart. Journ. Eng. Geol.* **20**, 59–73.

Milanovic, P. (2000) *Geological Engineering in Karst*. Zebra, Belgrade, 345pp.

NCB (1975) *Subsidence Engineers' Handbook*. National Coal Board, London, 111pp.

NCB (1982) *The Treatment of Disused Mine Shafts and Adits*. National Coal Board, London, 88pp.

Poland, J. F. (ed.) (1984) Guidebook to studies of land subsidence due to groundwater withdrawal. *Studies and Reports in Hydrology* **40**, UNESCO, Paris, 323pp.

Prokopovich, N. P. (1986) Origin and treatment of hydrocompaction on the San Joaquin Valley, CA, USA. *Int. Assoc. Hydrol. Sciences Publ.* **151**, 537–546.

Skempton, A. W. (1970) The consolidation of clays by gravitational compaction. *Quarterly Journal of Geological Society* **125**, 373–411.

Stephens, J. C. and Stewart, E. H. (1977) Effect of climate on organic soil subsidence. *Int. Assoc. Hydrol. Sciences Publ.* **121**, 647–655.

Tinsley, J. C., Youd, T. L., Perkins, D. M. and Chen, A. T. F. (1985) Evaluating liquefaction potential. *US Geological Survey Professional Paper* 1360, 263–315.

Waltham, A. C. (1989) *Ground Subsidence.* Blackie, Glasgow, 202pp.

Waltham, T. (2000) Peat subsidence at the Holme Post. *Mercian Geologist* **15**, 49–51.

Waltham, T. (2002) Sinking cities. *Geology Today* **18**, 95–99.

Waltham, A. C. and Swift, G. M. (2004) Bearing capacity of rock over mined cavities in Nottingham. *Engineering Geology* **75**, 15-31

Waltham, T., Bell, F. and Culshaw, M. (2005) *Sinkholes and Subsidence.* Springer-Verlag, Berlin and New York, 382pp.

Whittaker, B. N. and Reddish, D. J. (1989) *Subsidence.* Elsevier, Oxford, 528pp.

Further reading

Carbognin, L. and Gatto, P. (1986) An overview of the subsidence of Venice. *Int. Ass. Hydrol. Sci. Publ.* **151**, 321–328.

Holzer, T. L. (1991) Nontectonic subsidence. *Geological Society of America Centennial Special Volume* **3**, 219–232.

Stephens, J. C., Allen, L. H. and Chen, E. (1984) Organic soil subsidence. *Geol. Soc. Am. Reviews in Eng. Geol.* **6**, 107–122.

Waltham, T. (2001) *Foundations of Engineering Geology*, Second Edition. Spon, London, 92pp.

Part III

Environments and Landscapes

13. Glacial Environments

Lewis A. Owen and Edward Derbyshire

13.1 Introduction

At present, glaciers cover about 10% of the Earth's surface and create some of the most complex environments on the planet. During the last few million years the area covered by glaciers has fluctuated between times of more extensive glaciation (glacials) when glaciers covered as much as one-third of the Earth's surface, and times of more limited glaciation (interglacials) similar to today. Thus, glaciers exert an important influence upon most high-latitude and high-altitude environments, both now and in the past. Furthermore, glaciers affect adjacent non-glaciated regions by controlling the nature of stream systems, lakes, coastal environments and wind systems. The importance of glaciation in mid- and high-latitude regions may be gauged by the dominant influence on the landscape exerted by the extensive cover of glacial and proglacial sediments and landforms readily seen on maps of surface geology (Eyles and Dearman, 1981; Eyles *et al.*, 1983a). Most engineering activities in such regions are, therefore, bound to have to deal with sediments and landforms arising directly or indirectly from glaciation. It is important to note also that the need for engineers to understand the dynamics of glacial systems will increase as populations rise in such regions in response to continuing economic development. The overview of the nature and variability of glacial environments, glacial dynamics and glacial sediments and landforms presented here is designed to provide a groundwork of understanding for engineers working in glaciated or formerly glaciated terrains.

Glaciated environments are encountered in all latitudes (Broecker and Denton, 1990). However, latitude and altitude are fundamental in influencing the regional and local climate that, in turn, essentially controls the distribution, dynamics and morphology of glaciers. As such, in polar and sub-polar regions glaciers may extend to sea level, while in tropical and equatorial regions they are usually restricted to altitudes above $\sim 4000\,m$ above sea-level. In high latitude regions, the glacial environment is commonly bordered by tundra (a mixture of bare ground with varying degrees of lichen, moss and low bush cover), giving way distally to tundra woodland, and then to circumpolar taiga (coniferous forest), each of these regions being progressively less affected by perennially frozen ground, or permafrost, and a wide range of periglacial processes (Chapter 14). In mid-latitude, tropical and equatorial regions, glacier-bordering environments may include alpine meadows and forests. Topography is another strong influence on glaciation. For example, in deeply dissected terrain, glaciers tend to be confined to the valleys while, in low-relief landscapes, they may spread out to form ice sheets.

The morphological form of glaciers is highly variable, ranging from small cirque glaciers to extensive ice sheets (Figures 13.1 and 13.2). The type of glacier ice also varies geographically. In high-latitude areas, basal ice may be below the pressure melting point (cold-based ice); this diminishes the glacier's ability to slide on its bed with consequent reduction in erosion of the bed and entrainment of debris. In contrast, the basal ice in mid- and low-latitude glaciers is above the pressure melting point, the meltwater released at the glacier bed allowing the glacier to slide and readily erode the substrate, and entrain and deposit sediment. The last few million years have been characterised by major oscillations in climate that have varied on decadal to millennial timescales. As a consequence, glaciers have fluctuated in response to variations in temperature

A

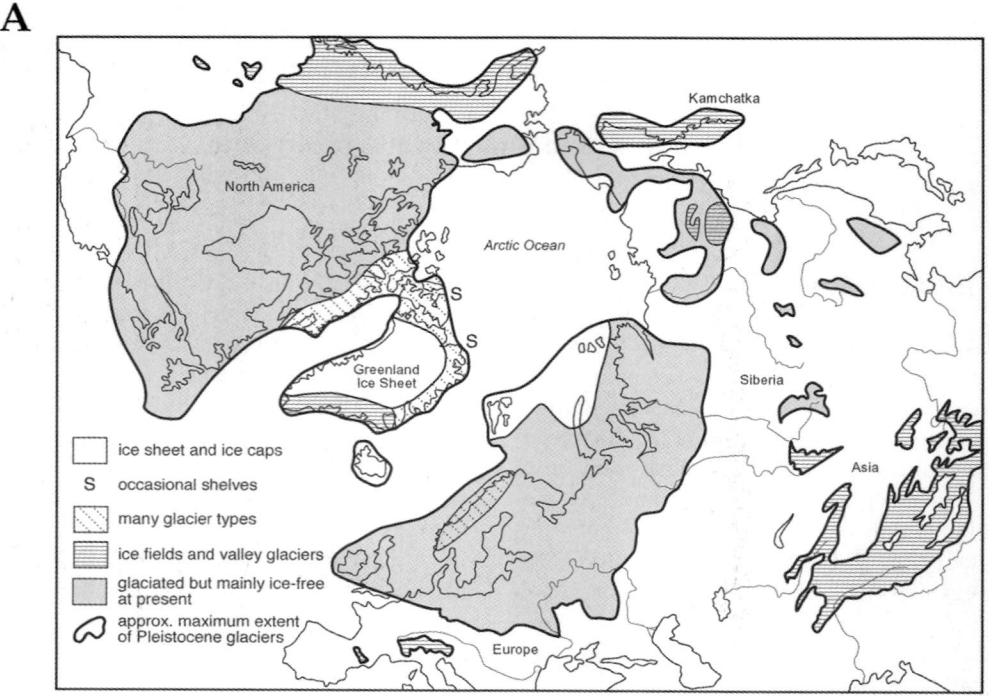

B

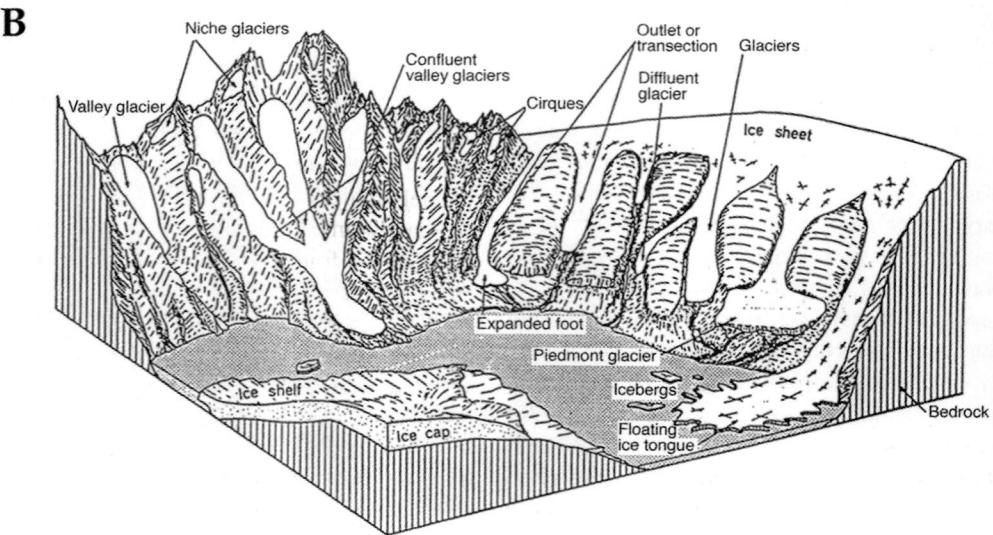

Figure 13.1   (*A*) The current distribution of major glacier types by morphology, within the areas affected by Pleistocene glaciation in the northern hemisphere (after Sugden and John, 1976); (*B*) The morphology of glaciers (after Sugden and John, 1976).

Figure 13.2a   Series of large nivation hollows, with well-developed cirque-like headwalls, facing north (left) along a sub-summit ridge in the Sierra de Peña Labra, north-western Spain.

Figure 13.2c   Two valley glaciers, with a cirque glacier between (centre, top), hanging above Wright Dry Valley. Wright Valley was formerly occupied by a transection glacier that drained the Antarctic ice sheet.

Figure 13.2b   Small cirque glacier adjacent to Wright Dry Valley, eastern Antarctica. The area of the glacier between the fresh snow and the boulder-topped moraine ridge (seen above the tents) is mantled with aeolian sand in this hyper-arid glacial environment.

and precipitation. This has further complicated the glacial environment, resulting in a wide variety of patterns and types of glacially eroded and depositional landforms and suites of sediments.

In order to aid understanding of the complexities of glacial environments, glacial geologists have resorted to developing facies and landsystem models for the various types of glacial settings. A facies is an individual sedimentary deposit having a distinct combination of properties that distinguish it from its neighboring sediments (Reading, 1986; Walker, 1992). The term lithofacies is used when referring only to the physical characteristics of a facies with no reference to the origin of the deposit. A genetic facies provides reference to the specific mode of formation. Facies can be grouped into associations and successions and can be used to help characterise a particular depositional environment (Walker, 1992). Study of contemporary environments allows these facies associations and successions to be compared with ancient sediments. Thus facies models facilitate the interpretation of sedimentary deposits. On a larger scale, the facies associations may be grouped together into a depositional system or landsystem model (Boulton and Eyles, 1979; Eyles, 1983; Evans, 2003). Most glacial geologists rely on facies and landsystem models to help them interpret ancient sediments and landforms. This approach has proved useful in providing a framework for predicting the spatial and temporal variability of glacial landforms and sediments that, in turn, can be used to help quantify the variability of materials likely to be encountered during an engineering project (Fookes, 1997). The main glacial landsystems are described in this chapter. However, because a full appreciation of the dynamics and products of the glacial system requires some understanding of the general physics of glaciers, glacial erosion, entrainment of debris and the main depositional processes and characteristics of glacial sediments, these topics are considered first.

13.2 Physics of glaciers

Glaciers tend to form in regions where snow layers persist year after year. The transformation of perennial snow into glacier ice is dependent on the temperature and precipitation regimes within a region. The successive accumulation of snow layers eventually leads to transformation of the snow into glacier ice as the volume of air-filled pores is reduced and the material increases in density from ~ 50–$200\,\text{kg/m}^3$ to 830–$910\,\text{kg/m}^3$ (Paterson, 1994). It is usual for glacial ice to form initially within small depressions, known as nivation hollows, where snow has preferentially accumulated, or on the tops of high mountains or plateaus (Figure 13.2a). If the ice continues to accumulate as a consequence of climate change, small cirque glaciers (Figure 13.2b) will eventually form from the perennial snow patches in nivation hollows, and ice caps will develop from accumulating ice on mountain peaks and plateaus. If climate change is sufficiently severe these glaciers may enlarge to form extensive valley glacier systems or ice sheets (Figure 13.2c).

The snow accumulation rate and atmospheric temperature of a region determine the glacial ice temperature. This, together with the thickness of the glacier, the geothermal heat flux and the frictional heat generated by ice flow determines whether the basal ice of the glacier is above or below the pressure melting point (warm- or cold-based glaciers). Under the influence of gravity, glacier ice moves downslope. This movement occurs mainly by creep involving plastic deformation within or between individual ice crystals, added to which movement may occur by fracturing when brittle deformation occurs (Weertman, 1983; Alley, 1992; Paterson, 1994). In warm-based glaciers, movement may also be accommodated by deformation in rocks or sediment beneath the glacier sole (Boulton and Jones, 1979; Boulton and Hindmarsh, 1987) and basal sliding where slippage occurs over a water layer (Echelmeyer and Wang, 1987).

The velocity of ice flow is a function of the shape of the glacial basin, the temperature of the ice and the mass balance characteristics. Mass balance is a measure of the difference between total accumulation and gross ablation (the sum of melting and sublimation) on a glacier. Mass balance may be measured at a particular point, over an area or for the glacier as whole, and the measured period of time may vary. Measurement of the annual mass balance of a whole glacier provides an indication of whether a glacier will expand or contract. A mass balance that is persistently negative (ablation > accumulation) will result in glacier contraction. Conversely, when the mass balance is persistently positive (ablation < accumulation) the glacier will expand. A zero mass balance (ablation = accumulation) results in a stable condition with no substantial advance or retreat of the glacier. Figure 13.3a–c shows the distribution of accumulation and ablation zones, and ice-flow characteristics for different types of glaciers.

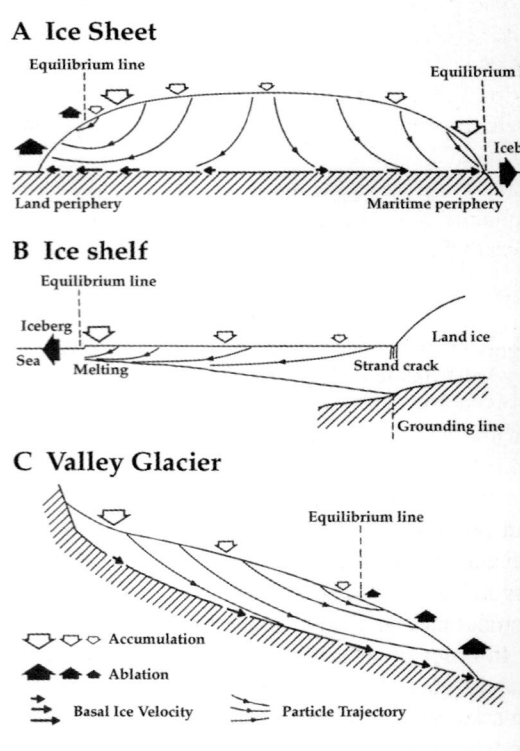

Figure 13.3 Models of (A) ice sheet (B) ice shelf and (C) valley glacier, showing the distribution of accumulation and ablation and related flow characteristics (after Sugden and John, 1976).

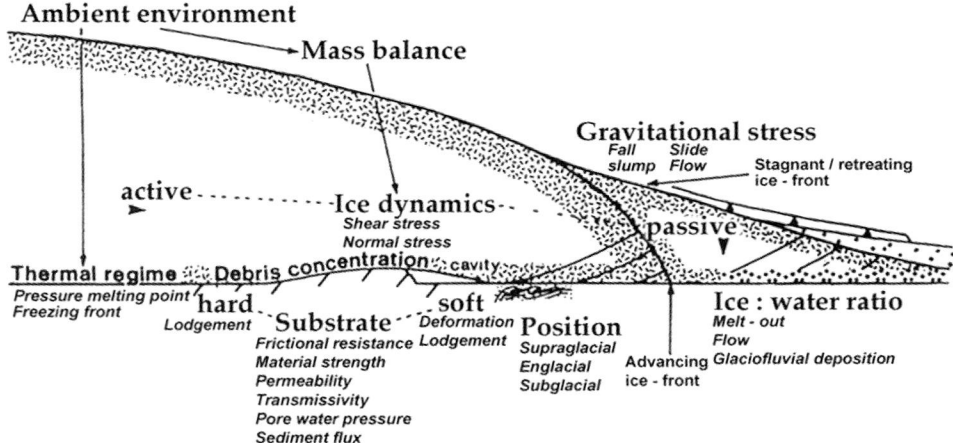

Figure 13.4   Major factors, associated variables and processes responsible for terrestrial glaciogenic deposition (after Whiteman, 1996).

The physics of glacier movement and mass balance is complex because of the highly variable nature of the geology and topography, climate and hydrology within any one region. This makes modelling and predicting the dynamics of glacial systems difficult. Nevertheless, a substantial amount of glaciological research on a wide range of glacier types has resulted in notable advances in predicting glacier movement.

13.3 Glacial erosion and entrainment of sediment

Erosion, deposition and deformation of rock and sediment occur as a consequence of the continuous movement of a glacier, and glacier advances or retreats (Figure 13.4). Glacial erosion occurs as a direct consequence of melting and refreezing of glacial ice on the underlying substrate, abrasion by particles carried in the ice and subglacial meltwater stream processes (Alley *et al.*, 1997). Glacially eroded forms vary in scale from single mm- and cm-size striae and friction cracks, m-size bedforms such as channels, depressions, *roche moutonnées* (glacially eroded streamlined bedrock knolls), to large scale forms such as cirque basins, glacial troughs and valleys that dominate the landscape (Benn and Evans, 1998). Further denudation

may result when glaciers retreat and leave behind undercut or unsupported slopes that may fail as a consequence of stress release (Bjerrum and Jorstad, 1968; Ballantyne and Benn, 1996).

The debris produced by subglacial erosion may become incorporated into the glacial ice at or near the base of the glacier when subglacial meltwaters refreeze to form regelation ice (Figures 13.4 and 13.5). Debris entrained within the ice is referred to as englacial debris. If such entrainment does not occur, it may form a subglacial debris layer beneath the glacier; even so, entrainment may eventually occur owing to subsequent deformation of the sediment with the advance of a glacier. Streams beneath glaciers may erode subglacial debris and redeposit it within or beyond the glacial system. Rock fall and avalanche debris deposited on the surface of the glacier may also become entrained within the ice by movement down open crevasses, or by incorporation by progressive burial under new snowfalls. Some of the surface (supraglacial) debris will be transported to the glacier's terminus without being incorporated into the ice. During its journey, supraglacial streams, as well as mass movement and aeolian processes may rework the supraglacial debris (Owen *et al.*, 2003).

When glaciers advance they may override sediments deposited earlier at points near or adjacent to the ice margin. Such proglacial sediments are

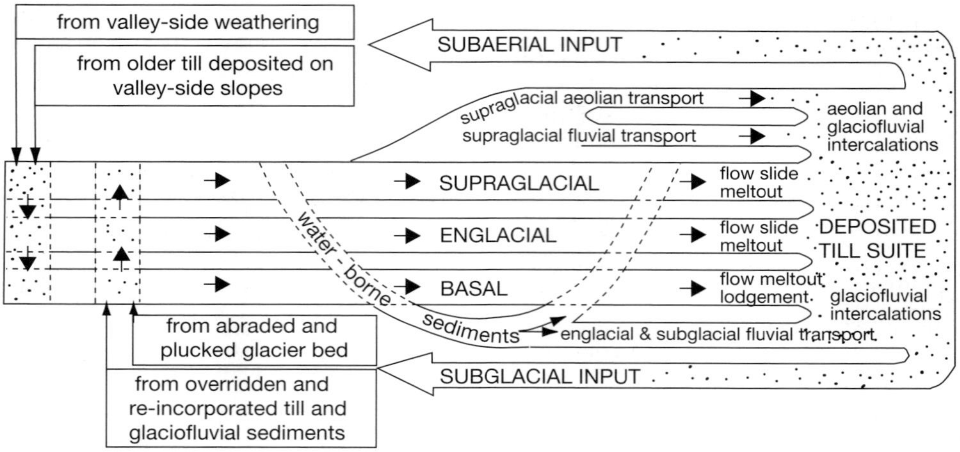

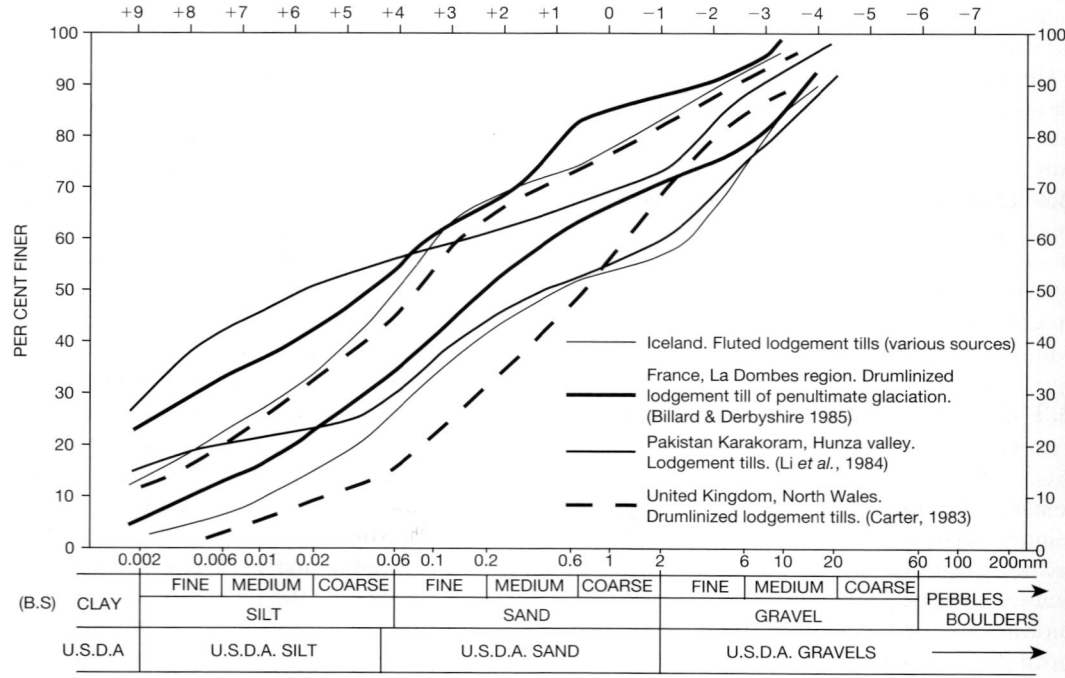

Figure 13.5   Schematic diagram representing glacial sediment systems (after Derbyshire *et al.*, 1979).

Figure 13.6   Particle size distribution envelopes for selected subglacial tills from low plateau (Iceland), piedmont (France), low maritime mountains (North Wales, UK) and high continental mountains (Karakoram Mountains, northern Pakistan).

deformed by the overlying ice and may become entrained within the glacier.

The sediment produced by glacial erosion and transport characteristically displays a very broad grain size range, being an important member of the group of multi-modal particle size sediments commonly referred to as diamicts or diamictons. Dreimanis and Vagners (1971) showed that, for

a particular source rock, there is a limiting particle size to which rock will be ground down. They
referred to this as the terminal grade, meaning
that, irrespective of the amount of time involved
or the distance over which the glacial sediment is
transported, the particles cannot be ground down
below a distinctive terminal diameter. The larger
particles, however, become progressively smaller
as the processing time and the distance of glacial
transport increase. Figure 13.6 shows the range of
variability of particle size distribution curves
(expressed as envelopes) for subglacial tills from
selected environments. Benn and Ballantyne
(1993; 1994) showed how particle shape is indicative of the mode of transport and deposition.
Reworking of sediments by ice-marginal or
proglacial processes following deposition may
continue to modify the size distribution and shape
characteristics of a sediment. Despite these general
relationships, it may prove very difficult to predict the characteristic properties of a given till.

13.4 Depositional processes and classification of tills

Entrained sediment will eventually be deposited
beneath, adjacent to or beyond the glacier margin.
Sediment deposited directly by ice, without any
appreciable meltwater action, is referred to as till
and the landforms so produced are known as
moraines. Numerous attempts have been made to
classify tills (Goldthwait and Matsch, 1989;
Trenter, 1999), but the generally accepted classifications are based on the mode of deposition. Till
may be deposited by lodgement, meltout, slide,
flow, sublimation and deformation. The interrelationships between the main factors and parameters influencing subglacial erosion and deposition
are shown in Figure 13.7.

Lodgement occurs when debris in the basal
layers of a sliding glacier is plastered on to a rigid
or semi-rigid bed (Dreimanis, 1989). Sediment
accumulated by the lodgement processes is
referred to as lodgement till. However, when
subsequently overridden, lodgement till behaves
as a deforming bed (Murray, 1997; Hindmarsh,
1997) beneath a glacier, and is then referred to as

deformation till. These tills are usually fine
grained, have high bulk densities and are generally over-consolidated. Comminution till, a term
rarely used today, is also a product of subglacial
deformation; composed of crushed and powdered
local bedrock, it has moderately high bulk densities and shear strengths (Elson, 1989), but may be
difficult to distinguish from deformation and
lodgement tills. When rocks and sediments are
subglacially sheared, the resulting *mélange* is
referred to as a glaciotectonite (Banham, 1977).

The process of meltout is one of direct deposition of sediments by melting of stagnant or very
slowly moving debris-rich ice. This may occur
below, within, or above the glacier and sediments
deposited by these processes are known as subglacial, englacial and supraglacial meltout tills,
respectively. During the meltout process, the sediment mass may fail resulting in debris slides or
flows, materials sometimes referred to as flow tills.
However, the term debris flows is more appropriate
because it refers to the true mode of deposition
(Lawson, 1981). When debris-rich ice sublimates,
the resultant sublimation till is formed and may
inherit the foliated structure of the ice (Shaw, 1977).
These tills usually have low bulk densities, are
poorly indurated and have coarser particular size
distributions than deformation and lodgement tills.

The sedimentary characteristics of each type
of till are shown in Table 13.1. However, given the
complexity of the multiple processes summarised
above, the origin of a till cannot always be determined unequivocally. Describing tills in terms of
their lithologies provides a more objective
approach. Miall (1978) and Eyles *et al.* (1983b)
provide a convenient framework for describing
the lithologies of glaciogenic sediments, known
as lithofacies coding (Table 13.2).

Different types of till may be deposited on top
of each other. For example, it is common for
supraglacial till to be deposited upon deformation
till. Furthermore, other sediments may be inter-
bedded into a glacial deposit such as glaciofluvial
and glaciolacustrine sediments that may be laid
down subglacially, englacially or supraglacially.
These sediments may be overridden by a glacier
and incorporated into a deformation till. The
different till types have different geotechnical

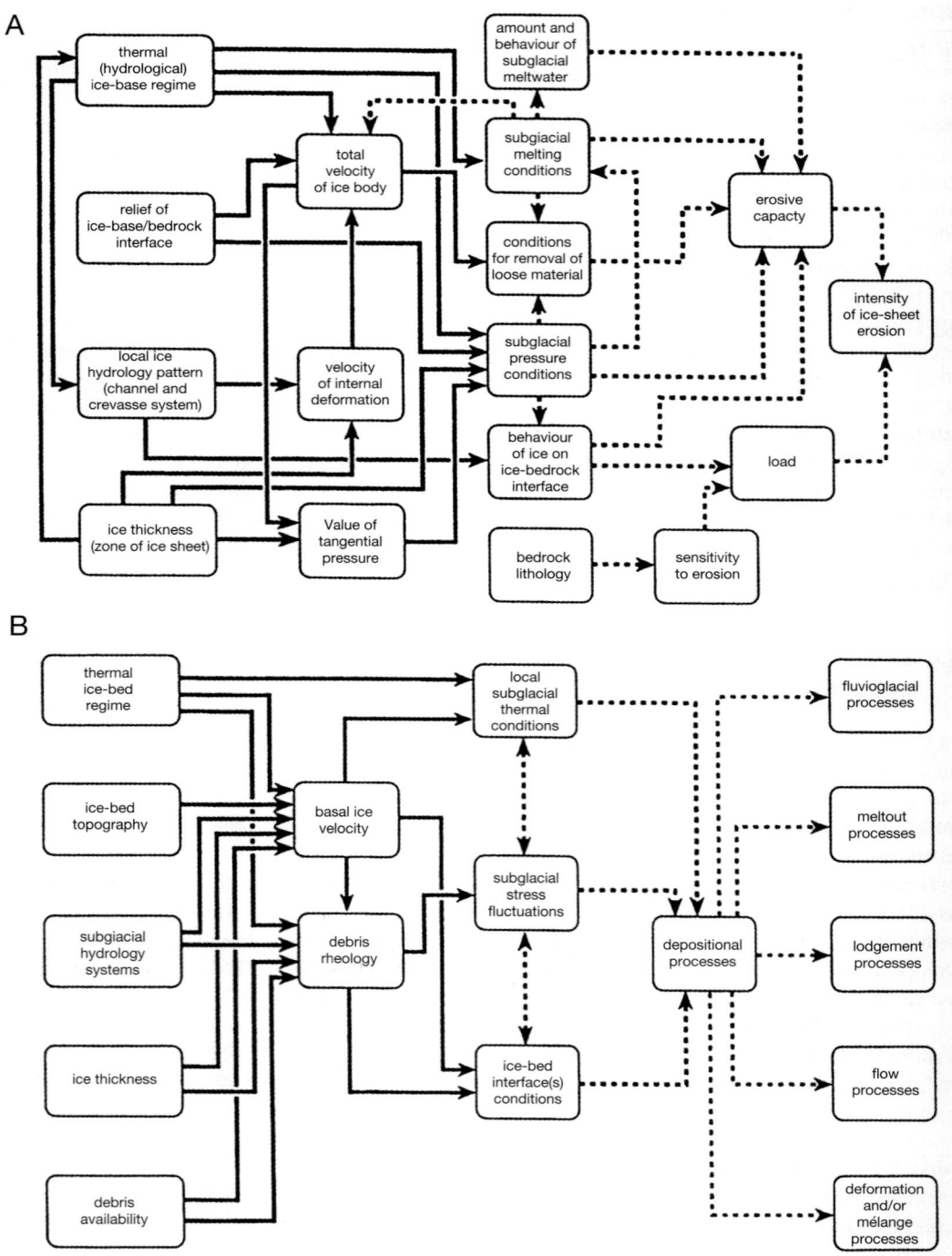

Figure 13.7   Models showing the interrelationship between main parameters influencing subglacial processes and (*A*) erosion and (*B*) deposition (after Menzies and Shilts, 1996).

Table 13.1   Some characteristics and properties of different types of till.

Till type	Genesis	Particle size distribution	Mesofabric	Microfabric	Bulk density	Shear strength
Glaciotectonite	Subglacially sheared sediment and bedrock	Poorly sorted	Moderate	Microshears	Moderate	Low
Comminution	Subglacially crushed and powdered local bedrock	Poorly sorted skewed towards fine	Moderate	Microshears	Moderate	Moderate
Lodgement	Subglacially plastered glacial debris on a rigid or semi-rigid bed	Poorly sorted	Strong up-valley dip	Microshears	High	High
Deformation	Subglacially deformed glacial sediment	Poorly sorted	Strong	Microshears	Moderate to high	Moderate to high
Meltout	Glacial sediment deposited directly from melting ice	Poorly sorted	May be stratified	Microshears	Low	Low
Sublimation	Glacial sediment deposited directly from sublimated ice	Poorly sorted	Preserves ice foliation	Microshears	Low	Low
Flow till*	Sediment deposited off the ice by debris flow processes	Poorly sorted	Downslope	Few	Low	Low

* This is classed as a debris flow rather than a till, but it is included here for comparison.

properties that are essentially a function of the depositional environment. In particular, the plasticity and particle size of tills are critically dependent on the mode of deposition and the nature of the terrain over which the glacier or ice sheet was active (Trenter, 1999). Predicting the vertical and lateral variability of the geotechnical properties is extremely difficult when a glacial deposit consists of several types of till, as may be seen from the study of modern ice margins (Figure 13.8).

13.5 Contemporary glacial environments: landsystem models

Ice sheets

Ice sheets extended over much of the high and mid-latitude regions of North America and Europe during the glaciations of the Quaternary (Figure 13.1a). As a result, characteristic suites of glacial landforms and sediments dominate extensive areas of these landscapes. Studies of the ice sheets in Greenland and Antarctica have provided some understanding of the nature of ice-sheet dynamics, although it is important to recognise that these modern ice masses are not perfectly analogous to the larger Quaternary ice sheets that formerly covered much of the northern hemisphere.

The variability in the environments produced by the Quaternary ice sheets is illustrated in the landsystems and lithofacies models shown in Figures 13.9 to 13.12. Many of these landforms were produced subglacially beneath several kilometres of glacial ice. Subglacial deposition and deformation produce elongated streamlined landforms (Figures 13.9 and 13.10). These range

Table 13.2 Simplified lithofacies codes (adapted from Miall, 1978 and Eyles *et al.*, 1983b).

Code	Description
Diamictons	Very poorly sorted admixture of wide particle size range
Dmm	Matrix-supported, massive
Dcm	Clast-supported, massive
Dcs	Clast-supported stratified
Dms	Matrix-supported stratified
Dml	Matrix-supported laminated
- - - (c)	Evidence of current reworking
- - - (r)	Evidence of resedimentation
- - - (s)	Sheared
Boulders	Particles >256 mm (b-axis)
Bms	Matrix-supported, massive
Bmg	Matrix-supported, graded
Bcm	Clast-supported, massive
Bcg	Matrix-supported, graded
Gravels	Particles of 8–256 mm
Gms	Matrix-supported, massive
Gm	Clast-supported, massive
Gh	Horizontally bedded
Gt	Trough cross-bedded
Gp	Planar cross-bedded
Granules	Particles 2–8 mm
GRcl	Massive with clay laminae
GRch	Massive and infilling channels
GRh	Horizontally bedded
GRt	Trough cross-bedded
GRcu	Upward coarsening
GRfu	Upward fining
Sands	Particles of 0.063–2 mm
St	Medium to very coarse and trough cross-bedded
Sp	Medium to very coarse and planar cross-bedded
Sm	Massive
Sr	Ripple cross-laminated
Sh	Very fine to very coarse horizontally/plane bedded or low angle cross-lamination
Sc	Steeply dipping planar cross-bedding
Sd	Deformed bedding
Suc	Upward coarsening
Suf	Upward fining
Srg	Graded cross-laminations
Silts & clays	Particles of <0.063 mm
Fl	Fine lamination often with minor fine sand and very small ripples
Fm	Massive
Flv	Fine lamination with rhymites or varves
Frg	Graded and climbing ripple cross-laminations

in size from a few metres wide and long (flutes) to tens of metres across and high, and hundreds of metres long (drumlins) to many hundreds of metres across and tens of metres high and kilometres long (megaflutes). Their internal structure and composition can be complex, comprising tills, glaciofluvial and glaciolacustrine sediments. Sinuous ridges (eskers), tens of metres high and many kilometres in length represent subglacial glaciofluvial channels that were infilled with sediments and left behind as the ice sheets melted. The internal sediments and structures within these landforms are highly complex, often including a whole range of lithologies, bedding types and deformation structures. Stagnating ice left during glacial retreat also often gives rise to a suite of complex landforms and sediment types that include hummocky moraines, kettle holes and kame terraces (Figure 13.11). Large proglacial lakes are also associated with ice sheets. Again, the sediments and landforms in these lacustrine environments may be very complex. Large supraglacial lakes may develop, particularly when the ice has retreated into mountainous terrain. A landsystems model for such a supraglacial lake is shown in Figure 13.12. The landforms along former ice margins may also be further complicated by deformation during glacial advances. The advancing glacier may override or push previously deposited sediment so as to form folded and faulted strata. Predicting the internal structure of these landforms is extremely difficult because of their complexity. In the first stages of analysis, therefore, each individual landform should be considered as a unique entity.

Valley glaciers

Valley glacier landsystems are highly variable and may include glaciers with limited debris cover, debris-mantled glaciers, rock glaciers and hanging glaciers (Benn *et al.*, 2003). Boulton and Eyles (1979) provide a model to describe a valley glacier landsystem for a glacier with moderate amounts of debris cover. However, this model cannot be fully applied to valley glaciers having substantial amounts of surface debris. Such debris-mantled glaciers produce large lateral-frontal 'dump' moraines, as well as ice-contact fans and ramps

Figure 13.8 Complex interrelationship between sedimentary facies at the terminus of the 59 km long Batura Glacier, Karakoram Mountains, Northern Pakistan. 'GFL' indicates glaciofluvial and 'GL' glaciolacustrine facies. The scale bars indicate metres. After Li *et al.* (1984).

(Small, 1983; Owen and Derbyshire, 1989; 1993; Owen, 1994). These glacial environments are characterised by the incorporation of ice-marginal, supraglacial, subglacial, proglacial, periglacial and paraglacial sediments and landforms, recording the juxtaposition and migration of very different depositional environments. When a rock glacier becomes progressively buried by debris, it may eventually form a glacial rock glacier comprising a core of deforming ice beneath thick debris (Benn and Evans, 1998). The resultant form is a steep sided tongue-like or series of lobate masses with ridges, furrows of coarse debris that advance downslope as a consequence of internal deformation of the ice at its core. Hanging glaciers occur in steep terrain in which the glacier tongues are perched in hanging fashion high above valley floors. Benn *et al.* (2003) provide models showing the details of glaciated valley landsystems, and highlight the importance of the strong influence of topography on glacier morphology, sediment transport paths and depositional basins, as well as the importance of debris from supraglacial sources in the glacial sediment budget.

Glaciomarine environments

The glaciomarine environment is one of the most complex, involving subglacial, subaerial and marine sedimentation. A schematic figure showing the major glaciomarine enviroments is presented in Figure 13.13. Here, sediment is transferred from supraglacial and subglacial pathways into the marine environment by density currents, mass movements and icebergs. These sediments may be deformed by grounded and advancing ice. Study of glacial processes within glaciomarine environments is clearly difficult because of their inaccessibility. However, the geometry and structure of offshore glaciomarine sediments are somewhat easier to study than terrestrial deposits because geotechnical techniques such as seismic reflection profiling can be more readily applied in the marine realm. Two main glaciomarine settings can be recognised (Hambrey, 1994; Trenter, 1999). The first is within fjords where sediments are supplied from glaciers in the form of ice-contact deposits, glaciofluvial deposits and ice-rafted debris. Clastic and biogenic marine sediments are supplied together with fluvial, mass movement and aeolian deposits from the terrestrial environment. The second setting comprises continental-shelf and open-ocean environments where sedimentation is dominated by grounded ice-margins, floating ice-tongues, ice-shelves and marine processes. Sediments are supplied mainly from subglacial sources when debris is exposed at grounding lines with minor

A

B

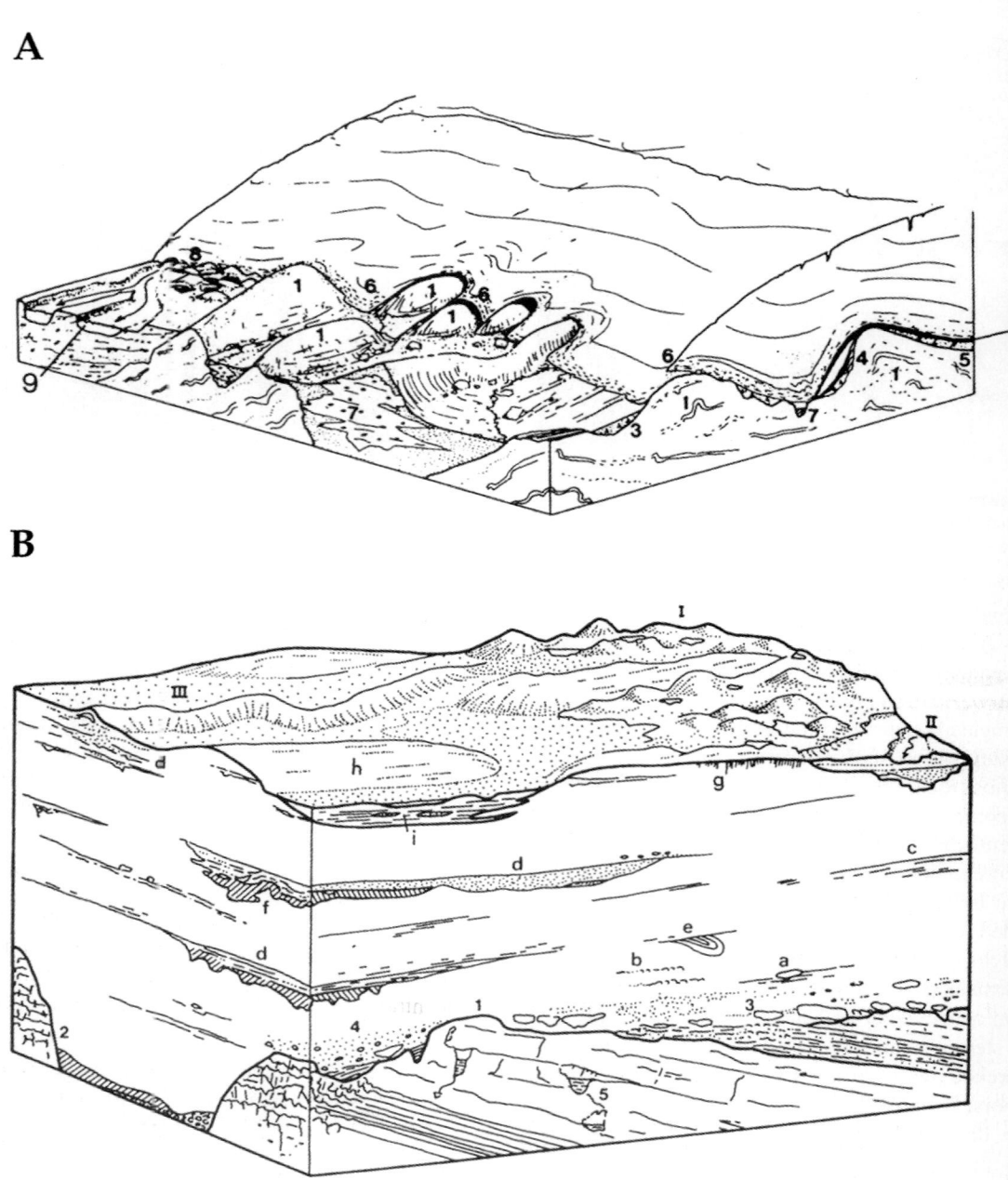

Figure 13.9 (*A*) Subglacial landsystem in an area of hard substrate (after Eyles, 1983). (1) abraded and stream-lined rock knobs; (2) Basal debris; (3) lodgement till on low-relief rock surface; (4) lee-side cavity fill; (5) basal melt-out till; (6) debris melting out at ice surface and dumped by gravity on freshly exposed subglacial surface; (7) subglacial esker with gravel core; (8) hummocky or kettled outwash surface produced by the melt-out of ice buried by outwash fans; (9) the proglacial stream carries subglacial abrasion products. (*B*) The subglacial landsystem in an area with low-relief limestone terrain and where multiple stacked till units have been deposited during a single glaciation (after Eyles, 1983; Benn and Evans, 1998). This subglacial landsystem is reworked by: (I) hummocky kame-and-kettle topography; (II) outwash cut into the till surface and comprising stratified sands and

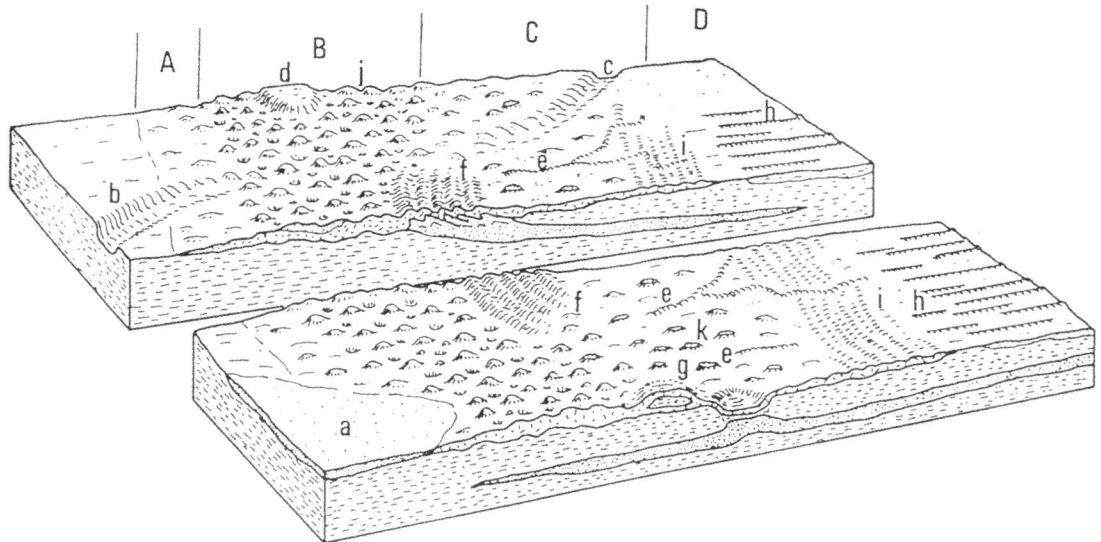

Figure 13.10   Idealised zones of glacial landforms and sediments on the prairies of North Dakota. (*A*) Proglacial suite; (*B*) supraglacial suite; (*C*) transitional (submarginal) suite; (*D*) subglacial suite. (a) Proglacial lake; (b) proglacial meltwater channel; (c) subglacial meltwater channel or tunnel valley; (d) ice-walled lake plain; (e) esker; (f) transverse thrust moraines cupola hills; (g) prairie mound; (h) flutings; (i) transverse recessional moraines; (j) hummocky moraine; (k) isolated kames. (Modified from Clayton and Moran, 1974).

inputs from supraglacial debris and debris trans-ferred from the ice to the marine environment, by way of ice shelves, ice cliffs, outlet glaciers and ice streams. At the terrestrial-marine boundary, a distinct set of ice-marginal landforms and deposits develop that are easily recognisable as glaciogenic in origin. These include stratified diamict (waterlain till), as well as laminated sediments both with and without ice-rafted debris, depending on proximity to the ice front when deposition took place. Beyond

this limit, the sediments are not easily recognisable as glaciogenic in origin and may comprise lami-nated sediment with only occasional dropstones, and debris flow dominated turbidites.

Glaciofluvial

Water drains through the glacial system in three main ways: supraglacially, englacially and subglacially. This glaciofluvial water is usually concentrated within channels that are thermally

gravels; (III) esker deposited during ice wastage and therefore not truncated by subglacial tills like other channel fills in the subglacial landsystem. The base of the subglacial landsystem is characterised by: (1) striated rock head; (2) buried channel/valley with a fill of subglacial sands and gravels and till; (3) glacitectonised rockhead, with rock rafts and boulder pavements; (4) lowermost till, comprising local lithologies, which thickens in the lee sides of rock protuberances as lee-side cavity fills; (5) cold-water karst. The sediments of the subglacial landsystem are charac-terised by: (a) predominantly preferentially aligned, faceted clasts; (b) crude shear lamination produced by the smearing of soft lithologies (deformation till/glaciotectonite); (c) slickensided edding planes (fissility) produced by glacitectonic shear; (d) stratified gravels, sands and clays deposited in subglacial cavities, pipes or canals and trun-cated by overlying till (the base of each till unit may be fluted) — they constitute lenses which are elongated in the direction of ice flow and typically internally disturbed by folding and faulting due to post-depositional deformation by glacier/till overriding; (e) folded and sheared-off channel fill; (f) diapiric intrusion of till squeezed up into sub-glacial cavity; (g) vertical joints produced by post-depositional pedogenic processes; (h) drumlinised surface of upper till sheet; (i) inter-drumlin depressions filled with post-glacial solifluction debris and peat. The horizontal scale may range from 10 m to 10 km and the vertical scale may be 10 cm to 100 m high.

A

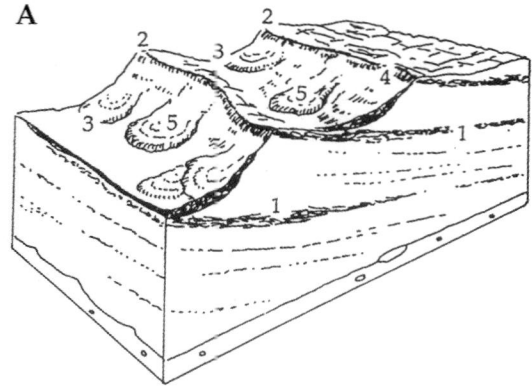

B

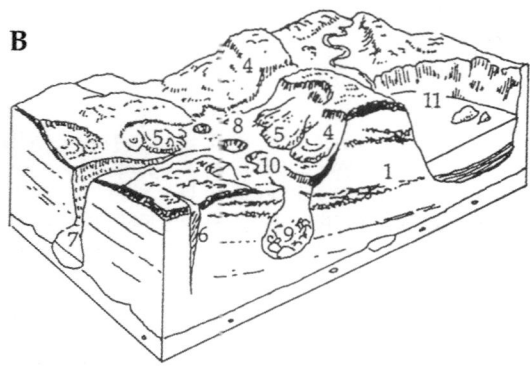

C

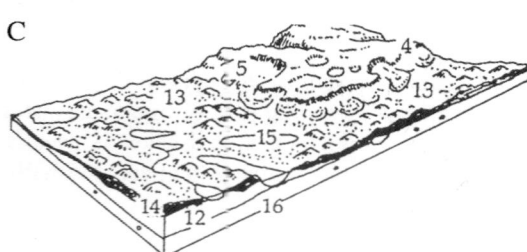

Figure 13.11 Evolution of glacier karst on an ablating debris-mantled glacier. (*A*) Young stage; (*B*) mature stage; (*C*) old stage. (1) Debris bands in glacier ice; (2) ice-cored ridge; (3) depositional trough; (4) back-wasting slope of exposed ice; (5) debris flow; (6) enlargement of crevasse by ice melt; (7) subglacial conduit; (8) sink holes; (9) collapsed tunnel roof; (10) enlargement of sink-hole by ice melt and collapse; (11) lake with back-wasting margins; (12) dead ice; (13) ice-free hummocky terrain; (14) deposits of supraglacial environments; (15) lakes; (16) deposits of subglacial environments. (Redrawn from Krüger, 1994).

eroded into the ice, commonly opening and closing in response to the stresses associated with flowing ice. This makes difficult the prediction of the points where meltwaters will exit from the glacier, a fundamental problem in reconstructing the glacial hydrology. Some subglacial channels, however, may be eroded into basal till or bedrock to form deep channels that may fill with glaciofluvial sediment. The waters that eventually exit the glacier are usually characterised by highly variable discharges that fluctuate diurnally and seasonally in response to variations in the glacier's mass balance. Glaciofluvial waters usually have high sediment loads, typically containing abundant silt, which is transported and held in suspension, as well as boulders and gravel in the traction load. Much of this sediment is carried to the proglacial environment where it is deposited in braided channel systems that may form extensive outwash fans (tens of kilometres in extent), referred to as sandurs. Some glaciofluvial sediment, however, may be deposited within the glacier as metre-size channel fills associated with subglacial tills. Such deformed channel-fill material sometimes forms part of deformation tills. When glaciofluvial sediment accumulates so as to fill large subglacial or englacial ice channels, long (100s to 1000s of metres) sinuous ridges of coarse gravels and boulders with steep cross bedding may be preserved as the ice retreats. These distinctive landforms, known as eskers, are often an important aggregate resource. Glaciofluvial sediments may also be deposited on or along the margins of the ice and they may form impressive terraces when the ice eventually melts. The bedding within these accumulations may be deformed and slumped when ice within or beneath melts out, leading to bowl-shaped, collapse depressions called kettle holes. Numerous lithofacies models have been produced to describe the range of sediment types and associations found in glaciofluvial terrains, a good review being that by Maizels (1996).

Glaciolacustrine

Several different types of lakes are commonly associated with glacial environments, occurring in subglacial, supraglacial and proglacial settings. Supraglacial lakes are often found on glaciers that

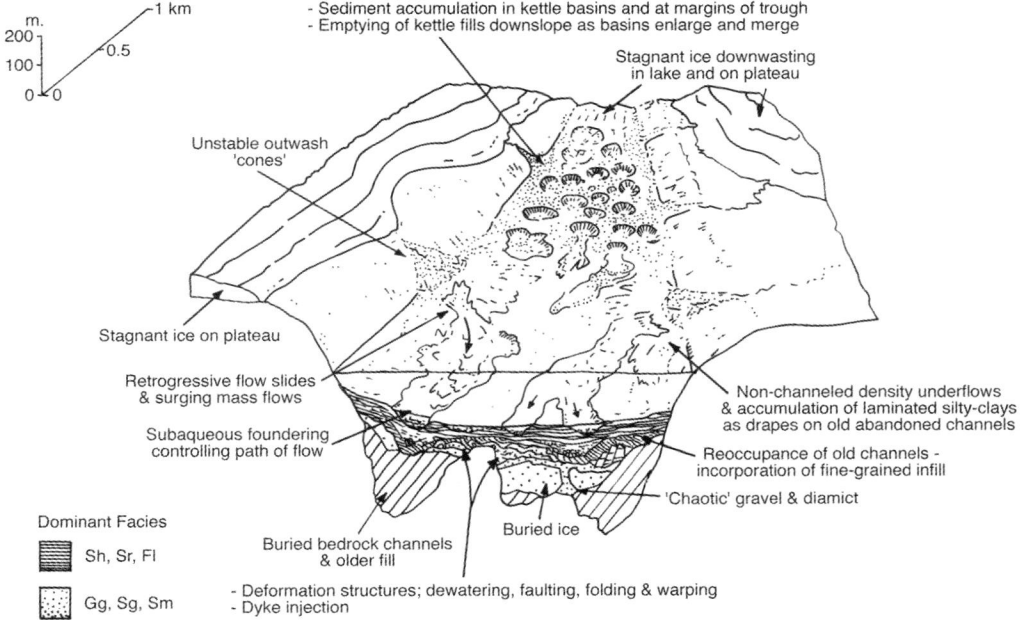

Figure 13.12   Model for sedimentation in a supraglacial lake system in an area of moderate to high relief, based on the stratigraphy of the Fraser River valley, British Columbia, Canada. (From Eyles *et al.*, 1987. Reproduced by permission of Blackwell Ltd.)

are in retreat. These may drain catastrophically as the ice thins and as channels open within the glacier. Subglacial lakes are common under ice sheets, surging glaciers, and where glaciers overlie volcanic centres such as in Iceland. These lakes may also drain catastrophically as new drainage conduits open and lead, through the glacier, to the ice margin. The high discharges result in the resedimentation of large quantities of glaciogenic sediment to form flood deposits in the proglacial environment. Proglacial lakes form within moraines and behind moraine ridges, within ice-marginal valleys, and within bedrock depressions (Figure 13.14). In all these lakes a complex assemblage of sediments may be deposited including fan deltas, turbidites, ice-rafted debris and laminated deposits. Deformation structures are common within many of these sediments, particularly if buried ice or ice adjacent to the lake melts out. This causes slumping and faulting. Furthermore, sediments of contrasting densities, deposited on top of each other, frequently

result in abundant load structures and slumping. All these lakes and the resultant sediments are intimately associated with glaciofluvial systems; in fact, extreme lateral and vertical variation may be observed within glaciolacustrine successions. Ashley (1996) provides a summary of the main environments and lithofacies associated with glaciolacustrine environments.

Glacioaeolian

Strong winds are associated with glacial environments because of the high-pressure systems that develop over glaciated regions. Severe cooling of air above a large valley glacier may give rise to air drainage closely constrained by high and steep valley walls. Such katabatic winds sometimes reach very high velocities and thus are capable of removing surface sands and silts well beyond the glacier snout. Thus, the finer fractions of the abundant debris produced by glacial abrasion and cryogenic (freeze–thaw) weathering contribute to aeolian sedimentation at a range of scales from local sand and

Figure 13.13   Schematic figure illustrating the major environments for glaciomarine sedimentation in a cool temperate climate, based on the present north-north-west Gulf of Alaska (after Powell and Domack, 1996). (1) Tidewater terminus; (2) side drainage with gravel pocket beach and talus fan on fjord floor; (3) bergstone mud on glacial fjord floor; (4) sandur plain in marine outwash fjord; (5) sediment gravity flow channels in marine outwash mud; (6) entrance sill capped with morainal banks and grounding-line fan; (7) continuation of outwash mud; (8) rocky shore with gravel pocket beaches; (9) tectonically uplifted banks of older rocks exposed by winnowing on the shelf; (10) extensive tidal mudflat in estuary; (11) recurved and cuspate spits from alongshore transport; (12) marine outwash mud on the continental shelf; (13) relict sand deposits; (14) thick temperate rain forest of spruce/hemlock with muskeg swamps; (15) raised marine terrace; (16) aeolian dunes fed from spit at wave-dominated delta; (17) large slide/slump areas on the continental shelf; (18) ice-dammed lake with sublacustrine grounding-line fan and lake laminites with iceberg-rafted debris; (19) terminal moraine and gravelly beach; (20) sandur/delta system with small moraines and gravelly beach; (21) modern littoral sand; (22) bank on continental shelf of relict glacier-marginal sediment, currently being winnowed; (23) tidewater terminus just at sea level with short-headed stream, fan deltas; (24) river-dominated estuary with stunted barrier island system offshore; (25) sea valley on continental shelf being infilled with marine outwash mud.

silt drapes on moraines to extensive mantles of regional extent (typically grading with distance from the front of an ice sheet from sands to progressively finer silts known as loess: Chapter 25). Derbyshire and Owen (1996) describe these processes and highlight the polygenetic nature of such glacioaeolian sediments. Three main glacioaeolian settings exist. The first occurs adjacent to contemporary glaciers and within the approximate maximum limit of Pleistocene glaciation. This is characterised by landforms that have been eroded by aeolian processes, have deflated surfaces, and on which are found ventifacts and sand dunes. The second setting is generally beyond, but adjacent to the Pleistocene glacial limit, and is characterised by extensive sheets of sands often referred to as cover-sands. The third setting is often a considerable distance from the Pleistocene glacial limit and is characterised by massive loess deposits. Many of the silts that comprise loess are thought to derive from glacial flour (silts produced by glacial grinding and carried and deposited by glaciofluvial

Figure 13.14   Partially floating tongue of Breidamer-kurjökull, an outlet glacier of the Vatnajökull ice sheet, south-east Iceland, showing large ice-front crevasses, calving, and the resulting icebergs in the proglacial lake Jökulsárlon.

streams) and then blown considerable distances beyond the ice margins. The loess covering the plains of Northern Europe and the northern USA are thought to be largely of this origin. However, the silt making up the loess in some other regions of the world, such as North Africa, may be derived from non-glacial sources. The great loess deposits of northern China, although undoubtedly related to the waxing and waning of the great Pleistocene ice sheets, are a dryland periphery accumulation of complex origin.

13.6 Engineering in glacial environments

From an engineering point of view, it is useful to distinguish between glacierised terrain (those occupied by active glaciers) and glaciated landscapes (those bearing the marks of past glaciation). The problems posed by the presence of glaciers differ in certain obvious ways from those arising from engineering works on glacially eroded rock surfaces and the associated and often diverse range of ancient glacial sediments. In the case of terrain close to the margins of glaciers and ice sheets, the dynamic behaviour of glaciers in response to several forcing factors, notably climatic variations, results in relatively short-term mass motion (glacier advance and retreat), as well

as a suite of hydrological changes arising from movement and storage of glacial meltwaters. In formerly glaciated terrain, factors of major significance include the bulk properties of the principal sedimentary facies (glaciogenic, glacio-fluvial, glaciolacustrine, glaciomarine, glacioa-eolian), the degree of alteration by processes including weathering and mass movements, the hydrological characteristics of the deposits, and the erosional and depositional landforms.

Glacial advance and retreat, shifting meltwater river regimes, and glacial lake outburst floods: hazards and mitigation

Tufnell (1984) regarded the three main types of glacial hazards as expanding glaciers, glacial lake outburst floods, and ice that breaks off from glacier snouts and avalanches into valleys, destroying all artificial objects in its path.

The advance and retreat of glaciers often constitutes a substantial hazard, especially in mountain regions with low, but locally concentrated human populations. Advancing and retreating glaciers may directly cause important landscape changes by re-shaping mountain slopes, depositing large morainic accumulations, and diverting and damming natural drainage. Indirect effects of ice advance and retreat include important changes in the local and regional climate and ecology, which, in turn, have a bearing on hydrology and slope stability. Changes in the position of ice fronts may cause several types of direct and substantial damage to important human structures including highways (Derbyshire et al., 2001), settlements, hydro-electric plants, irrigation channels and fields. Well-known examples include the advance of the Mer de Glace in Chamoix, Switzerland, between 1644 and 1919 (threatening the village of Les Bois) and the advance of the Pasu Glacier in the Hunza valley (Karakoram Mountains of northern Pakistan) that threatened Pasu village in 1913 (Figure 13.15; Tufnell, 1984).

Fluctuating meltwater rivers on glacial forelands can cause considerable damage to constructions such as bridges, roads and cultivated fields. Many historical changes of this type have been documented for the Icelandic glaciers including, for example, the foreland of Hoffellsjökull Glacier

Figure 13.15 The debris covered snout of the Pasu Glacier (upper right centre), Karakoram Mountains of northern Pakistan, in July 1988. The proglacial lake, fed by a subglacial stream (see exit tunnel: right centre) is held back by a complex set of end-moraine ridges (centre) left in 1981 as the glacier retreated. The moraine complex is being eroded by the lake-outlet stream (left centre). The village of Pasu is 2 km downslope of this site.

(Tufnell, 1984). The destructive effect of shifting meltwater channels also occurs on a smaller scale, though often with a major impact on routeways and settlements. For example, the main outlet stream of the Ghulkin Glacier, Karakoram Mountains, breached the Karakoram Highway in the summer of 1980, burying it under a new gravel fan *c.* 0.5 km across (Derbyshire and Miller, 1981; Jones *et al.*, 1983).

Glaciers and their moraines frequently provide temporary meltwater dams in mountain valleys. Two styles of glacier-dammed lakes and their associated breaching mechanisms are shown in

Figure 13.16*a* (Walder and Costa, 1996). Glaciers carrying large volumes of supraglacial debris, such as those in parts of the Himalaya (e.g. Owen *et al.*, 2003), pose a particular threat to human communities because such debris is usually coarse and so may be breached only a short time after formation of the lake, the process sometimes being catastrophic. The Pokhara basin in Nepal was catastrophically flooded by a glacier lake outburst flood through a breached ice-cored moraine dam about 450 years ago, the resulting lake attaining an area of 10 km² (Yamada, 1993). Rapid release of about 5 million m³ of rock debris into the proglacial Damenlahai Lake, Tibet, in 1964 triggered a breach of the end moraine, the resulting flood being estimated at over 3500 million m³; the debris carried by this flood buried twelve houses and over 2 km of the Sichuan–Tibet highway (Derbyshire, 1990). Richardson and Reynolds (2000) have recently provided a schematic diagram of a hazardous moraine-dammed glacial lake based on examples from Bhutan (Figure 13.16*b*; Table 13.3). Glaciers with a thick debris cover often develop supraglacial ponds and small lakes (Figure 13.17); in exposing bare ice cliffs by enhanced melting, these supraglacial water bodies coalesce, leading to rapid growth of ice marginal lakes (Kirkbride, 1993; Benn *et al.*, 2000). Large glaciers not in equilibrium with the present climate may constitute a heightened risk if the current warming trend continues (Naito, 2000). Methods of hazard assessment relevant to such situations have been applied relatively rarely in major mountain ranges such as the Himalaya, however. Mitigation measures, including progressive lowering of lake levels by siphoning, have met with some success in Nepal and Peru (Reynolds, 1998).

Building on glacial deposits: foundation problems both onshore and offshore

Any construction on glacial deposits requires a consideration of the geotechnical properties, hydrology and variability of the sediments. This is particularly important for large structures such as dams, oil and gas platforms and industrial complexes. Derbyshire (1992) provided examples of some of the geomorphological and engineering considerations encountered during the site

A

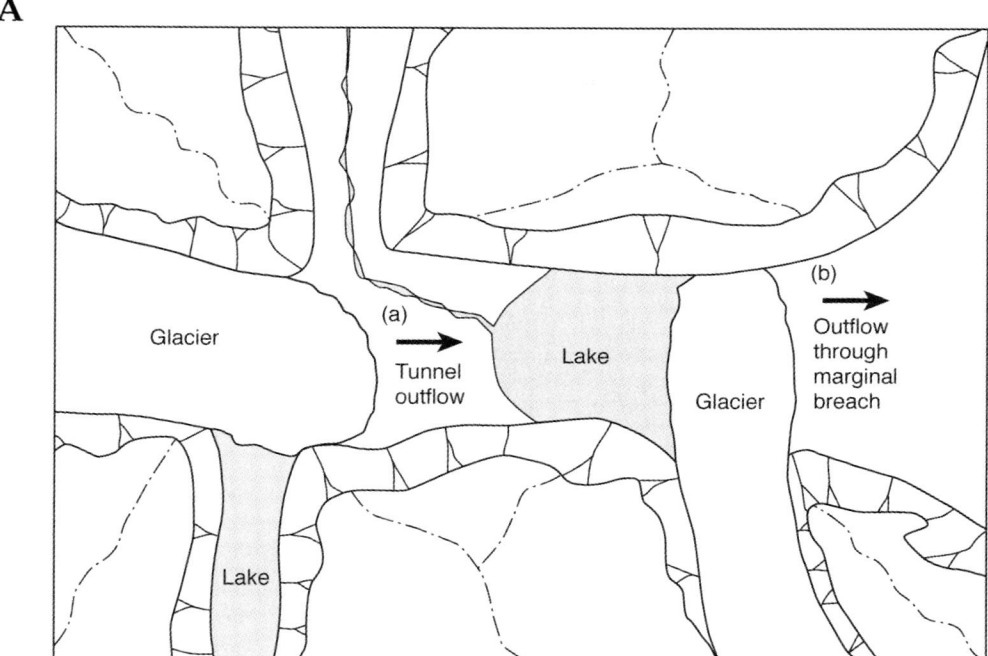

Figure 13.16   (*A*) Two styles of ice-dammed lakes and associated breach mechanisms (after Walder and Costa, 1996).

B

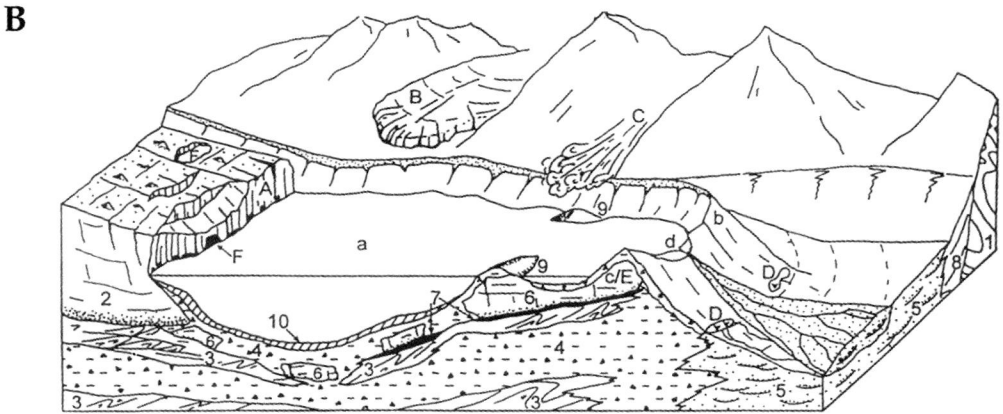

Figure 13.16   (*B*) Schematic diagram of a hazardous moraine-dammed glacial lake (after Richardson and Reynolds, 2000). The geologic successions include (1) bedrock; (2) receding glaciers; (3) stacked subglacial tills and (4) supraglacial tills representing multiple glacial advances; (5) complex inter-bedded glaciofluvial sediments, supraglacial tills and gravity flow diamicts in terminal and lateral moraine ridges; (6) stagnant glacier ice with; (7) basal meltout till; (8) valley-side fan deposits; (9) hummocky moraine resulting from melt of ice-cores (thermokarst), and (10) lake sediments. Factors contributing to the hazard risk include (a) large lake volume, (b) narrow and high moraine dam, (c) stagnant glacier ice within the dam, (d) limited freeboard between the lake level and crest of the moraine ridge. Potential outburst flood tiggers include avalanche displacement waves from (A) calving glaciers (B) hanging glaciers, and (C) rock falls; (D) settlement or piping within the dam (due to progressive seepage or seismic activity); (E) melting ice-core; (F) catastrophic glacial drainage into the lake from sub and englacial channels or supraglacial lakes.

Table 13.3   Types of glacial and glacially related hazards (adapted from Richardson and Reynolds, 2000).

Category	Hazard event	Description	Time scale
Glacier hazards	Avalanche	Slide or fall of large mass of snow, ice and/or rock	Minutes
	Glacier outburst	Catastrophic discharge of water under pressure from a glacier	Hours
	Jökulhlaup	Glacier outburst associated with sub-glacial volcanic activity	Hours–days
	Glacier surge	Rapid increase in rate of glacier flow	Months–years
	Glacier fluctuations	Variations in ice front positions due to climatic change, etc.	Years–decades
Glacier hazards as above, plus:	Glacier lake outburst Floods (GLOFs)	Catastrophic outburst from a proglacial lake, typically moraine dammed	Hours
	Débâcle	Outburst from a proglacial lake (French)	Hours
Related hazards	Lahars	Catastrophic debris flow associated with volcanic activity and snow fields	Hours
	Water resource problems	Water supply shortages, particularly during low flow conditions, associated with wasting glaciers and climate change etc.	Decades

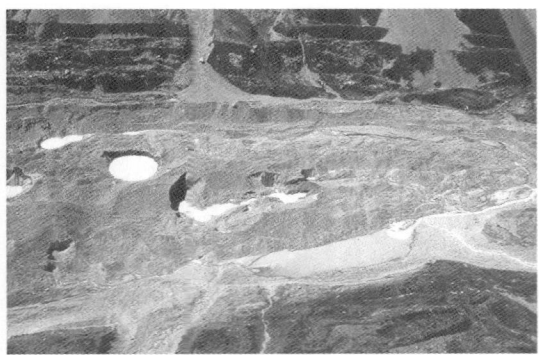

Figure 13.17   Aerial view northwards across the terminal zone of the Hooker Glacier, the south island of New Zealand, in 1974. The debris-covered glacier shows evidence of a complex hydrology. This includes several supraglacial lakes (each sealed off from the next, as indicated by the range of water levels); ice-contact (proglacial) lake (foreground); supraglacial streams, englacial streams (lower left of photograph); subglacial and proglacial streams (right of photograph). Parallel lateral moraine series on both side slopes of the glacier (top and bottom of photograph) show varying degrees of eroded incision with associated debris cone/fan development, one fan constraining the limits of the proglacial lake (lower right of photograph).

exploration stages of the Brenig dam site on the drumlins in the Denbigh Moors in North Wales and the Kielder dam site on tills along the North Tyne River, in northern England. He emphasised the variability of sediment types and geotechnical properties within the proposed dam sites. Furthermore, he showed that microstructure of tills may greatly influence the mechanical properties and loading characteristics (e.g. Boulton and Paul 1976; McGown and Derbyshire 1977).

The geotechnical properties of glacial sediments are usually a function of the depositional environment (Table 13.1). Farrell et al. (1995) showed that weathering may also alter the geotechnical properties of tills (Figure 13.18). In particular, weathering increases the fines content and the plasticity of till, reducing its density and shear strength. The typical ranges of geotechnical properties for different tills are illustrated in Tables 13.4 and 13.5 and Figure 13.19. Hird et al. (1991) and Lehane and Faulkner (1998) provide useful studies of the stiffness and strength characteristics of tills.

Depositional environment and pre- and post-depositional process history are important

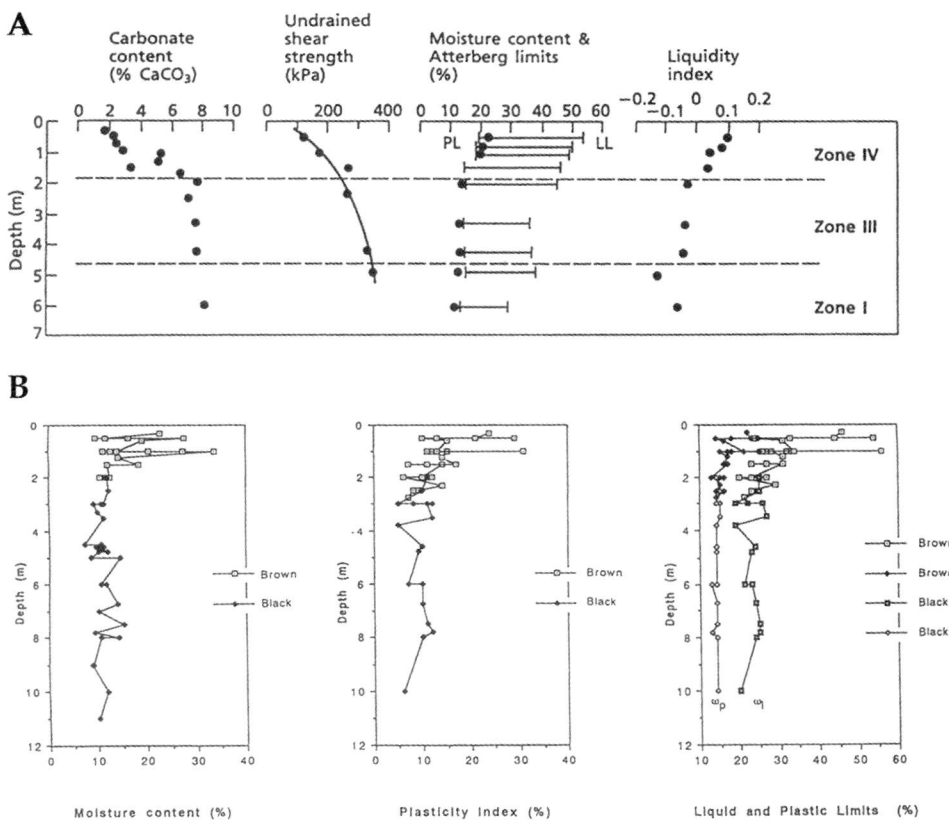

Figure 13.18   Variation in the geotechnical properties through (*A*) a lodgement till from Northumberland, England (after Eyles and Sladen, 1981) and (*B*) a weathered till (after Farrell *et al.*, 1995). See Table 13.5 for details of the weathering zones.

influences on the geotechnical properties of glacial sediments. Many glacially deposited sediment types have a multimodal or bimodal grain size distribution that includes fines (notably silts but often with some clay grade material: Figure 13.6). These fines, whether making up the matrix of tills or as relatively uniform fine-grained beds such as glaciolacustrine deposits, transfer stresses applied by glacial and other sources of loading, a process often intricately preserved in their microfabric properties. Examples of some common microfabric types are shown in Figures 13.20, 13.21 and 13.22.

Trenter (1999) emphasised several problems encountered when excavating in till for an engineering development or construction project.

These include: the misidentification of rockhead; the presence of large boulders; the presence of glaciolacustrine laminated silty clays; water-bearing silts and sands and water-bearing bedrock. These features may increase the risk of slope failure and subsidence and they can often substantially increase the cost of protective measures to assure the safety and longevity of the construction or project being undertaken. Furthermore, Trenter (1999) highlighted the following characteristics of glacial sediments that influence both foundation design and construction:

(1) spatial variations in soil type, exhibited as changes in strength, compressibility and consolidation properties

Table 13.4　Strength of tills of north Norfolk and Holderness (UK) (from Bell, 1993).

	Unconfined compressive strength (kPa)		
	Intact	Remoulded	Sensitivity
North Norfolk			
1. Hunstanton till (Holkham)			
Max.	184	164	1.22 (L)
Min.	152	128	1.18 (L)
Mean	158	134	1.19 (L)
2. Marly drift (Weybourne)			
Max.	120	94	1.49 (L)
Min.	104	70	1.28 (L)
Mean	110	81	1.34 (L)
3. Contorted drift (Trimingham)			
Max.	180	168	1.67 (L)
Min.	124	76	1.08 (L)
Mean	160	136	1.23 (L)
4. Cromer till (Happisburgh)			
Max.	224	188	1.19 (L)
Min.	154	140	1.10 (L)
Mean	176	156	1.13 (L)
Holderness			
1. Hessle till (Dimlington, Hornsea)			
Max.	138	116	1.31 (L)
Min.	96	74	1.10 (L)
Mean	106	96	1.19 (L)
2. Withernsea till (Dimlington)			
Max.	172	148	1.18 (L)
Min.	140	122	1.15 (L)
Mean	160	136	1.16 (L)
3. Skipsea till (Dimlington)			
Max.	194	168	1.15 (L)
Min.	182	154	1.08 (L)
Mean	186	164	1.13 (L)
4. Basement till (Dimlington)			
Max.	212	168	1.27 (L)
Min.	163	140	1.19 (L)
Mean	186	156	1.21 (L)

L = low sensitivity.

(2) variation in depth to rockhead, with consequential variation in thickness of overlying compressible soils
(3) the presence of groundwater, the spatial variation in sediment type, and the variation to rockhead (and hence compressible thickness).

Table 13.5　Typical geotechnical properties for Northumberland (UK) lodgement tills (after Eyles and Sladen, 1981). See Figure 13.8b.

Property	Weathering Zones	
	I	III & IV
Bulk density (kg/m³)	2150–2300	1900–2200
Natural moisture content (%)	10–15	12–25
Liquid limit (%)	25–40	35–60
Plastic limit (%)	12–20	15–25
Plasticity index	0–20	15–40
Liquidity index	−0.20 to −0.05	III −0.15 to +0.05 IV 0 to +30
Grading of fine (< 2 mm) fraction		
% clay	20–35	30–50
% silt	30–40	30–50
% sand	30–50	10–25
Average activity	0.64	0.68
c (kPa)	0–15	0–25
Φ (degrees)	32–37	27–35
Φ_r (degrees)	30–32	15–32

c = cohesion.
Φ = angle of internal friction.
Φ_r = angle of internal friction (residual).

These difficulties arise from:

(1) the presence of several different sediment types in the tunnel face
(2) variations in the thickness of sediments and the depth to rockhead
(3) the possibility that nests of cobbles and boulders may be present
(4) varying groundwater conditions, including artesian pressures.

When undertaking any substantial engineering project in glaciated or formerly glaciated terrain, it is advisable to undertake a comprehensive site investigation. Boone *et al.* (1998) emphasised that one of the major problems in underground excavations and piling in tills is the prediction of the likely size and frequency of the occurrence of boulders on the basis of the very limited sampling of the ground that takes place in a conventional site investigation. Preliminary investigations should,

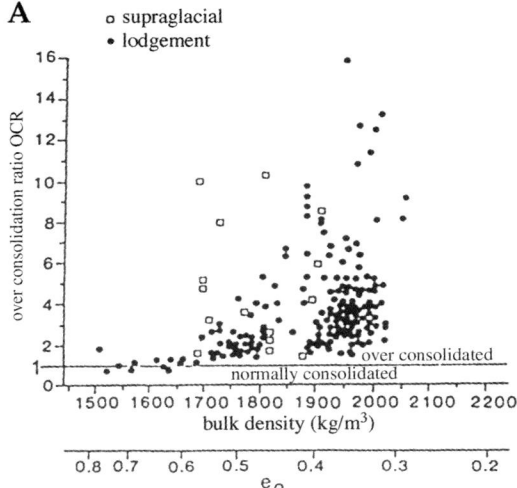

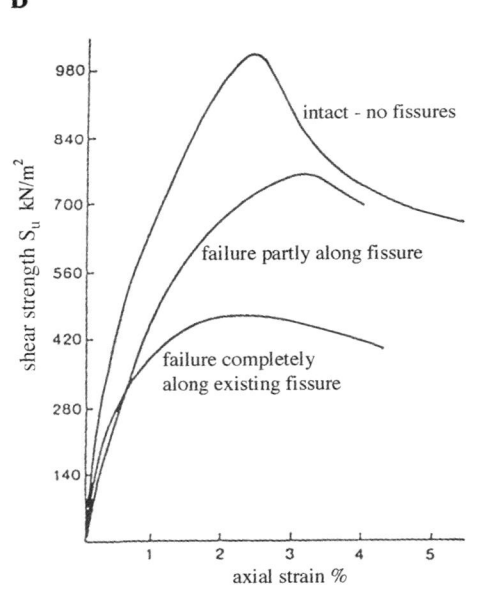

Figure 13.19 (*A*) Some bulk properties of subglacial and supraglacial tills (after Lutenegger *et al.*, 1983). (*B*) Effects of fissuring on shear strength (after McGown *et al.*, 1974).

therefore, include strata definition to characterise the vertical and lateral variation in sediment type. This should include geomorphic mapping the excavation of trial pits or the examination of natu-

ral exposures, and cable percussion or rotary core boring methods. The different sedimentary units should be sampled once the variability of sediment types has been determined. Sampling methods include tube sampling, rotary methods and bulk and block sampling. The extracted samples can be used to determine the geotechnical properties of the sediments including bulk density, moisture content, particle size distribution, liquid and plastic limits, compressive strength, coefficient of consolidation, and shear strength. More comprehensive data profiles include the addition of *in situ* tests, which are particularly valuable because they are more representative of true site conditions. They may include standard and cone penetration tests, permeability tests and pressure-metre tests.

Fookes (1997) has stressed the importance of using systematic site investigation to construct landsystem models. Not only does this provide a means of minimising uncertainties concerning ground conditions in engineering sites, it is likely to yield improved basic data and enhanced design quality. Given the state of knowledge of glacial landsystems, landsystem models are capable of further development, according to Fookes, by organising information on a site as a geological environment matrix, consisting of tabulated geological data susceptible to computer manipulation from which to derive predictive models of ground conditions. However, a number of steps have yet to be taken to render such data readily applicable as part of a standard approach.

Water resources and pollution in glacial sediments

An understanding of the nature of Quaternary sediments is important for assessing the potential for groundwater resources and predicting the nature of groundwater pollution. Most aquifers within Quaternary glacial deposits occur within proglacial sands and gravels. The geometry of these aquifers is a function of the depositional setting and the landforms. Buried glaciofluvial channels and eskers, for example, form substantial elongated aquifers. The lithology and structure of the glacial deposit affects its ability to transmit and store water and pollutants. Daly (1992) provided a useful classification of pollution vulnerability rating for

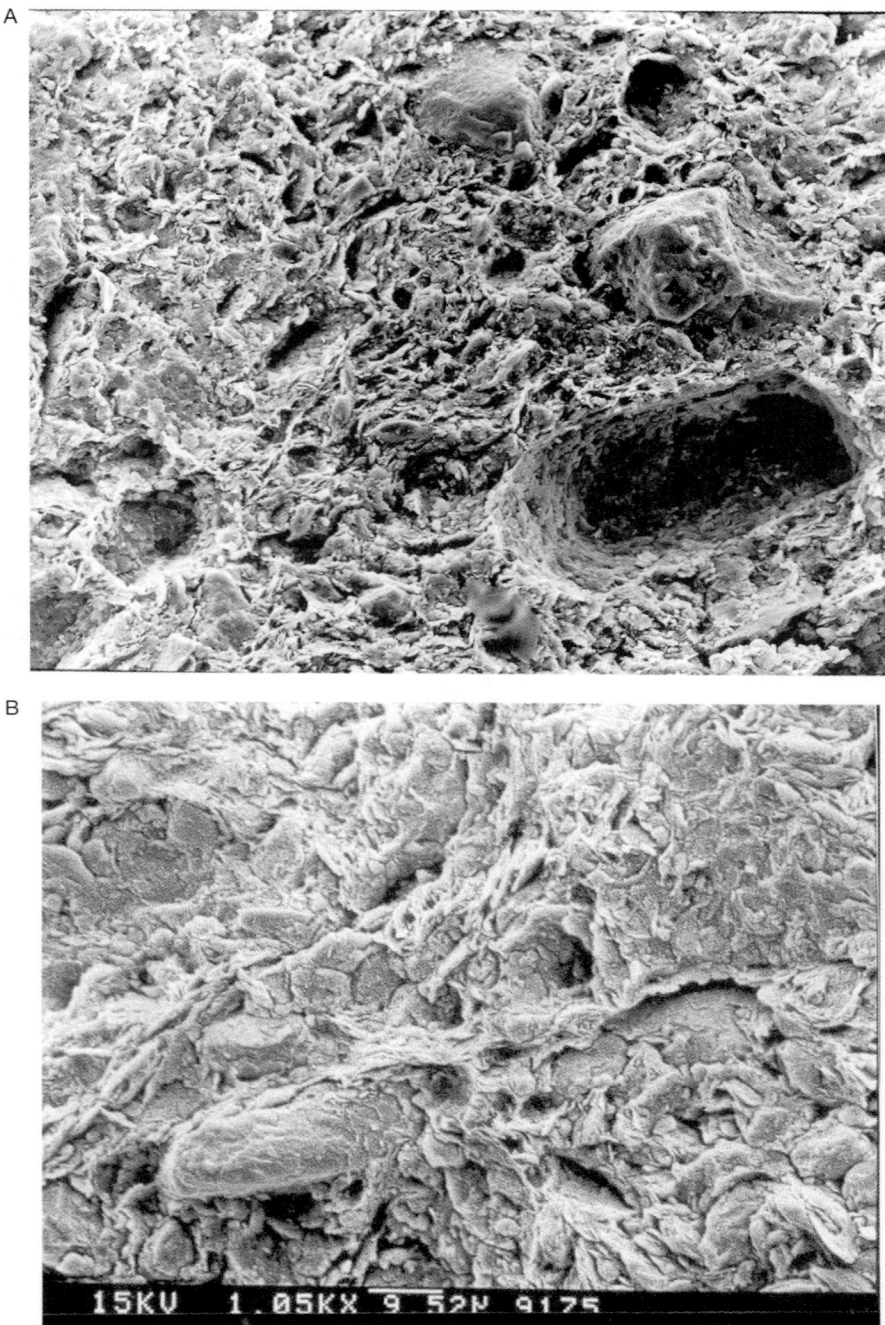

Figure 13.20   (*A*) Mildly over-consolidated clay-rich Upper Pleistocene lodgement till from Cowden, Holderness, north-east England, showing compact fabric (bulk density *c*. 2100 kg/m³). Ice moved from left to right. Field of view is 300 μm wide. (*B*) Over-consolidated clay-rich till of the penultimate Pleistocene glaciation at Boulton Moor, near Derby, northern England. The dense matrix fabric is cut by shear systems, rising from left to right (in the same sense as the glacier thrust).

Figure 13.21 (*A*) Pleistocene subglacial meltout till of the penultimate European glaciation, from Pouilleux, Rhône valley, south-east France. Relatively high voids ratio is evident from the loose fabric of this coarse-grained matrix. The coarser silts and sand grains are often coated with fine clays as a result of meltwater eluviation. (*B*) Modern meltout till from beneath the margin of Blåisen, southern Norway, released from the glacier in 1965. The loose and fragile fabric is a product of observed meltwater eluviation. Some of the silt grains show a capping of very fine clay grade material, attributed to the same process. Scale bar is *c.* 50 μm.

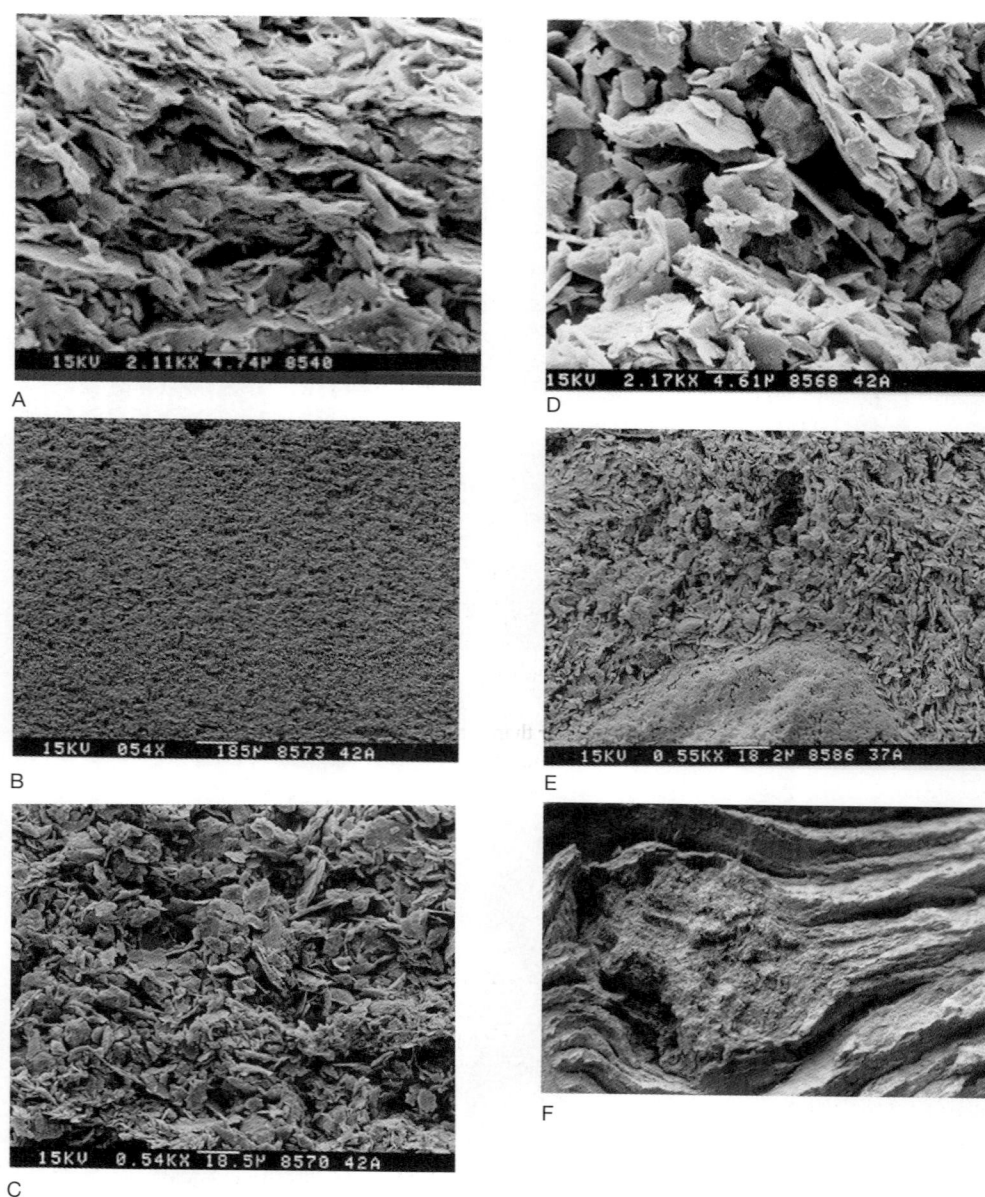

Figure 13.22 Micrographs of two British Pleistocene glaciolacustrine sediments. (*A*) Sub-horizontal fabric in silts rich in micas from a depth of *c.* 47 m within a finely laminated deposit. (*B*) Very uniform grain size of apparently massive silts with few, barely perceptible laminae: depth 42 m. (*C*) and (*D*) Enlarged views of silts shown in *B*, revealing an intimate mixture of medium and fine silts, scattered throughout. Fairly compact fabric, with variable grain anisotropy owing to abundant clusters of edge-to-face fabric elements in the medium silts. (*E*) Sand grain 'dropstone', showing evidence of disruption of the silt fabric (indicated by vertically aligned fabric zone above and to the right of the sand grain). Width of photograph is 1 cm. (*F*) Dropstone consisting of a small pod of till matrix, disrupting the laminae in ice-proximal lacustrine silts: depth 12 m.

Micrographs *A–E* are from a core section from Pentre, Shropshire, England. Micrograph *F* is from a core at Jarrow, northern England.

Table 13.6   Vulnerability rating for typical Irish geohydrological settings (after Daly, 1992).

Vulnerability Rating	Geohydrological setting
Extreme	1. Outcropping bedrock aquifers (particularly karst limestone) or where overlain by shallow (< 3 m*) subsoil 2. Sand and gravel aquifers with a shallow (< 3 m*) unsaturated zone 3. Areas near karst features such as sink holes
High	1. Bedrock — major, minor and poor aquifers and non-aquifers — overlain by 3 m + sand and gravel or 3–10 m sandy till or 3–5 m low permeability clayey tills or clays 2. Unconfined sand and gravel aquifer with 3 m+ unsaturated zone
Moderate	1. Bedrock — major, minor and poor aquifers and non-aquifers — overlain by 10 m + sandy till or 5–10 m clayey till, clay or peat 2. Sand and gravel aquifers overlain by 10 m + sandy till or 5–10 m clayey till, clay or peat
Low	1. Confined bedrock aquifers overlain by 10 m + clayey till or clay or low permeability bedrock such as shale 2. Non-aquifers and poor aquifers overlain by 10 m + clayey till or clay 3. Confined gravel aquifers overlain by 10 m + clayey till or clay

* Note: less than 1 m subsoil or unsaturated zone beneath a development rather than 3 m could be the cut-off depth for the 'extreme' rating. However, taking a thickness of 3 m rather than 1 m is regarded as more practical and useful for the following reasons:

1. The base of many developments — septic tank systems or farmyard effluence holding tanks for instance — are 1–3 m below ground level.
2. In preparing a vulnerability map the general rather than the site-specific situation must be taken into account.
3. A 3 m cut-off depth allows for lateral variations and often provides a safety margin. Obviously if the base of a potentially polluting development is >3 m deep, the rating classification may be affected.

Quaternary sediments in typical Irish hydrological settings that can be applied to other glaciated environments (Table 13.6). He emphasised the need for detailed, sedimentological and geotechnical studies of Quaternary sediments in arriving at a comprehensive assessment of the nature of groundwater resources and vulnerability to pollution.

Aggregate exploitation

Glacial and proglacial sediments can provide the engineer with a useful source of aggregates. The highly variable nature of these sediments offers an opportunity to exploit a variety of sediment types. An understanding of the landform and sediment associations is useful for efficiently exploiting these resources, including predicting the likely extent, geometry and distribution of the sediment. Particularly important aggregate resources include eskers, sandurs (outwash plains) and kames (dominated by ice contact glaciofluvial and glaciolacustrine sediments). The landsystems and lithofacies models described above, therefore, provide a working framework designed to aid in the efficient and effective exploration and exploitation of glacial sediments for aggregate.

Mass movements

Mass movement is particularly common in glaciated terrain because of the abundance of unconsolidated sediment and the generally steep slopes. Even in glaciated landscapes where slopes are gentle, such as the glaciated plains of North America, mass movement is common particularly because sensitive clays (a common component of sediments such as those deposited in glaciolacustrine environments) frequently play a major role in initiating landsliding. In a study of 8835 inland landslides in the UK, 20% were found to lie in tills, with most (32%) being complex (involving more than one mechanism), followed closely behind by debris flows (28%) and planar slides (26%) (DoE, 1994; Trenter, 1999). In glaciated terrain, mass movements may be both natural and human-induced (Vaughan and Walbancke, 1975; Chandler, 1984). Natural mass movements may occur as the sediments and landforms are forming or after glaciation as the landscape readjusts to the new environmental conditions. Eisbacher and Clague (1984) provided a useful overview of glacier-related mass movements in high mountains and suggested means of mitigating these hazards. Human-induced mass movements are particularly common when excavations are made in glacial sediments, as occurs during highway construction, for example. In a study of the hazards along the Karakoram Highway in Northern Pakistan, Derbyshire *et al.* (2001) have recently illustrated the variability of mass movements in glaciated terrain.

Trenter (1999) suggests that remediation strategies for landslides in glaciated terrains should include adequate drainage and the modification of slope profiles, as well as retaining and restraining structures. He stresses that the methods available for use in glaciated terrain do not differ in principle from those available for application to landslides in non-glaciated regions. However, when selecting an approach on a glaciated site, due account should be taken of the spatial variation in the substrate and the depth to rockhead, the presence of glaciofluvial and glaciolacustrine deposits, and other local factors such as climate, vegetation and land use.

13.7 Conclusions: recommendations to engineers working in glacial or glaciated environments

When undertaking any engineering activity on glaciated or formerly glaciated terrain it is important to understand the complex distribution and diverse properties of glacial sediments and landforms. The above descriptions of glacial environments and sediments illustrate some of the typical characteristics and the range of variability found within glacial terrain. However, predicting the nature of the substrate and processes within such landscapes is notoriously difficult. Thus, it is advisable to consider each engineering site within

glacial terrain as unique, at least in the earlier stages of site investigation. Nevertheless, a basic knowledge of the types and sediments associated with each of the landsystems described can provide the investigator with a valuable framework with which to build up an understanding of the reasons for the considerable diversity found in so many glacial terrains.

References

Alley, R. B. (1992) Flow-law hypotheses for ice-sheet modeling. *Journal of Glaciology* **35**, 108–139.

Alley, R. B., Cuffey, K. M., Evenson, E. B., Strasser, J. C., Lawson, D. E. and Larson, G. J. (1997) How glaciers entrain and transport basal sediment: physical constraints. *Quaternary Science Reviews* **16**, 1017–1038.

Ashley, G. M. (1996) Glaciolacustrine environments. In Menzies, J. (ed.) *Modern Glacial Environments: Processes, Dynamics and Sediments*, Vol. 1. Butterworth-Heinemann, Oxford, 417–444.

Ballantyne, C. K. and Benn, D. I. (1996) Paraglacial slope adjustment during recent deglaciation: implications for slope evolution in formerly glaciated terrain. In Brooks, S. and Anderson, M. G. (eds) *Advances in Hillslope Processes*. Wiley, Chichester, 1173–1195.

Banham, P. H. (1977) Glacitectonites in till stratigraphy. *Boreas* **6**, 101–105.

Bell, F. G. (1993) *Engineering Geology*. Blackwell Scientific Publications, Oxford, 359pp.

Benn, D. I. and Ballantyne, C. K. (1993) The description and representation of particle shape. *Earth Surface Processes and Landforms* **18**, 665–672.

Benn, D. I. and Ballantyne, C. K. (1994) Reconstructing the transport history of glacigenic sediments: a new approach based on the co-variance of clast shape indices. *Sedimentary Geology* **91**, 215–227.

Benn, D. I. and Evans, D. J. A. (1998) *Glaciers and Glaciation*. Arnold, London, 734pp.

Benn, D. I., Wiseman, S. and Warren, C. R. (2000) Rapid growth of a supraglacial lake, Ngozumpa Glacier, Khumbu Himal, Nepal. In Debris-covered glaciers. IAHS (International Association of Hydrological Sciences) Publication 264: 177–185.

Benn, D. I., Kirkbride, M. P., Owen, L. A. and Brazier, V. (2003) Glaciated valley landsystems. In Evans, D. J. (ed.) *Glacial Landsystems*. Arnold, London, 372–406.

Billard, A. and Derbyshire, E. (1985) Pleistocene stratigraphy and morphogenesis of La Dombes: an alternative hypothesis. *Bulletin de l'Association francaise pour l'etude du Quaternaire* **22**, 85–96

Bjerrum, L. and Jorstad, F. (1968) Stability of rock slopes in Norway. *Norwegian Geotechnical Institute Publication* **79**, 1–11.

Boone, S. J., Westland, J., Busbridge, J. R. and Garrod, B. (1998) Prediction of boulder obstructions. Negro Jr and Ferreira (eds) *Tunnels and Metropolises.* Balkema, Rotterdam, Vol. 2, 817–822.

Boulton, G. S. and Eyles, N. (1979) Sedimentation by valley glaciers: a model and genetic classification. In Schluchter, C. (ed.) *Moraines and Varves.* Balkema, Rotterdam, 11–23.

Boulton, G. S. and Hindmarsh, R. C. A. (1987) Sediment deformation beneath glaciers: rheology and sedimentological consequences. *Journal of Geophysical Research* **92**, (B9), 9059–9082.

Boulton, G. S. and Jones, A. S. (1979) Stability of temperate ice caps and ice sheets resting on beds of deformable sediment. *Journal of Glaciology* **24**, 29–43.

Boulton, G. S. and Paul, M. A. (1976) The influence of genetic processes on some geotechnical properties of glacial tills. *Quarterly Journal of Engineering Geology* **9**, 159–194.

Broecker, W. S. and Denton, G. H. (1990) What drives glacial cycles? *Scientific American* **January**, 43–50.

Carter, T. G. (1983) *Site investigation and engineering characterization of glacial and glaciolacustrine materials.* Unpublished PhD, University of Surrey.

Chandler, R. J. (1984) Recent European experience of landslides in over-consolidated clays and soft rocks. *International Symposium on Landslides* **4**, Toronto, Sept. 1984. Vol. 1.

Clayton, L. and Moran, S. R. (1974) A glacial process-form model. In Coates, D. R. (ed.) *Glacial Geomorphology.* State University of New York, USA, 89–119.

Daly, D. (1992) Quaternary deposits and groundwater pollution. *Quaternary Proceedings* **2**, 79–89.

Department of the Environment (DoE) (1994) *Landsliding in Great Britain.* Her Majesty's Stationery Office, London.

Derbyshire, E. (1990) Environment: understanding and transforming the physical environment. In Cannon, T. and Jenkins, A. (eds) *The Geography of Contemporary China.* Routledge, London and New York, 80–101.

Derbyshire, E. (1992) Engineering in Quaternary sediments: case studies from Western Europe and Eastern Asia. *Quaternary International* **2**, 33–48.

Derbyshire, E. and Miller, K. J. (1981) Highway beneath the Ghulkin. *Geographical Magazine* **53**, 626–635.

Derbyshire, E. and Owen, L. A. (1996) Glacioaeolian processes, sediments and landforms. In Menzies, J. (ed.) *Modern Glacial Environments: sediments Forms and Techniques*, Vol. 2. Butterworth-Heinemann, Oxford, 213–238.

Derbyshire, E., Fort, M. and Owen, L. A. (2001) Geomorphological hazards along the Karakoram Highway: Khunjerab Pass to the Gilgit River, Northernmost Pakistan. Erdkunde, Band **55**, 49–71.

Derbyshire, E., Gregory, K. J. and Hails, J. R. (1979) *Geomorphological Processes.* Dawson, Westview, 312pp.

Dreimanis, A. (1989) Tills, their genetic terminology and classification. In Goldthwait, R. P. and Matsch, C. L. (eds) *Genetic Classification of Glacigenic Deposits.* Balkema, Rotterdam, 17–84.

Dreimanis, A. and Vagners, U. J. (1971) Biomodal distribution of rock and mineral fragments in basal tills. In Goldthwait, R. P. (ed.) *Till: A Symposium.* Ohio State University Press, Columbus, 237–250.

Echelmeyer, K. and Wang, Z. (1987) Direct observation of basal sliding and deformation of basal drift at sub-freezing temperatures. *Journal of Glaciology* **33**, 83–98.

Eisbacher, G. H. and Clague, J. J. (1984) Destructive mass movements in high mountains, hazard and management. *Geological Survey of Canada* Paper 84–16.

Elson, J. A. (1989) Comment on glacitectonite, deformation till, and comminution till. In: Goldthwait, R. P. and Matsch, C. L. (eds) *Genetic Classification of Glacigenic Deposits.* Balkema, Rotterdam, 85–88.

Evans, D. J. (2004) *Glacial Landsystems.* Arnold, London, 535pp.

Eyles, N. (1983) *Glacial Geology.* Pergamon, Oxford, 409pp.

Eyles, N. and Dearman, W. R. (1981) A glacial terrain map of Britain for engineering purposes. *Bulletin of the International Association of Engineering Geology* **24**, 173–184.

Eyles, N. and Sladen, J. A. (1981) Stratigraphy and geotechnical properties of weathered lodgement till in Northumberland, England. *Quarterly Journal of Engineering Geology* **14**, 129–142.

Eyles, N., Dearman, W. R. and Douglas, T. D. (1983a) The distribution of glacial landsystems in Britain and North America. In Eyles, N. (ed.) *Glacial Geology.* Pergamon, Oxford, 213–228.

Eyles, N., Eyles, C. H. and Miall, A. D. (1983b) Lithofacies types and vertical profile models, an alternative approach to the description and environmental interpretation of glacial diamict and diamictite sequences. *Sedimentology* **30**, 393–410.

Eyles, N., Clark, B. M. and Clague, J. J. (1987) Coarse-grained sediment–gravity flow facies in a large supraglacial lake. *Sedimentology* **34**, 193–216.

Farrell, E. R., Coxon, P., Doff, D. H. and Pried'homme, L. (1995) The genesis of the brown boulder clay of Dublin. *Quarterly Journal of Engineering Geology* **28**, 143–152.

Fookes, P. G. (1997) Geology for engineers: the geological model, prediction and performance. *Quarterly Journal of Engineering Geology* **30**, 293–424.

Goldthwait, R. P. and Matsch, C. L. (eds) (1989) *Genetic Classification of Glacigenic Deposits*. Balkema, Rotterdam.

Hambrey, M. J. (1994) *Glacial Environments*. UCL Press, London, 296pp.

Hindmarsh, R. (1997) Deforming beds: viscous and plastic scales of deformation. *Quaternary Science Reviews* **16**, 1039–1056.

Hird, C., Powell, J. J. M. and Yung, P. C. Y. (1991) Investigations of the stiffness of a glacial clay till. *Proc. 10th ECSMFE*, Firenze **1**, 107–110.

Jones, D. K. C., Brunsden, D. and Goudie, A. S. (1983) A preliminary geomorphological assessment of part of the Karakoram Highway. *Quarterly Journal of Engineering Geology* **16**, 331–355.

Kirkbride, M. P. (1993) The temporal significance of transitions from melting to calving termini at glaciers in the central Southern Alps of New Zealand. *The Holocene* **3**, 232–240.

Krüger, J. (1994) Glacial processes, sediments, landforms and stratigraphy in the terminus region of Myrdalsjökull, south Iceland. *Folia Geographica Danica* **21**, 1–233.

Lawson, D. (1981) *Sedimentological Characteristics and Classification of Depositional Processes and Deposits in the Glacial Environment*. Cold Regions Research and Engineering Laboratory, Report 81–27, Hanover, New Hampshire.

Lehane, B and Faulkner, A. (1998) Stiffness and strength characteristics of a hard lodgement till. In Evangelista, A. and Picarelli, L. (eds) *The Geotechnics of Hard Soils–Soft Rocks*. Balkema, Rotterdam, 637–646.

Li, J. J., Derbyshire, E., Street-Perrott, F. A., Xu, S. Y. and Waters, R. S. (1984) Glacial and paraglacial pediments of the Hunza Valley, north-west Karakoram, Pakistan: a preliminary analysis. In Miller, K. J. (ed.) The International Karakoram Project, Vol. 2, Cambridge University Press, Cambridge, 496–535.

Lutenegger, A. T., Kemmis, T. J. and Hailberg, G. R. (1983) Origin and properties of glacial till and diamictons. In *Geological Environments and Soil Properties*. Special Publication American Society of Civil Engineers, Geotechnical Engineering Division, 310–410.

Maizels, J. (1996) Sediments and landforms of modern proglacial terrestrial environments. In Menzies, J. (ed.) *Modern Glacial Environments: Processes, Dynamics and Sediments*, Vol. 1. Butterworth-Heinemann, Oxford, 365–416.

McGown, A. and Derbyshire, E. (1977) Genetic influences on the properties of tills. *Quarterly Journal of Engineering Geology* **10**, 389–410.

McGown, A., Sali, A. and Radwan, A. M. (1974) Fissure patterns and slope failures in boulder clay at Hurlford, Ayrshire. *Quaternary Journal of Engineering Geology* **7**, 1–26.

Menzies, J. and Shilts, W. W. (1996) Subglacial environments. In Menzies, J. (ed.) *Modern Glacial Environments: Processes, Dynamics and Sediments*, Vol. 1. Butterworth-Heinemann, Oxford, 15–136.

Miall, A. D. (1978) Lithofacies types and vertical profile models in braided river deposits: a summary. In Miall, A. D. (ed.) *Fluvial Sedimentology*. Canadian Society of Petroleum Geologists, Memoir **5**, 597–604.

Murray, T. (1997) Assessing the paradigm shift: deformable glacier beds. *Quaternary Science Reviews* **16**, 995–1016.

Naito, N., Nakawo, M., Kadota, T. and Raymond, C. F. (2000) Numerical simulation of recent shrinkage of Khumbu Glacier, Nepal Himalayas. In *Debris-covered Glaciers*. IAMS (International Association of Hydrological Sciences) publication 264, 245–254.

Owen, L. A. (1994) Glacial and non-glacial diamictons in the Karakoram Mountains and Western Himalayas. In Warren, W. P. and Croot, D. G. (eds) *The Formation and Deformation of Glacial Deposits*. Balkema, Rotterdam, 9–28.

Owen, L. A. and Derbyshire, E. (1989) The Karakoram glacial depositional system. *Zeitschrift für Geomorphologie Supp.-Bd.* **76**, 33–73.

Owen, L. A. and Derbyshire, E. (1993) Quaternary and Holocene intermontane basin sedimentation in the Karakoram Mountains. In Shroder, J.F. (ed.) *Himalaya to the Sea*. Routledge, London, 108–131.

Owen, L. A., Derbyshire, E. and Scott, C. H. (2003) Contemporary sediment production and transfer in high-altitude glaciers. *Sedimentary Geology* **155**, 13–36.

Paterson, W. S. B. (1994) *The Physics of Glaciers*, 3rd edition. Pergamon, Oxford, 480pp.

Powell, R. and Domack, E. (1996) Modern glaciomarine environments. In Menzies, J. (ed.) *Modern Glacial Environments: Processes, Dynamics and Sediments*, Vol. 1, 445–486.

Reading, H. G. (1986) Facies. In Reading, H. G. (ed.) *Sedimentary environments and Facies*, 2nd edition. Blackwell, Oxford, 4–19.

Reynolds, J. M. (1998) High-altitude glacial lake hazard assessment and mitigation: a Himalayan perspective. In Maund, J. G. and Edleston, M. (eds) *Geohazards in Engineering Geology*. Geological Society, London, Engineering Geology Special Publications **15**, 25–34.

Richardson, S. D. and Reynolds, J. M. (2000) An overview of glacial hazards in the Himalayas. *Quaternary International* **65/66**, 31–47.

Shaw, J. (1997) Till deposited in arid polar environments. *Canadian Journal of Earth Science* **14**, 1239–1245.

Small, R. J. (1983) Lateral moraines of Glacier De Tsidjiore Nouve: form, development and implications. *Journal of Glaciology* **29**, 250–259.

Sugden, D. E. and John, B. S. (1976) *Glaciers and Landscape*. Arnold, London.

Trenter, N. A. (1999) Engineering in Glacial Tills. Construction Industry Research and Information Association (CIRIA) Report C504, London, 259pp.

Tufnell, L. (1984) *Glacial Hazards*. Longman, London, 97pp.

Vaughan, P. R. and Walbancke, H. J. (1975) The stability of cut and fill slopes in boulder clay. *Proc. Symp. Engineering Behaviour of Glacial Materials*. Midland Soil Mechanics and Foundation Engineering Society, Birmingham, UK, 209–219.

Walder, J. S. and Costa, J. E. (1996) Outburst floods from glacier-dammed lakes: the effect of mode of lake drainage on flood magnitude. *Earth Surface Processes and Landforms* **21**, 701–723.

Walker, R. G. (1992) Facies, facies models and modern stratigraphic concepts. In Walker, R. G. and James, N. P. (eds) *Facies Models: Response to Sea-level Change*. Geological Association of Canada, Toronto, 1–14.

Weertman, J. (1983) Creep deformation of ice. *Annual Review of Earth and Planetary Sciences* **11**, 215–240.

Whiteman, C. A. (1996) Processes of terrestrial deposition. In Menzies, J. (ed.) *Modern Glacial Environments: Processes, Dynamics and Sediments*, Vol. 1. Butterworth-Heinemann, Oxford, 293–308.

Yamada, T. (1993) *Glacier Lakes and their Outburst Floods in the Nepal Himalaya*. Water and Energy Commission Secretariat, Kathmandu, Nepal.

14. Periglacial Forms and Processes

H. Jesse Walker

14.1 Introduction

In 1909 Lozínski (Jahn, 1975; 1985) introduced the term 'periglacial' in order to provide a way of designating those areas affected by the cold conditions that bordered Pleistocene ice sheets. Today the term is used in reference to a much broader area than either etymology or original concept suggest. It is now generally accepted that periglacial environments possess temperature regimes that include alternations between terrestrial freezing and thawing. Some 35% of the earth's land surface is affected by frost action (cryogenic) processes (Williams and Smith, 1989). Because such temperature regimes occur at high elevations, as well as at high latitudes, periglacial processes also operate in alpine areas of low latitudes even including those straddling the Equator. At times in the past, conditions that favour the development of periglacial forms were more extensive than at present and, therefore, their remnants are often encountered in areas that today enjoy more temperate conditions. The inheritance of periglacial features can have significant implications for engineering works.

Because both high latitude and high altitude environments have generally remained beyond major human concern, scientific research on their landforms has been neglected and engineering experience has been limited. However, the realisation of the strategic importance of some periglacial areas and the recent push to exploit the mineral wealth they possess are providing an increasing number of opportunities for geomorphological and other research and for the development of appropriate engineering principles by scientists and engineers. This research is not only leading to a better understanding of present-day periglacial processes and environments but also is helping geomorphologists and engineers in their interpretation of the fossil forms encountered in areas formerly subjected to the same types of processes.

Emphasis is placed on the Arctic and its unique environmental conditions of which, from the standpoint of engineering, permafrost is of major importance (Ferrians *et al.*, 1969).

14.2 Permafrost distribution

Present-day permafrost, which underlies some 20% of the world's land area, is concentrated in the Northern Hemisphere and the ice-free portions of Antarctica. Although permafrost has been regionalised in a variety of ways, it is now usually displayed under four categories: sub-sea permafrost, continuous permafrost, discontinuous permafrost, and alpine permafrost (Figure 14.1). Continuous and sub-sea permafrost are presently limited to high latitudes; alpine permafrost, on the other hand, is found in both middle and low latitudes. The lower elevation of alpine permafrost gradually increases with decreasing latitude. Non-polar alpine permafrost is affected by a variety of factors that differ from those prevailing at high latitudes. Harris (1988) notes, for example, that

insulation and reradiation are particularly great at high altitudes . . . cold air drainage and inversions are very important, wind speeds are much higher, differences in aspect are accentuated, . . . [and] numbers of freeze–thaw cycles in the ground are much higher.

During parts of the Pleistocene, permafrost extended much further towards the equator than at present. By combining the present-day and Quaternary distributions of glacial ice and

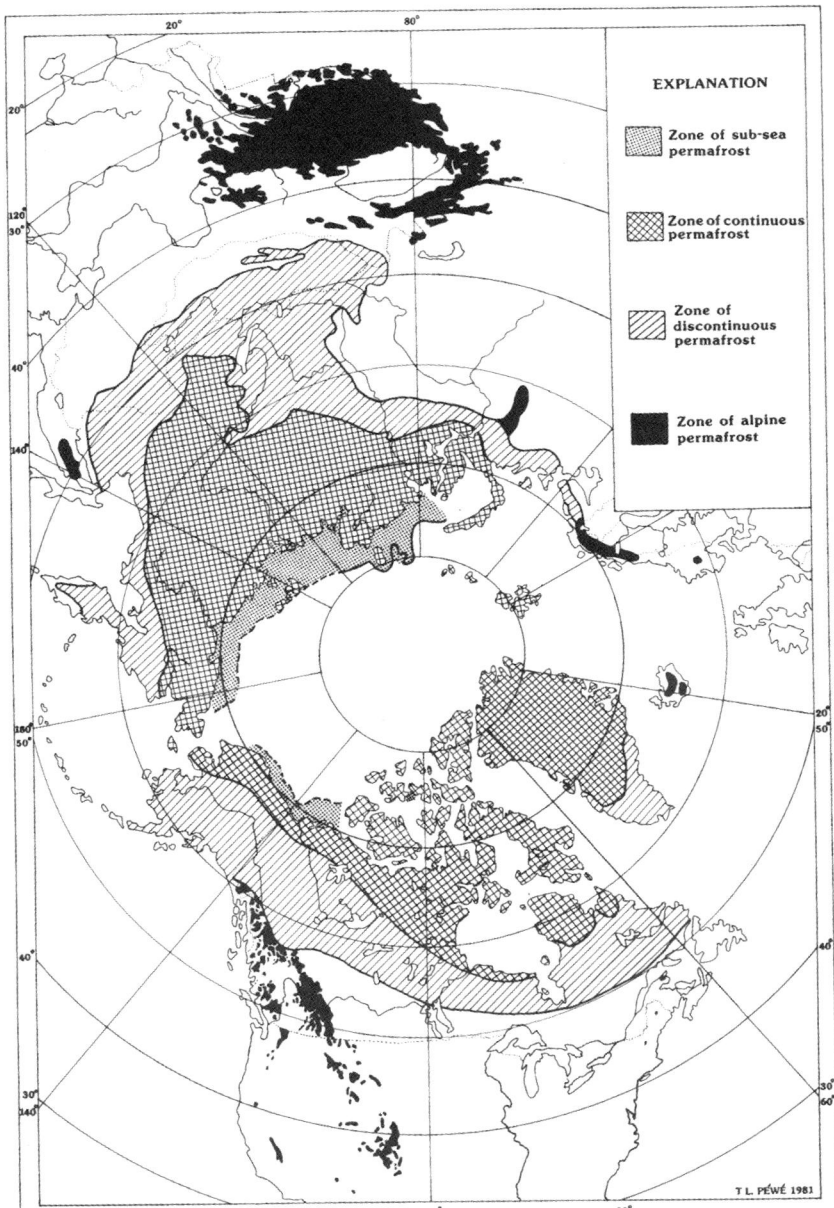

Figure 14.1 Permafrost in the Northern Hemisphere (after Péwé, 1983).

permafrost (Figure 14.2), a notion of the area in which periglacial processes have operated at times in the past and are presently operating can be obtained. Over much of the Northern Hemisphere, permafrost is relatively young, having only developed since the last glaciation. In the Arctic the areas with the oldest (and deepest) permafrost are those that were not glaciated during the Pleistocene, such as the North Slope of Alaska and Eastern Siberia.

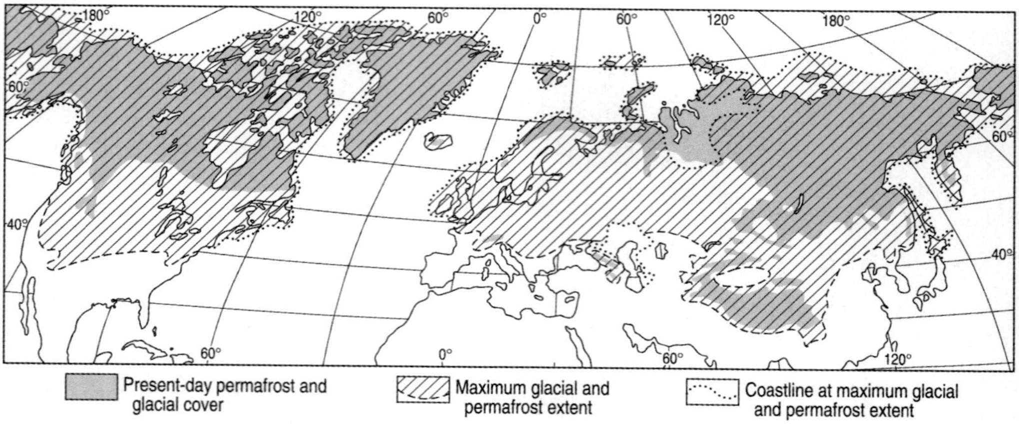
180°
120°
60°
0°
60°
120°
180°
60°
40°
60°
0°
60°
120°
60°
40°
Present-day permafrost and glacial cover
Maximum glacial and permafrost extent
Coastline at maximum glacial and permafrost extent

Figure 14.2 Permafrost and glacial cover in the Northern Hemisphere through time (modified from Nekrasov, 1983).

14.3 Periglacial processes

The permafrost landscape is an amalgam of forms, some of which are created by processes that are uniquely periglacial; others are the product of processes more commonly associated with other parts of the world. However, even they are usually modified to some extent within the permafrost environment. Forms created by freezing and thawing, mass wasting, nivation, wind, water, and biota are all in evidence.

The importance of freeze–thaw oscillations in geomorphology derives especially from the unique property changes of water as it freezes and thaws. Solidification on freezing leads to the formation of ground ice and expansion of the soil. Liquefaction and contraction on thawing lead to solifluction and subsidence.

Frost action

Frost action (the process of freezing and thawing) includes such specific actions as frost wedging (shattering, riving, scaling, splitting)*, frost heaving, frost creeping, frost sorting, nivation, and solifluction (gelifluction).

Frost wedging, considered by many to be the main physical weathering process in cold regions (Washburn, 1985), is the fracturing of rock as water freezes in its cracks and pores. The rate of fracture and the size and shape of the end product depend on the type and character of rock as well as the intensity, frequency, and rate of freezing and the quantity and availability of water (e.g. subsurface aquifers and melting snow). The end product of frost wedging ranges in size from large blocks (that are likely to be tabular in shape when derived from slate or schist) to silt (Figure 14.3). Frost wedging leads to the production of talus which accumulates at the base of rocky cliffs. When associated with a permanent snowfield it becomes a protalus rampart (see Figure 14.4).

In extremely arid periglacial zones exfoliation is common and may be more important than frost wedging; in the better vegetated zones (e.g. wet tundra) chemical and biochemical weathering also occur. However, unless they are derived from mudstones, most periglacial soils contain minimal amounts of clay, partly because of the relative unimportance of chemical weathering.

Frost heaving is the displacement of rock (soil) particles or other included materials by ice as it forms in a freezing soil. Although there is 9% expansion of pore water within the soil on freezing, the most important aspect in frost heaving is the production of segregation ice that forms as unfrozen water flows to the freezing front and freezes into lenses and flakes of various thicknesses. Although displacement caused by segregated ice is normally upward it may also be lateral

* Terms in parentheses are frequently found in the literature and are basically synonymous with those used here.

Figure 14.3   Angular debris produced by frost shattering and ice wedging of limestone in the English Pennines. Photograph courtesy of Tony Waltham Geophotos.

(sometimes referred to as frost thrusting). Frost heaving, which is irregular (except possibly in completely homogeneous material), is responsible for the production of surface forms such as earth

hummocks (thufur) and for the displacement upwards of buried rocks and other objects (including power poles and house pilings). It is especially common in fine-grained soils when moisture content is high. In clayey soils, on the other hand, differential freezing is accompanied by deflocculation and particle disaggregation. Upon thawing such soils are virtually cohesionless and flow easily. Frost heave and ice segregation cause some of the most severe problems facing engineers in the Arctic. They have to be considered in the construction of buildings, road beds, and pipelines.

Frost sorting and frost creeping are two other results of the freezing and thawing of soils. Size-sorting can lead to the formation of various types of patterned ground. Frost creeping results when frost-expanded soil thaws and settles downslope from where freezing occurred (Harris, 1986).

Nivation (snow-patch erosion), which occurs mainly because of the frost wedging that

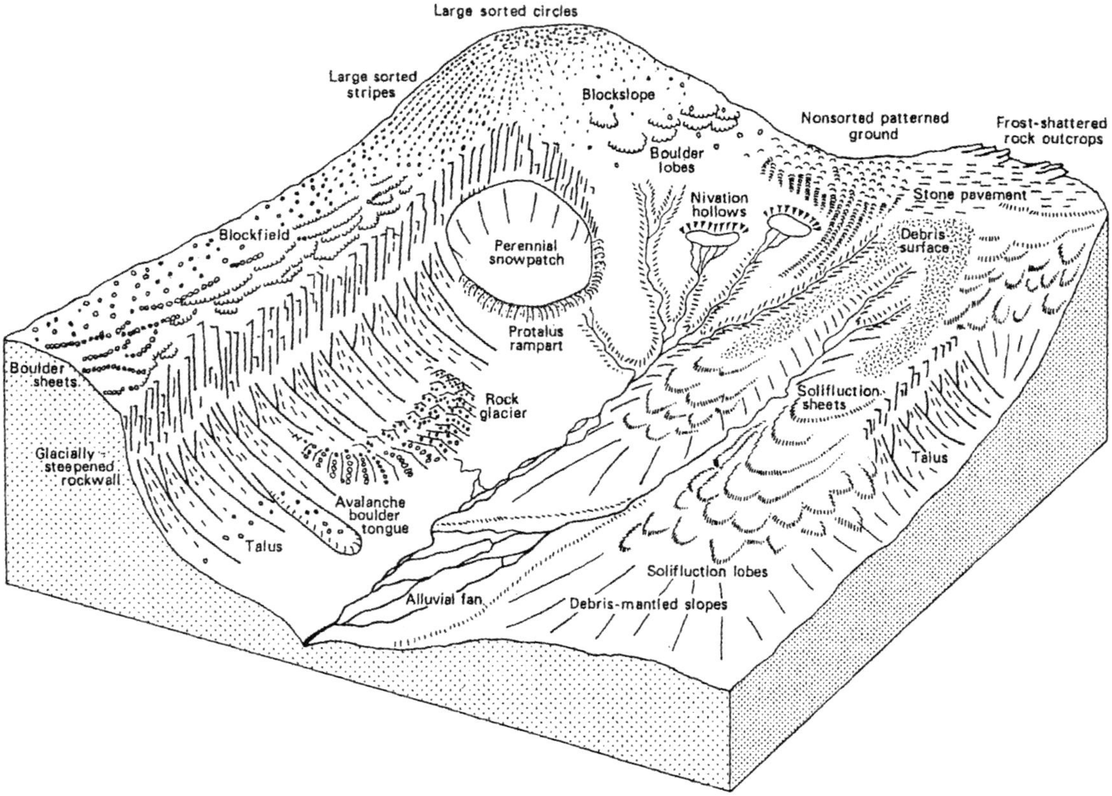

Large sorted circles

Large sorted stripes

Blockslope

Nonsorted patterned ground

Frost-shattered rock outcrops

Boulder lobes

Nivation hollows

Stone pavement

Blockfield

Perennial snowpatch

Debris surface

Boulder sheets

Protalus rampart

Rock glacier

Solifluction sheets

Glacially steepened rockwall

Talus

Avalanche boulder tongue

Talus

Solifluction lobes

Alluvial fan

Debris-mantled slopes

Figure 14.4   Schematic of Late Devensian periglacial features in Scotland. Adapted from Ballantyne and Harris, 1994.

accompanies freezing and thawing at the base and edge of snowbanks, also includes chemical action (Thorn, 1988). Snow has a high carbon dioxide content and is chemically active especially in limestone terrain (Higginbottom and Fookes, 1970). The frequency and intensity of nivation depend not only on rock type but also on the thickness and duration of snow cover. Snow, being a good insulator, tends to reduce the number and intensity of freeze–thaw cycles and therefore the effectiveness of mechanical weathering.

Solifluction (gelifluction)

It has been stated that the '. . . predominant form of periglacial mass movement in both the present Arctic and, in relic form, in Britain' is periglacial solifluction (Hutchinson, 1991, p. 285). In the most recent review of solifluction (Matsuoka, 2001), it is noted that the term has yet to be defined unequivocally (p. 108). Many definitions of the term have occurred since it was first proposed in 1906 by Anderson. Matsuoka (2001) states that solifluction, as understood today,

> . . . involves several components: (1) needle ice creep and diurnal frost creep originating from diurnal freeze–thaw action; (2) annual frost creep, gelifluction and plug-like flow originating from annual freeze–thaw action; and (3) retrograde movement caused by soil cohesion.

Although not limited to areas with frozen ground, solifluction is most active there, because, even on low slopes, the frozen subsurface ensures a saturated upper layer by preventing the water from melting snow and thawing surface soil from percolating downward. This concentration also may lead to excess pore pressures in the thawing soil which promotes slope movements (Lewkowicz, 1988). Rates are affected by degree of slope, soil texture, depth of thaw, water content, and vegetation composition.

In those permafrost zones where the terrain is sloping, active-layer failure is common. Although most solifluction is slow, relatively rapid failures can also occur. Known as skin flows (skin slides), detachment slides, and mudbursts, they appear to be caused by over-saturation at the permafrost table (Barsch, 1993). As potential

hazards they represent a challenge to the periglacial engineer.

As frozen soil thaws, the liquid water that does not run off accumulates increasing the pore pressure in the soil which in turn causes a reduction in soil strength often leading to soil flow. On the other hand, the runoff of melt water, combined with settling of the soil particles as the ice melts, results in thaw consolidation which affects the stability of the soil layer (Williams and Smith, 1989).

Solifluction was sufficiently widespread during the Pleistocene that it affected all of the British Isles where its effects today are relic. Hutchinson (1991) writes that:

> From an engineering viewpoint, relic solifluction materials may be regarded as having been loosened by frost heave, to some extent reconsolidated during and after thawing, and displaced downslope. For foundation purposes, they are likely to have a lower bearing capacity and a higher compressibility than the underlying, parent stratum.

Frost cracking

Frost cracking, (ground cracking, thermal contraction cracking) occurs as frozen ground becomes colder and contracts. The rate at which the temperature drops appears to be more important than the actual temperature at the time of cracking (Lachenbruch, 1962). Such thermal contraction is similar in many respects to the contraction that occurs with desiccation, some chemical reactions, and phase changes. The contraction polygons that may be produced are highly variable in size and shape, a variability that is controlled by the rheological nature of the cracking material as well as temperature conditions. In periglacial areas, thermal-contraction cracks range up to several millimetres in width, several metres in depth, and several tens of metres in separation and serve as voids into which melt water, snow, and sediment can collect.

Other processes

Although the interrelated processes (ground cracking, mass wasting, nivation, etc.) described above are those that are most distinctly periglacial, other, more familiar types of processes such as

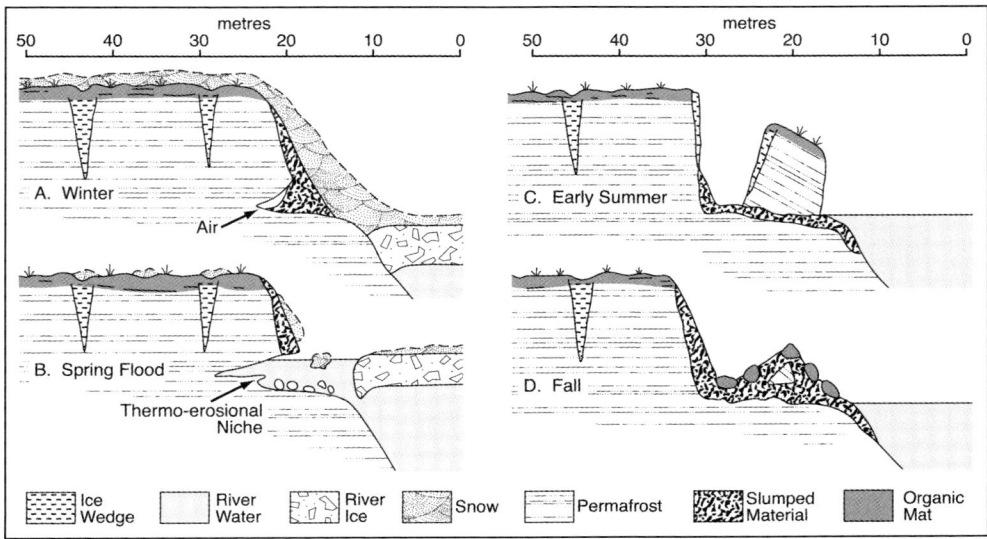

Figure 14.5   Riverbank erosion along the Nechelik channel of the Colville River.

wind and running water are also active in periglacial regions.

Wind is regionally highly variable in periglacial environments; its effectiveness depends upon many conditions other than the velocity, direction, and duration of the wind itself. During much of the year (usually that part of the year when wind velocities are highest) most of the surface in the Arctic is protected with a snow cover, albeit usually thin. On irregular surfaces snow accumulates in drifts, some of which may last through much of the summer. Even on level surfaces snow ridges (sastrugi) form on the wind-packed snow surface. Snow, with a hardness that may equal that of quartz grains at low winter temperatures, is easily moved and becomes an effective erosive tool. During snow-free periods, wind erosion, transport, and deposition are common (Walker, 1967). River bars, outwash plains, and other non-vegetated areas are sources of much sediment and help account for the wide distribution of loess in both present-day and former periglacial areas (Chapter 25).

Running water, a phenomenon of the frost-free period of the year in the Arctic, is especially abundant in the snow-melt season. It originates from melting snow, thawing ground, and rainfall. During the early stages of the snow-melt season,

when the ground surface is still frozen, water accumulates between the many hummocks that characterise much of the surface. Soon thereafter, however, it flows across the surface and concentrates in streams. The flooding caused by melting snow accounts for most of the annual discharge in many streams in periglacial areas. For example, the Colville River (the largest river on the North Slope of Alaska) carries about half of its total annual flow during a 3–4 week period (Arnborg *et al.*, 1966). It is during such flooding that much of the fluvial activity in periglacial areas occurs. As Clark (1988) notes, it includes 'water and sediment inputs, discharge and sediment transport, and the resulting hydrological and geomorphological outputs' (p. 415). Stream erosion may be extreme (Figure 14.5). Relatively warm water flowing against snow-covered and frozen banks is responsible for thermal erosion (Figure 14.6), which in turn may lead to block collapse (Figure 14.7; Walker and Arnborg, 1966). Similar results are found along the shorelines of lakes, lagoons, and the open coast.

Ice jams are the frequent cause of flooding on arctic rivers. For example, in May 2001, ice jams on the Lena River downstream from Yakutsk threatened to flood that city with water that accumulated

Figure 14.6   Thermoerosional undercutting of frozen stream bank with subsequent collapse, on the Arctic Plain of Alaska. Photograph courtesy of Tony Waltham Geophotos.

Figure 14.7   Block collapse along ice wedges into a small river channel on the Arctic Plain, Alaska. Photograph courtesy of Tony Waltham Geophotos.

upstream from the ice dam. Explosives dropped on the ice jams from helicopters breached the ice and reduced flooding. The 2001 ice-jam flooding, the worst in a century, was aggravated by exceptionally thick ice, heavier than usual snowfall, and an unusually warm spring (Kriner, 2001).

14.4 Ground ice and permafrost

Although temperature variations are a requisite for periglacial processes, it is the presence of ice at the surface and in the ground that renders these processes especially meaningful. Freeze and thaw, whether in river, lake, or sea, or within the ground, are responsible for most of the engineering

problems in periglacial environments. It is, as stated by Williams (1995), the '. . . formation (aggradation) and disappearance (degradation) of permafrost that are geotechnically the most important consequences of temperature change' (p. 348).

Engineering experience in coping with surface ice during freeze-up and break-up and in the period of time between them in temperate latitudes has been substantial. Many of the publications of CRREL (Cold Regions Research and Environmental Laboratory — see reference list for web site) treat these topics in detail.

Ground ice

Ground ice, a frost phenomenon, is nearly ubiquitous in periglacial areas where water is present. It may 'occur as thin fibers, grains, veinlets, large vertical wedges, horizontal sheets or in irregular masses of all sizes, shapes and colour . . . it may be prismatic or granular, contain numerous air bubbles which are either oriented or without orientation and have horizontal or vertical planes' (Hamelin and Cook, 1967, p. 17).

A number of classifications of ground ice have been proposed. Two of the most widely referenced (and modified) are those of Shumskiy and Vtyurin (1966) and Mackay (1972) (Figure 14.8). In addition to those that form *in situ* (pore ice, ice wedges, pingo ice) there are those ice masses that form above the surface and are buried (e.g. snow, icings, and glacial, river, lake, and sea ice). Although such buried ice may be significant locally, it generally is of small areal extent.

The most widely distributed of ground-ice types is pore ice (interstitial ice, ice cement) which bonds soil grains together. As long as there is no excess water involved, this type of soil is subjected to minimal disturbance with phase change. However, phase changes do bring about impermeability and strengthening with freezing and reduction of cohesiveness with thaw (Williams and Smith, 1989).

When moisture is present in the surface layers of the soil, freezing may produce nearly pure crystals that grow perpendicular to the surface. These crystals, known as needle ice (pipkrake), are capable of heaving materials above the ground surface. They can develop in soils with textures ranging

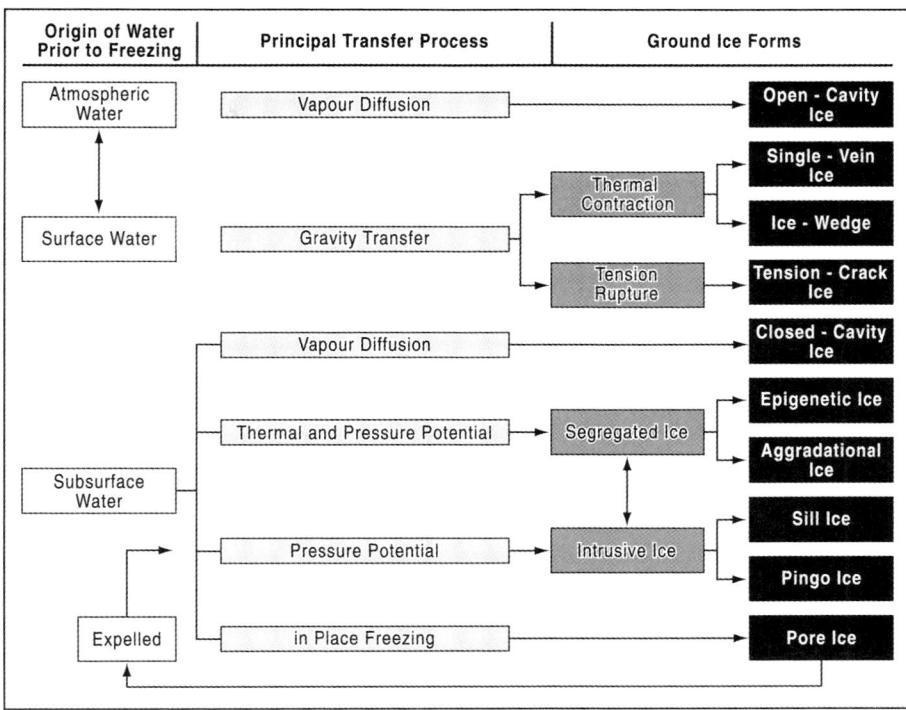

Origin of Water Prior to Freezing	Principal Transfer Process	Ground Ice Forms

Figure 14.8 Classification of underground ice (modified from Mackay, 1972).

from clay to coarse sand although they are best developed in silt. If that surface is generally level, soil stirring will be the major result; if, on the other hand, the ground is sloping (even if less than 5°), the lifted material will settle downslope as the needle ice melts resulting in one type of frost creep.

Segregated ice (taber ice) develops in supersaturated soils. It is 'epigenetic' when it predates the surrounding sediments and 'aggradational' when forming in conjunction with the material enclosing it. This enclosing material may be organic or mineral and it may accumulate slowly as with soil creep or more rapidly during river flooding, mudflows, or wind storms. Both types of segregated ice occur as lenses, some of which grow to be several metres thick. Such growth is responsible for ground heave. Upon melting the excess water that characterises these lenses is lost and subsidence occurs.

Intrusion ice (injection ice) develops where water is forced between sediment layers. Although less important quantitatively than pore or segregated ice, it produces surface forms that are more conspicuous than either of them, and is usually more pure and granular. When the injected water freezes in a tabular form it becomes sill ice; when it occurs as the core of a large mound it is pingo ice. Small mounds with injected ice are known as frost mounds.

Vein ice and wedge ice form in the contraction cracks that develop in frozen ground usually after rapid temperature drops in winter. Because these cracks fill with any material that drops into them, veins and wedges vary in composition. However, the most common ingredient appears to be the water that fills them during snow-melt season. In Antarctica, where the quantity of liquid water is minimal, the filling material is almost always fine-grained sediment which produces sand wedges instead of ice wedges.

An ice vein is the first step in the formation of wedge ice which develops when sequential cracking occurs along the vein (Figure 14.9). Ice wedges may grow to be several metres wide at their tops and occupy as much as half of the upper several

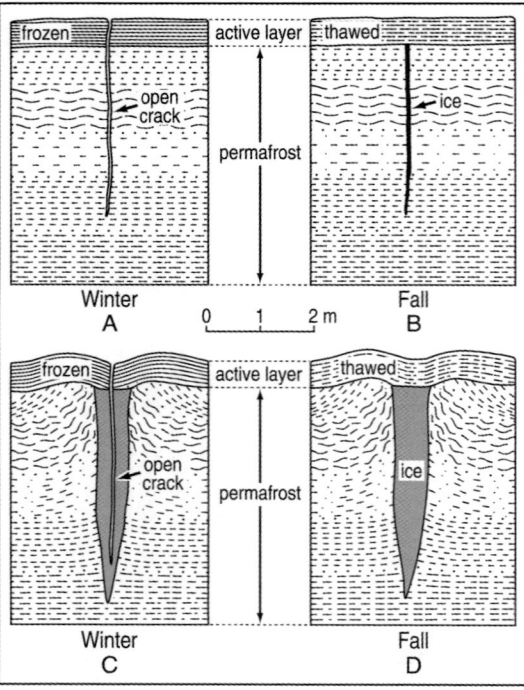

Figure 14.9 Ice wedge formation (after Lachenbruch, 1962).

Figure 14.10 Ice wedge exposed in the Fox Tunnel cut into frozen silts near Fairbanks, Alaska; the ice appears dark in this photograph as it absorbs more light from the camera flash. Photograph courtesy of Tony Waltham Geophotos.

metres of ground in many areas in the Arctic. In North America ice wedges are usually less than 8 m deep and 3 m wide; in Siberia, on the other hand, some wedges are much larger (Harry, 1988). Such sizeable growth occurs under extremely cold conditions, in fine-textured and peaty soils (Figure 14.10), and in locations with aggrading alluvial surfaces (Harry, 1988). Although wedge ice can develop in sandy soils (even sand dunes) it is generally a narrow vertical band of ice and seldom takes the classical wedge shape.

Permafrost and the active layer

The optimum expression of periglacial forms and processes occurs in those areas where the subsurface is perennially in a frozen state. The general term for such a condition is permafrost. It is defined as being a naturally occurring earth material in which the temperature has been below 0 °C for two or more years (Muller, 1947). It is initiated when a ground layer freezes in winter, but does not thaw in summer, and grows as freezing progresses downward. The definition has nothing to do with the presence of water. If water is absent, as in solid rock, the material is called dry permafrost. When water is present, it is wet permafrost, the type that is of major concern to engineers because thawing causes subsidence and may cause erosion. There are situations where water retains its liquid form even though temperature may render the term permafrost appropriate. One example is sub-sea permafrost where saline water fills the pores of the sediment; another is in clay soil when the freezing temperature is depressed below 0 °C.

The upper surface of permafrost (i.e. the level below which the temperature remains below 0 °C even during summer) is known as the permafrost table. Above this table is the active layer, the layer that freezes in winter and thaws in summer.

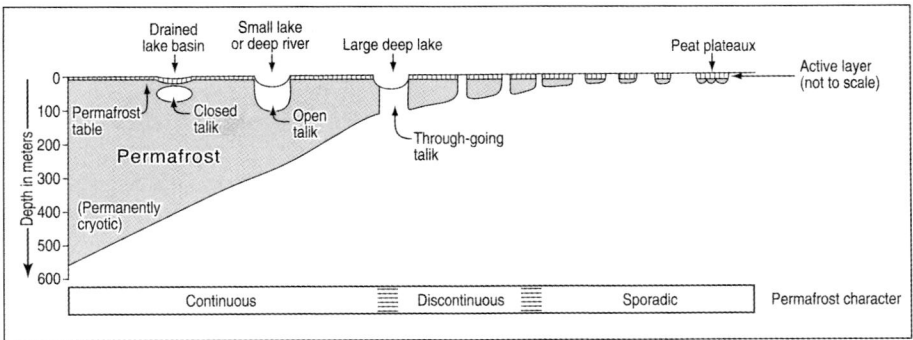

Figure 14.11   Permafrost characteristics from north to south across Canada. Adapted from Ballantyne and Harris, 1994.

Normally the bottom of the active layer is coextensive with the permafrost table. However, during some seasons or in some locations there is a zone between these two where the temperature remains above 0 °C. Such unfrozen layers are called taliks. Although taliks may be surrounded by permafrost some are open at the top as for example beneath lakes or river channels that are sufficiently deep not to freeze to the bottom (Figure 14.11).

Permafrost extends to depths of several hundred metres in the high Arctic. Depths of more than 600 m in North America and nearly 1500 m in Siberia have been recorded. Permafrost thickness is controlled by mean surface temperature and geothermal heat flow. Low values for each favour permafrost. Surface temperatures, and therefore permafrost growth, vary with surface character (Figure 14.12). Water bodies (lakes, rivers, lagoons, the ocean) are major modifiers of regional permafrost development. Water bodies that are more than about 2 m deep do not freeze to the bottom and, even though surface ice on them usually lasts more than half a year, the water beneath the ice cover ensures that bottom temperatures remain above freezing. In some cases, as when a river meanders across the land surface, thaw occurs beneath the river and a thaw-bulb develops. Where deep lakes and rivers are sufficiently large, the unfrozen zone (talik) may extend completely through the regional permafrost (Figures 14.11 and 14.12). Even shallow lakes and streams (i.e. those that freeze to the bottom in winter) influence permafrost development

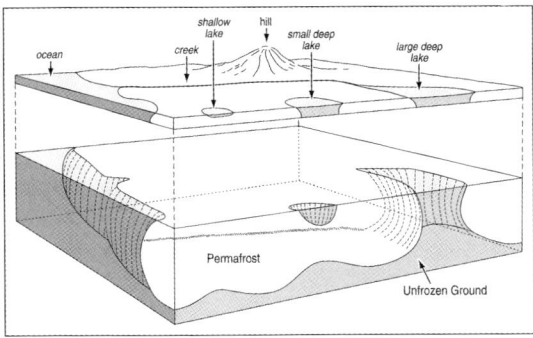

Figure 14.12   Surface features and the distribution of permafrost (after Lachenbruch, 1968).

because they possess mean annual bottom temperatures that are higher than those of the surrounding sub-aerial ground surface. In the case of the ocean, the same general characteristics prevail even though sea water freezes at temperatures below 0 °C (-1.9 °C at a salinity of 35‰).

Other factors influencing the average ground temperature include snow cover (thickness, duration), micro-relief, vegetation type, and soil characteristics (French, 1988). These factors have a major influence on the development of the active layer (Figures 14.13 and 14.14). This layer ranges in thickness from a few centimetres under dense vegetal mats to more than 2 m in non-vegetated, coarse textured, southerly exposed sand dunes. Thickness may vary greatly over short distances but basically follows the contours of the surface. In the zone of continuous permafrost, the active layer will normally be at or near saturation throughout the summer. Sand dunes and alluvial fans are often exceptions because of the permeability of their coarse grained texture. Although the total annual precipitation is low in the Arctic, high water content is possible in the active layer because the permafrost table prohibits percolation

and low gradient slopes suppress horizontal groundwater flow.

14.5 Periglacial landforms

Periglacial landforms are highly varied in size, shape, and distribution. Those most typical of periglacial regions are found in zones of permafrost. Many, such as solifluction stripes and ice-wedge polygons, are relatively small features but because they are highly repetitive in occurrence, they often extend over large areas (Walker, 1983). Still others, such as pingos, are fewer in number and often isolated.

Altiplanation terraces, block fields, and rock glaciers

Altiplanation terraces (cryoplanation terraces) are benches formed by frost action on hillsides. Formed in bedrock, these terraces usually support a thin layer of frost-shattered rock. They often occur in steps and may encircle individual hill crests. In contrast, block fields (Felsenmeere) are thick accumulations of frost-shattered rocks of different sizes and shapes. There is relatively little movement of individual blocks as most block fields tend to be level or have low gradients. Rock glaciers consist of masses of poorly sorted angular blocks that are consolidated by interstitial ice. Nonetheless, controlled by slope and ice content, movement, often several

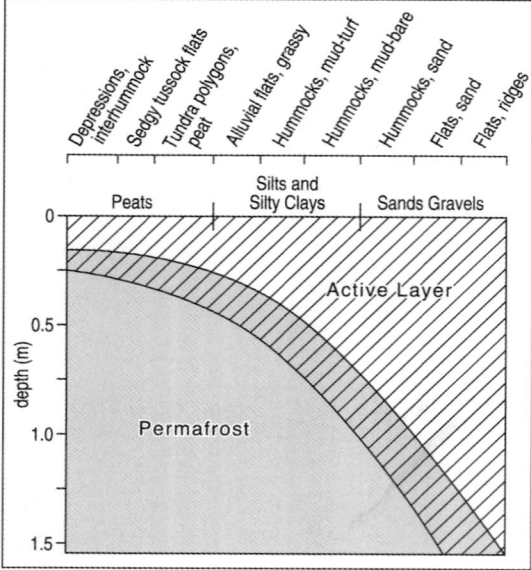

Figure 14.13 Active layer thickness in various periglacial terrains (after Mackay, 1970, in Washburn, 1980).

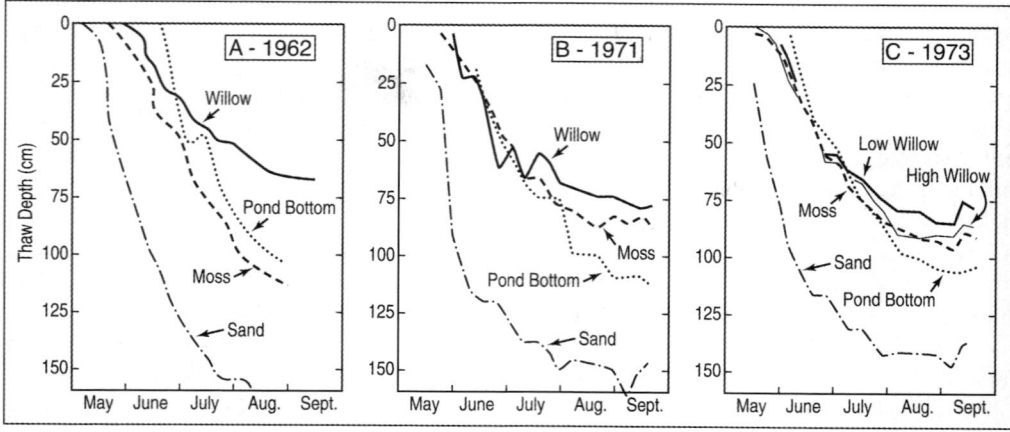

Figure 14.14 Active layer development in sand under different types of vegetation, Colville River delta.

metres per year, is characteristic of rock glaciers during their active stage (Barsch, 1988). Upon descending relatively steep slopes they may progress across more gently sloping terrains. Rock glaciers have shapes that usually range from elongated tongues to gently curving lobes. These are features that are especially common in mountains such as those in Scotland (Figure 14.4).

Patterned ground

Although patterned ground is not restricted to periglacial regions, it reaches its optimum development there, especially in zones of continuous permafrost. Patterned ground is one of the most distinctive features to be formed in freeze–thaw climates.

The surface forms in patterned ground are highly varied in shape, size, composition, degree of sorting, and rate of development. Basic patterns include circles, nets, polygons, steps, and stripes. Although the shapes are often geometrically representative, there are frequent gradations among them. The first three types are most common on low-gradient surfaces, the latter two are related to slopes.

Circles, which may be sorted or unsorted (non-sorted) and are widespread in periglacial environments, occur singly or, more commonly, in groups, and are generally small (< 3 m in diameter). There are different types of circles, e.g. among sorted circles some have stone borders that reach to the permafrost table whereas others have borders that are surface features only. Nets have patterns that are intermediate between those of circles and polygons.

Polygons (Figure 14.15), the most common of patterned ground types, range from a few centimetres to over 100 m in diameter. As in the case of circles and nets, the degree of sorting is diagnostic. Sorted polygons, which may be more than 10 m across, have borders comparable to those of sorted circles and do not need permafrost to form.

Unsorted polygons, which are also widespread, may be composed of peat, silt, sand, and

Figure 14.15   Sorted stone polygons in Alaska. Photograph courtesy of Tony Waltham Geophotos.

even gravel. The borders between unsorted poly-gons are often cracks which contain water and support a type of vegetation that differs from that of the centres. The most distinctive type of unsorted polygon in periglacial regions is the ice-wedge (tundra) polygon (Figure 14.16). It only forms in areas with permafrost and is repet-itive across thousands of square kilometres in the Arctic. Counterparts in ice-free Antarctica are the polygons that result from sand-wedge formation.

Ice-wedge polygons are usually classified as either low-centred or high-centred depending on the relative heights of their borders and centres. Low-centred polygons are indicative of actively growing ice-wedges, usually contain water in their centres during summer, and serve as ideal locations for peat formation. High-centred poly-gons reflect degradation and often have water in their troughs. In both types, surface irregularity is great although total relief is low.

Lachenbruch (1962) divided ice-wedge poly-gons into groups based on the way ground cracking occurs. When most of the cracks are at right angles, they are labelled orthogonal; when most are about 120°, non-orthogonal. Orthogonal polygons, which typically form in inhomogeneous material, may be random or oriented. The formation of oriented sys-tems is favoured by gradual recession of water from a surface as occurs during river meandering and lake drainage.

Steps (lobate soils) and stripes (solifluction stripes) occur on sloping terrain as the result of the gravitational migration of material downslope. If the downslope border of the step is composed of coarser material than occurs upslope, sorting has occurred. In contrast, unsorted steps are bordered by vegetation. In essence, stripes have the same distinctions as steps and differ from them mainly in that their orientation is down the slope (Figure 14.17) instead of across it. Both usually occur in groups although they can form singly.

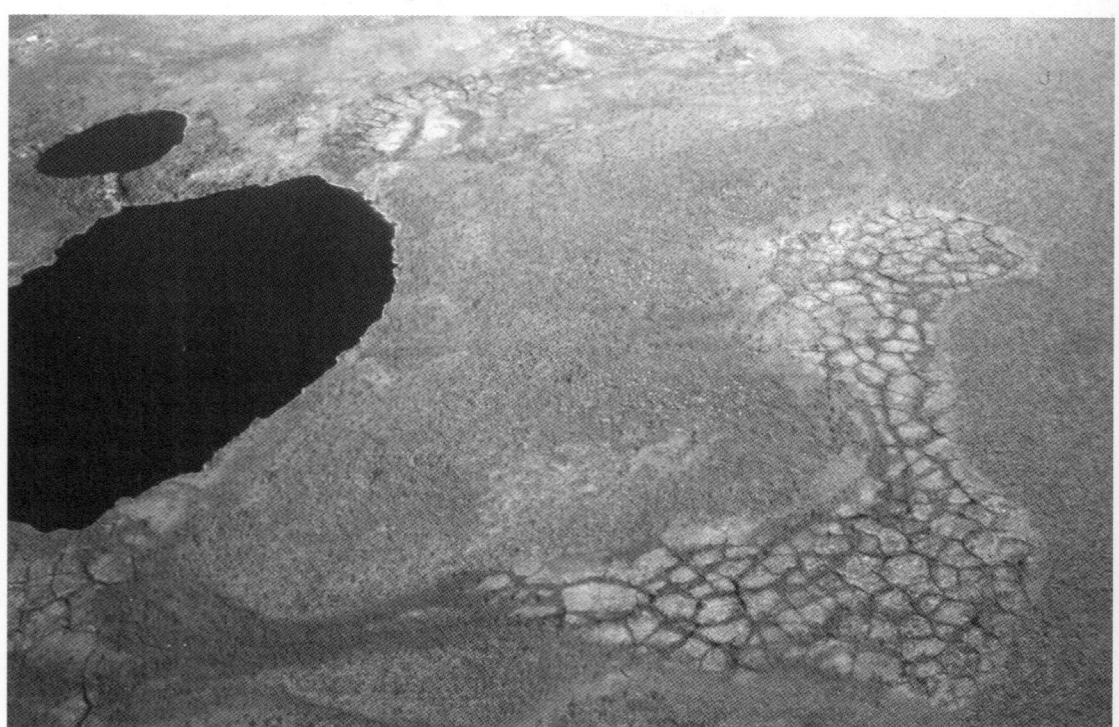

Figure 14.16   Ice-wedge polygons in old lake sediments in the Mackenzie Delta, Canada. Photograph courtesty of Tony Waltham Geophotos.

Figure 14.17   Solifluction stripes created by slope movement of a sorted soil in permafrost, on Disko Island, western Greenland. Photograph courtesy of Tony Waltham Geophotos.

Figure 14.18   Inverted relief formed as polygonal ice wedges melt in the summer at Prudhoe Bay, Alaska. Photograph courtesy of Tony Waltham Geophotos.

Frost mounds and pingos

The increase in volume that occurs with the freezing of water is responsible for the formation of rather symmetrical mounds some of which reach heights of more than 50 m. Such large forms are known as pingos. Although there are different ways in which pingos form, the majority develop when a talik (as would be present beneath a small lake) begins to freeze from all sides. Progressive expansion within the subsoil forces surface layers upward. Smaller features (frost mounds, palsas) form in shallow water bodies such as low-centred polygons. Although frost mounds, palsas, and pingos are perennial features there are others, including icing mounds, icing blisters, and frost blisters, that are seasonal (Pollard and van Everdingen, 1992).

Thermokarst and associated melt forms

If the mounds described above are considered to be positive forms (i.e. heave) then the development of thermokarst is negative (i.e. subsidence). When the ground ice within permafrost thaws, volume is decreased and the surface subsides. Such subsidence often leads to the creation of thaw lakes, some of which are oriented and which abound in permafrost regions. Melting ice-wedge junctions can lead to the formation of a distinctive drainage pattern known as beaded drainage. The melting of ice-wedges creates conspicuous troughs (Figure 14.18) bringing about an inversion of relief.

Thermokarst is considered by many to be one of the 'most dynamic process suites modifying arctic and subarctic landscapes' (Harry, 1988, p. 132). Although the development of thermokarst occurs naturally, it can also be initiated and aggravated by human activities. It provides periglacial engineers with some of their more serious problems in arctic development. Any future warming of the climate that may occur will aggravate thermokarst development.

Other landforms

Within periglacial regions, river valleys, deltas, sand dunes, glaciated terrain, beaches, and other more familiar landforms are also present. However, even these features are affected by periglacial processes and most have many of the forms discussed above superimposed on them. For example, the Colville River delta, Alaska, is a typical delta in most respects but also associated with its bars, flats, dunes, distributaries, and lakes are such periglacial forms as ice wedges, ice-wedge polygons, frost mounds, and pingos.

14.6 Periglacial soils

Soils formed in periglacial environments are like those that develop in other environments in that their morphological, physical, and chemical characteristics are dependent upon a number of factors including lithology, climate, hydrology, topography, biota, and time. In periglacial areas the relative importance of these factors varies greatly. In those regions with little or no biotic component (such as is common in Antarctica, the high Arctic, and at high elevations), soil development is almost exclusively the result of physico-chemical processes near the surface (Tedrow and Ugolini, 1966). Such areas also tend to be extremely arid. The soils are usually coarse grained.

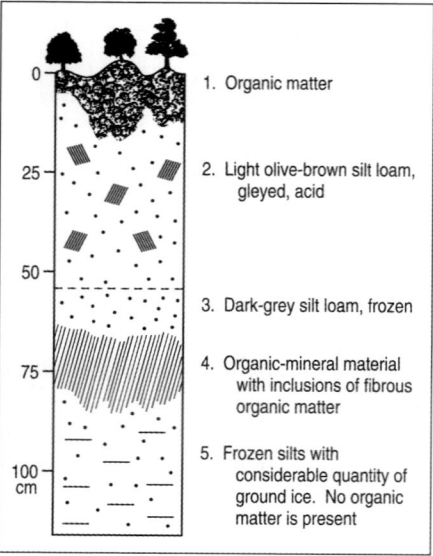

Figure 14.19 Tundra soil profile. Dashed line approximates permafrost table (after Tedrow, 1966).

Contrasting with these ahumic (i.e. without humus) and arid zones are those periglacial areas where biota flourish and liquid water (even if for only a part of the year) is abundant at the surface and in the active layer. Gentle slopes and impeded groundwater circulation, both of which occur over much of the Arctic, aggravate waterlogging, and lead to 'gleization' (development of a sticky, compact, blue-grey layer) and to the production of tundra and bog soils (Figure 14.19).

Vegetation affects the development of soils through the insulation it provides, through root binding, and through its slow decay rate (Washburn, 1980). Peat, which is extensive over much of the Arctic, has a lower thermal conductivity than mineral soil and thus different freezing characteristics.

One of the major variables in periglacial soil development is time. In areas only recently exposed because of glacial retreat and rebound (isostatic uplift), bare bedrock or freshly exposed deposits are little modified. Further, in areas of deposition (deltas, river valleys, sand dunes, loess deposits), the upward movement of the permafrost table limits the length of time during which soil development can occur.

Soil morphology is complicated by the 'soil-destroying' processes associated with frost action including solifluction (Figure 14.17) because, according to Tedrow (1977), these processes operate too fast for soil morphology to adjust (Figure 14.20).

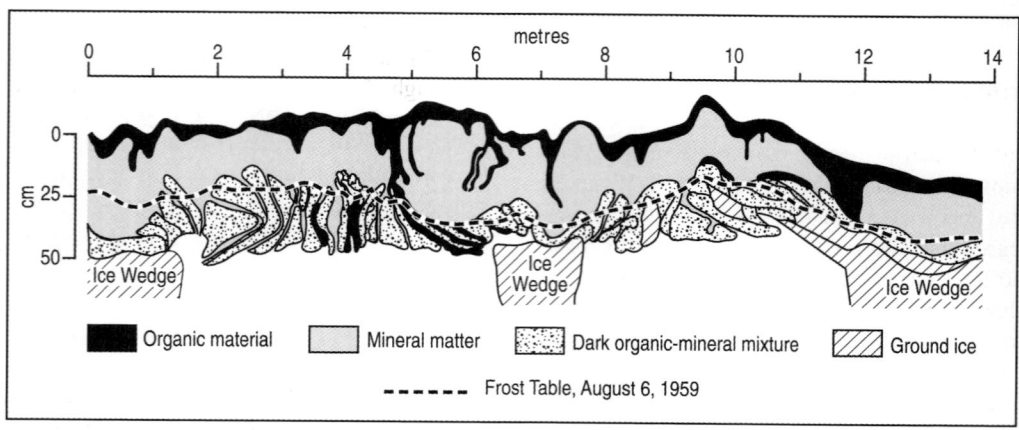

Figure 14.20   Cross-section of tundra soil illustrating complicated structure (after Tedrow, 1977).

PATTERNED GROUND	Rock land	Lithosols	Regosols	Arctic brown shallow phase	Podzol-like	Arctic brown	Upland tundra	Meadow tundra	Soils of the solifluction slopes	Bog
Sorted nets	X		X	X						
Non-sorted nets, stone-centred	X	X								
Sorted polygons			X	X	X					
Non-sorted ice-wedge polygons							X			X
Sorted circles	X		X							
Peat rings							X	X		X
Frost scars							X	X		
Frost scar mounds							X	X	X	
Non-sorted steps		X	X	X	X		X		X	
Non-sorted stripes		X	X				X		X	
Solifluction mounds							X	X	X	
Hummocks									X	
Peaty mounds										X

Figure 14.21   Soils and related patterned ground features (after Tedrow, 1977 as compiled from Brown, 1966).

Soil types in periglacial environments often reflect the type of patterned ground with which they are associated. Brown (1966), for example, found for one valley in arctic Alaska that there were no fewer than ten soil types associated with thirteen types of patterned ground (Figure 14.21). Two of the most important characteristics of soil in the Arctic is that it is frozen — seasonally, in the case of the active layer — and that frozen soil is usually stronger than either ice or unfrozen soil (Williams and Smith, 1989).

14.7 River ice, lake ice, and sea ice

One of the major components of the periglacial regions of the Northern Hemisphere is water, in the liquid state for part of the year and frozen (at least on the surface) for the rest of the year. Lakes and rivers, many of which enter the periglacial environment from more southerly latitudes, are abundant and their shorelines, along with those of the Arctic Ocean and its bordering seas, serve as the most important locations for human settlements. Just as the temperature regime is responsible for the formation and distribution of permafrost, it is also responsible for the lengthy period of the year during which rivers, lakes, and the sea are covered with ice, a cover that is usually topped by snow as is the surrounding land. Snow cover, over most of the Arctic, is so complete that at some locations it is difficult to distinguish visually between land and sea during winter.

Arctic rivers are highly variable in length, discharge, and dates of freeze-up and break-up. In the periglacial areas of the Northern Hemisphere, rivers are of two basic types: (1) those that originate outside the zone of continuous permafrost, and (2) those whose basins are confined to it. Although flow characteristics vary among the rivers that flow into the Arctic Ocean, break-up and the flooding that accompanies it is critical in all of them. The timing and intensity of break-up affects many human activities in the river's floodplains and deltas. Much of the annual erosion and deposition by rivers occurs in the relatively short period of break-up and the flooding that accompanies and follows it.

Sea ice, which dominates the Arctic Ocean (even in summer), is shore-fast along the coastline for varying lengths of time ranging from only a few months to a full year; along some coasts, for

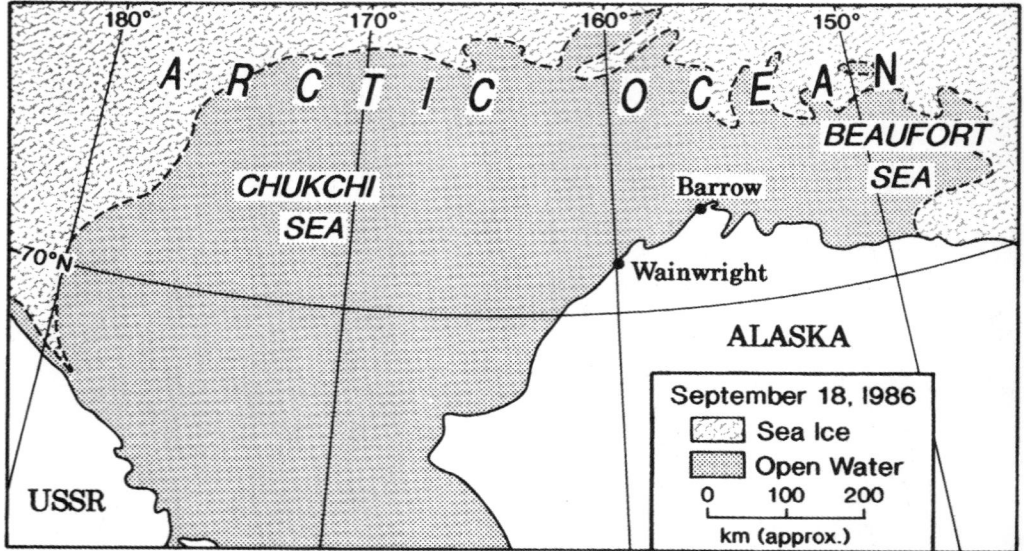

Figure 14.22   The Chukchi and Beaufort Seas showing the length of fetch during September 1986.

some years. It is during break-up and freeze-up and the time in between (i.e. the relatively ice-free summer period) that most of the shoreline is impacted. Although ice is an eroding, depositing, and transporting agent, especially during periods of freeze-up and break-up, it is the waves that develop during storms that are the major coastal modifiers. The most adverse conditions occur when the ice front retreats some distance from the shore thereby providing a long fetch over which storm waves can develop (Figure 14.22). At such times beach and bluff erosion may be severe endangering coastal settlements. To date few coastal engineering projects (such as sea walls and beach renourishment) have been undertaken to prevent or control coastal erosion (Walker, 1991).

14.8 Engineering and fossil periglacial forms

Even in those vast areas no longer subjected to periglacial conditions (Figure 14.2), surface and subsurface forms and soil characteristics frequently reflect former periglacial processes and forms. Evidence of such activity is well displayed in a variety of forms such as involutions, solifluction, valley bulging, and cambering in Britain, peat deposits in former pingo craters in The Netherlands, ice-wedge casts in Poland, and crop (ice-wedge) polygons in Denmark (Figure 14.23). Of these numerous forms and processes, Hutchinson (1991) writes that in Britain the

> reactivation of relic clayey solifluction mantles by ill-advised earthworks probably constitutes at present the most frequent and costly type of failure . . . having a periglacial origin.

Among the many micro-relief forms present in many parts of north-western Europe are the remnants of ice-wedge polygons. They often become conspicuous because of differences in the rates at which crops grow. Aerial photographs of these patterns have been made in Denmark, Sweden, Germany, and England (Christensen, 1978).

Several hundred locations of fossil ice wedges in Denmark have been identified (Figure 14.23). They tend to be of two types: those where the wedges are filled with fine-grained materials and those with sediments coarser than the surrounding material

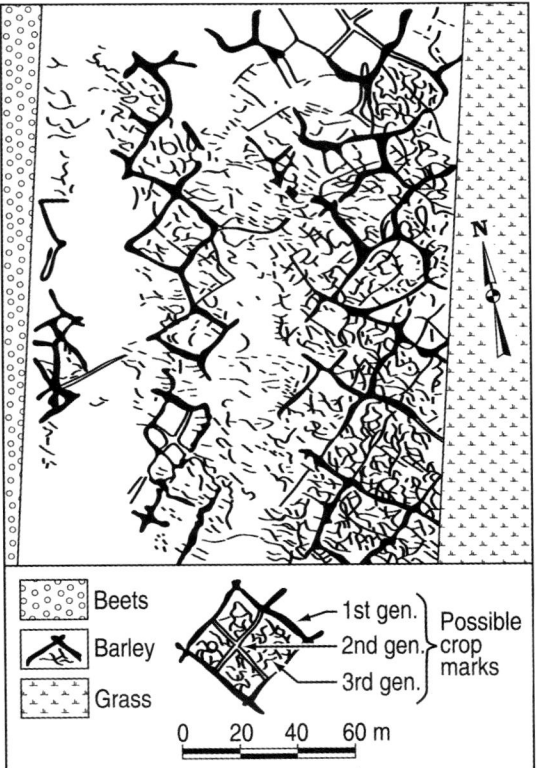

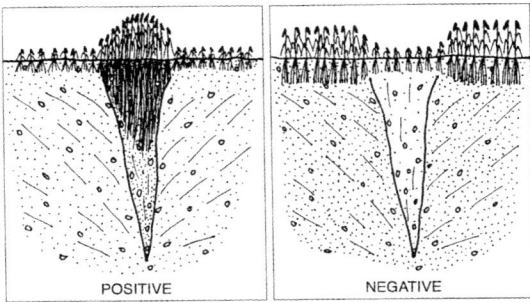

Figure 14.24   Illustrations of the relative effect of ice-wedge casts on cereal growth at maturity (July–August). Adapted from Christensen (1978, p. 259).

Figure 14.23   Crop polygons above fossil ice-wedge polygons in barley field, Denmark (after Christensen, 1978).

Figure 14.25   Fossil wedge from central Poland (Jahn, 1975).

(Christensen, 1978). These differences are reflected in variations of growth rates which may be negative or positive (Figure 14.24).

The occurrence of ice wedges, involutions, frost mounds, and pingos in the fossil form are major concerns in engineering because they may represent a '. . . sudden and usually unexpected replacement of one material by another, often with inferior geotechnical properties' (Higginbottom and Fookes, 1970, p. 95). They (1970) also note that the problem is especially important in connection with shallow foundations, roads and runways. A diagram by Alfred Jahn (Figure 14.25) illustrates the conditions one might encounter *vis-à-vis* ice wedges during construction in a former permafrost environment. Morgan (1971) notes that in the English Midlands fossil wedges affected trenches

in two ways:

Firstly, the trench was subject to local flooding by water running from the wedges. Secondly, the walls of the trench collapsed, either because the wedges were running parallel to, but behind, the trench wall or because the sand and gravel at the base of the till sheet was washed out.

Although such forms as fossil ice wedges may be unexpected, other fossil forms, including some pingo remnants which are common in northern Europe, are rather conspicuous and thus can be avoided by engineering works (Hutchinson, 1991).

Cambering and valley bulging are mass movements that may have a periglacial connection. Cambering (the downward displacement of strata) and valley bulging (the upward displacement of strata) are considered as related phenomena and are both responsible for a number of engineering problems. One of the major aspects associated with cambering are the gulls (filled fractures between blocks) that can cause leakage affecting foundation stability (Petley, 1991).

Present over much of former permafrost areas are vast deposits of loess which is considered by Eyles and Paul as 'the most widespread of periglacial sediments' (as reported in Culshaw et al., 1991, p. 14). Loess (Chapter 25) presents problems to the engineer primarily because of their tendency to collapse when wetted (Culshaw et al., 1991).

14.9 Engineering and present-day permafrost environments

The challenges that face engineers in present-day permafrost environments include those found elsewhere plus many that are unique; solutions are equally as diverse and complex (Morgenstern, 1981). Nearly every human activity in the Arctic vis-à-vis the environment is affected by periglacial conditions. Agriculture, even though generally limited to zones of discontinuous permafrost may, nonetheless, be affected. Large masses of residual ice in the ground when disturbed (e.g. by clearing the natural vegetation cover) is likely to result in thermokarst topography and field abandonment. In the case of mining, permafrost is usually a negative factor for open pit operations. However, permafrost, because of its stability (as long as it remains in the frozen state), can be a positive factor in underground operations.

Although agriculture and mining are affected by periglacial conditions, more important from the standpoint of engineering in the north is the influence of these conditions on the construction and maintenance of the many material structures (buildings, roads, sewers, etc.) that comprise modern culture. Engineering problems are especially acute where poorly drained, fine-grained sediment predominates. Péwé (1998, p.758), noting these difficulties, wrote that 'engineering problems in the north are of four fundamental types:

1. those involving thawing of ice-rich permafrost and subsequent subsidence of the surface under unheated structures such as roads and airfields
2. those involving subsidence under heated structures
3. those resulting from frost action, generally intensified by poor drainage caused by permafrost
4. those involved only with the temperature of permafrost that causes buried sewer, water, and oil lines to freeze'.

Because of low population density, roads and airfields are relatively few in number in the Arctic and railways are generally restricted to the sub-Arctic. Nonetheless, virtually all undergo frost heaving in winter and subsidence in summer. Maintenance is a major and almost continuous problem. The major challenge is the establishment of a bed that prevents permafrost degradation and thaw settlement (Williams, 1995). Such fills may have to be as much as 2–3 m thick to ensure sub-fill permafrost stability. Insulation materials including synthetic fabrics (Figure 14.26) help reduce the thickness of fill needed (Johnston, 1981). Another major concern in the construction of roads, railways, and airfields is drainage. Because natural drainage is poor any disruption of the surface can aggravate ponding and permafrost thaw.

Buildings, because they are heated, have an intensified degradational effect on the underlying permafrost. A variety of techniques has been devised to assist in preserving the integrity of

Figure 14.26   Oil tank built on a gravel pad in Inuvik, northern Canada; the tube vents within the gravel allow cold air to blow through in winter, ensuring the conservation of permafrost. Photograph courtesy of Tony Waltham Geophotos.

permafrost and therefore of the buildings. One of the earliest, other than thick gravel pads, was to build on pilings thus preventing heat from the building being transferred to the ground beneath. Of course pilings are subject to frost jacking which can damage bridge pilings as well as those used in buildings or other structures. A more recent technique for maintaining the integrity of the permafrost has been the utilisation of thermal piles (thermosyphons, thermotubes) which are passive heat transfer devices that operate when the air temperature is lower than that of the ground.

The problems of supplying utilities to the residents of centres of population in the Arctic and sub-Arctic are difficult and expensive undertakings. The common practice is the use of utilidors (Figure 14.27) which contain fuel distribution lines, electrical cables, sewer lines, and hot water pipes. They may be buried in or beneath the active layer or placed above ground and may vary in size up to those that can be walked through.

14.10 Oil industry

The relatively recent expansion of the oil industry in the Arctic — especially since the establishment of the Prudhoe Bay fields in the late 1960s — has led to much research on exploration, drilling, storage, and distribution of oil and gas in areas with permafrost. The challenges posed in the development of the resource involves those concerned with preserving the integrity of the environment including not only land, water, and air but also flora and fauna.

Many of the problems facing the construction of roads, airports, and buildings are aggravated in hydrocarbon exploitation (Ferrians *et al.*, 1969). These difficulties are possibly best illustrated in connection with the 1285 km long, 1.22 m diameter

Figure 14.27   Utilidors linking houses raised clear of the ground on pile foundations to conserve the permafrost in Inuvik, a modern Inuit town in northern Canada. Photograph courtesy of Tony Waltham Geophotos.

pipeline that delivers oil across Alaska from Prudhoe Bay to the Gulf of Alaska. The line crosses many different types of landscape including those with continuous permafrost, buried ice wedges, fine to coarse-grained sediments, bogs and swamps, and a variety of vegetation types. Because the oil is hot (70–80 °C), thawing of permafrost would occur if the pipeline was buried. Thus about half of the Alaskan pipeline is above ground. Frost jacking of the supporting pilings is prevented by the use of thermosyphons (Figure 14.28).

Exploitation of petroleum is being pursued offshore as well as onshore. Additional problems associated with exploration, drilling, and production involve the construction of platforms and offshore islands that can withstand sea ice and wave action. Further, in addition to these oceanographic hazards, sub-sea permafrost (Figure 14.1) must be considered in drilling and the placement of pipelines.

14.11 Summary and future challenges

The periglacial landscape presents the engineer with unique and serious problems: problems that can be solved only by the use of special techniques, procedures, and materials. Most of these problems are related to a thermal regime that includes alternating freeze and thaw. When freeze and thaw cycles occur in areas of poorly drained fine-grained sediments, periglacial processes are especially effective and surface forms such as patterned ground, pingos, and solifluction lobes are common. Changes in this thermal regime, which can be affected by many of man's activities, may lead to subsidence and frost-heave (Committee on Permafrost, 1983). Most of the challenges facing the periglacial and permafrost engineer are related to such changes and they are likely to increase in number and intensity in the near

Figure 14.28   The trans-Alaskan pipeline with thermosyphons. Photograph courtesy of Tony Waltham Geophotos.

future. General Circulation Models (GCMs) indicate that not only will atmospheric temperature rise around the globe but that it is likely to rise two to four times as much in the Arctic as elsewhere. High latitude periglacial regions, because of the unique characteristics discussed above, are especially sensitive to changes in temperature and other climatic variables and are likely to be altered more than most other regions (Maxwell, 1992). There are already indications that such changes are occurring in parts of the Arctic. In some locations, for example, the active layer is thickening, permafrost temperatures are rising, thermokarst development is accelerating, sea ice is thinning, snow fall amounts are changing as are albedos (reflected sunlight), and the dates of freeze-up are later and break-up earlier. All such changes will tax the ingenuity and expertise of those engineers called upon to work in the area.

References

Anderson, J. G. (1906) Solifluction, a component of subaerial denudation, *Journal of Geology* **14**, 91–112.

Arnborg, L., Walker, H. and Peippo, J. (1966) Water discharge in the Colville River Alaska, 1962. *Geografiska Annaler*, Series A **48**, 195–210.

Ballantyne, C. K. and Harris, C. (1994) *The Periglaciation of Great Britain*. Cambridge University Press, Cambridge.

Barsch, D. (1988) Rockglaciers. In Clark, M. J. (ed.) *Advances in Periglacial Geomorphology*. John Wiley & Sons, Chichester, 69–90.

Barsch, D. (1993) Periglacial geomorphology in the 21st century. In Vitek, J. D. and Giardino, J. R. (eds) *Geomorphology: The Research Frontier and Beyond*. Elsevier, Amsterdam, 141–163.

Brown, J. (1966) *Soils of the Okpilak River Region, Alaska*. CRREL Research Report 188, US Army, Hanover.

Christensen, I. (1978) Waterstress conditions in cereals used in recognizing fossil ice-wedge polygonal patterns in Denmark and northern Germany. In *Proceedings, Third International Permafrost Conference*. National Academy of Sciences, Washington, DC, 254–261.

Clark, M. J. (1988) Periglacial hydrology. In Clark, M. J. (ed.) *Advances in Periglacial Geomorphology*. John Wiley & Sons, Chichester, 415–462.

Committee on Permafrost (1983) *Permafrost Research: an Assessment of Future Needs*. National Academy Press, Washington, DC.

CRREL. Cold Regions Research and Environmental Laboratory. http://www. crrel.usace.army.mil/

Culshaw, M. G., Cripps, J. C., Bell, F. G. and Moon, C. F. (1991) Engineering geology of Quaternary soils: 1. Processes and properties. In Forster, A., Culshaw, M. G., Cripps, J. C., Little, J. A. and Moon, C. F. (eds). *Quaternary Engineering Geology*. The Geological Society, London, 3–38.

French, H. M. (1988) Active layer processes. In Clark, M. J. (ed.) *Advances in Periglacial Geomorphology*. John Wiley & Sons, Chichester, 151–177.

Ferrians, O. J., Jr, Kachadoorian, R. and Greene, G. W. (1969) *Permafrost and Related Engineering Problems in Alaska*. Geological Survey Professional Paper 67, Washington, D C.

Hamelin, I. and Cook, F. (1967) *Illustrated Glossary of Periglacial Phenomena*. Laval University Press, Quebec.

Harris, S. A. (1986) *The Permafrost Environment*. Barnes and Noble, Totowa.

Harris, S. A. (1988) The Alpine periglacial zone. In Clark, M. J. (ed.) *Advances in Periglacial Geomorphology*. John Wiley & Sons, Chichester, 369–413.

Harry, D. G. (1988) Ground ice and permafrost. In: Clark, M. J. (ed.) *Advances in Periglacial Geomorphology*. John Wiley & Sons, Chichester, 113–149.

Higginbottom, I. and Fookes, P. (1970) Engineering aspects of periglacial features in Britain. *Quarterly Journal Engineering Geology* 3, 85–117.

Hutchinson, J. N. (1991) Periglacial and slope processes. In Forster, A., Culshaw, M. G., Cripps, J. C., Little, J. A. and Moon, C. F. (eds) *Quaternary Engineering Geology*. London, The Geological Society, 283–331.

Jahn, A. (1975) *Problems of the Periglacial Zone*. Polish Scientific Publishers, Warsaw.

Jahn, A. (1985) Experimental observations of periglacial processes in the Arctic. In Church, M. and Slaymaker, O. (eds) *Field and Theory, Lectures in Geocryology*. Univ. of British Columbia Press, Vancouver, 17–35.

Johnston, G. H. (ed.) (1981) *Permafrost: Engineering Design and Construction*. John Wiley & Sons, Chichester.

Kriner, S. (2001) *Ice bombings bring reprieve from Siberian flood. http://www.redcross.org/news/in/flood/010522siberia.html*

Lachenbruch, A. (1962) *Mechanics of Thermal Contraction Cracks and Ice-wedge Polygons in Permafrost*. Geological Society of America, Special Publication No. 70, New York.

Lachenbruch, A. (1968) Permafrost. In Fairbridge, R. (ed.) *The Encyclopedia of Geomorphology*. Reinhold, New York, 833–839.

Lewkowicz, A. G. (1988) Slope processes. In: Clark, M. J. (ed.) *Advances in Periglacial Geomorphology*. John Wiley & Sons, Chichester, 325–368.

Mackay, J. R. (1970) Disturbances to the tundra and forest tundra of the western arctic. *Canadian Geotechnical Journal* 7, 420–432.

Mackay, J. R. (1972) The world of underground ice. *Annals of the Association of American Geographers* 62, 1–22.

Matsuoka, N. (2001) Solifluction rates, processes and landforms: a global review. *Earth Science Reviews* 55 (1-2), 107–134.

Maxwell, B. (1992) Arctic climate: potential for change under global warming. In Chapin, F. III, Jefferies, R., Reynolds, J., Shaver, G., Sroboda, J. and Chu, E. (eds) *Arctic Ecosystems in a Changing Climate: An Ecophysiological Perspective*. Academic Press, Inc., New York, 11–34.

Morgan, A. V. (1971) Engineering problems caused by fossil permafrost features in the English Midlands. *Quarterly Journal Engineering Geology* 4, 111–114.

Morgenstern, N. R. (1981) Geotechnical engineering and frontier resource development. *Geotechnique* 31 (3), 303–366.

Muller, S. W. (1947) *Permafrost or Perennially Frozen Ground and Related Engineering Problems*. Edwards Brothers, Ann Arbor.

Nekrasov, I. (1983) Dynamics of the cryolithozone in the northern hemisphere during the Pleistocene. In *Proceedings, Fourth International Permafrost Conference*. National Academy of Sciences, Washington, DC, 903–906.

Petley, D. J. (1991) Report on Session 2b. In Forster, A., Culshaw, M. G., Cripps, J. C., Little, J. A. and Moon, C. F. (eds) *Quaternary Engineering Geology*. The Geological Society, London, 409–414.

Péwé, T. L. (1983) Alpine permafrost in the contiguous United States: a review. *Arctic and Alpine Research* 15, 145–156.

Péwé, T. L. (1998) Permafrost. *Encyclopedia Britannica* 20, 752–759.

Pollard, W. H. and van Everdingen, R. O. (1992) Formation of seasonal ice bodies. In Dixon, J. C. and Abrahams, A. D. (eds) *Periglacial Geomorphology*. John Wiley & Sons, Chichester, 281–304.

Shumskiy, P. and Vtyurin, B. (1966) Underground ice. In *Proceedings*, *First International Permafrost Conference*. National Academy of Sciences, Washington, DC, 108–113.

Tedrow, J. (1966) Arctic soils. In *Proceedings, First International Permafrost Conference*. National Academy of Sciences, Washington, DC, 10–55.

Tedrow, J. (1977) *Soils of the Polar Landscapes*. Rutgers University Press, New Brunswick.

Tedrow, J. and Ugolini, F. (1966) Antarctic soils. In Tedrow, J. (ed.) *Antarctic Soils and Soil Forming Processes*. National Academy of Sciences, Washington, DC, 161–177.

Thorn, C. (1988) Nivation: a geomorphic chimera. In Clark, M. J. (ed.) *Advances in Periglacial Geomorphology*. John Wiley & Sons, Chichester, 3–31.

Walker, H. J. (1967) Riverbank dunes in the Colville delta, Alaska. *Coastal Studies Bulletin* **1**, 7–14.

Walker, H. J. (1983) E Pluribus Unum: small landforms and the arctic landscape. In Gardner, R. and Scoging, H. (ed.) *Mega-Geomorphology*. Clarendon, Oxford, 39–55.

Walker, H. J. (1991) Bluff erosion at Barrow and Wainwright, Arctic Alaska. *Zeitschrift für Geomorphologie,* Supplementband **81**, 53–61.

Walker, H. J. and Arnborg, L. (1966) Permafrost and ice-wedge effect on riverbank erosion. In *Proceedings, First International Permafrost Conference*, NAS-NRC, Publ. 1287, National Academy of Science, Washington, DC, 164–171.

Washburn, A. L. (1980) *Geocryology. Survey of Periglacial Processes and Environments*. Edward Arnold, London.

Washburn, A. L. (1985) Periglacial problems. In Church, M. and Slaymaker, O. (eds) *Field and Theory, Lectures in Geocryology*. Univ. of British Columbia Press, Vancouver.

Williams, P. J. (1995) Permafrost and climate change: geotechnical implications. *Philosophical Transactions* **352** (1699), 347–358.

Williams, P. J. and Smith, M. W. (1989) *The Frozen Earth*. Cambridge University Press, Cambridge.

Further reading

Church, M. and Slaymaker, O. (eds) (1985) *Field and Theory: Lectures in Geocryology*. University of British Columbia Press, Vancouver.

Davis, N. (2001) *Permafrost: A Guide to Frozen Ground in Transition*. University of Alaska Press, Fairbanks.

Dixon, J. C. and Abrahams, A. D. (eds) (1992) *Periglacial Geomorphology*. John Wiley & Sons, Chichester.

French, H. M. (1996) *The Periglacial Environment*. Second Edition. Addiser Wesley Longman, London and New York.

John, B. and Sugden, D. (1975) Coastal geomorphology of high latitudes. *Progress in Geography* **7**, 53–132.

King, C. (ed.) (1976) *Periglacial Processes*. Benchmark Papers in Geology 27. Dowden, Hutchinson & Ross, Inc., Stroudsburg, Pennsylvania.

Linell, K. and Johnston, G. (1973) Engineering design and construction in permafrost regions: a review. In *Proceedings, Second International Permafrost Conference*. National Academy of Sciences, Washington, DC, 553–575.

Linell, K. and Tedrow, J. (1981) *Soil and Permafrost Surveys in the Arctic*. Clarendon, Oxford.

Péwé, T. L. (ed.) (1969) *The Periglacial Environment: Past and Present*. McGill-Queens University, Montreal.

Williams, P. J. (1986) *Pipelines & Permafrost*. Carleton University Press, Ontario.

Woo, M-K. and Gregor, O. J. (eds) (1991) *Arctic Environment: Past, Present & Future*. McMaster University, Hamilton.

Walker, H. J. (1973) The morphology of the North Slope. In Britton, M. (ed.) *Alaskan Arctic Tundra*. Arctic Institute of North America, Washington, DC, 49–92.

15. Temperate Environments

Kenneth J. Gregory

Temperate landscapes have often been visualised as those that remain when other, more distinctive world environments have been identified. The types of temperate area (Figure 15.1) are listed with process characteristics, engineering problems and geohazards (Table 15.1). These areas include some of the most densely populated parts of the earth's surface, have presented great challenges for engineering, and have stimulated research which has given much of the foundation for geomorphological understanding. They are the areas which inspired the normal cycle of erosion, the importance of the drainage basin, and the basis for contemporary hydrological understanding of runoff generation.

However, with hindsight, for temperate areas to provide the norm against which other areas should be studied is inappropriate because the rates of operation of present processes (rivers, Chapter 10; coasts, Chapter 21; slopes, Chapter 11) are not typical; processes and environments have been greatly affected by human activity especially through removal of the original vegetation, and temperate landscapes contain much evidence of landscape change during former climates, so that the majority of landscape features and their deposits do not now relate to contemporary environmental processes. The range of temperate environments (Figure 15.1) can be thought of as the domain of rain and rivers operating on landscapes which often contain the record of a variety of past environmental processes. Process domains have been identified (Thornes, 1979; 1983) according to the spatial distribution of work done by several processes, often typified by some particular environmental parameter such as rainfall intensity. Domains therefore represent equilibrium relationships between processes as related to the controlling parameters such as climate,

infiltration rate, or vegetation or land use cover density. The temperate domain can be introduced by the deposits, sediments and the landscape (15.1 Deposits, sediments and the landscape), the movement of water through the landscape system (15.2 Water movement), leading to the significance of human impact for engineering (15.3 Human activity), and thence to the key elements for an engineering geomorphology approach (15.4 Summary).

15.1 Deposits, sediments and the landscape

Thin deposits over the surface of many temperate areas are often relict from former climatic or environmental conditions. It is therefore difficult for the engineer to deduce how extensive a particular deposit is and how much variability occurs within it because, as the deposits originated under conditions which no longer exist, there is no simple basis for estimation. The range of deposits encountered in temperate landscapes includes not only weathering, colluvial and fluvial materials originating from contemporary conditions, but also marine, aeolian, glacial, periglacial, and tropical deposits, which can still be significant amounts of material although they originated during earlier phases of landscape development. The major problems associated with the interpretation of each type of deposit are indicated in Table 15.2.

It is often necessary to have some indication of the lateral and vertical extent of a deposit; whereas some deposits blanket the landscape, as a glacial deposit may completely infill a former valley with little or no indication on the surface (Figure 15.2), in other cases the deposits are associated with particular features or distinctive

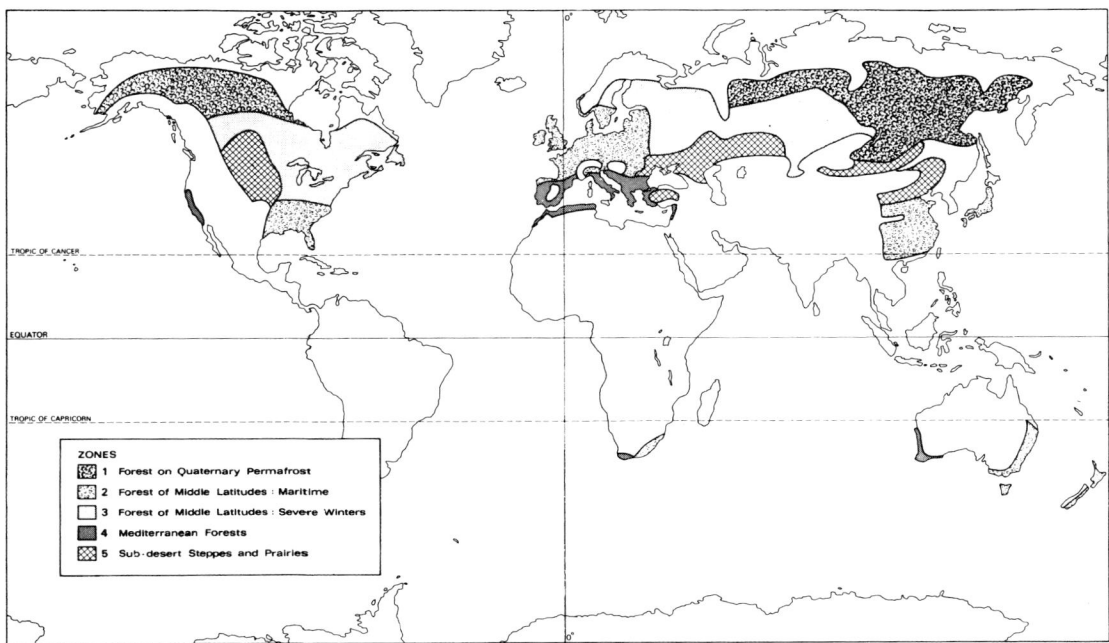

Figure 15.1 Temperate landscape systems. Based upon Tricart and Cailleux (1972). Details for each system including possible geohazards are given in Table 15.1.

landforms. Anomalous channels or valleys which are steep-sided and flat-floored and which at present contain a small stream or no drainage at all, were the lines along which glacial melt-water flowed. Such channels commonly have glacial till or periglacial deposits on their floor and perhaps also on their sides, and the floor may have been infilled by peat (Figure 15.3). When building construction takes place in areas where such channels occur the different properties of each deposit in the sequence dictate that site location must be planned very carefully and that engineering structures must often be anchored to the underlying bedrock. Deposits over the landscape may be changed significantly by various types of human activity as summarised together with attendant problems that may arise in Table 15.3.

Climatic changes

Four orders of change each having had a significant effect on the deposits and landforms of temperate environments can be envisaged (Brown, 1991) namely:

1. First order: caused by major cyclic, and probably Milankovitch mechanism, variations in climate involving large fluctuations in the water balance. Such changes included changes from glacial to interglacial climate conditions which occurred more than twenty times during the Quaternary.

2. Second order: caused by shorter climatic variations such as interstadials with a periodicity of either 40 000 or 25 000 years.

3. Third order: due to climate or geomorphology (e.g. earth movement in Japan, Turkey or New Zealand) or human activity (e.g. deforestation).

4. Fourth order: effects of individual years or events that may persist for many years and may be important thresholds to new states. The effects of a major flood could have long-term effects upon the landscape or lead to a threshold being crossed in order to move to a new state.

Table 15.1 Subdivisions of the temperate environment (Based upon Tricart, 1957; details of zones described in Tricart and Cailleux, 1972). Under natural conditions zones 1–4 would be forested but have been substantially transformed by human activity. Zones 1 and 5 are transitional to other zones.

Zone (see Figure 15.1)	Climate	Processes and features	Major hazards which may occur (using list in Alexander, 1999)	Problems relevant to engineering
(1) Forest on Quaternary permafrost	Severe winters, may be associated with periglacial zone	Permanently frozen ground beneath land surface may be continuous or discontinuous, is residual from the Quaternary and not forming at present	Wildfires Frost or ice storm Snowstorm Subsidence Windstorm	Modification of surface affects thermal regime and can lead to thermokarst features with surface collapse
(2) Forested zone of middle latitudes	Maritime without severe winters. No large seasonal variations in temperature or humidity	Chemical erosion limited by moderate temperatures, some frost action but penetration rarely reaches bedrock. High angle slopes can be stable where still covered by forest	Accelerated erosion Avalanches Soil heave and collapse Floods Landslides Coastal erosion	Landsliding on devegetated slopes Many ancient deposits over landscape Flooding may increase downstream of vegetation changes
(3) Forested zone of middle latitudes	With severe winters and seasonally distributed precipitation	Heavy showers and snowmelt can produce higher streamflow rates than in zone 2, mechanical processes more important as frost penetration is great and can reach bedrock. Chemical erosion limited by winter frost but slopes can be steep, up to 20° to 35°, and rectilinear	Drought Severe thunderstorms Hailstorms Snowstorms	Landslides when vegetation removed Downstream flooding increases when vegetation changed and other catchment characteristics altered
(4) Mediterranean forested zone of middle latitudes	Seasonal precipitation, mild winters, warm/hot summers. Frost uncommon at low elevations	Alternation of wet and dry conditions induce major influence such as an incidence of landslides. Seasonal stream flow regime can give high seasonal discharges which elevate course debris and rapid dissection and gullying where vegetation removed or degraded	Soil erosion Floods, high spatial and temporal variability High sediment yields along rivers Earthquakes Volcanic eruptions Landslides	Sheet erosion where vegetation removed Increased flooding downstream, and gully development may occur
(5) Subdesert steppes and prairies	Summer rainstorms, dry cold, severe winters	Transitional to temperate deserts with some frost action in winter. Wind action, occasional sheet wash and gullying	Drought Tornadoes Soil erosion	Deflation encouraged by removal of vegetation Gullying where land ploughed

Table 15.2   Types of deposit which may occur in temperate areas.

Type of deposit developing in temperate areas at present but also in the past	Characteristics	Associated problems
Weathering and colluvial	Medium grain size, not usually deep, may have developed on top of fossil deposits, podzolized soils on surface	Some lateral variation due to subsurface drainage
Fluvial	Range of grain sizes present and often incorporate remnants of fossil deposits — frequent lateral and vertical changes	Small changes in river position can release new exposures of relict sediments
Marine	Range of grain sizes present, may incorporate fossil deposits from cliff deposits or from off-shore sediments	Deposits may also occur above present sea level marking former shorelines
Aeolian	Fine silts common in zone 5 and also in other areas where vegetation removed and then field boundaries removed. Under former conditions had loess or water-sorted loess (brick earth) cover	May be mantle over surface and not related to deposits of very different character below
Relict from former conditions		
Glacial	Till deposits, lake clays, gravels and sands	Distribution and character not easily deduced; rapid variations in thickness
Periglacial	Angular scree deposits, unsorted slope deposits, fine wind-blown deposits	Distribution localised but character reflects locally available rock types — may be on slopes which are relict and unstable when modified
Tropical	Clays which are remnants of deep weathering	Occur localised, often on plateau sites, to considerable depth in pockets

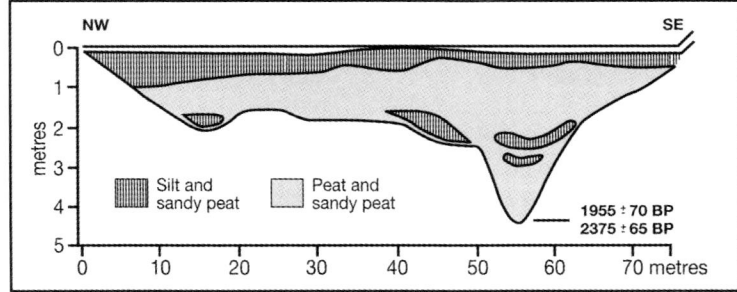

Figure 15.2   Section across an infilled channel palaeomeander at Czmoniec, Poland (redrawn from Kozarski, 1983) showing how there can be great differences between the form of the land surface (here virtually flat) and the form of the subsurface profile.

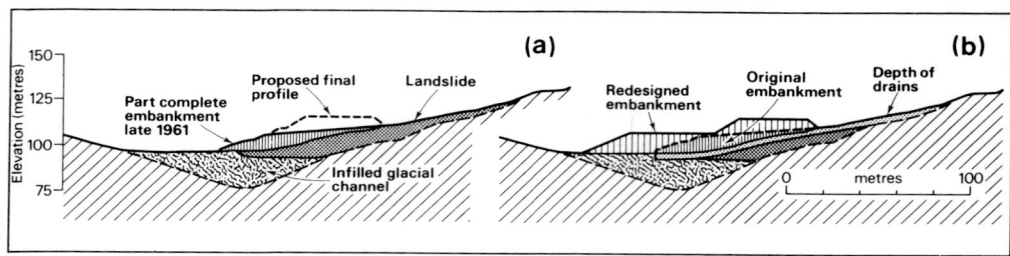

Figure 15.3　Significance of deposits in relation to road construction. In (*a*) the early stages of motorway embankment construction above an infilled glacial drainage channel at Walton's Wood Staffordshire are indicated. Failure of the embankment necessitated reconstructions and remedial measures as shown in (*b*). After Chandler (1977) and based upon Early and Skempton (1972).

Table 15.3　Examples of problems following changes in deposits.

Change due to:	Consequences
Weathering	Progressive reduction in shear strength may give slope failure in natural or manmade slopes
Drainage	Contraction of deposits can give substantial lowering of surface elevation, and so increase flooding. Desiccation cracks can provide location for slumps or for gully development
Groundwater changes	Groundwater rise, for example due to creation of reservoir, can increase pore pressures and increase possibility of slope failure
Loading	Uneven amounts of settlement because of variations in deposits and in layers of deposits
Piping	Localisation of infiltration on surface by land use practices can lead to increase of subsurface piping and this can subsequently lead to collapse and gully development
Expansive sediments and rocks	Swelling especially of soils with appreciable montmorillorite clay content can occur and with a slight increase of moisture content can have differential effect on surface

Temperate areas are now covered by several types of deposit, many of which originated under past process regimes. Similarly the landscape features and landforms owe their origins to the impact of a succession of climate changes. Deposits and landforms of temperate landscapes were once parts of one of the succession of dynamic systems that were arrested leaving material stored over the landscape. Analysis of this succession of changes has been undertaken using landscape stability and instability, identification of critical thresholds, and linkages between different sediment stores. From a sediment point of view the temperate landscape is the result of the intersection of different domains, with sediments from past systems still retained in the landscape environment. It has been suggested (Büdel, 1982; Lewin, 1987) that 95% of the temperate landscape may be composed of landforms inherited from pre-Holocene conditions. Such inherited landscapes include features produced by glacial processes, others which include evidence of periglacial processes (Chapter 14) and some reminiscent of contemporary loess lands. As the climate and the associated erosion systems in the past changed from glacial to periglacial to temperate and back to glacial again, so a considerable range of deposits and landform features can occur in a single area (Figure 15.4).

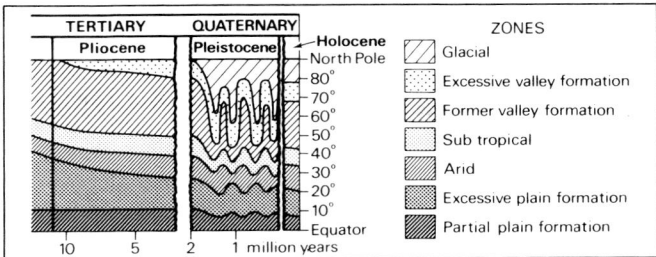

Figure 15.4 Fluctuations of world landscape zones in the late Tertiary and the Quaternary. The zone of former valley formation and the subtropical zone which are the principal temperate areas have varied in position during the glacial and interglacial phases of the Quaternary. This diagram based on Büdel (1982) presents a very simplified picture which is now known to be much more complex and composed of many oxygen isotope stages (Chapter 6).

Such changes, which occurred in a time sequence, were accompanied by changes in sea level, which in turn had an influence on landforms because valley deepening was associated with low sea levels leaving former valley floors as river terraces. The reverse development, aggradation, a building up of alluvium and alluvial deposits, accompanied higher sea levels of the past. Therefore the temperate landscape may be thought of as broadly composed of three major components (Figures 15.5 and 15.6) as listed below.

1. Plateau areas are where many features were initially produced by pre-Quaternary erosion. These are the remnants of landscapes produced under Tertiary conditions, which may have resulted in planation surfaces generated under tropical or subtropical conditions, later affected by erosional and depositional processes of glacial or periglacial conditions.

2. Major valleys are the product of the erosional and depositional phases of the last two million years of the Quaternary, in which the alternation of glaciations and interglacials was accompanied by major sea level fluctuations.

3. The detailed development of the environments along the present river courses and valley floors are the products of the last 10 000 years of the Holocene, when the climate changed from glacial to interglacial. The impact of human activity has become increasingly pronounced,

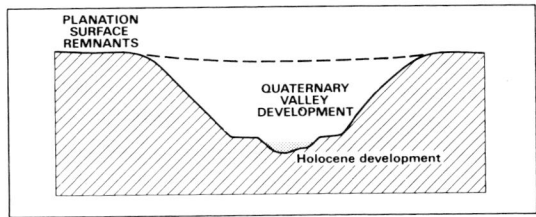

Figure 15.5 Generalised section across an upland valley in the temperate zone (compare with Figure 15.6). Three elements in the cross-section are (1) planation surface remnants which are the remnants of old land surfaces often produced in the Pliocene or earlier; (2) valley development which occurred mainly during the Quaternary and (3) Holocene erosion during the last 10 000 years often confined to areas closest to rivers and streams.

and some changes in slope or river dynamics may have been concentrated in short periods of time.

Viewed along the line of major valleys, this sequence has culminated in the stages illustrated in Figure 15.7. The present temperate landscape must therefore be viewed against the background of up to 10 million years of landscape development under the successive influence of a sequence of contrasting erosion systems so that a profile across a major temperate valley (Figure 15.8) can reflect many of the stages of Quaternary development (Chapter 6).

A composite section across the valley of the Wislowka in southern Poland is shown in

Figure 15.6 Headwaters of River Exe basin, Somerset, UK. The snow-covered plateau surfaces are remnants of land surfaces produced more than 2 million years ago. The valley sides, covered by woodland and plantations, are the result of dissection during the Quaternary.

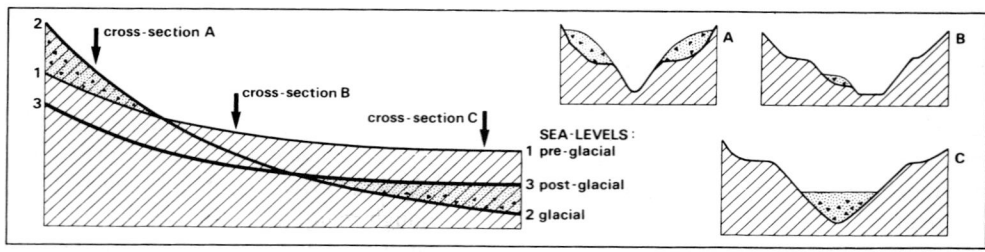

Figure 15.7 Diagrammatic representation of stages of Quaternary valley development. These stages occurred as valley development took place as shown in Figure 15.5. Diagram based on Butzer (1976) after Dury (1970). Phases of erosion and downcutting occurred when sea levels were relatively low and deposition occurred when sea levels were higher. In addition to the influence of sea level, changes of climate also affected river regime and sediment transport.

Figure 15.8. The deposits in that section have accumulated in the last 40 000 years and many of them during the last 11 000 years. In temperate areas the alluvial chronology may be reconstructed as the stages of development through which the valleys have evolved during the Quaternary (Figure 15.5). Such alluvial chronologies are based largely upon the deposits and dateable material which have survived and five major influences upon the alluvial chronology in any

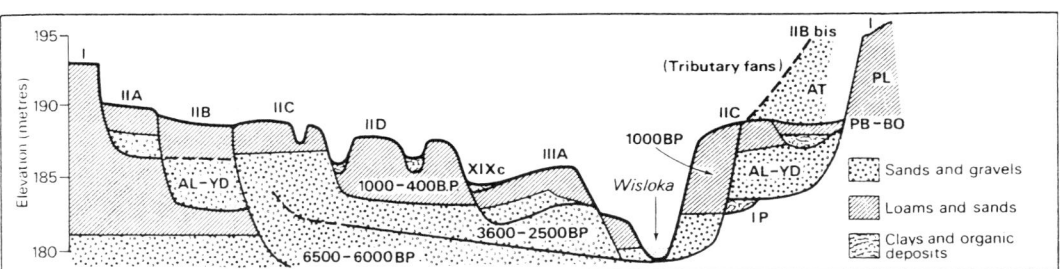

Figure 15.8 Detailed valley cross-section across Wisloka valley, southern Poland showing the complex sequence of deposits produced during the last 6000 years (after Starkel, 1983).

area may be recognised. First is the response to changes of sea level, as indicated in Figure 15.7, as there will be phases of deposition of river and marine deposits, and of erosion of such deposits. Secondly, there are effects of changes of climate which will be indicated by the character of the ground slope and the fluvial deposits. Thirdly, there will be influences from other areas, when, for example, ice-transported deposits or wind blown sediments are added to the alluvial sequence. Fourthly, in tectonically active areas like Japan, Turkey or New Zealand, there may be earth movements which displace the sediments. Finally, the influence of human activity may have affected not only the amount of particular deposits but also the type of sediment produced. Such deposits have been dated in a variety of ways including reference to archaeological evidence and by use of radionuclides (e.g. Stokes and Walling, 2003). The course of valley development in a temperate landscape can result in some or all of these five major contributions being blended together to make up cross-profiles such as the one illustrated in Figure 15.8.

Advances in the techniques available for the interpretation of the environment and the dating of its evolutionary stages have enabled a more refined chronology to be determined for the Cainozoic; this has been accompanied by an enhanced understanding of the way in which processes operate in temperate landscapes and of the ways in which those processes differ over the temperate environment and have changed in recent decades especially as a result of human activity. A change has therefore occurred in the

way in which the past evolution of temperate landscapes has been interpreted in that a more retrospective approach has been employed using knowledge of contemporary process domains to extrapolate back into the past. In this way palaeohydrology, which has been defined as 'the science of the waters of the earth, their composition, distribution and movement on ancient landscapes from the beginning of the first rainfall to the beginning of continuous hydrological records' (Gregory, 1983, p. 10; 1996, p. 2), has utilised an approach retrospective from current processes. Reconstruction of environmental change (Gregory, 2000) in temperate landscapes has come to depend upon an increasingly multidisciplinary approach and one which endeavours to understand the way in which contemporary process–response systems are at variance with those of the past as a consequence of human activity.

Sediment transfer and model hillside

Although many deposits in the temperate landscape may be inherited from past environmental conditions, two types of changes of deposit can occur under contemporary conditions. First, weathering processes (Chapter 2) can lead to a reduction in material strength and hence to slope failure. Secondly, transfers of sediments in the landscape can be achieved either by gravitational processes operating on hillslopes or by fluid transport.

The temperate hillslope can be regarded as composed of any combination of nine hypothetical landsurface units (Dalrymple et al., 1968) and

Table 15.4 Units of the hypothetical nine-unit landsurface model (Figure 15.9) based upon Dalrymple *et al.* (1968).

Unit	Predominant present geomorphic processes
(1) Interfluve	Pedogenetic processes associated with vertical subsurface soil water movement
(2) Seepage slope	Mechanical and chemical eluviation by lateral subsurface water movement
(3) Convex creep slope	Soil creep, terracette formation
(4) Fall face	Fall; slide; chemical and physical weathering
(5) Transportational midslope	Transportation of material by mass movement (flow, slide, slump, creep); terracette formation; surface and subsurface water action
(6) Colluvial footslope	Redeposition of material by mass movement and some surface wash. Fan formation. Transportation of material; creep; subsurface water action
(7) Alluvial toeslope	Alluvial deposition; processes resulting from subsurface water movement
(8) Channel wall	Corrosion, slumping, fall
(9) Channel bed	Transportation of material downvalley by surface water action; periodic aggradation and corrosion

each of these has certain properties and therefore a certain significance in relation to engineering development (Table 15.4). This hypothetical model (Figure 15.9), has been developed in relation to soil profile development and is particularly related to the mode of water flow. In addition it is necessary to envisage the ways in which several types of mass movement may take place and also to estimate where and when a particular type of movement will occur. The locations may be envisaged from the scars of previous mass movements but in general the incidence of specific failures may arise as a consequence of the reduction of the material shear strength occurring as a result of a number of significant variables (Table 15.5). In some areas such as California the sequence of intense rainstorms can produce mass movements in slopes over the landscape (Figure 15.10) and slope instability is also found in areas where recent uplift has occurred in Mediterranean areas and New Zealand.

Sediment is also transferred in the temperate landscape by water flow, over the slopes as sheet flow, and along river courses, where it is transported as bedload rolled or jumped along the bed, or as suspended load, carried in suspension

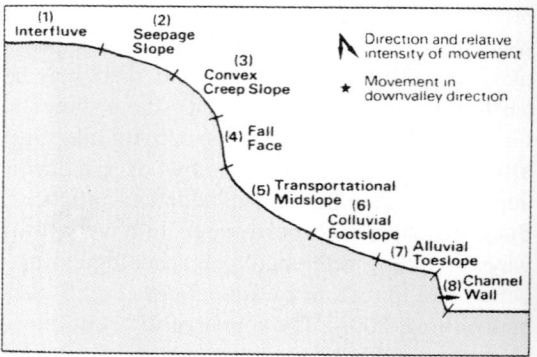

Figure 15.9 The hypothetical nine-unit landsurface model (after Dalrymple *et al.*, 1968). Details of the units are given in Table 15.4.

together with material carried as solutes or dissolved load (Chapter 10).

Location and storage of sediment

Although in temperate areas there are no very high rates of erosion, unless there is profound disturbance by human activity, two aspects of sediment transport are particularly relevant to engineering. First is the fact that in many temperate areas the

Table 15.5 Variables which may cause reduction in shear strength of materials and land-surface.

High intensity precipitation	The temperate landscape is adjusted to events which occur on average every few years so that precipitation falls which are rare (e.g. 1 in 100 years) may reduce material shear strengths in specific locations
Vegetation removal	Increases amount of moisture infiltrating and so can increase porewater pressures and decrease shear strengths
	Removal of root network removes binding influence upon material
Ploughing	In grassland areas breaks down the plant—soil structure where a mat is provided by the perennial grasses

(see also Tables 15.3, 15.8 for rate of groundwater changes, drainage, weathering).

Figure 15.10   Coast ranges of California showing extent of landslide scars produced by heavy rainfall January–March 1980. The rainstorms succeeded a wet autumn so that debris slides occurred extensively over the landscape and disrupted the chapparal (Mediterranean type) vegetation.

sources of sediment are restricted so that the movement of sediment through the temperate system is less than would be expected and takes place at sub-capacity levels. If unlimited amounts of sediment were to be available then it would be possible to deduce amounts of transport more easily from values of water discharge or of water velocity using hydraulic equations. However, in temperate areas most sediment sources occur near stream channels so that as little as 10% of a drainage basin will be contributing to sediment transfer. Erosion rates in temperate areas tend to be fairly low unless the erosion regime is very seasonal and characterised by rainfall intensities greater than 25 mm per hour. If stream or river discharge is reduced, for example as a result of the construction of a dam or reservoir, then the river channel may be armoured by the selective concentration of a coarse layer on the channel bed which reduces the availability of sediment to the stream channel. Conversely if the discharge is increased new sediment sources may be eroded, and wherever the landsurface is modified by human activity then the areas most likely to influence the sediment supply must be treated with great care. This applies particularly to the riparian vegetation along water courses which should be removed only if absolutely necessary; removal means that protective engineering measures are then required. This is one reason why strategies for the conservation

of aquatic ecosystems, for example in the United States, emphasise the protection and restoration of stream side riparian forests. However the issue of whether the banks of streams and rivers are better left to revert to forest or not is complicated because there is a myriad of interrelated variables involved which can include sediment supply, size and lithology; magnitude and frequency of water discharge; nature of overbank materials; presence and type of vegetation on the banks; types of channel and its history of disturbance; as well as the drainage basin context (Montgomery, 1997) — all affecting whether streambank trees can stabilise or erode the banks on which they grow.

A second feature affecting sediment transfer in temperate areas is the considerable amount of storage in the system. Not only does this apply on the long time scale in the sense of the various forms of deposition referred to earlier (climatic changes), but on a much shorter timescale of decades, there can be storage on slopes, in flood plains, and along valley floors (e.g. Figure 15.11) which can all be released if the hydrological system is changed substantially. In many temperate areas floodplains store the sediment that was released when the vegetation, usually forest, over the catchment areas was cleared by human activity which could have begun some 4000 years ago. In the Severn and Wye catchments it was suggested (Brown, 1987)

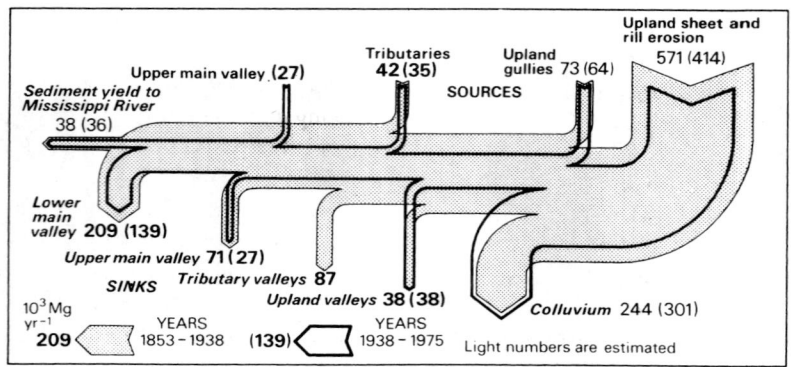

Figure 15.11 Sediment storage for Coon Creek, Wisconsin (after Trimble, 1983). Numbers are annual averages of sediment supply in 10^3 Mg year and show the importance of different sediment sources and also compare the storage and supply before conservation measures (1853–1938) with the subsequent situation (1938–1975). The storage for 1975–1993 is reduced significantly (Trimble, 1999).

that one might expect much of the Severn's flood-plain sediment, rapidly deposited around 2000–3000 years ago, to have been remobilised by increased fluvial action due to increases in flood magnitude and frequency and decreased slope erosion. In fact, the channelization and regulation of British lowland channels prevented this occurring by artificially lengthening the residence time of sediment in floodplain storage. Embanking had also led to a reduction in flooding and in flood-plain accumulation rates, but to an increase in within-channel sedimentation especially in lower reaches. The quantities of stored sediment in low-land basins such as that of the Severn suggested that human activity altered the relationships between the erosion, transport and storage processes during the mid and late Holocene. Consequently as much sediment went into storage as has left the Severn and Wye basins during the Holocene. The sediment storage component is therefore not spatially uniform and in particular it is likely that the majority of hillslope erosion went into proximal colluvial storage. The complexity of alluvial sediment accretion is illustrated by the agricultural Coon Creek basin, Wisconsin where sediment budgets with conservation measures 1938–1975 were a fraction of those 1853–1938 (Figure 15.11) and in 1975–1993 were only about 6% of the rate that occurred in the 1930s (Trimble, 1999). Adjustments in sediment storage that arise from specific structures are indicated in Table 15.6, together with indications of the increase in sediment production that can arise from human activity.

Some parts of the temperate zone have been characterised by conditions in which sediment supply has been greatly increased, usually as a consequence of human activity. Soil erosion over the catchment area of the Huang He (Yellow River) basin of China is indicated in the saying 'once the skin is gone where can the hair grow?' It has been responsible for the largest sediment load of any world river, carrying nearly 10% of all the sediment transported to the oceans from the surface of the globe (Walling, 1981) and averaging 1.69 billion tonnes annual sediment production 1919–1996 (Liu Changming, 2000). As a result of the high sediment load the lower

Table 15.6 Changes in sediment storage due to specific structures.

Type of structure	Effect on sediment storage
River dam with reservoir	Sediment trapped upstream of dam; accelerated erosion downstream of dam. May have deltaic accumulation of sediment where rivers enter reservoir
Channelization, flood prevention schemes, urban drainage systems	Sediment sources covered. Low rates of sediment transport until sources become available downstream. In-channel storage on riffles and bars also precluded
Slope control measures by grading or drainage	Render sediment unavailable to surface runoff

reaches of the river channel have silted and many sections of the river bed have risen. The 'suspended river' means that the river bed in the lower reaches is generally 3–5 m higher than the floodplain behind the levees and in some sections the height above the floodplain is 10 m (Liu Changming, 2000). Major problems in the basin include severe soil erosion in the middle reaches, flood hazard in the lower reaches, degradation of ecosystems throughout the basin as a result of deforestation, grassland degradation, desertification, saline and alkalisation, and water pollution throughout the basin for which a management strategy has been developed. The river has ceased flowing in its lower reaches on twenty-one occasions in the period from 1972 to 1998 and in the 1990s the river's lower reaches dried up every year for increasingly longer periods. Problems arising from the drying up of the river include increased flood hazard (as a result of the suspended channel), serious ecological effects in the delta area, increased saline intrusions, and water supply problems.

Deforestation in the Mediterranean basin produced accelerated erosion which has been reflected in a series of stages recorded in the alluvial chronology.

15.2 Water movement

Temperate areas are classically regarded as the domain of rain and rivers. As problems have arisen in relation to some parts of the transfer of water over the land surface, it has been discovered that an engineering solution, necessary to solve a particular problem in one area, may often create a further problem elsewhere. Thus a flood prevention scheme involving channelization in one area may lead to increased flooding downstream. Such experience has influenced the way in which engineering geomorphology has evolved (15.4 Summary).

Drainage basins and water transfer

Studies within temperate areas are often based upon the drainage basin (Figure 15.9) or catchment which is the unit bounded by a watershed or divide within which precipitation is collected and conveyed to rivers (Gregory and Walling, 1973). Such a unit is necessary to analyse the rain and river budget of temperate areas. The unit can be used for the calculation of sediment yield from an area or for the solution of the water balance equation which relates runoff (R in mm), precipitation (P), losses by evapotranspiration (E_t) and changes in surface soil or groundwater storage (ΔS) in the form:

$$R = P - E_t \pm \Delta S$$

This indicates why the drainage basin is often used as the basis for the management and planning of temperate areas.

Water transferred through drainage basins in temperate areas, as part of the hydrological cycle, follows a series of possible routes (Figure 15.12) and three aspects of this series of routes are potentially important. First is the relative importance of quick flow and delayed flow in the basin. The amount of delayed flow is important in relation to groundwater recharge and therefore to the provision of ground water supplies. Human activity tends to increase the amount of quick flow. Second is the fact that in a temperate landscape the network of stream channels changes in short periods of time. A network is composed of perennial stream channels that flow all the time and will

be fed by groundwater or delayed or base flow during dry periods; of intermittent streams that flow only when the water table is seasonally high, and of ephemeral streams that will flow only during particularly heavy rainstorms. Thus under natural conditions water will flow along a network of stream channels which is constantly changing in relation to the hydrograph expressing variations of discharge along a river against time (Figure 15.13). The drainage network will be most extensive and composed of perennial, intermittent and ephemeral streams just before times of peak

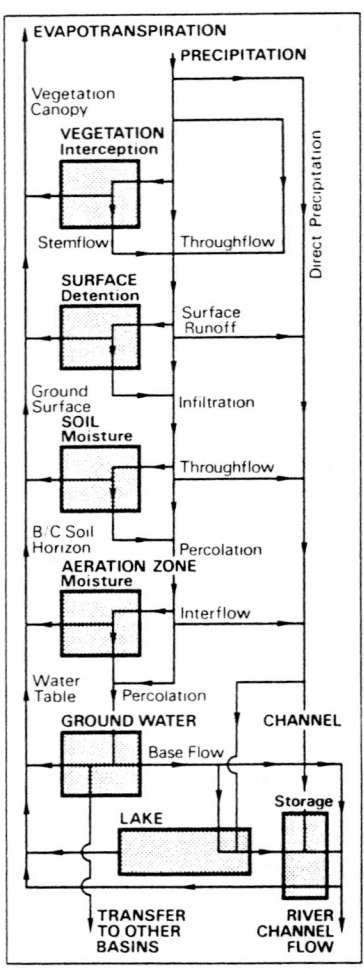

Figure 15.12   Routes taken by water in the land-based part of the hydrological system (after Derbyshire *et al.*, 1979).

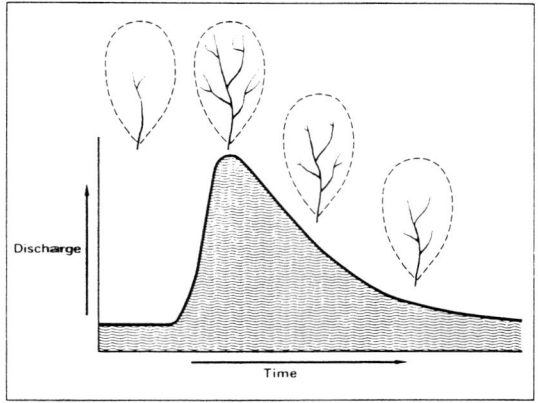

Figure 15.13   The relationship of a discharge hydrograph to the extent of the stream network. As the drainage network extends with more streams flowing so the rate of water flow through the drainage basin takes place more rapidly, affecting the shape of the discharge hydrograph.

Table 15.7   Examples of velocities of water flow in drainage basins. Based upon data collected in Table 2.2 of Derbyshire *et al.* (1979).

Type of flow	Velocity of water flow
Overland flow	3–15 cm/s on slope of 0.40 Less than 0.1 cm/s on low slopes with thick vegetation cover
Vertical percolation	Less than 7.5 cm/day in Georgia
Saturated throughflow	20 cm/h (0.2–37.2 cm/h saturated hydraulic conductivity values collected from various field measurements)
Throughflow	80 cm/day in B horizon in Somerset; 50 cm/day in B/C horizon, 10–20 cm/s in Central Wales
Stream channel flow	Average 45 cm/s

stream flow, whereas when the discharge is low it will be the less extensive network of permanent streams which will contribute to river flow. These three modes of water flow through a drainage basin are the end members of a continuum. At the head of the ephemeral streams there may be natural pipes which can be up to 0.5 m in diameter and there will also be flow through the soil, either matrix flow or diffuse flow. Although there is a continuum of water transfers through the basin, ranging from water flow through soil pores at one extreme, to open channel flow in a large river channel at the other, there is a contrast in the velocity of channel flow and other types of water flow (Table 15.7). If the characteristics of the flow routes are changed inadvertently, it is possible that water will move through the drainage basin more rapidly, so that an increased incidence of flooding may be an inevitable consequence further downstream.

A third aspect of the variety of flow routes is that they are often extended, modified or changed in temperate areas, and such changes can induce flooding as illustrated in Table 15.8. Particular concern has arisen from the modification of river courses by channelization. Measures of stream renovation or stream restoration have been proposed

which altogether work with the river rather than against it, and endeavour to imitate the natural features of river channels rather than to eliminate them from the drainage basin (Figure 15.14).

Wood debris in channels

Prior to the influence of human activity many temperate areas were forested, and the forest influenced the way in which environmental processes operated. A characteristic feature was that not only did the forest inhibit mass movement on slopes but it also contributed debris to the stream and river channels. In the headwaters of drainage basins accumulations of coarse woody debris were much more commonly encountered than at the present time. Such debris accumulations may occupy less than 2% of stream channel length, but can be responsible for half the total flow resistance, can account for 4% of the vertical drop in a channel long profile, and for 70% of the sediment stored in the channel. Such debris dams can be very densely distributed with average spacing as little as every 2.8 m and so they significantly affect stream channel processes. In central Europe the great phase of deforestation (Darby,

Table 15.8 Significance of changes of water flow routes. Left-hand column shows factors affecting the flow velocities indicated in Table 15.7.

Increase due to:	Consequences include:
Increased overland flow velocities due to reduction of surface roughness for example by removal of vegetation or of surface irregularities	Possible increased erosion of surface. Runoff production time reduced and possible greater peak discharges downstream with more frequent flooding
Land drainage whereby channel flow replaces overland flow	Decreased time of travel of water through drainage basin which can give higher flood peaks and increased flooding downstream
Channelization of river by decreased roughness and decreased flow resistance can give higher stream velocities	Flooding may be increased and erosion may be accelerated downstream
New areas of impervious surface such as roads allowing surface flow where infiltration occurred previously	Higher runoff rates, and possible localised erosion of sediment adjacent to roads

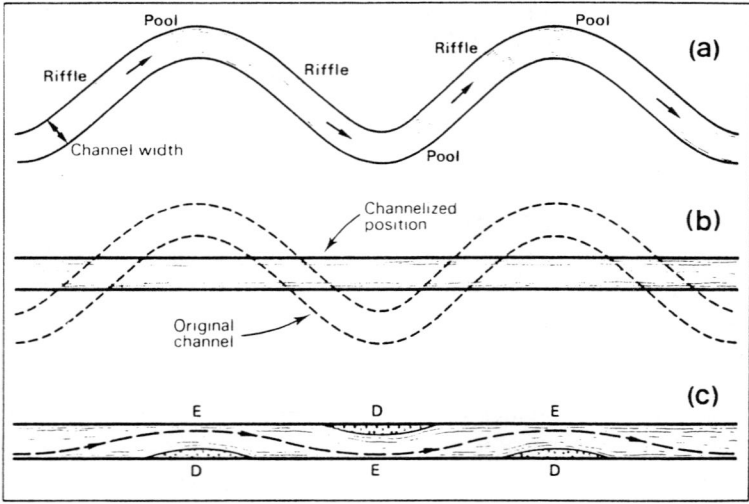

Figure 15.14 Effects of channelization (after Coates, 1982). The natural meandering stream (*a*) contrasts with the straightened engineered channel (*b*) and subsequently (*c*) the channelized stream begins to adjust by becoming sinuous with developing point bars.

1956) that occurred for 200 years after 1050 AD must have had a significant influence upon the supply of wood and upon the dynamic influence that it had on rivers. Riparian woodland persisted despite deforestation, although it is estimated that in North America and Europe more than 80% of riparian corridors have disappeared in the last 200 years (Hupp, 1999).

15.3 Human activity

References have already been made to deforestation which began in the Mesolithic (*c.* 9000 years ago) and the Neolithic (5000 years ago). Large tracts of land were deforested in Britain before the Romans arrived in the first century BC, and there was a great wave of deforestation in western and central Europe in mediaeval times. In mediterranean lands the natural vegetation was mixed evergreen and deciduous forest of oaks, pine, beech and cedars; deforestation began as early as 4600 years ago. Temperate North America which was wooded from the Atlantic coast as far west as the Mississippi when the first Europeans arrived, lost more woodland in the following 200 years than Europe had lost in the previous 2000. Deforestation, followed by conversion of land to grazing and arable land, or use for urbanisation affects much of the temperate zone and such changes influence the nature and rate of operation of geomorphological processes. The consequences for water and sediment movement are most pronounced in cities (Chapter 27), but throughout temperate areas there have been many effects which have become apparent only long after the changes due to human activity. It has been proposed that the rate of change of a landscape system is analogous to the decay of radioactive isotopes so that a rate law of landscape change may be envisaged (Graf, 1977) whereby there may be a reaction time before a change begins, and then a relaxation time which is necessary before the effects of the change are fully incorporated into the system (Figure 15.15). In most parts of the temperate zone the energy available to environmental processes is less than in other world areas, but the equivalent of a great increase in energy can be released by human activity although the effects may not always be immediately apparent for the reasons listed in Table 15.9.

The need for engineering geomorphology

It is the intersection of extensive human activity with the various characteristics of temperate landscapes that has prompted the need for

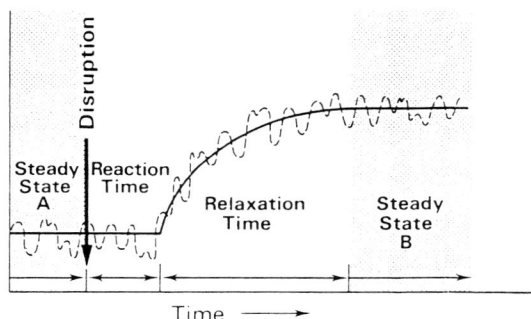

Figure 15.15   Response of a system to disruption (after Graf, 1977). After a disruption the reaction time is when the system accommodates the effect of the disruption and is followed by the relaxation time which sees the development of a new equilibrium (B). The *y* axis can be any parameter such as stream channel size, density of stream channels, or nature of channel pattern.

Table 15.9 Reasons for delays in landscape adjustment.

In Figure 15.15 there is a delay between the disturbance and the response of the landscape. That delay can be due to:

gradual changes in materials	see Tables 15.3, 15.5.
gradual changes in landscape	for example where a series of events or a certain amount of change is needed before adjustment will occur.
incidence of specific events	storm events of a particular size may be necessary to initiate change or floods of a particular size along a river valley.

engineering solutions to problems, often to prevent problems arising. Mass movement, slope instability and river flooding are examples of major problems that have arisen in many areas and engineering solutions have focused on structures to modify processes, as in the case of slopes prone to landslip, and structures to preclude process events affecting human activities as exemplified by river channelization and flood protection schemes. Particularly important was the analysis of time series of processes to establish the likely occurrence of hazards of a particular frequency and the design of structures to guard against an event of a particular frequency such as the 100 year flood. Also important has been the investigation of landscape sensitivity (Downs and Gregory, 1995) because it is the areas near the threshold between stability and instability that are most sensitive and so require the most careful management. Whereas initial solutions to landscape problems of mass movement or flooding employed have come to be thought of as 'hard engineering' methods, subsequent approaches have been developed to be more concerned with maintaining the maximum integrity and balance of the total land–water ecosystem as it relates to landforms, surface materials and processes, an approach which is called geomorphic engineering (Coates, 1976).

Restoration and key approaches for a geomorphological foundation for engineering

High population densities in many parts of the temperate zone mean that the consequences of human impact have to be minimised and protection has to be given against natural hazards, which may include earthquakes, floods, avalanches, soil erosion, landslides, subsidence and dam disasters (Alexander, 1999). Since 1990 there has been increasing pressure for the management of temperate environments to ensure not only that they utilise the approach of geomorphic engineering but also involve hard engineering methods only where absolutely necessary. This has therefore meant that 'soft engineering' approaches have been sought and in some cases restoration, the complete structural and functional return to a pre-disturbance state (Gregory, 2000), has been undertaken. Some rivers that were channelized, including straightening, by hard engineering methods have now been restored to a meandering condition (Brookes and Shields, 1996). It has been suggested (Brown, 1998) that multiple-channel systems characterised north-west European floodplains prior to deforestation and river channelization, so that it can be argued that multiple-channel systems which involve regular flooding of parts of the floodplain, should now be restored. Such an approach would significantly alter the way in which river channels are restored (e.g. Figure 15.14) and temperate landscapes are managed.

This example shows how a spectrum of issues needs to be considered in environmental management. There are areas where hard engineering provides the only possible solution but there are others where softer approaches are feasible. Issues which have to be considered in the geomorphic engineering of temperate environments are increasingly important in view of the effects of global change, and the issues may be summarised as:

1. Some unchanged landscapes should be maintained for future generations to appreciate — including Sites of Special Scientific Interest (SSSIs), heritage sites, wilderness areas and museums.
2. Hard engineering methods should be avoided except where absolutely essential.
3. Any engineering solution for the problems of a specific area should be designed in relation to adjacent areas — an approach described as total catchment management or a holistic approach.
4. Management strategies should include an awareness of stability/instability, sensitivity, and risk and uncertainty.
5. Knowledge of past environmental systems should inform decision–making about future management; acknowledging that the environment has a memory and that it is essential that design is undertaken with an awareness of past environmental change.
6. Restoration in a general sense should be attempted. Wherever possible landscapes and environments should be restored to more 'natural' conditions. To achieve restoration, a variety of overlapping approaches can be envisioned which have been described as recovery, re-establishment, enhancement, rehabilitation, reinstatement, creation and naturalisation as well as restoration.
7. In determining the management or restoration policy for a particular location, community views should be taken into consideration.

15.4 Summary

Temperate areas include ample evidence of development under former climates and often have been thought to be the norm against which other landscapes should be considered. There are several distinctive types of temperate area. Deposits over the landscape may be comparatively thin, although many are often inherited from earlier climatic conditions, and may locally be thick. It is often difficult to deduce quickly the character, extent, and nature of superficial deposits. Sediment transfers are fairly slow and storage of sediment on slopes and in flood plains is common, except where human activity, especially by deforestation, has accelerated change. Stream runoff is dependent upon the extension and contraction of drainage networks; most runoff is generated close to the drainage lines, and most fluvial sediment sources are also found close to streams and rivers. Human activity is very extensive and, whereas hard engineering solutions to problems are still necessary in some situations, in others softer methods sometimes involving restoration of more natural conditions are increasingly attempted.

References

Alexander, D. (1999) Natural hazards. In Alexander, D. E. and Fairbridge, R. W. (eds) *Encyclopedia of Environmental Science*. Kluwer Academic Publishers, Dordrecht, 421–425.

Brookes, A. and Shields, F. D. (eds) (1996) *River Channel Restoration: Guiding Principles for Sustainable Projects*. Wiley, Chichester.

Brown, A. G. (1987) Long-term sediment storage in the Severn and Wye catchments. In Gregory, K. J., Lewin, J. and Thornes, J. B. (eds) *Palaeohydrology in Practice*. Wiley, Chichester, 307–332.

Brown, A. G. (1991) Hydrogeomorphological changes in the Severn Basin during the last 15,000 years: orders of change in a maritime catchment. In Starkel, L., Gregory, K. J. and Thornes, J. B. (eds) *Temperate Palaeohydrology*. Wiley, Chichester, 147–170.

Brown, A. G. (1998) The maintenance of diversity in multiple channel floodplains. In Bailey, R. G., Jose, P. V. and Sherwood, B. R. (eds) *United Kingdom Floodplains*. Westbury, Otley, 83–92.

Büdel, J. (1982) *Climatic Geomorphology* (tr. L. Fischer and D. Busche). Princeton University Press, Princeton.

Butzer, K. W. (1976) *Environment and Archaeology*. Methuen, London.

Chandler, R. J. (1977) The application of soil mechanics methods to the study of slopes. In Hails, J. R. (ed.) *Applied Geomorphology*. Elsevier, Amsterdam, 157–180.

Coates, D. R. (1976) Geomorphic engineering. In Coates, D. R. (ed.) *Geomorphology and Engineering*. Dowden, Hutchinson and Ross, Stroudsburg, 3–21.

Coates, D. R. (1982) *Environmental Geology*. Wiley, Chichester.

Dalrymple, J. B., Conacher, A. J. and Blong, R. J. (1968) A hypothetical nine-unit landsurface model. *Z. Geomorphologie* **12**, 60–76.

Darby, H. C. (1956) The clearing of the woodland in Europe. In Thomas, W. L. (ed.) *Man's Role in Changing the Face of the Earth*. University of Chicago Press, Chicago, 183–216.

Derbyshire, E., Gregory, K. J. and Hails, J. R. (1979) *Geomorphological Processes*. Dawson, Folkestone.

Downs, P. W. and Gregory, K. J. (1995) Approaches to river channel sensitivity. *Professional Geographer* **47**, 168–175.

Dury, G. H. (1970) *The Face of the Earth*. Penguin, Harmondsworth.

Early, K. R. and Skempton, A. H. (1972) Investigations of the landslide at Walton's Wood, Staffordshire. *Q. J. Eng. Geol.* **5**, 19–41.

Graf, W. L. (1977) The rate law in fluvial geomorphology. *Am. J. Sci.* **277**, 178–191.

Gregory, K. J. (ed.) (1983) *Background to Palaeohydrology*. Wiley, Chichester.

Gregory, K. J. (1996) Introduction. In Branson, J., Brown, A. G. and Gregory, K. J. (eds) *Global Continental Changes: the Context of Palaeohydrology*. Geological Society Special Publication **115**, 1–8.

Gregory, K. J. (2000) *The Changing Nature of Physical Geography*. Arnold, London.

Gregory, K. J. and Walling, D. E. (1973) *Drainage Basin Form and Process*. Edward Arnold, London.

Hupp, C. R. (1999) Relations among riparian vegetation, channel incision processes and forms, and large woody debris. In Darby, S. E. and Simon, A. (eds) *Incised River Channels*. Wiley, Chichester, 219–245.

Kozarski, S. (1983) River channel adjustment to climatic change in west central Poland. In Gregory, K. J. (ed.) *Background to Palaeohydrology*. Wiley, Chichester.

Lewin, J. (1987) Stable and unstable environments — the example of the temperate zone. In Clark, M. J., Gregory, K. J. and Gurnell, A. M. (eds) *Horizons in Physical Geography*. Macmillan, Basingstoke, 200–212.

Liu Changming (2000) A remarkable event of human impacts on the ecosystems: The Yellow River drained dry. Paper presented to the 29th International Geographical Congress, Seoul, Korea, 17th August 2000.

Montgomery, D. R. (1997) What's best on the banks? *Nature* **388**, 328–329.

Starkel, L. (1983) The reflection of hydrological change in the fluvial environment of the temperate zone during the last 15,000 years. In Gregory, K. J. (ed.) *Background to Palaeohydrology.* Wiley, Chichester, 213–235.

Stokes, S. and Walling, D. E. (2003) Radiogenic and isotopic methods for the direct dating of fluvial sediments. In Kondolf, G. M. and Piegay, H. (eds) *Tools in Fluvial Geomorphology.* Wiley, Chichester, 233–267.

Thornes, J. B. (1979) Processes and interrelationships, rates and changes. In Embleton, C. and Thornes, J. B. (eds) *Process in Geomorphology.* Arnold, London, 378–387.

Thornes, J. B. (1983) Evolutionary geomorphology. *Geography* **68**, 225–235.

Tricart, J. (1957) Application du concept de zonalité à la géomorphologie. *Tijdschrift van het koninklijk Nederlandsch Aardrijkskunddig Genootschap,* Amsterdam **422**, 34.

Tricart, J. and Cailleux, A. (1972) *Introduction to Climatic Geomorphology* (tr. C. J. Kiewiet de Jonge). Longman, London.

Trimble, S. W. (1983) A sediment budget for Coon Creek basin in the driftless area, Wisconsin 1853–1977. *Am. J. Sci.* **283**, 454–474.

Trimble, S. W. (1999) Decreased rates of alluvial sediment storage in the Coon Creek basin, Wisconsin, 1975–1993. *Science* **285**, 123–124.

Walling, D. E. (1981) Yellow River which never runs clear. *Geographical Magazine* **53**, 568–575.

Further reading

Butzer, K. W. (1976) *Geomorphology from the Earth.* Harper Row, New York, London.

Gregory, K. J. and Benito, G. (2003) *Paleohydrology. Understanding Global Change.* Wiley, Chichester, 61–164.

Lewin, J. (1987) Stable and unstable environments – the example of the temperate zone. In Clark, M. J., Gregory, K. J. and Gurnell, A. M. (eds) *Horizons in Physical Geography.* MacMillan, Basingstroke, 200–212.

16. Hot Drylands

Mark Lee and Peter Fookes

16.1 Introduction

Hot drylands cover around one third of the earth's surface (Figure 16.1) and can be distinguished through the aridity index P/ETP (where P is the annual precipitation and ETP is the mean annual potential evapotranspiration, based on the Penman formula). Aridity increases as the P/ETP ratio declines, defining four zones of aridity (UNESCO, 1979):

1. sub-humid zone $P/ETP = 0.50–0.75$
2. semi-arid zone $P/ETP = 0.20–0.50$
3. arid zone $P/ETP = 0.03–0.20$
4. hyper-arid zone $P/ETP = {<}0.03$.

Most drylands are centred on the tropics where the stable, descending air of the Sub Tropical High Pressure Zone maintains aridity throughout the year. Among the other important factors in controlling the distribution of arid zones are the presence of large land masses which disrupt the zonal pattern of the global pressure systems (e.g. the Sahara), mountain barriers that generate rain-shadow areas in their lee (e.g. the extra-tropical deserts of North and South America) and the influence of cold currents on the eastern margins of the oceans.

Arid areas experience the highest average deviation in annual rainfall (Morales, 1977). This deviation tends to increase as the mean annual rainfall decreases. For example, a 35 year rainfall series for Elat in the Sinai desert (mean annual rainfall of 25.3 mm), includes 9 years with a quarter of the mean, 5 years with double the mean and 1 year with treble the mean (Schick, 1987). 'Cloudbursts' (associated with convection cells) are a characteristic feature of the climate, associated with almost instantaneous peaks in rainfall followed by a prolonged tail of low-intensity rain.

Rainfall can also be subject to extreme spatial variability. Yair and Lavee (1985) note that local variations in rainfall intensity in the Sinai can be considerable, with some areas within the same storm receiving twenty times more rainfall than areas less than a few kilometres away, during a particular rainfall event.

16.2 Desert landscapes

The popular notion is that barren sand dune-covered landscapes dominate hot drylands. The reality is somewhat different and much more complex. The magnitude, frequency, duration and timing of rainfall and the scale and significance of wind activity vary between arid zones. The range of bedrocks and geological histories and the availability of mobile sediments all add to the variety of desert environments. In general, however, there are two main types of desert landscape (Figure 16.2):

1. shield and platform deserts, including inselberg–pediment landscapes and canyon–scarp–pediment landscapes
2. basin and range deserts.

It is important to bear in mind that much of the current desert landscape detail is probably the product of the major climatic changes experienced throughout the Quaternary (i.e. the last 2–3 million years), during which there were many phases of alternating semi-arid or humid and dry (arid) conditions. Gerson (1982) and Bowman *et al.* (1986) have suggested that recent (i.e. the last 100 000 years) landform development has involved responses to varying climatic conditions:

Semi-arid conditions: these conditions are believed to have promoted the development of

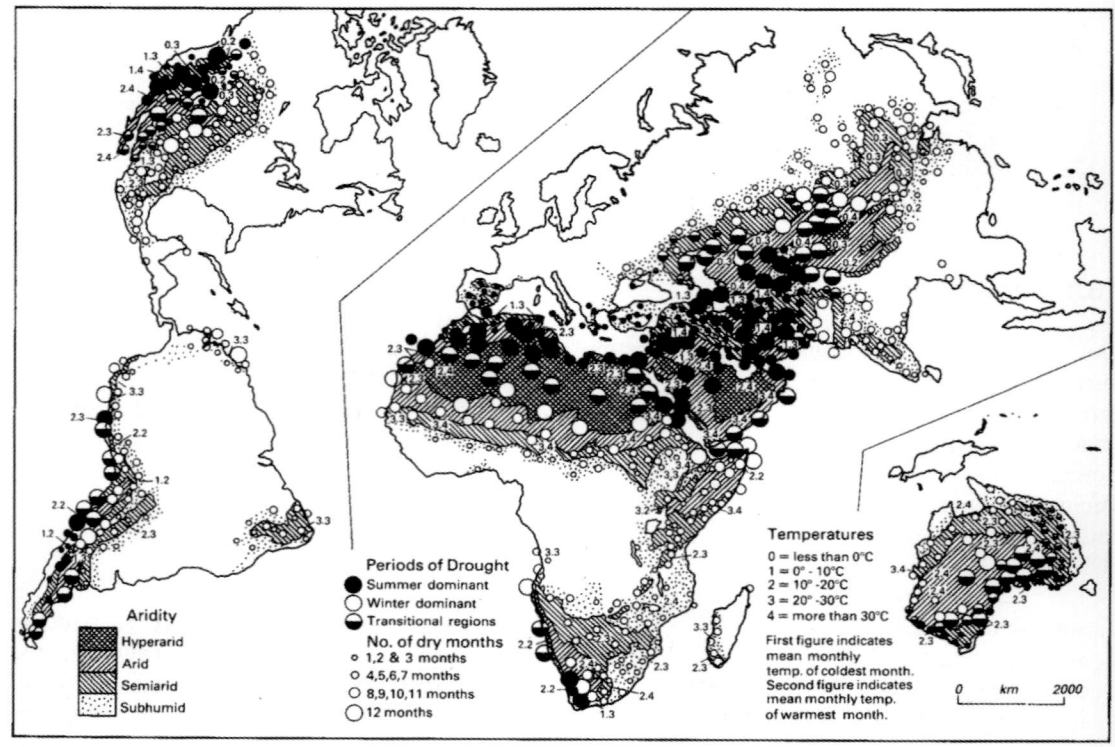

Figure 16.1   World distribution of deserts (from UNESCO, 1977).

debris flows off hills and scarps, creating relatively smooth talus aprons at the slope foot (Gerson, 1982) (i.e. a phase of talus apron creation). During this period, colluvial (e.g. debris flow textured) and alluvial sediments are deposited within wide, braided floodplains.

Arid and extremely arid conditions: on hillslopes and scarps, this phase would have been characterised by slope wash and gullying, with the dissection of the talus aprons producing 'flatirons' (i.e. a phase of talus apron destruction). As debris flow activity declines, so wadi channels transport less sediment of smaller size and tend to be 'misfits' within the relict floodplains. Later, when the talus aprons have been largely removed, combination of flash floods with limited sediment load promotes incision across the floodplains.

Lake formation occurred in enclosed basins and low-lying areas, during pluvial (wet) or humid phases, and was accompanied by the deposition of muds, usually carbonate. For example,

Fontes *et al.* (1985) and Gasse *et al.* (1987) identified relict lake bed deposits on the northern margins of the Grand Erg Occidental, Algeria. These lakes are believed to have developed in enclosed depressions (possibly wadi channels blocked by dune movement) and were present between 9300–3000 BP, when groundwater levels were probably close to the surface (nowadays they are around −50 m).

However, an important characteristic of the desert environment is the marked contrast between the long-term morphological stability of many landscape features (e.g. the scarps and the pediments) and the dynamic geomorphological activity of the alluvial plains, sand seas and alluvial fans.

Most desert soils are granular, with the grading related to the distance from the uplands. Fine sediments are transported by water or wind and stored in topographic lows or on base-level plains. Here, the presence of high and saline

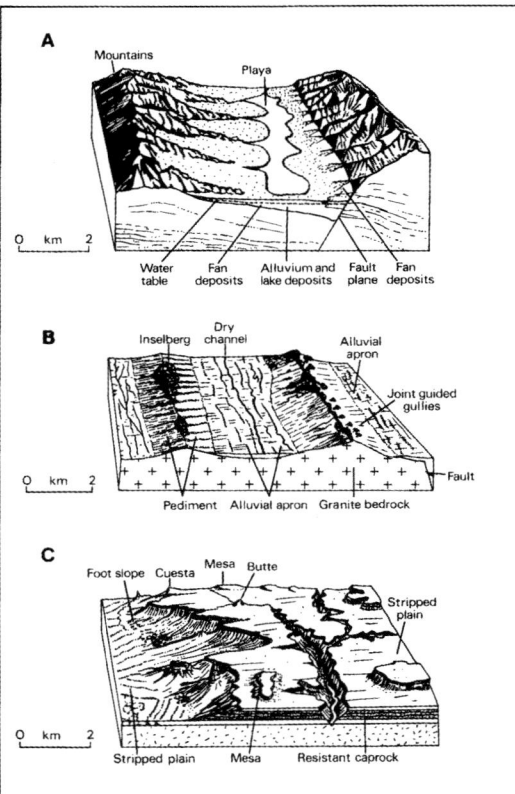

Figure 16.2 Characteristic desert landscapes: (*A*) basin and range topography with alluvial fans and playas; (*B*) inselbergs and pediment landscape and (*C*) canyon and scarp country (from Goudie and Wilkinson, 1978).

groundwater tables will create salt weathering problems. Wind-blown dust (*loess*) has accumulated across many desert landscapes. For example, in the Middle East the material is believed to have derived from dust-laden storms from North Africa between 20 000–80 000 years BP. This loess material is potentially metastable (i.e. will collapse on wetting, when under a higher load than it previously supported).

For engineering geomorphological purposes, the key hot dryland issues can be related to a number of distinctive terrain units that are characteristic of many desert environments (albeit in varying combinations). Four main zones can be recognised from uplands to base level plains, each

with typical surface features and with different engineering behaviour (Figure 16.3; Fookes and Knill, 1969).

A variety of issues can be encountered in drylands, including:

1. surface erosion and instability
2. difficult excavation
3. behaviour of desert soils
4. availability of aggregates
5. water and sediment movement problems
6. wind blown sand
7. aggressive salty ground.

These issues are considered briefly after an outline of the main terrain zones.

16.3 Zone I: the uplands

Mechanical and chemical weathering are active processes in uplands. However, these areas are often characterised by bare rock and boulder-strewn slopes as the rate of weathering is generally less than the rate of removal by mass movement or surface erosion. Many uplands comprise a relatively simple landscape of slowly retreating scarps with detached mesas and buttes, developed in near horizontal sedimentary rocks or tabular volcanic rocks. Elsewhere, as in the Zagros Mountains of Iran, there may be complex assemblages of cuestas, hogs-backs and strike valleys developed in folded or faulted rocks. Despite these variations it is possible to identify a number of broad terrain sub-zones (Figures 16.4 and 16.5).

Zone Ia: backslopes
Backslopes are where the caprock, often a duricrust (Chapter 2) is exposed as a low-relief stripped plain ('slickrock' slope) that is typically boulder or rock strewn, with little or no soil cover. Away from the scarp face the backslope may grade into a pediment fronting another scarp slope (see Zone IIa below).

Zone Ib: scarp slope and free face
These are often 'composite scarps' with a resistant caprock underlain by more erodible strata. This zone is prone to rockfalls and so-called

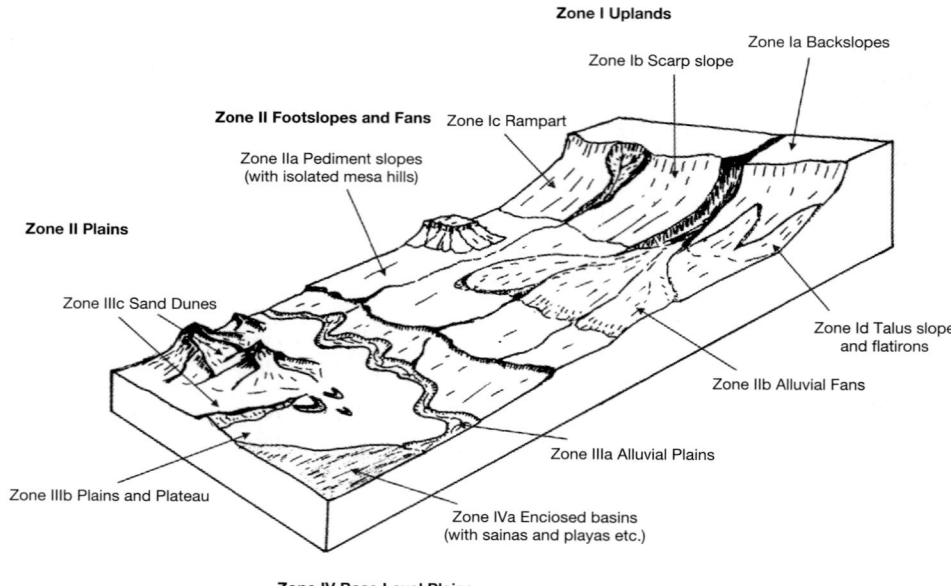

Figure 16.3   A simplified desert terrain model (based on Fookes and Knill, 1969; Fookes, 1976, 1978).

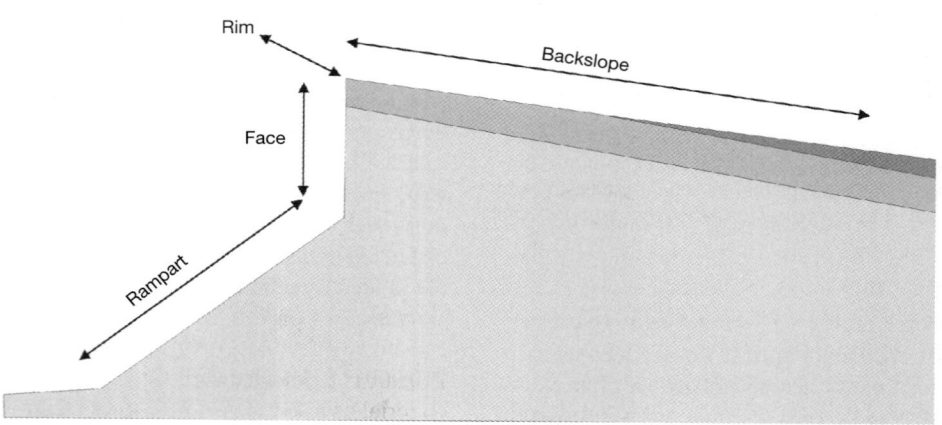

Figure 16.4   Components of a scarp–backslope terrain (from Howard and Selby, 1994).

block-by-block undermining and collapse. Major deep-seated landslides may be present, but these are often an inheritance from previous wetter climates (e.g. Figure 16.6). Where scarp faces comprise impermeable strata over massive permeable rocks they are often subject to basal undermining by seepage erosion. This leads to cave formation and ultimately collapse. Estimated scarp retreat rates in the order of 0.1–10 m/1000 years may be anticipated, depending on the setting. Yair and Gerson (1974) estimated that the average rate of scarp retreat in reef limestones in the Sinai was probably around 1.2–2 mm/year over the past 250 000 years. Higher average rates of up to 6.7 mm/year have been suggested for shales in the Grand Canyon (Lucchitta, 1975).

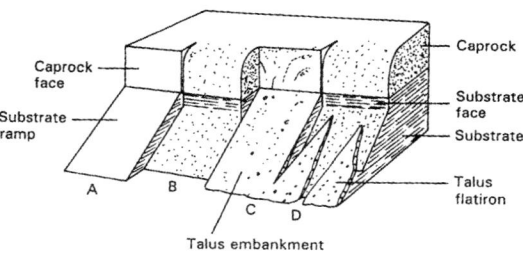

Figure 16.5 Scarp development model, highlighting different terrain elements. (A) Generalised form of compound scarp composed of caprock face and substrate ramp. (B) Erosion of caprock brow and substrate ramp, with downward cliff extension into substrate. (C) Failure of substrate face and collapse of unsupported caprock, producing talus embankment on substrate ramp. (D) Erosion of caprock brow, talus embankment, and substrate ramp; erosion of talus embankment produces talus flatirons; downward cliff extension into substrate approaches threshold for next collapse (from Oberlander, 1997).

Zone Ic: rampart

A rampart is an inclined rock surface at the scarp base, often mantled with rockfall debris. These slopes are believed to be formed at the characteristic friction angle of the debris, under conditions where debris removal just balances production (i.e. 'Richter' slopes; Figure 16.7).

Zone Id: talus slopes and flatirons

Many upland slopes are flanked by talus aprons or screes, produced by scarp face collapse. These slopes are prone to gullying and debris flow activity, producing so-called flatirons (i.e. talus remnants with marked upslope margins; Figure 16.5).

Considerable hydrological research has been undertaken in and around the uplands. Among the characteristic features of this environment are convective cell cloudbursts; very high runoff coefficients; destructive flood events; large sediment yields, and temporary dam formation.

Convective cell cloudbursts can have rainfall intensities of up to 90 mm/hour and considerable spatial variability. Much erosion of channel banks and beds, and deposition is achieved in rare storm events. For example, Schick (1977) estimated that 99% of erosive work during a 10 year period at the Nahal Yael catchment, Sinai occurred in 5 days and by only 7 runoff events that constituted 20% of the mean annual rainfall.

Runoff coefficients on the bare rock and debris mantled (scree) slopes can be very high, generally in the range 15–40% (Yair and Lavee, 1976). Relatively frequent rainfall events will generate runoff and overland flow on almost every slope. The lag from cloudburst to peak flow can be as little as 1 hour. However, up to 60–70% of runoff generated in small catchments may not reach a main channel because of infiltration losses (Yair and Lavee, 1985).

The probabilities of large, destructive flood events are relatively high and the frequency of historical events suggest that such events can be expected to occur within the main mountain catchments once every 25–50 years.

Widespread availability of loose, coarse sediment on the mountain slopes and the absence of vegetation can lead to very large sediment yields in large flood events.

Temporary dams across main stream channels (wadis) can be formed as a result of debris flows or flash floods from tributary wadis. The subsequent breach of these dams in storms can give rise to near instantaneous rises in wadi flow, generating large bores or 'walls of water'.

An example of the large, destructive floods that occur in the mountains is provided by the Wadi Mikeimin (a tributary of Wadi Watir, immediately south of Nuweiba) flood of January 1971 in Sinai (Schick and Lekach, 1987). Detailed reconstruction of the event indicated that the flow peaked at 91.9 m³/s, with a total flow in excess of 100 000 m³. The flow deposited 6200 m³ of coarse material as a fan at the junction with Wadi Watir. The fan dammed the main wadi for 21 months during which time a lake, up to 400 m long, formed. The first subsequent flood breached it in November 1972, resulting in large flood bores.

16.4 Zone II: footslopes and fans

The terrain surrounding the uplands may comprise two contrasting terrain units: gently sloping rock pediments and alluvial fans.

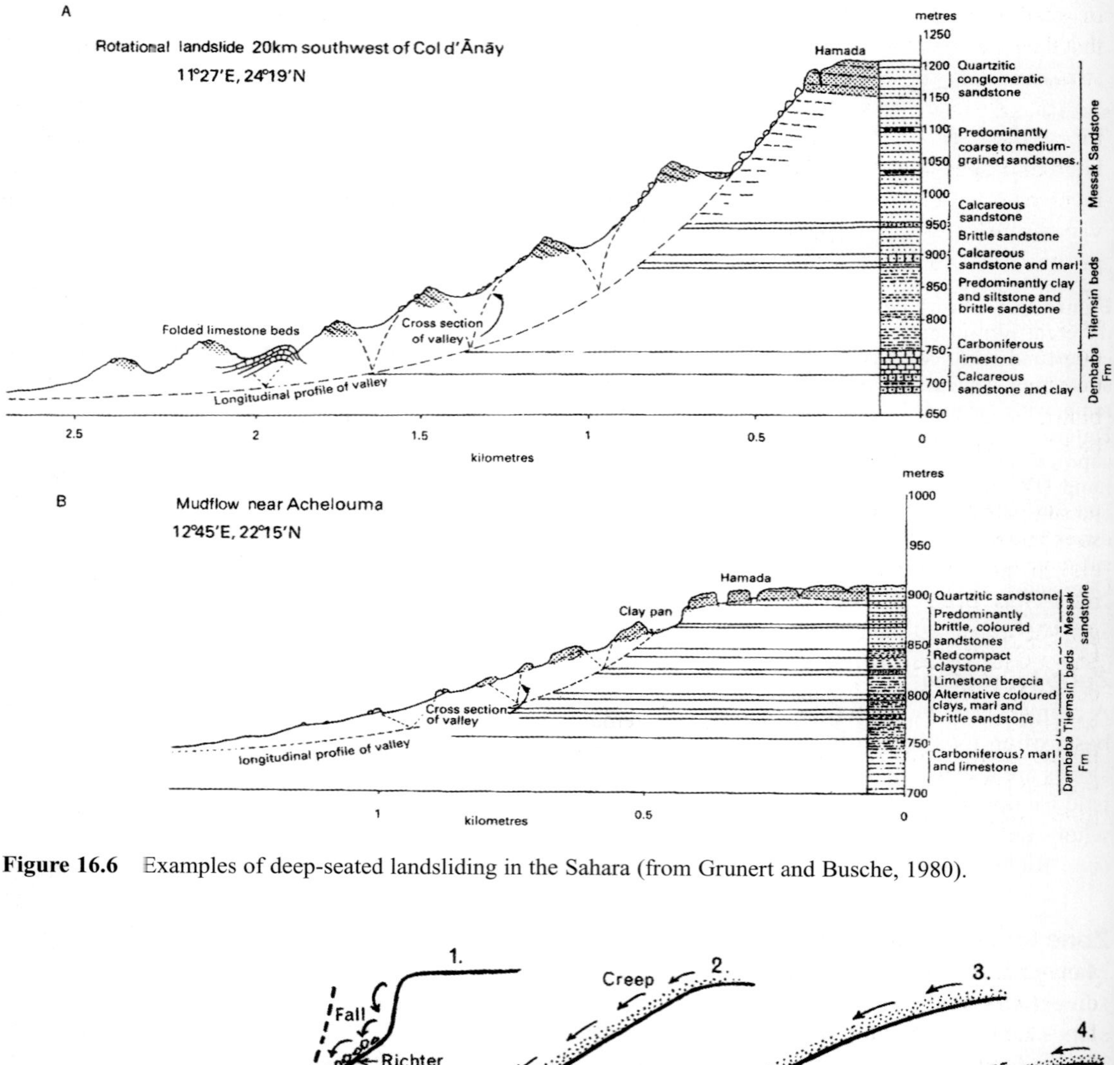

A
Rotational landslide 20km southwest of Col d'Ānāy
11°27'E, 24°19'N
Hamada
metres
1250
1200
1150
1100
1050
1000
950
900
850
800
750
700
650
Quartzitic conglomeratic sandstone
Predominantly coarse to medium-grained sandstones.
Calcareous sandstone
Brittle sandstone
Calcareous sandstone and marl
Predominantly clay and siltstone and brittle sandstone
Carboniferous limestone
Calcareous sandstone and clay
Messak Sandstone
Tilemsin beds
Dembaba Fm
Folded limestone beds
Cross section of valley
Longitudinal profile of valley
2.5
2
1.5
1
0.5
0
kilometres
B
Mudflow near Achelouma
12°45'E, 22°15'N
metres
1000
950
900
850
800
750
700
Hamada
Clay pan
Quartzitic sandstone
Predominantly brittle, coloured sandstones
Red compact claystone
Limestone breccia
Alternative coloured clays, marl and brittle sandstone
Carboniferous? marl and limestone
Messak sandstone
Tilemsin beds
Dembaba Fm
Cross section of valley
longitudinal profile of valley
1
0.5
0
kilometres

Figure 16.6 Examples of deep-seated landsliding in the Sahara (from Grunert and Busche, 1980).

1.
2.
3.
4.
Fall
Richter slope
Roll
River
Creep
wash

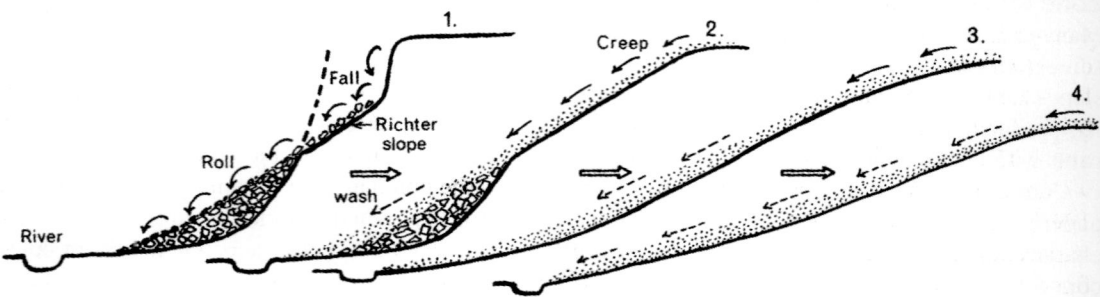

Figure 16.7 Development of talus and Richter slopes (from Selby, 1982).

Zone IIa: pediments

Pediments are gently sloping surfaces developed in bedrock, often cutting across a range of lithologies. On occasions there may be no change in lithology between the pediment and the adjacent uplands. Pediments are believed to be formed by a combination of lateral planation of mountain front streams and sheetwash (inter-rill) and rill erosion of weathered materials. It has also been suggested that they are relict forms, created by the progressive stripping of deeply weathered old tropical soils.

The surface may have irregular micro-topography, with low knolls, hills (inselbergs) or bedrock ridges that stand above the general pedi-ment level and a network of bedrock-lined and allu-vial channels (wadis). Stream dissection tends to be greatest close to the mountain front. Un-dissected areas often have shallow, ill-defined drainage chan-nels with low indistinct interfluves. The regolith is dominated by weathered residual soils, often man-tled by a veneer of superficial deposits, including blown sands, loess and alluvium (including infilled bedrock channels). Duricrusts may be present.

Erosion and debris flow activity from the upland areas and adjacent steep flanking slopes supplies sediment (predominantly sand and grav-els) to the surrounding plains. Where the sedi-ment has been transported further, extensive gravel spreads (piedmont plains) can developed. Sediment can be moved across pediments and deposited in Zone III because the velocity of water flows and, hence, stream power may remain in the same order of magnitude on both the uplands and plains. For example, with reference to the Manning equation (Chapter 11):

$$v = \frac{r^{2/3} S^{1/2}}{n}$$

where r is the hydraulic radius, S is the slope gra-dient (m/m) and n is the Manning roughness coef-ficient, if the gradient on the uplands is 0.8 and 0.1 on the pediment, and $n = 0.06$ on the uplands and 0.02 on the pediment, the velocity of flow would remain almost constant (Cooke *et al.*, 1993). This means that this Zone can experience significant sediment transport and erosion as flood flow velocities remain high across the rela-tively smooth sloping surfaces.

The key engineering issues in this zone include potentially difficult excavation of bedrock at or near the ground surface, flood hazard and chan-nel instability. However, foundation conditions away from drainage channels are likely to be good, though dependent upon the local bedrock type. The groundwater table is likely to be deep.

Zone IIb: alluvial fans

Alluvial fans are cones of poorly sorted coarse sediment (boulders, cobbles, gravels and some sands) laid down where a channel emerges from an upland area onto a plain. When a confined stream leaves the uplands it spreads out, with the increase in width accompanied by a reduction in flow depth and velocity, resulting in sedimenta-tion. In addition, as the stream crosses permeable fan sediments the discharge will decrease, increasing the sediment concentration and fur-ther enhancing deposition. Not all channels con-struct alluvial fans. Large perennial rivers are capable of maintaining channel banks and flow competency.

The main morphological features of a fan are (Figure 16.8):

1. the drainage basin, typically with steep, debris-covered, unvegetated or partly vegetated slopes
2. a feeder channel, the main stream that delivers water and sediment to the fan—some fans may have multiple feeder channels
3. a fan apex, the point at which the feeder chan-nel emerges from the uplands
4. an incised channel, the downslope extension of the feeder channel across the fan
5. an intersection point, where the incised chan-nel merges into the fan slope
6. an active deposition lobe, below the intersec-tion point, flows leave the channel at this point and spread out across the fan forming a zone of active deposition.

Fans often comprise a complex sequence of coa-lescing or segmented sections, often of different dates (Figure 16.9; Table 16.1).

The dominant sedimentary processes are sum-marised in Figure 16.10 and can involve both primary construction or aggradation and secondary fan erosion or degradation, especially by gullying or wind erosion. An important feature of fan develop-ment is the constantly changing pattern of channels. This is caused by the progressive infilling and over-flows of channels, often as a result of a blockage by boulders or mudflows. The primary processes are:

1. debris flows, transporting sediment supplied by a variety of landslide processes including

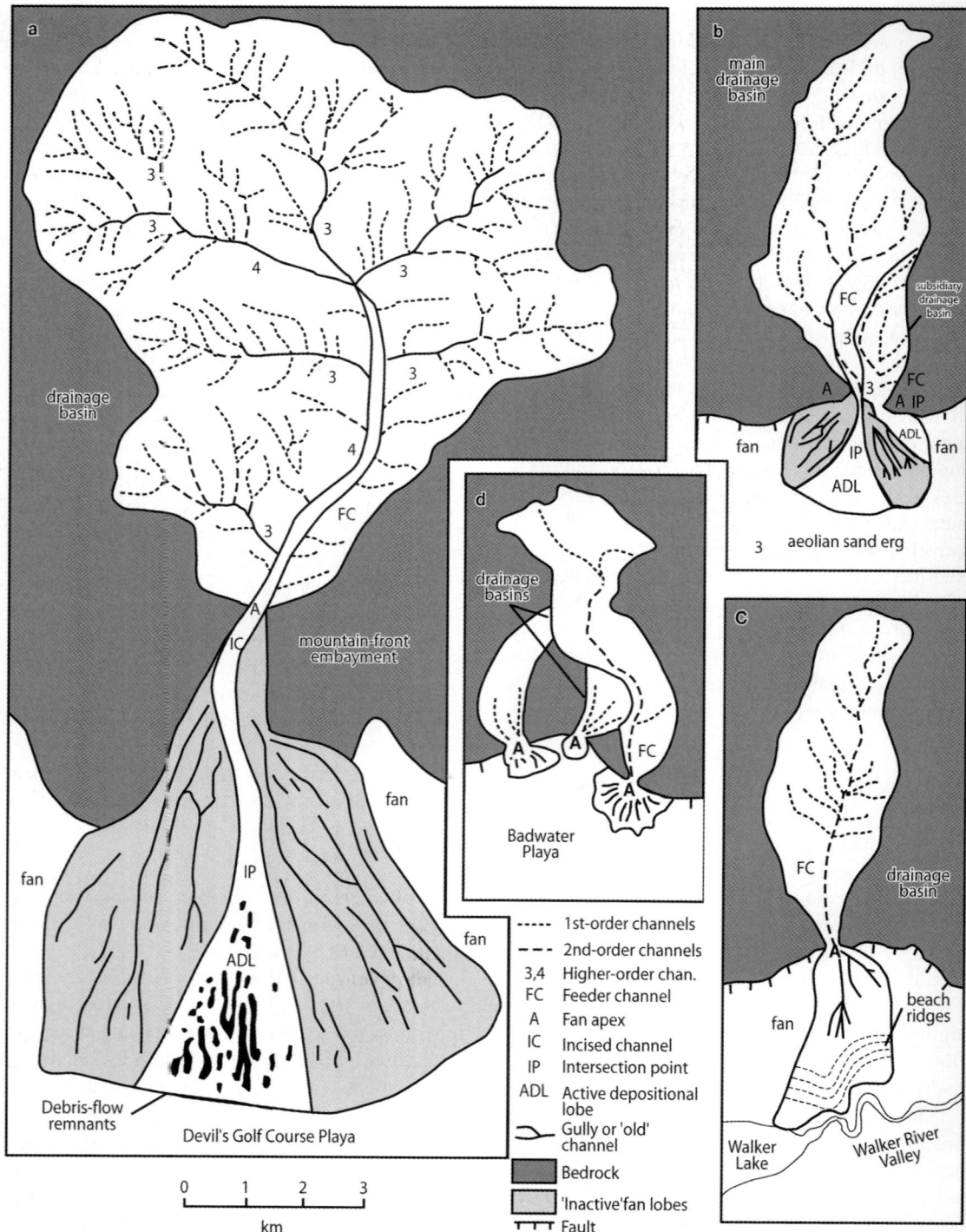

Figure 16.8 Principal features of alluvial fans (redrawn from Blair and McPherson, 1994).

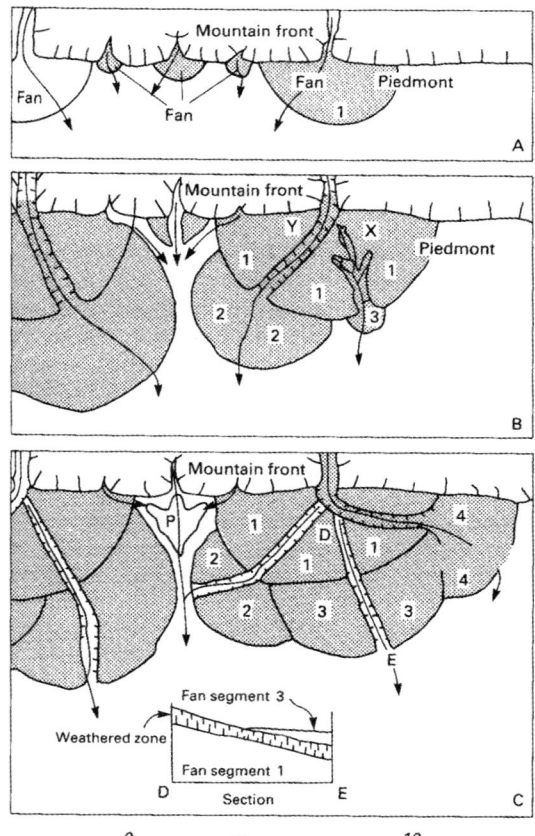

Figure 16.9   The development of a segmented fan complex (from Denny, 1967).

rockfalls, rock avalanches and slides (Chapter 8; Costa, 1988); debris flows may range from predominantly boulders or cobbles (clast-rich) or muds (clast-poor mudflows)

2. sheetfloods (unconfined flows), extensive but short-duration events produced by very intense rainstorms

3. incised channel flows, generally depositing only coarse sediments because of the high flow velocities created by the confinement by near-vertical channel walls. However, some finer sediments may accumulate along the channel during the waning flood stage or during low flow events.

Although many fans have a dense network of braided distributaries, these are secondary features not believed to be related to fan construction. Fan deposits do not generally show planar or trough cross-bedding typically associated with braided channel deposits.

Figure 16.11 indicates the relationship between average fan slope, the dominant processes and sediment size. Where one process is dominant (e.g. debris flows) or there is a uniform available sediment size, there may be a constant fan angle. In contrast fans with a progressive decrease in sediment size may be characterised by concave profiles.

Two main fan types can be recognised (Blair and McPherson, 1994):

Type 1: developed by cohesive clast-rich and clast-poor debris flows. Fan morphology include debris flow lobes and levees, and a constant 5–15° slope. Distal sand-skirts are absent. Because of the low frequency of debris-flow events the fan surface is often marked by the effects of secondary processes, especially gullies, rills and boulder lags created by the removal of fines by wind action. The fan sediments comprise stacked sequences of debris-flow deposits separated by gravel lags produced by secondary erosion. This type of fan is associated with catchments underlain by bedrocks that weather to produce boulders, cobbles, silt and clay, but little sand.

Type 2: developed by flash floods, especially sheetfloods. A prominent sand-skirt forms the distal fan margin. The fan surface is smooth with average slope angles of 2–8° with a progressive downslope decline in angle. Secondary processes are effective because of the smaller fan sediment sizes. This type of sand fan is often associated with catchments underlain by fractured or jointed granitic plutons, gneiss or friable sandstones.

Fan development may proceed from a relatively steep (10–25°) ramp abutting the mountain front (Stage 1), primarily through landslide processes. The emergence of debris flows and sheetfloods as the dominant processes results in the creation of a gentler fan surface (Stage 2) (Figure 16.11). Over time fan enlargement may involve the development of segmented fans in which secondary fans are built on or beyond the original segments (Figure 16.9) or telescope fans where successive fans occur within the boundaries of the initial fan (Figure 16.12).

Table 16.1   Characteristics of alluvial fans of different ages (from Christenson and Purcell, 1985).

Characteristic	Young	Intermediate	Old
Drainage pattern	Distributary; anastomosing or braided	Tributary; dendritic	Tributary; dendritic or parallel
Depth of incision	< 1 m	Variable (1–10 m)	> 10 m
Fan surface morphology	Bar-and-channel	Variable, generally smooth and flat	Ridge and valley, most of surface slopes
Preservation of fan surface	Presently active	Incised, but well preserved wide flat divides	Basically destroyed, locally preserved on narrow divides
Desert pavement	None to weakly developed	None to strongly developed	None (surface destroyed) to strongly developed (surface preserved)
Desert varnish	None to weakly developed	None to strongly developed	None (surface destroyed) to strongly developed (surface preserved)
Calcic horizon	None to weakly developed; carbonates disseminated throughout profile	Weakly to strongly developed	None, carbonate rubble on surface (surface destroyed) to strongly developed hardpan (surface preserved)

The key engineering issues in this zone include debris flows, flood hazard and channel instability. Foundation conditions on old fan surfaces away from drainage channels are likely to be good, although rapid horizontal and vertical changes in engineering properties may occur. In general, fan materials are a good source of 'borrow materials' for rockfills, gabions and aggregates, although their chemistry needs to be checked for use in concretes.

16.5 Zone III: plains

This is a zone of net deposition of sediments supplied by the erosion of the uplands and footslopes, comprising alluvial (Zone IIIa) and wind-blown sediments (Zone IIIc). Elsewhere stone-covered rocky surfaces can dominate (Zone IIIb). These gently sloping plains may comprise considerable thicknesses of water-borne materials. Two types commonly occur: sandy stony and silty stony desert (Fookes, 1978). Fluvial and aeolian processes

generate the main geohazards and, hence, dominate the engineering issues in this Zone.

Zone IIIa: alluvial plains

The main fluvial processes and forms are related to sediment transport and deposition within alluvial channels. Many desert plains support a complex dendritic drainage network, with numerous braided stream channels (wadis) and extensive floodplains; for the most part, these channels are relict forms, created during periods of higher rainfall, although they may be sustained by contemporary processes. For example, much of the gross form of the Sahara was probably fashioned during the early to mid-Tertiary, over 66–25 M years ago, prior to the establishment of full desert conditions (e.g. Williams and Faure, 1980).

Braided streams tend to be the dominant channel form, reflecting the interaction between slope gradient, sediment availability, erodible banks and highly variable flows. In floods, the flow occupies the entire channel system between low terraces.

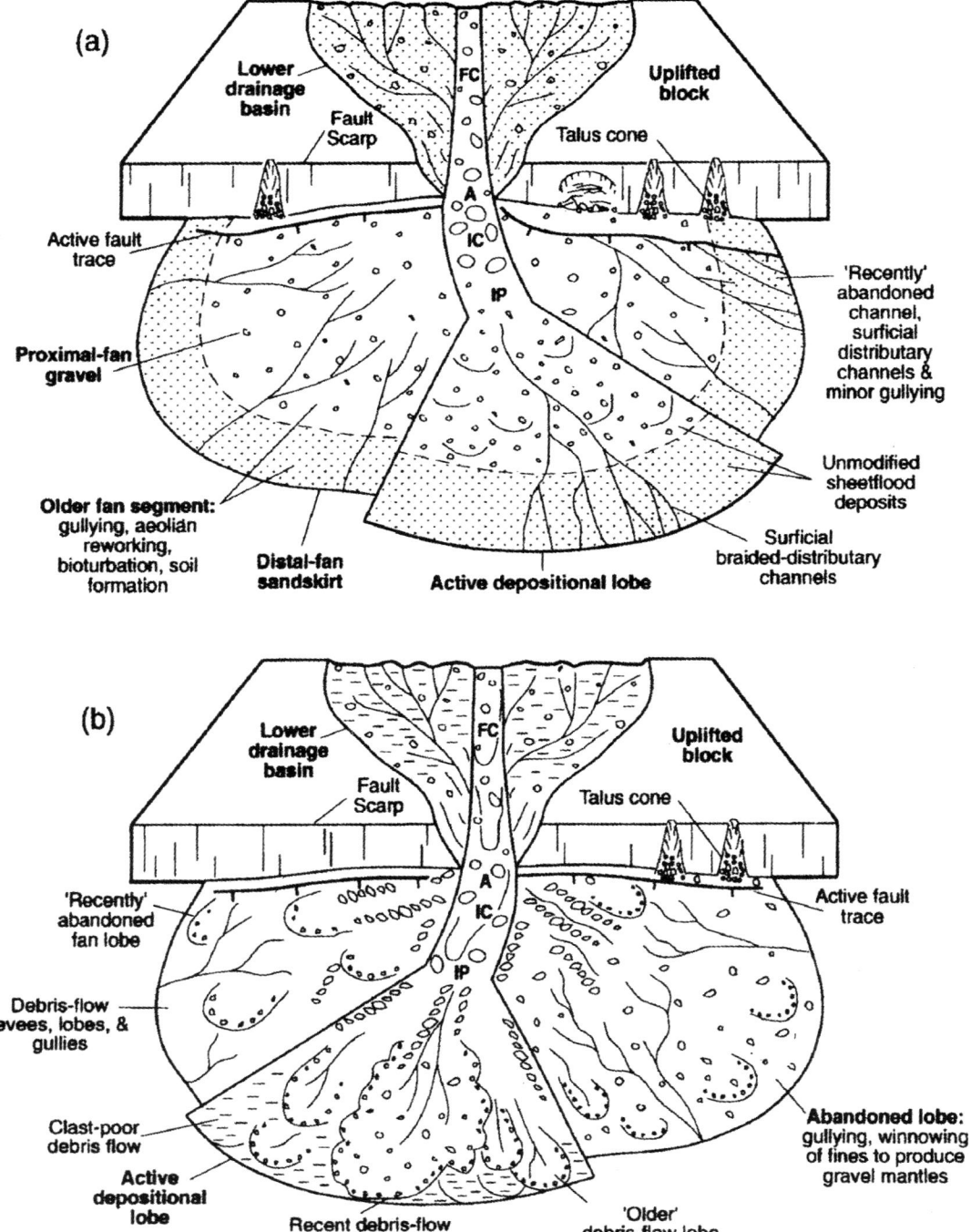

Figure 16.10   The dominant processes operating on alluvial fans: (*A*) fans dominated by water flows (*B*) fans dominated by debris flows (from Blair and McPherson, 1994).

Dryland streams with permanent low flows may develop compound forms. The low forms occupy a single meandering channel whilst high flows spread across a wider braided channel. When the meandering channel is near bankfull discharge it can transport large volumes of sediment and erode its banks. At higher flows the system becomes braided and flows more directly downslope along the sub-channels. The whole compound system may be occupied by very high

flows, during which bed materials are completely mobilised. After the flood has past a new dominant channel will develop.

Dryland flows are ephemeral (i.e. the channel is dry between runoff producing storms). Large floods do occur. For example, the El Arish flood of 1975 which resulted from a 48 hour storm across much of Sinai, with total rainfall of up to 73 mm at Santa Katharina Monastery (Gilead, 1975). A discharge of 1650 m³/s was recorded 30 km south of El Arish. The flood wave needed 39 hours to cross 250 km from mid-Sinai where it was first identified. A 1 m high wall of water was observed. The floodwaters destroyed a railway bridge and deposited a delta 500 by 300 m on the Mediterranean shoreline. This is believed to have been a 1 in 50–100 year event (Schick, 1987). Major floods have also been reported in the western Sahara (e.g. the Gur–Saoura–Messaoud catchment floods of 1890 and 1915 (Mabbutt, 1977; Figure 16.13).

Sheetfloods occur across the ground surface and when the channel system capacity is exceeded.

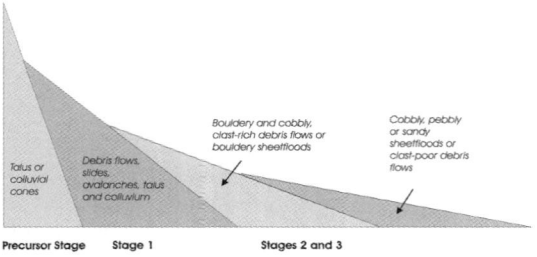

Figure 16.11 The association between fan form, processes and slope angle, drawn with a 2 × vertical exaggeration (from Blair and McPherson, 1994).

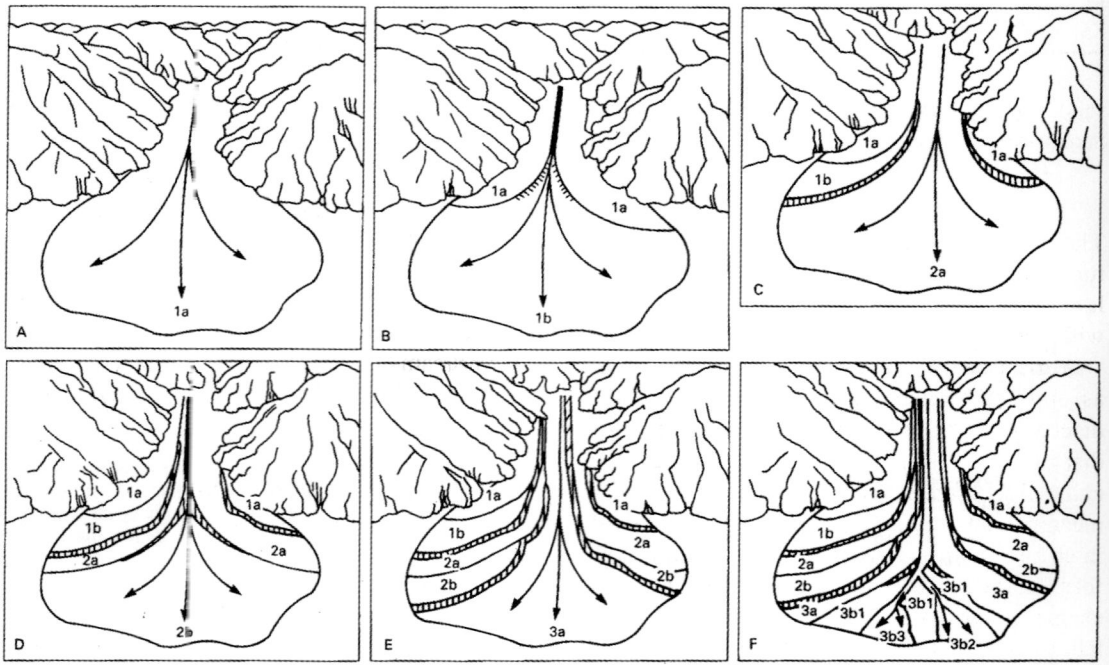

Figure 16.12 The evolution of telescopic fans in Death Valley, California (from Dorn, 1988).

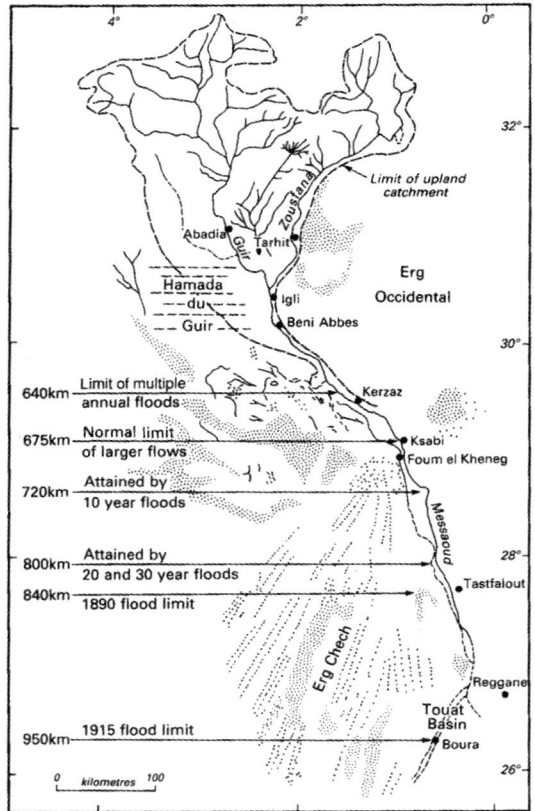

Figure 16.13 The limits of flooding in the Gur–Saoura–Messaoud catchment, Algeria (from Mabbutt, 1977).

These floods comprise a sheet of unconfined turbulent water moving across the alluvial plain; large events are often highly erosive and laden with debris.

Large desert floods tend to be formative events, as opposed to temperate rivers where bankfull discharge is believed to be the most effective discharge (Chapter 10), triggering significant channel changes, including the switch from meandering to braiding with the resulting increase in channel width (Table 16.2). Occasionally planform changes are accompanied by incision. For example, during a flood in 1896 the Fremont River, Utah entrenched 7 m and a 1.5 km wide braided channel replaced a 30 m wide meandering channel (Graf, 1983). Meandering channels may

Table 16.2 Changes in channel width along rivers in the western USA (from Graf, 1988).

River	Change	Time
Canadian River, Oklahoma	0.8–3.2 km	Flood in 1906
Rio Salado, New Mexico	15–168 m	1882–1918
Red River, Texas–Oklahoma	No change	1874–1937
Red River, Texas–Oklahoma	1.2–0.8 km	1937–1953
Cimarron River, Kansas	15–366 m	1874–1942
Cimarron River, Kansas	366–168 m	1942–1954
Platte River, Nebraska	1.16 km–111 m	1860–1979
S.Platte River, Colorado	790–60 m	1897–1959
N.Platte River, Wyoming	1.2 km–60 m	1890–1977
Gila River, Arizona	45–90 m	1875–1903
	90–610 m	1903–1917
	610–61 m	1917–1964
Salt River, Arizona	No change	1868–1980
Fremont River, Utah	30–400 m	Flood of 1896

become re-established over time as braided sub-channels infill with sediment, bars and islands expand and become attached to the banks. A new floodplain will develop between the banks of the braided system, across which the meandering channel will migrate.

A notable feature of dryland flows is the 'transmission losses' due to infiltration through the channel bed. Thus, discharge may decrease or dry up downstream (the reverse is true in temperate rivers) and larger flows will extend further than smaller floods (Figure 16.13).

Zone IIIb: plains and plateau surfaces

Extensive, apparently monotonous flat plains are a feature of many deserts. Often the plateaux surfaces are bounded by major escarpments, believed to be the product of a combination of differential erosion and structural controls (Zone Ib). A distinctive feature of many desert plains is a surface pavement of closely packed stones (*reg* or *sarir* in the Sahara, *gobi* in Asia and gibber plain in Australia), often in close association with boulder-strewn rock outcrops with a scatter of stones (*hamada*). The pavements are often a lag deposit left after wind deflation. In some regions, they are now believed to be the result of accretion of wind-blown silt and dust. The gradual surface accumulation of fine material results in the upward growth of the soil profile beneath the stone layer which settles evenly on the surface. In this way the uniform input of wind-blown material over a broad area can lead to the development of monotonous level surfaces with relatively consistent thicknesses of silty or fine sandy soils.

Streamlined wind-sculptured bedrock hills (*yardangs*) often occur on desert plains and plateaux, along with oddly shaped and fluted hillocks (*demoiselles* or *hoodoos*). Yardangs tend to be associated with unidirectional wind regimes and most features are aligned parallel to these winds. They range from small hillocks cut in

former lake bed deposits ('mud lions') to features extending several kilometres in length, carved in harder rocks such as the Nubian Sandstone or granites (Breed *et al.*, 1997).

Hydrocompaction can be a significant problem on some desert alluvial soils. Ground fissuring and subsidence of up to 3 m can result from compaction as clay bonds supporting the voids within the soil are weakened by wetting (e.g. during floods or after the onset of irrigation).

Zone IIIc: sand dunes

Many desert plains contain areas of wind-blown sand accumulation, either as thin sand sheets, dune fields (a collection of dunes in an area less than 30 000 km²) or sand seas (ergs, areas of dunes of varying forms and sizes extending over 30 000 km²; Figure 16.14). In general, dunes accumulate where concentrated sand flows converge or where the wind energy declines, as in topographic lows away from upland areas.

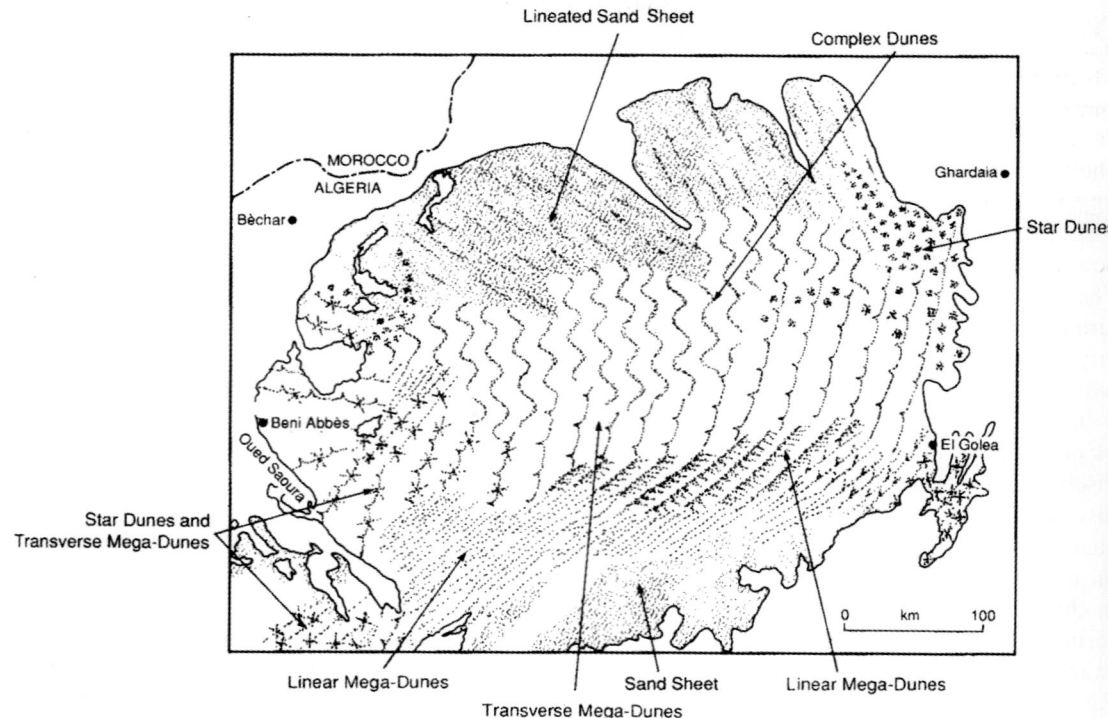

Figure 16.14   The Great Western Sand Sea, Algeria: dune assemblages (from Cooke *et al.*, 1993).

Most dune material is quartz, although some dune areas are dominated by calcium carbonate (e.g. the south Wahiba Sands, Oman), gypsum (e.g. White Sands, New Mexico), sodium chloride (e.g. southern Tunisia) or clay. Dune sands commonly have median grain diameters between 0.2–0.4 mm and range between 0.1–0.7 mm. Dunes are usually well graded, with a single dune having particles of similar size and roundedness. Dunes have poor load-bearing capacities (loose to medium dense) and can prove difficult to compact as fill.

Dunes exhibit a variety of forms, including barchans, crescentic ridges, linear, star, reversing and parabolic (Figure 16.15; Lancaster, 1994; Thomas, 1997). Transverse and barchan dunes tend to be migratory forms (see Table 16.3 and following discussion on due mobility), whereas linear dunes are extending or sand-passing forms; star dunes are accumulating forms. 'Zibar plains' are rolling sand plains without distinct dune forms.

Dune types appear to be related to the volume of sand and the wind direction variability, as expressed by the ratio between the total sand drift potential (DP) and the resultant drift potential RDP (i.e. the dominant drift direction). Barchans occur where there is little sand-moving wind variability and limited sand supply (Figure 16.16). Linear dunes are associated with restricted sand supply but variable wind direction; star dunes occur where sand is abundant with an unstable wind direction.

In sand seas there is a hierarchy of superimposed forms: megadunes: complex dune systems (*draa*); mesodunes: individual simple dunes superimposed on the megadunes; ephemeral dunes: reaching around 1 m in height and developed on constantly changing slopes—they may be

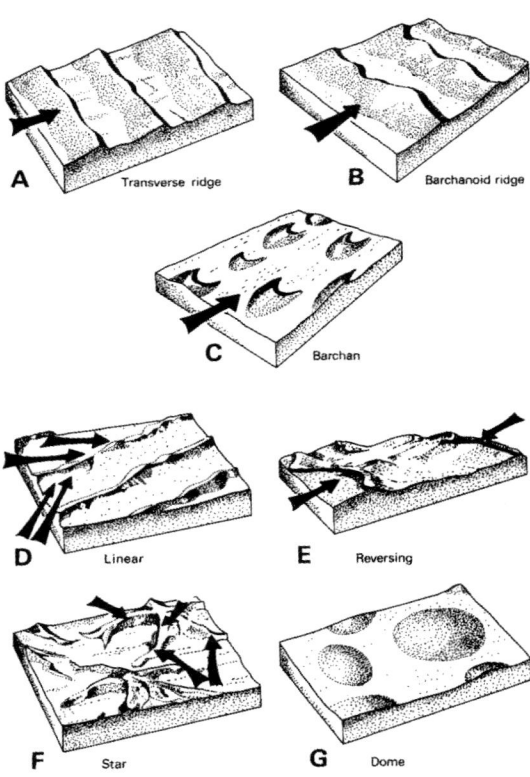

Figure 16.15 Morphological classification of dunes (from McKee, 1979).

Table 16.3 Dune activity rates (from Thomas, 1992).

Dune Type	Location	Dune height (m)	Net migration rate (m/year)
Barchan	Pampa de la Joya, Peru	1–7	32–9
	Pampa de la Joya, Peru	2.1–3.9	28–16
	Algodones, California	6	20
	Salton Sand Sea, California	3.1–8.2	27–14
	Abu Moharic, Egypt	6–14	8–5
	El-Arish, Sinai	1.9–4	13.1–6.2
	Mauritania	3–17	18–63
	Jafurah Sand Sea, Saudi Arabia	< 30	15
Transverse	Namib Sand Sea	2.5	5
	Erg Oriental	35	0.3
	Erg Oriental	240	0.16

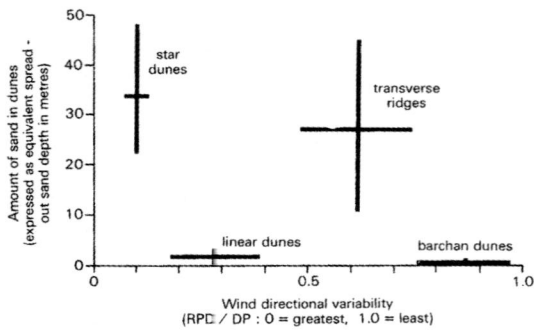

lines show x̄ ± 1 standard deviation for each axis

Figure 16.16   The occurrence of major dune types in relation to sand volume and wind direction variability (from Wasson and Hyde, 1983).

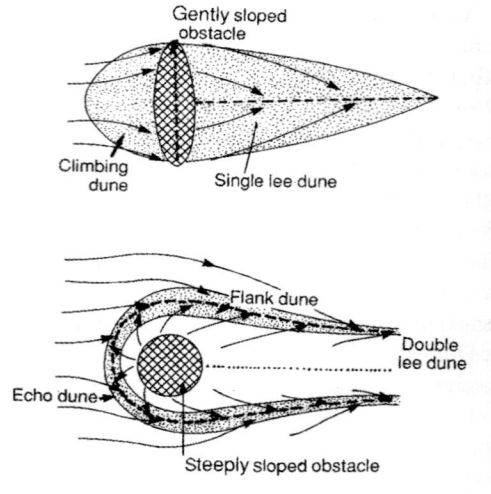

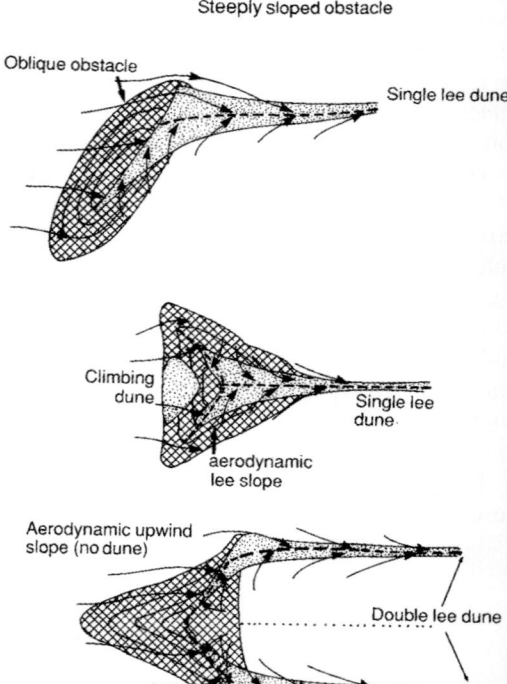

Figure 16.17   Types of anchored dunes (from Howard, 1985).

destroyed and rebuilt as the wind changes, and wind ripples. Each form is a response to an element of the wind regime, megadunes being related to long-term wind patterns (over 1000s of years), ripples to hourly or daily events.

Most sand seas and dune fields have sharply defined margins. Often this is due to topographic barriers such as scarps or rivers. Elsewhere the sharpness is explained by 'sand-shepherding', with sand moving quicker over stony plains than sandy dune surfaces. Sand entering a dune field suddenly decelerates and may be deposited, whereas sand leaving the dunes accelerates.

Other important dune forms include: anchored dunes, which are fixed by topographic obstacles (e.g. scarp slopes, hills) or plants (Figure 16.17). 'Climbing dunes' bank up against a slope as a sandy ramp. 'Echo dunes' form in response to the wind vortices set up in front of and around steep obstacles. In areas of high hills there can be a 1 km wide dune-free corridor separating the echo dunes from the high ground. 'Nebkha mounds' are formed as individual desert plants trap mobile sand—they have been reported as reaching in excess of 10 m in height. Stabilised dunes are immobilised by cementation or vegetation after their formation. The cementation may be by clays, creating 'lunettes' (often found on the downwind margins of ephemeral lakes); carbonates, forming aeolinite (calcarenite) or gypsum.

16.6 Zone IV: base level plains

This zone is dominated by the effects of groundwater and salt accumulation, often with soils dominated by wind-blown or water-lain silts and sands.

In areas of high saline groundwaters the fine-grained soils have restricted load bearing and other engineering performance characteristics. Dewatering of excavations may be needed (e.g. by using a well point system). Filter protection against migration of fines (e.g. geotextiles) may be necessary in underground and surface drainage systems. However, over time these can get clogged with salts. Calcrete duricrusts (carbonate cemented sands or clays) may be present in layered sediments, especially in sabkhas. The local salt regime may be very complex, varying with the seasons. Salts will contaminate fine and coarse aggregates and lead to salt attack on roads, structures and buildings. Each site needs to be investigated and tested separately.

Zone IVa: enclosed basins

The lowest areas within enclosed basins are characterised by almost flat vegetation-free ephemeral lakes (playas, salar or pans; Figure 16.18). Similar features may occur in topographic lows in both pediment and alluvial plains (salinas) or on coastal plains (sabkhas). Playas are net accumulation zones of both fine-grained clastic and non-clastic sediments with inputs received from aeolian and fluvial processes and through groundwater exchanges

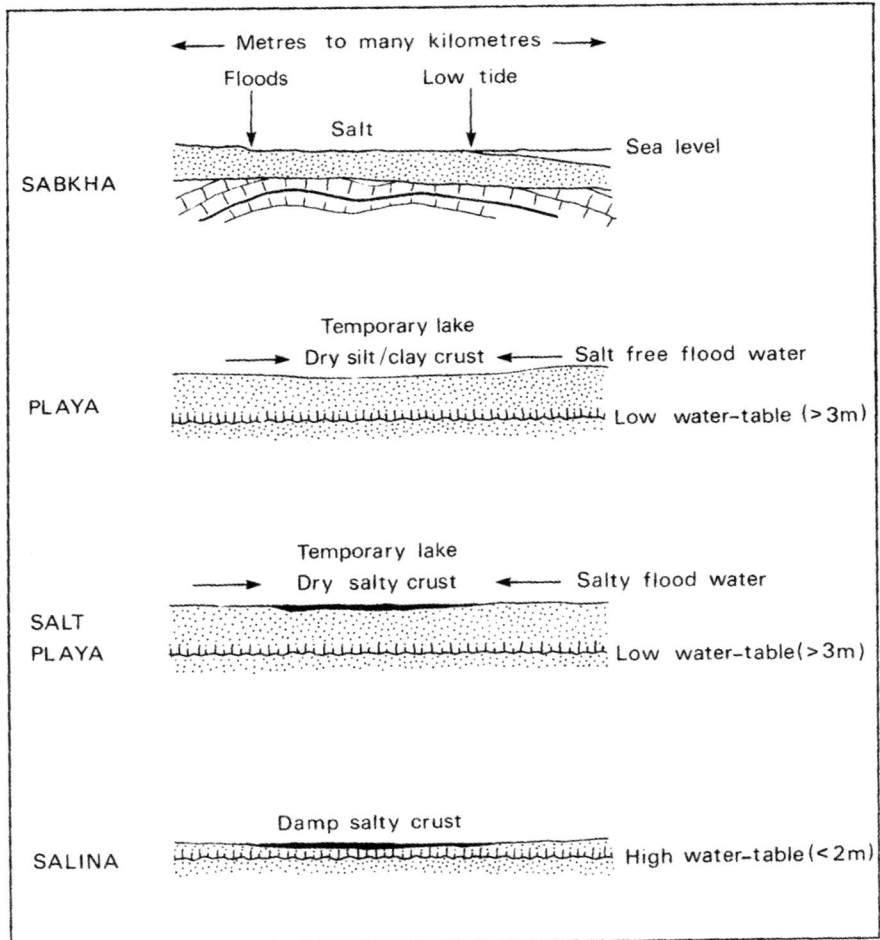

Figure 16.18a   Idealised cross sections through sabkha, playa, salt playa and salinas for engineering purposes (from Fookes, 1976).

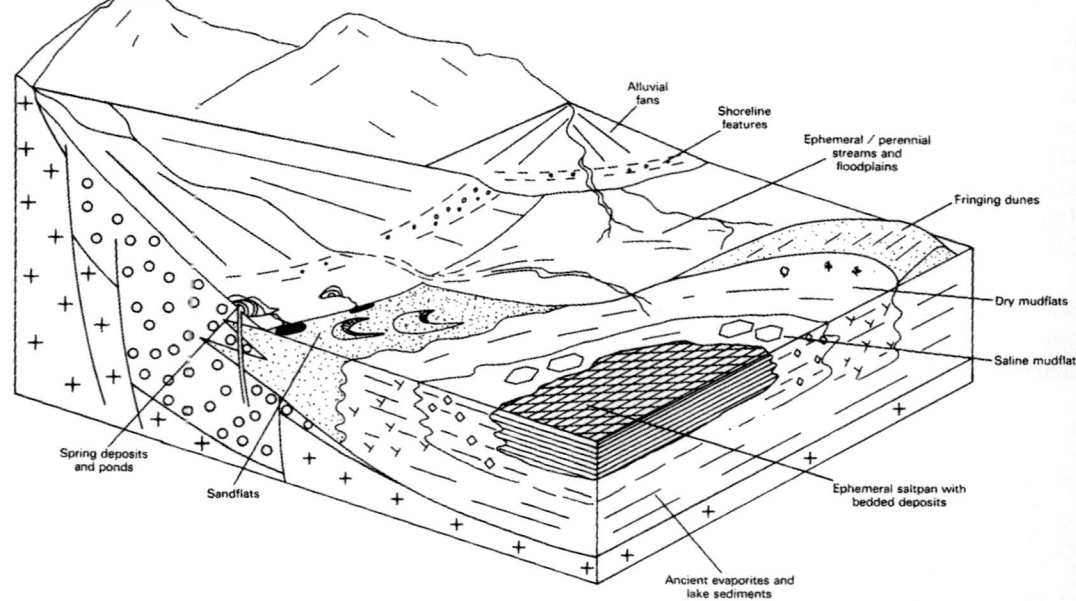

Figure 16.18b Simple playa basin model (from Hardie *et al.*, 1978).

(capillary rise or from through-flow). Groundwaters are often saline because of evaporite salts dissolved from the local or regional bedrock or from sea water intrusion. The predominance of silts and clays means that playa surfaces tend to be impermeable, encouraging the accumulation of surface runoff. They are also prone to desiccation phenomena, including raised-rim polygons, 'boils' of deliquescent salts, drying cracks and salt crusts.

Five principal types of playa can be recognised:

1. *Surface water discharging playas*: the playa surface is above the capillary limit and tends to be hard, dry and composed of fine-grained sediments.
2. *Capillary movement playas*: the surface tends to be saline, soft, permeable and loosely compacted, often with micro-relief of up to 15 cm. Where there is capillary discharge, 'self-rising' ground and salt-thrust polygons can develop.
3. *Direct groundwater playas*: these features are characterised by thin salt crusts of halite or gypsum of thicker salt pavements. The surfaces can be wet, soft and sticky. Solution pits and sinkholes may be present.
4. *Phreatophyte discharge playas*: where brackish groundwaters are close to the surface, phreatophytes and other plants may flourish and form large mounds. Ring fissures and localised surface subsidence may develop as a result of the groundwater being lowered beneath the plants.
5. *Spring discharge playas*: mounds can develop around saline springs, rising up to 30 m high (theoretically they can grow up to the level of the piezometric surface). Where artesian water emerges onto a playa surface large spring pools can develop, up to 4 m deep and 5 m wide, with near-vertical sides. These features probably form through a combination of surface solution and subsidence.

Sabkhas are saline coastal flats, noted for their high concentrations of carbonates and evaporites. They often form part of a complex assemblage of coastal landforms, including beach ridges, dunes, barriers and lagoons (Figure 16.19). Flooding may occur at high tides and during large runoff events. Various types of salts accumulate as a result of capillary rise from the saline intrusion.

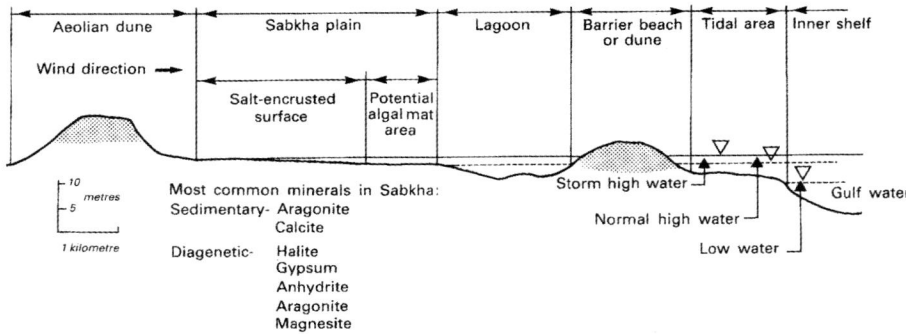

Figure 16.19   Generalised section through an Arabian sabkha (from Akili and Torrence, 1981).

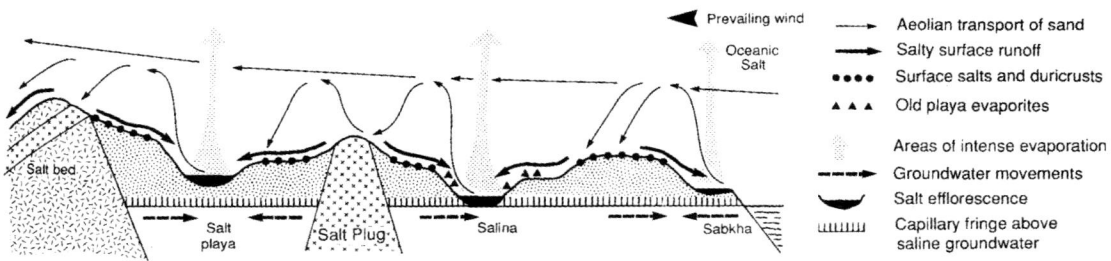

Figure 16.20   The salt cycle in deserts (from Cooke *et al.*, 1993).

The concentration of salts in playas, salinas and sabkhas can be viewed as part of a 'salt-cycle' (Figure 16.20). Salts are dissolved from rocks in the uplands and transported by water to depositional basins where they precipitate out. Wind erosion results in their removal by deflation and transport either back to the upper catchment or to an adjacent basin.

Zone IVb: depressions

Broad, low-lying depressions, bounded by steep scarps or terrace sequences, occur in many deserts (e.g. the Qattara, Siwa and Jaghbub depressions in the eastern Sahara). Although these features have been viewed as the product of extensive and prolonged wind deflation to below the groundwater table, they probably reflect a variety of factors working in combination. The Qattara depression (the lowest point is 134 m below sea level) in Egypt is believed to have been excavated as a stream valley and subsequently modified by solution weathering of the adjacent limestone

plateaux, and further extended by mass movement, deflation and surface water erosion (Albritton *et al.*, 1990). Wind erosion may have been significant during the Quaternary arid phases, with 20 000 km³ of removed debris providing the source of sand for the downwind sand seas.

16.7 Surface erosion and instability

A single layer of angular or sub-rounded gravels immediately above stone-free material covers many desert pediments and plains. Although stone pavements provide excellent surface protection against wind and water erosion they are easily destroyed by vehicle traffic or during road and pipeline construction. The resulting accelerated surface erosion leads to increased dust and sedimentation problems and creates notable environmental impacts. Once disrupted, pavement recovery may be very slow. Little is known of the rate of pavement development in different desert environments, although evidence from stone pavements on sequences of alluvial terraces in Israel suggests that a near-continuous pavement (i.e. stones touch each other) may take over 14 000 years to form (Amit and Gerson, 1986). Elsewhere, off-road vehicle tracks in the Mohave Desert, USA have recovered within decades (Webb and Wilshire, 1983).

16.8 Excavation

The presence of bedrock at or close to the ground surface on upland ramparts and pediments may come as a surprise if, for example, talus or alluvium was anticipated. However, it is the occurrence of duricrusts that can create a challenge to planning the method and rate of excavations (the need to bring in heavier equipment than originally expected, or to use blasting, can prove costly and cause delays). Duricrusts (Chapter 2) are generally the product of strong upward leaching in response to an excess of evapotranspiration over rainfall (e.g. Dixon, 1994; Watson and Nash, 1997). Desert duricrusts include silcrete, calcrete and weak gypsum crusts.

Silcrete is formed through the accumulation of silica, creating a brittle, intensely indurated material comprising quartz clasts cemented by crystallised or amorphous silica. Well-developed silcrete horizons are between 0.5 and 3 m thick, though multiple crusts may exceed 5–7 m. Silcretes are believed to be ancient features and are relatively common in Australia, southern Africa, parts of the Sahara and the Arabian Gulf (e.g. Milnes and Thiry, 1992).

Calcrete is formed through the accumulation of calcium carbonate. Calcretes may range from calcified soils, calcium carbonate nodules or concretions, honeycomb calcretes (coalesced nodules) to mature hardpans (Netterberg, 1980). Hardpans are typically 0.3–0.5 m thick, although superimposition of sequences of calcretes can create thick, complex profiles. Calcrete can form relatively quickly (over the 100–1000 year timescale) and is a common feature of deserts with extensive carbonate-rich bedrocks.

Weak gypsum crusts are found in warm deserts, where mean annual rainfall is less than 250 mm (Watson, 1985). The gypsum accumulations may reach 5 m thick, ranging from bedded crusts, bands of large crystals (desert roses—these can reach up to 5 m high) or surface crusts of alabastine gypsum.

An important distinction needs to be made between duricrusts developed within the soil profile (i.e. they form thin indurated layers or pans which 'float' within the soil profile) and those formed by the enrichment of the bedrock surface (i.e. 'caprocks'). Floating duricrusts can generally be ripped out by conventional excavation equipment. Well-developed caprock duricrusts can be strong and massive and, hence, present significant problems for excavation. They often require heavy ripping with hydraulic breaking or even blasting, especially if they are unjointed. As excavation can be a major cost driver in pipeline projects it is prudent to establish the nature and extent of duricrusts at an early stage in project planning (e.g. Fookes et al., 2001).

16.9 Behaviour of desert soils

In general the soil grading, in situ density and Casagrande classification are good guides to the potential engineering performance of desert soils. However there are important exceptions to this in that soils may often appear to be of high strength due to very low moisture contents (high suctions in fine-grained soils) or cementing due to clay or salt bonds between larger particles. Whereas in a temperate climate coarse-grained materials having less than 12% fines will generally be free draining, in deserts some sands with up to 30% fines of low plasticity may behave as free-draining soils (Fookes and Gahir, 1995).

Unless truly cemented, such soils can either collapse on wetting and loading if they are initially loose (e.g. aeolian deposits), or swell substantially on wetting if comprising high-plasticity clay. There is a need in classifying arid soils to identify those that are truly cemented and those whose strength can decrease or volume change dramatically on wetting (e.g. Fookes and Parry, 1994).

A simple slaking test whereby samples are left in water for a minimum of an hour is useful in distinguishing between real and apparent cementation. Swelling potential of clays can be estimated from the difference in behaviour of samples in water and carbon tetrachloride (Sridharan, 1998). Once the effects of cementation have been taken into account, normal engineering classifications are a good guide—the potential collapse of sandy and silty soil is indicated by low relative density and uniform grading, while the swelling potential of clay is indicated by high plasticity and low *in situ* moisture content.

Some desert soils may be prone to collapse or swelling affecting foundations, roads and pipelines. These problems can be the result of changes in moisture content associated with, for example, broken drains, irrigation of gardens, or failure to deal correctly with surface water, air conditioner or roof drainage. In other situations the problems may simply be the response to the presence of a new building changing the balance between capillary rise and surface evaporation.

Where potential problems of collapse or swelling are identified, special precautions may have to be taken with building foundations. These may range from attempts at prevention of the problem (e.g. double-sleeving of water service connections to buildings) to the use of deep foundations, the design of which must take account of the expected soil movements.

The shortage of water in deserts is often a problem in applying normal compaction methods. However it has been found that many soils have a second maximum in their water content/dry density curves at near zero water content, giving a dry density close to that achieved at the normal optimum moisture content. Dry compaction is often, therefore, a viable option. However, it must be recognised that such compacted materials have relatively high air voids and may be prone to collapse settlements on wetting. Special construction techniques may therefore be necessary to prevent wetting of the material in, for example, road embankments during periods of heavy rainfall.

16.10 Aggregates

The principal aggregate sources in drylands (Fookes and Higginbottom, 1980) are shown in Figure 16.21 and Table 16.4 and include alluvial sands and gravels, coastal beach sands, dune sands, and crushed rock aggregate.

Alluvial sands and gravels are sourced from fans and alluvial channels in Zones II and III. Leaching by floodwaters ensures that the chloride and sulphate levels are relatively low. However, the deposits may be cemented by carbonate or gypsum and, hence, require suitable processing. It is usually essential to screen and wash fan gravels carefully before use.

Coastal beach sands are often carbonate grains and may have a narrower and finer grading than that preferred for concretes. Shells may need to be screened out to avoid adverse effects on concrete structure or density. Often chlorides and sulphates may be unacceptably high, especially where borrow pits are worked close to the groundwater table and where salts can accumulate through evaporation.

Dune sands—wind-blown sands of Zone IIIc—have narrow, fine grading (c. 50–600 μm) with silt contents of around 10% and, hence, are not ideal sources, but can be used. The rounded nature of the grains gives them poor binding qualities. Experience has shown that it is possible to identify dunes or parts of dunes with better grading or to improve the resource by blending with other material.

Crushed rock aggregate can be obtained from the uplands (Zone I) sources or from duricrusts in Zones IIa and III. In some areas, such as the Gulf States, crushed calcrete has been an important source of aggregate (Fookes and Higginbottom, 1980). However, dolomite-rich rocks may be liable to alkali/carbonate reaction. Bedrock inselbergs in Zone IIa may prove excellent sources of suitable material.

16.11 Water and sediment problems

Major flood events and the associated erosion and deposition are rare and, hence, are often unexpected. Predictions of discharge and sediment

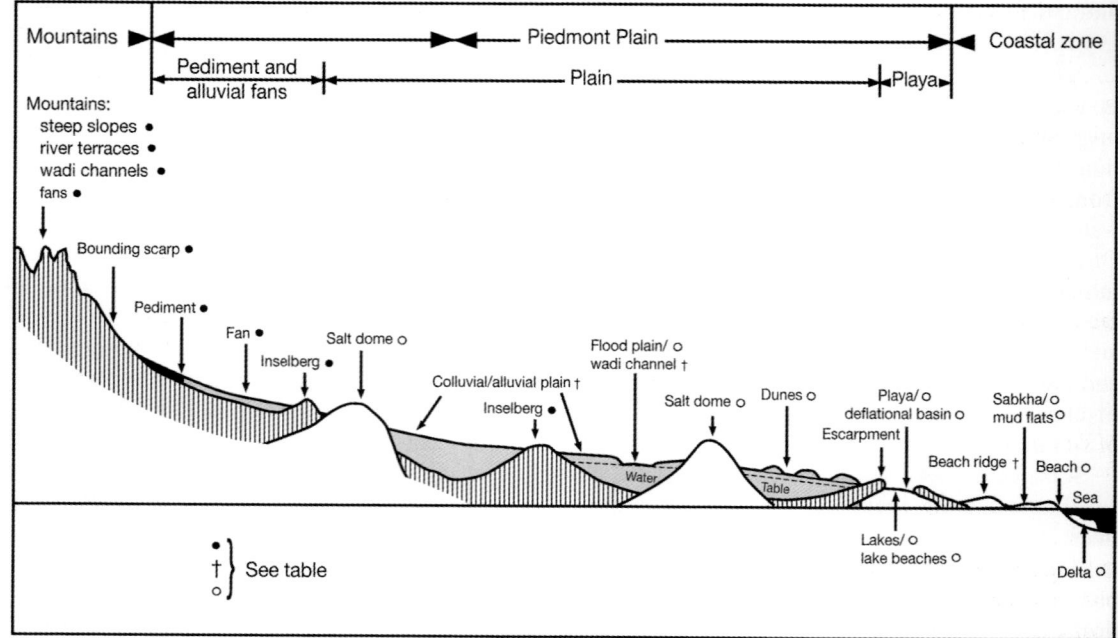

Major landforms as aggregate resources in drylands

Mountains
Including peaks, ridges, plateau surfaces, steep (excluding precipitous) slopes,* deep valleys and canyons, wadis, river terraces* and alluvial fans,* bounding scarp slopes.* Forms vary with rock type and the evolutionary history of the area.

Pediments and Alluvial Fans
Rock pediment,[†] fan* and bajada,[†] with occasionally inselbergs* or salt domes[0] forming locally high ground.

Plains
Occur downslope of pediments or alluvial fans without a distinct boundary and may include a whole variety of features including: alluvial[†] and colluvial plains,[†] wadi channels and flood plains, dune fields,[0] salt domes,[0] inselbergs,* and extensive stone pavement surfaces.[0]

Playa Basins
Enclosed depressions receiving surface runoff from internal catchments or within escarpment zones.[†] They frequently contain lakes (either temporary or permanent), lake beaches, evaporite deposits[0] and may be strongly influenced by aeolian, fluvial and salt processes in their baseland zones.

Coastal Zones
These include beach ridges[†] (formed at periods of higher sea-level or during exceptional storms), Sabkhas,[0] mud flats,[0] beach[0] and foreshore,[0] estuaries[0] and deltas.[0]

---

*Normally a major source of aggregate, conditional on suitable mineralogy
[†]May be a reasonable source, depending on specific characteristics
[0]Normally should *not* be used for aggregates.

Figure 16.21   Dryland landforms and aggregate sources (redrawn from Cooke *et al.*, 1982).

yield can be severely hindered by the absence of meteorological or hydrological records. In addition, the assessment of runoff based on catchment area (Chapter 11) are less reliable in drylands than temperate areas because most desert drainage networks are relict forms, only partly active even in the largest floods. The limited spatial extent of high intensity storms may mean that only part of a catchment is actively generating runoff at any one time.

Table 16.4   Landforms and aggregate potential (adapted from Fookes and Higginbottom, 1980; Cooke *et al.*, 1982)

Feature	Aggregate Type	Nature of Material	Engineering Properties/Fills	Potential Volume
Bedrock mountains	Crushed rock suitable for all types of aggregate	Angular, clean and rough texture	Depends on rock type and processing. Good fills	Very extensive
Duricrust	Road base and sub-base	Often contains salts. Needs crushing and processing	May be self-cementing with time. Quite good rock fill	Often only small deposits of good quality
Upper alluvial fan deposits	Concrete and when crushed for road base	When crushed and screened angular – and clean. Otherwise sometimes dirty and often rounded	Good compaction as fill	Often very extensive but good quality material may only be in small deposits
Middle alluvial fan deposits	May make road base	Often high fines content	Often good compaction as fill. Good bearing	Small to extensive
Lower alluvial fan deposits	Generally not useful	High fines content. Needs processing	Poor bearing and as fill	Small to extensive
Other piedmont plain alluvium	Variable, locally good concrete aggregate	Dirty, rounded and well graded. May contain salts	Good bearing capacity (dense sediments) and as fill. Locally poor due to clay and silt layers	Very extensive but good quality material in small deposits
Old river deposits	Concrete	Variable	Difficult to locate in field	Often deposits patchy and thin
Dunes	Generally not good	Usually too fine and rounded	Fills of poor compaction	Locally extensive
Interdunes	Fine aggregate	Coarse to fine, angular. Needs processing	Fills of poor compaction	Very localised
Salt playas and sabkhas	Not suitable	Very salty and aggressive	Poor. Special random fill	Locally extensive
Coastal dunes	Generally not suitable	Generally too fine and rounded	Special fills often poor compaction	Sometimes extensive
Storm beach	Fine aggregate may be sharp and/or salty	Sufficiently coarse for concrete sand Clean after processing	Random fill	Sometimes extensive
Foreshore	Generally not suitable	Fine, rounded sand.	Salt contaminated, but might make random fill	Locally extensive

In the absence of hydrological records, it is necessary to use simple field observations to provide an approximation of the flooding and scour hazard. Estimates of historical flood depths and velocities can be made from the theoretical relationships between channel form and bed grain size. Richards *et al.* (1987) have suggested a method for estimating wadi discharge through the reconstruction of credible flood events from field observations of channel-form (i.e. cross-section area, bed roughness and estimated water depths). The method involves the following steps:

1. Field mapping of the channel cross-section: a field sketch or surveyed profile (e.g. Figure 16.22) is required at representative channel cross-sections, defining separate morphological units within each cross-section (e.g. the active

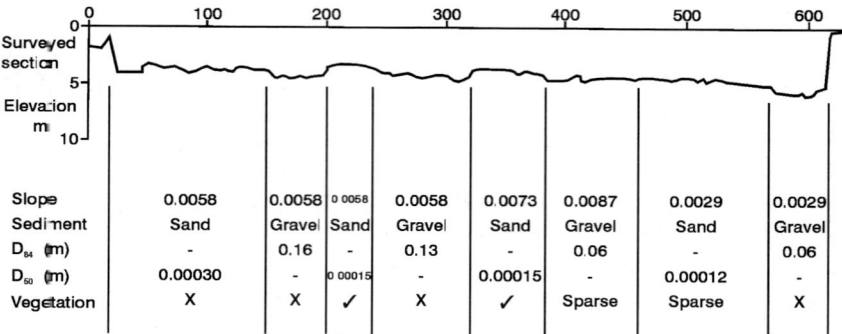

Figure 16.22   Sample cross section through Wadi Dhamad, Saudi Arabia illustrating the field data needed for assessment of flows (redrawn from Richards *et al.*, 1987).

channel, terraces, point bars etc.). For each unit, estimates need to be made of slope gradient, unit width, bed sediment size (D_{50} = median grain size, and D_{84} = the 84th percentile grain size) and the depth of recent and possible maximum flows through the channel section.

2. Estimation of the discharge passing through a unit at a particular flood depth (i.e. stage), from the mean depth and channel width and estimates of the mean velocity. This involves calculating velocity of flow through a unit at a particular flood depth using the Darcy-Weisbach equation:

$$v = \sqrt{8gR\,s/f}$$

where g is the acceleration due to gravity; R is the hydraulic radius (m); s is the slope gradient and f is the friction factor. The friction factor (f) can be calculated using the empirical relationship of Limerinos (1969):

$$1/\sqrt{f} = 1.16 + 2\log\left(R/D_{84}\right)$$

As a check on the results, the friction factor can also be calculated from:

$$n = 0.0151 D_{50}^{1/6}$$

where n is Manning's roughness coefficient, and

$$f = 8\,g\,n^2/R^{1/3}$$

where R is the hydraulic radius of the channel (i.e. cross section area/wetted perimeter). The velocity of flow through a unit at a particular flood depth can be calculated using either or both estimates of the friction factor. Discharge (Q) through each unit at a particular flood depth

is then calculated from the continuity equation:

$$Q = wRv$$

where w is the channel width.

3. Estimation of the overall discharge at a particular flood stage: this involves summing the discharges for each unit at a particular flood depth to estimate an overall discharge in a particular event.

4. Development of a flood stage-discharge relationship: this can be achieved by assessing the discharge at a range of credible event sizes (i.e. events that have left a morphological imprint on the channel).

An indication of the maximum flood depth experienced in a channel can be estimated from the largest mobile bed material. Baker and Ritter's (1975) method involves an empirical relationship between the largest grain size considered mobile (D_{max}, mm) and the bed shear stress (τ_0):

$$\tau_0 = 0.00044 D_{max}^{1.85}$$

Given the channel slope (S) the bed shear stress can be related to the channel hydraulic radius R by:

$$\tau_0 = \gamma RS$$

where γ is the specific gravity of water. Alternatively the critical velocity (V_c) needed to move the largest bed material can be estimated from:

$$V_c = 0.18 D_{max}^{0.49}$$

Careful morphological mapping can also be used to develop flood hazard maps (e.g.

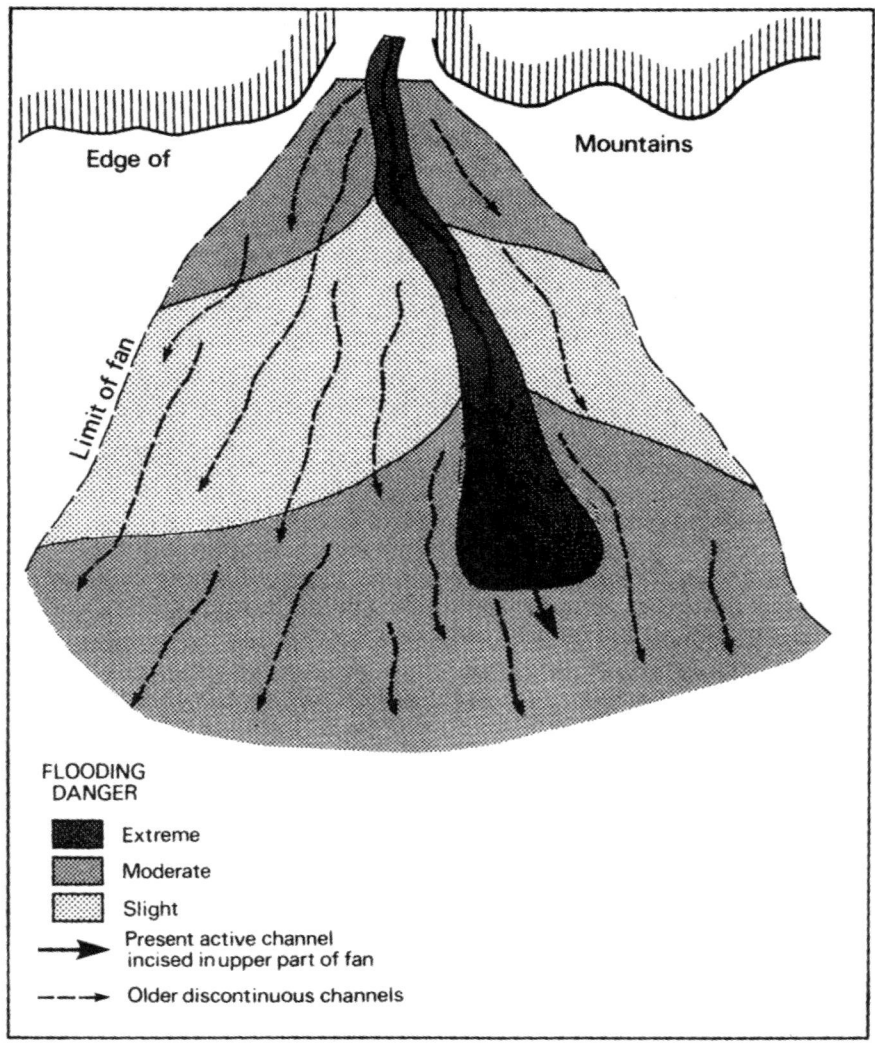

Figure 16.23   Flood hazard across an alluvial fan (from Kesseli and Beaty, 1959).

Figures 16.23 and 16.24). Useful strategies for managing flood risk include avoidance, flood warning systems (e.g. Figure 16.25) or adopting route alignments that minimise channel-crossing problems (e.g. Figure 16.26). A technique sometimes used to protect minor roads crossing areas prone to flash flooding is to have most of the road on low embankment, but have sections at intervals at or slightly below normal surface level which act as inverted bridges (referred to by engineers as 'Irish bridges') and allow flood water to pass over the road over controlled lengths (dependent on the likely maximum flows). Although the road may be closed during the actual flood event, damage to the road is localised and even prevented if the road is armoured over the low lengths.

16.12 Dune mobility

Mobile dunes may blow over roads or accumulate against buildings and structures. The migration

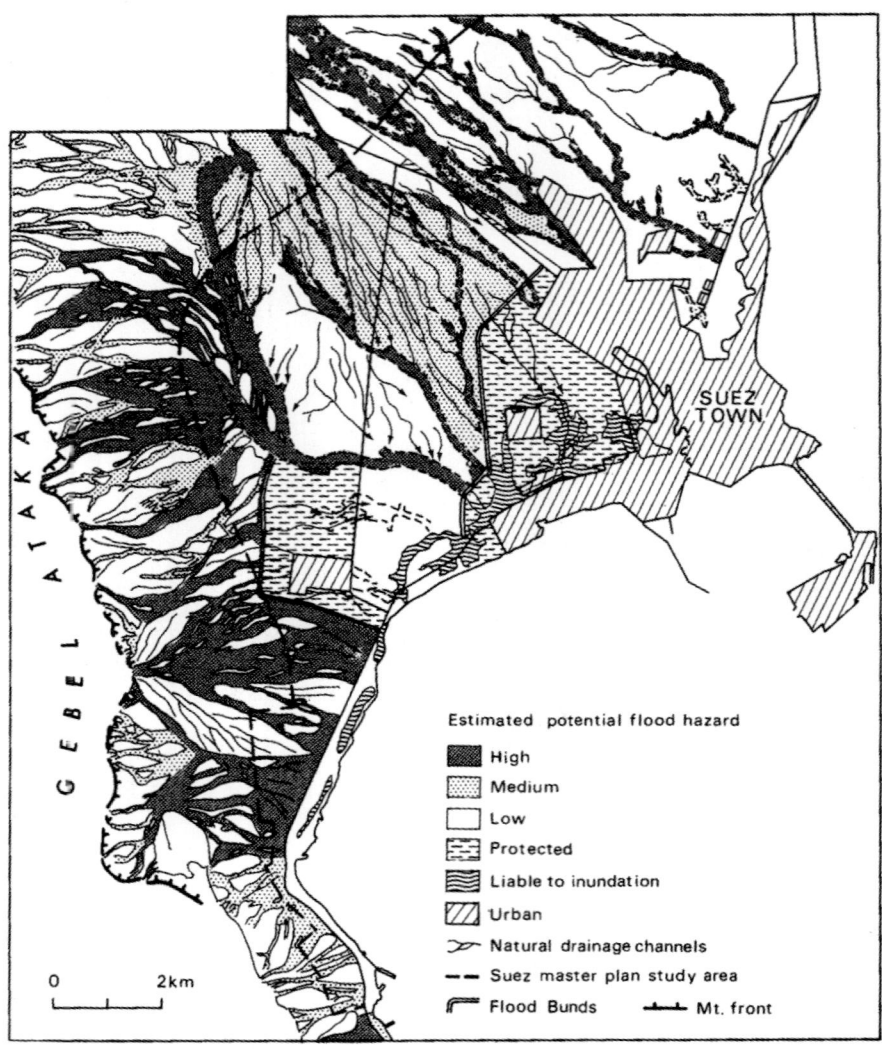

Figure 16.24   Flood hazard map of Suez City, Egypt (from Cooke *et al.*, 1982).

rate is inversely proportional to the dune wavelength and height (Simons *et al.*, 1965):

Migration Rate =

$$\frac{\text{Sand Transport Rate}}{(\text{Cross Section/Wavelength} \times \text{Height}) \times \text{Height} \times \text{Bulk Density}}$$

Thus, the bigger the dune, the less mobile it is (Figure 16.27; Table 16.3). However, above a certain size dune mobility appears to be fairly uniform.

Alternatively, the change in dune location on sequential aerial photographs or satellite imagery can provide an indication of the actual migration rates during that period.

Control measures include surface stabilisation and the use of fences. Surface stabilisation of source areas can be achieved with gravel, gypsum, saline water or a chemical or oil spray. Oil has been used successfully to stabilise large areas at low cost (Kerr and Nigra, 1952). Permanent stabilisation

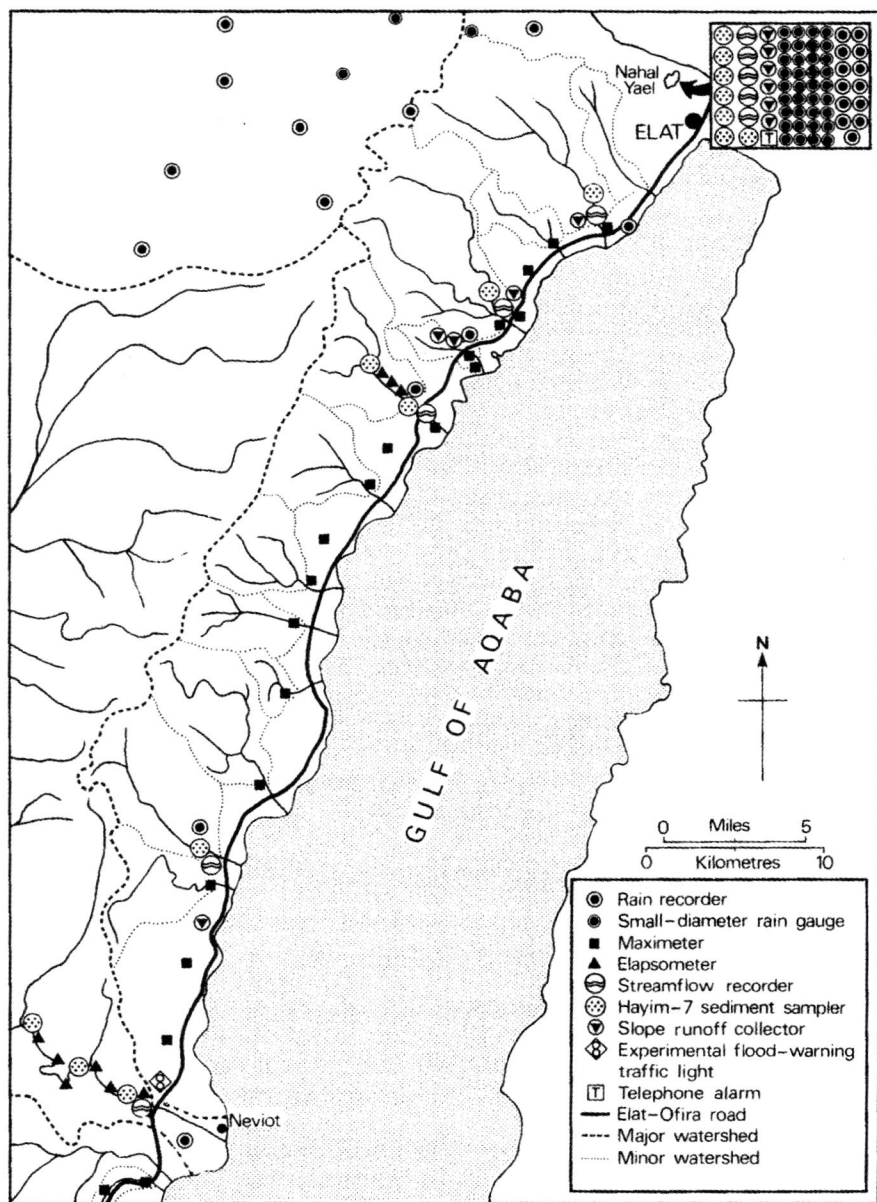

Figure 16.25 Flood warning system on the Gulf of Aqaba shoreline, Egypt (from Schick and Sharon, 1974).

can often only be achieved through the establishment of a vegetation cover. However, in many drylands this is not a straightforward task and requires careful consideration of suitable plant types and management. Control of movement can also be achieved through transport pathways with fences.

A frequently used method of aeolian sand control is the construction of panels or fences to divert or stop sand transport reaching a vulnerable site (e.g. a village). It is common practice to use a system of three parallel fences to provide effective protection to a site (Figure 16.28). Sand accumulates mainly

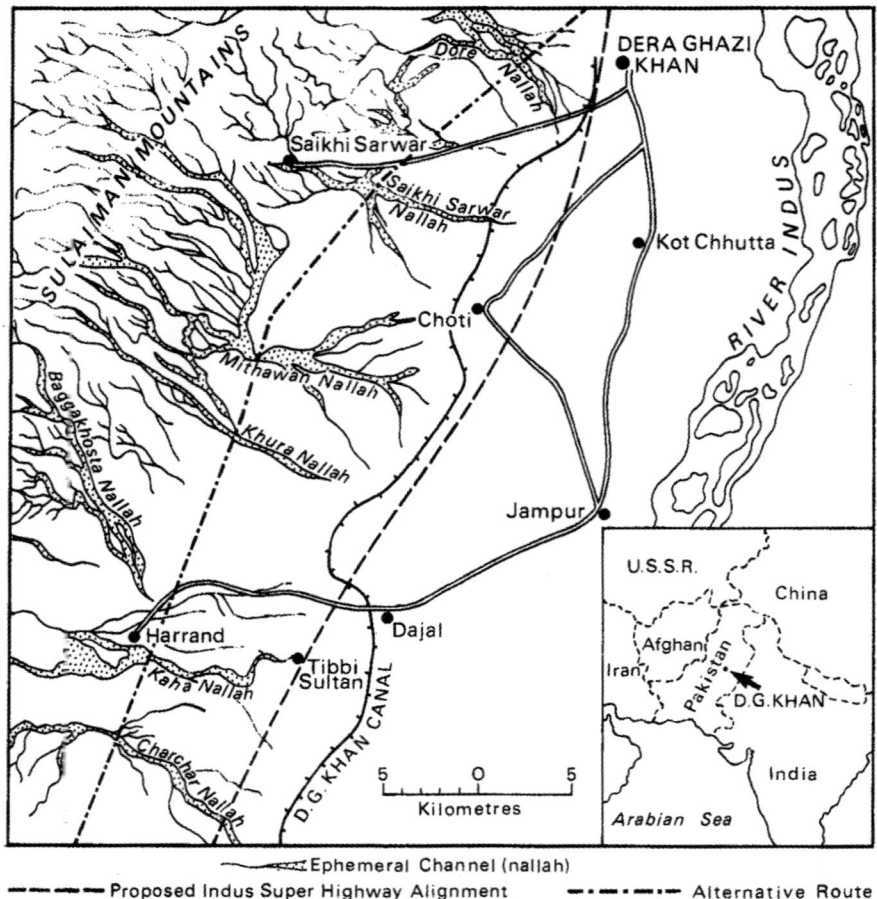

Figure 16.26   Proposed alignments of the Super-Indus highway, Pakistan. Note that the alternative route minimises channel crossings by crossing fans towards the apex (from Cooke *et al.*, 1982).

and first around fence 1; as the effectiveness of the first fence declines, sand accumulates increasingly at fence 2 and so on (Cooke *et al.*, 1982). Once the fence is covered it will no longer act as a sand trap, as new sand arriving from upwind will simply sweep across its smooth surface and be carried downwind (i.e. no reduction in transport capacity). Increasing the height will restore the trap efficiency.

16.13 Aggressive salty ground

The presence of saline groundwater close to the ground surface in topographic lows, base level

plains (Zone IV) and coastal margins creates a range of engineering hazards related to salt weathering (Figure 16.29). These problems largely occur as a result of post-construction accumulation of salts due to capillary rise of groundwater. The impact of chlorides and sulphates on concrete and roads is perhaps the most significant problem (e.g. Fookes and Collis, 1975a, 1975b, 1976; Fookes and French, 1977; French and Poole, 1976; Fookes *et al.*, 1985).

Figure 16.30 provides a simple model of the soil-groundwater conditions. The upper surface of the capillary fringe (limit of capillary fringe) can vary from being at depth (below foundation level) to potentially above ground level

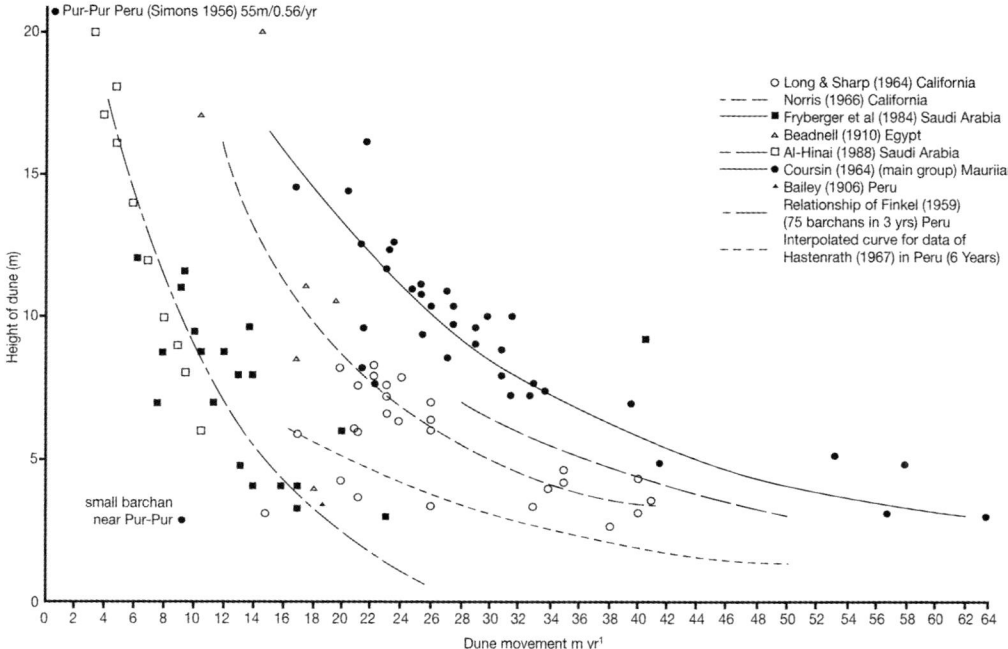

Figure 16.27   The relationship between dune size and mobility (redrawn from Cooke *et al.*, 1993).

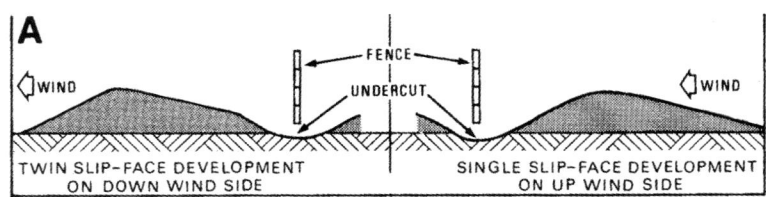

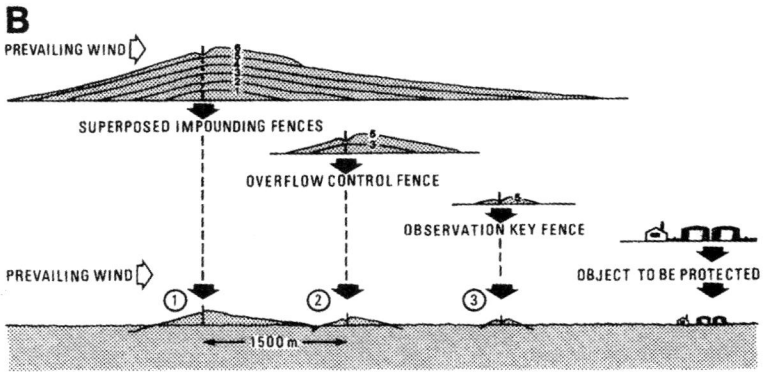

Figure 16.28   The use of porous fences as sand traps (from Kerr and Nigra, 1952).

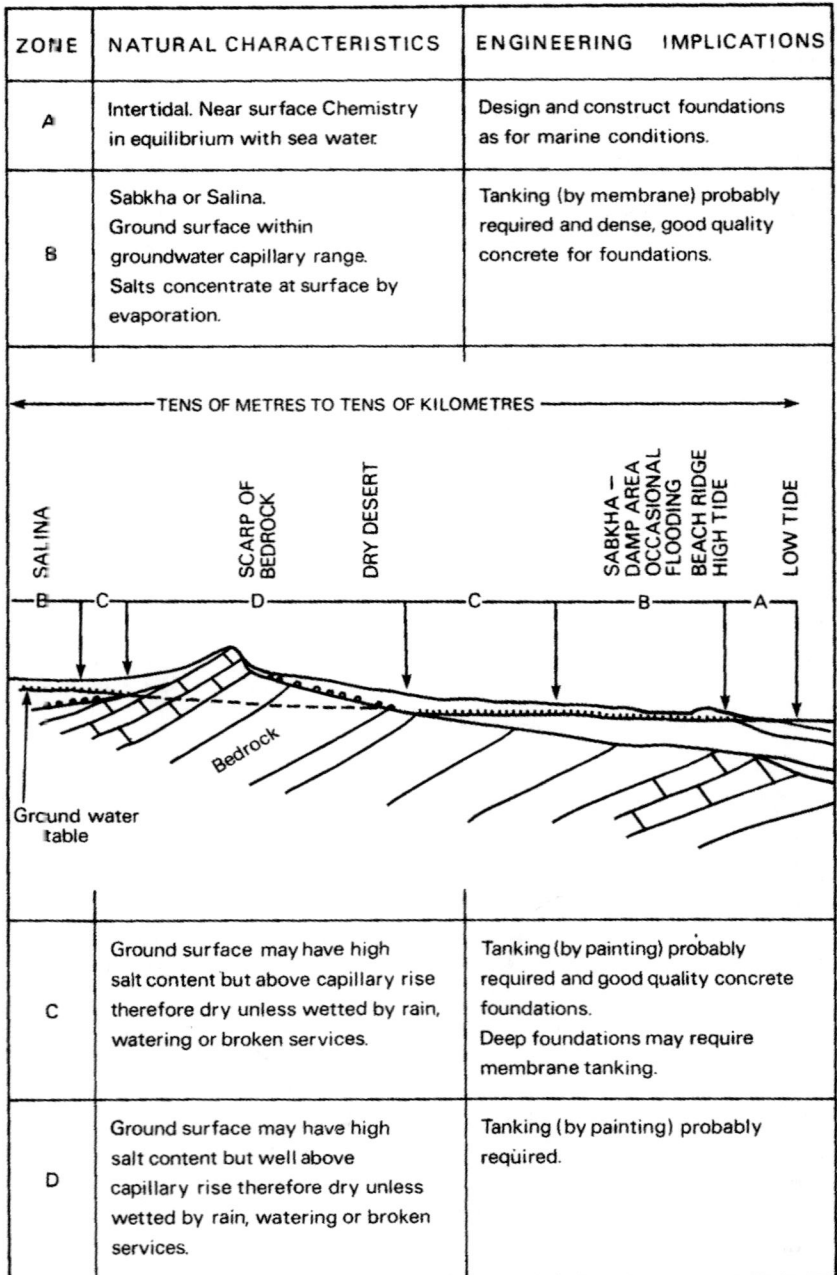

ZONE	NATURAL CHARACTERISTICS	ENGINEERING IMPLICATIONS
A	Intertidal. Near surface Chemistry in equilibrium with sea water	Design and construct foundations as for marine conditions.
B	Sabkha or Salina. Ground surface within groundwater capillary range. Salts concentrate at surface by evaporation.	Tanking (by membrane) probably required and dense, good quality concrete for foundations.

TENS OF METRES TO TENS OF KILOMETRES

SALINA — SCARP OF BEDROCK — DRY DESERT — SABKHA – DAMP AREA OCCASIONAL FLOODING — BEACH RIDGE HIGH TIDE — LOW TIDE

B — C — D — C — B — A

Bedrock

Ground water table

| C | Ground surface may have high salt content but above capillary rise therefore dry unless wetted by rain, watering or broken services. | Tanking (by painting) probably required and good quality concrete foundations. Deep foundations may require membrane tanking. |
| D | Ground surface may have high salt content but well above capillary rise therefore dry unless wetted by rain, watering or broken services. | Tanking (by painting) probably required. |

Figure 16.29   Salt problems in drylands (from Fookes and Collis, 1976).

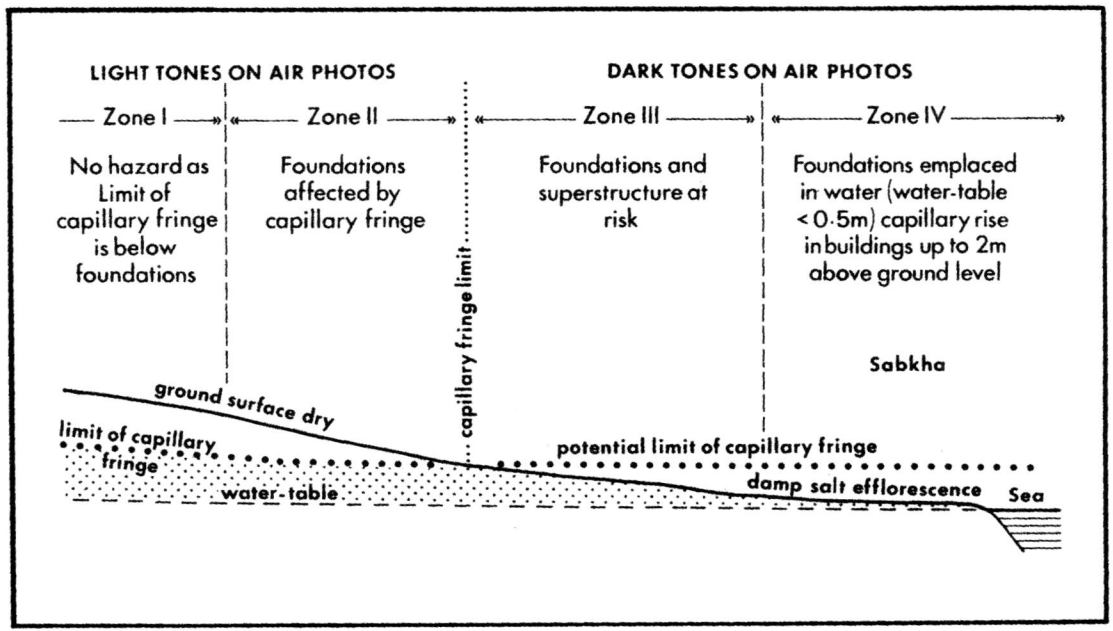

Figure 16.30 A simple soil-groundwater model for drylands (from Jones, 1980; after Fookes and French, 1977).

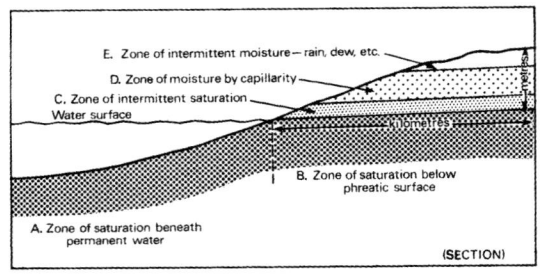

Figure 16.31 The interrelationships between soil moisture zones in drylands (from Fookes and French, 1977).

(resulting in salt crusts on the surface). Buildings or structures can allow capillary rise to extend to its full height, leading to damp-stained walls and the deposition of salts in the fabric. The capillary fringe limit is the location where the capillary fringe intersects the rising ground surface and marks the boundary between areas where only foundations are at risk and areas where both foundations and the superstructure may be attacked.

Figure 16.31 shows how the aggressive ground hazard can be related to observable soil moisture zones. This approach can be useful in

route selection and the evaluation of road maintenance priorities. An alternative approach is to convert the simple soil-groundwater model of Figure 16.30 into hazard classes:

1. Zone I: no hazard from groundwater as the limit of the capillary fringe is below foundation depth. Protection may be needed in some cases.
2. Zone II: the limit of the capillary fringe is below ground surface but sufficiently close to it to affect foundations. Concrete and structures will need protection.
3. Zone III: the limit of the capillary fringe is potentially above ground level so that both foundations and susperstructures are at risk. Use full protective measures if construction is necessary.
4. Zone IV: the water-table is within 0.5 m of the ground surface for most of the year so that foundations are emplaced in water and there is potential for capillary rise to well above ground level. Avoid these areas.

Figure 16.32 presents a hazard zone map of Suez City, Egypt, developed using this method.

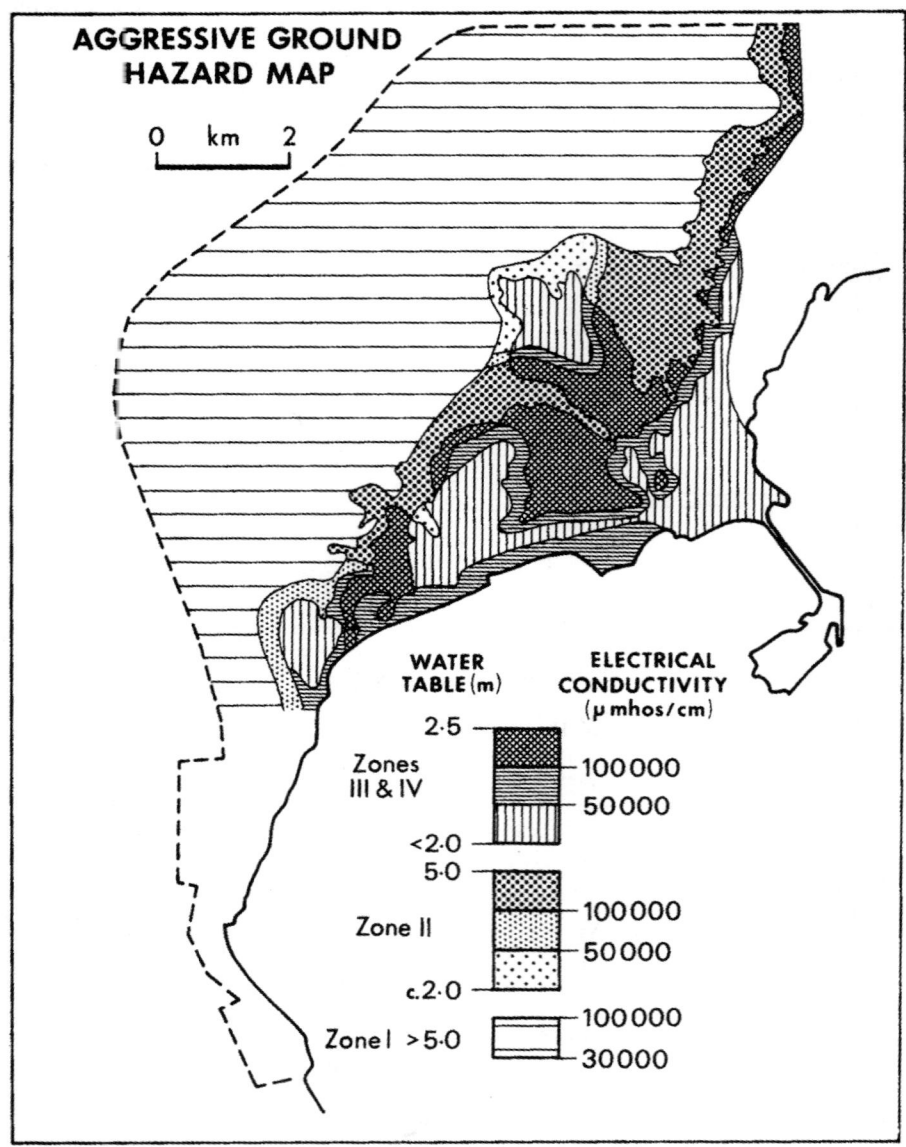

Figure 16.32 Aggressive ground hazard map, Suez City Egypt (from Cooke *et al.*, 1985).

16.14 Summary

Although desert landscapes show considerable variability, the simple classification of the terrain into uplands, footslopes and fans, plains and base-level plains provides a framework for identifying and evaluating engineering hazards and assessing resources. Indeed, deserts present a range of unique problems to engineers, including surface instability following disruption of stone pavements, difficult excavation of strong and massive duricrusts, unfamiliar (to western engineers) behaviour of the soils, restricted availability of aggregates, ephemeral flood activity, dune mobility and aggressive ground. In addition, active slope instability on upland slopes, hydro-compaction and dust generation are issues that need to be considered in many desert landscapes.

References

Akili, W. and Torrence, J. K. (1981) The development and geotechnical problems of sabkha, with preliminary experiments on the static penetration resistance of cemented sands. *Quarterly Journal of Engineering Geology* **14**, 59–73.

Albritton, C. C., Brooks, J. E., Issawi, B. and Swedan, A. (1990) Origin of the Qattara Depression, Egypt. *Bulletin of the Geological Society of America* **102**, 952–960.

Amit, R. and Gerson, R. (1986) The evolution of Holocene reg (gravely) soils in deserts—an example from the Dead Sea region. *Catena* **13**, 59–79.

Baker, V. R. and Ritter, D. F. (1975) Competence of rivers to transport coarse bedload material. *Geological Society of America Bulletin* **86**, 975–978.

Blair, T. C. and McPherson, J. G. (1994) Alluvial fan processes and forms. In Abrahams, A. D. and Parsons, A. J. (eds) *Geomorphology of Desert Environments*. Chapman and Hall, London, 354–402.

Bowman, D., Karnieli, A., Issar, A. and Bruins, H. J. (1986) Residual colluvio-aeolian aprons in the Negev Highlands (Israel) as a paleoclimatic indicator. *Palaeography, Palaeoclimatology, Palaeoecology* **56**, 89–101.

Breed, C. S., McCauley, J. F., Whitney, M. I., Tchakarian, V. P. and Laity, J. E. (1997) Wind erosion in drylands. In Thomas, D. S. G. (ed.) *Arid Zone Geomorphology*. Wiley, Chichester, 437–464.

Christenson, G. E. and Purcell, C. (1985) Correlation and age of Quaternary alluvial-fan sequences, Basin and Range Province, southwestern United States. *Geological Society of America Special Paper* **203**, 115–122.

Cooke, R. U., Brunsden, D., Doornkamp, J. C. and Jones, D. K. C. (1982) *Urban Geomorphology in Drylands*. Oxford University Press.

Cooke, R. U., Brunsden, D., Doornkamp, J. C. and Jones, D. K. C. (1985) *Geomorphological Dimensions of Land Development in Deserts—with special reference to Saudi Arabia*. Nottingham Monographs in Applied Geography No. 4, University of Nottingham.

Cooke, R. U., Warren, A. and Goudie, A. S. (1993) *Desert Geomorphology*. UCL Press.

Costa, J. E. (1988) Rheologic, geomorphic and sedimentalogic differentiation of water floods, hyperconcentrated flows and debris flows. In Baker, V. R., Kochel, R. C. and Patton, P. C. (eds) *Flood Geomorphology*. Wiley, New York, 113–122.

Denny, C. S. (1967) Fans and pediments. *American Journal of Science* **265**, 81–105.

Dixon, J. C. (1994) Duricrusts. In Abrahams, A. D. and Parsons, A. J. (eds) *Geomorphology of Desert Environments*. Chapman and Hall, London, 82–105.

Dorn, R. I. (1988) A rock varnish interpretation of alluvial fan development in Death Valley, California. *National Geographic Research* **4**, 56–73.

Fontes, J. Ch., Gasse, F., Callot, Y., Plaziat, J. -C., Carbonell, P., Dupeuble, P. A. and Kaczmarska, I. (1985) Freshwater to marine-like environments from Holocene lakes in northern Sahara. *Nature* **317**, 608–610.

Fookes, P. G. (1976) Road geotechnics in hot deserts. *Highway Engineer* **23**, 11–29.

Fookes, P. G. (1978) Middle East—inherent ground problems. *Quarterly Journal of Engineering Geology* **11**, 33–49.

Fookes, P. G. and Collis, L. (1975a) Problems in the Middle East. *Concrete* **9**, 12–17.

Fookes, P. G. and Collis, L. (1975b) Aggregates and the Middle East. *Concrete* **9**, 14–19.

Fookes, P. G. and Collis, L. (1976) Cracking and the Middle East. *Concrete* **10**, 14–19.

Fookes, P. G. and French, W. J. (1977) Soluble salt damage to surfaced roads in the Middle East. *Journal of the Institution of Highway Engineers* **24**, 10–20.

Fookes, P. G. and Higginbottom, I. E. (1980) Some problems of construction aggregates in desert areas, with particular reference to the Arabian peninsula 1: Occurrence and special characteristics. *Proceedings of the Institution of Civil Engineers, Part 1* **68**, 39–67.

Fookes, P. G. and Knill, J. L. (1969) The application of engineering geology to the regional development of Northern and Central Iran. *Engineering Geology* **3**, 81–120.

Fookes, P. G. and Parry, R. H. G. (eds) (1994) *Engineering Characteristics of Arid Soils*. Balkema, Rotterdam.

Fookes, P. G. and Gahir, J. S. (1995) Engineering performance of some coarse-grained arid soils in the Libyan Fezzan. *Quarterly Journal of Engineering Geology* **28**, 105–130.

Fookes, P. G., French, W. J. and Rice, S. M. (1985) The influence of ground and groundwater geochemistry on construction in the Middle East. *Quarterly Journal of Engineering Geology* **18**, 101–128.

Fookes, P. G., Lee, E. M. and Sweeney, M. (2001) Pipeline route selection and characterisation, Algeria. In Griffiths, J. S. (ed.) *Land Surface Evaluation for Engineering Practice*. Geol. Soc. Engineering Group Special Publication 18, 115–121.

French, W. J. and Poole, A. B. (1976) Alkali aggressive aggregates and the Middle East. *Concrete* **10**, 18–20.

Gasse, F., Fontes J. C., Plaziat, J. C., Carbonel, P., Kaczmarska, I., De Deckker, P., Soulie-Marsche, I., Callot, Y. and Dupeuble, P. A. (1987) Biological remains, geochemistry and stable isotopes for the reconstruction of environmental and hydrological changes in the Holocene lakes from North Sahara. *Paleaogeography, Paleaoclimatology, Paleaoecology* **60**, 1–46.

Gerson, R. (1982) Talus relicts in deserts. *Israel Journal of Earth Sciences* **31**, 123–132.

Gilead, D. (1975) *A Preliminary Hydrological Appraisal of the Wadi El-Arish Flood 1975.* Mimeogr. Rep. Israel Hydrological Service, Jerusalem.

Goudie, A. S. and Wilkinson, J. (1978) *The Warm Desert Environment.* Cambridge University Press.

Graf, W. L. (1983) Flood-related change in an arid region river. *Earth Surface Processes and Landforms* **8**, 125–139.

Graf, W. L. (1988) *Fluvial Processes in Dryland Rivers.* Springer-Verlag, Berlin.

Grunert, J. and Busche, D. (1980) Large-scale fossil landslides at the Msak Mallat and Hamadat Manghini escarpment. In Salem, M. J. and Busrewil, T. (eds) *Geology of Libya.* Academic Press, London.

Hardie, I. A., Smoot, J. P. and Eugster, H. P. (1978) Saline lakes and their deposits: a sedimentalogical approach. In Matter, A. and Tucker, M. (eds) *Modern and Ancient Lake Sediments.* International Association of Sedimentologists Special Publication 2, 7–41.

Howard, A. D. (1985) Interaction of sand transport with topography and local winds in the northern Peruvian coastal desert. In Barndorff-Nielson, O. E., Moller, J. -T., Rasmussen, K. and Willetts, B. B. (eds) *International Workshop on the Physics of Blown Sand.* University of Aarhus Memoir 8, 511–544.

Howard, A. D. and Selby, M. J. (1994) Rock slopes. In Abrahams, A. D. and Parsons, A. J. (eds) *Geomorphology of Desert Environments.* Chapman and Hall, London, 123–172.

Jones, D. K. C. (1980) British applied geomorphology: an appraisal. *Zeitschrift für Geomorphologie* Suppl. **36**, 48–73.

Kerr, R. C. and Nigra, J. O. (1952) Eolian sand control. *American Association of Petroleum Geologists, Bulletin* **36**, 1541–1573.

Kesseli, J. E. and Beaty, C. B. (1959) *Desert Flood Conditions in the White Mountains of California and Nevada.* US Army Quartermaster Research and Engineering Center, Technical Report, EP-108.

Lancaster, N. (1994) Dunes. In Abrahams, A. D. and Parsons, A. J. (eds) *Geomorphology of Desert Environments* Chapman and Hall, London, 474–505.

Limerinos, J. T. (1969) Relation of the Manning coefficient to measured bed roughness in stable natural channels. *US Geological Survey Professional Paper* 650-D, 215–221.

Lucchitta, I. (1975) Application of ERTS images and image processing to regional geologic problems and geologic mapping in Northern Arizona: Part IVB, The Shivwits Plateau. *National Aeronautical and Space Administration Technical Report* 32-1597, 41–72.

Mabbutt, J. A. (1977) *Desert Landforms.* MIT Press, Cambridge, Massachusetts.

Mckee, E. D. (ed.) (1979) A Study of Global Sand Seas. *US Geological Survey Professional Paper* 1052.

Milnes, A. R. and Thiry, M. (1992) Silcretes. In Martini, I. P. and Chesworth, W. (eds) *Weathering, Soils and Paleosols.* Developments in Earth Surface Processes 2. Elsevier, Amsterdam, 349–377.

Morales, C. (1977) Rainfall variability—a natural phenomenon. *Ambio* **6**, 30–33.

Netterberg, F. (1980) Geology of southern African calcretes. I: Terminology, description, macro-features and classifications. *Transactions of the Geological Society of South Africa* **83**, 255–283.

Oberlander, T. M. (1997) Slope and pediment systems. In Thomas, D. S. G. (ed.) *Arid Zone Geomorphology.* Wiley, Chichester, 135–163.

Richards, K. S., Brunsden, D., Jones, D. K. C. and McCaig, M. (1987) Applied fluvial geomorphology: river engineering project appraisal in its geomorphological context. In Richards, K. S. (ed.) *River Channels: Environment and Process.* Blackwell, Oxford, 348–382.

Selby, M. J. (1982) *Hillslope Materials and Processes.* Oxford University Press.

Schick, A. P. (1977) A tentative sediment budget for an extremely arid watershed in the southern Negev. In Doehring, D. O. (ed.) *Geomorphology in Arid Regions.* Allen & Unwin, London, 139–163.

Schick, A. P. (1987) Hydrologic aspects of floods in extreme arid environments. In Baker, V. R., Kochel, R. C. and Patton, P. C. (eds) *Flood Geomorphology.* Wiley, New York, 189–203.

Schick, A. P. and Lekach, J. (1987). A high magnitude flood in the Sinai desert. In Mayer, L. and Nash, D. (eds) *Catastrophic Flooding.* Allen & Unwin, London, 381–410.

Schick, A. P. and Sharon, D. (1974) *Geomorphology and Climatology of Arid Watersheds.* Mimeo Report Dept. Geography, Hebrew University, Jerusalem.

Simons, D. B., Richardson, E. V. and Nordin Jr, C. F. (1965) Bedload equation for ripples and dunes. *US Geological Survey Professional Paper* 462H, H1–H9.

Sridharan, A. (1998) Volume change behaviour of expansive soils. In Eiji Yanagisawa, Nobuchika Moroto and Toshiyuki Mitachi (eds) *Problematic Soils.* Balkema, Rotterdam.

Thomas, D. S. G. (1992) Desert dune activity: concepts and significance. *Journal of Arid Environments* **22**, 31–38.

Thomas, D. S. G. (1997) Sand seas and aeolian bedforms. In Thomas, D. S. G. (ed.) *Arid Zone Geomorphology*. Wiley, Chichester, 373–412.

UNESCO (1979) *World Distribution of Arid Regions*. Laboratoire de Cartographie Thématique du CERCG, CNRS, Paris.

Wasson, R. J. and Hyde, R. (1983) A test of granulometric control of desert dune geometry. *Earth Surface Processes and Landforms* **8**, 301–312.

Watson, A. (1985) Structure, chemistry and origins of gypsum crusts in southern Tunisia and the central Namib Desert. *Sedimentology* **32**, 855–875.

Watson, A. and Nash, D. J. (1997) Desert crusts and varnishes. In Thomas, D. S. G. (ed.) *Arid Zone Geomorphology*. Wiley, Chichester, 69–108.

Webb, R. H. and Wilshire, H. G. (eds) (1983) *Environmental Effects of Off-road Vehicles*. Springer, New York.

Williams, M. A. J. and Faure, H. (1980) *The Sahara and the Nile*. Balkema, Rotterdam.

Yair, A. and Gerson, R. (1974) Mode and rate of escarpment retreat in an extremely arid environment (Sharm el Sheikh, southern Sinai Peninsula). *Zeitscrift für Geomorphologie* **21**, 106–121.

Yair, A. and Lavee, H. (1976) Runoff generative process and runoff yield from arid talus mantled slopes. *Earth Surface Processes* **1**, 235–247.

Yair, A. and Lavee, H. (1985) Runoff generation in arid and semi-arid zones. In Anderson, M. G. and Burt, T. P. (eds) *Hydrological Forecasting*. Wiley, Chichester, 183–200.

17. Savanna

Michael F. Thomas

17.1 Introduction to savanna environments

The nature and status of 'savanna'

Savanna environments are generally regarded as typifying the seasonal tropics, beyond the range of the tropical forests, but excluding the extremities of desert and steppe. The present-day vegetation consists of a variety of open deciduous woodlands, woodland/grassland mosaics and areas of open grassland. In Brazil this vegetation is known as 'cerrado' and drier thorn woodlands as 'caatinga'. The transition to tropical rainforest occurs where rainfalls approach *c.* 1500 mm/year and the dry season is less than 4 months in duration. Where dry season conditions become more extreme and rainfall declines below 800 mm/year the deciduous broadleaved species give way to *Acacia* thorn woodland and semi-arid steppe, which includes the 'Sahel' of northern Africa, and the 'Sertao' of north-east Brazil. Internally the so-called savanna zone can be highly complex, the physiognomy of the vegetation reflecting stress factors such as seasonal soil moisture deficits, natural and man-made fires, grazing and other land-use pressures. Because much of this variation is between the content of trees, shrubs and grasses in the ground cover, and this does not reflect in a simple manner the rainfall amount, the status of the savannas is uncertain. Savanna woodland vegetation, dominated by *Brachystegia spp.*, covers a broad crescent-shaped zone of Africa, stretching from Senegal in the north-west to Moçambique in the south-east. In South America the Llanos is an island of savanna (La Grand Sabana) dominating parts of Venezuela and adjacent countries, while the cerrado and caatinga stretch from north-east Brazil to Chile, covering important areas of Argentina and Paraguay. In Australia, *Eucalyptus* savannas cover a transitional zone between the 'wet tropics' of the north and north-east and the interior steppe and desert.

Geomorphology in the savanna zone

The largest areas of savanna coincide with the extensive plateau surfaces of the southern, Gondwana continents, usually attributed to planation in the late Mesozoic or early Cenozoic eras. These are largely cratonic areas underlain by Archaean rocks and by undeformed platform sediments, though important areas of volcanic rocks occur, notably on the Deccan of India and in the east African rift zone. The ancient land surfaces are dominated by leached, nutrient-poor soils (oxisols and ultisols), and are commonly underlain by deep saprolites (see Fookes, 1997), which may be capped by duricrusts. They are also characterised by groups of 'inselbergs', prominent rocky hills, often of granite and usually domed in profile (Figure 17.1). The combination of open

Figure 17.1 Monolithic granite domes at Dombashawa, Zimbabwe, typical of the 'older granite' plutons within the cratonic landscapes of former Gondwanaland, associated with boulder strewn lower slopes derived from former weathering.

vegetation, extensive plains and steep, isolated hills led many geomorphologists to ascribe these features specifically to a climatic control that reflected the strong seasonal contrasts in moisture supply (Tricart, 1965; Büdel, 1982). In turn this led to the argument that a suite of landforms characteristic of savanna could be defined within a scheme of climatic geomorphology. There are, however, serious objections to this approach. The idea that there is a unique combination of features that characterise savanna areas is misleading; many landforms and deposits extend into both drier and wetter climatic zones. This is a result of changing global climates and plate positions since the break-up of Gondwanaland in the late Mesozoic. The nature and extent of the savanna, therefore, often reflects the geology and history of the landscape, more than the operation of recent processes. Nonetheless, when the details of local landscapes are considered, the imprint of Neogene dissection, Quaternary climate changes and current processes all become increasingly important for a full understanding of the regolith and its chemical and physical properties.

Climate history and the distribution of savanna

The climates of areas that now support deciduous woodland and savanna have been both wetter and drier in the geological past. Humid tropical conditions were more widespread during the late Mesozoic and Palaeogene. The growth of the Antarctic ice sheet during the Miocene appears to have been associated with a drying out of climate in many parts of the southern hemisphere tropics and sub-tropics. According to studies in South America, the weathering systems were effectively 'switched off' by this event (Alpers and Brimhall, 1988; 1989; Vasconcelos et al., 1992; 1994). The Miocene aridity may also help explain why sedimentary formations such as the Kalahari Sands and Continent Terminal in Africa, and possibly the Barreras Formation in Brazil, are found today in wetter savanna and forest areas. The prolonged humidity of the Mesozoic and Palaeogene had left a legacy of widespread, very deep saprolites across the continents. Declining rainfall during the Miocene may have led to the erosion and

redistribution of these saprolites. The renewal of weathering and saprolite formation in the later Neogene and throughout the Quaternary has been demonstrated more recently from work in Queensland (Feng and Vasconcelos, 2001; Li and Vasconcelos, 2002; Vasconcelos and Conroy, 2003). These finding are likely to apply to other continents, indicating that weathering systems have fluctuated in their effectiveness, responding to regional climatic and hydrological controls over long time periods.

During the Quaternary (the last 2 Ma) major fluctuations of rainfall accompanied the temperature variations corresponding to the ice ages of higher latitudes. During the Last Glacial Maximum (LGM), at 22–21 BP, much of the tropics experienced a dry phase lasting several millennia, when rainfall may have been reduced by 30%–60% compared to present-day levels. The semi-deciduous forests were almost certainly converted to open savanna during the late Quaternary dry periods, such changes possibly affecting a 200–400 km depth of the rainforest. On the drier margins of the savanna these climates, although not always in phase, were sufficiently arid in places to convert the landscape to desert. The widespread residual and alluvial sands became subject to aeolian processes and now form linear or other dune systems, which are extensive in the drier savanna regions of Africa and Australia.

17.2 The nature of the regolith in savanna areas

This climatic history has created many complexities in the altered substrates of savanna landscapes. In the first place, the zone experienced prolonged weathering, often to great depths, not only beneath older planation surfaces, but also in the growing mountain chains of the Andes (Alpers and Brimhall, 1988; 1989). Across the surviving plateaux a weathered mantle, in places more than 100 m deep, remains in place. But the geochemical and physical attributes of this mantle vary greatly, not only according to the parent rocks, but also as a result of a complex hydrological history.

Climate and weathering products

Despite the complexities of climate change, it has been shown that within the rainfall belts between 750–1500 mm/year, kaolinite clays predominate in well-drained surface soils (Pedro, 1968). However, smectites become increasingly abundant at rainfalls below 1000 mm/year, especially in receiving sites for surface and groundwater flows. A 60 km west to east transect in southern India reported by Bourgeon (1991) and discussed by Pedro (1997) demonstrates a transition from highly evolved, kaolinised saprolites (French: *alterites*) in the wet zone (>2500 mm/year), towards friable sandy materials (French: *arènes*; German: *grus*) in the dry zone (<750 mm/year) (see Figure 17.2). The transition from 'massive' weathering to 'limited' weathering (Bourgeon, 1991) depends on the availability of drainage water. According to this study, plagioclase will remain intact at depth in the driest zone, whereas biotite and amphibole will alter to low-charge 2 : 1 phyllosilicates (smectite clays), which retain Calcium. As weathering intensifies in the friable zone, so alteration of plagioclase, and the formation of high-charge 2 : 1 phyllosilicates (sericite, illite), leads to the export of calcium and the retention of Potassium. In

the wetter areas, alkali-poor 1 : 1 clays (kaolins) progressively dominate the saprolite.

The chemical composition of saprolite is of great significance to its behaviour as an engineering material. The presence of smectite in significant quantities leads to marked increases in swelling in the presence of moisture. Where Ca^{2+} is the dominant cation, the figure ranges from 45–145%. In Na^+ rich clays it rises to $>1400\%$, and such sodic clays are highly dispersible. Kaolinite, as a 1:1 lattice clay, absorbs less water and may expand by less than 10%. Both smectite and kaolinite commonly bond with free iron oxides, which are produced in the weathering of biotites and other dark minerals. The Fe_2O_3 acts as a weak cement that binds kaolinite and other clay minerals, leading to their appearance in the silt fraction of the saprolite, and effectively reducing the apparent clay content of the material. The delicate aggregates formed in this way exhibit open, 'cardhouse' microfabrics. These produce collapsible soils, which react to excess moisture or loading by rearrangement of the platy clay particles with a reduction of void space and loss of volume. Where the Fe_2O_3 is amorphous rather than crystalline it can be remobilised under saturated (reducing) conditions. Iron behaviour is complex, however, and Fe_2O_3 will crystallise, commonly as goethite or hemetite, to form segregations (mottles) and pisolitic nodules, usually within a few metres of the ground surface. Where the iron content is high enough (20–40% +) individual mottles or pisoliths may aggregate to form laterite (or ferricrete), which is resistant to erosion.

In drier areas and beneath more recent or dissected land surfaces, sandy regolith low in clay content, similar to the 'grus' found in colder climates predominates. In contrast to clay-rich saprolites that have 20–50% clay content, these less weathered materials typically contain $<15\%$ clay. According to engineering weathering classifications, the grus generally corresponds with Zone III, grading up into more clay-rich materials in Zones IV and V (Geological Society, 1990; Migoń and Thomas, 2002).

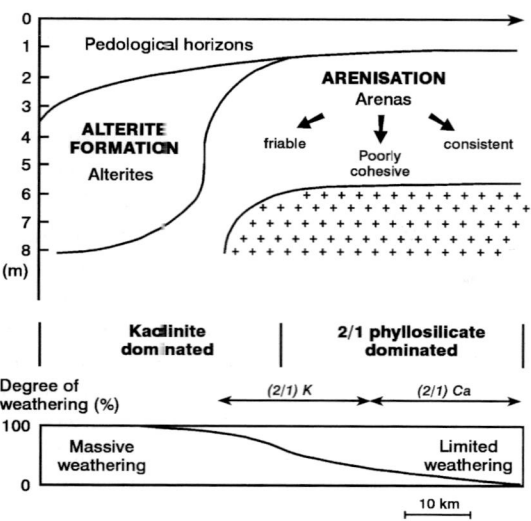

Figure 17.2 Geochemical and mineralogical characteristics of the weathering zone along a transect in southern India (redrawn from Bourgeon, 1991).

Duricrusts

The regolith in savanna areas is often characterised as having a widespread cover of surface or

Table 17.1   The occurrence of common duricrusts in relation to humidity of climate.

Humid	
↓	Alucrete or 'bauxite' (Al_2O_3 as gibbsite and boehmite)
↓	Ferricrete or 'laterite' (Fe_2O_3 as goethite and hemetite)
↓*	Silcrete or 'porcellanite' (SiO_2, silica)
↓*	Calcrete or 'caliche' ($CaCO_3$, with Mg as, dolocrete)
↓	Gypcrete ($CaSO_4$, gypsum)
Arid	

* Calcrete and silcrete do not follow a clear order of formation in relation to present-day climate.

near-surface, resistant duricrust. Duricrusts are exposed and hardened horizons first formed in saprolites or sediments, by the precipitation of hydrated oxides of Aluminium, Iron and Silicon, or of Calcium as the carbonate or sulphate (Geological Society, 1990; Fookes, 1997). As a broad generalisation, the formation of these materials follows a humidity gradient, as indicated in Table 17.1.

This sequence has been the subject of debate and the true situation is complex. However, most alucrete (or bauxite) deposits are found in rainforest environments, while it is possible that many ferricretes formed first as soft, mottled clay (often termed laterite) horizons under forests and have become indurated, with the formation of crystalline goethite and hemetite skeletons, as a result of dehydration. This has been due both to climate change and to the lowering of the water-table during the fluvial dissection of landscapes. Laterite in this sense was formed as part of a weathering profile, but ferricrete also develops in transported sediments, particularly in savanna regions, wherever seasonal water-table fluctuations occur near the surface. The distal ends of piedmont slopes, which are often alluvial fans or amorphous colluvium, are commonly observed to contain shallow ferricrete. Alluvial deposits, especially coarse sands and gravels are also readily cemented by iron in floodplains and river terraces. Such detrital ferricrete can later appear as a resistant caprock as a result of relief inversion. The many varieties of Fe-duricrust have led to continuing

disputes about the definition of 'laterite' (Taylor and Eggleton, 2001). The behaviour of iron duricrusts, however, depends largely on the crystallinity and continuity of the cementing oxides, and the amount of clay. But in some detrital ferricretes the clastic host sediment is an important component. Duricrusts that overlie clay-rich saprolite are often subject to marginal collapse as the clay is washed out by underground seepage. Pseudo-karst hollows are also found in duricrusted terrain, and appear to result from volume loss in the underlying saprolite, due in part to the formation of underground drainage channels.

Both calcrete and silcrete deposits are found widely in the drier savanna, where precipitation falls below c. 800 mm/year. Calcium liberated during the weathering of plagioclase, or derived from carbonate rocks initially travels as the dissolved bicarbonate in groundwater, but rapid degassing occurs in shallow, aerated rivers, leading to precipitation of tufa deposits. $CaCO_3$ precipitation also takes place as pods and domes of calcrete (also known as caliche or kunkar) in floodplain environments, probably due to water-table fluctuations. The precipitation of secondary silica to form silcrete can arise from different processes: by the loss of Al_2O_3 from kaolinised saprolite, in which case it forms within the profile, or by release of silica from rocks and saprolites and its subsequent precipitation from silica-rich groundwater. This may possibly take place under conditions of impeded drainage, where seasonal desiccation prevents the export of the silica. But long-distance transport of SiO_2 has also been proposed to explain the occurrence of silcrete around fluviatile and past lacustrine basins, as in central Australia (Stephens, 1971; Taylor and Eggleton, 2001).

The removal of iron, silicon and even aluminium from saprolites that already display advanced weathering increases the void ratio, and collapsible fabrics beneath well-drained sites can result. Elsewhere, the precipitation of ions in transport and their accumulation as smectite clays, creates environments where swelling clays can present problems for both agriculture and road foundations. On and beneath some ancient surfaces, karstic and pseudo-karst (silica) features have developed. These can lead to cavernous macro-voids and to surface features ranging from micro-forms such as rillenkaren (surface channels or corrugations) through sinkholes to macro-tower karst. The rocks affected range from ultra-basic igneous and metamorphic rocks to silica-cemented sandstones and conglomerates, in addition to the true karst of carbonate rocks.

Climate change and regolith transformation

The mineralogical changes brought about by the climatic transitions induced by plate-tectonic movement and global climate changes have been documented by Tardy and Roquin (1998). Weathering profiles developed during the Mesozoic have since experienced both geochemical destabilisation and physical dismantling as a consequence of altered hydrology and increasing dissection of the continents during the last 100 Ma. Profiles have been truncated by surface erosion (Butt, 1987) and the products of dismantling redistributed across the landscape by wind action and sometimes by water. The Kalahari Sands probably resulted from these processes, and 'sandplains' are encountered widely across the other continents (Fairbridge and Finkl, 1984). Those profiles that survived intact have experienced an altered hydrology, usually involving a fall of water-tables, sporadically through this 100 million years (Thomas, 1989; 1994; Twidale, 1991). This has been a necessary consequence of fluvial dissection of the continental margins, as well as a result of drier climates. Bauxitisation may have occurred in wetter areas, by de-silicification of kaolinitic weathering mantles while, in less humid areas (Valeton et al., 1991), induration of Iron-rich zones led to the formation of Fe-duricrusts. In a similar way silcretes now form cappings to low hills formed in saprolite, particularly in the interior of Australia.

Quaternary climate changes affected regoliths mainly by the impacts of periodic aridity and accelerated surface erosion, a result of reduced vegetation cover. Some weathering profiles became truncated, exposing horizons formed at depth to the desiccating effects of surface environments. Duricrusts became strongly indurated; pisolitic gravels developed from iron-rich soil mottles, and surface erosion removed fine

sediment downslope to form valley fills. These processes contributed to the formation of surface, or near-surface, 'lag' gravels.

Regolith materials and landforms

The savanna zone is, therefore, associated with the widespread occurrence of duricrusts that form tabular residual hills and bench-lands in valleys (Figure 17.3). In many cases these hills represent an inversion of relief, and the duricrusts commonly contain transported gravels. In fact, many ferricretes appear unconnected with a 'laterite profile' and demonstrate iron cementation of a transported regolith (Ollier and Galloway, 1990). Eroded and redistributed regolith materials accumulate to form residual sands and gravels on uplands (Figure 17.4) and valley-side slopes, while fine sands and clays accumulate in valley floors. This can occur on the scale of small, shallow valleys known as dambos in Africa (Figure 17.5; Clark, 1974; Thomas and Goudie, 1985) and also on the much broader scale of large alluvial basins, as in the middle Nile valley (Williams et al., 1998). The seasonal drying of soils and regoliths to considerable depths beneath elevated sites is largely responsible for the formation of Fe-Mn pisoliths, while the strong evaporation from floodplains and valley floors leads to the incorporation of divalent Ca^{2+} ions to form vertisols containing smectite clays.

Strong catenary relationships or toposequences of soils and slopes were first described in soils from the seasonal tropics, reflecting the strong

Figure 17.3   Duricrust (ferricrete) 'breakaway' in northern Nigeria.

leads to the break-up of the ferricrete and the redistribution of derived gravels across the lower slopes of valleys and around isolated hills (mesas). Subsequently, secondary or 'footslope' ferricretes may form within these deposits (Dowling, 1966; Figures 17.6 and 17.7).

It is, however, mistaken to think that the savanna zone is blanketed by ferricrete. In most areas these deposits occur as discrete caps or have formed around the margins of larger hills. On the other hand surface gravels are common on convex summits and these materials contain duricrust fragments, unrolled vein quartz and, commonly, rolled pebbles of fluvial origin (Figure 17.4). These gravels appear to mark the complex history and lowering of the land surface over long time periods. Related stonelines of gravel also appear within soil profiles, usually within 1–2 m of the surface, and these have led to considerable debate concerning their origins (Figures 17.8; 17.9; Alexandre and Symoens, 1989; Thomas, 1994). Such stonelines can create hydraulic discontinuities in the profile, encourage sub-surface water flows and also precipitation of Fe^{3+} to form ferricrete. While thin stringers of stones may arise from simple downslope processes, many stonelines are more complex, commonly exceeding 50 cm in thickness beneath surfaces of gentle slope. They are often viewed as lag deposits, dating from episodes of aridity and powerful sheetwash, which have been followed by accumulation of finer sediment and soil, often by bioturbation (especially by termites) and soil development. Others may be degraded river terrace gravels.

The rate of topsoil formation by termites and other soil fauna has been quoted as 0.05–0.5 mm/year (see Thomas, 1994), which might imply that a topsoil of 2 m depth could have formed during the Holocene (10 ka, at 0.2 mm/year). Large termitaria (e.g. *Macrotermes natalensis*) can reach 10 m in height and 30 m in diameter, with densities of 2–5 per hectare (Figure 17.10; Aloni, 1975). The material brought to the surface is fine grained (< 2 mm, *c.* 20% clay) and is washed over the intervening areas by rainsplash erosion, but very large mounds may have an age of at least 700 years (Watson, 1974). The action of termites has a major impact on

Figure 17.4   Residual gravels forming a surface 'lag' deposit in northern Sierra Leone. The gravels contain fragments of ferricrete, vein quartz and quartzite pebbles.

seasonal contrasts in slope and groundwater hydrology (Milne, 1935). Such contrasts are heightened by the dissection of ancient saprolites, leading to the formation of residual hills with the exposure of different weathering zones across the slope. The exposure of duricrusts results from the loss of topsoil and dissection, which creates low cliffs or 'breakaways' (Figure 17.3). Water penetrating the duricrust via numerous cavities and cracks will find subterranean pathways towards nearby valleys. This can create a form of lateritic pseudo-karst and water will issue from tunnels below the duricrust capping, often leading to its collapse. Closed depressions can form below summit areas, where removal of material from beneath the capping has led to sagging of the crust. The undermining and erosion of the breakaways

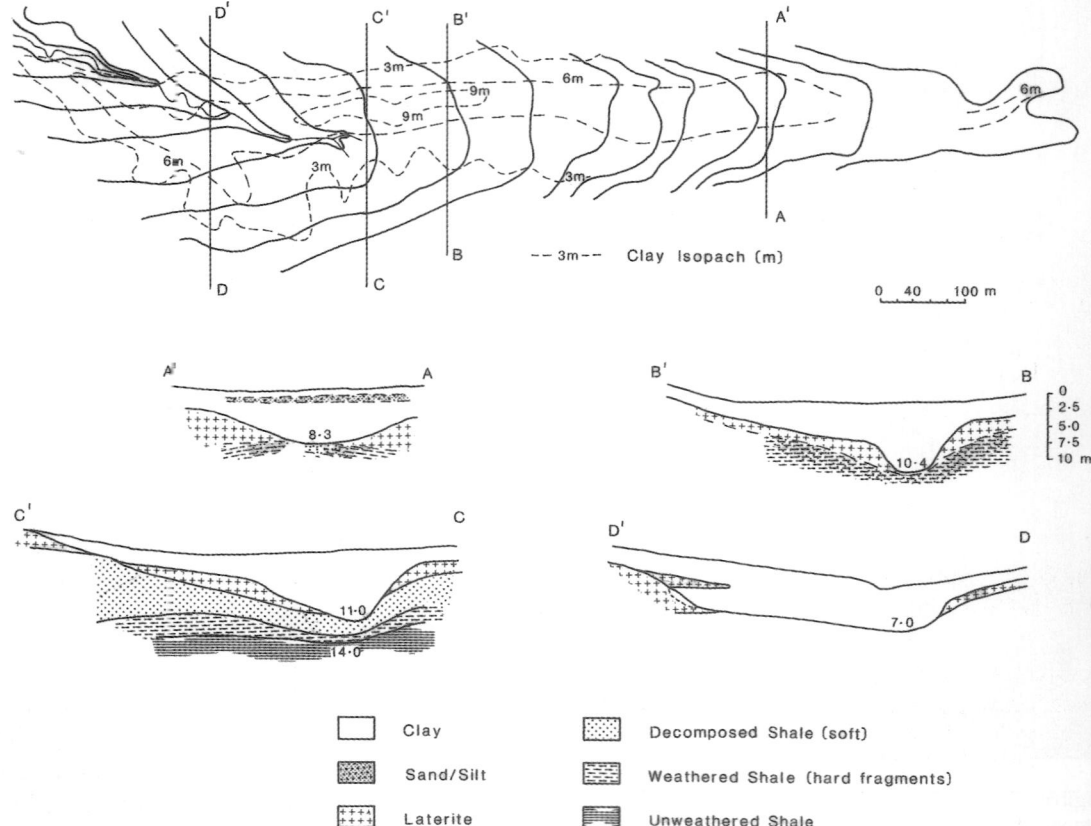

Figure 17.5   Dambo valley at Kankamo, Zambia, showing the flat floor and sedimentary fill.

sub-surface conditions, contributing to the partitioning of the soil profile (Figure 17.11), with the penetration of chambers and channels continuing to depths of many metres. The contribution of small *Microtermes* species to the total rate of soil accumulation is important.

17.3 Landform assemblages and surface materials

Regolith features and landforms

Although savanna is associated with the broad plateaux and lowlands of the ancient fragments of Gondwanaland, there are of course grasslands and grassland–tree mosaics present in montane environments. But they are seldom described as, or grouped with, savanna. For this reason, a range of forms and deposits that characterise the 'old lands' of the seasonal tropics will be emphasised here (see Table 17.2; Figure 17.12).

Regional weathering and landform evolution

In essence Table 17.2 lists residual and transported materials derived from the weathering of bedrock, and regarded as 'regolith' (Ollier and Pain, 1997; Taylor and Eggleton, 2001). Bedrock forms are ignored. These materials co-exist within regional landscapes as quazi-regular patterns. But before these are discussed, the pattern of weathering of the mantle itself deserves

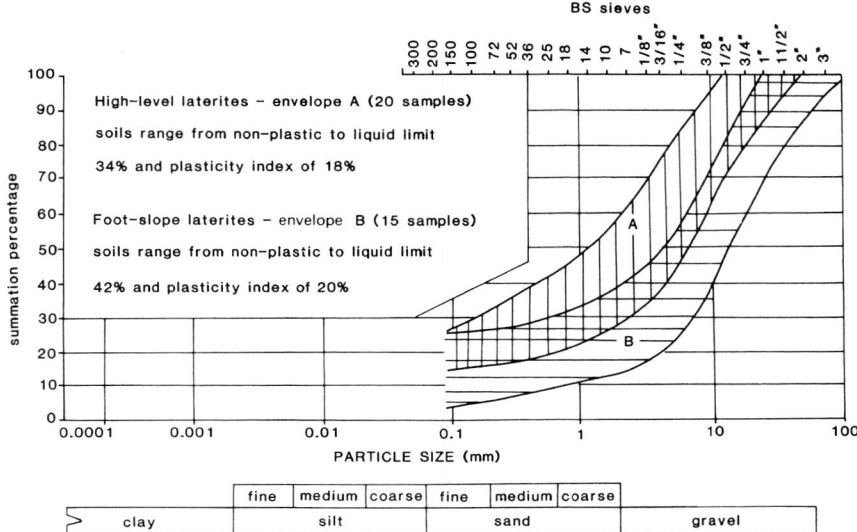

Figure 17.6   Particle-size distributions and plasticity data for high-level and footslope laterites from the Fika-Nafada area, Nigeria (after Newill and Dowling, 1968).

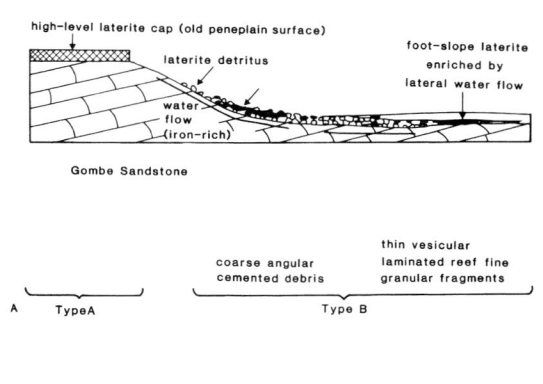

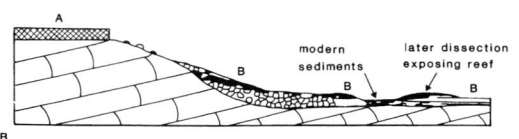

Figure 17.7   (*A*) Formation of footslope laterite (ferricrete) and relation to high-level laterites and (*B*) distribution of ferricretes (laterites) in the present landscape (after Newill and Dowling, 1968).

Figure 17.8   Colluvium overlying stoneline gravel above a granite saprolite, adjacent to a granite outcrop on the Jos Plateau, Nigeria.

comment. The base of the weathered zone is sometimes abrupt and referred to as the weathering front or basal weathering surface, and it may lie more than 30 m (>100 m in special circumstances) below the ground surface (Figures 17.13; 17.14). Many studies have indicated regional and local variations in the depth of the weathered mantle. On a regional scale, greater depths and differentiation correspond with the survival of ancient land surfaces of relatively low relief. Many parts of the central African plateau (or 'High Veld'), and significant sections of other cratonic areas, such as east

Figure 17.9 Thick stoneline gravels below a planate land surface in eastern Zambia.

Figure 17.10 Large termitarium (of *Macrotermes natalensis*) within a deciduous forest in Shaba Province, Republic of the Congo.

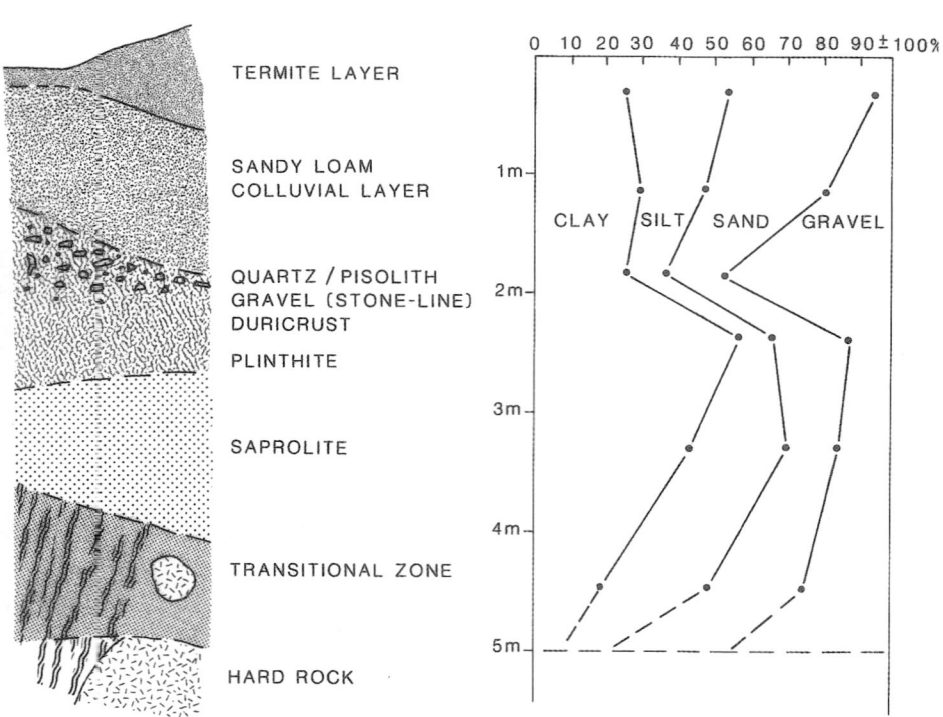

Figure 17.11 Composite soil–saprolite profile showing typical combination of termite layer, colluvium, stoneline, plinthite (incipient laterite), saprolite and transitional (saprock) zone, with hypothetical grading curves. The 5 m depth may be much greater than shown, up to 15 m or more.

central Brazil and the Yilgarn Block in western Australia, fall within this category. Radioisotope and other studies converge to indicate that many of these regoliths date from the Mesozoic and most contain weathering products older than 10 Ma (Vasconcelos *et al.*, 1992; 1994).

During the ensuing period, partial dismantling of these ancient regoliths has occurred, and led to

Table 17.2 Common associations of landforms and materials in the seasonal tropics. This list excludes bedrock surfaces, and does not specifically list Landforms characteristic of erosion (modified from Thomas, 1994, after Thomas, 1974 and Butt, 1987).

A. Residual materials

I. *Interfluve and plateau residuum*
  1. Residual soil on rocks, duricrusts and saprolites
  2. Duricrust of weathering profile (tabular summits and plateaux) including ferricrete, silcrete, and calcrete cappings (alucrete rare in savanna)
  3. Duricrust derived rubble (e.g. from *in situ* breakdown of fericrete)
  4. Blockfields of rock cores (residuals from weathering)
  5. Eluvial 'white sands', resulting from collapse of weathering profile

II. *Hillslope residuum*
  6. Duricrust cliffs — cliffs around 1. — see also B I
  7. Upper zones of weathering profile
  8. Lower zones of weathering profile (saprolite and corestones on steep and lower slopes)
  9. Boulder-controlled slopes; with *tors* as residual hills (convex summits and steep slopes $> 20°$)

B. Transported materials

I. *Interfluve and plateau deposits*
  1. Duricrusts containing transported clasts (tabular summits) including ferricrete, silcrete cappings (often impregnating sediments)
  2. Gravel layers (exposed or as stone-lines) comprised of duricrust fragments, rolled and unrolled quartz
  3. Sandplains: aeolian sand sheets, alluvial and colluvial sands (possibly from eluvial sands)

II. *Hillslope deposits*
  4. Colluvium: often in hillslope hollows or former gullies
  5. Landslide debris, including: boulder accumulations from rock falls or slides; earth flow debris from slumps and translational slides, and debris flow and mudflow sediments, often channelised

III. *Lower hillslope and piedmont zone deposits*
  6. Coarse talus and debris fan deposits, including: clast-supported talus and alluvial fans, and matrix supported debris flow deposits
  7. Colluvium, including: mudflow and sheetflood sediments; laminated sands and silts, and undifferentiated colluvium including hillfoot 'pedi-sediment'
  8. Cemented deposits, including: ferricrete benches in colluvial and alluvial deposits; silcreted sands and gravels, and calcrete layers and benches

IV. *Valley floor deposits*
  9. Alluvial deposits, including: floodplain facies from meandering streams; deep channel fills; flood sediments; shallow braided valley sediments; clay fills in channelless *dambo* valleys, and clay pans
  10. Lacustrine deposits, often associated with alluvial plains
  11. Aeolian deposits, including: linear dunes marking former extensions of desert conditions; sandplains, see also 3, and lunette dunes associated with deflation pans
  12. Cemented deposits, as in 8.

the range of surface deposits listed in Table 17.2. Degrees of regolith stripping vary and the landscapes produced have been described as varieties of etchplain (see Thomas, 1994). Although brought about by fluvial dissection, outcrop patterns become accentuated during stripping, because of the differential response of various rocks to chemical weathering processes through time. A sequence of scale-dependent changes is indicated in schematic terms in Table 17.3.

The results of this evolution are reflected in the detail of surface deposits. Many saprolite profiles deepen away from stream channels under duricrust-capped hills and plateaux. Renewed

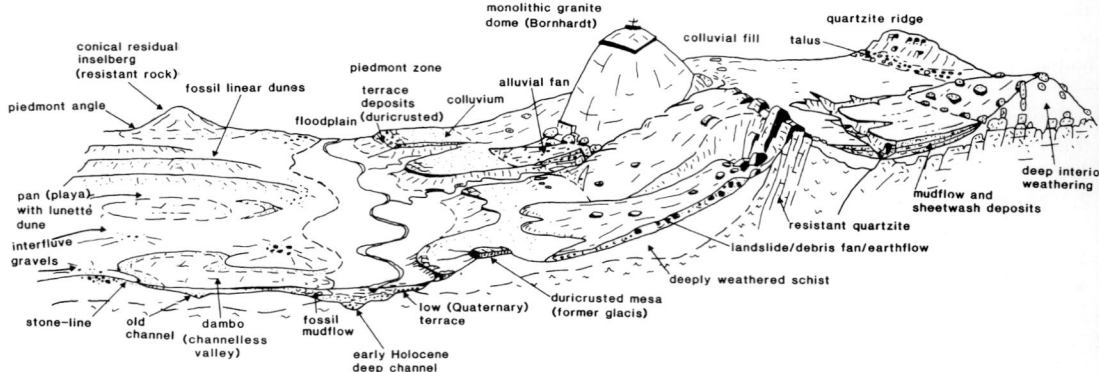

Figure 17.12   Composite block diagram to show assemblage of typical savanna landforms and deposits.

Figure 17.13   Core boulders and a sharply defined weathering front in granite, Jos Plateau, Nigeria.

weathering has led to the formation of troughs and basins of weathering along shatter zones or has been influenced by orthogonal joint sets and variations in mineralogy or rock fabric (Figure 17.14). The juxtaposition of outcrops and deep regoliths is a function of all these factors. The rapid variations in sub-surface conditions also make a detailed ground investigation difficult using conventional boreholes or geophysics. The best approach is usually to undertake visual inspections and block sampling in trial trenches or shafts of sufficient diameter to allow man entry (with appropriate safety measures).

Sediment stores are found as plateau sands and gravels; hillslope debris, which may be cemented, and gully fills; piedmont fans and colluvium, and

floodplain sediments. In drier areas, past desiccation has led to the formation of deflation pans and lunette dunes, even to the formation of linear dunes across large areas. Many such deposits are functions of Quaternary climatic fluctuations and they may not be in equilibrium with the balance of present-day processes (Figures 17.14; 17.15).

Implications for foundations and earthworks

In general, conditions for shallow foundations in savanna areas may be reasonably good; allowable bearing capacities tend to decrease with increasing clay content. Problems that may be encountered in foundation engineering include excessive settlement of collapsible soils, or heave of expansive soils. There is now a considerable literature related to these soils, and useful introductions are provided by ASCE (1982), Blight (1997) and Fookes (1997). Undermining by erosion of duricrust layers provides the possibility of local subsidence (methods of combating this are discussed in Chapters 12 and 24). Piling for heavy structures can be problematic, due to the variability of the ground both horizontally and vertically, with the potential for zones of fairly strong material to be underlain by more weathered and weaker ground. Expansive or collapsible soils also provide problems for road construction, but duricrusts and pisolith gravels can provide good quality materials for road bases and surfacing.

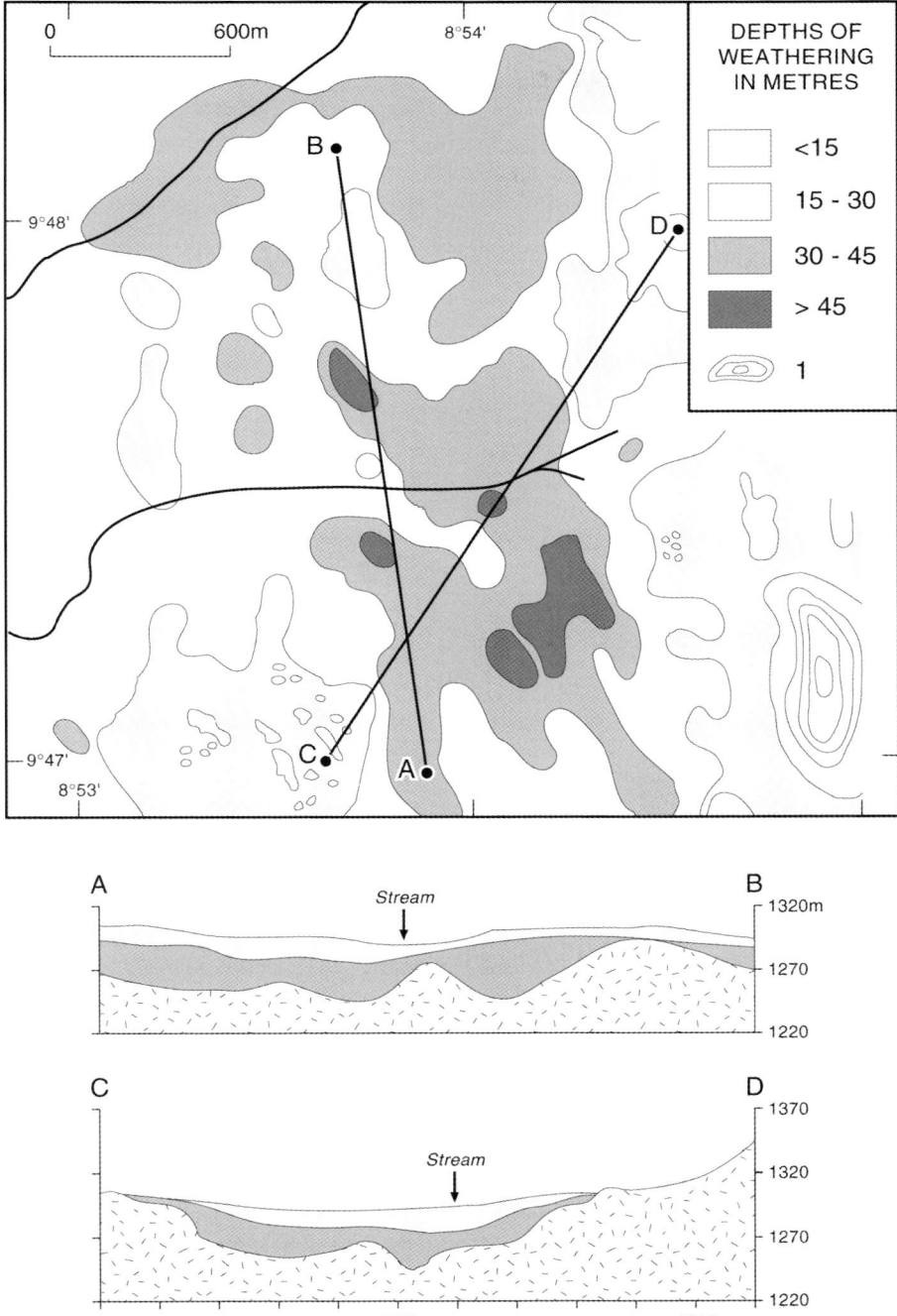

Figure 17.14 Deep weathering patterns in fine-grained biotite granite near Jos, Nigeria. Sections A–B, C–D show colluvial mantle (stipple) over saprolite (diagonal rule).

Table 17.3   Schematic representation of the evolution of deeply weathered plateaux through time.

Geologic evolution	Landform development
Regional cratonic surfaces (plateau surfaces—Mesozoic or pre- Mesozoic in origin)	Warped and faulted during late Mesozoic–Cenozoic plate tectonic movement
↓	
Mid-late Cenozoic dissection (along continental margins; within interior basins)	100 Ma of increasing relief; local planation within major catchments
↓	
Geochemical differention in relief patterns during dissection (relief sensitive to outcrop geology)	Duricrusted hills and ranges; elevated inselbergs; weathered plains, and alluvial valleys
↓	
Neogene weathering and erosion (renewed weathering within relief compartments)	Penetration of weathering beneath residual hills; weathering of new lavas and sediments; new ferricretes
↓	
Quaternary redistribution of regolith (formation of present mosaic of surface deposits)	Formation of transported regolith stores on hillslopes, benchlands and valley plains

17.4 Surface processes and geohazards in the savanna

Runoff and surface erosion

Tropical savanna areas share rainfall characteristics with more humid zones but experience these processes for a shorter period, usually less than 6 months in the year. Rainfall frequently occurs as intense convectional downpours, possibly delivering more than 100 mm in a single storm lasting 2–3 hours, with intensities exceeding 100 mm/hour over periods of 15–30 minutes. Prolonged heavy rainfall lasting many days, characteristic of monsoon climates worldwide, occurs less often in drier areas. Because seasonal moisture stress limits the density of the vegetation it is often argued that these characteristics lead to high runoff ratios and sediment yields. But there are problems with this argument because, under natural conditions, long grass provides a protective ground cover, in the wetter savanna at least. If the grass cover is removed by frequent fires, which are usually followed by grazing or cropping, then the impact of the intense rainfall can be very severe. Sediment will be released from most well-drained, exposed soils after 15–20 minutes of rainfall exceeding only 25 mm/hour (Hudson, 1995). The impact of

the heavy rainfall is increased where soils are developed on sandy colluvium, and severe soil erosion can result on slopes as low as 2°. Recent work on the soil erosion history of central Tanzania has shown (Eriksson et al., 1999) that 'modern' soil erosion in this region is no more than 900 years old, but a major episode of colluviation took place between 14.5 ka and 11.4 ka, a period of rapidly changing climate in the late Pleistocene. However two other studies, from eastern Zambia and from central Tanzania have indicated a more continuous history of punctuated colluviation spanning most of the last glacial cycle (Sørensen et al., 2001; Thomas and Murray, 2001).

Slope failures

On the other hand slope failures are commonly restricted to shallow debris slides and flows that occur on steeper hillslopes during heavier storms. The restricted amounts of groundwater present across much of the savanna zone appears to limit the potential for deep-seated landsliding, which would otherwise be favoured by the occurrence of deep, clay-rich saprolites. Nonetheless, the impact of large intense storms can be devastating. In February 1970, more than 100 mm of rain fell on the Mgeta area of Tanzania in less than 3 hours,

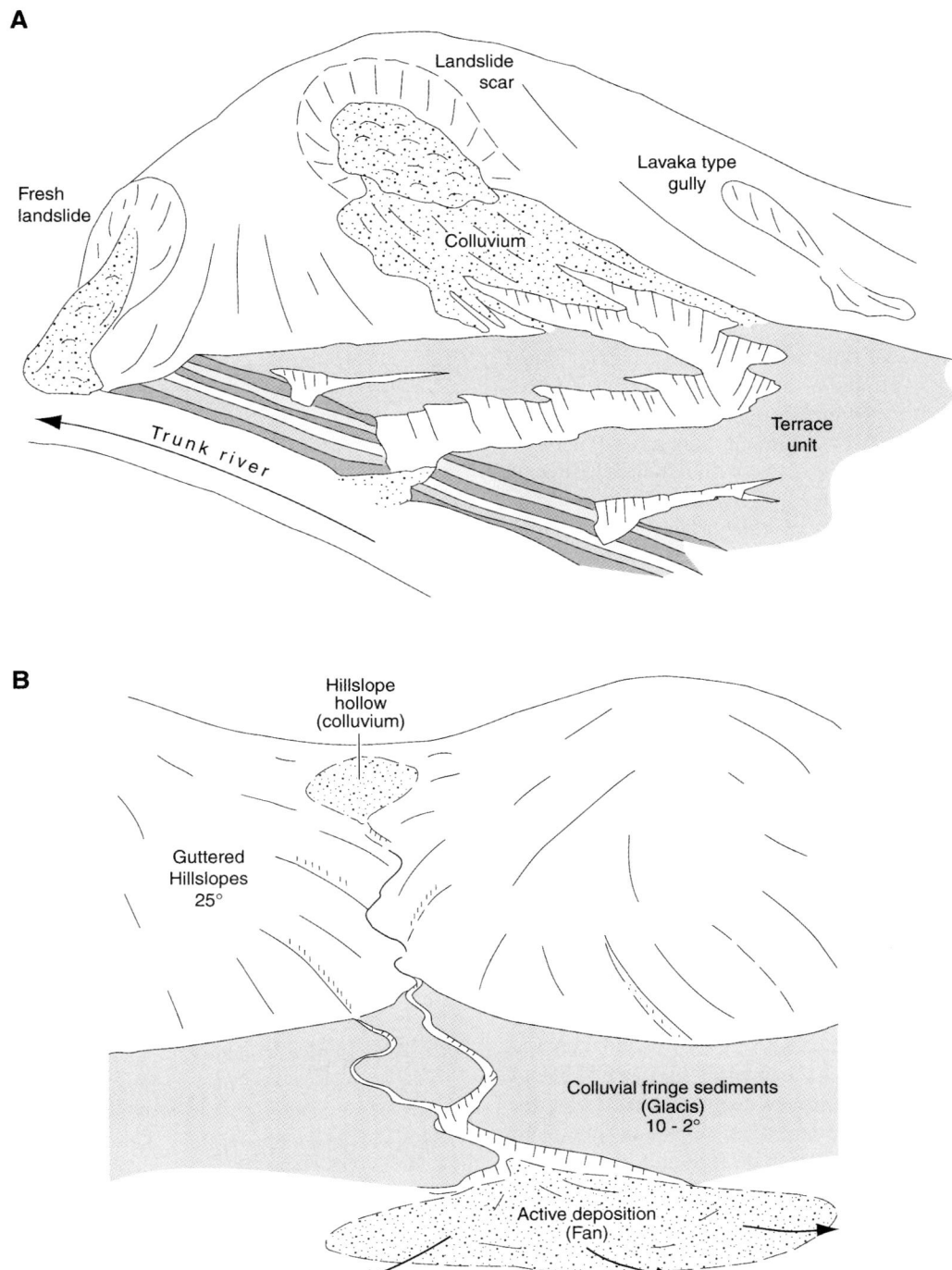

Figure 17.15 Erosion and deposition features common in savanna areas: (*A*) landslides and 'lavaka' type gullies supply colluvium to lower slopes, here comprised of river terrace alluvium, sensitive to active gullying; (*B*) amorphous colluvium and more definite fan deposits occur widely in piedmont zones in savanna areas.

leading to more than 1000, mainly small, debris slides, which transformed into mudflows downs-lope. Most slides were 5–20 m wide and their average depths were only 1–1.5 m. More than 90% of the slides occurred beneath open grassland or cultivated fields (Temple and Rapp, 1972).

Some larger landslides found in the savanna may be relict features from periods of more humid climate during the Pleistocene. Well-dated land-slide deposits from the tropics are rare. A few scat-tered dates suggest that landslides may have been favoured by the early Holocene pluvial conditions (c. 10–8 ka). But, recent work in Zambia, where rainfall is around 1100 mm/year, has revealed OSL (Optically Stimulated Luminescence) dates of > 200 ka for sediments associated with an episode of widespread rotational sliding around Chipata in the Eastern Province (Thomas, 1999; Thomas and Murray, 2001). On the other hand, large landslides may be triggered in monsoon and cyclonic areas, even when total rainfalls are insuf-ficient for rainforest. They are probably rare events in the drier savanna, unless caused by exceptional ground conditions.

Gully erosion

Gullies commonly develop in deep colluvium, as observed above. At certain sites, as at St Michael's Mission in Zimbabwe, colluvial sediments are exposed in modern gullies and several episodes of deposition and erosion have clearly taken place at this site, old gully floors being marked by bands of coarse sediment (Figure 17.16; Stocking, 1981). The current phase of gully development appears to have been triggered by cattle tracks. But there is also an interaction between the sediments, soil development and erosion. The colluvium contains clays with a high exchangeable sodium percentage (ESP), and is subject to deep cracking during the dry season. As in many soils of the savanna zone there is also a clay-rich or argillic horizon of low permeability, found within 50 cm of the surface, leading to sub-surface water flows and piping (Figures 17.17; 17.18).

Once gully incision has taken place, the hydraulic gradient is greatly increased at the head-cut, and groundwater penetration via surrounding cracks is rapid. If sodic clays are present, they are

Figure 17.16 Section through gully-fill sediments near St Michael's Mission, Zimbabwe. Deep vertical cracks become enlarged into 'pipes' due to the presence of dispersible Sodium-rich clays.

Figure 17.17 Partly collapsed 'tunnel' gully devel-oping above an impermeable clay horizon in alluvial sediments, Luangwa Valley, Zambia.

Figure 17.18 Shallow gullying resulting from the extension of the process described in Figure 17.18, Luangwa Valley, Zambia.

highly dispersible and enter into suspension, rapidly enlarging the soil macropores into pipes (Figure 17.19). Flow through the pipes progressively enlarges the cavities until roof collapse becomes common, thus extending the gully system. This destabilisation of valley fills can lead to a rapid propagation of lateral gullies, fed partly by groundwater flows. The extent of gullying will be limited by the distribution of the colluvium, but other factors operate. Away from the valley floors the smectite clays may be replaced by more stable kaolinite within a more sandy sediment. In the absence of deep cracks and pipes, groundwater movement becomes more diffuse and less aggressive, and more water is likely to enter the gully by surface flow. As the gully system extends, contributary slope catchments become reduced in area and runoff lengths leading to the gully-heads become shorter. These factors can reduce effective water flows in individual gullies, which may then evolve by wall collapse and stabilisation of debris by vegetation growth (Figure 17.20).

The trigger mechanisms leading to gully incision may also be complex. Some gullies develop from river banks, where the hydraulic gradient is controlled by the seasonal rise and fall of the river level (possibly > 10 m). Normally, adjacent floodplain sediments will be stabilised by riparian forest vegetation, but this can be breached by natural bank erosion and by animal tracks. Large game animals such as buffalo, which travel in large herds, and hippopotamus, which move frequently

Figure 17.19 Deep vertical fissure-like pipes developed in the colluvium at St Michael's Mission, Zimbabwe.

Figure 17.20 Inactive gully on the Jos Plateau, Nigeria, now evolving by wall collapse and vegetation.

in and out of water, are instrumental in causing serious bank erosion. Trampling by these animals also leads to the collapse of soil pipes, accelerating the formation of open gullies. Intensive grazing of

Figure 17.21 Development of deep gullies in Quaternary colluvium on the Jos Plateau, Nigeria. The gully is extending partly as a result of runoff via the road culvert.

floodplain silts by domestic animals can have devastating effects, as can be seen in resettlement areas downstream of the Kariba dam in Zambia.

Road culverts and ditches designed for the dispersal of storm runoff from sealed road surfaces are also liable to cause or exacerbate gully development, and the tendency to lead road runoff into depressions and stream heads, without adequate protection is indefensible. Examples abound, including sealed roads built in the 1970s on the Jos Plateau of Nigeria (Figure 17.21). Spillways from river impoundments, using low earth-fill dams also have the potential to cause downstream erosion during storms.

Although the role of colluvium in the extension of gullying has been emphasised, the *in situ* saprolite is also subject to slope failure and gully formation. The *lavaka* gullies of Madagascar with their distinctive 'hourglass' shape were thought by Wells *et al.* (1991) to start from bare patches on hillslopes underlain by ferrallitic saprolites. In this case the development of the gullies by overland flow on overgrazed slopes is the likely mechanism (Morgan, 1986). Similar, and even more aggressive gullying, can be seen in north-east Brazil, where convex relief compartments entirely comprised of granite saprolite have become carved into deep canyons.

Experience in a humid tropical area around Bananal, eastern Brazil led De Oliveira (1990) to propose a model for the two major types of gully formation: by seepage erosion, and by overland flow (Figure 17.22). Where these become

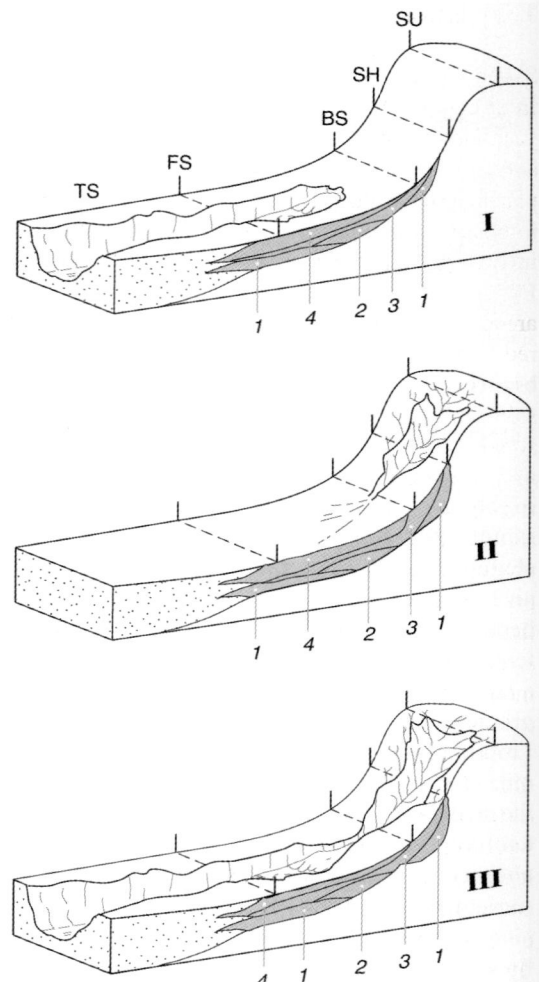

Figure 17.22 Model for gully development on concave slopes, based on conditions near Bananal, São Paulo, Brazil. I: Gullies connected to the drainage network, extending by seepage erosion along discontinuities in colluvial sediments. II: Gullies not connected to the drainage network, but forming on steepest slopes by runoff erosion. III: The combination of types I and II: connecting upper and lower slopes. SU = summit; SH = shoulder; BS = backslopes; FS = footslope; 1–4 superposes colluvial sheets (after De Oliveira, 1990).

connected the potential for serious land degradation is obvious. This model is equally relevant to the savanna areas of the tropics, but the occurrence of deeply weathered relief compartments will become limited in the drier areas to hills protected by duricrust.

17.5 Summary

Savanna regions receive annual rainfalls varying from below 800 mm/year to as much as 1600 mm/year and comprise mixtures of deciduous woodland and grass. Their association with the plains and plateaux of the ancient continents of Gondwanaland implies a long and complex history of development, involving climate changes on different timescales. Extensive mantles of weathered rock and saprolite are legacies of past eras of humid climate, and more recent history has involved the erosion and redistribution of the regolith to form a variety of sediment stores in the landscape. Limited leaching in the seasonal climates has retained much of the silicon, iron, calcium, and magnesium liberated from rock minerals during weathering, either as duricrusts or as 2:1 phyllosilicates (mainly smectite clays). The properties of these materials, when subject to the prevailing regime of heavy tropical rainfall, and their occurrence within Quaternary sediments that accumulated under a different climatic environment, have created conditions for serious and rapid extension of gully erosion.

Savanna environments can be viewed as transitional between the humid tropics and the drylands of the deserts and their borders. Active processes work on inherited materials that can be particularly sensitive to destabilisation, and foundation conditions reflect the complexity of geologic and environmental changes over long time periods. The deceptive simplicity of many of the great savanna plains when viewed from afar, can, therefore, be seriously misleading.

References

Alexandre, J. and Symoens, J. J. (eds) (1989) Stone-Lines. Journée d'Étude, Bruxelles, 24 mars 1987. *Académie Royale des Sciences D'Outre-Mer, Geo-Eco-Trop.* **11**, 239pp.

Aloni, J. (1975) Le sol et l'évolution morphologiques des ermitières géantes du Haut-Shaba (Rép. Zaire). *Pedologie* **25**, 25–39.

Alpers, C. N. and Brimhall, G. H. (1988) Middle Miocene climatic change in the Atacama Desert, Northern Chile: evidence from supergene mineralisation at La Escondida. *Geological Society of America Bulletin* **100**, 1640–1656.

Alpers, C. N. and Brimhall, G. H. (1989) Paleohydrologic evolution and geochemical dynamics of cumulative supergene metal enrichment at La Escondida, Atacama Desert, northern Chile. *Economic Geology* **84**, 229–255.

ASCE (1982) *Engineering and Construction in Tropical and Residual Soils.* ASCE, New York.

Blight, G. E. (ed.) (1997) *Mechanics of Residual Soils.* Balkema, Rotterdam.

Bourgeon, G. (1991) See Pedro (1997) for a published version of this thesis.

Büdel, J. (1982) *Climatic Geomorphology* (trans. L. Fischer and D. Busche). Princeton University Press, 443pp.

Butt, C. R. M. (1987) A basis for geochemical exploration models for tropical terrains. *Chemical Geology* **60**, 5–16.

Clark, D. A. (1974) *The Kankoma Clay Deposit.* Economic Report 49, Geological Survey, Zambia.

De Oliveira, M. A. T. (1990) Slope geometry and gully erosion development: Bananal, São Paulo, Brazil. *Zeitschrift für Geomorphology N.F.* **34**, 423–434.

Dowling, J. W. F. (1966) The mode of occurrence of laterites in northern Nigeria and their appearance in aerial photography. *Engineering Geology* **1**, 221–223.

Eriksson, M. G., Olley, J. O. and Payton, R. W. (1999) Late pleistocene colluvial deposits in central Tanzania: erosional response to climate change? *GFF* **121**, 198–201.

Fairbridge, R. W. and Finkl, Jr, C. W. (1984) Tropical stone lines and podzolised sand plains as palaeoclimatic indicators for weathered cratons. *Quaternary Science Reviews* **3**, 41–72.

Feng, Y. X. and Vasconcelos, P. M. (2001) Quaternary continental weathering geochronology by laser heating $^{40}Ar/^{39}Ar$ analysis of supergene cryptomelane. *Geology* **29**, 635–638.

Fookes, P. G. (ed.) (1997) *Tropical Residual Soils.* The Geological Society, London, 184pp.

Geological Society (1990) Engineering Group Working Party Report: Tropical residual soils. *Quarterly Journal of Engineering Geology* **23**, 1–101.

Hudson, N. W. (1995) *Soil Conservation* (3rd edn). Batsford, London, 391pp.

Li, J. W. and Vasconcelos, P. (2002) Cenozoic continental weathering and its implications for palaeoclimate: Evidence from $^{40}Ar/^{39}Ar$ geochronology of supergene K-Mn oxides in Mt Tabor, central Queensland, Australia. *Earth and Planetary Science Letters* **200**, 223–239.

Marsh, G. P. (1864) *Man and Nature.* Scribner, New York.

Migoń, P. and Thomas, M. F. (2002) Grus weathering mantles—problems of interpretation. *Catena* **49**, 5–24.

Milne, G. M. (1935) Some suggested units of classification and mapping particularly for East African soils. *Soil Research* **4**, 183–198.

Morgan, R. P. C. (1986) *Soil Erosion* (2nd edn). Longman, New York, 298pp.

Newill, D. and Dowling, J. W. F. (1968) Laterites in west Malaysia and northern Nigeria. In *Engineering Properties of Laterite Soils*, Proceedings ICS MFE VII, 133–150.

Ollier, C. D. and Galloway, R. W. (1990) The laterite profile, ferricrete and unconformity. *Catena* 17, 97–109.

Ollier, C. D. and Pain, C. F. (1997) *Regolith, Soils and Landforms*. Wiley, Chichester, 316pp.

Pedro, G. (1968) Distributions des principaux types d'altération chimique à la surface du globe. *Revue de Géographie Physique et Géologie Dynamique* X, 457–470.

Pedro, G. (1997) Clay minerals in weathered rock materials and soils. In Paquet, H. and Clauer, N. (eds) *Soils and Sediments*. Springer, Berlin, 1–20.

Sørensen, R., Murray, A. S., Kaaya, A. K. and Kilasara, M. (2001) Stratigraphy and formation of late Pleistocene colluvial apron in Morogoro district, central Tanzania. *Palaeoecology of Africa* 27, 95–116.

Stephens, C. G. (1971) Laterite and silcrete in Australia: a study of the genetic relationship of laterite and silcrete and their companion materials, and their collective significance in the weathered mantle, soils, relief and drainage of the Australian continent. *Geoderma* 5, 5–52.

Stocking, M. A. (1981) Causes and prediction of the advance of gullier. *Proceedings of the South-East Asian Regional Symposium on Problems of Soil Erosion and Sedimentation* 1981, Bangkok, 37–47.

Tardy, Y. and Roquin, C. (1998) *Dérive des Continents Paléoclimats et Alterations Tropicales*. editions BRGM, Orléans, France, 473pp.

Taylor, G. and Eggleton, R. A. (2001) *Regolith Geology and Geomorphology*. Wiley, Chichester, 375pp.

Temple, P. H. and Rapp, A. (1972) Landslides in the Mgeta area, western Uluguru Mountains, Tanzania. *Geografiska Annaler* 54A, 157–193.

Thomas, M. F. (1974) *Tropical Geomorphology*. Macmillan, Basingstoke, pp 332.

Thomas, M. F. (1989) The role of etch processes in landform development II. Etching and the formation of relief. *Zeitschrift für Geomorphologie N.F.* 33, 257–274.

Thomas, M. F. (1994) *Geomorphology in the Tropics*. Wiley, Chichester, 460pp.

Thomas, M. F. (1999) Evidence for high energy landforming events on the central African plateau: Eastern Province, Zambia. *Zeitschrift für Geomorphologie N.F.* 43, 273–297.

Thomas, M. F. and Goudie, A. S. (eds) (1985) Dambos: small channelless valleys in the tropics. *Zeitschrift für Geomorphologie, Supplementband* 52, 222pp.

Thomas, M. F. and Murray, A. (2001) On the age and significance of Quaternary colluvium in eastern Zambia. *Palaeoecology of Africa* 27, 117–133.

Tricart, J. (1965) *The Landforms of the Humid Tropics, Forests and Savannas* (English trans. C. J. KiewietdeJonge, 1972). Longman, London, 306pp.

Twidale, C. R. (1991) A model of landscape evolution involving increased and increasing relief amplitude. *Zeitschrift für Geomorphologie, N.F.* 35, 85–109.

Valeton, I., Beissner, H. and Carvalho, A. (1991) The Tertiary Bauxite Belt on Tectonic Uplift Areas in the Serra da Mantiqueira, South-East Brazil. *Contributions to Sedimentology* 17, Schweizerbart, Stuttgart, 101pp.

Vasconcelos, P. M. and Conroy, M. (2003) Geochronology of weathering and landscape evolution, Dugald River valley NW Queensland, Australia. *Geochimica aet Cosmochimica Acta* 67, 2913–2930.

Vasconcelos, P. M., Renne, P. R., Brimhall, G. H. and Becker, T. A. (1992) Age and duration of weathering by $^{40}Ar/^{40}Ar$ and $^{40}Ar/^{39}Ar$ analysis of potassium manganese oxides. *Science* 258, 451–455.

Vasconcelos, P. M., Brimhall, G. H., Becker, T. A. and Renne, P. R. (1994) $^{40}Ar/^{39}Ar$ analysis of supergene jarosite and alunite: implications to the weathering history of the western USA and west Africa. *Geochimica et Cosmochimica Acta* 58, 401–420.

Watson, J. P. (1974) Termites in relation to soil formation, groundwater, and geochemical prospecting. *Soils and Fertilizers* 37, 111–114.

Wells, N. A., Andriamihaja, B. and Rakotovololona, H. F. S. (1991) Patterns of development of lavaka, Madagascar's unusual gullies. *Earth Surface Processes and Landforms* 16, 189–206.

Williams, D., Dunkerley, D., De Dekker, P., Kershaw, P. and Chappell, J. (1998) *Quaternary Environments* (2nd edn). Arnold, London, 329pp.

Further reading

Chorley, R. J., Schumm, S. A. and Sugden, D. E. (1984) *Geomorphology*. Methuen, London, 606pp.

Douglas, I. and Spencer, T. (eds) (1985) *Environmental Change and Tropical Geomorphology*. George Allen & Unwin, London, 378pp.

Eggleton, R. A. (ed.) (2001) *The Regolith Glossary*. CRCLEME, CSIRO Mining, Canberra, ACT, 144pp.

Goudie, A. S. and Pye, K. (eds) (1983) *Chemical Sediments and Geomorphology*. Academic Press, London, 439pp.

Summerfield, M. A. (1993) *Global Geomorphology*. Longman, Harlow, UK, 537pp.

Acknowledgements

The author wishes to thank George Milligan for additional material, and William Jamieson for cartographic assistance and for compiling the figures for this chapter.

18. Hot Wetlands

Ian Douglas

18.1 The tropical rainforest environment

The high temperatures and humidity of hot wetlands support a dense, diverse forest cover where biological activity is continuous. The apparent uniformity of the humid tropical rain forest is deceptive. There are wide variations in forest type and in the engineering soils, rocks and landforms beneath the forest (Figure 18.1). This diversity of humid tropical landforms must be appreciated, as engineering experience in one part of the hot wetlands is not necessarily transferable to another.

The hot wetlands embrace landscapes as diverse as the tectonically active mountains of

equatorial parts of the Pacific 'rim of fire', the volcanic landscapes of Java and central America, the old plateaux of the northern part of the Brazilian shield and the flooded forests and freshwater swamp forests of the lower parts of the Amazon and Congo Basins and of eastern Sumatra and southern Borneo. The tectonically active areas tend to have younger rocks that weather and break-up more rapidly than rocks of more stable areas. Given the same general rock type, topography, runoff and temperature, rates of chemical weathering of younger igneous rocks, such as those in island arcs, are roughly twice rates in old cratonic settings (Stallard, 2000). In tectonically active areas, earth movements frequently trigger landslides and thus remove the

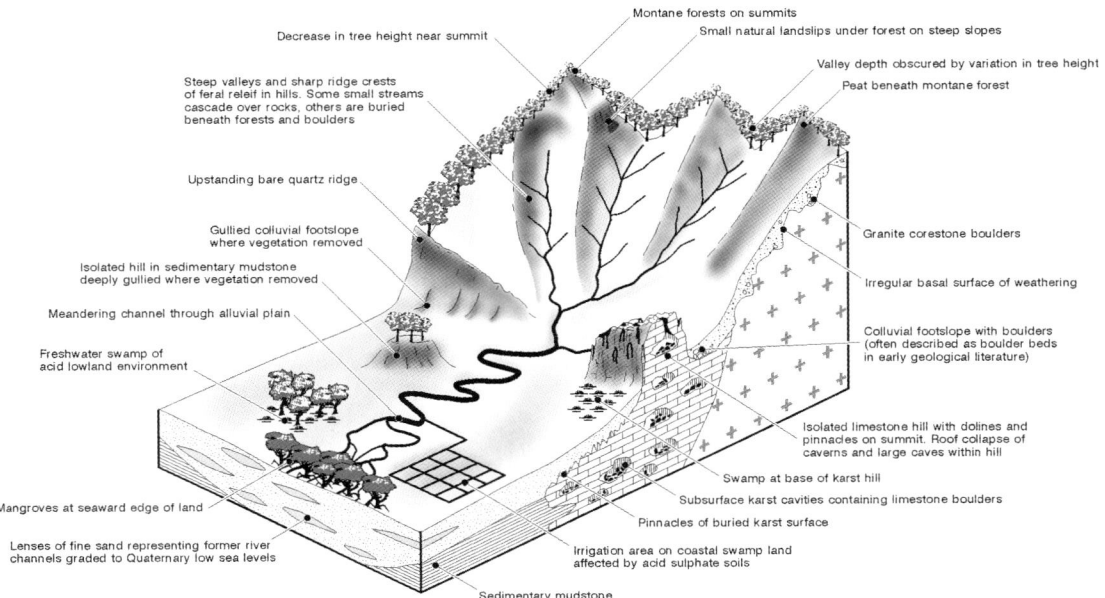

Figure 18.1 Block diagram of humid tropical terrain to show associations between vegetation, weathering, sedimentary deposits and geomorphology.

decomposing rock, supplying large quantities of sediments to rivers.

The rate of chemical weathering in these hot wetlands, and thus the preparation of rock material for detachment and transport by eroding agents, is largely determined by the combination of tectonics, lithology, climate, relief and vegetation. The resulting weathering environment can be viewed in terms of transport-limited and weathering-limited erosion regimes (Johnsson, 2000; see Chapter 11). Material loosened by weathering is carried downslope by transport processes such as landslides and soil creep. If these slope erosion processes work faster than the weathering processes, then erosion is limited by the rate at which the rock is weathered. However, if the rate of weathering exceeds the rate at which detachment and transport processes operate, erosion is limited by the efficiency of the transport process and a deeper weathering profile develops (Johnsson, 2000). Vegetation cover is often the regulator of the efficiency of transport processes, preventing weathered particles from being detached and trapping particles carried downslope. When tree-fall or landslides expose bare weathered material, detachment and downslope transport are rapid. The occurrence of erosion in areas of natural forest is often related to irregular disturbance of small parts of the forest (see also Chapter 19).

Weathering in well-drained sites thus tends to be highly efficient, reducing clay minerals to oxides of iron and aluminium. However, where the water-table is at or near the surface for most of the year, the breakdown of clays characteristic of well-drained sites does not occur, and smectite, swelling and cracking clays may accumulate. In tectonically active areas mass movements are often frequent enough to prevent the accumulation of more than a metre or so of soil and weathered material. In stable areas, deep weathering profiles of 40 m or more occur.

The nature of the rainforest

The problems of engineering soils and geomorphology cannot be separated from the vegetation. The rain forest plays a critical role in the hydrological cycle, in pedogenesis and in landform stability. Intense rainfall is a feature of the equatorial tropics. Mean annual rainfall erosivity may be 25 times greater than in humid temperate latitudes (for example in the northeast of peninsular Malaysia mean annual erosivity is over $25 \, kJ \, m^{-2}$, while in western Britain it is just over $1.3 \, kJ \, m^{-2}$ (Morgan, 1986)).

As elsewhere, rainfall varies with elevation, the highest mean annual rainfall total recorded in the hot wetlands of northeast Queensland, Australia, being 7664 mm (maximum annual total 11 521 mm in 1979) at the summit station on Mt. Bellenden Ker, whereas at the eastern side of the mountain, Babinda the annual mean is $4092 \, mm \, y^{-1}$.

In the areas prone to tropical cyclones, high rainfall intensities can occur for several hours (Table 18.1). Such extreme falls have great geomorphological significance. Shorter return

Table 18.1 Examples of recorded hot wetland rainfall intensities for different storm durations.

Time period	Rainfall (mm)	Intensity mm h^{-1}	Location	Date
1 minute	33	1980	Barot, Guadeloupe	26 Nov 1970
3 minutes	63	1260	Porto Bello, Panama	29 Nov 1911
15 minutes	198	792	Plumb Point, Jamaica	12 May 1916
9 hours	1087	121	Belouve, La Reunion	28 Feb 1964
12 hours	1340	112	Belouve, La Reunion	28 Feb 1964
18.5 hours	1689	94	Belouve, La Reunion	28 Feb 1964
24 hours	1870	78	Cilaos, La Reunion	15–16 Mar 1952
48 hours	2500	52	Cilaos, La Reunion	15–17 Mar 1952
72 hours	3240	45	Cilaos, La Reunion	15–18 Mar 1952
120 hours	3854	32	Cilaos, La Reunion	13–18 Mar 1952
144 hours	4055	28	Cilaos, La Reunion	13–19 Mar 1952

Table 18.2   2-year return period rainfall intensities for selected tropical cities in mm/h.

Place	Continent	5 min	15 min	30 min
Niamey	Africa	160	110	79
Kinshasa	Africa		126	93
Kuala Lumpur	Asia	250	148	110
Babinda	Australia	129	107	84
Paranaibo	South America		124	88
Manaus	South America	186	112	76
Belém	South America	167	106	72

period rainfall intensities (Table 18.2) indicate that 30 minute rainfalls of 80 to 100 mm h^{-1} are widespread in hot wetlands. In the Danum Valley area of Sabah, Malaysian Borneo, 5-minute rainfall intensities rarely exceed 100 mm h^{-1} (Bidin, 2001). In the period May 1995 to April 1998, 5-minute intensities of over 100 mm h^{-1} only occurred on 26 occasions and on only two of these days was an intensity of over 50 mm h^{-1} sustained for over 25 minutes. These figures are within the general 2-year return period ranges shown in Table 18.2.

Depending on the type of forest, some 11 to 25 per cent of the annual rainfall may be intercepted by the foliage and returned to the atmosphere. Transpiration may account for 40 to 65 per cent of rainfall. Although the high canopy breaks the fall of raindrops, the coalescing of drops on leaves may create large drops that can reach their terminal velocity before hitting the ground. Localised splash erosion of patches of bare soil in the forest illustrates the erosivity of raindrops below the canopy. Overall, the forest retains water in the biomass, soils and weathering profile. Surface runoff is rare, save during the most intense and prolonged storms. Although small streams rise and fall rapidly during short duration thunderstorms, the vegetation reduces the height of storm runoff peaks.

Most tropical forest soils are acid, low in organic matter and available plant nutrients (i.e. oxisols and ultisols). Most of the nutrients that support the forest are in the decomposing leaves and other plant material on the soil surface. Once this organic material is removed, new plant cover is difficult to re-establish, but cover is essential if erosion is to be avoided.

Natural forest disturbance and sediment yields

Overall, the forest operates as a pulse-driven system (Table 18.3). Rainstorms can initiate geomorphological changes, such as small mass movements, sudden stream rises and pulses of sediment moving downstream. The bigger the rain event, the larger is the work done. However other pulses are partly driven by the biota. Not only are there animals that create tracks and wallows where erosion occurs, but tree falls knock down other plants and create gaps in the canopy that allow light to reach the forest floor. In the streams, falling trees create debris dams that temporarily store up both plant material and sediment until either biotic decay or lifting of the wooden matter in a flood, releases a pulse of sediment to be swept downstream.

Disturbances of this type reach their greatest frequency on slopes affected by earth quake-induced mass movements, in forests affected by volcanic activity and on hillsides exposed to the full force of tropical cyclones making their landfall after a long path across the ocean. In such environments, the frequency and magnitude of disturbance of the forest and soil may be so great that high natural rates of erosion occur (see also Chapter 11).

Considering combinations of tectonic and lithology, it is possible to develop a schema of how sediment yields will vary under natural conditions in the humid tropics that helps to explain the broad inter-regional differences in sediment yields in the humid tropics (Table 18.4). The highest sediment yields in the humid tropics occur in tectonically active areas, such as the mountains of New Guinea and Taiwan, experiencing sediment yields of the order of 10 000 t km^2 y^{-1} (Pickup et al., 1981; Shimen Reservoir Authority, 1975). The lowest sediment yields are on old land surfaces or sedimentary basins of low relief. Areas like the Congo Basin in Africa and eastern South America (Orinoco and Negro basins) have sediment yields of the order of 100 t km^2 y^{-1} (Milliman and Meade, 1983).

Quaternary environmental change in the humid tropics

Despite claims for their ancient heritage, tropical rainforests have changed greatly in extent and

Table 18.3 Principal types of hot wetland (evergreen) rainforests.

Forest type	Forest characteristics (after Prance, 2002)	Engineering geomorphological significance
Lowland rain forest	Multi-layered, many trees exceeding 30 m in height, closed canopy with sparse undergrowth, large lianas frequent, epiphytes relatively scarce	Predominant forest type in areas subject to agricultural, urban and industrial development; removal leads to accelerated erosion, slope instability and increased runoff unless protective measures are undertaken.
Montane rain forest	Few trees exceeding 30 m, palms and tree ferns common in undergrowth, ground layer rich in herbs and mosses, epiphytes common; often subdivided into lower and upper montane forest	Often on steep slopes with considerable surface organic matter, due to slower decomposition in cooler conditions, but with shallow soils and shallow root systems. In exposed areas subject to windthrow or cyclone damage. Variable foundation conditions.
Cloud forest	High altitude closed forest with many gaps, trees rarely over 20 m, often gnarled, numerous lianas, tree ferns, herbs, shrubs, epiphytes and mosses.	Often associated with upland peat soils that take hundreds of years to develop and with shallow depths to unweathered bedrock. Highly susceptible to instability when earthworks undertaken. Greatly sensitive to expansion of temperate crop cultivation and hill resorts and golf courses. Rapid erosion follows cloud forest removal.
Alluvial forest	Forest growing in seasonally inundated areas along river banks, multi-layered, closed forest with numerous gaps, buttresses and stilt roots, palms, herbaceous undergrowth, epiphytes and lianas common	Widespread in Amazonia: in white water river areas known as *várzea* forest, in black and clear water areas as *igapó;* seasonal flooding disrupts mobility and restricts use of heavy machinery; complex soil shrink-swell problems or surface loading issues may arise; major impacts on aquatic life.
Swamp forest	Forest on permanently wet areas, buttresses, stilt roots and pneumatophores common, palms, ferns and herbs abundant	Often traversed by blackwater streams with pH as low as 3.0 which may be corrosive to concrete and metals; these forests often overlie former distributary channel systems whose coarse sands and gravels may pose problems of stability and water retention when canals and embankments are constructed.
Peat forest	Forest over deep peat deposits in nutrient poor soils. Rarely above 20 m in height, ground cover mostly of ferns	On drainage these lowland peats are likely to shrink and develop acid sulphate soil conditions and cracking clay, shrink-swell phenomena. Low pH conditions may result in corrosion of structures.
Mangrove forest	Single layered forests up to 30 m in height growing in intertidal zones in salt water throughout tropics and subtropics, evergreen, stilt roots and pneumatophores common, little ground vegetation, few epiphytes	Associated with marine clays and silts, but often conceal lenses of sand derived from former river channels. Greatly modified for aquaculture, port development and tourist resorts. Major installations usually require deep piling, unless substantial layers of stiff clay occur. Potential for disruption of natural sediment fluxes and natural coastal protection causing harbour siltation or coastal erosion.

Table 18.4 Schema of regional variations in sediment yields with tectonic situation in the humid tropics.

Tectonic setting	Major channel characteristics	Dominant sediment source	Estimated erosion rate (range of annual sediment yield)/(t km^{-2} a^{-1})	General depth of soil on slope	Example
Active plate margin rift or half graben edge	Braided channels, abundant gravel	Mass movement often triggered by earthquakes	up to 10000	thin	Markham River, Papua New Guinea
Active volcanic areas of recent lava flows	Braided channels	Volcanic debris unstable ash deposits	up to 10000	skeletal	Toto Amarillo, Costa Rica
Tectonically active mountain areas	Deep gorge sections where river usually occupies whole valley floor in flood; some braiding, but lateral movement of river restricted	Valley wall failure, landsliding associated with seismic activity	7000–10000	thin	Upper Fly River, Papua New Guinea
Late Tertiary tectonic activity and weak mudrocks	Wide channel with gravel; bare, frequent undercut banks and grassed flood deposits	Bank erosion in main channel and tributaries, erosion by saturated overland flow in streamhead hollows	ca. 1000	ca. 1m	Segama River Sabah, Borneo
Passive margin with relief of 2000 m in equatorial climate	In uplands, boulder strewn channels, with mature trees right up to water's edge; abundant quartz sand between boulders	Continuation of boulders, surface wash on slopes, bank erosion	50–100	up to 30 m	Gombak, Malaysia
Passive margin with relief of 2000 m in tropical cyclone zone	Upland boulders strewn or rock-cut channels, with areas of exposed rock or boulders at low flows; channel capacities much greater than previous case	Disintegration of boulders, surface wash on slopes, bank erosion	50–200	up to 30 m	Babinda and Behana Creeks, North Queensland
Ancient craton with relief of 2000 m	Upland channels guided by ancient structural lineaments, resistant angular blocks and some derived gravel; monsoonal climates produce large broad lowland valleys	Little sediment supply except through river action on rock of channel wall	up to 50	up to 30 m	Mahaweli Ganya System, Sri Lanka
Ancient craton with erosion surface	Wide sandy channels, little incision; occasional rock bars which suffer little erosion	Wash from etchplain surface	up to 30	up to 30 m	Zaire, Africa
Sedimentary basin on ancient craton	Wide anastomising channels, often with legacies of past phases of fluvial erosion	Little locally derived sediment, except from bank erosion	up to 30	up to 30 m	Amazon basin

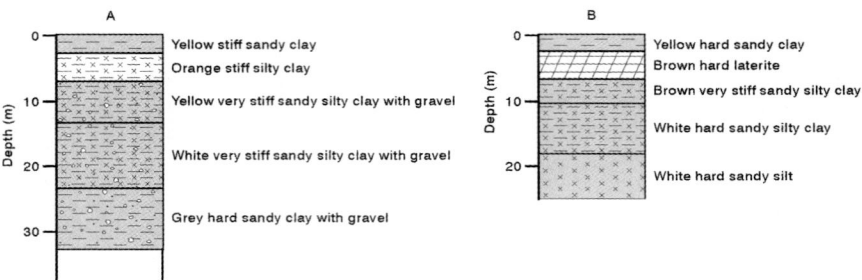

Figure 18.2   Typical boreholes on weathered granite in hot, wet lowland environments.

altitudinal range during the Quaternary as sea
levels and climatic conditions changed (Stanley
et al., 2002). Much of the area now covered by
tropical rain forest became more arid during
cooler periods with lower sea levels, with a moderate extension towards the equator of the
savanna, semi-arid and arid areas (Haberle, 1997).
In many places soil formations and river channels
characteristic of these more seasonal, less wet,
climates are found. However, new palynological
evidence after 1990 has suggested that the
hypothesis of the Amazon rainforest being
reduced to several discrete 'refuges' may be no
longer tenable. (Colinvaux et al., 2000, 2001)
suggest that the lowland Amazon forests persisted
through glacial cycles, with some reassortment of
species as temperature fluctuated from the Late
Glacial Minimum to the present.

Many formations, from the Older Alluvium of
Malaysia and Singapore (Gupta and Pitts, 1992)
and the widespread ferricretes (the iron-oxide-
rich duricrust often termed laterite, Chapter 2)
under some rainforests in SE Asia and Africa are
more likely to have been formed under more
seasonally wet climates than those prevailing at
present. The laterite in the granite residual soil
borehole (Figure 18.2b) is probably a relic from
such seasonally dry quaternary conditions. The
Older Alluvium has depositional features typical
of large braided stream channels that probably
flowed across the exposed floor of the Sunda Sea
to a sea level well below the present. Not surprisingly, therefore, boreholes through the marine
clays of coastal plains often encounter beds of
sand with some gravel that represent former

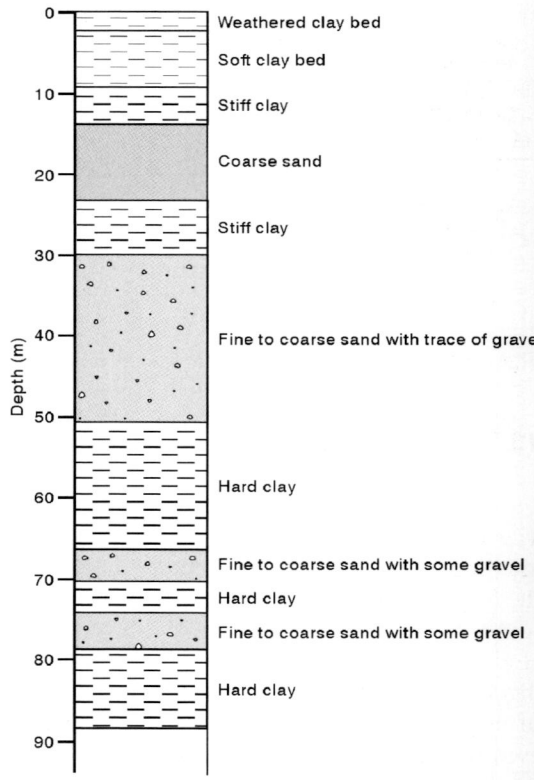

Figure 18.3   Typical borehole in a humid tropical
coastal plain environment with evidence of former river
channels (sand and gravel beds).

stream channels (Figure 18.3). In such former
channels, groundwater movement is relatively
rapid, and clay tends to be washed, i.e. piped,
between particles.

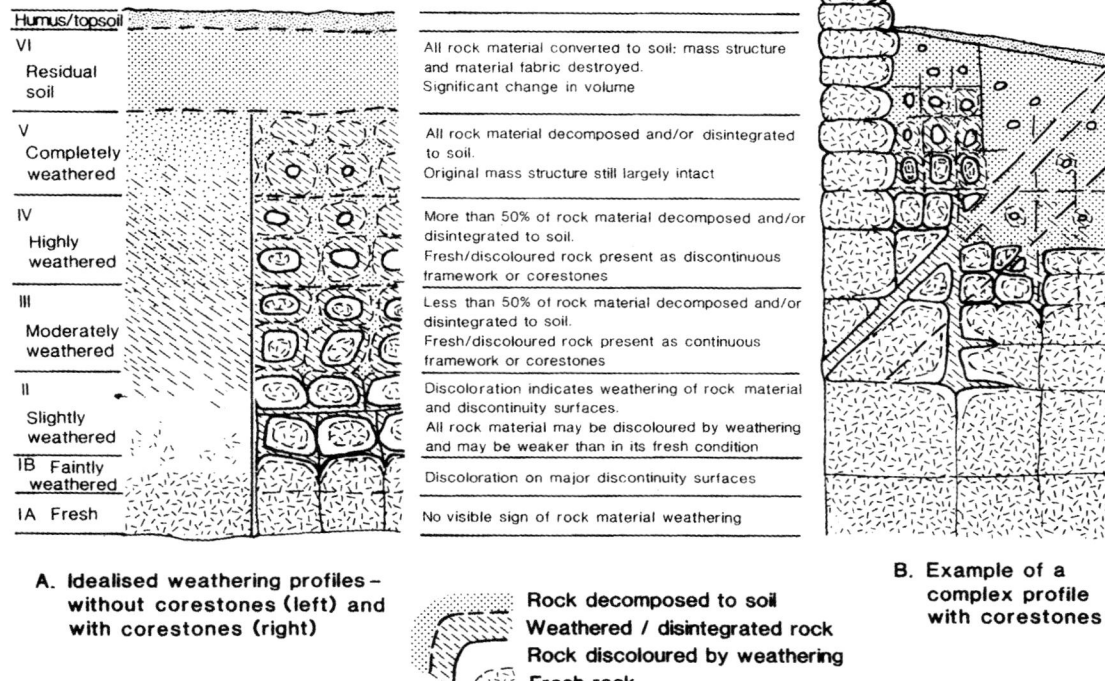

Humus/topsoil		
VI Residual soil	All rock material converted to soil: mass structure and material fabric destroyed. Significant change in volume	
V Completely weathered	All rock material decomposed and/or disintegrated to soil. Original mass structure still largely intact	
IV Highly weathered	More than 50% of rock material decomposed and/or disintegrated to soil. Fresh/discoloured rock present as discontinuous framework or corestones	
III Moderately weathered	Less than 50% of rock material decomposed and/or disintegrated to soil. Fresh/discoloured rock present as continuous framework or corestones	
II Slightly weathered	Discoloration indicates weathering of rock material and discontinuity surfaces. All rock material may be discoloured by weathering and may be weaker than in its fresh condition	
IB Faintly weathered	Discoloration on major discontinuity surfaces	
IA Fresh	No visible sign of rock material weathering	

A. Idealised weathering profiles –
without corestones (left) and
with corestones (right)

Rock decomposed to soil
Weathered / disintegrated rock
Rock discoloured by weathering
Fresh rock

B. Example of a
complex profile
with corestones

Figure 18.4   Classification of tropical weathering grades (from Fookes, 1997a).

18.2 Weathering and soil characteristics

The deep weathering profiles of the humid tropics are probably the most well-known aspect of the engineering geomorphology of hot wetlands (e.g. Lovegrove and Fookes, 1972; Fookes, 1997b). However, while deep weathering is commonly thought to be the norm and does prevail in much of the area covered by rainforest, it is variable, even sometimes absent. Parent material, rainfall, drainage and the age of the landsurface are the prime factors determining the character of weathering profiles.

The combination of high temperature and rainfall produces the conditions for intense tropical weathering causing decomposition of the rock forming minerals and a general decrease in rock intact strength and interparticle bonding. This weathered material has not been transported—it is in its original position (*in situ*) and has only

undergone a decompositional change to a soil type material (residual soil). Weathering is initiated at the surface and from discontinuities in the rock that allow the ingress of water to greater depths. The weathering profiles in Figures 18.4 and 18.5 are schematic representations of a classification for the various stages of decomposition (Tables 18.5 and 18.6). The result is a weakened 'skin' of decomposed material which can be tens of metres thick. In Hong Kong, for example, the rocks have been deeply weathered, giving rise to depths of up to 60 m of silty-sand residual soil (Grades IV–VI) with large corestones in the matrix or exposed on the surface.

The lower sections of many mountain slopes are likely to be mantled by varying thicknesses of colluvium (i.e. transported material '... *heterogeneous, generally structureless materials of soil/rock material and sometimes organic matter, deposited on and at the base of natural slopes by predominantly mass-wasting*' Evans *et al.*, 1997). Various forms of

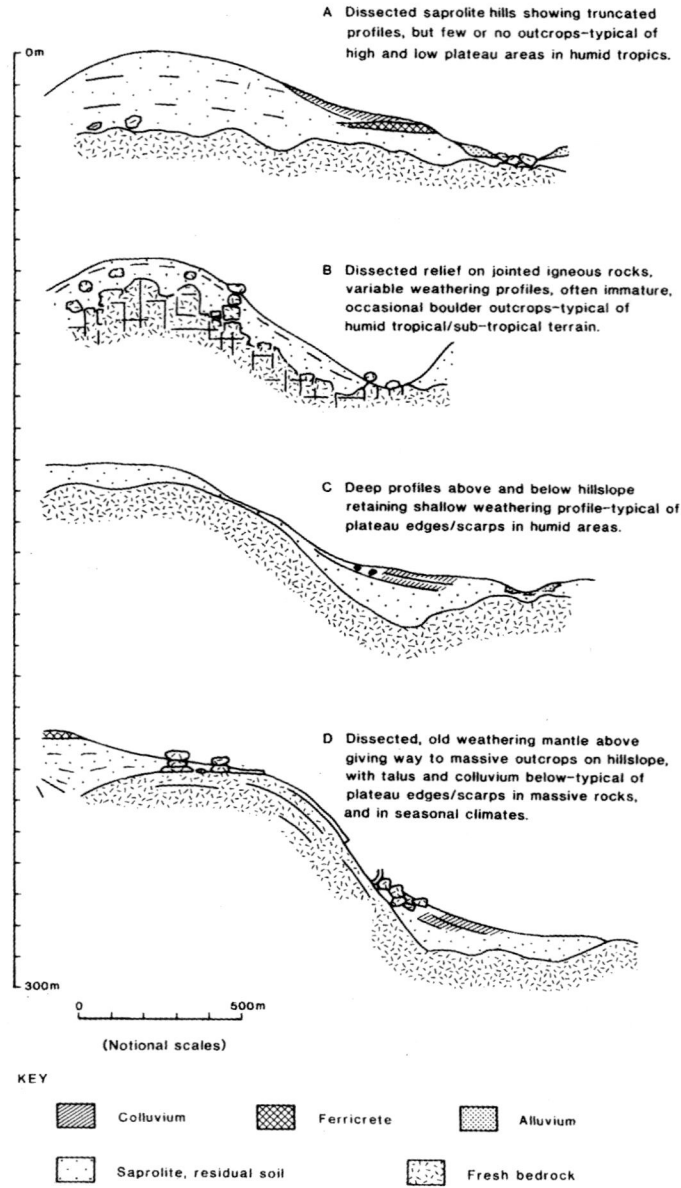

0m

A Dissected saprolite hills showing truncated profiles, but few or no outcrops–typical of high and low plateau areas in humid tropics.

B Dissected relief on jointed igneous rocks, variable weathering profiles, often immature, occasional boulder outcrops–typical of humid tropical/sub-tropical terrain.

C Deep profiles above and below hillslope retaining shallow weathering profile–typical of plateau edges/scarps in humid areas.

D Dissected, old weathering mantle above giving way to massive outcrops on hillslope, with talus and colluvium below–typical of plateau edges/scarps in massive rocks, and in seasonal climates.

300m

0 ————————— 500m

(Notional scales)

KEY

Colluvium   Ferricrete   Alluvium

Saprolite, residual soil   Fresh bedrock

Figure 18.5 Examples of hillslope profiles illustrating common sequences of weathering and landforms (from Fookes, 1997a).

colluvium are to be expected, relating to different generic processes, including the products of soil creep, rock falls, and landslide activity.

Parent material character often produces abrupt changes in the depth of weathering. In a metamorphosed sedimentary sequence containing quartzite

and phyllite, for example, the depth of weathering may be as little as 0.1 m on the quartzite and as much as 15 m on the phyllite. Quartzite is resistant to weathering but other common rock minerals, particularly feldspars, decompose rapidly in the high humidity, high temperatures and high organic

Table 18.5 Scale of weathering grades of rock mass (from Fookes, 1997a).

Term	Description	Grade
Fresh	No visible sign of rock material weathering; perhaps a slight discolouration on major discontinuity surfaces.	I
Slightly weathered	Discolouration indicates weathering of rock material and discontinuity surfaces. All the rock material may be discoloured by weathering.	II
Moderately weathered	Less than half of the rock material is decomposed or disintegrated to a soil. Fresh or discoloured rock is present either as a continuous framework or as corestones.	III
Highly weathered	More than half of the rock material is decomposed or disintegrated to a soil. Fresh or discoloured rock is present either as a discontinuous framework or as corestones.	IV
Completely weathered	All rock material is decomposed and or disintegrated to soil. The original mass structure is still largely intact.	V
Residual soil	All rock material is converted to soil. The mass structure and material fabric are destroyed. There is a large change in volume, but the soil has not been significantly transported.	VI

activity of rainforest areas. Where parent rocks contain silicate minerals with a high proportion of more soluble elements, such as calcium, magnesium and potassium, then decomposition is most effective. However, even silica eventually becomes mobile, leaving behind oxides of iron and aluminium that eventually appear as nodules or concretions in the weathering profile, and, finally, as massive hardpan layers, or as duricrusts, termed ferricrete and bauxite respectively. The types of clay found in the weathering profile also reflect the parent material. Although the trend in hot wetlands is to the formation of kaolinitic clays (Table 18.7), such clays are by no means found everywhere in such regions. While soils developing on granites and pegmatites tend to be rich in kaolinite, those on slate and shales, perhaps because of the character of the sediments from which these rocks were formed, are usually dominated by illite.

Rainfall influences weathering through both quantity and seasonality. In general, the proportion of kaolinite in the clay increases with greater humidity and greater rainfall. In the wettest areas with high rainfall throughout the year, kaolinite is broken down and gibbsite and haematite become more prominent. Weathering profiles tend to be deeper in the wetter areas, but as the highest rainfalls tend to occur on steep mountainsides, excessive water there may cause episodic mass

movements and relatively high soil loss, preventing a deep weathering profile from developing.

Topography and drainage influence weathering-profile characteristics. On well-drained, upper and middle sections of slopes on suitable rocks, kaolinite will be the dominant clay mineral, but downslope it will tend to be replaced by illite or montmorillonite (Table 18.7). Where drainage is impeded, montmorillonite becomes the dominant clay mineral. Thus quite different clay conditions will be encountered in flat valley floors and on adjacent spurs and hillslopes. Materials mobilised by weathering on the well-drained sectors of slopes may accumulate in the valley floors to such an extent that secondary ferricretes can be found where the iron mobilised upslope is redeposited in the fluctuating water-table zone of the floodplain.

The age of the landsurface is of major significance for the depth of weathering. Many parts of the hot wetlands lie in the zone of present day tectonic mountain building, the tectogene tropics (Table 18.4). Other parts belong to the stable, older cratonic landsurfaces of the cratogene tropics. On the latter surfaces, weathering profiles may be old and carry features inherited from past climates. In the tectogene tropics, frequent tectonic activity may trigger earthquakes that induce so many mass movements that deep weathering

Table 18.6   Summary of Duchaufour tropical soil phases, location and climate (from Fookes, 1997b).

Factor/conditions: Soil phase	Mineralogy	Climate needed to reach the phase	Typical locations of the phase	FAO/UNESCO equivalents (USA — Soil survey)
1. Fersiallitic	Upper soils undergo decalcification and weathering of primary minerals. Quartz, alkali feldspars and muscovite not affected. Free iron usually >60% of total iron. Main clay mineral formed is 2:1 smectite; 1:1 kaolinite may appear in older well drained surfaces. With recent volcanic ashes porous andosol soils formed which are eventually replaced by 1:1 halloysites.	Mean annual temperature (°C) 13–20 Annual rainfall (m) 0.5–1.0 Dry season — Yes	Mediterranean, subtropical	Cambisols, calcisols, luvisols, alisols, andosols (alfisols, inceptisols)
2. Ferruginous (ferrisols-transitional)	More strongly weathered soils form but orthoclase and muscovite typically remain unaltered. Kaolinite is the dominant clay mineral; 2:1 minerals are subordinate and gibbsite usually absent. On older land surfaces and more permeable and base rich parent material, ferrisols transitional to phase 3. Partial alteration to gibbsite may occur.	Mean annual temperature (°C) 20–25 Annual rainfall (m) 1.0–1.5 Dry season — Sometimes	Subtropical	Luvisols, nitosols, alisols, acrisols, lixisols, plintha sols (alfisols, ultisols, oxisols)
3. Ferrallitic	All primary minerals except quartz are weathered by hydrolysis and much of the silica and bases removed by solution. Remaining silica combines to form kaolinite but with excess aluminium gibbsite is usually formed. Depending on the balance between iron and aluminium, iron oxide or aluminium oxide will predominate. Soils currently take 104 or more years to form.	Mean annual temperature (°C) >25 Annual rainfall (m) >1.5 Dry season — No	Tropical Can occur in modern savanna from previous wetter climate. Conversely, some currently hot wet areas are still only in the ferruginous phase (e.g. by climate change or by rejuvenation of slopes).	

Table 18.7 Clay-sized weathering products and humid tropical geomorphology.

Weathering product (in order of stage of weathering)	Character and composition	Presence in hot wetlands	Location and geomorphic significance
1. *Gypsum*, halite, etc.	Gypsum, $CaSO_4 \cdot 2H_2O$: evaporite from brine and salt solutions.	Rarely found, being characteristic of arid regions.	
2. *Calcite*, dolomite, aragonite	Calcite, $CaCO_3$; dolomite, $CaCO_3 \cdot MgCO_3$; aragonite, $CaCO_3$ (less stable form than calcite).	Widespread only in karst environments — readily weathered from igneous rocks.	Primarily as a secondary deposit in karst cavities or on tufa in rivers draining karst areas.
3. *Olivine—hornblende*	Olivine: magnesium silicate; hornblende: calcium silicate.	Minerals readily weathered from igneous rocks with loss of magnesium, calcium and silica in solution.	Found on fresh rock exposures only. Rapidly decomposed.
4. *Biotite*, chlorite	Ferromagnesian mica, $KFe_3 (AlSi_3 O_{10}) Mg_3 (OH, F)_2$; chlorite is a clay mineral formed from weathering of hornblende, augite and hypersthene rock.	Minerals readily weathered from magmatic rocks due to solubility of potassium and magnesium.	Found on fresh rock exposures only, biotite granites likely to weather readily.
5. *Albite*	Sodium silicate, $Na (AlSi_3 O_8)$; primary mineral in acid igneous and metamorphic rocks.	Readily weathered in decomposition of feldspars in igneous rocks.	
6. *Quartz*	Silica, SiO_2: quartz is a major rock constituent and quartzite veins often occur in metamorphosed rock.	Less readily weathered than preceding minerals.	Primary mineral which is more resistant than other common minerals in weathering profile. May comprise a high proportion of the sand fraction in many tropical deep weathering profiles. Often the major component of sand size bed material in tropical rivers. Quartzite forms upstanding bare ridges or hills in hot wetlands.
7. *Illite*, muscovite, sericite	Clay minerals: secondary micas. Muscovite, $K_2O . 3Al_2O_3 . 6SiO_2 . 2H_2O$: Structure is platy with potassium ions sandwiched between alumino-silicate layers.	Produced by breakdown of primary micas in igneous rocks and tend to be weathered themselves by loss of potassium.	
8. *Hydrous mica intermediates*, vermiculites	Similar layer lattice structured clay minerals, often with similar composition to illite and muscovite but with less potassium and more hydroxyl (H_3O^+).	Usually found as an intermediate stage in the breakdown of primary micas in igneous rocks but can be found as minerals of neoformation in swamp environments.	

profiles do not develop or persist on steep slopes. Irregular regolith depth and instability of the weathering mantle is likely to be the norm in these tectogene areas. In Papua New Guinea, for example, most weathering profiles are shallow, immature and kaolinitic (Löffler, 1977). However, rates of chemical denudation may be higher in the tectogene tropics than in the cratogene tropics, because of their association with high rates of physical erosion (Riebe *et al.*, 2001). Deep weathering profiles may thus indicate landscape stability, age and duration of weathering much more than rate of weathering.

On the cratogene land surfaces, legacies from drier, more seasonal Quaternary climates can be found. While some basalts, shales and similar rocks have iron accumulations in the form of gravels in the upper part of the present-day weathering profile (Figure 18.6), such accumulations may well be broken fragments from a ferricrete formed under an earlier, seasonally dry climate. Duricrusts develop in some hot wetland environments at the present time, such as bauxites on well-drained basic rocks in Sarawak, or ferruginous concretions found in the weathering profiles on basalts and andesites (Figure 18.6). It is wrong to think that a particular type of duricrust is an indicator of a specific palaeoclimate. Topography, drainage, parent material and time all play a role in weathering and duricrust formation.

The combination of tropical climate, deeply weathered rocks and steep slopes generate a number of active surface processes that are of importance to engineering in hot wetlands (e.g. Fookes 1997a; Figure 18.7). These are:

1. *Widespread landslide activity*. In Hong Kong, for example the combination of widespread landslide activity, dense development of a hilly terrain, highly variable seasonal rainfall regime and a large number of potentially substandard man-made slopes (mostly formed before the 1970s without proper geotechnical input and control) have created acute slope safety problems. Landslides have caused over 470 deaths since 1940.

On average, some two to three hundred slope failures are reported in Hong Kong every year. This equates approximately to an average density of one landslide per year for every square kilometre of the developed area. Almost all of the landslides are triggered by heavy rainfall. Rainfall events that trigger landsliding at medium densities ($1–10/km^2$) occur every 2 years at the local scale (i.e. any given site), and 5-times/year for the region (i.e. Hong Kong as a whole), as indicated in Table 18.8.

Most of the landslides occur on man-made slopes (i.e. cut slopes and retaining walls) and are relatively small ($<50\,m^3$), although some can be $5000\,m^3$ or more. Until recently attention has focused on the stability of man-made slopes. However, landslides on natural terrain also present a significant hazard, as highlighted by:

- the 1.2 km long debris flow on the eastern slopes of Tsing Shan, above Tuen Mun New Town, that occurred in 1990 and reached a planned development site (King, 1996);
- numerous debris flows on the northern slopes of Lantau Island, near Tung Chung in 1992/1993 (Franks, 1996), which would have presented a threat to the North Lantau Expressway had it been opened at the time.

Concerns for the safety of new development sites led the Geotechnical Engineering Office to undertake a programme of region-wide landslide hazard mapping – the Natural Terrain Landslide Study (Evans and King, 1998; Evans *et al.*, 1997; King, 1997, 1999). As a first step, an inventory of landslide features and areas of intense gullying was compiled from the interpretation of high-level aerial photographs (1:20 000 to 1:40 000 scales) taken in 1945, 1964 and annually from 1972–1994 (excluding 1977). Most parts of Hong Kong appeared on between 20–23 sets of photographs.

A total of 26 870 natural terrain landslides were identified within an area of 640 km². Of this total, 8,804 landslides were described as having occurred within the last 50 years (i.e. 'recent'), with evidence for 17 976 slides over 50 years old. Most failures were debris slides

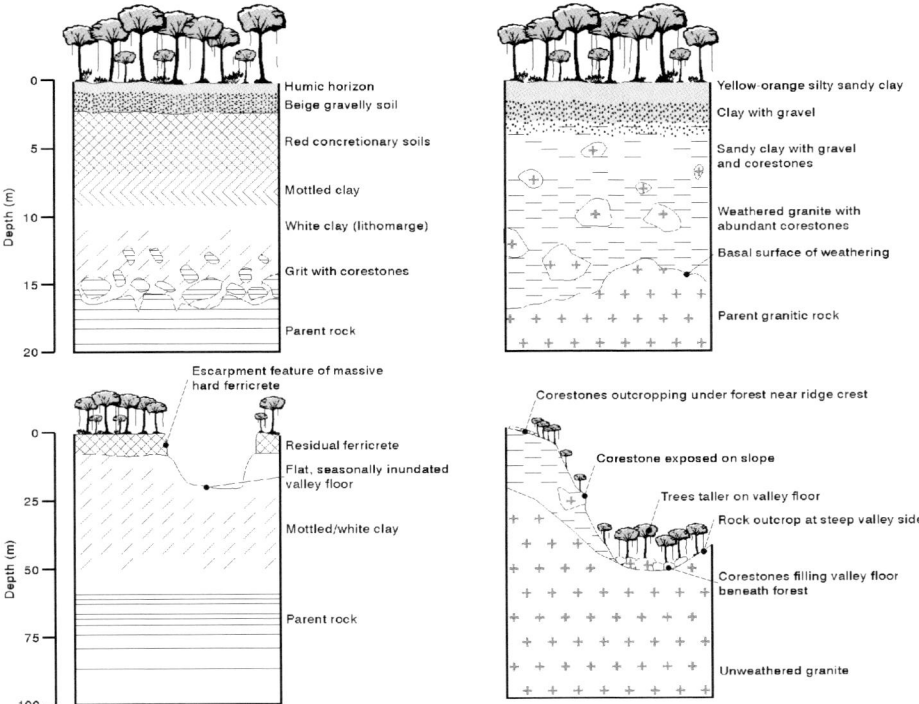

Figure 18.6   Examples of weathering profiles and topographic relationships in humid tropical terrain.

or debris avalanches within weathered rock (i.e. residual soil) or overlying colluvium. Mapped landslide frequencies on different geological units range from 0.3/km^2 per year to 2.8/km^2 per year, with an overall average of 0.8/km^2 per year for all landslides. The average density of *recent* slides is 0.275/km^2 per year. A clear association between landslide density, geological unit and slope angle was identified and provided the basis for a regional hazard assessment.

2. Areas of severe soil erosion hazard, especially after vegetation has been removed prior to construction. Soil erosion can lead to major land degradation and off-site sedimentation problems (e.g. decline in reservoir water quality, increased flood risk because of siltation within river channels etc.) and have significant visual impact (Chapter 11).

3. The potential for metastable collapse of residual soils, leading to a combination of erosion

(e.g. gullying) and landslide activity. Note that large run-out flow slides can develop where relatively saturated metastable soils fail under undrained conditions (e.g. during excavation of cuttings).

18.3 Soil characteristics on different rock types

Granitic rocks contain quartz, feldspars and micas which weather to produce a soil with a bimodal grain size distribution, a clay fraction from the decomposition of the feldspars and micas, and a coarse sand fraction of relatively unweathered quartz grains. Such weathering profiles are likely to contain boulders, or corestones, of unweathered rock one or more metres in diameter. Such corestones are often exposed as outcrops on hillsides (Figure 18.6) whence they may eventually roll on to the valley floor. Such corestones

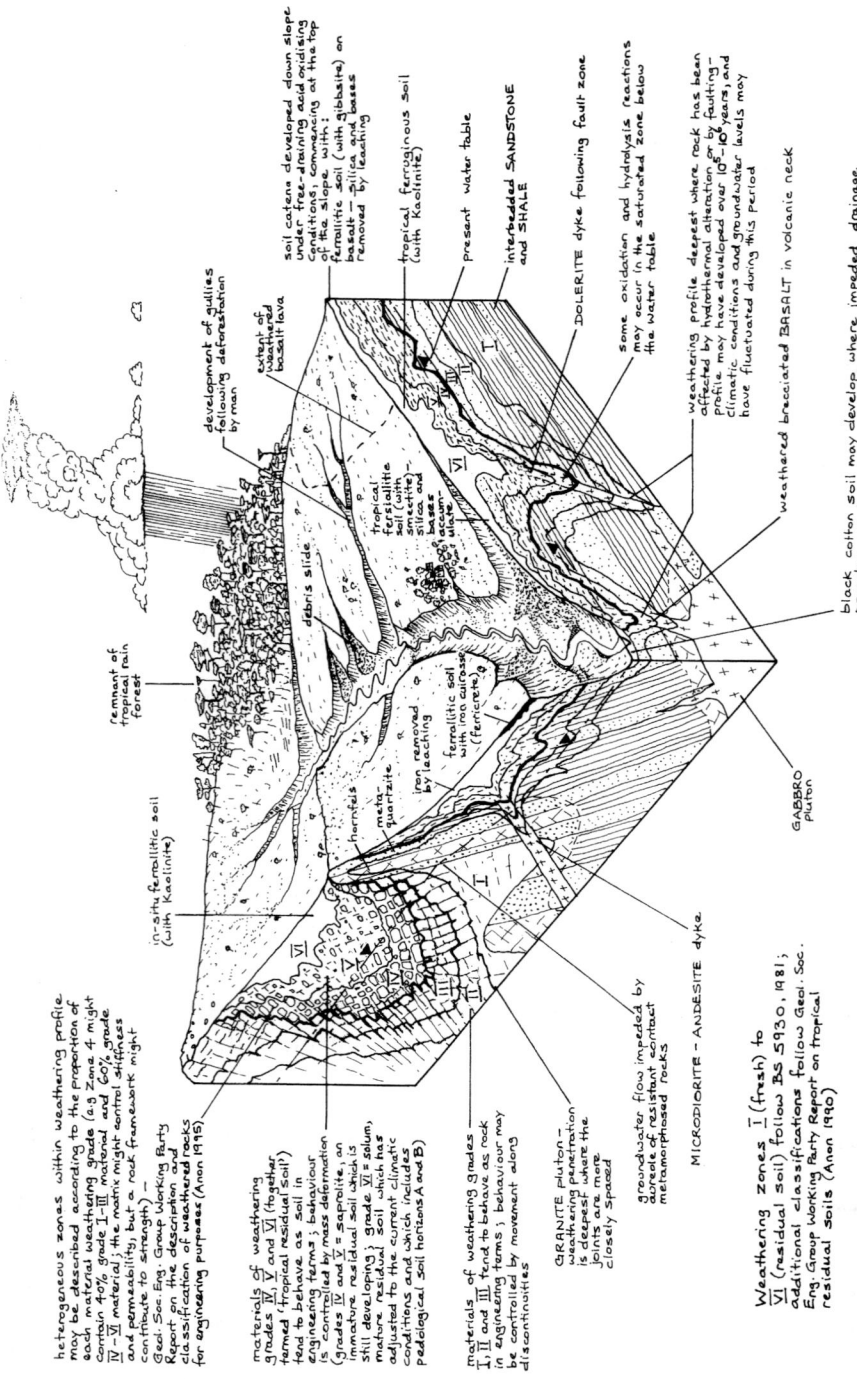

Figure 18.7 Wetland weathering and landscape model (from Fookes, 1997b).

Table 18.8   Rainfall thresholds for landslide activity in Hong Kong (Evans, 1997; Evans *et al.*, 1997)

	Threshold I—start of landsliding		Threshold II—start of landsliding at medium densities ($1–10/km^2$)		Threshold III—start of landsliding at high densities ($>10/km^2$)	
	Rainfall	Return period	Rainfall	Return period	Rainfall	Return period
Local	60–70 mm	2.5 times/year	180–220 mm	Every 2 years	380–450 mm	Every 20 years
Regional		25 times per year		5 times per year		Every 2 years

can cause problems in excavations and in the stabilisation of road cuts. Boreholes can hit corestones, giving the impression that solid rock has been reached.

Similar corestones are likely to be found in sandstone and conglomerate weathering profiles, but in mudstones, shales and schists, the transition between weathered and unweathered material is gradual. Weathered shale materials retain the lineaments and hues of the parent rock but are friable and loose to touch. Where such stratified rocks contain vein quartz or quartz grains in a coarse sand horizon, quartz fragments can form a characteristic stone-line in the soil. Some shales have ferruginous gravel in the upper part of the weathering profile.

Basalt weathers into deep red-brown, uniform soil, but under some circumstances rounded basalt boulders may form. Although usually smaller than the granitic corestones, such boulders can be inconvenient when excavating. Being rich in basic minerals, basalt regoliths tend to develop bauxitic concretions in the wettest, well-drained areas, iron-rich concretions in slightly less wet and more seasonal conditions and montmorillonitic clays generally. Some basaltic soils have marked shrink-swell characteristics. Depth of weathering and regolith development on such volcanic rocks vary with the age of the lava flows, as well as with other environmental characteristics.

Unlike most other rocks in tropics, save for quartz and quartzite, limestone appears in bare vertical cliffs protruding from rainforests. The solution process on calcareous rocks is so effective that the only soil material is the insoluble residue

that accumulates at the base of dolines or cockpits in the karst surface (see Chapter 24). Clays are washed through surface openings and accumulate in the cavern systems inside karst hills (see below). In high rainfall areas a steep, sharp edge pinnacle or needle karst develops on pure limestones, producing an almost impenetrable knife-edge terrain (see also Chapter 24).

18.4 Rivers and alluvial landforms

Most water falling on the rainforest floor infiltrates, then is either transpired by the plants or finds its way by subsurface routes to the watertable or to stream channels (see Chapter 11). Streams emerge as seepages or small springs in stream-head hollows, which may be choked with boulders, depending on the local terrain and parent material. Under normal conditions, a concentrated stream will begin some distance down the channel from the stream head, but after prolonged rain, with a return period of once a year or less frequently, concentrated surface runoff may occur along every declivity where slopes converge above the hollow, producing a cascade of water into the stream head hollow. Along the banks of such small headwater streams, several natural pipe outlets may be active after a major storm, indicating that quick subsurface return flow operates.

The heads of streams tend to be the most sensitive part of the landscape, being likely to erode rapidly if surface runoff above it is accelerated by human activity. Gullies frequently develop above stream head hollows following removal or severe disturbance of the forest, becoming the prime pathway into the stream for the eroded material as land is developed in these hot, humid regions. Disturbance can also lead to the enlargement of natural pipes and their collapse to form gullies, especially in the colluvial fill of stream-head hollows.

Often, in extremely wet conditions, the walls of stream-head hollows are so saturated that fragments of weathered and colluvial material fall to the floor of the hollow and are carried downstream. Along small channels, rivers erode their banks, undercutting trees and scouring around logs that have fallen in the river. Some of the logs trap other branches and plant fragments, developing debris dams which may eventually form a small step in the channel long profile with sand and gravel accumulating upstream of the blockage. Logging and land clearance operations can increase the number of these temporary sediment stores. A high flow following a major storm can

lift the jammed logs, setting them and all the sediment behind them in motion, sometimes provoking bank scour and often leading to the floating timber building up against bridges or other structures further downstream. Careless tree removal and grading operations close to rivers can lead to extremely serious downstream consequences.

Usually the concentrations of dissolved solids in tropical rainforest streams are low. There is plenty of rain to dilute ground and soil water. However, the waters tend to corrode both natural rocks and industrial materials in rivers. Appropriate concrete is needed for bridge abutments and similar structures if corrosion is not to become a serious problem in a few years. Boulders of crystalline rocks develop irregular surfaces as the more soluble minerals are broken down and the resistant minerals protrude from the remaining boulder (Figure 18.8). In many small streams emerging from hilly terrain, boulders

Figure 18.8   Boulders in a tributary of the Sungai Langat, Selangor, Malaysia. The stream bed here is largely quartz sand, the suspended sediment load mainly kaolinite clay, reflecting the bimodal grain size of the weathered granite regolith in the area.

do not persist far below the last rock outcrop and the sand in the stream bed gradually decreases in grain size. Larger rivers, however, can carry boulders, sand and gravel considerable distances, as demonstrated by the in-stream gravel abstraction close to many lowland towns in Asian rainforest regions. Although movement of bed material may contribute a relatively small part of the total sediment transport by equatorial rivers, it is a powerful agent that needs to be considered whenever structures are built along or across rivers.

Large flood events such as those of a 1 in 50 recurrence interval (or less frequent) can lead to massive avulsion, river terrace dissection and channel change. Such events can disrupt communications, especially bridges and their approaches. Often the available hydrological data have gaps caused by recorder damage during floods, while many remote areas only have rainfall information for villages on valley floors. Detailed hydro-geomorphological surveys are often the only

means of establishing the positions of former channels and evidence of past flood heights (see Chapter 10).

Removal of vegetation and shifting of earth alter rainfall: runoff: sediment yield relationships. Erosion plot experiments suggest that, depending on parent material and slope, soil loss under forest may be of the order of $5–100\,t\,km^{-2}\,y^{-1}$, but around $2400–4000\,t\,km^{-2}\,y^{-1}$ from bare soil. This 100-fold increase in the erosion rate may exaggerate what happens over larger areas, but nonetheless indicates the potential severity of the sediment problems created by ground clearance. Road building and the use of unsealed roads with inadequate drainage in rural and forest areas are major sources of sediment, particularly when steep road-cuts and embankments along roads fail (Figure 18.9). Stabilisation of cuttings and embankments by turf or spraying grass seed and the provision of adequate cross drains and safe disposal of roadside runoff should be part of the construction brief of all

Figure 18.9   Stabilising a fill slope in the Kinabalu Highlands.

roads, even those for temporary use in forestry and similar operations.

High sediment yields often last for a relatively short period during road building and clearance. The disturbance usually shifts, in activities like forestry and plantation development or the building of new residential areas to another part of the catchment. Around a large city, the outward movement of bare ground, cleared for new construction, can be readily traced by remote sensing (Figure 18.10). Eventually a whole catchment could be come less prone to erosion as the ground is covered by forest regrowth, or new forms of vegetation, or by paved areas. However, extreme events can always trigger the collapse of some apparently stabilised slopes. Failure of abandoned forest roads, partly through the decay of hollow log culverts and the blocking of subsurface drainage by cut and fill, following rainstorms of 1 in 5 year, or less frequent, recurrence interval, can set in train almost as much sediment movement into channels as the original road building and logging operations (Douglas et al., 1999).

These changes in land cover and the irregular pulsed nature of major sediment inputs to rivers complicate the estimation of sediment yields to reservoirs and peak flows in rivers. In places, lowland channel siltation affects irrigation channels in ricefields such that there is no longer sufficient head for the water to reach all the planted area. Water intakes for urban supplies have become silted. Engineering design and maintenance has to cope with the consequences of changing land cover and land use.

Rivers in alluvial lowlands

In the lowlands, rivers often meander dramatically, or develop anastomosing channels, through freshwater swamps and brackish water mangrove systems (Chapter 10). The várzea and igapó forests of the Amazon probably represent the most complex such environment with changes of several metres in water levels between wet and dry seasons. Traditional farmers have many ways of adapting to the seasonal flooding, but modern settlements and transportation systems often require careful estimation not only of the foundation problems and ground settling when land

surfaces are raised, but also of the probability of shift in channel position.

In the deltaic areas, the problems become even more complex as changed sediment and solute loads coming from the upper catchment interact with the brackish water from the sea, as modified by canal building, flood defences and irrigation works (Chapter 20). Flooding can last several months in major river deltas such as the Mekong. River channel change occurs through erosion and accretion. Shifts of river banks in the Mekong delta of up to 1000 m laterally have occurred over reaches of up to 6 km in length in the period 1966–1999 (Viet et al., 2000). Sand and gravel islands in the delta rivers change position, altering conditions for navigation. These naturally unstable areas show constant adjustment to present day conditions, but also overlie complex sedimentary sequences of former channels and backswamps sloping down to former shorelines, associated with lower sea levels during the Quaternary. In the dry season salt water penetration becomes a problem and changes in land use and reservoir capacity upstream may alter low flows in major rivers such that salinity begins to affect new areas further inland.

When rivers reach the coast, their dry season flow may be insufficient to prevent the virtual closure of the mouths by beach ridges developed by longshore drift. Many humid tropical shorelines have a series of parallel sand beach ridges with intervening peat filled backswamps. The latter play important flood storage roles but are often reclaimed as ports, industries and coastal settlements expand.

18.6 Volcanic terrain

The varied character of the volcanoes dominating many humid tropical landscapes produces a variety of landforms. Active volcanoes create new landforms, burying older landscapes, often blocking watercourses, and creating temporary lakes which can overflow and cause major floods. Many volcanoes have irregular periods of activity between which people occupy the fertile soils that develop on their flanks. The density of agricultural

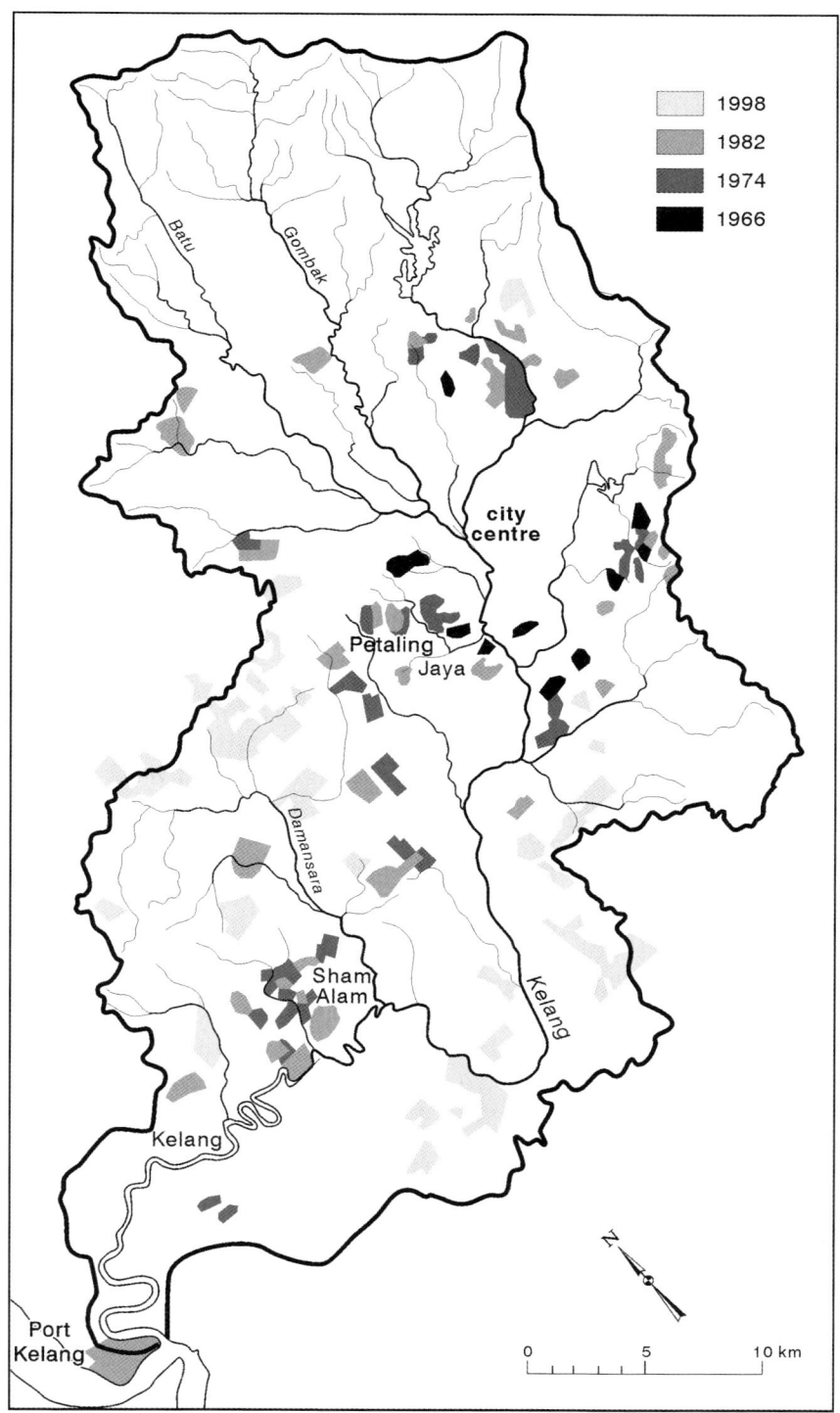

Figure 18.10   Map to show the outward shift of bare ground (mainly on construction sites) since the 1960s in the Kelang Valley of Malaysia.

settlements often places people at risk from future lava flows or nuées ardentes, as on the flanks of Mt. Merapi in Java, Indonesia (see Chapter 23). The burning of the vegetation that follows showers of *nuées ardentes* exposes the soil to erosion.

Deep gulleys on the volcano flanks provide pathways for the lahars that develop when water is caught up with molten lava. The lahars move rapidly down valley with the power to obliterate everything in their path. Large blocks several metres in diameter are moved by these flows. Concrete dams to slow lahar movement have been built in many channels on the blanks of Mt. Merapi, but Engineering works further downvalley need to take account of the lahar risk.

Bandung, Java provides a good example of a humid tropical city with complex volcano-related geomorphological problems. The city municipality and district cover some 2200 km², with nearly 6 million people living at a density of over 17 500 per km², overlying important aquifers of volcanic deposits or fluvial sediments (Dam, 1994). The city essentially is built on a fan of volcanic material from Gunong Tangkubanparahu to the north that becomes inter-fingered with alluvial material and is eventually overlain with lacustrine deposits that formed in the Pleistocene Bandung Lake. This sequence of volcanic and alluvial deposits gives rise to several good aquifers yielding high quality water. However, that water supply is being threatened by pollutants from the expanding urban area, siltation from erosion and the runoff of excess agricultural chemicals from surrounding farmland.

For further discussion of the engineering properties of volcanic soils and their role in earthworks and foundations, see Chapter 23.

18.7 Karst terrain

Limestone landscapes in the humid tropics take a variety of forms (Table 18.9), all reflecting combinations of lithology, structure and climate (see Chapter 24). The most striking general features of karst in these hot, wet environments are the great vertical development of the residual hills and karst pinnacles, and the sheer size of many underground caverns (Figure 18.11). High rainfall areas may have large underground river systems, often discharging through major cave openings or springs. The density of enclosed depressions in pyramid and doline and cockpit karst may be as high as 12 per km², making the centripetally focusing polygonal karst drainage net probably the most efficient natural drainage system (Williams, 1978). Solution rates often exceed $300\,t\,km^{-2}\,y^{-1}$ (Laverty, 1980), and are associated with decay of the limestone surface, particularly at the base of dolines. Thus partial collapse of cavern roofs often change landforms leaving piles of boulders choking many parts of tower karst caves. Residual soils frequently accumulate on the floor of enclosed depressions, supporting a denser, taller forest vegetation than the adjacent limestone slopes.

Around the karst hills, karst plains have often developed, probably by lateral corrosion by rivers. Sometimes outcrops of limestone occur at the surface of the ground. In other places, the limestone surface has been buried by alluvium and is only exposed at the base of excavations such as tin mines or gravel pits. This buried karst is a major engineering problem in many parts of the tropics, particularly where pinnacle karst surfaces have been buried or where large cavern systems lie underground. Boreholes sunk into the buried karst may penetrate a succession of voids in the limestone (Figures 18.1 and 18.12). Some of these voids have a clay infill, others contain so little material that borehole loggers record that the drill bit 'fell under its own weight'. These irregular cavities may only affect part of a site and in the past have been missed where exploratory boreholes have been widely spaced. Details of the siting of several major buildings in cities like Kuala Lumpur have had to be changed because of problems associated with buried karst. Grouting techniques to fill such cavities are not always successful, as material may be lost into deeper cave systems. As the sub-surface water circulation is not well understood, it is possible that solution and cavern formation are continuing and that sub-surface cavern collapse may lead to subsidence of surface structures.

Table 18.9 Karst landscapes of the hot wetlands.

Type	Description	Problems	Typical localities
1. Polygonal	General term for any system of closed depression tropical karst where divides between depressions form a cellular network in plan. Subdivided into five major categories.	Complex drainage likely to be cut by excavations. Fill material may be washed into underground cavern systems.	New Guinea.
a. Pitted undulating surfaces	Surfaces incised by close-set depressions with gully-like channels leading to roughly central stream sinks.	Rapid runoff in gullies after rain.	
b. Pyramid and doline	Bowl-shaped depressions and concave pyramidal hills.	Steep-hillsides, irregular depressions, chaotic relief.	Lake Kutubu area, New Guinea.
c. Cockpit (cone or kegal)	Conical or hemispheroidal residual hills with closed depressions between them. Red earths or rendzinas in depression floors. Forest covering craggy slopes.	Water-logging of depressions in wet season. Minimal surface drainage. Underground water connections usually unknown.	Gunung Sewu, Java, Indonesia; Puerto Rico; Jamaica; Kikori, New Guinea.
d. Tower	Steep-sided, vertical or overhanging residual hills, surrounding deep enclosed depressions or rising from karst margin plains, often covered with alluvium, cliff-foot caves or notches.	Massive caves in hills, complex subsurface water circulation, collapse debris and secondary formations often filling caves.	Porema, New Guinea (Löffler, 1977).
e. Arête and pinnacle	Spectacular, knife-edge ridge country consisting of nearly vertical limestone ridges with sharp crests surrounding enclosed depressions.	Extremely difficult country. Limestone edges so sharp they can cut boots! Vegetation usually obscures true height of pinnacles.	Gunung Mulu, Sarawak; Mt Kaijende, New Guinea.

Table 18.9 (*continued*).

Type	Description	Problems	Typical localities
2. Karst margin plains	Flat to gently undulating plains developed along river channels and coastal margins, usually covered with alluvium, often with isolated upstanding residual tower karst hills. Buried karst plains often have cavern systems in the underlying limestone.	Massive caves in residual hills, pinnacles beneath adjacent alluvial plain. Summits of hills often form cockpit karst. Hills frequently exploited for limestone aggregate. Foundation problems on plains acute due to difficulty of locating buried karst pinnacles and partially filled caverns. Virtually nothing known of deep karstic water circulation below buried plains.	Kinta Valley and Kuala Lumpur areas of Malaysia; Sulawesi; Cuba; Vietnam; southern China; Sarawak.
3. Corridor and crevice	Flat-topped plateaux with narrow, often discontinuous corridors cut in the limestone preferentially along joints and faults and other lines of weakness. Crevice karst is usually at more closely dissected scale, akin to the grikes of northern England.	Irregular spacing of corridors. Some may be obscured by vegetation.	Kikori area, New Guinea (Löffler, 1977).
4. Doline	Enclosed depressions pitting a plateau surface. Relief akin to that of limestone areas of Europe, with occasional enclosed depressions and dry valleys.	Dolines often filled with alluvium. Extremely variable in depth. Connections to subsurface drainage complex.	Müller Plateau, New Guinea (Löffler, 1977).

A good example of the problems in these buried karst plains is provided by the building of a 4 km tunnel for the Light Rail Transit (LRT) System in Kuala Lumpur. The metamorphosed limestone has a complex series of buried pinnacles and cavities that pose tremendous challenges to engineers. Overlying the limestone in places is the Kenny Hill Formation consisting of interbedded shales and sandstone of upper Silurian-Devonian age. Alluvial deposits mainly consisting of loose silty sand and gravels were deposited over these formations during the Quaternary. The building of a tunnel through these diverse materials saw the collapse of thin sandy alluvium over the weathered Kenny Hill Formation with excessive settlement and sinkhole development near Masjid Jamek Station. When a large limestone pinnacle 33 m below ground level was encountered, high rates of water seepage occurred through re-cemented weathered limestone material of high permeability than the adjacent hard rock. Surface investigations failed to show many of the karst features encountered or indicate the presence of variable strengths of rock.

Figure 18.11   Solutionally rounded ceilings distinguish the large passages of the Niah Caves in Sarawak's Subis Hills. Photograph courtesy of Tony Waltham Geophotos.

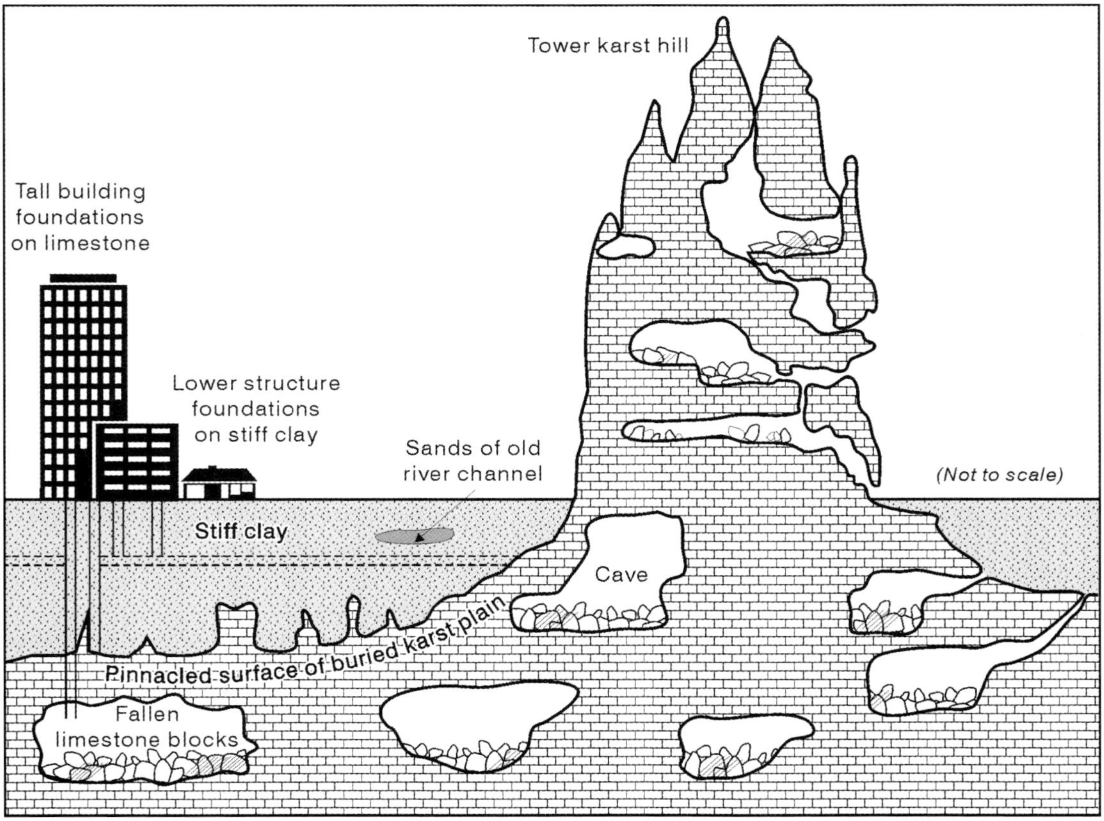

Tower karst hill

Tall building foundations on limestone

Lower structure foundations on stiff clay

Sands of old river channel

(Not to scale)

Stiff clay

Cave

Pinnacled surface of buried karst plain

Fallen limestone blocks

Figure 18.12   Typical borehole in buried karst terrain overlain with alluvium in the humid tropics.

Awareness of the problems of buried karst and its complex relationship to the overlying formations was highly important in this case (see Chapter 24 for discussion of problems of site investigation and foundation engineering in Karst terrain).

18.8 Slope instability and landsliding

Any hill, or steepland area, is difficult to develop without having a risk of causing environmental damage. Throughout the humid tropics granite is often deeply weathered and can be liable to mass movement, through soil slips or landslides, if vegetation is removed and if excavations are made into steep slopes. The depth of unconsolidated material on such slopes may be well over 30 metres. While the bulk of this profile may be undisturbed weathered rock, some of it is likely to be colluvial and taluvial material, weathered rock and soil material that has been washed down from higher upslope. The presence of these materials above weathered rock can create a zone of lateral water flow at the discontinuity. This lateral movement may facilitate soil slips and landslides. The colluvial material usually found only on the lower parts of slopes is a problem when construction involves cut and fill techniques.

In the areas subject to tropical cyclones (typhoons), landslides triggered by extreme rainfalls are always a threat to urban areas. Throughout the Philippines, landslides often occur, whether caused by typhoons or earthquakes. Roads, railways, and other key transport systems can be rendered unusable, hampering rescue and relief operations and isolating disaster areas from the rest of the community. In January 1999 at least 200 000 people were made homeless by landslides in and around the towns of Viga, Bagamanoc, Payo, Caramoran and San Miguel on the island of Catanduanes. In August 1999, the Manila area experienced the heaviest rainfalls for 25 years. A torrent of mud engulfed the Cherry Hills housing estate in Antipolo, 36 people being lost. Elsewhere in the area 46 people were killed in other floods and landslides. The most vulnerable areas were the shanty towns which have grown up along rivers and on steep hillsides (Gittings, 1999).

Hong Kong has had to cope with similar problems. In the 1950s and 1960s the Hong Kong Government built resettlement estates for immigrants. Unfortunately, some of the earthworks were not designed and constructed well enough to cope with severe wet season rainstorms. The result was frequent failures of artificial slopes, culminating in 1972 in two major disasters on the same day. On 18 June 1972 in Sau Mau Ping Estate in Kowloon, a 40 m high road embankment collapsed, killing 71 people. This was followed a few hours later by the collapse of the hillside above a steep temporary excavation on Po Shan Road in the Mid-Levels area of Hong Kong Island which triggered a landslide that demolished a 12-storey residential building and killed 67 people. Four years later, another severe rainstorm hit Hong Kong and brought down three fill slopes in Sau Mau Ping Estate which again were constructed without proper compaction. The resulting landslides killed 18 people.

After these tragic events the Hong Kong Government developed a policy on landslide risk management entitled 'Slope Safety for All' aimed at meeting the highest standards of slope safety. The Slope Safety System is managed by the Geotechnical Engineering Office (GEO) of the Civil Engineering Department (CED), with the overall target of achieving reduction in risk to the whole community (e.g. GEO, 1984). The System is based on seven key result areas (Table 18.10).

18.9 Extreme events and major geomorphological problems in hot wetlands

The humid tropics are often classified as an extreme or risky environment. This view emphasises the climatic hazards associated with heavy rainfalls, tropical cyclones, coastal storm surges and El Niño Southern Oscillation (ENSO, see Chapter 2) related floods and droughts. In terms of engineering geomorphology it is important to recognise what these climatic hazards mean for earth surface processes. Wet and drying of the

Table 18.10 Strategies and action to cope with landslides in Hong Kong.

	Seven point strategy	Other measures
Standards and regulations	Improve slope safety standards, technology, and administrative and regulatory frameworks	Publish geotechnical standards
Implementing standards	Ensure safety standards of new slopes	Catalogue some 54 000 sizeable man-made slopes and retaining walls in Hong Kong
Improving slope stability	Rectify sub-standard Government slopes	Identify the maintenance responsibility of all the catalogued slopes
Maintenance and warning	Maintain all Government man-made slopes	Extensive network of automatically recording raingauges: real time rainfall data for emergency warnings
Public responsibility	Ensure that owners take responsibility for slope safety	24-hour year-round emergency service
Awareness and help	Promote public awareness and response in slope safety through public education, publicity, information services and public warnings	Investigation of serious landslides
Good design	Enhance the appearance and aesthetics of engineered slopes	Auditing the design and supervision of construction of all new slopes
Prevention		Landslip Preventive Measures (LPM) Programme
Squatter settlements		Inspect squatter villages on steep hillsides: guidance to residents; clearance if necessary

land surface takes place rapidly. Most of the time the climate above the tropical rainforest canopy is like that of a desert. In the fierce sunlight, the humidity is low and temperatures are high. Once the forest is removed, the surface of the ground is subjected to this daytime aridity.

When the rain starts, conditions change rapidly, ground water tables in stream head and below valley floors respond rapidly to even a few millimetres of rain. Infiltrating water moves rapidly to streams which begin to rise quickly; a river only 4 m wide can rise 1 metre in 15 minutes. Heavy rains lead rivers to flow into riverine vegetation. Channels often do not have a well-defined bankfull stage of precise recurrence interval. Flood height prediction from morphological features is not always easy.

Much of the sediment load is carried in a short period of time. Close observations on the Segama River in Sabah (Douglas *et al.*, 1999) show that the sediment load carried by a 1 in 10 year flow on a single day may exceed that carried in a whole year (Figure 18.13). Attention to records of past floods, extreme rainfalls, and mass movements is of great importance in developing plans for new infrastructure or urban developments.

Foundation conditions are complicated not only by the energy of the present-day environment, but also by the legacies of Pleistocene climatic and sea-level changes. Understanding of Quaternary events in the humid tropics has advanced greatly since 1980, but there are still uncertainities about the extent of climate and vegetation changes. Buried former river channel and karst features are however a practical issue for anyone concerned with water supplies, major infrastructure developments, city sewerage schemes or construction of high rise buildings.

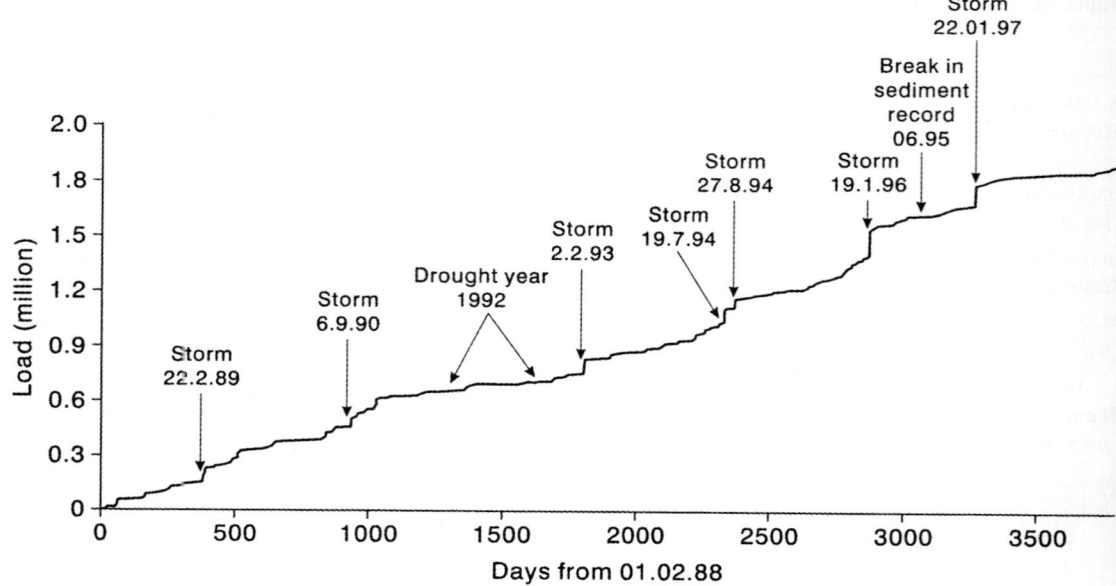

Figure 18.13   Cumulative plot of sediment yield over time for the Segama River at the Danum Valley Field Centre, Sabah, Malaysian Borneo emphasizing the days on which extreme events occurred.

18.10 Summary

Most hot wetlands, with their green covering of tropical rain forest, appear similar to each other. However, the jungle conceals a wide variety of forest types, engineering soils, rocks and landforms. Engineering experience in one part of the hot wetlands is not necessarily transferable to another. High mountain areas were subject to shifts in climatic and vegetational belts during the Quaternary. Lowland areas were greatly affected by Quaternary sea-level changes. Soil formations, relict river channels and deposits of sand and gravel typical of more seasonally, less wet environments are frequently encountered. Large river flows, major mass movements, and rockfalls tend to happen more frequently than elsewhere on similar terrain, just because of the amount, intensity and magnitude of rain events. Many humid tropical landsurfaces are old, with deep weathering profiles. Others are young, steep and unstable, with relatively thin regoliths. Particular problems are associated with individual rock types, such as the buried karst plains of limestone areas. The legacies of the past are still less well understood in the humid tropics than in most other places and planned new developments require careful and cautious investigation.

The humid tropics are an environment to be treated with care and caution. In particular, the downslope, downstream changes that result from changes in runoff and sediment yield must always be borne in mind. Good foundation conditions and effective stabilization of one site may mean the creation of off-site problems for someone else.

References

Asdak, Chay., Jarvis, Paul G., Gardingen, Paul V. and Fraser, A. (1998) Rainfall interception loss in unlogged and logged forest areas of Central Kalimantan, Indonesia. *Journal of Hydrology* **206**, 237–244.

Bidin, K. (2001) *Spatio-temporal variability in rainfall and wet-canopy evaporation within a small catchment recovering from selective tropical forestry*. Unpublished PhD. thesis, University of Lancaster.

Colinvaux, P. A., Irion, G., Rasanen, M. E. and Bush, M. B. and De Mello, J. (2001) A paradigm to be discarded: geological and paleoecological data falsify the Haffer & Prance refuge hypothesis of Amazonian speciation. *Amazoniana* **16**, 609–646.

Colinvaux, P. A., De Oliveira, P. E. and Bush, M. B. (2000) Amazon and Neotropical plant communities on glacial time scales: the failure of the aridity and refuge hypotheses. *Quaternary Science Reviews* **19**, 141–169.

Dam, M. A. C. (1994) *The Late Quaternary evolution of the Bandung Basin, West java, Indonesia*. Ph.D. thesis, Vrije Universiteit, Amsterdam.

Douglas, I., Kawi Bidin, G., Balamurugan, N. A., Chappell, R. P. D., Walsh, T., Greer and Sinun, W. (1999) The role of extreme events in the impacts of selective tropical forestry on erosion during harvesting and recovery phases at Danum Valley, Sabah. *Philosophical Transactions of the Royal Society of London* B, **354**, 1749–1761.

Evans, N. C. and King, J. P. (1997) *The Natural Terrain Landslide Study: Debris Avalanche Susceptibility*. GEO Technical Note TN 1/98.

Evans, N. C., Hung, S. W. and King, J. P. (1997) *The Natural Terrain Landslide Study: Phase III*. GEO Special Project Report SPR 5/97.

Fookes, P. G. (ed.) (1997a) *Tropical Residual Soils*. Geological Society Engineering Group Working Party Revised Report. Geological Society Publishing, London.

Fookes, P. G. (1997b) First Glossop Lecture: Geology for Engineers: the geological model, prediction and performance. *Quarterly Journal of Engineering Geology* **30**, 290–424.

Franks, C. A. M. (1996) *Study of Rainfall Induced Landslides on Natural Slopes in the Vicinity of Tung Chung New Town, Lantau Island*. GEO Special Project Report SPR 4/96.

GEO (1984) *Geotechnical Manual for Slopes*. Geotechnical Control Office, Hong Kong Government.

Gittings, J. (1999) Floods engulf East Asia. *The Guardian*, August 5th 1999, 13.

Gue See Sew and Muhinder Singh (2000) Design & Construction of A LRT Tunnel in Kuala Lumpur, *Seminar on Tunnelling, IEM, Kuala Lumpur, 28 February, 2000*.

Gupta, A. and Pitts, J. (eds) (1992) *The Singapore story: Physical adjustments in a changing*. Singapore: Singapore University Press.

Haberle, S. (1997) Upper Quaternary vegetation and climate history of the Amazon Basin: correlating marine and terrestrial pollen records. In Flood R. D., Piper, D. J. W., Klaus, R. and Peterson, L. C. (1997) (eds) *Proceedings of the Ocean Drilling Programme, Scientific Results* **155**, 381–396.

Johnsson, M. (2000) Chemical weathering and soils. In Ernst, W. G. (ed.) *Earth Systems: Processes and Issues*. Cambridge University Press, Cambridge, 119–132.

King, J. P. (1996) *The Tsing Shan Debris flow*. GEO Special Report SPR 6/96.

King, J. P. (1997) *Natural Terrain Landslide Study: The Natural Terrain Landslide Inventroy*. GEO Report No. 74.

King, J. P. (1999) *Natural Terrain Landslide Study: The Natural Terrain Landslide Inventory*. GEO Technical Note TN 1/97.

Laverty, M. (1980) Water chemistry in the Gunung Mulu National Park including problems of interpretation and use. *Geographical Journal* **146**, 232–245.

Löffler, E. (1977) *Geomorphology of Papua New Guinea*. ANU Press, Canberra.

Lovegrove, G. W. and Fookes, P. G. (1972) The planning and implementation of a site investigation for a highway in topical conditions in Fiji. *Quarterly Journal of Engineering Geology* **5**, 43–68.

Milliman, J. D. and Meade, R. H. (1983) World-wide delivery of river sediment to the oceans. *Journal of Geology* **91**, 1–21.

Morgan, R. P. C. (1986) *Soil Erosion & Conservation*. Harlow: Longman.

Pickup, G., Higgins, R. J. and Warner, R. F. (1981) Erosion and sediment yield in the Fly River drainage basin, Papua New Guinea. *International Association Hydrological Sciences Publication* **132**, 436–456.

Prance, G. T. (2002) Tropical Forests. In Mooney, H. A. and Canadell, J. G. (eds) *The Earth System: biological and ecological dimensions of global environmental change*. (Vol. 2 of Encyclopaedia of Global Environmental Change (ed.) Munn R. E.) Chichester, Wiley, 582–586.

Riebe, C. S., Kirchner, J. W., Granger, D. E. and Finkel, R. C. (2001) Strong tectonic and weak climatic control of long term chemical weathering rates. *Geology* **298**, 511–514.

Shimen Reservoir Authority (1975) *Shimen Reservoir Catchment Management Work Report*. Taipei: The Authority (in Chinese).

Stallard, R. F. (2000) Erosion. In Hancock, P. L. and Skinner, B. J. (eds) *The Oxford Companion to the Earth*. Oxford University Press, Oxford, 314–318.

Stanley, S., Chopra, P. and De Deckker, P. (2002) Sea level changes in SE Asia over the last 150 000 years. Implications for geographical settings and oceanic currents. In Chen-Tun, A. C. (ed.) *Marine Environment: The Past, Present and Future.* Sueichan Press, Taipei, 76–78.

Viet, Pham Bach Nguyen, Lam Dao and Duan, Ho Dinh (2002) Using remotely sensed data to detect changes of riverbank in Mekong River, Vietnam. *www.gisdevelopment.net/application/natural_hazards/floods/nhcy0009.htm.*

Williams, P. W. (1978) Interpretations of Australasian karst. In Davies, J. L. and Williams, M. A. J. *Landform evolution in Australasia.* ANU Press, Canberra, 259–286.

Further reading

Douglas, I. and Spencer, T. (1985) *Environmental Change and Tropical Geomorphology.* Allen & Unwin, London.

Reading, A. J., Thompson, R. D. and Millington, A. C. (1995) *Humid Tropical Environments.* Blackwell, Oxford.

Styles, K. A. and Hansen, A. (1989) *Geotechnical Area Studies Programme: Territory of Hong Kong.* GASP Report XII, Geotechnical Control Office, Hong Kong.

Thomas, M. F. (1994) *Geomorphology in the Tropics: A Study of Weathering and Denudation in Low Latitudes.* Wiley, Chichester.

19. Mountain Environments

John Charman and Mark Lee

19.1 Introduction

It has been estimated that around 36% of the land surface of the earth is covered by mountains, highlands and hills (Fairbridge, 1968). Around 10% of the world's population live in mountain regions and more than 40% are believed to be dependent on mountain resources (Messerli, 1983). They are extremely dynamic and sensitive environments, reflecting the combination of steep slopes, high altitudes and relative relief, together with the presence of numerous relict landforms often inherited from previous phases of glaciation. This landscape sensitivity is easily disrupted by deforestation, land use change and construction projects, often leading to dramatic impacts on the scale and intensity of landslide activity and soil erosion.

A great variety of classification systems have been developed to try and capture the distinctiveness of mountain environments (e.g. Table 19.1). Perhaps the most pragmatic are the simple divisions based on relative relief:

1. hill: $< 700\,\text{m}$
2. highland and upland: $700-1000\,\text{m}$
3. mountain: $> 1000\,\text{m}$.

Alternatively, absolute height can be used:

1. low mountain: $< 1000\,\text{m}$
2. intermediate mountain: $1000-1500\,\text{m}$
3. high mountain: $> 1500\,\text{m}$.

The locations of the world's major mountain belts are shown in Figure 19.1. Many have been formed in the relatively recent geological past, originating in the Tertiary Alpine orogeny but in many areas, for example the Himalaya, uplift is still occurring. The main active mountain belts are the Alpine–Himalayan belt, running from Borneo through northern India into Iran, Turkey and through southern Europe, and the Circum–Pacific belt, encompassing the Andes of south America and the Rocky Mountains of the USA, and encompassing the island arcs of the west Pacific. These mountain belts are located close to tectonic plate boundaries and continuing seismic events are a firm indication that mountain building is still active. Their relative immaturity is reflected in high relative relief, steep slopes, folded and fractured rocks and high rates of weathering and erosion.

Mountain belts formed in earlier orogenies, principally the Caledonian and the Variscan

Table 19.1 Classification of mountain types (from Fairbridge, 1968).

Type	Sub-type
1. Structural, tectonic or constructional forms	A: volcanic mountains B: fold or nappe mountains C: block mountains D: dome mountains E: erosional uplift or outlier mountains F: structural outlier or klippe mountains G: polycyclic tectonic mountains (Alpinotype) H: epigene mountains
2. Denudational, subsequent, destructional or sequential forms	A: differential erosion or relict mountains B: exhumed mountains C: igneous (plutonic) and metamorphic complexes D: polycyclic denudational forms

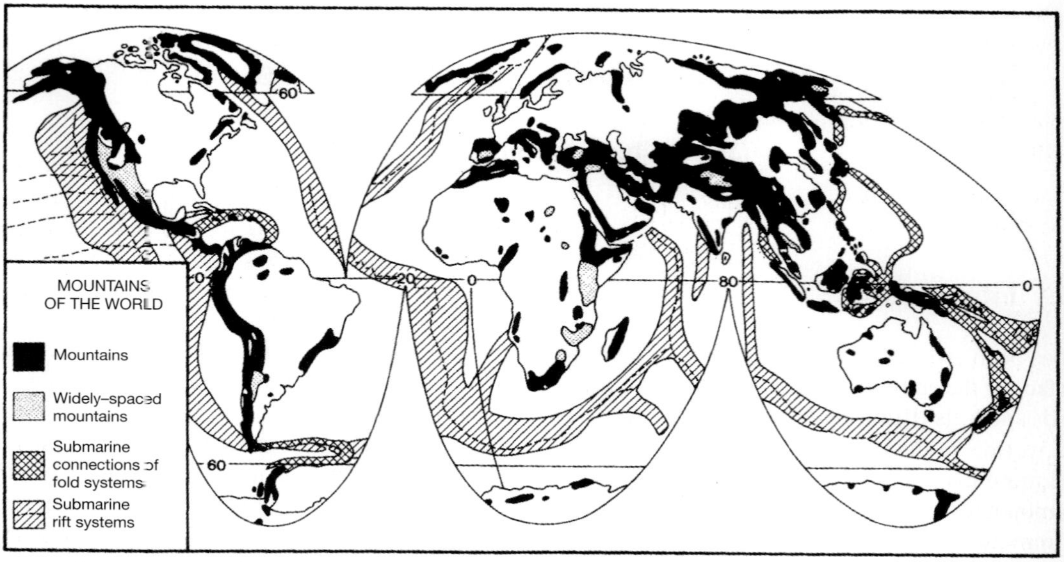

Figure 19.1   Distribution and types of mountain (from Gerrard, 1990).

(Hercynian), exist elsewhere but a longer history of weathering and erosion have worn them down and the extreme relative relief, characteristic of the Alpine belts, is no longer present.

19.2 Rocks, climate and weathering

In young fold mountains the landscape is characterised by the interaction of several factors, including tectonic activity, geology, topography and climate. The most important geotechnical problems relate to the nature of the near surface rocks and soils, slope instability and drainage. In tectonically active zones rapid rates of uplift lead to over-steepened slopes (Chapter 9). In active mountain belts major shear zones and faults are initiated and continue to develop. Accompanying metamorphism produces rocks with preferred structural trends, for example, slaty cleavage and schistocity. These structural planes exhibit lower shear strength than the intact rock.

The influence of geology is reflected both in rock type and rock structure. Mountain belts contain a variety of rock types which vary in their degree of resistance to weathering and erosion. Near the core of the mountain belt both volcanic (fine-grained) and deep-seated (coarse-grained) igneous rocks tend to form. Intense metamorphism also occurs, yielding slates, schists and gneisses. The result is a suite of minerals that formed in conditions far removed from present day atmospheric conditions and they are highly susceptible to weathering (Chapter 2).

In the outer regions sedimentary rocks, sometimes lightly metamorphosed, are more dominant. These rocks are often buried beneath igneous and metamorphic rocks transported from the core zone by gravity slide and nappe structures.

Mountain slopes are often controlled by differential erosion of the various rock types. However, lithological variability can lead to a complex pattern of weathering and soil formation, with marked changes in landslide and surface erosion potential. An indication of the relative susceptibility of some of the more common minerals is given in Table 19.2 (Chapters 3 and 4).

Table 19.3 gives an indication of the relative weathering resistance of the main rock types in relation to their intact rock properties.

Table 19.2   Susceptibility to chemical weathering of common rock minerals.

Fine-grained in minerals sedimentary rocks	Weathering susceptibility	Minerals in igneous rocks
Primary minerals	Most	*Primary minerals*
Gypsum	↑	Olivine
Calcite		Ca-Plagioclase
Olivine,		feldspar
Amphiboles		Na-Plagioclase
Biotite		feldspar
Alkali feldspar		Biotite
		Alkali feldspar
Secondary minerals		
Quartz		
Illite	↑	
Hydrated mica		
Montmorillonite		
Hydrated aluminium oxide		
Hydrated iron oxide	Least	

Igneous rocks

The texture of igneous rocks depend on the rate at which the magma cools. Granites and gabbros are coarsely crystalline because they are emplaced below the earth's surface and cool relatively slowly. Basalts are finely crystalline because they are ejected onto the earth's surface and cool quickly. The coarser grained varieties, such as gabbros, weather more quickly than the finer grained varieties, such as basalts, because they possess a higher porosity.

Sedimentary rocks

Sedimentary rocks have a texture that depend on the mode and distance of sediment transport and the conditions under which they were deposited and subsequently buried. Such rocks may be loosely compacted and voided, densely compacted with a range of grain sizes or cemented with a secondary constituent. Sedimentary mudrocks such as clays and shales also contain clay minerals but weather less quickly. Carbonate-rich rocks such as limestones and gypsum-rich rocks such as evaporites tend to dissolve easily.

Metamorphic rocks

Metamorphic rocks possess a texture that depends on the character of the original rock and the particular conditions of temperature and pressure under which it has been modified. For example, rocks that have been modified under high temperatures and pressures during mountain building episodes are often coarsely crystalline, such as gneisses.

Rock fabric

The fabric of a rock is the spatial arrangement of the textural features. The rock structure is the result of processes that have impacted on the rock during and after deposition. Igneous rocks may contain flow bands, sedimentary deposits may contain alternating beds of differing grain size and metamorphic rocks may contain a preferential mineral orientation as a result of the dominant stress pattern during formation. Major faults and joints result from post-depositional processes and are a major factor in controlling the mass stability of the rock mass. The major geological structural trends affect the major valley profiles, the mass stability mechanisms active on the slope and the depth to which weathering will penetrate.

Climate

Climate varies considerably in mountainous regions because of the effects of altitude and slope orientation. The active mountain belts straddle the main climatic zones and weathering is at its most intense in the tropical and equatorial regions. Altitude is also an important climatic influence and over relatively short distances a traverse across an area of high relative relief can experience several climatic zones. Landscape forming processes depend on the interaction of temperature and rainfall and on the relative interaction of weathering (*in situ* decomposition) and erosion (sediment transport and mass wasting).

Temperature, both seasonal and daily, together with rainfall influences the rate and type of weathering. Mechanical weathering may cause breakage of rock into more closely fractured components while

Table 19.3   Resistance to weathering related to rock properties (modified from Cooke and Doornkamp, 1990).

Rock properties	Physical weathering (disintegration)		Chemical weathering (decomposition)	
	Resistant	Non-resistant	Resistant	Non-resistant
Mineral composition	High feldspar content Calcium plagioclase Low quartz content Ca CO$_3$ Homogeneous composition	High quartz content Sodium plagioclase Heterogeneous composition	Uniform mineral composition High silica content (quartz, stable feldspars) Low metal ion content (Fe–Mg) Low biotite High aluminium ion content	Mixes/variable mineral composition High CaCO$_3$ content Low quartz content High calcic plagioclase High olivine Unstable primary igneous minerals
Texture	Fine-grained Uniform texture Crystalline or tightly packed clastics Gneissic Fine-grained silicates	Coarse-grained Variable texture Schistose Coarse-grained silicates	Fine-grained dense rock Uniform texture Crystalline Clastics Gneissic	Coarse-grained igneous Variable texture (porphyritic) Schistose
Porosity	Low porosity Free-draining Low internal surface area Large pore diameter permitting free drainage after saturation	High porosity Poorly draining High internal surface area Small pore diameter hindering free drainage after saturation	Large pore size Low permeability Free-draining Low internal surface area	Small pore size High permeability Poorly draining High internal surface area
Bulk properties	Low absorption High strength, elasticity Fresh rock Hard	High absorption Low strength Partially weathered rock Soft	Low absorption High compressive, tensile strength Fresh rock Hard	High absorption Low strength Partially weathered rock Soft
Structure	Minimal foliation Clastics Massive formations Thin-bedded sediments	Foliated Fractured, cracked Mixed soluble, insoluble mineral component Thin-bedded sediments	Strongly cemented Dense grain packing Siliceous cement Massive	Poorly cemented Calcareous cement Thick-bedded Fractured, cracked Mixed soluble, insoluble mineral component
Representative rocks	Fine-grained granites Some limestones Diabases, gabbros Coarse-grained granites Rhyolites Quartzites Strongly cemented sandstones Slates Granitic gneisses	Coarse-grained granites Dolomites, marbles Many basalts Soft sedimentary rocks Schists	Acidic igneous varieties Crystalline rocks Rhyolites, granites Quartzite Granitic gneisses Metamorphic rocks	Basic igneous varieties Limestones Marbles, dolomites Poorly cemented sandstones Slates Carbonates Schists

chemical weathering causes decomposition of the rock and the disaggregation of minerals into a soil comprising a collection of discrete particles.

Given the role of weathering in producing a residual mantle of potentially erodible disaggregated particles, rainfall is probably the most important climatic factor governing whether this mantle is subject to soil erosion or landsliding. While annual rainfall totals have some influence the greater role is provided by seasonal rainfall patterns, particularly when the rainy season is populated by short intense storms which can produce catastrophic slope erosion. The onset of intense periods of rainfall provides the medium to transport the weathered materials. In temperate and colder climates the rate of weathering is considerably slower so that significant thicknesses of weathered materials do not form. In these regions transported soils are more prevalent.

With such a diversity of topography, geology and climate a variety of issues can be encountered in mountain regions, including:

1. access and routing
2. large landslides
3. debris flows and torrents
4. snow melt and flash floods
5. landslide dams
6. glacial lake outburst floods
7. snow avalanches
8. ice avalanches
9. suitable sources for construction materials.

These issues are considered briefly after an outline of the main terrain zones.

19.3 Mountain terrain model

In active young fold mountain belts, cycles of relatively rapid uplift initiate a period of intense erosion as rivers cut down to lower base levels and produce steep-sided valleys. Intervening more dormant periods allow weathering agencies to dominate and cause rock decomposition, and the reduction in shear strength causes landslide activity in the valley sides. Meanwhile, periods of intense rainfall initiate high erosion

rates. In this dynamic environment any rural management programme or new engineering project, such as a road or a hill irrigation canal benefit from a careful evaluation of landslide and erosion hazard, allowing them to be planned accordingly.

The cyclic nature of mountain development provides the basis for developing a model in which the system can be divided into land units (zones) and further sub-divided into land facets. Figure 19.2 is based on a mountain system classification developed in Nepal (Fookes *et al.*, 1985). The land units and sub-units (facets) are described in Table 19.4.

The cycles of high tectonic activity lead to the forming of narrow incised valleys. The steep slopes of these valleys, immediately bordering the main rivers, are often very unstable, depending on the underlying geological structure, and can be areas of high landslide hazard. These are designated as land unit 4, characterised by slopes steeper than 35°, often steeper than 45°, and actively degrading to shallower slope angles.

In periods of lower activity and relatively slow uplift, continuing landslide activity eventually produces shallower and more stable slopes. These less active areas are subject to a longer period of chemical weathering and because erosion is less intense a mantle of weathered residual soil develops. These are designated as land unit 3, characterised by slopes shallower than 35° and chemically weathered to produce friable and easily erodible soils.

During these periods the river may begin to widen the valley floor and deposit alluvium. The alluvial areas are designated as land unit 5, characterised by flat tracts of granular material, the higher, older terraces having steep frontal slopes, and the tops of the terraces being subjected to chemical weathering. The next phase of high activity initiates another cycle in which the river cuts down through the alluvium, which is left as a depositional terrace above newer steep slopes of land unit 4 dropping down to the new river level.

When considering an engineering project, particularly linear projects which traverse several land units, the use of the mountain zone

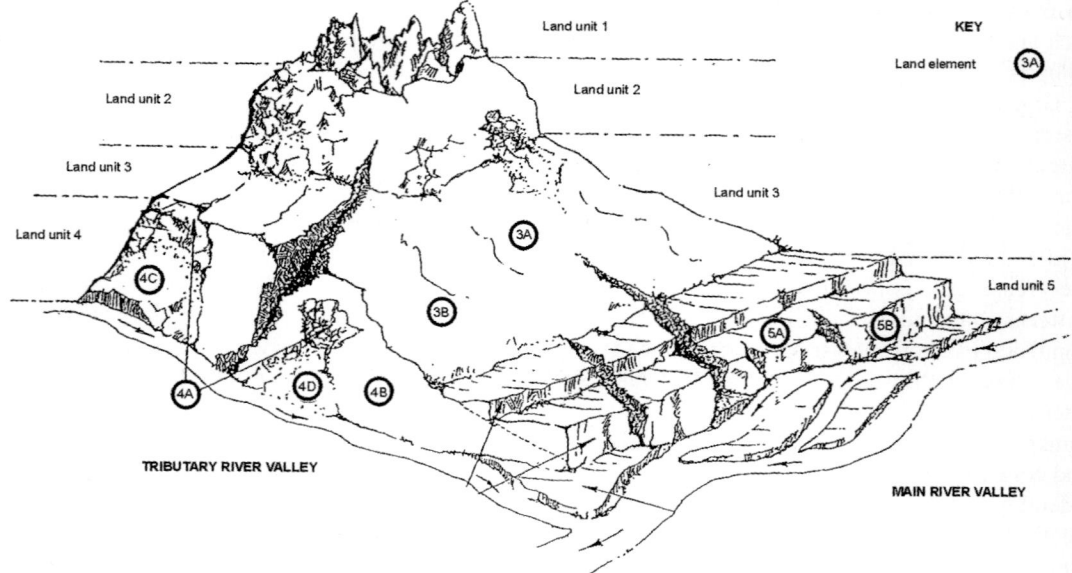

Figure 19.2   The mountain terrain model (from Fookes *et al.*, 1985).

classification to differentiate individual land units and facets is invaluable.

Terrain maps based on this system, with knowledge of the distribution of land elements and typical engineering approaches in each, provides engineers with the information to prioritise the location of individual sites or route alignments.

19.4 Access and routing

Steep, dissected terrain presents extreme difficulties in providing safe, maintained access to a site and associated borrow or spoil disposal sites throughout the year. On road and pipeline sites it is often impossible to provide temporary access, parallel to the alignment, without endangering the integrity of the route. Construction materials and other essential supplies must be transported along the right-of-way, potentially causing disruption to operations.

Pipeline routes through mountainous terrain should avoid sidelong ground and the crossing of deeply incised gullys and stream valleys. The 'straight-line' approach to pipeline alignment is rarely feasible in rugged terrain. The most efficient routing approach to minimising risk in such areas involves adopting 'ridge and spine' or 'ridge and spur' alignments, wherever possible. These are the topographic positions where landslide and erosion problems are considerably lower than on the adjacent hillslopes or valley side-slopes. The most suitable ridgelines are those which are generally aligned with the desired alignment direction. Where it is not possible to use spurs or ridgelines, it is desirable to climb or descend steep slopes normal to the contours, minimising the exposure to landslide hazard.

The principles of road alignment through mountainous terrain have been set out in Fookes *et al.* (1985), based on their experience on the Dharan–Dhankutra Road, Nepal. The ideal corridor is along the mature slopes, stale terraces and plateaux of land unit (zone) 3. Where it is necessary to traverse all the zones, the key principle is make distance in land unit (zone) 3 and make height in land unit (zone) 4, avoiding extended runs of the line across land unit (zone) 4 slopes

Table 19.4 A mountain system classification: description of terrain units (see Figure 19.2).

Land Unit		Land Facet	
No.	Description	No.	Description
1	High altitude glacial and periglacial areas subject to glacial erosion, mechanical weathering, rock and snow instability and solifluction movements with thin rocky soil, boulder fields, glaciers, bare rock slopes, talus development and debris fans.		
2	Free rock face and associated steep debris slopes subject to chemical and mechanical weathering, mass movement, talus creep, freeze-thaw, and debris fan accumulation.		
3	Degraded middle slopes and ancient valley floors forming shallow eroded surfaces subject to chemical weathering, soil creep, sheetflow, rill and gully development and stream incision.	3A	Ancient eroded terraces covered with a weathered residual soil mantle generally up to 3 m thick. Slope angle generally < 35° and stable. Often farmer terraced. Highly susceptible to water erosion.
		3B	Degraded colluvium comprising landslide debris of gravel, cobbles and boulders in a matrix of silt and clay. Slope angle < 35°. Relatively stable. Often farmer terraced. Variable permeability.
4	Steep active lower slopes with chemical and mechanical weathering, large-scale mass movement, gullying, undercutting at base and accumulation of debris fans and flows of marginal stability.	4A	Bare rock slopes. Steep slope angles > 60°. Stability dependent on orientation of discontinuities, such as joints and bedding planes.
		4B	Rock slopes with mantle of residual soil usually < 2 m thick. Steep slope angles > 45°. Prone to extensive shallow debris slides. Deeper instability as for 4A.
		4C	Active colluvium. Thick landslide debris often at base of slope and subject to active river erosion. Slope angle > 35°. Highly unstable, particularly during wet season.
		4D	Degraded colluvium. Thick landslide debris. Slope angle < 35°. Marginally stable and susceptible to gradual downslope creep during wet season.
5	Valley floors associated with fast flowing, sediment laden rivers, and populated by sequences of river terraces.	5A	Top of old alluvial terraces above present river level. Generally flat to shallow, < 10°. Coarse granular and permeable soils. May be covered by a less permeable residual soil mantle.
		5B	Front scarp face of old alluvial terraces. Steep slope angle > 65°, but subject to sudden collapse when cementation breaks down under weathering or when subject to toe erosion.

(Figure 19.3). In addition, experience suggests the following principles:

1. Avoid structures and rock cuts, both of which are costly, unless the alternative is clearly more expensive. Rock cuts will be unavoidable on slopes over 45°.
2. Retaining structures will be needed on slopes greater than 30°.
3. Earthworks problems will arise when the alignment generates a surplus or deficit of material. On such slopes the location of borrow and spoil disposal sites should be given as much attention as the alignment itself.

19.5 Large landslides

Mountains are extreme, high-energy environments and probably the most landslide prone landscape in the world. This is because of the combination of steep slopes, high relative relief, active river incision, widespread structural weaknesses in the rock masses, intense physical weathering, harsh climates and, in neotectonic regions, seismic activity. The full variety of landslide forms described in Chapter 8 can be found in mountain regions, however it is worth highlighting a number of landslide forms that are perhaps unique to this type of environment.

Rock spreading and sagging

Rock spreading and sagging involves the large-scale (volumes can exceed 1 M m^3), extremely slow deformation of mountain ridges and scarps. This often results in the development of double ridges, trenches (grabens) and uphill facing scarplets. Hutchinson (1988) suggested that the features represent the early stages of major landsliding. A number of different mechanisms can be recognised: lateral spreading in homogeneous rocks and brittle rocks overlying ductile materials, and rock flows.

Lateral spreading in homogeneous rocks (Figure 19.4a) can involve ridge core subsidence and downward movement along low-angled shear surfaces, causing rock spreading at the ridge

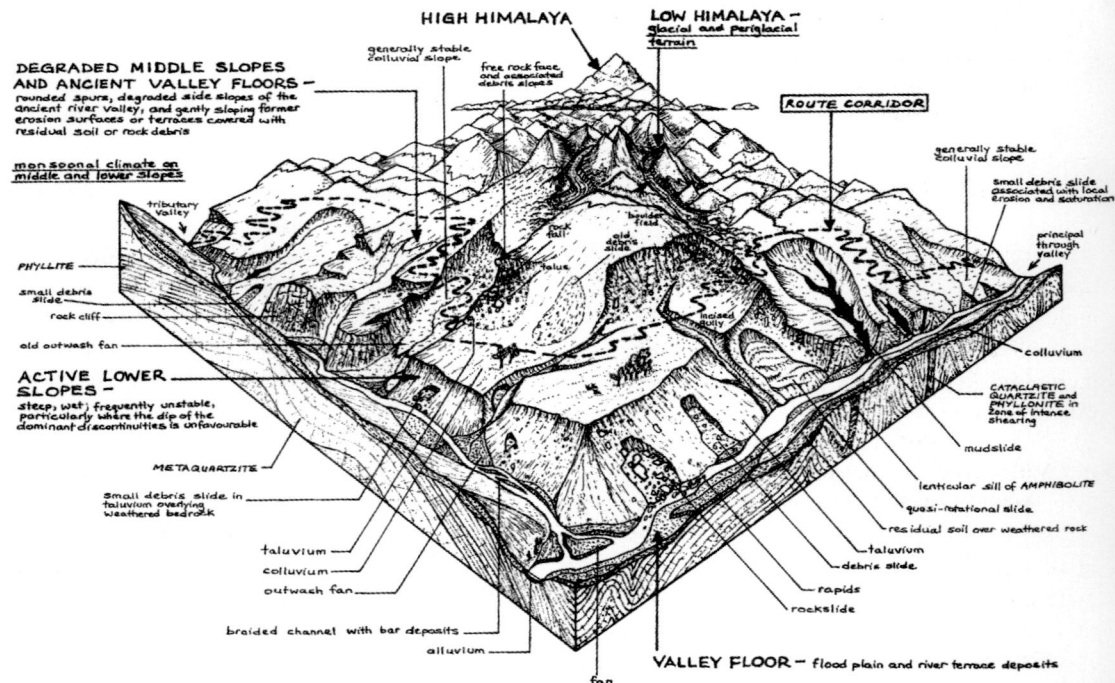

Figure 19.3   Route selection through mountainous terrain (from Fookes, 1997).

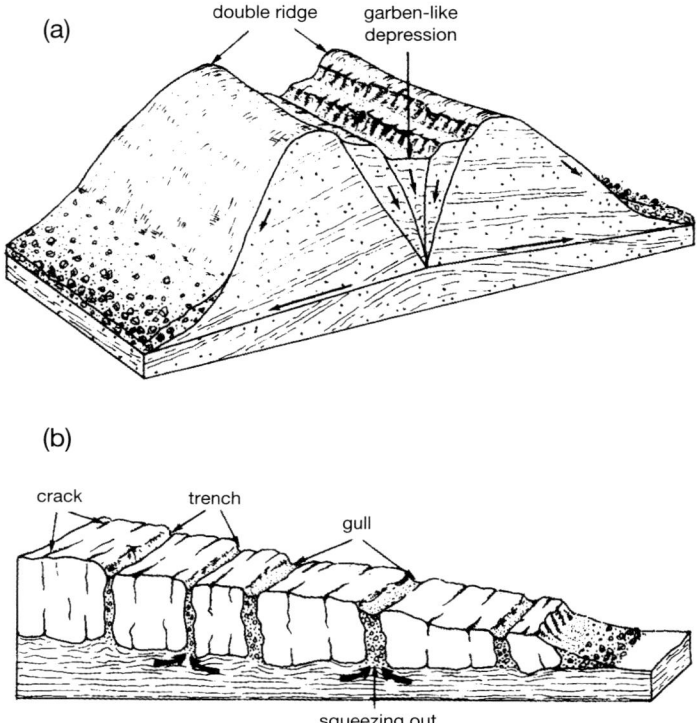

Figure 19.4 Lateral spreading (*A*) in homogeneous rocks and (*B*) in a brittle formation over a ductile layer (from Pasuto and Soldati, 1996).

crest. Examples have been described in the Carpathians, USA, New Zealand and Venezuela (Pasuto and Soldati, 1996).

Lateral spreading in brittle rocks overlying ductile materials (Figure 19.4b) leads to the development of trenches, gulls, grabens and pseudo-karst depressions in the competent rocks and bulges in the softer materials. Movements may extend for several kilometres back from a plateau edge (the zone of decompression). Figure 19.5 illustrates the development of this style of movement which involves progressive visco-plastic deformation of the ductile material along a system of partial shear surfaces or a band of softened material. Over time these partial shear surfaces may combine to generate distinct block slides (Záruba and Mencl, 1982).

Rock flows (sackung) are creeping flow-type, deep-seated deformations affecting densely jointed or stratified hard rock masses (Bisci *et al.*, 1996). Typical features include trench-like grabens, double ridges and counterscarps, with bulges and other compressional features on the lower slopes (Figures 19.6 and 19.7). It is believed that the rock mass deforms through viscous flow because the confining pressures are high and the deviator stresses too low to produce shearing. On the upper and lower slopes where the confining pressures are low, the rock mass moves along shear surfaces and deforms in a dilatant mass.

As rock spreading and flow movements tend to be extremely slow, they do not generally present a significant problem. However, as development encroaches onto mountain regions there is a need to ensure that dams or hydro-electric power (HEP) stations, for example, are not located at the toe of these features. In some instances the creeping

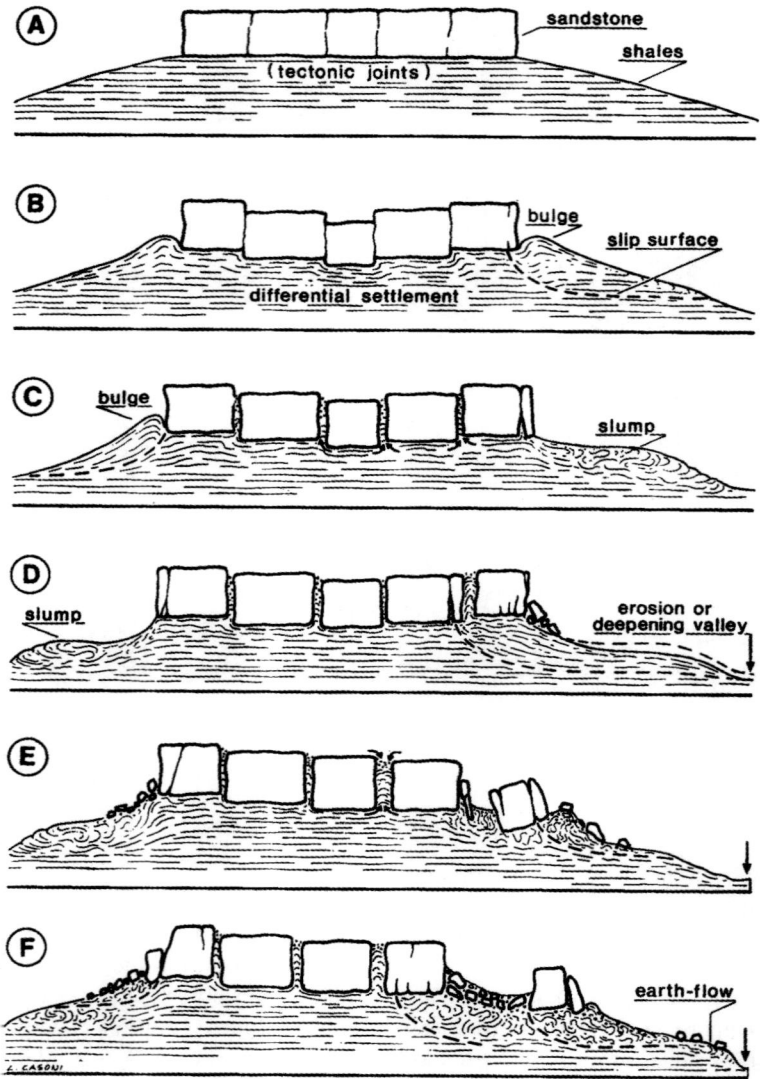

Figure 19.5   Schematic sequence of rock spreading in brittle and ductile rocks (from Cancelli and Pellegrini, 1987).

deformations may be indicative of the potential for the development of a catastrophic landslide that could cause widespread damage and loss of life. Field investigations are needed to assess the risk associated with these sites. Monitoring of the displacements through the use of GPS, geodetic or topographic surveys or crack monitoring can be an important tool in managing the risks. However, because of their scale and the depth of movement,

long-term remedial measures are unlikely to be effective.

Catastrophic landslides

Many high-impact landslide events in mountainous regions involve rapid displacement and long run-out of the debris (Chapter 8, Table 8.1; see Voight, 1978, for an indication of the enormous scale of some historical rock avalanches). The

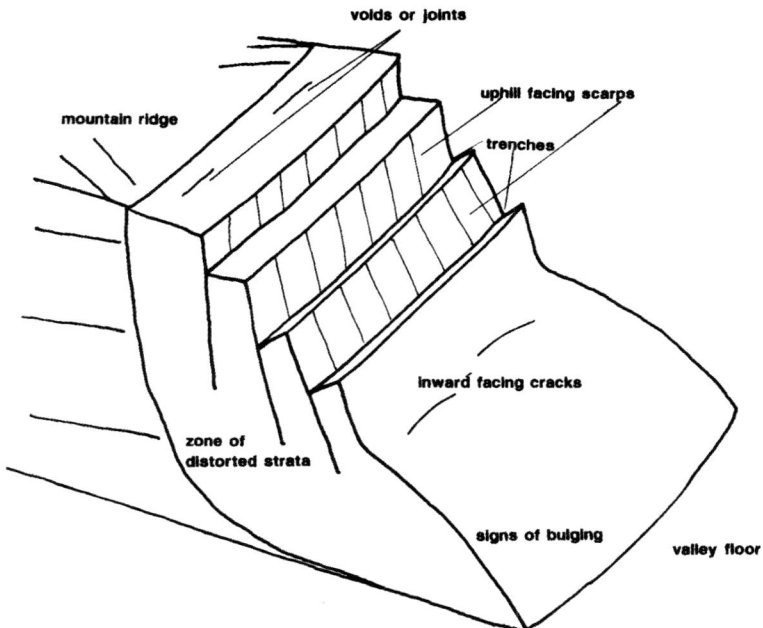

Figure 19.6   Block model of a rock flow (sackung) (from Bisci *et al.*, 1996).

1970 Huascaràn disaster, for example, destroyed the town of Yungay, Peru, killing between 15 000–20 000 people. An offshore earthquake (M = 7.7) triggered a massive rock and snow avalanche from the overhanging face of the mountain peak. The resulting turbulent flow of mud and boulders (estimated at 50–100 × 10⁶ m³) passed down the Rio Shacsha and Santa valleys as a 30 m high wave travelling at an average speed of 270–360 km/hour in the upper 9 km of its path. The landslide run-out buried the towns of Yungay and Ranrahirca beneath 10 m of debris (Plafker and Ericksen, 1978).

The Bairaman landslide, Papua New Guinea, in May 1985 in the Nakanai Mountains of New Britain was 200 m thick and covered 1 km², with an estimated volume of 180 × 10⁶ m³. The slide probably began as a down-dip failure of saturated weak biospartite. The mass then broke down into a debris avalanche flow that travelled distances of 2 km upstream and 1.5 km downstream, damming the Bairaman River (King *et al.*, 1999).

The 1987 Mount Zandila rock avalanche in the Valtellina of the central Italian Alps destroyed the village of Morignone (Govi, 1999). The front of the rock mass continued up the opposite side of the valley, where the runup was 300 m above the valley floor. Part of the avalanche fell into a small lake and caused a 95 m high wave of muddy water and debris to travel upstream, flattening the villages of Poz, San Antonio and Tirindre. Twenty-seven people were killed when the mud wave hit Aqmlone 2.1 km upstream. The main avalanche continued downstream, travelling 1.4 km and forming a landslide dam across the valley.

An earthquake in 1987 triggered landslides in the north-eastern Ecuador Andes which resulted in the destruction or local severance of nearly 70 km of the trans-Ecuadorian oil pipeline and the only highway from Quito to the eastern rain forests and oil fields (Nieto and Schuster, 1999). Economic losses were estimated at $1B, with between 1000–2000 deaths caused by the landslides and associated flooding.

In the 1991 rock avalanche off the top of Mount Cook in the Southern Alps of New Zealand (McSaveney *et al.*, 1999), an estimated 14 × 10⁶ m³ of rock cascaded downslope, with an

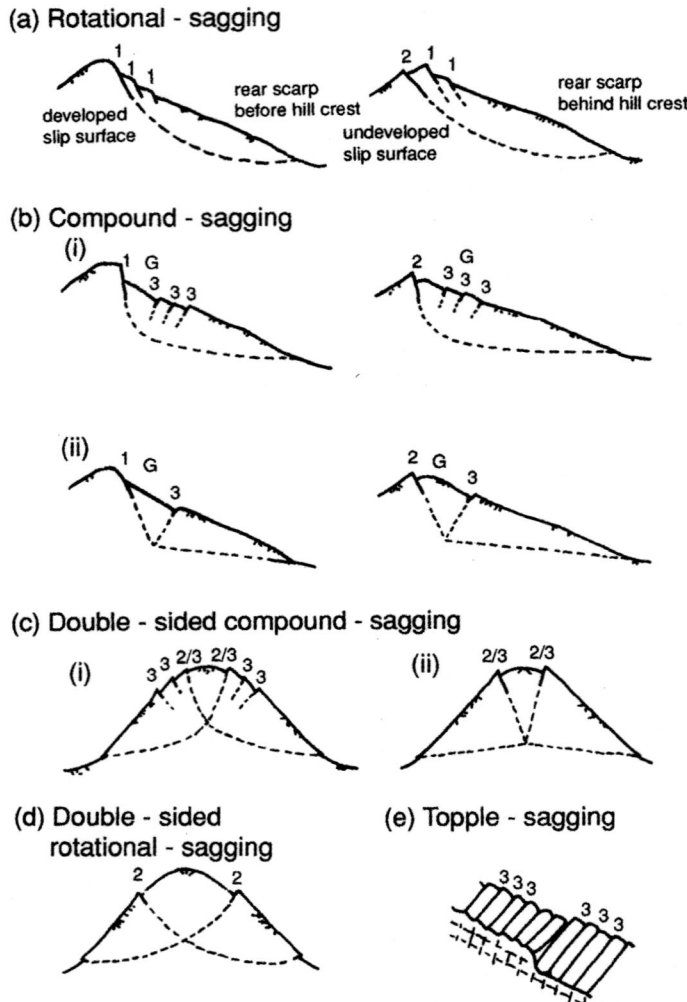

Figure 19.7   The main types of rock flow (from Hutchinson, 1988). Notes 1 = normal, down movement scarps; 2 = upslope, down movement scarps; 3 = upslope, up movement scarps; G = graben.

initial fall of 1500 m and an average speed of 300 km/hour. Part of the avalanche overtopped an adjacent ridge, the remainder was deflected down the Tasman Glacier. The runup at the far side of the glacier was 70 m.

Large run-out landslides are a regular hazard in Japan. For example, the 1995 Nikawa landslide, in the Kobe area destroyed eleven houses and buried thirty-four people. The landslide was triggered by the Hyogoken–Nanbu earthquake — 7.2 Richter magnitude (Sassa *et al.*, 1999). The

1997 rapid Harihara debris slide-flow in Izumi City killed twenty-one people and destroyed nineteen homes and eleven non-residential buildings. The slide was triggered by heavy rainfall during the rainy season and involved $1.3 \times 10^5 \, \text{m}^3$ of debris (Sassa *et al.*, 1998).

The threat posed by such events has led to considerable research being undertaken to explain the speed of movement. A failure mechanism was proposed by Sassa (1996) by which 'sliding-surface liquefaction' occurs when a shear surface develops in sandy soils and the grains are crushed or comminuted in the shear zone. The resulting volume reduction causes excess pore pressure generation which continues until the effective stress becomes small enough that no further grain crushing occurs. This mechanism is common in earthquake-triggered debris slides and flow. The mechanism can also take place in rain-induced landslides if the amount of grain crushing in the shear zone is high, as in andesitic debris.

Large runout, rapid landslides generally present a very low probability, high consequence threat to mountain societies. Landslide hazard assessment and avoidance of critical settings are likely to prove the most effective means of managing the risk, along with monitoring of vulnerable slopes linked to early warning and evacuation procedures (Chapter 8). In Japan, Sassa (1996) has suggested that those slopes that are prone to sliding surface liquefaction are potential sites of large runout landslides. The key slope conditions are the presence of a saturated layer within the slope and soils susceptible to sliding surface liquefaction, such as weathered granitic soils, volcanic deposits and sandy river terrace deposits.

19.6 Debris flows and torrents

Debris flow (debris torrents) comprise a slurry of fine (sand, silt and clay) and coarse material (cobbles and boulders) mixed with varying quantities of water — natural debris flows may have viscosities of over 1000 poise and densities of 2–2.4 g/cm³, whereas water has a viscosity of 0.01 poise and a density of around 1.0 g/cm³ (Costa, 1984). The mixture moves downslope, often along pre-existing drainage paths, in surges induced by gravity and channel bank collapses. They are a common and highly destructive feature in mountain environments, where earthquakes, heavy rainfall or snowmelt can mobilise surface debris and the thin soil cover and incorporate it into a flow (Costa and Wieczorek, 1987). Observed velocities are in the range 0.5–20 m/s.

There is much damage associated with debris flow activity for example in the 1988 Morin debris

flow disaster in the mountainous region of Serra do Mar, south-eastern Brazil (Ogura and Filho, 1999), heavy rainfall triggered a rockslide on the slopes above Petropolis City, which slid from approximately 500 m above street level. The slide moved downslope as a debris flow, mobilising further colluvial material. The flow destroyed or damaged housing at the slope foot and killed six people.

In 1988 an earthquake (magnitude 5.5) southwest of Dushanbe, Tajikistan triggered a series of landslides in loess country (Chapter 25). The slides turned into a massive mudflow (20×10^6 m³) which travelled around 2 km across an almost flat plain (Ishihara, 1999). It buried more than 100 houses in Gissal Village under 5 m of debris; 270 villagers were killed or reported missing.

In 1993 heavy rainfall in Kagoshima Prefecture, Japan caused widespread debris flows in the Ryugamizu District (Sasahara and Tsunaki, 1999). Soil from shallow hillside failures transformed into channelled debris flows that seriously damaged a hospital, houses, roads and railways. The flows killed 20 people; 2500 were cut off from their homes.

In 1994 an earthquake (magnitude 6.4) in the Paez Valley, south-west Colombia triggered landslides and debris flows over an area of around 250 km² (Martinez et al., 1999). Some debris flows travelled some 120 km downstream, destroying villages, six bridges and 100 km of road before being dumped into a reservoir. The maximum debris flow reached 10–40 m above the pre-earthquake river level, with a velocity of 8–12 m/s.

Table 19.5 summarises the main protection measures that can be use to mitigate debris flows hazards (Hungr et al., 1987); all are reliant upon a thorough investigation of the problems (Chapter 8). So-called 'passive' measures either involve avoiding the risk areas or minimising the impact of debris flows through the use of early warning systems and evoking evacuation procedures. 'Active' measures can be directed towards the different parts of a debris flow catchment: the source areas, transport zones and deposition zones.

Source areas: check dams can be installed along debris flow channel sections to prevent scour and remobilisation of the bed and bank materials.

Channel linings (e.g. grouted riprap) can also be used to stabilise the channel bed (Figure 19.8).

Transport zone: as the blockage of a debris flow channel can lead to avulsion (uncontrolled channel branching) or flooding, measures can be used to improve the ability of the channel to pass the debris surges downstream. Options include debris chutes (Figure 19.9), channel diversions or bridges designed to allow the free passage of flows (e.g. Figure 19.10). Sacrificial bridges can be used that can pass minor surges but will fail in large flows: the bridge can be anchored at one abutment with a cable to ensure that it does not join the flow, adding to its destructive power.

Deposition zone: debris sheds or galleries can be used to protect transportation routes where they cross debris flow paths. Open basins bounded by dykes can be used to divert and store debris, preventing the accumulation in high-risk areas (Figure 19.11). Closed debris barriers and storage basins (Figure 19.12) can be constructed to store design flows while passing water discharges; spillways can act as debris overflows in an emergency.

19.7 Snow melt and flash floods

Mountain streams and rivers are strongly influenced by the seasonal impact of snowmelt and glacier melt. When combined with steep, bare rocky and boulder strewn catchments, these processes tend to give rise to flash floods and compressed annual hydrographs. The discharge characteristics of the Upper Hunza River in the Karakorum Mountains illustrate this point (Figure 19.13). Between November and February, the river is fed by groundwater flows. As snowmelt

Table 19.5 Debris flow protection measures (from Hungr *et al.*, 1987).

Measure	Purpose
Passive measures	
Hazard mapping and zoning	Restrict use of endangered areas
Warning systems: advance, during event or post-event	Facilitate evacuation at times of danger
Active measures: source area	
Reforestation/controlled harvest	Reduce debris production due to logging or natural loss of forest cover
Forest road construction	Eliminate unstable cuts and fills that could be debris sources
Stabilisation of debris sources (channel linings or check dams)	Stabilise channel bed and side slopes in source reaches
Active measures: transport zone	
Training by chutes, channels, deflecting walls or dykes	Ensure passage of debris surges, without blockage or overflowing
Channel diversion	Change path of flow away from endangered area
Bridges designed for passage	Protect traffic and prevent blockage due to bridge obstruction
'Sacrificial' bridges	Prevent channel blockage due to bridge obstruction
Bypass tunnels beneath creek bed	Protect transport route without modifying stream channel
Active measures: deposition zone	
Open debris basins, dykes and walls	Control extent of natural deposition area by shaping and dyking
Closed retention basins and barriers — full or partial volume	Create a controlled deposition area fronted by a barrier and spillway
Bridges or other structures designed for burial	Prevent damage during burial by a flow
Debris sheds, galleries or cut and cover tunnels	Place transport route beneath deposition area

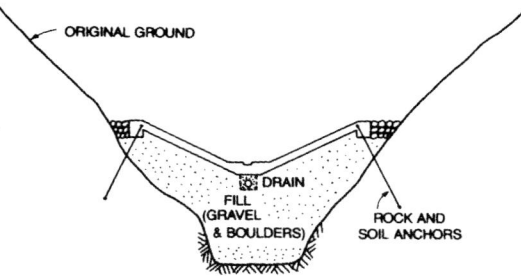

Figure 19.8 Channel stabilisation measures, Alberta Creek, British Columbia (from Hungr *et al.*, 1987).

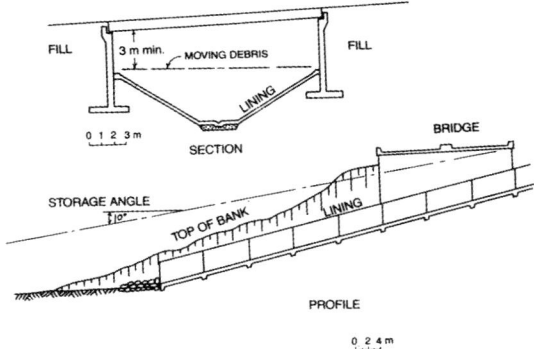

Figure 19.10 Bridge on Coquihalla Highway, British Columbia, designed to allow debris flow passage (from Hungr *et al.*, 1987).

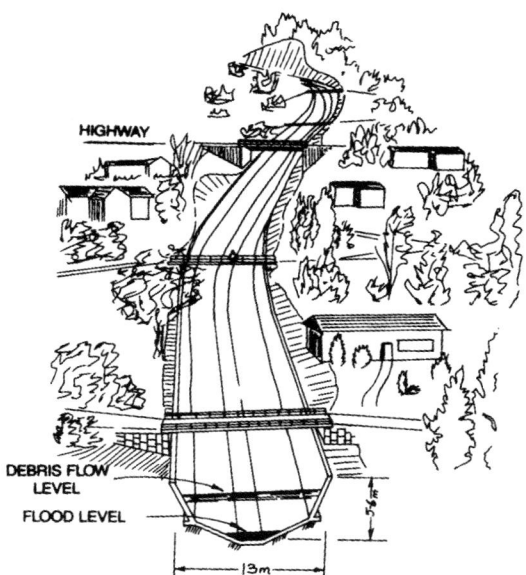

Figure 19.9 Debris chute on Alberta Creek, British Columbia (from Hungr *et al.*, 1987).

begins in the lower parts of the basin in March, so the discharge increases. Between 40–70% of the total annual discharge occurs in July and August, where the flows can be up to forty times those earlier in the year.

Marked diurnal fluctuations in discharge are common in mountain streams, reflecting both day–night variations in snowmelt and interruptions in melting conditions during periods of cloud cover or during snowfalls (Figure 19.14). In Miller Creek,

British Columbia, the maximum discharge occurs around 19:00–20:00 hours, with a minimum around 09:00–10:00 hours (Slaymaker, 1974). There can also be pronounced year-to-year fluctuations in discharge as a result of annual variations in snow cover and glacier mass balance.

Rainfall-triggered flash floods are common, especially during the latter part of the snowmelt season. Flash floods are characterised by the rapid rise and fall of the floodwaters, with peak flows often occurring within hours of the onset of heavy rain. The combination of steep terrain, very high runoff, small catchments and episodic heavy rainfall creates the unique flash flood problems of upland areas. The rapid rise of floodwaters is a reflection of the speed at which water is delivered to the stream channel because of the very high runoff rates and the fact that water arrives at the channel at nearly the same time throughout the small catchment (i.e. a short time of concentration: Chapter 11). The absence of significant floodplain areas adjacent to many mountain streams tends to intensify the flood conditions as floodwaters are not 'stored' in relatively slow-flowing backwaters.

The Highlands of Scotland are noted for frequent flash flooding (Table 19.6); in one event, in 1829, the River Findhorn rose 15 m above its normal level during the course of a storm (Nairne, 1895). In many cases the impact of flash floods may extend beyond the inundation of communities

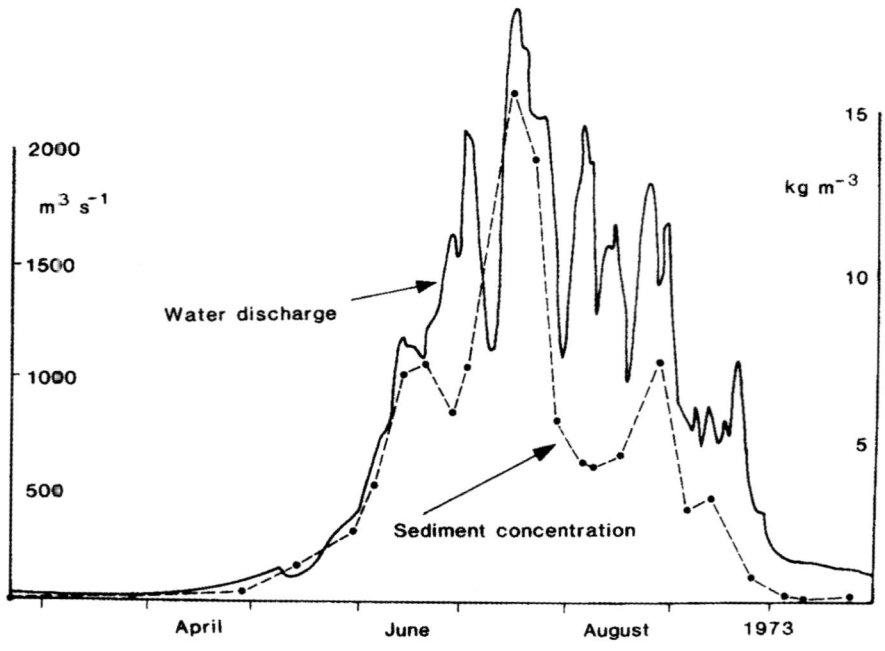

Figure 19.13 Discharge characteristics of the Upper Hunza River, Karakorum Mountains (from Ferguson, 1984).

form high dams because they tend to be high-velocity events that can cause complete stream or river blockage. Tables 19.7 and 19.8 provide an indication of the scale of some recent landslide dams and a simple classification, respectively.

Landslide dams give rise to two important flood hazards: upstream or backwater flooding, and downstream flooding.

Upstream or backwater flooding occurs as a result of the relatively slow impounding of water behind the dam. In 1983, the small town of Thistle, Utah was inundated by a 200 m long, 50–60 m deep lake that formed behind a landslide dam on the Spanish Fork River (Figure 19.15; Kaliser and Fleming, 1986). A lake that had formed behind a rock avalanche dam in 1513 led to the droning of the Swiss hamlets of Malvaglia and Semione (Eisbacher and Clague, 1984).

Downstream flooding occurs in response to the failure of the landslide dam. The most frequent failure modes are overtopping because of the lack of a natural spillway, or breaching due to erosion. Failure of the poorly consolidated landslide debris

generally occurs within a year of dam formation (Figure 19.16). The effects of the resultant flooding can be devastating, as indicated by the following examples.

In the winter of 1840–41 a spur of the Nanga Parbat Massif failed during an earthquake, completely blocking the River Indus and causing the impoundment of a 60–65 km long lake. The dam, which had been up to 200 m high, breached in early June 1841. The lake emptied in 24 hours causing what has been called 'the Great Indus flood' during which hundred of villages and towns were swept away. A Sikh army encamped close to the river about 420 km downstream was overwhelmed by a flood of mud and water estimated to be 25 m high; about 500 soldiers were killed (Mason, 1929).

Failure of the Deixi landslide dam on the Min River, China in 1933 resulted in a wall of water that was 60 m high 3 km downstream (Figure 19.17). The floodwaters had an average velocity of 30 km/hour and reached the town of Maowen (58 km downstream) in 2 hours. The total length

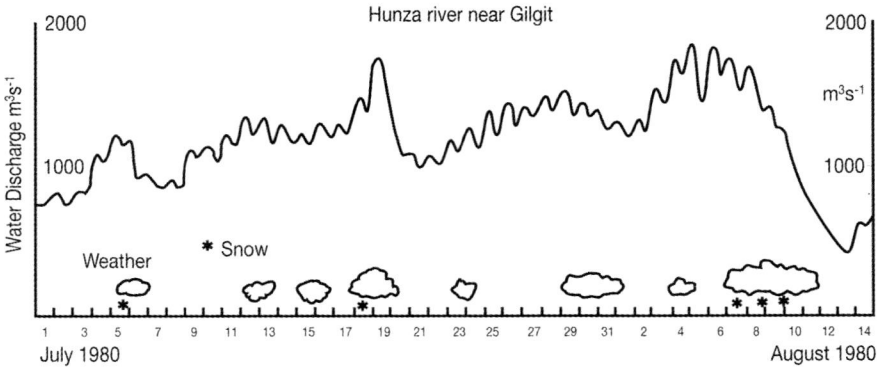

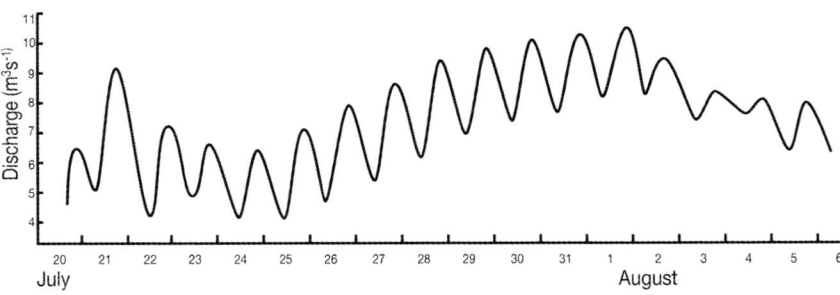

Figure 19.14   Daily variations in discharge of glacial fed rivers (redrawn from Ferguson, 1984; Slaymaker, 1974).

of the flood was 253 km and at least 2423 people were killed (Li *et al.*, 1986).

Between thirteen and forty people were killed in the Canadian Cordillera in 1891, when a landslide dam burst engulfing an Indian village near the North Pacific Cannery on Inverness Passage (Evans, 1986).

As indicated in Figure 19.18, the peak discharge associated with the failure of landslide dams is lower than the equivalent man-made structures. This is because full breach tends to develop slower in landslide debris and, hence, the discharge is spread over a longer time period.

There is an urgent need to undertake a rapid risk assessment once a landslide dam has formed, because of their temporary nature and the potentially devastating consequences of downstream flooding. Engineering options generally involve controlling the dam discharge and thus preventing overtopping, either through the construction of a spillway or draining the lake waters through pipes or tunnels. Although many dams have been successfully controlled, in some cases overtopping has occurred before stabilisation measures were fully implemented.

19.9 Glacial lake outburst floods

Proglacial lakes impounded behind moraines or an ice dam can present a major flood risk to downstream communities. Failure of the moraine or ice dam can lead to a catastrophic outburst flood (*aluvión*, *débâcles* or glacier lake outburst floods) with the discharge reaching up to several thousand cubic metres per second (Reynolds, 1998), often accompanied by large volumes of mobile sediment. Perhaps the greatest impacts have occurred in the Cordillera

Table 19.6 Examples of flash flood events in the Scottish Highlands (from Nairne, 1895 and later sources).

Date	Location	Comment
3–4 August 1829	Moray district	Around 90 mm of rain; the R. Findhorn rose 15 m above its normal level causing immense damage; bridges swept away, crops and farms destroyed or ruined by deposited gravels. Numerous families left destitute and damage estimated at £20 000 (1829 prices). Severe floods on the Nairn and Spey; 'great landslips' occurred, farms swept away. In Spey valley damages estimated at over £37 000, plus countless livestock and several lives lost.
27 August 1829	Inverness district	Considerable flood damage in Inverness; crops flattened, numerous bridges lost, mills and homes damaged. Estimated as several thousand pounds damage.
24–26 January 1849	Inverness district	The 'Inverness Flood': most disastrous flood in the NW Highlands. Bridges lost at Aberchalder and Forst Augustus, Caledonian Canal breached. In Inverness the stone bridge was lost and a third of the town flooded by the combined waters of the Ness and the canal. Immense damage but no loss of life.
30 January – 1 February 1868	Inverness district	Farms damaged, crops lost, bridges swept away throughout district. Inverness flooded with extensive damage to property.
29 January 1892	Strathglass, Strathspey	Great, extensive flood following unpredicted snow falls for ten days. Damage extensive especially in Strathglass, Bonar-Bridge and Strathspey, but no fatalities. Railways washed away on Skye.
25 May 1953	Lochaber, Appin and Benderloch	Road bridges destroyed, disruption to road and rail traffic; extensive damage to forestry property through floods and landslips. In Argyllshire damage to roads estimated at £130 000.
30 July 1956	Cairngorm and Moray	Flooded houses and bridges damaged throughout region, especially around Forres. Main railway line from Inverness washed away. Livestock swept away. Extensive erosion and deposition of gravels on agricultural land. 72 hr maximum rainfall of 250 mm.
17–18 August 1970	Moray	72 hr maximum rainfall of 150 mm. Extensive damage to roads and bridges, agricultural land flooded and covered by gravels.

Blanca of Peru, where there are some 722 glaciers covering an area of 723 km^2 (Lliboutry *et al.*, 1977). An outburst flood in 1725 killed between 1500 and 2000 people in the village of Ancash. In 1941 an outburst from Laguna de Palcacocha destroyed part of the city of Huaraz and killed 6000. In 1945 the ancient town of Chavin de Huantar was destroyed by an outburst from the Quebrada Huachescsa. Similar floods have been widely reported in the Himalayas (Table 19.9; Reynolds, 1998), the Canadian Cordillera (Clague and Evans, 1994; an outburst in 1978 damaged the Canadian Pacific Railroad and the Trans-Canadian Highway and derailed a freight train; Jackson, 1979), Washington State and Alaska in the USA (O'Conner and Costa, 1993; Stone, 1963) and the Alps (e.g. Tufnell, 1984), where the Vernagt glacier has caused serious *débâcles* in 1600, 1678, 1680, 1773, 1845, 1847 and 1848.

Failure can result from a surge of the glacier or an ice avalanche into the lake, causing waves that either overtop or erode the dam. Moraine dams can be affected by piping of the heterogeneous materials, weakening the feature until it fails (e.g. during an earthquake). If the moraine has a core of stagnant ice, its melting can result in the gradual lowering of the crest height until the lake can

Table 19.7 A selection of historic landslide dams (from Schuster, 1986 and Sassa, 1999).

Landslide	Year	Dammed river	Landslide volume (m³)	Dam height (m)	Dam width (m)	Lake length (km)	Lake volume (m³)	Dam failure
Slumgullion Earth Flow, USA	1200–1300	Lake Fork, Gunnison River	50–100 × 10⁶	40	1700	3		No
Usay landslide, Tadzhikistan Partial	1911	Murgab River	2–2.5 × 10⁶	300–550	1000	53		
Lower Gros Ventre landslide, USA	1925	Gros Ventre River	38 × 10⁶	70	2400	6.5	80 × 10⁶	Yes
Deixi landslide, China	1933	Min River	150 × 10⁶	255	1300	17	400 × 10⁶	Yes
Tsao-Ling rockslide, Taiwan	1941–42	Chin-Shui-Chi River	250 × 10⁶	217	2000		157 × 10⁶	Yes
Cerro Codor Sencca rockslide, Peru	1945	Mantaro River	5.5 × 10⁶	100	580	21	300 × 10⁶	Yes
Madison Canyon rockslide, USA	1959	Madison River	21 × 10⁶	60–70	1600	10		No
Tanggudong debris slide, China	1967	Yalong River	68 × 10⁶	175	3000	53	680 × 10⁶	Yes
Mayunmarca rock slide, Peru	1974	Mantaro River	1.6 × 10⁹	170	3800	31	670 × 10⁶	Yes
Gupis debris flow, Pakistan	1980	Ghizar River		30	300	5		No
Polallie Creek debris flow, USA	1980	East Fork Hood River	70–100 × 10³	11	230		105 × 10³	Yes
Thistle earth slide, USA	1983	Spanish Fork River	22 × 10⁶	60	600	5	78 × 10⁶	No
Pisque River landslide, Partial Ecuador	1990	Pisque River	3.6 × 10⁶	56	60	2.6	3.6 × 10⁶	
Tunawaea landslide, New Zealand	1991	Tunawaea Stream	4 × 10⁶	70	80	0.9	9 × 10⁶	Yes
Rio Torro landslide, Costa Rica	1992	Rio Torro	3 ×10⁶	100	75	1.2	0.5 × 10⁶	Yes
La Josefina rockslide, Partial Ecuador	1993	Paute River	20–44 × 10⁶	100+	500	10	177 × 10⁶	

Table 19.8   Classification of landslide dams (after Swanson *et al.*, 1986).

Speed of movement	Landslide type	Comment
Slow landslides	A1 Basal shear surface emerges at ground surface upslope of channel A2 Basal shear surface emerges in channel	Slides does not impinge on channel, but delivery of colluvium may occur by secondary failures. A common setting for landslide dams. Potential for damming can be represented by: $$\frac{\text{Annual}}{\text{constriction ratio}} = \frac{\text{Speed of landslide toe}}{\text{Channel width}}$$ Ratios in excess of 100 appear to be required for development of lakes.
	A3 Basal shear zone emerges on far side of channel	Landslide movement can result in an upstream back-water effect from upward heave of the streambed. Limited storage capacity behind these dams.
Rapid landslides	B1 Landslides small in relation to width of valley floor B2 Landslide deposits spanning the valley floor B3 Landslide deposits covering valley floor over valley length much greater than the width	Typically cause only minor damming, although channel can be diverted. Typically produces a single impoundment which may flood broad areas upstream. Deposits may dam both main channel and tributaries, tributaries, forming multiple lakes. Volumes of impounded water may be very large where dams are high.

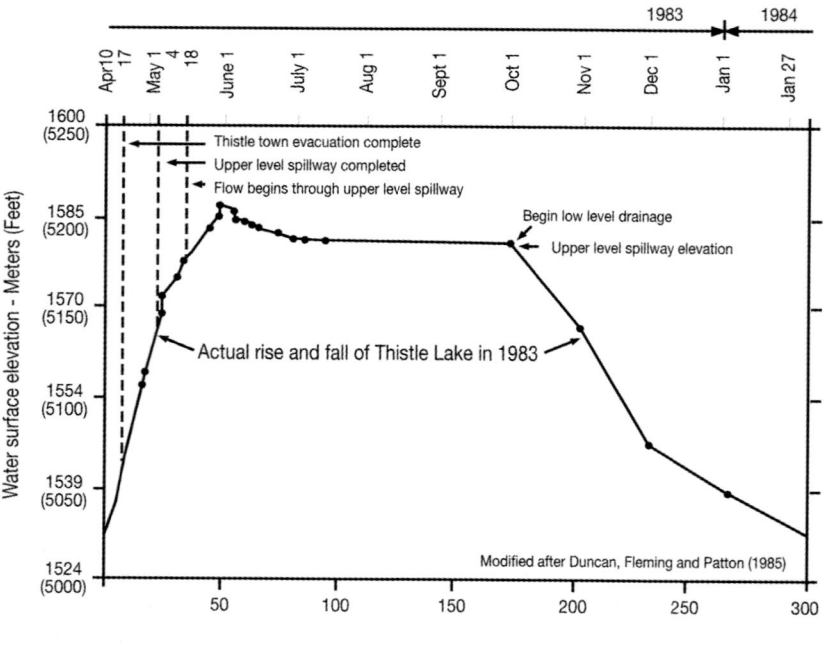

Figure 19.15   Key events in the rise and fall of the Thistle landslide dam lake, 1983 (redrawn from Kaliser and Fleming, 1986).

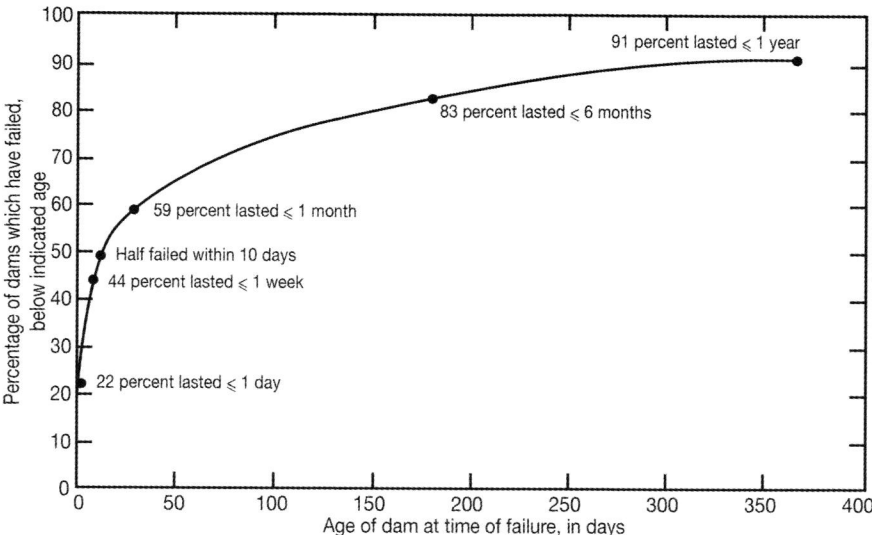

Figure 19.16   Survival time of landslide dams (redrawn from Schuster and Costa, 1986).

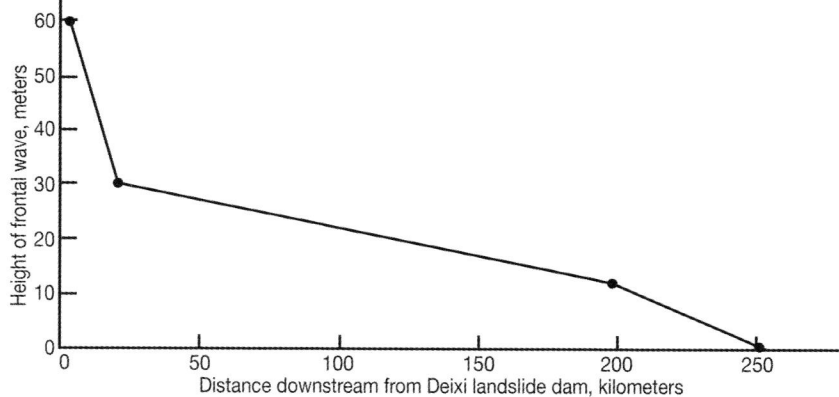

Figure 19.17   Downstream heights of the frontal wave of the 1933 Min River flood caused by the breaching of the Deixi landslide dam, China (redrawn from Li *et al.*, 1986).

overtop the dam. Climate change and glacier fluctuations are key controls on the occurrence of outburst floods and it is worth stressing that Reynolds (1998) notes that the number and size of high altitude lakes in both the Andes and Himalayas has increased in recent decades because of climate change.

According to Reynolds (1992; 1998) and Reynolds *et al.* (1998), mitigation of the flood hazard can involve: the excavation of open cuts in the moraine dam to lower the lake levels; the construction of siphons, as used to drain the Tsho Rolpha lake, Nepal in 1994; tunnelling through bedrock beneath or beside a moraine dam into the lake to drain it (Figure 19.19); and the construction or restitution of the natural dam by installing a culvert or spillway across the moraine. The need for ongoing monitoring of the whole glacier/lake system is essential for successful hazard mitigation.

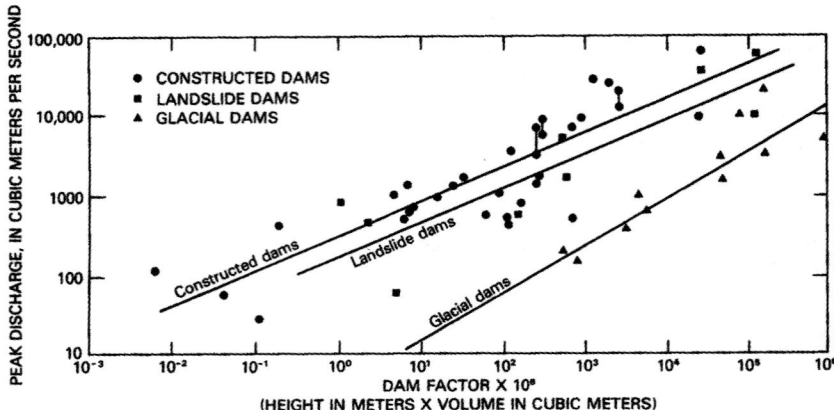

Figure 19.18   Flood peak discharge versus dam factor (height × volume × 10^6) for constructed, landslide and glacial dams (from Schuster and Costa, 1986).

Table 19.9   Recent glacier lake outburst floods in Nepal (from Reynolds, 1998).

Year	Location	River system	Principal damage
1964	Gelhaipco Lake	Arun/Pumqu	End moraine collapsed due to ice avalanche into the lake; road damaged and 12 trucks lost
1964		Arun/Pumqu	Damage to forest, bridges and trucks
1964	Longda Glacier	Trisuli River	No data
1964	Zhangzangbo Glacier	Sun Kosi/Poiqu	Moraine collapsed due to seepage
1968	Aycio Lake	Arun/Pumqu	Damage to roads and bridges
1969	Aycio Lake	Arun/Pumqu	Damage to roads and bridges
1970	Aycio Lake	Arun/Pumqu	Damage to roads and bridges
1977	Nare Glacier Lake	Dudh Kosi	Ice-cored moraine collapsed. Damage to mini-HEP station, road, bridges, fields etc.
1980	Phuchan Glacier Lake	Sapta Kosi	Damage to forest and river bed etc.
1981	Zhangzangbo Glacier	Sun Kosi/Poiqu	Moraine collapse due to glacier front calving. Damage to Arniko Highway, bridges, Sun Koshi HEP station, fields, killed livestock and caused casualties
1982	Jinco Lake	Arun/Pumqu	Moraine collapsed due to glacier tongue sliding into lake. Damage to eight villages, livestock killed, fields roads and bridges damaged
1985	Dig Tsho	Dudh Kosi	Moraine collapse after rock avalanche. Destroyed Namche HEP plant, damaged roads, bridges, fields, houses and caused casualties

19.10 Snow avalanches

The rapid movement of detached masses of snow presents a serious hazard in many mountain regions, often causing loss of life to both local inhabitants and tourists. Three main types of avalanche motion are recognised:

1. Powder avalanches, in which an aerosol of fine snow is carried in the air as a snow cloud. At the front of the avalanche the snow may reach 20–70 m/s and victims can die by inhaling the snow particles.
2. Dry-flowing avalanches, involving dry snow travelling over steep irregular terrain with

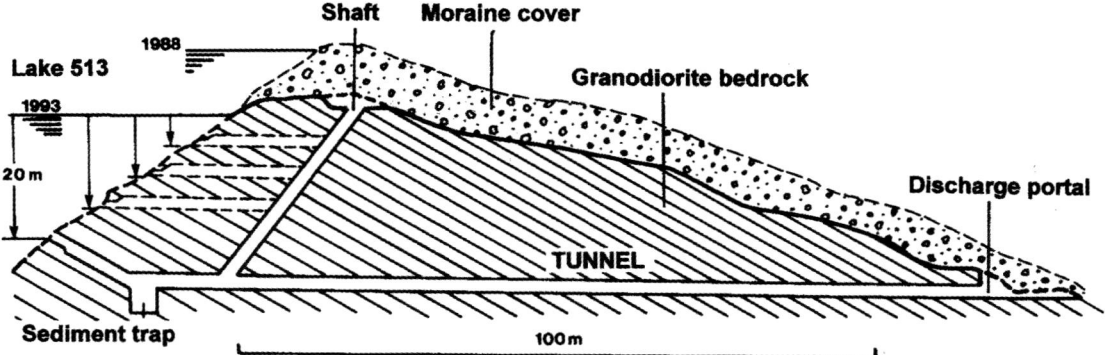

Figure 19.19   The tunnel design for draining the glacial lake at Hualcán, Cordillera Blanca, Peru (from Reynolds *et al.*, 1998).

particles ranging from powder grains to snow blocks. These events tend to follow well defined channels such as gullys. Typical speeds range from 15–60 m/s up to 120 m/s while travelling through the air.

3. Wet flowing avalanches, composed of wet snow or a mass of sludge. These features tend to flow in stream channels. Wet avalanches tend to be relatively slow moving (5–30 m/s) and can have a high density (300–400 kg/m³, compared with 50–150 kg/m³ for dry flows). As they support large volumes of debris and boulders they can cause considerable erosion along the avalanche track.

Avalanche impacts can exert high external loadings on structures in the path (typically in the range 5–50 t/m², but up to 100 t/m²) and, hence, can cause considerable damage (Table 19.10). In addition to the direct impact, avalanches may also exert both upwards and downwards forces which have lifted vehicles and buildings.

Although the nature and conditions vary between avalanches (see Figures 19.20 and 19.21 for avalanche classifications), all are associated with the weight of the snow exceeding the frictional resistance of the underlying surface. Snow layers can sustain large density changes; a layer deposited with an original density of 100 kg/m³ may densify to 400 kg/m³ during the course of a winter largely due to the weight of overlying snow. Shear strength, however, decreases, as the temperature rises towards 0 °C. As temperature

Table 19.10   Impact pressures and potential damage from snow avalanches (from Perla and Martinelli, 1976).

Impact pressure (t/m²)	Potential damage
0.1	Break windows
0.5	Push in doors
3.0	Destroy wood-frame houses
10.0	Uproot mature trees
100.0	Move reinforced concrete structures

rises further and meltwater appears within the snow mass, so the risk of failure increases.

Avalanches tend to be triggered by heavy snowfall, rain, thaw or an increase in dynamic loading such as skiers moving across the slope. Failure tends to start at a fracture point where there is high tensile stress, as at a break of slope or an overhang, and involves either: loose snow avalanche, which develops in cohesionless snow where the intergranular bonding is very weak giving rise to behaviour similar to dry sand, or slab avalanche, where a cohesive snow layer breaks away from a weaker underlying layer (the 'lubricating layer'). The initial slab may be up to 10 000 m² and involve a 10 m thick slab. When a slab breaks loose it may mobilise additional snow, with the avalanche increasing in volume by as much as 100 times.

Slope angle is a key factor in avalanche susceptibility. Most events occur on slopes between 30–45°; slopes less than 20° tend to be too slack for failure to occur, whereas slopes above

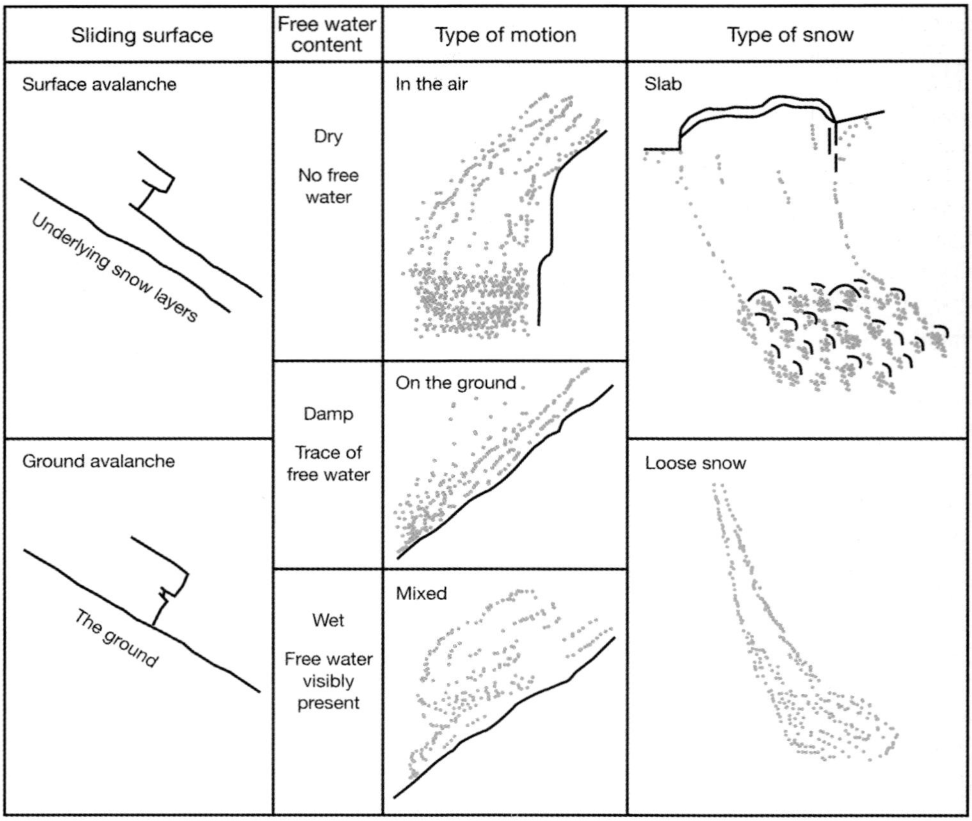

Sliding surface
Free water content
Type of motion
Type of snow
Surface avalanche
Underlying snow layers
Dry
No free water
In the air
Slab
Damp
Trace of free water
On the ground
Ground avalanche
The ground
Loose snow
Wet
Free water visibly present
Mixed

Figure 19.20 Classification of avalanches (redrawn from Embleton and Thornes, 1979).

From single point
From large area leaving wall
Loose-snow avalanche
Slab avalanche
Whole snow cover involved
Some top strata only involved
Full depth avalanche
Surface avalanche
Dry
Wet
Dry-snow avalanche
Wet-snow avalanche
Open slope
In gully
Unconfined avalanche
Channelled avalanche
Through air
Along ground
Airborne-powder avalanche
Flowing avalanche

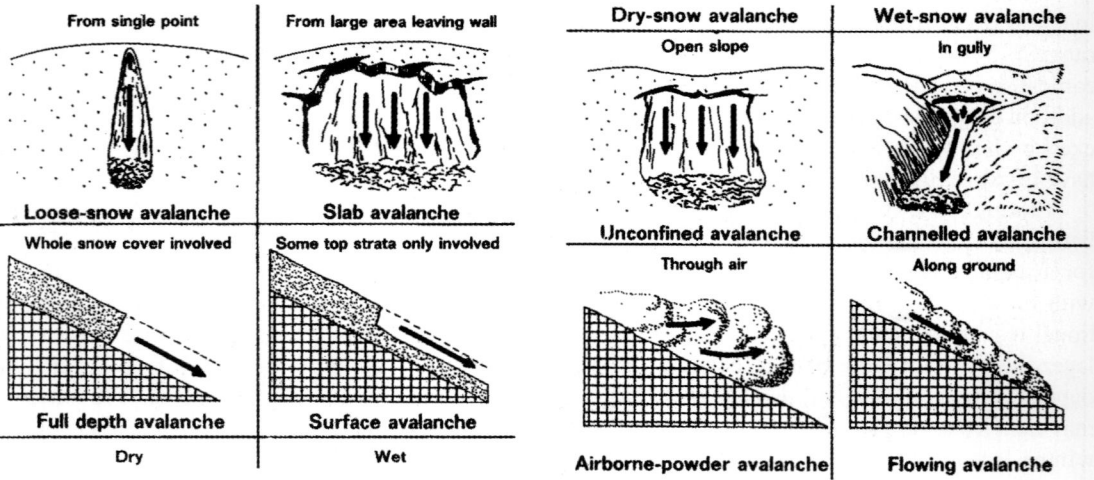

Figure 19.21 Avalanche classification (from Verstappen, 1983).

50° rarely accumulate sufficient snow to present a hazard. However, dry snow can avalanche on slopes as low as 25° and wet snow on slopes as low as 10° (Mellor, 1978). As avalanches tend to recur at the same sites, the potential threat can often be readily defined by the mapping of past avalanche tracks and runout zones (e.g. from breaks of slope, eroded channels, damaged vegetation or different aged trees in forests).

A variety of methods of avalanche mitigation are available, including the construction of defence structures (Figure 19.22; La Chapelle, 1977a; 1977b):

1. snow fences above the starting zone to trap and hold up snow that otherwise would have fallen onto the avalanche slope
2. supporting structures such as snow rakes to provide external support for the snowpack — these can stop small avalanches before they gain destructive momentum
3. deflectors (e.g. wedges pointing upslope) or retarding devices (e.g. earth mounds) built within the avalanche track or runout zone — the most successful deflectors have been those that steer the avalanche by less than 15–20°
4. direct protection structures such as snow sheds and galleries designed to allow the flow over railway line or roads (Figure 19.23) — in France buildings in avalanche-prone communities are oriented to allow the snow to pass between them (De Crecy, 1980).

Avalanches can also be managed by artificial release, through the use of explosives or weapons. This allows the unstable snowpack to be released in a series of small manageable failures rather than a single major event. Also the release can be timed to occur when the ski runs and highways are closed. For example, Parks Canada and the Canadian Armed Forces trigger avalanches in the Rogers Pass where the Canadian Pacific railway and the Trans-Canadian highway pass through the Selkirk Mountains (Smith, 1992).

19.11 Ice avalanches

Unstable ice on steep mountain slopes can lead to the initiation of *ice avalanches* through the calving and free fall of ice from a hanging glacier or the

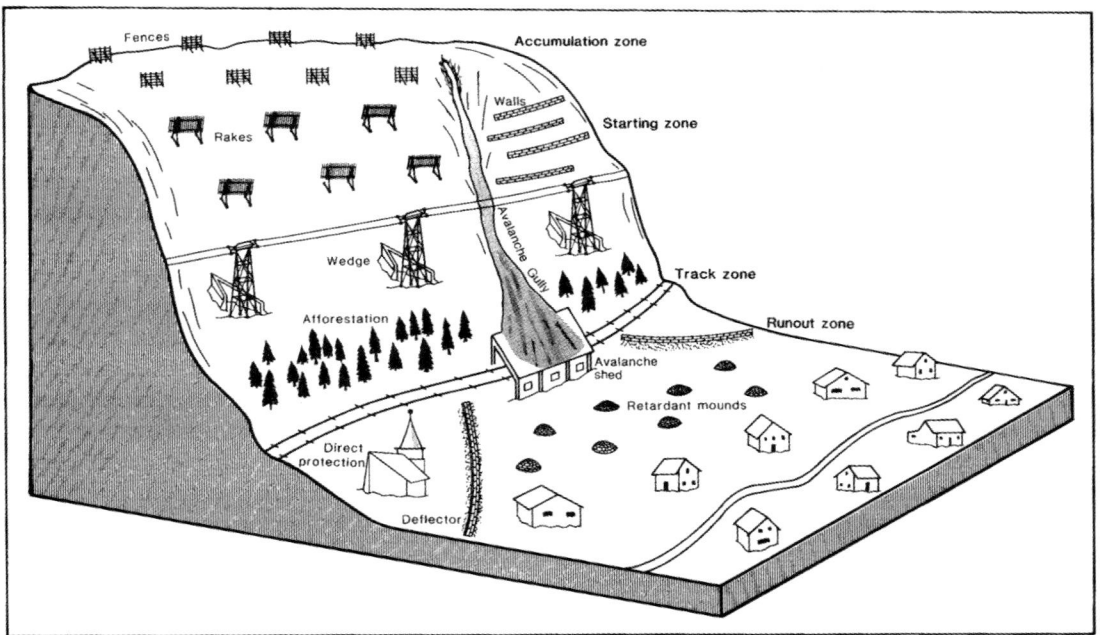

Figure 19.22   Avalanche mitigation measures (from Smith, 1992).

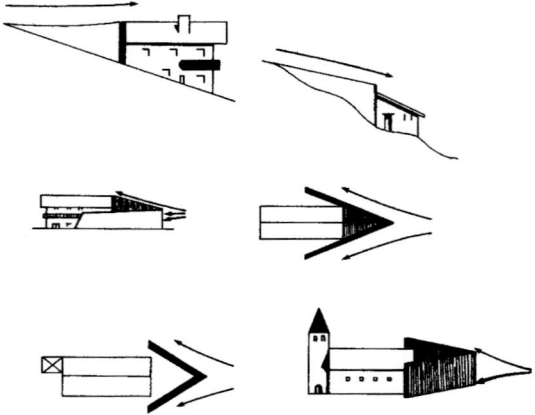

Figure 19.23   Various approaches to avalanche protection (from Verstappen, 1983).

detachment and sliding of tabular masses of ice. Often such events fail to extend beyond the immediate glacier margins and, hence, have little impact on society. However, when the avalanche travels further the results can be devastating. The 1962 Huascaràn avalanche in the Santa Valley, Peru began as a displacement of up to 3 million m³ of ice. As the ice moved downhill it was transformed into a mudflow of some 13 million m³ which attained a maximum velocity of over 100 km/hour (McDowell, 1962; Morales, 1966). The avalanche travelled 16 km from its source and descended a

vertical elevation of 4000 m. The scale of the devastation was immense: 4000 people were killed, 6 villages were destroyed and 3 partly destroyed; 600 ha of agricultural land was lost as well as many thousands of livestock; 10 flour mills and 4 bridges were destroyed (Morales, 1966).

In most instances the greatest threat is to climbers and trekkers; an avalanche off the Le Tour glacier, France killed six walkers in 1949, and in 1981 part of a glacier in the Swiss Alps swept down the mountain side killing six climbers (Tufnell, 1984).

Little can be done to control the hazard other than avoidance of avalanche-prone areas of the mountains or the monitoring of glacial flow patterns and crevasse development to forecast potential events.

19.12 Construction materials

The low and moderate grade metamorphic rocks that dominate many mountain areas present a series of problems for construction (Fookes and Marsh, 1981a; 1981b). Typically these rocks show considerable variations in strength and durability with orientation. Natural rock fragments and crushed aggregates often have flaky and elongated, shapes reflecting the metamorphic fabric (Table 19.11). The poor shape characteristics severely limits the

Table 19.11   Excavated rock characteristics from the Lower Himalayas (from Fookes and Marsh, 1981b).

Rock Type	Shape characteristics of slightly to highly weathered rock					Soundness of slightly to highly weathered rock	10% fines: slightly to highly weathered rock
	Axial Ratio			Numerical flakiness	Numerical elongation	Loss, % passing (Na$_2$SO$_4$)	Average value (kN)
	a/b	a/c	b/c				
Middle Siwalik sandstone	1.4	2.3	1.6	33	18	1.7–29.9	150
Sangure Series phyllite	1.7	4.3	2.5	69	40	0.2–22.7	87
Sangure Series metasandstones	1.6	3.0	1.9	52	23	2.0–6.9	123
Sangure Series white quartzite	1.5	2.4	1.6	35	13	1.5–3.1	188
Sangure Series purple quartzite	1.6	2.6	1.6	32	28	< 1.0	353
Lower Himalayan Unit plyllonite and schist	1.8	5.9	3.3	95	43	0.5–14.2	70
Lower Himalayan Unit gneisses	1.6	2.7	1.7	39	21	0–5.2	79

quantities of well shaped stone for masonry, gabion and rockfill; reduces the effectiveness of normal compaction procedures to produce dense, stable layers of granular material; reduces the effective area of adhesion of bitumen to aggregate in surface dressing, increasing the binder requirement, and substantially increases the water requirement in concrete mixes for a given degree of workability.

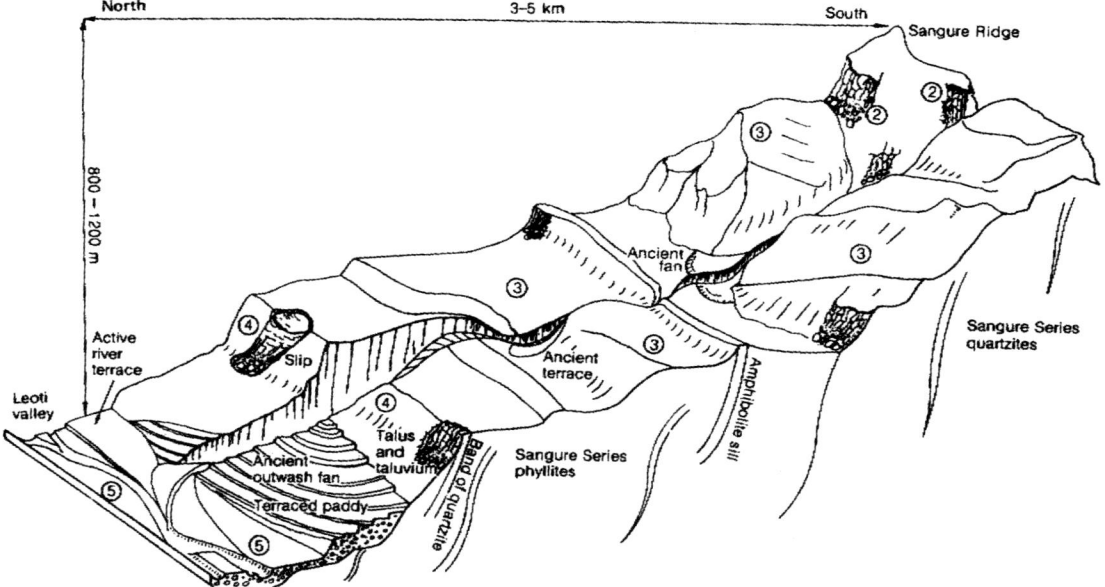

Figure 19.24   Idealised block model of the Leoti Valley to Sangure Ridge, Lower Himalayas — see Table 19.12 for the accompanying legend (from Fookes and Marsh, 1981a).

Table 19.12   Mountain zones between the Leoti Valley and Sangure Ridge, Lower Himalayas — see Figure 19.24 (from Fookes and Marsh, 1981a).

Mountain zone	5. Valley floor	4. Active lower slopes	3. Degraded middle slopes and ancient valley floors	2. Free rock faces and associated debris slopes
Topography	Low angle slope <15°	Steep undercut slopes at 25–45°	Undulating stepped slopes at moderate to low angles <35°	Steep and very steep slopes >25°
Foundation soils	Sands and gravels, silty sand paddy	Coarse taluvium and colluvium soils	Coarse and finer colluvium and residual soils with some taluvium and unweathered sandy gravel terraces and fans	Coarse talus, taluvium and some colluvium over residual soils and unweathered rock
Drainage	Good	Often poor with springs and seepage	High rate of runoff and generally poor subsoil drainage	High rate of runoff and good subsoil drainage
Available sub-base material	Sands and gravel		Sands and gravel from ancient terraces and fans	Naturally well graded quartzitic soils
Available base material	Crushed gravels and boulders, uncrushed gravels	Crushed rock	Crushed and uncrushed gravels occasionally crushed rock	Crushed rock

Table 19.13   Classification of common engineering soil types in the Lower Himalayas (from Fookes and Marsh, 1981a).

Soil type	Description	Weathering state of coarse fragments	Typical slope angles (degrees)
Residual	Material remaining *in situ* during the weathering of rock	Highly to completely weathered	
Colluvium	An accumulation of usually well graded soil from the weathering of rock which has been transported downslope mainly by gravity and slope wash processes, destroying the original rock structure; fine to coarsely graded	Moderately through to completely weathered	<35
Taluvium	Soil formed by the chemical and mechanical weathering of scree; often medium to coarsely graded or gap-graded	Moderately to completely weathered	45–25
Scree or talus	An accumulation of angular rock fragments formed by weathering and rockfall processes on exposed rock faces; often medium to coarsely graded	Slightly to moderately weathered	38–33
Gully infill	An accumulation of sub-angular debris in stable gullies and natural depressions which has been transported a short distance by water; fine to coarsely graded	Slightly to completely weathered	<25
Alluvium	An accumulation of water transported weathering products in either primary or reworked river deposits; fine to coarsely graded	Slightly to completely weathered	<15

Poor particle shape can hinder the achievement of a dense, fully compacted, fully interlocking base layer in road construction (Fookes and Marsh, 1981b). The flaky and elongated particles tend to rotate to lie parallel to the surface during rolling and greatly reduce the effectiveness of binding techniques in filling voids between the crushed stone. In a compacted, well graded, wet mix or crusher run base, flakes lying on the surface cannot be adequately bound into the compacted mass and are easily worked out by subsequent construction traffic.

The potential for quarry development is often limited because of the very steep natural slope angles which would lead to unworkable quarry faces. Local practice often involves exploiting river boulder deposits for construction stone and crushed rock. However, working these deposits can be wearing on mechanical plant. Seasonal floods will also restrict the safe period of working.

Widespread slope instability and erosion tends to produce extensive fan and terrace deposits within mountains and around their margins (Figure 19.24; Tables 19.12 and 19.13). These generally offer the best sources of mechanically stable fills, natural sub-base and bases for roads and backfill for pipelines. Grading is related to the source rocks and the position within the catchment. Boulders are common within all the transported soils and can present problems during excavation. The equipment used to exploit these sources should be capable of handling very large boulders. Screening uncrushed river gravels from coarsely graded river deposits will result in a relatively low overall yield of processed aggregates; two-stage crushing with suitable equipment is more efficient.

19.13 Summary

Mountains are one of the most dynamic environments in the world and, as a result, can present a range of major challenges to engineering projects. Many of the hazards faced in these areas are complex, with earthquakes, snowmelt or heavy rainfall being the trigger for potentially massive events involving both landslides, debris flows and flooding that could cause widespread damage. The simple classification of the terrain into a series of distinct mountain zones provides a framework for identifying and evaluating engineering hazards and assessing resources. Figure 19.25 illustrates

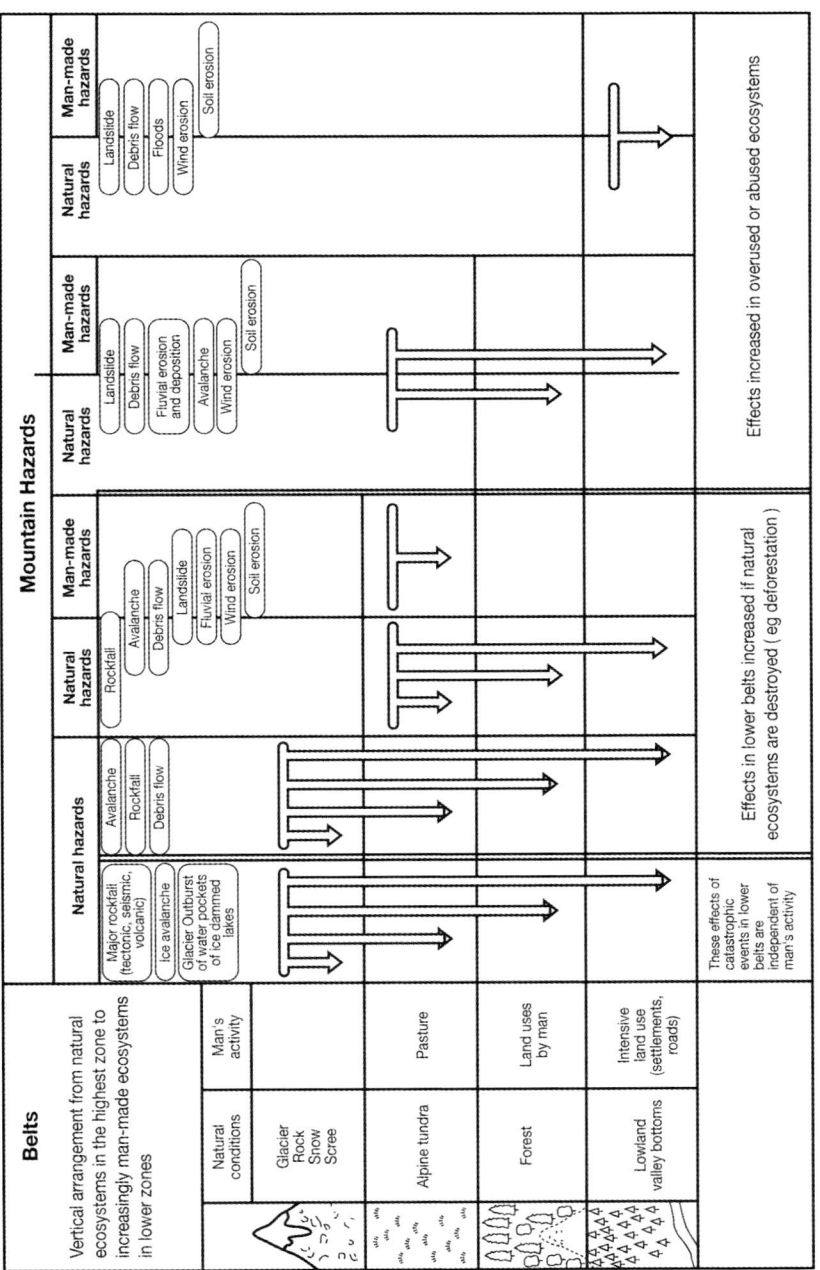

Figure 19.25   Classification of mountain hazards (redrawn from Messerli and Ives, 1984).

how many hazards correspond with different zones and emphasises the interaction between natural processes and human activity.

References

Baldwin II, J. E., Donley, H. F. and Howard, T. R. (1987) On debris flow/avalanche mitigation and control, San Francisco area, California. In Costa, J. E. and Wieczorek, G. F. (eds) *Debris Flows/avalanches: Processes, Recognition and Mitigation*. Geological Society of America, Reviews in Engineering Geology VII. Boulder, Colorado, 223–236.

Bisci, C., Dramis, F. and Sorriso-Valvo, M. (1996) Rock flow (sackung). In Dikau, R., Brunsden, D., Schrott, L. and Ibsen, M.-L. *Landslide Recognition: Identification, Movement and Causes*. Wiley, Chichester, 150–160.

Cancelli, A. and Pellegrini, M. (1987) Deep-seated gravitational deformations in the Northern Apennines, Italy. *Proceedings 5th ICFL, Australia and New Zealand*, 1–8.

Clague, G. J. J. and Evans, S. G. (1994) Formation and failure of natural dams in the Canadian Cordillera. Geological Survey of Canada Bulletin 464.

Cooke, R. U. and Doornkamp, J. C. (1990) *Geomorphology in Environmental Management*. Oxford University Press.

Costa, J. E. (1984) Physical geomorphology of debris flows. In Costa, J. E. and Fleisher, P. L. (eds) *Development and Applications of Geomorphology*. Springer–Verlag, Berlin 268–317.

Costa, J. E. and Wieczorek, G. F. (eds) (1987) Debris flows/avalanches: processes, recognition and mitigation. *Geological Society of America*, Reviews in Engineering Geology VII. Boulder, Colorado.

De Crecy, L. (1980) Avalanche zoning in France — regulation and technical bases. *Journal of Glaciology* 26, 325–330.

Eisbacher, G. H. and Clague, J. J. (1984) *Destructive Mass Movements in High Mountains: Hazard and Management*. Geological Survey of Canada Paper 84–16, 230pp.

Embleton, C. and Thornes, J. B. (eds) (1979) *Process in Geomorphology*. Arnold, London.

Evans, S. G. (1986) Landslide damming in the Cordillera of Western Canada. In Schuster, R. L. (ed.) *Landslide Dams: Processes, Risk and Mitigation*. Geotechnical Special Publication No. 3, American Society of Civil Engineers, 111–130.

Fairbridge, R.W. (1968) Mountain and lilly terrain, mountain systems: mountain types. In Fairbridge, R.W. (ed.) *Encyclopedia of Geomorphology*. Reinhold, New York, 745–761.

Ferguson, R. I. (1984) Sediment load of the Hunza River. In Miller, K. J. (ed.) *International Karakorum Project* Vol. II. Cambridge University Press, 581–598.

Fookes, P. G. (1997) Geology for engineers: the geological model, prediction and performance. *Quarterly Journal of Engineering Geology* 30, 293–424.

Fookes, P. G. and Marsh, A. H. (1981a) Some characteristics of construction materials in the low to moderate metamorphic grade rocks of the Lower Himalayas of East Nepal. 1: Occurrence and geological features. *Proceedings of the Institution of Civil Engineers*, Part 1 70, 123–138.

Fookes, P. G. and Marsh, A. H. (1981b) Some characteristics of construction materials in the low to moderate metamorphic grade rocks of the Lower Himalayas of East Nepal. 2: Engineering characteristics. *Proceedings of the Institution of Civil Engineers*, Part 1 70, 139–162.

Fookes, P. G., Sweeney, M., Manby, C. N. D. and Martin, R. P. (1985) Geological and geotechnical engineering aspects of low-cost roads in mountainous terrain. *Engineering Geology* 21, 1–152, Elsevier, Amsterdam.

Gerrard, A. J. (1990) *Mountain Environments: an Examination of the Physical Geography of Mountains*. Belhaven Press, London.

Govi, M. (1999) The 1987 landslide on Mount Zandila in the Valtellina, Northern Italy. In Sassa, K. (ed.) *Landslides of the World*. Kyoto University Press, 47–50.

Hungr, O., Morgan, G. C., VanDine, D. F. and Lister, D. R. (1987) Debris flow defences in British Columbia. In Costa, J. E. and Wieczorek, G. F. (eds) *Debris Flows/avalanches: Processes, Recognition and Mitigation*. Geological Society of America, Reviews in Engineering Geology VII. Boulder, Colorado, 201–222.

Hutchinson, J. N. (1988) General report: morphological and geotechnical parameters of landslides in relation to geology and hydrogeology. In Bonnard, C. (ed.) *Landslides*. Balkema, Rotherdam, Vol. 1, 3–35.

Ishihara, K. (1999) Liquefaction-induced landslide and debris flow in Tajikistan. In Sassa, K. (ed.) *Landslides of the World*. Kyoto University Press, 224–226.

Jackson, L. E. (1979) A catastrophic glacial outburst flood (jokulhlaup) mechanism for debris flow generation in the Spiral Tunnels, Kicking Horse River Basin, British Columbia. *Canadian Geotechnical Journal* 16, 806–813.

Kaliser, B. and Fleming, R. W. (1986) The 1983 landslide dam at Thistle, Utah. In Schuster, R. L. (ed.) *Landslide Dams: Processes, Risk and Mitigation*. Geotechnical Special Publication No. 3, American Society of Civil Engineers, 59–83.

King, J. P., Loveday, I. and Schuster, R. L. (1999) The Bairaman river landslide and natural dam, Papua New Guinea. In Sassa, K. (ed.) *Landslides of the World*. Kyoto University Press, 51–53.

La Chapelle, E. R. (1977a) Snow avalanches: a review of current research and applications. *Journal of Glaciology* **19**, 313–324.

La Chapelle, E. R. (1977b) Alternative methods for the artificial release of snow avalanches. *Journal of Glaciology* **19**, 389–397.

Li Tianchi, Schuster, R. L. and Wu Jishan (1986) Landslide dams in south-central China. In Schuster, R. L. (ed.) *Landslide Dams: Processes, Risk and Mitigation*. Geotechnical Special Publication No. 3, American Society of Civil Engineers, 146–162.

Lliboutry, L., Arnao, B., Morales, A., Pautre, A. and Scneider, B. (1977) Glaciological problems set in the control of dangerous lakes in the Cordillera Blanca, Peru. Part 1 Historical failures of moraine dams, their causes and prevention. *Journal of Glaciology* **18** (79), 239–254.

McDowell, B. (1962) Avalanche! *National Geographic* **121**, 855–880.

McSaveney, M. J., Chinn, T. J. and Hancox, G. T. (1999) Mount Cook rock avalanche of 14 December 1991, New Zealand. In Sassa, K. (ed.) *Landslides of the World*. Kyoto University Press, 56–59.

Martinez, J. M., Avila, G., Agudelo, A., Schuster, R. L., Casadevall, T. J. and Scott, K. M. (1999) Landslides and debris flows triggered by the 6 June 1994 Paez earthquake, southwestern Colombia. In Sassa, K. (ed.) *Landslides of the World*. Kyoto University Press, 227–229.

Mason, K. (1929) *Indus floods and Shyock glaciers: the Himalayan Journal*. Records of the Himalayan Club, Calcutta, Vol. 1, 10–29.

Mellor, M. (1978) Dynamics of snow avalanches. In Voight, B. (ed.) *Rockslides and Avalanches*. Elsevier, Amsterdam, 753–792.

Messerli, B. (1983) Stability and instability of mountain ecosystems: introduction to the workshop. *Mountain Research and Development* **3**, 81–94.

Messerli, B. and Ives, J. D. (1984) Gongga Shan (7556 m) and Yulongxue Shan (5596 m). Geoecological observations in the Hengduan Mountains of Southwestern China. In Lauer, W. (ed.) *Natural Environments and Man in Tropical Mountain Ecosystems*. Springer-Verlag, Berlin, 55–77.

Morales, B. (1966) The Huascaràn avalanche in the Santa valley, Peru. In *International Symposium on Scientific Aspects of Snow and Ice Avalanches*. International Association of Hydrological Sciences Publication 69, 304–315.

Nairne, D. (1895) *Memorable Floods in the Highlands during the Nineteenth Century*. Northern Counties Printing and Publishing.

Nieto, A. S. and Schuster, R. L. (1999) Mass wasting and flooding induced by the 5 March 1987 Ecuador earthquakes. In Sassa, K. (ed.) *Landslides of the World*. Kyoto University Press, 220–223.

O'Conner, J. E. and Costa, J. E. (1993) Geologic and hydrologic hazards in glaciated basins in North America resulting from 19th and 20th century global warming. *Natural Hazards* **8**, 121–140.

Ogura, A. T. and Filho, O. A. (1999) The Morin debris-flow disaster at Petropolis City, Rio de Janeiro State, Brazil. In Sassa, K. (ed.) *Landslides of the World*. Kyoto University Press, 289–292.

Pasuto, A. and Soldati, M. (1996) Rock spreading. In Dikau, R., Brunsden, D., Schrott, L. and Ibsen, M.-L. *Landslide Recognition: Identification, Movement and Causes*. Wiley, Chichester, 122–136.

Perla, R. I. and Martinelli, M. Jr (1976) *Avalanche Handbook*. Agriculture Handbook 489 USDA (Forest Service), Washington DC.

Plafker, G. and Ericksen, G. E. (1978) Nevados Huascaràn avalanches, Peru. In Voight, B. (ed.) *Rockslides and Avalanches*, 1 Natural Phenomena. Elsevier, Amsterdam, 277–314.

Reynolds, J. M. (1992) The identification and mitigation of glacier-related hazards: examples from the Cordillera Blanca, Peru. In McCall, G. J. H. Laming, D. J. C. and Scott, S. C. (eds) *Geohazards*. Chapman and Hall, London, 143–157.

Reynolds, J. M. (1998) High altitude glacial lake hazard assessment and mitigation: a Himalayan perspective. In Maund, J. G. and Eddleston, M. (eds) *Geohazards in Engineering Geology*. Geological Society, London Engineering Geology Special Publication, 15, 25–34.

Reynolds, J. M., Dolecki, A. and Portocarrero, C. (1998) The construction of a drainage tunnel as part of glacial lake hazard mitigation at Hualcan, Cordillera Blanca, Peru. In Maund, J. G. and Eddleston, M. (eds) *Geohazards in Engineering Geology*. Geological Society, London Engineering Geology Special Publication, 15, 41–48.

Sasahara, K. and Tsunaki, R. (1999) Landslide disasters triggered by heavy rainfall in Kagoshima Prefecture, July to September 1993. In Sassa, K. (ed.) *Landslides of the World*. Kyoto University Press, 253–256.

Sassa, K. (1996) Prediction of earthquake induced landslides. In Senneset, K. (ed.) *Landslides*. Balkema, Rotterdam, 115–132.

Sassa, K. (1999) Geotechnical classification of landslides. In Sassa, K. (ed.) *Landslides of the World*. Kyoto University Press, 139–145.

Sassa, K., Fukuoka, H. and Wang, F. W. (1998) Mechanisms of rapid long run-out motion in the May 1997 Sumikawa reactivated landslide in Akita Province and the July 1997 Harihara landslide-debris flow, Kagoshima Prefecture, Japan. *Journal of Japan Landslide Society*, **35**, 29–37.

Sassa, K., Fukuoka, H. and Sakamoto, T. (1999) The rapid and disastrous Nikawa landslide. In Sassa, K. (ed.) *Landslides of the World*. Kyoto University Press, 27–31.

Schick, A. P. and Lekach, J. (1987) A high magnitude flood in the Sinai desert. In Mayer, L. and Nash, D. (eds) *Catastrophic Flooding*. Allen & Unwin, 381–410.

Schuster, R. L. (ed.) (1986) *Landslide Dams: Processes, Risk and Mitigation*. Geotechnical Special Publication No. 3, American Society of Civil Engineers.

Schuster, R. L. and Costa, J. E. (1986) A perspective on landslide dams. In Schuster, R. L. (ed.) *Landslide Dams: Processes, Risk and Mitigation*. Geotechnical Special Publication No. 3, American Society of Civil Engineers, 1–20.

Slaymaker, O. (1974) Alpine hydrology. In Ives, J. D. and Barry, R. G. (eds) *Arctic and Alpine Environments*. Methuen, London, 134–155.

Smith, K. (1992) *Environmental Hazards: Assessing Risk and Reducing Disaster*. Routledge, London.

Stone, K. H. (1963) Alaskan ice-dammed lakes. *Annals of the Association of American Geographers* **53**, 332–349.

Swanson, F. J., Oyagi, N. and Tominaga, M. (1986) Landslide dams in Japan. In Schuster, R. L. (ed.) *Landslide Dams: Processes, Risk and Mitigation*. Geotechnical Special Publication No. 3, American Society of Civil Engineers, 131–145.

Tufnell, L. (1984) *Glacier Hazards*. Longman, London.

Verstappen, H. Th. (1983) *Applied Geomorphology*. Elsevier, Amsterdam.

Voight, B. (ed.) (1978) *Rockslides and Avalanches*, 1 Natural Phenomena. Elsevier, Amsterdam.

Záruba, Q. and Mencl, V. (1982) Slope movements caused by squeezing out of soft rocks. In Záruba, Q. and Mencl, V. (eds) *Landslides and their Control*. Elsevier, Amsterdam, 110–120.

20. Estuaries and Deltas

Warren E. Grabau, H. Jesse Walker and Mark Lee

20.1 Estuaries and deltas: introduction

The very beginnings of human civilizations were nurtured by deltas and estuaries, and that relationship has continued unbroken through the ages up to and including the present time. Many of the world's major cities are sheltered by them. Their waterways carry the cargoes of the world, and in some cases whole nations depend upon their fortunate combinations of fertile soil and abundant water for their food supply. Given such a concentration of interests, it is hardly surprising that mankind, through combinations of engineering works and management, has tried throughout the ages to bend the natural environments better to suit the needs of their human populations.

Estuaries and deltas are surprisingly difficult to define, in part because they so often occur as components of a single compound feature which consists of an association of a sub-aerial and sub-aqueous sedimentary structure (the delta) with a semi-enclosed sub-marine structure filled with salt water (the estuary). However, it is entirely possible for an estuary to occur without an accompanying delta, and a delta may occur without an associated estuary. Thus, while the two features are commonly closely associated, and perhaps inextricably linked conceptually within the human mind, such an association is not a requirement.

Estuaries and deltas are usually defined as features that form where freshwater rivers meet the salt sea, ignoring the obvious fact that features similar in almost every way, save for the presence of salt water, form on the margins of freshwater lakes. Indeed, features analogous in many ways to deltas can form even without the presence of a water body, as witness the eerie Okavango Swamp in northern Botswana, which has formed on the margins of an enormous dry depression.

The matter is still further confused by the fact that some definitions incorporate tides as an essential component of both estuaries and deltas, thus focusing on the ceaseless combat between freshwater streams and saltwater tides. However, that view seems too restrictive because the ancient Nile delta, which gave its name to the landform, formed at the margin of the nearly tideless Mediterranean. Accordingly, the following discussion will be confined to those features, both sedimentary and hydrologic, occurring where freshwater rivers meet the saltwater sea.

20.2 Estuaries

The definitions of estuary are many and varied, but nearly all reflect one element of commonality, namely the concept that an estuary is a semi-enclosed body of water which has at least one free connection with the sea, as well as a supply of fresh water, thus implying the presence of features consequent to the mixing of saline and fresh water. Some extend the definition to include not only the arm of the sea but all portions of the related freshwater streams that are affected by tides. However, the implication that tides are essential would eliminate consideration of the many river-meets-the-sea situations around the nearly tideless Mediterranean. To avoid such an obvious imbalance, the zero-tide examples are included as end-members of a continuous series ranging from tideless to very high tides indeed. Other definitions reflect in some degree the varying points of view of the various scientific disciplines of those who study them, and thus focus on the complexity of the interactions of hydrological, geological, chemical and biological processes that operate within them.

Basic morphology

In most cases sedimentation pushes the shoreline seaward from the head of the estuary, creating a positive fill estuary. There are a few exceptions, in which the estuaries receive the bulk of their sediment from offshore sources, thus creating negative fill estuaries. There are even a few neutral estuaries in which the delta front remains in equilibrium because the rate of sedimentation is matched by the rate of flushing (transport of the sub-aqueous portion of the delta to the open sea).

The rate at which the sub-aerial delta margin is pushed towards the sea varies greatly from estuary to estuary. The most significant factors are the rate at which sediment is delivered to the delta by the stream, the volume of the estuary which must be filled and the flushing rate.

Many of the glacial-era valleys, such as those of the Mississippi and the Nile, have already been filled and overfilled, so that the modern delta extends well beyond the mouth of the original estuary. If sea level remains essentially unchanged, the ends of all estuaries are foreordained: all will eventually fill completely to the mouth, and most will overfill, the exceptions being those where waves and currents are able to sweep sediments away as rapidly as the river deposits them.

Estuaries vary enormously in size, shape, and environmental setting, with much of the variability dependent on the mode of formation. In principle, all estuaries are drowned valleys. Most existing estuaries, but by no means all, are artefacts of the depressed sea level during the last episode of the Pleistocene glaciation and the resurgence of the sea level during postglacial times. During the long glacial period, many streams carved deep valleys or gorges in the margins of the continents and formed deltas at their mouths at the margins of a sea that was perhaps as much as 160 metres lower than the modern level. The rising sea that accompanied the melting of the glaciers flooded those ancient deltas as well as the glacial-era valleys and gorges upstream, producing the countless estuaries that notch the coasts of all of the continents.

The modern sea level was reached about 6000 years ago, and since that time the drowned valleys have been modified in four principal ways, thus producing four clearly recognisable types (Figure 20.1):

1. unmodified drowned river valleys
2. barred (drowned river valleys which have been partially walled off from the sea by bars built across their mouths)
3. fjords and fiards (glacier-scoured valleys invaded by the sea after the melting of the ice)
4. tectonic (tectonic structures, not necessarily originally associated with a river, that has been flooded by the sea).

Two additional forms are sometimes recognised:

5. riverine estuaries
6. compound estuaries.

Unmodified drowned river valleys (Figure 20.1) are especially common on temperate zone low-relief coastlines and are accordingly often referred to as coastal plain estuaries, even though they are also often found on coastlines bordering low plateaux. In cross-section they tend to be shaped like an open V, like the original river valley, and normally widen and deepen as they approach the open sea. Because the invading sea usually covered the river floodplain as well as the river channel, the deepest part of the estuary is commonly the sinuous channel of the pre-flooding river. However, such is not always the case, because sedimentation after the drowning sometimes obscures the old channel. The planimetric shapes of this type of estuary are often very complex because tributary valleys were also often flooded. Depths at the mouth vary widely, in some instances reaching in excess of 150 metres. Much shallower ones are very common, especially in those estuaries where sedimentation has been rapid.

Barred estuaries (Figure 20.1), which are often referred to as bar-built estuaries, or depositional barrier estuaries, can form on any gently shelving coast that is subjected to even modest wave attack. In such cases, wave action mobilises sediment (usually sand) from the sea floor and constructs a

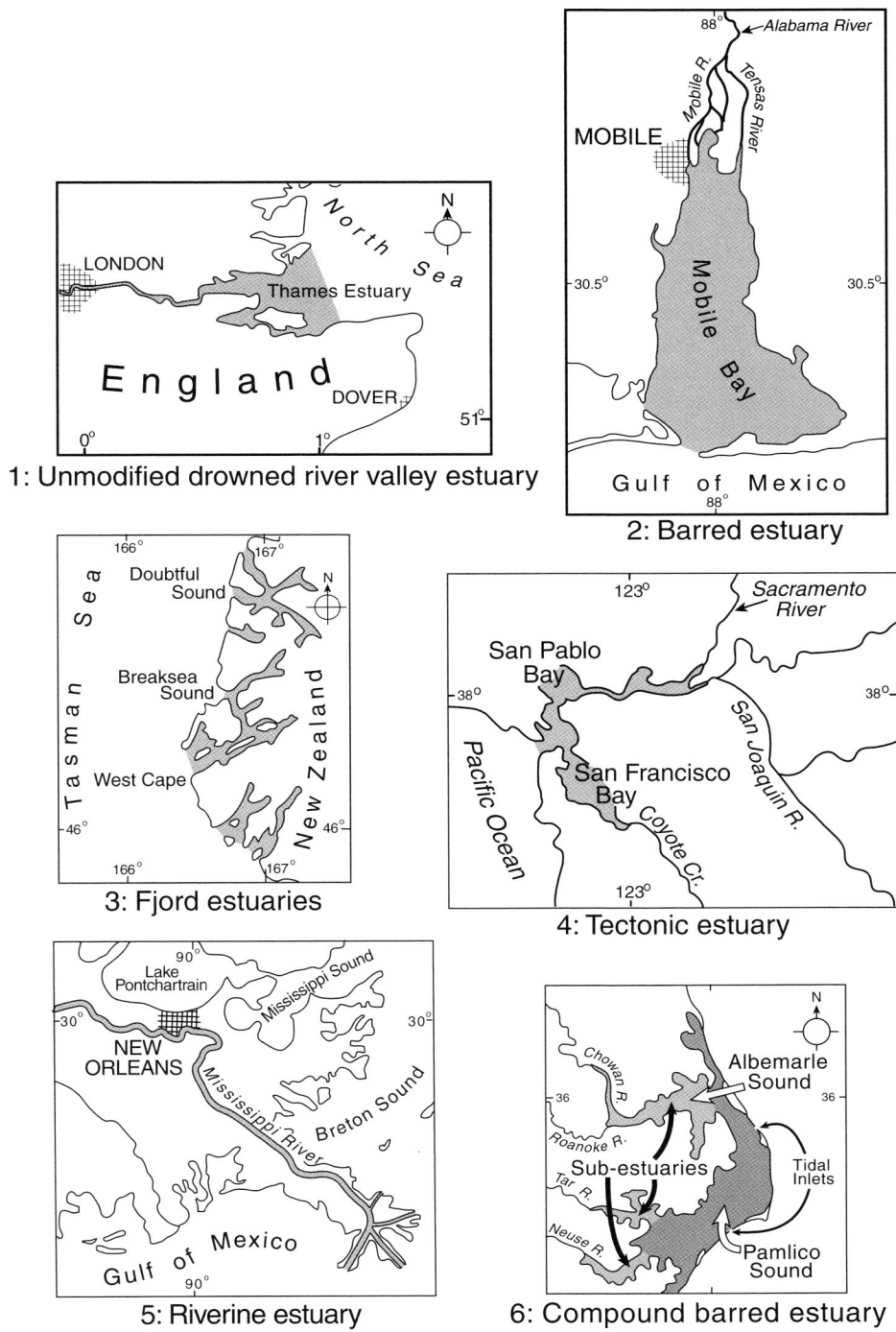

Figure 20.1   Types of estuaries.

barrier beach island across the mouth of the stream valley, so that the estuary communicates with the sea only through one or more tidal inlets which penetrate the barrier. However, a variant form can also develop in drowned valleys whose mouth is exposed to along-shore currents strong enough to mobilise and transport sediments, creating a bay-mouth bar. Such estuaries also communicate with the open sea through tidal inlets. Not uncommonly, the barrier beach islands form so far offshore that several river valleys are blocked off by a single extensive barrier, so that in effect the mouth of the individual estuaries open, not into the open sea, but into a lagoon or sound, the outer limit of which is the enclosing barrier beach island. This often has profound effects on the dynamics of the component estuaries, because such marine influences as salinity, wave action, tidal variations and along-shore currents may be strongly moderated in the lagoon behind the barrier islands. Indeed, it is not unknown for the tidal inlets to seal over completely from time to time and for extended periods, during which intervals the river estuaries are discharging into a brackish, tideless lake instead of the open sea. Such estuaries, having formed on gently shelving coasts, are usually very shallow (often less than 10 metres), one consequence being the absence of high-energy waves at their mouths. Nearly the entire Atlantic coast of North America south of the mouth of the Hudson River is characterised by barred estuaries, as is much of the coast of the Gulf of Mexico

Fjords (Figure 20.1) typically have U-shaped cross-sections resulting from glacial scour prior to glacial melting and subsequent flooding by the sea. They are typically steep-sided, narrow in relation to length, and straight or slightly sinuous. They are often very deep, up to as much as 800 metres, but depth is usually not continuous. The bottoms of many, and perhaps most, are characterised by rock sills (transverse rock bars) which often rise nearly to the surface, especially at or near their mouths. The bottoms of many fjords consist of linear basins strung end to end and separated from each other by rock sills which may nearly reach the surface. Fjords are the glacially scoured courses of rivers,

so they typically have substantial streams discharging into them at or near their head. Fiards, however, are glacially scoured depressions which are not related to pre-glacial stream valleys. Most of them do not have steep sides and they are normally much shorter, broader and shallower than fjords.

Tectonic estuaries (Figure 20.1) are created when diastrophism drops a segment of the Earth's crust below sea level in a position such that the sea can flood the depression. Down-dropped fault blocks (grabens) and fold valleys are the most common causes, although volcanic activity and general crustal subsidence are also responsible for some estuaries. Because tectonic forces do not follow the rules of normal weathering and erosion processes, there are no preferred shapes or features, with the possible exception being that the shorelines of those formed in grabens and structural troughs are often relatively simple and regular. The same cannot be said for those formed by volcanic action or general subsidence.

Although the classic estuary is formed when the sea enters a pre-existing depression, some investigators recognise another way in which estuaries may form. As rivers create deltas which extend into the open sea (e.g. through the overfilling of the original flooded valley), they flow on their own sedimentary deposits seaward of the mouth of the original valley. Such river channels are called riverine (Figure 20.1), or deltaic distributary estuaries. The classic examples are, of course, the Nile and the Mississippi, although there are many others.

Most estuaries are characterised by a single large freshwater stream entering an arm of the sea, but there are many examples of situations in which several streams of approximately equal size enter a lagoon (or sound) which has been created by the development of an offshore barrier bar. In such cases, the 'mouth' of the estuary is the tidal inlet that pierces the barrier bar. However, each of the streams entering the lagoon is itself an independent sub-estuary which may behave quite differently than its neighbours because of local conditions. Albemarle Sound, in North Carolina, is of this kind. It is also possible for a single complex river system to be have been flooded in such a way that several major tributaries coalesce just before passing through a gorge or narrow valley into the open sea. An example of this variety is Chesapeake Bay, in which the bay mouth comprises the mouth of the primary estuary, and each major tributary, such as the James, York, Potomac, and so on is a sub-estuary. Such systems may be called compound estuaries (not illustrated).

Estuarine classifications

Given the extraordinary complexity of the estuarine system, it is perhaps to be expected that classifications of estuaries are also highly varied. However, most classifications reflect, to some degree, the fact that an estuary is a semi-confined body of water in which sea water is mixed with salt water. The potential variations in size and shape are almost limitless in number (as described in Basic morphology, above). Some are affected by enormous tidal ranges, while others have little or none. Some are conditioned by great ranges of meteorological conditions, while others rest in regions having climates that vary little from day to day or season to season. Some are in the tropics, and some are in the Arctic. Some have abundant biotas, while others are impoverished. All have been used at one time or another to classify estuaries.

Despite the complexities, there is more or less general agreement that useful characterisations can be achieved by describing transverse and longitudinal components (Figure 20.2). In this scheme the transverse components describe the configuration of the estuarine basin, not the water, while the longitudinal components describe, somewhat indirectly, the salinity of the water. Thus, the generally recognised transverse components are supratidal (the part of the basin above tide level), intertidal (the part of the basin alternately flooded and drained during the tidal cycle) and subtidal (continuously below water level). Similarly, the commonly recognised longitudinal zones are upper (freshwater dominated, or river dominated), middle (the primary zone of mixing of saline and fresh waters) and lower (saltwater dominated, or ocean dominated). Profound disagreements arise when attempts are made to define the longitudinal components quantitatively,

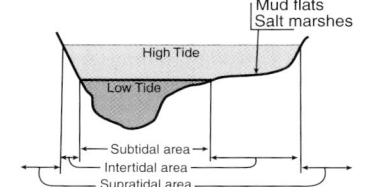

A = Transverse components (Cross-section)

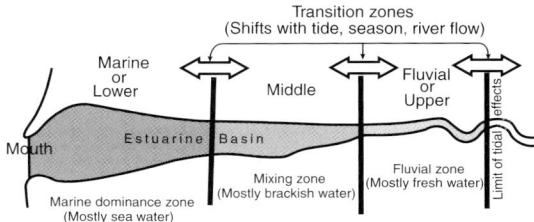

B = Longitudinal components (Planimetric view)

Figure 20.2   Estuarine components (modified from Fairbridge, 1980).

in part because the mixing processes are so complex as to defy ready description.

Every individual estuary displays marked temporal changes in the position and extent of the transitions between the longitudinal categories. Further, depending on the criteria being applied and the methods of measurement or detection being used, it is often difficult for two independent investigators to agree on the location of the transition zones at any arbitrary time. It is also worthy of note that not all longitudinal components are necessarily present at all times. For example, a river flood period may force both the middle and lower components seaward past the mouth of the estuary. Conversely, a desert estuary with an intermittent stream may permit the entire estuary to be ocean dominated for an extended period. Further, an estuary at the margin of a tideless sea, such as the Mediterranean, will have no intertidal zone.

Estuarine dynamics

The common feature of all estuaries on marine littorals is the interaction of sea and river. It has been said that sea and river are in an eternal conflict that neither can wholly win. Of all the factors that are germane to the struggle, the most important are relative water densities, tidal action, river flow cycles, the size and three-dimensional shape of the estuarine basin, and wave climate, because those factors collectively control both the mixing of fresh and salt water as well as the erosive and depositional processes. All these factors — and more — are linked in exquisitely complex interactions within the estuary.

While there is no universal system for classifying the characteristics of the tidal cycle, three categories are often used to describe amplitude: microtidal (0–2 m), mesotidal (2–4 m), and macrotidal (> 4 m), and the temporal characteristics are classified as: diurnal (one high and one low water per day), semidiurnal (two high and two low waters per day with small diurnal inequality), and mixed (two high and two low waters per day with one high water very significantly higher than the other).

The beds of streams (especially large ones) approaching an estuary are very commonly below sea level for significant distances upstream. As a result of the differences in specific gravity (the ratio of the specific gravity of fresh water to sea water is approximately 1.0/1.03), a salt wedge often forms because the fresh water tends to override the salt water, with the result that within the inflowing stream, the water is layered, or stratified, with fresh water on top flowing seaward and the salt water at the bottom, producing a vertically stratified estuary (Figure 20.3). This effect persists even when the fresh water is carrying significant amounts of materials (chiefly silt and clay) in suspension.

In principle such a salt wedge can extend as far upstream as the point at which the bed of the river rises to an elevation above sea level. The effect is most pronounced in estuaries subject to relatively modest tidal oscillations (i.e. microtidal situations). For example, the salt wedge in the Mississippi River may extend 150 km upstream. In practice such an extreme rarely occurs because the rush of fresh water tends to push the end of the saltwater wedge towards the sea.

At the contact between the fresh and salt water (Figure 20.3), turbulence develops because of the differences in velocity, and as a result a layer of brackish water develops between the saltwater wedge and the freshwater outflow, and the

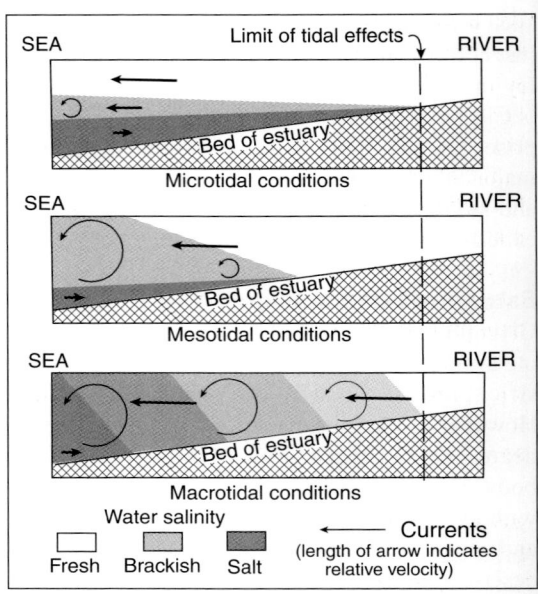

Figure 20.3   Fresh water/salt water mixing modes in estuaries (modifed from Biggs, 1982).

brackish-water layer also moves seaward. To replenish the water lost in the mixing process, seawater flows up the wedge, so that the result is an upstream bottom current of salt water, and a downstream primary current of fresh water. Clearly any sediment being transported downstream by the freshwater flow, either in saltation or suspended as bedload, will strike the point at which the currents reverse direction, come out of suspension and begin the construction of a bar transverse to the bed of the stream. Eventually the rise in elevation of the stream bed generates a ramp which directs the freshwater flow upward, which creates significant amounts of turbulence, which in turn enhances the mixing of the fresh and salt waters. The consequence of this seems to be that the head of the saltwater wedge is driven downstream at the same time that the zone of mixing is greatly enlarged.

As tidal oscillations increase, salt wedge formation is reduced because the increasingly rapid tidal currents increase turbulence, which enhances vertical mixing. Estuaries exposed to mesotidal ranges are often characterised as partially mixed estuaries (Figure 20.3) because there tends to be a relatively smooth vertical salinity gradient, saltiest at the base (but not as salty as sea water) and brackish to nearly fresh at the top. In these estuaries, the transition zone between the upper and middle estuary is both indistinct and highly mobile, moving upstream at low tide and downstream at high. Estuaries with macrotidal ranges may have no perceptible salt wedge formation, because the very rapid tidal currents thoroughly mix the entire water column, thus producing a vertically homogeneous estuary (Figure 20.3). The only detectable salinity gradient in this type of estuary is longitudinal, and the transition zones between fresh and salt water are so indistinct as to be undefinable.

Adjustments between processes and forms

'Neither the channel morphology nor the tidal properties . . . can be explained solely in terms of the other, though the two are mutually interdependent.' (Wright *et al.*, 1973).

Because estuaries are links between a river and the sea, the reversing flows of fresh and salt water interact in tidal estuaries to produce unique depositional environments. The three-dimensional configuration of the estuary is of critical importance as it may profoundly modify the shape of the tidal surge as it enters the relatively shallow and confined estuary basin. In the open ocean the tide may be thought of as two waves, one sublunar and the other anti-lunar, which travel around the earth in approximately 24.8 hours. Thus, at the equator, the wavelength (the distance from crest to crest) is approximately 20 000 km, and a tidal limb (the distance from crest to trough) is about 10 000 km, because the waves are essentially symmetrical.

However, as the tidal wave approaches the shore of an island or continent, the front of the wave (the rising limb) commonly steepens as the water becomes more shallow, and accordingly at the shoreline the tide often requires less time to go from trough to crest than from crest to trough: that is, the tidal wave has become asymmetrical. Complex configurations of the shoreline and ocean floor may magnify this effect enormously.

For example, estuaries with wide, unobstructed, deep mouths, but gradually shoaling and constricting channels, like the Severn Estuary, often create a strongly asymmetrical tidal surge. This is because the gradually narrowing and shoaling channel allows the crest of the tidal wave to travel faster than the more slowly moving water in the trough of the tidal wave. This produces a markedly asymmetric tidal wave, with a steep rising limb of short duration, while the ebb duration increases. In extreme cases this combination of conditions may produce a bore, a breaking wave marking the front of the tidal surge as it moves up the channel. In such cases the rising tide is a period of high current velocities and extreme turbulence, while the long ebb-tide period is accompanied by relatively modest turbulence and current velocities. Such estuaries are almost invariably characterised by a cross-section such that the average water depth is greatest at high tide.

When such flood dominant estuaries are associated with a net input of sediment from the sea, as they often are, the estuary becomes a sediment sink with rapid deposition rates on the intertidal areas. Thus, in such estuaries, the elevations of intertidal areas gradually increase.

At the other extreme are those estuaries in which the cross-section is such that the mean water depth is greater at low tide than at high tide. This apparent contradiction results because the high tide cross-section is dominated by shallow water over intertidal flats, whilst the low tide cross-section is confined to the deeper subtidal channel. In this condition, the asymmetry of the tidal wave is reversed, with the passage of the tidal wave crest being slower than the trough (Dronkers, 1986). In such ebb dominant estuaries, the intertidal areas are often actively eroded by the high current velocities of the ebbing tide, and the net movement of sediment tends to be out to sea.

It is at least conceptually possible for a single estuary to alternate between flood dominant and ebb dominant over long periods of time. However, many estuaries are characterised by a long-term balance between tidal processes and form. That is, deposition within the estuary during the period of rising tide is more or less precisely balanced by erosion and sediment removal during the period of ebb tide. In other words, these are neutral estuaries, in which sedimentation rate is equal to flushing rate. Pethick (1996) hypothesised that the system works somewhat as follows:

Estuary depth and tidal wave length: in the open ocean the tidal wave propagates in response to extraterrestrial (especially lunar) gravitational attractions. However, as it approaches the shore, other factors such as decreasing water depths and shoreline configuration, decouple the tidal wave from its lunar influence. After the wave enters the mouth of an estuary, its velocity slows in proportion to the shallowing of the estuary from mouth to head. This reduction in velocity is accompanied by deposition which, in turn, reduces the depth of water. If, and when, the deposition rate balances the flushing rate, the estuary can remain in a relatively static status for a long period of time. The precise mechanisms of this process are not well understood, but it appears that deposition on the intertidal areas eventually reaches elevations such that they are flooded only at spring tides (maximum high tide, which occurs once per lunar month), thus effectively reducing the area of the intertidal zone, and at the same time modifying the cross-sectional shape such that rising tide and falling tide current regimes are approximately equal.

Estuary mouth and tidal prism: in every estuary a relationship exists between the size and shape of the cross-section of the estuary mouth, and the volume of the tidal prism (defined as the total volume of water injected into, and drained from, the estuary during each tidal cycle). It should be noted that the tidal prism is never a constant, because its value changes (in some instances very significantly) over the course of each approximately 28-day lunar cycle, with a minimum value at neap tide and a maximum value at spring tide.

If the estuary mouth is formed of unconsolidated materials, any long-continued period of environmental stability will produce a cross-sectional shape and area for the estuary's mouth that is adjusted to the tidal prism, although it will exhibit small-scale variations produced by the difference between neap-tide and spring-tide prisms. It should be noted that the cross-sectional area alone is not a sufficient measure, because (all other conditions being equal) a broad and shallow cross-sectional shape will not transmit as much water as a narrower but deeper channel. The wetted perimeter, defined as the contact length (in a cross-section) between the water and the channel bed, is a useful shape parameter because maximum hydraulic efficiency occurs when it reaches a minimum for a given cross-sectional area.

The shape and size adjustment of the mouth occurs because the current velocity through the opening will be a function of the tidal prism (the total amount of water which flows through the opening) and the cross-sectional size and shape of the opening. Note that the tidal prism is also a function of the magnitude of the tide. If the opening is too small, the current velocity may be high enough to mobilise sediments in the opening, thus changing the shape or enlarging the opening until the maximum current velocities are no longer capable of mobilising the sediments. Of course the contrary is also true: if the opening is too large, then the tidal currents flowing through it will not only fail to mobilise the sediments already there, they will also deposit sediments (chiefly the coarser fractions) which are already entrained within the tidal waters until the available current will pass through the opening at velocities capable of carrying the pre-existing sedimentary load but incapable of mobilising new material from the bed.

However, it is important to note that any apparent stability is dynamic. During the period of rising tide the currents in most estuaries will be flowing inland, and the size and shape of the opening will adjust to the maximum current velocities achieved during some (usually brief) part of the rising limb of the tide. Following that period, sedimentation will again occur as the current through the mouth slows to a complete stop during high tide, at which time the highest sedimentation rate will occur within the mouth. When the tide begins to ebb, gradually accelerating current velocities through the mouth will again begin to mobilise materials within the mouth and again reach a maximum at some point on the falling limb of the tidal cycle. Thus, if the estuary is exposed to semidiurnal tides, the cross-sectional area and shape of the mouth will actually go through four periods of expansion and contraction during each 24-hour period. Added to the normal daily cycle is the 28-day lunar cycle of extreme high and extreme low tides, which of course exacerbates the effect because the tidal prisms are respectively larger and smaller than normal. This effect may be very limited in many estuaries with very wide mouths, but it is present all the same.

Nevertheless, certain kinds of estuaries, notably the unmodified drowned river valley type, have apparently evolved relatively stable relationships between the cross-sectional area of the mouth and the volume of the tidal prism. For example, thirteen estuaries along the east coast of England demonstrate such a relationship even though the estuaries are widely different in size and configuration (Pethick, 1996) (Figure 20.4). However, even in estuaries characterised by such broadly similar environmental settings, variations occur. For example, two of the estuaries exhibit mouth cross-sections larger than necessary to accommodate the tidal prism, suggesting that rapid accumulation of sediment within the estuary, coupled perhaps with extensive human reclamation of the intertidal flats over the last 500 years, has significantly reduced the tidal prism, and that this reduction has not yet been accompanied by compensatory reduction in the cross-sectional area of the mouth.

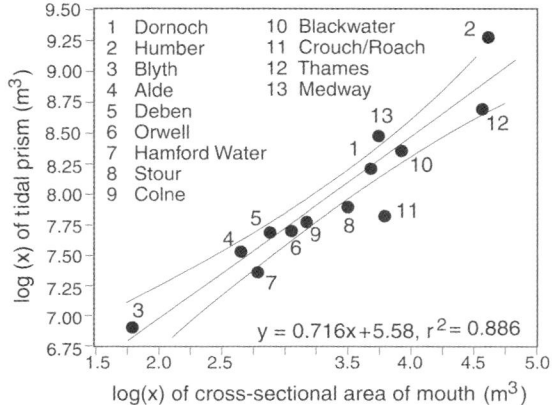

Figure 20.4 Best-fit logarithmic relationship between tidal prism and cross-sectional area of mouth for British east coast estuaries (modified from Pethick, 1996).

Other types of estuary, notably those with markedly constricted openings, such as the tidal inlets of most barred estuaries, obey very different rules. Such restricted openings often produce secondary bottom structures called channel-mouth bars or tidal deltas resulting from the drop in current velocity as the tidal flow passes through the constricted opening and emerges into either the open sea (during the ebbing limb of the tidal cycle) or the interior of the estuarine basin (during the rising limb of the tidal cycle). Thus, while it seems counter-intuitive, channel-mouth bars commonly form both inside (flood tide deltas) and outside (ebb tide deltas) the mouths of estuaries with restricted openings. The mouth of St Johns River estuary in north-east Florida is an example of this condition, although in this case the interior bar is less well developed than the exterior bar because the current of St Johns River is strong enough to inhibit its full development.

As a general principle, any substantial modification of the three-dimensional geometry of the estuary may result in a cascade of secondary and tertiary effects. Geometric changes may be caused by increased sedimentation, land reclamation of intertidal areas, enhancement or degradation of tidal deltas, or dredging and channelling.

Sedimentation within an estuary results in a gradual reduction of the tidal prism, which creates a complex chain of responses. As previously noted, the reduced water flow through the mouth (which may be thought of as an inlet jet) may result in changes in cross-sectional shape and reduction of cross-sectional area of the estuary mouth, as well as modifications to the tidal deltas. For example, if the estuarine system provides an abundant supply of sediment that can be transported by the reduced velocity of the inlet jet, the ebb tide delta may actually increase in size because the reduced inlet jet velocity permits an increase in sedimentation as it spreads and dissipates in the sea. Alternatively, a well-developed flood tide delta implies the existence of a high-velocity inlet jet and a copious supply of sediment derived from marine deposits outside the estuary. If the process continues, the flood tide delta may grow so large that it significantly reduces the tidal prism, which will in turn reduce the velocity of the inlet jet. The mouths of many estuaries are restricted by the growth of spits which are created usually on the seaward side of the inlet by combinations of strong along-shore currents, wave action and wind. Because the restricted opening is incapable of transmitting the necessary volume of water to either fill or empty the estuary within a tidal cycle, the water level on the upstream side rises and creates a hydraulic 'head' which converts the inlet to a short stretch of stream with a very high gradient and an accordingly high current velocity, which in turn affects the morphology of both the ebb-tide and flood-tide deltas, as well as the cross-sectional size and shape of the inlet.

Land reclamation within an estuary can result in dramatic changes throughout the estuarine system, including within the estuary, at the mouth, and even along the adjacent coastline. The interactions are complex and not fully understood. However, it is conjectured that the large changes in the tidal prism often associated with reclamation of the intertidal areas may result in greatly reduced tidal current velocities through the mouth, as well as within the estuary itself, thus reducing the capacities of those currents to mobilise and transport sediments. The reduced sediment quantities delivered to the ebb-tide delta thus results in such impoverishment that wave action and along-shore currents are capable of removing and rearranging the deltaic materials faster than they are replenished. This process may result in the partial closure of the inlet by features such as spits, bars, and dune systems. The reduction in sediment transport out of the mouth of the estuary implicit in a reduced current velocity may result in accelerated coastal erosion, because there may no longer be a supply of sediments large enough to replenish the losses along the nearby seashore due to wave and along-shore current action (Figure 20.5). For example, it has been postulated that this sequence was a major factor in the events in the vicinity of the former port of Dunwich, Suffolk (Pethick, personal communication). The concept is that reclamation in the Blyth and Dunwich estuary on the coast of England north-east of the mouth of the Thames, triggered the southwards growth of a shingle spit across the mouth and the collapse of the ebb tide delta. This was followed by accelerated coastal erosion, culminating in the loss of much of Dunwich village (Gardner, 1754; Bacon and Bacon, 1988).

Estuarine sedimentation

The nature of the mixing processes has important consequences for the sedimentology of the estuary, especially if the suspended load in the freshwater stream contains large amounts of clay. The carrying capacity (whether in suspension, bedload or saltation) of both salt and fresh water is dependent on current velocity. Because the current velocity of a freshwater stream entering an estuary is reduced as the estuary broadens toward the sea, the sedimentary deposits within the estuary tend to exhibit a marked longitudinal grain-size gradient, from large upstream to small toward the sea. This tendency is especially marked in estuaries with little tidal action. Increasingly large tidal oscillations progressively destroy the grain-size gradient, in part because tidal currents may be rapid enough to mobilise and carry even large particle sizes.

Moreover, the depositional dynamics are also altered by the electrochemical action of salt water, which produces flocculation (aggregation) of clay particles. Because the flocs have densities significantly higher than water, they come out of suspension as soon as the current velocities fall below a critical value, and in consequence they settle out in the seaward parts of the estuary as a low-density clay depositional layer. All estuaries experience this phenomenon, the intensity being a function of the proportion of clay minerals in suspension in the freshwater stream and the degree to which the water column is mixed. Because very little of the fresh water is mixed with salt water in vertically stratified estuaries (usually those with little tidal action), such estuaries may send plumes of fine-grained sediment well beyond the mouth of the estuary, while at the same time accumulating nearly all of the coarse-grained sediments within the estuary itself. Progressively more of the clay-sized materials are deposited within the estuary as tidal ranges increase, because the increasingly well-mixed fresh and salt water results in higher flocculation rates. Thus, a vertically homogeneous estuary can be a highly efficient sediment trap for all grain sizes. However, it

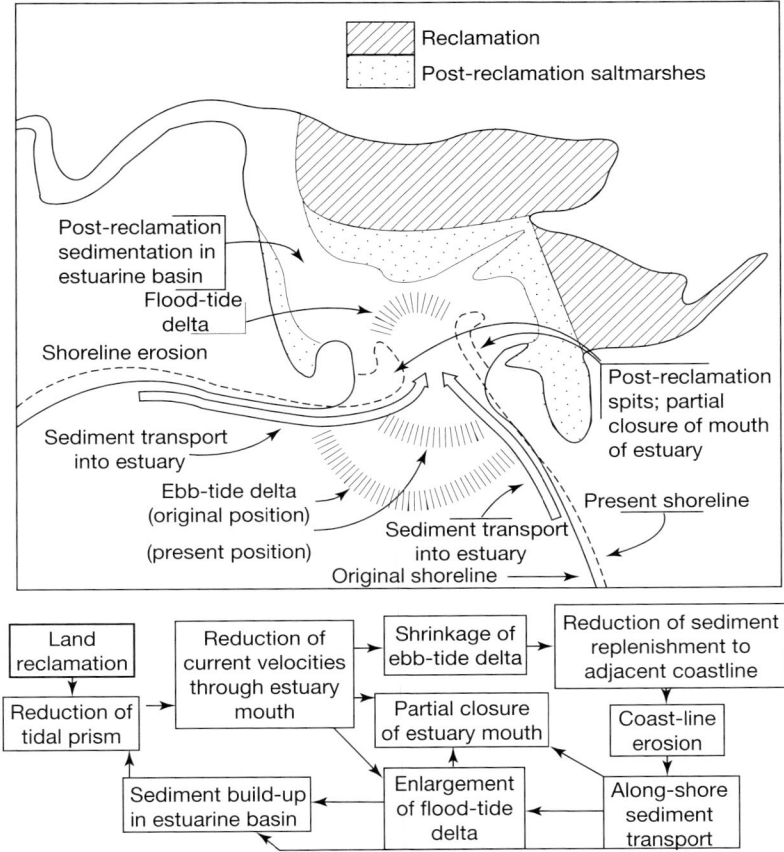

Figure 20.5   Concept for explaining the evolution of meso/macrotidal estuarine environments along the east coast of Britain (modified from Carter, 1992).

must be emphasised that this tendency may be interrupted or even reversed if tidal currents are very rapid, because the high-velocity currents flush the low-density clay-rich sediments accumulated because of flocculation out to sea.

Estuaries are found along most coastlines of the world and are therefore subjected to the full spectrum of tidal cycles and ranges. In general, the greater the tidal range, and to a somewhat lesser extent the greater the tendency toward semidiurnal tides, the greater the amount of energy available for mixing. Thus, estuaries exposed to semidiurnal macrotides tend to be the best mixed.

Tidal currents are often strong enough to act as agents of erosion as well as transport and deposition, and because the currents are fastest during the rising and falling stages and slowest during the periods of high and low water, the largest grain sizes, as well as the largest amounts of suspended sediments, tend to be moved during the rising and falling limbs of the cycle. Because the current velocities of flood tides are usually greater than those of ebb tides, tides sometimes cumulatively move sediment upstream, a process which produces an inverse-fill estuary.

Tidal effects are greatly complicated by the size and three-dimensional shape of the estuary, because complex shapes tend to produce greater turbulence and therefore enhanced mixing. Thus, estuaries with extensive intratidal zones tend to be better mixed than those in which intratidal zones are small or absent. In funnel-shaped estuaries (unmodified drowned river valleys) the tidal range is frequently accentuated from mouth to head. In some estuaries the development of a bore greatly enhances mixing. Tidal enhancement of mixing is normally minimal in microtidal estuaries, but such estuaries usually fill rapidly with coarse-grained sediment because tidal currents are inadequate to transport such materials to the mouth. As tidal ranges rise through mesotidal to macrotidal, ever larger areas of intertidal zones are affected, and tidal currents increasingly dominate circulation, so that current pulses (generated by rising and falling periods of the tide) transport pulses of sediment through the estuarine channels and out to sea. Because the high-tide period is a time of near-zero current velocities, fine-grained sediment is deposited in the intertidal zone, especially on mud flats and marshes. Such tidal marshes and flats are very extensive in estuaries such as the barred estuaries of the Atlantic coast of the United States. Tidal effects tend to be minimal in fjords, which are usually very deep and steep-sided, because the intertidal zones are very restricted or even absent.

Barred estuaries constitute a special case with respect to tidal influence, because in those instances in which the lagoon behind the barrier bar is large, the small tidal inlets may so restrict the entrance of the tidal surge that the tide is scarcely felt. In effect, regardless of the amplitude of the tides in the open sea beyond the barrier bar, the estuary may become microtidal.

The flow cycles of the rivers which feed fresh water into the estuaries are at least as variable and far more irregular and unpredictable than the tides. Streams may actually cease to flow in estuaries along arid and arctic coasts, in which case the estuary temporarily becomes nothing more than an arm of the sea. On the other hand, the flows of some rivers are so powerful, especially during flood periods, that the mixing zone is pushed well out to sea beyond the mouth of the estuary. A notable example is the Amazon. All possible intermediate conditions may occur, depending on the local conditions within any given estuary.

Sediment entrained in estuarine waters come from a wide variety of sources. As modified from

McDowell and O'Conner (1977), the following sources have been identified (Figure 20.6):

1. stream erosion (sediment carried by streams entering the estuary SR)
2. littoral drift (sediment carried by along-shore currents SL)
3. bank (derived from the banks of streams within the estuary SB)
4. shore erosion (derived from erosion of shores of the estuary SS)
5. erosion of near-shore ocean bed (SO)
6. aeolian sources (wind erosion of coastal dunes and intertidal shoals SW)
7. biological sources (decomposition and excretions of aquatic biota SA)
8. human activities (derived from domestic and industrial effluents and solid wastes SP)
9. dredged spoil (derived from re-mobilisation of dredged materials SD).

The relative importance of these sediments varies with the size and nature of the estuarine drainage basin (the combined drainage basins of the inflowing streams, plus that of the estuary itself), the nature of the associated marine environments and the degree and nature of human development of the estuarine drainage basin, as well as with the estuarine dynamics discussed above. Indeed, especially in large estuaries, the sources of sediments may vary significantly within various parts of a single estuary. For example, in Chesapeake Bay (Figure 20.7), a very large compound estuary, there is a large contrast in the relative contributions of inorganic and organic materials between the upper and middle parts of the estuarine system.

Typically the depositional facies gradually change from marine (usually sand, and occasionally with marine-derived gravels) through estuarine (combinations of detrital and organically derived materials) to fluvial (usually fine-grained clastics) from mouth to head (Figure 20.8). If tidal flats exist, they are normally covered with silts and clay and often mixed with substantial amounts of organic material. However, there is normally much overlapping and inter-fingering of sediment types because the upper, middle, and lower estuarine zones shift up or downstream

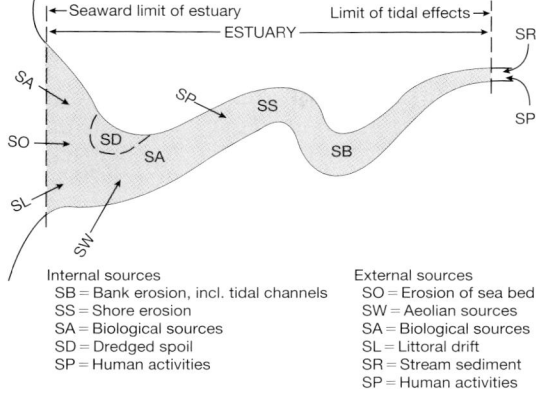

Figure 20.6 Sources of sediment in an estuary (modified from McDowell and O'Connor, 1977).

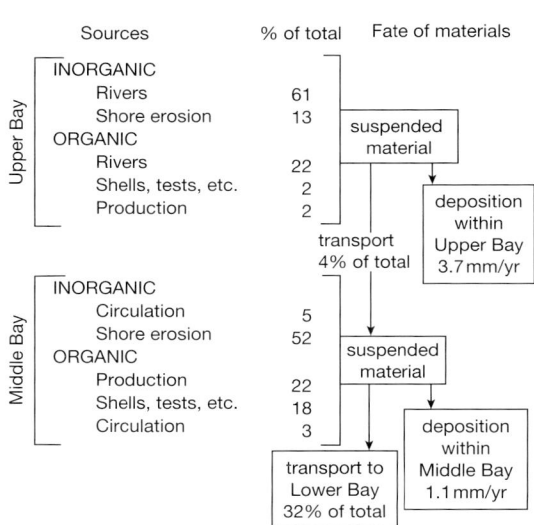

Figure 20.7 Differences in sources of suspended sediments between upper and Middle Chesapeake Bay. (modified from McLusky, 1981).

with every change of tide or river flow. Because the number of combinations and permutations is nearly infinite, the depositional characteristics of each estuary are unique and often surprising. For example (Figure 20.9), the Tay estuary on the east coast of Scotland is a long, narrow, macrotidal estuary with extensive intertidal areas, complex tidal channels and a mouth constricted in both width and depth. Contrary to expectations, the maximum suspended sediment load and minimum salinity occurs within the estuarine basin (i.e. the deep basin within the constricted mouth) at low tide. As the tide ebbs, fine sediment is entrained from the mudflats and tidal channels and concentrated in the estuarine basin, from which it cannot readily escape because of the constricted mouth and relatively high sill at the entrance. During flood tide, the suspended load is diluted by inflowing sea water and dispersed over the mudflats, where some of it is deposited during the slack-water high-tide period.

Mudflats (horizontal or near-horizontal intertidal surfaces devoid of vegetation and composed of fine-grained sediments) and saltmarshes (similar to mudflats, but covered with salt-tolerant reeds and grasses) are formed by complex alternations between deposition and erosion of fine sediments (Pethick, 1992a; 1996). While many exceptions occur, there is a normal sequence from open water to mudflat to saltmarsh within most estuarine intertidal zones.

Mudflat surfaces are steepened by storm surges, probably because sediment near the open water edges of the surface is mobilised by the extreme turbulence generated by the breaking storm waves, and the sediment thus entrained is deposited farther inland because water turbulence decreases as the wave moves inland and loses energy. The original nearly horizontal surface may be restored during the long intervals between storms by erosion of the high parts and re-deposition of the materials on the low parts. The mechanisms are not well understood, but among those postulated is wave planation by wind-generated waves in the shallow water covering the mudflat during periods of high tide. Thus, both erosive and depositional processes are fundamental to their long-term stability.

Saltmarshes are much less subject to gradient changes in their surfaces, presumably because the surfaces are 'armoured' by the root-masses of the vegetation, but their surfaces nevertheless rise and fall in response to local conditions. Investigations, primarily in small estuaries along the east coast of England, have demonstrated that a major storm event will significantly reduce the average elevation of a saltmarsh (Pethick, 1992b),

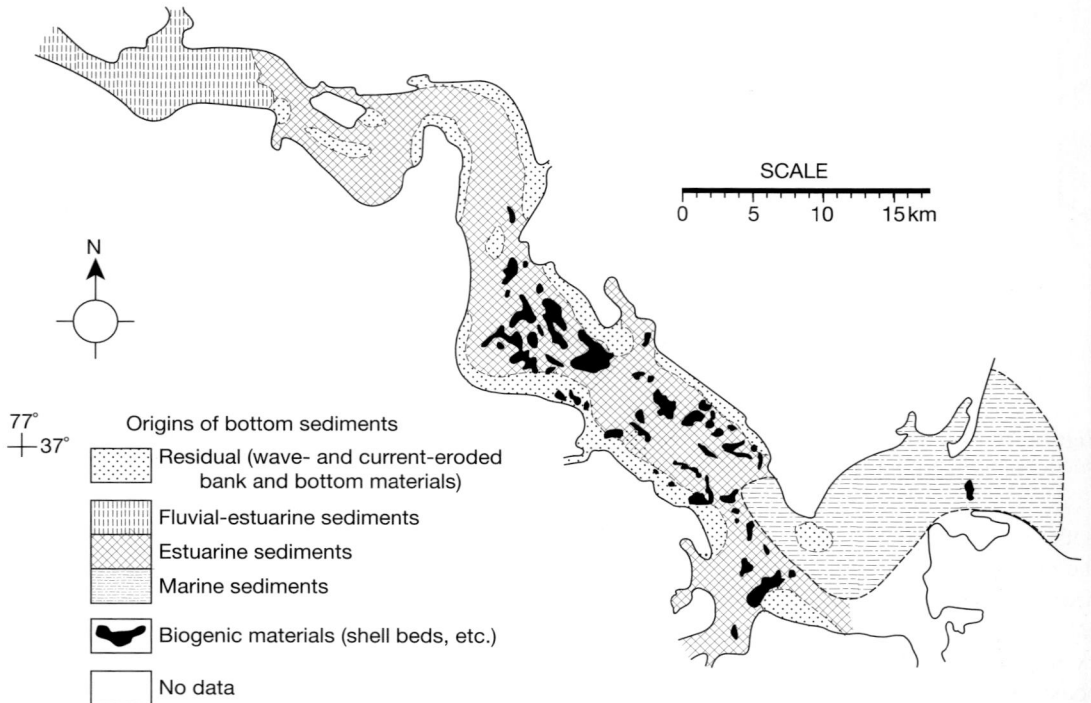

Figure 20.8   Distribution of depositional facies in James estuary (modified from Nichols, 1972).

but that it will recover and in some cases exceed the original surface over a period of several years if the recovery is not interrupted by a second storm event. The reduction resulting from the storm is apparently the product of the extreme water turbulence at the water–soil contact during the period of storm-wave activity. Despite the protection afforded by the vegetation, sediment particles are mobilised and removed during the progress of the storm. The surface of the saltmarsh begins to rise again as soon as the storm has passed, probably because the water after the storm is slightly deeper than before. If entrained sediment is evenly distributed through a water column, then the deeper the water, the greater the amount of sediment available for deposition.

Curiously, while the surfaces of the salt-marshes decline during a storm event, the surfaces of mudflats actually rise, probably because the materials entrained by the storm waves are re-deposited on the adjacent mudflats because the water is deeper, and therefore less turbulent, than

over the neighbouring saltmarshes. In brief, high-energy events cause saltmarsh erosion, and the entrained sediments are transported to and deposited upon the fronting mudflats, resulting in mudflat accretion (Figure 20.10), with the process reversed during intervals of calm.

As with all sedimentation processes, mudflat sedimentation is controlled primarily by two sets of factors: the concentration and grain sizes of the suspended sediment, and the velocity of the water in contact with the mudflat surface. Grain size is important because, all other things being equal, large particle sizes settle more rapidly than small sizes. Current velocity is important because, all other things being equal, the faster the current, the greater the quantity of material that can be kept in suspension and the larger the size of the particle that can be mobilised. However, it is not the mass velocity of the water (a measure of the total amount of water passing through a given cross-section per unit of time) that is important, but rather the velocities of the individual packets of

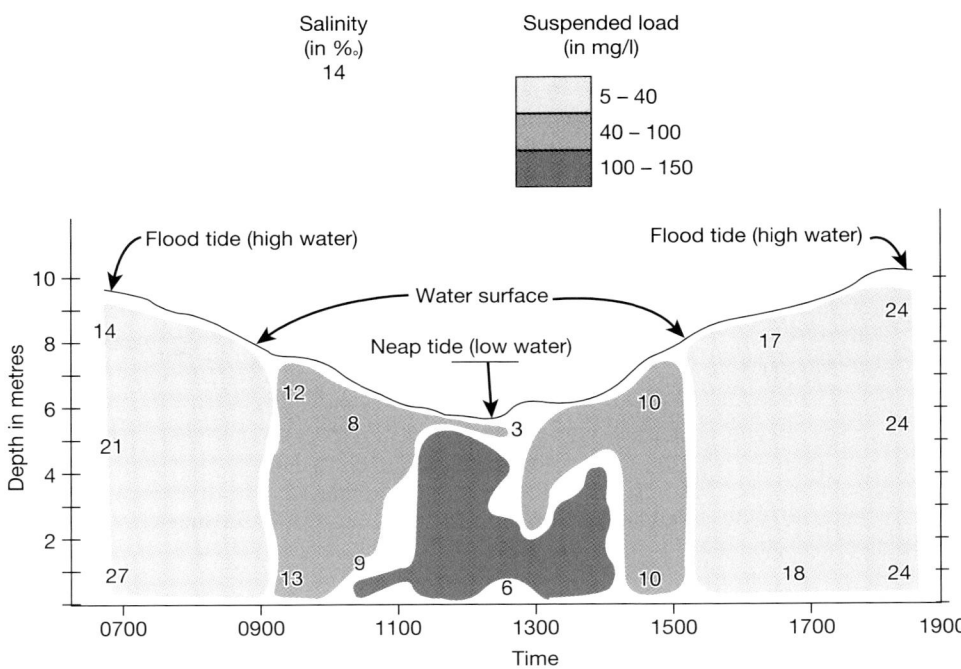

Figure 20.9   Variations in salinity and suspended load in the Tay estuary, Scotland, during one tidal cycle (after Sholkovitz, 1979 as modified from McLusky, 1981).

water surrounding the individual sediment grains. Thus laminar flow can mobilise particles of sizes appropriate to the flow velocity, but transport is confined to saltation (particles tumbling along the channel floor), because the mobilised particles settle out of the laminar flow as soon as they are no longer in contact with the bottom. However, if water having the same mass velocity is also turbulent, the carrying capacity, as well as the particle mobilisation capability, will be a function of the mean velocities of the innumerable packets of water in the turbulence vortices. A particle of appropriate size injected into such a flow may be suspended indefinitely because it will be carried along a path that is closely analogous to the random Brownian motion of a particle suspended in a gas.

True laminar flow is rare, because irregularities in bottom configuration, the presence of vegetation, inequalities due to density variations, wind at the surface, and many other factors create

turbulence. For the most part all of these factors reinforce each other in such a way that there is a marked tendency for turbulence to increase with mass velocity.

As a rapidly moving tidal current slows (and becomes less turbulent) as it flows across the progressively shallower water of a mudflat, first the larger grains, followed by progressively smaller ones will settle out. Sometimes the rates are phenomenal, especially during storm events. Centimetres of sedimentation have been recorded in the course of a single tide cycle. However, the average rate of sedimentation for most mudflats is unlikely to be greater than about a centimetre per year. This value may reflect a balance between erosion and deposition, both of which occur at timescales reflecting both the local tide cycle and the local storm events. For example, most of the sediment deposited on the mudflats during high tide may be remobilised on the succeeding ebb tide and carried toward the sea until the low-tide

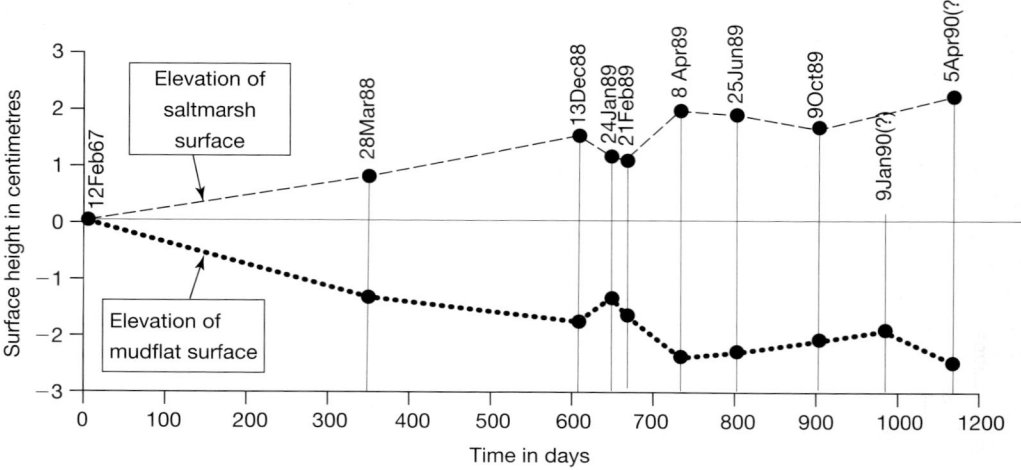

Figure 20.10 Changes in mudflat and saltmarsh surfaces over a 2.9-year period. Extreme storm event 20 December 1988 (modified from Pethick, 1992a).

slack water occurs, at which time much of it will be deposited. The currents of the next incoming tide may then remobilise much of it and carry it back to the mudflats, with the process repeating endlessly. In stable or accreting estuaries, some of the material is lost by long-term deposition on the mudflats and saltmarshes, and some lost by being carried out to sea, with the loss being made up by new sediment added to the estuarine system by the inflowing freshwater stream, or from the sea carried in through the mouth by strong high-tide currents.

As the mudflats grow in height and extent, they eventually assume the form of a wide inter-tidal 'beach' which is capable of dissipating the wave energy which is imposed on it. The reduction in wave energy caused by these wide intertidal flats allows rapid sedimentation at their upper, shore-ward extremity where only infrequent high energy storm waves act to remobilise the deposited material. As their elevation increases, their tidal inundation period is lowered until they may be covered by tidal water for only relatively short periods during the 12.4-hour tidal cycle. Under these conditions, vegetation may develop on these upper mudflats, thus forming embryo saltmarshes.

The mechanisms of formation and evolution of saltmarshes is strongly dependent on tidal range

and exposure to waves originating in the open sea. Thus saltmarshes in mesotidal and macrotidal estuaries with relatively broad mouths, such as those along the eastern shore of England, develop primarily on the landward edges of mudflats, as described above. However, the very extensive saltmarshes which commonly develop within microtidal estuaries with narrow mouths, espe-cially such barred estuaries as those along the south-eastern coast of the United States, develop simply by the gradual accumulation of sediments on previously shallow portions of the estuarine basin. Lacking both strong tidal currents and energetic wave action, sedimentation simply raises the general floor until the process slows and stops as the sedimentary surface rises to a level such that it is exposed to the atmosphere for a sig-nificant length of each tidal cycle. Colonisation by salt-tolerant vegetation usually occurs well before that stage is reached, so the saltmarshes originate and evolve without the prior develop-ment of mudflats.

Once a saltmarsh is established it acts as an efficient sediment trap, because the vegetation stems reduce the velocities of any currents flow-ing across them. This process can be so efficient that the surface can be raised to approximately the same level as that of spring tides, at which point they are often colonised by salt-tolerant shrubs and trees. This evolution of saltmarsh to tree island occurs chiefly in microtidal environments lacking frequent high-energy wave events.

Conversely, especially in mesotidal and macrotidal estuaries with wide mouths, the marsh can act as a reservoir of sediment, as previously described. The extreme turbulences and high current velocities produced by storm waves are capable of mobilising sediment despite the inhibiting effect of the vegetation. During such events the sediments derived from the saltmarsh surfaces are distributed over the adjacent mudflats as well as the estuary basin as a whole, thus effecting an overall widening and flattening of the intertidal zone.

Estuaries and engineering

Estuaries are so complex and dynamic that it is impossible to generalise on engineering requirements within them. Current velocities and directions change in unpredictable patterns because of chance combinations of river flows, tidal cycles and meteorological conditions. Near random variations in water salinity create extraordinarily harsh environments for many structural materials. Foundation conditions may change both vertically and laterally over astonishingly short distances as a result of the complex dynamics of the system. Even more troubling is the fact that even relatively small changes in the configuration of the estuary, such as those which might be created by the emplacement of piers or retaining walls, or the dredging of new channels, may result in dramatic large-scale changes in estuarine dynamics, with consequences that may not be entirely predictable. Channel creation such as dredging and jetty construction, can result in important modifications to the natural regime. The hydraulic effect of deepening or otherwise constraining the flow of a subtidal channel is dependent on the relative size of the channel to the tidal prism (if any), as well as to the overall geometry of the estuarine basin. For example, if a deep longitudinal channel is dredged in the bed of a wide and shallow estuary, one effect will be to enhance the velocities of both the flood tide and ebb tide currents within the newly created channel. If the result is

to produce ebb asymmetry (i.e. faster ebb-tide than flood-tide currents), then there is a high probability that active erosion of the intertidal zone will occur, resulting in a net loss of sediment from the estuarine basin, a slowly enlarging tidal prism, and marked changes in the size and configuration of the estuary mouth. A channel confined by jetties may have similar effects. Without the enhanced hydraulic efficiency of the new channel, sediment would be more likely to be contained within the estuarine system, thus reducing the tidal prism, with all the secondary effects that can flow therefrom.

On the other hand, if the deepened or confined channel produces flood asymmetry, there is a reasonable probability that sediment will be brought into the estuarine basin by the enhanced flood-tide currents, resulting in a long-term reduction of the tidal prism, followed by increased sedimentation within the estuarine basin. A classic example is provided by a study of the Thames estuary (Inglis and Allen, 1957), which suggested that the removal of large volumes of dredged sediment to the outer estuary merely resulted in its immediate transfer back into the estuarine basin. This consequence was avoided by transporting dredged spoil entirely out of the estuary, with the result that dredging (required to maintain an adequate ship channel) was reduced by a factor of twelve.

Even such seemingly innocuous things as boat wakes, if repeated over a long span of time, can affect sediment deposition patterns, current dynamics and salinity variations. Upstream (even outwith the estuary) engineering projects such as river diversion, reclamation, dredging, navigation channel maintenance and dam construction may result in profound changes in the behaviour of estuaries, because they induce changes in river flows and sediment loads. To summarise, an engineering project within an estuary may require the services of a very broad spectrum of engineering specialisations.

Estuaries also have a profound significance to conservation and wildlife management. It is clear that any inhibition or enhancement of sediment movement within the intertidal zone will affect the stability of both the mudflat and saltmarsh forms and, consequently, the habitats which they support. In this context, fixed flood defences may prevent the free adjustment of the saltmarsh and mudflat balance and, hence, may lead to the degradation of these features.

The complex of subtidal, intertidal, and terrestrial habitats characteristic of estuaries often support unique indigenous communities of plants, invertebrates, birds and animals, including fish, amphibians and reptiles. Mudflats and saltmarshes are rich habitats for invertebrates and plants which supply food and shelter for other animals (Davidson *et al.*, 1991). Many estuaries are also of major national and international importance as sites for rookeries, as well as for resting places for migrant and wintering waterfowl. In recognition of their importance to wildlife, many estuaries have been designated or otherwise identified under a variety of national and international measures, including: Ramsar Sites, which are wetlands of international importance designated under the 1971 Ramsar Convention; Special Protection Areas (SPAs) designated under the EC Birds Directive (79/409/EEC Conservation of Wild Birds); Special Areas of Conservation (SACs) designated under the EC Habitats Directive (92/43/EEC Conservation of Natural Habitats and of Wild Fauna and Flora).

20.3 Deltas

Deltas are accumulations of sediment, primarily of fluvial origin, which are created where a stream discharges into a receiving basin such as an ocean, inland sea or lake. While virtually identical expressions form at the margins of both fresh and saltwater bodies, the following discussion will be restricted to those at the margins of salt water. Both the materials and the morphology of deltaic systems are functions of the complex interactions of the hydraulic regime and sedimentary load of the river including: the waves, tides and coastal currents of the sea; the temperature, wind, and precipitation regimes of the local climate; the morphology and tectonic stability of the earth's crust over the area occupied by the delta. Indeed, conditions well beyond

the region of the delta may have a significant impact, the most obvious example being the climatic and geologic conditions over the drainage basin of the stream, because those factors control both the grain-size distribution, mineralogy and the quantities of the sedimentary load carried by the stream. Because the number of combinations and permutations of variables is so very large (Figure 20.11), deltas display wide variations in all aspects of their composition and morphology.

Deltas occur in two very broad contexts. Estuarine, or bay-head deltas are those which form where a stream enters a drowned valley of some type (see above). Continental shelf deltas are those which are constructing a sedimentary structure out onto the continental shelf beyond the mouths of bays, fjords, or other coastal re-entrants. While the dynamics of delta formation and growth are very similar, the lateral restrictions imposed by a drowned valley, of whatever form, result in a very significant simplification of the processes. Consequently, the following discussion will focus primarily on the phenomena associated with classic deltas, which are those that extend beyond the normal continental shoreline onto the continental shelf.

Delta components

The components of a delta system are easy to conceptualise, but extraordinarily difficult to identify in the field because the various parts grade smoothly into each other, without any sharp lines of demarcation. The delta includes the sub-aerial deltaic plain and a sub-marine sub-aqueous delta (Figure 20.12). The deltaic plain commonly consists of three components: a) the upper deltaic plain, which is normally an older part of the deltaic structure, inland from and topographically higher than the remainder of the delta, as well as above the limit of tidal influences; b) the lower deltaic plain, which lies within the realm of river–marine interaction and normally extends seaward of the limit of tidal effects and c) the marginal deltaic plain, which is a somewhat indeterminate zone that normally lies on the flanks of the lower deltaic plain.

The upper deltaic plain may include both active and abandoned channels and is comprised

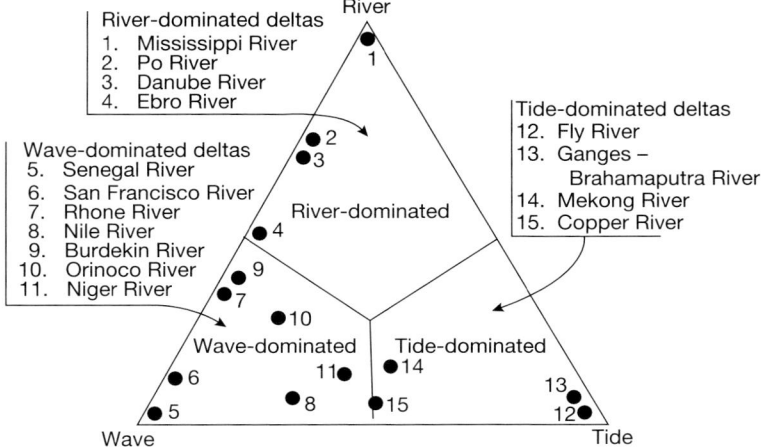

1: Qualitative ternary diagram illustrating how deltas may be categorized in terms of predominant processes operating at the delta front (modified from Galloway, 1975 and Elliott, 1978b).

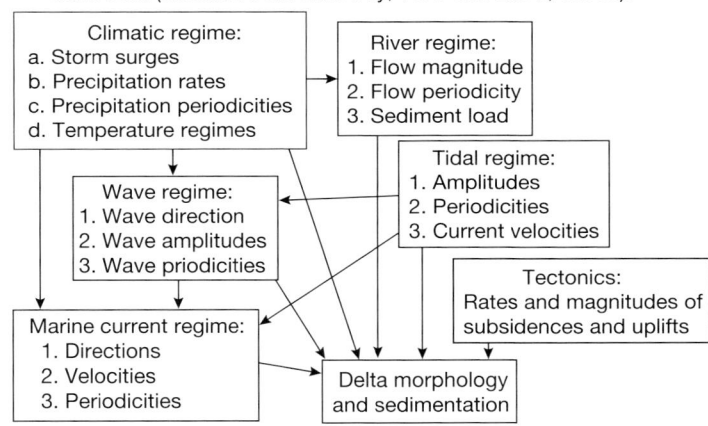

2: Qualitative illustration of the interactions among major factors affecting delta morphology and sedimentation (modifed from Elliott, 1978a).

Figure 20.11 Illustrations of the complexities of interactions among enviromental factors affecting the morphology and sedimentation of deltas on marine shorelines.

of sediments and landforms resulting from the migrations or abandonments of both main and distributary channels, overbank flooding and periodic crevasses (breaks) in the natural levees that characteristically line both banks of both main channel and distributaries. These processes distribute river sediments into adjacent inter-distributary basins (topographic basins of very low relief between the natural levees of active or abandoned channels). The depositional

environments are normally very complex as well as gradational, and include braided channels, point-bars in meander belts, lakes, backswamps and marshes.

The lower deltaic plain is normally even more complex than the upper plain, in part because it is directly affected by all of the vagaries of the associated marine environment, as well as those of the river. Channels are usually numerous and form complex bifurcating and anastomosing patterns,

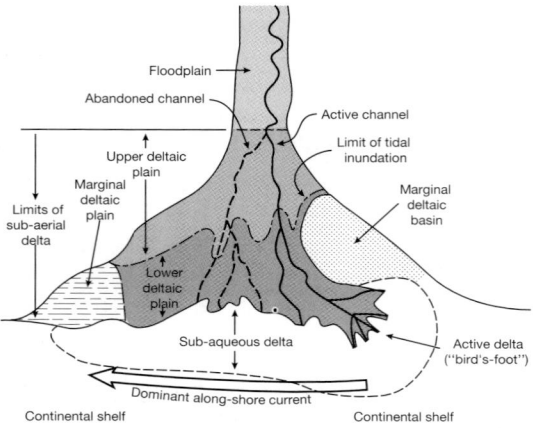

Figure 20.12   Components of a delta (modified from Coleman and Prior, 1982).

with each channel paralleled by natural levees of varying degrees of development. As a direct consequence of the natural levees, there are normally extensive inter-distributary basins, some of which are flooded by the sea to form inter-distributary bays, while others form marshes, swamps and lakes. Within these basins and bays there are complex, intergrading and overlapping sedimentary deposits called 'overbank splays' and 'crevasse splays'.

The marginal deltaic plains are features which form chiefly under the influence of the sea, and thus consist primarily of extensive saltmarshes cut by complex patterns of tidal channels, and interrupted by active or abandoned wave-created beach ridges. In some places the beach ridges form offshore of the marsh edge, creating semi-landlocked salt bays and lagoons between the open sea and the saltmarshes. Occasionally the rapid growth of a delta lobe will almost completely cut off a large area adjacent to the lower deltaic plain, thus creating extensive brackish-water lakes in areas that would otherwise develop the more characteristic saltmarsh. An example of this development is Lake Pontchartrain, adjacent to the Mississippi River on the Gulf Coast of the United States.

The sub-aqueous delta is that portion of the delta that lies below sea level, but nevertheless actively receives fluvial sediments. In principle,

it consists of two quite different depositional environments. The delta front is characteristically a relatively steep slope dropping into deep water from the seaward margin of the sub-aerial delta. However, the 'steep' slope may exhibit an inclination of 0.5°, or even less, especially in front of very large deltas such as that of the Mississippi River. In general, the sediments tend to be coarse-grained near the mouths of the distributaries, becoming finer with greater distance from the shore. Beyond the relatively steep delta front slope, the bottom slopes flatten still further. This pro-delta area, sometimes called bottomset beds, receives only the very finest grain sizes, including fine silt and flocculated clays.

Finally, there is a highly specialised component of the sub-aqueous delta known as the channel mouth bars, which form as highly localised arcuate sea-bed structures just offshore of the mouths of all riverine channels reaching the sea, including both the main channel as well as the distributaries. In general they consist of ridges of the most coarse-grained sediments transported by the streams.

Delta dynamics

Deltas will build seaward from stream mouths only if two conditions are met. First, the supply of sediment delivered by the stream at its mouth must exceed the amount that is mobilised and removed by the waves and currents of the offshore marine environment. Second, the supply of sediment to the sub-aerial portion of the delta must more than compensate for any subsidence caused by crustal down-warping or compaction of previously deposited deltaic sediments. The process of progradation (compensating for subsidence by the progressive superimposition of younger sedimentary layers on older deposits) can be extraordinarily complex. For example, cores from the modern Mississippi delta often reveal complex sequences of strata, each characteristic of a unique depositional environment, and often exhibiting shifts from a shoreline environment to a delta front, back to a shoreline and ending with a modern marsh veneer (Figure 20.13). The continually shifting geometry of channels, natural levees, inter-distributary bays, crevasse splays, and all of the other depositional and dynamic features so characteristic of the deltaic environment results in considerable three-dimensional variability in the sequences (Figure 20.14). In fact, it may be said with some confidence that no two cores taken 50 metres apart will display precisely the same sequence of deposition.

The proclivity for rapid spatial variability within a delta is even further exacerbated by a process known as delta switching (Figure 20.15). It is essentially a consequence of the process of progradation. As the delta front moves seaward, the efficiency of the discharge channels is reduced because their thalweg (i.e the planimetric expression of the thread of maximum current velocity) length is increased and their gradient is decreased. This reduces the current velocity in the channels, which increases the flood heights, which results in a more frequent inundation of the delta surface, which results in higher sedimentation rates. Eventually the river is flowing down the crest of a low ridge which projects into the sea. The end result is that at some arbitrary point upstream, the gradient of a distributary which departs at an angle from the main channel is steeper than the gradient of the main channel because the distance to the sea is shorter, at which time the river switches to the distributary channel, and

the original channel is abandoned. A new delta lobe is then initiated at the point where the new main channel reaches the sea. The Mississippi provides a text-book example of the process. Over the past 7000 years the river has switched seven times, each delta remaining active only about 1000 to 1500 years. The last switch, which began the formation of the modern Balize delta, began only about 600 to 800 years ago (Figure 20.15).

With the abandonment of each delta lobe, regional subsidence, which in the Mississippi delta is caused by a combination of crustal warping and sediment compaction, results in marine transgressions. Thus, a coastal plain formed by delta switching will exhibit an orderly repetition of depositional events characterised by interfingering and overlapping deltaic sequences separated by shallow-water marine facies.

Channel-mouth bars are formed where the confined current of the stream, whether main channel or distributary, slows as it spreads into the sea beyond the mouth. The rapid drop in current velocity results in rapid deposition of the coarsest components of the transported sediments closest to the river mouth, with progressively finer sediments being deposited with increasing distance from the mouth. Theory predicts a smoothly arcuate ridge concave to the mouth of

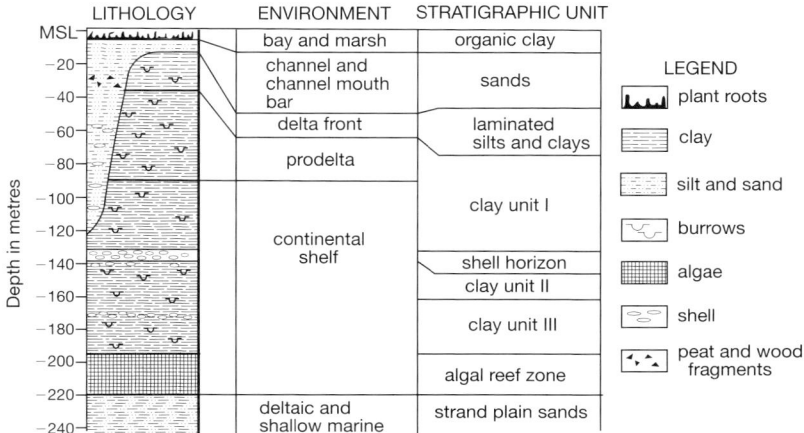

Figure 20.13   Vertical sediment sequence in modern Mississippi River delta (modified from Coleman and Prior, 1980).

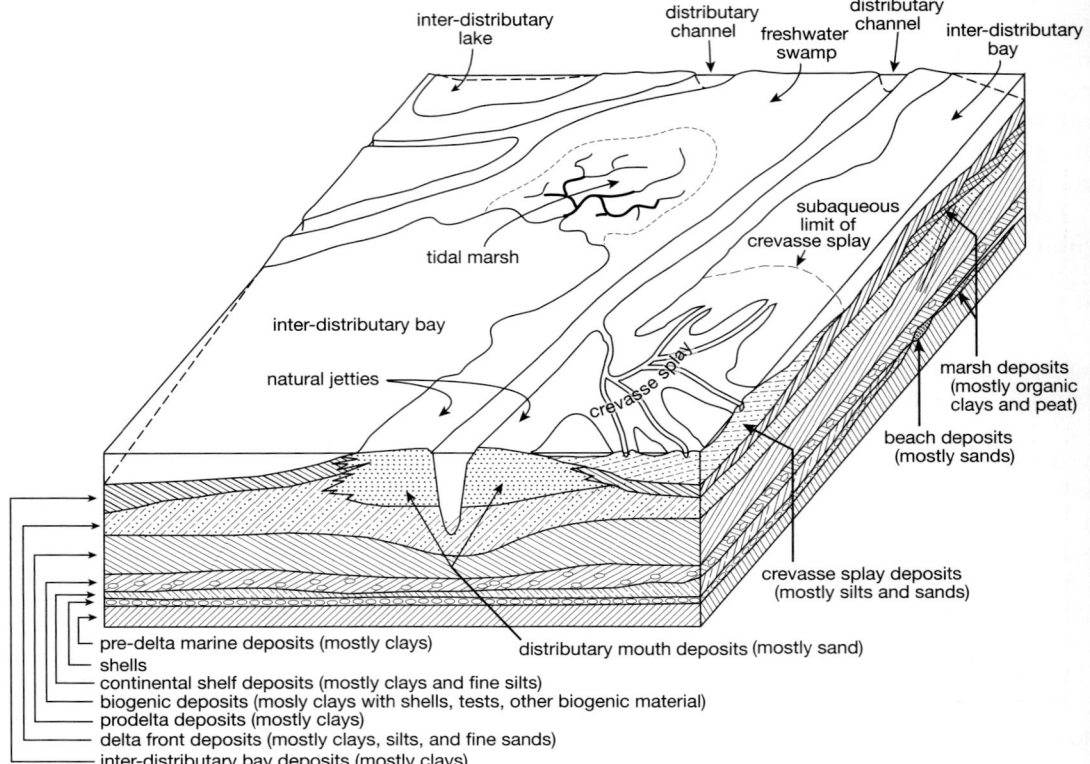

Figure 20.14   Isometric block diagram illustrating rapid vertical and lateral facies changes in deltaic deposits (modified from Coleman and Prior, 1980).

the stream, but the actual form is greatly modified by such factors as wave climate, tidal range, and along-shore currents. As the margin of the sub-aerial delta moves seaward, the river discharge channels are extended, and in the process a channel is progressively scoured through the channel-mouth bar, while at the same time new deposition on the bar extends further toward the sea. The forms remain, but the entire structure gradually moves seaward.

20.4 Depositional environments

Inter-distributary bays

Inter-distributary bays, which form between the natural levees paralleling the main channel or the distributaries of the river, are normally shallow,

open bodies of water, bounded on the seaward side by open sea and on the landward side by saltmarsh penetrated by bifurcating and anastomosing tidal channels and by reaches of natural levee. While the water is usually saline, or nearly so, brackish periods may occur, especially when the river is in flood, or when a crevasse in a natural levee permits large amounts of river water to enter the bay. Such breaks in the natural levees initiate bay-filling episodes, because the sediment-laden river water creates a fan-shaped sub-delta known as a crevasse splay, which itself develops a radial system of bifurcating channels. On a large scale, entire delta systems may enlarge by sequences of bay-filling episodes (Figure 20.16).

Typically, each crevasse splay evolves from an initial break in the natural levee during a flood

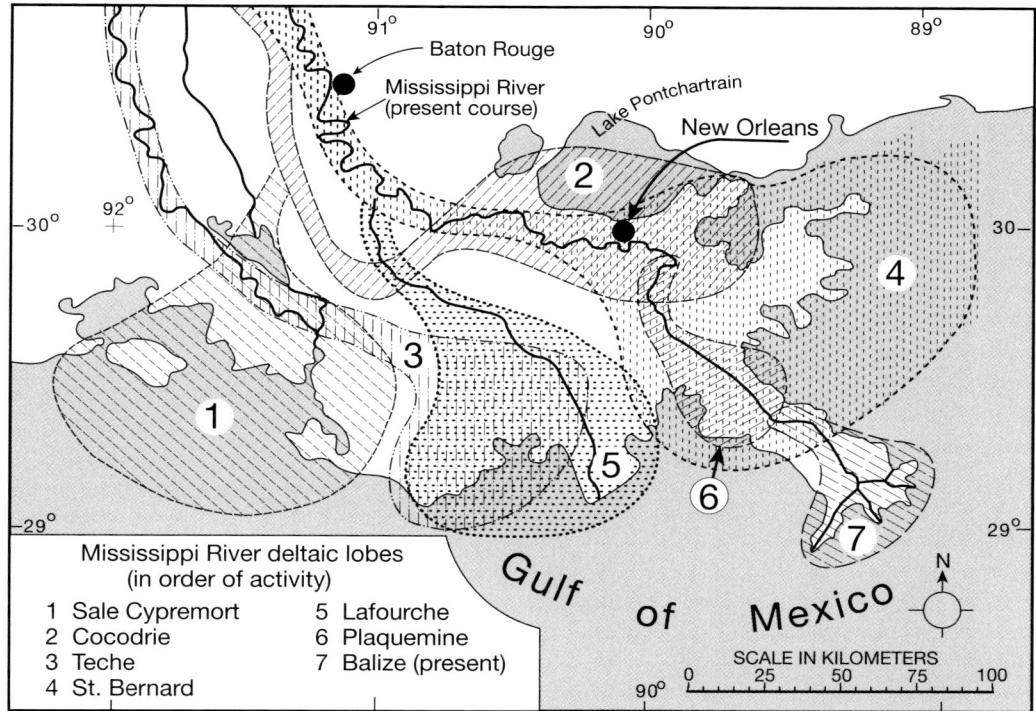

Figure 20.15 Delta switching in the Mississippi River delta (modified from Kolb and Van Lopik, 1966).

episode. If conditions are propitious, the crevasse enlarges during successive flood episodes, and sediment deposition in the resultant splay proceeds rapidly until the distributary channels in the splay offer no hydraulic advantage over the main channel. Thereafter water volumes and sediment transport decrease, until at last the system becomes inactive. The evolution of such crevasse splays may be nearly as complex as that of a primary delta (Figure 20.17). If regional subsidence is significant, the entire sub-delta may eventually be lowered below sea level, thus returning the area to an approximation of the original configuration and condition. This may set the stage for another episode of sub-delta formation in the same inter-distributary bay, as illustrated in Figure 20.16.

Bay-filling episodes are usually readily identifiable in the stratigraphic record, because they result in a coarsening upward sequence of deposits (Figure 20.18). The lowest member is

commonly a highly organic peaty clay or more rarely, a thick peat, which presumably formed while the area was a saltmarsh, prior to the formation of the crevasse splay. The organic layer is then overlain by alternating silts and silty clays which are extensively burrowed by marine organisms. These may be interpreted as pro-delta sediments deposited on the floor of the inter-distributary bay in advance of the sub-delta front. These beds are in turn followed by layers of coarser silt and fine sand which are only slightly disturbed by burrowing marine organisms. As the sub-delta front advances into the bay, sands and silts form graded beds and small-scale climbing ripples may be preserved. The capping unit usually consists of either clay which has been extensively burrowed and reworked, or highly organic peaty clays. If the former, the implication is that regional subsidence dropped the sequence below sea level, where it remained without further sedimentation for a considerable time; if the latter, the implication is that

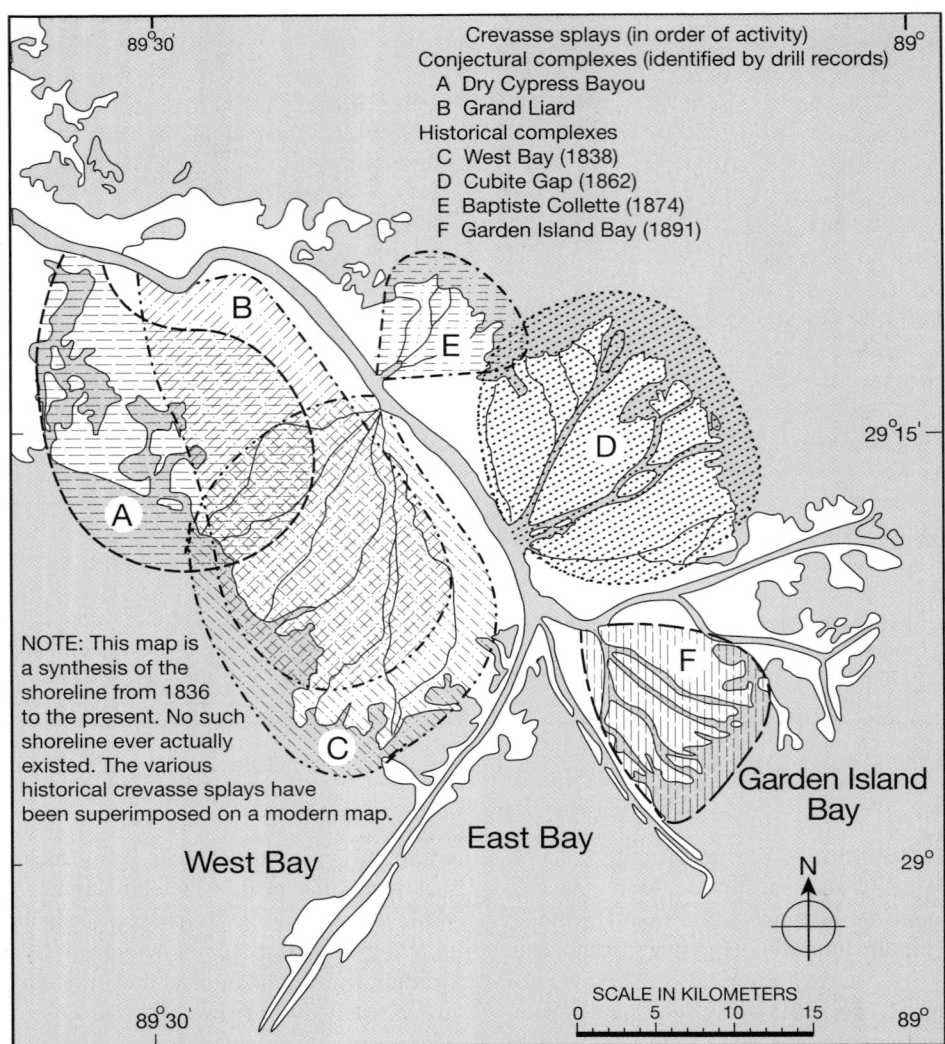

Figure 20.16   Crevasse splays in the modern Mississippi River delta (modified from Coleman and Prior, 1980).

it remained at or near the surface, and thus provided the base for a tidal saltmarsh.

Engineering works in inter-distributary bay sediments and environments must be undertaken only with full awareness that such deposits consists of materials that have extremely high porosities and permeabilities, and that their structures are partially water supported. The combination makes them extremely prone to shrinkage on draining or drying. This is especially true of the peats and highly organic clays. In addition,

because the sedimentary structure is partially water supported, inter-distributary bay deposits are universally in the process of compaction, which results in continuous surface subsidence. The only variable is the *rate* of subsidence, which may vary over a broad range, depending on local circumstances.

Abandoned distributaries

Distributary channels are the conduits for sediment transport within deltaic systems. They rarely

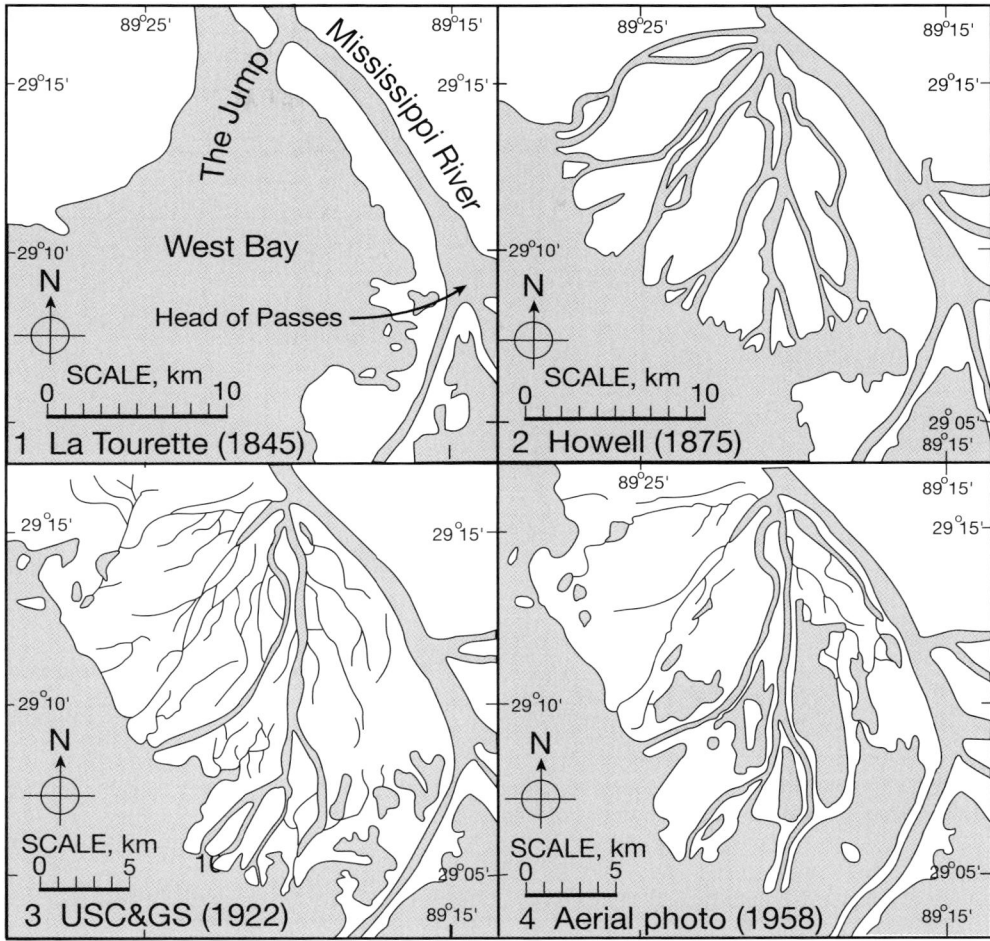

Figure 20.17   Development of the West Bay crevasse splay (bay-filling episode), 1845 to 1958 (modified from Coleman and Prior, 1982).

exhibit point-bar or meander-belt deposits because they do not usually migrate laterally. The reasons remain obscure. The reasons for the abandonment of such channels is equally uncertain, and it is likely that local circumstances dictate individual instances. The following reasons have been suggested for certain Mississippi delta distributaries: log jams blocking the entrances may become so impervious that flow is prevented; the thalweg gradient of the distributary may become less than that of the main channel; changes in the course of the main channel may isolate the head of the distributary; the channel efficiency of the

distributary may be so reduced by such obstructions as tree growth that the river blocks off the head of the channel by the construction of a bar, in a process analogous to the creation of ox-bow lakes.

Deprived of its river source, the water flow becomes very slow and may cease entirely during certain seasons. Once abandoned, the channel begins to fill with mostly locally derived sediments, although significant amounts are usually derived from overbank flood flows in the main channel. Because all sediment sources, including the overbank flood flows, consist of fine-grained materials,

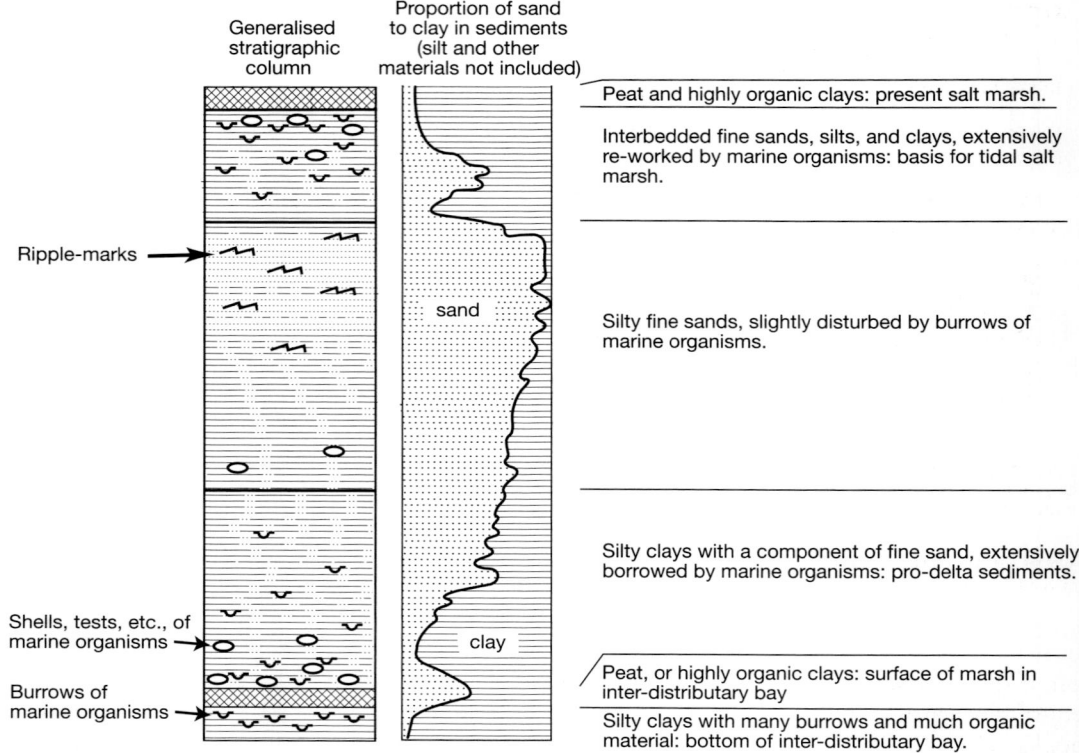

Figure 20.18   Characteristic sediment sequence in a crevasse splay (bay-filling episode). Sample from West Bay episode, Mississippi River delta (modified from Coleman and Prior, 1982).

the channels eventually fill with fine silts and clays. Without significant currents, the infilling process is hastened by water-tolerant trees and shrubs which colonise the shallow edges and act as highly efficient sediment traps, gradually constricting the channel to scarcely more than a thread. Significant amounts of wood and other vegetal debris are sometimes incorporated, and in some instances the final depositional layer is peat or highly organic clay.

Thus, the abandoned distributary channels are represented by channel plugs consisting almost entirely of fine-grained silts and clays, often slightly coarser near the base, incorporating vegetal material and often capped with peat or highly organic clay (Figure 20.19).

Channel-mouth bars

Channel-mouth bars are an integral part of the stratigraphic record of the sub-aqueous delta. As

the edge of the sub-aerial delta advances seaward, the channel is defined by natural jetties which are extensions of the natural levees which normally border the river channel as it crosses the sub-aerial delta. However, the function of the natural levee changes subtly as the river builds out into open water. As long as the land on either side is above sea level, the natural levees serve to guide and constrain the current only during high water stages, but as soon as the sedimentary surface recedes below sea level, the natural levee, now a natural jetty, serves to constrain and direct the current during all water stages. The natural jetties are continued as sub-aqueous forms as the river current escapes from confinement and spreads out, and therefore reduces in velocity. This mechanism is responsible for the 'toes' of the 'bird's-foot deltas' that are so prominent at the mouth of the Mississippi and other deltas. The sands which

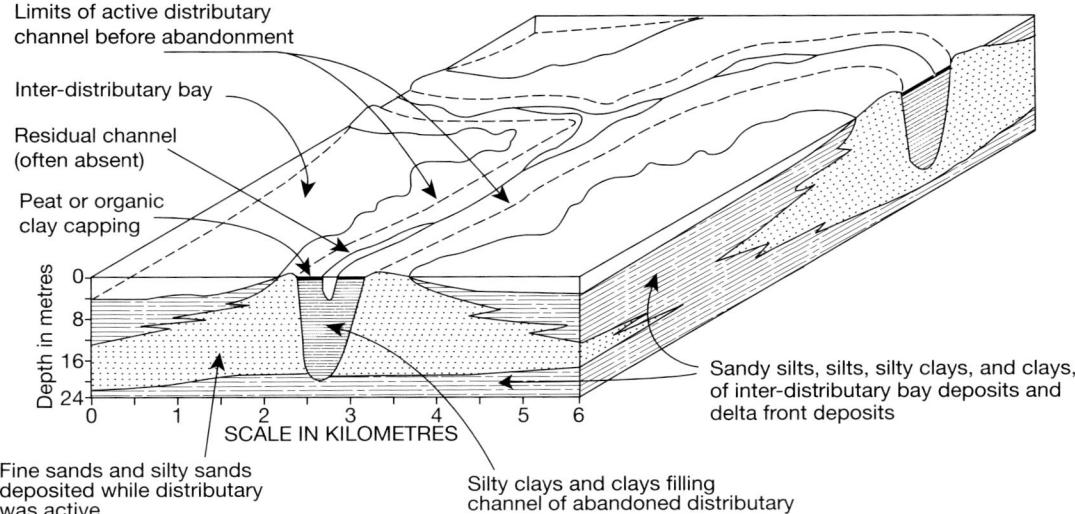

Figure 20.19   Isometric block diagram of deposition in channels of abandoned distributaries in the lower deltaic plain of the Mississippi River (modified from Coleman and Prior, 1982).

comprise the dominant materials of the natural jetties are deposited over, and generally become indistinguishable from the sub-aqueous channel-mouth bars. Some of the sand comprising the bar itself is usually mobilised and transported in all directions away from the river mouth by wave action, along-shore currents and tidal currents, the extent and direction being a function of the local conditions. As the channel mouth advances sea-ward, the sandy bar and the natural jetty materials are emplaced over the silts and clays of the delta slope (Figure 20.20). The result is a feather-edged 'tongue' of sandy sediments that gradually pro-grades, bounded by feather edges as it merges with the more fine-grained inter-distributary bay sediments on either flank.

The bar sediments typically show a coarsening upward sequence, with pro-delta silty clays at the base. Sequentially above these are strata of increasingly silty cross-laminated sand and silt layers with burrows and shell fragments, followed still higher in the sequence by fine sand structures such as ripples caused by the oscillatory currents generated by waves, and cross-laminae and cut-and-fill features resulting from the unidirectional tidal and along-shore currents which more or less continually sweep the upper portions of the channel mouth bars. Bar deposits are usually capped by the silts and silty clays of inter-distributary bay sedi-ments, which normally include sand grains, shells and shell fragments and organic debris.

Engineering problems associated with channel-mouth bars are in most instances only those related to the properties of unconsolidated sands and silty sands. However, local diapirism around the channel mouths of the Mississippi River can produce extensive and relatively rapid ground motions, including both uplift and horizontal dis-placements which can rapidly destroy buildings, wharfs and pipelines. The cores of the diapirs are the infamous mudlumps almost unique to the Mississippi River.

Sub-aqueous slopes

Rapid sedimentation on the sub-aqueous delta slopes, including both the delta front and the pro-delta, results in soft silts and clays with low strengths. For reasons not fully understood, the sediments are often rich in interstitial methane gas and can have extremely high pore fluid pressures (Figure 20.21). Under these conditions, even slopes as low as 0.5° are subject to a number of different

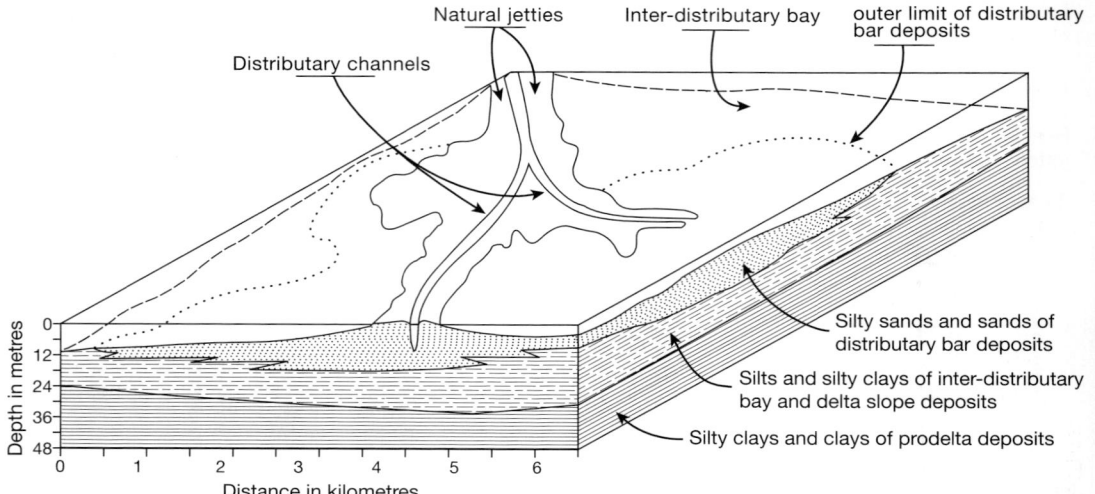

Figure 20.20   Isometric block diagram illustrating characterisitic sediment sequence in distributary mouth bar deposits (modified from Coleman and Prior, 1982).

kinds of slope instabilities which collectively move large quantities of sediment downslope into deeper water (Figure 20.22). Sub-aqueous landslide varieties include: shallow rotational sliding, which occurs most commonly near the seaward edges of channel mouth bars; collapse depressions resulting from the subsidence of inter-distributary bay sediments; elongated mudflows or mudslides which may carry very large quantities of sediment downslope for very long distances, resulting in mudflow lobes or debris lobes that prograde across the continental shelf, or even into the offshore deep sea basin. The causes of the initiation of sediment movement are not fully understood, but they are known to include the complex interaction of several processes (Figure 20.23), including excessive pore pressures, which are enhanced by the generation of methane gas within the sediments themselves. Surface waves, especially if amplified by storm surges during great storms, create high-frequency cyclic loadings which produce instabilities. Typically delta-front sediments have such high water contents (often in excess of 50%) that they are nearly always near the plastic limit, and frequently exceed the liquid limit.

The development of oil and gas resources in sub-aqueous deltas has encountered problems of design, installation and maintenance of platforms and pipelines. The instabilities characteristic of sub-aqueous deltaic sediments pose particular difficulties because they are especially prone to both vertical subsidence and lateral displacement. Many existing structures have been damaged, and some have been lost due to some combination of these factors. Modern geological and geotechnical capabilities, particularly the mathematical modelling of tide and wave interactions with the sea bottom, can be used to provide useful engineering solutions.

20.5 Summary

Estuaries and deltas comprise only a small percentage of the total area of the coastal zone of the world. Yet, because of their attractiveness to humans, they are among the most intensively utilised landscape on the planet. Part of the reason they are so desirable relates to the biological, hydrological, climatological, sedimentological and morphological dynamics that results from their position astride the boundary of sea and land. Many of the ways in which the estuaries and deltas have been used, or are proposed for use, are

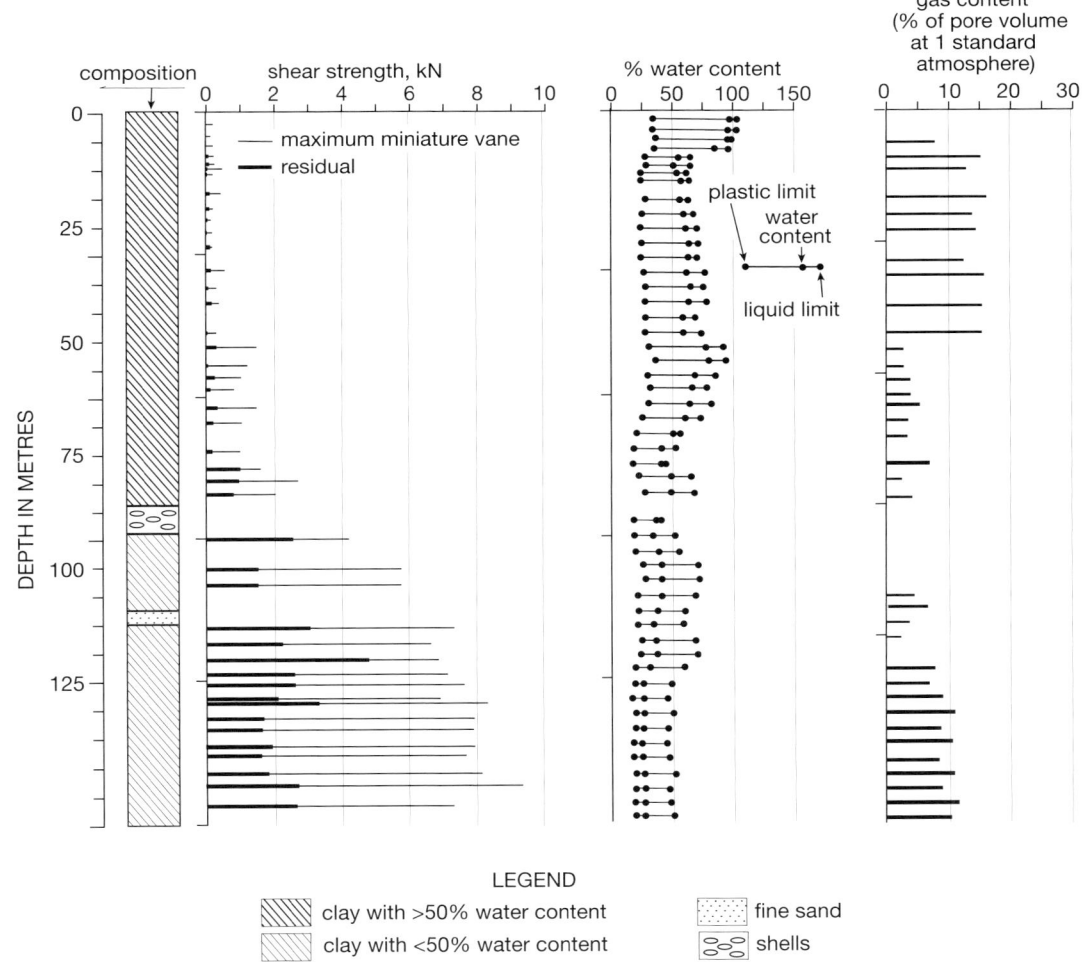

Figure 20.21   Soil characteristics on the delta front, Mississippi River delta (modified from Bea and Audibert, 1980).

mutually incompatible, and methods of reconciling such diversity in the extraordinarily complex environments of the land–sea interface are likely to provide unique challenges to the engineer for the foreseeable future.

References

Bacon, J. and Bacon, S. (1988) *Dunwich Suffolk.* Segment Publications, Colchester.

Bea, R. G. and Audibert, J. M. E. (1980) Offshore platforms and pipelines in the Mississippi river delta. *Proc. Am. Soc. Civil Engrs.* **106** (GT 8), 853–869.

Biggs, R. B. (1982) Estuaries. In Schwartz, M. L. (ed.) *The Encyclopedia of Beaches and Coastal Environments.* Hutchinson Ross, Stroudsburg, Pa., 393–397.

Carter, R. W. G. (1992) Coastal conservation. In Barrett, M. G. (ed.) *Coastal Zone Planning and Management.* Thomas Telford, 21–48.

Coleman, J. M. and Prior, D. B. (1980) *Deltaic Sand Bodies.* Am. Assn. Petrol. Geol. Continuing Education Series, No.15, Tulsa, Oklahoma.

Coleman, J. M. and Prior, D. B. (1982) Deltaic environments of deposition. In Scholle, P. A. and Spearing, D. (eds) *Sandstone Depositional Environments.* Am. Assn. Petrol. Geol., Memoir 31, 139–178.

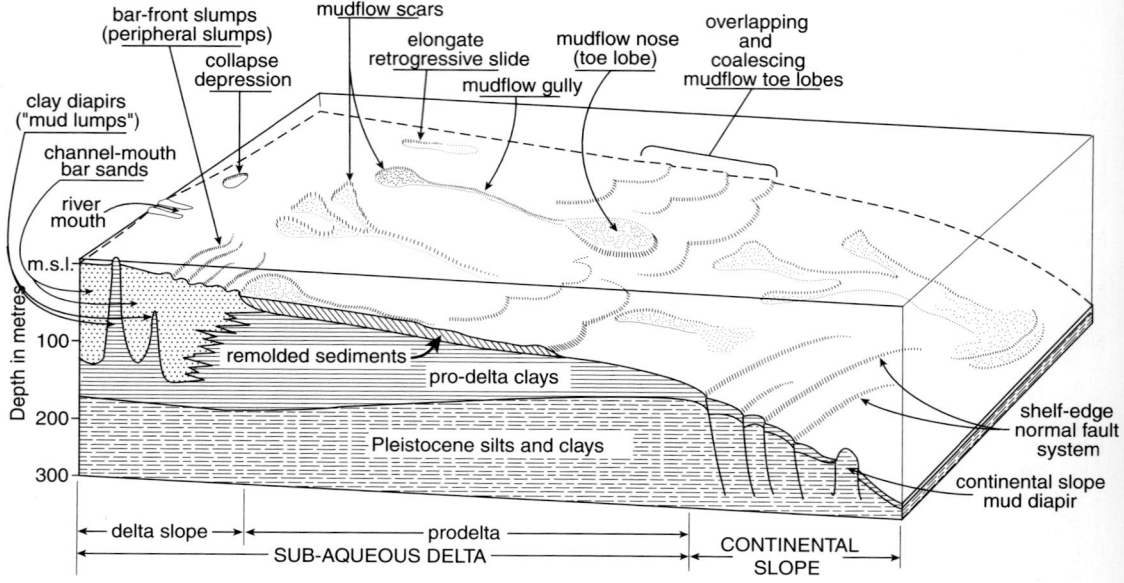

Figure 20.22   Block diagram illustrating soil failure modes on the sub-aqueous delta of Mississippi River (modifed from Coleman *et al.*, 1998).

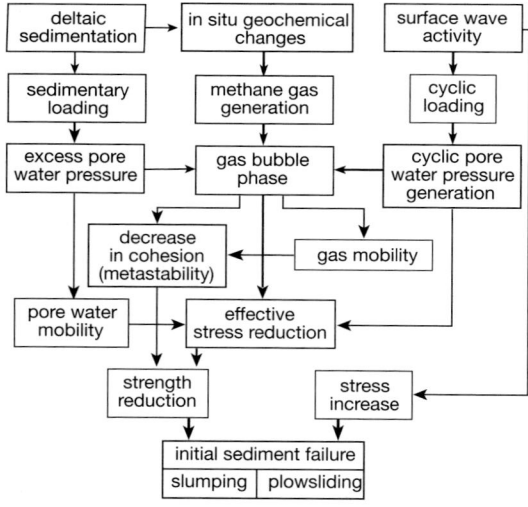

Figure 20.23   Summary of factors contributing to failure of marine sediments on low slopes on the Mississippi River delta (modifed from Prior and Coleman, 1982).

Coleman, J. M., Roberts, H. and Stone, G. (1998) Mississippi River delta: An overview. *Journal of Coastal Research* **14** (3), 698–716.

Davidson, N. C., d'A. Laffoley, D., Doody, J. P., Way, L. S., Gordon, J., Key, R., Drake, C. M., Pienkowski, M. W., Mitchell, R. and Duff, K. L. (1991) *Nature Conservation and Estuaries in Great Britain*. Nature Conservancy Council, Peterborough.

Dronkers, J. (1986) Tidal asymmetry and estuarine morphology. *Netherlands Journal of Sea Research* **20**, 117–131.

Elliott, T. (1978a) Deltas. In Reading, H. G. (ed.), *Sedimentary Environments and Facies*. Oxford University Press, 97–142.

Elliott, T. (1978b) Clastic shorelines. In Reading, H. G. (ed.), *Sedimentary Environments and Facies*. Oxford University Press, 143–77.

Fairbridge, R. W. (1980) The estuary: its definition and geodynamic cycle. In Clausson, E. and Cato, I. (eds) *Chemistry and Biogeochemistry of Estuaries*. Wiley, New York, 1–31.

Galloway, W. E. (1975) Process framework for describing one morphologic and stratigraphic evolution of the deltaic depositional system. In Broussard, M. L. (ed.) *Deltas, Models for Exploration*. Houston Geol. Soc., Houston, Texas, 87–98.

Gardner, T. (1754) *Historical Notes on Dunwich*. Blythburgh and Southwold.

Inglis, C. and Allen, F. (1957) The regimen of the Thames estuary as affected by currents, salinities and river flow, *Proceedings of the Institution of Civil Engineers* **7**, 827–868.

Kolb, C. R. and Van Lopik, J. R. (1966) Depositional environments of the Mississippi river deltaic plain, southeastern Louisiana. In Shirley, M. L. and Ragsdale, J. A. (eds) *Deltas.* Houston Geol. Soc., Houston, Texas, I 7–62.

McDowell, D. M. and O'Connor, B. A. (1977) *Hydraulic Behaviour of Estuaries.* Wiley, New York.

McLusky, D. S. (1981) *The Estuarine Ecosystem.* Blackie, Glasgow and London; Wiley, New York.

Nichols, M. M. (1972) Sediments of James River estuary. In Nelson, B. W. (ed.) *Environmental Framework of Coastal Plain Estuaries.* Geol. Soc. of America, Inc., Boulder, Colorado, 169–212.

Pethick, J. (1992a) Salt marsh geomorphology. In Allen, J. R .L. and Pye, K. (eds) *Saltmarshes: Morphodynamics, Conservation and Engineering Significance.* Cambridge University Press, 41–62.

Pethick, J. (1992b) Natural change. In Barrett, M. G. (ed.) *Coastal Zone Planning and Management.* 49–63. Thomas Telford, London.

Pethick, J. (1996) The geomorphology of mudflats. In Nordstrorn, K. F. and Roman, C. T. (eds) *Estuarine Shores: Evolution, Environments and Human Alterations.* Wiley.

Prior, D. B. and Coleman, J. M. (1982) Active slides and flows in underconsolidated marine sediments on the slopes of the Mississippi delta. In Saxov, S. and Nieuwenhuis, P. K. (eds) *Marine Slides and other Mass Movements.* Plenum, New York, 21–49.

Sholkovitz, E. R. (1979) Chemical and physical processes controlling the chemical composition of suspended material in the River Tay estuary. *Estuarine and Coastal Marine Science* **8**, 523–545.

Wright, L. D., Coleman, J. M. and Thom, B. G. (1973) Processes of channel development in a high-tide range environment: Cambridge Gulf-Ord River Delta, W. Australia. *Journal of Geology* **81**, 15–41.

Further reading

Bowden, K. F. (1967) Circulation and diffusion. In Lauff, G. H. (ed.) *Estuaries.* American Association for Advancement of Science, Washington DC, 15–36.

Coleman, J. M. (1988) Dynamic changes and processes in the Mississippi River delta. *Geological Society of America Bulletin* **100**, 999–1015.

Dronkers, J. and van Leussen, W. (eds) *Physical Processes in Estuaries.* Springer-Verlag, Berlin.

Dyer, K. R. (1986) *Coastal and Estuarine Sediment Dynamics.* Wiley, Chichester.

French, P. W. (1997) *Coastal and Estuarine Management.* Routledge, London.

Hobbie, J. E. (ed.) (2000) *Estuarine Science: a Synthetic Approach to Research and Practice.* Island Press, Washington DC.

Ippen, A. T. (1966) *Estuary and Coastline Hydrodynamics.* McGraw-Hill, New York.

Lauff, G. H. (1967) *Estuaries.* American Association for the Advancement of Science, Washington DC.

Leeder, M. R. (1982) *Sedimentology, Process and Product.* George Allen & Unwin, London.

National Research Council (1983) *Fundamental Research on Estuaries.* National Academy Press, Washington DC.

Nelson, B. W. (ed.) (1972) *Environmental Framework of Coastal Plain Estuaries.* Geol. Soc. of America, Boulder, Colorado.

Perillo, G. M. E. (ed.) (1995) *Geomorphology and Sedimentology of Estuaries.* Elsevier, Amsterdam.

Shirley, M. L. and Ragsdale, J. A. (eds) (1966) *Deltas.* Houston Geol. Soc. Houston, Texas.

Stanley, D. J. (1997) Mediterranean deltas: subsidence as a major control of relative sea-level rise. *Bulletin de l'Institut Océanographique*, Monaco, Numero special **18**, 35–62.

Stone, G. W. and Donley, J. C. (eds) (1998) The World Deltas Conference. *Journal of Coastal Research* **3** (14), 695–915.

van Westen, C. J. and Scheele, R. J. (1996) *Planning Estuaries.* Plenum Press, New York.

21. Coastal Environments

Julian Orford

21.1 Introduction

This chapter is about how engineers can use geo-morphology in support of their activities in the coastal zone. There is a range of texts (e.g. Komar, 1998; Horikawa, 1988) for engineers working on coastal problems that identify empirical and theo-retical approaches to wave and tidal processes and associated sediment transport. Considerable under-standing of the geomorphological basis of coasts can also be found (e.g. Davies, 1972; Carter, 1988; Trenhaile, 1997). There are texts that attempt to straddle both approaches (van Rijn, 1998), however there is a scarcity of material that specifically iden-tifies the geomorphological approaches and diffi-culties that engineers need to identify when dealing with coastal problems in terms of causes as well as symptoms. This chapter outlines a framework in which engineers might consider the nature and solution of coastal problems with which they deal. There is little space available to cover the variation and complexity of the world's coastlines, so this chapter will emphasise the features of mid-latitude coasts where engineering intervention is at its most complex and advanced.

The seaward boundary of the coastal zone (CZ) is given, for convenience, as the depth of the effective wave base related to extreme storms, while the landward boundary is often well land-ward of coastal deposition, in that terrestrial sources of sediment and water masses need to be included in the CZ system. The importance and role of energy and mass exchanges from terres-trial geomorphological systems into the CZ, as well as exchanges between it and offshore envir-onments should be recognised as part of the background to understanding the dynamics of the CZ. Engineering geomorphology of the CZ con-cerns the actual shoreline (i.e. the beach and its

spatially contiguous geomorphological compon-ents) more than it does the terrestrial element that theoretically defines the landward zone of the CZ. The shoreline is the pivotal position with respect to hazardous contact between the human and physical worlds, hence maximal engineering intervention at this position.

The importance of the CZ in the future should not be underestimated, as the physical challenges to societies adjusting to CZ changes accelerated by climate change, will occupy a central position in future environmental agendas. It has been esti-mated that 37% of the world's population lives within 100 km of the coastline (Cohen *et al.*, 1997) and is growing at a rate greater than the overall population rate. This fact alone serves to emphasise the strategic need to be able to deal with the range and magnitude of the impacts of future human occupancy of the coastline.

21.2 Perspectives affecting our understanding of the coastal zone

Human interventions in the coastal zone

The perils posed by living next to a hazardous and unstable coastal zone have not prevented a tradi-tion of positive intervention into the coastal zone by most coastal societies. The success of inter-vention has been constrained by the technical effi-ciency of the moment. Lack of ability did not prevent past societies from taking risks in estab-lishing a presence in hazardous positions with respect to the CZ. But prior to the Industrial Revolution, there was a more fatalistic acceptance that nature could not be dominated, such that soci-ety had to live (and die) with the impact of extremes of coastal hazards. This attitude still has echoes in some contemporary societies that are striving to achieve an economic transformation of their CZ.

The nature of human intervention in the CZ reflects the cultural imperatives of each society. In western-orientated societies, technological advances in engineering design and material behaviour over the last two centuries support the view that CZ vicissitudes could be contained. For most of the twentieth century a persistent culture of living by the seashore was not balanced by a recognition of the limits of technology, nor by an understanding of how coastal systems were changed by human intervention. However, experience over the last fifty years, whereby engineering intervention has generally proven to be unsustainable *per se*, over-expensive for the benefits derived, and incapable of withstanding the extremes of coastal hazards, has led to the recognition that intervention carries a sustainability cost to the coastal environment. Whilst it is likely that debates over the realities and costs of engineered coastal living will continue into the future (e.g. the conflicts on the value of sea walls between Kraus, 1988 and Pilkey and Wright, 1988), most coastal engineers accept that intervention into the CZ is at a price. Once started, engineering needs constant support as it rarely creates a mutually balancing cultural–physical system. As Doyle *et al.* (1985) commented in their 'Laws of the Coast', once you start intervening in the coast you can never stop. Whether this cost is acceptable, depends on the perspective and relative valuation of natural and human environments.

To protect or allow change?

The scale of perceived damage to coastal systems by human intervention has generated tension. Some engineers believe that their mission is foremost to protect the landward investment while others would argue that such protection has been at the cost of the seaward environments. Rather than continue this debate, it is now time for both sides to reflect on the best ways to achieve a sustainable accommodation of human coastal usage. It has become increasingly recognised that human coastal activity has to be conducted within self-regulating coastal systems so that the sustainability

of natural resources can be assured. This means that societies will have to work with the coast and not block or deflect coastal processes and responses. This idealistic approach is unlikely to happen overnight, if at all, given the history of human investment in the CZ. The pace of natural coastal change is, however, speeding up as globally accelerated climate change is translated into rising sea levels and transgressive shifts of coastal environments onshore. This means that there is a rising risk associated with the shoreline as a point of hazardous interaction between natural and cultural activities.

Scenarios for future intervention in the coastal zone

There are four scenarios that societies can explore in response to these shoreline pressures:

1. Hold the line by trying to prevent all potential changes.
2. Do nothing and let society take a progressive loss of function and value due to coastal change.
3. Accommodate the pressures by spatially mixing holding the line and doing nothing in proportion to the risk and utility of maintaining specific coastal functions.
4. Undertake managed retreat by shoreline realignment.

The last approach is to 'give up' marginal coastal areas by allowing them to form the accommodation space for coastal environments associated with the present (and future?) shoreline, and by so doing, allow an increasing revision to a natural and self-sustaining coastal system. Although this is an ideal state and in theory allows for the retention or expansion of natural habitats that have come under severe pressure in the twentieth century, there has to be a realistic sense of the likelihood of society allowing sufficient release of coastal space to keep pace with the landward shift of coastal environments under the predicted eustatic sea-level changes of the twenty-first century (Houghton, 1997). This means that engineering will be found in the future, but that its mode is diversifying from outright resistance to subtle

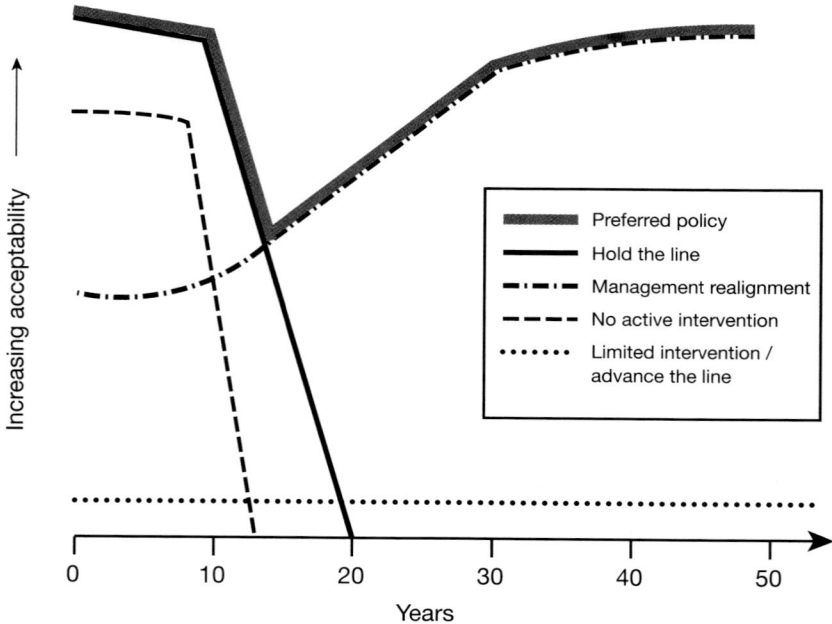

Figure 21.1   The most acceptable coastal management policy pathway for the next half century in the UK. Note the switch from holding the line to managed realignment.

reinforcement of ongoing processes. Figure 21.1 shows how the acceptance of these modes of response varies over time, and that the selected priority method is likely to change due to changing society and coastal circumstances.

Coastal engineering as a function of geomorphology

The reasons for engineering in the CZ constrain the type of geomorphological information that is required. In outline, engineers design coastal structures to the needs of two main generic situations (S):

1. S1: to prevent loss of both coastal land and built infrastructure by erosion and coastal flooding.
2. S2: to support new developments in the CZ by reclaiming land from areas that are active geomorphological units of coastal environments.

These two requirements can be reduced to geomorphological terms to identify the problems

created when society intrudes into highly mobile and dynamic natural coastal systems. In this sense, S1 reflects the need to mitigate, deflect or stop a natural tendency of the system that is perceived to be causing human problems; S2 reflects on controlling the consequences of intervening into an existing self-regulating natural coastal system that is likely to be thrown out of balance by intervention. To work with these system perspectives requires an understanding of the controls and especially the timescales by which the processes and responses are structured.

Time and space scales in the coastal zone

Geomorphic elements of the coastal zone respond differentially to process timescales. Individual sand grains (mm scale) move in response to water current activity that is driven by periodicities of seconds to hours (waves and tides). Aggregate sand grain morphology ($>10\,m$) responds more slowly and may require seasonal, annual or annual plus event time frames before change is noticeable. In this way, time and space scales of

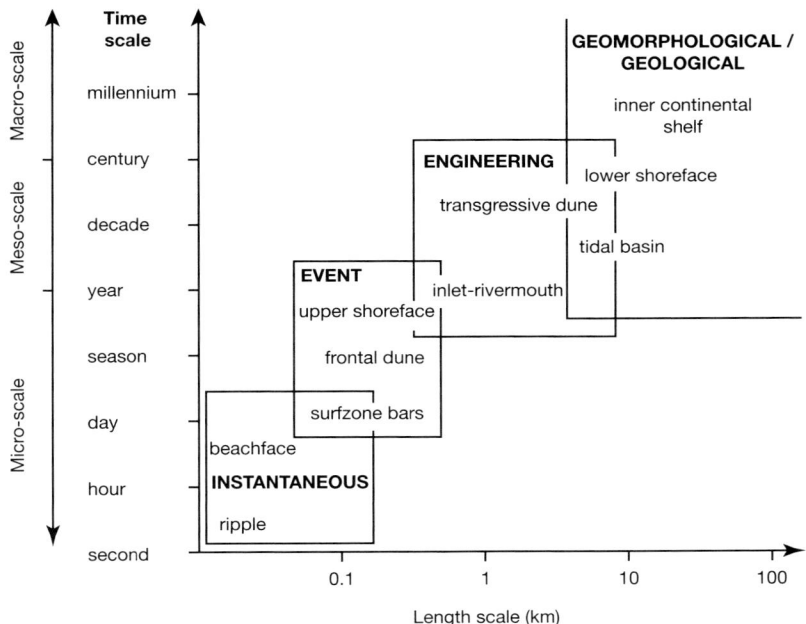

Figure 21.2   Time and space scales controlling coastal response (after Cowell and Thom, 1994).

morphology are positively associated (Cowell and Thom, 1994; Figure 21.2) such that geomorphic analysis needs to be set in the correct time frame. Engineering responses are often critically short of observation time, yet estimates of long-term change need to be central to the study. Engineers should be aware of the potential for time-dependent change and variability of natural systems.

A principal element of most coastal problems is insufficient longshore sediment supply (cf. situation 1 above), yet information on the natural annual variation of the supply is rarely available, while investigations into the longer-term (decade, century or even millennium) sediment budget context are usually absent. It is a fallacy to consider that all coastal situations are in a steady state or dynamic equilibrium (Chapter 1) such that instantaneous change rates are representative of long-term change. Such situations are rare given that a stationary process set has only a short-term context (hours at best). The forcing of coastal processes is dependent on atmospheric and oceanographic conditions that are highly vulnerable to non-stationary changes over time. The recent

emphasis on the effects of accelerated climate-induced change has alerted coastal geomorphologists to potential timescales of coastal change that had been previously ignored. In particular, the consideration of meso-scale processes (decades to centuries) (De Vriend et al., 1993) characteristic of large spatial scales (km), have proven fruitful in providing a framework for the analysis of coastal changes (e.g. beach profile volumes along the central Dutch coast: Wijnberger and Terwindt, 1995).

Coasts are dynamic, with extreme events of one to two days' duration affecting morphology that may remain in evidence for decades to centuries. Many events of longer-term processes, e.g. sea-level change and sediment supply, define the boundary conditions in which problems of a smaller time span, for example storms and their products, are considered critical by coastal engineers. Coastal problems exercising engineers rarely involve just short-term or spatially isolated processes. An understanding of geomorphological constraints on coastal evolution is an essential prerequisite for successful engineering in the CZ.

Sensitivity and response in the coastal zone

As a further refinement on process–response timescales, engineers need to be aware of the sensitivity issues of processes related to change (Thornes and Brunsden, 1977). The magnitude of the forcing process has an associated reaction time, which is the timescale before the morphology being forced starts to change, and a relaxation time, over which the morphology still responds to the forcing event's occurrence even after the forcing has ceased.

Extreme storms are likely candidates for forcing coastal changes that have long meso-scale relaxation times (Orford et al., 1999a). Coastal changes in relaxation phases are rarely linear in response over time. This can lead to false perspectives on the rate and route of coastal change, if such changes are measured over short timescales (< 5 years). The sensitivity of geomorphic units can be characterised by the reaction times. Saltmarshes and sand dunes are likely to respond almost instantaneously to direct wave attack carried by storm surges and would be defined as highly sensitive environments, while the same storm may have no impact on a hard rock cliff (Pethick and Crooks, 2000).

A further dimension of sensitivity can be seen in the concept of coastal vulnerability, which is the balance between coastal susceptibility and coastal resilience. Whereas sensitivity reflects a temporal dimension to change, vulnerability carries a value that any sensitive change is sudden and detrimental in a human context.

1. Vulnerability is a function of coastal susceptibility and resilience.
2. Susceptibility is an indication of the forcing conditions to be experienced.
3. Resilience is the ability of coastal environments to absorb forcing change and return to the existing state once the forcing relaxation phase is over.

A vulnerable coastal environment is one that shows a low and diminishing resilience in the face of consistent forcing, often associated with persistent coastal change, and as a consequence presents a problem for associated human activities. Low coastal resilience is usually related to falling beach face sediment volumes due to failing supply. Environments with low resilience may show a major shift in the structure of the coast to a new domain in which the coastal sediments are self-organised into new stable configurations (Forbes et al., 1995; Orford et al., 1996). Such reorganisations on low-lying coasts are often associated with transgressive movements of geomorphic units. This can lead to severe impact problems when any potential accommodation spaces are already occupied by human activity — hence the issue of 'coastal squeeze', whereby coastal defences hold a stable line in the face of a rising sea level and as a consequence the natural on–offshore sequences of geomorphic environments are differentially squeezed spatially.

An added problem in understanding coastal sensitivity is the deflection of natural process–response timescales by human intervention. Reaction and relaxation times to dis-equilibria created by man's interventions can range from the instantaneous to well beyond the design life of the structure installed. In general, open coast changes as a function of changes in the landward elements (e.g. estuary reclamation and wetland draining) take longer to appear (> 100 years) than do the changes incurred by intervention on the beach face per se. For example, a cessation of beach sediment supply by sea-wall protection of eroding cliffs may show an impact on beach volume within a year.

21.3 Process and response controls in the coastal zone

Relative sea-level change

Relative sea-level change (RSLC) concerns long-term variation (usually annual) in the vertical position of mean water level relative to the land. Such change refers to the apparent rise and fall of mean sea level due to geodetic, sea volume (eustatic) and land surface (isostatic/tectonic variation) changes. As these water and land movements are independent, the final resolution of sea-level position and any change per se, can only be relative, hence Relative SLC. Determination of annual RSLC rate over decades provides a measure of the long-term shift in a statistical position of sea level that is the effective datum for all periodic activity at the shoreline. The movement of this position defines whether the coastal zone responds transgressively (movement upwards) or regressively (downwards) over the long term. As this posture will influence the long-term morphological context of the CZ, it is essential to know RSLC before attempting engineering intervention. As RSLC is a statistical concept, its value is open to movement due to a variety of water level forcing influences at radically different time periods: seconds (waves); hours to years (tides); seasonal to inter-annual (oceanic circulation, e.g. up-welling and seasonal oceanic density changes due to temperature variation). There is a general consensus that accelerating atmospheric–oceanic heat exchange accounts for 50% of eustatic change in the late twentieth century, the rest relating to increased release of ice melt (Raper *et al.*, 2000). RSLC is statistically de-trended to remove the periodic effects of tidal activity, but inter-annual variation related to atmospheric and oceanographic conditions can persist (Figure 21.3), which in the short term may be the cause of shoreline behaviour out of step with the general decadal trend. The superimposition of differing coastal responses at different timescales is a problem in understanding coastal behaviour, especially when relaxation phases may overlap.

The importance of RSLC for engineers in tectonically mobile coastal zones is limited, though seismic disturbance of water elevation in the form of tsunami is a sizable threat (Dawson, 1996). In tectonically stable areas, the rate of RSLC over a structure's design life is important. Southern England is currently experiencing a RSLC of +1.5 mm/year, which increases the risk of estuary flooding and structures overtopping during surges. Barkham *et al.* (1992) estimated the change in return period of the present 100-year extreme water elevation (storm plus surge) for ports around the UK, given projected rises in RSL by AD 2040 of *c.* 50 cm. The return period falls to 1 in 45 (Glasgow) as one of the least, while the potential reductions in return period of 1 in 3.5 for Milford Haven and less than 1 in 10 for other southern British ports indicates the problems that lie ahead for assets in areas at risk from flooding.

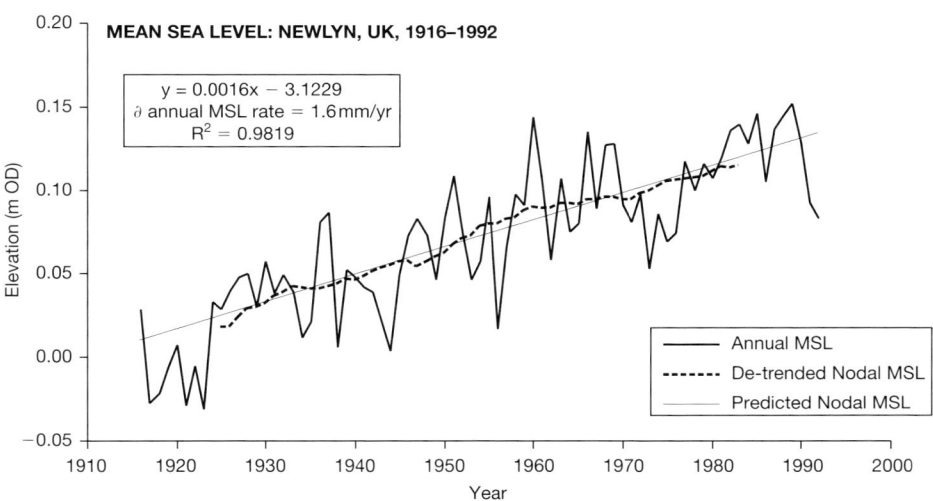

Figure 21.3 Inter-annual to decadal scales of variation in residuals of annual mean sea level from the empirical secular trend in sea level based on the nodal tidal cycle (18.6 years). Clustering of residuals relates to other oceanographic forcing factors. Annual MSL data (1916–92) are from the Newlyn (UK) tide gauge.

A rising sea level is a passive factor *per se*; its importance is that it transmits onshore energetic geomorphological processes at positions shorewards of a minimum lateral position due to inundation alone. Bruun (1962; 1983) has attempted to relate this horizontal shoreline displacement to RSLC. He empirically related coastal retreat to sea-level rise, weighted by the depth offshore at which sediment transport under prevailing local conditions starts (Figure 21.4). Bruun's Rule is a mantra for planners along the heavily developed eastern USA barrier islands where future progressive shoreline erosion as a result of RSLC has been predicted (Leatherman, 2000). Despite the relationship between RSLC and shoreline erosion being equivocal (SCOR, 1991), the debate on Bruun underlines the role of RSLC as a major coastal control element. Since the last glacially related sea-level minimum, RSLC has caused a substantial volume of unconsolidated sediment to be reworked in the CZ.

Generally during the middle and late Holocene (last 6000 years) there has been a deceleration in RSL rise rates, such that reworking has led to reduced sediment volumes on contemporary beaches, which allows greater potential for increased geomorphic sensitivity in the CZ (e.g. Carter and Wilson, 1993).

In areas with near-stationary RSL, the reworking and reduction of non-renewable sediment volume has also led to increased longshore erosion of existing coastal sediment (cannibalisation). Many of the current coastal engineering problems result from prevailing beach sediment scarcity related to low RSL change over the last two millennia.

The IPCC (Inter-governmental panel on climate change) predicts (Houghton, 1997) that future eustatic rise rates may be up to 8 mm/year. During the middle Holocene a similar variation in RSL rise may have led to the discontinuous rolling onshore of beaches and barriers (Figure 21.5: Jennings *et al.*, 1998). There is a suggestion that some barriers can be overstepped and 'drowned' during periods of fast rise rates (Forbes *et al.*, 1991). Shoreline deposition may not be able to react coherently to the speed of future sea-level change, which opens up the issue of coastal

domain changes (Orford *et al.*, 1996) that could spell dramatic instability for the CZ of the twenty-first century. The potential for increased sediment in the intertidal zone with accelerated RSLC predicted for the next century, is dependent on the degree to which new terrestrial sources are already protected from this future erosion.

Tidal activity

Tidal action concerns the effects of diurnal and semi-diurnal periodic vertical changes in still water level due to the gravitational effects of the Moon and Sun on the Earth's water bodies (Pethick, 1984). Tidal range translated into horizontal displacement of the breaking wave on coastal slopes is important in designing structures. Tidal range controls the rate of wave energy applied per unit area of beach face. In semi-diurnal conditions the high and low neap tidal positions are occupied for the longest times (Pugh and Faull, 1983). Tides are not entirely passive, but on the beach face their value lies in the elevation of waves and surges, and as such, joint tide and wave plus surge probabilities are important in the design of structures. Severe storms superimposed on a rare high-tide level can have more significance than frequent lower-energy events at neap-tide positions (e.g. the 1953 East Coast Flood in England).

Tidally induced movement of water masses generates currents capable of sediment transport that are enhanced in shallow seas. Such currents show a prevailing bi-directional structure related to flood and ebb directions. The net velocity difference in these currents is mirrored in an asymmetrical morphology of offshore and estuary sand bodies (Chapter 22). The position of shoals may offer shoreline coastal protection from wave attack, as well as being a source and sink of littoral mobile sediment. The influence of tidal currents on coastal morphology appears maximal for beaches with a tidal range between 2–4 m (Hayes, 1979), especially on offshore barrier beaches rather than fixed/fringing beaches (see below). On beaches with high tidal ranges (> 5 m) there is a likelihood of tidal currents interacting with wave-generated currents to provide greater transport potential around the low

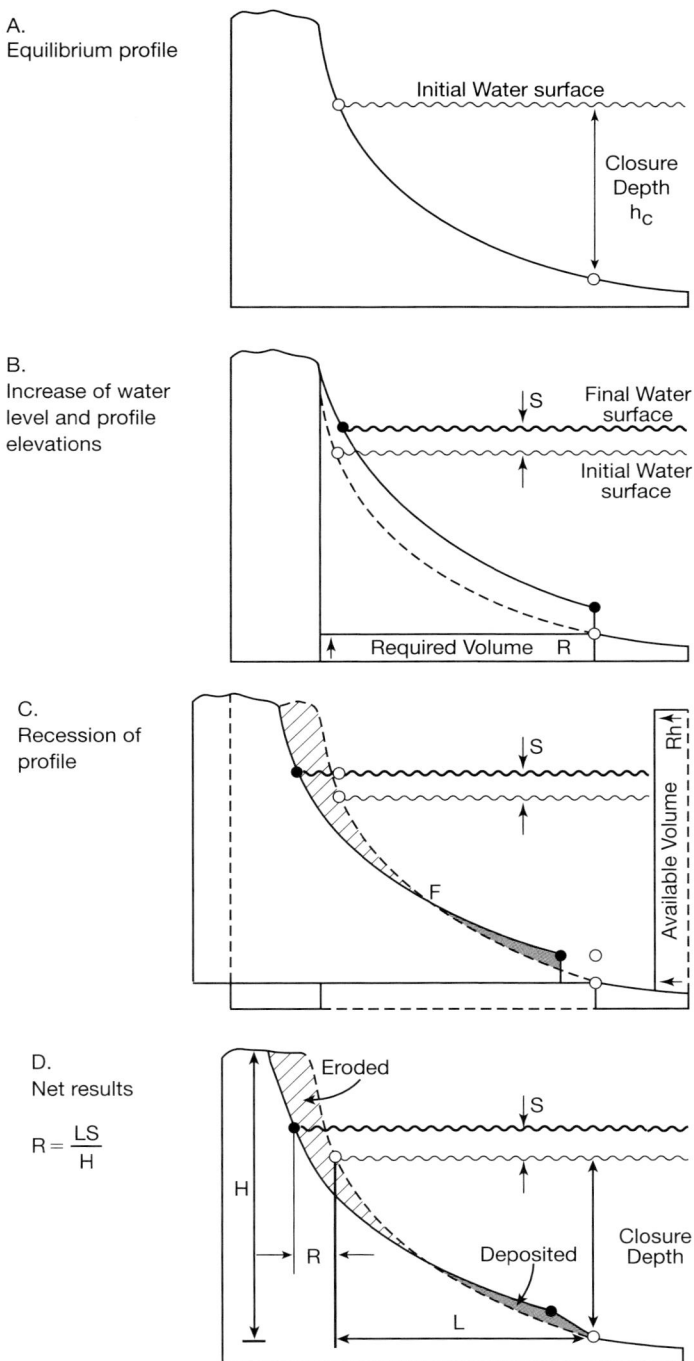

A.
Equilibrium profile

Initial Water surface

Closure
Depth
h_C

B.
Increase of water level and profile elevations

↓S   Final Water surface

↑   Initial Water surface

↑   Required Volume   R

C.
Recession of profile

↓S

Available Volume

Rh

F

D.
Net results

$R = \dfrac{LS}{H}$

Eroded

↓S

H

R

Deposited

Closure
Depth

L

Figure 21.4   Schematic development of shoreline recession as a function of rising sea level (after Bruun, 1962). Equilibrium beach profile (*A*) moves upwards (*B*) and onshore (*C*) with a rise in sea level. Eroded shoreline sediment is required to build up the beach face to maintain an equilibrium beach profile. Recession is a function of length of profile times sea-level rise increment divided by height of profile (*D*).

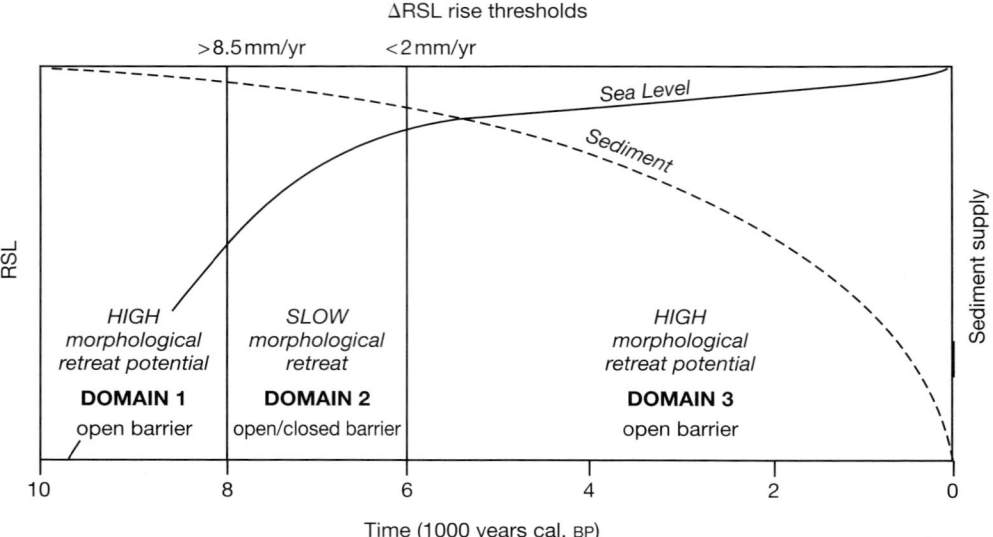

Figure 21.5   Domains of relative instability and stability of a gravel barrier as a function of RSLR over the last 10 k years (after Jennings *et al.*, 1998). Barrier stability depends on longshore sediment supply, which is best with high RSLR. However high RSLR discourages barrier spatial stability due to rapid barrier rollover. Barriers appear to have been most stable under medium RSLR, although intermittent breaches in the barrier are likely.

tide position. Flood–ebb flow asymmetry depends on the relative balance of an estuary's tidal prism retention capacity over the tidal period, and dictates the relative size of any sub-aqueous deltas at the estuary mouth (Chapter 20). Ebb-tide deltas influence the consistency of longshore sediment transport in the near-shore zone. The lack of wave activity and the dominance of fine sediment deposition in the upper reaches of tidal estuaries encourage saltmarsh growth.

History has tended to view tidal marshes as being of little value, allowing them to became waste sites for growing estuary communities. In the post-1850s, many marsh sites became potential sites for reclamation, culminating in the Dutch agricultural expansion as a by-product of the impounding of the Zuider Zee for coastal protection purposes. The east coast of the USA has seen considerable reclamation in the post-1950s of back-barrier marshes for urban expansion (Lins, 1980).

Reclamation has had a long history with respect to agricultural uses of estuaries by the rapid enclosure of shallow areas (polderisation).

The impact of empoldering was a reduction of the tidal volume in the estuary (prism) and associated disturbance of the flood–ebb fluid flow asymmetry at the estuary mouth or tidal exit, as reclamation especially reduced the intertidal friction generating areas of the estuary. The cross-section of the inlet mouth is proportional to the prism (O'Brien, 1969). As the prism decreases so does the inlet width, and by association the size of the ebb-delta diminishes, as does its protection of open-coast flanking (Nordstrom *et al.*, 1986). Major post-seventeenth-century marsh reclamation around the Wash and along the north Norfolk barrier coast has been considered responsible for ongoing coastal changes (Orford *et al.*, 2000). Post-nineteenth-century marsh reclamation in north-west Lancashire has been thought to account for substantial open-coast erosion (cf. Pethick, 2001), originally thought to be a function of the engineering protection of Victorian holiday facilities. The problem with this reclamation trend is the sub-century timescale by which the open-coast erosion problems became apparent (Orford, 1988). The wave of nineteenth-century reclamation optimism never had to deal with its twentieth-century impact.

Wave energy and storm activity

Breaking wave height (H_b) is the driving force of coastal zone morphological variation as wave energy is proportional to wave height squared. Longshore gradients of H_b can be related to variation in beach ridge height and beach sediment size. On gravel beaches, H_b correlates positively with mean sediment size. Engineers should, in particular, be aware of two related aspects of wave activity: the range of high-magnitude, low-frequency storms that occur, and the temporal and spatial distribution of wave steepness (wave height over wave length).

Storms dominate CZ geomorphological development. Figure 21.6 shows which coasts are affected by winter storms. Most offshore sediment losses from the CZ are due to return flows from breaking storm waves. Littoral sediment drift rates are proportional to $H_b^2 \times$ cosine of the breaker crest angle with the shoreline, and are several orders of magnitude greater in storms than under fair-weather swell, as the wave types of the latter have longer associated periods ($>12\,s$)

which refract more and hence have smaller breaker approach angles. The majority of the annual longshore sediment transport in southern California is estimated to occur in $c.\,10\%$ of the time indicating the importance of storms in the coastal climate (Seymour and Castel, 1985).

Onshore storms also generate a rise in water level above the tidal prediction (i.e. storm surge). This rise is both a response to the low barometric pressure accompanying storm centres and the geostrophic effect of wind activity over the sea surface. Elevations of $0.5\,m$ are common with westerly-dominated depressions in the UK, with supra-elevations of several metres being recorded in severe storms. Surges are amplified when depressions drive into enclosed basins and can be major flood hazards for low-lying coastal margins, as witnessed by the destruction associated with the 1953 storm surge around the southern North Sea basin and the catastrophic succession of cyclone surges along the Orissa and Bangladesh coastlines of the Bay of Bengal (Murty et al., 1986). The risk, scale and economic impact of such flooding can account for major protective measures like the Thames Barrier scheme to prevent extreme surges flooding central

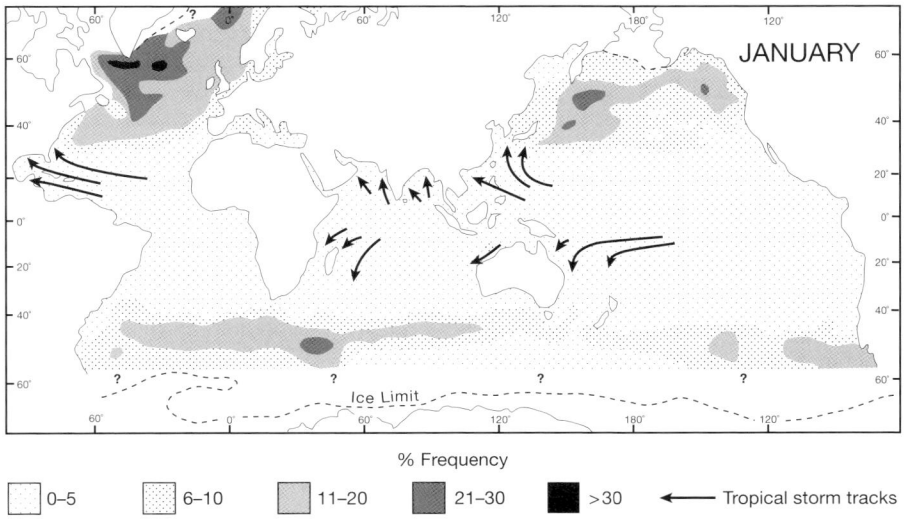

Figure 21.6 Storm wave zones: frequency of ocean gales (Force 8 winds) in January. This is the basis of storm wave environments ($> 20\%$ gales). Note the tropical storm tracks that generate short-term severe storm wave activity in otherwise swell wave dominated mid- and low-latitude coasts (after Davies, 1972).

London. Such massive engineering structures are rarely viable options for economically developing countries of the low latitudes where greater magnitude surges are common as tropical storms (hurricanes, cyclones and typhoons) impinge on the CZ (Figure 21.6).

Storm-induced surges are often the principal marine agent in back-beach and back-barrier dynamics and deposition. Beach crest heights on barriers (ignoring any windblown element) are related to storm wave plus surge elevations, though the highest combinations may overwash the crest and produce remnant crest heights. Therefore design of elevated coastal structures to resist flooding requires a confidence band of water elevations above existing beach crest heights. Prohibition on building within a set-back zone that recognises the back beach spatial limits to severe surges may need to become mandatory in some settings, so as to free the accommodation space for geomorphic activity associated with extreme events. Maximum surge penetration may be spatially progressive along a shallow sea coastline, when a storm moves in conjunction with the peak of a moving tidal wave (Orford et al., 1999b). Surges may show a rhythmic preferential longshore penetration (Dolan and Hayden, 1983), though over time such positions appear to be self-balancing. Artificial breaches on the beach crest due to vehicle, pedestrian and services' access can be critical to accelerating erosion positions during storms.

Wave steepness has been used as an indicator of the likelihood of a beach eroding. Steep waves associated with storms tend to erode sediment from the beach face and transport it into temporary sinks in the sub-tidal zone, often in longshore bar form. Waves of low steepness attributed to fair-weather swell tend to be constructional and rebuild the beach, in particular creating step-like berms that act as reservoirs of finer sediment that may be deflated into back-beach dunes. The critical value of wave steepness that discriminates between step and bar profiles also depends on sediment size, wave length and beach slope (Figure 21.7; Sunamura, 1975). The value of such predictive relationships for long-term planning is however being questioned and they should only be used as guides to beach behaviour.

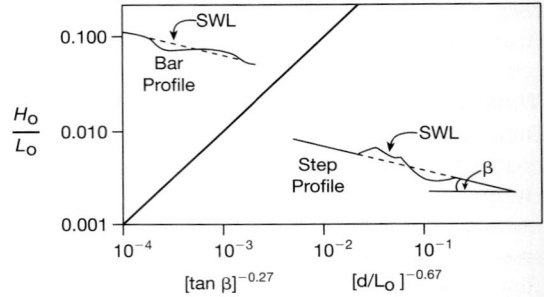

Figure 21.7   Beach profile response to wave steepness (H_o/L_o). Step (constructional) profiles appear with low wave steepness and bar (destructional) profiles appear with high wave steepness (after Sunamura, 1975).

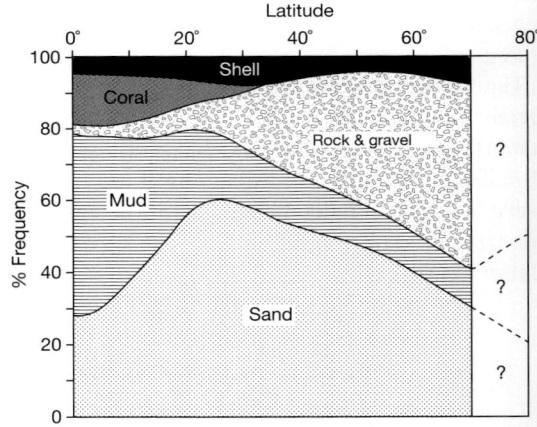

Figure 21.8   Relative textural composition of continental shelf sediment (after Hayes, 1967).

21.4 Sediments in the coastal zone

Modern coastal zone sediment distributions are related to products both of Pleistocene glacial events (Chapter 6) and of current terrestrial weathering zones (Chapter 2). Figure 21.8 (Hayes, 1967) shows proportions of material types on the world's continental shelves. The substantial element of rock and gravel in the mid to high latitudes is the result of Pleistocene glacigenic activity. Bioclastics (coral) are restricted to equatorial warm waters, though carbonate material (shell) is equally present at all latitudes. Low-latitude mud results from fluvially transported weathering products of equatorial chemical weathering regimes.

Sediments in the CZ range from boulders to mud. The coarsest material exists in the most energetic zones unless it is a lag deposit in situations with sufficient energy only to remove smaller sediment sizes. Silts and clays (i.e. mud) are rarely found at the shoreline of wave-dominated coasts. Mud dominates in peri-coastal (marsh and estuary) settings where wave energy is expended seawards and tidal current activity is maximised, as well as offshore beneath the effective wave-base. The continental shelves' sediment zones are observed to a large extent at the adjacent coastline (Figure 21.8). Figure 21.9 shows the importance of pebbles on a world scale as a beach-forming material and reflects the direct impact of upper to mid latitude glacial deposits. This should not underestimate sand-sized material as the dominant sediment by volume of beach material at almost all latitudes.

Zonal patterns of sediment transport pathways are disturbed at the local scale by the textural variation of the eroding terrestrial cover, as well as wave and current energy assemblages. These assemblages help to sort and transport initial heterogeneous influxes of sediment into more texturally selective environments. McLaren and Bowles (1985) have shown how spatial residual sediment deposition can be seen as parts of sediment pathways that link coastal geomorphic units, although as Carter (1988) shows, similar coastal sediment pathways may emerge from different energy gradients, such that uni-dimensional sediment selection is not a consistent mechanism.

21.5 Coastal organisation

One of the difficulties with assessing human intervention in the coastal zone is the lack of recognition given to the linkage of coastal geomorphic units and the cumulative effect that intervention can have spreading across spatially related units. Determination of linkage, therefore, needs to be a central theme of coastal engineering. In

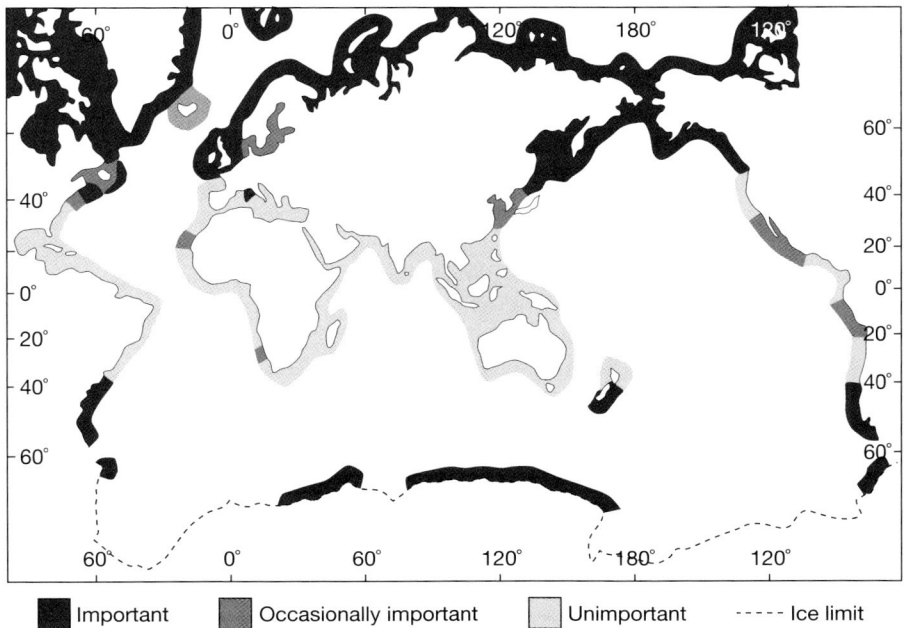

Figure 21.9 Relative importance of gravel (pebbles) to world shoreline sediment (after Davies, 1972).

this sense there is a difference between beach and coast scales. The former is concerned with short-term geomorphic analysis dominated by within-unit activity (i.e. how does the specific cliff or beach or dune system change?). Coastal scale is more concerned with inter-unit linkages that respond in the meso-scale.

There are three basic methodologies that focus on coastal linkage:

1. facies analysis
2. coastal cell analysis
3. coastal behaviour systems.

Facies analysis is a traditional sedimentological method by which vertical sequences of sediment-ary deposits are interpreted into horizontal suc-cessions of environments. The facies model is an attempt to reconstruct the energy gradients and sediment flows (process and response) connect-ing this spatial sequence (Reading, 1996; Relson, 1979 for barriers). There is nothing unique to the coastal zone in facies modelling *per se* but as its outcome is to interpret vertical successions as reflecting contemporary spatial sequences, its usefulness is beyond the requirement to establish linkages. Coastal cell analysis is a methodology specific to the CZ and concentrates on the wave–sediment linkages of the shoreline and as such is now a standard perspective from which to understand coastal organisation and shoreline behaviour.

Coastal cells

Coastal cells are defined by: sediment inputs to the shoreline (source); sediment movement along the shoreline (transport corridor or pathway); sediment loss (sink) from the shoreline. The con-centration on beach sediment in cell analysis reflects the importance of beach width and height as a control on terrestrial coastal erosion. Cell analysis, therefore, is a major tool for coastal zone management with respect to set-back and other controls on human activity, as cell delimitation identifies areas of potential coastal erosion and deposition. Given the importance of sediment to coastal cells, as well as breaking wave activity, coastal cells are sometimes referred to as wave–sediment cells (and also 'littoral' cells or coastal process units).

Most coastal cells rely on coastal erosion of terrestrial sediment by waves as the contributing source. Where sand is the dominant sediment type then the cell sinks may be either onshore sand dunes or offshore shoals. A beach represents a transport corridor or sediment conveyor. They are not necessarily the most efficient conveyor as maximum transport rates may occur only for a low proportion of the time. The persistence of a beach is either dependent on consistent up-drift sources of sediment, or on less efficient longshore transport conditions that leave sediment behind. A beach's formation and deposition can therefore be viewed as temporary, representing a hiatus or waiting stage in the long-term transport of sedi-ment alongshore.

Adjacent cells can be linked longshore by a constant drift direction, especially where geologic-ally variable (crenellated or closed) coastlines occur (see below). Directional changes of drift in adjacent cells will lead to either progradation or erosion boundaries (Figure 21.10). The volume of sediment moving along the transport corridor is reflected in beach morphology: single beach ridges with a concave plan view form, open to the coast, suggest limited longshore drift volumes, while multiple ridge systems suggest high trans-port volumes. Pulses of sediment moving wave-like along the shore can indicate irregular temporal episodes of cliff erosion by major storms. These features should not be confused with the regular shoreline perturbations generated by reflected wave energy (see below).

The recognition of cell element boundaries is crucial in addressing S1-type problems (see above). Cells are delimited by positions where the value of longshore directed wave power (P_l) is zero. This occurs where there is no breaking wave or the angle of breaker approach to the shoreline is zero, regardless of the height of the breaking wave (H_b). Longshore variations in both H_b and approach angle are caused by wave refraction off-shore (Komar, 1998). Wave refraction is the process by which deepwater waves differentially adjust to a varying bathymetric topography as the wave moves onshore. The effect of offshore rises

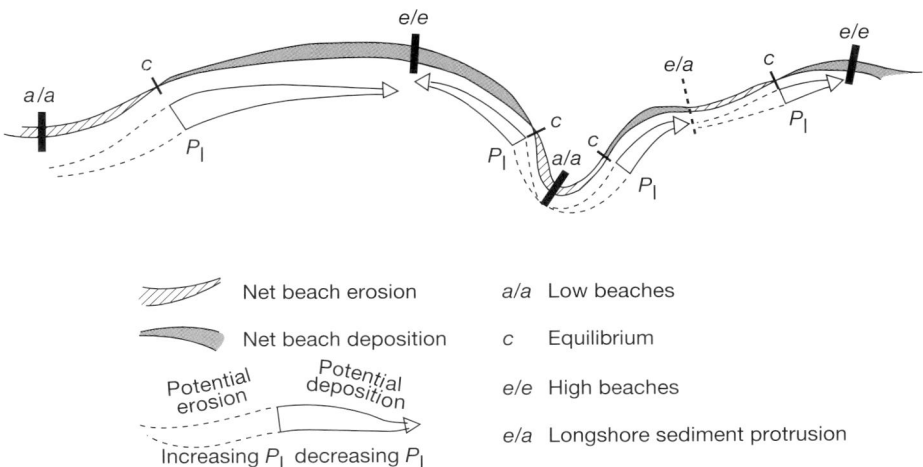

Figure 21.10   Relative structure of coastal cells as a function of longshore variation in wave power: see text for development (after May and Tanner, 1973).

and hollows under different points of the wave crest causes a variable response in wave crest deformation along the crest length. Acceleration and retardation of different elements of the crest length cause it to deform when viewed in plan, thus varying the angle of breaker approach to the shoreline. Wave crest distortion leads to bunching and stretching of the crest length, which translates into a variable longshore H_b and P_l.

Cell delimitation requires increasing or decreasing P_l values in the down-drift direction (May and Tanner, 1973). Boundary a is the up-drift cell limit where erosion starts, while e is the down-drift cell limit where no further longshore transport occurs (Figure 21.10). Between a and e is c where P_l values are insufficient to maintain the sediment supply already being transported, thus c marks the start of net deposition. The transport corridor stretches between the source and sink zones. Two contiguous cells with a common a/a boundary results in an erosion node with low or non-existent beaches, due to drift moving away from the boundary in both shore parallel directions. A common e/e boundary results in a depositional node associated with high-volume beaches. An e/a boundary marks a sub-cell boundary of a major drift cell with a consistent drift direction. A budgetary amalgamation of P_l results, for cells

defined by all wave directions, leads to a residual cell structure (usually for a year). Figure 21.11 shows this analysis for part of Northern Ireland's Irish Sea coast (Bowden and Orford, 1984). Note that multiple drift reversals can be seen; such refraction-generated reversals are common on irregular or crenellated coasts like this one. Bray *et al.* (1995) offer further evidence of cell delimitation exercises at differing scales. Empirical formulae for the calculation of P_l and its connections to longshore sediment drift rates abound. That of Komar (1998) has been widely used, though Horikawa (1988) cites alternative equations that underline the very imprecise predictive nature of drift calculations. It is best to use non-calibrated coastal cell analysis only as a long-term indication of what might be expected with respect to cell structure.

Cells and coastal geomorphology

Usually the development of cells can be seen in the geomorphological structure and disposition of the cliff, beach, barriers and dunes. Figure 21.12 shows the structure and relationship between elements of a sediment cell on a sand and gravel fringing beach system that is common to UK shorelines. It depends on cliff sediment sources

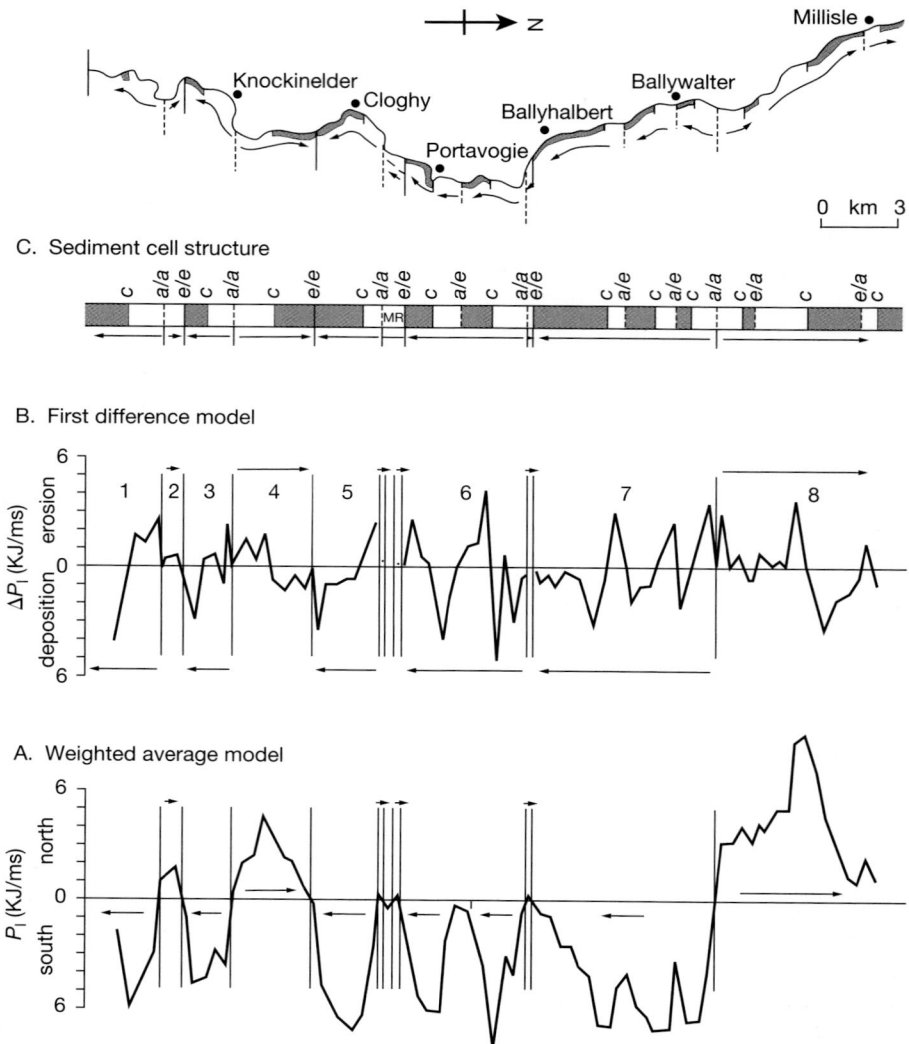

Figure 21.11   Coastal cell modelling along part of the Northern Ireland's Irish Sea coast (after Bowden and Orford, 1984). A budgetary analysis (based on 400 m coastal units) of annual longshore wave power experienced at the coast (*A*) is reduced to adjacent unit differences in wave power (*B*), which allow residual coastal cells (source and sink elements) and boundaries to be identified (*C*).

and reflects the separation of gravel and sand on the beach and the development of dunes as the down-drift sink. This typology is conditional on sediment volume and in most UK cases the source areas are insufficient to maintain the cells as coastal defences incrementally seal the source areas off, such that beaches are moving into deficit sediment volume and thereby reduce in

area and sediment size. This process has compounded an ongoing reducing beach sediment availability due to the reduction in the rate of relative sea-level rise over the last two millennia.

Figure 21.13 shows an example of cell development from the high-energy west coast of Ireland, where it is possible to recognise the cell elements from the geomorphology of the beach.

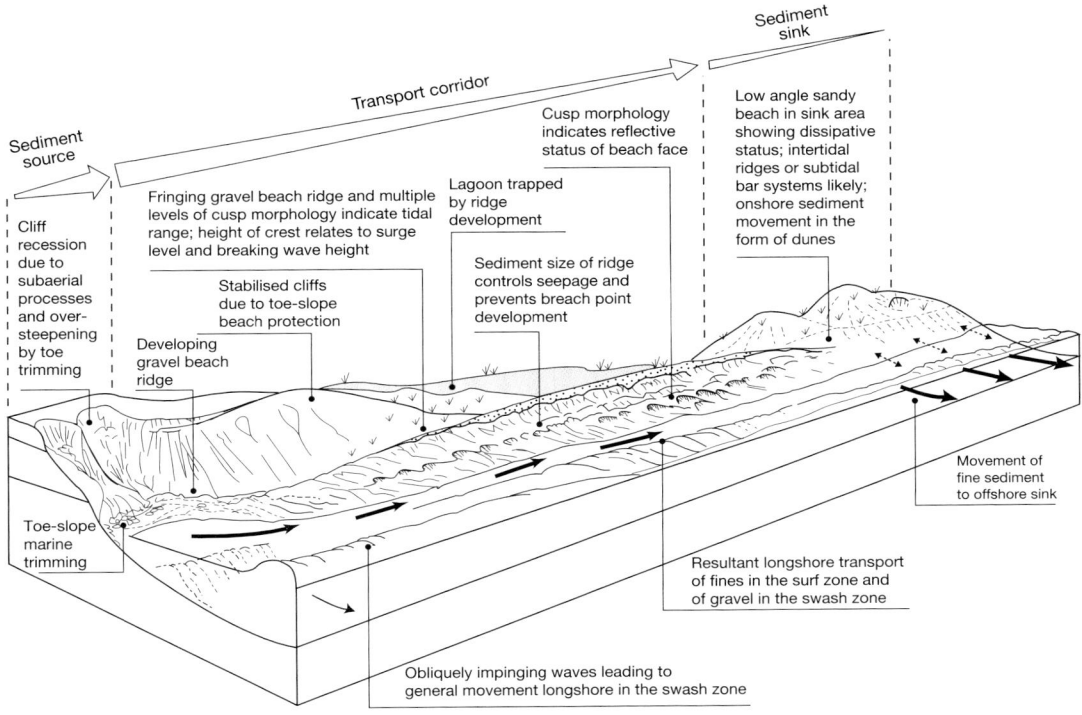

Figure 21.12 Generalised coastal cell typical of the British Isles.

Exposure to Atlantic storm waves has established a net longshore transport corridor (A) from the source cliffs south of the village of Lahinch (B), to the sediment sinks of the dune field north of Lahinch (C) and to the nearshore shoals in front of the dunes (C). Sediment source depletion has resulted in a variable volume of beach sediment along the shoreline. The higher beach face in front of the dunes (D) as opposed to the lower beach face in front of the village (E) reflects the net deposition of the sink and the net erosion of the source. This difference in beach height is picked out by the extent of draining water on the lower beach face (E). To avoid storm flooding as well as erosion of the village foundations, a continuous sea wall with all the attendant difficulties of arranging access to the beach (steps and slipways) has been constructed. The rockfill (G) at the base of the wall is designed to prevent wave reflection off the wall, which can cause beach scouring and undermining of coastal defences.

Depletion and disturbance of this rubble mound is common under severe Atlantic storms, and such material has to be renewed to maintain the defence. The depletion in beach sediment is now progressing along in front of the dunes and its associated golf course where further coastal defences (late 1990s) have now been constructed to prevent storm erosion of the dune given the decreasing sediment on the beach. The high cost of coastal protection for the village, given the static and massive structure involved, generated much debate over its construction.

Coastal cells and relative sea-level change

One of the unresolved issues about coastal cells is their stability and relationship to rapid sea-level change. Forbes *et al.* (1995) have indicated that a rising sea level will by accelerated erosion generate more beach face sediment than a stationary sea level. Some form of equilibrium will evolve

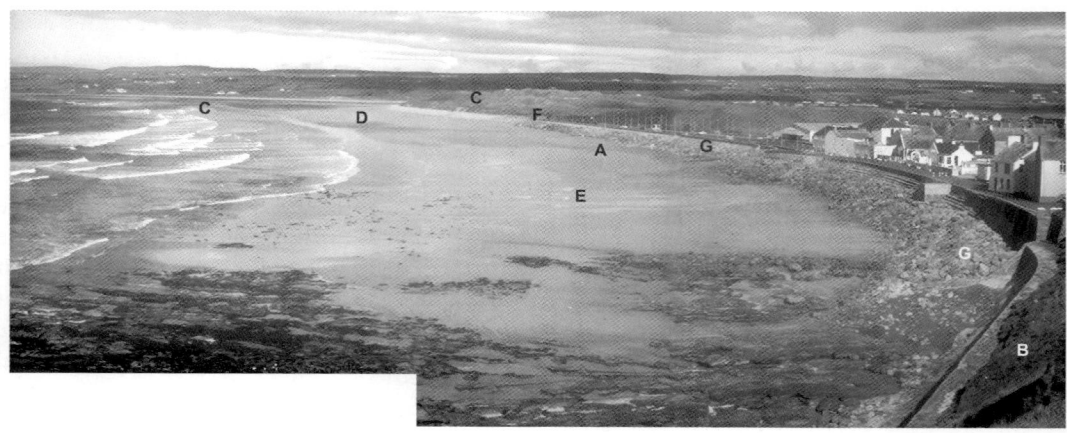

Figure 21.13   Coastal cell elements observed at Lahinch, western Ireland; positions A to G are identified in the text.

between the number of source areas and the beach volume that is being dispersed alongshore, which can maintain protection against further erosion. Fast relative sea-level rise (RSLR) is likely to be associated with smaller longshore cells due to multiple erosion sources producing higher volumes of sediment than can be dispersed by the beach conveyor. As the rate of RSLR decelerates, beach volumes are likely to reduce and greater longshore cell elongation occurs. This is because as beach volume falls there is a reorganisation of the beach face away from the drift-aligned high-volume beaches to swash-aligned low-volume beaches characterised by a zero-net longshore transport rate (Orford *et al.*, 2001). Swash-aligned beaches tend to attach themselves to any differentially resistant coastal feature; therefore their persistence depends on the sensitivity of the feature to which they are attached.

During periods of low RSLR in the late Holocene around the English coast, the sediment supplied to the shoreline during the main Holocene transgression would have been reworked by long-shore transport. In general, apart from limited beach progradation associated locally with fluvial inputs, most beach progradation was at the cost of adjacent longshore erosion of cliff-supplied sediment or of existing beaches (Lee, 2001). Hard rock cliffs provided little sediment volume for beach replenishment. The principal terrestrial sources of

beach sediment were materials supplied by coastal cliff and slope erosion of Pleistocene glacigenic deposits plus unlithified Tertiary deposits in the south and south-east. Much of the fine sediment from cliff erosion would have been transported off-shore while long-term sediment attrition would have depleted the beach-retained sediment. Little fresh sediment would have moved ashore in the storm wave dominated conditions. The retained beach volume was quickly dispersed longshore, but it was rarely sufficient to maintain other than swash-aligned beaches and in some cases insufficient to do even that. The English south coast (specifically west Sussex) reflects a chequered history of depleted cells where *e/a* positions should have set a template for swash-alignment beaches, but for human intervention whereby beach drift-alignment has been maintained over the last century, originally by groyning, now by beach re-profiling and nourishment (Gifford Associated Consultants, 1997). This is a somewhat inefficient approach, as a controlled movement to swash-alignment (cf. 21.2 Perspectives affecting our understanding of the coastal zone) would reduce the requirement for contemporary sediment maintenance, which will only accelerate in the light of future predicted RSLR.

The west Sussex coast exemplifies a lack of understanding of cell analysis in coastal management. Although the concept was taken up as the basis for shoreline management planning in England and Wales (MAFF 1995: Figure 21.14), the delimitation of cells was based on a regional analysis of major headland controls (macro-geology) and major estuary sinks (Motyka and Brampton, 1993). This delimitation can be criticised as being too coarse to offer specification at the local scale, given that only eleven cells cover *c.* 1800 km of highly variable coastline. This coarseness dissuaded a furtherance of the proactive use of cells as an instrument of positive shoreline planning, though the human dislocation in realigning the Sussex coast would have created a political barrier to its implementation at even the local scale. A further problem of cell definition relates to the movement of substantial coastal sediment offshore of the beach and beyond the nominated down-drift cell boundary. This is evident along eastern England where major volumes of mud derive from the Holderness coastline, move south past several cells into the southern North Sea and potentially end up in the Wadden Sea of northern Holland and Germany. It could be argued that this negates the value of coastal cells, though it does not disallow the use of shoreline cells wherever the issue of coastal erosion and fluctuating beach sediment volumes has to be considered.

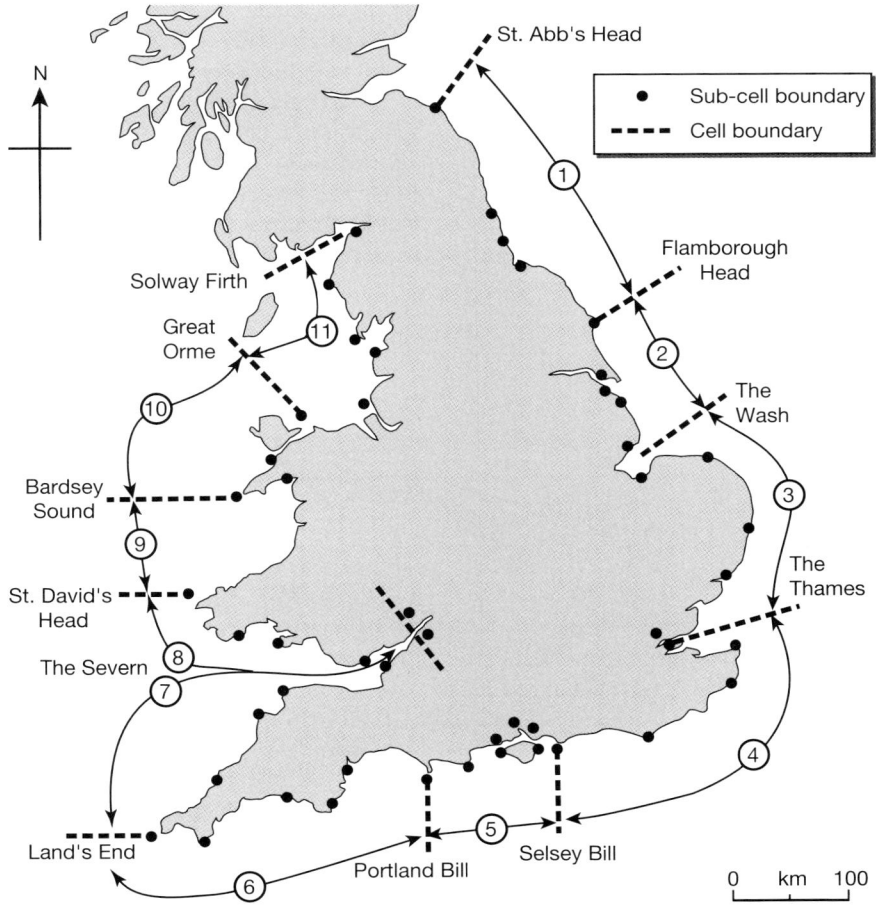

Figure 21.14 Macro-scale coastal cells identified by Motyka and Brampton (1993) and used as the basis for Shoreline Management Plans by MAFF during the 1990s.

Coastal behaviour systems

Almost by definition coast cell analysis concentrates on beach and shoreline activities, which should not be surprising given beaches' crucial position as an indicator of likely coastal erosion problems. However the beaches' behavioural character may well be related to activity across the whole range of the coastal zone as well as to regional-scaled activities that transcend single cells. Coastal behaviour systems (CBS) is a form of analysis which attempts to broaden the investigator's perspective beyond the day-by-day activity and to use meso-scale plus timeframes in order to set a geomorphological context to the problems. There is also a requirement to consider behaviour as a linkage between what might be considered an ensemble of geomorphic units. Although the activity of a beach in terms of particle movement is a legitimate study, it is the aggregate behaviour of a beach that is the required scale for CBS. Establishing the controls on aggregate behaviour clearly lift the analysis away from the instantaneous scale and into contexts identified under 'Time and space scales in the coastal zone' and 'Sensitivity and response in the coastal zone',

above. The specification of behaviour at the meso-scale allows attempts to parameterise processes for predictive estimation at a similar scale (Terwindt and Kroon, 1993). Such parameters are often statistical statements of forcing and though they have no physical manifestation, they do have a characterising power. An example is the use of a surge-forcing coefficient (Wheeler et al., 1999) to reflect annual storminess.

The value of CBS is enhanced when the coastal system is variable in form and type. Figure 21.15 exemplifies schematically the coastal system of the north Norfolk coast (UK) consisting of eroding and accreting sand-dominated barriers, dune fields, tidal inlets, ebb-tide deltas and back-barrier marshes in a meso-tidal environment, which all interact in terms of sediment pathways. Cell analysis will work at the shoreline scale, but in this case is clearly limited to supporting a structural analysis of the barrier beaches *per se* and ignores the relationships between RSLR, historical sediment supply variations, barriers, tidal inlets and associated sedimentation, tidal prisms and back-barrier marshes: areas that are essential in understanding Norfolk barrier extension and

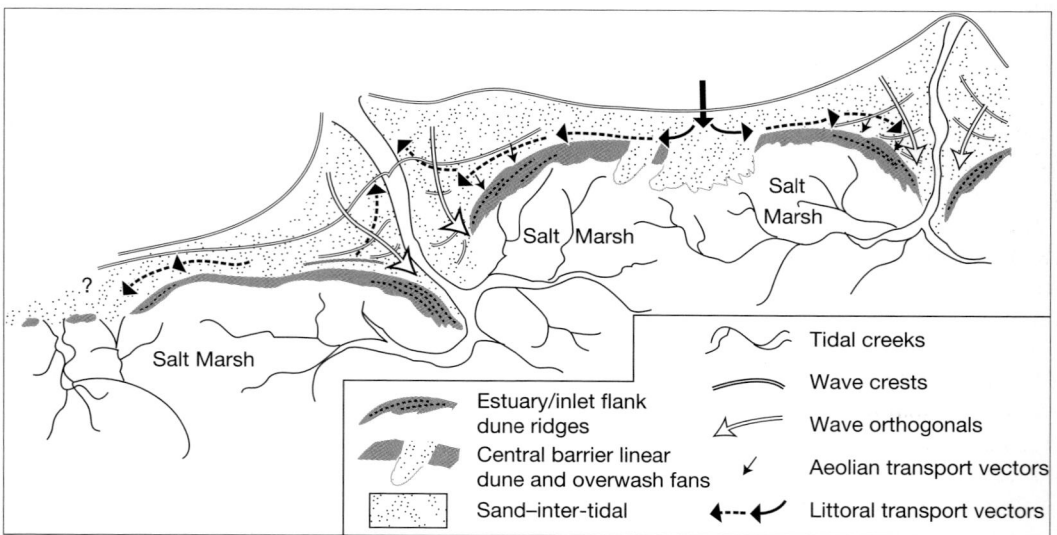

Figure 21.15   Schematic structure of the north Norfolk barrier coast to show strong interlinked relationship between open-coast and tidal-dominated passes and back-barrier areas, which control sediment pathway and coastal evolution. Coastal cell analysis understates the intimate connection of open coast and estuary (after Orford *et al.*, 2000).

dune emplacement (Orford *et al.*, 2000). The CBS scenario should provide the holistic context in which coastal cell development and contribution can be understood for a range of time and space scales.

21.6 Effects of morphology on coastal zone structure

Terrestrial morphology

Further constraints on coastal zone morphology are imposed by the terrestrial morphology that the CZ borders. A coastline which is geologically complex, with folding and faulting paralleling beach dip (discordant), and which exposes a wide range of lithologies of varying durability to wave attack, is likely to be a crenellated one with numerous fixed or fringing beaches with geological structures differentiating coastal cells. Bray *et al.* (1995) have identified how the discordant coastal lithologies along the Devon and Dorset south coast (UK) have contributed to uneven cell development as a function of the episodic retreat of coastal cliffs. Accordant (open) coastlines, especially those cut in unconsolidated glacigenic deposits, can show elongated cells unless facies variation (Chapter 7) offers initial perturbations that generate small disturbances to the macro-cell structure. Orford *et al.* (1996) stress how drumlin density along the Atlantic Nova Scotia coast sets the focus of coastal structure as the drumlins provide the point sources for barrier feeding, while the inter-drumlin areas provide the accommodation space in which the barriers move under RSLR. Irregularities in the offshore bathymetry can cause wave focusing leading to coastal cell drift reversals.

Wave shoaling slope

The offshore shoaling slope of beaches affects the translation of sea-level rise, tidal and surge amplification and rates of wave attenuation. The beach slope *per se* can be viewed as the final part of the shoaling slope, whose importance in final wave transformation grows as the tidal range increases. Macro-tidal ($>4\,\mathrm{m}$) sand beaches can have intertidal widths of hundreds of metres and

at high tide act as influential extensions of the shoaling slope. Shallow slopes and wide offshore zones dampen wave energy, so that relatively high wave-lain beach ridges on the upper beach are unlikely except with extreme events (Aagaard *et al.*, 1997). Low shoaling slopes are often associated with sediment sinks, and subsequent deflation of the intertidal zone leads to back-beach dunes.

Differentiation between the slope of the intertidal beach and the adjacent offshore shoaling slope is needed as they can reflect different circumstances. The steepness of the intertidal beach face needs to be assessed against the width of the area immediately landward of the high-tide position. Low-angle beach faces with no back-beach volume can be indicative of reduced beach volume, while a low beach face and a wide back-beach area (where foredunes may be developing) is indicative of a radically different condition of positive sediment supply to the beach. It is usual to find with sand beaches, holding both waves and tidal range constant, that steeper beaches are associated with reducing sediment supply. As the tidal range expands, the likelihood of a single slope across the beach face reduces such that an upper steeper beach slope relative to a lower low-angle slope appears.

The nearshore slope strongly controls the rates and modes of breaking wave deformation. Beaches can be placed on a continuum between total reflected wave energy (steep beaches dominated by gravel) and total dissipated wave energy (low-angle sand beach). Incident wave energy can generate a range of secondary water movements that have a range of morphological responses in the type and spatial arrangement of inter and subtidal sand bodies. Figure 21.16 shows this morphological variation as a function of the surf-similarity index (Wright and Short, 1984) that characterises the beach as a statement of wave structure relative to slope. Pertinent to engineers from this type of surf-zone analysis is the likelihood of longshore, spatially periodic, high and low beach morphology. Beach cusps on reflective gravel beaches and rip-current spacing on sandy dissipative beaches control topographic cross-beach lows which may act as avenues for breaker access that overtop

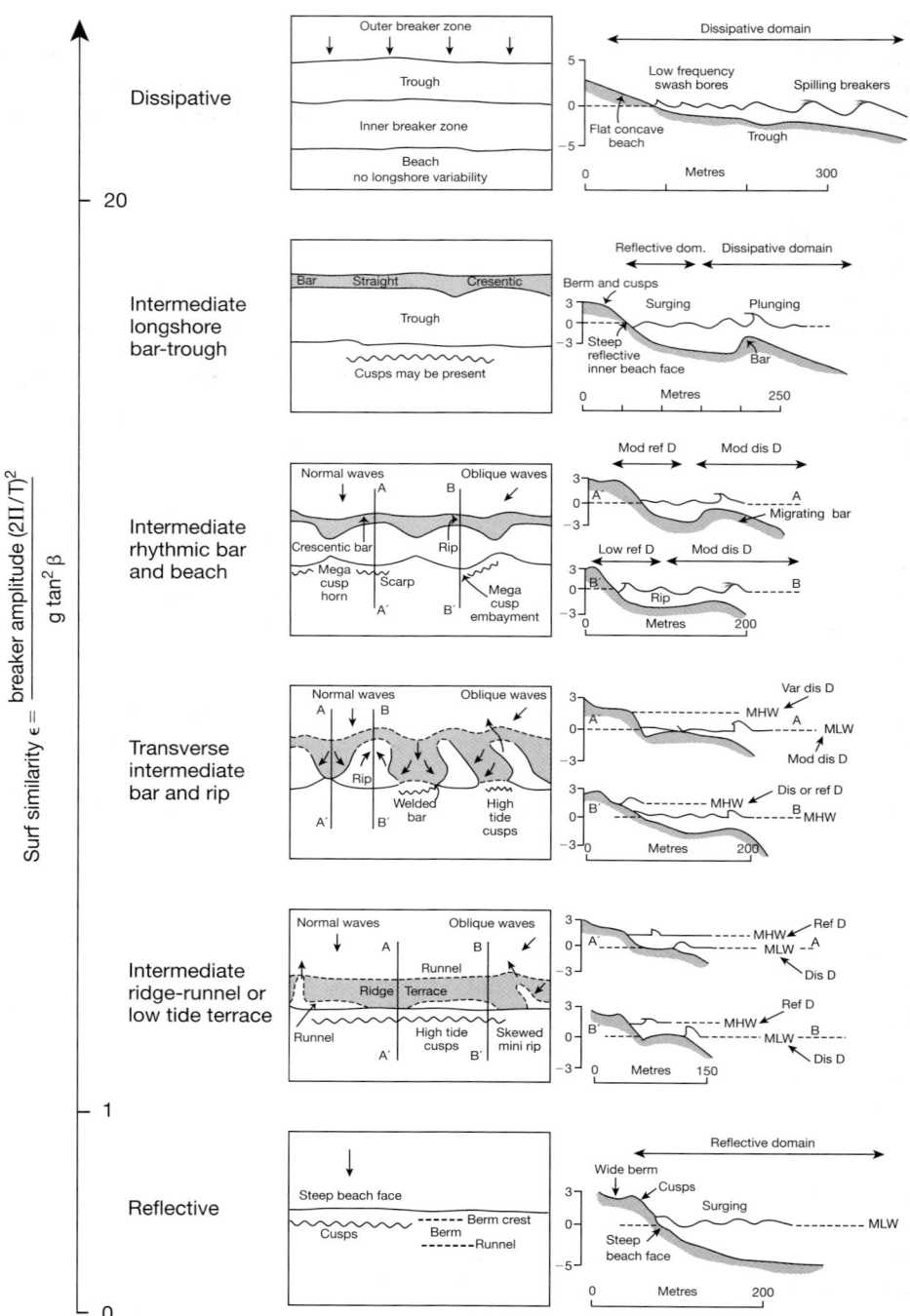

Figure 21.16   Integration of beach face morphological variation through the concept of surf similarity (after Wright and Short, 1984).

beach crests. Otherwise temporal changes in beach-face morphology affect rates of longshore sediment transport and longshore H_b gradients. These morphologies are variable as a function of wave climate, however less variation is found with beaches that are at the continuum end positions (reflective and ultra-dissipative: Wright and Short, 1984). Wave climate generally refers to the distribution of wave heights and periods that can be expected over a year at any one site. While regional wave climate refers to offshore conditions, more specific wave climates apply at different positions along a coastline when wave transformation through shoaling and refraction would have taken place. It is this distribution that will specify the beach morphological range, controlled by available sediment size and offshore slope. At a meso-scale level it is important to recognise that offshore, intertidal and back-beach slopes are rarely in step due to the variable sediment volume working through a coastal system. These slopes would only become balanced when sediment input was either zero, or when sediment input balanced output across the coast, and process conditions were stationary. The improbability of these conditions means that slopes are usually out of step such that any one slope is not a safe guide for characterising the whole coastal system.

21.7 Controls on coastal morphology

The morphology of coastal environments is too varied for discrimination on other than a multivariate basis. Coastal morpho-sedimentary environments reflect five interacting factors:

1. Mean sea level acts as the datum for all activity, while its relative change rate provides the tempo control on change and sediment availability.
2. Wave climate controls the retained sediment size as a function of wave height. Climate also identifies the extreme events that control the basic depositional framework.
3. Sediment source, supply and size range: excess sediment leads to progradation (usually in the form of multiple beach ridges), while

sediment deficit leads to an erosive pseudo-transgressive morphology (usually in the form of a single beach ridge dominated by crest overtopping as a function of extreme events, hence 'pseudo-transgressive'). The rate of transgression is controlled by RSL rise while progradation can be conditional on RSL fall (forced regression) or on sediment deposition pushing the shoreline seawards (normal regression).

4. Terrestrial basement morphology upon which the coastal environments are formed: in particular the degree of coastal crenellation or irregularity indicates the likelihood of open (wave dominated) and protected (tide dominated) depositional environments. However, note that protection is often associated with the ability of processes to use the available sediment to form wave-lain structures that offer seaward protection to areas where wave energy is reduced or negligible and tidal currents enhanced.
5. Tidal range, which has the effect of enhancing dislocation of longshore beach continuity as it increases.

The combination of these variables leads to several basic beach forms that, once established, control the spatial disposition of other environments dependent on beach morphology for protection from wave energy.

21.8 Key clastic coastal depositional morphology

Mid-latitude coastal morphology dominated by clastic sediments varies from fringing beaches (Figure 21.12) to beaches found on barriers standing seaward of the terrestrial shoreline (Figure 21.17). In general, fringing or fixed beaches will result from:

1. higher median wave energy (maximised in storm wave environments: Davies, 1972)
2. higher tidal range (meso–macro)
3. low sediment availability
4. steeper offshore slope.

Figure 21.17   A natural coastal barrier at Long Island, New York, USA (photograph courtesy of J. Allen, US Parks).

Coastal barriers are more likely with the reverse of the above conditions. The origin of coastal barriers is still debated (Leatherman, 1982) though a single origin is now considered unlikely. Although their presence is limited to about 15% of the world's coastline, they are the physical foundation to some of the world's most valuable real estate: property worth $> \$5 \times 10^9$ can be found along the barrier coast of the eastern USA alone. Even barriers in economically developing countries like Brazil can attract a disproportional share of accumulated wealth into coastal real estate (Figure 21.18) and thus create the potential for major protection issues in the future.

Fringing or fixed beaches

Fringing or fixed beaches are a common feature of meso- to macro-tidal (> 3 m), storm-wave ($H_b > 1.5$ m) dominated coasts (e.g. north-west Europe). This mid-latitude zone tends to have experienced a decelerating rate of relative sea-level rise since the mid-Holocene. These beaches,

Figure 21.18   Urbanised barrier west of Rio de Janeiro. There is a future issue of beach face protection to be faced as well as an existing pollution problem in the lagoon separating the two phases of urbanisation on the seaward and landward barrier ridges.

particularly those that are gravel-dominated (shingle), are often remnants of a rolling-onshore process associated with a weak transgressive shoreline. The growth and diminishment of fixed or fringing beaches is strongly controlled by longshore supply. Around the UK coastline much of that supply comes from cliffs formed from Quaternary glacigenic deposits and unlithified Tertiary sands and clay that have been weakened by Quaternary periglacial activity. The episodic failure of such cliffs as a function of toe-slope erosion has been noted as a formative process of coastal scenery (Brunsden and Jones, 1980). Lee (2001) offers an illuminating perspective of how cliff supply fluctuations of the last two millennia have contributed to the strength or weakness of contemporary fringing beaches along the coast of south central England. Sediment reductions in the face of reducing RSLR identify a period of beach consolidation, characterised by a single beach ridge that has to act as the basic coastal protective bulwark to flooding and to wave energy along the coast. There is little coastal zone related deposition landward of these fixed units, though limited wetland development may form at the rear of such beaches where seaward terrestrial drainage is impeded. Extensive wetland areas are limited to major depositional sinks in crenellated and estuary coasts (e.g. The Wash, outer Thames estuary), or are remnants

linked to the rear of once-larger bay-wide barrier beaches controlled by a hinge-point terrestrial morphology (e.g. Start Bay, Devon). Most of these latter beach and barrier forms were transgressive and have now overridden the once back-barrier wetland deposits, generating an unstable unconsolidated sub-beach stratigraphy. Gravel barriers driven onshore quickly lose their coherence when pushed against headlands of a crenellated coast or when RSLR slows and longshore sediment supply diminishes (Jennings *et al.*, 1998). Headlands act as refraction hinge-points, upon which the remnant barrier pivots, to be left stranded in the bay head as a fixed beach.

On most contemporary fringing beaches around the UK, reducing longshore sediment supply and longshore transport specify beach budgets. Few beaches show sediment progradation as most are in deficit mode, so that the best depositional coast and beach morphology is usually exhibited at the down-drift cell boundary (e.g. sand dunes, Figure 21.12). However, many dune systems around the British Isles show a lack of modern foredune development reflecting a contemporary reduction in beach volume. Fringing beaches show marked spatial separation of foreshore sand and rear beach gravel elements, with the highest wave-lain ridges found on gravel-dominated beaches. The beach crest height is governed by extreme events (i.e. related to the vertical limit of swash run-up in storms). Run-up can overtop the beach crest by which sediment can be incrementally added to the ridge top, while overwashing of crests in severe storms can remove crest sediment to a back-beach position (hence beach ridge rollover). The rate of rollover retreat for gravel-dominated ridges has been related to RSLR rate: *c.* 1 m barrier retreat per 1 mm of RSLR per year (Orford *et al.*, 1995). It is not a continuous process as the interplay of overtopping and overwashing is dependent on the size and spacing of storms. Overtopping can also occur under swell wave conditions, as the high reflective status of gravel beaches can generate secondary water motions that can amplify swell run-up sufficiently to overtop. Most fringing beaches are still quasi-transgressive due to overtopping, despite longshore-dominated transport on the lower beach.

Beach face constructional berm height and beach slope angle are a function of H_b, wave period and grain size (Figure 21.12). Mature gravel ridges down-drift of the source area can show alongshore sediment-size grading related to longshore H_b gradients. This is best observed on beaches without sediment recharge, and points to the difficulty of nourishing gravel beaches without the correct specification of depleted sediment size elements. On less energetic gravel beaches, cross-beach particle-form grading occurs, related to size. Beach-face sand size depends on wave energy and sediment renewal. A typical median size for high-energy beaches ranges from 0.15 mm to 0.18 mm, with the best sediment sorting found on high-energy beaches. Sedimentation depths of modern fringing beaches are small, with the sweep zone (difference between the lowest and highest beach profile at the same position over time) far less than the absolute beach crest height. Stratification of beach ridges is common, with wave-lain sand-based ridges usually incremented by blown sand. Such aeolian development has been termed dune decoration. Fringing gravel beaches contain predominant seaward dipping sedimentary units with limited landward dipping ones at the rear of the crest related to the rollover process. Gravel-dominated remnants of bay-mouth barriers now stranded can show major landward dipping, planar and lenticular stratification (Figure 21.19) with only superficial surface seaward dipping units. Back-beach stratification formed by storm washover (< 5 cm thick), show a coarsening upwards of grain size, and have been used as groundwater or storm water seepage corridors.

Beach ridge breaching by fluvial sources depends on the sediment-controlled percolation rates (Figures 21.20 and 21.21). A transition from seepage to open-channel flow (breach) for gravel can occur when percolation rises in excess of 1–2.5 m³/s per metre beach width, and for sand beaches around 0.3–0.8 m³/s per metre beach width. Fluvially derived breaches usually occur at cell divides (*a/a* point) or at low beach positions due to low H_b values. The absence of distinctive breaches in gravel ridges means that overtopping by storms leads to a constant onshore movement of sediment, as there is no feasible sediment return to the near-shore, as happens with tidal

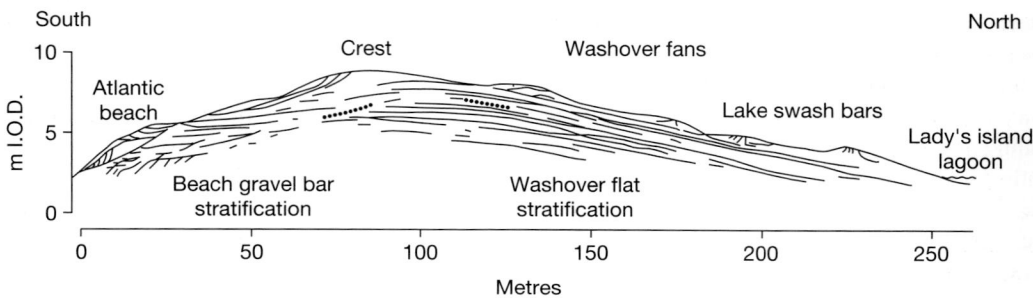

Figure 21.19   Internal structure of a gravel-dominated barrier in southern Ireland ('m I.O.D.' refers to metres above the Irish Ordinance Datum, the local levelling datum in Ireland, approximately mean sea level).

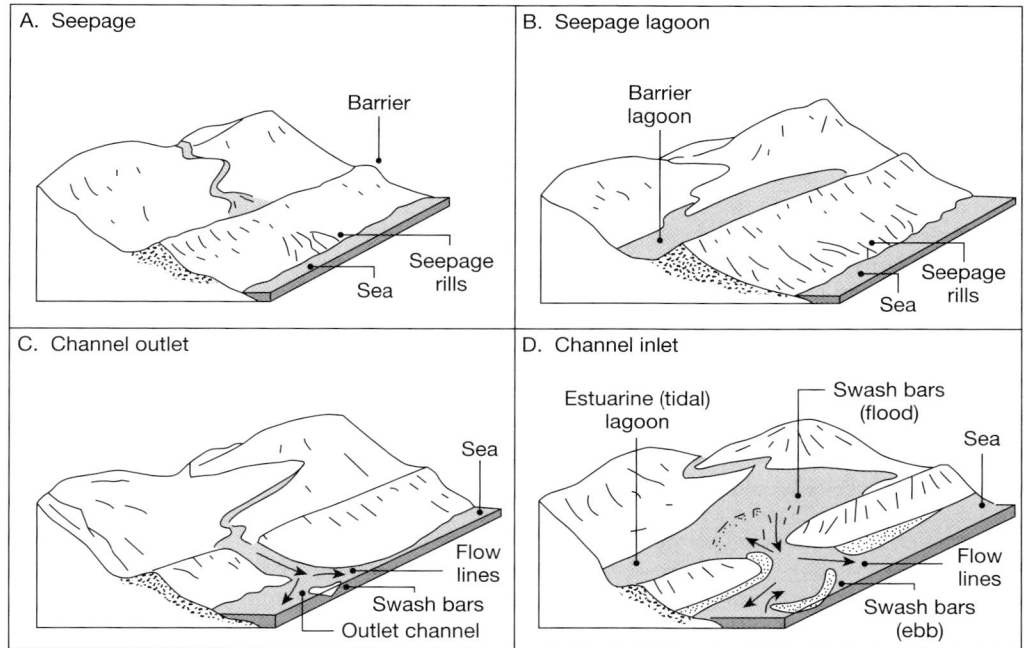

Figure 21.20 The development of fluvially generated breaches in coastal barriers as a function of seepage (*A*); back-barrier freshwater lagoon development (*B*); overtopping in storms cutting channels enlarged or maintained by fluvial flow (*C*), and substantial maintenance by tidal action forming a brackish lagoon (*D*) (after Carter *et al.*, 1984).

passes on sandy barriers. Seaward dip of fringing beaches stops excessive landward seepage under storms, while a lack of distinctive seepage points on the landward side of beach ridges reduces the possibility of incipient breach points and landward flooding.

Engineering intervention on beaches

Engineering intervention should be assessed in the light of the sediment balance in the coastal cell. Successful attempts at controlling cliff erosion may well cause sediment supply problems for down-drift beaches, though cliffs of a high clay content are unlikely to be the source for down-drift sandy beaches. The requests for beach engineering defences are usually prompted because of a lack of beach volume that can act as either a natural adjusting buffer to breaking wave energy, or an elevation to prevent storm overwash. Such sediment losses need to be carefully assessed, as seasonal differences in beach-face volume should not be taken as the only indication

of major beach sediment depletion. Dunes and beaches may also show rhythmic variations in volumes related to seasonal wave climate variation. If there is long-term sediment loss, then resorting to primary engineering structures (i.e. walls and groynes) will at best only reduce rates of loss in the short term, while at worst accelerate erosion at other points. The uses of structures that *de facto* prevent further sediment passage to the beach (walls, bulkheads, revetments), or interrupt longshore sediment transport (groynes), have been commonplace in the engineer's armoury. However useful these have been in reducing landward erosion rates, they have serious detrimental consequences for the beach sediment system, as these interventions in longshore sediment supply only cause further depletion down-drift. The role of such structures should now be more a question of regional perspective than local need.

The priority of engineering solutions has in the past two decades given way to the use of a 'greener' solution, that of beach renourishment. It is usually

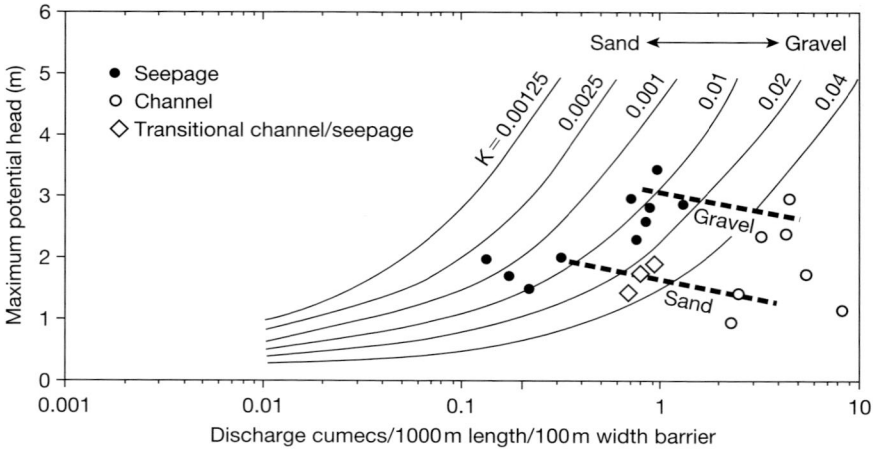

Figure 21.21 Seepage flows required to form channels in sand and gravel barriers (after Carter *et al.*, 1984). ('Discharge cumecs' is a measure of discharge in cubic metres per second.)

nourishment at point sites rather than nourishment for the whole beach conveyor *per se*, given that the volumes required for the latter are prohibitive. One-off exercises at beach rebuilding with sand have been restricted to large sites (km) to ensure economies of scale (e.g. Bournemouth, southern England), as obtaining sand coarser than the original (for beach retention) is environmentally difficult. England also exemplifies the undertaking of extensive gravel nourishment, notably the west Sussex coast. The problem of attempting to match sediment in terms of size and shape is reduced given the anisotropic dominant beach lithology — flint — that supports a low level of shape selectivity in beach sedimentation. This issue gains almost national importance at Dungeness where a nuclear power station needs protection from longshore gravel depletion, with down-drift sediment being transported by vehicle to an up-drift re-injection site. Cynics debate whether the recharged sediment or vehicles reach the down-drift sink sooner! When sediment recharging and profile remodelling is undertaken to raise beach crest levels against storm flooding, it is inevitable that upper beach stratification will be lost. Raising beach crests in this way means that upper-beach backwash cannot dissipate via the subsurface, such that sediment starts to move down-beach under even constructive swell conditions, and by a pseudo-headward erosion

distinctive channels can cut back into the beach crest and become early potential avenues for over-washing and back-barrier flooding. Engineers should aim at preserving the natural pattern of beach form and material in the transport corridor of fringing beaches — though this is virtually impossible to do. The disturbance of a beach structure by cross-beach trenching will, despite backfilling, also lead to localised seepage disturbances.

Barriers

The geomorphic attributes of fringing beaches can be applied equally to beaches on barriers. However, the engineer has to widen the reference frame when working with barriers by considering the barrier's regional setting and its position relative to the terrestrial shoreline, back-barrier water bodies (lagoons or bays) and water connections between the open sea and lagoons. Barrier types can be represented as a continuum from free-standing offshore to fixed position onshore, but are best differentiated on the basis of tidal range and sediment size. Although there are numerous gravel-dominated barriers in the mid to high latitudes (Orford *et al.*, 2001), they pose substantially fewer engineering problems than sand barriers of the mid-latitudes

The best freestanding and continuous sand barriers are in micro-tidal conditions, with medium to low wave energy ($H_b < 0.5$ m) punctuated by peaks of high wave energy in tropical storms and hurricanes, sand dominated, and on a low-angle shoaling slope (e.g. Gulf of Mexico). Barriers are absent from the waters of north-west Europe, except where substantial fine sediment and low-angle shoaling slopes coincide (e.g. the southern North Sea: FitzGerald *et al.*, 1984). Hayes (1979) and Leatherman (1982) indicate basic controls on barrier type. Leatherman (1982) divides barriers on the basis of their regressive or transgressive status (in terms of sediment supply) and micro- or meso-tidal range (Figure 21.22). As the tidal range increases towards meso-tidal, then tidal inlets that have a strong spatial stability segment the barrier's length. Barriers in micro-tidal environments tend to be wave dominated, long, narrow, low lying and continuous. Such inlets as do occur are generated initially as washover sites during severe storms (usually hurricanes) and then enlarged into outlets by out-flowing storm waters trapped in the lagoon. The inlets are often unstable, as they can migrate rapidly in the face of high-volume longshore drift, or seal up quickly given there is no tidal exchange to keep them open. Sediment washed over the low barrier by a storm surge may return to the beach face via temporary outlets, flowing out with the storm water return. Fine sediment can also return to the beach face by aeolian action. Lack of lagoon tidal action means little marsh growth.

Meso-tidal (2–4 m) environments show more stable barrier islands fronting substantial wetlands caused by tidal exchanges of water and sediment into the lagoon from the open sea via relatively fixed inlets (Figure 21.22). Inlets are more common, with barriers readily defined as islands. Most inlets are linked with flood and ebb-tide deltas as a consequence of cross-barrier tidal flows. Ebb deltas are important for the open coast as their morphology influences nearshore wave refraction, and hence beach development at the island–inlet boundary. Switches in ebb-delta deposition may change the wave refraction focus positions and move erosion or deposition alongshore; it is not uncommon for these zones to jump spatially longshore. Multiple beach ridges at island longshore ends are common in regressive and transgressive islands and reflect swash bar accretion of longshore drifted sand moulded by wave refraction around the ebb-tide deltas. Although island positions tend to be stable, the erosion of centre beach material for barrier growth at terminal positions means that the islands appear to oscillate in place. Oblique wave approaches (e.g. east coast swell from the south-east along the US barriers) cause island elongation and down-drift thickening, leading to islands' 'drumstick'-shaped plan view. Storm-generated overwash is not as geomorphologically important on meso-tidal as on micro-tidal barriers, as barriers are wider and sediment can be moved to back-barrier positions via inlet activity.

Engineering intervention on barriers

The US east coast barriers have had a major history of engineering intervention during the twentieth century. Intervention on barriers has followed the pressure of human recreational occupancy. There has been a strong tradition of building right at the shoreline edge to take advantage of the view, but also to maximise the effect of onshore coastal breezes during otherwise stifling and humid summers experienced on the eastern seaboard. The density of barrier recreational use is a result of the proximity of the main urban centres of the north-east with the northern barriers of New Jersey (Nordstrom *et al.*, 1986) and New York having a long history of traditional coastal defences: walls, revetments and groynes. This is a micro-tidal area with long transgressive barriers that experience severe pressure for defences, as they are both low and narrow. The amount of natural dune activity — that might elevate the barriers — is limited, and attempts to maintain any dune line as a natural defence have had a varied history of success using mixtures of dune fences, vegetation seeding and fixing, as well as the near-universal dune access control. The concept of shoreline-parallel boardwalks for recreational promenading has been extended to major structures overarching the dunes to prevent dune trampling, but allow beach access (Figure 21.23).

The main engineering issue is about maintaining beach width in the face of barrier rollover and

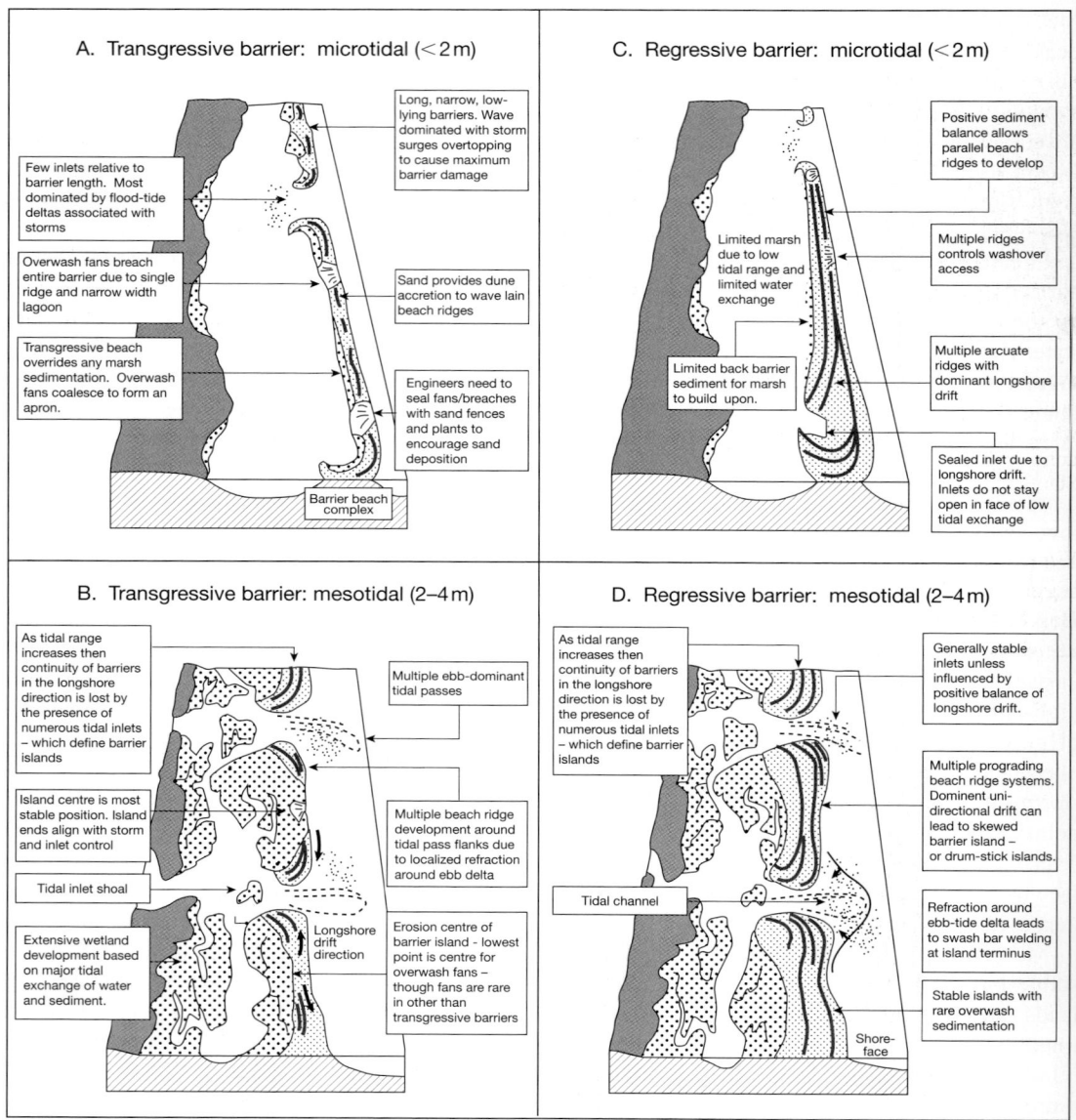

A. Transgressive barrier: microtidal (<2m)

Few inlets relative to barrier length. Most dominated by flood-tide deltas associated with storms

Overwash fans breach entire barrier due to single ridge and narrow width lagoon

Transgressive beach overrides any marsh sedimentation. Overwash fans coalesce to form an apron.

Long, narrow, low-lying barriers. Wave dominated with storm surges overtopping to cause maximum barrier damage

Sand provides dune accretion to wave lain beach ridges

Engineers need to seal fans/breaches with sand fences and plants to encourage sand deposition

Barrier beach complex

C. Regressive barrier: microtidal (<2m)

Limited marsh due to low tidal range and limited water exchange

Limited back barrier sediment for marsh to build upon.

Positive sediment balance allows parallel beach ridges to develop

Multiple ridges controls washover access

Multiple arcuate ridges with dominant longshore drift

Sealed inlet due to longshore drift. Inlets do not stay open in face of low tidal exchange

B. Transgressive barrier: mesotidal (2–4m)

As tidal range increases then continuity of barriers in the longshore direction is lost by the presence of numerous tidal inlets – which define barrier islands

Island centre is most stable position. Island ends align with storm and inlet control

Tidal inlet shoal

Extensive wetland development based on major tidal exchange of water and sediment.

Multiple ebb-dominant tidal passes

Multiple beach ridge development around tidal pass flanks due to localized refraction around ebb delta

Longshore drift direction

Erosion centre of barrier island - lowest point is centre for overwash fans – though fans are rare in other than transgressive barriers

D. Regressive barrier: mesotidal (2–4m)

As tidal range increases then continuity of barriers in the longshore direction is lost by the presence of numerous tidal inlets – which define barrier islands

Tidal channel

Generally stable inlets unless influenced by positive balance of longshore drift.

Multiple prograding beach ridge systems. Dominent uni-directional drift can lead to skewed barrier island – or drum-stick islands.

Refraction around ebb-tide delta leads to swash bar welding at island terminus

Stable islands with rare overwash sedimentation

Shore-face

Figure 21.22   Structure of micro-tidal and meso-tidal barriers along the US east coast (after Leatherman, 1982).

predominant longshore transport. The 'northern states' beaches are relatively free from hurricanes but 'Northeasters' (intense, fast moving storms) are more than capable of generating overwash. Not only is overwashing the only viable means for coherent barrier retreat, but it is also seen as the major disturbance factor to human barrier activity, so that attempts to build up the beach ridge

elevation and prevent overwashing are actively employed. Re-profiling (bulldozing) has been a common method of raising beach ridge elevation as a defence against storm surge. Beach sediment renourishment has also been widely employed as a means of expanding the beach width to act as a wave buffer, though its efficiency in the long term has been widely debated in recent years.

Figure 21.23 Possible misplaced concern for barrier dune protection shown by over-elaborate boardwalks when the broader issue of urbanisation and its effect on dune evolution has been ignored. Fire Island, New York State, USA.

Figure 21.24 shows the changing state of the biggest USA renourishment project at Miami Beach, Florida: (*A*) shows the human investment at risk in 1972 due to the loss of the natural beach,

underlining the need for beach nourishment; (B) the initial 16 km beach replenishment costing 60×10^6 in 1980 and (C) the state of the beach over a decade later. The sheer size of this scheme and the lack of any hurricane activity on Miami Beach probably accounts for its persistence, although it has been prone to some longshore reorganisation. However, other US experiences show that few beach renourishment projects last for longer than 10 years with nearly 50% of projects failing within 5 years. This high failure rate confirms that continual renewal of beach sediment is required and identifies how ineffective nourishment can be if the offshore extension of re-profiling is not fully undertaken. Even then a sequence of stormy events may reduce nourished volumes to a position of non-seasonal recovery.

The use of beach nourishment has received considerable support from conservationists, relative to the alternatives of fixed engineering and is still seen by many as the only effective sustainable control on coastal erosion despite its cost (160×10^6

Figure 21.24 The use of beach nourishment as a means of coastal defence at Miami Beach Florida. Miami before nourishment (*A*) relied on sea walls and groynes. The threat to major real estate investment generated the then (late 1970s) biggest US renourishment scheme designed to act as a barrier to hurricanes (*B*). The scheme in 1992 (*C*) shows a substantial element of the beach in place, however Miami Beach has not been tested by a hurricane as yet. Photograph (A) Courtesy of RWG Carter; photogaph (B) Courtesy of S Royle.

for US national renourishment schemes). However, it is important that this approach should not be seen as a one-off exercise and that remedial renourishing will be needed, despite the fact that the US federal authorities are reducing their support of this cost. The presence of nourishment is often noticeable by the down-drift structural containment barrier that marks an administrative or property boundary, rather than a natural sediment cell boundary (Figure 21.25).

Figure 21.26 shows Long Beach, mid New Jersey, an example of a transgressive micro-tidal barrier island that has been extensively urbanised for recreational purposes. Property development has been a continuing aspect of the island's history since the late nineteenth century, and represents the main incentive for Federal-funded coastal defences to counter shoreline recession (> 2 m/year prior to groynes) as a function of RSLR of +2–3 mm/year. Rock groynes have greatly reduced longshore drift rates, but the terminal scour problem necessitates their presence along the entire island's shoreline. The narrow beach indicates attempts to control shoreline recession. Attempts to recreate a dune ridge by erecting sand fences proved

unsuccessful given the lack of sediment volume currently in the transport corridor and the virtual absence of a drying area for sediment deflation. The threat of 'Northeasters' is not only an ocean-side problem, the associated surges in the tidal lagoon can flood and erode bay-side developments, many of which are built on dredge spoil. The hurricane that intermittently reaches this far north further compounds this bay surge problem. Evacuation of the island's population in the face of a hurricane landfall is a real possibility for many US barriers. The limited access to and from barriers (one bridge for Long Beach's 31 km shoreline) presents a bottleneck for all evacuation plans. The lack of hurricane-proof buildings on the barriers offers little hope for those people caught on them.

A major threat to barrier stability and human use arises with hurricane landfall on barriers. The threat of hurricane landfalls rises in central and southern east coast states and the risk of severe storm-induced surges (> 2–3 m) strongly influences engineering responses. The emphasis is on hurricane proofing properties rather than hurricane proofing barriers, as the latter is near impossible in view of the accumulated experience over

Figure 21.25   Beach renourishment has become a major element of US coastal protection, but issues of federal support relative to private and public beach access inevitably leads to piecemeal beach nourishment that leads to a distortion of beach widths given longshore redistribution of sediment. Public and private access along St Petersburg's (Florida) shoreline can account for inconsistencies in beach width post nourishment. Source of photograph unknown.

Figure 21.26   The scale of urbanisation on barriers is exemplified on Long Island Beach, New Jersey, USA. Controlling the loss of beach by multiple phases of groyning can be seen, though the success is negligible given that the loss is due to human attempts to prevent beach movement onshore with RSLR. The loss of beach width makes the attempts to form dunes by dune fencing (at the high water positions) a pointless exercise, indulged in as a means of morale building more than beach building.

the twentieth-century, while the massive urban development of vulnerable barriers underlines the need for the former. Further emphasis is placed on forecasting hurricane landfall positions in order to give time-dependent warnings for evacuation. The difficulty of obtaining precise landfall positions other than within 12–24 hours of the event and the limited evacuation routes means that warnings are not spatially specific and require substantial population movement that might not be justified in light of the actual landfall. People resist calls for evacuation and so instead of horizontal evacuation, some areas legislate for vertical evacuation by proofing buildings to resist the force of surge flows and elevating dwelling positions above predicted surge levels. Coastal zone management on the eastern US seaboard is among the most proactive in the world with emphasis now on planning preventative measures and educating the population to be more responsive and flexible to the hazards of the CZ rather than expecting engineers to defend them totally. This is not to say that more defences are

to be built, but that more strategic zoning and better building regulations are established and a better awareness of normal coastal behaviour is encouraged.

The level of urban use of gravel-dominated barriers is virtually negligible in comparison to sand barriers. The main problems posed by gravel barriers are related to a reducing sediment supply and the retention of spatial position under rising RSL. Gravel-based barriers in the UK that have acted in the past as flood barriers are now being rolled landwards under rising sea levels. The associated problems are not always with the barriers *per se*, but rather with the wetlands (both fresh and brackish) that they protect, which form bird habitats and sites of special interest with respect to bio-diversity. As the barriers migrate, the wetlands are caught in the ensuing coastal squeeze, and often their legal designation means conservationists have to argue for replacement areas ('no net-loss' policy). The difficulty in finding effective replacement areas for the loss of old habitats

Further reading

The following are some standard texts that identify the range of approaches to coastal environments
and their management. Clearly some sources are
dated, but scanning the historical development of
procedures can tell us about pitfalls, rather than
experiencing them directly!

Carter, R. W. G. (1988) *Coastal Environments*. Academic
    Press, London.
    A substantial statement about coastal processes,
    responses and management. Inevitably somewhat
    dated but still a sound geomorphological synthesis
    of coastal problems.
Davies, J. L. (1972) *Geographical Variation in Coastal
    Development*. Oliver and Boyd, Edinburgh.
    Still one of the most readable geomorphological
    accounts of worldwide variation in coastline devel-
    opments.
DEFRA (ex MAFF) *http://www.defra.gov.uk*; English
    Nature *http://www.English-nature.org.uk*, and the
    UK Environment Agency *http://www.environment-
    agency.gov.uk* all provide full web sites where issues
    of coastal protection and management in the UK are
    explored with differing perspectives. For compari-
    son of a US perspective try *http://www.ocrm.nos.
    noaa.gov/czm*.
Horikawa, K. (ed.) (1988) *Nearshore Dynamics and
    Coastal Processes*. University of Tokyo Press, Japan.
    Modern numerical review of processes in the
    nearshore and on the beach. Pertinent overview of
    Japanese contributions to CZ engineering.
Institution of Civil Engineers (1983) *Shoreline
    Protection*. Thomas Telford, London.
    A useful UK view of shoreline protection of fixed
    or fringing beaches, probably somewhat dated now
    but a reflection of how coastal defence policy is
    changing.

Kaufman, W. and Pilkey, O. (1979) *The Beaches are
    Moving*. Doubleday, New York.
    A stimulating, but partisan account of coastal engi-
    neering disasters along the USA shoreline. A more
    up-to-date statement can be found in Dean, C.
    (1999) *Against the Tide: The Battle of America's
    Beaches*. Columbia University Press, New York,
    278pp.
Komar, P. D. (1998) *Beach Processes and Sedimentation*.
    (2nd edn). Prentice-Hall. Englewood Cliffs, New Jersey.
    A major authority on approaches to wave mechan-
    ics and sediment transport. It is less developed
    with respect to longer-term approaches to beach
    development.
Komar, P. D. (ed.) (1983) *Coastal Erosion Handbook*.
    CRC Press, Boca Raton, Florida.
    A selection of mainly US-orientated papers con-
    cerned with the then modern approaches to beach
    and coastal dynamics that still act as the mainstay of
    today's approaches.
Silvester, R. (1972) *Coastal Engineering*. Elsevier,
    Amsterdam, 2 vols.
    An engineer's approach to the understanding of
    wave dynamics. A good statement of a theoretical
    approach to coastal engineering that has not always
    been successful in the light of empirical evidence.
Trenhaile, A. (1997) *Coastal Dynamics and Landforms*.
    Clarendon Press, Oxford.
    A good systematic review of coastal environments,
    especially in terms of geomorphology.
US Army (1977) *Shore Protection Manual* (3rd edn).
    Coastal Eng. Res. Centre, Fort Belvoir, Virginia,
    3 vols.
    The standard reference for US coastal engineering
    design, though now updated and issued as software
    that can be accessed by means of web site
    http://www.veritechinc.com.
van Rijn, L. C. (1998) *Principles of Coastal Morphology*.
    Aqua Publications, Amsterdam.
    A comprehensive approach to coastal morphology
    based on numeric and dynamic principles.

22. Continental Shelves

Colin Jago

22.1 Introduction

The continental shelves do not constitute a quantitatively impressive proportion of the oceans. In geological and morphological senses, the shelves are not even part of the ocean basins, being essentially the drowned margins of the continents. But in economic and social terms, the shelves are the most important parts of the oceans. They are the primary source of fish and shellfish, and a major source of hydrocarbons, hard minerals, sand and gravel. They are sites of increasing navigation, engineering activity, waste disposal, and recreation.

A major concern of the coastal and offshore engineer is the nature of the shelf floor — its morphology and its sediments — and its response to both natural processes (waves and currents) and human interference (emplacement of structures). To this end, it is crucial to understand not only the distribution and properties of shelf sediments, but also the dynamics of sediment entrainment, transport and deposition. The shelf floor is a complex physico-chemico-biological environment: the shelf hydraulic regime drives the sediment system; the sediment flux carries nutrients and contaminants, affects the composition of the benthic community and determines the suitability of the sea floor for building and dumping. However, the response of the shelf floor to the hydraulic regime is partly determined by chemical and biological processes within the surficial sediments.

Inasmuch as the shelves are the recently drowned margins of continents, they carry sediments that were deposited in a range of coastal environments — deltas, estuaries, barriers and lagoons. These sediments, modified by modern processes, lie either on the shelf floor or just below a thin veneer of modern sediments. The geotechnical properties of the substrate may then be dramatically different from the properties of the surface. It is essential that the often complex three-dimensional array of shelf deposits be thoroughly explored in any site investigation. The engineer should consider the geological, as well as the oceanographic, controls of continental shelf sedimentation.

22.2 Morphology of continental margins

The continental margin consists of three segments: shelf (of average gradient 7 minutes), slope (average 4°) and rise (less than 1.5°). The shelf break between shelf and slope lies at an average water depth of 132 m (Figure 22.1). Together these three segments amount to 21% of the sea surface, but they hold 73% of total marine sediments. It has been estimated that there are $150 \times 10^6 \, km^3$ of sedimentary material on the continental margins; this compares with $125 \times 10^6 \, km^3$ on the continents. Given the mineral wealth associated with sediments, the potential economic value of the margins is clear, although as yet almost all engineering activity has been confined to the shelves. As a result of tectonic controls (Chapter 4), there is a wide physiographic diversity among margins, but a small number of basic categories can be identified (Inman and Nordstrom, 1971).

Tectonically active margins
Tectonically active margins occur at destructive plate boundaries and subduction zones, and at conservative plate boundaries and transform faults, and are thus subject to volcanism and

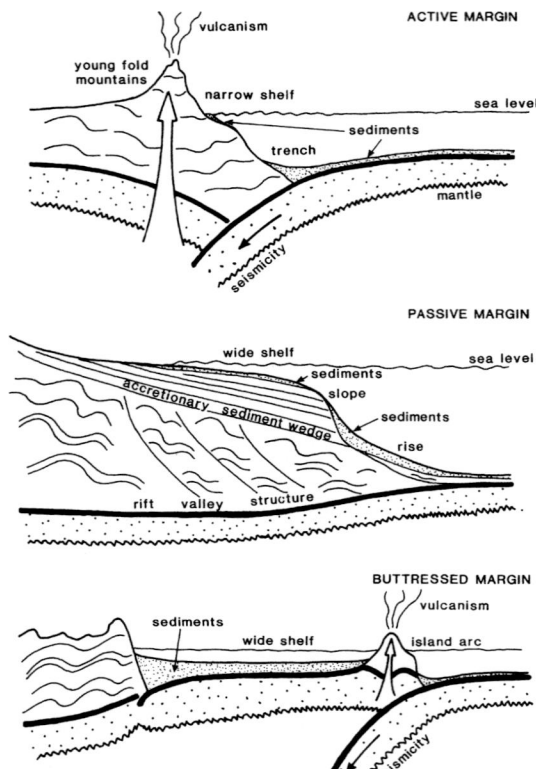

Figure 22.1 Morphotectonic classification of continental margins. After Hayes, 1967a.

seismicity (Chapter 23). They are characterised by mountainous hinterlands, narrow shelves, steep fault-defined slopes and often deep-sea trenches (Figures 22.1 and 22.2). They frequently constitute the leading edges of continents, and are thus adjacent to mountain belts (e.g. Andes, Rockies). Large numbers of high-gradient streams provide sediments (often sandy) to the narrow shelves. A combination of rapid sedimentation, steep gradients and seismic activity produces an unstable substrate prone to failure. Tectonic forces give rise to rapid vertical movements of the margin; subsidence rates of 300 cm/1000 years have been measured. Examples of active margins are the west coast of the Americas, much of the Mediterranean, South East Australia and New Zealand. These are often described as Pacific-type margins.

Tectonically passive margins

Tectonically passive margins are mainly the trailing edges of continents. Their continental slopes mark the original points of rifting at constructive plate boundaries. Most of these margins are now distant from spreading centres and are seismically passive. They border coastal plains of subdued relief, and are characterised by wide shelves, continental slopes of low gradient and extensive continental rise wedges of sediment (Figures 22.1 and 22.2).

Although the major mountain belts, and hence sources of sediment, form on the leading edges of the continents, some major rivers and sediment suppliers (e.g. the Amazon) disgorge on the trailing edges, because of the asymmetry of the watersheds. As the trailing edges move away from spreading centres, cooling and isostatic sinking occur, although subsidence is of the order of 100 mm/1000 years, much less than on active margins. As a consequence of sedimentation on a stable sea floor coupled with slow subsidence, such shelves consist of a seaward-thickening wedge of sedimentary material. Examples of passive margins are the east coast of the Americas, North West Europe ('Amero' subtype) and much of Africa ('Afro' subtype). Juvenile forms, where spreading was initiated in geologically recent times, are the Red Sea and Gulf of California ('Neo' subtype). Passive margins are described as Atlantic-type margins.

Tectonically buttressed margins

Tectonically buttressed margins (Figure 22.1) are associated with destructive plate boundaries behind island arcs. They are less clearly categorised than the other types, but frequently consist of wide and shallow shelves (e.g. Yellow Sea, East China Sea, South China Sea, although the shelf is narrow at Taiwan). These Asian marginal seas are supplied by two of the major sediment suppliers to the oceans (Huanghe and Changjiang rivers) and sedimentation rates are high. Coupled with the seismic activity that results from plate subduction, such shelves may be high-risk areas from the engineering point of view, with sediments liable to liquefaction. Good examples are most of the margins of the Western Pacific and the Gulf of Mexico.

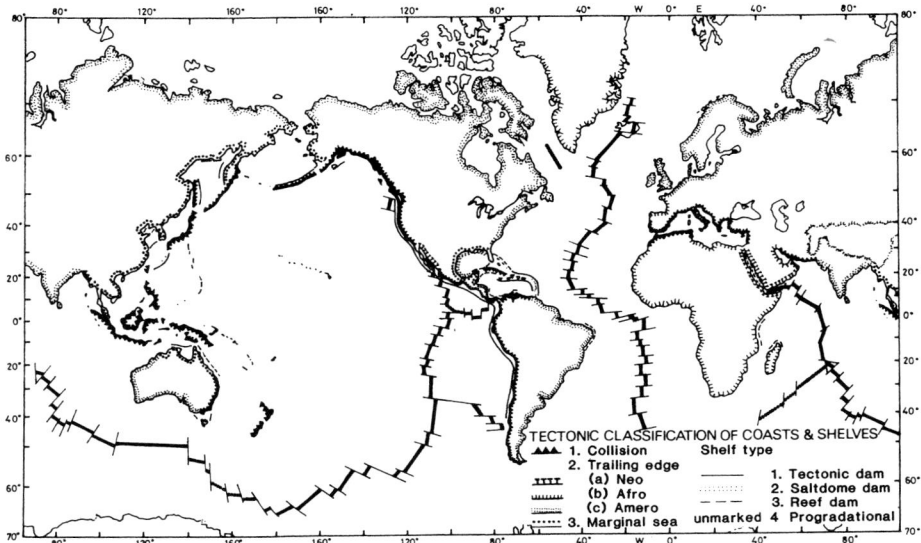

Figure 22.2   Worldwide distribution of margin types. From Inman and Nordstrom (1971).

Many shelves are cut by deep valleys (submarine canyons) which can transfer large amounts of sediment from the shelf to the rise. Shelf sediments, moving with a shore-parallel component, tumble into the canyons; after a period of temporary storage, the sediments are dislodged (by earthquakes or storm waves) and travel at high speeds in turbidity currents which carry enormous quantities of sediment to the rise. Large submarine fans develop at the canyon mouths. Canyons are especially important in shelf sedimentation on Pacific-type margins where the canyon heads can be very close to the shore, thereby intercepting longshore drift. The canyon slopes are steep, and the sediments deposited on them are gravitationally unstable and of low shear strength. Failure of the sediments is most likely on active margins, and on margins where the rate of sedimentation is high.

22.3 Major controls of shelf sedimentation

The nature of shelf sediments and sedimentation is exceedingly varied as a result of the interaction of a number of processes, some of which occur on a larger than basinal scale (tectonics, climate, sea-level change), some of which are restricted to a particular region (tide- or storm-driven hydrodynamics).

Those controls which operate on a scale that is larger than a localised area of a continental shelf are often designated as external controls. These controls are tectonics, climate and sea-level change.

Tectonics

Global tectonics are the fundamental control of shelf sedimentation. Terrigenous sediments are ultimately derived from erosion of continental areas, primarily areas of elevated topography, and mountain building is driven by movement of lithospheric plates. As a result, the continental shelves which receive most terrigenous sediments are those which are adjacent to young mountain belts — the faster the rate of uplift, the faster the rate of supply of sediments to the shoreline and shelf. So the shelves of South East Asia, fed by rivers from the Himalayas and from mountainous island arcs, receive large quantities of modern sediments while the shelves of North West Europe and the Atlantic shelf of North America are sediment starved.

Climate

Climate exerts a profound influence on the weathering of the continents which constitute the
fundamental source of supply of sediments
(Chapter 2). It further determines the mode of erosion by wind, water or ice of the weathered material. Climate accordingly controls the rate and type
of terrigenous sediment supply to the shelves and
produces broad Latitude zones of shelf sediment
types. Polar climates discourage chemical weathering and so production of clay minerals is minimal; production of rock fragments by mechanical
weathering due to ice action is paramount. Land
ice and floating ice are effective transporters of a
range of sediment grain sizes, including gravels
too coarse for transport by most water currents.
High-latitude shelves are therefore characterised
by maximum concentrations of gravels and minimum concentrations of muds (Figure 22.3).
Tropical rainy climates, however, accelerate the
rate of chemical weathering and the production of
clay minerals. Low-latitude shelves are therefore
areas of maximum mud concentrations.

A high proportion of sand and the maximum
yield of sediment are produced where the mean
annual precipitation is about 300 mm/year. On both
high- and low-latitude shelves, seasonal climatic
variations will give rise to corresponding fluctuations in the rates of sediment supply. The yield of

sand is increased under climates that have an
intense dry season rather than a uniformly distributed rainfall. So the sediment supply is high along
coasts with monsoon climates (e.g. South Vietnam,
India). A combination of climate and mountainous
hinterlands puts the Ganges-Brahmaputra and
Mekong among the major sediment carriers.

Climate also affects the rate of production and
distribution of biogenic, chiefly carbonate, sediments on the shelves. Carbonate sedimentation
is encouraged by a low influx of terrigenous
material and high water temperatures (the latter
reduces carbonate solubility — organisms can
then more readily extract carbonate to build their
shells, and the shells of dead organisms dissolve
more slowly). The temperature control means that
there is an increase in carbonate sedimentation in
low latitudes; furthermore, since the westward-
flowing waters of the equatorial currents are
progressively heated before turning polewards at
the western margins of the oceans, the highest
temperatures and most abundant carbonates are
found at these western margins in equatorial
latitudes (Figure 22.4).

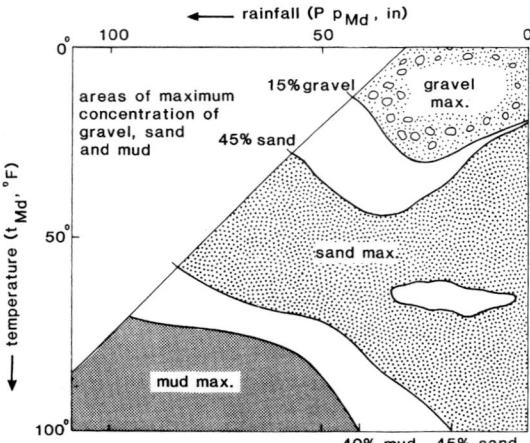

Figure 22.3   Climate control of shelf sediment distribution. After Hayes (1967a).

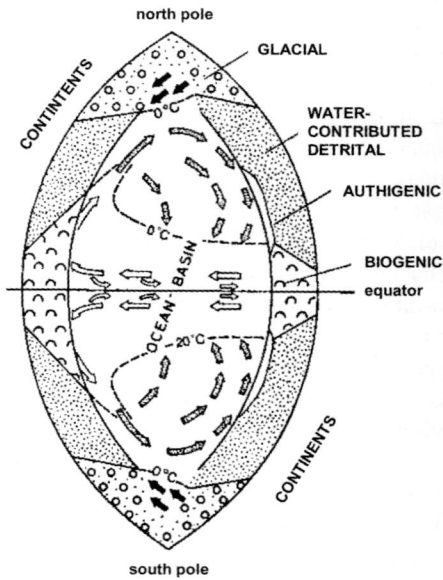

Figure 22.4   Idealised distribution of shelf sediment
types. White arrows—warm water; dotted arrows—
upwelling water; black arrows—cold water. From
Reineck and Singh (1975).

On trailing-edge shelves in low latitudes, carbonates are deposited as the initial shoulder of the rifting coast (e.g. Red Sea). Off-lying banks, born early in the rifting process, drift close to, but detached from, the margin; continuous deposition of carbonates on the banks perpetuates shallow-water conditions despite the subsidence that results from spreading (e.g. the Bahamas). Peninsulas and adjacent shelves distant from terrigenous sources have also maintained a long history of carbonate deposition (e.g. Florida shelf, Yucatan shelf). Carbonates also occur on some mid-latitude shelves where the terrigenous supply is low (e.g. off southern Australia, western Ireland, western Scotland). The nature of carbonate sediments is often largely determined by the supplying organisms rather than by the hydraulic regime — coarse-grained carbonates can occur on low energy shelves.

Sea-level change

Eustatic sea-level changes occur in response to climatic variations (e.g. glaciations) and to tectonic processes (e.g. sea-floor spreading and the formation of mid-ocean ridges). Local changes can occur as a result of isostatic adjustments (e.g. because of accumulation and subsequent melting of ice sheets) and tectonic uplift or downwarping.

Such changes obviously control the position of the shoreline; most of the continental shelves were exposed to sub-aerial processes at various times during the Pleistocene (Chapter 6) when a complex mosaic of glacial, fluvial and lacustrine sediments was deposited on them. Sea-level changes, especially when rapid, alter river-mouth morphology and equilibration which in turn control the supply of terrigenous sediment to the shelf. Thus modern river mouths are predominantly estuarine, because of the Flandrian transgression, and are effective sediment traps (Chapter 20). Much of the shelf is therefore deprived of a supply of terrigenous material. Sea level determines the water depth and hence available hydraulic energy at the sea bed, and many shelf sediments are now at depths too great to be frequently affected by modern shelf processes.

The foregoing factors combine to determine the flux of sediment from land to shelf. There are gross geographical disparities in this flux: the shelves of South East Asia receive large quantities of terrigenous sediment while the shelves of North West Europe and the Atlantic shelf of North America are starved of sediment. Although comparable with respect to their dominant tidal regimes, the Yellow and North Seas are remarkably different in sedimentary regimes because of gross differences in sediment supply ($> 2000 \times 10^6$ and $> 10 \times 10^6$ tonnes/year, respectively). The well-supplied shelves, rather than the starved shelves, present most engineering problems since they are floored with rapidly deposited, under-consolidated muds prone to failure when dynamically loaded by waves, earthquakes or man-made structures. Tectonics and sea-level change determine the size and morphology of continental shelves. Shelves are open (e.g. Atlantic and Pacific shelves of North America) or semi-enclosed (e.g. North Sea, Yellow Sea). Of the major sediment suppliers, only the Ganges, Brahmaputra, Indus and Amazon supply open ocean shelves.

22.4 Shelf hydrodynamic regime

Shelf sediments are subjected to various water currents acting alone or in concert (Figure 22.5). Of these, meteorological currents and tidal currents are the most important on most shelves, but others, especially oceanic currents, can affect some shelves.

Oceanic currents

Differential solar illumination gives rise to a heat flow from the equator to the poles, accomplished by shallow wind-driven currents and deep thermohaline currents. The major geostrophic currents hug the continental margins, flowing mostly parallel to the shelf break. These currents undoubtedly deposit sediments on the continental rise — thinly bedded fine-grained contourites — but they can also intrude onto the shelf as a result of eddy shedding and large-scale meandering. The resulting flows can transport suspended sediment and sometimes cause scour (e.g. the Gulf Stream over the Blake Plateau on the US Atlantic shelf. The Canaries Current generates megaripples on the

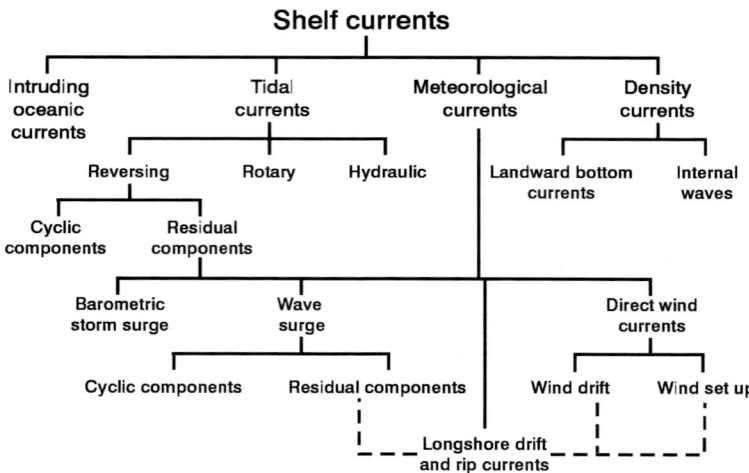

Figure 22.5   Summary of the main physical processess of the shelf hydraulic regime. Redrawn from Swift *et al.*, 1971.

outer Saharan shelf, and the Agulhas Current produces spectacular bed forms (sand ribbons, sand waves) on the South East African shelf. The Agulhas Current flows at up to 2.5 m/s at the surface, but oceanic currents usually flow at lower velocities. The slower flows may not in themselves be competent to transport much sediment, but they may be significant when superimposed on other shelf currents.

Meteorological currents

Meteorological forces generate a range of currents that become the paramount energy input on some shelves. Wind drift currents arise from shear stresses at the air–sea boundary. Such currents deepen with time, as turbulent mixing induces a transfer of momentum. The subsurface currents are not, however, parallel to the wind direction, since the Coriolis effect and the Ekman spiral cause a deviation of the surface and subsurface flows respectively. Surface wind drift current velocities are some 2–5% of the surface wind velocities. Bottom currents so produced are much less, though flows up to 80 cm/s have been recorded at depths of 50–80 m.

Currents are also produced by waves whose motions are accompanied by an orbital motion of water particles. The circular orbits have a diameter equal to the wave height at the surface but this diameter decreases with depth. At depths greater than about half the length of the wave, the movement of water particles ceases. At shallower depths, the particle motion extends to the sea floor, which then modifies the orbits from circular to elliptical and absorbs energy through friction. The orbits become increasingly elliptical with depth until they are linear oscillations at the sea floor of 10 cm/s or more. These oscillations constitute wave surge currents which may have sufficient energy to entrain sediment. But wave surge has little direct effect on sediment transport since the movement is only back and forth.

However, sediments set into motion by wave surge may then be moved by residual components of the velocity field which by themselves would be below the sediment movement threshold. Moreover, percolation of water through the porous sediment induced by the pressure field of passing storm wave crests and troughs induces cyclic loading, liquefaction, and loss of sediment strength. This may lead to the collapse of engineering structures placed on the sea floor.

In shallow water, the surge currents exhibit time–velocity asymmetry; the shoreward oscillation is stronger and shorter than the seaward, and the water particle orbits no longer close. This residual is the wave drift current which produces a net shoreward movement of sediment on the shore face under some conditions.

The capability of waves to entrain sediment on the shelf floor depends on their energy (Figure 22.6). The largest waves are generated in mid to high latitudes by winds associated with mid-latitude cyclones. These waves travel to the north-east till they break in storm-wave and west-coast swell environments. They also move along some east-coast shelves where they are augmented by local summer sea-breeze waves. The average annual wave energy input is therefore high on storm-wave shelves and west-coast swell shelves where wave surge currents augmenting other currents can stir sea-floor sediments down to water depths of 200 m or more, rather less on east-coast swell shelves. In the subtropics, the trade winds and seasonal monsoons create moderate energy waves, while in the tropics wave energy further diminishes (though punctuated by occasional hurricanes). Polar shelves are protected from waves by ice; shelves in semi-enclosed seas are protected by the surrounding land masses and limited fetch.

On all shelves, wave energy is attenuated as the waves approach land and water depth decreases; the wider and shallower the shelf, the greater the attenuation. Wave attenuation is therefore greatest on Atlantic-type margins, least on Pacific-type. Thus the energy loss across the wide shelf of the South East United States is up to 85%, compared to a loss across the narrow shelf of South East Australia of only 4%.

Storm surge, due to reduced barometric pressure or high wind stress, produces a long-period oscillation of water level that may not only flood coastal regions but may also induce sediment transport on the shelf. Currents of 1 m/s, both onshore and offshore, can be generated by hurricanes and the resulting transport of sediments, especially offshore, may be intense. Storm surges are especially spectacular on the trailing edges of continents which have low-lying constructional coasts.

Tidal currents

Tides, the result of gravitational attractions between Moon and Earth, and Sun and Earth, occur as a wave of long period propagates on to the shelf from the adjacent ocean. The tidal wave

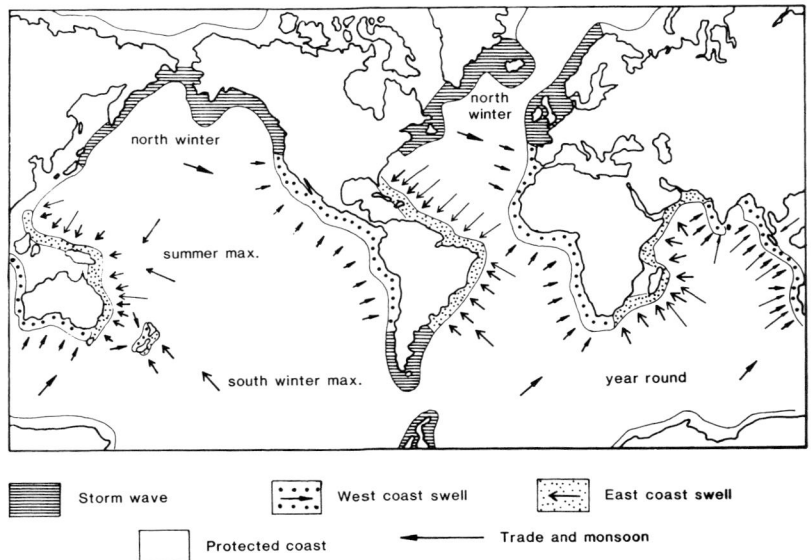

Figure 22.6 Wave energy input to the continental shelves.

represents an equilibrium attained by the sea surface between the centripetal and gravitational forces of the Earth–Moon system and, to a lesser extent, the Earth–Sun system. This equilibrium gives rise to spring tides, when the gravitational forces of the two systems are additive, and to neap tides when they are in opposition: this is a fortnightly cycle. Further variations arise from the eccentricity of the lunar orbit (the highest tides occurring when the perigee falls at either new or full Moon), and from variations in the Earth–Sun distance in summer and winter.

The tide is much modified by the distribution of the continents and the shape of the ocean basins. If a sea is semi-enclosed with only a small connection to the open ocean, propagation of the tidal wave is hampered; the sea is microtidal, like the Mediterranean (Figure 22.7). If the length of the basin is such that its natural period is close to that of the principal tide-producing force, then resonance occurs and a large tidal range and strong tidal currents result, as in the Southern Bight of the North Sea. Such macrotidal shelves are to be found in the Gulf of Korea, the Gulf of California, the Arabian Gulf, the Gulf of Maine and elsewhere. The highest tides of all are in the Bay of Fundy (Canada) whose resonance period is 12.58 hours, close to the lunar semidiurnal period of 12.42 hours.

The principal tidal components are reversing so that, in the first approximation, there is no net flow of water — but there are invariably residuals caused by basin topography. In coastal waters, a time–velocity asymmetry develops because of the shoaling effect of the sea bed and constriction of the tidal wave by the enclosing coastal regions. Fast flows of 1–2 m/s can then be generated. The resulting net transport of water may be accompanied by a net transport of sediment. On the open shelf and in wide bays, the Coriolis effect brings about a constant change of direction of the tidal currents which are accordingly rotary. Close to shore, the tide becomes strongly elliptical to rectilinear. Macrotidal shelves are high-energy shelves in terms of sediment transport since sediment entrainment can occur twice daily.

Density currents

Density currents occur in response to differential densities of sea water of varying salinity, temperature or suspended sediment load. The density stratification characteristic of estuarine circulation (Chapter 20) can continue across the shelf, with

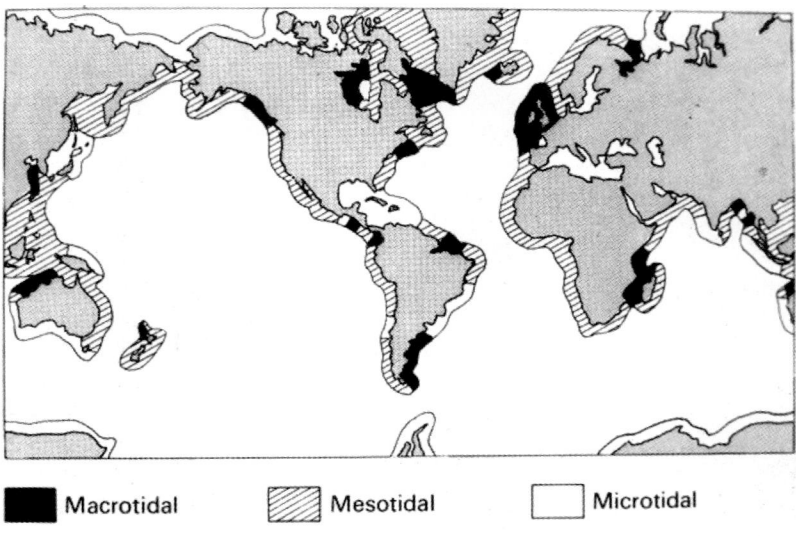

■ Macrotidal   ▨ Mesotidal   □ Microtidal

Figure 22.7 Tidal ranges: microtidal (<2 m), mesotidal (2–4 m) and macrotidal (>4 m). After Davies (1964).

surface water flowing offshore and bottom water onshore. The bottom drift is slow and is not generally able to move bedload, but it will affect sediment already entrained by other stronger currents. Stratification is destroyed by the turbulence from high-velocity tidal currents.

22.5 Properties of shelf sediments

Shelf sediments may be genetically classified as:

1. detrital (i.e. supplied by rivers, coastal erosion, aeolian activity or glacial activity)
2. biogenic (i.e. whole and fragmented shells; faecal pellets)
3. authigenic (i.e. formed, like phosphorite, within a host sediment — typical in areas of upwelling where cold, nutrient-rich water from depth rises to the near surface at mid-latitudes on the eastern sides of the oceans: Figure 22.4)
4. volcanic (predominantly at active margins)
5. residual (i.e. from *in-situ* weathering of rock outcrops).

While geologists have traditionally classified sediments in terms of texture (mainly grain size) and mineralogy, a more useful classification for oceanographers and engineers, who are concerned with the properties and behaviour of sediments, is based on sediment cohesion. Non-cohesive sediments are those in which there is no effective cohesion between particles. Sands and gravels are non-cohesive; they are composed of silicates (primarily quartz), carbonates (calcite and aragonite) and rock fragments. Individual grains are, by definition, coarser than 63 μm diameter. Cohesive sediments are those in which there is significant cohesion between adjacent particles: muds are cohesive. Cohesion is dependent on electrochemical forces, van der Waals forces and stickiness imparted by organic compounds created by marine organisms (mostly algae and bacteria). Muds, with particles finer than 63 μm diameter, are composed typically of clay minerals (complex alumino-silicates), organic matter and water.

Non-cohesive siliciclastic sediments

The dynamics of non-cohesive sands and gravels have been extensively researched over past decades. The transport of non-cohesive sediments is controlled by inertia forces dependent on grain size. Both the critical (or threshold) conditions for initial transport and the transport rate can be parameterised in terms of the bed shear stress (usually expressed as τ_0 in Pascals), the shear velocity (expressed as U_* in m/s or cm/s, where $U_* = \sqrt{\tau/\rho}$), or the flow speed expressed as U_1 in m/s or cm/s, the flow speed 1 m above the bed). The shear velocity is proportional to the flow speed. Reasonably accurate estimates (to within a factor or two) can be made of the critical condition for transport using the grain diameter as the sole descriptor of the bed sediment for both unidirectional currents (Miller *et al.*, 1977) and oscillatory currents (Komar and Miller, 1975). Simple equations of the form:

$$q = k\left[(U_1 - U_{1c})/U_{1c}\right]^n \qquad (22.1)$$

have been derived (Bagnold, 1963; Hardisty, 1983; Jago and Mahamod, 1999) for sand transport by unidirectional currents where q is sand transport rate, c indicates critical values for transport, and k and n are dependent on grain diameter and bed shear stress. It is generally agreed that $n = 3$ for bedload transport, increasing to 4 or 5 for total load transport (which includes suspension). Such equations are not very accurate but will predict transport to within an order of magnitude. The major problem is the non-linear dependence of q on U_* or U_1 so small errors in the latter produce large errors in q.

Further difficulties arise when steady currents are enhanced by waves. Heathershaw (1981) showed that the sand transport rate increases many fold when even quite small waves are superimposed on steady currents. Waves, with some help from the current, entrain the sediment and the current transports it. The combined bed stress is not simply additive but involves non-linear interactions. There are several formulations which consider sediment transport by combined waves and currents including the following given by Soulsby (1997).

$$q_t = A_s \bar{U} \left[\left(U^2 + \frac{0.018}{C_D} U_{rms}^2 \right)^{0.5} - \bar{U}_c \right]^{2.4}$$
$$\times (1 - 1.6 \tan \beta) \tag{22.2}$$

where $A_{sb} = [0.005h(d/h)^{1.2}]/[(s-1)gd]^{1.2}$, $A_{ss} = [0.012dD*]^{-0.6}$ and $A_s = A_{sb} + A_{ss}$. In equation 22.2, q_t is total sand transport rate, $\bar{U}$ is depth-averaged current velocity, U_{rms} is root-mean-square wave orbital velocity, C_D is a drag coefficient due to the current alone, U_c is the critical current velocity required for transport, β is the slope of the seabed, h is water depth, d is the median grain diameter and $D*$ depends on the relative density of the sediment and the kinematic viscosity of the water. The term A_{sb} gives bedload transport and term A_{ss} gives suspended load transport.

The sand transport rate predicted using an equation of the form of equation 22.2 is orders of magnitude greater than the rate predicted using a simple power relationship of the form of equation 22.1. This suggests that sand transport on shelves is dominated by extreme events and on many shelves most sand transport does occur during storms. However, on shelves with large tides, extreme events are less significant: moderate waves superimposed on average tidal currents occur frequently, while storm waves superimposed on peak spring tidal currents occur rarely, and both combinations may make comparable contributions to long-term sand transport.

Carbonate sediments

Although nominally non-cohesive, carbonate sands and gravels differ from their siliciclastic counterparts in both hydrodynamic and geotechnical properties. This is because carbonate grains, which are essentially biological in origin, have irregular shape and variable effective density, both properties being determined by the nature of the organism that produced the grain. Carbonate grains are whole shells, fragments of shells, tests, skeletons, etc. and these may contain chambers and holes. Furthermore, due to diagenesis in the upper centimetres of the seabed, carbonate grains are often cemented, varying from slightly cemented to strong calcarenite rock. As a result, settling velocity, critical shear stress and transport rates are not accurately parameterised by grain diameter alone. Therefore, equations 22.1 and 22.2 are likely to be in error when applied to carbonates.

There are also important differences in the geotechnical properties of siliciclastic and carbonate sediments. Up until the late 1980s, information on the geotechnical properties of carbonates was scarce. However exploitation of the continental shelf off North West Australia and in the Bass Strait required installation of offshore production platforms in deep carbonate deposits. Major problems were encountered with the installation of piles for the foundations of the North Rankin A platform, with piles penetrating many metres under their own weight and generating negligible capacity in either end bearing or skin friction. These failures led to important programmes of research and a dedicated conference on engineering with carbonate sediments (Jewell and Andrews, 1988). The proceedings of this conference include useful reviews of the engineering geology of carbonates by Fookes (1988) and their geotechnical properties by Semple (1988). A second conference followed in 1999 (Al-Shafei, 1999). Ongoing research has sought to define the mechanics of carbonate sediments and rocks in relation to critical-state frameworks, see for example Coop (1990), Coop and Atkinson (1993), Lagioia and Nova (1995) and Cuccovilli and Coop (1997).

Uncemented carbonate sands generally display much higher critical-state friction angles than silica sands, around 40° compared with 33° to 35°, due to the interlocking of the shelly particles. However the shear strength decreases with increase in stress more rapidly than for silica sands due to particle crushing. The *in-situ* void ratio may be very high in comparison with silica sands due to the very angular and plate-like shape of the particles and to a lesser extent the presence of intra-particle voids. Initial compressibility is also very high — this is the reason for very low confining stresses (and hence skin friction) being developed on piles after driving.

For slightly cemented materials the effects of cementing are difficult to separate from those of particle breakage. In general, three phases of behaviour are observed: an initial elastic phase where the material behaves essentially as a rock and the stiffness is related to the degree of cementation rather than the porosity; a yield or 'destructuration' stage when rapid compression occurs at constant stress and the behaviour changes from rock-like to soil-like; the final stage when the behaviour is the same as for an initially uncemented material.

Cohesive sediments

The foregoing equations cannot be applied to cohesive sediment transport. While non-cohesive sediment transport is dominated by inertia forces, cohesive sediment transport is controlled by cohesion. Cohesion depends on many factors including water content (i.e. compaction), mineralogy, temperature, salinity and organic mucus content. Grain size *per se* is not important (see also Chapters 3 and 7). This means that it is difficult to parameterise and predict cohesive sediment properties and transport. A fundamental property of cohesive sediments is their propensity to form aggregates — weakly bound flocs of clays, organic matter and water — which form in suspension under the influence of turbulence, differential settling and organic binding. Aggregation varies with the concentration of suspended matter and aggregates are easily broken by turbulent shear. Consequently, the size and behaviour of aggregated matter change rapidly on short time and length scales and it is difficult to model and predict settling and resuspension fluxes. The difficulty is compounded by biophysical interactions particularly during plankton blooms. Blooms give rise to phytodetritus which, by combining with cohesive matter in the water column, settles rapidly to the seabed as a layer of benthic fluff (Jago *et al.*, 1993). Large aggregates exceeding 600 μm are common in shelf seas. These have low density but fast settling velocity, typically faster than fine non-cohesive sand grains (settling velocities of 5 mm/s are not uncommon: Jago *et al.*, 1993).

Deposition of cohesive particles from low concentrations occurs under low shear stress conditions, typically when $\tau_0 < 0.1$ Pa ($u_1 < c$.

0.15 m/s), but deposition from very high concentrations may occur under much faster currents with τ_0 up to 1 Pa. A thin region of quasi-laminar (non-turbulent) flow develops at the seabed if bed roughness is small and particles that fall into this region will be deposited (McCave, 1971). Mud deposition generally occurs in low energy parts of the shelf below wave base but deposition can occur under higher energy conditions if there is sufficient material in suspension.

Once deposited on the bed, the structure of aggregates changes over time due to compaction and the erosion stress likewise changes over time. The early hypothesis of Krone (1963) that critical erosion stress is controlled by the strength of aggregated matter on the surface of the seabed remains the most plausible physical model. While a freshly deposited mud with high water content may be readily eroded, a compacted mud will resist erosion; compacted mud may require a substantially greater erosion stress than sand. A freshly deposited mud may erode at rates greater than 1 mm/hour but a compacted mud, under the same excess shear stress, may erode at less than 1 mm/day. Critical erosion conditions and transport rate have been shown to be related empirically to yield strength (Migniot, 1968) and density (Delo, 1988) but universally-applicable algorithms have not been developed (Vanoni, 1975; Thorn and Parsons, 1980; Mehta *et al.*, 1982; Villaret and Paulic, 1986, and Mehta, 1988).

Other geotechnical problems

Large-scale offshore development, initially on the continental shelf but progressively moving into deeper water, has been a driving force behind research on many areas of soil mechanics and geotechnical engineering of sediments. Examples have been the behaviour of sediments under cyclic loading due to wind and waves, the undrained behaviour of sand under very large platform foundations experiencing short-term loading and design and installation methods for very large piles.

Following loss and damage involving oil platforms in the Gulf of Mexico due to the underwater mud slides triggered by Hurricane Camille (Sterling and Strohbeck, 1973), attention was also turned to the potential of wave loading to cause massive slides of very soft sediments on even very shallow slopes following initial work by Henkel (1970). As offshore developments extend across the shelf break into deeper water, the problems of underconsolidated sediments on the relatively steep gradients of the continental slopes will become paramount. The continental margins are characterised by hydrodynamic and sedimentological regimes that differ considerably from those of the continental shelves: these present geotechnical problems and engineering scenarios are beyond the scope of this chapter.

22.6 Organism–sediment interaction

This is a factor that is too often overlooked by oceanographers and engineers, perhaps because it is notoriously difficult to quantify. However, the marine biomass is at a maximum on the shelf where infauna and epifauna continually modify the upper 10–15 cm of the substrate (Figures 22.8 and 22.9). Organism activities alter the strength and stability of the sediments, while the strength and stability of the sediments control the diversity and density of the organisms.

Benthic organisms may secrete organic films, produce dense fields of vertical tubes and form calcareous crusts, all of which activities may stabilise the host sediment. Others, or the same, organisms effect grain agglomeration and flocculation, burrow through and track the surface of the sediment and produce faecal pellets; as a result the water content of the sediment is increased, shear strength decreased and scour by marine currents accelerated (see Rhoads and Boyer, 1982, for a review of animal–sediment interactions).

The magnitude of such biological processes should not be underestimated: a typical shelf sediment contains 150–500 g/m^2 of biomass in its upper few centimetres and, for example, mussels in the Wadden Sea generate some 150 000 metric tonnes of sediment (i.e. faecal pellets) per year. Biological alteration of the sediment's geotechnical properties may become critical on shelves with inherent sediment instability due to earthquakes, storms or steep sea-floor slopes.

22.7 Patterns of shelf sedimentation

The patterns of sedimentation on modern shelves have been established in varying measure by the sea-level oscillations of the recent past, by the quantity and quality of sediment supply and by the hydrodynamic regime. Shelves were periodically exposed to sub-aerial processes, then crossed by transgressing shorelines and finally swept by shelf currents and waves. Shelves adjacent to the

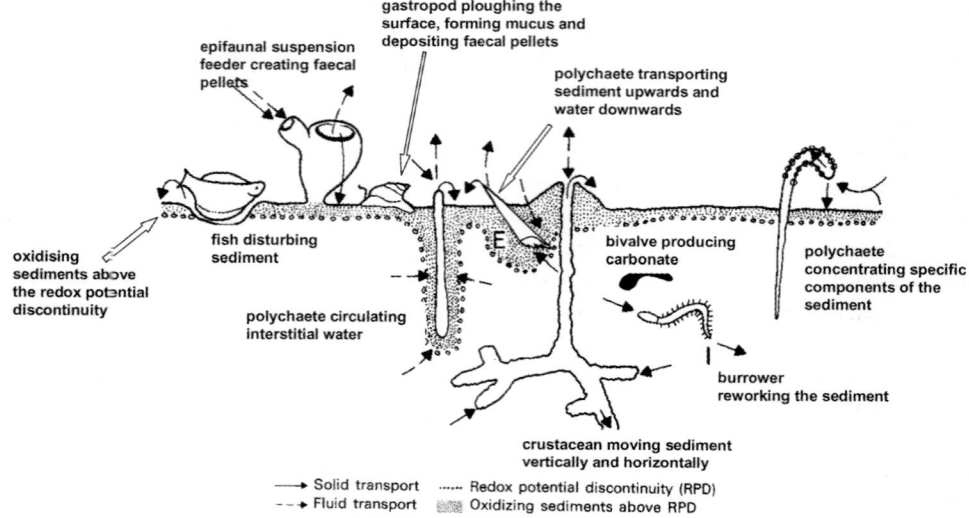

Figure 22.8   Organism–sediment interaction. From Webb *et al.* (1976).

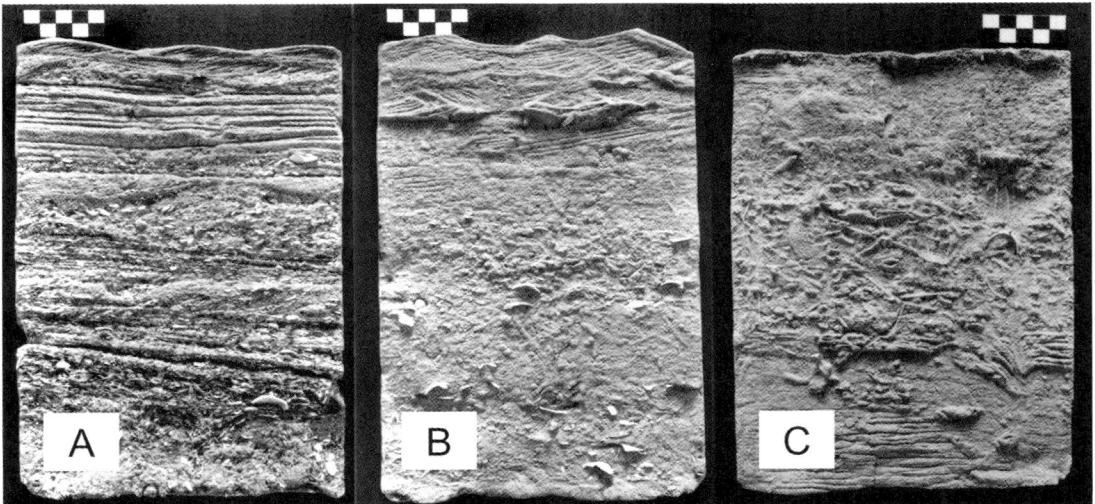

Figure 22.9 Bioturbation (sediment reworking by organisms) in intertidal sands. Primary sedimentary structures (cross-bedding and plane bedding) progressively destroyed by burrowing tellinids and worms. Note that surface bed forms and structures may not be affected. Scale in cm. From Jago and Hardisty (1984).

mouths of major sediment-carrying rivers receive large supplies of sediments, while shelves decoupled from any terrestrial source of sediment are effectively sediment starved. There are large temporal and spatial variabilities of controlling factors, so shelf morphologies and sediment distributions are complex and variable.

Most shelf sediments are not in equilibrium with present-day conditions because they were emplaced during periods of low sea level. Many shelves have sediment distributions that owe more to past processes than to modern processes. Sediments that are now in deep water may have been emplaced in glacial, fluvio-glacial, fluvial or coastal environments. Thus coarse sands occur on outer shelves. Such sediments are designated as relict (i.e. remnant from a different earlier environment and now in disequilibrium with the shelf hydraulic regime). On high energy shelves, much of this relict material has been reworked to form palimpsest sediments (i.e. possessing vestiges of a former environment but with the imprint of the present environment).

Where the shelf is subjected to a high-energy hydraulic regime, the relict sub-aerial and nearshore morphologies and sediments become progressively modified by modern waves and currents. How this

is achieved, and the nature of the resulting patterns of sedimentation, depend on the rate and type of sediment supply (primarily whether it is cohesive or non-cohesive) and the nature of the hydraulic regime (primarily whether it is tide dominated or storm dominated). A genetic classification of shelf sedimentary regimes has been developed, based on these controls, which differentiates muddy and sandy shelves, while sandy shelves are differentiated into tide- and storm-dominated types.

22.8 The shoreface and coastal zone bypassing

Under most conditions, the inner shelf hydraulic regime pins sands against the shoreline and maintains the nearshore sand prism. During major storms, the role of wind-driven currents becomes important. More sand is entrained because of increased wave energy. Winds drive surface water shorewards, creating downwelling and offshore bottom flow; massive amounts of sand may then be transported seawards. This loss of sand may be greater than can be recovered by the prevailing onshore wave drift currents. The seawards escape

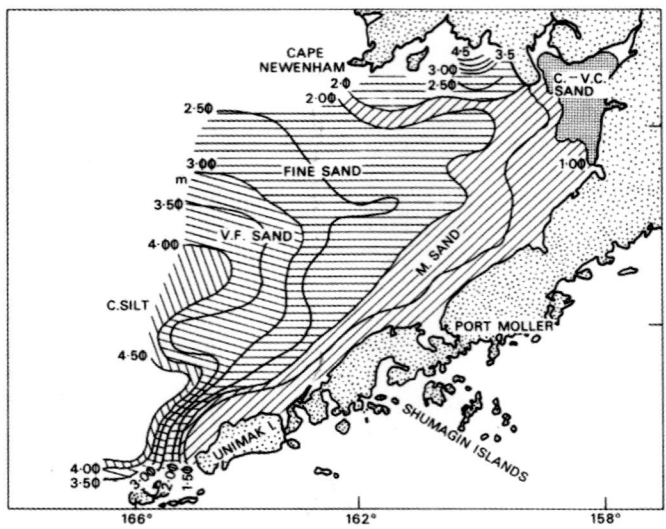

**V.C. SAND = 1–2 mm, C.SAND = 0.5–1.0 mm, M.SAND = 0.5–0.25 mm,
F.SAND = 0.25–0.125 mm, V.F. SAND = 0.063–0.125 mm, C.SILT = 0.031–0.063 mm**

Figure 22.10   Seaward-fining textural gradient due to waves, in Bristol Bay, southern Bering Sea. $\phi = -\log_2 D$ (where D is grain diameter in mm). After Sharma *et al.* (1972).

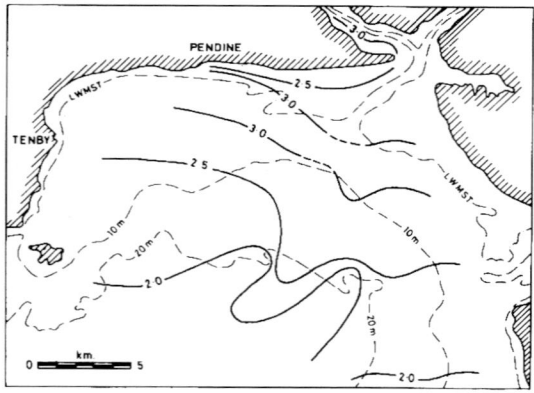

Figure 22.11   Seaward-coarsening textural gradient in tide-dominated Carmarthen Bay, Bristol Channel. $\phi = -\log_2 D$ (where D is grain diameter in mm).

of sand becomes permanent, especially during a marine transgression when sea level is rising. The shoreface then serves as a zone of sediment bypassing as it erodes and retreats. Such shoreface bypassing (Swift, 1976a, b) leaves a thin sand sheet in the wake of the transgression.

The fine-grained sediments escape from the coasts more easily. Suspended material disperses seawards and is deposited as mud, usually beyond the shoreface where wave surge currents stir the bottom only rarely. This route becomes dominant where a river is supplying large amounts of material to the coast. Sediment passes through the littoral energy fence via jets associated with flood stages of the river flow and ebbing tides. River-mouth bypassing (Swift, 1976a, b) of modern sediments produces a mud blanket that steadily grows towards the shelf break, gradually covering older sediments; few shelves have yet reached this equilibrium situation because significant river-mouth bypassing occurs in only a few areas.

The inner shelf is affected at most times by the passage of fairweather waves and hence by turbulence. On most shelves this turbulence prevents the permanent deposition of fine-grained sediments, so that the innermost shelf is clothed by sand and muds are deposited in deeper water. If, however, the supply of fine-grained sediment is very large — as off the Amazon — then muds may prevail even in shallow water. Usually, though, sands are deposited on the intertidal foreshore and the subtidal shoreface. The shoreface consists of an upper part which is subject to a wave drift current and a lower part which is

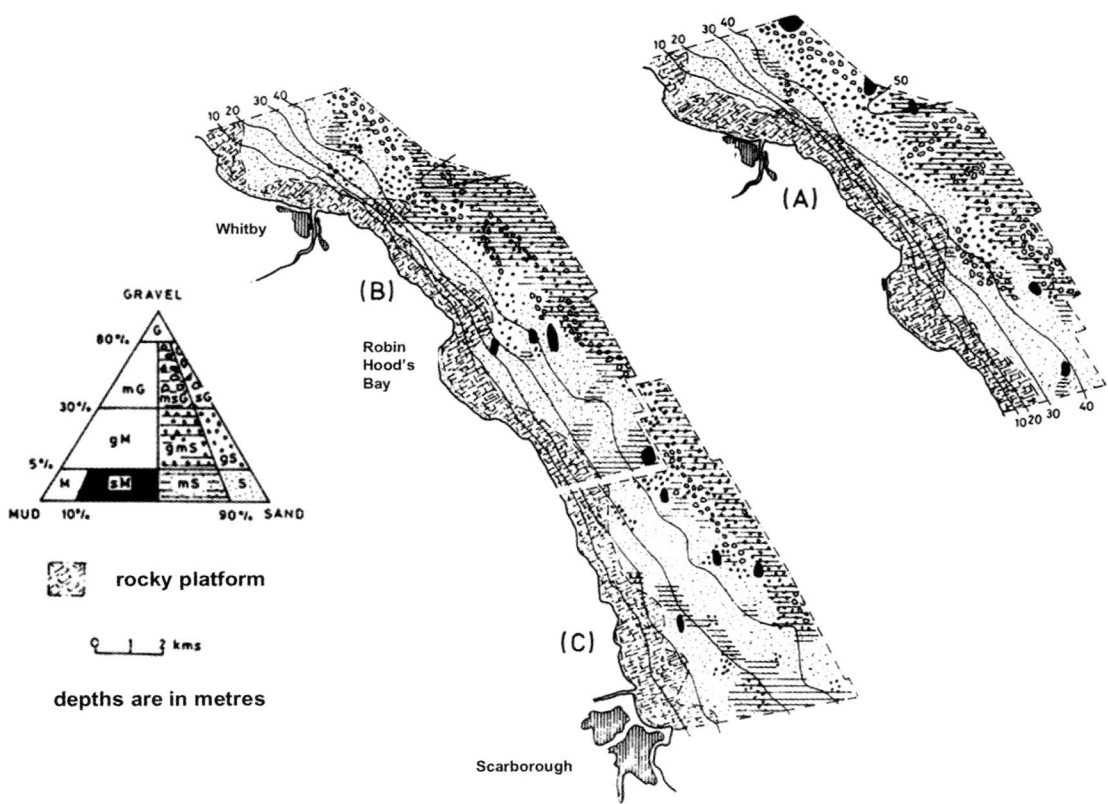

Figure 22.12   Sediment-free shoreface, North Yorkshire shelf, North Sea. From Jago (1981).

agitated by wave surge. Depths of 10–15 m mark the division between these two parts, though this depends on the rigour of the wave climate. Sand in shallow water is pushed shorewards by wave drift currents and seawards by rip currents, so that much of the shoreface is the dynamic surface of a nearshore sand prism (Chapter 21), sand moving shorewards during fairweather conditions and seawards during storms.

On many shorefaces, there is a seaward-fining textural gradient in response to seaward diminishing sea bed shear stresses and to rip current fallout (Figure 22.10). However, where tidal currents augment wave surge currents, there may be a seaward coarsening of sands that corresponds to increasing tidal flows in deeper water (Jago, 1981 and Figure 22.11). On inner shelves where there is a paucity of sand, much of the shoreface

may be swept free of sand altogether so that a wide wave-cut platform extends seawards from a rocky shoreline (Figure 22.12).

22.9 Muddy shelves

Muds dominate the shelf sediment regime in regions where major rivers discharge large quantities of cohesive sediments. The sedimentary regime of these shelves is autochthonous. The pattern of mud deposition depends on the rate of supply of sediment. Where this is relatively low, mud deposition occurs on the lower shoreface (typically at water depths of 5–20 m) while sands are deposited on the upper shoreface. The mud belt is replaced by relict or palimpsest sands on the outer shelf. Examples are the North Sea shelf of the

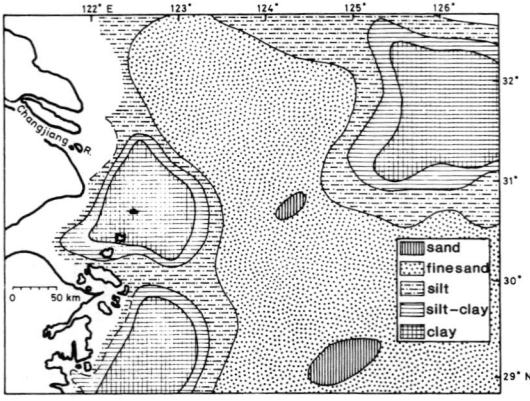

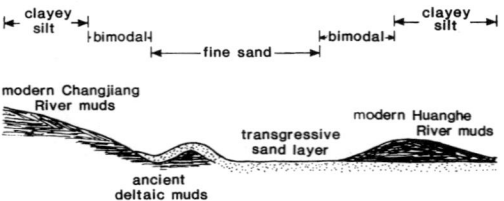

Figure 22.13 Allochthonous sedimentation showing autochthonous sand 'windows', East China Sea. From Nittrouer et al. (1984).

Netherlands and Belgium and the inner German Bight (Dorjes et al., 1970), the Adriatic shelf south of the Po, parts of the north-west Gulf of Mexico (Curray, 1960) and the Washington shelf of the USA (Nittrouer and Sternberg, 1981) the shelf north of the Zaire (Congo) river (Giresse and Kouyoumontzakis, 1973) and the Golfe du Lion adjacent to the Rhone (Jago and Barusseau, 1981). Generally, only the inner shelf sands and shelf muds are zones of active sedimentation. Wave energy is the most critical factor: in parts of the German Bight of the North Sea, mud is accumulating at a rate of 15.5 cm/100 years even though near-bottom tidal currents reach 0.3 m/s, but wave effectiveness is low. The seaward extension of the mud belt depends on the concentration of suspended sediment (which determines the deposition rate) relative to wave- or current-generated bed shear stress (which determines the resuspension rate); the greater the wave energy, the further seaward the boundary between inner shelf sand prism

and shelf mud blanket (McCave, 1972). The outer shelf relict sand blanket escapes mud deposition because the rate of supply of mud is small relative to the resuspension capacity of tides, internal tides and internal waves (Huthnance, 1981).

Where the supply of fine sediment from rivers is high, the mud blanket progrades towards the outer shelf. This occurs on parts of the north-west Gulf of Mexico (Curray, 1960; van Andel, 1960) and the Oregon shelf (Kulm et al., 1975) of the USA. On the shelf of the East China Sea, where there is a large supply of fine sediment from the Changjiang and Hwanghe, a lens of mud is rapidly accumulating on the inner shelf, which thins seawards towards the mid shelf where it is replaced by relict/palimpsest sands. Because of the semi-enclosed morphology of this shelf, more muds supplied by the Hwanghe encroach on the outer shelf (Butenko et al., 1985). The resulting distribution of sediments is essentially an autochthonous regime (modern muds) with autochthonous windows (relict sands) (Figure 22.13). The inner shelf mud deposit is accumulating at rates up to 5cm/year and is dispersed by a southward-flowing coastal current during the stormy winter months (DeMaster et al., 1985). These deposits have high water contents and porosities, with average values of 111% and 72%, respectively; shear strength is as low as 1.5 kPa (Keller and Yincan, 1985). The maximum mud accumulation rate on the outer shelf is 0.3 cm/year (DeMaster et al., 1985).

The most dramatic examples of muddy shelves are those adjacent to major fluvial sediment sources such as the Amazon shelf (Nittrouer and DeMaster, 1986), Ganges–Brahmaputra system (Kuehl et al., 1990) and the Yellow Sea shelf adjacent to the Hwanghe river (Alexander et al., 1991). For obvious reasons, it is difficult to quantify the sediment loads of large rivers but it is estimated that it is up to 1300×10^6 tonnes/year (Meade et al., 1985) for the Amazon, so two orders of magnitude greater than the sediment supply to the North Sea. In the case of the Amazon, more than 80% of the drainage basin is low-lying rainforest and accounts for most of the water discharge, but more than 80% of the sediment is derived from the Andes; up to 95% of the sediment is silt and clay-size material carried in suspension.

In these systems, muds are deposited on the shelf faster than they are resuspended even in shallow water stirred by waves. Aggregation and settling of suspended sediment occurs as the rivers reach the sea so suspended sediment concentrations decline seawards. For example, on the Amazon shelf surface suspended sediment concentrations decrease from 400 mg/l near the coast to 10 mg/l within 200 km of the river mouth (Curtis and Legeckis, 1986) though higher concentrations near the seabed extend further offshore. The seabed, made up of very fluid muds, effectively attenuates wave energy and net accumulation occurs as a mud blanket that extends from the shore towards the outer shelf. The muds form large sub-aqueous deltas defined by gently dipping topset beds, more steeply inclined foreset beds and thin bottomset beds (Figures 22.14 and 22.15) which can be identified through high-resolution seismic studies. Sedimentation rates vary from small or negligible on the topset and bottomset beds to a maximum on the foresets (e.g. 10 cm/year and 8 cm/year for the Amazon and Ganges–Brahmaputra systems, respectively). Since most deposition occurs on the foresets, the deltas remain sub-aqueous and prograde seawards over older transgressive sand sheets.

Although dominated by fine sediments, these shelves are not low energy environments. Thus on the Amazon shelf, the best documented of these systems, tides are energetic and current speeds up to 2 m/s occur within 2 m of the seabed near the river mouth (Nittrouer et al., 1986). The large discharge of freshwater produces an estuarine-like circulation on the inner shelf, with offshore flows in surface waters and onshore flows in bottom waters. An intruding oceanic current, the South Equatorial Current, gives rise to the North Brazilian Coastal Current, which sweeps the shelf in a north-westward direction with surface speeds in excess of 0.75 m/s. Combination of this current and tidal currents produces net advection towards the north-west with superimposed onshore–offshore tidal excursions (Nittrouer et al., 1986). A combination of the buoyancy-forced circulation, particle aggregation and settling ensures that most sediment does not escape from the shelf and a high concentration belt is maintained along the

coast. Muds dominate the shelf out to a water depth of 60 m and are then replaced in deeper water by palimpsest sands, relict sands reworked into a transgressive sand sheet as sea level rose (Nittrouer et al., 1983). The delta is defined by topset beds between shore and 40 m isobath, foreset beds between 40 m and 60 m isobaths, and thin bottomset beds in deeper water. The beds are defined acoustically by sandy interbeds which pinch out towards the north-west along the transport path.

Rapid settling of aggregates and frequent resuspension by tidal currents and waves generate a very turbid layer at the seabed above the mud blanket. The bottom turbid layer occurs with increasing thickness shoreward of the 30 m isobath until it fills the entire water column shoreward of the 10 m isobath. The bottom turbid layer grades exponentially into a fluid mud on the seabed. Sediment remains on the seabed due to a combination of drag reduction by high suspended sediment concentrations, yield stress development during periods of slow flow speeds, and shear thickening of the fluid mud (Faas, 1986). Intensely reworked mud forms a surficial layer up to 2 m thick at the 15 m isobath (Kuehl et al., 1986). Limited sands occur seaward of the river mouth where interlaminated mud and sand is present (Kuehl et al., 1982; Nittrouer et al., 1983).

Much of the north-westward advection of fine sediment appears to occur in a belt shoreward of the 10 m isobath (Nittrouer et al., 1986). This gives rise to a 30 km wide and tens of metres thick mud deposit on the coastline and inner shelf of the Guianas. The Guiana shoreline has been prograding seawards at rates up to 250 m/year in association with growth of oblique, shoreface mud banks which are typically 5 m high, 10–20 km wide, and 50–60 km long (Rine and Ginsburg, 1985). Sediment is eroded from the upstream side and deposited on the downstream side of the mudbanks resulting in a migration rate of 1.5 km/year. The downstream muds are exceedingly fluid with bulk densities of 1.03–1.22 g/cm^3 and concentrations exceeding 10^4 mg/litre (Wells and Coleman, 1981). These fluid muds attenuate surface gravity wave energy so that, although some resupension occurs on the shoreface, the shoreline is protected

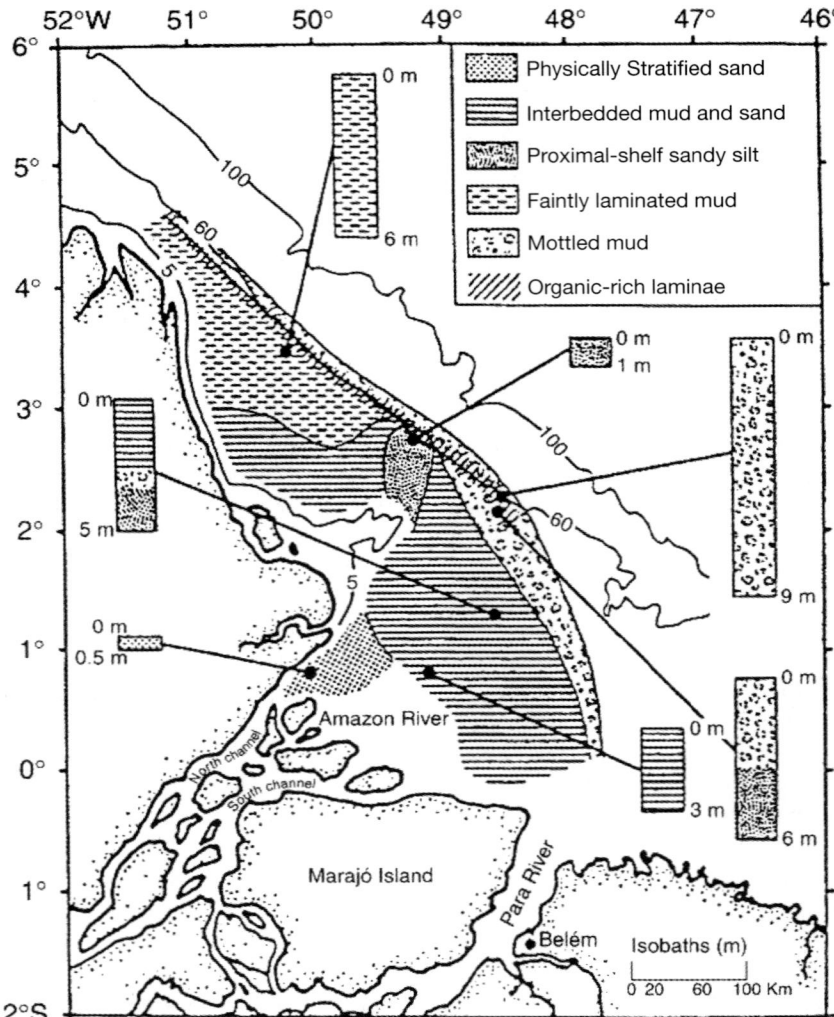

Figure 22.14   Surficial sediment distribution on the sub-aqueous delta of the Amazon shelf (from Kuehl *et al.*, 1986).

and is able to prograde. By contrast, the upstream side of the mudbanks, and inter-mudbank areas, contain overconsolidated muds so that wave energy attenuation by the seabed decreases and erosion of the adjacent shoreline occurs.

While along-shelf advection of fine sediment is important on the Amazon shelf due to the influence of shelf edge processes on this open shelf, on a more enclosed shelf, such as the Yellow Sea, the dispersal system has important shore-normal

as well as shore-parallel components. High suspended sediment concentrations, up to 10 kg/m³ during normal conditions, move alongshore due to tidal currents up to 1 m/s and plunge down the delta front beneath the ambient waters as gravity-driven hyperpycnal underflows (Wright *et al.*, 1990); storm resuspension of delta front sediments can increase the suspended sediment concentration of underflows by an order of magnitude. The paths of the underflows are

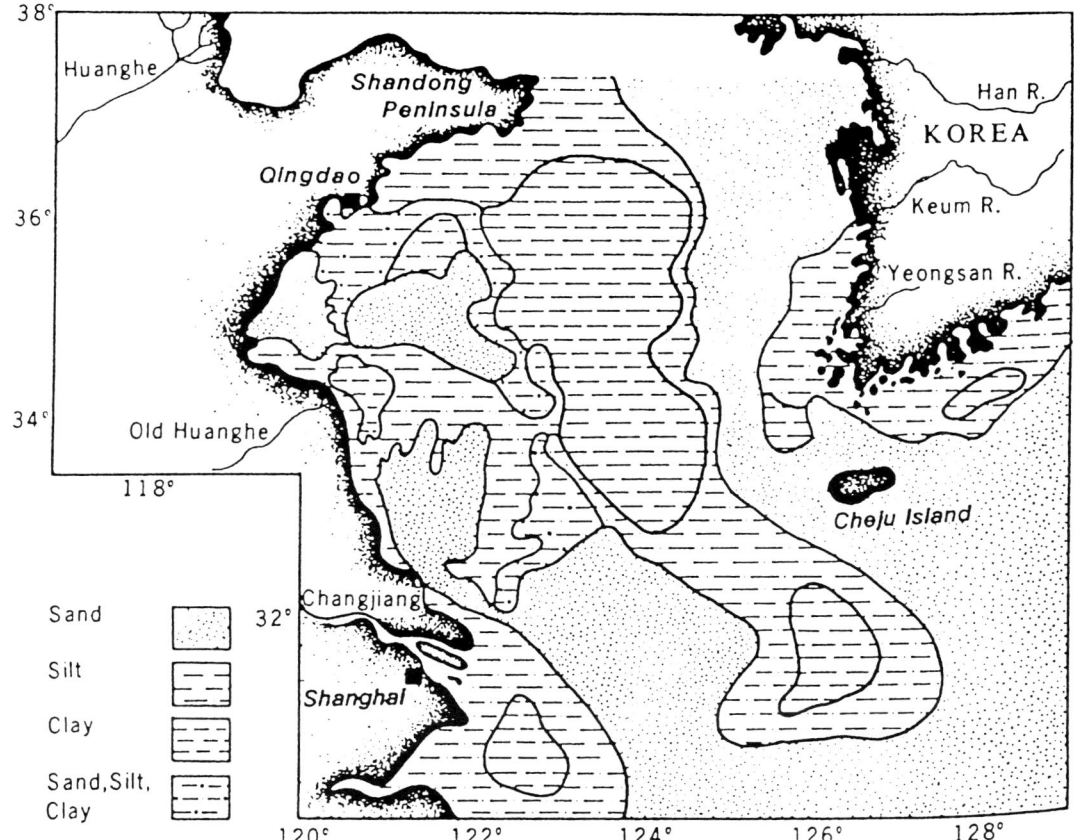

Figure 22.15 Surficial sediment distribution in the Yellow Sea (from Alexander *et al.*, 1991).

curved. They decay rapidly due to entrainment of ambient mass so that rapid deposition occurs on the delta front. The geometries of the deltas are controlled by the nature of the dispersal system. In the pericontinental Amazon and Ganges–Brahmaputra systems, since sediment dispersal follows the dominant shelf circulation along the shelf perpendicular to the progradation direction, the bottomset deposits extend less than 50 km across the shelf. By contrast, in the epicontinental Yellow Sea, the dispersal system is parallel to the progradation direction and both axially and laterally extensive bottomset deposits have developed (Alexander *et al.*, 1991).

Characteristic features of these muddy shelves include rapid sedimentation and frequent resuspension which give rise to turbid waters and a substrate with a high water content. The inner shelf muds of the Amazon shelf contain a sparse benthic community due to the unstable nature of the seabed and poor nutrient supply from the overlying turbid waters (Aller and Aller, 1986). Allied to intense physical reworking, this results in a deposit in which primary sedimentary structures prevail. At water depths greater than 30 m, the influence of bioturbation increases. Along the shelf, in the net sediment transport direction, sands become less common so the deposits change from interbedded sands and muds to laminated muds (Kuehl *et al.*, 1986). The sandier sediments near the river mouth have low water contents, high shear strength and normal to over consolidation as opposed to the high water contents, low shear strength and normal to under consolidation of the muds (Faas, 1986). The foreset regions, with seabed gradients up to 1°, are potential sites for sediment failure although there is no evidence of slumping or mass movement (Adams *et al.*, 1986). Tide-driven, asymmetric megaripples occur on a sandy shoal at the mouth of the Amazon; the outer shelf sands have bedforms but these appear to be generally moribund.

Most allochthonous muddy shelves, although constructional in nature, tend to be smooth and featureless in areas of mud accumulation due to deposition of sediments too fine grained for the formation of large-scale bedforms. Such shelves are in many respects sedimentologically and geomorphologically less complex than autochthonous shelves, but in engineering terms they are likely to be more problematical. This is partly due to the high rates of sedimentation. Allochthonous shelves on tectonically active margins are prone to earthquakes, the combination of rapid sedimentation and seismic activity produces a geotechnically difficult substrate. The liquefaction potential of the sediments further increases on shelves attacked by storm waves and swept by fast tidal currents.

22.10 Tide-dominated sandy shelves

The most intensely studied shelf in its response to modern hydraulic processes is the tide-swept shelf around the British Isles (Stride, 1982). Parts of this shelf support intense activity by the oil and offshore industries, as well as some of the world's busiest shipping lanes. Active sediment transport on much of the shelf produces scour around pipelines, platforms and previously buried wrecks, while deposition on shifting shoals presents a constant hazard to shipping.

This shelf is exposed to frequent storms and surges but much more work is done on the sea floor by semidiurnal tidal currents which regularly exceed 0.5 m/s over vast areas and locally 2.0 m/s (surface currents). In areas where such

tidal currents coincide with large storm waves, the resulting sediment transport is intense. With such a high rate of energy expenditure, morphological and textural patterns inherited from the retreating nearshore zone have been effectively erased.

Although tidal currents are reversing, a sediment transport asymmetry develops from several causes: flood and ebb currents that are unequal in speed and duration; flood and ebb currents that follow mutually evasive transport paths; lag effects resulting from rotary currents; the enhancement of the flood or the ebb currents by other currents. Well-defined transport paths are the result, with sand streams which diverge from 'bed load partings' and flow down a hydraulic gradient until either the shelf break or a 'bed load convergence' and sand accumulations are reached (Figure 22.16).

Each stream is defined by a characteristic progression of bottom morphologies and sediment textures (Figure 22.17). This begins in high velocity zones (maximum surface currents exceeding 1.5 m/s) where the sea floor is scoured down to bedrock locally and thinly veneered with lags of shells and gravel. Further down the transport path, at maximum surface velocities of 1.3–1.5 m/s, sand ribbons become the dominant bedform. These are longitudinal bedforms, up to 15 km long and 200 m wide, though usually less than 1 m high. They consist of sands in transit over a pavement of shell and gravel and they can be identified using side-scan sonar by their textural contrast with the underlying gravel. They originate in secondary flows generated by variable turbulence over a substrate of variable roughness (McLean, 1981). Sand ribbon morphology varies somewhat depending on the current velocity and the sand availability. Beyond the sand ribbons are sand waves where maximum surface velocities are 0.5–1.0 m/s. Sand waves are transverse bed forms with straight crests and well-defined avalanche faces, wavelength 30–500 m, and height < 1–20 m. There is disagreement about whether all sand waves, regardless of their size, are formed in the same way and some authorities classify megaripples (which have wavelengths up to 2 m) as separate bedforms

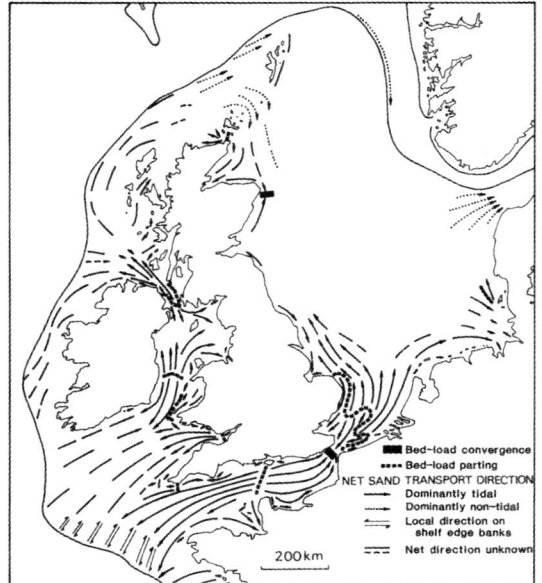

Figure 22.16 Sand transport paths on the UK continental shelf. From Stride (1982).

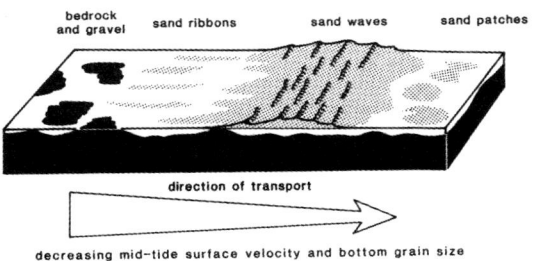

Figure 22.17 Schematic model of sand transport path down a hydraulic gradient which is matched by the textural gradient. In Swift (1976b).

(Belderson *et al.*, 1982) while others view them as essentially the same as the larger sand waves (Allen, 1980). Although sand waves are usually considered as actively migrating features responding to the present-day hydraulic regime, data on their migration rates are lacking. Next down the velocity gradient are sand patches; these are sheets of fine sand and muddy fine sand lacking bedforms other than ripples and supporting a varied infauna. They seem to be essentially the product of suspensive transport of sediment that

has outrun the bedload stream. The sands may be as thick as 10 m, but where they do not continue into mud, they break up into irregular patches less than 2 m thick on a gravel substrate. Many transport paths are terminated by mud zones formed by deposition of the suspensive load in areas where suspended sediment concentrations are high and bottom stirring by waves infrequent. The most extensive mud areas are at water depths greater than 30 m. Wave effectiveness, rather than tidal current intensity, is a critical control of mud deposition (McCave, 1971). Mud is absent from large areas of the northern North Sea because of increased wave activity.

The transport paths are predominantly parallel to the tidal streams and to the coastline, up to 400 km long though usually incomplete. Areas of bed-load convergence are sites of deposition marked by fields of sand waves. Areas of bed-load parting are usually erosive, although some deposition may occur where there is a local influx of sand from the shoreline. The transport paths imply that the shelf is moving toward a state of equilibrium with its tidal regime.

Bedforms larger than sand ribbons and sand waves occur in the Southern Bight of the North Sea (e.g. the Norfolk Banks and Flemish Banks, Figure 22.18) and in the Celtic Sea. These are sand ridges or sand banks. In the North Sea they are up to 40 m high, 6 km wide and 60 km long, with their crests spaced at 4–12 km. Water depth between the ridges is 30–50 m but over their crests only 3–13 m; they clearly present a navigational hazard. They have their long axes orientated almost parallel to the direction of the strongest tidal currents, which suggests that they are at least in part related to present-day hydraulic conditions. It is not clear however whether they were initiated by modern processes or whether they are modified relict features responding to modern processes.

The discovery of a slight obliquity (of usually 7°–15°) of sandbank axes to peak tidal current direction led to a model for sand-ridge genesis by a rectilinear tidal flow (Huthnance, 1982). This gives a residual flow due to the effects of the earth's rotation and bottom friction. Given an initial bed waviness of small amplitude with contours making a small angle with the peak current direction, a series of bed forms should grow in amplitude. Once formed, helical circulations may develop in conformity with the ridge axes, and these may assist in building up the banks further (Figure 22.19). The establishment of mutually evasive flood and ebb channel systems could then aid the growth and proliferation of sand ridges (Caston, 1972; Figure 22.20).

It was at one time thought that the lateral sequences of bedforms seen in the North Sea were diagnostic of a tide-driven dispersal system. But Flemming (1980) identified very similar sequences on the shelf of South East Africa

where an intruding oceanic current, the Agulhas Current, has sufficient energy to generate large sand waves and sand ribbons in palimpsest carbonate sands moving over a gravel pavement. Field *et al.* (1981) identified large sand waves formed by a steady northerly coastal current on the northern Bering Sea shelf.

22.11 Wave and storm-dominated sandy shelves

Wave dominance *per se* is generally confined to the shoreface during fair-weather conditions but

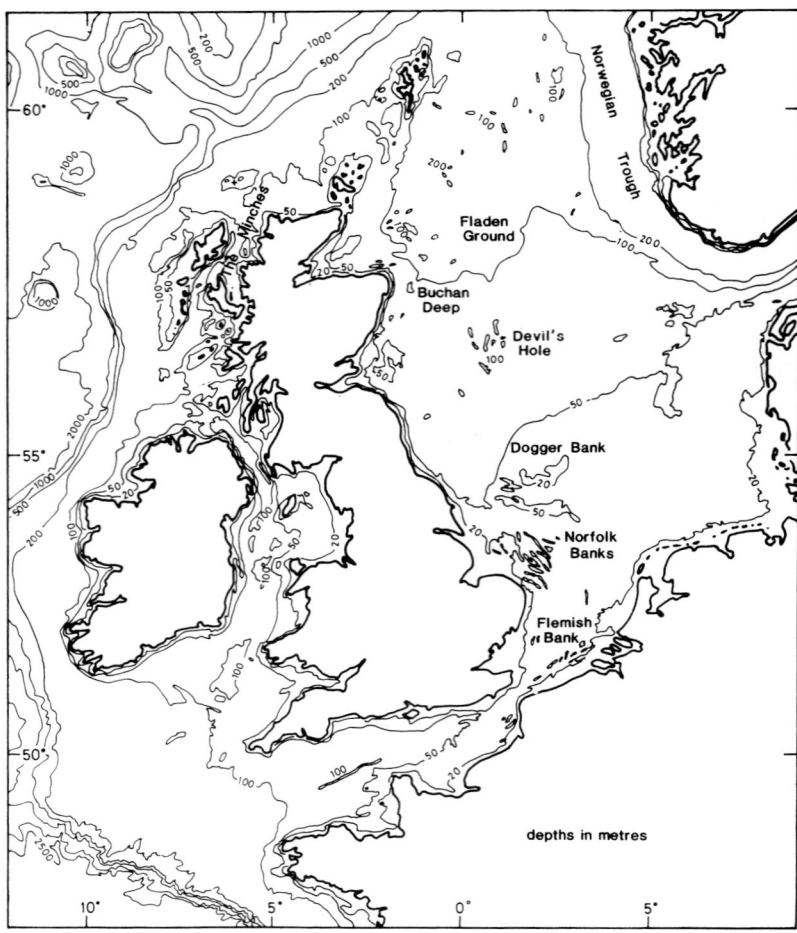

Figure 22.18   Bathymetry, erosional features and constructional features of the north-west European shelf.

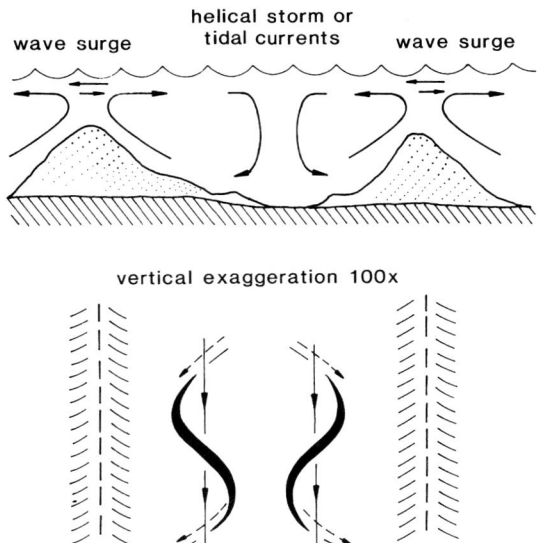

Figure 22.19   Schematic representation of helical flow structure believed to be associated with sand ridges.

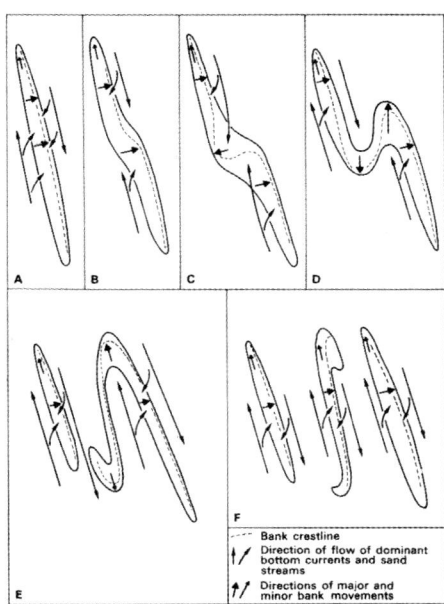

Figure 22.20   Model for growth and development of linear tidal sand ridges. (*A*) Development of mutually evasive ebb and flood channels. (*B, C*) Inequality of secondary cross-shoal currents leads to destruction of straight crest line. (*D*) Double curve develops into incipient ebb and flood channels. (*E*) Lengthening of channels so that centre ridge becomes parallel to adjacent ridges. (*F*) Continuation of cycle. From Caston (1972).

can affect the whole shelf during storms. Wave action on the sandy inner shelf of the Golfe du Lion in the western Mediterranean generates a strongly defined seaward-fining textural gradient between the beach and the start of the shelf mud blanket (Jago and Barusseau, 1981). Such a textural gradient also occurs on the sandy, open shelf of the Bering Sea (Figure 22.10) but here it extends into deeper water due to the more intense wave action; note the absence of muddy sediments on this high latitude shelf (Sharma *et al.*, 1972). On sandy shorefaces, waves generate a shore-normal sequence of bedforms which reflect the decreasing bed shear stresses seawards of the breaker zone. On high energy shorefaces of open shelves, the bedforms include asymmetric lunate dunes and upper flow regime plane beds (Clifton *et al.*, 1971). On low energy shorefaces, the bedforms are small asymmetric ripples. Modified shoreface bedforms are exposed on intertidal foreshores in macrotidal regions (Jago and Hardisty, 1984).

The best-documented sandy shelf dominated by storms is the Middle Atlantic Bight of the North American Atlantic shelf (summarised by Swift, 1976a, b) but comparable conditions occur elsewhere — for example the Gulf of Mexico (Hayes, 1967b) and the Bering Sea (Nelson, 1982). Under fair-weather conditions, the prevailing movement of shoreface sand is probably onshore, but during storms a downwelling jet can move large quantities of sand offshore. Storm wind stress leads to storm set-up on the coast which in turn gives rise to a compensatory return flow at the seabed. This geostrophic flow has a large shore-parallel component as well as a shore-normal component. The shoreface is characterised by sand ridges which are oblique to the shoreline (Duane *et al.*, 1972). These form under the opposing forces of the wind-driven coastal current (moving sand seawards) and wave drift currents (moving sand shorewards). The process of sediment convergence is analogous to the process which forms sand ridges in tidal seas. When the storm set-up abates, the offshore flow is augmented by a further pulse as the piled-up water relaxes. Under these conditions, fast nearshore currents in excess of 1 m/s are generated (Morton, 1981) which can rapidly move large quantities of sand (up to 10^3 tonnes/m per day) over distances of several kilometres. The sand is deposited as a sand sheet across the shelf. This mechanism is an extreme form of shoreface bypassing which allows sand to escape the nearshore wave drift. The sand sheet is characterised by small dunes (< 0.3 m high) covered by small ripples (Swift et al., 1979). Formed during winter storms, these tend to be flattened during summer.

22.12 Preservation of relict features

The preservation of relict features is most likely where the influx of detrital sediment is small and the speed of the transgression is rapid. In such cases, coastal sediments enter the shelf system by shoreface bypassing; since these are of local origin, the sedimentary regime is described as autochthonous. The preservation potential of relict features is low if the detrital sediment influx from rivers is high (so that river-mouth bypassing is enhanced) and the speed of transgression is slow; this is an allochthonous regime. The proportion of relict sediment decreases, and of palimpsest sediment increases, if the shelf is transgressed slowly or exposed to a high-energy hydraulic regime.

Sub-aerial morphologies have survived on autochthonous shelves. On high-latitude shelves, relief may exceed 200 m, a consequence of fluvial and glacial erosion of bedrock. A good example is the North American Atlantic shelf where the Fall Line (the demarcation between piedmont and coastal plain provinces) intersects the shoreline at New Jersey and runs across the shelf. Sub-aerial cuesta have been either cut off (e.g. Long Island) or drowned (e.g. Georges Banks) by the transgression. The ridges are separated by drowned

basins (e.g. Long Island Sound, the Gulf of Maine).

Shelves of lower relief consist of flat plateau divided by shelf valleys cut during low stands of the sea. Thus glacier tongues excavated former river valleys to produce the lochs and fjords of Scotland and Norway, and their seaward extensions in the Minches, the Sea of Hebrides and the Norwegian Trough (Figure 22.18). Glacial meltwater, flowing into an island lake located over the Fladen Ground of the North Sea, cut long deep channels known as tunnel valleys which remain as local basins (e.g. the Buchan Deep and Devil's Hole: Figure 22.18).

In addition, there are many examples of drowned river valleys, surface and subsurface channels, such as in the English Channel and the Middle Atlantic Bight of North America, where the valleys terminate in low-stand deltas whose fronts are seaward-bulging shelf-edge scarps (Figure 22.21). Buried river channels that were formed and filled during the last low stand of sea level occur below the Holocene sediment sequence on the shelf of the East China Sea (Butenko et al., 1985). These constitute a major geological hazard for seafloor engineering such as emplacement of drilling platforms.

Depositional features are also preserved on glaciated shelves; in the northern North Sea glacial moraines were deposited by successive ice sheets and these survive as prominent banks (e.g. the Dogger Bank and the Jutland Bank). Smaller-scale features seem in general to have been destroyed by erosive retreat of the shore face, so that most shelves are somewhat featureless when compared to adjacent coastal plains, unless subsequently covered by complex sand ridges that are largely the result of marine processes initiated after the transgression.

Shelves covered with biogenic carbonates are exceptional in that relict features are more readily preserved owing to the early lithification of carbonate sediments. Thus, on the Yucatan Shelf of the Gulf of Mexico, reefs have developed on the cemented remnants of low-stand coastal ridges or dunes.

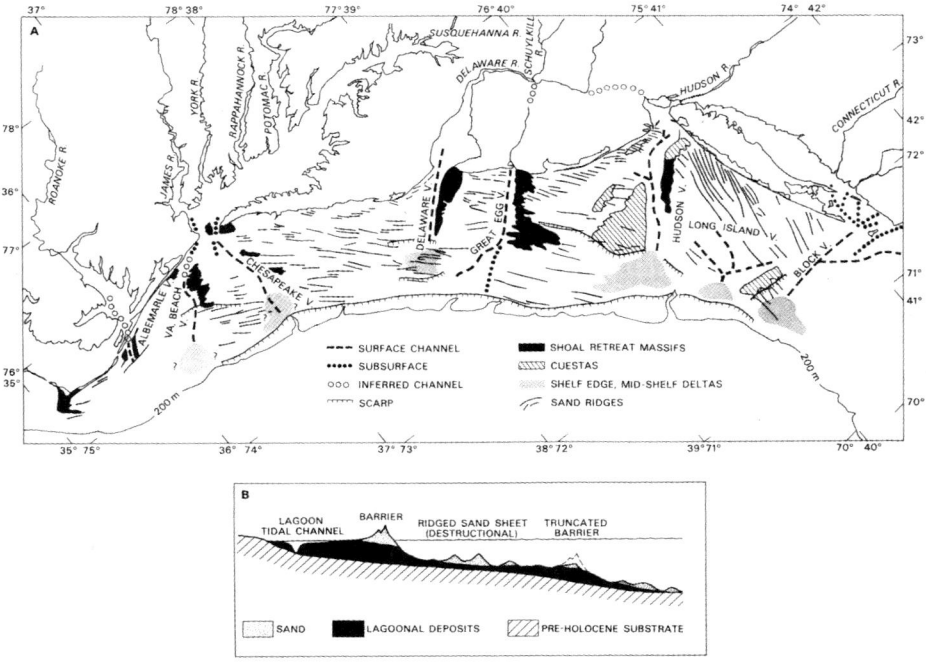

Figure 22.21 (*A*) Main morphological features of the Middle Atlantic Bight, USA. (*B*) Schematic profile across the shelf showing transgressive sheet sand overlying lagooned and pre-Holocene deposits. After Swift (1976b).

22.13 Preservation of nearshore marine patterns

As the Flandrian coasts transgressed the shelves, relict morphologies were modified and coastal sand bodies deposited. The most characteristic feature of shelves in an autochthonous regime is a discontinuous sand sheet up to 10 m thick that was laid down during the erosive retreat of the shoreface by a seaward transfer of sand during storms. On shelves bordering low, constructional coasts (such as the USA Atlantic shelf), this sand sheet overlies back barrier sands and lagoonal muds. Pauses in the general rise of sea level enabled shorelines to temporarily stabilise, so that shoreface profiles built upwards by means of upper shoreface and barrier aggradation, rather than retreating landwards. With the resumption of the transgression, the coasts resumed their migration through shoreface erosion and storm washover; the short hiatuses were

commemorated by scarps 10 m or more in height (Figure 22.21).

Superimposed on the sand sheet of the USA Atlantic shelf are various shoals formed in zones of nearshore sand storage associated with river mouths and headlands. These form shoal-retreat massifs (Figure 22.21). The first type, called inlet-associated bars, consist of seaward convex bars at the mouths of major estuaries and tidal inlets. They extend seawards for several kilometres. Formed as ebb tide deltas, they were dissected by mutually evasive flood and ebb channels flaring outwards from the estuary mouths, passing seawards into storm-dominated sand ridges. Coastal retreat during transgression lessened tidal dominance and increased reworking by storms, so that shoal-retreat massifs remain reworked with superimposed linear sand ridges moulded by unidirectional storm-generated currents (Figure 22.22). The second type are cape-associated bars: these are adjacent to prominent

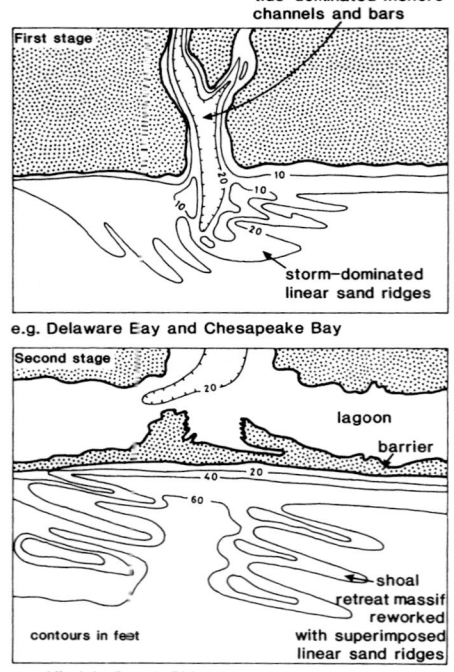

Figure 22.22 Schematic model for evolution of inlet-associated bars into offshore linear sand ridges during a marine transgression From Swift *et al.* (1972).

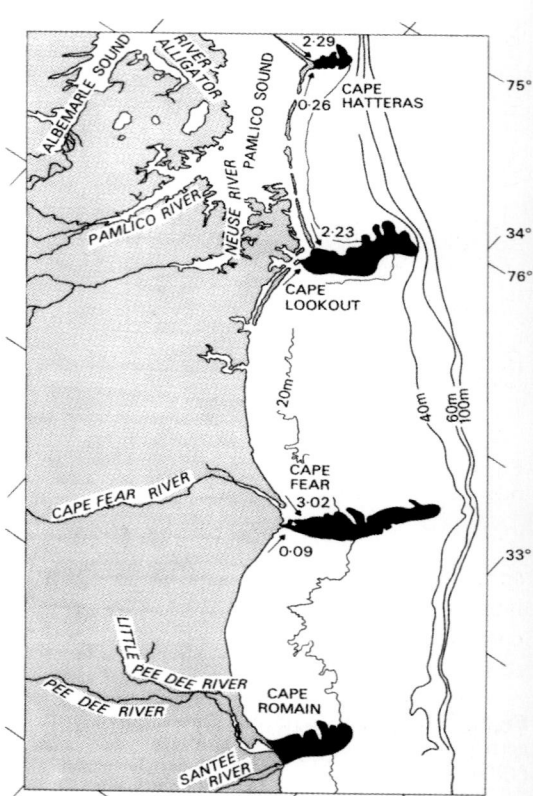

Figure 22.23 Cape-associated sand bars forming at sites of littoral drift convergence (sand transport rates in yd/y $\times$ 10^{-3}). From Swift (1976b).

headlands and possess a distinctive hammer-head shape (Figure 22.23). Common on the shelf south of the Middle Atlantic Bight of North America, they are the response of the shoreface regime to a wave climate that includes a high directional variance. Waves approach from both the north-east and south-east and cuspate forelands arise by littoral drift convergence; this creates shoals which then maintain patterns of wave refraction that further drive littoral drift convergence. Probably initiated at low-stand deltas, tidal inlets or estuary mouths, the cape-associated bars were self-maintaining features throughout the transgression, with subsequent modification by storm-generated currents which developed superimposed sand ridges and troughs. The shoal-retreat massif therefore marks the retreat path of the drift convergence.

In the Middle Atlantic Bight, shoreface sand ridges were detached from the coast by sea-level rise and then maintained by helical currents (Figure 22.19). It is unlikely that the abundant mid-shelf ridges are simply inherited shoreface features. Once formed, the ridges are maintained by the shelf hydrodynamic regime which may even create secondary ridges (Figure 22.24) during intense winter storms. Evidence of sand transport on the ridges is provided by current ripples, sand ribbons and occasionally sand waves.

22.14 Summary

The pattern of shelf sedimentation and the properties of the shelf substrate vary in time and

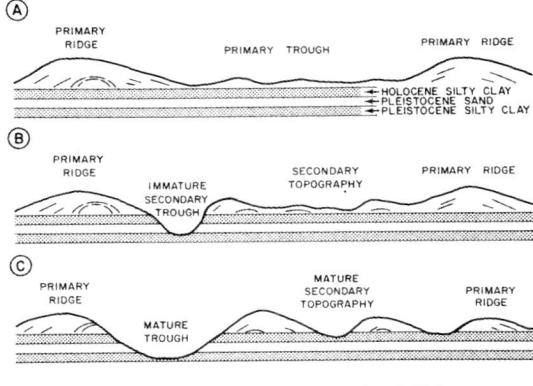

DEVELOPMENT OF RIDGE TOPOGRAPHY

Figure 22.24 Evolution of storm-generated sand ridges. From Stubblefield and Swift (1976).

space. Young mountains are in places providing large quantities of terrigenous sediments to the coast and shelf via rivers carrying large quantities of suspended fine sediment. However, as a result of sea-level changes, many river mouths are now estuarine, trapping sediments from their rivers and poaching sediments from the shelves. Many shelves are therefore sediment starved. Most shelves have inherited relict morphologies and a mosaic of sediments deposited in a range of coastal environments. Allochthonous shelves, dominated by fine cohesive sediments supplied by rivers, exhibit very different properties and dynamics compared to autochthonous shelves, characterised by relict and palimpsest non-cohesive sands. Muddy shelves, on which sedimentation rates are high, develop sub-aqueous deltas which are prograding across the shelf; the sediments have high water content and low shear strength. Rapid sedimentation on shelves on seismically active margins gives rise to a substrate with a high liquefaction potential. Reasonably accurate estimates of the transport of sands can be made but it is not yet possible to quantify the dynamics of cohesive sediment transport. The surface veneer of sediments on some shelves may be atypical of the subsurface sediments and it is therefore vital that the three-dimensional geometry of the shelf deposits be thoroughly investigated prior to any engineering activity.

References

Adams, C. E., Wells, J. T. and Coleman J. M. (1986) Transverse bedforms on the Amazon shelf. *Continental Shelf Res.* **6**, 175–187.

Alexander, C. R., DeMaster, D. J. and Nittrouer, C. A. (1991) Sediment accumulation in a modern epicontinental-shelf setting: the Yellow Sea. *Mar. Geol.* **98**, 51–72.

Allen, J. R. L. (1980) Sand waves: a model of origin and internal structure. *Sediment. Geol.* **26**, 281–328.

Aller, J. Y. and Aller, R. C. (1986) General characteristics of benthic faunas on the Amazon inner continental shelf with comparison to the shelf of the Changjiang River, East China Sea. *Continental Shelf Res.* **6**, 291–310.

Al-Shafei (ed.) (1999) *Engineering for Calcareous Sediments*. Balkema, Rotterdam.

Andel, T. H. van (1960) Sources and dispersion of Holocene sediments, northern Gulf of Mexico. In Shepard, F. P., Phleger, F. B. and van Andel, T. H. (eds) *Recent Sediments, Northwest Gulf of Meixco.* Amer. Ass.Petroleum Geol., Tulsa, 44–55.

Bagnold, R. A. (1963) Mechanics of marine sedimentation. In Hill, M. N. (ed.) *The Sea,* 3. Wiley Interscience, New York, pp. 507–582.

Belderson, R. H., Johnson, M. A. and Kenyon, N. H. (1982) Bedforms. In Stride, A. H. (ed.) *Offshore Tidal Sands.* Chapman and Hall, London, 27–57.

Butenko, J., Milliman, J. D. and Yincan, Ye. (1985) Geomorphology, shallow structure, and geological hazards in the East China Sea. *Continental Shelf Res.* **4**, 121–141.

Caston, V. N. D. (1972) Linear sand banks in the southern North Sea. *Sedimentology* **18**, 63–78.

Clifton, H. E., Hunter, R. E. and Phillips, R. L. (1971) Depositional structures and processes in the non-barred high energy nearshore. *J. Sedim. Petrol.* **41**, 651–670.

Coop, M. R. (1990) The mechanics of uncemented carbonate sands. *Geotechnique* **40**, No. 4, 607–626.

Coop, M. R. and Atkinson, J. H. (1993) The mechanics of cemented carbonate sands. *Geotechnique* **43**, No. 1, 53–67.

Cuccovilli, T. and Coop, M. (1997) Yielding and pre-failure deformation of structured sands. *Geotechnique* **47**, No. 3, 491–508.

Curray, J. R. (1960) Sediments and history of Holocene transgression, continental shelf, northwest Gulf of Mexico. In Shepard, F. P., Phleger, F. B. and can Andel, T. H. (eds) *Recent Sediments, Northwest Gulf of Meixco.* Amer. Ass. Petroleum Geol., Tulsa, 221–266.

Curtis, T. B. and Legeckis, R. V. (1986) Physical observations in the plume region of the Amazon river during peak discharge — 1. Surface variability. *Continental Shelf Res.* **6**, 31–51.

Davies, J. L. (1964) A morphogenetic approach to world shorelines. *Z. Geomorph.* **8**, (Sp. No.) 127–142.

Delo, A. (1988) Estuarine muds manual. *HR Wallingford, England, Report No.SR164.*

DeMaster, D. J., McKee, B. A., Nittrouer, C. A., Jiangchu, Q. and Guodong, C. (1985) Rates of sediment accumulation and particle reworking based on radiochemical measurements from continental shelf deposits in the East China Sea. *Continental Shelf Res.* **4**, 143–158.

Dorjes, J., Gladow, S., Reineck, H.-E. and Singh, I. B. (1970) Sedimentologie und Makrobenthos der Nordergrunde und der Aussenjade (Nordsee). *Senckenberg. Mar.* **2**, 31–59.

Duane, D. B., Field, M. E., Meisburger, E. P., Swift, D. J. P. and Williams, S. J. (1972) Linear shoals on the Atlantic inner continental shelf, Florida to Long Island. In Swift, D. J. P., Duane, D. B. and Pilkey, O. H. (eds) *Shelf Sediment Transport: Process and Pattern.* Dowden, Hutchinson and Ross, Stroudsburg, 447–499.

Faas, R. W. (1986) Mass physical and geotechnical properties of surficial sediments and dense nearbed sediment suspensions on the Amazon continental shelf. *Continental Shelf Res.* **6**, 189–208.

Field, M. E., Nelson, C. H., Caccione, D. A. and Drake, D. E. (1981) Sand waves on an epicontinental shelf: northern Bering Sea. *Mar. Geol.* **42**, 233–258.

Flemming, B. W. (1980) Sand transport and bedform patterns on the continental shelf between Durban and Port Elizabeth (southeastern African continental margin). *Sediment. Geol.* **26**, 179–205.

Fookes, P. G. (1988) The geology of carbonate soils and rocks and their engineering characteristics and description. *Proc. Int. Conf. on Calcareous Sediments* Perth, Australia, March 1988.

Giresse, P. and Kouyoumontzakis, G. (1973) Cartographie sédimentologique des plateaux continentals sud du Gabon, du Congo, du Cabinda et du Zaire. *Cah. ORSTOM, Ser. Geol.* **5**, 235–257.

Hardisty, J. (1983) An assessment and calibration of formulations for Bagnold's bedload equation. *J. Sediment.Petrol.* **53**, 1007–1010.

Hayes, M. O. (1967a) Relationship between coastal climate and bottom sediment type on the inner continental shelf. *Mar. Geol.* **5**, 111–132.

Hayes, M. O. (1967b) Hurricanes as geological agents, south Texas coast. *Bull. Am. Ass. Petrol. Geol.* **51**, 937–942.

Heathershaw, A. D. (1981) Comparisons of measured and predicted sediment transport rates in tidal currents. *Mar. Geol.* **42**, 75–104.

Henkel, D. J. (1970) The role of waves in causing submarine landslides. *Geotechnique* **20**, No. 1, 75–80.

Huthnance, J. M. (1981) Waves and currents near the continental shelf edge. *Prog. Oceanog.* **10**, 193–226.

Huthnance, J. M. (1982) On one mechanism forming linear sand banks. *Estuarine Coast. Mar. Sci.* **14**, 79–99.

Inman, D. L. and Nordstrom, C. E. (1971) On the tectonic and morphologic classification of coasts. *J. Geol.* **79**, 1–21.

Jago, C. F. (1981) Sediment response to waves and currents, North Yorkshire Shelf, North Sea. In Nio, S. D., Schuttenhelm R.T.E. and van Weering, T.C.E. (eds) *Holocene Marine Sedimentation in the North Sea Basin.* I.A.S. Spec. Publ. **5**, 283–301.

Jago, C. F., Bale, A. J., Green, M. O., Howarth, M. J., Jones, S. E., McCave, I. N., Millward, G. E., Morris, A. W., Rowden, A. A. and Williams, J. J. (1993) Resuspension processes and seston dynamics. *Phil. Trans. Royal Soc.* **343** (1669), 475–491.

Jago, C. F. and Barusseau, J.-P. (1981) Sediment entrainment on a wave graded shelf, Roussillon, France. *Mar. Geol.* **42**, 279–299.

Jago, C. F. and Hardisty, J. (1984) Sedimentology and morphodynamics of a macrotidal beach, Pendine Sands, SW Wales. *Mar. Geol.* **60**, 123–154.

Jago, C. F. and Mahamod, T. (1999) A total load algorithm for sand transport by fast steady currents. *Estuar. Coast. Shelf Sci.* **48**, 93–99.

Jewell, R. J. and Andrews, D. C. (eds) (1998) *Proc. Int. Conf. on Calcareous Sediments* Perth, Australia, March 1988.

Keller, G. H. and Yincan, Y. (1985) Geotechnical properties of surface and near-surface deposits in the East China Sea. *Continental Shelf Res.* **4**, 159–174.

Komar, P. D. and Miller, M. C. (1975) On the comparison of the threshold of sediment motion under waves and unidirectional currents. *J. Sediment. Petrol.* **45**, 362–367.

Krone, R. B. (1963) A study of rheologic properties of estuarial sediment. *Sanit. engrg. Res. Lab. Rept.* University of California, Berkeley, 63–8, 91 pp.

Kuehl, S. A., Hariu, T. M. and Moore, W. S. (1990) Shelf sedimentation off the Ganges–Brahmaputra river system: evidence for bypassing to the Bengal Fan. *Geology* **17**, 1132–1135.

Kuehl, S. A., Nittrouer, C. A. and DeMaster, D. J. (1982) Modern sediment accumulation and strata formation on the Amazon continental shelf. *Mar. Geol.* **49**, 279–300.

Kuehl, S. A., Nittrouer, C. A. and DeMaster, D. J. (1986) Distribution of sedimentary structures in the Amazon subaqeous delta. *Continental Shelf Res.* **6**, 311–336.

Kulm, L. D., Roush, R. C., Harlett, J. C., Neudeck, R. H., Chambers, D. M. and Runge, E. (1975) Oregon continental shelf sedimentation: inter-relationships of facies distribution and sedimentary processes. *J. Geol.* **83**, 145–175.

Lagioia, R. and Nova, R. (1995) An experimental and theoretical study of the behaviour of a calcarenite in triaxial compression. *Geotechnique* **45**, No. 4, 633–648.

McCave, I. N. (1971) Wave-effectiveness at the sea-bed and its relationship to bed-forms and deposition of mud. *J. Sedim. Petrol.* **41**, 89–96.

McCave, I. N. (1972) Transport and escape of fine grained sediment from shelf areas. In Swift, D. J. P., Duane, D. B. and Pilkey, O. H. (eds), *Shelf Sediment Transport: Process and Pattern.* Dowden, Hutchinson and Ross, Stroudsburg, 225–248.

McLean, S. R. (1981) The role of non-uniform roughness in the formation of sand ribbons. *Mar. Geol.* **42**, 49–74.

Meade, R. H., Dunne, T., Richey, J. E., de M.Santos, U. and Salati, E. (1985) Storage and remobilisation of suspended sediment in the lower Amazon river of Brazil. *Science* **228**, 488–490.

Mehta, A. J. (1988) Laboratory studies on cohesive sediment deposition and erosion. In Dronkers, J. and van Leussen, W. (eds) *Physical Processes in Estuaries.* Springer, Berlin, Heidelberg, New York, pp. 427–445.

Mehta, A. J., Parchure, T. M., Dixit, J. G. and Ariathurai, R. (1982) Resuspension potential of deposited cohesive sediment beds. In Kennedy, V. (ed.) *Estuarine Comparisons.* Academic Press, New York, pp. 591–609.

Migniot, C. (1968) Étude des propriétés physiques de différents sediments très fins de leur comportement sous des actions hydrodynamiques. *Houille blanche* **7**, 591–620.

Miller, M. C., McCave, I. N. and Komar, P. D. (1977) Threshold of sediment motion under unidirectional currents. *Sedimentology* **24**, 507–527.

Morton, R. A. (1981) Formation of storm deposits by wind-forced currents in the Gulf of Mexico and the North Sea. In Nio, S. D., Schüttenhelm, R. T. E., and van Weering, T. C. E. (eds) *Holocene Marine Sedimentation in the North Sea Basin.* I.A.S. Spec. Publ. **5**, 385–396.

Nelson, C. H. (1982) Modern shallow water graded sand layers from storm surges, Bering shelf: a mimic of Bouma sequences and turbidite systems. *J. Sedim. Petrol.* **52**, 537–545.

Nittrouer, C. A. and DeMaster, D. J. (1986). Sedimentary processes on the Amazon continental shelf: past, present and future research. *Continental Shelf Res.* **6**, 5–30.

Nittrouer, C. A. and Sternberg, R. W. (1981) The formation of sedimentary strata in an allochthonous shelf environment: the Washington continental shelf. *Mar. Geol.* **42**, 210–232.

Nittrouer, C. A., Sharara, M. T. and DeMaster, D. J. (1983) Variations of sediment texture on the Amazon continental shelf. *J. Sediment. Petrol.* **53**, 179–191.

Nittrouer, C. A., DeMaster, D. J., McKee, B. A., Guodong, C. and Jiangchu, Q. (1984) Formation of sedimentary strata in the East China Sea. In Luo Yuru (ed.) *Proc. Int. Symp. on Sedimentation on the Continental Shelf with Special Reference to the East China Sea.* Beijing, pp. 696–704.

Nittrouer, C. A., Sharara, M. T. and DeMaster, D. J. (1986) Concentration and flux of suspended sediment on the Amazon continental shelf. *Continental Shelf Res.* **6**, 151–174.

Reineck, H. E. and Singh, I. B. (1975) *Depositional Sedimentary Environments.* Springer-Verlag, Berlin.

Rhoads, D. C. and Boyer, L. F. (1982) The effects of marine benthos on physical properties of sediments: A successional perspective. In McCall, P. L. and Tevez, M. J. S (eds) *Animal–Sediment Relationships.* Plenum, New York, 3–52.

Rine, J. M. and Ginsburg, R. N. (1985) Depositional facies of a mud shoreface in Suriname, South America — a mud analogue to sandy shallow marine deposits. *J. Sediment. Petrol.* **55**, 633–652.

Semple, R. (1988) State of the art report on engineering properties of carbonate soils. *Proc. Int. Conf. on Calcareous Sediments* Perth, Australia, March 1988.

Sharma, G. D., Naidu, A. S. and Hood, D. W. (1972) Bristol Bay: a model contemporary graded shelf. *Bull. Am. Ass. Petrol. Geol.* **56**, 2000–2012.

Soulsby, R. L. (1997) Dynamics of marine sands: A manual for practical applications. *HR Wallingford Report No. 14*, 85 pp.

Sterling, G. H. and Strohbeck, E. E. (1973) The failure of the South Pass 'B' Platform in Hurricane Camille. *Proc. 5th Offshore Technology Conf. Houston* **2**, 719–730.

Stride, A. H. (1982) *Offshore Tidal Sands.* Chapman and Hall, London.

Stubblefield, W. L. and Swift, D. J. P. (1976) Ridge development as revealed by sub-bottom profiles on the central New Jersey shelf. *Mar. Geol.* **20**, 315–334.

Swift, D. J. P. (1976a) Coastal sedimentation. In Stanley, D. J. and Swift, D.J.P. (eds), *Marine Sediment Transport and Environmental Management.* Wiley, New York, 255–310.

Swift, D. J. P. (1976b) Continental shelf sedimentation. In Stanley, D. J. and Swift, D. J. P. (eds), *Marine Sediment Transport and Environmental Management.* Wiley, New York, 311–352.

Swift, D. J. P., Stanley, D. J. and Curray, J. R. (1971) Relict sediments on continental shelves: a reconsideration. *J. Geol.* **79**, 322–346.

Swift, D. J. P., Freeland, G. L. and Young, R. A. (1979) Time and space distribution of megaripples and associated bedforms, Middle Atlantic Bight, North American Atlantic shelf. *Sedimentology* **26**, 389–406.

Swift, D. J. P., Kofoed, J. W., Saulsbury, F. P. and Sears, P. (1972) Holocene evolution of the shelf surface, central and southern Atlantic shelf of North America. In Swift, D. J. P., Duane, D. B. and Pilkey O. H. (eds) *Shelf Sediment Transport: Process and Pattern.* Dowden, Hutchinson and Ross, Stroudsburg, 499–574.

Thorn, M. F. C. and Parsons, J. G. (1980) Erosion and cohesive sediments in estuaries: an engineering guide. *Proc. 3rd Int. Symp. Dredging Technol.* Bordeaux, 349–358.

Vanoni, V. A. (1975) Sedimentation engineering. *ASCE Task Comm.* New York, 745 pp.

Villaret, C. and Paulic, M. (1986) Experiments on the erosion of deposited and placed cohesive sediments in an annular flume and a rocking flume. *Rep. UFL/COEL–86/007 Coast. Oceanogr. Eng. Dep.* University of Florida, Gainesville.

Webb, J. E., Dorjes, D. J., Gray, J. S., Hessler, R. R., van Andel, T. H., Rhoads, D. C., Werner, F., Wolff, T. and Zijlstra, J. J. (1976) Organism sediment relationships. In McCave, I. N. (ed.) *The Benthic Boundary Layer.* Plenum, New York, 273–295.

Wells, J. T. and Coleman, J. M. (1981) Physical processes and fine grained sediment dynamics, coast of Surinam, South America. *J. Sediment. Petrol.* **51**, 1053–1068.

Wright, L. D., Wiseman, W. J., Yang, Z.-S., Bornhold, B. D., Keller, G. H., Prior, D. B. and Suhayda, J. N. (1990) Processes of marine dispersal and deposition of suspended silts off the modern mouth of the Huanghe (Yellow River). *Continental Shelf Res.* **10**, 1–40.

Further reading

Dalrymple, R. W. (1992) Tidal depositional systems. In (eds) Walker, R. G. and James, N. P. *Facies Models: Response to Sea Level Change.* Geological Association of Canada, Waterloo, Ontario, 195–218. New information on sediments in tidal environments.

Flemming, B. W. and Bartholoma, A. (1995) *Tidal Signatures in Modern and Ancient Sediments.* Special Publication of International Association of Sedimentology, **24**, 358pp. Blackwell Scientific Publications, Oxford. Diagnostic features of sediment deposits in tidal environments.

Milliman, J. D. and Qingming, J. (1985) *Sediment Dynamics of the Changjiang Estuary and the Adjacent East China Sea. Continental Shelf Research* **4** (1/2), 251pp. A collection of research papers documenting various aspects of shelf sedimentation in the East China Sea which receives large quantities of fine sediments from the Changjiang and Hwanghe rivers.

Nio, S. D., Schuttenhelm, R. T. E. and van Weering (eds) (1981) *Holocene Marine Sedimentation in the North Sea Basin.* Inter. Ass. Sedimento-logists, Sp. Pub. **5**. A collection of research papers which covers important aspects of North Sea sediments.

Nittrouer, C. A. (ed.) (1981) *Sedimentary Dynamics of Continental Shelves* (Developments in Sedimentology **32**). Elsevier, Amsterdam. Mostly research papers that describe developments in instrumentation, techniques, and scientific rationale.

Nittrouer, C. A. and Demaster, D. (eds) (1986) *Sedimentary Processes on the Amazon Continental Shelf. Continental Shelf Research* **6** (1/2). A collection of research papers providing detailed information on Amazon shelf sediments, properties and dynamics.

Reading, H. G. (ed.) (1996) *Sedimentary Environments and Facies* (3rd edition). Blackwell, Oxford. An excellent review of modern (and ancient) sedimentation in both continental and marine environments.

Stanley, D. J. and Swift, D. J. P. (eds) (1976) *Marine Sediment Transport and Environmental Management.* Wiley, New York. A collection of papers on the physical oceanography and sedimentology of continental margins. Includes papers on problems facing the offshore engineer.

Stride, A. H. (ed.) (1982) *Offshore Tidal Sands.* Chapman and Hall, New York. An excellent synthesis of our knowledge of the tidal shelves of north-west Europe.

Swift, D. J. P., Duane, D. B. and Pilkey, O. H. (eds) (1972) *Shelf Sediment Transport: Process and Pattern.* Dowden, Hutchinson and Ross, Stroudsburg. Collection of research and review papers on sediment dynamics and sedimentation on continental shelves.

Swift, D. J. P., Oertel, G. F., Tillman, R. W. and Thorne, J. A. (1991) (eds) *Shelf Sand and Sandstone Bodies: Geometry, Facies and Sequence Stratigraphy.* Special Publication of International Association of Sedimentology **14**, 532pp. Blackwell Scientific Publications, Oxford. A good synthesis of information on sandy sediments in shelf seas.

Walker, R. G. and Plint, A. G. (1992) Wave- and storm-dominated shallow marine systems. In (eds) Walker, R. G. and James, N. P. *Facies Models: Response to Sea Level Change.* Geological Association of Canada, Waterloo, Ontario, 219–238. A facies approach to shallow marine sediments forced by waves and storms.

23. Volcanic Landscapes

Tony Waltham

23.1 Plate tectonics and volcanoes

The popular concept of a volcano as a tall conical mountain, such as Fujiyama, Vesuvius or Popacatapetl, belies the variety of terrain styles that are created by different types of volcanic activity. High symmetrical cones are the conspicuous product of viscous magmas and explosive activity (Figure 23.1), but equally important are the effusive volcanoes that produce fluid and basaltic lavas to build the gentle profiles of shield volcanoes, typified by the island mountains of Hawaii. Add to these the great sheets of flood basalts, notably India's Deccan Traps, then the ignimbrite eruptions of caldera collapses, epitomised by Yellowstone, and finally the many small scoria cones, to appreciate the full range of volcanic landscapes. Each type has its own implications for the engineering geomorphologist.

Viscosity of the parent magmas accounts for the main contrasts between volcanoes and owes its properties to origins in different plate tectonic environments. Plate boundaries are the loci of most active, dormant and recently active volcanoes (Figure 23.2), notably over hot spots. Away from the active plate margins, and a few intra-plate sites, there are no true volcanic landscapes. Old volcanic rocks lose their shapes and create high ground, merely on account of their resistance to erosion.

Volcanic materials fall into four main groups:

1. *Lava*: rock crystallised by direct cooling of a liquid (also known as lava once it emerges above ground and therefore ceases to be known as magma), including bubbly scoria and frothy pumice.
2. *Tephra*: all forms of exploded and partially or totally cooled volcanic debris, including airfall

Figure 23.1 The active andestic volcano of Koryaksky forms a classic cone to 3456 m high on Russia's Kamchatka peninsula. In the foreground, a dome of dark andesitic lava rises almost to the rim of the summit crater of Avacha.

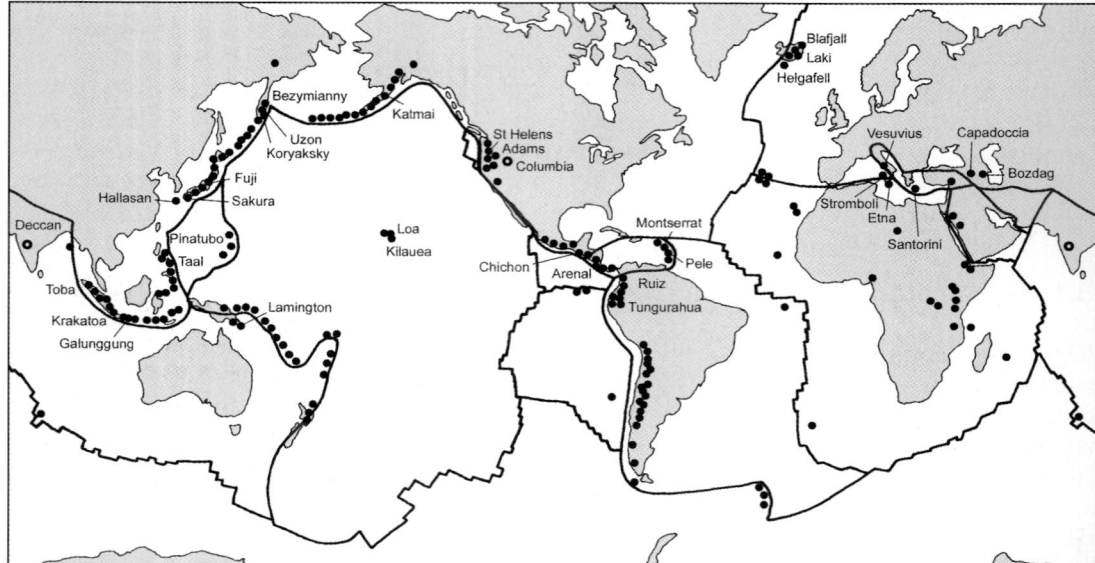

Figure 23.2   The major volcanic belts of the world in relation to the active plate boundaries. Named volcanoes are referred to in the text.

ash, cinder cones and tuffs of various grain sizes, and individual volcanic bombs.

3. *Pyroclastic flows*: the hot gas and debris mixtures that flow down a volcano's flanks, and the deposits left by these.

4. *Fluvial deposits*: mudflows and lahars rich in volcanic debris, and jökulhlaup deposits from subglacial volcanic meltwater floods.

The proportions of these materials on any one volcano depend largely on its explosivity, and therefore largely on the chemistry of its parent magma (Table 23.1).

Plate divergence and basaltic volcanoes

The crustal construction that takes place on a divergent plate boundary, such as the Mid-Atlantic Ridge (Figure 23.2), is almost entirely the emplacement of new igneous material derived from the uprising mantle convection flow that lies beneath. Mantle fractionation produces basaltic magma that is low in silica and therefore very fluid. Most of it is emplaced as dyke swarms that infill tension fissures and form most of the new lower crust as the sheeted dolerites and coalesced gabbros of ophiolite sequences. Magma that emerges above its dyke feeders becomes the basalt lava of mild and effusive

Table 23.1   A classification of the eight most important types of igneous rocks (named in the shaded box) related to their magma properties, chemistries, morphologies, eruption styles and appearances. Granite and basalt are the most abundant rocks as the dominant morphologies are determined by the contrasting viscosities of the acidic and basic magmas.

Classification	Acid	Intermediate	Basic
Lavas	rhyolite	andesite	**basalt**
Dykes	porphyry		dolerite
Batholiths	**granite**	diorite	gabbro
Magma source	continent	mountain chain	ocean floor
Silica content	70%	60%	50%
Viscosity	very viscous	viscous	fluid
Dominant form	batholiths	pyroclastics	lavas
Eruption style	violent	explosive	effusive
Iron content	3%	8%	10%
Lava colour	pale	grey	black

volcanoes. Gases escape from the fluid magma before pressures can build up and there is, therefore, little scope for explosive eruption activity. Most of these volcanoes are on mid-ocean floor

Figure 23.3 Mount Etna, Sicily, with an ash plume blown to the east by the wind. The upper slopes of the volcano are bare rock and tephra, but the lower, slightly more gently graded slopes are cultivated between many villages and towns. In the foreground, Giardini-Naxos stands on limestone just clear of the basalt.

sites and their morphology has little direct impact on the human world. Where lava builds up above sea level it creates shield volcanoes over the divergent plate boundaries; the prime example is Iceland, which lies above an over-productive hot spot within the divergent boundary.

Basalt lavas may also be produced where subducted ocean floor melts and rises through the crust on a convergent plate boundary. Mt Etna in the Mediterranean (Figure 23.3) and Mt Adams in the Cascades are two examples of these basaltic volcanoes amid lines of andesitic cones.

Plate convergence and explosive volcanoes

Within the orogenic belt of any convergent plate boundary a major process is the downward movement of crustal material. Some is truly destroyed when it is subducted into the mantle but much is melted and rises back to the surface. Because this material originates wholly or partly from continental crust, its composition is normally close to that of andesite or dacite. The high silica content ensures a magma viscosity that is high enough to minimise gas escape. Close to the surface gas pressures can rise enormously, and the eruption activity is typically violent. Lavas are produced, but are normally outweighed by pyroclastic flows

and airfall tephra. Together with giant explosions, lateral blasts and flank collapses, these eruptions create an extremely dangerous environment.

Convergent plate boundaries form very narrow zones across the Earth's crust, and these explosive volcanoes are all within them, notably around the Pacific Ocean's 'Ring of Fire'. The sites of all the famous and recent seriously destructive eruptions — Krakatoa, Mt St Helens, Vesuvius, Katmai, Pinatubo, Mt Pele, Santorini and many more — are landmarks on the convergent plate boundaries (Figure 23.2).

The older convergent boundaries are also the sites of the major granites, which have the most extensive outcrops of the intrusive igneous rocks exposed by erosion (Chapter 4).

Intra-plate plumes and flood basalts

One of the world's most famous volcanoes, Kilauea on Hawaii, is conspicuously remote from any plate boundary. This is one of a group of volcanoes formed over hot spots within the mantle, where rising thermal plumes have burned holes through the overlying plates. Movement of the plates has then created lines of volcanoes with the activity continuing at the trailing end.

These hot spots are now recognised as representing the trailing tails of massive blobs of hot magma that have risen slowly through the mantle. When the main blobs of magma first melted through the crust, they produced enormous eruptions of extremely fluid lava — the flood basalts that now form massive lava plateaux. None of these is active at present, perhaps fortunately, as their impacts spread to world climates and the entire Earth environment. Plumes within plate boundary structures may account for centres of greater activity, notably Iceland within the Mid-Atlantic Ridge.

23.2 Basaltic volcanic landscapes

Eruption styles

Individual eruptions from basaltic volcanoes vary from purely effusive Hawaiian and Icelandic styles to regularly and mildly explosive Strombolian activity (Figure 23.4), both named after their type localities. More violently explosive events are not normally associated with fluid basaltic magmas, though submarine sites can create Surtseyan eruptions where steam-driven explosions are caused by the meeting of water and magma (Table 23.2).

Figure 23.4 A modest night-time explosion hurls glowing lumps of lava 100 m into the air, and some fallen blocks remain incandescent as they roll down the flanking tephra slopes. This is the summit crater of Stromboli, off the north coast of Sicily, the type-site of Strombolian activity.

Figure 23.5 Basaltic lava emerges from a small parasitic vent on Sicily's Mount Etna. The paler material crossing the top of the picture is bright red and moving at about 0.1 m/s. The dark material in the foreground forms a splendid pahoehoe flow about 2 m wide, whose surface was rolled into classic ropy structure. It has cooled and solidified, as it flowed from the vent towards the camera the previous day.

Eruptions may be from the summit vents of volcanoes, but often occur where the magma finds an exit at lower altitude. Flank eruptions may be from single parasitic vents or from long fissure vents. A variation in the Hawaiian eruption style is the lava fountain, where molten lava is driven by hydrostatic pressure into fountains over 100 m high, but these rarely last more than a few minutes or hours.

Lava flows

Basaltic lava that emerges from either a summit crater or a flank vent (Figure 23.5) flows

Table 23.2   Classification of volcanic eruptions (based largely on Francis, 1993). The Volcanic Explositivity Index (VEI) is a relative scale based on the volume of material ejected into the air (thereby excluding lava volumes), the scale of the explosions and the height of the eruptive cloud.

Type	Style	VEI	Product	Magma
Icelandic	Outpouring of lava from fissure vents	0	Thick flood lava flows building large plateaus	basalt
Hawaiian	Outpouring of lava from central vent	0	Thin lava flows building shield volcanoes	basalt
Strombolian	Periodic mild ejections of pasty lava	1–2	Spatter, cinders and bombs building small steep cones	basalt
Surtseyan	Larger explosions over subaqueous vent	2	Tephra and glass building flat cone up to water level	basalt
Vulcanian	Repeated explosions blasting debris high in the air	2–3	Airfall ash, some pyroclastic flows and lava, building composite volcanoes	basalt/andesite
Vesuvian	Strong Vulcanian events (*term not widely used*)	4	More airfall ash and pyroclastic flows, building composite volcanoes	andesite
Plinian	Continuous powerful jets blasting out debris	4–5	Pyroclastic flows and airfall ash, building composite volcanoes	andesite
Pelean	Magma dome squeezed out of vent	5	Magma dome or spine, with nuees ardentes and pyroclastic flows	andesite/rhyolite
Krakatoan	Multiple giant blasts and caldera collapse	6	Airfall ash and pyroclastic flows, blast crater or collapse caldera	andesite/rhyolite
Super-volcano	Cataclysmic eruptions and giant caldera collapse	7–8	Thick and extensive ignimbrite sheets, collapse of large calderas	rhyolite

downslope until it cools enough to preclude further movement as it solidifies. Flow rates may be tens or thousands of cubic metres per hour, but lava cannot flow far when it is exposed to atmospheric cooling. All long flows are tube-fed, in that new molten lava is supplied from the vent to the growth zone through tubes insulated and maintained within the older cooled lava. In some flows, lava streams that are typically 5–20m wide form levees of abandoned, chilled rock, and then roof over; normally as crusts grow by accretion, levees coalesce and rafted blocks jam against each other. Many flows are inflated by the injection of molten lava beneath an already solid crust (Walker, 1991), when extensive or braided tube systems develop. Inside its insulated tube, the lava remains hot, continues to flow and may entrench itself by thermal erosion of the tube floor (Figure 23.6). The gradient on a shield volcano can allow a tube to drain out at the end of an eruption, so that it remains as a cave tunnel down the core of its host flow. Access to these tubes is generally via skylights that never fully roofed over, or via collapse holes that failed subsequently.

A lava lake can develop within a vent crater, until it is drained either by overflow or back into the vent, or cools into the top of a vent plug. Equally, lava can pond within a valley or depression, and the cooling area of flat lava may collapse or be uplifted as the balance of flow changes beneath it.

Figure 23.6   A skylight in an active lava flow from the Kupianaha vent of Hawaii's Kilauea volcano. The skylight is about 3 m across, and through it the yellow-hot lava can be seen flowing in a tube over 4 m high and wide. The surface of the flow is smooth and ropy pahoehoe, much of which formed as mini-flows downslope of breakouts from the tube when lava filled it under pressure.

Tephra falls

Lava (and included vent wallrocks) blasted into the air, and cooled before landing back on the ground, is collectively known as either tephra or pyroclastic material; only the fine-grained component is labelled volcanic ash.

Basaltic volcanoes commonly have a minor component of pyroclastic production. A cascade of lava from an active fountain maintains enough heat to weld together when it lands as a pile of spatter. The same material thrown higher by the fountain and blown further by the wind may land as a fan of loose ash just downwind of the vent. Lava ejected in cooler Strombolian explosions may partially weld on landing, but most drapes the volcano as a debris slope of loose tephra, isolated bombs and rolled blocks; this builds up to form cones as steep as, but smaller than, andesitic volcanoes. Many eruption events that initially produce lava flows progressively lose energy and heat, so that a cone of scoria is created in a dying phase of mild explosions as gas escapes from the solidifying lava.

Distal airfall ash is minimal from basaltic eruptions, which normally lack the explosive power to blast debris high enough for redistribution by the wind. Similarly, gas-rich pyroclastic flows are not a component of basaltic eruptions.

Figure 23.7   A shower of basaltic tephra on the town of Zafferana Etnea, on the downwind flank of Sicily's Mount Etna. This was produced a few hours earlier by a single fountaining eruption, which lasted just 15 minutes, from Etna's summit vent, 12 km away.

Eruption hazards

A basaltic eruption presents almost no threat to human life. A lava flow never advances faster than walking pace, and tephra falls are rarely more than a minor inconvenience (Figure 23.7). Gas is only an occasional hazard. These eruptions are generally safe enough to be tourist attractions, with positive economic impact; Costa Rica's Arenal has become almost ringed by hotels since going into steady Strombolian mode (even though Arenal's lava is a basaltic andesite with the potential for more violent eruptions that could render the tourist development most inappropriate).

Lavas can however totally destroy landscapes and built structures. Vent production is unstoppable, but there is scope for controlling or diverting lava flows that threaten houses and villages. Lava

Figure 23.8   Diversion of a lava flow, in 1983, around the Rifugio Sapienza on the southern slope of Sicily's Mount Etna. The dark material is the basaltic lava that flowed across the site a few weeks before the picture was taken. The bank of lighter material on the right is old lava and tephra bulldozed into place to protect the main Sapienza building, whose top storey is visible beyond the bank.

flows gravitate to valley floors, but the gentle topography of shield volcano flanks commonly allows scope for diverting lava away from its natural line. Some basalt lavas are so fluid that they can flow through the windows and doorways of a house, but more viscous lavas can push down structural walls (Bolt *et al.*, 1975).

Some basaltic lava flows have been successfully managed by engineered diversion methods. Bulldozed banks of earth and lava debris, built obliquely across slopes have proved very effective on Hawaii and Etna (Figure 23.8); they cannot dam lava flows, but can divert them towards less developed land. Permanent barriers have been proposed above Hilo (below Hawaii's Mauna Loa), but successes to date have been with smaller banks built immediately in front of wayward flows. Diversion can also be achieved by cooling one flank of a flow, and a massive water-spraying operation contributed to keeping the lava out of the Icelandic town of Heimay during the 1973 flank eruption of Helgafell (Williams and Moore, 1973). Attempts at blasting holes in flow levees, and blocking tubes with dropped concrete blocks,

both on Etna in 1983, had debatable impact on the lava's progress.

A special eruption hazard is generated by subglacial volcanoes where huge volumes of meltwater are created over erupting lava. Eventually water pressure floats the ice, so that the water escapes under the glacier to emerge as a sudden and potentially catastrophic flood perhaps far from the causative volcano. These jökulhlaup are common in Iceland (where the name means 'glacier flood'). Comparable melting of summit icecaps on high volcanic cones create lahars, usually with less water and more debris.

Predictions of volcanic eruptions rely largely on comparison with recorded data from previous events and are therefore reliable for Kilauea, the world's most monitored volcano on Hawaii. There and elsewhere, inflation of the volcano and increased numbers of micro-earthquakes, both caused by rising magma inside, are the best indicators of impending eruption. But they can only allow broad predictions of increasing activity with minimal practical use. Even when an eruption is in progress, day-to-day predictions remain impossible.

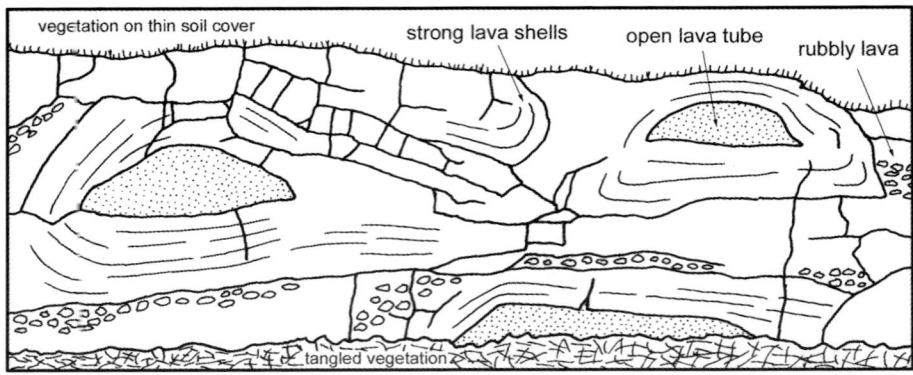

Figure 23.9   Typical structure in pahoehoe lava, with small open tubes in the cores of lava lobes that are now shells of strong and minimally fractured basalt, with more broken lava between the lobes. Based on a road-cut exposure in Hawaii.

Figure 23.10   The advancing front of a small aa flow, about 4 m thick, on Sicily's Mount Etna. The pale patch in the centre of the picture is red lava, not yet chilled, where it was exposed by blocks tumbling off the moving face.

Lava landforms

There are three main types of small-scale basalt surface morphology — 'pahoehoe', 'aa' (both terms are adopted from the Hawaiian language, so pronounced 'pa-hoy-hoy' and 'are-are') and 'pillow'.

Fluid lava develops a smooth pahoehoe surface as it forms a thin, viscous crust. This may be stretched until it ruptures as a new pahoehoe toe emerges from it. Much of the stretched skin is rolled into ropy pahoehoe by plastic movement of the hot lava beneath (Figure 23.5). Successive pulses of lava build the pahoehoe into a welded mass with a lenticular structure. Where it lies on almost flat ground, inflow of molten lava continues beneath the cooled crust, so that the whole flow can swell; the interior lava then cools with a massive structure, which is typically more vesicular towards its top. A lava pile of pahoehoe can be uniform and strong, but it can contain open lava tubes and variable fracture patterns (Figure 23.9).

More viscous lava may continue to flow on a steeper slope or where it is driven by a high rate of vent outpouring. It is continually ruptured by shear within the stressed flow, so that it breaks into aa (rubble of 100–500 mm size, also known as scoria). The front of an aa flow appears as an advancing rubble pile (Figure 23.10), and the molten interior of the lava advances over this rubble. The final cooled structure of the flow is a solid interior between a top and base of loose or partly welded aa (Figure 23.11). The resultant lava pile has alternating layers of strong, massive, basalt rock and weak, loose, aa scoria. Sequences of both aa and pahoehoe can be interrupted by weaker horizons of weathered soil, which may be lateritic in warm environments, or tephra or both.

Whether basalt cools as pahoehoe or aa depends on the original chemistry, viscosity and gas content, but eruption rate is the key background

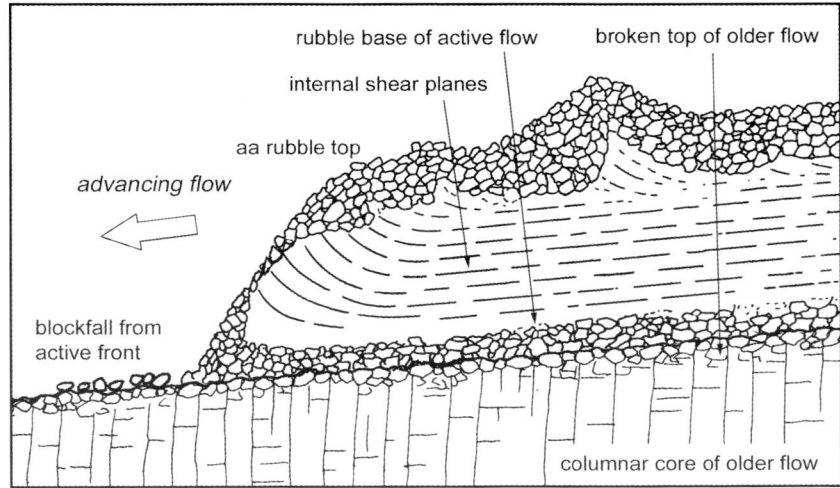

Figure 23.11 The internal structure of aa lava flows. The upper flow has a sheared but relatively solid core between a rubble top and a rubble base, and is advancing over an older flow with a core that has poor columnar jointing.

Figure 23.12 Pillow lavas exposed just north of San Francisco, USA. The pillows create a crude bedding, which dips to the right due to subsequent folding.

factor. A flow rate of > 5–$10\,m^3/s$ generally forms aa, while lower effusion rates permit flow without internal shear and thereby create pahoehoe (Rowland and Walker, 1990). Within a single flow, ground slope and degree of cooling influence the lava style. Many pahoehoe lava flows develop into aa, as they cool downflow while movement is maintained down a hillside, but aa rarely reverts to pahoehoe. The surfaces of some large basalt flows can break up into blocks larger than those of aa, thereby approaching the morphology of blocky andesites.

Pillow lavas are a larger scale of pahoehoe toes caused by the stretching of the lava crusts cooled during underwater eruptions. Individual pillows are less than a metre across, and a mass of pillow lavas has a bulbous lenticular structure (Figure 23.12).

Shield volcanoes

The low slopes necessary to maintain flows in fluid basalt lava ensure that basaltic volcanoes typically develop with the low profile of a shield (Figure 23.13); Mauna Loa is the type example (Figure 23.14), with slopes nowhere steeper than about 8°. Only a slight increase in lava viscosity, and the proportion of ejected tephra and spatter, creates the steeper profile of a very flat cone. Etna is a fine example of a composite volcano with its steeper summit standing on a flatter shield base. Large shields are generally dotted with cones and craters on parasitic vents: Hallasan, on Korea's Cheju Island has 360 parasitic vents.

In a sub-glacial eruption, lava is chilled and broken into glassy fragments by steam explosions

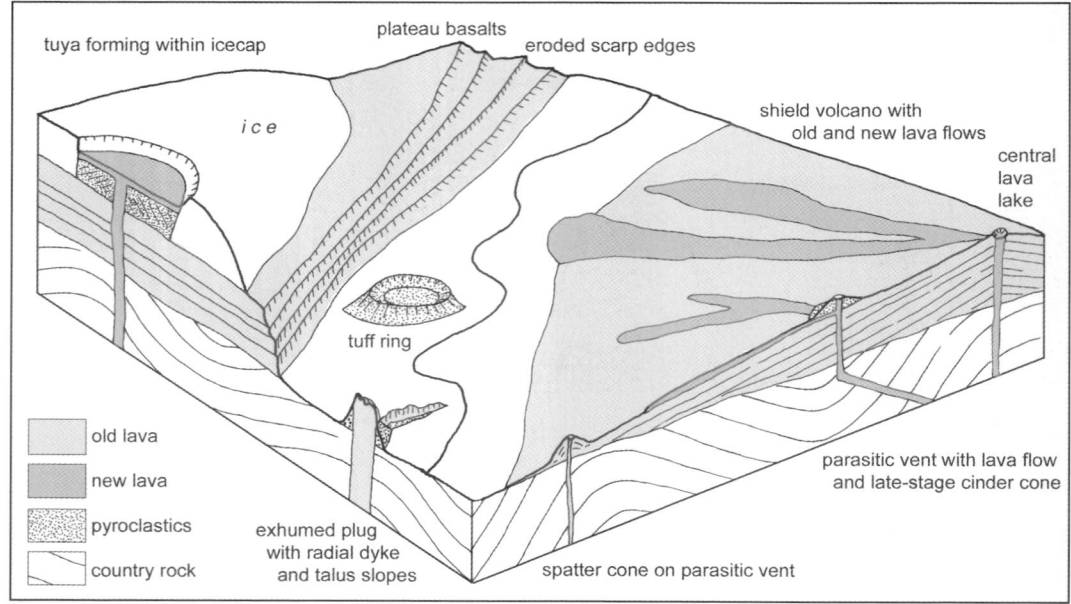

Figure 23.13   Block diagram showing various landforms associated with basaltic volcanoes.

Figure 23.14   The basaltic shield volcano of Mauna Loa, on Hawaii's Big Island. The darker lava flows are the more recent, and there is almost no tephra on it. The paler vegetated area in the foreground is the moss-covered weathered slope of the older Mauna Kea shield, with many small cinder cones that were formed in the dying stages of parasitic events.

Figure 23.15   The tuya of Blafjall, a Pleistocene sub-glacial volcano just south of Myvatn, Iceland.

on contact with ice and meltwater. This accumulates within a glacial lake of its own making, and is eventually capped by lava when the pile of ejecta rises above lake level. The lava cap creates the table mountain profile of the tuya, in place of a shield; tuyas have been exposed by post-Pleistocene ice retreat in Iceland (Figure 23.15).

Vent landforms

Volcanic craters are either excavated as debris is blasted out from the vent, or are created as ejecta accumulates around the vent. Summit craters are commonly blast features, maybe hundreds of metres deep, but they may be left as lava drains back to a lower outlet; all of them suffer subsequent wall collapse that ultimately leaves them as debris-floored bowls.

Proximal ejecta from individual vents, central or parasitic, create symmetrical cones of loose rock, known as tephra, scoria or cinder cones (Figure 23.16). Typically, these have side slopes of about 33°, and most are produced in single eruptions that last just a few days or weeks. More gassy eruptions leave larger craters within the cones, in extreme cases creating open tuff rings (Figure 23.13), of which Hawaii's Diamond Head is the classic example appreciated best from above by passengers flying into Honolulu. Spatter cones are smaller features with steeper slopes in the welded spatter (Figure 23.16).

Many vents are finally filled by solidified lava. This forms very strong plugs in the cores of the extinct volcanoes which are frequently exhumed by differential erosion to create rocky crags or towers with steep sides (Figures 23.16 and 23.17).

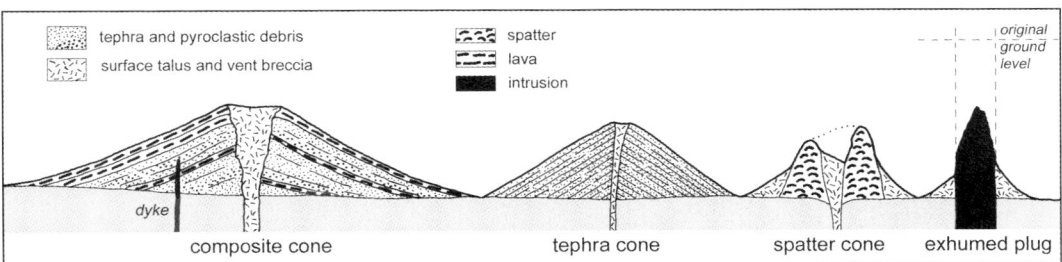

Figure 23.16   Cross-section through contrasting landforms developed on minor vents of basaltic volcanoes. Each feature can be of any height between a few metres and a few hundred metres. The composite cone is similar to a small andesitic strato-volcano except that its slopes are less steep due to the fluidity of its lava. Based on coastal and quarried exposures in Sicily, Costa Rica, Idaho and Arizona.

Figure 23.17 Mount Popa, in Burma, a volcanic plug exhumed by erosive removal of its original tephra cone, and now standing high above the Ayerwadi lowlands with a group of temples on its summit.

Plateau basalts

Continental flood basalts represent enormous outpourings of lava. Single events may last for many months and total thousands of cubic kilometres. The source vents are fissures, commonly associated with plate rifting, and the lava is so fluid that it spreads into almost horizontal sheets (Self *et al.*, 1996). The lava first spreads out as a thin sheet, which develops a chilled crust and then grows by sheet flow and injection beneath this protective insulation. Individual flows may inflate to over 80 m thick. The hot buried lava is extremely fluid, and large flow sheets are almost horizontal; their cores do not drain out, and they contain no tubes. None has been active in recorded history, and Tertiary examples have been stripped by erosion to stand as huge plateaus, notably the Deccan of western India and in the Columbia River basin, USA (Figure 23.18). During the Laki eruption of 1783 in Iceland, flows were partly tube-fed on sloping ground, but they spread out and inflated on level ground to create a small-scale example of flood basalt.

Figure 23.18 The thick sequence of flood basalts of the Columbia River basin form the cliffs of Cove Palisades in Oregon, USA. The skyline cone of Mt Jefferson is not related, as it is one of the andesitic volcanoes of the Cascades.

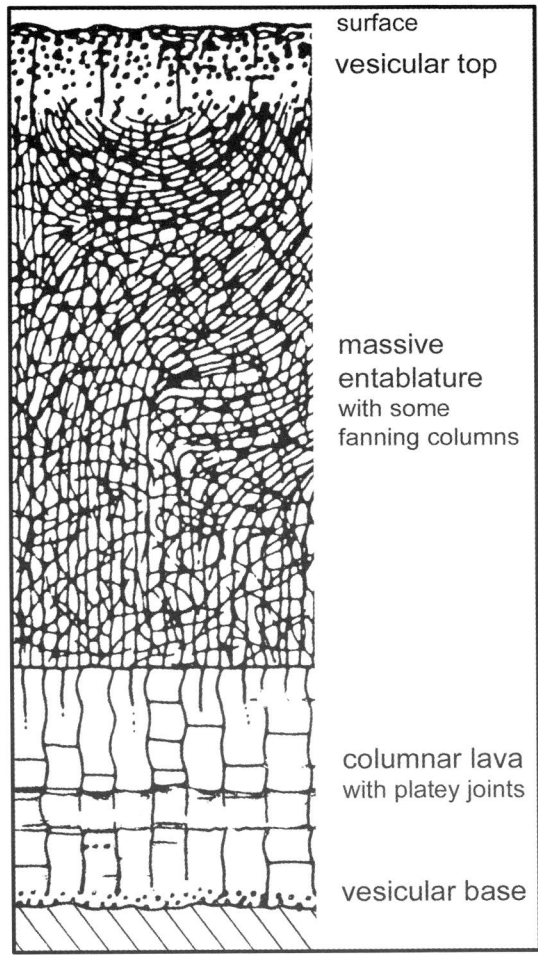

Figure 23.19   Profile through a typical sheet of flood basalt, with a total thickness of 10–40 m.

Figure 23.20   Basalt columns, each about 600 mm across, forming the Devil's Postpile in California. The flow is not extensive, but the columns are very well developed, and their fallen neighbours form the debris slope below the cliff.

The thick, horizontal flows of the plateau lavas contain some of the finest of the columnar jointing that is so well known in basalt (Figure 23.19). No two flows are the same, but most have an interior entablature that is either massive or with curved fractures and poor columns (Figure 23.20). The columns grow by extension of polygonal contraction fractures as a cooling front migrates from the base of the flow, or from its top. Some flows are entirely columnar with two sets of unequally sized columns meeting in the flow core. Each column is terminated and broken across its length by concave or convex joints that are also formed by cooling contraction.

Engineering implications

Intact basalt is a very strong rock with a compressive strength normally > 200 MPa. It commonly has a fracture spacing of < 1 m, and its rock mass strength is typically less than that of more massive granite. Basaltic lava can form strong ground with a safe bearing capacity of 6–10 MPa, except where horizons of scoria or open tubes create structural weakness. A sequence of aa flows, each with a rubble top and a solid core, creates ground that is notoriously difficult to excavate or tunnel through (Figure 23.21). Excavation of a water-collection tunnel into the flank of Mt Etna

Figure 23.21 Very mixed ground exposed in a road cutting 5 m high on Hawaii. The rock is all basalt, but the solid cores of two lava flows are separated by partly weathered aa rubble, and the thin lower flow has a small open tube with only a thin roof of solid rock.

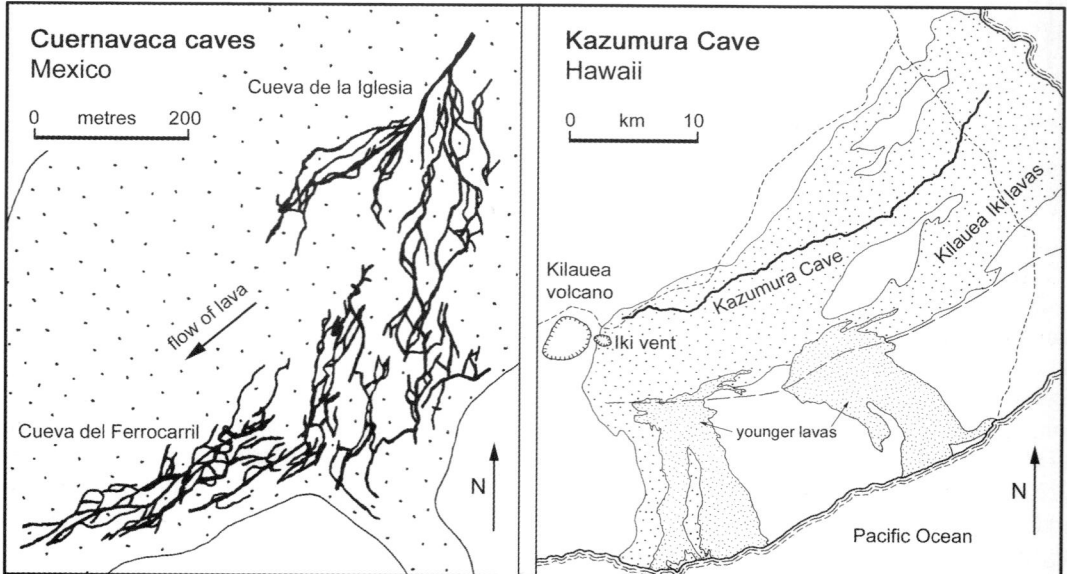

Figure 23.22 The extent of the open lava tubes in two basalt flows: the complex branching system of lava caves at Cuernavaca, Mexico; and the single very long tube of Kazumura Cave in an old lava from Kilauea, Hawaii.

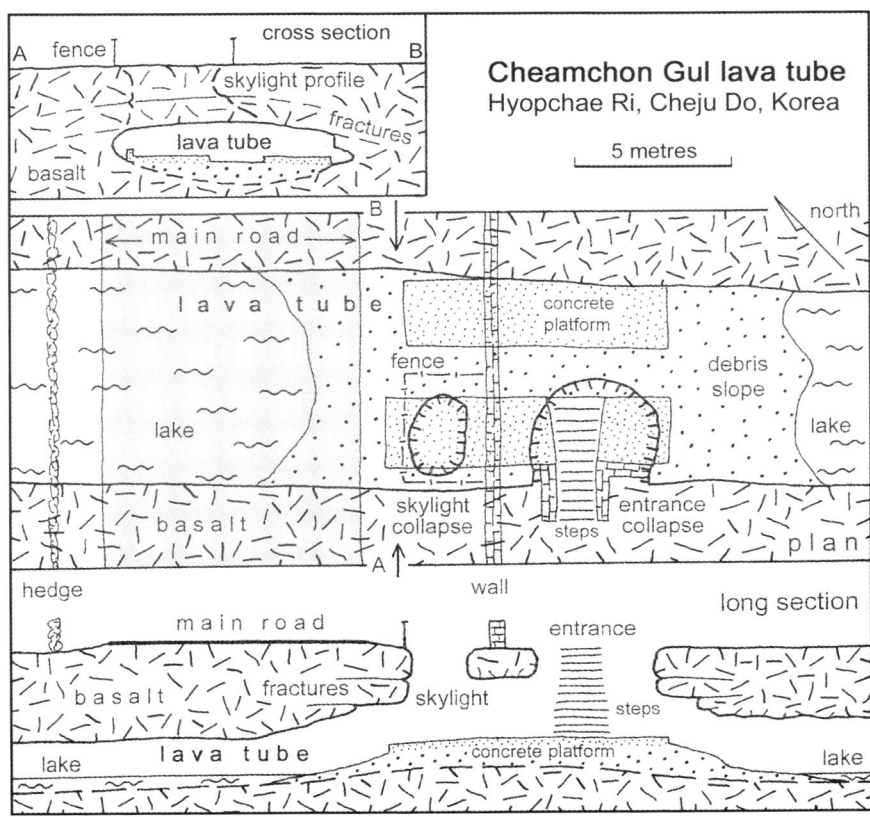

Figure 23.23   A lava tube in Korea with a main road safely standing on a thin basalt roof adjacent to two small natural collapses (from Waltham and Park, 2002).

was abandoned when blasting of the solid basalt lava cores caused unmanageable overbreak and drill-hole collapse in the rubbly lava of the flow tops and bases. Outcrops of pahoehoe basalts can have well-defined platy structure, and may include areas of heavily disturbed ground, typically up to 100 m across, caused by crustal decline and deflation where underlying pools of lava drained out.

Lava tubes can be a foundation hazard on shield volcanoes; they are not found in level plateau basalts. Drained flows on very gently sloping shield flanks can contain multiple braided tubes; the Cuernavaca lavas in Mexico have more than 10 km of braided and distributary tubes, each 2–10 m in diameter, in an 800 m long segment of a single flow 600 m wide (Figure 23.22). In contrast, a flow on the east slope of Kilauea contains a single continuous tube 41 km long with relatively few

branches (Figure 23.22). The special danger of lava tubes is that many lie close to the surface, beneath thin crusts with unknown fracture patterns; injection tubes can have thick and very massive roofs, while tubes formed by the crusting of lava streams commonly have thin roofs of more broken rock. Basalt can create a strong rock arch: many roads and light structures rest on less than 2 m of sound rock over voids 5 m or more wide (Figure 23.23). Different roof origins create very variable ground conditions, and the basalt above any tube should be individually assessed before load is imposed over it.

The only lava tube roof collapses yet recorded have been of very heavy machines crashing through during construction. Precautionary engineering may be required where a road unavoidably crosses a tube. Reinforced concrete slabs have been placed over some lava caves on Cheju Island,

Korea, and cave roofs have been broken in to create a stable rock pile within tubes on Hawaii. Heavy buildings demand a safer thickness/width ratio, and the usual precaution is to avoid the lava tubes in selecting the site. This demands careful ground investigation, with direct underground mapping, systematic drilling or microgravity surveys as appropriate in the light of the frequency of observed tube occurrences around a particular site.

Basaltic pyroclastic material behaves as a loose granular soil, while tephra cones are commonly quarried for hard granular aggregate. Inter-bedded lava and tephra sequences can create unstable slopes where masses of strong lava fail over weaker tephra; deglaciated valley sides in Iceland have some very large rotational slides that occurred soon after ice support melted away. Basalts tend to weather to smectite-rich soils that provide excellent arable land, but can create engineering instability through their high shrink–swell capability: the unstable ferrisiallitic soils of Australia are a notable example (known as black cotton soils after their colour and widespread crops).

Basalt lava sequences are very permeable and surface water retention can be difficult on them. Work on the reservoirs of the Columbia River basin, USA, has shown that permeabilities can be very high, variable and unpredictable within the layers of rafted crustal slabs and rubbly lava between the flow cores; columnar jointing within the cores is generally tight.

23.3 Explosive volcanoes

Eruption styles

While all the powerful explosions owe their origin to accumulated gas pressure in viscous magmas, variations in the eruptions occur as the style of gas release varies. Vulcanian eruptions have brief explosions that blast debris to heights of many kilometres. Plinian events last longer with powerful jets spewing out vast amounts of debris. As a variant, Pelean eruptions are those with *nuées ardentes*, the most lethal of the descending clouds of gas and debris (TABLE 23.3). The names are from the type localities of Vulcano and Pele, and from Pliny's description of the ad 79 eruption of Vesuvius. Even larger events are described as ultraplinian or as super-eruptions, causing total destruction on national or continental scales. The last eruption of this size was that of Toba, on Sumatra, 73 500 years ago. There is no engineering defence against super-volcanoes, and the rarity of events makes response programmes in civil logistics impractical for the foreseeable future.

Because these eruptions are so explosive, little magma emerges as liquid lava. When it does, as andesite, dacite, rhyolite or obsidian, it is so viscous that much of it remains as almost hemispherical domes or upheaved plugs. True lava flows can only move on steep slopes, and then for no more than a few kilometres. Slopes of $> 40°$ are normal on angular blocky andesitic lava, and also survive on

Table 23.3   Relationships between volcanic landforms, magma types and eruption scales. VEI is Volcanic Explosivity Index (see also Table 23.2); eruption types are in italics.

Magma type	VEI	Quantity of magma produced small → → → → → → large or repetitive → → → → → → very large		
fluid basic	0	lava flows		flood basalts — *Icelandic*
↓			shield volcanoes — *Hawaiian*	
↓	1–2	scoria cones	— *Strombolian*	
↓	3	tephra cones	composite volcanoes — *Vulcanian*	
↓	4–5		— *Plinian*	
↓	5	tephra maars	plug domes — *Pelean*	
	6		calderas — *Krakatauan*	
acid viscous	7–8			ignimbrite sheets

some conical volcanoes armoured by lava flows. Modest andesitic flows have been observed on some of the Pacific rim volcanoes, but no lavas of the other materials have been observed within recorded history.

Pyroclastic flows

Explosive volcanoes eject mixtures of hot gas and incandescent lava fragments high into the atmosphere. As the blast energy is dissipated, the tephra of lava fragments is blown by the wind and cascades back to the ground as an airborne deposit. While a metre of coarse fragments may accumulate close to the volcano, a thin layer of fine ash may reach 100 km from the vent.

These eruption columns are intrinsically unstable, with their high content of solids supported only by the blast and heat energy. Many therefore collapse — into surges of hot, turbulent clouds that roll down the volcano flanks under their own weight. These are one type of pyroclastic flow. Others are produced when expanding lava domes collapse and thereby release debris-charged gas clouds, and yet others are the direct products of vent explosions that did not have the energy to rise high into the atmosphere. Individual pyroclastic flows may be limited in size, but they can occur in a rapid series of explosions that can cover the volcano flanks. In AD 79, Pompeii (Figure 23.24) was completely buried within about seven hours by a steady rain of pumiceous tephra from the Plinian eruption of Vesuvius, but most of the people died in the hot gases of a series of short-lived surges that interrupted the steady fall of pumice.

The morphology of pyroclastic flow deposits varies considerably. A typical flow unit, 5–20 m thick, has fine ashes at both base and top, with a core that has particles graded by density (not mass) so that lithic fragments near its base give way upwards to pumice fragments (Figure 23.25). Little sediment remains from the short-lived, low-density surges whose hot gasses are so lethal. *Nuées ardentes* are glowing clouds with a higher proportion of solid debris than the more common surges. Pumice flows, and the ignimbrites that they form, are the product of unrestrained lava frothing and even more dense eruption clouds.

Figure 23.24   The ruins of Pompeii excavated from the pyroclastic debris of the AD 79 eruption of Vesuvius in southern Italy. In the background, Monte Somma, on the extreme right horizon is a part of the rim of the crater formed in that catastrophic event, and the higher summit cone of Vesuvius has grown almost entirely since then.

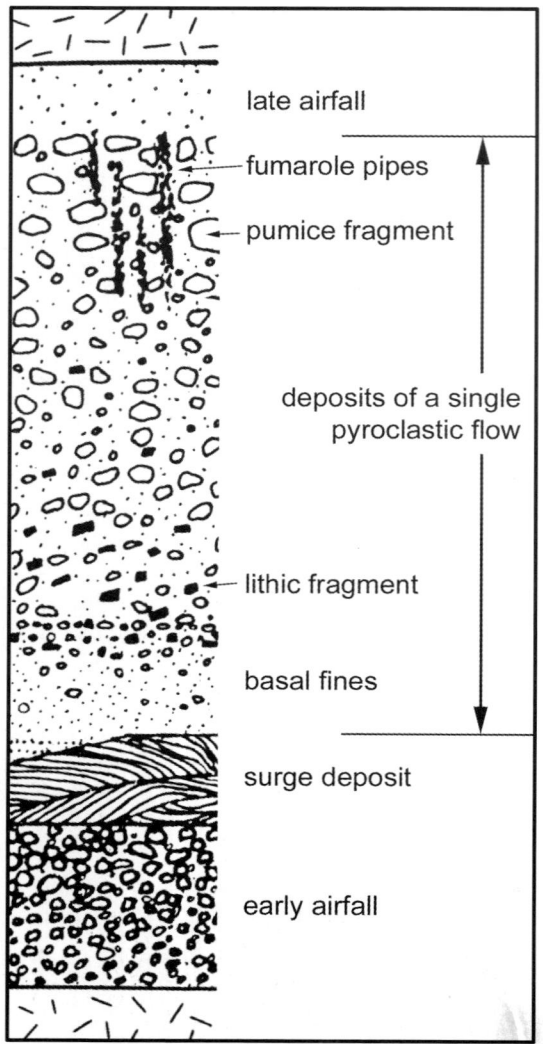

late airfall

fumarole pipes

pumice fragment

deposits of a single
pyroclastic flow

lithic fragment

basal fines

surge deposit

early airfall

Figure 23.25 Diagrammatic profile through the elements of a pumiceous pyroclastic flow that emerged during an explosive eruption.

Avalanches and lahars

Viscous lava and proximal pyroclastics from explosive eruptions build steep conical mountains whose slopes progressively become unstable. Collapse is a natural event on conical volcanoes. Massive landslides may occur as part of an eruption, as famously at Mt St Helens in 1980 (Figure 23.26), and the debris avalanches from these constitute the most voluminous of volcanic sediments. Hot debris avalanches from dome collapses create *nuées ardentes*, and a spectrum of intermediate materials can be produced within complex eruptions.

Lahars are mudflows and saturated debris flows that descend the lower flanks of erupting volcanoes. Their water content is derived from crater lakes, from ice caps melted by volcanic heat, from eruption-induced thunderstorms or from flank lakes and rivers. Their solid fractions derive from debris avalanches, pyroclastic flows and airfall pyroclastics scoured from the flanks. Once mobilised, lahars can travel many kilometres down river valleys (Figures 23.26 and 23.27) before their sediment settles, and they evolve into very muddy flood pulses. They can become the most destructive of an eruption's elements when they are channelled into populated valleys. Over 20 000 people died in Armero, Colombia, when Nevado del Ruiz erupted 50 km away in 1985; they died in lahars, without ever seeing either lava or a pyroclastic event (Williams, 1990).

Eruption hazards

Large explosive eruptions are massively destructive. Once an eruption is in progress, the only safe response is to be somewhere else. Consequently, predictions of events become extremely valuable, as they may allow timely evacuation of potential danger zones. Predictions are based largely on inflation of the volcano and increased seismic activity. Both can be monitored, but the accuracy of a prediction generally depends on having recorded data for comparison and this is rarely available for explosive volcanoes that typically have eruption intervals of 100 years or more. Monitoring of vent gases and ground temperatures can support earlier predictions, but suffer even more from the lack of comparative databases.

A volcano is described as 'dormant' through the times in between active periods of eruptions. It is only considered 'inactive', 'dead' or 'extinct' when there are no historical records of activity. There are no clear boundaries between active, dormant and extinct, and use of the terms is discouraged as they can mask the real situation. Poorly documented volcanoes that were thought to be extinct have suddenly erupted, notably Lamington (Papua New Guinea) in 1951 and

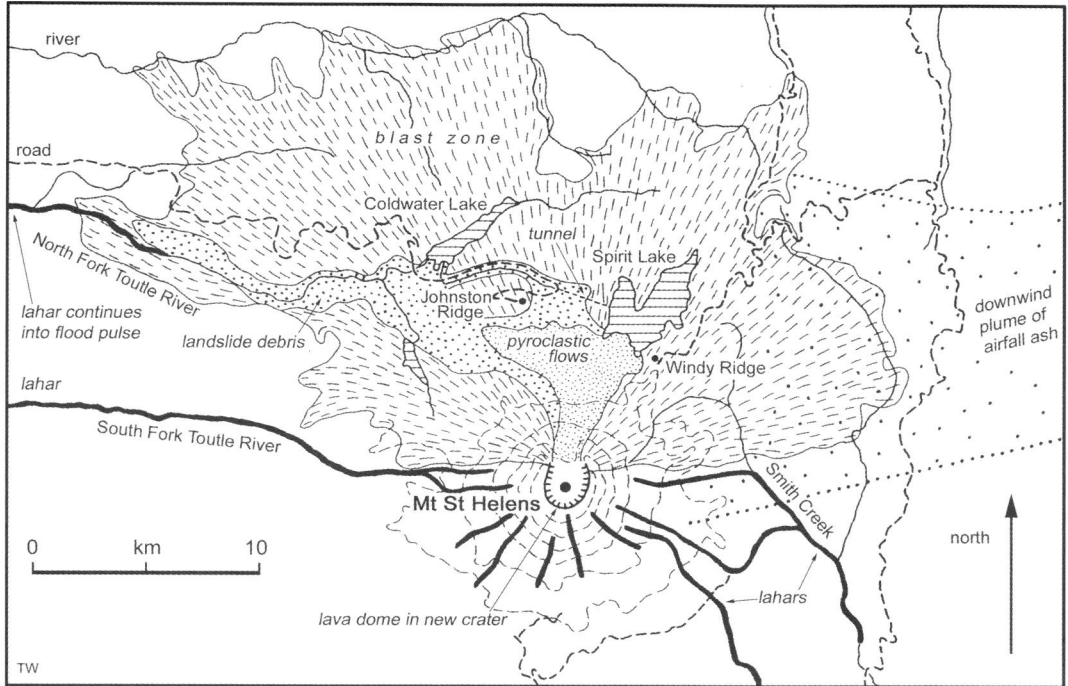

Figure 23.26   Features of the 1980 eruption of Mt St Helens, USA. The landslide debris from the failure of the northern flank fell into the valley of the Toutle North Fork. The debris flow dammed, and created lakes in, three tributary valleys and then developed into a lahar which extended west of the map and evolved downstream into a muddy flood pulse. Spirit Lake existed before the eruption, but was raised in level. The lateral blast overtook the landslide and destroyed the forest over hills and valleys alike. Deposits from pyroclastic flows created the Pumice Plain and also fed smaller lahars on the south flank, while the main airfall ash landed 250 km downwind to the east.

Figure 23.27   The debris of mud, tephra, soil and broken trees remaining from the lahar that swept down the Smith Creek valley from America's Mt St Helens during the 1980 eruption.

Bezymianny (Kamchatka, Russia) in 1956. The island volcano of Taal, in the Philippines, erupted in 1911 and killed nearly all the 500 people who had resettled on the island during a century of dormancy, when memories and folklore of earlier massive eruptions were long forgotten. Such is less likely to happen today due to the more intense monitoring and documentation of all volcanoes.

Evacuation of any area around a volcano is often not practically possible until people can perceive a risk, or unless they have seen other recent, destructive events. Fortunately, nearly all large and dangerous, explosive eruptions are preceded by days or weeks of milder activity, which provide this perception. Evacuation is likely to proceed when advice from an official source is backed by the sight of a tall ash column or the first pyroclastic flows high on the volcano slopes. Increased steam plumes and seismic activity on Ecuador's volcano of Tungurahua prompted evacuation of 22 000 people from the town of Banos in mid-September 1999. Only in mid-October did the first major explosive eruption occur. Pyroclastic flows and lahars were generated, but Banos was only impacted by air-fall ash, and most inhabitants had returned to their homes and fields by the following August. The evacuation could have been judged as unnecessary, but was a wise precaution.

It is significant that the main destructive elements from an explosive eruption are directed largely down valleys. Pyroclastic flows can descend a volcano at over 200 km/hour, but they rarely travel far from the edifice flanks (Figure 23.28). Though many are confined to valleys, they are so mobile that they can override interfluves and large base surges spread radially from the eruption column over ridges and valleys alike. Lahars cannot escape from their valleys, but travel at 50 km/hour or more and can be extremely destructive up to 100 km from their source volcano. Hazard zoning around a volcano is therefore influenced by topography, with the absolute distances determined by evidence from previous events and recognisable deposits. Open-framed barriers in ravines work well in ameliorating lahars on Sakurajima and other Japanese volcanoes.

Symmetrically concentric hazard zones can be rendered useless by a flank collapse and subsequent lateral blast. The 1980 event at Mt St Helens was the first to be observed and interpreted (Lipman and Mullineaux, 1981), but subsequent research has revealed that similar

Figure 23.28   A small pyroclastic flow hurtles down a hillside on Montserrat. It originated as a gas blast released by a partial collapse of the chilled crust over the expanding lava dome.

destructive events have taken place on dozens of the world's volcanoes (Siebert, 1984; Crandell, 1989). A rising magma dome, offset from the central vent, caused the disturbance and subsequent landslide failure of the flank, which allowed the initial blast to escape sideways instead of upwards (Figure 23.26). Greater attention to the shape of the inflation is now routine in volcano monitoring. St Helens' landslide scar was modified into a new crater by subsequent explosive action, and this now contains a new lava dome that has risen within the vent (Figure 23.29). This dome temporarily blocks the vent, and the mechanism of its eventual failure will determine the scale of the next eruption.

Airfall ash is rarely more than an inconvenience, unless it is as thick as that which blanketed villages and rice paddies below Galunggung, in Java, in 1982. With respect to roof loading, the density of airfall tephra is typically about five times that of snow. High eruption clouds of finely dispersed ash can seriously impede aircraft jet engines and volcanoes below air routes of the western Pacific are now carefully monitored for even modest upward blasts. Very large explosive eruptions can inject so much ash into the upper atmosphere that they can influence global climates. There were measurable short-term impacts from the eruptions of Krakatoa and El Chichon, and realistic debate continues unresolved on the impacts of much larger prehistoric events.

Volcanic landforms

Accumulations of alternating lava and tephra build the classic composite cones (Figure 23.1), also known as 'strato-volcanoes', whose simple, symmetrical profiles belie a complexity of internal structure (Figure 23.30). Variations in cone and crater dimensions relate largely to the explosivity of eruptions (Table 23.3).

Viscous lavas, of andesite, dacite, rhyolite or obsidian only flow on steep slopes, where they break into chaotic masses of blocks that are typically 1–5 m across; this is auto-brecciation on a grand scale. Like giant aa, these flows can have more solid cores but their surfaces create extraordinarily inhospitable terrain. They form lobes down the volcano flanks, some elongated with high levee margins, and others more bulbous with crescent pressure ridges like super-giant ropy pahoehoe. The inter-bedded tephra range from coarse lahar deposits to laminated airfall ashes. These fill valleys or blanket terrains and may be loose sediment or may gain strength where indurated or partly welded.

Figure 23.29 The 1980 crater of Mt St Helens, USA, seen from its rim. Essentially a blast-modified landslide scar, the break in its far wall is where the landslide and lateral blast headed out northwards. The valley beyond is floored by pale pyroclastic flows on top of the landslide debris. This dams Spirit Lake, the dark patch below Mt Rainier, another andesitic cone 80 km away. The lava dome has grown in the crater since 1980, and is now surrounded by dust and debris from the crumbling crater walls, inter-bedded with winter snow layers.

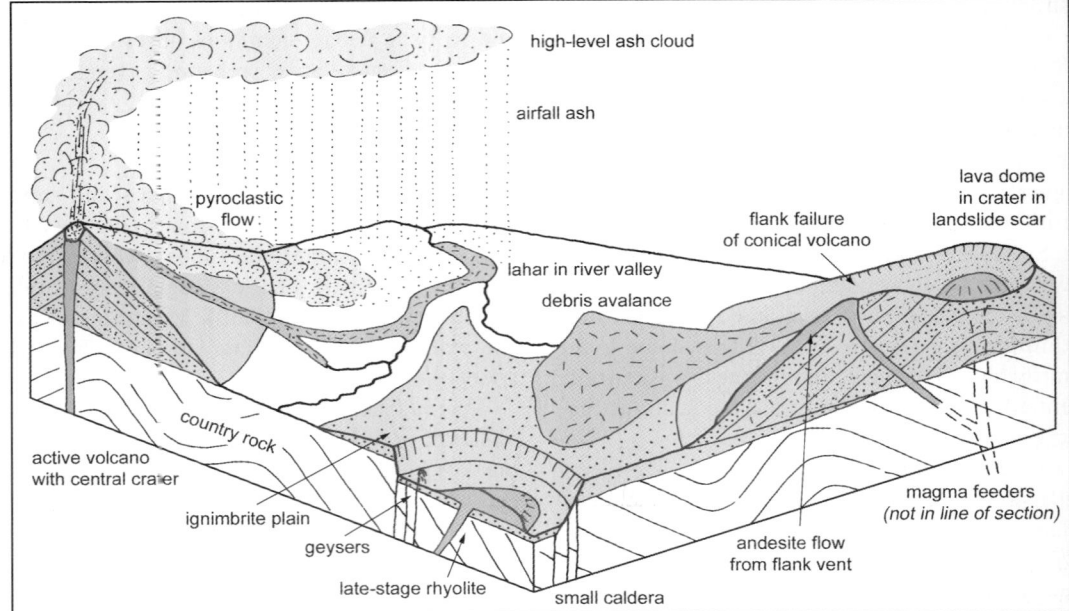

Figure 23.30   Block diagram showing various landforms associated with explosive volcanoes.

The craters of explosive volcanoes are rarely deep pits, as they tend to be backfilled by slumped tephra to create only shallow bowls. Many vents are plugged by lava, which may have risen in the late stage of an eruption to create a short-lived spine (as at Mt Pele until it collapsed) or the longer-lived dome at Mt St Helens since 1980 (Figure 23.29).

Calderas and ignimbrites

In the most massive volcanic eruptions, vast amounts of gas-rich magma emerge so fast from multiple vents that magma chambers are drained and the land above collapses into a caldera. The best analogy is with a frothing champagne bottle. Liquid lava flows from the vents as rhyolite or obsidian during the milder phases of the eruption. These are overshadowed by the climax events when the extremely mobile froth emerges as a fluid mass of rhyolitic pumice fragments supported by hot expanding gas. On settling, this forms an ignimbrite (Sparks *et al.*, 1973); some of it may weld into a very strong, glassy rock, but most remains as a partly indurated, granular sand. Evacuation of the magma chamber then causes

ground collapse, creating a deep caldera ringed by steep walls along marginal faults (Figure 23.31).

The only historical eruption of this style was of Mt Katmai, Alaska, in 1912 (Hildreth, 1983). More than 10 km³ of pumice emerged from a parasitic vent (Figure 23.32) to fill the Valley of Ten Thousand Smokes to a depth of over 200 m (Figure 23.33), while the summit of Katmai collapsed into a caldera 600 m deep and 2000 m across. It was a tiny event in comparison with ancient eruptions from super-volcanoes, which were two or three orders of magnitude greater (Table 23.2). The planet's most recent super-eruption was at Toba, in Sumatra, 73 500 years ago (Chesner and Rose, 1991). It formed a caldera 100 km long, when 2800 km³ of pyroclastic debris were ejected. Sulphuric acid aerosols created from the sulphurous gas, together with the ash in the high atmosphere, appear to have caused the worldwide cooling of 3–5 °C that is recognised as Stage 4 in the oxygen isotope record from Greenland ice cores.

Volcanic activity of this style is rare; the worldwide frequency of super-eruptions is less than once every 100 000 years. This still represents a

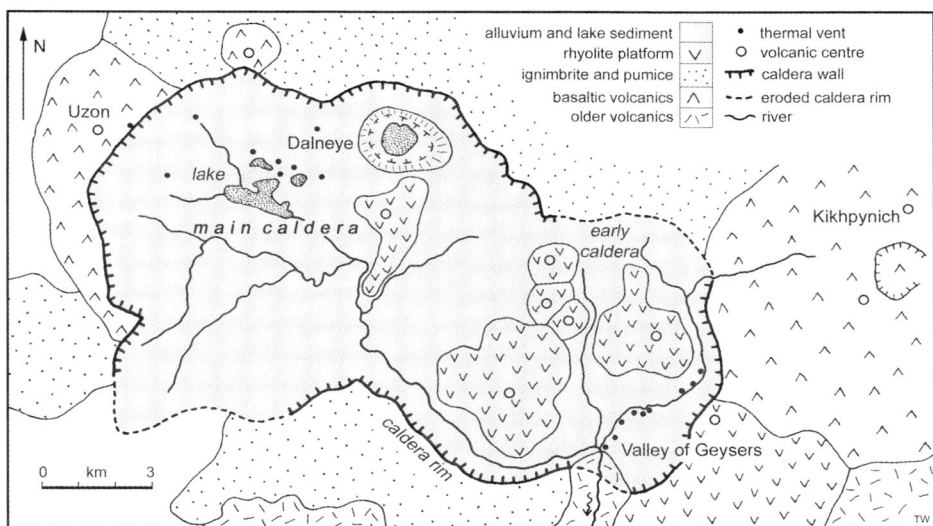

Figure 23.31   Geological map of the Uzon caldera, Kamchatka, Russia. Ignimbrites outside the caldera were formed during its succession of collapses, starting 40 000 years ago. The rhyolite platforms, Belaya lava dome and Dalneye tuff ring all lie inside, and post-date, the last caldera collapse. The Valley of Geysers lies close to the caldera's marginal faults, where groundwater circulation and geothermal heat are both at their maximum.

significant annual probability, but is very difficult to relate to daily, individual or personal hazards. These eruptions can influence world climates and they would severely impact human life and civilisation on a continental scale. They also create entire new landscapes. The rhyolites survive as high plateaus, and the caldera walls remain as circular cliffs until steadily degraded by erosion. The weakly indurated ignimbrites are commonly gullied into badland topography, of which the dissected plateaus and remnant pillars of Capadoccia form a splendid example in eastern Turkey (Figure 23.34).

Engineering implications

The explosive volcanics create notoriously variable ground conditions. Very strong rhyolitic lavas and welded ignimbrites alternate with loose tephra and coarse pyroclastic flow deposits containing large lava blocks in a fine matrix. Steep initial dips in loose tephra can create slope instability. At the volcanic centres, vertical vent structures, lobate and lenticular lavas, infilling tephra and hard intrusive dykes can create a three-dimensional complexity that makes interpretation

and prediction of ground conditions very difficult for the engineering geologist.

Under warm and wet weathering conditions, basic volcanic rocks weather to produce swelling smectite clays, creating moisture-sensitive, compressible and subsidence-prone soils within adjacent basins (Chapter 18). Andesites can weather to unstable andisols rich in halloysite clay minerals. Acid volcanics tend to weather to kaolinite clays, which are relatively stable and engineering-friendly. However, many volcanic soil sequences are notoriously variable, due to their accumulation as layers of volcanic debris from events at different times at varying distances, with periods of weathering and erosion between and subsequent to the volcanic events. This complexity can cause significant problems in ground investigation and for all forms of construction.

Perhaps the most important feature of many residual soils derived from volcanic material is that they do not fit well into the pattern of behaviour established for alluvial soils (Wesley, 1998; Shimizu, 1998). This is mainly because of the porosity and crushability of the soil grains in the silt, sand sizes and the presence of particular clay

Figure 23.32 The lava dome inside the Novarupta vent, Alaska, from which emerged 10 km^3 of pumice during the 1912 eruption, when the summit of Mount Katmai (on the horizon), collapsed into a small caldera.

Figure 23.33 The Valley of Ten Thousand Smokes, Alaska, filled to a depth of 200 m with pumiceous ignimbrite in the 1912 eruption, and since entrenched by meltwater rivers. Novarupta, the 1912 vent, is off to the right; Mt Griggs, in the background, is a dormant volcano.

minerals such as allophane. *In situ* water content and liquid limit tend to be much higher than usual, due to the predominance of hollow or porous particles that hold water within the particles. Particle crushing gives rise to a number of features:

1. investigation by cone testing shows very little difference between dense and loose states, the behaviour in both cases being similar to that for loose quartz sand
2. the reduction in friction angle with increasing pressure is greater than for quartz sands, and it is therefore more important to ensure that triaxial strength tests are carried out at an appropriate confining stress
3. during consolidation, the void ratio has a linear (not logarithmic) relationship with pressure, so that compressibility is described by a constant value of the coefficient of volume change (m_v)
4. during compaction, soils may soften so that no peak is obtained on the curve of dry density against moisture content; the soil is easily damaged by earth-moving machinery leading to loss of trafficability and reduced strength (Tonogaito *et al.*, 1998), however the drying of the soil leads to improved performance, which is not reversible, while addition of quicklime is effective in improving stability.

Low particle densities can make volcanic soils very susceptible to surface erosion, by water or wind, which can be prevented by rapidly establishing vegetation or by placing a surface layer of soil-cement. Internal erosion can be equally serious where water is able to flow through embankments, though the hazard is reduced by adequate drainage measures. Liquefaction of loose volcanic soils during earthquakes is another potential hazard.

Some fine-grained volcanic soils in Indonesia have been observed to have much higher strengths than expected in the steep terraces of the rice fields. Their strength does not appear to be dependent on plasticity in the manner of alluvial soils. Peak friction angles may be as high as 37–40°, while back analysis of slopes suggests that values of effective cohesion (c') must be above 10–15 kPa.

Many sequences of primary pyroclastics are well bedded, with variable strengths and steep initial dips (Figure 23.35), but some large areas of ignimbrite can offer very simple ground conditions. The thick and extensive Neapolitan Tuff that underlies the western part of Naples, Italy, is partly indurated, so that it is easily excavated yet has a bearing capacity of over 1 MPa, can be cut back to stable vertical faces and can also be cut

Figure 23.34   Gully erosion of weakly indurated ignimbrites has created the towers and pinnacles of Capaddocia, Turkey. Houses and churches carved inside some of the pinnacles are exposed where they have partly collapsed.

Figure 23.35   A sequence of coarse, poorly sorted deposits from four small pyroclastic flows are separated by thinner beds of pale, fine-grained surge deposits. The exposure is a roadcut about 8 m high on the entirely volcanic island of Lipari, off the north coast of Sicily.

into building blocks. The main geohazard of the Naples tuff is now the collapses over the hundreds of ancient tunnels and mines which were left with thin, unstable roofs prone to piping failure. At the same time, some ignimbrites are notable for their deep open fissuring; these include the fissured rhyolitic flow deposits that initiated the failure of the dam at Teton Falls, Idaho, in 1976.

Hydrothermal alteration can occur anywhere, but is particularly common under and around water-rich explosive volcanoes, including those no longer active. Its main effect is the hydration of silicates, producing clay minerals that inevitably reduce the strength of the rocks. The alteration is commonly along relatively narrow zones associated with fissure lines that were pathways for the hot and corrosive escaping fluids.

23.4 Mud volcanoes

Mud volcanoes occur where mud (and some larger clasts), liquids and gasses erupt from the ground surface, so that the sediment forms downslope mudflows. Sediment emerging from

one or more vents accumulates over time to form conical or gently domed edifices. Many mud volcanoes are associated with magmatic volcanoes, but others are cold and are driven by gas that emerges from diapiric intrusions of underconsolidated mudrocks. The main gases are methane and carbon dioxide.

A typical mud volcano has a summit source area that comprises one or more vents in a circular or elongated crater. Individual vents may be cones just a few metres high or bubbling mud pools just a few metres across. They may form cones, shields, maars or basins comparable to lava features but on a smaller scale. Ground rupturing is a common feature of eruption activity and fissures may extend for hundreds of metres. The body of a mud volcano may be a steep-sided cone of viscous sediment that may reach 400 m high. More fluid muds create low-profile shields with base widths that commonly exceed a kilometre. Mudflows commonly reach over 100 m from the source crater (Figure 23.36), and more voluminous eruptions may mantle the entire cone in a single event. The Otman Bozdag mud volcano in Azerbaijan has a mudflow 2900 m long.

Sediment ultimately accumulates as a series of overlapping, low-angled (generally <5°) flow lobes, with deformational structures that match those in basalt lavas. Recent lobes tend to be unvegetated and dark grey with distinct morphologies. Over time the lobes degrade, the surface weathers to a lighter grey and plants colonise the site. The extent of this zone of mudflow lobes is controlled largely by the magnitude of the eruptions and the water content of the emitted material.

Though mud volcanoes are recorded in many countries, the world's greatest assemblage is in Azerbaijan, where over 1800 recorded eruptions have occurred within the last 190 years. The eruptions fall into four types:

1. Type I: eruptions of large volumes of mud breccia with numerous rock fragments, accompanied by explosions of varying strength, the emission of a powerful gas jet, with or without combustion, and the formation of fissures. Flame heights can exceed 500 m. In the 1902 eruption of Bozdag-Gezdeg, 6 men and 2000 sheep were burned to death.
2. Type II: explosions of gas, and the formation of large fissures, without the emission of flowing mud. These approach the characteristics of the more violent mud-laden eruptions of some of New Zealand's geysers.
3. Type III: relatively small outflows of mud, without intense gas emission.
4. Type IV: extrusion of sediment, with negligible gas emission. These match the style of mud volcanoes on sedimentary basins in Sicily (Figure 23.36) and Alaska, that are in a state of almost continuous mild eruption with modest rates of mud emission.

Figure 23.36   One of the Macalube mud volcanoes near Agrigento, in Sicily, with flows of cold liquid mud radiating from a small vent pool.

that are potentially catastrophic, only broadly predictable and barely controllable. While some lava flows may be diverted, many eruptions must simply be avoided. Inactive volcanic landscapes are more widespread; they offer no dramatic hazard, but can provide difficult ground conditions where areas and horizons of very strong rock mask loose and weak materials.

23.5 Summary

An engineer's appreciation of a volcano starts with recognising its type and the potential form of its eruptions. The hazard contrast between the mild basaltic eruptions and the dangerous andesitic or rhyolitic explosions is significant (Table 23.4). Active volcanic landscapes provide geohazards

References

Bolt, B. A., Horn, W. L., Macdonald, G. A. and Scott, R. F. (1975) *Geological Hazards*. Springer-Verlag, New York.
Chesner, C. A. and Rose, W. I. (1991) Stratigraphy of the Toba Tuffs and evolution of the Toba caldera complex, Sumatra, Indonesia. *Bulletin of Volcanology* **53**, 343–356.

Table 23.4   A summary of the general features of volcanoes and volcanic landscapes. The two main types are best described as basaltic (or effusive) and explosive (or mainly andesitic). There is some overlap with respect to features and landforms, and some volcanoes fit in between these two broad categories.

Magma type	Basaltic	Andesitic to rhyolitic
Activity style	Effusive	Explosive
Plate environment	Divergent boundaries	Convergent boundaries
Rock types	Basalt	Andesite
	Basaltic andesite	Dacite
		Rhyolite
		Obsidian
Eruption events	Hawaiian: fountain or flow	Vulcanian: brief explosions
	Strombolian: mild explosions	Vesuvian: powerful jets
	Surtseyan: phreatic explosions	Plinian: larger Vesuvian
		Pelean: very large explosions
		Krakatoan: giant explosion
		Super-eruptions
Duration of eruption	Strombolian: minutes	Hours to days
	Hawaiian: hours to years	
Eruption interval	Strombolian: minutes to hours	1–>100 years
	Hawaiian: weeks to years	Super-eruptions: > 100 000 years
Lava flows	Aa: rubbly surfaces	Blocky surfaces
	Pahoehoe: smooth surfaces	
	Flood basalt: large volumes	
Volcano profiles	Shield: low profile	Conical strato-volcanoes
	Lava plateau	Caldera collapses
	Tuya: table mountain	
Explosive activity	Airfall ash: minor	Airfall ash: major
	Tephra fountain	Vertical eruption column
	Spatter fountain	Lateral blast
		Pumice frothing
Pyroclastic deposits	Airfall tephra: proximal	Airfall tephra: proximal and distal
		Pumice
		Ignimbrite: loose or welded
Small landforms	Tephra cone: loose debris	Lava dome
	Spatter cone: welded debris	Lava spine (temporary)
	Tuff ring: thrown loose debris	
Slopes	Lava flows: 3–8°	Lava flows: 40°
	Flood basalts: stepped profile	Tephra: 30–33°
	Tephra: 33°	Welded ignimbrite: 90°
Ground conditions	Strong lava cores	Strong lavas
	Rubbly lava tops	Granular pyroclastics
	Loose tephra	Strong welded ignimbrites
Rock density (Mg/m³)	2.9	2.7 (pumice: 0.9)
Intact rock strength (MPa) (UCS)	Lava: 200–300	Lava: 100–300
		Tephra: 0–1
		Welded ignimbrite: 5–200
Typical permeability (m/day)	Solid lava cores: < 0.01	Fractured lavas: 0.1–5
	Rubbly lava tops: 10–100	Tephra: 0.1–50
Ground hazards	Inter-beds of loose lava rubble	Hydrothermal alteration
	Lava tubes	
Eruption hazards	Invasion by lava flows	Major explosions
	Proximal airfall tephra: modest	Pyroclastic flows and surges
		Nuées ardentes
		Lateral blasts
		Distal airfall ash: major

Crandell, D. R. (1989) Gigantic debris avalanche of Pleistocene age from ancestral Mount Shasta volcano, California, and debris avalanche hazard zonation. *U. S. Geological Survey Bulletin* **1861**, 32 pp.

Francis, P. (1993) *Volcanoes: A Planetary Perspective.* Oxford University Press.

Hildreth, W. (1983) The compositionally zoned eruption of 1912 in the Valley of Ten Thousand Smokes, Katmai National Park, Alaska. *Journal of Volcanological and Geothermal Research* **18**, 1–56.

Lipman, P. W. and Mullineaux, D. R. (eds) (1981) The 1980 eruptions of Mt. St. Helens, Washington. *U. S. Geological Survey Professional Paper* **1250**, 843 pp.

Rowland, S. K. and Walker, G. P. L. (1990) Pahoehoe and aa in Hawaii: volumetric flow rate controls the lava structure. *Bulletin of Volcanology* **52**, 615–628.

Self, S., Thordarson, Th., Keszthelyi, L., Walker, G. P. L., Hon, K., Murphy, M. T., Long, P. and Finnemore, S. (1996) A new model for the emplacement of Columbia River basalts as large inflated pahoehoe lava flow fields. *Geophysical Research Letters* **23**, 2689–2692.

Shimizu, M. (1998) Geotechnical features of volcanic-ash soils in Japan. In Yanagisawa, E., Moroto, N. and Mitachi, T. (eds) *Problematic Soils.* Balkema, Rotterdam, 907–927.

Siebert, L. (1984) Large volcanic debris avalanches: characteristics of source areas, deposits and associated eruptions. *Journal of Volcanological and Geothermal Research* **22**, 163–197.

Sparks, R. S. J., Self, S. and Walker, G. P. L. (1973) Products of ignimbrite eruptions. *Geology* **1**, 115–118.

Tonogaito, M., Mishima, N. and Kawai, Y. (1998) Design and construction of volcanic soil embankments on expressways. In Yanagisawa, E., Moroto, N. and Mitachi, T. (eds) *Problematic Soils.* Balkema, Rotterdam, 929–944.

Walker, G. P. L. (1991) Structure and origin by injection under surface crust of tumuli, lava rises, lava-rise pits and lava-inflation clefts in Hawaii. *Bulletin of Volcanology* **53**, 546–558.

Waltham, A. C. and Park, H. D. (2002) Roads over lava tubes in Cheju Island, South Korea. *Engineering Geology* **66**, 53–64.

Wesley, L. D. (1998) Some lessons from geotechnical engineering in volcanic soils. In Yanagisawa, E., Moroto, N. and Mitachi, T. (eds) *Problematic Soils.* Balkema, Rotterdam, 851–863.

Williams, R. S. and Moore, J. G. (1973) Iceland chills a lava flow. Geotimes **18** (8), 14–17.

Williams, S. N. (ed.) (1990) Nevado del Ruiz volcano, Colombia. *Journal of Volcanological and Geothermal Research* **42**.

Further reading

Chester, D. K., Duncan, A. M., Guest, J. E. and Kilburn, C. R. J. (1985) *Mount Etna: The Anatomy of a Volcano.* Chapman & Hall, London.

Fisher, R. V. and Schmincke, H. U. (1984) *Pyroclastic Rocks.* Springer-Verlag, New York.

Green, J. and Short, N. M. (eds) (1971) *Volcanic Landforms and Surface Features.* Springer-Verlag, New York.

Kauahikaua, J., Cushman, K. V., Mattox, T. N., Heliker, C. C., Hon, K. A., Mangan, M. T. and Thornber, C. R. (1998) Observations on basaltic lava streams in tubes from Kilauea volcano, island of Hawaii. *Journal of Geophysical Research* **103** (B11), 27 303–27 323.

Kilburn, C. R. J. and Guest, J. E. (1993) Aa lavas of Mount Etna, Sicily. In Kilburn, C. R. J. and Luongo, G. (eds) *Active Lavas: Monitoring and Modelling.* UCL Press, London.

Lipman, P. W., Self, S. and Hieken, G. (1984). Calderas and associated igneous rocks. *Journal of Geophysical Research* **89** (B10), 819–841.

Pyle, D. M. (1989) The thickness, volume and grain size of tephra fall deposits. *Bulletin of Volcanology* **51**, 1–15.

24. Karst Terrains

Tony Waltham

24.1 Soluble rocks terrains

Karst

Limestone, and a few other rock types that are almost completely soluble in natural waters, may be almost entirely removed in the weathering process, leaving only very small insoluble residues. The effect of this on the ground surface is the development of a distinctive assemblage of landforms that create a range of characteristic management problems. These processes contribute minimal detrital sediment to the river systems and soils are therefore thin and very slow to accumulate. The effect underground is that the ubiquitous fractures are enlarged by chemical dissolution by slow-moving groundwater as the fracture walls are removed in solution. Over time, the initial cracks in soluble rocks are enlarged into wide fissures and then into open caves that are capable of carrying all natural drainage underground. This widening process is impossible in insoluble rocks: surface tension prevents water moving fast enough to erode mechanically in fractures that are initially narrow.

A karst terrain (karst is the German form of a Slovene word meaning bare stony ground) is defined as one with underground drainage and also a distinctive landform that is the consequence of this (Figure 24.1). Caves, closed depressions and sinkholes are therefore essential components of a karst terrain; dry valleys, isolated hills and bare rock outcrops are also characteristic but are not present in all karsts. For the engineer, karst provides unique styles of difficult ground conditions with open voids, potential collapse and subsiding soil cover, and also a special case of hydrological condition where surface water is difficult to retain and underground water is difficult to exploit.

Karst occurs worldwide, virtually wherever limestone is at outcrop. It is therefore less common in the Precambrian blocks, such as Northern Canada and much of Africa, and is notably abundant in south-east China, south-east Asia, the Balkans and south-east USA, but karst can create difficult ground for engineers on even the smallest outcrop of limestone.

Soluble rocks

By far the most widespread soluble rocks, that form most of the world's karst terrain, are the carbonates. They are soluble in natural waters that contain carbon dioxide, with which they combine to form the soluble bicarbonate. Soils are the main source of groundwater carbon dioxide, as they contain up to 3% of the gas, compared with only 0.03% in normal atmosphere. The main karst rock is, therefore, limestone. The largest caves and the most rugged surface karst landforms are formed in the older, stronger limestones. Compact, well-lithified, strong limestone has negligible permeability in the intact rock, so that all groundwater flow and dissolution effort is concentrated in the fissures, creating large caves and efficient drainage paths. Marble may be similarly cavernous and karstic; there are large cave systems just behind the marble quarries in Italy's famous Carrara district.

Dissolution effort is dispersed within the more porous, softer limestones, which therefore contain numerous small cavities in place of fewer large ones, though their bulk permeabilities are comparable. Chalk is the softest limestone and generally has only small caves beneath a distinctive karst landscape with few rock outcrops on account of its low strength and rapid degradation.

Dolomite, with magnesium replacing half the calcium to form the mineral dolomite in place of

Figure 24.1   Bare limestone deeply fretted by dissolution so that steep runnels drain into deep fissures, forming the pinnacle karst of a stone forest at Shilin, China. Comparable deep fretting of a buried rockhead provides the difficult ground conditions typical of karst.

calcite, along with partially replaced dolomitic limestones, are also soluble, but less so than pure limestone. They can support karst landforms but they are generally more resistant, therefore forming crags and residual hills and containing only smaller caves.

The main evaporite rocks, gypsum ($CaSO_4.2H_2O$) and halite (NaCl, rock salt), are both much more soluble than limestones. Gypsum may contain extensive caves within fine karstic terrains, and creates special engineering problems by way of its rapid dissolution and its low mechanical strength. Anhydrite is also soluble in water but only occurs at depth because it converts to gypsum at depths less than about 100 m. Halite is so rapidly dissolved in normal weathering that it cannot survive at outcrop except in deserts. Its surface landforms are therefore rare, and caves are never large because halite is so ductile that it will deform and close voids within erosion timescales. The main features of salt karst are

subsidence depressions over areas of rockhead dissolution.

Quartzite, composed almost entirely of silica, is so insoluble that only tiny amounts of siliceous cement are dissolved over very long periods of time. However, this can then permit piping failure of the disaggregated quartz grains — which is the origin of the pseudo-karst of deep fissure caves and giant sinkholes in the quartzite plateaux (the tepuis) of Venezuela and just a few other sites.

24.2 Surface landforms of karst

Karren and rockhead

Dissolution of strong and massive limestone is highly selective as fractures are etched out and enlarged into open fissures, while the intervening rock is minimally eroded and retains its high strength. The normal result is a rock surface fretted by open fissures and entrenched runnels, both of which are known as types of karren (Figure 24.2). Rillenkarren are channels only 10–20 mm wide with rounded floors between sharp interfluve ridges; they are formed by sub-aerial rainwater runoff which deepens the channel floors by dissolution. Rundkarren are channels (or runnels) 100–400 mm wide with rounded floors and interfluves; they are formed beneath a cover of soil or vegetation that supplies corrosive water to the whole surface. Kluftkarren are generally straight fissures which have been etched out along tectonic fractures by sub-aerial or sub-soil dissolution. A surface fretted by karren may also be known as lapies (or lapiaz).

The sizes of the karren features on an outcrop increase with age, rainfall supply and water aggressiveness (which increases with biogenic carbon dioxide in a plant-rich environment). Bare alpine terrains have only small rillenkarren excavated by snowmelt water. Rainforest terrains have giant karren runnels cutting back into steep slopes and draining into wide fissures, which can create a terrain that is extremely difficult to traverse.

Soil water is normally rich in biogenic carbon dioxide and continues the dissolution attack of the rockhead, except in rare situations beneath totally impermeable clay soils. Consequently, rockhead on limestone is typically highly irregular and may contain wide buried fissures, creating difficult ground conditions for heavy foundations that have to bear on rock. Karst in tropical regions may develop pinnacled rockheads with tens of metres of buried relief.

Sinkholes and dolines

Surface hollows or closed depressions are the ubiquitous landform of a karst terrain, as they can only survive (outside deserts) where underground drainage prevents them filling with rainwater to become lakes. They may be 1 m or many kilometres across and 1–300 m deep with sides that are gentle slopes or rocky cliffs. They may contain

Figure 24.2   A limestone pavement on the Astraka plateau of northern Greece, which was glaciated during the Pleistocene. Deep kluftkarren fissures have been opened by postglacial dissolution, and are fretted with small rillenkarren formed by modern snowmelt drainage.

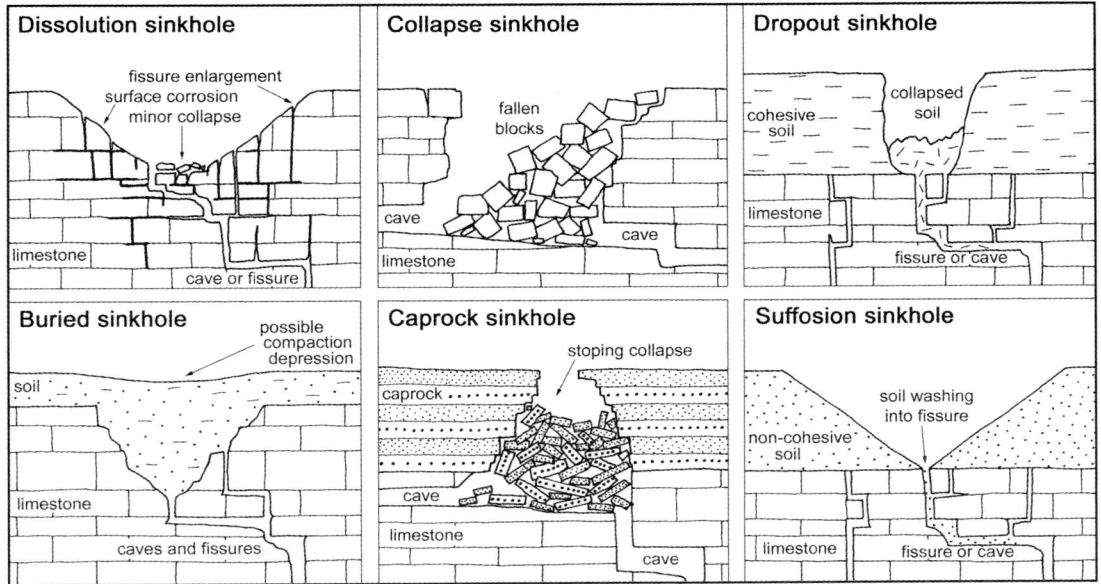

Figure 24.3   Profiles through the six types of sinkholes (or dolines) formed in karst. The dropout and suffosion sinkholes may be known collectively as subsidence sinkholes.

sizeable sinking streams or rivers, or may absorb all rainfall by percolation and fissure flow. To geomorphologists, these closed hollows are known as dolines, but most engineers refer to them as sinkholes (regardless of whether or not they contain sinking streams).

There are six types of sinkholes (Waltham *et al.*, 2005), classified on the processes by which they develop (Figure 24.3); they have been given various other names in the past (Table 24.1). Suffosion and dropout sinkholes both form entirely within the soil profile; infiltrating rainfall washes soil down into pre-existing rockhead fissures at rates that can be significant to engineered structures; they are known collectively as subsidence sinkholes. The slow slumping of a non-cohesive sandy soil produces suffosion sinkholes that may damage structures but are not life threatening. In a cohesive clay soil, cavitation initiates immediately above a rockhead fissure and grows slowly beneath an arched soil roof. It propagates upward until the surface collapses instantly and without warning — such a dropout sinkhole can be a major engineering hazard in a soil-covered karst, particularly if it is large. There are many dropout sinkholes more

than 50 m across in the karst lowlands of central Florida.

Dissolution, collapse and caprock sinkholes are all dependent on processes that mature over geological timescales in limestone; although they form significant and abundant landforms, their collapse events are so rare that they present only a very minor hazard to engineering projects. Halite may be dissolved so rapidly at a rockhead beneath a permeable soil that dissolution sinkholes can form within engineering timescales; along with areal and linear dissolution at rockhead, this may create a significant form of ground subsidence in salt regions (Chapter 12). Gypsum dissolution is too slow to create new sinkholes by rockhead lowering, but rapid corrosion by a flowing cave stream can undermine gypsum fast enough to create collapse sinkholes along with the dropout failure of any soil cover (Figure 24.4). Buried sinkholes are features of an irregular karst rockhead that may be relevant to foundation conditions (see below).

Valleys are entrenched in karst terrains only where and when their formative rivers cannot sink underground. This may occur before significant

Table 24.1 Nomenclature of sinkhole types with past and present classifications compared.

Lowe & Waltham 2002 Waltham & Fookes 2003 Waltham et al., 2005		Ford & Williams 1989	White 1988	Jennings 1985	Bogli 1980	Sweeting 1972	Culshaw & Waltham 1987	Beck & Sinclair 1986	other terms in use
dissolution		solution	solution	solution	solution	solution	solution	solution	
collapse		collapse	collapse	collapse	collapse (fast) *or* subsidence (slow)	collapse	collapse	collapse	
caprock			-	subjacent collapse		solution subsidence	-		
dropout	**subsidence**	suffosion	cover collapse	subsidence	alluvial	alluvial	subsidence	cover collapse	ravelled, shakehole
suffosion			cover subsidence					cover subsidence	
buried		-	-	-	-	-	buried	-	filled, palaeo-

Figure 24.4   A ground failure that destroyed buildings in Ripon, Yorkshire (UK), was due to a series of sinkhole processes. Rapid dissolution of a gypsum bed caused collapse of a cave within it, collapse of the overlying cover rocks and then a dropout sinkhole in the soil cover, so that the ultimate ground failure was an instantaneous event. Sinkholes had been recorded at the site before the houses were built, and part of the failure was due to collapse of the fill within a buried sinkhole.

karstic permeability is developed, but underground capture of the rivers is inevitable with maturation of the karst. Dry valleys are then left as abandoned fluvial features. By subsequent karstic modification, these may evolve into linear systems of dolines, as are commonly found in mature tropical karsts (see below). Dry valleys may also originate from periglacial interludes when groundwater flow was temporarily prevented by permafrost. Surface rivers do exist where major flows of allogenic drainage are just too large to sink into available caves, especially where low profiles create minimal hydraulic gradient through the rock. Ultimately many of these do find underground routes, but the timescale of cave enlargement generally restricts very large cave rivers to the tropical karsts (see below).

Deep and spectacular gorges are common features in karst terrains. None of the world's larger gorges originated as a collapsed cave. They were entrenched by rivers whose incision was far more rapid than was slope degradation, where surface water is minimal because it sinks underground. Many were cut, at least in part, when the ground was sealed by permafrost and others were meltwater channels from Pleistocene glaciers. The largest gorges are cut by large allogenic rivers that flow from outcrops of insoluble rock and remain on the surface throughout the maturation of tropical karst.

Dissolution planation by hillside undercutting is concentrated at valley floor level by surface water on an alluvial plain. An alluviated closed depression may, therefore, widen into a polje, with a flat floor and abrupt margins to steep confining slopes and an outlet cave drain whose evolving morphology controls the level of the polje floor. The many large poljes in the Dinaric karst, in the Balkans, are partly fault controlled, with their levels constrained by underground drainage over impermeable basements in the adjacent fault blocks. Poljes in the tropical karsts expand at local base level regardless of the basement position.

Tufa and travertine

These two materials are essentially similar, both formed by calcite deposition along stream courses. They are the sub-aerial equivalent of cave stalagmite, except that their precipitation is normally due to algae extracting the carbon dioxide from carbonate-saturated water. The same algae create the open porous texture typical of tufa, while massive banded material is generally described as travertine. Algal growth is stimulated in warmer environments, where travertine may build into extensive terraces, as along the southern coast of Turkey. In cooler climates, tufa is generally restricted to small stream dams and thin crusts. Engineering problems can be encountered where a buried layer of strong travertine, or tufa-cemented gravel, is mistaken for rockhead when it conceals underlying sediment of low bearing capacity or high permeability.

Rapid deposition builds travertine dams, typically in long flights that hold back a staircase of small lakes. These tend to create rather beautiful tourist sites (Figure 24.5), but most cases owe their origin to sources of geothermal water with greater potential for carbonate dissolution and precipitation.

Engineering implications

Karren runnels are merely attractive landform details in bare rock outcrops but they create difficult ground conditions where structures have to be founded on a very irregular rockhead. Pinnacled rockhead is the extreme form, found

Figure 24.5 Travertine dams create a staircase of small lakes at Pammukale in western Turkey.

largely in tropical karsts, where the depth to rockhead beneath level ground may vary by tens of metres (Figure 24.6). A construction project in Kuala Lumpur, Malaya, had to place end-bearing piles through soft sediment and found rockhead at depths of 5 m and 80 m at adjacent pile sites only few metres apart. Costs become high and unpredictable in such karst terrain. Rafts or mattresses may bear on the tops of pinnacles, except on those that are undercut to become large unstable boulders within the lower soil profile.

Buried sinkholes are an added component of rockhead relief. The only indication of their scale is from local exposures and drilling records, but they may be up to 100 m wide and deep. Soft fills can cause differential settlement in structures built across their margins. In the karst of South Africa, buried sinkholes over 100 m wide and nearly as deep have been found to be the sites of reactivated subsidence and new dropout sinkholes. The sediment-filled pipes that are common in chalk are buried sinkholes that may be almost cylindrical in shape and extend downwards into choked cave passages. Their sediment fills commonly

settle to cause subsidence depressions when they are disturbed by loading or drainage changes within construction projects. Electrical contrast between the chalk and the pipe fills make it possible to detect these features reliably by the resistivity techniques of geophysics.

Subsidence sinkholes are the most frequent engineering hazard in karst terrains, especially the rapidly formed dropouts. The scale of potential hazard is best assessed from local records of previous ground failures. The sites of new failures are impossible to predict, except that zones of water input across the limestone boundaries are more at risk, especially where soil cover is within the range 1–15 m thick (though ground collapses have occurred where the cavernous limestone lies beneath more than 100 m of soil cover). Desk studies of recorded ground failures and a field geomorphological survey may identify areas of greater or lesser risk that may be used to plan site development. Any amount of further ground investigation cannot reliably indicate future sites of sinkhole failure in undisturbed ground. Dynamic compaction with drop weights can collapse existing

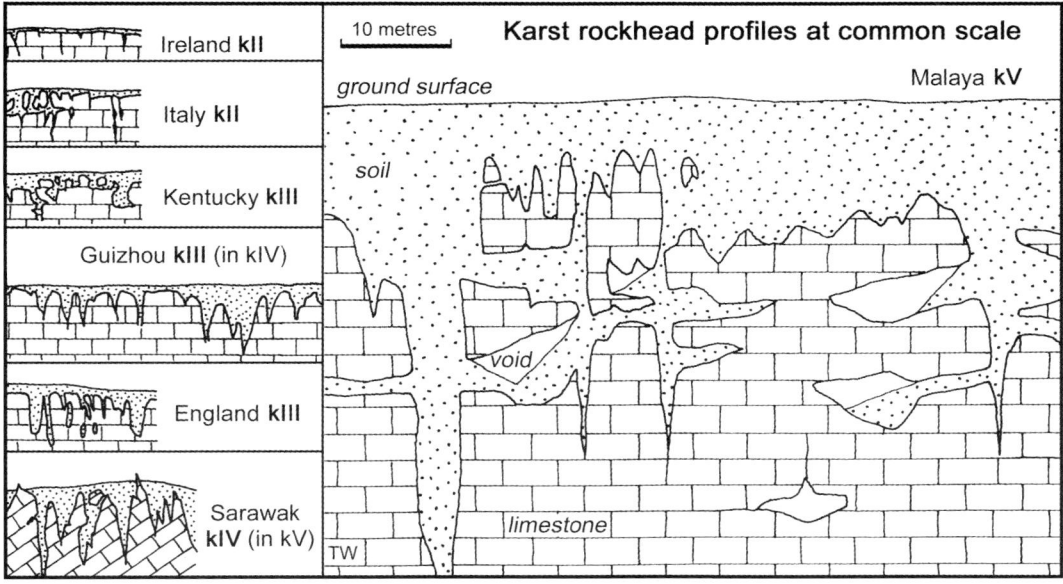

Figure 24.6 Rockhead profiles beneath soil cover in karst terrains, all drawn to the same scale. These range from minimal dissolution fissuring to pinnacled conditions. The codes kI–kV refer to the engineering classification of karst (see Table 24.4).

soil cavities at shallow depths but does little to preclude future void development.

The better engineering response to sinkhole risk is to design structures that will survive any undermining by the size of sinkhole collapse that is likely to occur within the area, as judged from existing features and past records. Roads can be effectively protected by underlayers of geogrid; these will span a new cavity for long enough to recognise the hazard by a sag in the road before total collapse occurs (Villard *et al.*, 2000). Buildings may be protected by reinforced strip footings or rafts that will span a potential new failure, and bridges can be designed so that they will survive the loss of any single pier footing.

Dropouts are caused by water flow, and therefore occur most frequently during rainstorms, or where drainage paths have been modified or water tables have declined due to over-abstraction. Most sinkhole collapses are induced by engineering activity and are therefore avoidable (Newton, 1987). Proper drainage control is critical in karst terrains, as it is uncontrolled drainage that causes the great majority of new sinkhole collapses and the reactivation of existing sinkholes. The same applies to sinkholes in gypsum karst. The Dserzhinsk region of Belarus averages more than four new sinkholes per km^2 per year, over both limestone and gypsum: this is a very high rate, but it does include many construction-induced collapses. An entire machine factory was lost in a sinkhole in 1992. When it was built, four adjacent sinkholes were filled in and the factory floor had at least four small failures before the 1992 event, but all the warning signs had been ignored.

Sinkholes in drift soils over limestone may be remediated by excavation to rockhead, then choking the open fissures with coarse rock fill. The covering fill should then be stabilised with geotextiles or by appropriate grading, so that drainage can still pass through to the cave below, as a seal could divert water to form new sinkholes over nearby fissures (Sowers, 1996). Remediation of sinkhole-prone sites (as opposed to individual collapsed features) may also be achieved by compaction grouting of the soils immediately above the rockhead. Repair and filling of sinkholes over gypsum can be difficult; any diverted drainage

flow may create new voids within timescales of only a few years, and then cause renewed ground failure adjacent to the original site.

24.3 Underground features of karst

Caves

Underground drainage through a karstic limestone is largely through conduits that have been enlarged by dissolution; those large enough to be entered by man are caves, which are interconnected with smaller un-enterable fissures. Caves may form anywhere within a limestone mass, wherever there is, or has been, through-drainage of groundwater. They may form above, below or at a conceptual base-level water table. Individual passages may have freely draining flow beneath an air surface or may be filled by water under hydrostatic pressure. Vadose caves, formed by free flowing streams, are underground canyons interrupted by waterfall shafts along continuously descending profiles. Phreatic caves, formed beneath a local or regional water table, generally have tubular cross-sections, and their long profiles may switchback up and down. Both types may include rift fissures opened on joints or faults, larger chambers formed by coalesced passages modified by roof breakdown, dendritic tributary systems and passage loops that are either braided channels or maze networks. The dimensions of caves can be spectacular: ten of the world's caves each have more than 100 km of interconnected passages and 72 caves reach depths of 1000 m. Passage widths are typically 1–20 m, but the largest cave chamber (in the Mulu karst of Sarawak) is 700 m long and 300 m wide.

Cave systems can also be of spectacular complexity. Surface lowering and valley entrenchment over long periods of time mean that most limestone masses have evolved through an earlier phase when they were saturated beneath a water table and a subsequent phase when they were largely free-draining into adjacent valleys. Most cave systems are therefore multi-phase, with an early network of tubular phreatic caves modified and entrenched by later phases of vadose canyon caves. The older passages are generally modified by roof breakdown debris, allogenic clastic sediments and also stalactites and stalagmites formed by calcite preciptation from saturated percolation waters that drip or seep into the cave. Patterns, shapes and profiles of caves are guided by structural and stratigraphic features of the host limestone (Figure 24.7). Though this can be recognised in all mapped caves, locations of unknown caves cannot be predicted, except in the broadest of terms. Limestones have too many structural elements to consider and fissures may

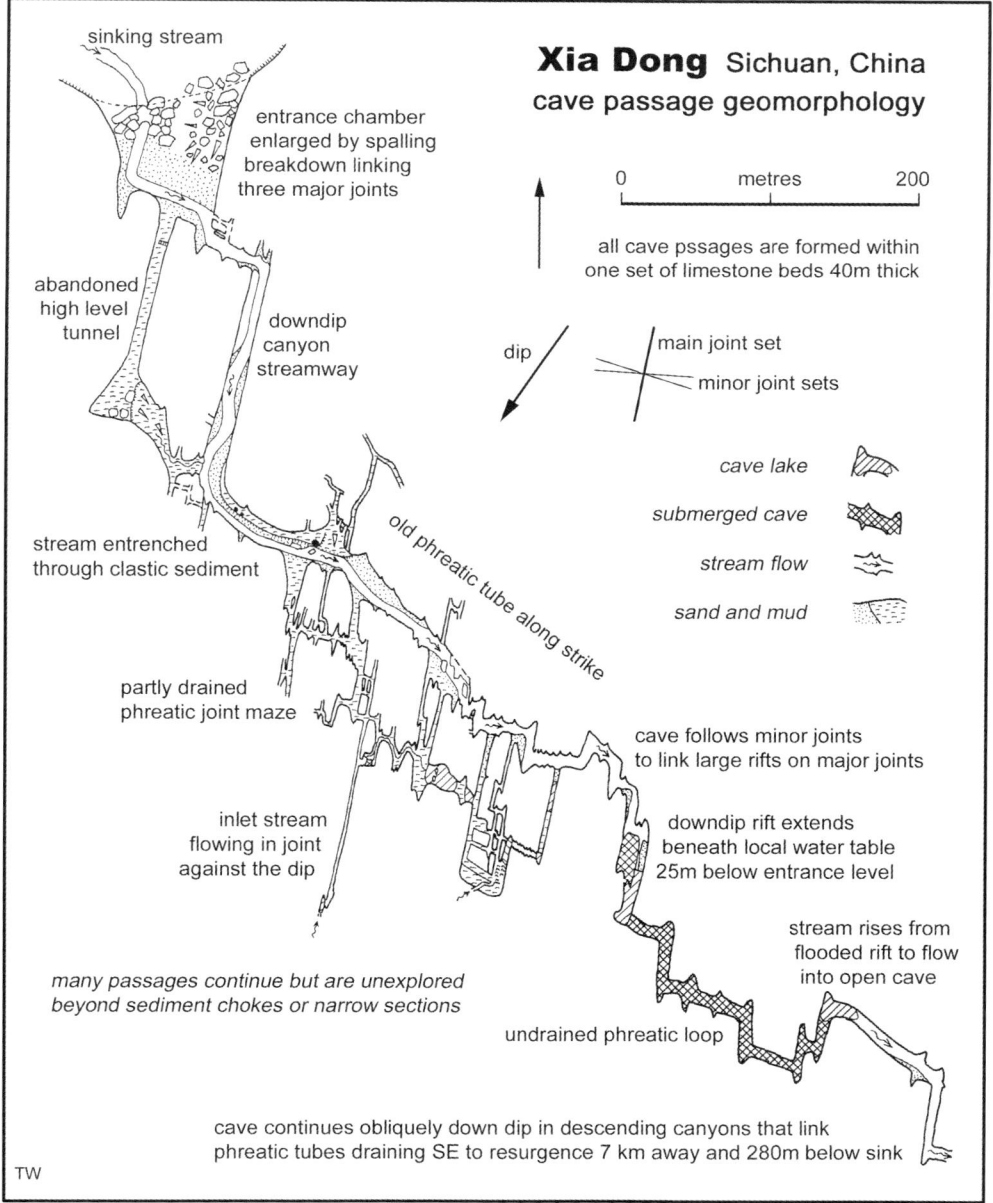

Figure 24.7   The pattern of passages in a cave in China reveals their guidance by features of the limestone geology.

develop on any or all of them, and there are too many choices for subsequent cave development from only some of the fissures.

The initiation, or inception stage of cave formation is dependent on water that is moving extremely slowly through very narrow fractures or bedding planes within tight limestone. The opening of the first millimetre of a fissure width may take a million years, but exactly where it occurs influences the pattern of the cave systems that subsequently develop. The slightest chemical contrasts in the rock or groundwater cause enhanced dissolution related to carbon dioxide. In addition, dissolution by sulphuric acid, derived from connate waters and pyrite breakdown, is probably significant in the initial stages of most caves' origins. Some caves also owe their subsequent enlargement to sulphuric acid migrating into a karst limestone from adjacent sedimentary basins. These include the well-known Carlsbad Caverns, in New Mexico, whose huge banks of gypsum are a by-product of the sulphate reactions.

Once groundwater flow is established, cave passage enlargement is mainly by dissolution in drainage waters containing carbon dioxide derived from soil air and, to a lesser extent, directly from the atmosphere; there is an element of mechanical abrasion and clastic removal in larger river caves. A cave passage 1 m in diameter may be formed from an initial fissure within about 5000 years. The rate increases with large drainage flows and also with waters that are more aggressive due to higher carbon dioxide contents collected from basins with lush vegetation. Cave passages are commonly 50–100 m high and wide in limestone karst beneath tropical rainforests (Figure 24.8), where their expansion has also been aided by long periods of uninterrupted dissolution attack. In temperate environments, where Pleistocene evolution of karstic landforms was constrained by glacial and periglacial interludes, caves are rarely more than 15 m wide.

Extensive cave systems can also form in gypsum. Rapid dissolution of the rock means that networks of fractures are widened simultaneously at shallow depth beneath the water table creating very complex maze caves in areas of low relief. Inputs of allogenic water can create individual phreatic tubes and open stream passages. Passage widths rarely exceed about 10 m before roof spans in the weak rock collapse and the cave is choked by breakdown and trapped sediment.

Engineering implications

Caves constitute an engineering hazard in that they are the ultimate form of zones of weak ground within terrains of strong rock. Construction over a cave relies on the integrity of its rock roof under imposed load. A broad guideline is that a cave roof in karstic limestone is stable under engineering loading where the thickness of rock roof is greater than the cave width. This concept is conservative in many cases, and a cover/width ratio of 0.7 is generally more realistic (Waltham et al., 2005). Strong beds of intact limestone are stable in very thin spans, but the degree of fracturing and fissuring must be assessed for each site and inspection of a cave roof may indicate variance from the guide ratio. The Dublin–Sligo railway in Ireland stood for many years on limestone only 2.5 m thick over a cave more than 6 m wide. When the single track was replaced by a main road around 1960, a ground-slab of reinforced concrete effectively carried the road over the cave in order to minimise the perceived risk. Most caves lie at depths far greater than their width, where the guideline indicates that they offer no danger to ground loading.

Near-surface caves are a ubiquitous hazard (Figure 24.9), and locations of any without surface entries are almost impossible to predict. Cavernous karst warrants a clear philosophy of ground investigation, but this may have to lean heavily on local knowledge and experience of karst conditions. Without either of these, engineers may waste their investigation budgets or may rely on inadequate assessments of the ground's bearing capacity. Adequate investigation may demand extensive and expensive exploration by drilling to prove the absence of caves. The number of exploratory drill-holes can only be defined in terms of the known local site conditions (including its geomorphological history), the sensitivity of the structure to be built and the results achieved as the investigation proceeds in stages. Belgium's Remouchamps Viaduct provided a classic example of the unpredictable nature of karst (Waltham et al., 1986): 31

Figure 24.8 Gruta de Janaleo, in Brazil, is an indication of the size that cave passages can reach where a river of chemically aggressive water has drained through a limestone karst from a tropical forest environment for over a million years. Note the person standing on the sandbank for scale.

boreholes on the initial investigation found no caves; excavation of the pier footings found 2 unknown caves; 308 boreholes in a second investigation found no more caves.

Local data is the only guide to local cave passage widths and also to the extent of caves with respect to the statistical chance of one lying under a given point. Site-specific risk assessment can then indicate whether a typical or a maximum cave width is used to determine drill hole probing depths beneath foundation sites.

A similar approach may have to be adopted beneath the bases of end-bearing piles, by subdrilling to prove sound rock at whatever depth they

Figure 24.9   A cave passage lies only a few metres below ground surface in the Yorkshire Dales (UK). Its roof contains tectonic fractures and fissures widened by dissolution which weaken the limestone roof beam and make its bearing capacity difficult to assess.

terminate (Foose and Humphreville, 1979). In this respect, shallow caves are a component of any pinnacled rockhead where the fissures and caves are components of a single three-dimensional void network in an eroded limestone. Additionally, inter-pinnacle soils and soft cave sediments extracted in excavation may be unsuitable for use in a balancing fill operation.

Detection of caves by geophysical surveys would be a welcome engineering aid, but the science is not easy and no single technique has proven totally reliable. Useful case studies and experience with electrical, seismic, radar and gravity surveys are reviewed in Beck *et al.* (1999) and Beck and Herring (2001). Micro-gravity surveys provide the most consistently useful data, as negative anomalies always indicate the absence of rock and therefore some potential hazard. They may be due to open cave passages, fissure networks, filled sinkholes, sediment-choked caves or water-filled caves. With sufficient collected data, an analysis of anomaly amplitude and wavelength may deduce the size and depth of the 'missing rock', but drilling is normally required to ascertain the details of any detected karstic

voids. Three-dimensional cross-hole seismic tomography (3dT) is a new technique that exploits improved computer analyses of massive banks of data to image cavities by their seismic shadows. With data points in boreholes or a tunnel it can produce spectacular results, but is limited in application to surface investigations of flat ground. Any geophysical survey is best used only as a guide to ground conditions. Exploratory borehole programmes are more efficient and instructive when targeted on geophysical anomalies, and every anomaly should routinely be verified by drilling (Waltham *et al.*, 2005).

Caves are notoriously unpredictable. A single cave was found, purely by chance during routine maintenance, just a few metres beneath the main runway of Palermo airport, on Sicily. It was 25 m wide, and though there were no signs of breakdown, the consequence of even partial failure was so severe that it was filled with concrete (Jappelli and Liguori, 1979). The site is on a coastal platform of young limestone, where wide cavities are notably prone to development by dissolution at the interface between salt and fresh water at either current or past sea levels.

Caves in gypsum can be a greater hazard than they are in limestone. The lower rock strength requires greater thickness of rock roof, but there is little available data on loading over gypsum caves. Concrete filling of caves and fissures in gypsum may not be a safe option where diverted groundwater flow or cave streams could excavate an alternative route through the gypsum within the lifetime of an engineered structure. Remediation of subsiding control structures on the River Neckar, in Germany, required filling voids in the gypsum with concrete and also placing an upstream grout curtain to prevent continuing groundwater flow and subsequent renewed dissolution (Wittke and Hermening, 1997).

Groundwater flow

The hydrology of karst aquifers is primarily distinguished by extreme variability between intact rock and large open caves. Traditional concepts of diffuse groundwater flow in uniform aquifers cannot be applied to karst terrains drained by caves. Ox Bel Ha is one of a series of very long cave systems that drain Mexico's Yucatan peninsula (Figure 24.10). The map of 70 km of its flooded tubes, mostly 2–20 m wide and entirely below the water table, is a fine demonstration of conduit flow in a karst aquifer. If the scale is changed, it is representative of many other karst aquifers which are not accessible for comparable mapping by divers.

Limestone and gypsum may transmit groundwater by all types of rock permeability (Table 24.2). Young poorly lithified limestones, including chalk, have high inter-granular porosity, which is replaced in older limestones by fracture porosity. It is the fractures that are enlarged by groundwater flow to create enhanced secondary permeability. Networks of fractures and fissures may be so dense that they provide diffuse flow,

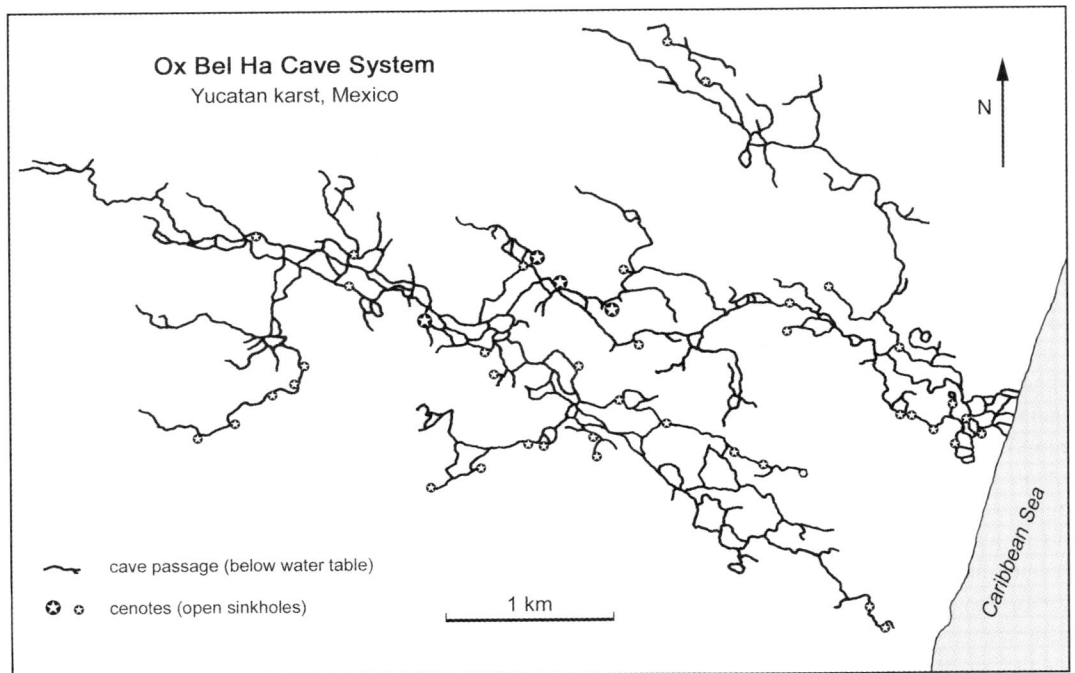

Figure 24.10 Outline map of the Ox Bel Ha cave with its system of flooded cave conduits that drain the Yucatan karst in Mexico. The cenotes are collapse sinkholes that provide access to the cave passages. Passage widths and cenote sizes are exaggerated to make them visible at this scale (after maps by the cave divers of Grupo de Exploracion Ox Bel Ha).

Table 24.2   Typical hydrogeological properties of the different groundwater elements that occur within a limestone karst.

Cavity type	Typical size	Morphology	Hydrogeology
Inter-granular	0.001–0.1 mm	Pores between grains Skeletal voids in reef limestones	Porosity up to 40% Slow laminar flow in small openings Diffuse permeability
Fractures	0.01–10 mm	Joints, faults and bedding planes	Network of planar voids Effectively diffuse flow Porosity < 1% (fractures only)
Fissures	10–100 mm	Fractures widened by dissolution	Greatly increased permeability Porosity may be up to 3% (fissures only) Flow is both diffuse and conduit Rapid laminar or turbulent flow
Conduits/caves	100 mm–10 m	Maze and network caves Dendritic tube and canyon caves	Underground rivers or random flow through maze cave Porosity 1–4% (higher only in small zones) Turbulent flow

except when viewed on the smallest of scales. Increasing drainage through fissure systems is normally accompanied by local variations in permeability as a result of flow concentrations and subsequent enlargement by dissolution along favourable routes. In a mature karst, the hydrology evolves towards efficient conduit flow in large open caves — many caves in the wet tropics carry base flows of > 1 m³/s. It is significant that cave drainage patterns are normally related to geological structure and may have no relationship to topographic divides and surface basins (Figure 24.11).

Typical values for overall porosity and permeability of limestones are indicated in Figure 24.12. Values for hydraulic conductivity and flow rates cannot be assigned to cavernous limestones because they vary enormously with cave morphology. Completely vadose cave streamways may transmit water at many kilometres per day. Flow rates are much lower where water is ponded in flooded networks of phreatic caves, but flood pulses are transmitted instantly through a flooded conduit. Most drainage paths through karst limestone are a mixture of stream canyons, underground lakes and phreatic loops, with hydrological properties unique to each and only determined by observation.

The hydrology of gypsum karst is analogous to that of limestone. Deeply buried salt beds are almost impermeable, and groundwater normally impacts on salt only at its boundaries with permeable cover rocks or sediments. Very high secondary permeability is then created by rapid dissolution, notably in breccia zones of collapsed insoluble inter-beds that evolve over a salt rockhead and are the prime sites of ground removal and surface subsidence (Chapter 12).

Engineering implications

The large groundwater flows through discrete cave conduits in karst are difficult either to constrain or exploit. Chalk and other young porous limestones have significant diffuse permeability that make them excellent productive aquifers, though enhanced flows in fissures in chalk are well known. The major hand-dug wells of Victorian London were sunk to just below the water table. Adits were then driven at a depth where they could be kept dry by pumping. Each was driven just as far as a chance encounter with a major fissure, where inflows increased to over 0.1 m³/s, an order of magnitude greater than seepage inflows from un-fissured chalk.

A cavernous limestone has minimal diffuse flow, and wells and boreholes are only productive

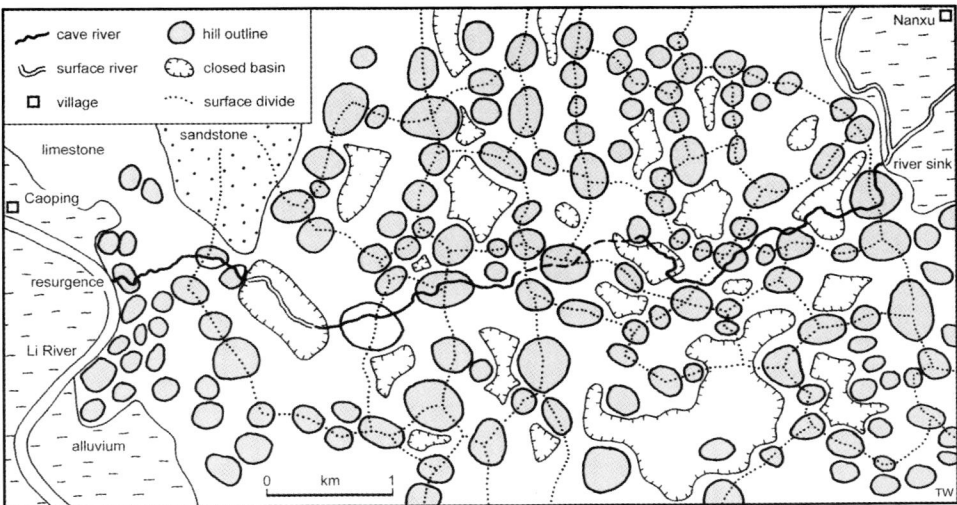

Figure 24.11 The main stream passage of the Guanyan cave system carries drainage through the fengcong karst beside the Li River in Guangxi China. The cave is unrelated to the surface topography, as it passes beneath the polygonal basins of internal drainage between the conical hills. Geological influences on the cones, basins and caves are no longer recognisable in the mature karst.

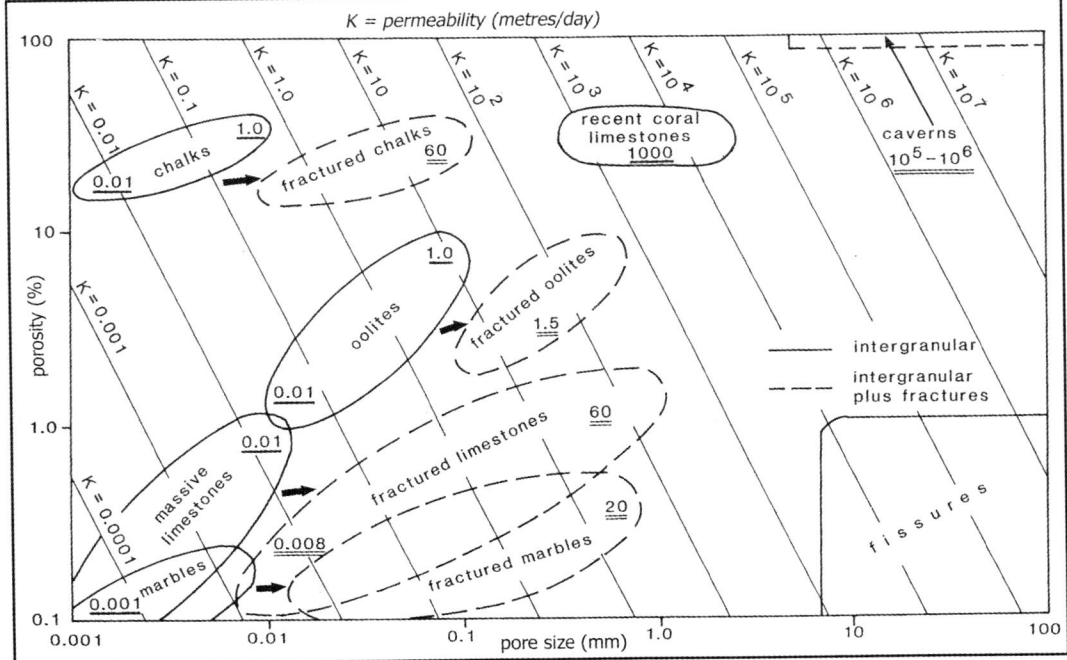

Figure 24.12 Broad relationships between porosity, pore size and permeability in contrasting types of karst aquifer. The permeability contours (in m/day) are theoretical values based on laminar flow in parallel straight tubes, while the numbers indicate typical values found in each lithology.

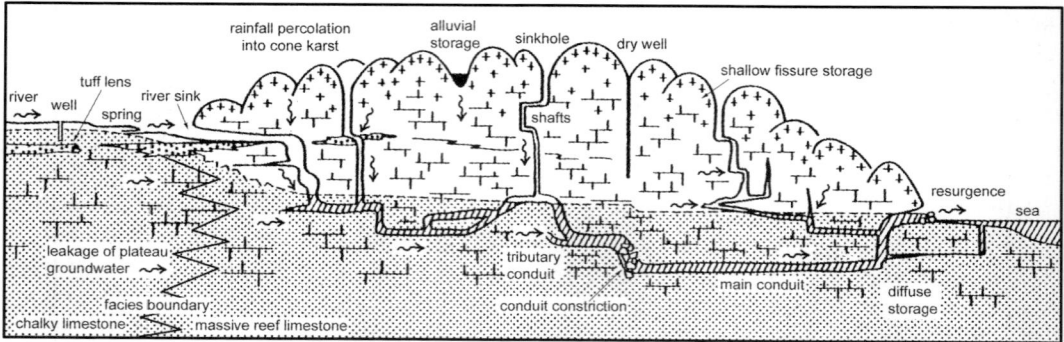

Figure 24.13   Elements of groundwater flow and storage in the limestone of Gunung Sewu, an area of fengcong cone karst in Java.

if they penetrate a cave conduit. Vadose fissure storage is small in karstic limestone, and flows into boreholes below the regional water table may be limited where fissures are poorly interconnected or are choked with sediment (Figure 24.13). Well yields are small, and systematic exploitation of a cavernous aquifer is difficult unless it targets conduits and resources identified by underground exploration and mapping (Waltham *et al.*, 1985).

Discharges from karstic springs can measure tens of cubic metres per second. Water may cascade out of open cave mouths or pour from smaller fissures against an impermeable boundary. Alternatively, water may rise under pressure from deep phreatic loops at sites known as 'vauclusian risings', named after the Fontaine de Vaucluse in southern France. Hydrographs of karst springs generally have high peaks when flood pulses are transmitted rapidly through the efficient cave conduits. Percolation water provides an element of base flow, but storage is largely within the soil cover and shallow fissures from where most of it drains rapidly into the cave conduits.

Fissures and caves have almost no filtration capability. Pollution is, therefore, transmitted easily and rapidly through a karst aquifer once it has escaped through or round any soil horizon. Disposal of waste and sewage requires considerable care on karst. Underground drainage paths are directly determined by programmes of dye

tracing. These have to be designed so that they identify flow paths that may be active only at flood stage, when water (and pollutants) may cross over within a simpler pattern of flow paths that are active at low stage. The town of Horse Cave, Kentucky, has long had problems with contamination of the karst aquifer and the caves beneath the town emanating from its own sewage plant and industrial wastes. The pollution has been transmitted to eight springs along the Green River, but these have unpolluted springs lying between them fed by separate discrete conduits within the limestone (Quinlan and Ewers, 1989).

Abstraction of karst water requires careful monitoring and control of potential pollution sources. Water resources from caves under Kentucky's Sinkhole Plain are susceptible to fuel pollution from roads across the source area; however, point spillages drain into discrete conduits and therefore usually contaminate only one resource point, which is predictable now that the drainage paths have been traced (Quinlan and Ewers, 1989). Similar problems can be created by fuel draining into sinkholes from roads across the chalk outcrop, where there is adequate fissure flow to transmit pollutants many kilometres underground (Atkinson and Smith, 1974).

Reservoir impoundment on a karst limestone may be appropriate where limestone can provide strong foundations for a concrete dam. Successful dams have been built, but extensive grout curtains are generally required and these can be difficult to place where large caverns can absorb immense volumes of concrete. Drilling at the Hales Bar dam site in Tennessee found 8411 cavities in 2000 boreholes (Schmidt, 1943), some of which were then exposed in the cut-off trenches (Figure 24.14). Repeated grouting programmes failed to seal the leakage and the dam was subsequently demolished. Leakage is self-increasing where sediment is washed out of previously choked fissures, and the same sediment may prevent effective grouting. The best dam sites in karst are where the geological structure and adjacent insoluble beds create natural hydrological barriers to which shorter grout curtains can be connected; alluvial fills are not adequate seals as they are soon lost in sinkholes beneath impounded water. China has over 5000 reservoirs impounded on karst in the provinces of Guangxi, Guizhou and Hunan alone, and about a third of these suffer significant leakage (Yuan, 1991). The Chinese have also impounded groundwater flow by building dams in caves; some have succeeded, but others have failed, mainly due to leakage through unrecognised sediment chokes in branch cave passages (Waltham and Smart, 1988).

Reservoirs on gypsum are generally undesirable. Any fissures that carry leakage are widened by dissolution within the lifetime of an impound-ment. Many countries have dams on gypsum that have never retained water.

24.4 Types of karst terrain

Any unit of soluble rock at or close to the ground surface develops a range of karst features that mature with time. Immature karst may have few caves, some degree of remnant surface drainage and landforms dictated largely by resistance of the rocks to surface denudation. A very old limestone outcrop may be reduced to a karst plain with thick soils and only some of its drainage underground in phreatic loops. Between these end stages, a mature karst has large cave systems within terrains of high, irregular and spectacular relief.

Most of the world's karst terrains have been maturing throughout Pleistocene times, and the main factor that determines their morphology is the history of their climatic environments (Table 24.3). Hot and wet tropical (and equatorial) zones are covered in lush vegetation and have developed the most mature karst landforms. Temperate regions have mature karsts, generally with less spectacular landforms. Glaciokarst occurs at higher latitudes or altitudes, has a history interrupted by glacial or periglacial conditions during Pleistocene cold stages and still bears the scars of

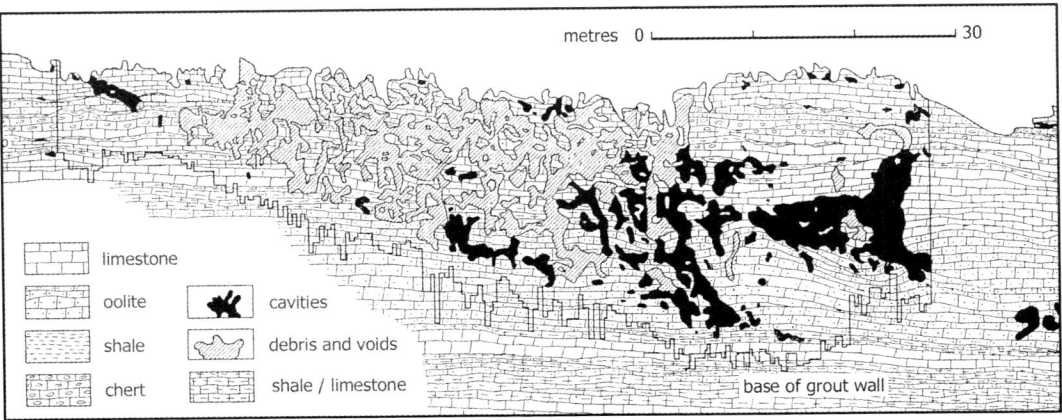

Figure 24.14   Dissolution cavities exposed in the limestone beneath the Hales Bar Dam, Tennessee (after Schmidt, 1943).

Table 24.3   Representative features of the five major types of karst terrain.

	Glaciokarst	Fluviokarst	Doline karst	Fengcong karst	Fenglin karst
Positive landforms	rock scars	interfluve ridges	polygonal ridges	cones	towers
Hill distribution	–	–	–	clustered	isolated
Relief	inherited alpine	up to 300 m	less than 100 m	up to 1000 m	up to 300 m
Negative landforms	stream sinks	dry valleys	dolines	dolines	alluviated plains
Hill slopes	stepped	20–30°	20–30°	30–45°	45–90°
Soil cover	bare rock, pavements	thin	thin	patchy	in fissures, thick on plains
Rockhead	exposed	fissured	fissured	pinnacled	pinnacled
Cave locations	anywhere	within ridges	under dolines	anywhere	within towers
Cave passage width	1–5 m	1–5 m	1–10 m	1–30 m	1–30 m
Surface drainage	flood streams	ephemeral streams	minimal	none	on alluviated plains
Main origins	ice scouring	periglacial streams	pure karst	tropical	tropical, with uplift
Examples	Yorkshire Dales (UK) Calcareous Alps (Austria)	Peak District (UK) Causses (France) Chalk Downs (UK)	Sinkhole Plain (Kentucky) Kras (Slovenia) Yucatan (Mexico)	Guizhou (China) Sewu (Java) Cockpits (Jamaica)	Guangxi (China) Halong Bay (Vietnam)

ice cover. Dissolution processes virtually cease in both desert and glacial environments, where the few karst features are generally inherited from Tertiary climates that were warmer and wetter. These may include the very large but very shallow sediment-floored depressions known as 'dayas' in the North African deserts; these appear to represent very old buried features but their depths and floor structures are largely unknown. Even older palaeokarst can survive beneath lithified sediments to create difficult ground. Such was encountered in Devon (UK) where road excavations were hampered by a deeply eroded and locally pinnacled surface on Devonian limestone, found infilled and buried by Cretaceous sands (Fookes and Hawkins, 1988).

Glaciokarst

In many high mountain terrains of limestone, the larger landforms relate to glacial erosion during the Pleistocene cold stages, while the smaller landforms comprise a glaciokarst, also known as

an alpine karst. Bare rock outcrops are conspicuous and include terraced scars left by ice plucking. Limestone pavements fretted with small karren are the result of glacial scouring followed by minimal soil formation and slow karren evolution (Figure 24.2). Typically small vadose cave passages carry underground drainage efficiently on steep gradients in the mountainous terrains, and generally intersect old, dry cave passages abandoned as a result of multiple glacial rejuvenations. Gorges were cut by proglacial meltwater rivers, at least partly when the bedrock was frozen and the caves inactive, but many are dry now that more modest rain-fed drainage has sunk underground.

Temperate karst

The natural landscape on limestone in a temperate environment is doline karst, of which Kentucky's Sinkhole Plain is a prime example. The entire surface is pitted by dolines (also known as sinkholes) and where the doline rims coalesce, a polygonal karst forms having almost no flat ground in a terrain of interfluve networks. All drainage is underground into mature cave systems, except for short lengths of temporary surface flow feeding to sinks in the floor of the larger dolines. Most gypsum karsts fall into this category. A variety of doline karst is the cenote karst of the Yucatan, in Mexico — there is minimal surface erosion on the dry plains but numerous cenotes are collapse sinkholes into long caves that carry drainage from inland mountains (Figure 24.10).

Fluviokarst is dominated by dendritic systems of dry valleys, as in the Derbyshire Pennines of England (Figure 24.15). Some valleys are blind where they terminate at old stream sinks. Many valleys have sinkholes along their floors, fed by short lengths of surface streams active only in storm events. Extensive cave systems are developed within the interfluve ridges and also pass beneath the valleys. Most of the valleys were cut as fluvial features when the caves were sealed by ice in the permafrost of Pleistocene periglacial stages, but others are relics from erosion before the limestone developed secondary permeability. Chalk karst is a type of fluviokarst, where the low strength of the host chalk precludes development of bare rock scars along the sides of the dry valleys; the typical result is the downland of eastern England.

Tropical karst

Massive production of carbon dioxide in the hot-house environment of the wet tropics provides the acidic waters that create the most spectacular karst landforms. The natural landscape of mature tropical karst has an 'egg-box' topography of conical or hemispherical hills separated and drained by dolines (Figure 24.11). Known in the western world as cone karst (or cockpit karst), it is better described as fengcong karst, after the Chinese landscape fengcong which has clusters of hills with little flat land between them (Figure 24.16). Fenglin karst has isolated hills with steep sides undercut by planation at the level of the intervening alluviated plain across which the limestone towers are scattered; it is essentially equivalent to the western concept of 'tower karst' (Figure 24.17). The Chinese terms should be used since the huge karst areas of Guangxi and Guizhou have been recognised as critical to the understanding of karst processes and landforms (Table 24.3).

There is an evolutonary progression from fluviokarst to doline karst to fengcong karst to

Figure 24.15   The dry valley of Cressbrook Dale in the Peak District fluviokarst of the southern Pennines. Streams re-occupy stretches of the thalweg after heavy rainstorms and drain through the thin soils on the valley floor into limestone fissures which are connected to deeper cave conduits.

Figure 24.16   Fengcong karst of conical hills in Guizhou, China.

Figure 24.17   Fenglin karst with isolated limestone towers around Yangshuo in Guangxi, China.

fenglin karst to karst plain. Valleys are broken into lines of dolines, and as these deepen the remnant intervening cones become the dominant landforms; traces of the early valleys are recognisable through many fengcong karsts, though they are unrelated to the subsequent underground drainage. As fengcong karst degrades, the hills diminish while flats and poljes expand at base level between them until a karst plain evolves with few remnant hills; the intervening stage may be described by the

Chinese as fenglin with cones. True fenglin, with tall narrow tower hills, only breaks into the evolutionary path where tectonic uplift is matched by planation and alluviation at the declining base level and is matched by undercutting at the margins of the emerging towers. The long evolution of fenglin karst is matched by major surface lowering, and it only matures in limestone sequences thick enough to not be entirely removed in the process. This and the necessary coincidence of various parameters accounts for the almost complete absence of fenglin karst outside China and south-east Asia.

Pinnacle karst and shilin are very similar terrain types with sharply pointed limestone pinnacles 10–30 m high. These are effectively giant karren ridges created in massive limestones in wet tropical regions. The pinnacles are micro-landforms that may form the main surfaces over fenglin towers. Shilin is the Chinese variety, whose name translates as 'stone forest' (Figure 24.1).

Though these extreme forms of karst landscape are largely confined to the tropical regions, geological structure can locally influence landforms to mimic the tropical features. In the Nahanni karst of northern Canada, remnant towers amid polje basins owe their slope profiles to control by strong beds of horizontal limestone. Geology is also influential in that towers rarely develop in dolomite, where conical hills dominate even in fenglin terrains.

Engineering implications

Ground conditions on karst can be extremely difficult for engineers, and each site may warrant a prolonged ground investigation. Bearing capacity and potential instability of karstic ground are only fully appreciated after careful consideration of rock structure, drainage conditions and geomorphic evolution. This does benefit from a thorough understanding of the local karst conditions, preferably by an engineering geologist with karst experience. Engineering difficulties broadly increase in warmer and wetter climatic zones where rock dissolution is more rapid and karst landforms are more mature. Relict landforms can occur, whereby a large cave in a glaciated terrain may be inherited from an ancient phase of warmer conditions, but such features are statistically rare.

An engineering classification of ground conditions on limestone karst identifies five classes of terrain (Waltham and Fookes, 2003). These cover the considerable range of karstic conditions, with increasing engineering difficulties created by more mature karstic erosion, cave and landform evolution (Figure 24.18), and can be delimited by broad groups of terrain features and landform dimensions (Table 24.4). However, karst features do not all evolve in strict unison, and the

classification can only define a broad suite of ground conditions. A full assessment of any site should have numerical values for critical parameters appended to the karst class description. The important quantifiable parameters include:

1. amplitude of rockhead relief
2. typical and maximum cave passage widths
3. frequency of new sinkhole collapses.

Even with these parameters defined, the complexity of karstic ground must be recognised. The approach to terrain assessment by the construction of geological models (Fookes, 1997) is beneficial in karst because it directs the engineer's thoughts to the various possible conditions that can occur on cavernous limestone. It is however notable that a large site, such as a transport corridor, may generate many 'sub-models' where variations are created by the many factors involved in karstic evolution. Perhaps the most successful approach to engineering on karst is an observational method, where there is full scope for adaptation to ground conditions that cannot fully be appreciated in advance. A safe bearing pressure for strong un-fissured limestone may be 4 MPa, but this is normally found only in footings that have been excavated to below all rockhead pinnacle and fissures, and must still be proven by sub-drilling to ensure sound rock with no caves. More typically, foundations on karst require innovative designs adapted to the specific site in order to successfully transfer loads to stable rock (Figure 24.19).

Karst on gypsum can have all the difficulties of a limestone karst. The absolute size of features is generally smaller, but rapid dissolution of gypsum may allow new cavities to form within engineering timescales. Salt karst has few caves and the engineering hazards are created by ongoing dissolution and ground subsidence (Chapter 12).

24.5 Summary

Karst terrains are distinguished by landforms that are unique to soluble rocks and by engineering difficulties that are unique to those landforms. No other terrain has such a marked contrast in bearing

Table 24.4 The engineering classification of karst. The table provides outline descriptions of selected parameters, which are not mutually exclusive, and give only broad indications of likely ground conditions that can show enormous variation in local detail. It should be viewed in conjunction with Figure 24.18 that shows some of the typical morphological features. The comments on ground investigation and foundations are only broad guidelines to good practice in the various classes of karst. NSH is the rate of formation of new sinkholes per km^2 per year.

Karst class	Locations	Sinkholes	Rockhead	Fissuring	Caves	Ground investigation	Foundations
kI Juvenile	Only likely in deserts and periglacial zones, or on impure carbonates	Rare NSH <0.001	Almost uniform; minor fissures	Minimal; low secondary permeability	Rare and small; some isolated relict features	Conventional	Conventional
kII Youthful	The minimum in temperate regions	Small suffusion or dropout sinkholes; open stream sinks NSH 0.001–0.05	Many small fissures and depressions	Widespread in the few metres nearest surface	Many small caves; most <3 m across	Mainly conventional, probe rock to 3 m, check rockhead fissures	Grout open fissures; control drainage
kIII Mature	Common in temperate regions; the minimum in the wet tropics	Many suffosion and dropout sinkholes; large dissolution sinkholes NSH 0.05–1.0	Extensive fissuring; relief of <5 m; some loose blocks in cover soil	Extensive secondary opening of most fissures	Many <5 m across at multiple levels	Probe to rockhead, probe rock to 4 m, micro-gravity survey	Rafts or ground beams, consider geogrids, driven piles to rockhead; control drainage
kIV Complex	Localised in temperate regions; normal in tropical regions	Many large dissolution and numerous subsidence sinkholes NSH 0.5–2.0	Pinnacled; relief of 5–20 m; loose pillars	Extensive large dissolution openings, on and away from major fissures	Many >5 m across at multiple levels	Probe to rockhead, Prove rock to 5 m with splayed probes, micro-gravity survey	Bored piles to rockhead, or cap grouting at rockhead; control drainage and abstraction
kV Extreme	Only in wet tropics	Very large sinkholes of all types; remanent arches; soil compaction in large buried sinkholes NSH ≫1	Tall pinnacles; relief of >20 m; loose pillars undercut between deep soil fissures	Very complex dissolution cavities	Complex 3-D cave systems, with galleries and chambers >15 m across	Make individual ground investigation for every pile site	Bear in soils with geogrid, load on proven pinnacles, or on deep bored piles; control all drainage and control abstraction

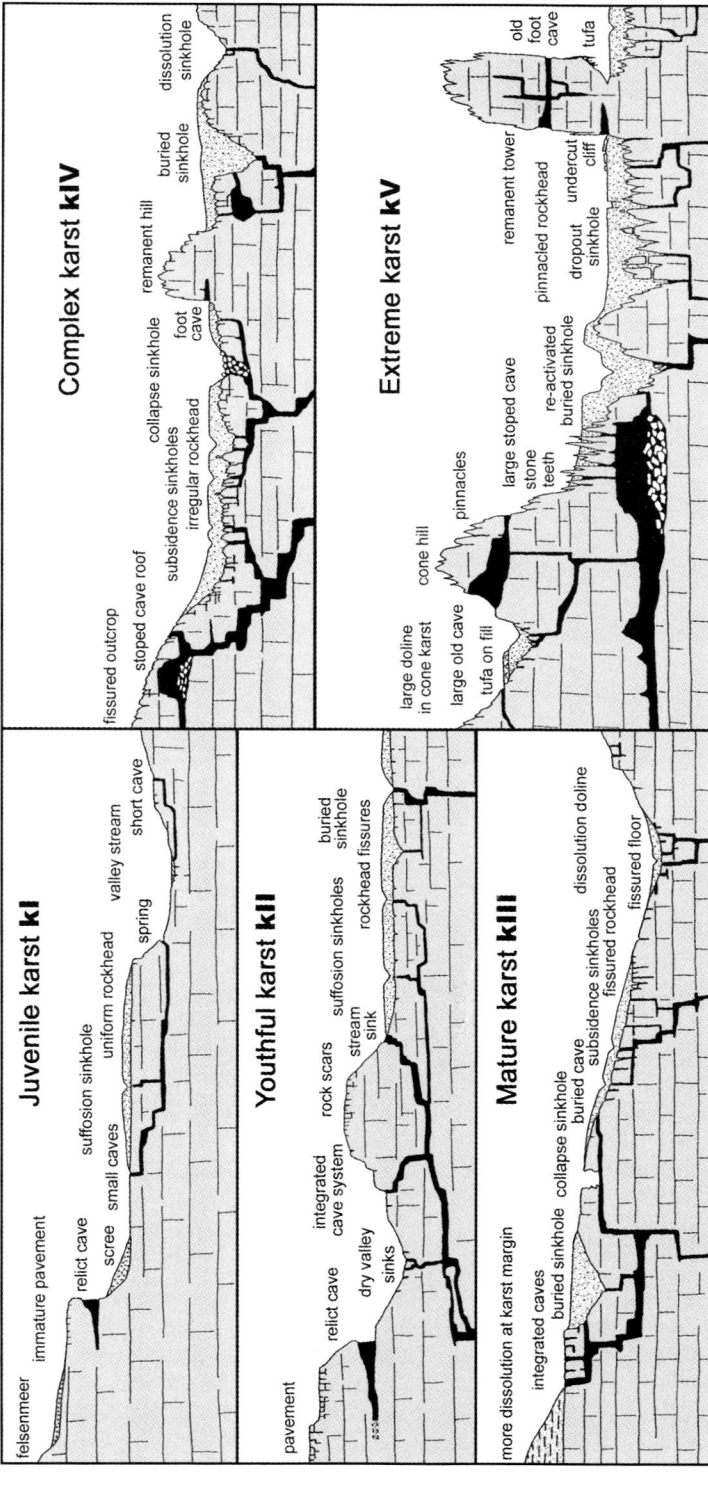

Figure 24.18   Profiles through the five classes within the engineering classification of karst ground conditions. The five classes provide guideline subdivisions within a continuous spectrum of evolving karst terrains, and not all the landforms shown appear in all areas of each karst class.

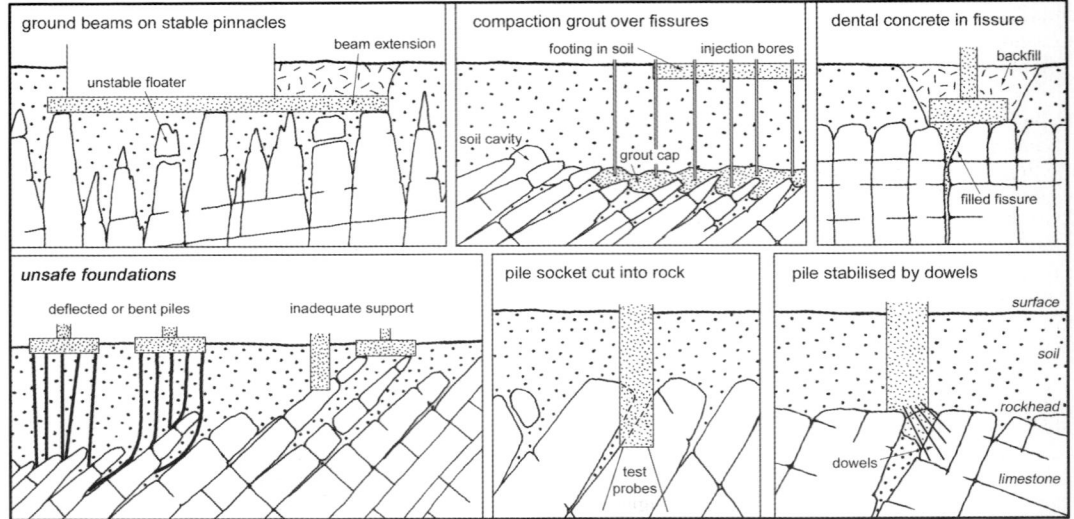

Figure 24.19   A variety of design options appropriate for foundations on karst (based largely on Sowers, 1996).

capacity — between strong rock and open caves. Soil collapses over unseen ground voids add to the problems, and every site has to be investigated as if it was unique. The hydrology of karst is also unlike any other, in that high groundwater flows are concentrated in conduits that make control and abstraction difficult.

References

Atkinson, T. C. and Smith, D. I. (1974) Rapid groundwater flow in fissures in the chalk: an example from south Hampshire. *Quarterly Journal of Engineering Geology* **7**, 197–205.

Beck, B. F. and Sinclair, W. C. (1986) *Sinkholes in Florida: an Introduction*. Florida Sinkhole Research Institute Report 85-86-4, 16pp.

Beck, B. F. and Herring, J. G. (eds) (2001) *Geotechnical and Environmental Applications of Karst Geology and Hydrology*. Balkema, Rotterdam, 437pp.

Beck, B. F., Pettit, A. J. and Herring, J. G. (eds) (1999) *Hydrology and Engineering Geology of Sinkholes and Karst*. Balkema, Rotterdam, 477pp.

Bögli, A. (1980) *Karst Hydrology and Physical Speleology*. Springer-Verlag, Berlin, 284pp.

Culshaw, M. G. and Waltham, A. C. (1987) Natural and artificial cavities as ground engineering hazards. *Quarterly Journal of Engineering Geology* **20**, 139–150.

Fookes, P. G. (1997) Geology for engineers: the geological model, prediction and performance. *Quarterly Journal of Engineering Geology* **30**, 293–431.

Fookes, P. G. and Hawkins, A. B. (1988) Limestone weathering: its engineering significance and a proposed classification scheme. *Quarterly Journal of Engineering Geology* **21**, 7–13.

Foose, R. M. and Humphreville, J. A. (1979) Engineering geological approaches to foundations in the karst terrain of the Hershey Valley. *Bulletin of the Association of Engineering Geologists* **16**, 355–381.

Ford, D. C. and Williams, P. F. (1989) *Karst Geomorphology and Hydrology*. Unwin Hyman, London, 601pp.

Jappelli, R. and Liguori, V. (1979) An unusually complex underground cavity. *Proceedings of International Symposium on Geotechnics of Structurally Complex Formations,* Associazione Geotecnica Italiana, Rome **2**, 79–90.

Jennings, J. N. (1985) *Karst Geomorphology*. Blackwell, Oxford, 293pp.

Lowe, D. and Waltham, T. (2002) A dictionary of karst and caves (2nd edn), *British Cave Research Association Cave Studies* **10**, 1–40.

Newton, J. G. (1987) Development of sinkholes resulting from man's activities in the eastern United States. *US Geological Survey Circular* **968**, 1–54.

Quinlan, J. F. and Ewers, R. O. (1989) Subsurface drainage in the Mammoth Cave area. In White, W. B. and White, E. L. (eds) *Karst Hydrology: Concepts from the Mammoth Cave Area*. Van Norstrand Reinhold, New York, 65–103.

Schmidt, L. A. (1943) Flowing water in underground channels, Hales Bar dam, Tennessee. *Proceedings of the American Society of Civil Engineering* **69** (9), 1417–1446.

Sowers, G. F. (1996) *Building on Sinkholes*. American Society of Civil Engineers, New York, 202pp.

Sweeting, M. M. (1972) *Karst Landforms*. Macmillan, London, 362pp.

Villard, P., Gourc, J. P. and Giraud, H. (2000) A geosynthetic reinforcement solution to prevent the formation of localized sinkholes. *Canadian Geotechnical Journal* **37**, 987–999.

Waltham, A. C. and Fookes, P. G. (2003) Engineering classification of karst ground conditions. *Quarterly Journal of Engineering Geology Hydrogeology* **36**, 101–118.

Waltham, T, Bell, F. and Culshaw, M. (2005) *Sinkholes and Subsidence*. Springer-Verlag, Berlin and New York, 382pp.

Waltham, A. C. and Smart, P. L. (1988) Civil engineering difficulties in the karst of China. *Quarterly Journal of Engineering Geology* **21**, 2–6.

Waltham, A. C., Smart, P. L., Friederich, H. and Atkinson, T. C. (1985) Exploration of caves for rural water supplies in the Gunung Sewu karst, Java. *Annales de la Société Géologique de Belgique* **108**, 27–31.

Waltham, A. C., Vandenven, G. and Ek, C. M. (1986) Site investigations on cavernous limestone for the Remouchamps Viaduct, Belgium. *Ground Engineering* **19** (8), 16–18.

Wittke, W. and Hermening, H. (1997) Grouting of cavernous gypsum rock underneath the foundations of the weir, locks and powerhouse at Hessigheim on the River Neckar. *Proceedings of 19th Congres, Florence*, Commission Internationale des Grands Barrages, 613–626.

White, W. B. (1988) *Geomorphology and Hydrology of Karst Terrains*. Oxford University Press, 464pp.

Yuan, D. (ed.) (1991) *Karst of China*. Geological Publishing House, Beijing, 232pp.

Further reading

Gunn, J. (ed.) (2003) *Encyclopedia of Caves and Karst Science*. Routledge, New York, 950pp.

Palmer, A. N. (1991) Origin and morphology of limestone caves. *Geological Society of America Bulletin* **103**, 1–21.

Waltham, A. C. (1989) *Ground Subsidence*. Blackie, Glasgow, 202pp.

Wilson, W. L. and Beck, B. F. (1988) Evaluating sinkhole hazard in mantled karst terrane. *American Society of Civil Engineers Geotechnical Special Publication* **14**, 1–24.

25. Loess

Edward Derbyshire and Xingmin Meng

25.1 Introduction

Loess is a clastic sediment formed from wind-lain deposits predominantly of silt-size particles that have subsequently been altered to varying degrees by diagenetic processes, principally weathering and pedogenesis (soil formation).

The dominant mineral in most of the world's loess is quartz (~50–70%), with a significant content of feldpars and mica, as well as minority carbonates and clay minerals. In some loess regions of the world, however, distinctive source materials yield quite different mineralogical populations. This is notably the case adjacent to active volcanic belts, which inject rock fragments and glass shards into accumulating loess, as in Argentina and New Zealand for example.

The quartz grains making up the loess 'skeleton' are mainly angular to sub-angular and blade-like in form. Such characteristics, taken together with the characteristic airfall depositional process, are conducive to a rather loosely packed, moderate to poorly sorted deposit lacking any visible stratification, and with distinctive geotechnical behaviour.

Surface detachment, entrainment and transport of mineral dust by the wind are a function of several variables, notably critical wind speed, particle exposure (for example when freshly deposited — seasonally, and when a vegetation cover is lacking as in dryland regions) and particle size. Dust storms vary in scale between local vortices ('e.g. dust devils', 'willy-willies', etc.) only a few metres in diameter that raise dust for a few seconds to a few hours, and extensive atmospheric turbulence associated with air-mass frontal systems that may transport mineral dust for great distances. Air-mass frontal systems associated with the hemispherical wind regimes, notably the upper Westerlies, frequently transport terrestrial dust across oceans including the Atlantic and Pacific, with deposition occurring some two or three weeks after initial entrainment (Pye, 1987). It is important to discriminate between regional and extra-regional events (Pye, 1995), as the former occur at lower atmospheric levels and transport coarser silts while the latter carry finer fractions at high levels within the troposphere and are readily observed on earth-orbiting satellite images (Figure 25.1). In North China, for example, the driving force is the north-westerly hemispherical wind system controlled by the Mongolian and Siberian high pressure anticyclonic systems and the Aleutian low pressure cell. The regional transport path (source proximal) in this case is broadly north-west to south-east, as demonstrated by the systematic fining of loess grain size across the Chinese Loess Plateau from sandy loess in the north-west to 'typical' loess and then, in the south-east, to clayey loess (Figure 25.2). The long distance, high altitude transport path (source distal) is one in which central Asian dusts are carried by the upper Westerlies sometimes as far as the western United States. A number of factors influence the location and rate of loess deposition, including trapping of dust by vegetation or surface water bodies, a reduction in wind speed and turbulence, formation of agglomerations caused by increase of atmospheric moisture and washing out of particles by precipitation (Pye, 1995), the latter process being well characterised by the ancient Chinese descriptor 'loess rain'. There is strong support for the view that long-term vertical accretion of dust at rates sufficiently high to form loess is likely when there is some surface vegetation cover (Tsoar and Pye, 1987; Pye and Tsoar, 1987).

The process relationship between wind-blown dust transport and loess deposition was recognised at least 2000 years ago in China (Liu, 1985), although this link did not become widely accepted

Figure 25.1 SeaWiFS (sea-viewing Wide Field-of-view Sensor) image, showing a large dust storm over and the leeward of the Chinese Loess Plateau (lower centre of photograph) on 16 April 1998. The outline of Chinese coastline, and the Bohai Gulf and East China Sea (largely covered by broken clouds) may be clearly seen in the lower right quadrant. The storm is driven by a cold front, with its outburst of cold air from the 'Mongolian High', that can be traced by the leading (eastern) edge of the dust cloud and its continuation as a zone of white cloud running northeast (towards the upper right of the photograph). Driven by a vigorous depression (low pressure system), some of the finer component dusts crossed the Pacific Ocean to reach North America on 25 April. (The SeaWiFS Project (Code 970.2) and the Goddard Earth Sciences Data and Information Services Center/Distributed Active Archive Center (Code 902) at the Goddard Space Flight Center, Greenbelt, MD 20771, USA, are gratefully acknowledged for the production and distribution of these data, respectively, within NASA's Earth Science Enterprise.)

in Europe and elsewhere until the publication of von Richthofen's work on the Chinese loess (Pye, 1995; Richthofen, 1877–85; 1882). Loess drapes, including the classic Chinese deposits, have been attributed to numerous other processes, including lacustrine and alluvial deposition. In addition, there is a strong body of opinion in eastern Europe and middle Asia favouring attribution of the term loess to aeolian silts only after they have been modified by *in situ* weathering (the process known as loessification: e.g. Lozek, 1965; cf. Smalley, 1971). However, a number of landscape and compositional characteristics leave no doubt about the wind-blown origin of loess, as Pye (1987; 1995), among others, has shown. Typically, loess mantles the landscape as a drape, and includes a variety of

relief forms extending over a wide altitude range. Both the thickness of this mantle and the loess mean grain size decrease progressively with distance from the silt sources. Tracts of loess and associated deposits such as wind-blown sands frequently bear a clear geographical relationship to present or former prevailing wind tracks. Loess is unstratified and lacks the pebble stringers so common in water-lain deposits. The mineralogical composition of loess is often quite different from that of the subjacent rocks or sediments, making an *in situ* origin unlikely. Accretion of loess by deposition of wind-blown dust can be observed in some parts of the world today (Figure 25.3), detailed study of individual, hemispherical-scale dust storm events being possible with the advent

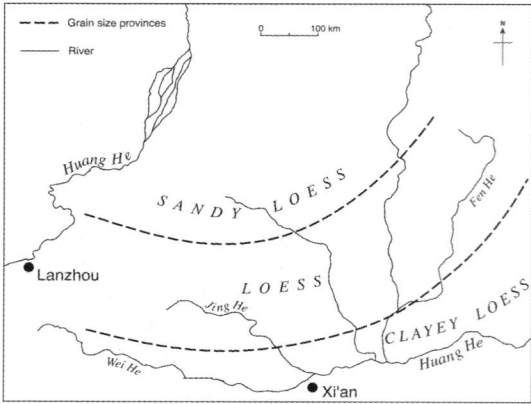

Figure 25.2 Gradational change in the particle size population from north-west to south-east across the Loess Plateau of China, expressive of the two major factors of dominant regional winds and the humidity gradient from arid in the north-west to sub-humid to humid in the south-east (after Liu *et al.*, 1964).

of high quality satellite imagery (Figure 25.1). Finally, the particle size distributions of typical loess and modern, source-adjacent aeolian dusts are identical (Pye, 1987).

Siltstones are found throughout the geological column, but conditions in the Quaternary (approximately the last 2 Myr) appear to have been particularly conducive to silt generation and its accumulation on the land surface, especially as alluvial silts and as loess. Although aeolian silts have been found on all the world's landmasses, large concentrations of loess are associated with particular regions of the globe. Of the estimated 10% of the earth mantled by loess, by far the greatest volumes are to be found in the great accumulations making up the Eurasian (western and central Europe, central Asian and north China), Siberian, North American and South American loess regions (Figure 25.4). Less extensive, although regionally important deposits are also found in New Zealand and Alaska. Also less extensive, but more variable in particle size composition, are the silt-rich deposits of fine sands associated with the world's sandy deserts, notably around the Sahara (Nigeria, Tunisia, Libya, Israel, Iran), frequently described as loess in the literature. In addition, thin, discontinuous loess

drapes are found in many mountain regions of the world, especially in High Asia (e.g. the Karakoram (Owen *et al.*, 1992) and the Anyemaqen Mountains of north-eastern Tibet) but also including loess in sub-Andean montane basins as in north-west Argentina (e.g. Sayago, 1995; Iriondo, 1997). Finally, a loessic silt component has been detected in some surface soils outside such generally recognised loess regions (e.g. Catt, 1978; 2001).

Numerous processes contribute to loess formation, but provenance (including particle generation as well as the geology of the source regions), transportation by the wind, deposition and post-depositional modification, especially by the climatically driven processes of weathering and pedogenesis, are of particular importance. Opinion varies on the question of the relative effects of these processes on some of the bulk properties of silt deposits. Wentworth (1933) favoured the transport mechanism as the major determinant of the grain size. More recently, attention has been directed particularly to wind transportation as a sorting mechanism (e.g. Tsoar and Pye, 1987) and a case has been made for regarding particle formation as at least as important a determinant of the grain size of silts as the transportation process (Assalay *et al.*, 1998). Fookes and Best (1969) favoured processes operating during deposition as the principal determinant of the engineering properties of loess. In extreme cases of modification of a unit of aeolian silt by advanced pedogenesis, the dominant control on bulk properties may be largely attributable to post-depositional events (see below). It follows, therefore, that careful account should be taken of the process history of individual loess units during site assessment.

When loess accumulates to a thickness that effectively masks some or all of the morphology of the underlying terrain, it assumes a series of terrain types that reflect its distinctive composition, structure and behavioural properties, especially in response to water and seismic shock.

25.2 Loess Terrain

The great loess deposits of the Earth mantle extensive areas of variegated terrain, but thicken

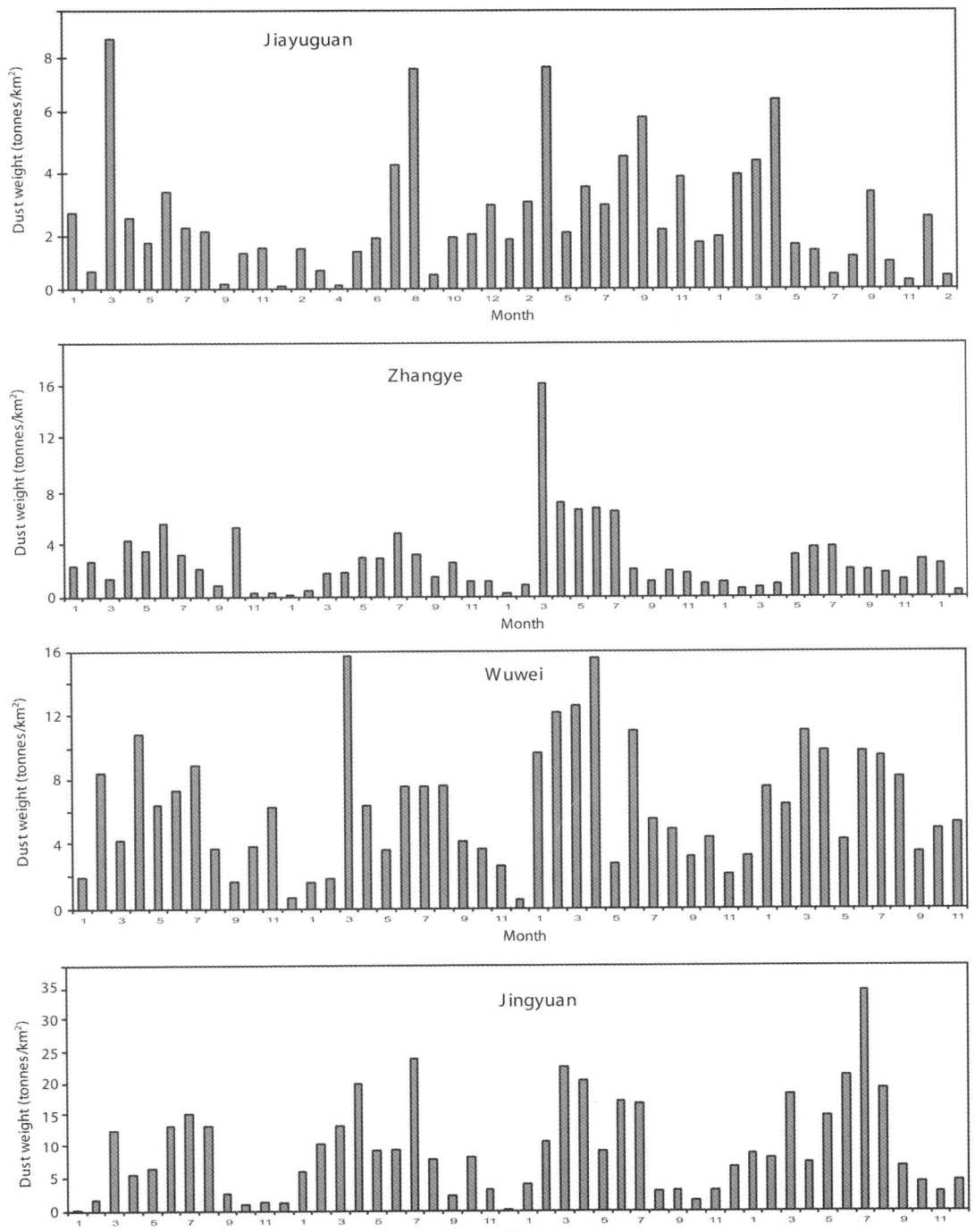

Figure 25.3 The year-round pattern of dust falls at four stations from WNW (top) to ESE (bottom) along the Hexi Corridor (west of the Chinese Loess Plateau) for the four-year period 1988–1991 (after Derbyshire *et al.*, 1998).

significantly in broad basins and on gently sloping plains and plateaus. The thickest loess on Earth has accumulated in the continental interior basins of Eurasia, as part of the great loess belt that stretches from the Atlantic coast of Western Europe to the Yellow Sea (Figure 25.4). Within this 10 000 km long tract, the Chinese loess reaches a thickness of *c*. 500 m towards the western margin of the Loess Plateau, and the central Asian loess is about 200 m thick in the Tajikistan–Uzbekistan region.

At the macroscopic scale, the geomorphology of loess terrain is relatively simple, consisting of a limited number of elements. Unlike the distinctive suite of depositional landforms associated with aeolian sand, for example, the surface forms of loess landscapes rarely reflect dominant wind direction. Thin loess (a few metres in thickness) commonly drapes rocky footslopes and terrace surfaces, locally masking irregularities in the underlying bedrock. Thick loess (tens to hundreds of metres thick) is found on plains and within basins where it may completely mask the subjacent relief; such plateau-like loess has extensive planar surfaces, interrupted to varying degrees by the action of water, and other processes including tectonics. The major morphological features found in thick loess terrain are thus plateaus, fluvially cut valleys and interfluvial ridges and, with more advanced degradation, linear ridges and conical hills. Long slopes in loess exposed by rapid river incision, faulting and related mechanisms are vulnerable to degradation by landslides, subsidence, mass flowage and associated processes, especially in environments with high-magnitude, low-frequency rainfall events and seismic activity. Complexity is introduced in loess terrain by certain ground and surface water regimes, giving rise to distinctive

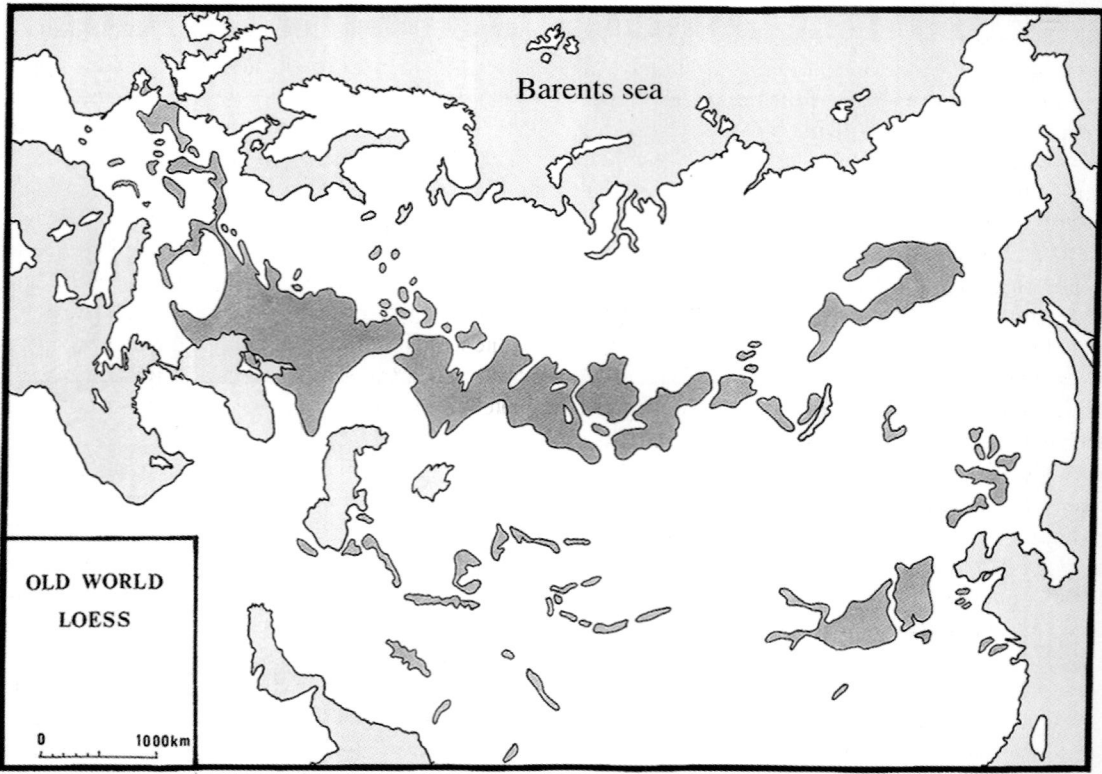

Figure 25.4 The distribution of thick loess in Eurasia, indicated by the darker shading (abstracted and simplified from the map in Rózycki, 1991).

and sometimes complex morphology at the field scale. In general, complex loess morphology is a function of the bulk properties of loess, and it is this relationship that is emphasised here.

The regional-scale morphology of loess is best illustrated by the thick loess terrains in central and eastern Asia. The Chinese Loess Plateau is a classic region exemplifying most of the characteristics arising from the fundamental relationship between the bulk properties of loess and the morphology of loess at all scales of enquiry. The macroscopic morphology of this region is traditionally characterised using the descriptive native terms 'yuan', 'liang' and 'mao'. *Yuan* are flat plateaus in thick loess, best seen in southern Shaanxi and western Gansu Provinces (Figure 25.5*a*). *Liang* are ridges, common in northern Shaanxi and Shaanxi Provinces (Figure 25.5*b*). A *mao* is a rounded loess hill; *mao* hills often occur in liang country, and dominate the landscape only east of the Yellow River (Hwang He) in Shanxi Province. Although many view this set of forms as an erosive sequence, neotectonics has certainly played a role in its development. A fourth morphological type, termed 'cone and plain', was recognised on the western edge of the Loess Plateau north of Lanzhou (Derbyshire, 1983a). Broadly similar landform suites occur in other regions of thick loess, including central Asia and North America, albeit often on a smaller scale.

Lohnes and Handy (1968) proposed a general model for slope form and type in loess terrain that takes into account the bulk behaviour of loess and the surface and sub-surface processes, notably the hydrology. Fluvial incision by gully systems leads to two common slope elements. First, there is a steep upper facet with gradients of 70–85° that reflects sets of stress relief joints induced by tensional stresses near the ground surface. Retreat of this upper cliff is largely by slab fall and sliding. The variously disaggregated slabs of loess mantle the basal shear plane sloping at 51–59°, depending on the values of φ' and c'. Accumulation of these slabs and blocks may obscure the lower shear slope at, or close to their angle of repose (38°), explaining the large number of observed slopes in the range 34–38°. Many active gullies in thick loess terrain maintain a rectilinear cross-profile as a result of an irregular annual cycle of flood discharges in the summer wet season followed by a long winter dry season in which gully-side slopes relax and readjust (Figure 25.6), in some regions with the help of additional processes such as freeze–thaw and snowmelt. More stable loess slopes in areas protected from active undercutting show varying degrees of departure from the model slope set.

The dominant slope facet on many stable *liang* and *mao* forms in North China frequently lies in the range 35–45°. Aside from the flat-topped (*yuan*) surfaces, many slopes in this region thus fall into the third category of Lohnes and Handy. Derbyshire *et al.* (2000), working on the very thick loess in the western part of the Loess Plateau, showed that the slope population is almost normal in distribution, with a modal category at 36–43°, and a principal mode of 39°. There are conflicting views on the question of the optimal gradient for cut slopes in loess. Vertical slopes, believed by some to minimise erosion by water, are subject to shear failure of their upper parts in uniform, thick loess (> 10 m) as an unloading response not necessarily related to rainfall events (see 'Mass movement types' below). In North China, slopes are commonly cut at grades of 1 : 1.2 to 1 : 0.25, depending on age, slope height, local mean annual precipitation and loessic facies (notably primary air-fall loess and reworked material — loessic alluvium). The commonest grade used by engineers in the Loess Plateau of China is 1 : 0.75 (*c.* 53°, which approximates the basal shear plane angle of Lohnes and Handy, 1968).

The relationship between formative wind systems and loess landscapes is subtle and, in detail, quite complex. A distinctive zoning in the particle size of loess from source regions to the most distal loess-covered terrain has been demonstrated in many loess regions of the world, including China and Tajikistan–Uzbekistan. In China, the north-west to south-east gradient in climate from the desert basins west and north-west of the Yellow River, and the dominant north-westerly winds in winter and spring, are reflected in a progressive fining of the mean grain size of the surface aeolian deposits from sandy silts near the sources to

Figure 25.5a Thick loess mantling the landscape to form a plateau (*yuan*), and dissected to varying degrees to produce deep valleys. Zhenyuan, Gansu Province (reproduced from Wang and Zhang, 1980).

Figure 25.5b Ridges (*liang*) in the thick loess near Ansai, Shaanxi Province. Note the alternation of groups of thick palaeosols (dark) between loess units, the deep incision and the parallel valley-side gullies (reproduced from Wang and Zhang, 1980).

Figure 25.6 Active erosion of loess by a tributary of the Yellow River (Hwang He), *c.* 50 km SW of Lanzhou city, western Loess Plateau. The channel is characteristically flat-floored, with classic 'upper cliff and lower debris slope' morphology on the channel sides (centre and right). There has been notable re-entrainment of loessic silt by fluvial action, as can be seen in the bedded silts in the bluff ('loessic alluvium': left of photograph).

wind-lain silts ('typical' loess) to progressively more clay-rich loess towards the south-east (Liu, 1985). Similar systematic changes in sedimentary properties with distance from source have been demonstrated for loess units of several ages back to *c.* 7.7 Myr (Ding *et al.*, 1998). In addition to this effect of differential sorting during transport, enrichment with clay minerals is progressively important south-eastwards as rates of weathering and pedogenesis increase as climatic conditions become warmer and wetter.

Considerable volumes of silt enter the geomorphological system from the seasonal melting cycle in both high, glaciated mountains and lower, seasonally snow-covered hills. Using northern China as a basis for a general model (Figure 25.7),

source areas are characterised by a mixed suite of landforms including sand plains and dunes, extensive (up to 50 km in diameter) silt-rich alluvial fans and silt river terraces and plains. Mixed gravels, sands and silts are commonly differentially eroded by the wind, frequently leaving a protective and stable surface lag known by the Mongolian word *gobi* (stony desert). The regional gradient in mean particle size is also reflected in a progressive change in landform types. On some piedmont fans within or close to silt source areas, locally reworked and alluvially re-deposited loessic silts are found inter-bedded with sand and gravel units. In rare instances, relatively thin drapes of sandy loess show linear forms parallel to dominant wind directions, and some small barchan-like forms

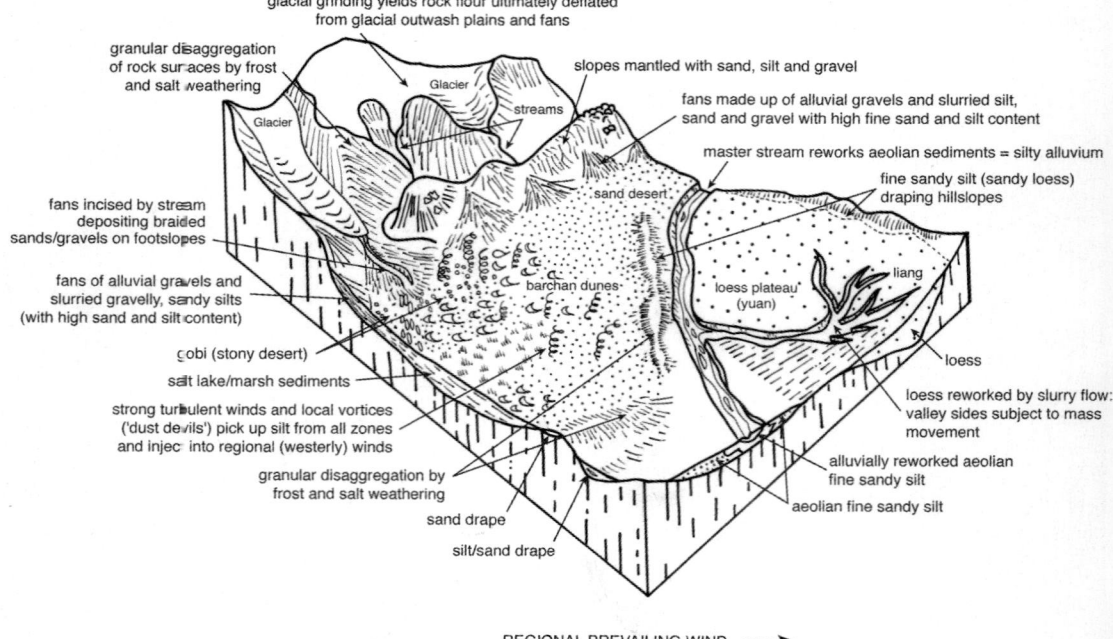

Figure 25.7   A generalised model of a mid-latitude loess landscape, based on northern China, showing glacial, nival and seasonal alluvial sources, entrainment, the relationship to sand dune desert and plateau-like accumulation of thick loess and its dissection.

have been reported from the mountains of central Asia (Keš, 1972). Some thin (< 1 m) sandy loess drapes mantle steep (> 20°) rock slopes near source areas, as in the eastern Kunlun Shan in Tibet, attesting to its direct aeolian accretion (Figure 25.8). With increasing distance from the sources, typical loess may accumulate to great thicknesses. However, loess thickness is affected significantly by the underlying geomorphology in the depositional zone. The thicker loess areas in northern China, for example, have been correlated with broad, usually structurally defined bedrock basins (Liu, 1985; Zhu and Ding, 1994). It is in such areas that the thick loess-palaeosol sequences, as seen in Asia, are cut by river systems, exposing long slopes subject to several mass failure processes, so completing the model landform sequence (Figure 25.7) Outside such favourable regions, as in plateau and mountainous bedrock terrain, the deposition–erosion balance may be more variable, leading to thinner loess and greater fluvial reworking, but often adding earthquake shock as

a trigger causing failure of loess slopes (Dijkstra *et al.*, 1993).

Given the high rainfall intensities required to saturate and mobilise the surface of sparsely vegetated thick loess (Muxart *et al.*, 1994), slopes arising from mass slurrying of loess are relatively rare, but rilling and gullying become progressively more notable in the finer, clay-rich (leeward) loess zone, as exemplified by the southern and eastern Loess Plateau of North China. The combination of joint enlargement by water, suffosion, development of sinkholes and major systems of large pipes and associated subsidence, together with gravitational failure by toppling, falling and sliding (as thin loess zones above clay-enriched bedrock or palaeosols give rise to localised liquefaction of the base of the loess), are the major land-forming processes in loess terrain. Rivers that drain thick loess landscapes are among the most turbid on earth.

The tendency to collapsibility in loess affects plain and plateau surfaces well away from

Figure 25.8   Accretion of loess of uniform thickness (25–40 cm) draping a rectilinear rock slope in the eastern Kunlun Mountains, Tibet. Sub-parallel rills eroding the loess drape can be seen on the slopes above and to the left of the Kirghiz herdsmen.

hillslopes. Movement along active faults certainly plays a role in generating large and potentially destructive fissures in flat loess terrain, although severe seasonal rainfall has similar capability as shown by fissures > 1 km long and up to 14 m wide in China's Shaanxi province. Such fissure systems, often enhanced by extraction of groundwater for domestic and industrial purposes, are a continuing threat to the great historical city of Xi'an, where subsidence in the period 1962–83 totalled 777 mm, with measured annual extreme values of up to 123 mm/yr (Sun, 1988).

25.3 Engineering properties of loess

The behavioural properties of loess are influenced by several factors, past and present climate being particularly important. Often regarded as a homogeneous material, the bulk properties of loess may vary in important respects with age, location, and site. Determination of the depositional and weathering history of loess underlying a site may thus provide a deeper appreciation of the range of bulk properties and the likely behaviour of specific loess units.

The history and type of source rock (e.g. granites, gneisses, vein-quartz) dictates the concentration of defects within quartz, the dominant mineral in much of the world's loess. These 'Moss defects' may control the minimum size attainable during natural crushing and abrasion (the 'comminution limit') of quartz (Moss, 1966; Moss and Green, 1975), thought to lie in the range 10–30 μm. This may explain differences between loess derived from the Laurentian and Baltic shields, and loess adjacent to the Himalaya and Tibet. Other regional distinctions arise from different process histories, such as the dominance of glacial comminution as a source in the mid-western loess of North America.

The *adobe* of the desert south-west of the United States (Rogers and Smalley, 1996) is an example of a non-glacial loess. The combination of a suite of processes including tectonic crushing, frost and salt weathering and glacial grinding, as in the High Asia region including Tibet and the Himalayan tract, is the source of the greatest piles of loessic silts on Earth.

Particle shape

The quartz grains making up the skeletal framework of much of the world's loess show a mixture of forms including equant, tabular and blade shapes, with the latter two being dominant (Smalley, 1966a; Krinsley and Smalley, 1973; Derbyshire, 1988). Rogers and Smalley (1993) obtained probabilities of 1% cubic, 26% tetragonal/disc and 73% blade shapes, the axial ratios of the latter category being 8 : 5 : 2. Such grain morphology is typical of loess in Asia and parts of eastern Europe, for example, the occasional rounded or irregular forms probably being attributable to partial weathering of non-quartz components and the varying extent of cementation or clay minerals (Derbyshire *et al.*, 1988). Most loess particles are transported in suspension and thus lack the edge-rounding and 'frosted' surface texture so characteristic of aeolian sand grains.

Particle size

The term 'loess' has been applied rather loosely to a broad range of silt-rich sediments. Many of these are calcareous, with a broad particle size envelope in which clay-size content varies between *c*. 8 and more than 30%, and sand may exceed 60%. Published mean grain sizes of material described as loess range from at least 5.2 to 8 *phi*, with sorting coefficients (Folk and Ward, 1957) between 3 and 1.3. Views on the mean or modal particle size of loess have been greatly influenced by regional factors. For example, in regions dominated by ice sheet glaciation in the Quaternary (notably Western Europe and North America) the modal grain size of loess appears to average about 30 μm (5 *phi*). In central Asia and North China (where loess often occurs along sand desert margins), in contrast, the modal grain size

is 15–20 μm (6 *phi*). In northern Africa, aeolian silts are much coarser than the Asian average, being dominated by the fine sand size fraction, as in Tunisia (e.g. Coudé-Gaussen, 1987) and Libya (Assallay *et al.*, 1996), while they are finer than the central Asian–Chinese average in northern Nigeria (McTainsh, 1987). Assalay *et al.* (1998) provide a detailed discussion of the possible role of silt formation processes in contributing to such diversity in silt sizes from region to region.

The range of variation in the particle size envelopes of loess from central China, Tajikistan, Uzbekistan, Italy, France and North Africa is considerable (Derbyshire *et al.*, 2000; Figure 25.9*a*; Figure 25.9*b*). Fine aeolian deposits referred to as loess from some parts of the climatically diverse region of Western Europe and North Africa may contain up to 70% fine sand grade and only *c*. 20% silt (Figure 25.9*c*); elsewhere in Europe some deposits regarded as loess contain 50% clay grade (< 2 μm) material. Inter-regional comparisons of grading envelopes for loess from China,

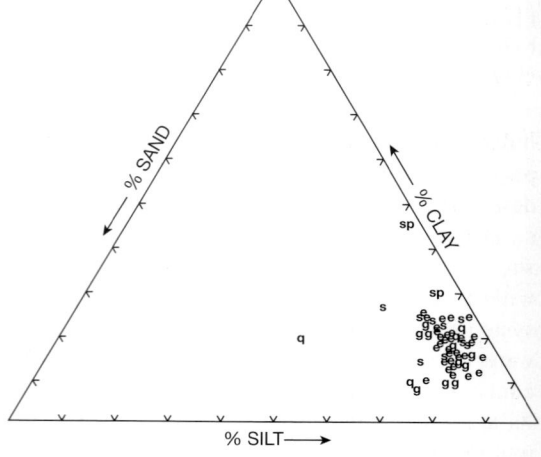

Gansu, Shaanxi and Qinghai

g and e :   Lanzhou loess (Gansu)
s :   Luochuan loess (Shaanxi)
sp :   Luochuan palaeosol
q :   Qinghai silts
no. = 55

Figure 25.9*a*   Ternary diagram showing typical closely clustered envelope for the Chinese Loess Plateau (with some divergence in Qinghai, Tibet) (after Derbyshire *et al.*, 2000).

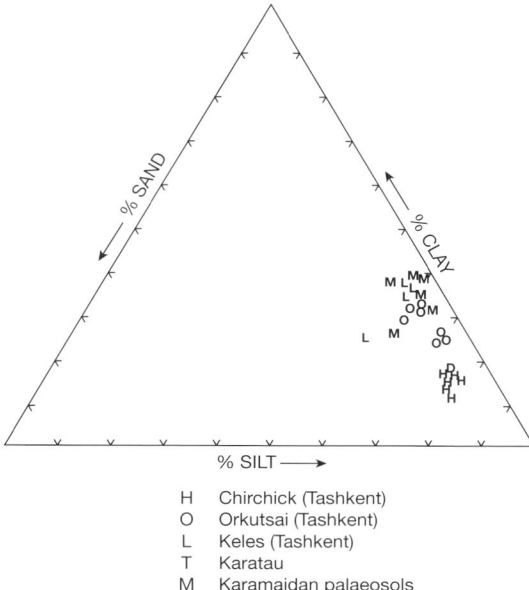

H   Chirchick (Tashkent)
O   Orkutsai (Tashkent)
L   Keles (Tashkent)
T   Karatau
M   Karamaidan palaeosols
D   Karadarya (Samarkand)
no. = 26

Figure 25.9*b*   Ternary diagram showing consistent, but less closely clustered envelope for the loess of Tajikistan–Uzbekistan (after Derbyshire *et al.*, 2000).

Western Europe and North Africa present some interesting contrasts (Figure 25.10).

Most loess is poorly to very poorly sorted when measured in the laboratory using conventional indices such as those of Folk and Ward (1957) (cf. Derbyshire *et al.*, 1993), an apparently anomalous outcome in view of the apparent effectiveness of wind as a sorting agent. Once deposited, many of these silt-sized aggregates are broken down and redistributed (dispersed or redeposited in higher concentrations) in response to overburden pressures, shearing, leaching and diverse reactions during *in situ* weathering. Thus, standard granulometric methods present the constituent rather than the dynamic grain size of loess (cf. Derbyshire *et al.*, 1980, p.185).

Particle size distribution also varies with age, i.e. the depth-down profile. The mean grain size of the North China loess lies in the range 10–60 μm (*phi*-value range 5.0–8.0; Derbyshire *et al.*, 2000: cf. 25 μm mode of Jefferson and Smalley, 1995). The older loess formations are clearly finer than

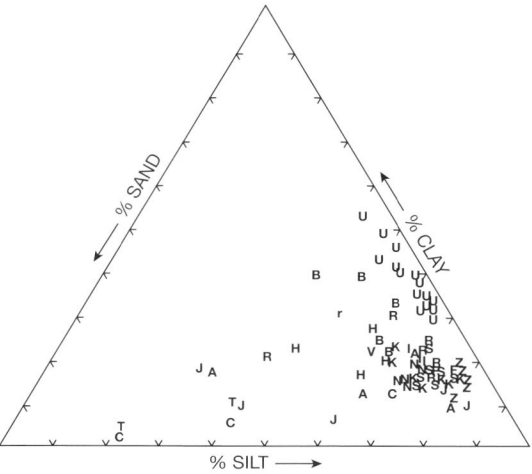

W. Europe and Tunisia

B   Barlassina (Italy)
C   Copreno (Italy)
V   Bella Vista (Italy)
R   Rocourt (Belgium) (r = Rocourt palaeosol)
H   Harmignies (Belgium)
K   Kesselt (Belgium)
N   Nuth (Netherlands)
S   Saint Roman (Normandy)
I   Iville (Normandy)
F   Saint Pierre les Elbeuf (Normandy)
J   Jersey (U.K.)
A   Kent (U.K.)
U   Primarette (SE France)
Z   Saint Prim (SE France)
T   Matmata Plateau (Tunisia)
no. = 72

Figure 25.9*c*   Ternary diagram showing the much more divergent particle size pattern found in Western Europe and Tunisia (after Derbyshire *et al.*, 2000).

the Late Pleistocene loess (Figure 25.11*a*). Mean versus sorting co-plots reveal several loess groups ranging from better-sorted and coarse sandy loess, through the 'typical' loess (the 'D1' population at 25–30 μm of Jefferson and Smalley, 1995: cf. Browzin, 1985), to fine-grained and very poorly sorted palaeosols (Figure 25.11*b*).

Soil fabric

The granular skeleton of silt-sized particles usually show variable amounts of finer particles occurring as both loosely adhering clusters and cements, as buttresses and bridges linking the skeletal grains, and as occasional silt-sized aggregations. Fabric refers to all directional

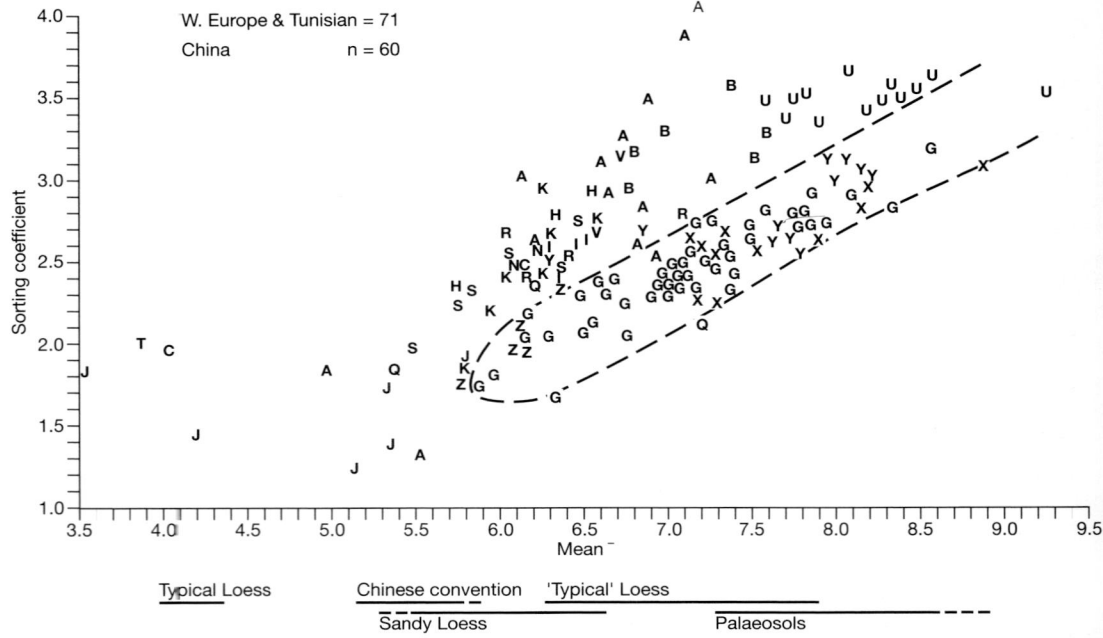

Figure 25.10   Plot of mean particle size versus sorting coefficient for samples of loess and palaeosols from China and Western Europe. The consistency of the envelope for Gansu province ('G'), China, contrasts with the rather diffuse Western European envelope (site nomenclature as in Figure 25.9c). The grain size median zone for 'typical' Western European loess (left, horizontal axis) and western Chinese Loess Plateau loess (bold print) are also shown.

properties of a sediment (Derbyshire *et al.*, 1976), although particle orientation and packing (Rogers *et al.*, 1994a) have attracted most attention in geological and geotechnical studies of loess. Fabric is influenced by several factors including grain shape, particle size distribution and the stress history of the material.

Following deposition of wind-blown mineral dust as a loosely packed agglomeration of angular, platy quartz grains and silt-sized aggregates of fine silt and clay-sized material, several surface processes progressively affect changes in the deposit. Wetting and drying induce flocculation of constituent fines, as cationic concentration increases with any decline in porewater content (Grim, 1953). The fine silts and clay-sized grains making up aggregates are drawn to pore margins by porewater menisci, yielding fine-particle bridges, buttresses and adhering aggregates (Derbyshire, 1984). Soluble salts become redistributed to form inter-particle cements and silt

grain coatings, and redistribution of clay-sized particles often results in aggregation and clay coatings on larger grains (cf. Grabowska-Olszewska, 1988; Derbyshire and Mellors, 1988; Billard *et al.*, 1993).

The sequence dispersion — redistribution–cementation increases fabric compaction and strengthens inter-particle bonds in some horizons. In addition, the process of normal consolidation is associated with inter-granular shearing. Such disruption of aggregates in pore fillings may further strengthen loess structure. Increasing normal loads or porewater pressures lead to irreversible collapse of the characteristic edge-to-face and face-to-face fabrics considered syngenetic in undisturbed loess (Derbyshire and Mellors, 1988; Gao, 1983; Derbyshire *et al.*, 1988). Eluviation of the finer particles tends to break down the characteristic clay bridges, which do not reform in the horizons below. Loess fabrics thus show both intra- and inter-site variation.

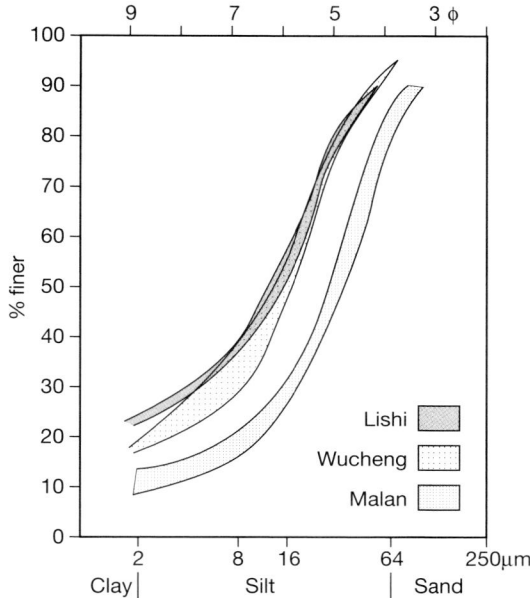

Figure 25.11a   Particle size variation with age and degree of soil development in the loess of the western Loess Plateau of China: particle size envelopes for Malan (youngest), Lishi and Wucheng (oldest) loess units (after Derbyshire and Mellors, 1988).

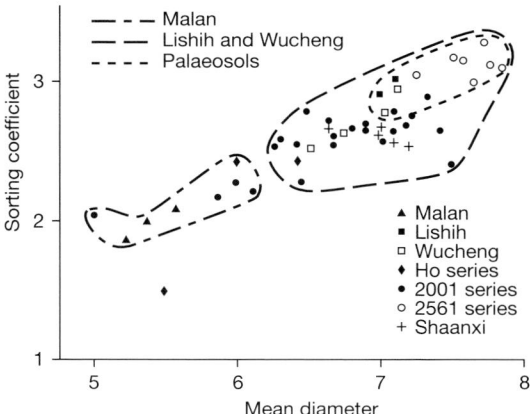

Figure 25.11b   Discrimination, using a co-plot of mean versus sorting coefficient, between older loess with palaeosols, and the younger loess (after Derbyshire and Mellors, 1988).

The progression of fabric changes in the loess and intercalated palaeosols from the drier western margins of the Chinese Loess Plateau to the warmer and more humid east have been described and illustrated using both optical thin sections (micromorphology) and scanning electron microscopy (Derbyshire *et al.*, 1995a).

The loess of Western Europe also varies with age and geographical location. Loosely packed fabrics in coarse to medium angular silts, with little clay mineral content and limited, often localised cementation (comparable to much Asian loess) can be found at many sites in continental Europe (Figure 25.12). However, fabric varies with climatic diversity and past variations, shifting soil water regimes from saturation to desiccation (including freeze-drying associated with cryoturbation: van Vliet-Lanoë *et al.*, 1984), bioturbation, leaching and re-deposition, snow meltwater infiltration, mineral weathering, natural loading and unloading, and reworking by running water and mass movements. Particle packing, the nature and degree of cementation, and the content of clay-grade material provide a short but useful list of comparative indices. An additional variable in loess fabrics is preferred fabric trends (anisotropic fabrics). These vary from visible lamination (as in alluvially re-deposited loessic silts: Derbyshire 1983b) to strongly parallel particle fabrics, such as those generated by cyclic freezing and thawing (Figure 25.12). The quartz grains in the loess of Normandy, France and the nearby Channel Islands appear rounded because of an almost complete clay encrustation rich in Si, Al and Fe (Figure 25.13a). *Limon à doublets*, a loess facies that

Figure 25.12   (*Continued*).

Figure 25.12 (*a*) Scanning electron micrograph of Last Glacial loess from Göllersdorf, *c.* 30 km NW of Vienna, Austria, showing a fabric skeleton made up of very angular quartz silt particles relatively poor in clay minerals. The rather coarse grain size, angularity of grains and generally 'clean' appearance are a product of proximity to a source characterised by a grain crushing geosystem (glacier and ground ice in the nearby Alps), as well as a freely draining fabric. Sample provided by David Boardman. (*b* and *c*) Fabrics in loess produced by the development of ground ice. Both micrographs are from samples located 1 m below the surface at Kesselt, Belgium. This calcareous loess dates from the upper part of the Last Glacial. The distinctive fabric produced by segregated ground ice development is shown. Sub-horizontal lens-like structural units are delineated by overlapping, slightly curved, sub-horizontal discontinuitities that mark the surfaces of former segregated ground ice lenses. (*c*) In this detailed micrograph (scale bar 20 μm), the size sorting affected by ice segregation is clearly evident, with coarser silts at the top and bottom of the micrograph. In between (lower centre), fine silts densely packed by ice-growth compression, and the characteristic thin clay film draping the top of this compact fine silt unit are clearly visible. Sample supplied by Sanda Balescu.

occurs across Europe as far as the Russian Plain, consists of thin, gently dipping alternations of clay-rich brownish laminae and grey, clay-poor layers between 1 mm and > 1 cm in thickness (Figure 25.13*b*). The grey layers often have high void ratios and lack clay bridges between silt grains (Figure 25.14), while both clay bridges at silt grain contacts and clay-coated silt grains are common in the brown layers, and sometimes more concentrated at the base (Figure 25.14). This distinctive loess is considered to be a periglacial facies.

The dominant feature of the macrofabric of loess is a system of sub-vertical joint sets. These arise from several processes including tectonic shock and dilation (associated with erosive unloading, natural and artificial), and are of importance because of the effects on permeability and bulk strength. At shallow depths, plant roots may directly affect vertical permeability and, in giving rise to sub-vertical contraction cracks, they may constitute an indirect factor. The relative importance of these factors varies regionally in

Figure 25.13*a* Loess from Green Island, Island of Jersey (UK), off the coast of Normandy. Although the particle size is similar to other lowland loess in Western Europe, the fabric is distinctive. The silts appear sub-rounded, but this arises from the presence of widespread surface mantle on the silt grains rich in Si, Al and Fe (as can clearly be seen in the upper right centre, where the coating has been broken during sample preparation). The absence of any scattered or aggregated fine silt and clay particles/aggregates indicates severe alteration of this shoreline loess.

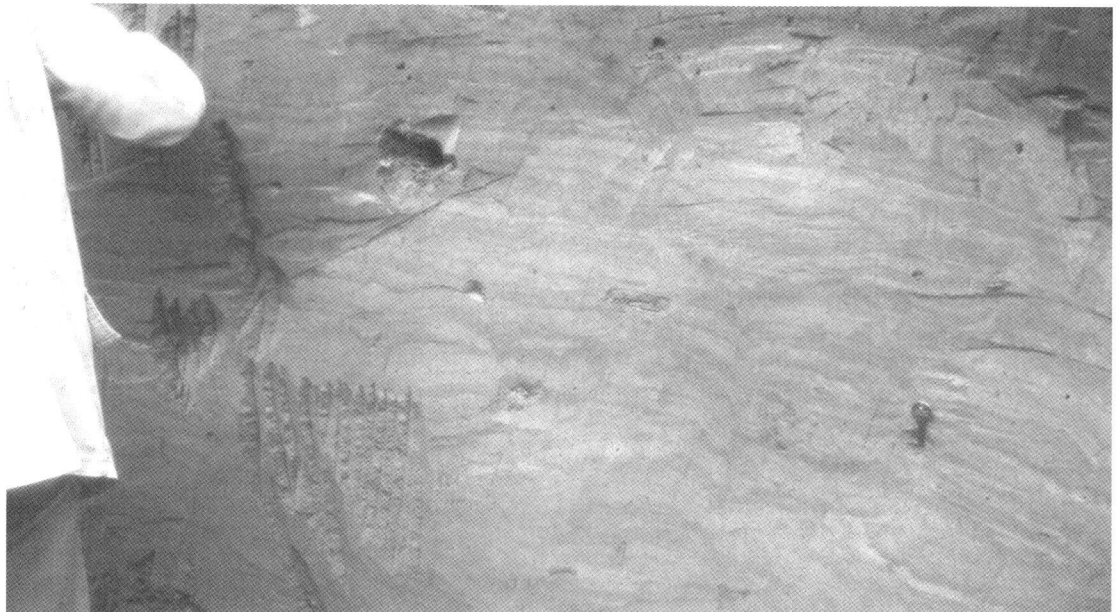

Figure 25.13b   An exposure of the distinctively banded *limons à doublets* at Port Racine, Normandy, France.

Figure 25.14   *Limon à doublets* from Belcroute, southern Jersey, UK Channel Islands: (*a*) dense lamina of the finer, more compact, clay-rich 'brown doublet' (lower part of photograph) compared to the more open fabric of the 'grey doublet' above it; (*b*) Sharp contact between grey doublet (top of photograph) and brown doublet. Some cemented bridges occur in both types, but the greater void ratios in the grey layers are clearly evident.

response to climate, seismic conditions and the erosion regime. In general, rates of vertical permeability (Kv) in loess are higher than horizontal permeability (Kh) rates; reported differences are high in relatively thin loess in which there is tree growth, but differences in thick, massive loess, such as that found in parts of dryland Asia decline rapidly with depth. In North China, Kv : Kh ratios decline with depth, there being little difference below 100 m (Derbyshire *et al.*, 2000). At the same time, marked variations in both Kv and Kh values and ratios arise from the presence of sub-horizontal discontinuities in the loess, notably intercalated palaeosols.

Mineralogy

In addition to quartz, a range of subsidiary minerals is present in loess, notably feldspars, micas, calcium carbonate and gypsum. Various heavy minerals (density > 2.7) also occur, but often constitute only a few percent by weight of the total mass. Calcite is an important constituent of much loess and certain horizons within loessic palaeosols, occurring as both clastic and secondary types. Secondary carbonate precipitates,

Figure 25.15   These two micrographs (from Longpré-les Corps Saint, Picardy, France) have a common centre point. They provide a good example of a Western European calcareous loess. It consists of a skeleton of coarse angular to sub-angular silt grains with calcite cementation, a moderately open structure, and well-developed, carbonate-strengthened inter-silt buttresses and bridges made up of aggregates of very fine silt and clay grade particles. Sample supplied by Sanda Balescu.

occurring as concretions ('loess dolls'), pore linings, encrustations and inter-granular cements, are an important source of strength in loess (Figure 25.15). Variations in calcium carbonate content in the loess of the western Loess Plateau of China are greatest in the palaeosol horizons. Billard and Muxart (2000) have shown that drying of the palaeosols within loess successions leads to supersaturation of the soil solution and precipitation of carbonate components and cementation bonds. This strengthens the loess skeleton and the aggregates of finer particles in particular. Leaching of such cements, and any associated clay minerals, is inhibited in semi-arid conditions, so preserving the openwork fabric. Carbonate contents also tend to rise with increasing loess age (Derbyshire and Mellors, 1988; Derbyshire et al., 1991). Such 'strength stratigraphy' is revealed in some cone penetrometer profiles (Figure 25.16). Pedogenic gypsum occasionally occurs as asymmetrical void fills (Kemp et al., 2001) and sometimes as small concretions (Derbyshire, 1983a) in the semi-arid parts of the western Loess Plateau of China.

Distinctive mineralogy is found in loess enriched with volcanic tephra, as in parts of South America. Near Córdoba, close to the western margin of the Argentine *pampa*, the loess has the following mineralogical composition: quartz 27%, volcanic glass 20%, altered feldspar 20%, plagioclase 14%, lithic fragments 10%, potassic feldspar 7% and biotite 2% (Iriondo and Kröhling, 1995). Loess situated at greater distances from volcanic centres, such as in the sub-Andean basins of the north-west of the country for example, contain only 3–6% of volcanic fragments and glass shards (Ovejero, 1980).

Dissolution of salts and carbonate precipitates during eluviation by percolating groundwater considerably reduces soil strength, a factor affecting the known high sensitivity values of loess (Derbyshire et al., 1993). In the dry state, carbonates act to strengthen loess, but at higher moisture contents they may be leached out, eventually contributing to collapse of the loess skeleton (Wang and Derbyshire, 1994).

Quartz is also frequently the principal mineral in the finest (< 2 μm) loess fractions, occurring

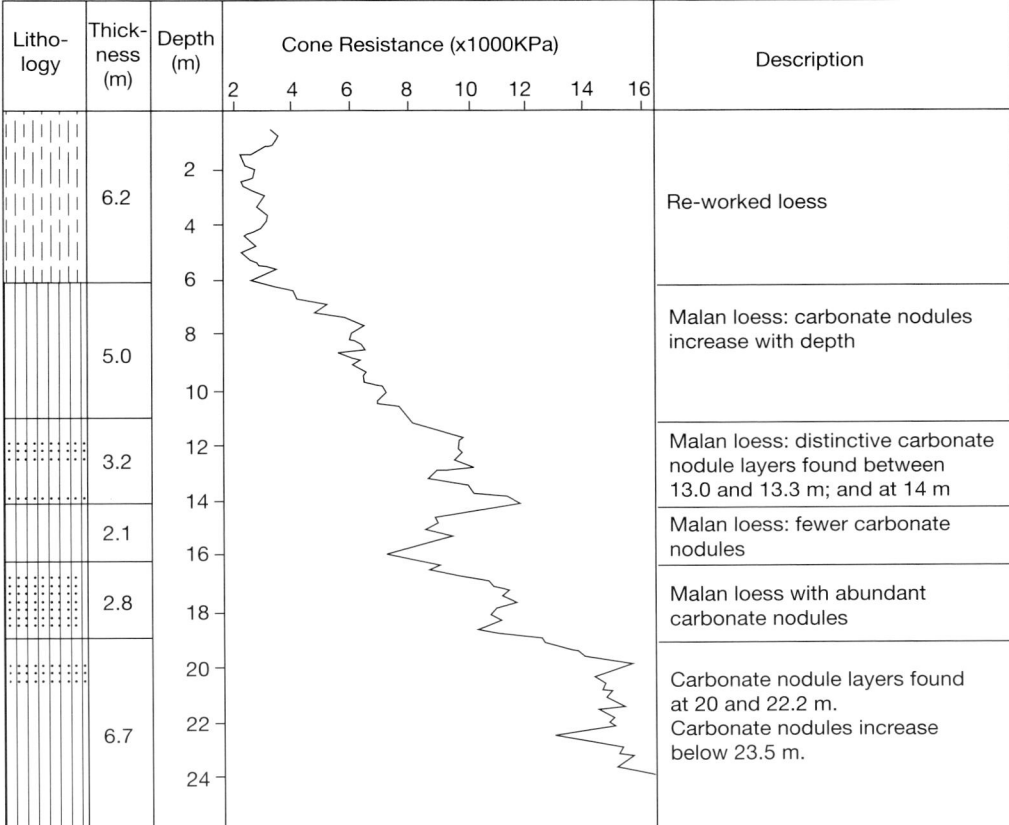

Figure 25.16 Penetrometer profile to a depth of 24 m in the Late Pleistocene loess near Lanzhou city, western Chinese Loess Plateau. The general increase of cone resistance with depth is clear, much of the variance arising from the presence of carbonate-rich horizons in the intercalated palaeosol units.

in a range from lithic fragments to biogenic opal. Kaolinite and illite are the most frequently occurring clay minerals, with chlorite, vermiculite, smectite and several mixed layer clays, but there are some significant regional variations arising from the dual influence of the rock composition in the dust source area and the effects of post-depositional weathering. The pioneering work of Frye *et al.* (1962) showed that the amount of montmorillonite in the Peoria (late Last Glacial) loess in the United States varies systematically depending on the different sources provided by different tributaries of the Mississippi River, rising to as much as 70% of all clay minerals present in the loess. The range of variation is generally less than this on the Chinese Loess

Plateau, although the present and past north-west to south-east climatic gradients in China are broadly reflected in the clay mineralogy (Derbyshire, 1983a; 1983b). In Argentina, the dominant clay mineral appears to be illite in both the Pampean (Karlson *et al.*, 1993) and the northwestern loess regions, with both kaolinite and smectite being minority minerals (Sayago, 1995). Smectite/illite is prominent in the loess of Poland (50–80%, being highest in the older loess units), with illite (up to 40%) and minor kaolinite (2–5%: Grabowska-Olszewska, 1988).

Determination of the mineralogy of the clay size fraction is an important precursor to geotechnical testing. For example, the tensile strength of loess is significantly affected by the presence of clay particles in response to a complex of variables. These include soil moisture content and the clay cohesion (including the degree to which clay mineral particles adhere both to each other and to the silt-sized primary mineral particles making up the silt skeleton: Smalley, 1966b; Derbyshire, 1983b; Tan, 1988; Derbyshire et al., 1994).

Natural moisture content

Ambient natural moisture content in loess is broadly controlled by the climatic regime, and specifically the annual water balance, but it is also affected to varying degrees by the detailed geomorphology, and the loess sedimentary characteristics including particle size, percentage clay content, degree of consolidation (age and depth) and the surface and sub-surface hydrology. Because of the complexity of patterns of hydraulic conductivity in some loess, the relationship between natural moisture content and loess thickness is not a simple linear function. Loess of different ages and different degrees of consolidation may have widely different mean natural moisture contents. For example, Lin (1995) and Derbyshire et al. (2000) provide mean values of 7.4–20.1% for the Loess Plateau of China; Phien-wej et al. (1992) quotes 8–12% for the 'fine sandy silt' of Thailand; Assallay et al. (1996) cites 2–6% for North African loess (Libya); Feda (1966) gives 14.8–17.0% for the Czech Republic; Lautridou (1993) cites 17–25% for Normandy (France), and Grabowska-Olszewska (1988) quotes 3–26% for the diverse loess deposits of Poland.

Bulk density

The specific gravity of loess particles varies according to the mineralogy of the source materials, much lying in the range 2.5–2.8. Bulk density values increase with age, and so with several other factors, most notably degree of consolidation. Values for the loess of the western Loess Plateau of China are $1.38 \, \text{Mg/m}^3$ (Upper Pleistocene), 1.56 (Middle and the upper part of the Lower Pleistocene) and 1.71 (Lower Pleistocene: Derbyshire et al., 2000). Porosity values show the same trend (0.53, 0.47 and 0.43, respectively). Palaeosols have the highest bulk densities (c. 1.70 ± 0.05). Similar trends occur elsewhere:

$1.48–1.69 \, \text{Mg/m}^3$ in Kent, England (Derbyshire and Mellors, 1988), 1.57–1.77 in south-west central Iran (Fookes and Knill, 1969), and 1.36–1.46 in Libya (Assallay et al., 1996). In New Zealand, bulk densities range more widely in response to the variable content of allophane, halloysite, glass and other volcanic materials in the loess, as well as to different climatic environments (Stevens, 1988; Eden and Hammond, 2003).

Atterberg limits

The range of liquid and plastic limits in loess reflects the clay content, the highest values in most regions occurring in the palaeosols. In the western Chinese Loess Plateau, the lowest mean liquid (LL) and plastic limits (PL) (28.4% and 18.9% respectively) are found in the youngest, and coarsest, loess. Values for LL and PL in the older and rather finer-grained units are c. 27% and 18% respectively, with values in the more clay-rich palaeosols being c. 30% and 20%, respectively (Derbyshire et al., 2000). In more humid parts of the world, including the central and south-eastern Chinese Loess Plateau, values are higher (LL 29.9–31.1%; PL 18.6–19.4%: Lin, 1995). Reported values of LL for the Vicksburg loess of the lower Mississippi valley (USA) fall in the range 28–43%, with a PL range of 25–29% (Krinitzsky and Turnbull, 1967). The range of LL and PL values for loess in the Mississippi catchment is 23–58% and 17–29%. In contrast, the loesses of Alaska and Washington State are of very low plasticity (Sheeler, 1968). LL and PL ranges in the loess of south-east England are 30–46% and 17–23%, respectively.

To make a general case, the dominance of silt-sized primary minerals and the generally low percentage of clay minerals in many loesses, yields LL values of between 25 and 35% and low plasticity index (PI) values. PI values for loess sampled in many parts of the world lie in the range 10–25, encompassing the traditional descriptive classification, based on particle size, of sandy loess, 'typical' loess and clayey loess (Figure 25.17; Gibbs and Holland, 1960).

Permeability and infiltration

Given the tendency of loess to develop vertical joint systems, it is important to distinguish

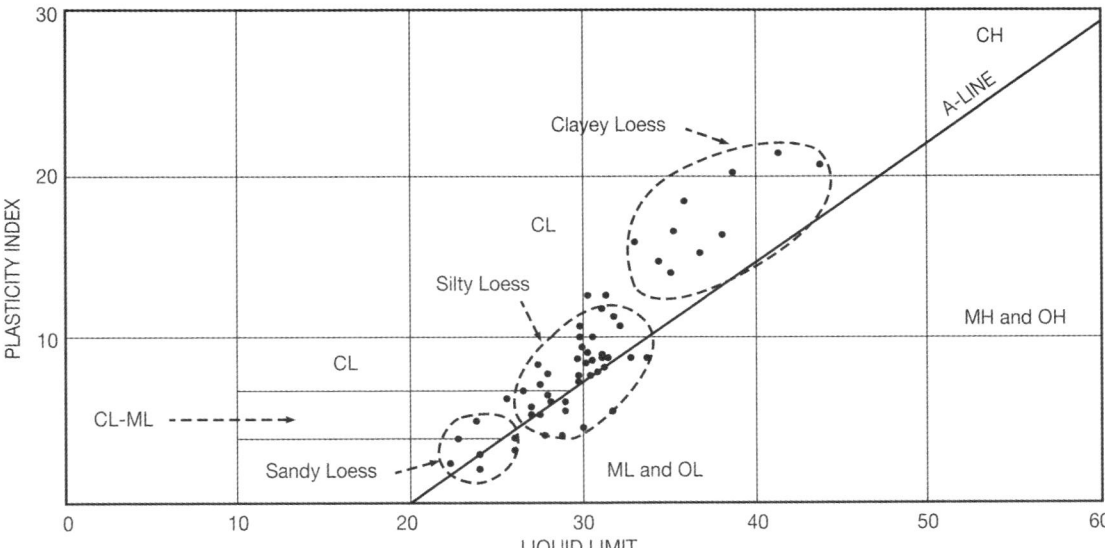

Figure 25.17a   Plasticity chart characterising the range of variation in the major loess deposits in the Missouri River basin, USA. The largest cluster is found between the zone demarcated by PI values of 5 to 12%, and LL values between 25 and 35% ('typical' or silty loess). The two other groups, sandy loess and clayey loess are minority populations. Based on eleven sites, redrawn after Gibbs and Holland (1960).

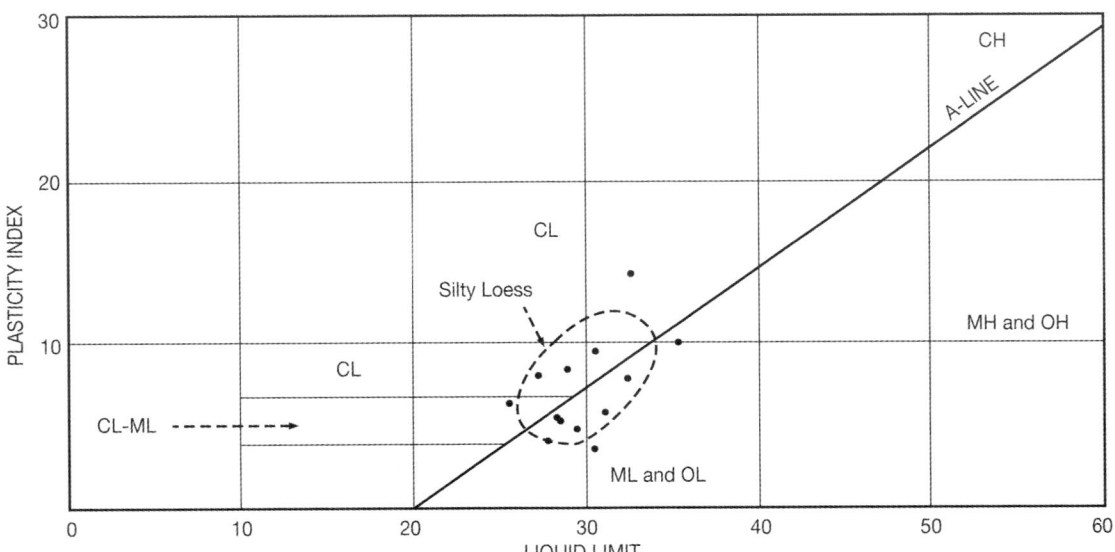

Figure 25.17b   Plasticity chart for samples taken from a single site (Dawan, 20 km WSW of Lanzhou in the semiarid western part of the Chinese Loess Plateau), for comparison with the 'silty loess' category in Figure 25.17a.

between intact permeability and bulk permeability that includes bypass drainage by way of joints and fissures. Coefficients of permeability in intact loess are $1.6–3 \times 10^{-8}$ m/s and 8×10^{-9} m/s for vertical and horizontal cases, respectively. Rates of permeability involving bypass drainage, however, may be orders of magnitude higher. Rates generally decrease with depth below the surface in loess with few or closed joint systems, any modification of the latter increasing the variance in values and sometimes locally reversing the gradient of rate with depth. Well-developed palaeosols, being more clay rich, are also a complicating factor. In addition, various types of structural anisotropy may lead to the development of preferred routes for sub-surface water movement. At the field scale, for example, the presence within thick loess of more compact or more plastic horizontal layers, including palaeosols, may influence the geometry of water-enlarged joint sets (piping: see 'Loess hydrology and pseudo-karst' below), a process that is thought to contribute to destabilisation of some loess slopes. The tendency for piping to develop appears to be true of the microfabric as well as the field scale in Chinese loess, as shown by artificial eluviation experiments (Figures 25.18a and 25.18b: Muxart et al., 1995).

Collapsibility

When loess under a load reaches critical moisture contents close to saturation, it displays collapse behaviour sometimes referred to as hydrocompaction or hydroconsolidation. Collapsibility at the field scale is a visible characteristic of loess, and one with important implications for engineering works. Collapsibility has been quantified using a number of criteria. Denisov (quoted in Feda, 1966) used void ratio above a critical value (e_c) and at the liquid limit (e_l), while Feda (1966) also used critical and liquid limit void ratios but added void ratio at the plastic limit (e_p), in the following relationship:

$$e_c = 0.85\,e_l + 0.15\,e_p$$

Widespread engineering problems involving collapse in thick loess terrain in China have given

rise to criteria taking account of several variables, in the following form:

$$\delta_c = (h_p - h_p')/h_o = (e_p - e_p')/(1 + e_o)$$

where δ_c is the coefficient of collapsibility; h_p and e_p are the sample height and voids ratio, respectively, under a loading pressure p; h_p' and e_p' are the sample height and voids ratio under pressure p in a saturated condition; h_o and e_o are the original sample height and voids ratio (CSCCC,

Figure 25.18a   Undisturbed Upper Pleistocene loess, showing typical intact fabric (right) adjacent to a large natural drainage channel (left).

Figure 25.18b   Locally compacted inter-silt fabric and large, ovoid drainage pipes (both vertical and transverse) developed during 240 hours of controlled laboratory leaching of Lanzhou loess. For details, see Muxart et al. (1995).

(Collapsibility Standard for Chinese Civil Construction) 1978, cited in Derbyshire et al., 1995b).

It has been pointed out by Lutenegger (1981) that loess, consisting predominantly of silt, would not be regarded as a sensitive soil in the classic Terzaghi sense, because the ratio of undisturbed to remoulded strength at a constant moisture content is approximately 3, placing loess in the medium sensitivity class only. Using the inference of Feda (1966) that sensitivity can be expressed by the ratio between undisturbed and saturated strength (in unconfined compression), many loess soils would be classified as 'quick'. Classic quick-clays, however, are saturated, whereas some loess displays collapse behaviour when wetted to some critical but unsaturated level and beneath a load, as shown in Feda's oedometer tests. Handy (1973) noted that it is the loess soils deficient in clay minerals that tend to collapse in Iowa (USA). He went on to argue that, because of the increase in liquid limit and decline in saturation moisture content with distance from the loess source regions, the extent of collapsible loess in Iowa may be defined at a point when the saturation moisture content equals or exceeds the liquid limit.

The collapsibility of loess varies geographically in response to climate and related factors, and also with depth below the surface (i.e. age). Lin and Liang (1982) and Qian et al. (1985) show how collapsibility in the loess of northern China varies with climatic region and mean thickness of collapsible loess (Figure 25.19). Tests on the loess at Lanzhou show that, below a void ratio of c. 0.95, coefficient of collapsibility rapidly approximates zero. The commonly observed spread of data above the voids ratio threshold indicates the role played by other factors such as cementation, shape and distribution of pores, particle size distribution, and zones of collapse involving inter-granular compaction, introducing complex post-collapse permeability patterns (Derbyshire et al., 2000). In more humid climates, collapsibility may show greater variability depending on factors such as clay content, natural moisture content, modification by reworking in very different climates of the past, and geomorphological location and history.

Shear strength

In general, shear strength rises with a decline in moisture content and a rise in bulk density. Moisture content exerts a fundamental influence upon the shear strength of loess. In laboratory tests on Chinese loess, the internal friction angle rose as moisture content increased from zero to 5%, after which internal friction angle declined with rise in the moisture content (Dijkstra et al., 1994). The range of internal friction angles in the essentially uncemented loess of North China is 25–40°. The primary cohesion is dominated by particle size, mineral composition and material density, and rises with percentage clay content. Consolidation cohesion occurs when chemical cementation is present, common cementing agents in loess being calcium carbonate, magnesium sulphate, gypsum and sodium chloride. These minerals are taken up and re-precipitated by soil moisture solutions (e.g. Muxart et al., 1995), so that consolidation cohesion is frequently more marked in the older loess units.

Rogers et al. (1994b) discussed the relationship between cementation and clay mineral bonding in loess. The presence of significant amounts of clay mineral allows deformation of loess fabric without any accompanying decrease in strength because the clay bonds are long range in type. However, the percentage of clay mineral is low in most loess, so that it is dominated by short-range bonds, has low plastic limits and is of low plasticity (I_p range 6–8%). At the same time, varying degrees of cementation bonding are commonly present so that loess may achieve greater yield strengths even though these may dissipate rapidly with relatively modest amounts of deformation. When loess fails, it approximates a liquefaction state either because of saturation or, if dry, as a result of shear failure in cementation bonds (Rogers et al., 1994b; Mellors, 1995), a condition that persists until pore pressures dissipate and effective stresses recover. The resultant material, sometimes referred to as 'secondary loess', has closer particle packing and diminished cemented strength. In freshly deposited loess, with its characteristically very high void ratios, relatively little moisture is required to mobilise the cementing agents that are re-deposited as bridges between

aggregates and mechanical eluviation of solids. These processes, added to liquefaction of thin layers of loess along joint walls, result in joint enlargement. In this way, an essentially homogeneous material following deposition is transformed into a quasi-coherent sediment divided into blocks by a dense network of pipes, cavities and channels due to the process of infiltration of water, dissolution of soluble material and piping, to create the characteristic macroporosity.

The presence of the macroporosity, known as pseudo-karst, is important for two reasons. First, notwithstanding the results of the mechanical–eluviation experiments summarised above, effective permeability values determined in field situations are very much higher than those derived from measurements using blocks of intact loess in the laboratory. Second, the joint and pipe systems conduct water to the loess–bedrock interface zone, which may then reach saturation even though the loess above remains relatively dry. As this process is time dependent, the proportion of a given rainfall amount infiltrating through the pipe system increases with time, at least until localised collapse diverts the conduits and the process recommences in whole or in part. The metastability of some parts of the loess always tends to increase either progressively or abruptly. The waters may also partly infiltrate into bedrock joints and bedding planes, a process that may be enhanced in its efficiency by the concentration of waters within the pseuodo-karst conduits. The loess–bedrock contact remains a particularly critical zone where the water may concentrate during long or heavy rainy periods or following several such episodes.

The role of bypass drainage varies according to local circumstances. Many systems are highly integrated and drain freely for long periods giving rise to slope-foot springs. Sometimes roof falls occur, the result of which is a coincident distribution of sinkholes and shallow but dry gully networks on some loess slopes. Some sub-surface distributary pipes develop and lead down into reduced-diameter pipes or even dead ends. In such situations, water pressure potentials may rise significantly, leading to local saturation and creating temporary perched water tables; liquefaction may then be followed by collapse of the superjacent loess mass (Figure 25.21b). In such circumstances loess hydrology and erosion may involve more tunneling, collapse and mass sliding than rilling and gullying. Erosion can be particularly severe around some types of artificial structures, notably bridge foundations.

Mass movements in loess

Slope failure is manifested by a range of landscape features of considerable diversity and involving a number of processes. These processes include fall, toppling, flow and spreading, in addition to sliding. While these and other processes are commonly grouped together in referring to failed masses on slopes as 'landslides', the term mass movement is more comprehensive in that it emphasises mass transport (Brunsden, 1984). The general classification of mass movements is not considered in detail here, as important contributions to this topic are readily available in the work of Varnes (1978), Brunsden and Prior (1984), Hutchinson (1988), Cruden (1991) and Dikau *et al.* (1996), for example.

Landslides in loess are very diverse as a result of the broad range of field conditions in which landsliding occurs. The mode of failure of loess, as an undisturbed, partially cemented mass of silt, is greatly influenced by the presence of joint systems; in this, it has some similarities with failure on rock slopes. At the other extreme, as a cohesionless particulate body, loess may fail as a flowslide given certain moisture levels at depth, and as a mudflow when saturated. An intermediate type of failure in loess involves extensional or translated slides, or lateral spreads (Figure 25.23). Creep-like behaviour also occurs.

Despite the extensive literature on landslides and landslide classification, mass movements in loess have received relatively little attention, not appearing at all in some classifications. The large and catastrophic loess landslide in Gansu province, China, in 1920 (Zhang, 1995) is the only loess mass failure mentioned (and referred to as 'typical') in the Varnes (1958) classification, for example. Some more recent classification schemes (e.g. Sassa, 1989) underestimate the special geotechnical characteristics of loess

Figure 25.23   Detail of widespread tension cracks in laterally spreading, slow, rainfall-induced landslide at Tala, *c*. 40 km SW of Lanzhou city, western Loess Plateau of China, in August 1988. For context, see Figure 25.25*b*.

and the distinctive failure behaviour of massive silts (Derbyshire *et al.*, 1990). Suggestions such as that the 1920 Gansu slide might have been prevented if the loess had been artificially hydro-consolidated (Veder, 1981) arise from a failure to appreciate the implications for the study of mass movement types in loess of structure and fabric, moisture content and the geomorphology, particularly the slope steepness. Attention has been drawn to the similarities of behaviour between loess slopes and those in classic 'quick-clays' (Smalley and Derbyshire, 1991; Derbyshire *et al.*, 1995b). Both loess and quick-clays are sensitive materials and, although most of the latter have higher sensitivities (16 as against *c*. 10 for loess), both have openwork fabrics, the obvious difference between the two being the order of magnitude difference in scale (loess $\pm 50\,\mu m$ versus ± 2–$5\,\mu m$ for quickclays). The behaviour of both materials is dependent upon the presence of short-range bonds.

Mass movement types
Although the greatest potential database for loess landslides probably lies in the thick loess terrain of northern China and central Asia, local conditions

and land use practices vary to such a degree that no single basis for landslide classification has yet emerged from this huge region. However, a number of commonly agreed criteria have been proposed by practising engineers and geologists as a basis for damage limitation strategies in China, namely:

1. the material contents of the sliding mass
2. the location of the failure plane in relation to the lithology.

Additional practical considerations are the geometry of the slide and the type of movement.

Landslide types based on slide mass composition and location of failure plane
In a recent review of literature and practice in China, Meng and Derbyshire (1998) described three major types of landslide defined in terms of the composition of the slide mass and the situation of the failure plane, named 'bedrock contact landslides', 'palaeosol contact landslides' and 'mixed landslides', to which they added a fourth (the special case of 'terrace landslides') (Figure 25.24).

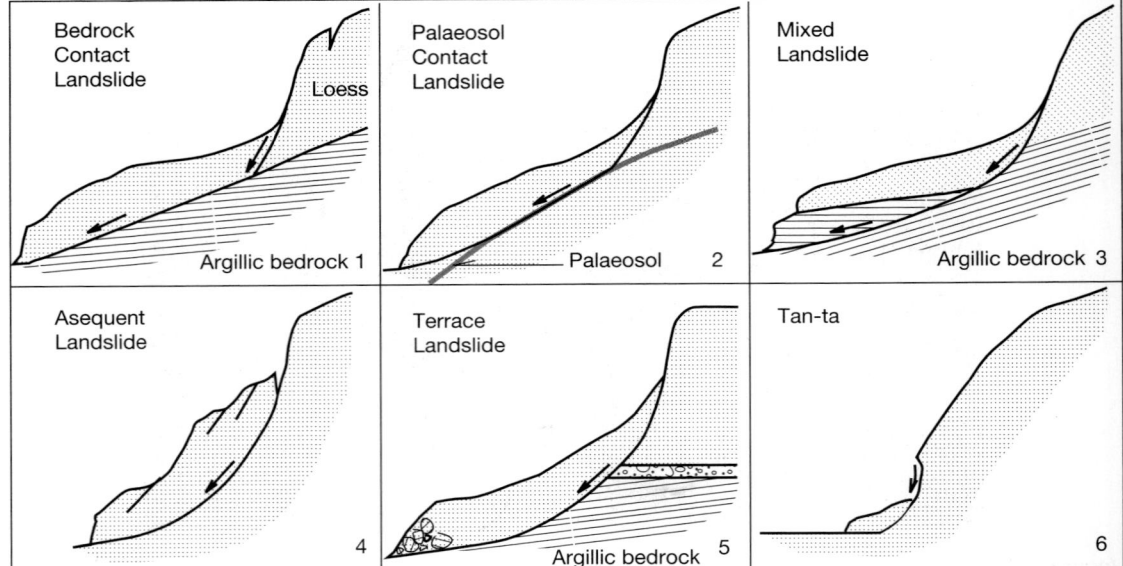

Figure 25.24   Main loess landslide types on the western Chinese Loess Plateau (after Meng and Derbyshire, 1998).

Bedrock contact landslides, as the name implies, involve the sliding of a mass of loess along a bedrock surface. This is the commonest type throughout the Chinese Loess Plateau. The underlying bedrock is generally of low to very low permeability (commonly argillites or mudstones). The resulting impedance of groundwater movement results in a zone of saturation at the loess–bedrock interface. This situation is frequently exacerbated by the presence of a weathered zone on the bedrock surface that contains smectitic clays (Derbyshire and Meng, 1995). Loess landslides of this type include some of the largest loess mass failures known (Figures 25.25a and 25.25b).

Palaeosol contact landslides involve the sliding of a mass of loess on or within the numerous palaeosols found within the Chinese loess. Water circulation is impeded by such relatively compact, clay-enriched horizons, causing localised rises in soil moisture and consequent decline in soil strength. The probability of failure is enhanced when the basal parts of the loess slopes are undercut by river incision or artificial excavation. This slide type is not well documented.

Mixed landslides consist of mixtures of loess and bedrock fragments sliding along bedding planes or joints in the bedrock. Frequent situations are those in which argillites or mudstones underlie the loess drape, or where shallow faults or clay-lined shear zones occur within the bedrock. A number of cases in which intermittent sliding of mixed material occurs on slopes of loess overlying Cretaceous sandstone with intercalated thin mudstones are documented in the Chinese literature.

Terrace landslides are a common and important type in China. Many river terraces consist of a planed rock surface with fluvial or colluvial sediments above, upon which rests loess (both primary and reworked) up to thicknesses of 100 m or more. Given the much lower permeability of most of the common bedrock types, the fluvial or colluvial gravels serve as an aquifer between rock and loess. Surface efflorescence of salts is common along seepage lines, and eluviation of salts and finer particles in moist conditions causes softening of the lower parts of the loess cover, which respond by deforming under the overburden stresses. If the strength of the bedrock falls to a critical value, cracks develop throughout the loess, leading to failure. Many of the thicker loess landslides in northern China are found on terrace

Figure 25.25*a*   Landslide consisting of mass movement of loess on the sloping bedrock beneath a *liang*, in Ningxia province, North China. The slide mass reached the valley floor and has since largely been removed by fluvial erosion. From an original photograph by the late Wang Jingtai.

Figure 25.25*b*   Part of an extensive (*c.* 4 km²) spreading slide at the loess–bedrock contact at Tala, August 1988 (cf. Figure 25.23). This was a slow landslide induced by prolonged monsoon rain; drinking water wells and roads were all destroyed and there were some animal deaths, but no people died. The sloping contact between loess (pale) and underlying argillite bedrock can be clearly seen on the ridge face in the middle distance.

edges, some of the failures being rapid and occasionally catastrophic. Many cities, towns and villages have been built on such loess-draped river
terraces, urban concentrations served by railways,
roads, and industrial plants. Thus, even quite
small examples of this type of landslide have
proved serious, especially as, in the larger slide
masses, high pore pressures may build up during
sliding, so greatly extending the sliding distance.

Slides entirely within loess: large landslides
composed exclusively of loess are relatively
uncommon in the Chinese Loess Plateau, presumably because the only features within loess having
any potential to initiate slip surfaces (gravity-
induced joints and shear planes) are usually too
dry in this region, with typical ambient moisture
contents < 12% (Meng and Zhang, 1989). In
such conditions, the cementation bonds remain
unbroken though brittle, providing loess with sufficient shear strength to resist slope failure
(Dijkstra *et al.*, 1994). Nevertheless, landslides
entirely within loess have been quite widely
recorded, although special conditions (including
human action in undercutting slopes and raising
the local water table) are thought to have been
present in most cases. Throughout the loess
region of North China, abundant small-scale
displacements (~ 10 m diameter) with planar slip
surfaces from one to several metres deep show
high sliding velocities and rapid disintegration of
the slide mass (Derbyshire *et al.*, 1993). These
small features are known as *tan-ta* (both singular
and plural: Figure 25.26). The consistency of the
gradients on these features is in accordance with
long-term but episodic movement triggered by
precipitation and valley incision. Several factors
appear to be involved, including infiltration of
snowmelt and rainwater, slope undercutting and
natural earth tremors. Activities such as irrigation
and slope undercutting during road and railway
construction also trigger *tan-ta*. These small features are a recurrent nuisance, and occasionally
constitute a real hazard. Collapse of components
of the pseudo-karst on relatively thin loess slope
mantles may increase the volume of loess likely
to become saturated in subsequent rainfall
events (Figure 25.27). Shallow sliding may
disrupt antecedent sinkhole-and-pipe systems,

Figure 25.26 Small slide (*tan-ta*) entirely in loess:
Gaolanshan, Lanzhou, China.

exacerbating the 'closed pipe' problem mentioned above and leading to retrogressive failure
(Figure 25.27).

It is important to recognise that the hydrology
of loess in general and conditions at loess–water
interfaces may vary with the age of the loess unit
under consideration because of differences in
bulk properties, notably particle size and bulk
density. Loess-karst is more common in young
loess, which generally has a coarser mean particle
size and lower bulk densities. Landslides in
younger loess are influenced by the distinctive
physical properties of the loess but also by other
factors such as its occurrence as a relatively thin
drape that cuts unconformably across older, sub-
horizontal loess units on long, steep slopes. In
such cases, flowslide type failures associated with
saturated conditions on the basal slip surface

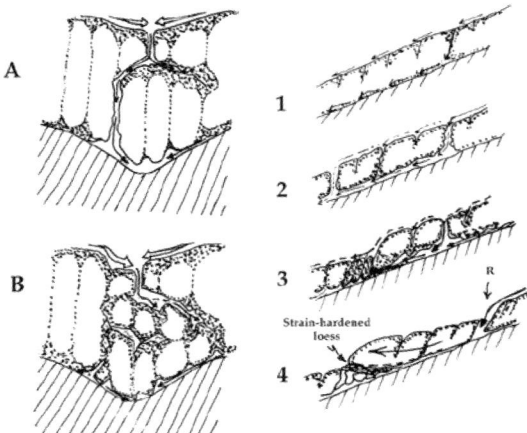

Figure 25.27 Pseudo-karst collapse in loess. The cartoon on the left shows a cross-section of loess in which the potentially saturated loess mass increases following collapse. The sketches on the right show the development of loess pseudo-karst on a slope, and associated sliding. This may be a retrogressive process, which may begin again above the point marked 'R'. Broken arrows indicate water flow (overland flow and pipe flow).

Figure 25.28 Scanning electron micrograph (vertical face) showing discrete, low-dipping shear planes and large secondary pores arising from high porewater pressures associated with rapid undrained shearing during sliding in the lowermost few centimetres of a thin slide mass of loess overlying argillite bedrock, near Lanzhou, northern China.

following prolonged rainfall may be a critical triggering factor. Failure planes develop on or close to the loess–bedrock interface where conditions favour porewater fluctuations and efficient throughflow, as shown by features including dewatering structures and microshearing zones in the fabric of loess within 0.5 m of slide planes on low permeability bedrock (Figure 25.28).

Landslide types based on their geometry and movement mechanism

Meng *et al.* (2000) have described the types of landslide types in the thick loess terrain of northern China, classified on the basis of their geometry and mechanism of movement. Included are flows and complex mass movements, large rotational mass movements, planar slides and debris and mudflows.

Flows and complex mass movements. When a thin layer at the base of a thick loess mantle on a sloping palaeo-relief approaches saturation, it is liable to suffer instantaneous collapse, leading to large-scale failure of the flowslide or 'spread' type (Figure 25.25*b*) in the classification of

Varnes (1958), together with lateral mass flowage comparable to mudflows.

Mass movements in loess are often complex. Block-type movement in the upper area of a failed mass may gradually change into a flow type of movement in the lower parts. In China, it is often the great extent of these flow lobes that causes the greatest damage to the human environment.

The best, and most fully documented case is the slide that occurred at 17:46 hours on 7 May 1983 at Saleshan, a rural area about 80 km south of Lanzhou (Cao, 1986; Zhang, 1989; Zhu, 1989; Mahaney and Hancock, 1990; Cao *et al.*, 1997). Here, about 150 m of loess overlies Neogene argillites, thin conglomerates and clayey limestones. Sustained high groundwater levels in the contact zone between loess and argillites caused the argillites to soften. A mass of loess, with an estimated volume of 31 million m^3, and measuring

1000 m in width and 170 m high, travelled a distance of 1600 m in less than one minute, killing 273 people (Figure 25.29a).

Large rotational mass movements. Large rotational landslides have frequently been triggered by major earthquakes, as a comparison of their distribution with the remarkable Chinese written records describing their impact on daily life clearly shows. The Tawa landslide (Figure 25.29b), situated about 15 km south-west of Lanzhou city, was triggered by an earthquake in 1125 AD. The slide shows a maximum vertical displacement along the slip surface of 150 m, the shape and gradient of its landslide scar suggesting a single major circular failure plane, and so quite unlike the mode of failure of the Saleshan slide. Little disintegration of the slide mass is evident, although a stream was displaced and continues to undercut part of the main slide mass. Some moderate-sized mass movements have occurred towards the toe of the slide because of river undercutting, and others have been induced by highway construction.

Planar slides involve the movement of an essentially rigid body over a shallow slip surface. Sliding of loess slabs up to about 5 m in thickness appears to take place on slip surfaces that roughly parallel the slope surface. Planar slides are defined by the shear strength conditions on the failure surface and the cohesion of the material. Such slides are usually restricted to the Late Pleistocene (Malan) loess, but some have also been recorded in slope deposits (reworked loess). Some degree of strain hardening may be involved so that, as a consequence, such planar slides are not usually associated with the same acceleration and release of energy typical of slides that develop in the older, more brittle loess units. Usually, therefore, they constitute only a minor hazard.

Debris and mudflows. Debris flows in loess terrain are usually found in locations where the loess is a thin drape over bedrock. The term 'debris flows' is probably a misnomer, however, because such silt-dominated flows are best described as mudflows. They commonly occur in association with slope failures of other types, being found with debris flows and debris floods in some areas. Since 1949, large-scale loessic mud and debris flows have occurred on at least 50 occasions in Gansu province, western China, killing more than 2100 people. More than 50 of Gansu's 82 counties and at least 7000 valleys have been affected.

Soil erosion and sediment yield

The distinctive sedimentary and geotechnical properties of loess explain the very high rates of erosion to be seen in many loess landscapes. Measurements of sediment yield in the youngest (Upper Pleistocene) and most friable loess in China by Gong and Jiang (1979) gave values of 14 200–34 500 t/km^2. They quote yields of 0.7–1.0 t/m^3 (48.7–61.5% by weight) during the heaviest summer rainstorms. These storm rates are 1.10–1.76 times greater than the measured rates of loss from the interfluves, indicating that the loess slope profiles maintained by seasonal undercutting in the south-central Loess Plateau of China represent a rate of abstraction some 1.10 to 1.76 times the loss rate from the interfluves. Thus, it appears that such slope profiles are maintained largely by periodic basal corrasion. It has been claimed by Dai (1987) that, in present conditions, the volume of loess lost by erosion from the Loess Plateau in one year is equivalent to 100–300 years of accumulation. This is graphically illustrated by comparison of some present-day features with the remarkable descriptions to be found in the Chinese written records. Near Xifeng, in eastern Gansu province, the part of the Loess Plateau surface known as Dungzhi Yuan was 32 km wide in the Tang Dynasty (618–907 AD). Thus yuan has been reduced in area from 1344 to 756 km^2 in little more than 1000 years, leaving parts of it now only 1 km wide (Figure 25.30).

When vegetation cover is sparse or absent, either because of a dry climate or intensive land use, as in many parts of China and central Asia, loess is highly susceptible to erosion by rainbeat, sheet erosion, gullying, mass movement and subsidence. These processes deliver pulses of sediment into rivers, material that may remain suspended for long distances until re-deposited on floodplains, deltas and in the coastal seas. The coincident distribution of major loess

Figure 25.29a   The Saleshan landslide, about 80 km south of Lanzhou city, western Loess Plateau, China. Photograph taken 5 years after the failure (7 May 1983). The massive area of run-out, clearly seen by its dark shade below the slide scar, covered three villages in less than one minute, blocked the Ba Xie River and ruined two reservoirs used for drinking water in this semi-arid area.

Figure 25.29b   The Tawa landslide (right half of photograph), its crest c. 380 m above the floor of the Xuangjia valley, SW of Lanzhou in the western Loess Plateau. This is an example of a large rotational slide triggered by an earthquake. Written records show that this occurred in the AD 1125 Lanzhou earthquake (estimated magnitude 7). Maximum vertical displacement along the slip surface was c. 150 m. In contrast to the Saleshan slide (above), all evidence here points to a single, major circular failure plane, with no evidence of widespread disintegration of the slide mass. However, the slide displaced the river, and subsequent undercutting has occurred, triggering the secondary slides, some of which still threaten the major roadway (bottom of picture). Photograph by Armelle Billard.

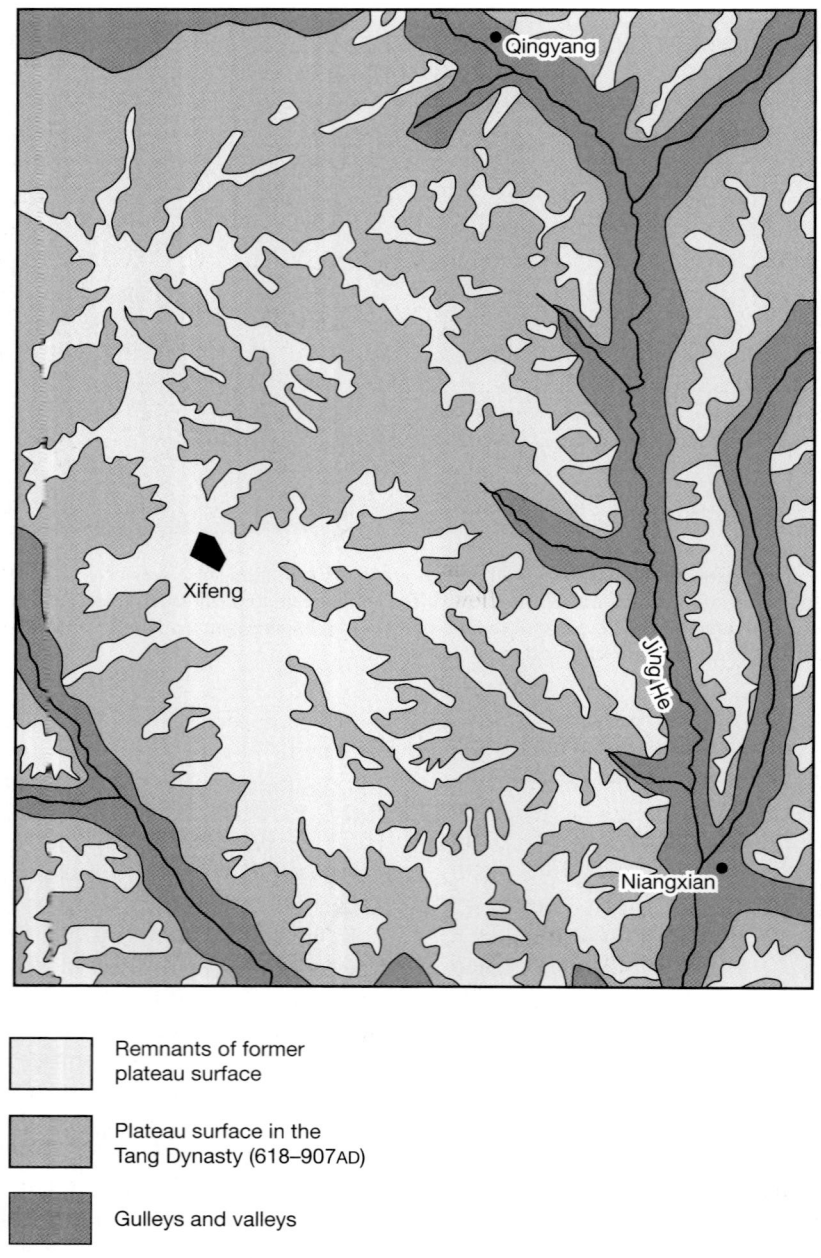

Figure 25.30 Documented progressive elimination of a loess yuan near Xifeng, eastern Gansu province (after Derbyshire, 2001).

deposits and great rivers, and the possible role played by the latter in the accumulation of the former, has been persuasively argued by Smalley (1972; 1995). Geologically recent crustal overlap has made High Asia a region of considerable geodynamic energy, combining tectonic, glacial, periglacial, river and other means of particle generation and transportation (Smalley and

Derbyshire, 1990). Some of the greatest rivers on Earth drain the loess lands of Asia, including the Yellow, the Yangtze, the Mekong, the Irrawadi, the Ganges-Brahmaputra, the Indus, the Amu and the Syr-Darya. They transport huge volumes of silt, the most turbid major fluvial system being the Yellow River and its tributaries, its suspended sediment concentration exceeding 30 kg/m³ (Ferguson, 1984). Such high silt contents are ultimately deposited as extensive alluvial plains, and as some of the world's biggest deltas. Also, there are important silt components in the deltas of the Mississippi, the Rhine and the Danube, and the role of river transport has been invoked in explanation of the complex silt and sand facies of the Argentine *pampa* (e.g. Zarate and Blasi, 1988, 1993; Rabassa, 1990; Iriondo, 1990).

The Yellow River (Hwang He), with more than 70 tributaries in its 750 000 km² catchment, flows through extensive drylands in its middle reaches. Erosive inputs of loessic soils to the river in the drylands of Ningxia and Inner Mongolia and, further downstream, in the Loess Plateau, reach rates of 25 000 t/km² per year, with some individual tributary catchments yielding up to 50 000 t/km² per year. It has been estimated that the Loess Plateau contributes some 75% of the suspended sediment load of the Yellow River (Ren and Shi, 1986). In dramatic contrast, the North China Plain and the Yellow River delta have gradients as low as 0.1–0.2%. The river, flowing across this plain in levées up to 10 m high, has regularly shifted channel course during recent historical time, often flooding up to 250 000 km², and posing a major hazard for more than 120 million people. At extreme suspended sediment concentrations exceeding 1300 kg/m³, the river enters the hyper-concentrated condition capable of blocking its own flow, as happened at Tongguan in 1977.

25.5 Conclusions

Typically consisting of accumulations of wind-transported and deposited silts, but sometimes containing a variable amount of clay and fine

sand, loess is light brown in field exposures, only weakly plastic, with a mean particle size in the coarse silt range. Quartz is the dominant mineralogy, apart from certain deposits influenced, for example, by volcanic sources. Exceptionally, some fine sandy aeolian deposits on dryland margins are referred to as loess. The airfall nature of loess gives it a distinctive sedimentary fabric, which is an important influence upon a range of bulk properties. The presence of inter-particle cement and variations in the generally low percentage of clay mineral in loess are particularly influential factors.

Loess terrain varies from plateaus to small hills, and is frequently cut by vertical fissures, especially near the surface, although fissures occur in loess of all ages. Loess permeability is dictated by these systems rather than by the porosity of the intact loess. This imposes a distinctive hydrology on thick loess terrain, involving enlargement of the joints by suffosion, leading to an integrated subterranean system of pipes and chambers. The appearance and general handling properties of most loess are those of a soft rock. However, though hard or firm when dry, much loess has a tendency to liquefy on approaching saturation, so that it may properly be regarded as metastable. Such hydrocompaction is common and often constitutes a major geological hazard.

Failure of loess slopes is expressed as several types of landslide, best documented from northern China. Some of the largest, most rapid and most destructive landslides are those found in areas where a thick loess mantles a hilly palaeo-relief consisting of low permeability rocks. These distinctive sedimentary, hydrological and geotechnical properties provide a key to the understanding of the bulk behaviour of loess, and the susceptibility of loess terrain to rapid degradation. The major loess landscapes of Asia are subject to some of the highest rates of erosion on Earth, their rivers being among the most turbid on the planet. Loess landscapes thus continue to be a serious threat to the lives and livelihoods of millions of people. They constitute a substantial challenge to developers, planners and engineers around the world.

References

Assallay, A. M., Rogers, C. D. F. and Smalley, I. J. (1996) Engineering properties of loess in Libya. *Journal of Arid Environments* **32**, 373–386.

Assalay, A. M., Rogers, C. D. F., Smalley, I. J. and Jefferson, I. F. (1998) Silt: 2–62 μm, 9–4φ. *Earth Science Reviews* **45**, 61–88.

Billard, A. and Muxart, T. (2000) Mineralogy, chemistry and elemental composition of loess and palaeosols. In Derbyshire, E. Meng, X. M. and Dijkstra, T. A. (eds) *Landslides in the Thick Loess Terrain of Northwest China*. Wiley, Chichester, 65–71.

Billard, A., Muxart, T., Derbyshire, E., Wang, J. T. and Dijkstra T. A. (1993) Landsliding and land use in the loess of Gansu Province, China. *Zeitshrift für Geomorphologie* Suppl. Band **87**, 117–131.

Browzin, B. S. (1985) Granular loess classification based on loessial fraction. *Bull. Assoc. Eng. Geol.* **22**, 217–227.

Brunsden, D. (1984) Mudslides. In Brunsden, D. and Prior, D. B. (eds) *Slope Instability*. Wiley, Chichester, 363–418.

Brunsden, D. and Prior, D. B. (eds) (1984) *Slope Instability*. Wiley, Chichester.

Cao, B. (1986) The geologic characteristics of the Sala Shan type of super landslide and a model for spatial prediction. *Proceedings of the 5th International IAEG Conference* **3**, 1989–97.

Cao, B., Yang, Z. and Zheng, X. (1997) Movement mechanism and disaster prediction of typical high speed landslide in Northwest China. In Sassa, K. (ed.) *International Symposium on Landslide Hazard Assessment*. Kyoto University, Japan, 247–260. ISBN 4-9900618-0-2.

Catt, J. A. (1978) The contributions of loess to soils in lowland Britain. In Limbrey, S. and Evans, J. G. (eds) The effect of man on the landscape: the lowland zone. *Council for British Archaeology Research Report* **21**, 12–20.

Catt, J. A. (2001) The agricultural importance of loess. *Earth Science Reviews* **54**, 213–229.

Coudé-Gaussen, G. (1987) The presaharan loess: sedimentological characterisation and palaeoclimatological significance. *GeoJournal*, **15**, 177–183.

Cruden, D. M. (1991) A simple definition of a landslide. *Bulletin International Association for Engineering Geology* **43**, 27–29.

Dai, Y. S. (1987) The engineering geological characteristics of the loess and soil erosion in the Middle Reaches of the Huanghe River. In Liu, T. S. (ed.) *Aspects of Loess Research*. China Ocean Press, Beijing, 432–436.

Derbyshire, E. (1983a) On the morphology, sediments and origin of the Loess Plateau of central China. In Gardner, R. and Scoging, H. (eds) *Megageomorphology*. Oxford University Press, 172–194.

Derbyshire, E. (1983b) Origin and characteristics of some Chinese loess at two locations in China. In Brookfield, M. E. and Ahlbrandt, T. S. (eds) *Eolian Sediments and Processes*. Elsevier, Amsterdam, 69–90.

Derbyshire, E. (1984) Granulometry and fabric of the loess at Jiuzhoutai, Lanzhou, People's Republic of China. In Pecsi, M. (ed.) *Lithology and Stratigraphy of Loess and Palaeosols*. Hungarian Academy of Science, Budapest, 95–103.

Derbyshire, E. (1988) Granulometry and fabric of Quaternary silts from Eastern Asia. *Proc. of Second Conference of the Palaeoenvironment of East-Asia from the Mid Tertiary* **1**, 215–245.

Derbyshire, E. (2001) Geological hazards in loess terrain, with particular reference to the loess regions of China. *Earth Science Reviews* **54**, 231–260.

Derbyshire, E. and Mellors, T. W. (1986) Loess. In Fookes, P. G. and Vaughan, P. R. (eds) *A Handbook of Engineering Geomorphology*. Surrey University Press and Blackie, Glasgow, 258–269.

Derbyshire, E. and Mellors, T. W. (1988) Geological and geotechnical characteristics of some loess and loessic soils from China and Britain: a comparison. *Engineering Geology* **25**, 135–175.

Derbyshire, E. and Meng, X. M. (1995) The landslide hazard in North China: characteristics and remedial measures at the Jiaoshuwan and Taishanmiao slides in Tian Shui city, Gansu Province. In McGregor, D. F. M. and Thompson, D. A. (eds) *Geomorphology and Land Management in a Changing Environment*. Wiley, Chichester, 89–104.

Derbyshire, E., McGown, A. and Radwan, A. (1976) 'Total' fabric of some till landforms. *Earth Surface Processes* **1**, 17–26.

Derbyshire, E., Gregory, K. J. and Hails, J. R. (1980) *Geomorphological Processes*. Dawson-Westview Press, Folkestone, UK and Boulder, USA, 312pp.

Derbyshire, E., Billard, A., van Vliet-Lanoë, B., Lautridou, J.-P. and Cremaschi, M. (1988) Loess and palaeoenvironment: some results of a European joint programme of research. *Journal of Quaternary Science* **3**, 147–169.

Derbyshire, E., Wang, J. T. and Smalley, I. J. (1990) Loess landslides and geotechnical classification. *Landslide News* **4**, 21–23.

Derbyshire, E., Wang, J. T., Jin, Z. X., Billard, A., Egels, Y., Kasser, M., Jones, D. K. C., Muxart, T. and Owen, L. (1991) Landslides in the Gansu loess of China. *Catena Supplement* **20**, 119–145.

Derbyshire, E., Dijkstra, T. A., Billard, A., Muxart, T., Smalley, I. J. and Li, Y. J. (1993) Thresholds in a sensitive landscape: the loess region of central China. In Thomas, D. S. G. and Allison, R. J. (eds) *Landscape Sensitivity*. Wiley, Chichester, 97–127.

Derbyshire, E., Dijkstra, T. A., Smalley, I. J., Rogers, C. D. F. and Li, Y. J. (1994) Failure mechanisms in loess and the effects of moisture content changes on remoulded strength. *Quaternary International* **24**, 5–15.

Derbyshire, E., Kemp, R. and Meng, X. M. (1995a) Variations in loess and palaeosol properties as indicators of palaeoclimatic gradients across the Loess Plateau of North China. *Quaternary Science Reviews* **14**, 691–699.

Derbyshire, E., Dijkstra, T. A. and Smalley, I. J. (eds) (1995b) *Genesis and Properties of Collapsible Soils.* NATO ASI Series C: Mathematical and Physical Sciences, Vol. 468, Kluwer, Dordrecht, 413pp.

Derbyshire, E., van Asch, T., Billard, A. and Meng, X. M. (1995c) Modelling the erosional susceptibility of landslide catchments in thick loess: Chinese variations on a theme by Jan de Ploey. *Catena* **25**, 315–331.

Derbyshire, E., Meng, X. M. and Kemp, R. A. (1998) Provenance, transport and characteristics of modern aeolian dust in western Gansu Province, China, and interpretation of the Quaternary loess record. *Journal of Arid Environments* **39**, 497–516.

Derbyshire, E., Meng, X. M. and Dijkstra, T. A. (eds) (2000) *Landslides in the Thick Loess Terrain of Northwest China.* Wiley, Chichester, 288pp.

Dijkstra, T. A., Derbyshire, E. and Meng, X. M. (1993). Neotectonics and mass movements in the loess of north central China. *Quaternary Proceedings* **3**, 93–110.

Dijkstra, T. A., Rogers, C. D. F., Smalley, I. J., Derbyshire, E., Li, Y. J. and Meng, X. M. (1994) The loess of north-central China: geotechnical properties and their relation to slope stability. *Engineering Geology* **36**, 153–171.

Dikau, R., Brunsden, D., Schott, L. and Ibsen, M.-L. (1996) *Landslide Recognition; Identification, Movement and Causes.* Wiley, Chichester, 251pp.

Ding, Z. L., Sun, J. M., Liu, T. S., Zhu, R. X., Yang, S. L. and Guo, B. (1998) Wind-blown origin of the Pliocene red clay formation in the central Loess Plateau, China. *Earth and Planetary Science Letters* **161**, 135–143.

Eden, D. N. and Hammond, A. P. (2003) Dust accumulation in the New Zealand region since the Last Glacial Maximum. *Quaternary Science Reviews*, **22**, 2037–2052.

Feda, J. (1966) Structural stability of subsident loess soil from Praha-Dejvice. *Engineering Geology* **1**, 201–219.

Ferguson, R. I. (1984) Sediment load of the Hunza River. In Miller, K. J. (ed) *International Karakoram Project.* Vol. 2, Cambridge University Press, 581–598.

Folk, R. L. and Ward, W. C. (1957) Brazos River bar. A study in the significance of grainsize parameters. *J. Sed. Pet.*, **27**, 3–27.

Fookes, P. G. and Best, R. (1969) Consolidation characteristics of some late Pleistocene periglacial metastable soils of east Kent. *Quart. J. Eng. Geol.* **2**, 103–128.

Fookes, P. G. and Knill, J. L. (1969) The application of engineering geology in the regional development of northern and central Iran. *Engineering Geology* **3**, 81–120.

Frye, J. C., Glass, H. D. and William, H. B. (1962) Stratigraphy and mineralogy of the Wisconsinan loesses of Illinois. Illinois State Geological Survey Circular 334, Urbana, Illinois, USA, 55pp.

Gao, G. (1983) Microstructure of loess soil in China relative to geological environment. *Geological Environment and Soil Properties.* Special Publication, ASCE Geotechnology, Engineering Division, Houston, Texas, 121–136.

Gibbs, H. J. and Holland, W. Y. (1960) Petrographic and engineering properties of loess. United States Department of the Interior, Bureau of Reclamation, Engineering Monograph No. 28, Denver, Colorado, 37pp.

Gong, S. and Jiang, D. (1979) Soil erosion and its control in small watershed of the loess plateau. *Scientia Sinica* **22**, 1302–1313.

Grabowska-Olszewska, B. (1988) Engineering–Geological problems of loess in Poland. *Engineering Geology* **25**, 177–199.

Grim, R. E. (1953) *Clay Mineralogy.* McGraw-Hill, New York.

Handy, R. L. (1973) Collapsible loess in Iowa. *Procs. Soil Sci. Soc. America* **37**, 281–284.

Hutchinson, J. N. (1988) Morphological and geotechnical parameters of landslides in relation to geology and hydrology, general report. In Bonnard, C. (ed.) *Landslides.* Fifth Int. Symposium on Landslides Vol. 1, 3–35.

Iriondo, M. H. (1990) A late Holocene dry period in the Argentine plains. In Rabassa, J. (ed.) *Quaternary of South America and Antarctic Peninsula.* Balkema, Rotterdam, 197–218.

Iriondo, M. H. (1997) Models of deposition of loess and loessoids in the Upper Quaternary of South America. *J. South American Earth Sciences* **10**, 71–79.

Iriondo, M. H. and Kröhling, D. M. (1995) *El Sistema Eólico Pampeano.* Com. Mus. Prov. Ca. Naturales (Nueva Serie 'Florentino Ameghino'), Santa Fe, Argentina (ISSN 0325-3856), 45pp.

Jefferson, I. and Smalley, I. J. (1995) Six definable particle types in engineering soils and their participation in collapse events: proposals and discussions. In Derbyshire, E., Dijkstra, T. A. Smalley, I. J. (eds) *Genesis and Properties of Collapsible Soils.* NATO ASI Series C: Mathematical and Physical Sciences, Vol. 468, Kluwer, Dordrecht, 19–32.

Karlson, A., Tauber, A. and Torres Anza, D. (1993) Tipificación mineralogical de paleosuelos del sur de la provincia de Córdoba. Procs. XIV Congreso Argentino de la Ciencia del Suelo, Mendoza, Argentina, 463–474.

Kemp, R. A., Derbyshire, E. and Meng, X. M. (2001) A high-resolution micromorphological record of changing landscapes and climates on the western Loess Plateau of China during oxygen isotope stage 5. *Palaeogeography, Palaeoclimatology, Palaeoecology* **170**, 157–169.

Keš, A. S. (1972) On spreading and forming loess relief in Europe. *Acta Geol. Acad. Sci Hungar.* **16**, 359–370.

Krinitzsky, E. L. and Turnbull, W. J. (1967) *Loess Deposits of Mississippi.* Geol. Soc. America, Special Papers No. 94, 64pp.

Krinsley, D. H. and Smalley, I. J. (1973) Shape and nature of small sedimentary quartz particles. *Science* **180**, 1277–1279.

Lautridou, J.-P. (1993) L'eau dans les loess de Normandie. *Quaternaire* **4**, 91–96.

Lin, Z. (1995) Variation in collapsibility and strength of loess with age. In Derbyshire, E., Dijkstra, T. A. and Smalley, I. J. (eds) *Genesis and Properties of Collapsible Soils.* NATO ASI Series C: Mathematical and Physical Sciences, Vol. 468, Kluwer, Dordrecht, 413pp.

Lin, Z. and Liang, W. (1982) Distribution and engineering properties of loess and loesslike soils in China. Schematic map of engineering geological zonation. *Bull. Int. Assoc. Eng. Geol.* No. **21**, 112–117.

Liu, T. S. (ed.) (1985) *Loess and the Environment.* Science Press, Beijing, 215pp.

Liu, T. S., Wang, T. M., Wang, K. L. and Wen, C. C. (1964) *Loess in the Middle Reaches of the Yellow River.* Science Press, Beijing, 234pp. (in Chinese).

Lohnes, R. A. and Handy, R. L. (1968) Slope angles in friable loess. *J. Geol* **76**, 247–258.

Lozek, V. (1965) Das Problem der Lössbildung und die Lössmollusken. *Eiszetalter und Gegenwart* **16**, 61–75.

Lutenegger, A. J. (1981) Stability of loess in light of the inactive particle theory. *Nature* **291**, 360.

Mahaney, W. C. and Hancock, R. G. V. (1990) Stratigraphy and paleosols in the Sale terrace loess section, northwestern China. *Catena* **17**, 357–367.

McTainsh, G. (1987) Desert loess in Northern Nigeria. *Zeitschrift für Geomorphologie* **31**, 145–165.

Mellors, T. (1995) The influence of the clay component in loess on collapse of the soil structure. In Derbyshire, E., Dijkstra, T. A. and Smalley, I. J. (eds) *Genesis and Properties of Collapsible Soils.* NATO ASI Series C: Mathematical and Physical Sciences, Vol. 468, Kluwer, Dordrecht, 207–216.

Meng, X. M. and Derbyshire, E. (1998) Landslides and their control in the Chinese Loess Plateau: models and case studies from Gansu Province, China. In Maund, J. D. and Eddleston, M. (eds) *Geohazards in Engineering Geology.* Geological Society, London, Engineering Geology Special Publications **15**, 141–153.

Meng, X. M. and Zhang, S. W. (1989) Investigation of dangerous loess slopes in Lanzhou region. Internal Report, 89pp. (in Chinese).

Meng, X. M., Dijkstra, T. A. and Derbyshire, E. (2000) Loess slope stability. In Derbyshire, E., Dijkstra, T. A. and Smalley, I. J. (eds) *Genesis and Properties of Collapsible Soils.* NATO ASI Series C: Mathematical and Physical Sciences, Vol. 468, Kluwer, Dordrecht, 173–202.

Moss, A. J. (1966) Origin, shaping and significance of quartz sand grains. *J. Geol. Soc. Australia* **13**, 97–136.

Moss, A. J. and Green, P. (1975) Sand and silt grains: predetermination of their formation and properties by micro fractures in quartz. *J. Geol. Soc. Australia* **22**, 485–495.

Muxart, T., Billard, A., Derbyshire, E. and Wang, J. T. (1994) Variation in runoff on steep unstable loess slopes near Lanzhou, China: initial results using rainfall simulation. In Kirkby, M. J. (ed.) *Process Models and Theoretical Geomorphology.* Wiley, Chichester, 337–355.

Muxart, T., Billard, A., Andrieu, A., Derbyshire, E. and Meng, X. M. (1995) Changes in water chemistry and loess porosity with leaching: implications for collapsibility in the loess of North China. In Derbyshire, E. Dijkstra, T. A. and Smalley, I. J. (eds) *Genesis and Properties of Collapsible Soils.* NATO ASI Series C: Mathematical and Physical Sciences, Vol. 468, Kluwer, Dordrecht, 313–331.

Ovejero, R. (1980) Geología y geomorfología de la cuenca del rio San Javier, Depto. Tafi (Tucumán – Argentina). Unpublished thesis, Facultad de Ciencias Naturales, Tucumán University, Argentina.

Owen, L. A., Derbyshire, E., White, B. J. and Rendell, H. (1992) Loessic silt deposits in the western Himalayas: their sedimentology, genesis and age. *Catena* **19**, 493–509.

Phien-wej, N., Pientong, T. and Balasubramaniam, A. S. (1992) Collapse and strength characteristics of loess in Thailand. *Engineering Geology* **32**, 59–72.

Pye, K. (1987) *Aeolian Dust and Dust Deposits.* Academic Press, London, 334pp.

Pye, K. (1995) The nature, origin and accumulation of loess. *Quaternary Science Reviews* **14**, 653–667.

Pye, K and Tsoar, H. (1987) The mechanics and geological implications of dust transport and deposition in deserts, with particular reference to loess formation and dune sand diagenesis in the northern Negev, Israel. In Frostick, L. E. and Reid, I. (eds) *Desert Sediments: Ancient and Modern.* Geol. Soc. Lond. Spec. Publ. 35, Blackwell, Oxford, 139–156.

Qian, H., Wang, J., Luo, Y., She, G., Shi, G. and Qi, W. (1985) *Foundations on Collapsible Loess.* Publishing House of the Chinese Architecture Industry, 470pp. (in Chinese).

Rabassa, J. (1990) Late Pleistocene and Holocene loess deposits in the upper Rio Sauce Grande basin, Sierra de la Ventana, Argentina. *International Symposium on Loess* (Expanded abstracts), Mar del Plata, Argentina, 84–88.

Ren, M. E. and Shi, Y. F. (1986) Sediment discharge of the Yellow River (China) and its effect on the sedimentation of the Bohai and Yellow Sea. *Continental Shelf Research* **6**, 785–810.

Richthofen, F. von (1877–85) *China. Ergebnisse Eigener Reisen und Daran Gegruendeter Studien.* Reimer, Berlin, 5 vols.

Richthofen, F. von (1882) On the origin of the loess. *Geological Magazine* **9**, 293–305.

Rogers, C. D. F. and Smalley, I. J. (1993) The shape of loess particles. *Naturwissenschaften* **80**, 461–462.

Rogers, C. D. F. and Smalley, I. J. (1996) The adobe reaction and the use of loess mud in construction. *Eng. Geol.* **40**, 137–138.

Rogers, C. D. F. Dijkstra, T. A. and Smalley, I. J. (1994a) Particle packing from an earth science viewpoint. *Earth Science Reviews* **36**, 59–82.

Rogers, C. D. F., Dijkstra, T. A. and Smalley, I. J. (1994b) The Teton Dam failure (Idaho, USA, 1976). In Fookes, P. G. and Parry, R. H. G. (eds) *Engineering Characteristics of Arid Soils*. Procs. of the First International Symposium. Balkjema, Rotterdam, 415–417.

Rózycki, S. Z. (1991) *Loess and Loess-like Deposits.* Polish Academy of Sciences Publishing House, Wroclaw, 187pp.

Sassa, K. (1989) Geotechnical classification of landslides. *Landslide News* **2** (June 1989).

Sayago, J. M. (1995) The Argentine neotropical loess. *Quaternary Science Reviews* **14**, 755–766.

Sheeler, J. B. (1968) Summarization and comparison of engineering properties of loess in the United States. *Highway Research Record* No. 212, 1–9.

Smalley, I. J. (1966a) Formation of quartz. *Nature* **211**, 476–479.

Smalley, I. J. (1966b) The properties of glacial loess and the formation of loess deposits. *Journal of Sedimentary Petrology* **36**, 669–676.

Smalley, I. J. (1971) 'In situ' theories of loess formation and the significance of the calcium carbonate content of loess. *Earth Science Reviews* **7**, 67–85.

Smalley, I. J. (1972) The interaction of great rivers and large deposits of primary loess. *Trans. New York Acad. Sci.* **34**, 534–542.

Smalley, I. J. (1995) Making the material: the formation of silt-sized primary mineral particles for loess deposits. *Quaternary Science Reviews* **14**, 645–651.

Smalley, I. J. and Derbyshire, E. (1990) The definition of 'ice sheet' and 'mountain' loess. *Area* **22**, 300–301.

Smalley, I. J. and Derbyshire, E. (1991) Flowslide-type ground failures in the airfall deposits of northern China and the shallow-marine postglacial clays of eastern Canada. In Jones, M. and Cosgrove, J. (eds) *Geology of Neotectonic Environments*. Bellhaven Press, Leichester, 202–219.

Stevens, K. F. (1988) A preliminary investigation of the physical and chemical properties of the Pahiatua Terrace loess beds, north-western Waisawapa, New Zealand. In Eden, D. N. and Furkert, R. J. (eds) Loess: Its Distribution, Geology and Soils. Balkema, Rotterdam, 193–199.

Sun, J. Z. (1988). Environmental geology in loess areas of China. *Environmental Geology and Water Science* **12**, 49–61.

Tan, T. K. (1988) Fundamental properties of loess from northwestern China. *Engineering Geology* **25**, 103–122.

Tsoar, H. and Pye, K. (1987) Dust transport and the question of desert loess formation. *Sedimentology* **34**, 139–153.

Varnes, D. J. (1958) Landslide types and processes. *Landslides and Engineering Practice*. Nat. Acad. Sci., Nat. Res. Council, HRB **544**, 20–47.

Varnes, D. J. (1978) Slope movements: type and processes. In Eckel, E. B. (ed.) *Landslide Analysis and Control*. Transport Res. Board, Special report 176.

Veder, C. (1981) *Landslides and their Stabilization*. Springer-Verlag, New York, 247pp.

Vliet-Lanoë, B. van, Coutard, J.-P. and Pissart, A. (1984) Structures caused by repeated freezing and thawing in various loamy sediments: a comparison of active, fossil and experimental data. *Earth Surface Processes and Landforms* **9**, 553–565.

Wang, J. and Derbyshire, E. (1994) Engineering soil classification of the loess of Gansu, China, based on sedimentary properties: relationships to geotechnical behaviour. In Fookes, P. G. and Parry, R. H. G. (eds) *Engineering Characteristics of Arid Soils*. Balkema, Rotterdam, 153–157.

Wang, Y. and Zhang, Z. (eds) (1980) *Loess in China*. Shaanxi People's Art Publishing House, Xi'an, China, n.p.

Wentworth, C. K. (1933) Fundamental limits of the sizes of classic grains. *Science* **77**, 633–634.

Zarate, M. and Blasi, A. (1988) Depositos loessicos del Pleistoceno tardio-Holoceno del flanco sudeste de Tandila. CADINQUA, 27pp.

Zarate, M. and Blasi, A. (1993) Late Pleistocene–Holocene eolian deposits of the southern Buenos Aires Province: a preliminary model. *Quaternary International* **17**, 15–20.

Zhang, L. (1989) The landslide history and Late Cenozoic environmental factors in Sale Shan area. In Zhang, L. and Siwei, S. (eds) *International Field Workshop on Loess Geomorphological Processes and Hazards*. Lanzhou, China, 81–93.

Zhang, Z. (1995) Geological disasters in loess areas during the 1920 Haiyuan earthquake, China. *GeoJournal* **36**, 269–274.

Zhu, H. (1989) Some types of seismic landslides in loess area in China. In Zhang, L. and Siwei, S. (eds) *International Field Workshop on Loess Geomorphological Processes and Hazards*. Lanzhou, China, 64–71.

Zhu, Z. Y. and Ding, Z. L. (1994) *The Climatic and Tectonic Evolution in the Loess Plateau of China during the Quaternary*. Beijing, China, Geological Publishing House. 226pp. (in Chinese).

Acknowledgements

The authors gratefully acknowledge the support provided by the Centre for Quaternary Research at Royal Holloway (University of London) and the cartographic help of Jenny Kynaston. We also acknowledge the provision of sample loess material by David Boardman (Figure 25.12*a*) and Sanda Balescu (Figures 25.12*b*, 25.12*c*, and 25.15), and original photographs by the late Professor Wang Jingtai of Lanzhou (Figure 25.25*a*) and Dr. Armelle Billard (Figure 25.29*b*).

26. Chalk Landscapes

Fred Bell and Martin Culshaw

26.1 Introduction

Chalk has been defined as a soft, fine textured, usually white or grey limestone of marine origin that consists primarily of calcite that was deposited during Upper Cretaceous and early Tertiary times. However, hard horizons do occur in the Chalk and it can have a reddish or greenish colour due to the presence of iron oxide or glauconite respectively. Most European chalks were formed by pelagic sedimentation in a temperate to tropical shelf sea environment with water depths of between 100 and 300 m.

Generally, chalk is a remarkably pure micritic limestone, which, excluding flints and marl bands, contains over 95% calcium carbonate, which can be divided into coarse and fine fractions. The Chalk Marl in the Lower Chalk is an exception, it containing an appreciable amount of silt and clay. The coarse fraction, which may constitute 20 to 30%, falls within the 40 to 100 μm range. This contains material derived from the mechanical breakdown of large shelled organisms and, to a lesser extent, from foraminifera. The fine fraction, which takes the form of calcite particles that may be less than one micron in size, is composed almost entirely of coccoliths and may form up to, and sometimes over, 80% of certain horizons. Flints are common, especially in the Upper Chalk.

Most of the material constituting the Chalk was deposited as low-magnesium calcite, with less than 5 mole per cent magnesium in the lattice. Although low-magnesium calcite is not uncommon in other ancient limestones, usually the carbonate was deposited as aragonite and high magnesium calcite that then, because it is less stable at ordinary temperature and pressure conditions, reverted to a low-magnesium form. This is

one of the reasons why most limestones are better indurated than chalks.

The Chalk in western Europe continues westwards from northern Germany, Denmark and the Netherlands, to France and the British Isles (Figure 26.1). An extensive part of the North Sea is floored by chalk where deposition continued into Lower Tertiary or Danian times. In central and southern Europe, Upper Cretaceous strata are represented by limestones within the Alpine mountain belt. Eastwards from Poland, the Chalk extends to the northern slopes of the Caucasus, with extensions to the Black Sea, Iraq, the Caspian Sea and south-western Siberia (Figure 26.1). It also occurs around the Mediterranean Sea from the Greek islands into Turkey, in Israel, Sinai, Egypt, Libya and Algeria. Accordingly, the Chalk of the European, Asian and African basins covers a vast area. Outside this region, chalk is found in North America. For example, the Niobrara Chalk is present in Kansas and Nebraska; the Selma Chalk in Alabama, Mississippi and Tennessee; the Austin Chalk in Texas, and the San Felipe Chalk in Mexico. Mortimore (1990) indicated that chalk also occurs in Indonesia, Western Australia and New Zealand and has been found in cores obtained by the Ocean Drilling Program (ODP) (Figure 26.1).

The Chalk of England is divided into three principal depositional and faunal provinces (Figure 26.2); further details can be found in Mortimore, 1983; Bristow et al., 1997; Mortimore et al., 2001. Table 26.1 shows the main bio- and litho-stratigraphic sub-divisions of the Cretaceous Chalk Group of England and the North Sea.

Chalk varies in hardness. Hard chalks may be associated with the formation of rhombic calcite crystals that had a range of sizes, pore filling microspar and cement overgrowths on coccoliths. On the other hand, the individual particles in soft chalk are

Figure 26.1 Worldwide distribution of chalks (after Mortimore, 1990). The main shaded areas are the known extent of chalk; the small circles are sites where the Ocean Drilling Program found Chalk in cores.

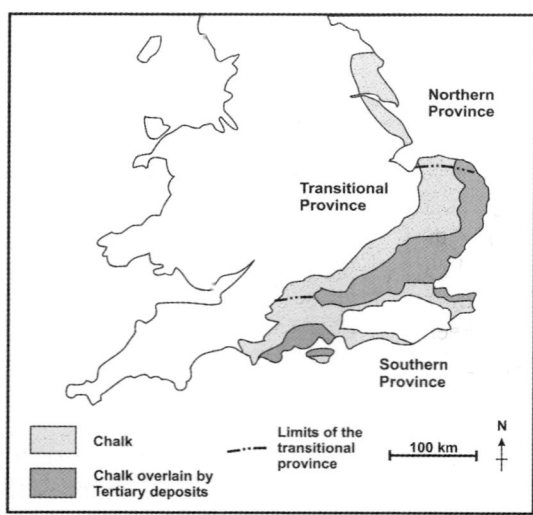

Figure 26.2 Distribution of the Chalk outcrop in England.

bound together at their points of contact by thin films of calcite. Such chalk contains only minute amounts of cement. Diagenesis, tectonic compaction, pressure solution and late-stage solution have all altered the properties of chalk from place to place.

The Alpine orogeny in Tertiary, primarily Miocene, times affected southern England, and thereby the Chalk, giving rise to structures with a dominant east-west trend (Figure 26.3). Tectonic compaction, pressure solution and late-stage solution have altered the engineering behaviour of chalk (Clayton, 1983). Stresses due to Alpine earth movements brought about mechanical disaggregation of the coccolith matrix, followed by reconsolidation, which then was followed by pressure solution and reprecipitation of the calcite under sustained pressures. Such action affected the strength and density of chalk. For instance, in areas where the Chalk dips at angles in excess of 30°, the density of some of the weakest chalks has been increased to that of the hardest hard-ground. Nonetheless, the density of tectonically modified chalks varies widely over relatively short distances.

26.2 Geomorphology of the Chalklands

The Chalk in England generally gives rise to a distinctive rolling country. Its outcrop normally forms a prominent escarpment with hard bands within the

Table 26.1   English Chalk stratigraphy (after Rawson *et al.*, 2001 and Mortimore *et al.*, 2001).

Old units	Stage	Sub-group	Southern Province		Northern Province	
			Formation	Member	Formation	Member
Upper Chalk	Campanian	White Chalk	Portsdown Chalk		Rowe Chalk	
			Culver Chalk	Spetisbury Chalk	Flamborough Chalk	
				Tarrant Chalk		
			Newhaven Chalk			
	Santonian		Seaford Chalk		Burnham Chalk	
	Coniacian					
Middle Chalk	Turonian		Lewes Nodular Chalk		Welton Chalk	
			New Pit Chalk			
			Holywell Nodular Chalk	Plenus Marls		Plenus Marls
Lower Chalk	Cenomanian	Grey Chalk	Zig/Zag Chalk		Ferriby Chalk	
			West Melbury Marly Chalk			

Chalk forming intermediate escarpments. The Chalk escarpment and dip slopes represent the surviving remnants of a sub-Palaeogene erosion surface that has been modified by late Tertiary and Quaternary geomorphological processes. It may be that the crest of the Chalk escarpment was nearing its present position by the end of Tertiary times and has undergone little further retreat subsequently.

In places, superficial deposits rest directly on the Chalk. For example, tills of Anglian and Devensian age are found on the west and east flanks of the Lincolnshire Wolds respectively. Similarly, in East Anglia much of the Chalk is covered with tills, primarily of Anglian age. These tills contain fragments of chalk of varying size, some of which may be several cubic metres in volume.

Another superficial deposit, which is associated with the Chalk of southern England, is clay-with-flints. These deposits tend to occur scattered in patches over the high ground of the Chalk, and at times are found in sinkholes and solution pipes.

Some patches of clay-with-flints may be over 10 m in thickness. As the name implies, flints and clay constitute the bulk of the material but some gravel, sand and silt may be present at some localities. Clay-with-flints generally has been considered as having been derived from the Chalk, it being a residual product left after chalk was dissolved. However, in some localities in the Weald of southeast England it would appear to have been derived from Tertiary beds. It tends to be brownish in colour and may darken with depth. The deposit was more widely distributed than it is at present and has been dissected by valleys that now are dry. Angular flint gravel occurs on the higher parts of the downlands, notably in south Dorset and the Isle of Wight. This consists of an accumulation of unworn and sometimes fragmentary flints in a matrix of quartz sand or chalk rubble. It is thought to have a similar derivation to that of clay-with-flints.

During Devensian times much of the Chalk outcrop was unglaciated and, as such, the surface

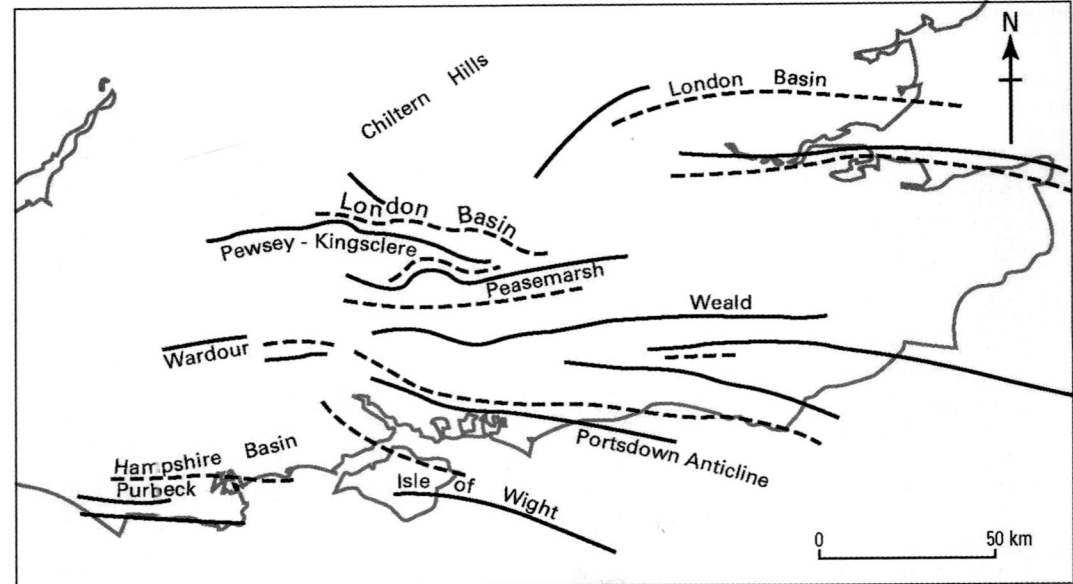

Figure 26.3   Structural trends in the southern England Chalk (after Evans and Stubblefield, 1929).

was subject to periglacial activity. Frost shattering of chalk and mass wasting of valley sides
occurred, and because of lower sea levels the
existing drainage system was rapidly incised and
accentuated. Some Tertiary deposits were eroded
from above the Chalk and some superficial
deposits were removed from valley slopes by hill-
wash and solifluction. These materials now tend
to occupy valley floors and the base of slopes in
chalkland areas. They may have been affected by
cryoturbation.

Drainage patterns

A trellis pattern of drainage is developed over
much of the chalklands, which is characteristic of
scarpland topography, with examples of river
piracy (where one stream has eroded to intersect
and capture another), obsequent streams (streams
that flow opposite to the original downslope direction following erosion at the head capturing part of
the original stream) and wind gaps occurring in
places. The Chalk escarpment of England is rarely
crossed by valleys from the west, the notable
exception being the Goring Gap through which
the River Thames flows into the London Basin.
A notable feature of chalkland scenery is the

occurrence of dry valleys and the scarp edge of the
Chalk is frequently notched by wind or dry gaps,
that is, dry valleys beheaded by the recession of
the scarp (Figure 26.4). Another significant breach
in the Chalk outcrop is made by the River Humber.

Dry valleys

Dry valleys probably represent the remnants of
a drainage system that was developed on the
Palaeogene sediments covering the dip slopes of
the Chalk. Birch and Griffiths (1996) suggested
that the position and orientation of dry valleys are
controlled by subtle structural features that have
been superimposed on the Chalk due to flexuring
and possibly faulting in older Mesozoic rocks
beneath. The form of dry valleys indicates that they
were excavated by flowing water. For instance,
typical chalk landscape consists of convex divides,
and reasonably steep-sided trough-shaped valleys.
However, there are numerous variations, valleys
may be wider and shallower or they may be asymmetrical in cross-section. At times the tributaries
of a system hang slightly above the main valleys,
whether wet or dry. A small rise in the water table
often is sufficient to provide the larger valleys
with streams, as is illustrated by bournes. These

Figure 26.4   The Devil's Dyke, Poynings, near Brighton, Sussex, England. The Dyke is a deep dry valley in the South Downs escarpment formed by the capture and rejuvenation of a dip-slope valley by a scarp-facing spring and stream (British Geological Survey photograph no. A13405, NERC (Natural Environment Research Council) copyright).

are intermittent streams that occur in larger dry valleys after heavy rainfall. Bourne flows may last for a few weeks or months. There frequently is a lag in time between the maximum heavy rainfall and the maximum flow of a bourne because of the time taken for the rainwater to percolate through the Chalk and thereby cause the water table to rise. The head of a bourne may extend upsteam after it makes its first appearance.

Three explanations have been advanced for the formation of dry valleys. As streams flow in some dry valleys after heavy rainfall, it has been suggested that dry valleys may have been carved by water during times when the climate was wetter and the water table therefore higher. Alternatively, it has been suggested that these valleys could have been formed when the ground was frozen during Pleistocene times. This would mean that the dispersal of precipitation, because it could barely infiltrate the frozen ground, was confined predominantly to runoff. The third explanation maintains that the height of the water table is influenced

to a large extent by the level of the discharge of springs at the foot of the escarpment and that the escarpment has receded as a result of erosion. The height of the escarpment increases as it retreats, which leads to a relative fall in the position of the water table; the latter also is affected by the surface of the clay beneath the Chalk being lowered.

Coombes

Coombes are another topographical feature characteristic of chalklands. These are bowl-shaped hollows that occur on the flanks of dry valleys. Some coombes that cut back into scarp slopes are similar to corries, except in size, and may have been formed by frost action working back into snow-filled hollows on a hillside during Pleistocene times. Other coombes resemble overflow channels in form but strike into hillsides and end blindly in a steep face. A number of such coombes often occur together, for example, in the South Downs, and may have been formed by spring sapping near the base of scarps under the periglacial conditions that existed in the south of England during the Pleistocene. Springs presumably flowed more vigorously as ice thawed and still-frozen ground meant that all meltwater was concentrated in runoff. The material eroded from coombes formed extensive fans of reworked chalk debris that accumulated downhill of the coombes and over the slopes at the base of the escarpments. This material is referred to as coombe rock.

Dissolution

Chalk being a carbonate rock, is subject to dissolution. The degree of aggressiveness of water to calcium carbonate can be assessed on the basis of the relationship between the dissolved carbonate content, the pH value and the temperature of the water. At any given pH value, the cooler the water the more aggressive it is. As solution continues its aggressiveness decreases and eventually ceases when saturation is reached. Hence, solution is greatest when the bicarbonate saturation is low. This occurs when water is circulating, and fresh supplies with low lime saturation are available continuously.

Calcium carbonate dissolves at a rate and in a manner which is influenced by its solubility ($0.015\,kg/m^3$ at $10\,^{\circ}C$ in pure water) and specific solution rate constant ($0.4\,m/s \times 10^3$ at $10\,^{\circ}C$ with a flow rate of $0.05\,m/s$). Aqueous dissolution of calcium carbonate introduces the carbonate ion into water (Weyl, 1959; Garrels et al., 1960). Further slight dissociation releases H^+ and OH^- ions, after which H^+, HCO_3^-, CO_3^{2-}, OH^- and Ca^{2+} become active. The relative distribution of H_2CO_3, H^+, HCO_3^- and CO_3^{2-} in solution as a function of pH is: below pH 5, H_2CO_3 dominates; from pH 6 to 10, HCO_3^- dominates; above pH 10, CO_3^{2-} dominates. The pH of rainwater is commonly 5 to 5.5 (although it may range from below 4 to more than 7), but in contact with $CaCO_3$, it rises to about 8 to 8.2. In the typical range of pH 7 to 9 of waters draining from carbonate material, the dominant anion associated with dissolved Ca^{2+} is HCO_3^-. Thus, $CaCO_3$ is dissolved slightly as $CaCO_3$ but predominantly as Ca^{2+} with HCO_3^-, and less as Ca^{2+} with CO_3^{2-}.

Both the rate of flow and the area of the material exposed to flowing water are significant controls on the process of dissolution. Although the porosity of chalk may be high, the pore size is small (see above). Consequently, most of the water that flows through chalk is concentrated along systems of discontinuities. Dissolution along discontinuities can give rise to the formation of solution pipes, which tend to develop at the intersection of discontinuites. Enlarged discontinuity systems mean that the flow of water through chalk increases. For instance, Atkinson and Smith (1974) mentioned a flow rate of 2 to 3 km per day through a well-developed system in chalk in Hampshire, and Pitman (1983) referred to transmissivities of approximately 1200 m per day in fissured zones in chalk in east Yorkshire.

Dissolution aids disintegration of chalk by weakening its fabric and by emphasising structural weaknesses, however slight. Any enlargement of pores brings about an increase in stress within the fabric of the rock thus reducing its strength and increasing stress corrosion. Accordingly, dissolution can weaken the bonding between grains, producing a weaker material. In fact, such breakdown ultimately can give rise to putty chalk.

Sinkholes

Sinkholes are a part of chalkland landscape (Figure 26.5), as well as areas where the Chalk occurs beneath shallow overlying deposits. Solution features, including cavities, occur across the outcrop of the Chalk. Although sinkhole formation is occurring now, Banks et al. (1995) suggested that it was probably more noteworthy under periglacial conditions towards the end of the Pleistocene.

Culshaw and Waltham (1987) described four types of sinkholes formed in more soluble rocks (solution, collapse, subsidence and buried) while Waltham and Fookes (2003) proposed a six-fold classification in which collapse sinkholes were subdivided into collapse, caprock and dropout sinkholes. In chalk, solution sinkholes of significant size are rare, and collapse sinkholes almost non-existent because of the absence of large caves. Large subsidence sinkholes are known in sand-covered chalk areas in Norfolk and Dorset (UK) but are inactive; active subsidence sinkholes are small and of limited distribution. Buried sinkholes are commonly found near the contact of the

Figure 26.5   Buried sinkholes in the Chalk of north Kent (UK) exposed in excavations for the Channel Tunnel Rail Link (photograph courtesy of Richard Ellison, British Geological Survey, NERC copyright).

Chalk and the overlying Tertiary and superficial deposits. This is largely due to the concentration of runoff at such locations. These sinkholes are usually small, and wide and shallow in shape, though deep narrow pipes occur in some areas (for example, north of London).

The enlargement of pipes or cavities by continuing dissolution can lead to sudden collapse with associated subsidence at the surface. Moreover, voids can migrate gradually upwards through chalk due to material collapsing. Collapsed material occupying a sinkhole may possess a metastable structure; for instance, a zone of loosely packed ground may be present if chalk, sand and gravel collapsed into a sinkhole. Lowering of the Chalk surface beneath overlying deposits due to solution can occur, disturbing the latter deposits and lowering their degree of packing. Hence, the Chalk surface may be extremely irregular in places.

Most sinkholes can be detected at the surface but if a cavity develops in clayey deposits overlying a sinkhole, then its presence may not be seen at the surface. Roof collapse of the cavity means that it migrates upwards and eventually it may occur at the surface without warning. When streams disappear into sinkholes, then the feature is called a swallow hole. The diameter of sinkholes varies from a few metres up to 20 m or more and occasionally can be 100 m across or even greater. On occasion numerous sinkholes may occur. For example, the Mimms Hall Brook in Hertfordshire flows over the London Clay and onto the Chalk going underground via sinks at Water End. The many sinkholes that have developed have merged into a single depression so that a temporary lake may form after heavy rainfall because the sinkholes have been choked by mud and cannot convey water away. As this example illustrates, a common location of sinkholes is near the boundaries of overlying relatively impervious London Clay or clay-with-flints. Streams flowing over these beds descend into sinkholes after reaching the Chalk, the latter often having an aligned occurrence. Sinkholes also may occur in valleys, on slopes or near the tops of hills. Conversely, they appear to be absent from areas of bare chalk such as found on Salisbury Plain. Some sinkholes have a smaller steeper depression within the normal smooth depression, which suggests that they have been reactivated.

As noted, sinkholes also may develop beneath a shallow cover of sand or gravel, which subsides as the sinkhole grows. These areas of sands and gravels commonly possess a heathland vegetation that gives rise to humic acid, which presumably aids the dissolution process. Exposures in quarry faces in the Chalk show that sinkholes narrow rapidly downward, holes of 6 m diameter or more at the surface tapering to less than a metre diameter within a shallow depth in the Chalk (Figure 26.5). Some of the older, better developed sinkholes may descend to over 15 m in the Chalk. Dissolution may have taken place along bedding planes crossed by such sinkholes. Normally, these sinkholes are occupied by gravelly, sandy or silty material, as are the associated enlarged bedding planes.

New sinkholes often appear at the surface after a period of heavy rainfall, and may form a circular shaft or pipe with steep sides a metre or two in diameter and a few metres deep (Figure 26.6). Indeed, West and Dumbleton (1972) suggested that pipes a metre in depth could form within 10 years where the infiltration of rainfall into the ground is concentrated. They further suggested that such a concentration could be due to local variations in the topography or permeability of the Chalk or shallow overburden, or could occur at the intersections of major joints in the Chalk. West and Dumbleton concluded that care should be taken to avoid the risk of pipe formation or the reactivation of sinkholes due to excessive concentration of water into chalk from hard surfaces such as major roads, soak-aways, leaking water mains or sewers, and that soak-aways also should be located with care. Wherever possible, engineering structures obviously should avoid areas affected by sinkholes. If they cannot be avoided, then the sinkholes will require treatment (see below). Sinkholes, their investigation, assessment and consequences for engineering are discussed by Waltham *et al.* (2005).

Cavern systems

True karstic cavern systems, which are accessible for exploration, have been recorded in the Chalk. For example, Banks *et al.* (1995) mentioned that

Figure 26.6 Solution pipes at Seaford Head, Sussex (UK) (photograph courtesy of Andy Farrant, British Geological Survey, NERC copyright).

such systems have been encountered in south-east England during sewer construction and well drilling. They also have been exposed in cliff sections, perhaps the most notable being at Beachy Head where 354 m of accessible passageway have been explored. Furthermore, Bradshaw *et al.* (1991) referred to cavern systems up to 2 km in length in the Chalk of the Paris Basin. Indeed, Lowe (1992) suggested that sub-anthropogenic caves (that is, caves too small to be explored by man) are probably much more common in the Chalk than previously have been appreciated.

Subsidence in chalk due to mining

The Chalk has been, and still is, extensively worked for lime. In the past many lime workings took the form of surface quarries and pits often with associated subsurface mining. These mine workings were located at shallow depths and, because they lay above the water table, were dry. Subsurface workings took the form of, and relied on the strength of, the Chalk to form self-supporting roofs.

In addition to being dug for lime, the Chalk also has been worked as a source of flint. The earliest flint mines in England date back to Neolithic times. The presence of such cavities, galleries and chambers at shallow depth beneath the ground, most of which are unrecorded, represents a hazard to surface development, as well as to existing property and, indeed, lives have been lost, owing to roof collapse into the workings. The timing of collapse generally is difficult or impossible to predict. However, factors that influence the possibility of roof collapse, together with the migration of the resultant void to the surface, include the width of the unsupported roof span, the character of the Chalk and any other cover materials, particularly their shear strengths and the incidence and geometry of any discontinuities, the dip of the beds, the height of the workings, the depth of overburden and the groundwater conditions.

A review of the geological conditions that may aid the collapse of mine workings in chalk has been provided by Smith and Rosenbaum (1993). Where weak superficial deposits occur above workings, they may 'flow' into voids that have reached rockhead, thereby giving rise to surface features, which may vary from gentle dish-shaped hollows to inverted cone-like depressions of significantly greater diameter than that of the associated void.

The height to which voids can migrate depends largely on the bulking factors of the roof rocks and the height of the workings. When a fall occurs, the material involved bulks so that there comes a point at which upward migration is arrested unless the workings are near the surface. If the bulking factor of chalk is around 1.3, then the height above the roof of the workings to which a void would migrate is approximately 2.5 to 3.5 times the height of the workings. However, void migrations in excess of this have been recorded but most of these are due to the near-surface weakening effects of periglacial action and of dissolution.

Old mine workings are entered by shafts, drifts or adits that run directly from the surface. Each mine could have more than one entrance. Unless filled correctly, these also represent hazards. In the case of shafts, many have been only partially filled, others may be capped and the cap covered with fill. With time, deterioration of the capping may lead to collapse, exposing the shaft. Such an event, although rare, can have serious consequences. Indeed, lives have been lost when individuals have fallen into shafts that opened beneath them. For instance, on 22 November 1967, a woman and child fell into a shaft, with a surface diameter of 5.5 m, which opened up in West Street, Frindsbury, near Rochester, Kent (Charman and Cooper 1987). The shaft was 38.7 m deep with water occurring at 23.8 m. The bodies were not recovered.

An additional factor that must be considered when dealing with subsidences of workings is that the Chalk, as mentioned above, is subject to dissolution. In fact, many chalk subsidence incidents have been man induced where flows of water into the ground have been concentrated into small areas by soak-aways, leaking drains and water supply pipes, and even over-zealous garden watering can have similar effects (Figure 26.7). Such concentrated flows accelerate dissolution, as well as reducing the strength of chalk. Urbanisation can make a significant contribution in causing subsidence in that it radically changes the way in which water enters the ground compared with an open, undeveloped site. By rendering 50 to 80%

Figure 26.7   Subsidence in the Chalk at Jacqueline Close, Bury St Edmunds, Suffolk, due to dissolution and thinning of the mine roof below leading to its collapse (photograph courtesy of the East Anglian Daily Times).

of the surface plan area relatively impermeable, infiltration and percolation of precipitation occurs only in the remaining areas of open ground. Furthermore, much of the water draining from built-over areas is concentrated into soak-aways. Backfilled service trenches also tend to act as preferred paths for water ingress.

Mine workings in the Chalk in south-east England are referred to as 'deneholes'. Some of the deneholes have ancient origins in that they are believed to have been excavated during pre-Roman times. Others are presumed to be of Roman origin whilst the remainder could have been excavated as late as the fourteenth century (Bell *et al.*, 1992). Commonly, a circular shaft, 10 to 20 m or so in depth, leads to two or more chambers. The most commonly occurring form is the six chambered or double trefoil type of dene-hole. Subsequent improvement in mining practice led to the development of the pillared denehole in

which the walls between the chambers were partially removed to create pillars (Figure 26.8). Other forms of workings in the Chalk include chalkwells, bell pits, chalkangles and pillared workings (Edmonds *et al.*, 1987). Pillared workings in the Chalk, as noted above, are the most recent and generally date from the beginning of the nineteenth century. The pattern of the workings varies, in particular the size of the pillars. The joint pattern and the presence of any solution pipes tended to influence the layout of the mines. The stalls were typically 2 to 3 m wide and about 9 m in height and were worked by benching. Subsidences resulting from the collapse of mines in the Chalk occasionally occur, one of the most notable being at Jacqueline Close, Bury St Edmunds, Suffolk, in 1966 (Figure 26.7). More recently, chalk mine workings have collapsed in Chislehurst, south-east London (1988 but with a recorded history of collapse going back to 1858),

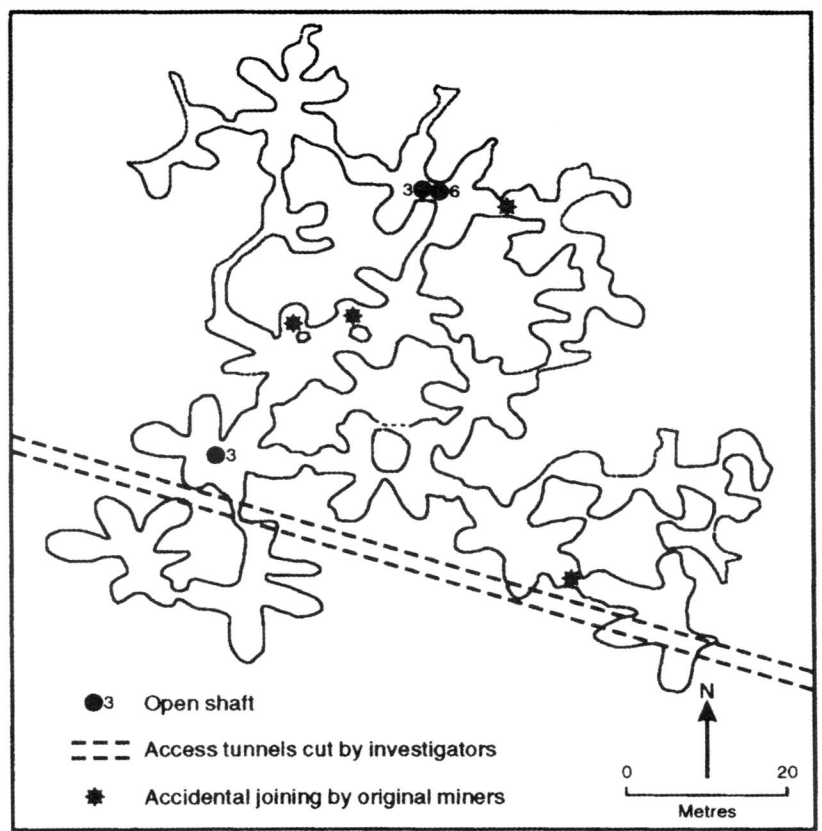

Open shaft

Access tunnels cut by investigators

Accidental joining by original miners

N

0   20
Metres

Figure 26.8   Linked deneholes at Hangman's Wood, Grays, Essex (after Le Gear, 1979).

Reading, Berkshire (2000) and Blackheath, in south-east London (2002).

Erosion and mass movement

The silty clay soils that are associated with areas of chalklands may be prone to soil erosion by water or wind when the ground lies fallow. The impacts of soil erosion are confined primarily to the degradation of agricultural land. Nonetheless, roads and ditches at times have to be cleaned, some siltation may occur in waterways and standing bodies of water, and surface water quality may be adversely affected (Boardman, 1990).

The Chalk gives rise to some spectacular cliffs where it meets the coast (Figure 26.9). Many of these are near vertical and may exceed 150 m in height. The cliffs are undergoing erosion and the

average rate of retreat may be between 0.05 and 0.10 m per year. Rockfalls initiated by weathering processes are the main agent of degradation (Figure 26.10). Nevertheless, landslides in the form of toppling, planar and wedge failures occur along the cliffs, the types of slide being controlled by the discontinuity pattern, and often are triggered by heavy rainfall or freeze–thaw action. The cliffs also are undermined by coastal erosion that ultimately leads to a near-vertical collapse, discontinuities again affecting the character of the failure.

Yet another cause of landslides in some parts of the Chalk has been attributed to groundwater trapped in stress relief fissures behind and sub-parallel to the cliff face. This stress causes shearing in the lower strength, less permeable marly

Figure 26.9   Steep chalk cliffs north of Flamborough Head, Yorkshire (photograph: F. G. Bell copyright).

horizons, especially when they occur near the base of the cliff. A notable example of a landslide-affected area is Folkestone Warren, which is a naturally unstable stretch of coastline, 3.2 km long, between Folkestone and Dover, in south-east England. A succession of slides affected the railway, located at the foot of the cliff, after it was built in the mid-nineteenth century but the most notable occurred in 1915 when a series of rotational slides and chalk falls from the cliffs blocked the line. One of these slides was transformed to a flow and had a runout to sea of some 400 m. Fortunately, a train was stopped by a signalman before it entered the section of the line affected by the slide but subsequent movements derailed the train. It took five years to rectify the damage. According to Hutchinson *et al.* (1980) the main causes of the movements were intense marine erosion along the toe of previously slid material and the development of high porewater pressures along potential sliding surfaces.

Occasionally, landslides occur on chalk scarplands. For example, several old landslide areas occur on the Chalk escarpment immediately north of the entrance to the Channel Tunnel and the possibility of their reactivation, particularly by construction operations, had to be investigated (Birch and Warren, 1996) (Figure 26.11). However, even engineered slopes in sub-horizontally bedded chalk are rarely subject to major instability and cuttings with 45° slopes in rubbly, blocky chalk rarely exhibit spalling (that is, the detachment usually of individual blocks commonly by freeze–thaw action). Nonetheless, Phipps and McGinnity (2001) concluded that a slope angle of 53° is the steepest that can be maintained in unweathered chalk before degradation processes present a risk of causing disturbance in cuttings.

As would be expected, landslides are more common in the superficial deposits. Phipps and McGinnity (2001) noted that, especially where slopes are steepened in superficial deposits for example by cuttings, debris flows, washouts and slumps may take place. They claimed that debris flows occur in deep heterogeneous superficial deposits that contain inter-layered clay soils and

Figure 26.10   Rockfall at Beachy Head, East Sussex, England. Left the Devil's Chimney in February 1999; right, the Devil's Chimney in August 2001 following the rockfall of 4 April 2001 (photographs courtesy of Peter Hobbs, British Geological Survey, NERC copyright).

gravels, and that they are triggered by extreme rainfall events. Indeed, Phipps and McGinnity further maintained that collapse of the soil fabric on saturation of these materials might lead to their liquefaction. Similarly, washouts are initiated by heavy rainfall affecting clayey sands and silts. Initially, slump failure occurs as a result of rotational or sub-planar movements in clay-rich deposits and may be followed by flow.

Springs

A spring line occurs at the foot of the scarp slope where the Chalk rests upon less permeable ground such as the Gault Clay, from which contact springs issue. Similarly, where the dip slope of the Chalk sinks beneath less permeable sediments, then another spring line occurs from which overflow springs may emerge.

Flooding

Flash floods have been associated with stream catchments located on the Chalk. Such floods often are due to a high water table caused by either intense storms or rapid snowmelt associated with frozen ground, both generating rapid runoff. An example of the first type of flash flood occurred at Louth on the margins of the Lincolnshire Wolds in 1922 when up to 153 mm of rain fell in the neighbouring small catchment on the Chalk. This gave rise to torrents in valleys that normally were dry and led to the level of the River Ludd rising by 5 m in 15 minutes, the flood wave moving through Louth at approximately 140 m³/s. Twenty-two people were drowned and 1250 made homeless, with buildings being demolished. The second type of example is provided by the Great Till flood that occurred in a chalk catchment on the downlands of Salisbury Plain. Melting snow and frozen ground meant that the River Till burst its banks at Shrewton and rose 2.5 m above its normal level. In this instance, three people were drowned, 72 houses destroyed and 200 individuals made homeless (Cross, 1967).

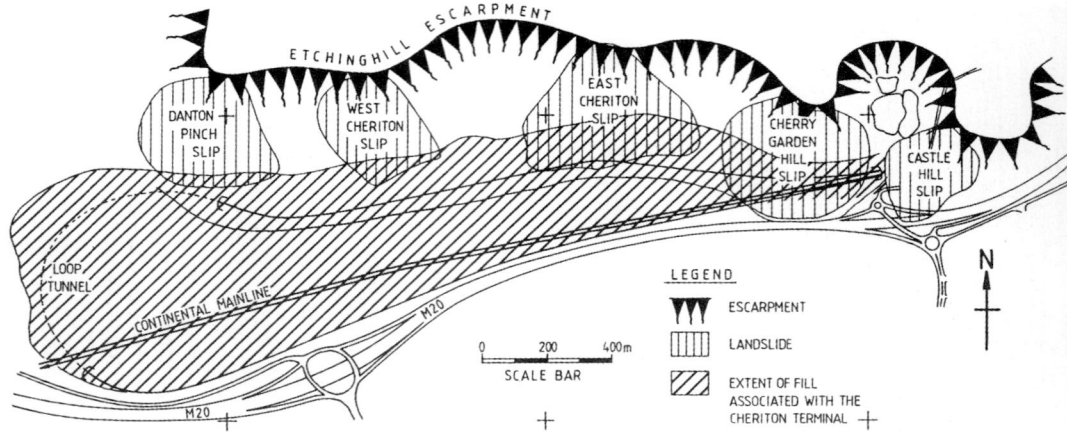

Figure 26.11   Landslides on the Chalk escarpment immediately north of the Channel Tunnel (after Birch and Griffiths, 1996).

Many of the features referred to above are illustrated in the idealized model of the chalk-lands given in Figure 26.12.

26.3 Chalk engineering

The Chalk has a range of distinctive engineering properties (Table 26.2), including significant variations in density (Higginbottom, 1965; Bell, 1977; Bell *et al.*, 1999, Clayton, 1983) and hardness (Mortimore and Fielding, 1990; Lord *et al.*, 1994, 2002). Generally, the Chalk is regarded as moderately strong, but occasionally is moderately weak to moderately strong (Anon., 1977). The unconfined compressive strength of chalk undergoes a marked reduction when it is saturated (Table 26.2). For instance, according to Bell *et al.* (1999) some samples of Upper Chalk from Kent suffer a dramatic loss on saturation amounting to almost 70%. Samples from the Lower and Middle Chalk may show a reduction in strength averaging over 50%. In fact, the weaker the rock, the greater the loss of strength appears to be on saturation. Porosities vary between less than 25% to greater than 40%. Saturation moisture content has been used as an indicator of hardness with a value of 15% being regarded as hard whilst those chalks for which it exceeds 25% are soft. The permeability of the

Chalk is governed by its discontinuity pattern rather than by inter-granular flow. Values of primary permeability range from roughly 1×10^{-11} to 1×10^{-8} m/s (Ineson, 1962; Bell *et al.*, 1999). Fissure permeability can be as much as 1×10^{-5} m/s, which accounts, in part, for why the Chalk is the most important aquifer in Britain, supplying approximately 40% of the groundwater abstracted (MacDonald and Allen, 2001).

A number of Chalk classifications have been developed (e.g. Ward *et al.*, 1968, see Table 26.3; Wakeling, 1970; Spink and Norbury, 1990, see Table 26.4); the most recent (Lord *et al.*, 1994) is based on the hardness, or more specifically dry density on the one hand and the nature of the discontinuities on the other (Figure 26.13). Chalk discontinuities are an important control on the mass hydrogeological and mechanical behaviour, ranging from microscopic grain boundaries and microfractures to major tectonic joint and fault structures (Figure 26.14; Table 26.5).

During cold weather chalk may suffer frost heave, ice lenses up to 25 mm in thickness being developed along bedding planes (Lewis and Croney, 1965). Higginbottom (1965) suggested that a probable volume increase of some 20 to 30% of the original thickness of the ground might ultimately result. The presence of permafrost in the surface layers of chalk during the late

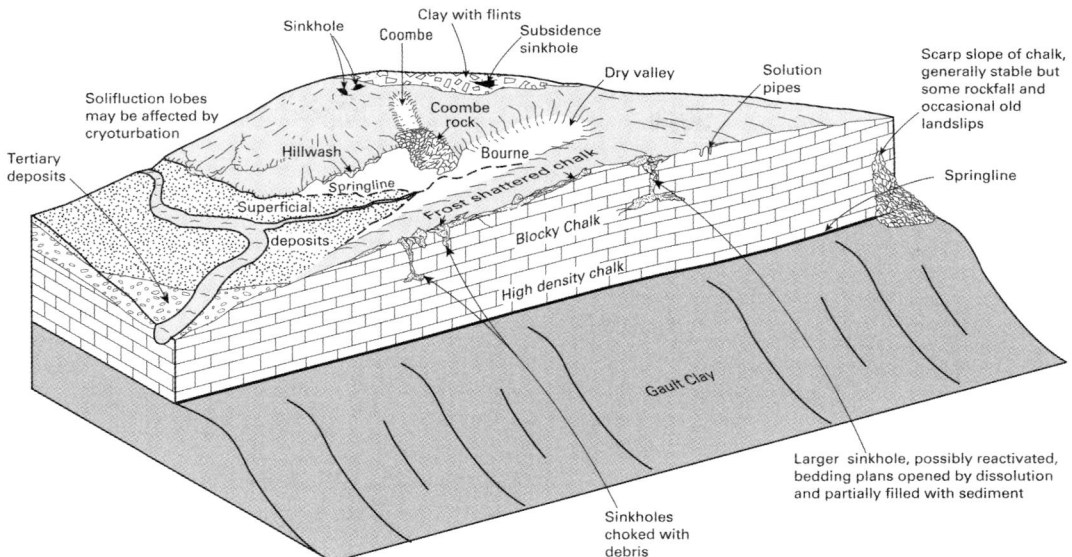

Figure 26.12   Conceptual block model of chalk landscape in the UK.

Pleistocene meant that it was subject to frost heave. In fact, most chalk possesses a surface zone of very closely jointed material in which fractures tend to be crazed, with curvi-planar surfaces typical of frost shattering. Frost shattering has been found to extend to depths of 30 m and may be concentrated along certain preferred planes (Figure 26.15). Frost churning has led to the obliteration of the macrostructure in the upper metre or so of the Chalk in many places and to the formation of a mass of pasty remoulded chalk (putty chalk) enclosing angular fragments of unaltered material, which increase in size with depth.

Chalk can present a number of problems to the engineer because of the presence of discontinuities; because of the reduction in mechanical properties such as density, strength and deformation modulus by weathering; because of the presence of dissolution features; because of the effects of periglaciation that has given rise, for example, to frost shattering, and because of past mining.

Foundations

Fresh chalk generally has proved a satisfactory foundation material for many buildings and structures. For instance, in a review of foundations on chalk, Lord (1990) maintained that shallow footings could be founded successfully in grades IV and V chalk, according to the Mundford classification, at bearing pressures of 250 and 300 kPa. In 1994 Lord et al. noted that the ultimate bearing capacity of chalk could range between 1.5 MPa for low-density chalk to in excess of 8 MPa for chalk of high density. They went on to point out that even with the lowest ultimate bearing capacity recorded and a factor of safety of 3, the minimum safe bearing capacity would be 500 kPa and that loads of this amount are rarely imposed by shallow foundation structures. Consequently, the design of shallow foundations in chalk is governed by settlement.

What is more, Burland et al. (1975) found that settlements of a five-storey building constructed on soft low-grade chalk at Reading were very small, as did the results of an investigation carried out at Basingstoke by Kee et al. (1975). Previously, Wakeling (1968) had indicated that the settlement of the piers carrying the bridge over the river Medway, Kent, amounted to some 28 and 35 mm for a mean effective foundation pressure of around 650 kPa.

Table 26.2 Some physical properties of chalk from Yorkshire, Norfolk and Kent (after Bell et al., 1999).

	Yorkshire*			Norfolk**				Kent***
	Lower	Middle	Upper	Lower	Melbourn Rock	Middle	Upper	Upper
Specific gravity								
Maximum	2.73	2.71	2.72	2.71	2.72	2.74	2.72	2.72
Minimum	2.65	2.69	2.67	2.66	2.68	2.68	2.70	2.65
Mean	2.71	2.70	2.70	2.68	2.70	2.70	2.71	2.69
Dry density (Mg/m³)								
Maximum	2.13 (L)	2.30 (M)	2.23 (M)	2.17 (L)	2.23 (M)	1.81 (L)	1.70 (VL)	1.61 (VL)
Minimum	1.85 (L)	1.76 (VL)	1.77 (VL)	1.71 (VL)	2.04 (L)	1.62 (VL)	1.54 (VL)	1.35 (VL)
Mean	2.08 (L)	2.14 (L)	2.06 (L)	1.99 (L)	2.17 (L)	1.76 (VL)	1.61 (VL)	1.44 (VL)
Saturated density (Mg/m³)								
Maximum	2.34	2.43	2.41	2.38	2.44	2.16	2.04	1.97
Minimum	2.16	2.11	2.05	2.06	2.27	2.01	1.93	1.82
Mean	2.31	2.34	2.29	2.24	2.34	2.14	2.00	1.91
Saturation moisture content (%) (saturation method)								
Maximum	17.2	19.9	22.9	21.5	11.4	24.8	28.0	37.2
Minimum	10.8	7.2	7.9	9.2	7.5	18.3	21.9	25.2
Mean	11.2	10.2	11.6	13.4	9.5	19.9	25.2	32.6
Effective porosity (%)								
Maximum	30.2 (VH)	35.0 (VH)	36.4 (VH)	34.4 (VH)	27.0 (H)	38.2 (VH)	43.2 (VH)	45.7 (VH)
Minimum	17.2 (H)	16.2 (H)	17.7 (H)	19.9 (H)	16.1 (H)	30.2 (VH)	34.3 (VH)	29.6 (H)
Mean	20.6 (H)	21.8 (H)	23.9 (H)	26.5 (H)	19.8 (H)	34.4 (VH)	39.9 (VH)	41.7 (VH)
Absolute porosity (%)								
Maximum	31.9	35.1	38.3	36.9	33.7	40.2	46.2	52.4
Minimum	20.5	18.2	17.7	19.9	18.8	33.2	37.3	38.7
Mean	26.8	24.8	24.4	26.6	29.2	35.1	40.6	47.0
Permeability (×10⁻⁹ m/s)								
Maximum	1.2					2.2		37.0
Minimum	0.3					0.5		13.9
Mean	0.9					1.4		27.7

Table 26.2 (*Continued*).

Dry unconfined compressive strength (MPa)								
Maximum	32.7 (MS)	36.4 (MS)	34.0 (MS)	30.5 (MS)	38.3 (MS)	25.1 (MS)	12.7 (MS)	6.2 (MW)
Minimum	19.1 (MS)	25.2 (MS)	18.1	14.2	22.1 (MS)	7.4 (MW)	6.9 (MW)	4.8 (MW)
Mean	26.4 (MS)	30.7 (MS)	25.6 (MS)	21.0 (MS)	29.1 (MS)	13.0 (MS)	9.5 (MW)	5.5 (MW)
Saturated unconfined compressive strength (MPa)								
Maximum	16.2	20.4	15.9	13.7	17.5	10.3	5.1	2.2
Minimum	8.6	11.7	7.4	6.2	8.9	3.1	2.8	1.4
Mean	13.7	16.8	11.9	10.7	14.3	5.8	3.6	1.7
Young's modulus (E_{t50} GPa) (tangent modulus at 50% failure)								
Maximum	18.4	21.7	17.1	14.1	18.9	10.4	8.2	4.6
Minimum	7.5	9.1	7.4	6.9	7.3	5.0	4.1	4.2
Mean	12.7	15.2	11.7	8.7	13.5	8.4	6.7	4.4

Dry density: VL = very low, less than 1.8 Mg/m³; L = low, 1.8 to 2.2 Mg/m³; M = moderate, over 2.2 Mg/m³.

Porosity: H = high, 15 to 30%; VH = very high, over 30% (Anon., 1979).

*Yorkshire: Lower — *H. subglobosus* (? = *S. gracile*) zone near Speeton; Middle — *T. lata* zone, Thornwick Bay; Upper — *M. coranguinum* zone, Selwicks Bay.

**Norfolk: Lower — *S. varians* (? = *M. mantelli*) zone, Hunstanton; Melbourn Rock and Middle — *T. lata* zone, Hillington; Upper — *M. coranguinum* zone, Burnham Market.

***Kent: Upper — *M. coranguinum* zone, Northfleet.

Unconfined compressive strength: W, weak, 1.25–5 MPa; MW, moderately weak, 5–12.5 MPa; MS, moderately strong, 12.5–50 MPa (Anon., 1977).

Table 26.3 Correlation between grades and the mechanical properties of Middle Chalk at Mundford (after Ward et al., 1968).

Grade	Description	Approx. range of Young's modulus E (MPa)	Approx. value of dynamic Young's modulus $F_{d,u}$ (MPa) (after Abbiss, 1979)	Range of compression wave velocities (km/s) (after Grainger et al., 1973)	Bearing pressure causing yield (kPa)	Creep properties	SPT N value (after Wakeling, 1970)**	Rock mass factor (after Burland and Lord, 1970)
V	Structureless melange. Unweathered and partly weathered angular chalk blocks and fragments set in a matrix of deeply weathered remoulded chalk. Bedding and jointing are absent.	Below 500	Below 500	0.65–0.75	Below 200	Exhibits significant creep	Below 15	0.1
IV	Friable to rubbly chalk. Unweathered or partially weathered chalk with bedding and jointing present. Joints and small fractures closely spaced, ranging from 10–60 mm apart.	500–1000	800	1.0–1.2	200–400	Exhibits significant creep	15–20	0.1–0.2
III	Rubbly to blocky chalk. Unweathered medium to hard chalk with joints 60–200 mm apart. Joints open up to 8 mm sometimes with secondary staining and fragmentary infillings.	1000–2000	4000	1.6–1.8	400–600	For pressures not exceeding 400 kPa creep is small and terminates in a few months	20–25	0.2–0.4
II	Medium hard chalk with widely spaced, closed joints. Joints more than 200 mm apart. Fractures irregularly when excavated, does not break along joints. Unweathered.	2000–5000	7000	2.2–2.3	600–1000	Negligible creep for pressure of at least 400 kPa	25–35	0.6–0.8
I	Hard, brittle chalk with widely spaced, closed joints. Unweathered.	Over 5000	Over 10 000	Over 2.3	Over 1000	Negligible creep for pressure of at least 400 kPa	Over 35	Over 0.8

* Ward et al. (1968) emphasised that their classification was specifically developed for the site at Mundford and hence its application elsewhere should be made with caution.

** The correlation between SPT (Standard Penetration Test) N value and grade may be different in the Upper Chalk (see Dennehy, 1976).

Table 26.4 The description and grading of Middle and Upper Chalk for engineering purposes (after Spink and Norbury, 1990).

Grade[1]	Structure	Colloquial description of grade[2]	Definitions of grade		Typical features of grade		Word order for descriptions
			% comminuted chalk matrix[3]	% coarser fragments[3]	Weathering of coarser fragments[8]	Strength of coarser fragments[9]	
VI	Structureless chalk, bedding and jointing absent	Putty chalk with small lumps	Greater than about 35%	Less than about 65%	Moderately, highly or completely weathered	Very weak or weak	Structureless[10] CHALK composed of: soil strength of matrix;[11] colour of matrix;[12] nature of matrix material; amount of fragments; presence and nature of flints; other features (Grade VI)
V		Chalk lumps in comminuted matrix	Less than about 35%	Greater than about 65%	Moderately or highly weathered	Very weak or weak	Structureless[10] CHALK composed of: angularity and size of fragments: colour of fragments: weathering of fragments;[14] strength of fragments;[15] amount of matrix;[13] nature of matrix material; presence and nature of flints; other features (Grade V)
			Fracture spacing (mm)[5,6]	Fracture width (mm)[5,6]	Material weathering[8]	Material strength[9]	Colour: rock material weathering;[14] CHALK:
IV		Rubbly chalk	Extremely closely to very closely spaced <60[7]	Open or infilled >5[7]	Moderately or highly weathered	Very weak or weak	rock material strength;[15] discontinuity type;[16] discontinuity spacing;[17] discontinuity width and nature of infill if appropriate; discontinuity orientations (*in situ* observations only);[18]
III	*In situ* structured chalk, with bedding and jointing	Rubbly to blocky chalk	Closely spaced 60–200	Open or infilled <3	Slightly or moderately weathered	Weak or moderately weak	
II		Blocky chalk	At least medium spaced >200	Tight and clean	Fresh or slightly weathered	Weak or moderately weak	presence and nature of flints; other features (Grade)
I		Brittle and massive chalk	At least medium spaced >200	Tight and clean	Fresh or slightly weathered	Moderately weak or moderately strong	

Notes:

1. Intermediate grades may be assigned to material that is intermediate between two grades e.g. III/IV. However V/IV or IV/V is not admissible.

2. Colloquial descriptive terms are undefined and should not be used in description.

3. The cohesive matrix is comminuted fragments of chalk typically of silt and sand size. The coarser fragments of chalk are typically gravel size or larger.

4. The 35/65% cut-off between Grades V and VI is approximate and dependent on assessed engineering behaviour. Grade VI is cohesive matrix dominated, Grade V is granular dominated by the coarser fragments of chalk.

5. The grading is determined on the assessed average fracture spacing (including both bedding fractures and joints) and the assessed typical fracture width or aperture.

6. Other combinations of fracture spacing and fracture width are possible. Any undefined gradings have to be assessed with care, noted and discussed.

7. The original 'Mundford' definition of Grade IV is fractures 10 to 60 mm spacing, open or infilled up to 20 mm. Original 'Mundford' terms are undefined.

8. Within each grade the degree of weathering varies, typical ranges are given. Original 'Mundford' terms are undefined.

9. Within each grade strengths vary, typical ranges are given. Original 'Mundford' terms are undefined. If the strength is exceptionally high or low compared with the typical range, and considered to alter the engineering behaviour, then the chalk may be upgraded or downgraded respectively, but this should be noted and discussed.

10. 'Structureless' may be replaced with 'Reworked' if evidence of reworking is present, such as included foreign gravel, or if the chalk is present within an alluvial or glacial sequence.

11. Soil strengths may be assessed where the matrix is dominant, but only for *in situ* observations.

12. Although Upper and Middle Chalk are white the matrix of Grades VI and V is often cream or light brown.

13. The proportion of secondary matrix or coarser fragments is quantified by the terms 'little', 'some', 'much', as defined in Table 1 of Spink and Norbury (1990).

14. The weathering of the chalk material or coarser fragments is described in accordance with Anon. (1972) The preparation of maps and plans in terms of engineering geology. *Quarterly Journal Engineering Geology* **5**, 293–382.

15. The strength of the chalk material or coarser fragments is described in accordance with BS 5930 (1999) *Code of Practice on Site Investigations*. British Standards Institution, London.

16. Discontinuity types are 'bedding fractures' when sub-horizontal, joints when non-horizontal, or 'fractures' to include both bedding fractures and joints.

17. The one dimensional fracture spacing scheme of BS 5930 (1999) is used. This should be supplemented by assessed minimum/average/maximum fracture spacing in millimetres for core and *in situ* observations.

18. Discontinuity set orientations are given as dip direction/dip e.g. 275°/10°. Ranges may be indicated e.g. 260°–290°/05°–15°. Major discontinuities are logged individually.

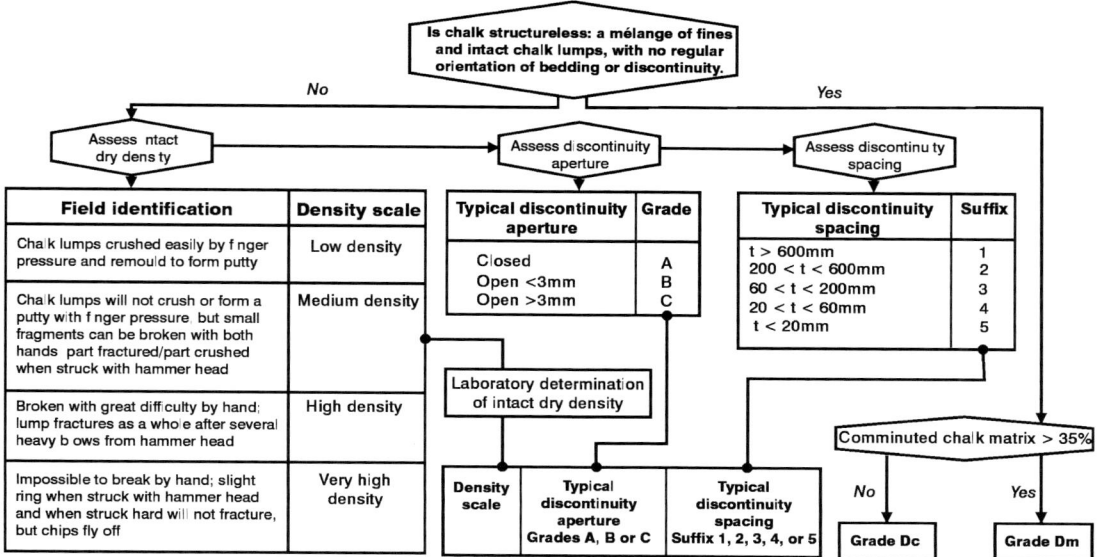

Figure 26.13   Engineering geological classification of chalk (redrawn from Lord *et al.*, 1994).
Examples: White, slightly weathered, CHALK, very weak, low density. Bedding fractures very closely spaced
(20–60 mm), open up to 20 mm, infilled with powdery chalk; discontinuities widely spaced (>600 mm), open
>3 mm: Low density Grade C4.
White, moderately to highly weathered, CHALK, very weak. Structureless mélange composed of 70% sub-angular to
rounded medium to coarse gravel size fragments set in 30% soft, light brown, sandy silt size chalk matrix: Grade Dc.

Lord (1990) concluded that the most effective method of estimating settlement was to carry out a series of 600 or 900 mm diameter plate load tests at 2.5 to 5.0 m intervals vertically into the Chalk. Subsequently, after an extensive study of data from plate load tests (Figure 26.16), Lord *et al.* (1994) concluded that the performance of chalk in shallow foundations could be related to the grades adopted by the CIRIA classification as shown in Figure 26.13. Plate load tests undertaken at Mundford by Burland and Lord (1970) showed that at low applied pressures even grade IV chalk behaved elastically. They pointed out that plate load tests for stresses up to 1.0 MPa demonstrated that grades IV and V chalk exhibited significant creep, and that long-term deflections may be appreciably larger than immediate deflections. Creep in grade III chalk is smaller and terminates more rapidly whilst grades II and I undergo negligible creep.

At higher pressures chalk exhibits yielding behaviour. Burland and Bayliss (1990) observed that the rate of settlement beneath four silos founded in the Upper Chalk at Bury St Edmunds, Suffolk, increased rapidly once the load exceeded the yield stress. Hence, if settlement is to be limited, then the stress applied by a foundation structure should not exceed the yield stress of the Chalk. Lord *et al.* (1994) suggested allowable bearing pressures for medium/high density, low density, and clast dominated structureless chalk of 300 kPa, 240 kPa and 225 kPa respectively. Allowable bearing pressure on matrix dominated structureless chalk should be assessed independently during a site investigation.

The use of pile foundations in chalk has been discussed in detail by Lord *et al.* (1994). The choice of pile and its diameter depends on the structure it has to support on the one hand and the ground conditions on the other. In the case of chalk, the latter includes its dry density, nature of the discontinuities, the presence of solution features and the position of the water table. If the blow count obtained from the standard penetration test is used to derive the ultimate base stress of piles in chalk, then Lord *et al.* suggested a range of 2000 to

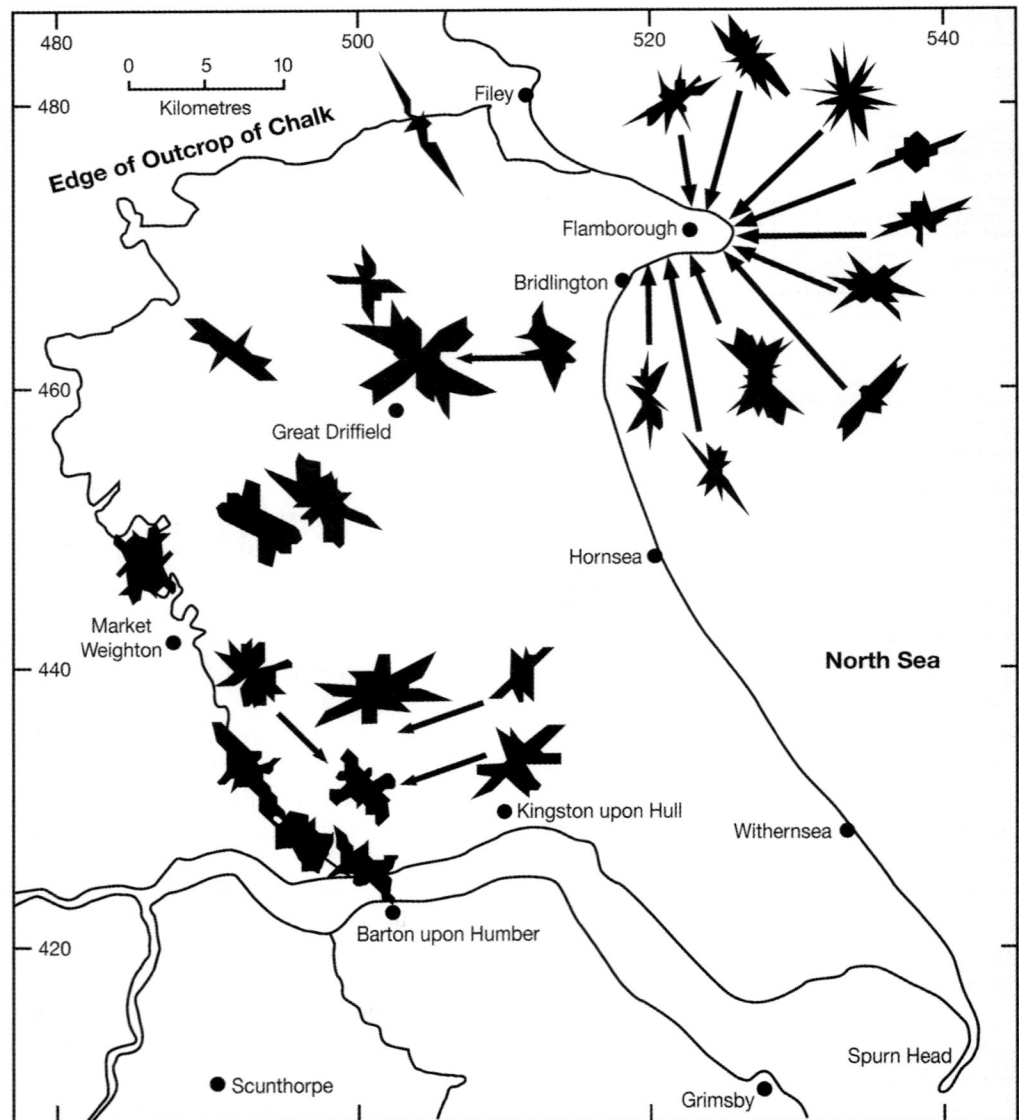

Figure 26.14   Joint orientation data in the Chalk of east Yorkshire (redrawn from Bell *et al.*, 1999).

3000 kPa for low density chalk, and from 4000 to 5000 kPa in medium/high density chalk.

Water can cause problems in engineering in chalk. For instance, a problem may occur in an excavation made in chalk that is well jointed and overlain by structureless chalk if the water table occurs in the latter. In other words, as excavation proceeds, then the hydrostatic head may exceed

the submerged density of the chalk leading to disturbance of the floor due to hydrostatic uplift. In such an instance, the groundwater pressure needs to be relieved by the installation of wells. In addition, water entering the ground from soak-aways, leaking sewers or leaking water supply pipes may give rise to accelerated dissolution of chalk, especially in relation to solution features or old mine workings,

Table 26.5 Discontinuity data and RQD (Rock Quality Designation) values derived from scanlines.

Scanline				Discontinuity data	
Location	Length (m)	Orientation	Number	Mean spacing (mm)	RQD (%)
Selswick Bay — north cliff	9.95	N 106° E	29	329	95.2
Selswick Bay — wave-cut platform	4.00	N 70° E	20	184	76.3
Selswick Bay — wave-cut platform	4.20	N 58° E	27	147	79.6
High Stacks — horizontal line	9.00	N 135° E	29	318	97.4
High Stacks — horizontal line	4.00	N 458 E	19	203	92.3

thereby leading to subsidence. Introduction of water into the ground or a fluctuating water table can disturb metastable deposits occupying sinkholes and pipes, and so cause these deposits to collapse.

When solution features can be located and occur at shallow depth, then, if occupied, the material can be removed and replaced by concrete or suitably compacted fill; they can be similarly filled if unoccupied. A raft may be used as a bridging structure to support a building if a void or disturbed ground is suspected to occur beneath deposits overlying chalk. Piles frequently are used in such situations, being founded below the suspect ground. However, their installation may adversely affect such suspect ground. Grout can be injected to fill voids that have been located or to stabilise loosely packed deposits in sinkholes. Known abandoned mines in chalk may be filled with bulk grouts, the processes involved being similar to those used to stabilise abandoned pillar and stall workings in old coal mines (Healy and Head, 1984).

Earthworks

According to Rat and Schaeffner (1990) the behaviour of chalk in earthworks depends on the production of a matrix of disturbed chalk due to it being crushed by construction equipment; on the possibility of blocks of chalk in fill being reduced in size by static or dynamic stresses, and on the susceptibility of chalk to frost activity. The excavation and compaction processes during earthworks partially break down chalk into a mixture of lumps and fines.

A wide range of equipment can be used quite successfully to compact chalk fill. Heavy deadweight rollers and pneumatic-tyred rollers are particularly satisfactory with hard chalk since it usually requires some degree of crushing during compaction. The softer varieties are usually broken down on excavation and can be compacted to a low air content by most types of plant. However, a danger often exists with soft chalk in that it may be over-compacted, thereby giving rise to high porewater pressures. Unfortunately, the fine material forms a slurry or putty chalk at higher moisture contents. Hence, fill material will be weak and unstable (that is, ruts are produced by construction traffic) if the proportion of putty chalk controls the behaviour of the fill, so that the stability of freshly placed chalk fill depends on both its fines content and moisture content.

Ingoldby and Parsons (1977) described the apparatus developed by the UK Transport and Road Research Laboratory to determine the moisture condition value and chalk crushing strength value, both of which parameters are used in a classification to avoid or minimise instability of chalk used as fill. The breakdown of blocks of chalk within fill can lead to settlement that can continue for several years after construction operations have ceased.

Tunnelling

The Chalk is a formation in which a number of tunnels have been constructed. Generally, the material is easily excavated by either hydraulic excavators or tunnel boring machines (TBMs) but is strong enough to stand unsupported in the short term. However, flints in the Chalk can cause difficulties for tunnelling machines, especially those of small diameter. They can produce high rates of

Figure 26.15 Upper surface of the Chalk showing a high degree of fracturing due to periglacial action and solution pipes, near Wells, Norfolk (UK) (photograph: F. G. Bell copyright).

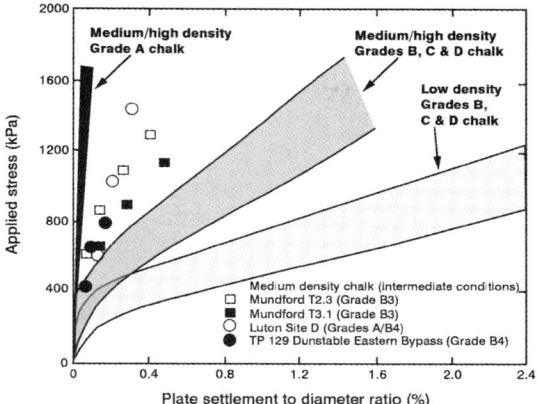

Figure 26.16 Plate load test data for chalk (redrawn from Lord *et al.*, 1994).

wear in cutting tools and larger pieces tend to clog up the mucking system.

A notable example is the Channel Tunnel, which follows the chalk syncline between England and France. For most of its length it is in chalk marl, the clay content being beneficial in reducing the permeability of the ground and reducing water inflow. At either end, but particularly near France, fault zones in harder chalk introduced problems related to high water inflow (Barthes *et al.*, 1994). The Channel Tunnel was constructed using TBMs and segmental lining.

The North Downs Tunnel, on the first stage of the rail link between the Channel Tunnel at Folkestone and London, was above the water table and was constructed by open face excavation using road-header machines and provided with a primary lining using 'shotcrete'. Progress was very rapid, the only problems being related to sinkholes in the Chalk reaching the tunnel crown at either end where the Chalk cover was relatively thin.

A much smaller tunnel constructed for the Barking Reach Power Station cooling water outlet to the River Thames, east of London, was also constructed using a TBM but considerable problems were encountered in both shaft sinking and tunnelling (Evans and Hodgkins, 1997). The TBM was able to work in either open or closed mode (as an earth pressure balance machine).

However, high water inflows were difficult to cope with in open mode and the disintegration of the excavated material into a slurry made control of muck disposal via the screw conveyor difficult in closed mode.

Frost susceptibility

Frost activity can lead to the development of segregated layers of ice forming near the ground surface in chalk that grow by drawing groundwater from below. This gives rise to frost heave adversely affecting embankments and roads, which may also be affected by subsequent thaw settlement. Frost-susceptible soils can be dealt with by adequate drainage, by replacement with gravels, by the addition of certain chemicals to the soil that reduce its capacity for water absorption or by the addition of a cement binder. D'Hem *et al.* (1990) described the use of cement, lime, pulverised fly ash and wet sand to enhance the engineering performance of chalk used in roadworks.

26.4 Summary and conclusions

Chalk is a usually soft, fine textured, often white or grey limestone of marine origin that consists primarily of calcite that was deposited during Upper Cretaceous and early Tertiary times. It is a remarkably pure limestone. For instance, excluding flints and marl bands, the calcium carbonate content of the English white chalk facies usually exceeds 98%. Chalk produces a distinctive 'rolling' countryside. Its outcrop normally forms a prominent escarpment with hard bands within the Chalk forming intermediate escarpments. The escarpment and dip slopes represent the surviving remnants of a sub-Palaeogene erosion surface that has been modified by late Tertiary and Quaternary geomorphological processes. These processes have produced a series of geomorphological features that characterise chalk landscapes. These include dry valleys, bournes, coombes, sinkholes, solution pipes and minor cave systems. The Chalk has also been excavated for lime and flints in some places, by means of surface quarries and pits and associated subsurface mining. Mining has left a legacy of entrances, galleries and chambers (known as deneholes) that are prone to collapse. Chalk slopes can be prone to surface erosion, landslides and, on steeper slopes, rockfalls. Flooding is not uncommon and can occur rapidly in response to intense rainfall because of the high permeability and porosity of the Chalk. Chalk has a distinctive range of engineering properties and can present foundation and earthworks problems.

References

Abbiss, C. P. (1979) A comparison of the stiffness of the Chalk at Mundford from a seismic survey and large scale test. *Geotechnique* **29**, 461–468.

Anon. (1972) The preparation of maps and plans in terms of engineering geology. Geological Society Engineering Group Working Party Report. *Quarterly Journal of Engineering Geology* **5**, 293–382.

Anon. (1977) The description of rock masses for engineering purposes. Geological Society Engineering Group Working Party Report. *Quarterly Journal of Engineering Geology* **10**, 355–388.

Anon. (1979) Classification of rocks and soils for engineering geological mapping. Part 1: rock and soil materials. *Bulletin of the International Association of Engineering Geology* **19**, 364–371.

Atkinson, T. C. and Smith, D. I. (1974) Rapid groundwater flow in fissures in chalk. *Quarterly Journal of Engineering Geology* **7**, 197–205.

Banks, D., Davies, C. and Davies, W. (1995) The Chalk as a karstic aquifer: evidence from a tracer test at Stanford Dingley, Berkshire, UK. *Quarterly Journal of Engineering Geology* **28**, Supp. 1, S31–S38.

Barthes, H., Bordas, D., Bouillot, M., Buzon, M., Dumont, P., Fermin, J., Landry, J. C., Larive, J.-P., Leblond, L., Merlot, J. J., Szypura, L., Vandebrouck, P. and Vielliard, B. (1994) Tunnels — geology. The Channel Tunnel, Part 3: French Section. *Proceedings of the Institution of Civil Engineers, Civil Engineering* **102**, Special Issue 1, 6–10.

Bell, F. G. (1977) A note on the geotechnical properties of the Chalk. *Engineering Geology* **11**, 27–225.

Bell, F. G., Culshaw, M. G., Moorlock, B. S. P. and Cripps, J. C. (1992) Subsidence and ground movements in chalk. *Bulletin of the International Association of Engineering Geology* **45**, 75–82.

Bell, F. G., Culshaw, M. G. and Cripps, J. C. (1999) A review of selected engineering geological characteristics of English chalk. *Engineering Geology* **54**, 237–269.

Birch, G. P. and Griffiths J. S. (1996) Engineering geomorphology. In Harris, C. S., Hart, M. B., Varley, P. M. and Warren, C. D. (Eds) *Engineering Geology of the Channel Tunnel*. Thomas Telford Press, London, 64–75.

Birch, G. P. and Warren, C. D. (1996) The cliffs behind the Channel Tunnel workings. In Harris, C. S., Hart, M. B., Varley, P. M. and Warren, C. D. (eds) *Engineering Geology of the Channel Tunnel*. Thomas Telford Press, London, 76–87.

Boardman, J. (1990) Soil erosion on the South Downs: a review. In Boardman, J., Foster, I. D. L. and Dearing, J. A. (eds) *Soil Erosion on Agricultural Land*. Wiley, Chichester.

Bradshaw, J., Caiger, N., Halpin, M., Le Gear, R., Pearce, A., Pearman, H., Reeve, T. and Sowan, P. (1991) *Kent and East Sussex Underground*. The Kent Underground Research Group, Meresborough Books, Rainham, Kent.

Bristow, C. R., Mortimore, R. N. and Wood, C. J. (1997) Lithostratigraphy for mapping the Chalk of southern England. *Proceedings of the Geologists' Association* **108**, 293–315.

BS 5930 (1999) *Code of Practice for Site Investigations*. British Standards Institution, London.

Burland, J. B. and Bayliss, F. V. S. (1990) Settlement and yielding of Upper Chalk supporting foundations for a silo complex. *Proceedings of the International Chalk Symposium*, Brighton. Thomas Telford Press, London, 365–374.

Burland, J. B. and Lord, J. A. (1970) The load-deformation behaviour of Middle Chalk at Mundford: a comparison between full-scale performance and *in situ* and laboratory measurements. *Proceedings of a Conference on In Situ Investigations in Soils and Rocks*. British Geotechnical Society, London, 3–15.

Burland, J. B., Kee, R. and Burford, D. (1975) Short term settlement of a five storey building on soft chalk. In *Settlement of Structures*. British Geotechnical Society, Pentech Press, London, 259–265.

Charman, J. H. and Cooper, C. G. (1987) The Frindsbury area, Rochester: a review of historical data and their implication on subsidence in an urban area. In Culshaw, M. G., Bell, F. G., Cripps, J. C. and O'Hara, M. (eds) *Planning and Engineering Geology, Engineering Geology Special Publication No 4*. The Geological Society, London, 115–124.

Clayton, C. R. I. (1983) The influence of diagenesis on some index properties of chalk in England. *Geotechnique* **33**, 225–241.

Cross, D. A. E. (1967) The Great Till floods of 1841. *Weather* **22**, 430–433.

Culshaw, M. G. and Waltham, A. C. (1987) Natural and artificial cavities as ground engineering hazards. *Quarterly Journal of Engineering Geology* **20**, 139–150.

D'Hem, P., Fevre, A., Hiernaux, R. and Holef, J. (1990) Stabilization of chalk in north-west France. *Proceedings of the International Chalk Symposium*, Brighton. Thomas Telford Press, London, 449–456.

Dennehy, J. P. (1976) Correlation of the SPT value with chalk grade for some zones of the Upper Chalk. *Geotechnique* **26**, 610–614.

Edmonds, C. N., Green, C. P. and Higginbottom, I. E. (1987) Subsidence hazard prediction for limestone terrains, as applied to the English Cretaceous Chalk. In Culshaw, M. G., Bell, F. G., Cripps, J. C. and O'Hara, M. (eds) *Planning and Engineering Geology, Engineering Geology Special Publication No 4.* The Geological Society, London, 283–294.

Evans, J. F. and Hodgkins, D. J. (1997) Barking Reach Power Station cooling water system. *Proceedings of the Institution of Civil Engineers, Civil Engineering* **120**, 15–26.

Evans, J. W. and Stubblefield, C. J. (1929) *Handbook of the Geology of Great Britain: a Compilative Work.* Thomas Murby and Co., London, 557pp.

Garrels, R. M., Thompson, M. E. and Siever, R. (1960) Stability of some carbonates at 25° and one atmosphere total pressure. *American Journal of Science* **258**, 402–418.

Grainger, P., McCann, D. M. and Gallois, R. W. (1973) The application of the seismic refraction technique to the study of the fracturing of the Middle Chalk at Mundford, Norfolk. *Geotechnique* **23**, 219–232.

Healy, P. R. and Head, J. M. (1984) *Construction over Abandoned Mine Workings.* Special Publication 23, Construction Industry Research and Information Association, London.

Higginbottom I. E. (1965) The engineering geology of the Chalk. *Proceedings of the Symposium on Chalk in Earthworks.* Institution of Civil Engineers, London, 1–14.

Hutchinson, J. N., Bromhead, E. N. and Lupini, J. F. (1980) Additional observations on the Folkestone Warren landslides. *Quarterly Journal Engineering Geology* **13**, 1–31.

Ineson, J. (1962) A hydrogeological study of the permeability of Chalk. *Journal of the Institution of Water Engineers* **16**, 43–63.

Ingoldby, H. C. and Parsons, A. W. (1977) *The Classification of Chalk for use as a Fill Material.* Laboratory Report 806, Transportation and Road Research Laboratory, Crowthorne, Berkshire.

Kee, R., Parker, A. S. and Wehale, J. E. C. (1975) Settlement of a twelve storey building on pile foundations in chalk at Basingstoke. In *Settlement of Structures.* British Geotechnical Society, Pentech Press, London, 275–282.

Le Gear, R. F. (1979) Deneholes, Part 2. *Records of the Chelsea Speleological Society* **10**, 1–116.

Lewis, W. A. and Croney, D. (1965) The properties of chalk in relation to road foundations and pavements. *Proceedings of the Symposium on Chalk in Earthworks.* Institution of Civil Engineers, London, 27–42.

Lord, J. A. (1990). Foundations in chalk. *Proceedings of the International Chalk Symposium,* Brighton. Thomas Telford Press, London, 301–326.

Lord, J. A., Twine, D. and Yeow, H. (1994) *Foundations in Chalk.* Project Report 11, Construction Industry Research and Information Association (CIRIA), London.

Lord, J. A. Clayton, C. R. I. and Mortimore, R. N. (2002) *Engineering in Chalk.* Publication C574. Construction Industry Research and Information Association (CIRIA), London.

Lowe, D. J. (1992) Chalk caves revisited. *Cave Science* **19**, 2, 55–58.

MacDonald, A. M. and Allen, D. J. (2001) Aquifer properties of the Chalk of England. *Quarterly Journal of Engineering Geology and Hydrogeology* **34**, 371–384.

Mortimore, R. N. (1983) The stratigraphy and sedimentation of the Toronian-Campanian in the Southern Province of England. *Zitteliana* **10**, 27–41.

Mortimore, R. N. (1990) Chalk or chalk? *Proceedings of the International Chalk Symposium,* Brighton. Thomas Telford Press, London, 15–45.

Mortimore, R. N. and Fielding, P. M. (1990) The relationship between texture, density, and strength of chalk. *Proceedings of the International Chalk Symposium,* Brighton. Thomas Telford Press, London, 109–132.

Mortimore, R. N., Wood, C. J. and Gallois, R. W. (2001) British Upper Cretaceous Stratigraphy. *Geological Conservation Review Series No. 23.* Nature Conservancy Committee, Peterborough, 558pp.

Phipps, P. J. and McGinnity, B. T. (2001) Classification and stability assessment for chalk cuttings: the Metropolitan Line case study. *Quarterly Journal of Engineering Geology and Hydrogeology* **34**, 353–370.

Pitman, J. I. (1983) Chemical weathering of the East Yorkshire chalk. In Paterson, K. and Sweeting, M. M. (eds) *New Directions in Karst.* Geo Books, Norwich, 77–113.

Rat, M. and Schaeffner, M. (1990) Classification of chalks and conditions of use in embankments. *Proceedings of the International Chalk Symposium,* Brighton. Thomas Telford Press, London, 425–428.

Rawson, P. F., Allen, P. and Gale, A. (2001) The Chalk Group — a revised lithostratigraphy. *Geoscientist* **11**, 1, 21.

Smith, G. J. and Rosenbaum M. S. (1993) Abandoned shallow mine workings in chalk: a review of the geological aspects leading to their destabilisation. *Bulletin of the International Association of Engineering Geology* **48**, 101–108.

Spink, T. W. and Norbury, D. R. (1990) The engineering geological description of chalk. *Proceedings of the International Chalk Symposium,* Brighton. Thomas Telford Press, London, 153–159.

Wakeling, T. R. M. (1968) Foundations on chalk. In *Chalk in Earthworks.* Institution of Civil Engineers, London, 15–23.

Wakeling, T. R. M. (1970) A comparison of the results of standard site investigation methods against the results of a detailed geotechnical investigation in Middle Chalk at Mundford, Norfolk. *Proceedings of a Conference on In Situ Investigations in Soils and Rocks.* British Geotechnical Society, London, 17–22.

Waltham, A. C. and Fookes, P. G. (2003) Engineering classification of karst ground conditions. *Quarterly Journal of Engineering Geology and Hydrogeology* **36**, 101–118.

Waltham, A. C., Bell, F. G. and Culshaw, M. G. (2005) *Sinkholes and subsidence: karst and cavernous rocks in engineering and construction.* Springer-Verlag, Berlin and Praxis Publishing Ltd, Chichester.

Ward, W. H., Burland, J. B. and Gallois, R. W. (1968) Geotechnical assessment of a site at Mundford, Norfolk, for a large proton accelerator. *Geotechnique* **18**, 399–431.

West, G. and Dumbleton, A. J. (1972) Some observations on swallow holes and mines in the Chalk. *Quarterly Journal of Engineering Geology* **5**, 171–178.

Weyl, P. K. (1959) The solution kinetics of calcite. *Journal of Geology* **66**, 163–176.

Acknowledgements

This paper is published with the permission of the Director of the British Geological Survey, Natural Environment Research Council (NERC).

27. Urban Geomorphology

Ian Douglas

27.1 Introduction

The rate of urban growth is accelerating throughout the world. In North America urban sprawl is seen as one of the greatest threats to biodiversity, conservation and sustainable development. The valley around Las Vegas, the fastest growing city in the U.S.A, is being developed commercially at the rate of about one square kilometre every four weeks. The suburban expansion of cities like Jakarta, Indonesia and Delhi in India is adding 5 to 15 per cent to the built-up area every year. The spectacular economic growth of China has led to even more rapid urban spread (Figure 27.1). This transformation of peri-urban areas is becoming one of the most challenging environmental problems of the twenty-first century. Even in countries where there is relatively little growth of the total population, urban areas are expanding, altering the way geomorphological processes work on landforms (Figure 27.2). Even though perhaps only 2% of the land surface of the Earth is built-up,

Population (millions)

Year	Population
1949	1.6
1959	3.2
1982	5.5
2000	7.6
2010	8.5 est
2025	11.0 est

by 1950
1951-1960
1961-1983
1984-1997

0   5 km

Figure 27.1   The urban growth of Beijing, China.

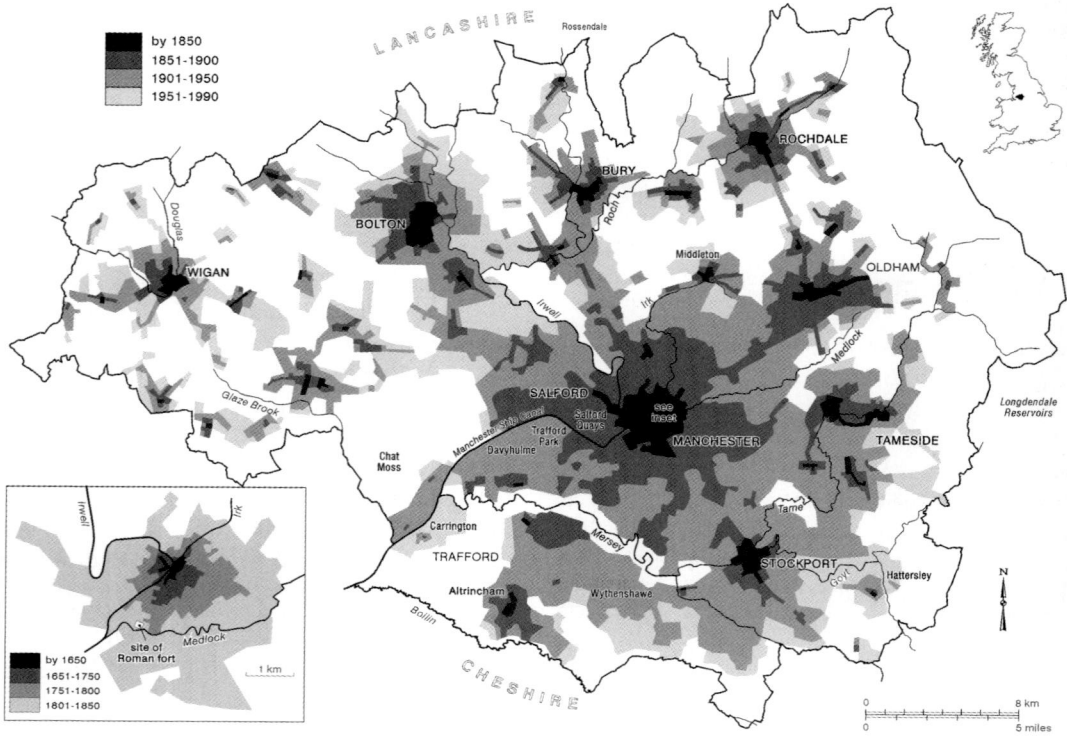

Figure 27.2   The urban growth of greater Manchester, England.

because over half the world's population is in urban areas, and because so much capital is invested in housing, commercial premises, public buildings and infrastructure, urban gemorphology is highly relevant to people's everyday lives and to national and regional economies.

27.2 Problems of siting and planning urban developments

Particular environments pose specific problems for urban development (Table 27.1, Figure 27.3) due to climate, geomorphology or geologic situation. In many environments, Quaternary climatic change has left a legacy of ground conditions that impose constraints on modern urban development. Thus detailed awareness of previous periglacial and glacial conditions and the resulting surface deposits is essential for the siting and planning of urban development over much of northern

America, northern Europe and northern Asia. Knowledge of both present-day process conditions and legacies of past geomorphological activity is needed in urban development planning. In older cities a further emerging need is understanding of past urban and industrial activities, including mining, quarrying and waste disposal, as urban regeneration is accelerating the re-use of old industrial land and other 'brownfield' sites.

Periglacial (see also Chapters 14, 24, 25)
Developments in active periglacial areas face four problems: a) permafrost or perennially frozen ground; b) frost heave problems; c) seasonal melt water flooding; d) river, lake and sea ice. One of the best examples of urban planning in a permafrost area with an active layer is the town of Inuvik, the capital of the Northern Territories in Canada. Inuvik had to be built with as little disturbance of the ground as possible. The key planning decisions were: a) to leave the natural

Table 27.1 Climatic, geomorphological and geological problems for urban development (based on data in Marker, 1996 and Bennett and Doyle, 1997).

Environment	Chief problems
A. Climatic	
Periglacial	Permanently frozen ground and overlying active layer require special types of construction and foundations for buildings and infrastructure.
Hot drylands	Water supply problems; wind erosion; dune migration; flash floods; possibility of salt weathering of building materials and foundations.
Hot wetlands	Rapid weathering and decomposition of building materials; deep, uneven weathering of most rocks in tectonically stable areas; frequent rain events causing rapid water erosion of exposed ground surfaces; frequent landsliding.
B. Geomorphological	
Mountainous	Unstable slopes, soil erosion, rockfalls, large run-out landslides, landslide dams debris flows and avalanches; potential for flash floods.
Flood plains and estuaries	Liable to periodic flooding; variable foundation conditions over former, buried river channels and alluvial deposits.
Coastal plains and deltas	Storm surge, tsunami and flooding risk likely to increase with rising sea levels; coastal erosion; complex ground conditions reflecting former shorelines and old drainage channels; saline intrusion; possible salt penetration in groundwater affecting foundations.
Coasts with weak rock cliffs	Liable to rapid coastal erosion, cliff undercutting and collapse; eroded debris often deposited in ports and harbours causing dredging expenditure.
Islands	Storm-surge, tsunami, rising sea-level and salt water penetration risks on low-lying atolls and coastal plains.
C. Geological	
Active plate margins	Major risks associated with coastal urban developments, especially on Pacific rim, special foundation requirements on filled areas, lake sediments and other unconsolidated materials; major earthquake triggered landslide hazards; volcanic debris and lahar risks requiring awareness of flow pathways on lower volcanic slopes likely to have urban settlements.
Shrink-swell clays	Cracking clay problems likely to be accentuated by climate change.
Karst	Buried karst a major problem for foundations of tall buildings and for sinkhole development; need for knowledge of buried karst plains and effects of lowered Quaternary sea levels.

moss covering intact to maintain its insulating values; b) to place all permanent structures on piles securely embedded in permafrost; c) to ban all road cuts and ditches to avoid permafrost degradation; d) to have culverts installed in gravel fill to drain seasonal surface water runoff; e) to build gravel pads on top of the natural vegetation to support all access roads and tracks and some temporary structures. When installing the piles, steps had to be taken to avoid disturbing the permafrost and to protect the mainly wooden piles against damage in the active layer. Piles were sunk to a depth twice that of the active layer. All

buildings had air spaces of about 1 metre beneath them and all utility pipes were encased in an insulated, raised utilidor (Cooke and Doornkamp, 1990).

Past periglacial conditions pose several, usually highly localised and variable problems, the most serious of which relate to the re-activation, during construction on slopes, of fossil slope failures and solifluction lobes (Table 27.2). Slopes on stiff fissured clays affected by periglacial solifluction some 10–13 000 years ago are widespread in England and Wales (e.g. Hutchinson, 1991; Jones and Lee, 1994). Although they may have

Figure 27.4   Example of a bridge that has suffered scour around its abutments and has required additional protective works: Pickre l Bridge, Nogales Wash, Arizona.

and, for a single storey property, incorporating a means of escape such as a dormer window into the building design.

Active floodplains are often subject to rapid lateral migration of river channels, particularly during major floods. In and around many urban areas, especially in semi-arid environments such as the American southwest (Figure 27.4) and North Africa, lateral migration of channels can wash away the approaches to bridges in a few hours. To avoid the scour of embankments and bridge approaches, bridge wingwalls or spur dikes may be constructed to deflect flows away from the channel banks and direct them underneath the bridge. Another alternative is to build additional bridge openings or spans, so widening the channel and increasing the flood capacity beneath the bridge.

Urban areas on alluvial fans

Urban settlements cn alluvial fans have long proved both attractive foci for habitation and geomorphologically problematic. Alluvial fans are composed of debris deposited by floods, mudflows or debris flows, often with minor channels running over them (see Chapter 16). Urban settlements tend to grow around the minor channels, ignoring the risk of the whole fan becoming active in an extreme event.

In most fans, the channel repeatedly branches down the fan, often through bifurcation associated with avulsion, conveying the water-sediment mixture to all parts of the fan surface. The return period for complete inundation of the fan may be several decades or more. In dry environments, the effects of a major rare erosional-depositional event may persist for tens of years.

Many ancient cities were cited on fans. The Roman city at Nimes, France was built on an upper Pleistocene-Holocene alluvial fan, the archaeological sedimentary record showing sequences of fluvial deposits and anthropogenic debris, indicating that the site suffered from repeated flooding and deposition across the fan (Fabre and Monteil, 2001). Some 53 significant floods have occurred since the 14th century, the majority producing some additional deposition on the fan. Now although the surface of the fan has been made virtually totally impermeable, major floods, like that of 3rd October 1988, still deposit sediment throughout the city and the main streets running down the fan surface become major torrents.

In Alpine regions, villages and towns are often situated on alluvial cones where streams emerge from the mountains into major valleys. The Alps has known several disasters, such as that of the

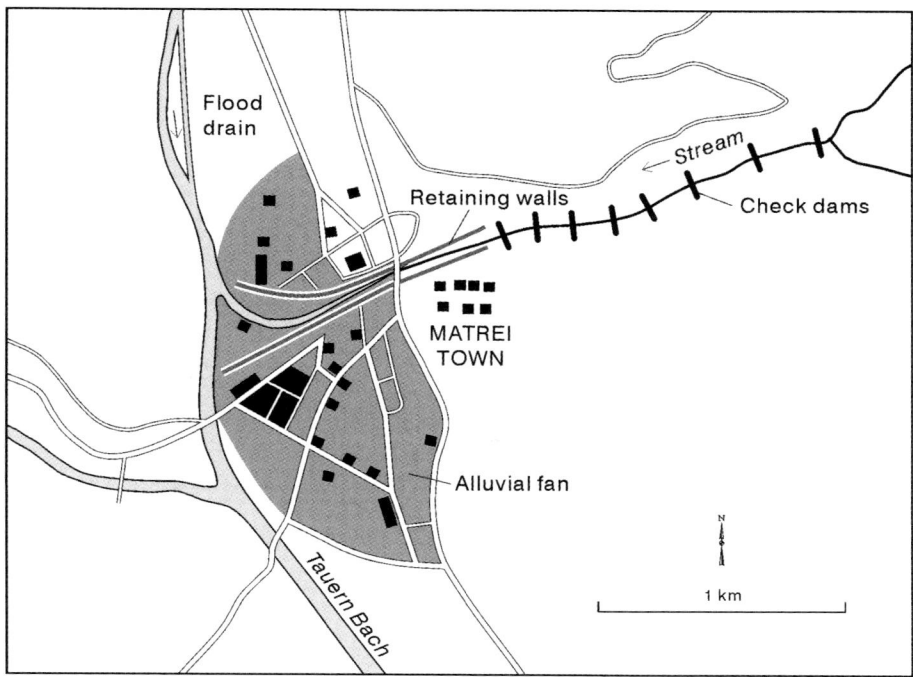

Figure 27.5 The alluvial fan of the Betterwand torrent (marked stream in the diagram) and the town of Matrei in the Austrian Tyrol.

Guil valley, where urban settlements were partly destroyed by debris flows (Tricart, 1961a, b). As tourist developments increase, with the growth of high altitude impermeable surfaces and compacted routes for surface runoff, the consequences of urbanization become increasingly severe, aggravating the impact of rapid melting of deep snow packs under warm spring rains, as in the French Alps in 1981.

Nimes has crucial lessons for the many modern desert cities that flourish on or adjacent to desert fans. In such cities, much of the partially or completely urbanised fan terrain has become impermeable, creating major urban drainage problems. Usually towns seek protection from floodwaters by building walls or diversion weirs to send the floodwater away from the built-up areas. However, on fans the floods usually bring large volumes of sediment, which tend to build up in the channels and reduce the flood conveyance capacity of the embanked waterways and diversionary channels. As in Nimes, the

water and sediment often invade the desert fan towns.

Flooding from the Betterwand torrent greatly affected the town of Matrei in the Austrian Tyrol, which is built on the alluvial fan created by the torrent as it discharges from the mountains towards the Tauern River (Figure 27.5). The torrent is now constrained by high retaining walls and numerous debris barriers in the channel upstream of the town trap sediment (Verstappen, 1983). The considerable engineering works offer protection from all but the most extreme floods.

Slope stability and mass movement

Slope failure is a natural process that in urban areas is often induced, accelerated or retarded by human actions, such as:

1. Cutting slopes for roads and other structures, quarrying, removal of retaining walls, and lowering of reservoirs.

2. Adding weight by landfills, stockpiles of ore or rock, waste piles and construction of heavy buildings and other structures.
3. Water leaking from pipelines, sewers, canals, and reservoirs.
4. Vibrations from explosions, machinery, road and air traffic.
5. Decrease of underlying support by mining, loss of strength or failure and/or squeezing out of underlying material.

The provision of and subsequent neglect of land drainage systems has had a significant effect on some slopes (e.g. Bromhead and Ibsen, 1996). One of the most serious effects is the artificial recharge of the groundwater table. Tebbutt (1998) highlights how average water usage per person has increased by 50% in the UK over the last 20 years. Water supply and drainage pipes can be vulnerable to ground movement, and it was a key factor in triggering landslides at Luccombe, Isle of Wight, in 1988 (Lee and Moore, 1989).

Pressures of land availability and cost leading increasingly to the building of houses on or close to potentially unstable slopes. A landslide of some 180 000 m³ occurred in March 2000 on a steep, 50 m-high, 40-year-old cut slope between two rows of residences in Millbrae, California (McCormick, 2002). The landslide left a 20 m head scarp and encroached upon 6 homes at the base of the slope, damaging 3 and knocking 1 off its foundations. The landslide is along the western margin of an old, large quarry cut into highly sheared and foliated shales and melange rocks. After quarrying ceased, houses were built at the top and the base of the 1.5 : 1 to 2 : 1 (horizontal to vertical) slopes.

The translational landslide failed as a block landslide (with minor rotation) along a basal failure surface dipping 5 to 10 degrees out of the slope. The lower portion failed as a semi-plastic flow above the toe of the slope. The failure was due to a gradient that was too steep for long-term stability of the materials, weathering of the bedrock over time and to localised surface instabilities and erosion that ultimately overcame the marginal stability of the existing cut slope (McCormick, 2002).

Piping

The development of subterranean channels by water moving through incoherent sediments is an important form of subsurface discharge of water and sediment in many areas (Jones, 1981; see Chapter 11). Piping regularly damages bridge abutments, piers, retaining walls and culvert facings. Where piping is extensive, highway subsidence or railway distortion can seriously disrupt traffic. The dominant type of piping causing damage to urban buildings and infrastructure is 'stress desiccation', in which surface runoff, including road surface water, drains into a desiccation crack and causes erosion of silts and fine sands, the collapse of channel banks and eventually surface settlement, creating an irregular topography. Other types of piping include the process of particle entrainment and high hydraulic gradients, and the development of subsurface passages due to variable lithological permeabilities. In some cases, large voids do not develop, but seepage erosion and 'running sand' cause morphologically similar surface phenomena.

In order to avoid piping problems, preconstruction surveys should carefully define materials and other environmental conditions that might favour piping. Remedial actions may include filters or filter drains, catchment drains with stable channels designed to carry water away from threatened structures, protection for abutments, and careful avoidance of concentration of drainage water where it infiltrates into natural sediments.

Salt (see also Chapter 16)

The breakdown of rock and concrete by salt weathering, found in many environments, is particularly serious in arid and semiarid areas. High rates of evapotranspiration concentrate salts in ground water hot, dry environments, while salt water penetration into aquifers occurs in coastal zones. Thus the risk of salt attack is often highest along arid coastlines. Salts tend to be precipitated in the upper layers of the soil. This salt causes accelerated weathering in four ways: a) by the growth of salt crystals in cracks and fissures in rocks and building materials; b) through thermal expansion of salt crystals under the diurnal heating cooling in hot deserts, leading to stresses in the building materials and rocks; c) breakage of rocks through expansion caused by hydration of salts; d) reactions of chloride and sulphate salts with cement. The areas where these salt problems are likely to be acute need to be avoided in developing new urban areas. Foundations can be protected by using salt resistant cement, careful screening and washing of aggregates to avoid salts, using dense low porosity concrete, and sealing foundations to prevent attack by groundwater (see Chapter 16).

Expansive soils (see also Chapter 7)

In many parts of the world, cracking clay soils are a major constraint on urban development. Many clay soils expand when wet and shrink and crack during long dry periods. Such 'shrink-swell' phenomena cause differential shifts of parts of the structure, such that floors tilt slightly and windows and doors no longer close properly. A high smectite clay content is particularly likely to raise the shrink-swell potential. The new Denver International Airport is built on clays with swell potentials as high as 15 per cent and averaging 8 per cent. These potentials had to be taken into account in the design of the terminal, parking garages and airport office building as well as the airfield pavements themselves. Consideration had to be given to the effects of heavy aircraft landing on the runways and the compaction and compression effects this would produce on the underlying clays.

The US Soil Conservation Service uses the coefficient of linear extensibility (COLE) as a measure of the shrink-swell potential of soils. The COLE represents an estimate of the vertical component of swelling in a natural soil clod. Soils with coefficients less than 3% are judged to rank low in soil expansion capability; those between 3 and 6% are judged to be moderate; and those with coefficients greater than 6% are judged to rank high in this capability. The higher the coefficient, the more severe is the shrink-swell behaviour of a soil, and the greater the risk of damage to buildings. Variations in the properties and potential of expansive clays soils are related to the substrate, water-table depth, antecedent soil conditions, amount of new water added to the soil, vegetation,

topography, drainage conditions and the way buildings are constructed. Drainage of such smectitic clays alters their swell potential.

Precautionary measures that can be undertaken include:

1. Adequate site investigations of both the soil and the substrate;
2. Location of vegetation away from the house;
3. Establishing strong foundations (building on a reinforced concrete platform ('raft') rather than brick piers);
4. Elevating the structure so that moisture cannot collect near or under it;
5. Replacing potential swelling soil with more stable earth material;
6. Providing physical barriers in the ground to prevent moisture from accumulating there;
7. Pre-wetting the terrain prior to construction to ensure that all stresses are relieved (Coates, 1985).

Coastal sites (see also Chapter 21)

Urban development on the coast alters the dynamics of beaches, cliffs, estuaries, mangroves and salt marshes. Examples abound of former ports that silted up and are now well inland, such as Rye in Kent, England and Sluis in West Flanders, Belgium. Other instances occur of shorelines needing protection and beaches requiring replenishment, as at Bournemouth, England (Spencer, 1999) and Knokke-le-Zoute, Belgium (Douglas, 1971). At Bournemouth, the sandy beach so essential for the tourist industry had beach replenishment activities in 1970, 1974–5 and 1988–90 with the next phase due in the autumn of 2005. At Knokke, annual replenishment of beach sand has been necessary since the late nineteen-fifties.

Coastal landslides affect a wide variety of urban settlements, some simply because houses are built in unsuitable places, others because slope stability has been altered by urban development. Undercutting of cliffs has affected residential areas from San Diego, California to the boulder clay coasts of eastern England.

Around the world, large sums have been invested in an attempt to resist the loss of cliff top properties, services and infrastructure and to mitigate the risk to public safety (Brunsden and Lee, 2004). In England, for example, there are some 860 km of coastal protection works, with over £20 million spent each year on maintaining and improving these defences, and providing new schemes. The Holderness coast, UK has retreated by around 2 km over the last 1000 years, leading to the loss of 26 villages listed in the Domesday survey of 1086. 75 Mm³ of land has been eroded in the last 100 years (Valentin, 1954; Pethick, 1996). Rapid recession has also caused severe problems on the Suffolk coast, most famously at Dunwich where much of the city has been lost over the last millennium (e.g. Bacon and Bacon, 1988). Gardner (1754) records that by 1328 the port was virtually useless and that 400 houses together with windmills, churches, shops and many other buildings were lost in one night in 1347.

Karst (see also Chapter 24)

Soluble rocks have subsurface conditions that are particularly acute in buried karst plains, such as those underlying Florida, parts of the Caribbean and parts of Southeast Asia. Around many of the prominent tower karst hills of northern Vietnam, southern Thailand, Malaysia and Indonesia, extensive alluvial deposits bury karst surfaces that are often pinnacled, encumbered with limestone blocks and pitted with sinkholes that developed during Pleistocene low sea levels. The buried karst now poses serious problems for civil engineering works (Bergado and Sebanayagan 1987). New high-rise buildings require deeper foundations than the low rise buildings that sufficed until the nineteen-seventies. In Kuala Lumpur, the low-rise structures had their foundations on the stiff clay layer within the alluvium. Taller multi-storey structures require piling into the underlying limestone. However, the irregularity of both the karst surface and the cavities within the buried karst means that foundation investigations have to be particularly careful. Drill holes may strike limestone, unaware as to whether it is buried rockfall material or a pinnacle, while a neighbouring hole might pass through several more metres of alluvium before hitting limestone. Analogous karst terrain in northern Vietnam exhibits a whole range of buried karst engineering problems that can only be overcome by careful geophysical investigation of subsurface geomorphology.

These foundations problems had to be taken into account during the preparations for building the Petronas Twin Towers, the world's tallest building at the time (Pelli *et al.*, 1997). Beneath the level surface of the site, the pinnacled surface of the karst plain sloped from 15 m depth under the northwestern corner to a depth of over 180 metres at the southeastern edge. In order to get a conventional level foundation for such a huge weight of concrete, the bedrock would have to be excavated at one edge, while at the other piles would have had to be sunk to a far greater depth than normal at the other. The risks and costs were so great that the building was moved 60 m to the southeast to allow for more than 55 m of alluvium beneath each tower. To do this successfully, a foundation designed to withstand movements with the alluvium was required.

The solution was to spread the load throughout the alluvium so that even movement could occur. A concrete mat was designed to spread the weight of the building over a set of drilled 1.3 m diameter piles. These piles would transfer the weight of the towers to the soil more gradually than a simple mat would. Friction between the surface of a pile and the surrounding soil would prevent the foundation supports from sinking. Settlement would then occur between the pile tips and the bedrock. Pile lengths were varied so that their tips were all at the same height above bedrock, so as to avoid any tilting of the foundations.

This technique introduced a new concern. The interlocking grains within the alluvium would cease to be locked together if excavation removed overlying material and allowed the alluvial mass to expand. The piles were therefore sunk from as near the ground surface as possible and the concrete mat was laid on top of them. The extremely careful investigations and detailed precautions taken during the planning of the foundations of the Petronas Twin Towers show just how important it is to have a full understanding of the ground and subsurface conditions of any major engineering structure.

Urban areas on peat

Many towns have expanded on to peat, building houses in areas where drainage has lead to ground surface lowering (Berry *et al.*, 1985). This is not a new phenomenon in human settlements. For example on the coastal plain of Flanders in north-eastern France and Belgium, inversion of relief has occurred in historic time, with the narrow ridges on sandy materials being the former beds of streams between which the drained peat has shrunk, with the result that the oldest settlements are on the low peat surface and younger villages on the ridges.

Industrial development and railway construction brought the classic engineering problems of crossing peat deposits, such as Chat Moss west of Manchester on the route of the 1829 construction of the Liverpool to Manchester railway. Now new housing developments on drained peat often require pre-loading of the peat to ensure that subsidence is artificially induced before any buildings are constructed.

The role of geomorphological mapping

Geomorphology in city planning. If geomorphological surveys are carried out sufficiently early in the planning of a new city, and their results are digested into planning discussions, some of the potential dissonance between environmental conditions and the city plan can be avoided (e.g. Bernknopf *et al.*, 1993).

Geomorphology in site development. Geomorphological contributions to site development include aid in rationalising site plans to avoid hazards, formulation of building and site choice adapted, and sampling sites for ground material investigations.

Geomorphology and post-construction management. The monitoring of geomorphological processes and landform changes after construction can play an important role in helping environmental managers to influence future planning policy and alleviate or avoid environmental hazards.

In the Las Vegas area, the US Geological Survey is engaged in the systematic production of general-purpose 1:24 000 and 1:100 000-scale geologic maps to provide a uniform data base for urban environmental planning and management. This involves: 1) Mapping the aerial distribution of permeable strata and fracture zones and effects on the migration of ground and surface water, and on pollution plumes in these waters; 2) Mapping, geophysical surveys, and hydrologic studies of hazards, such as expansive soils, flooding, subsidence due to ground water withdrawal, landslides and faulting; 3) Providing GIS-based information critical for planning and zoning and communication and outreach to the public through the Lake Mead-Lake Mojave National Recreation Areas; 4) Evaluation of the stratigraphic and structural setting of Las Vegas basin to determine ground water, petroleum, building materials, and mineral resource potential. The primary project objective is to determine the geologic, geophysical, and hydrologic history of Las Vegas Valley.

In Umbria, Italy Antonini *et al.* (2000) used aerial photographs taken over 50 years for a multi-temporal investigation of the slopes at 80 sites to map the most active landslide hazard areas. A simple vulnerability map showing houses, buildings, roads, railroads, power lines, and related infrastructure was derived from available 1:10 000 topographic maps and the most up-to-date aerial photographs at each site. A GIS intersecting the hazard and vulnerability maps was used to derive ranked levels of risk (Cardinali *et al.*, 2002).

Radar remote sensing has also been used to monitor landslides (e.g. Kimura and Yamaguchi, 2000). Differential interferometric synthetic aperture radar (DInSAR) allows for the measurement of small deformations of the terrain (at a fraction of the radar wavelength i.e. millimetres). DInSAR uses two different SAR images acquired at different times plus either a third image or a digital terrain model (DTM). The imagery and DTM are geocoded and after the removal of topographic and atmospheric effects, surface shift over the period spanning the image pair is calculated. DInSAR complements traditional ground survey measurements and has a wide range of applications, such as the monitoring of mining-induced subsidence, urban subsidence due to tunnel construction and/or groundwater extraction, landslides and mass movements.

In Italy, both DInSAR and multi-temporal SAR have been used to investigate slope movements at Caramanico (Refice *et al.*, 2000, 2001). Landsliding is the most frequently occurring geological hazard in the resort of Caramanico, in the south-central Apennines. Seasonal re-activation of several mass movements threatens the town's historic centre, cemetery and infrastructure. The large landslides, up to 1 km in maximum dimension, generally move slowly, and thus their changes are potentially detectable from SAR data. An approach similar to the so-called 'permanent scatterers' (PS) technique, originally proposed by Ferretti *et al.*, 1999, has also been attempted to monitor small displacements of objects characterised by high backscattering stability.

Kuching, Sarawak, provides a good example of the variety of foundation conditions experienced in a low-lying coastal city with large areas of peat (Figure 27.6). Detailed mapping enables four broad categories of foundation conditions to be recognised (Lam, 1992):

1. Very suitable areas underlain by residual deposits and terrace deposits or both; generally hilly or undulating terrain; tens of metres above flood level; less than 10 m above bedrock; no subsidence risk; little or no filling required (surface stripping usually required to level the ground).
2. Suitable areas underlain by shallow estuarine/deltaic deposits and riverine deposits less than 5 m thick; peat deposits less than 1 m thick; normally occurring as narrow belts around the residual deposits; at or below flood or highest tide levels; 10 to 15 m to bedrock; minor subsidence risk; some filling required.
3. Less suitable areas underlain by estuarine or deltaic deposits more than 5 m thick; below flood or high tide level; 15 to 30 m above bedrock; moderate amount of prolonged subsidence; large amount of fill required.
4. Least suitable areas underlain by deep peat of 5 m or more thickness; above flood level but always water-logged; very poor surface foundation (extremely soft and wet); 12–20 m to bedrock; severe subsidence risk; large amount of fill required.

Ideally, development in Kuching should be directed towards the very suitable and suitable areas and expansion on the less and least suitable areas should be controlled. Nevertheless, shortage of suitable land has meant that a few of the unfavourable areas have already begun to be used for development. Among the problems arising in the peat areas in the least suitable category is excessive settlement. Any increase in the load on the peat expels pore water and leads to compaction of the peat and lowering of the ground surface. Any attempt to drain the peat leads to further water expulsion and more settlement, as happened in the Petra Jaya housing estate close to Kuching where poorly supported fences collapsed (Lam, 1992). Houses built on piles on the peat tend to remain stable, while those with concrete floor slabs laid directly on the peat generally find that eventually the concrete cracks and breaks.

27.3 Impact of urban activities on geomorphic processes

Impacts during construction
During construction, the original vegetation is disturbed and the soil and weathering profile is exposed to the erosive agents of wind and rain. Road construction can lead to rapid erosion if cuts are too steep and roadside drains, with regular cross drains, are not established. Temporary access roads and building yards are particularly liable to surface erosion. Erosion on construction sites results in the deposition of about 80 million tonnes of sediment a year in US lakes (Harbor, 1999).

Many construction sites are levelled by cut and fill operations, with the less compacted fills being particularly liable to be eroded by rills and gullies if a protective cover is not provided quickly. If the fill lies on weathered rock or colluvium, the permeability of the fill is likely to be greater than that of the material below and thus water will tend to move laterally at the base of the fill and may induce sliding. Erosion of fill is likely to create deep gullies whose width will depend on the cohesiveness of the material, but normally is greater than the width of gullies in weathered rock

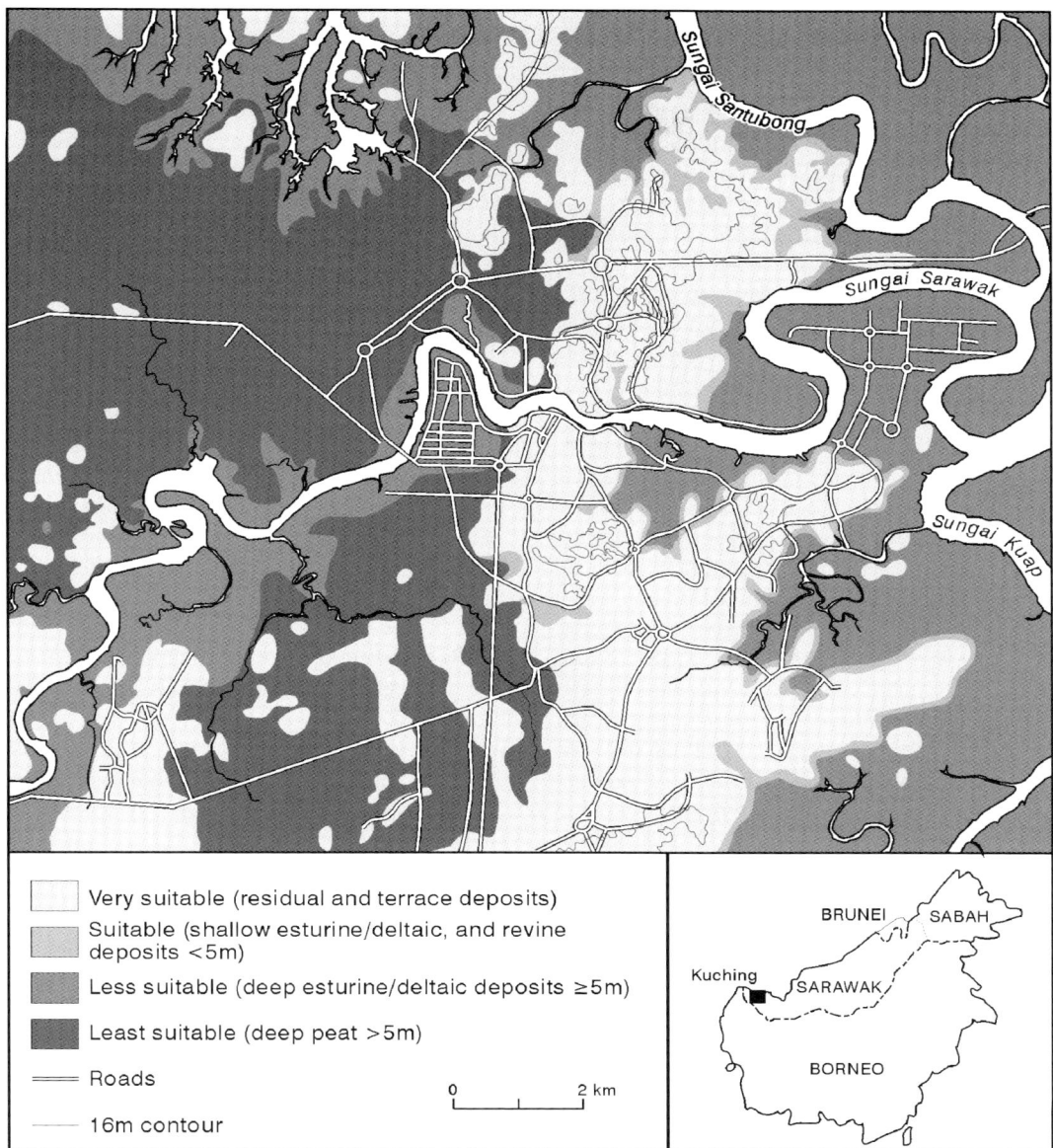

Figure 27.6   Urban land suitability classification for Kuching, Sarawak.

in situ. At places where throughflow emerges, overhanging rounded hollows, or shallow cavities develop in the gully walls.

Impacts after development is completed

Subsidence (see also Chapter 12). Many urban processes lead to geomorphological change, particularly those associated with the removal of underground resources, including ground water, oil, brines, salt, coal and other minerals. The extraction of resources necessary to sustain the industry and people of a city can lead to problems for sectors of the urban population. In north-eastern Phoenix, Arizona, groundwater abstraction resulted in up to

1.6 m of subsidence in the centre of a cone of depression of the groundwater table, caused by constant pumping between 1962 and 1982 (Péwé, 1990). Sometimes, earth fissures develop as a result of such subsidence. A 130 m-long fissure in a northeast Phoenix residential construction site was reported as the first known occurrence in an urbanising part of Arizona (Péwé, 1990), but it was expected that as cities grow, more and more such fissures will occur, endangering structures in the affected areas.

The Valley of Mexico is a hydrologically closed, graben structure, that had a series of shallow lakes in the early 16th century. Groundwater recharged in the mountains surrounding the Valley flows into the Basin of Mexico. Thick lacustrine clays cover the Valley floor and artesian conditions once prevailed. The water table is close to the surface over much the lowland around the ancient lakes. Relatively small rain events cause overland flow and flooding, problems always faced by communities in the Valley. Heavy pumping has caused drainage and consolidation of the lacustrine clays, and consequently land subsidence of up to 8 m in the central part of the city (Durazo and Farvolden, 1989). While in 1975 the recharge of the aquifer was $38 \, m^3 \, s^{-1}$ and abstractions were $30.3 \, m^3 s^{-1}$, by 1990 recharge had declined to $25 \, m^3 s^{-1}$ and abstractions had risen to about $53 \, m^3 s^{-1}$, drastically lowering the water table. In the centre of Mexico City, subsidence developed following the construction of over 50 deep wells. Subsidence averaged $8 \, cm y^{-1}$ from 1935 to 1947, $29 \, cm y^{-1}$ from 1947 to 1958, and $7.5 \, cm y^{-1}$ from 1959–1985. Since 1985 subsidence has probably continued at about $10 \, cm y^{-1}$. However, the slowing of subsidence was the result of controls on pumping in the central area. New wells were then sunk in the southern areas of Chalco, Tláhuac and Xochimilco. Since 1995, differential SAR interferometry has shown subsidence of around $40 \, cm y^{-1}$ in some of these areas, with a maximum of $47 \, cm y^{-1}$ southeast of the International Airport (Carnec et al., 2000). This case illustrates that measures successfully reducing geomorphological problems in one part of a large urban area may have to be repeated years later in another area.

Problems of former mine workings. Urban impacts of past activities may occur with long time lags. In southeast London on April 7th 2002, nearly 100 residents from council and private properties were urgently evacuated from properties on Blackheath Hill after subsidence caused a crater nine metres in diameter to appear in the road next to their homes. The hole caused both carriageways of the main A2 road to be closed. The crater is thought to have been caused by collapse of old, and possibly illegal or undocumented, chalk workings. Drilling revealed additional chalk workings, 15 metres and 30 metres below the surface of the road, at two locations. Most of the cavities had probably been filled with quarrying spoil or loose sand several centuries ago, as a result of earlier chalk tunnel roof failures. The Blackheath Hill area was nearest source of chalk for lime making to the centre of London and was already in use in 1666 when the Great Fire of London destroyed the City. Few controls on chalk removal existed at the time (Smith, 2002). Similar issues arise in most large urban areas. Planning departments need access to inventories of past land use and former mineral workings so that problems of this type can be avoided.

Many early mine workings were not recorded and only manifest themselves when a collapse occurs. The appearance of several holes alongside the UK East Coast main railway line to the southeast of Edinburgh was traced to the collapse of mined coal seams. Overcoming the problem required the realignment of the high-speed track onto ground, which had been treated by grouting at close centres to fill voids, with three sections of track carried on concrete slabs supported by mini-piles. A total of 17 000 t of grout and 162 km of drilling were required for the 1.8 km long diversion (New Civil Engineer, 2003).

Mass movements. Landsliding triggered by urban development is found from the polar to the equatorial regions. Changes to water flows caused by construction, paving and the weight of structures on the ground, disrupt patterns of pore water pressures and lead to slope failure. Lee et al. (2000) note the importance of human impact in the artificial recharge of groundwater, and also the re-activation of old slides by changes in loading conditions. Gutjahr et al. (2000) discuss the correlation between increase in urbanisation and

Table 27.3   The hydrologic impacts of urbanisation (after Leopold, 1968; see Chapter 10).

Impact	Effects
Change in total runoff	Impermeable surfaces alter balance between evaporation, transpiration, infiltration and surface flows.
Alteration of peak flow characteristics	More rapid rise of streamflow to peak storm discharge in totally or largely urbanised streams, especially in short duration, intense storms. More frequent moderately high flows in rivers flowing through cities.
Decline in water quality	Initially through increased sediment flows from construction sites followed by chemical and organic pollution from urban drainage and storm sewer overflows. In a growing city all these sources operate at the same time but in different parts of the urban areas.
Changes in the hydraulic behavior of streams	Hydrogeomorphic changes, including meandering to braided state when sediment yields and peak discharges are greatly increased. Channel widening downstream of culverts and bridges. Channelisation and embankments construction. Flood controls works. Channel change downstream of regulated reaches. Changes in channel vegetation.

Table 27.4   Sequence of fluvial geomorphological response to land use change: Sungai Anak Ayer Batu, Kuala Lumpur.

Land cover/land use	Channel condition
Forest	Narrow, meandering with low sediment load
Rubber plantation	Gullying during clear weeding; Peak discharge increased; Channel slightly widened; Later stabilized; Few cut-offs
Urban Construction	High sediment yield; High peak discharge; Metamorphosis to wider, steeper, shallower braided channel
Channelisation and stable urban built-up area	Higher peak discharge; Less sediment load; Channel enlargement downstream; Bank erosion, Minor channel incision; Loss of fine bed material by scour.
Post channelisation siltation	Where large quantities or organic debris enter concrete channels and are deposited, vegetation can become established and build up deposits that reduce channel capacity.

higher frequency of landslides, triggered by lesser rainfall events.

Kingston, Jamaica (Gupta and Ahmad, 1999) illustrates the interplay of biogeophysical and societal pressures in the slope instability: urban land use sequence. Deforestation and vegetation modification followed by construction by the affluent on the higher slopes and invasion of the lower slopes by the less wealthy leads to shallow slides and debris flows following heavy rains. Failed slopes recover and acquire new vegetation, but later are used as house sites with temporary access roads until the slope fails again in the next heavy storm. The sequence is then repeated. In the mountains of St. Andrew where much of

Kingston's new development is concentrated, 866 different slope failures were counted prior to intense rains in May 1991 that added 84 new failures and reactivated 540 others (Maharaj, 1993).

Fluvial channel change. Urbanisation alters the water balance of the affected areas and changes the local hydrology (Table 27.3). The changed water flows alter patterns of land surface and channel erosion. In extreme cases, increased peak runoff and sediment yields from construction sites lead to channel metamorphosis from meandering to braided state (Table 27.4, Figure 27.7). Bank protection and channelisation work often leads to renewed erosion further downstream, just as below the fixed abutments

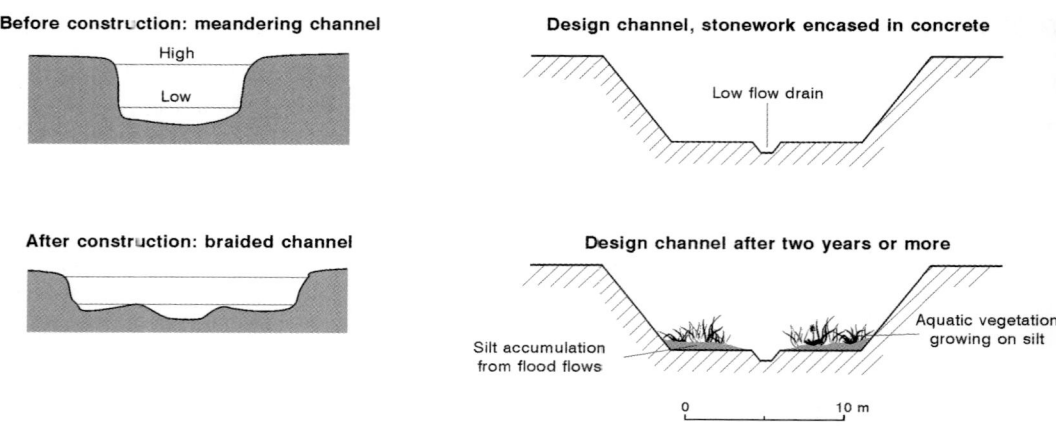

Figure 27.7   Channel changes during and after construction in the Sungai Anak Ayer Batu catchment, Kuala Lumpur.

of bridges, rivers tend to widen by scour of the banks. In many ways, work to protect river banks and to reduce flooding can serve to shift the flood, sediment and erosion problems further downstream. Thus many large cities have well protected, highly managed river reaches through their central commercial and smart residential districts, but eroding banks and degraded channels further downstream. Barbatic *et al.* (1998) describe measures involving flow velocity reduction, energy dissipation and increased river bed stability, required to prevent potentially catastrophic failures due to erosion of a channelled river. These channel adjustments can threaten buildings, bridges and utility pipelines. Channel aggradation can often raise stream beds above the level of adjacent urban areas and, hence, increase flood risk.

27.4 Landform creation by urbanisation

Landforms of accumulation

Much modern urban development involves land reclamation and major landform modification. In extreme cases, huge quantities of material are moved, for example in the development of major airport sites such as Kansai, Singapore and Hong Kong (Figure 27.8). At Kansai, the fill material

has caused some subsidence of the original seabed, with allowance having to be made for this in the operation and maintenance of the airport. The problems of subsidence of the second stage runway are expected to be more severe than in the first runway, with a prediction that after 50 years subsidence will have been 18 m compared to 11 m for the first stage (Kansai Airport Authority, personal communication 25 September 2000). Detailed analyses have been made of the way landing aircraft cause small temporary depressions in the runway that in turn affect the drag on aircraft moving along the runway (Endo, 2000).

As disposal of solid waste moves from landfill to land raise, new hills appear on the edges of floodplains, above former gravel pits and quarries and on offshore islands. From the huge dumps on the edges of cities like Istanbul and Manila to the managed disposal areas, such as Freshkills on Staten Island, New York, which has been taking nearly all the 17 000 tons of waste the city collects each day (Polan, 1992), local skylines are becoming dominated by landforms of rubbish accumulation, in the way that others were marked by coal waste tips in the early 20th century. As events at the Payatas tip in Manila have shown, some of these urban wastes mounds are unstable, prone to massive slumps and landslides. The loss of life and property that ensues is a challenge to the management of the waste disposal and the application of

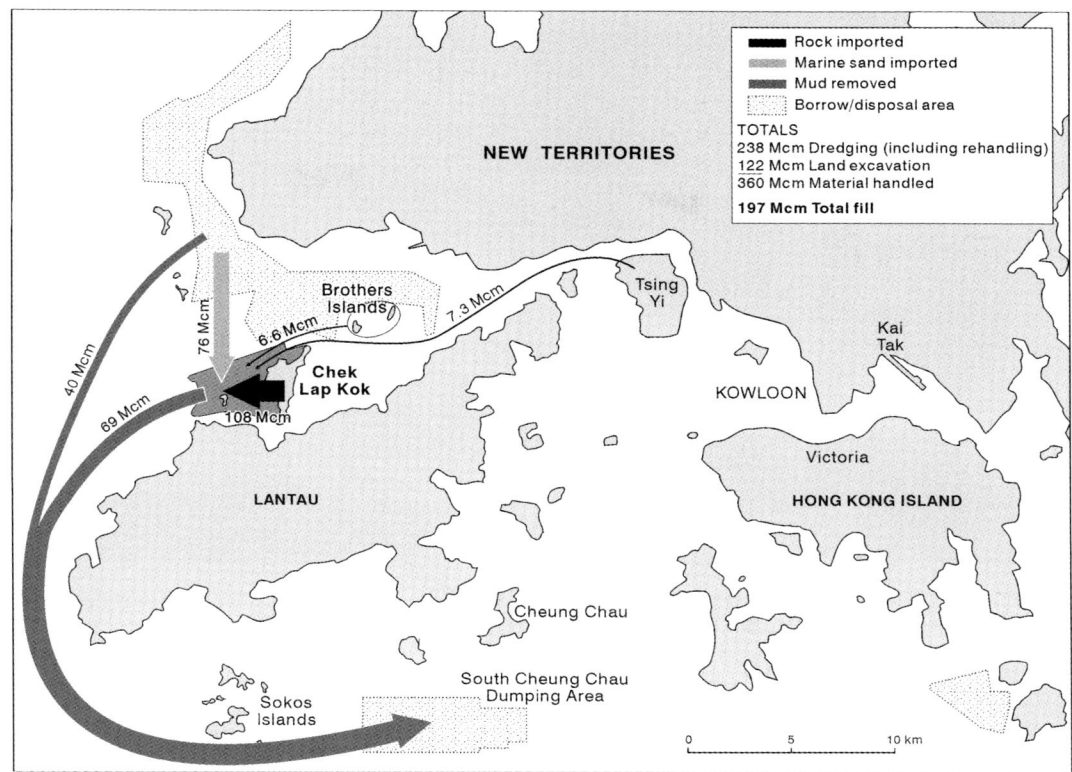

Figure 27.8   Transfers of earth materials involved in the creation of the new airport for Hong Kong.

geomorphology to the construction of land raise mounds.

In some parts of Britain, waste dumps are prominent features of the landscape. The older dumps are the result of coal, slate and china clay production. Modern land raise features dominate many low relief areas. Disposal to landfill of controlled industrial and municipal waste in Britain accounts for some 108 Mt per annum (DETR 1997). Whilst much of this waste is deposited in disused open pit mines and quarries, land raise mounds are probably the fastest growing artificial landforms in Britain today. The greatest geomorphological impact of landfill is in river valleys, sections of which are being filled, raising the height of the ground surface well above the former floodplain level. This effectively reduces the flood storage capacity of the floodplain, shifting the flood problems downstream.

Landforms of extraction

Urban demands for building materials, stone, gravel, sand, limestone and brick clays have led to the profound transformation of landforms around cities. Many areas of gravel extraction along river valleys and former floodplains have transformed large areas into wetlands (Figure 27.9). As cities have been extracting these materials for centuries, many former pits and quarries have been filled in and re-used for other purposes. Detection of filled quarries and pits may require the use of geophysical techniques (McDowell, 1975).

Sometimes there has been quarrying of clays from the surface and mining for chalk below ground at the same site, posing problems for later urban development. Near Reading, England, brickmaking occurred at the former Katesgrove clay pit from at least the 1600s. The resulting urban geomorphological hazards include

Figure 27.9   Development of the Trent Valley, Nottingham, UK. Gravel extraction has left numerous flooded pits. Photograph courtesy of Tony Waltham Geophotos.

significant thicknesses of made ground and remnant over-steepened clay pit faces. Houses built above the clay pit face have been so affected by slope instability that a number had to be domolished. When the clay pit floor was redeveloped for housing it was first necessary to stabilise the old clay pit face. However, when the developer started to put piles down into the underlying chalk for stronger foundations, a chalk mine was discovered below the site. The type of foundation had to be changed. However, as the mine was not filled-in, future potential risks for adjacent roads, car parks and landscaped areas remain.

In the clays, sands and gravels that overlie the chalk (known as the Lambeth Group deposits in Britain) and similar materials elsewhere, the following precautions should be taken during site investigations:

1. Always consider the potential for instability.
2. Carry out a desk study and look at historical maps for evidence of past quarrying in Lambeth Group deposits and similar formations. Some evidence may be old (1500s to 1700s).
3. Solution features commonly affect Lambeth Group and similar deposits – evidence is visible

in terms of anomalous, sudden lithology, colour and structure changes.
4. Information on the recorded occurrence of solution features and mining cavities can be obtained by searching national databases.

In the UK, a series of National Reviews of specific ground-related problems were commissioned by the Department of the Environment, including: Review of Landsliding (Geomorphological Services Ltd, 1986–87); Review of Mining Instability (Arup Geotechnics, 1991); Review of Natural Underground Cavities (Applied Geology (Central) Ltd, 1994); Review of Foundation Conditions (Wimpey Environmental Ltd, 1994); Review of Natural Contamination (British Geological Survey, 1994).

Often former quarries and pits are used as landfill sites before being redeveloped for other urban uses. Construction over landfill introduces another set of problems, unless great care is taken. Structures placed directly on a landfill are subject to settlement and to the escape of methane gas. A restaurant and shops built on a shallow, 8 metre thick, landfill south of Los Angeles, were placed on piles driven through the landfill to stop them from settling. The pathways around the buildings were not and a few years later had settled so much that they had to be raised up to the level of the adjacent buildings. The methane gas at the site was collected and used for gas lighting in the shopping precinct (Pipkin and Trent, 1994). A useful report relating to construction over landfill was written by Charles (1993).

27.5 Remediation and Prevention

Erosion avoidance, prevention and control

Construction site erosion control is important for the control of non-point source pollution requiring the combined efforts of design engineers, site engineers, geomorphologists and soil conservation experts (Harbor, 1999). The practical guidelines of the US Department of Agriculture and the New South Wales Department of Conservation are a good starting point for assessing erosion risk on construction sites.

Construction should be carried out in phases to avoid disturbing too much of the land at any one time (Table 27.5). No unnecessary clearing should be undertaken. Immediately below any cleared area, detention ponds should be constructed to retain any sediment washed of the site and to hold back stormwater runoff so that peak discharges in the stream below are not increased. Particular attention should be paid to the design of construction roads and later of permanent roads, following four basic principles and adopting erosion control measures (Table 27.6).

Landslide avoidance, prevention and control (see also Chapters 8, 9)

The Japanese Landslide Prevention Law permits the designation of a landslide-threatened area. In an urban area the landslide-threatened area to be designated must have a risk of causing damage to at least one of the following: a river of public

Table 27.5 Erosion and sediment control on construction sites (after Harbor, 1999).

Strategy	Implication
Fit the development to the site conditions	Fit development to the natural contours of land; use unstable and highly erodible areas as open space; avoid steep andfrequently wet areas.
Retain as much existing vegetation as possible	Avoid disturbing as much ground as possible; clear small parcels at a time and re-establish permanent plant cover as quickly as possible.
Minimise bare soil exposure by managing grading and construction timing	Delay clearing and earth moving until just before construction starts; clear large sites in stages; concentrate work in dry seasons or periods when erosion risk is least.
Establish temporary surface cover	Use mulches, temporary turf, shredded plant material, geotextiles or plastic sheeting as temporary ground cover.
Divert water away from disturbed areas	Use diversion channels to take water away from construction areas.
Minimise slope length and gradient	Break long slopes with intercept ditches; install cross drains on roadways.
Minimise runoff velocity and discharge	Divert unconcentrated runoff onto vegetated surfaces; line drainage channels with vegetation or rock to increase roughness; use check dams to slow flows.
Design drainage channels and outlets to be non-erosive	Ensure channels can cope with worst case flows; avoid risk of undercutting and erosion at channel outlets.
Trap sediment on site	Use silt fences, straw bale dikes and sediment traps, ponds and basins to hold back sediment.
Inspect and maintain	Ensure all erosion control works are in good condition; maintain sediment trapping structures, including controlled removal of accumulated sediment where necessary.

Table 27.6   Basic principles and erosion control measures for urban road construction.

Basic principles	Erosion control
Minimise the amount of disturbance caused by road construction by a) controlling the total mileage of roads, and b) reducing the area of disturbance on the roads that are built	Use as narrow a road as possible commensurate with traffic speed, safety requirements and erosion hazards.
Avoid construction in areas of high erosion hazard	If roads have to be built over steep and potentially unstable areas, and no alternative route is available, traffic restrictions, such as one way working along a narrow stretch, should be considered.
Minimise erosion on areas that are disturbed by road construction by a variety of practices designed to reduce erosion	Attempt to balance the volume of cut and fill material to minimise excavation. Use proper layer placement and compaction techniques wherever possible on fills to ensure their stability against mass movement or erosion. In potentially unstable materials, such as deeply weathered granite, full bench construction (with no fill slope) may be preferable. Haul excavated material to safe disposal areas. Stabilise the waste disposal areas as well as the road.
Minimise the off-site impacts of erosion	Allow and promote vegetation establishment on unpaved areas, especially the faces of cut and fill areas. Provide properly designed road surfacing to prevent roadway erosion and maintain a stable road. Provide adequate cross drains to prevent water erosion and gully development in ruts and drains. Drains should be located to cope with particular runoff and sediment problems. They must be adequately maintained and their outlets to natural channels designed to avoid downstream scour and siltation.

interest; railway tracks; prefectural and certain municipal roads; other important public facilities and buildings, such as government offices, schools and hospitals. A landslide-threatened area can also be designated in special circumstances so that houses at risk may be moved.

Under the Swedish Planning and Building Act of 1987, planning and granting of building permits must take into account the suitability of the ground on which development is to take place. This includes soil, rock and groundwater conditions and thus slope stability conditions. The responsibility for ensuring that urban development does not take place in areas of insufficient stability and for avoiding cumulative development eventually leading to increased slope instability rests firmly with the municipalities and partly with county administrative boards (Ahlberg *et al.*, 1988).

Landslide hazard mapping and assessment of landslide risks is widely used to identify the location and magnitude of potential slope instability

(e.g. Halcrow, 1986; Lee and Moore, 1991; Lee *et al.*, 2000). In France, production of maps of zones exposed to risks of soil movements (the ZERMOS system) have sometimes preceded the production of 1 : 5000 land occupation plans and have even led to 1 : 5000 plans of risk exposure (Brand, 1988). Since 1994, plans for the prevention of foreseeable risks have been required in France, involving public participation and the application of the principles of sustainable development (Dagorne and Dars, 1999). These developments give geomorphology a key role in the French urban planning and development control process.

27.6   Summary

Urban authorities and governments have to plan for multiple land uses, and sequential regeneration and re-use of the land. Increasingly building codes and planning regulations and guidance have to accommodate not only the natural geomorphological and surface geological conditions that may create hazards for urban activities, but the people-made geomorphology and surficial sediments from clay pits and landfills to made-ground and re-used industrial land (Leach and Goodger, 1991). In some cities, the present buildings are the fourth or fifth set of structures to occupy a particular site over the last 200 years. In urban areas, it is not only essential to 'know the ground you build on' in terms of geomorphology, geology and urban sedimentation, it is important to 'know the land use history and people-made geomorphological changes to the land you build on'. Building codes and planning guidance have to recognise the need for a geomorphology of brownfield sites, old quarries, landfills and gravel pits. In England and Wales, Planning Guidance note PPG14 on building on unstable ground (Department of Environment and Welsh Office, 1990, 1996) makes a start in this direction.

Research in the UK on land instability (for example Doornkamp, 1988; Lee and Moore, 1991; Thomson *et al.*, 1996; DETR, 1998) has integrated complex land instability, flooding and former industrial land use information to provide guidance on how planning conditions in specific localities, including Ripon, Rotherham, St Helens, Swansea and Torbay can incorporate geomorphological considerations. Guidance forms for conveyancing could include questions about land instability and former land uses. However, the questions are often general in nature, and solicitors may not always be aware of the full range of information to seek and of the full range of sources of information on ground conditions. Urban engineering geomorphology is thus more than identifying problems and finding solutions. Asking questions about geomorphology must be part of the total business of urban development and management.

References

Ahleberg, P., Stigler, B. and Viberg, L. (1988) Experiences of landslide risk considerations in land use planning in Sweden. In Bonnard, C. (ed.) *Landslides: Proceedings of the Fifth International Symposium on Landslides* Volume 2, Rotterdam: Balkema, 1091–1096.

Antonini, G., Ardizzone, F., Cacciano, M., Cardinali, M., Castellani, M., Galli, M., Guzzetti, F., Reichenbach, P. and Salvati, P. (2000) A geomorphological approach to estimate landslide hazard and risk in urban and sub-urban areas: examples from the Umbria region. *EGS 2000*, 25–29 April, Nice.

Applied Geology (Central) Limited (1994) *Review of Natural Underground Cavities*. Report to the Department of the Environment.

Arup Geotechnics (1991) *Review of Mining Instability*. Report to the Department of the Environment.

Bacon, J. and Bacon, S. (1988) *Dunwich Suffolk*. Segment Publications, Colchester.

Barbatic, Z., Langof, Z., Goluza, M., Steger, Z., Martinovic, D. and Lasic, M. (1998) *Geotechnical Hazards,* Balkema, Rotterdam.

Bennett, M. R. and Doyle, P. (1997) *Environmental geology:Geology and the Human Environment*, Chichester, Wiley.

Bergado, D. T. and Sebanayagan, A. N. (1987) Pile foundation problems in Kuala Lumpur limestone, Malaysia. *Quarterly Journal of Engineering Geology* **20**, 159–175.

Bernknopf, R. L., Brookshire, D. S., Soller, D. R., McKee, M. J., Sutter, J. F., Matti, J. C. and Campbell, R. H. (1993) Societal value of geologic maps. *U.S. Geological Survey Circular 1111*.

Berry, P. L., Illsley, D. and McKay, I. R. (1985) Settlement of two housing estates at St. Annes due to consolidation of a near surface peat stratum. *Proceedings Institution of Civil Engineers, Part 1* **77**, 111–136.

Brand, E.W. (1988) Special Lecture: Landslide risk assessment in Hong Kong. In Bonnard, C. (ed.) *Landslides: Proceedings of the Fifth International Symposium on Landslides* Volume 2, Rotterdam: Balkema, 1059–1074.

British Geological Survey (1984) *Review of Natural Contamination in Great Britain*. Report to the Department of the Environment.

Bromhead, E. N., and Ibsen, M.-L. (1998) Land use and climate-change impacts on landslide hazards in SE Britain. In Cruden, D. and Fells, R. (eds.) *Landslide Risk Assessment,* Balkema, 65–176.

Brunsden, D. and Lee, E. M. (2004) Behaviour of coastal landslide systems: an inter-disciplinary view. *Zeitschrift fur Geomorphologie* **134**, 1–112.

Cardinali, M., Reichenbach, P., Guzzetti, F., Ardizzone, F., Antonini, G., Galli, M., Cacciano, M., Castellani,, M. and Salvati, P. (2002) A geomorphological approach to the estimation of landslide hazards and risks in Umbria, Central Italy. *Natural Hazards and Earth System Sciences* **2**, 57–72.

Carnec, C., Huré A., Ledoux, E., Raucoules, D. and Rivera, A. (2000) Mapping and modelling of major urban subsidence on Mexico City from SAR interferometry, *EGS 2000*, 25–29 April, Nice.

Charles, J. A. (1993) *Building on fill: Geotechnical Aspects*. BRE Report.

Coates, D. R. (1985) *Geology and Society*. New York: Chapman & Hall.

Cooke, R. U. and Doornkamp, J. C. (1990) *Geomorphology in Environmental Management: a new introduction*. (Second Edition), Oxford: Oxford University Press.

Dagorne, A. and Dars, R. (1999) *Les risques naturels*. Paris: Presses Universitaires de France.

Department of the Environment, Welsh Office (1990) *Planning Policy Guidance: Development on Unstable Land* PPG 14.

DETR (Department of the Environment Transport and Regions) (1997) *Digest of Environmental Statistics* **19**, HMSO, London.

DETR (Department of the Environment Transport and Regions) (1998) *Land Instability, Flooding and Hazard Assessment in Ripon. Environmental Geology in Land-Use Planning Case Study No.8, http://www.symonds-group.com/services/ environ_ eng/publications/casestudies/casestudy8.pdf*

Doornkamp, J. C. (1988) Applied earth science background: Torbay. *GSL Publications*.

Durazo, J. and Farvolden, R. N. (1989) The groundwater regime of the Valley of Mexico from historic evidence and field observations. *Journal of Hydrology* **112**, 171–190.

Douglas, I. (1971) Dynamic equilibrium in applied geomorphology. *Earth Sciences* **5**, 29–35.

Endo, H. (2000) The behavior of a VLFS and an airplane during takeoff/landing run in wave condition. *Marine Structures* **13**, 477–491.

Fabre, G. and Monteil, M. (2001) Sur l'hydrogéomorphologie d'un espace à forte anthropisation urbaine: le site de Nîmes (Languedoc, France) du Pléistocène superior à l'Antiquité; impacts postérieurs. *Comptes Rendus de l'Académie des Sciences Series IIA— Earth and Planetary Sciences* **333**(8), 435–440.

Ferretti, A. Prati, C. and Rocca, F. (1999) Monitoring terrain deformations using multi-temporal SAR Images, *Proceedings of FRINGE '99, Liege Belgium, November 1999, http://esrin.esa.it/fringe99*

Gardner, T. (1754) Historical notes on Dunwich, Blythburgh and Southwold.

Geomorphological Services Limited (1986–1987) *Review of Landsliding*. Report to the Department of the Environment.

Gupta, A. and Ahmad, R. (1999) Geomorphology and the Urban tropics: building an interface between research and usage. *Geomorphology* **31**, 133–149.

Gutjahr, M. R., Tavares, R., Pereira, P. R. B. and Santora, J. (2000) Natural hazards vs man-made hazards: landslides in the escsarpment of Serro de Mar, SP, Brazil. In Bromhead, E., Dixon, N. and Ibsen, M-L. (eds) *Landslides in Research, Theory and Practice*, Proc. of 8th Int. Symp. on Landslides, Cardiff, 2000, Thomas Telford, London, **2**, 687–692.

Halcrow, W. (1986) *Rhondda Landslip Potential Assessment*. Department of the Environment and Welsh Office.

Harbor, J. (1999) Engineering geomorphology at the cutting edge of land disturbance: erosion and sediment control. *Geomorphology* **31**, 247–263.

Hutchinson, J. N. (1991) Periglacial and slope processes. In Forster, A., Culshaw, M. G., Cripps, J. C., Little, J. A. and Moon, C. F. (eds) *Quaternary Engineering Geology*, Geological Society Publishing, London, 283–334.

Jones, J. A. A. (1981) *The nature of soil piping: a review of research*. GeoBooks, Norwich.

Jones, K. D. C. and Lee, E. M. (1994) *Landsliding in Great Britain*. HMSO.

Kaye, C. A. (1976) Beacon Hill end moraines, Boston: New explanation of an important urban feature. In Coates, D. R. (ed.) *Urban geomorphology: Geological Society of America Special Paper* **174**, 7–20.

Kimura, H. and Yamaguchi, Y. (2000) Detection of landslide areas using satellite radar interferometry. *Photogrammetric Engineering and Remote Sensing* **66**(3), 337–344.

Lee, E. M. (1994) *The Investigation and Management of Erosion, Deposition and Flooding in Great Britain*. HMSO.

Lee, E. M. and Moore, R. (1989) *Landsliding in and around Luccombe Village*. HMSO.

Lee, E. M. and Moore, R. (1991) *Coastal landslip potential assessment: Isle of Wight Undercliff, Ventnor*. Department of the Environment.

Leopold, L. B. (1968) *Hydrology for urban land planning: A guidebook on the hydrologic effects of urban land use*. US Geological Survey Circular 554.

Maharaj, R. J. (1993) Landslide processes and landslide susceptibility analysis from an upland watershed: a case study from St. Andrew, Jamaica, West Indies. *Engineering Geology* **34**, 53–79.

Lam S. K. (1992) Progress report: Quaternary Geological Mapping of the Kuching City area, Sarawak. *Proceedings of the 23rd Geological Conference: Technical Papers*. Kuala Lumpur: Geological Survey of Malaysia, 96–107.

Leach, B. A. and Goodger, H. K. (1991) *Building on Derelict Land*, CIRIA Special Publication, CIRIA, London.

Lee, E. M., Jones, D. K. C. and Brunsden, D. (2000) The Landslide Environment of Great Britain. In *Landslides in Research, Theory and Practice*, Bromhead, E., Dixon, N. and Ibsen, M-L. (eds) Proc. of 8th Int. Symp. on Landslides, Cardiff, 2000, Thomas Telford, London, **2**, 911–916.

Marker, B. R. (1996) The role of the earth sciences in addressing urban resources and constraints. In McCall, G. J. H., De Mulder, E. F. J. and Marker, B. R. (eds) *Urban Geoscience*, Rotterdam, Balkema, 163–179.

McCormick, W. V. III. (2002) Urban landslide hazards, Millbrae Ca. Landslide: Case History. Geological Society of America, *Cordilleran Section - 98th Annual Meeting (May 13–15, 2002). http://gsa.confex.com/gsa/2002CD/finalprogram/abstract_34955.htm*

McDowell, P. W. (1975) Detection of clay filled sinkholes in the chalk by geophysical methods. *Quarterly Journal of Engineering Geology* **8**(4), 303–310

New Civil Engineer (2003) *Keep on running.* Emap Construct Ltd., 31–33.

Pelli, C., Thornton, C. and Joseph, L. (1997) The world's tallest buildings. *Scientific American* **277**(6), 64–73.

Pethick, J. (1996) Coastal slope development: temporal and spatial periodicity in the Holderness Cliff Recession. In Anderson, M. G. and Brooks, S. M. (eds.) *Advances in Hillslope Processes* **2**, 897–917.

Péwé, T. I. (1990) Land subsidence and earth-fissure formation caused by groundwater withdrawal in Arizona; A review. In Higgins, C. G. and Coates, D. R. (eds) *Groundwater geomorphology; The role of subsurface water in earth-surface processes and landforms. Boulder, Colorado, Geological Society of America Special Paper 252*, 219–233.

Pipkin, B. W. and Trent, D. D. (1994) *Geology and the Environment (2nd edn)*, Belmont, California, Wadsworth.

Polan, S.M. (1992) Solid waste problems in the Big Apple. *Forum for Applied Research and Public Policy* **7**, 62–64.

Refice, A., Bovenga, F., Wasowski, J. and Guerriero, L. (2000) Use of InSAR data for landslide monitoring: A case study from southern Italy, *Proceedings of IGARSS'2000, Honolulu, Hawaii USA, July 2000.*

Refice, A., Bovenga, F. and Wasowski, J. (2001) Monitoring landslide activity in a peri-urban area by SAR interferometry. EGS Nice. *http://www.copernicus.org/EGS/egsga/nice01/programme/abstracts/aai4893.pdf*

Skempton, A.W. and Weeks, A.G. (1976) The Quaternary history of the Lower Greensand escarpment and Weald Clay vale near Sevenoaks, Kent. *Philosophical Transactions of the Royal Society, London* **A283**, 493–526.

Smith, R. (2002) The Great Blackheath Hole. *Greenwich Industrial History* **5**(3), *http://gihs.gold.ac.uk/gihs25. html#blackhole*

Spencer, T. (1999) Coastal erosion. In Pacione, M. (ed.) *Applied Geography: Principles and Practice*, London, Routledge, 109–123.

Tebbutt, T. H. Y. (1998) *Principles of Water Quality Control.* Butterworth Heinemann, Oxford.

Thompson, A., Hine, P. D., Greig, J. R. and Peach, D. W. (1996) *Assessment of subsidence arising from gypsum solution.* Symonds Travers Morgan Report to the Department of the Environment.

Tricart, J. (1961a) Les modalités de la morphogénèse dans le lit du Guil au cours de la crue de la mi-juin 1957. *International Association of Scientific Hydrology Publication* **53**, 65–73.

Tricart, J. (1961b) L' evolution du lit du Guil au cours de la crue de juin 1957. *Bulletin de la Section Géographique, Comité de la Travaux Historiques et Scientifiques* **72**, 169–403.

USGS *Las Vegas Urban Corridor Geologic Mapping Project http://geology.wr.usgs.gov/wgmt/lasvegas/lvmap.html*

Valentin, H. (1954) Der landverlust in Holderness, Ostengland von 1852 bis 1952. *Die Erde* **6**, 296–315.

Vestappen, H. Th. (1983) *Applied Geomorphology*, Amsterdam. Elsevier.

Wimpey Environmental Limited (1994) *Review of Foundation Conditions.* Report to the Department of the Environment.

Further reading

Bennet, M. R. and Doyle, P. (1997) *Environmental Geology and the Human Environment.* Wiley, Chichester.

Douglas, I. (1983) *Urban Environments.* Edward Arnold, London.

Eyles, N. (1997) (Ed) *Environmental Geology of Urban Areas.* Geological Association of Canada, St. Johns.

Gupta, A. and Pitts, J. (1992) *Physical Adjustments in a Changing Landscape: The Singapore Story.* Singapore University Press, Singapore.

Appendices

Appendix A1: Visual Identification

Summary of contents

Table A1.1   Chart for estimating percentage cover or composition. From Gardiner and Dackombe (1983) after Hodgson (1974).

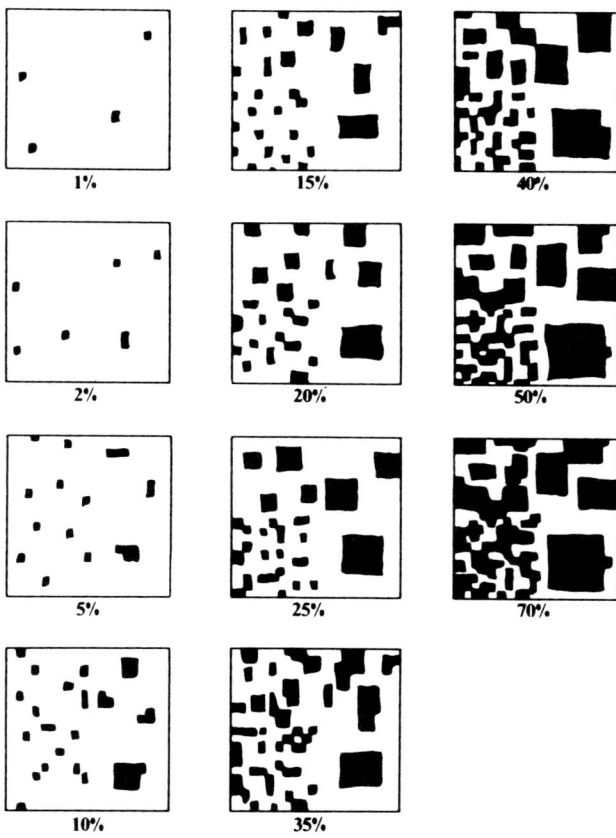

Table A1.2   Images for the visual assessment of pebble roundness. From Gardiner and Dackombe (1983), after Krumbein (1941).

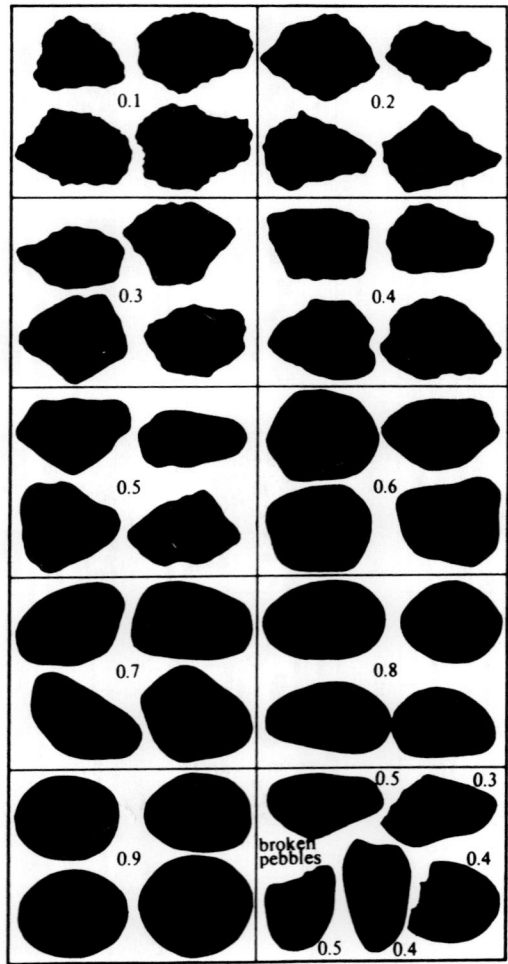

Table A1.3 The Beaufort scale of wind force.

Beaufort number	General description of wind	Approximate wind speeds (ms⁻¹)	Approximate wave height (m)	Specification	
				For coastal use	For inland use
0	Calm	0.0–0.2	0	Sea like a mirror	Smoke rises vertically
1	Light air	0.3–1.5	0.1–0.2	Ripples without appearance of scales; no foam crests	Wind direction shown by smoke drift but not by wind vanes
2	Light breeze	1.6–3.3	0.3–0.5	Small wavelets; crests have glassy appearance but do not break	Wind felt on face; leaves rustle, ordinary vane moved by wind
3	Gentle breeze	3.4–5.4	0.6–1.0	Large wavelets; crests begin to break; scattered white horses	Leaves and small twigs in constant motion; wind extends light flag
4	Moderate breeze	5.5–7.9	1.5	Small waves becoming longer; fairly frequent white horses	Raises dust and loose paper, small branches are moved
5	Fresh breeze	8.0–10.7	2.0	Moderate waves; many white horses and chance of some spray	Small trees in leaf begin to sway; crested wavelets form on inland water
6	Strong breeze	10.8–13.8	3.5	Large waves begin to form; white foam crests extensive everywhere and spray probable	Large branches in motion; whistling heard in telegraph wires; umbrellas used with difficulty
7	Near or moderate gale	13.9–17.1	5.0	Sea heaps up and white foam from breaking waves begins to be blown in streaks; spindrift begins to be seen	Whole trees in motion; inconvenience felt when walking against wind
8	Gale or fresh gale	17.2–20.7	7.5	Moderate high waves of greater length; foam blown in well marked streaks; edges of crests break into spindrift	Breaks twigs off trees; generally impedes progress
9	Strong gale	20.8–24.4	9.5	High waves; crests begin to topple and roll over; spray may affect visibility	Slight structural damage occurs; chimney pots and slates removed
10	Storm or whole gale	24.5–28.4	12.0	Very high waves; long overhanging crests, tumbling of sea becomes very heavy and shocking; sea	Seldom experienced inland; trees uprooted; considerable structural damage occurs

Table A1.3   (*Continued*).

Beaufort number	General description of wind	Approximate wind speeds (ms⁻¹)	Approximate wave height (m)	Specification	
				For coastal use	For inland use
				surface takes on white appearance as foam in large patches is blown in very dense streaks	
11	Violent storm	28.5–32.7	15.0	Exceptionally high waves, sea covered by long patches of foam; small and medium-sized ships might be lost to view behind waves for long periods	Very rarely experienced; widespread damage
12	Hurricane	>32.7	>15	Air filled with foam and spray; sea completely white with driving spray; visibility greatly reduced	

Table A1.4   Earthquake intensity (after Waltham, 1994).

Modified Mercalli intensity	Sensation	Damage	Area of influence, distances from epicentre	Equivalent Richter* Magnitude
I	Not Felt			
II	Felt at rest ⎫	Social		
III	Felt indoors ⎬	disturbance,	Limited	1 to 3
IV	Windows rattle ⎭	no damage		
V	Felt outdoors			
VI	Frightening	Slight damage	Slight damage to 10 km away	4
VII	Severe damage to adobe houses, little damage to reinforced concrete		Damage to 10 km away	5
VIII	Masonry damaged		Damage to 50 km away	
IX	Foundations damaged, severe damage to many buildings		Damage to 200 km away	6
X	Major damage to or destruction of most buildings		Severe damage to 20 km away	7
XI	Railways buckled			
XII	Total destruction			8

* Richter magnitude is given by $\log_{10}$ of the maximum wave amplitude in microns measured on a Wood Andersons seismograph 100 km from the epicentre.

Table A1.5 Roughness profiles and corresponding range of Joint Roughness Coefficient (JRC) associated with each (ISRM, 1981). (JRC is a measure of the relative roughness of surfaces in rock joints).

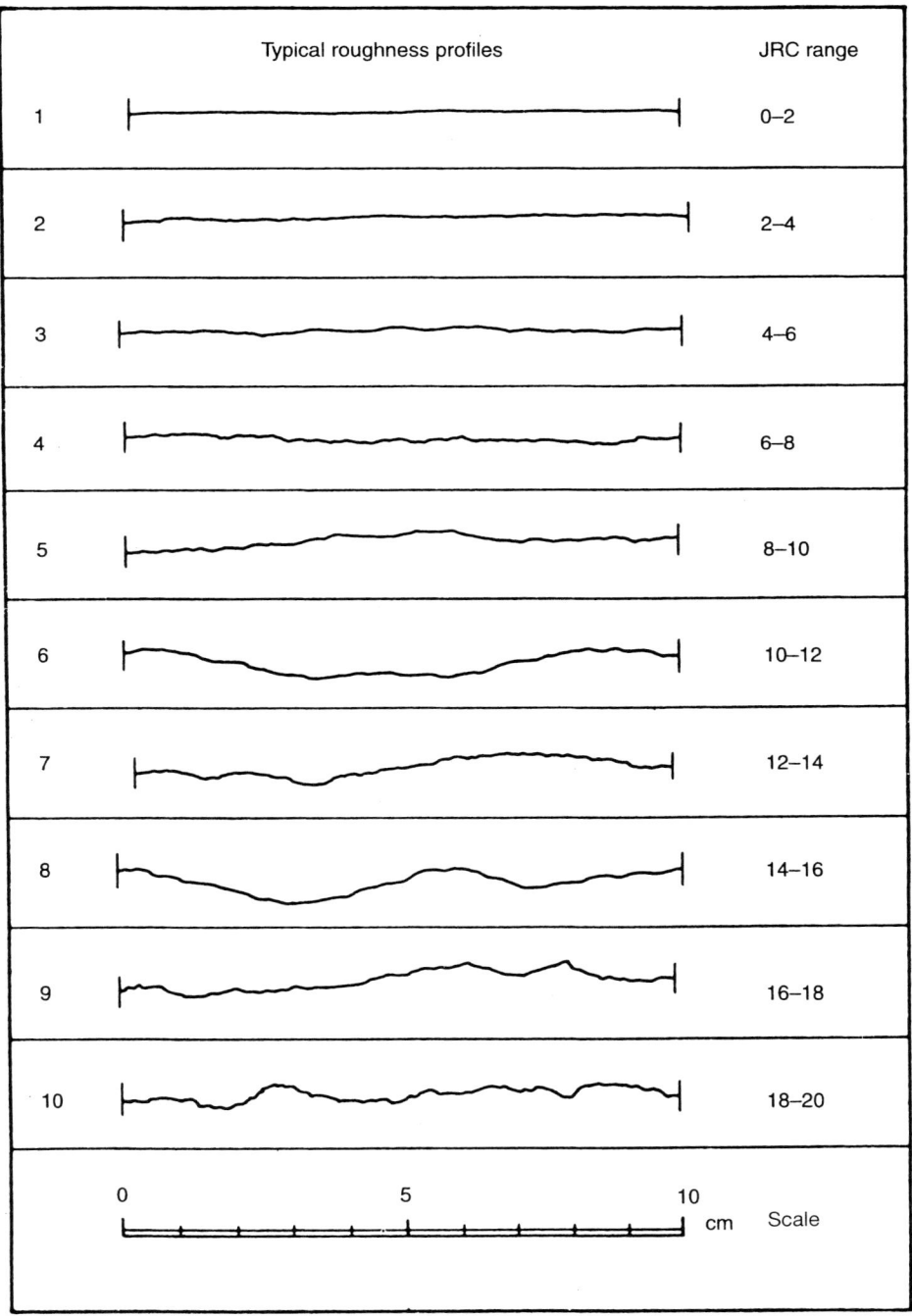

References

Gardiner, V. and Dackombe, R. V. (1983) *Geomorphological Field Manual.* George Allen and Unwin, London.

Hodgson, J.M. (ed.) (1974) *Soil Survey Field Handbook.* Soil Survey Technical Monograph 5, Rothamsted Experimental Station.

ISRM (International Society for Rock Mechanics) (1981) *Suggested Methods for the Quantitative Description of Discontinuities in Rock Masses.* (ed. E.T. Brown), Pergamon Press, Oxford.

Krumbein, W. C. (1941) Measurement and geological significance of shape and roundness of sedimentary partices. *J. Sed. Petrol.* **II**, 64–72.

Waltham, A. C. (1994) Foundations of Engineering Geology. Blackie, Glasgow, UK.

Appendix A2: Identification, description and classification of soils

Summary of contents

Table A2.1　Flowchart for identification and description of soils (BS5930: 1999).

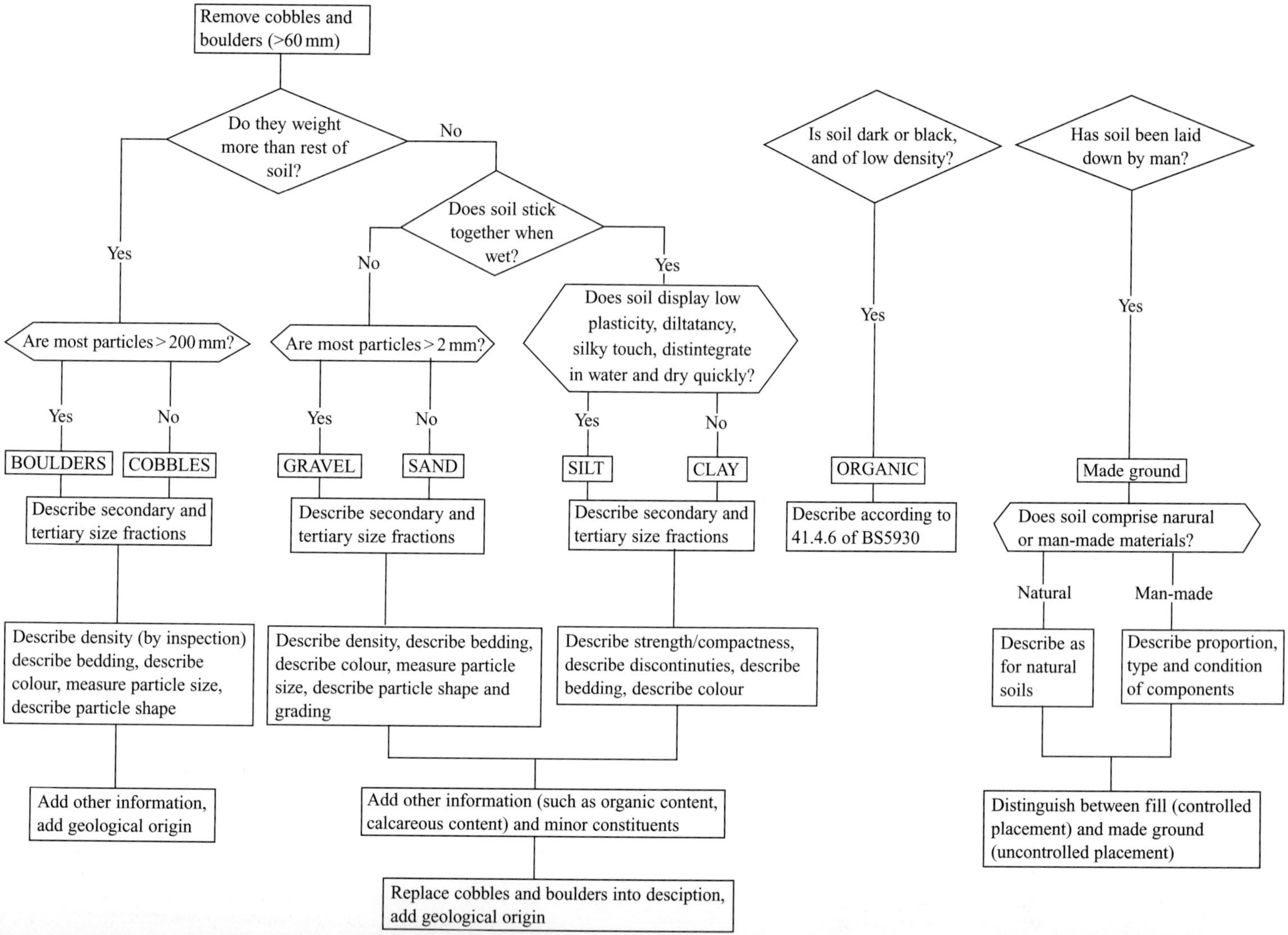

Remove cobbles and boulders (>60 mm)

Do they weight more than rest of soil?

Yes → Are most particles > 200 mm?

Yes → BOULDERS
No → COBBLES

Describe secondary and tertiary size fractions

Describe density (by inspection), describe bedding, describe colour, measure particle size, describe particle shape

Add other information, add geological origin

No → Does soil stick together when wet?

No → Are most particles > 2 mm?

Yes → GRAVEL
No → SAND

Describe secondary and tertiary size fractions

Describe density, describe bedding, describe colour, measure particle size, describe particle shape and grading

Yes → Does soil display low plasticity, dilatancy, silky touch, distintegrate in water and dry quickly?

Yes → SILT
No → CLAY

Describe secondary and tertiary size fractions

Describe strength/compactness, describe discontinuities, describe bedding, describe colour

Add other information (such as organic content, calcareous content) and minor constituents

Replace cobbles and boulders into description, add geological origin

Is soil dark or black, and of low density?

Yes → ORGANIC

Describe according to 41.4.6 of BS5930

Has soil been laid down by man?

Yes → Made ground

Does soil comprise natural or man-made materials?

Man-made → Describe proportion, type and condition of components

Natural → Describe as for natural soils

Distinguish between fill (controlled placement) and made ground (uncontrolled placement)

Table A2.2 Identification and description of soils (BS5930: 1999).

Soil group	Density/compactness/strength		Discontinuities		Bedding		Colour	Composite soil types (mixtures of basic soil types)		Particle shape	Particle size	Principal soil type
	Term	**Field test**	Scale of spacing of discontinuities		Scale of bedding thickness			For mixtures involving very coarse soils, see Table A2.3				
			Term	Mean spacing mm	**Term**	Mean thickness mm		**Term**	Approx, %[c] secondary			
Very coarse soils	Loose	By inspection of voids and particle packing					Red Orange Yellow Brown			Angular	−200	BUILDERS
	Dense										Coarse	COBBLES
										Sub-angular	−60	
	Borehole with SPT N-value		Very widely	Over 2000	Very thickly bedded	Over 2000	Green Blue White Cream	Slightly (sandy[d])	<5	Sub-rounded Rounded	Coarse	GRAVEL
	Very loose	0–4	Widely	2000 to 600	Thickly bedded	2000 to 600					−20	
										Flat	Medium	
	Loose	4–10	Medium	600 to 200	Medium bedded	600 to 200	Grey Black etc.	(sandy[d])	5 to 20[b]		−6	
	Medium dense	10–30	Closely	200 to 60	Thinly bedded	200 to 60				Tabular	Fine	
	Dense	30–50	Very closely	60 to 20	Very thinly bedded	60 to 20		Very (sandy[d])	>20[b]	Elongated	−2	
	Very dense	>50	Extremely closely	Under 20	Thickly laminated	20 to 6				Minor constituent type	Coarse	SAND
											−0.6	
					Thinly laminated	Under 6		SAND AND GRAVEL	about 50[b]	Calcareous, shelly, glauconitic, micaceous etc. using terms such as	Medium	
			Fissured	Breaks into blocks along unpolished discontinuities							−0.2	
Slightly cemented		Visual examination: pick removes soil in lumps which can be abraded					Light	**Term**	Approx. %[c] secondary	Slightly calcareous,	Fine	SILT
	Un-compact	Easily moulded or crushed in the fingers	Sheared	Breaks into blocks along polished discontinuities	Inter-bedded	Alternating layers of different types Prequalified by thickness term	Dark				−0.06	
						if in equal proportions. Otherwise thickness of and	Mottled	Slightly (sandy[e])	<35	calcareous,	Coarse	
Compact		Can be moulded or crushed by strong pressure in the fingers			Inter-laminated						−0.02	

Soil group: Coarse soils (over about 65% sand and gravel sizes)

Table A2.2 (*Continued*).

Soil group	Density/compactness/strength		Discontinuities	Bedding	Colour	Composite soil types (mixtures of basic soil types)		Particle shape	Particle size	PRINCIPAL SOIL TYPE
	Term	Field test								
Fine soils (over about 35% silt and clay sizes)	Very soft 0–20 (KPa)	Finger easily pushed in up to 25 mm	Spacing terms also used for distance between partings, isolated beds or laminae, dessication cracks, rootlets etc.	spacing between subordinate layers defined				very calcareous.	Medium	
	Soft 20–40	Finger pushed in up to 10 mm							−0.006	
	Firm 40–75	Thumb makes impression easily				(sandy[e])	35 to 65[a]		Fine	CLAY/SILT
	Stiff 75–150	Can be indented slightly by thumb						% defined on a site or material specific basis or subjective	−0.002	
	Very stiff 150–300	Can be indented by thumb nail				Very (sandy[f])	>65[a]			CLAY
	Hard (or very weak mudstone Cu>300 kPa)	Can be scratched by thumbnail see 41.2.2 of BS5930								

Soil group	Term	Field test	Discontinuities	Transported mixtures	Colour	PRINCIPAL SOIL TYPE
Organic soils	Firm	Fibres already compressed together	Fibrous	Plant remains recognizable and retains some strength — Slightly organic clay or silt / Slightly organic sand / Organic clay or silt / Organic sand	Grey as mineral / Dark grey / Dark grey	Contains finely divided or discrete particles of organic matter, often with distinctive smell, may oxidize rapidly. Describe as for inorganic soils using terminology above.
	Spongy	Very compressible and open structure	Pseudo-fibrous	Plant remains recognizable, strength lost — Very organic clay or silt / Very organic sand	Black / Black	
	Plastic	Can be moulded in hand and smears fingers	Amorphous	Recognizable plant remains absent — Accumulated *in situ* — Peat		Predominantly plant remains, usually dark brown or black in colour, distinctive smell, low bulk density. Can contain disseminated or discrete mineral soils

PRINCIPAL SOIL TYPE	Visual identification	Minor constituents	Example Stratum name	Example descriptions
BOULDERS	Only seen complete in pits or exposures			
COBBLES	Often difficult to recover whole from boreholes	Shell fragments, pockets of peat, gypsum crystals, flint gravel, fragments of brick, rootlets, plastic bags etc.		Loose brown very sandy sub-angular fine to coarse flint GRAVEL with small pockets (up to 30 mm) of clay. (TERRACE GRAVELS)
GRAVEL	Easily visible to naked eye; particle shape can be described; grading can be described	using terms such as: with rare with occasional with abundant/frequent/ numerous	RECENT DEPOSITS, ALLUVIUM, WEATHERED BRACKLESHAM CLAY,	Medium dense light brown gravelly clayey line SAND. Gravel is fine (GLACIAL DEPOSITS)
SAND	Visible to naked eye; no cohesion when dry; grading can be described.	% defined on a site or material specific basis or subjective	LIAS CLAY, EMBANKMENT FILL	Stiff very closely sheared orange mottled brown slightly gravelly CLAY. Gravel is fine and medium of rounded quartzite. (REWORKED WEATHERED LONDON CLAY)
SILT	Only coarse silt visible with hand lens; exhibits little plasticity and marked dilatancy; slightly granular or silky to the touch: disintegrates in water, lumps dry quickly; possesses cohesion but can be powdered easily between fingers		TOPSOIL, MADE GROUND OR GLACIAL DEPOSITS? etc.	
CLAY/ SILT	Intermediate in behaviour between clay and silt. Slightly dilatant			Firm thinly laminated grey CLAY with closely spaced thick laminae of sand. (ALLUVIUM)
CLAY	Dry lumps can be broken but not powdered between the fingers; they also disintegrate under water but more slowly than silt; smooth to the touch; exhibits plasticity but no dilatancy; sticks to the fingers and dries slowly; shrinks appreciably on drying usually showing cracks.			Plastic brown clayey amorphous PEAT. (RECENT DEPOSITS)

Notes:

[a] Or described as coarse soil depending on mass behaviour.
[b] Or described as fine soil depending on mass behaviour.
[c] % coarse or fine soil type assessed excluding cobbles and boulders.

[d] Gravelly or sandy and/or silty or clayey.
[e] Gravelly and/or sandy.
[f] Gravelly or sandy.

Table A2.3   Description of composite soil types for very coarse soils (BS5930: 1999)

		Main name	Estimated boulder or cobble content of very coarse fraction
	Over 50% of material is very coarse (>60 mm)	BOULDERS	Over 50% is of boulder size (>200 mm)
		COBBLES	Over 50% is of cobble size (200 mm to 60 mm)

Term	Composition
BOULDERS (or COBBLES) with a little finer material[a]	up to 5% finer material
BOULDERS (or COBBLES) with some finer material[a]	5% to 20% finer material
BOULDERS (or COBBLES) with much finer material[a]	20% to 50% finer material
FINER MATERIAL[a] with many boulders (or cobbles)	50% to 20% boulders (or cobbles)
FINER MATERIAL[a] with some boulders (or cobbles)	20% to 5% boulders (or cobbles)
FINER MATERIAL[a] with occasional boulders (or cobbles)	up to 5% boulders (or cobbles)

[a] The description of 'finer material' is made in accordance with 41.4.2 to 41.4.6 of BS5930, ignoring the very coarse fraction; the principal soil type name of the finer material may also be given in capital letters, e.g. sandy GRAVEL with occasional boulders, COBBLES with some sandy CLAY.

Term	Principal soil type	Approximate proportion of secondary constituent
slightly sandy or gravelly	SAND	up to 5%
sandy or gravelly	or	5% to 20%
very sandy or gravelly	GRAVEL	over 20%
	SAND and GRAVEL	about equal proportions

Table A 2.4 BS soil classification system (BS5930: 1981).

Soil groups		Sub-groups		Subdivisions		Liquid Limit (%)	Fines (%) < 0.06 mm
Slightly silty or clayey GRAVEL	G	Well graded GRAVEL	GW				0–5
		Poorly graded GRAVEL	GP	Uniformly graded	GPu		
				Gap graded	GPg		
Silty or clayey GRAVEL	G–F	Silty GRAVEL	G-M	Well graded	GWM		5–15
				Poorly graded	GPM		
		Clayey GRAVEL	G-C	Well graded	GWC		
				Poorly graded	CPC		
Very silty or clayey GRAVEL	GF	Very silty GRAVEL	GM	As for GC	GML etc.	As GC	
		Very clayey GRAVEL	GC	Low plasticity clay	GCL	<35	15–35
				Intermediate "	GCI	35–50	
				High "	GCH	50–70	
				Very high "	GCV	70–90	
				Extremely high "	GCE	>90	
				—			
Slightly silty or clayey SAND	S	Well graded SAND	SW				0–5
		Poorly graded SAND	SP	Uniformly graded	SPu		
				Gap graded	SPg		
Silty or clayey SAND	S–F	Silty SAND	S-M	Well graded	SWM		5–15
				Poorly graded	SPM		
		Clayey SAND	S-C	Well graded	SWC		
				Poorly graded	SPC		
Very silty or clayey SAND	SF	Very silty SAND	SM	As for SC	SML etc.	As SC	
		Very clayey SAND	SC	Low plasticity clay	SCL	<35	15–35
				Intermediate "	SCI	35–50	
				High "	SCH	50–70	
				Very high "	SCV	70–90	
				Extremely high "	SCE	>90	

GRAVELS (> 50% of coarse material is of gravel size >2 mm)

SANDS (> 50% of coarse material is of sand size < 2 mm)

Coarse soils (< 35% fines)

Table A 2.4 (*Continued*).

Soil groups	Sub-groups			Subdivisions		Liquid Limit (%)	Fines (%) < 0.06 mm
Gravelly or sandy SILTS and CLAYS (35–65% fines)	Gravelly SILT or gravelly CLAY*	FG	MG	Gravelly SILT	MLG etc.	As CG	
			CG	Gravelly CLAY	As for CG		
				Low plasticity "	CLG	<35	
				Intermediate "	CIG	35–50	
				High "	CHG	50–70	
				Very high "	CVG	70–90	
				Extremely high "	CEG	>90	
SILTS and CLAYS (>65% fines) — Sandy SILT or sandy CLAY*	Sandy SILT or sandy CLAY*	FS	MS	Sandy SILT	MLS etc.	As CG	
			CS	Sandy CLAY	As for CG	CLS etc.	As CG
Fine soils (>35% fines)	SILT or CLAY	F	M	SILT	ML etc.	As for C	As C
			C	CLAY		CL	Low plasticity	<35
				Intermediate "	CI	35–50	
				High "	CH	50–70	
				Very high "	CV	70–90	
				Extremely high	CE	>90	

ORGANIC SOILS   Letter 'O' suffixed to any group or sub-group symbol.
e.g. MHO = organic silt of high plasticity.

PEAT   Pt (Peat) soils consist predominantly of plant remains (fibrous or amorphous).

* GRAVELLY if > 50% coarse material is gravel sized; SANDY if > 50% is sand sized.

Notes:

[1] It should always be made clear whether the group classification is based on laboratory test results or simply on inspection of the soil.

[2] Silt (M-soil), M is material which plots below the A-line on a standard plasticity chart (see Table A2.6). Fine soils of this type include micaceous and diatomaceous soils, pumice and volcanic soils, and soils containing halloysite, in addition to silt-sized soils and rock flour. The alternative term 'M-soil' is sometimes used to describe this group of soils, to avoid confusion with purely silt-sized material.

[3] The soil classification applies to material of gravel, sand, silt and clay sizes. Larger particles must be removed before classification and their proportions recorded separately.

[4] Material is generally considered to be uniformly graded if it has a uniformity coefficient of less than 6 (see Table A2.6).

Table A2.5   Unified soil classification system (field observation), after Carter (1983).

Soil Groups		Description and Identification	Subgroups	Symbol
Coarse grained soils	Boulders and cobbles	Soils consisting chiefly of boulders (> 200 mm) or cobbles (75–200 m). Identifiable by visual inspection.	Boulder gravels.	
	Gravel and gravelly soils	Soils with an appreciable gravel content (2–75 mm). Generally easily identifiable by visual inspection. A medium to high dry strength indicates some clay is present. A negligible dry strength indicates the absence of clay.	Well-graded gravel–sand mixtures, little or no fines	GW
			Well-graded gravel–sands with small clay content	GC
			Uniform gravel with little or no fines	GU
			Poorly-graded gravel–sand mixtures with little or no fines	GP
			Gravel–sand mixtures with excess of lines	GF
	Sands and sandy soils	Soils with an appreciable sand content (0.06–2 mm). Majority of particles can be distinguished by eye. Feel gritty when rubbed between fingers. A medium to high dry strength indicates some clay is present.	Well-graded sands and gravelly sands, little or no fines	SW
			Well-graded sand with small clay content	SC
			Uniform sands with little or no fines	SU
			Poorly-graded sands, little or no fines	SP
			Sands with excess of fines	SF
Fine-grained soils	Fine-grained soils having low plasticity (silts)	Soils of low plasticity: cannot be readily rolled into threads when moist. Do not feel gritty. Exhibit dilatancy*.	Silts (inorganic), rock flour, silty fine sands with slight plasticity	ML
			Clayey silts (inorganic)	CL
			Organic silts of low plasticity	OL
	Fine-grained soils having medium plasticity	Can be readily rolled into threads when moist. Do not exhibit dilatancy*. Show some shrinkage on drying	Silty clays (inorganic) and sandy clays	MI
			Clays (inorganic) of medium plasticity	CI
			Organic clays of medium plasticity	OI
	Fine-grained soils having high plasticity	Can be readily rolled into threads when moist. Greasy to touch. Considerable shrinkage on drying. All highly compressible	Highly compressible micaceous or diatomaccous soils	MH
			Clays (inorganic) of high plasticity	CH
			Organic clays of high plasticity	OH
Fibrous organic soils with very high compressibility		Usually brown or black in colour. Very compressible. Easily identifiable visually.	Peat and other highly organic swamp soils.	Pt

* *Dilatancy*: If a pat of wet silt is shaken in the hand, water will appear on the surface and, if the pat is then pressed, the water will retreat into the silt, leaving a matt surface, because of an increase in volume. This property

Table A2.5   (*Continued*).

is known as dilatancy and is useful for distinguishing silt from clay. Fine sands also exhibit dilatancy but can be distinguished from silts by their particle size.

Group symbols

Main soil types		*Sub divisions*	
G	Gravel	W	Well graded
S	Sand	U	Uniformly graded
M	Silt	P	Poorly graded
C	Clay	C	With clay
O	Organic silts and clays	F	With fines
Pt	Peat	L	Low plasticity
		I	Intermediate plasticity
		H	High plasticity

Table A2.6 Unified soil classification system (laboratory test), after Carter (1983).

Soil groups		Laboratory classification criteria		Symbol
COARSE-GRAINED SOILS: more than half of material is coarser than 0.075 mm		Appreciable number of cobbles or boulders; coarser than 75 mm		–
	Gravels: more than half the coarse fraction is coarser than 2 mm	Less than 5% of whole sample finer than 0.075 mm	$U = \dfrac{D_{60}}{D_{10}} > 4$ and $S = \dfrac{(D_{30})^2}{D_{10} \times D_{60}}$ between 1 and 3	GW
			Does not meet both the above requirements. One size predominates	GU
		More than 12% of whole sample finer than 0.075 mm	One size lacking	GP
			Below 'A' line or PI < 4.	GF
			Above 'A' line and PI > 7. Above 'A' line and PI between 4 and 7 is borderline and requires dual symbols	GC
		Borderline soils (5–12% fines) require dual symbols		
	Sands: more than half the coarse fraction is finer than 2 mm	Less than 5% of whole sample finer than 0.075 mm	$U = \dfrac{D_{60}}{D_{10}} > 6$ and $S = \dfrac{(D_{30})^2}{D_{10} \times D_{60}}$ between 1 and 3	SW
			Does not meet both the above requirements. One size predominates	SU
		More than 12% of whole sample finer than 0.075 mm	One size lacking	SP
			Below 'A' line or PI < 4.	SF
			Above 'A' line and PI > 7. Above 'A' line and PI between 4 and 7 is borderline and requires dual symbols	SC

Table A2.6 (Continued).

Soil groups	Laboratory classification criteria	Symbol
FINE-GRAINED SOILS: more than half of material is finer than 0.075 mm.	Fine-grained soils are classified by reference to the plasticity chart, using the results of index tests carried out on the fraction finer than 0.425 mm. 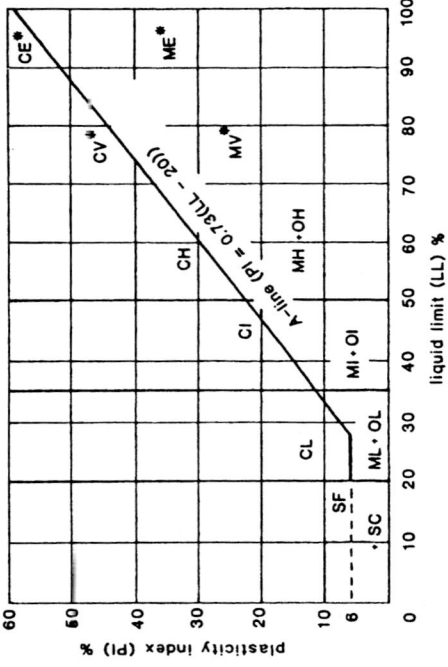	

* The division of the high plasticity range into very high and extremely high applies to the British Standard classification system only.

[1] D_{60} is the '60 per cent size' — the maximum size of the smallest 60% of the material. D_{30} and D_{10} are similarly defined. D_{10} is also called the 'effective size'.

[2] U is the uniformity coefficient; PI is the plasticity index.

[3] Borderline soils—those possessing characteristics of two groups—are designated by combinations of symbols. For example, GW-GC; well-graded gravel-sand mixture with clay binder.

[4] The soil classification applies to material of gravel, sand, silt and clays sizes. Large particles must be removed before classification and the proportions recorded separately.

Table A2.7 AASHTO/ASTM soil classification, after Carter (1983).

Soils are classified into seven groups, based on the results of grading and plasticity tests. The evaluation of soils within each group is made by means of a 'group index' which is a value calculated from an empirical formula. The group classification should be useful in determining the relative quality of the soil for use in earthwork structures, particularly embankments, subgrades, sub-bases and bases.

The classification and group index is obtained by using the test results and the tables and charts given in Table A2.8. The basic classification is obtained by using Table A2.8. If a more detailed classification is desired, a further subdivision of the groups may be made. Table A2.8 gives a classification with suggested subgroups.

The group index may be calculated using the empirical formula given below but is usually estimated using the figures in Table A2.8. The group index is given in parenthesis after the group symbol and is quoted to the nearest whole number (e.g. A-2-6 (3), A-4 (5)). The group number for A-I-a, A-I-b, A-3, A-2-4 and A-2-5 will always be zero so need not be calculated or quoted.

Descriptions and suitability of materials

Descriptions of the typical materials found in each group are given in Table A2.9.

In general, the lower the classification symbol and the lower the group index of a soil within a given classification, the higher will be the supporting value of the soil as a subgrade.

Classification procedure

With the required test data available, proceed from left to right in Table A2.8, and the correct group will be found by process of elimination. The first group from the left into which the test data will fit is the correct classification. For purposes of classification, all test values should be shown as whole numbers.

The group index is obtained by summing the two values obtained rounding to the nearest whole number.

Group Index

The group index is calculated from the following formula:

$$\text{Group index} = (F - 35)\,\{0.2 + 0.005\,(LL - 40)\} + 0.01\,(F - 15)\,(PI - 10)$$

where F is the percentage passing the 0.075 mm sieve, expressed as a whole number. This percentage is based only on the material passing the 75 mm sieve.
LL is the liquid limit.
PI is the plasticity index.

Notes:
[1] When the calculated group index is negative, the result is expressed as zero.
[2] The group index is reported to the nearest whole number.
[3] Table A2.8 may be used to estimate the group index.

Table A2.8 AASHTO/ASTM classification tables and group index charts, after AASHTO (1974).

General classification	Granular materials (35% or less passing 75 μm)			Silt-clay materials (more than 35% passing 75μm)			
1 Classification of soils into main groups							
General classification	A-1	A-3*	A-2	A-4	A-5	A-6	A-7
Sieve analysis, % passing:							
2 mm	–	–	–	–	–	–	–
0.425 mm	50 max	51 min	–	–	–	–	–
0.075 mm	25 max	10 max	35 max	36 min	36 min	36 min	36 min
Fraction passing 0.425 mm:							
Liquid limit	–	–	–	40 max	41 min	40 max	41 min
Plasticity index	6 max	N.P.	–	10 max	10 max	11 min	11 min
General rating as subgrade	Excellent to good			Fair to poor			

* The placing of A-3 before A-2 is necessary in the 'left to right elimination process' and does not indicate superiority of A-3 over A-2.

Table A2.8   *(Continued).*

General classification	Granular materials (35% or less passing 75 μm)								Silty-clay materials (>35% passing 75 μm)			

2 More detailed classification of soils into groups and subgroups

	A-1		A-3	A-2				A-4	A-5	A-6	A-7
Group classification	A-1-a	A-1-b		A-2-4	A-2-5	A-2-6	A-2-7				A-7-5 A-7-6
Sieve analysis, % passing:											
2 mm	50 max	–	–			–					
0.425 mm	30 max	50 max	51 max	–							
0.075 mm	10 max	25 max	10 max	35 max	35 max	35 max	35 max	36 min	36 min	36 min	36 min
Fraction passing 0.425 mm:											
Liquid limit	–			40 max	41 min	40 max	41 min	40 max	41 min	40 max	41 min
Plasticity index	6 max		N.P.	10 max	10 max	11 min	11 min	10 max	10 max	11 min	11 min
Usual types of significant constituents	Stone fragments, gravel, sand		Fine sand	Silty or clayey gravel and sand				Silty soils		Clayey soils	
General rating as subgrade	Excellent to good							Fair to poor			

* PI of A-7-5 subgroup < (LL–30); PI of A-7-6 subgroup > (LL–30).

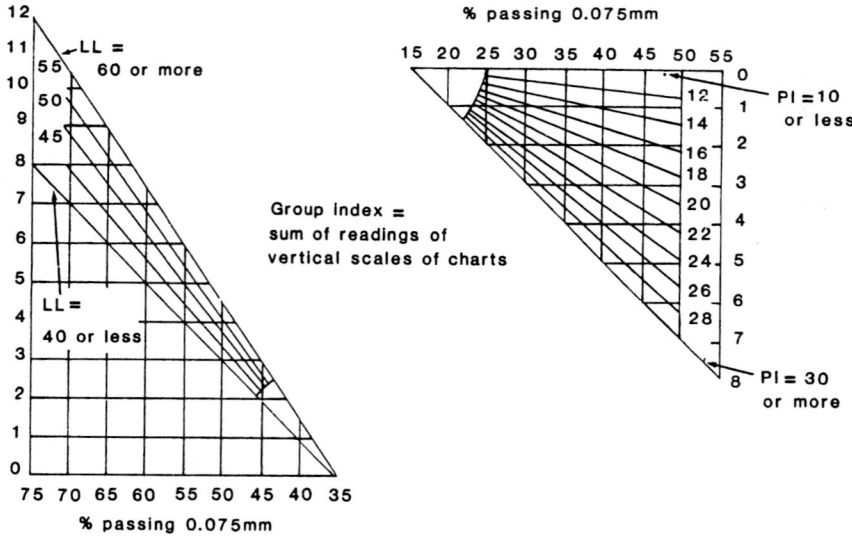

Classification A-8

Highly organic soils such as peat may be classified in an A-8 group. Classification of these materials is based on visual inspection and is not dependent on the grading or plasticity. The materials are composed primarily of decayed organic matter and have a fibrous texture, dark brown or back colour and a smell of decay.

They are unsuitable for use in embankments or subgrades.

Table A2.9   Description of soil types in AASHTO/ASTM groups after AASHTO (1974).

Classification of materials in the various groups applies only to the fraction passing the 75 mm sieve. The proportions of boulder and cobble sized particles should be recorded separately and any specification regarding the use of A-1, A-2 or A-3 materials in construction should state whether boulders are permitted.

Granular materials

Group A-1.   Typically a well-graded mixture of stone fragments or gravel, coarse to fine sand and a non-plastic or feebly plastic soil binder. However, this group also includes stone fragments, gravel, coarse sand, volcanic cinders, etc. without soil binder.

    Subgroup A-1-a is predominantly stone fragments or gravel, with or without binder.

    Subgroup A-1-b is predominantly coarse sand with or without binder.

Group A-3.   Typically fine beach sand or desert sand without silty or clayey fines or with a very small proportion of nonplastic silt. The group also includes stream-deposited mixtures of poorly graded fine sand with limited amounts of coarse sand and gravel.

Group A-2.   Includes a wide variety of 'granular' materials which are borderline between the granular A-1 and A-3 groups and the silty–clay materials of groups A-4 to A-7. It includes all materials with not more than 35% fines which are too plastic or have too many fines to be classified as A-1 or A-3.

    Subgroups A-2-4 and A-2-5 include various granular materials whose finer particles (0.425 mm down) have the characteristics of the A-4 and A-5 groups, respectively.

    Subgroups A-2-6 and A-2-7 are similar to those described above but whose finer particles have the characteristics of A-6 and A-7 groups, respectively.

Silty–clay materials

Group A-4.   Typically a non-plastic or moderately plastic silty soil usually with a high percentage passing the 0.075 mm sieve. The group also includes mixtures of silty fine sands and silty gravelly sands.

Group A-5.   Similar to material described under group A-4 except that it is usually diatomaceous or micaceous and may be elastic as indicated by the high liquid limit.

Group A-6.   Typically a plastic clay soil having a high percentage passing the 0.075 mm sieve. Also mixtures of clayey soil with sand and fine gravel. Materials in this group have a high volume change between wet and dry states.

Group A-7.   Similar to material described under group A-6 except that it has the high liquid limit characteristics of group A-5 and may be elastic as well as subject to volume change.

    Subgroup A-7-5 materials have moderate plasticity indices in relation to the liquid limits and may be highly elastic as well as subject to volume change.

    Subgroup A-7-6 material have high plasticity indices in relation to the liquid limits and are subject to extremely high volume change.

Group A-8.   Includes highly organic materials. Classification of these materials is based on visual inspection and is not related to grading or plasticity.

References

AASHTO (1974) Specification M145–73. The classification of soils and soil–aggregate mixtures for highway construction purposes. American Association of State Highway and Transport Officials, Washington.

BS5930: 1981 Code of practice for site investigations (formerly CP 2001). British Standards Institution, London.

BS5930: 1999 Code of practice for site investigations. British Standards Institution, London.

Carter, M. (1983) *Geotechnical Engineering Handbook.* Pentech Press.

Appendix A3: Identification, Description and Classification of rocks

Summary of contents

Quantitative description of rock masses		Table No.
1. Rock material description	a. Rock type	A3.1
	b. Wall strength	A3.2
	c. Weathering	A3.3
2. Discontinuity description	BS5930 checklist	A3.4
	d. Type	A3.5
	e. Orientation	A3.4
	f. Roughness	A3.4, A3.6, A1.5 (A1)
	g. Aperture	A3.4, A3.7
3. Infilling	h. Infilling type/width	A3.4, A3.8
4. Rock mass description	i. Spacing	A3.4, A3.9
	j. Persistence	A3.4, A3.10
	k. Number of sets	
	l. Block size and shape	A3.11
5. Groundwater	m. Seepage	A3.4, A3.12

Table numbers refer to the following tables which give scales and descriptive terms of each aspect of rock mass properties. In many cases the terms or numerical scales suggested by the International Society for Rock Mechanics (ISRM, 1981) and British Standard Code of Practice for Site Investigation (1999) differ. It should therefore always be made clear which system is being used, and as far as possible systems should not be mixed. Table A3.13 provides a rock mass strength classification system for geomorphic purposes.

Table A3.1   Aid to identification of rocks for engineering purposes (BS5930: 1999).

Grain size (mm)	Grain size description	Bedded rocks (mostly sedimentary) — siliceous	At least 50% of grains are of carbonate	At least 50% of grains are of grained volcanic rock	Saline / Carbonaceous
20 – 6 (Coarse, RUDACEOUS)		CONGLOMERATE — Rounded boulders, cobbles and gravel cemented in a finer matrix. Breccia Irregular rock fragments in a finer matrix	Calcirudite	Fragments of volcanic ejecta in a finer matrix. Rounded grains AGGLOMERATE. Angular grains VOLCANIC BRECCIA	SALINE ROCKS; HALITE; ANHYDRITE; GYPSUM
2 – 0.06 (Fine / Medium, ARENACEOUS)		SANDSTONE — Angular or rounded grains commonly cemented by clay, calcitic or iron minerals. Quartzite. Quartz grains and siliceous cement. Arkose Many feldspar grains. Greywacke Many rock chips	Calcarenite	Cemented volcanic ash; TUFF	
0.06 – 0.002 (ARGILLACEOUS)		SILTSTONE Mostly silt; MUD-STONE	Calcisiltite; Calcilutite; Calcareous mudstone; CHALK	Fine-grained TUFF; Very fine-grained TUFF	
Amorphous or crypto crystalline		Flint: occurs as bands of nodules in the Chalk. Chert: occurs as nodules and beds in lime stone and calcareous sandstone	Granular cemented - except amorphous rocks		COAL; LIGNITE
		SILICEOUS	CALCAREOUS	SILICEOUS	CARBONACEOUS

LIMESTONE and DOLOMITE (undifferentiated)

Grain size boundaries approximate

Sedimentary Rocks: Granular cemented rocks vary greatly in strength, some sandstones are stronger than many igneous rocks. Bedding may not show in hand specimens and is best seen in outcrop. Only sedimentary rocks, and some metamorphic rocks derived from them, contain fossils.

Calcareous rocks contain calcite (calcium carbonate) which effervesces with dilute hydrochloric acid.

Table A3.1b (*Continued*).

Igneous rocks: generally massive structure and crystalline texture

Grain size description	ACID Much quartz	INTERMEDIATE Some quartz	BASIC Little or no quartz	ULTRA BASIC
COARSE	GRANITE[1]	DIORITE[1,2]	GABBRO[1,2]	Pyroxenite
	These rocks are sometimes porphyritic and are then described, for example, as porphyritic granite			
MEDIUM	MICROGRANITE[1]	MICROIORITE[1,2]	DOLERITE[3,4]	Peridotite
	These rocks are sometimes porphyritic and are then described as porphyries			
FINE	RHYOLITE[4,5]	ANDESITE[4,5]	BASALT[5]	
	These rocks are sometimes porphyritic and are then described as porphyries			
Amorphous Crypto-crystalline	OBSIDIAN[5]	VOLCANIC GLASS		
	Colour: Pale ← → Dark			

increasing grain size (COARSE → MEDIUM → FINE)

Metamorphic Rocks

Foliated	Massive
GNEISS Well developed but often widely spaced foliation sometimes with schistose bands	MARBLE
	QUARTZITE
	GRANULITE
Migmatite Irregularly foliated; mixed schists and gneisses	HORNFELS
SCHIST Well developed undulose foliation; generally much mica	AMPHIBOLITE
	SERPENTINE
PHYLLITE Slightly undulose foliation; sometimes spotted	
SLATE Well developed plane cleavage (foliation)	
MYLONITE Found in fault zones, mainly in igneous and metamorphic areas	
CRYSTALLINE	
SILICEOUS	Mainly SILICEOUS

Igneous Rocks: Composed of closely interlocking mineral grains. Strong when fresh; not porous.

Mode of occurrence: 1. Batholiths, 2. Laccoliths, 3. Silts, 4. Dykes, 5. Lava flows, 6. Veins.

Metamorphic Rocks: Generally classified according to fabric and mineralogy rather than grain size. Most metamorphic rocks are distinguished by foliation which may impart fissility. Foliation in gneisses is best observed in outcrop. Non-foliated metamorphics are difficult to recognise except by association. Most fresh metamorphic rocks are strong although perhaps fissile.

Table A3.2a Descriptive terms for soil and rock strength (ISRM, 1981).

Grade	Description	Field identification	Approximate range of compressive strength	
			(MPa)	(psi)
R6	Extremely strong rock	Specimen can only be chipped with geological hammer	>250	(>36 000)
R5	Very strong rock	Specimen requires many blows of geological hammer to fracture it	100–250	(15 000–36 000)
R4	Strong rock	Specimen requires more than one blow with a geological hammer to fracture it	50–100	(7000–15 000)
R3	Medium weak rock	Cannot be scraped or peeled with a pocket knife; specimen can be fractured with single firm blow of geological hammer	25–50	(3500–7000)
R2	Weak rock	Can be peeled with a pocket knife; shallow indentations made by firm blow with point of geological hammer	5–25	(725–3500)
R1	Very weak rock	Crumbles under firm blows with point of geological hammer; can be peeled by a pocket knife	1–5	(150–725)
R0	Extremely weak rock	Indented by thumbnail	0.25–1	(35–150)
S6	Hard clay	Indented with difficulty by thumbnail	>0.5	(>70)
S5	Very stiff clay	Readily indented by thumbnail	0.25–0.5	(35–70)
S4	Stiff clay	Readily indented by thumb but penetrated only with great difficulty	0.1–0.25	(15–35)
S3	Firm clay	Can be penetrated several inches by thumb with moderate effort	0.05–0.1	(7–15)
S2	Soft clay	Easily penetrated several inches by thumb	0.025–0.05	(4–7)
S1	Very soft clay	Easily penetrated several inches by fist	<0.025	(<4)

Table A3.2b Descriptive terms for rock strength (BS5930: 1999).

Term	Field definition	Unconfined compressive strength (MN/m²)
Very weak	Gravel size lumps can be crushed between finger and thumb.	<1.25
Weak	Gravel size lumps can be broken in half by heavy hand pressure.	1.25 to 5
Moderately weak	Only thin slabs, corners or edges can be broken off with heavy hand pressure.	5 to 12.5
Moderately strong	When held in the hand, rock can be broken by hammer blows.	12.5 to 50
Strong	When resting on a solid surface, rock can be broken by hammer blows.	50 to 100
Very strong	Rock chipped by heavy hammer blows.	100 to 200
Extremely strong	Rock rings on hammer blows. Only broken by sledgehammer.	>200

Table A3.3a   Descriptive terms for weathering grades (ISRM, 1981).

Grade	Term	Description
W1	Fresh	Not broken easily by geological hammer; ringing sound if struck by geological hammer; no visible signs of decomposition
W2	Slightly weathered	Not broken easily by geological hammer; ringing sound if struck by geological hammer; fresh rock colours generally retained but stained near joint surfaces
W3	Moderately weathered	Cannot usually be broken by hand, easily broken by geological hammer; dull or slightly ringing sound if struck by geological hammer; completely stained throughout
W4	Highly weathered	Can be broken by hand into smaller pieces; dull sound if struck by geological hammer; not easily indented by point of geological pick; does not slake when immersed in water; completely discoloured compared with fresh rock
W5	Completely weathered	Original rock texture preserved; can be crumbled by hand and finger pressure into constituent grains; easily indented by point of geological pick; slakes when immersed in water; completely discoloured compared with fresh rock
W6	Residual soil	Original rock texture completely destroyed; can be crumbled by hand and finger pressure into constituent grains

Table A3.3b   Description and classification of weathered rock for engineering purpose (BS5930: 1999).

Approach 1: Factual Description of Weathering (Mandatory)

Standard descriptions should always include comments on the degree and nature of any weathering effects at material or mass scales. This may allow subsequent classification and provide information for separating rock into zones of like character. Typical indications of weathering includes:

- Changes in colour
- Reduction in strength

- Changes in fracture state
- Presence, character and extent of weathering products

These features should be described using standard terminology, quantified as appropriate, together with non-standard English descriptions as necessary to describe the results of weathering. At the mass scale the distribution and proportions of the variously weathered materials (e.g. corestones vs matrix) should be recorded.

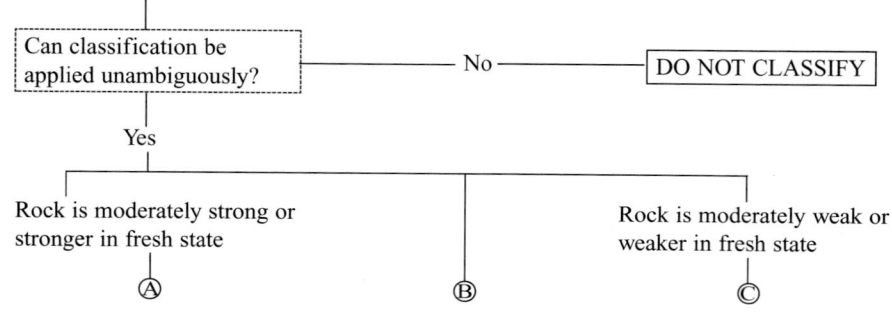

Table A3.3b *(Continued).*

Approach 2: Classification for Uniform Materials		
Grade	Classifier	Typical characteristics
I	Fresh	Unchanged from original state
II	Slightly weathered	Slight discoloration, slight weakening
III	Moderately weathered	Considerably weakened, penetrative discoloration. Large pieces cannot be broken by hand
IV	Highly weathered	Large pieces cannot be broken by hand. Does not readily disaggregate (slake) when dry sample immersed in water
V	Completely weathered	Considerably weakened. Slakes. Original texture apparent
VI	Residual soil	Soil derived by *in situ* weathering but retaining none of the original texture or fabric

Approach 4: Classification Incorporating Material and Mass Features		
Class	Classifier	Typical characteristics
A	Unweathered	Original strength, colour, fracture spacing
B	Partially weathered	Slightly reduced strength, slightly closer fracture spacing, weathering penetrating in from fractures, brown oxidation
C	Distinctly weathered	Further weakened, much closer fracture spacing grey reduction
D	Destructured	Greatly weakened, mottled, ordered lithorelics in matrix becoming weakened and disordered, bedding disturbed
E	Residual or reworked	Matrix with occasional altered random or 'apparent' lithorelics, bedding destroyed. Classed as reworked when foreign inclusions are present as a result of transportation

Is a zonal classification appropriate and is there enough information available? —No— Use Rock Mass Classification if appropriate

Yes

Approach 5: Special Cases

For rocks whose weathering state does not follow the other patterns indicate here, such as karst in carbonates and the particular effect of arid climates

Approach 3: Classification for Heterogenous Masses		
Zone	Proportions of material grades	Typical characteristics
1	% G I–III (not necessarily all fresh rock)	Behaves as rock: apply rock mechanics principles to mass assessment and design
2	> 90% G I–III < 10% G IV–VI	Weak materials along discontinuities. Shear strength stiffness and permeability affected
3	50% to 90% G I–III 10% to 50% G IV–VI	Rock framework still locked and controls strength and stiffness; matrix controls permeability
4	30% to 50% G I–III 50% to 70% G IV–VI	Rock framework contributes to strength; matrix or weathering products control stiffness and permeability
5	< 30% G I–III > 70%G IV–VI	Weak grades will control behaviour. Corestones may be significant for investigation and construction
6	100% G IV–VI (not necessarily residual soil)	May behave as soil although relict fabric may still be significant

Table A3.4 Terminology and checklist for discontinuity description (BS5930: 1999).

Spacing	Orientation	Persistence	Type of termination	Roughness	Wall strength	Aperture	Filling	Seepage	No. of sets
Extremely wide > 6 m	Dip amount only in cores	Discontinuous	Cannot normally be described	Small scale (cm) and intermediate scale (m)	Schmidt hammer	Cannot normally be described in cores	Clean	Cannot be described in cores	Cannot be described in cores
Very wide 2 to 6 m				Stepped — Rough, Smooth, Striated			Surface staining (colour)		
Wide 600 mm to 2 mm		Continuous in cores		Undulating — Rough, Smooth, Striated			Soil infilling (describe in accordance with 41)		
Medium 200 to 600 mm									
Close 60 to 200 mm	Take No. of readings, of dip direction/dip e.g. 015/08	Very high > 20 m	Termination x (outside exposure)	Planar — Rough, Smooth, Striated	Point load test	Very open > 10 mm	Mineral coatings (e.g. calcite, chlorite gypsum etc.)	Moisture on rock surface	
		High 10 to 20 m	r (within rock)		Other index tests	Open 2.5 mm to 10 mm		Dripping water	
Very close 20 to 60 mm		Medium 3 to 10 m	d (against discontinuity)			Moderately open 0.5 mm to 2.5 mm	Other—specify	Water flow measured per time unit on an individual discontinuity or set of discontinuities	Record spacing and orientation of sets to each other and all details for each set
Extremely close < 20 mm	Report as ranges and on stereo net if appropriate	Low 1 to 3 m		Large scale (dm) — Waviness, Curvature, Straightness	Visual assessment	Tight 0.1 mm to 0.5 mm	Record width and continuity of infill		
						Very tight < 0.1 mm		Small flow 0.05–0.5 l/s	
Take number of readings state min. average and max.		Very low < 1 m	Record also size of exposure	Measure amplitude and wavelength of feature		Take number of readings state min. average and max.		Medium flow 0.5–5.0 l/s	
								Strong flow > 5 l/s	

Table A3.5a   Types of discontinuities (ISRM, 1981).

Type	Term	Description
B	Bedding	The arrangement of a sedimentary rock in layers; stratification.
J	Joint	A break of geological origin in the continuity of a body of rock along which there has been no visible displacement. A group of parallel joints is called a set and joint sets intersect to form a joint system. Joints can be open, filled or healed.
SL	Slickenside	A fracture along which there has been recognisable displacement, from a few centimetres to a few kilometres in scale. The walls are often striated and polished (slickensided) resulting from the shear displacement.
F	Fault	A fracture zone along which there has been recognisable displacement, from a few centimetres to a few kilometres in scale. The walls are often striated and polished (slickensided) resulting from the shear displacement. Frequently rock on both sides of a fault is shattered and altered or weathered, resulting in fillings such as breccia and gouge. Fault widths may vary from millimetres to hundreds of metres.

Table A3.5b   Types of discontinuities (BS5930: 1999).

Type of discontinuity	Description
Joint	A discontinuity in the body of rock along which there has been no visible displacement.
Fault	A fracture or fracture zone along which there has been recognisable displacement.
Bedding fracture	A fracture along the bedding (bedding is a surface parallel to the plane of deposition).
Cleavage fracture	A fracture along a cleavage (cleavage is a set of parallel planes of weakness often associated with mineral realignment).
Induced fracture	A discontinuity of non-geological origin, e.g. brought about by coring, blasting, ripping etc.
Incipient fracture	A discontinuity which retains some tensile strength, which may not be fully developed or which may be partially cemented. Many incipient fractures are along bedding or cleavage.

Table A3.6   Descriptive terms for discontinuity roughness (ISRM, 1981).

Level	Description
I	Rough, stepped
II	Smooth, stepped
III	Slickensided, stepped
IV	Rough, undulating
V	Smooth, undulating
VI	Slickensided, undulating
VII	Rough, planar
VIII	Smooth, planar
IX	Slickensided, planar

Table A3.7   Descriptive terms for discontinuity aperture and thickness (ISRM, 1981).

Aperture (mm)	Description	
< 0.1	Very tight	
0.1–0.25	Tight	'Closed' features
0.25–0.5	Partly open	
0.5–2.5	Open	
2.5–10	Moderately wide	'Gapped' features
> 10	Wide	
10–100	Very wide	
100–1000	Extremely wide	'Open' features
> 1 m	Cavernous	

Table A3.13   Geomorphic rock mass strength classification and ratings, after Selby (1980).

	1 Very strong	2 Strong	3 Moderate	4 Weak	5 Very weak
Intact rock strength (N-type Schmidt hammer 'R')	100–60 Very strong $r = 20$	60–50 Strong $r = 18$	50–40 Moderate $r = 14$	40–35 Weak $r = 10$	35–10 Very weak $r = 5$
Weathering	Unweathered $r = 10$	Slightly weathered $r = 9$	Moderately weathered $r = 7$	Highly weathered $r = 5$	Completely weathered $r = 3$
Spacing of discontinuities	>3 m Solid $r = 30$	3–1 m Massive $r = 28$	1–0.3 m Blocky/seamy $r = 21$	300–50 mm Fractured $r = 15$	<50 mm Crushed or shattered $r = 8$
Joint orientations	Very favourable; steep dips into slope, cross joints interlock $r = 20$	Favourable; moderate dips into slope $r = 18$	Fair; horizontal dips, or nearly vertical (hard rocks only) $r = 14$	Unfavourable; moderate dips out of slope $r = 9$	Very unfavourable; steep dips out of slope $r = 5$
Width of joints	<0.1 mm $r = 7$	0.1 mm $r = 6$	1–5 mm $r = 5$	5–20 mm $r = 4$	>20 mm $r = 2$
Fracture continuity	None continuous $r = 7$	Few continuous $r = 6$	Continuous, no infill $r = 5$	Continuous, thin infill $r = 4$	Continuous, thick infill $r = 1$
Outflow of groundwater	None $r = 6$	Trace $r = 5$	Slight (<25 l/min from 10 m²) $r = 4$	Moderate (25–125 l/min from 10 m²) $r = 3$	Great (>125 l/min from 10 m²) $r = 1$
Total rating	100–91	90–71	70–51	50–26	<26

References

BS5930: 1999 Code of practice for site investigations. British Standards Institution, London.

Geological Society Engineering Group Working Party (1977) The description of rock masses for engineering purposes. *Q. J. Eng. Geol.* **10**, 355–388.

Selby, M. J. (1980) A rock mass strength classification for geomorphic purposes: with tests from Antarctica and New Zealand. *Z. Geomorph.* **24**, 31–51.

ISRM (International Society for Rock Mechanics) (1981) *Suggested Methods for the Quantitative Description of Discontinuities in Rock Masses.* (ed. E. T. Brown), Pergamon Press, Oxford.

Appendix A4: Geomorphological mapping

The Geological Society Working Party Report on maps and plans (Anon., 1972) identified examples of geomorphological mapping that could be of use to engineers. Lee (2001) presents a brief introduction to the preparation of engineering geomorphology maps. Stylised symbols are often used (see Demek and Embleton, 1978; Gardiner and Dakcombe, 1983; Cooke and Doornkamp, 1990).

The style and format of an engineering geomorphological map needs to reflect the nature of the environment and the problems to be addressed. Amongst the more common types of map are: regional surveys of terrain conditions, either to provide a framework for land use planning or as part of the baseline studies for environmental impact assessment; general assessments of resources or geohazards within an area or along a route corridor; specific-purpose surveys to delineate and characterise particular landforms, e.g. the investigation of pre-exising landslide problems in and around the Channel Tunnel terminal area, Folkestone (Griffiths *et al.*, 1995); the delineation of gypsum-related subsidence problems in the Ripon area of the UK (Thompson *et al.*, 1996).

Map scale is an important issue. It will not only determine the level of detail that should be sought, but it will also dictate what features should be recorded. For example, at a large scale (i.e. 1 : 500 to 1 : 5000) it will be practical to record individual landslides and their components (e.g. the head, track and accumulation zone of a mudslide), whereas at a medium scale (i.e. 1 : 10 000 to 1 : 50 000) it will generally only be possible to show the limits of larger landslides. At small scales (i.e. 1 : 100 000 and smaller) it may only be practical to note 'areas prone to landsliding' or to show the larger features by a stylised map symbol. Ideally map scale should reflect the

objective of the exercise and the complexity of the landscape; often, however, it is driven more by the availability of suitable base maps and the constraints of the fieldwork programme.

There is no single approach to geomorphological mapping. The method chosen will generally reflect the nature of the problem to be solved, the resources available and, not least, the training and experience of the mapper. However, all maps should seek to subdivide the landscape into units with similar surface form, materials and process characteristics. At the smaller scales these units will, inevitably, be terrain models or land systems (Figure A4.1; see Phipps, 2001). Individual landforms or terrain units (e.g. escarpments, dune ridges, river channels, landslides) might be recorded on medium-scale maps. Landform elements or geomorphological units (e.g. individual landslide blocks, within-channel bars, gullys) might be recorded on large-scale maps.

The method of recording geomorphological information may change with map scale, but the basic approach to data collection should remain the same: mapping of surface form, description of materials and recording evidence of process.

A key stage in map creation is the identification of suitable mapping units to reflect the scale of mapping and the objectives of the study. Ideally, each unit should have consistent geomorphological characteristics, although internal variability of materials or rate of process may be a feature of the unit (e.g. in areas mantled by glacial tills). Three broad categories of geomorphological unit can be recognised:

1. Units reflecting the control of the underlying geology (e.g. plateau surfaces, lithological benches)

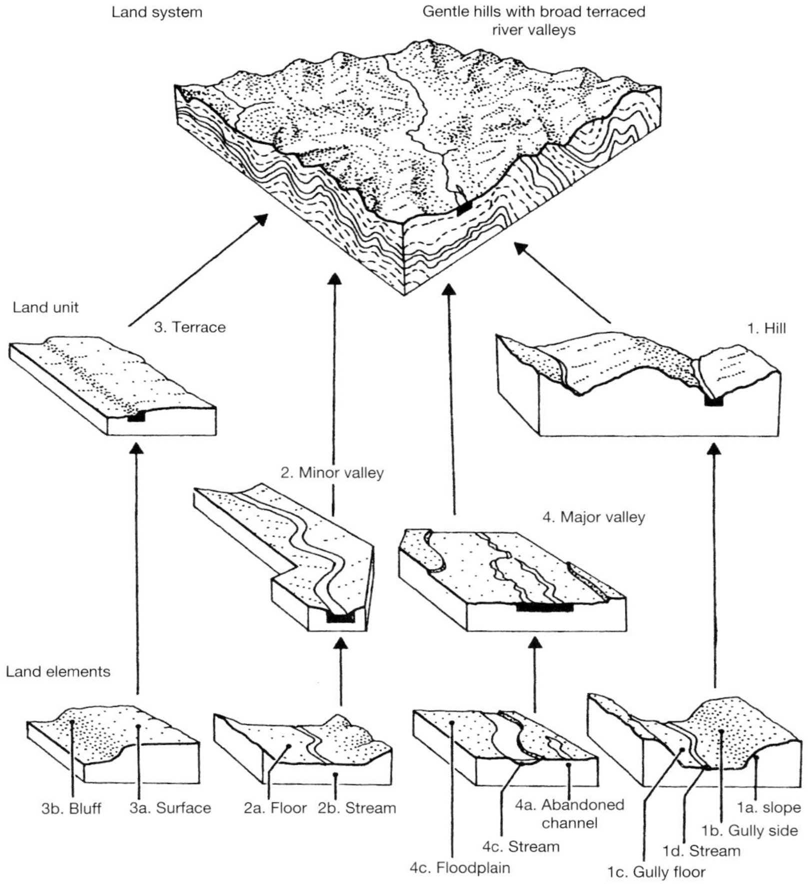

Figure A4.1 Diagram to show the relationship between land system, land facet (or land unit), and land element (after Lawrence, 1972).

2. Units reflecting the activity of active or relict surface processes (e.g. landslide, fluvial, aeolian features etc.)

3. Units reflecting modification of the landscape by man (e.g. areas of cut and fill, quarries, made ground etc.).

At the smaller scales, topographic map contours can be used to subdivide the landscape into units, based on slope steepness (i.e. from the spacing of contour lines) and slope form (i.e. from the shape of the contours). However, at larger scales contours reveal little about landforms and their assemblages. The technique of morphological mapping is the most convenient and efficient way of recording the surface morphology and allows later interpretation of form and process (Savigear, 1965; Waters, 1958). Breaks or changes of slope are identified from aerial photographs or in the field, and recorded using standard symbols (Figures A4.2 and A4.3).

Geomorphological map unit boundaries should generally follow morphological boundaries. Depending on the map scale these units can be portrayed in blocks of contrasting colour or shading, or as stylised symbols (Figures A4.4 to A4.16) or a combination of both.

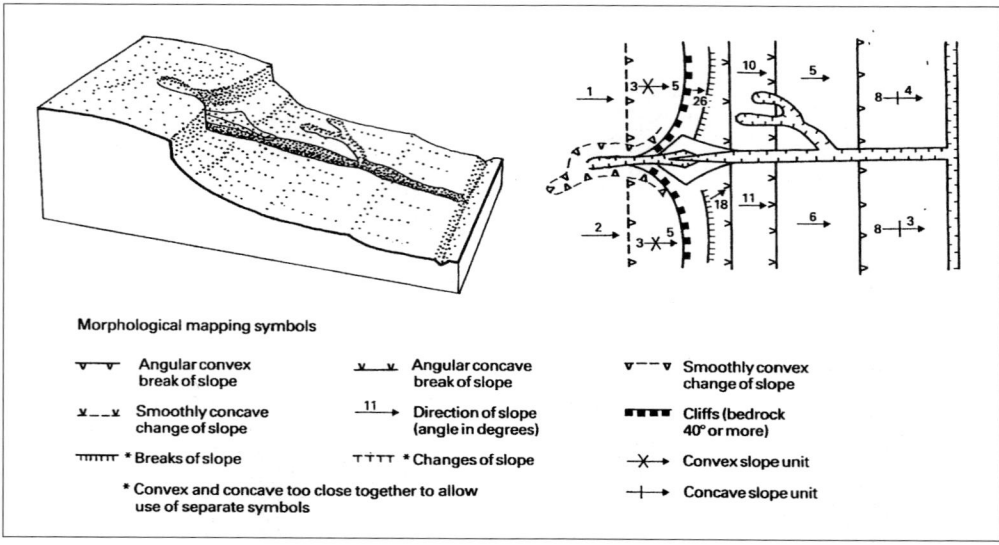

Morphological mapping symbols

v—v	Angular convex break of slope		x—x	Angular concave break of slope		v--v	Smoothly convex change of slope
x--x	Smoothly concave change of slope		—11→	Direction of slope (angle in degrees)		■■■■	Cliffs (bedrock 40° or more)
ттттт	* Breaks of slope		ттТТ	* Changes of slope		—X→	Convex slope unit
	* Convex and concave too close together to allow use of separate symbols					—I→	Concave slope unit

Figure A4.2   A morphological mapping system (after Savigear, 1965).

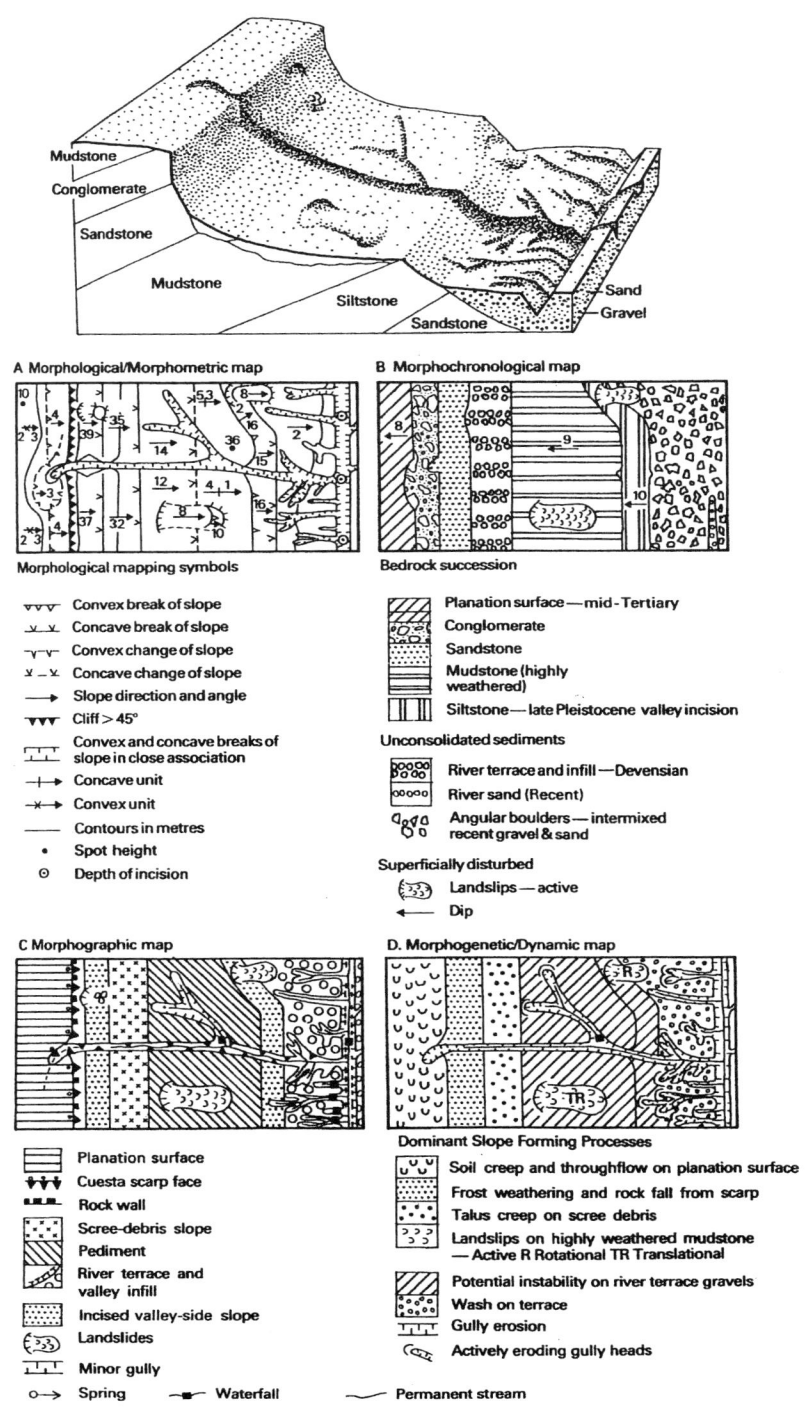

Figure A4.3 Different uses of geomorphological mapping data and styles of presentation for the same land surface (from Cooke and Doornkamp, 1990).

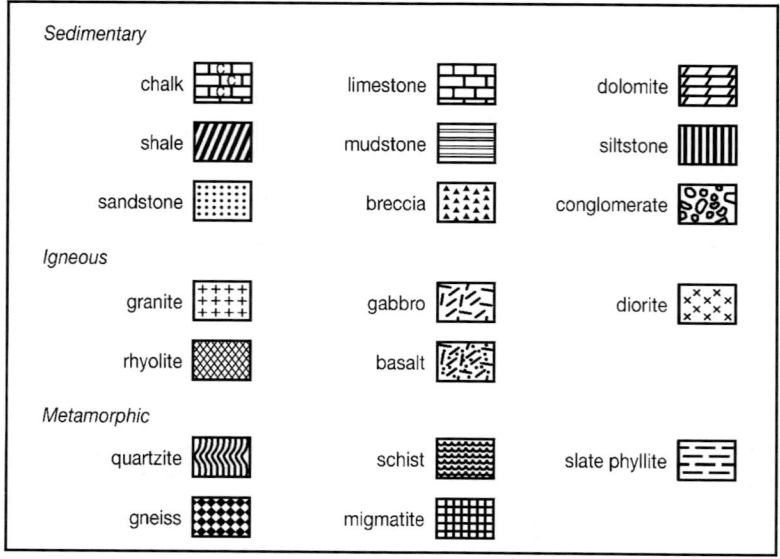

Figure A4.4 Bedrock lithology: geomorphological mapping symbols (redrawn from Cooke and Doornkamp, 1990).

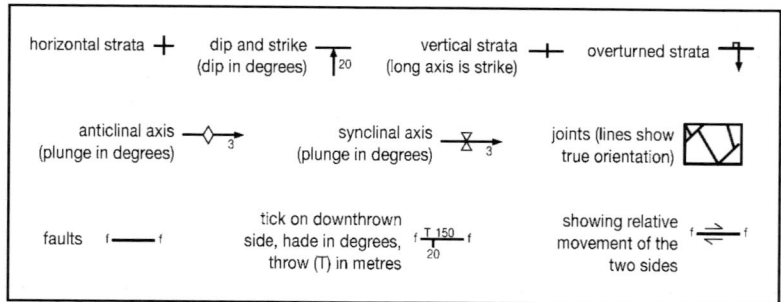

Figure A4.5 Geological structure: geomorphological mapping symbols (redrawn from Cooke and Doornkamp, 1990).

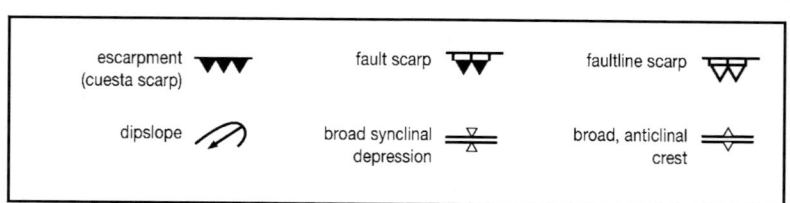

Figure A4.6 Features resulting from bedrock structure: geomorphological mapping symbols (redrawn from Cooke and Doornkamp, 1990).

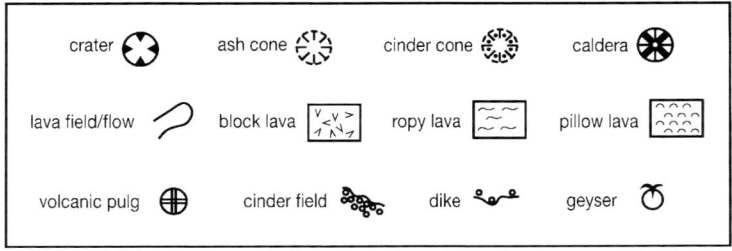

Figure A4.7   Features of volcanic origin: geomorphological mapping symbols (redrawn from Cooke and Doornkamp, 1990).

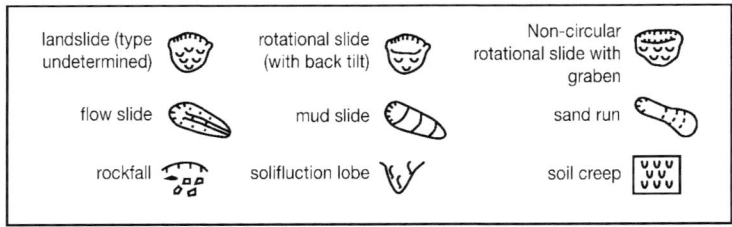

Figure A4.8   Superficial unconsolidated materials: geomorphological mapping symbols (redrawn from Cooke and Doornkamp, 1990).

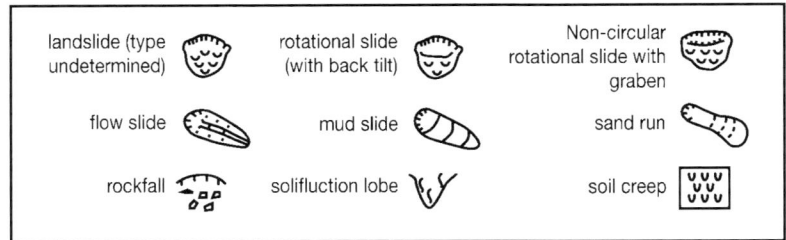

Figure A4.9   Slope instability features: geomorphological mapping symbols (redrawn from Cooke and Doornkamp, 1990).

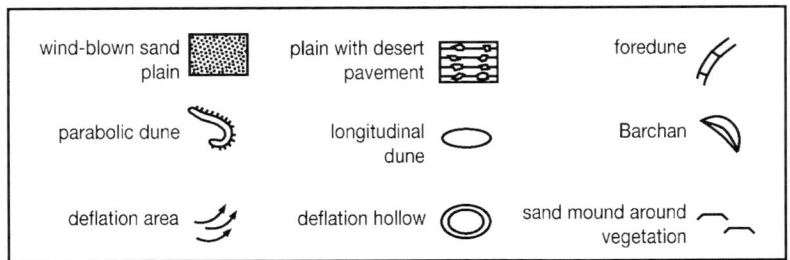

Figure A4.10   Aeolian features: geomorphological mapping symbols (redrawn from Cooke and Doornkamp, 1990).

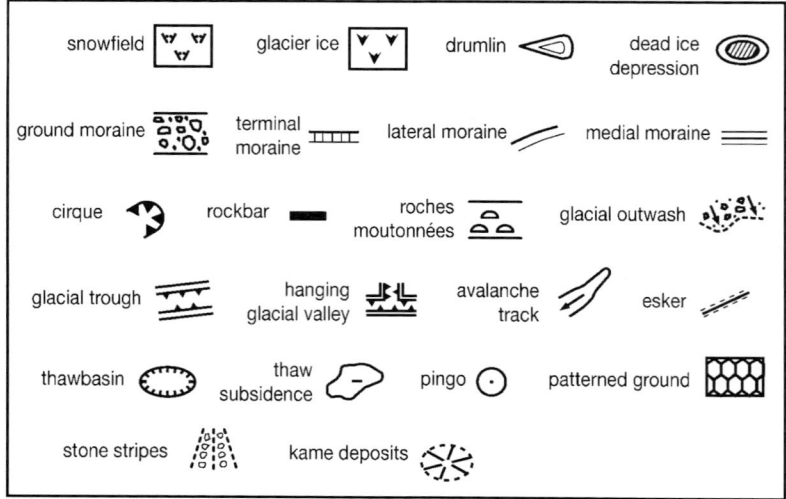

Figure A4.11   Coastal features: geomorphological mapping symbols. (redrawn from Cooke and Doornkamp, 1990).

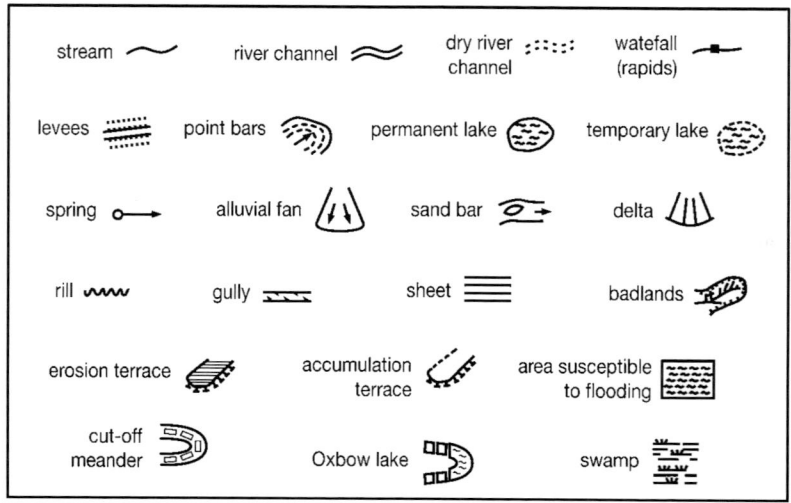

Figure A4.12   Forms of permafrost areas, glacial and periglacial features: geomorphological mapping symbols (redrawn from Cooke and Doornkamp, 1990).

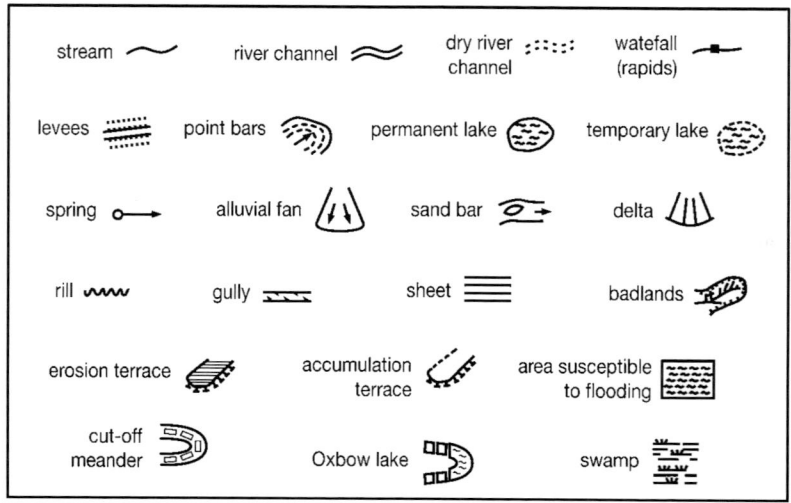

Figure A4.13   Forms of fluvial origin: geomorphological mapping symbols (redrawn from Cooke and Doornkamp, 1990).

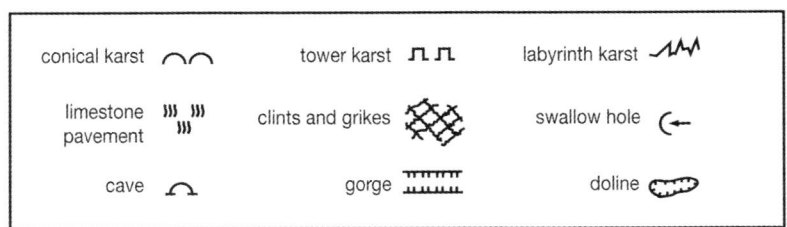

Figure A4.14   Karst landscape features: geomorphological mapping symbols (redrawn from Cooke and Doornkamp, 1990).

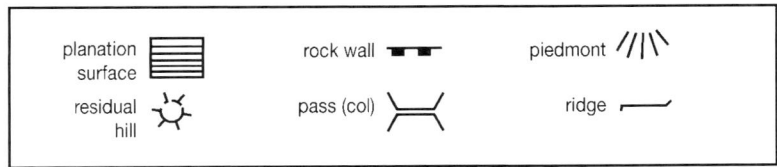

Figure A4.15   Major features not included in previous figures: geomorphological mapping symbols (redrawn from Cooke and Doornkamp, 1990).

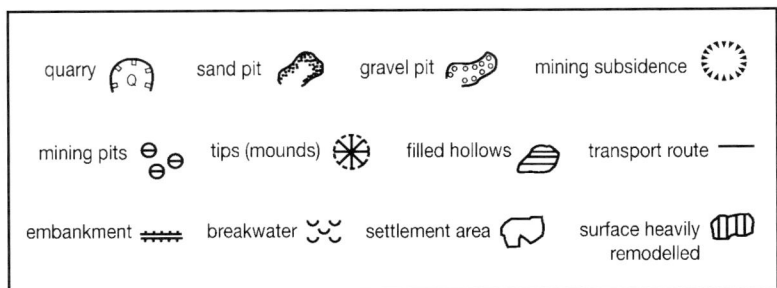

Figure A4.16   Man-made features: geomorphological mapping symbols (redrawn from Cooke and Doornkamp, 1990).

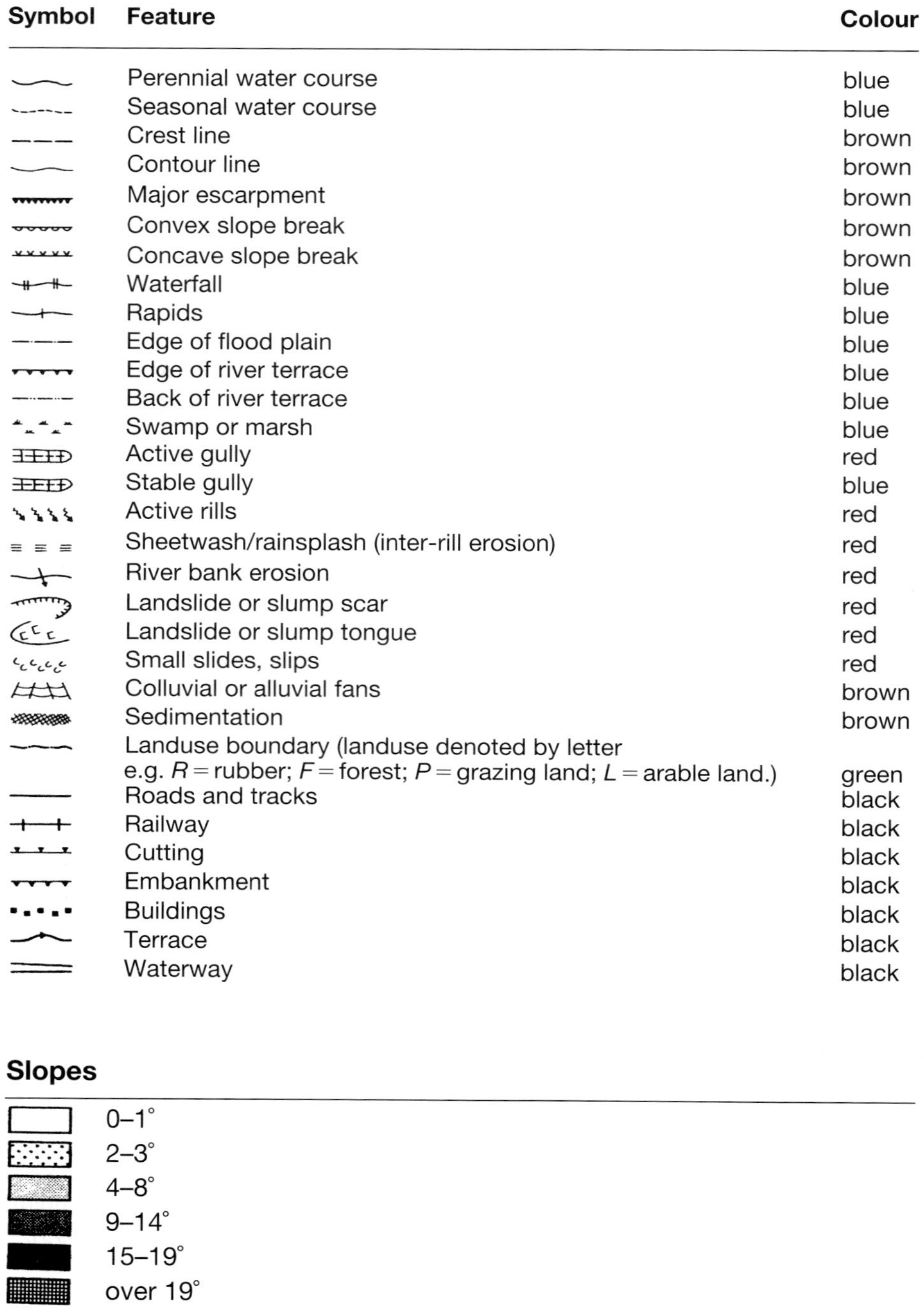

Symbol	Feature	Colour
~~	Perennial water course	blue
-.-.-.-	Seasonal water course	blue
– – –	Crest line	brown
~~	Contour line	brown
wwwww	Major escarpment	brown
ooooo	Convex slope break	brown
xxxxx	Concave slope break	brown
–#–#–	Waterfall	blue
–+–	Rapids	blue
– · –	Edge of flood plain	blue
vvvv	Edge of river terrace	blue
–·–·–	Back of river terrace	blue
·-·-·-	Swamp or marsh	blue
⊞⊞⊞⊃	Active gully	red
⊞⊞⊞⊃	Stable gully	blue
＼＼＼＼	Active rills	red
≡ ≡ ≡	Sheetwash/rainsplash (inter-rill erosion)	red
–+–	River bank erosion	red
⌒⌒⌒⌒⌐	Landslide or slump scar	red
(ccc_	Landslide or slump tongue	red
cccccc	Small slides, slips	red
/⊢⊣\	Colluvial or alluvial fans	brown
▨▨▨▨	Sedimentation	brown
–~~–	Landuse boundary (landuse denoted by letter e.g. R = rubber; F = forest; P = grazing land; L = arable land.)	green
——	Roads and tracks	black
+–+	Railway	black
·+·+·+	Cutting	black
vvvv	Embankment	black
■·■·■	Buildings	black
–⌃–	Terrace	black
═══	Waterway	black

Slopes

☐	0–1°
▦	2–3°
▨	4–8°
▓	9–14°
■	15–19°
▦	over 19°

Figure A5.1   Legend for mapping soil erosion (from Morgan, 1995).

Recorder :		Area :												FACET NO.	
Date :		Air photo no :													
Altitude :		Grid Reference :													

Present landuse															

Climate	Month	J	F	M	A	M	J	J	A	S	O	N	D	Erosivity	
	Rainfall (mm)														
	Mean temp (C)														
	Maximum intensity														

Vegetation	Type		% Ground cover		% Tree and shrub cover	

Slope	Position	Degree		Distance from crest	Shape	

Soil	Depth	Surface texture		Erodibility	
		Permeability	Clay fraction		

Erosion	

REMARKS	

EROSION CODE	0	½	1	2	3	4	5

Figure A5.2   Proforma for recording soil erosion in the field (from Morgan, 1995).

Table A5.1   Guide values for Manning's *n*.

Land use or cover	Manning's *n*
Bare soil	
roughness depth < 25 mm	0.010–0.030
roughness depth 25–50 mm	0.014–0.033
roughness depth 50–100 mm	0.023–0.038
roughness depth > 100 mm	0.045–0.049
Bermuda grass—sparse to good cover	
very short (> 50 mm)	0.015–0.040
short (50–100 mm)	0.030–0.060
medium (150–200 mm)	0.030–0.085
long (250–600 mm)	0.040–0.150
very long (> 600 mm)	0.060–0.200
Bermuda grass—dense cover	0.300–0.480
Other dense sod-forming grasses	0.390–0.630
Dense bunch grasses	0.150
Kudzu	0.070–0.230
Lespedeza	0.100
Natural rangeland	0.100–0.320
Clipped rangeland	0.020–0.240
Wheat straw mulch	
2.5 t/ha	0.050–0.060
5.0 t/ha	0.075–0.150
7.5 t/ha	0.100–0.200
10.0 t/ha	0.130–0.250
Chopped maize stalks	
2.5 t/ha	0.012–0.050
5.0 t/ha	0.020–0.075
10.0 t/ha	0.023–0.130
Cotton	0.070–0.090
Wheat	0.100–0.300
Sorghum	0.040–0.110
Concrete or asphalt	0.010–0.013
Gravelled surface	0.012–0.030
Chisel-ploughed soil	
< 0.6 t/ha residue	0.006–0.170
0.6–2.5 t/ha residue	0.070–0.340
2.5–7.5 t/ha residue	0.190–0.470
Disc-harrowed soil	
< 0.6 t/ha residue	0.008–0.410
0.6–2.5 t/ha residue	0.100–0.250
2.5–7.5 t/ha residue	0.140–0.530
No tillage	
< 0.6 t/ha residue	0.030–0.070
0.6–2.5 t/ha residue	0.010–0.130
2.5–7.5 t/ha residue	0.160–0.470
Bare mouldboard-ploughed soil	0.020–0.100
Bare soil tilled with coulter	0.050–0.130

After Petryk and Bosmajian (1975), Temple (1982) and Engman (1986).

Table A5.2   Values of Manning's *n* for major steams and streams with coarse bed materials or rock-cut channels (Chow, 1959).

	Manning's *n*
MAJOR STREAMS (bank full width > 30 m)	
(1) Regular sections with no boulders or brush	0.025–0.060
(2) Irregular and rough sections	0.035–0.100
CHANNELS WITH RELATIVELY COARSE BED MATERIALS AND ROCK-CUT CHANNELS	
gravel, 4–8 mm in diameter	0.019–0.020
gravel, 8–20 mm in diameter	0.020–0.022
gravel, 20–60 mm in diameter	0.022–0.027
pebbles and shingle, 60–110 mm in diameter	0.027–0.030
pebbles and shingle, 110–250 mm in diameter	0.030–0.035
smooth and uniform rock-cut channels	0.025–0.035–0.040
jagged and irregular rock-cut channels	0.035–0.040–0.050

Table A5.3   Manning's *n* for minor streams and flood plains (after Chow, 1959).

	Manning's *n*
MINOR STREAMS (bank full width <30 m)	
(a) *Streams on plain*	
(1) Clean, straight, full-stage, no rifts or deep pools	0.025–0.030–0.033
(2) As (1) but more stones and weeds	0.030–0.035–0.040
(3) Clean, winding, some pools and shoals	0.033–0.040–0.045
(4) As (3) but with some weeds and stones	0.035–0.045–0.050
(5) As (4) but lower stages, less efficient slopes and sections	0.040–0.048–0.055
(6) As (4) but with more stones	0.045–0.050–0.060
(7) Sluggish reaches, weedy, deep pools	0.050–0.070–0.080
(8) Very weedy reaches with deep pools	0.075–0.100–0.150
(9) Floodways with heavy stands of timber and vegetation	0.075–0.100–0.150
(b) *Mountain streams* with no vegetation in channel, usually with steep banks and bank vegetation submerged at high stages	0.030–0.040–0.050
(1) Bed of gravel, cobbles and a few boulders	
(2) Bed of cobbles with large boulders	0.040–0.050–0.070
FLOOD PLAINS	
(a) *Pasture, no brush*	
(1) Short grass	0.025–0.030–0.035
(2) Long grass	0.030–0.035–0.050
(b) *Cultivated areas*	
(1) No crops	0.020–0.030–0.040
(2) Mature row crops	0.025–0.035–0.045
(3) Mature field crops	0.030–0.040–0.050
(c) *Brush*	
(1) Scattered brush, heavy weeds	0.035–0.050–0.070
(2) Light brush and trees, in winter	0.035–0.050–0.060
(3) As (2), in summer	0.040–0.060–0.080
(4) Medium to dense brush, in winter	0.045–0.070–0.110
(5) As (4), in summer	0.070–0.100–0.160
(d) *Trees*	
(1) Dense straight willows, summer	0.110–0.150–0.200
(2) Cleared land with stumps, not sprouting	0.030–0.040–0.050
(3) As (2), but heavy sprout growth	0.050–0.060–0.080
(4) Heavy stands of timber, a few fallen trees, little undergrowth, flood stage below branches	0.080–0.100–0.120
(5) As (4), but floods reaching branches	0.100–0.120–0.160

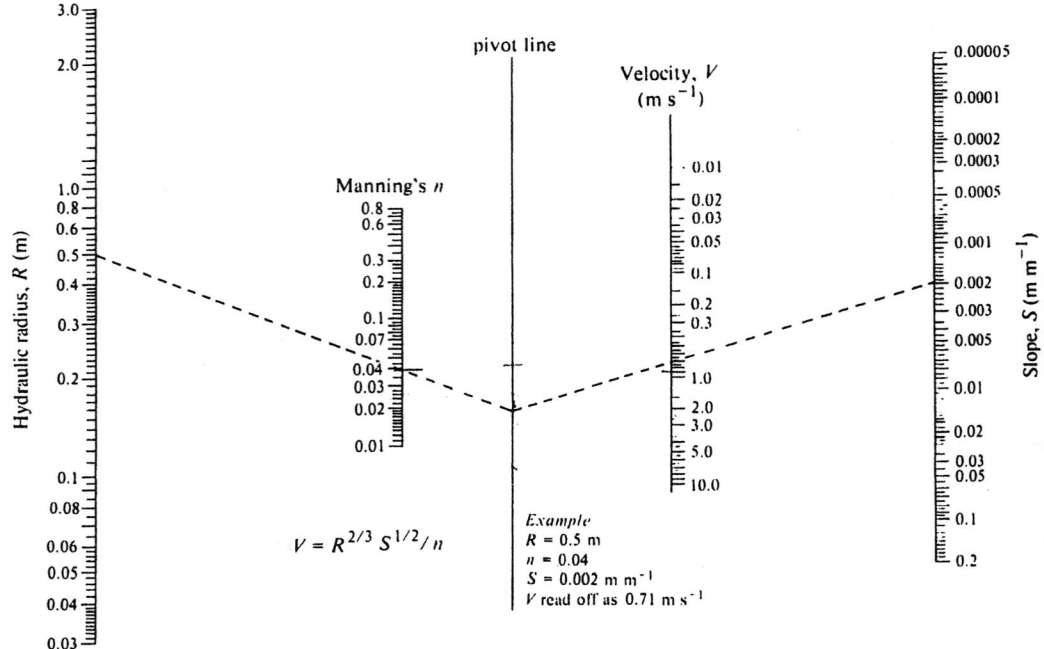

Figure A5.3   Nomogram for evaluation of the Manning equation (after Chow, 1959).

Table A5.4   C-factor values for the Universal Soil Loss Equation (from Morgan 1995).

Practice	Average annual C-factor
Bare soil	1.00
Forest or dense shrub, high mulch crops	0.001
Savanna or prairie grass in good condition	0.01
Overgrazed savanna or prairie grass	0.10
Maize, sorghum or millet: high productivity conventional tillage	0.20–0.55
Maize, sorghum or millet: high productivity conventional tillage	0.50–0.90
Maize, sorghum or millet: low productivity no or minimum tillage	0.02–0.10
Maize, sorghum or millet: high productivity chisel ploughing into residue	0.12–0.20
Maize, sorghum or millet: low productivity chisel ploughing into residue	0.30–0.45
Cotton	0.40–0.70
Meadow grass	0.01–0.025
Soya beans	0.20–0.50
Wheat	0.10–0.40
Rice	0.10–0.20

Table A5.4 (*Continued*).

Groundnuts	0.30–0.80
Palm trees, coffee, cocoa with crop cover	0.10–0.30
Pineapple on contour: residue removed	0.10–0.40
Pineapple on contour: with surface residue	0.01
Potatoes: rows downslope	0.20–0.50
Potatoes: rows across-slope	0.10–0.40
Cowpeas	0.30–0.40
Strawberries: with weed cover	0.27
Pomegranate: with weed cover	0.08
Pomegranate: clean-weeded	0.56
Ethiopian tef	0.25
Sugar cane	0.13–0.40
Yams	0.40–0.50
Pigeon peas	0.60–0.70
Mungbean	0.04
Chilli	0.33
Coffee: after first harvest	0.05
Plantains: after establishment	0.05–0.10
Papaya	0.21

Table A5.5 *P*-factor values for the Universal Soil Loss Equation (from Morgan 1995).

Erosion-control practice	*P*-factor value
Contouring: 0–1° slope	0.60*
Contouring: 2–5° slope	0.50*
Contouring: 6–7° slope	0.60*
Contouring: 8–9° slope	0.70*
Contouring: 10–11° slope	0.80*
Contouring: 12–14° slope	0.90*
Level bench terrace	0.14
Reverse-slope bench terrace	0.05
Outward-sloping bench terrace	0.35
Level retention bench terrace	0.01
Tied-ridging	0.10–0.20

* Use 50% of the value for contour bunds or if contour strip cropping is practised.

Table A5.6   Prediction of soil loss using the Universal Soil Loss Equation (from Morgan 1995).

Problem

Calculation of mean annual soil loss on a 100 m long slope of 7° on soils of the Rengam Series under maize cultivation with contour bunds spaced at 20 m intervals, near Kuala Lumpur.

Equation

Mean annual soil loss $= R \times K \times LS \times C \times P$

Estimating R (Rainfall erosion index)

Method 1:

Mean annual precipitation (P)	$= 2695$ mm
From Roose (1975), mean annual rainfall erosion index (R) in US units	$= 0.5\,P$
	$= 0.5 \times 2695$
	$= 1347.5$
Conversion to metric units	$= 1347.5 \times 1.73$
	$= \underline{2331.18}$

Method 2:

From Morgan (1974), mean annual erosivity ($KE > 25$)	$= 9.28\,P - 8838$
	$= (9.28 \times 2695) - 8838$
	$= 16\,171.6$ J/m^2
Multiply by I_{30} (use 75 mm/h; maximum value recommended by Wischmeier and Smith 1978)	$= 16171.6 \times 75$
	$= 1\,212\,870$
Divide by 1000 to give R value in metric units	$= \underline{1212.87}$

Method 3:

From Foster *et al.* (1981), mean annual EI_{30} (kg.m.mm)/(m^2.h)	$= 0.276\,P \times 75$
	$= 0.276 \times 2695 \times 75$
	$= 55\,786.5$
With these units, divide by 100 to give R value in metric units	$= \underline{557.9}$

Best estimate: discard result from Method 3 which is rather low. Take average value of Methods 1 and 2:

$$R = \underline{1772}$$

Estimating K (Soil erodibility index)

The soils have a 43% clay, 8% silt, 9% fine sand and 40% coarse sand; organic content about 3%.

Using the nomograph (Figures A5.3; 11.13) gives a first approxiamation K value $= \underline{0.05}$

Estimating LS (Slope factor)

For slope lenght (l) and slope steepness (s) in meters and per cent respectively,

$LS = (l/22)^{0.5}\,(0.065 + 0.045s + 0.0065s^2)$

With contour bunds at 20 m spacing, $l = 20$ m and $s = 12\%$ (approximation of 7°)

$LS = (20/22)^{0.5}\,(0.065 \times (0.045 \times 12) + (0.0065 \times 12^2)$

$LS = 0.95 \times 1.54$

$LS = \underline{1.46}$

Estimating C (Crop management factor)

According to Table A5.4, the C value for maize ranges between 0.2 and 0.9, depending on the productivity. For many tropical farming conditions, C for maize lies between 0.4 and 0.9 (Roose, 1975), depending on the cover.

During the three-month period from seeding to harvest, the cover is likely to vary from 9 to 45 per cent in the first month, to 55 to 93 per cent in the second month, and 45 to 57 per cent in the third month. Therefore, we might assume C values of 0.9, 0.4 and 0.7 for the three respective months.

Maize can be planted at any time of year in Malaysia but assume planting after the April rains, allowing growth, ripening and harvesting in June and July which are the driest months. Land is under dense secondary growth prior to planting (assume $C = 0.001$) and allowed to revert to the same after harvest (assume $C = 0.1$).

Of the mean annual precipitation, 32 per cent falls between January and April inclusive, 10 per cent in May, 6 per cent in June, 7 per cent in July, and 45 per cent between August and December. Assuming that erosivity is directly related to precipitation amount, these values can be used to describe the distribution of the R factor throughout the year.

From these data, the following table is constructed.

Months	C value	Adjustment factor (% R value)	Weighted C value (col. 2 × col. 3)
January–April	0.001	0.32	0.000 32
May	0.9	0.10	0.09
June	0.4	0.06	0.024
July	0.7	0.07	0.049
August–December	0.1	0.45	0.045
		Total	0.208 32

C factor value for the year $\qquad = \underline{0.028}$

Estimating P (Erosion-control practice factor)
From Table A5.5, P value for contour bunds $\qquad = \underline{0.3}$

Soil loss estimation
Mean annual soil loss $\qquad = 1772 \times 0.05 \times 1.46 \times 0.208 \times 0.3$
$\qquad = 8.07\,\text{t/ha}$

References

Chow, V. T. (1959) *Open-channel Hydraulics*. McGraw-Hill, New York.

Morgan, R. P. C. (1995) *Soil Erosion and Conservation*. Longman, London.

Engman, E. T. (1986) Roughness coefficients for routing surface runoff. *Transactions of American Society of Civil Engineers* **112**, 39–53.

Foster, G. R., Lane, L. J., Nowlin, J. D., Laflen, J. M. and Young, R. A. (1981) Estimating erosion and sediment yield on field-sized areas. *Transactions of American Society of Agricultural Engineers* **24**, 1253–1263.

Morgan, R. P. C. (1974) Estimating regional variations in soil erosion hazard in Peninsula Malaysia. *Malayan Nature Journal* **28**, 94–106.

Petryk, S. and Bosmajian, G. (1975) Analysis of flow through vegetation. *Transactions of American Society of Civil Engineers* **101**, 871–884.

Roose, E. J. (1975) *Erosion et ruissellement en Afrique de l'ouest: vingt années de measures en petites parcelles expérimentales*. Cyclo, Orstom, Ivory Coast.

Temple, D. M. (1982) Flow retardance of submerged grass channel linings. *Transactions of American Society of Agricultural Engineers* **25**, 1300–1303.

Wischmeier, W. H. and Smith, D. D. (1978) *Predicting rainfall erosion losses*. USDA Agricultural Research Service Handbook 537.

Appendix A6: Units, Scales and conversions

Table A6.1 Derived units in the SI and CGS systems (after Tennent, 1971).

Quantity and recommended symbol	Dimensions	SI unit	CGS unit	Ratio CGS: SI units
Mass, m	m	kilogram (kg)	gram (g)	10^{-3}
Length, l	l	metre (m)	centimetre (cm)	10^{-2}
Time, t	t	second (s)	second (s)	1
Area, A, S	l^2	m^2	cm^2	10^{-4}
Volume, V	l^3	m^3	cm^3	10^{-6}
Density, ρ	ml^{-3}	kg m^{-3}	g cm^{-3}	10^3
Velocity, u, v	lt^{-1}	m s^{-1}	cm s^{-1}	10^{-2}
Acceleration, G	lt^{-2}	m s^{-2}	gal	10^{-2}
Momentum, P	mlt^{-1}	kg m s^{-1}	g cm s^{-1}	10^{-3}
Moment of inertia, I, J	ml^2	kg m^2	g cm^2	10^{-7}
Angular momentum, L	$ml^2\,t^{-1}$	kg m^2 s^{-1}	g cm^2 s^{-1}	10^{-7}
Force, F	mlt^{-2}	newton (N)	dyne (dyn)	10^{-5}
Energy of work, E, W	ml^2t^{-2}	joule (J)	erg	10^{-7}
Power, P	ml^2t^{-3}	watt (W)	erg s^{-1}	10^{-7}
Pressure or stress, p	$ml^{-1}t^{-2}$	pascal (Pa)	dyn cm^{-2}	10^{-1}
Surface tension, γ	ml^{-2}	N m^{-1}	dyn cm^{-1}	10^{-3}
Viscosity, η	$ml^{-1}\,t^{-1}$	kg m^{-1} s^{-1}	poise	10^{-1}
Frequency, ν, f	t^{-1}	hertz (Hz)	s^{-1}	1

Table A6.2 Conversion factors.

To convert A to B multiply by	A	B	To convert B to A multiply by
Length			
2.54×10^{-2}	inch (in)	metre (m)	39.37
0.3048	foot (ft)	metre (m)	3.2468
0.9144	yard (yd)	metre (m)	1.0936
1.8288	fathom (fm)	metre (m)	0.5468
20.1168	chain (ch)	metre (m)	4.97×10^{-2}
201.168	furlong (fl)	metre (m)	4.97×10^{-3}
1609.34	mile (mi)	metre (m)	6.214×10^{-4}
1853.2	nautical mile (UK)	metre (m)	5.3961×10^{-4}
1852.0	nautical mile (Int.)	metre (m)	5.3996×10^{-4}

Table A6.2   (*Continued*).

To convert *A* to *B* multiply by	*A*	*B*	To convert *B* to *A* multiply by
1853.25	nautical mile (US)	metre (m)	5.3959×10^{-4}
100.0	cable	fathom (fm)	10^{-2}
1.6094	mile (mi)	kilometre (km)	0.6214
8.0	mile (mi)	furlong (fl)	0.125
1760.0	mile (mi)	yard (yd)	5.6818×10^{-4}
5280.0	mile (mi)	foot (ft)	1.8939×10^{-4}
Area			
6.4516×10^{-4}	sq. inch (in^2)	sq. metre (m^2)	1.55×10^{3}
9.2903×10^{-2}	sq. foot (ft^2)	sq. metre (m^2)	10.764
0.8361	sq. yard (yd^2)	sq. metre (m^2)	1.196
2 589 988.0	sq. mile (mi^2)	sq. metre (m^2)	3.861×10^{-7}
4046.856	acre	sq. metre (m^2)	2.4711×10^{-4}
2.590	sq. mile (mi^2)	sq. kilometre (km^2)	0.3861
0.4047	acre	hectare (ha)	2.471
640.0	sq. mile (mi^2)	acre	1.5625×10^{-3}
10.000	hectare (ha)	sq. metre (m^2)	10^{-4}
3.861×10^{-3}	hectare (ha)	sq. mile (mi^2)	259.0045
247.105	sq. kilometre (km^2)	acre	4.047×10^{-3}
Volume			
1.6387×10^{-5}	cubic inch (in^3)	cubic metre (m^3)	6.1024×10^{4}
16.387	cubic inch (in^3)	cubic centimetre (cm^3)	6.1024×10^{-2}
2.8317×10^{-2}	cubic foot (ft^3)	cubic metre (m^3)	35.314
28 316.8	cubic foot (ft^3)	cubic centimetre (cm^3)	3.5315×10^{-5}
28.3168	cubic foot (ft^3)	litre (l)	3.5315×10^{-2}
0.7646	cubic yard (yd^3)	cubic metre (m^3)	1.3079
3.785×10^{-3}	gallon (US)	cubic metre (m^3)	264.2
4.546×10^{-3}	gallon (UK)	cubic metre (m^3)	219.97
0.21998	litre (l)	gallon (UK)	4.546
0.26418	litre (l)	gallon (US)	3.7853
8.0	bushel	gallon (UK)	0.125
9.608	bushel	gallon (US)	0.104
231.0	gallon (UK)	cubic inch (in^3)	4.329×10^{-3}
1233.482	acre foot	cubic metre (m^3)	8.1071×10^{-4}
0.8326	gallon (US)	gallon (UK)	1.2011
2.8413×10^{-2}	fluid ounce (fl. oz)	litre (l)	35.195
Mass			
2.835×10^{-2}	ounce (oz)	kilogram (kg)	35.273
0.4536	pound (lb)	kilogram (kg)	2.2046
6.3503	stone (st)	kilogram (kg)	0.1575
50.8023	hundredweight (cwt)	kilogram (kg)	1.9684×10^{-2}
1016.04	ton	kilogram (kg)	9.8421×10^{-4}
907.20	short ton	kilogram (kg)	1.1023×10^{-3}
1.016	ton	metric tonne (t)	0.9842
2204.6	metric tonne (t)	pound (lb)	4.536×10^{-4}
14.5939	slug	kilogram (kg)	6.852×10^{-2}

Table A6.2　(*Continued*).

To convert *A* to *B* multiply by	*A*	*B*	To convert *B* to *A* multiply by
Weight			
4.448	pound force (lbf)	newton (N)	0.2248
0.1383	poundal (pdl)	newton (N)	7.233
10^5	newton (N)	dyne (dyn)	10^{-5}
32.17	pound-force (lbf)	poundal (pdl)	0.031 08
980.7	gram-force (lgf)	dyne (dyn)	1.0197×10^{-3}
Pressure or stress			
15.44×10^6	ton-force per sq. in	pascal (Pa)	6.4767×10^{-8}
157.47	ton-force per sq. in	kilogram-force per sq. cm	6.3504×10^{-3}
107.3×10^3	ton-force per sq. ft	pascal (Pa)	9.3197×10^{-6}
1.0936×10^4	ton-force per sq. ft	kilogram-force per sq. m	9.1441×10^{-5}
6.895×10^3	pound-force per sq. in	pascal (Pa)	1.4503×10^{-4}
7.03×10^{-2}	pound-force per sq. in	kilogram-force per sq. m	14.225
47.9	pound-force per sq. ft	pascal (Pa)	2.0877×10^{-2}
4.882	pound-force per sq. ft	kilogram-force per sq. m	0.204 83
101.325×10^3	standard atmosphere	pascal (Pa)	9.869×10^{-6}
1.033	standard atmosphere	kilogram-force per sq. m	9.6805
14.697	standard atmosphere	pound-force per sq. in	6.804×10^{-2}
760	standard atmosphere	millimetres of mercury (mm Hg)	1.315×10^{-3}
33.901	standard atmosphere	feet of water (ft H_2O)	2.9498×10^{-2}
3.05×10^{-2}	foot of water (ft H_2O)	kilogram-force per sq. cm	3.2787×10^{-3}
2.989×10^3	foot of water	pascal (Pa)	3.3456×10^{-4}
10^5	bar	pascal (Pa)	10^{-5}
133.322	millimetre of mercury	pascal (Pa)	7.5×10^{-3}
9.80665×14^4	kilogram-force per sq. cm	pascal (Pa)	1.0197×10^{-5}
Energy or work			
1.3558	foot pound-force	joule (J)	0.7376
1.3558×10^7	foot pound-force	erg	7.3757×10^{-8}
0.1383	foot pound-force	metre kilogram force	7.2307
4.2140×10^2	foot poundal	joule (J)	2.373×10^{-3}
4.2140×10^9	foot poundal	erg	2.373×10^{-10}
1.055×10^3	BTU	joule (J)	9.4787×10^{-4}
10^7	joule	erg	10^{-7}
4.1855	calorie at 15°C	joule (J)	0.2389
Power			
550	horse-power	foot pound-force per second	1.8182×10^{-3}
7.457×10^9	horse-power	erg per second	1.341×10^{-10}
7.457×10^2	horse-power	watt (W)	1.341×10^{-3}
1.3405	kilowatt (kW)	horse-power	0.74599
1.3558	foot pound-force per second	watt (W)	0.7376
Density			
16.019	pound per cubic foot	kilogram per cubic metre	6.243×10^{-2}
1.6019×10^{-2}	pound per cubic foot	gram per cubic centimetre	6.243×10^{-3}
1.0012	ounces per cubic foot	gram per litre	0.9988

Table A6.2 (*Continued*).

To convert *A* to *B* multiply by	*A*	*B*	To convert *B* to *A* multiply by
Unit weight			
16.019	pound-force per cubic foot	kilogram-force per cubic metre	6.243×10^{-2}
1.571×10^{2}	pound-force per cubic foot	newton per cubic metre	6.3654×10^{-3}
27.68	pound-force per cubic inch	gram-force per cubic centimetre	3.613×10^{-2}
271.4×10^{3}	pound-force per cubic inch	newton per cubic metre	3.6846×10^{-6}
Compressibility			
1.45×10^{-4}	sq. in per pound-force	sq. m per newton	6.897×10^{3}
14.22	sq. in per pound-force	sq. cm per kilogram-force	7.032×10^{-2}
9.324×10^{-6}	sq. foot per ton-force	sq. m per newton	1.0725×10^{5}
0.914	sq. foot per ton-force	sq. cm per kilogram-force	1.0941
Speed			
2.54	inch per second	centimetre per second	0.3937
30.48	foot per second	centimetre per second	3.281×10^{2}
0.447	mile per hour	metre per second	2.2371
0.5144	knot (Int.)	metre per second	1.944
1.0973	foot per second	kilometre per hour	0.9113
0.61818	foot per second	mile per hour	1.61765
0.5925	foot per second	knot (Int.)	1.6878
3.6	metre per second	kilometre per hour	0.2778
0.9659×10^{-8}	foot per year	metre per second	1.0353×10^{8}
Rate of flow and discharge			
2832	cubic foot per second	cubic centimetre per second	3.53×10^{-4}
2.832×10^{-2}	cubic foot per second	cubic metre per second	35.311
76 464	cubic yard per second	cubic centimetre per second	1.3078×10^{-5}
0.7646	cubic yard per second	cubic metre per second	1.3078
101.941	cubic foot per second	cubic metre per hour	9.8096×10^{-3}
2446.57	cubic foot per second	cubic metre per day	4.0874×10^{-4}
28.3161	cubic foot per second	litres per second	3.5315×10^{-2}
11.573 75	cubic metre per day	litres per second	8.6402×10^{-2}
0.408 735	cubic metre per day	cubic foot per second	2.4466
4.3813×10^{-2}	million gallons (US) per day	cubic metre per second	22.824
5.261×10^{-2}	million gallons (UK) per day	cubic metre per second	19.008
0.2713	acre-feet per day	million gallons (UK) per day	3.686
0.3259	acre-feet per day	million gallons (US) per day	3.0684
Yield			
0.699 725	cubic feet per acre	cubic metres per hectare	1.4291
1.120 85	pounds per acre	kilograms per hectare	0.892 18
Coefficient of consolidation			
0.1075	sq. in per minute	sq. cm per second	9.3023
1.075×10^{-5}	sq. in per minute	sq. m per second	9.3023×10^{2}
2.94×10^{-5}	sq. ft per year	sq. cm per second	3.4014×10^{4}
2.94×10^{-9}	sq. ft per year	sq. m per second	3.4014×10^{8}
Concentration			
1.0012×10^{-3}	ounce per cubic foot	gram per cubic centimetre	9.988×10^{2}
1.0012	ounce per cubic foot	kilogram per cubic metre	0.9988

Table A6.2   (*Continued*).

To convert *A* to *B* multiply by	*A*	*B*	To convert *B* to *A* multiply by
16.019×10^{-3}	pound per cubic foot	gram per cubic centimetre	62.426
16.019	pound per cubic foot	kilogram per cubic metre	6.2426
Dynamic viscosity			
47.8803	pound-seconds per square foot	newton-seconds per sq. metre	2.0885×10^{-2}
47.8803	slugs per foot-second	newton-seconds per sq. metre	2.0885×10^{-2}
10^{-3}	centipoise (cP)	newton-seconds per sq. metre	10^{3}
Kinematic viscosity			
$9.290\ 30 \times 10^{4}$	sq. foot per second	centistoke (cSt)	1.0764×10^{-5}
$9.290\ 30 \times 10^{-2}$	sq. foot per second	sq. metre per second	10.764
10^{-6}	centistoke (cSt)	sq. metre per second	10^{6}

Table A6.3   Conversion tables for slope data.

Degrees to percentages			Percentage to degrees		Natural tangent to degrees	
Degrees	**Minutes**		**Per cent**	**Degrees and minutes**	**Ratio (tan)[1]**	**Degrees (rounded)**
	0′	**30′**			1 : 0.25	76
0°	0	0.87	1	0°34′	1 : 0.5	63
1	1.75	2.62	2	1°09′	1 : 1	45
2	3.49	4.37	3	1°43′	1 : 1.5	34
3	5.24	6.12	4	2°18′	1 : 2	27
4	6.99	7.87	5	2°52′	1 : 2.5	22
5	8.75	9.63	6	3°26′	1 : 3	18
6	10.51	11.39	7	4°00′	1 : 3.5	16
7	12.28	13.17	8	4°34′	1: 4	14
8	14.05	14.95	9	5°09′		
9	15.84	16.73	10	5°43′		
10	17.63	18.53	11	6°17′		
11	19.44	20.35	12	6°51′		
12	21.26	22.17	13	7°24′		
13	23.09	24.01	14	7°58′		
14	24.93	25.86	15	8°32′		
15	26.80	27.73	16	9°05′		
16	28.68	29.62	17	9°39′		
17	30.57	31.53	18	10°12′		
18	32.49	33.46	19	10°45′		
19	34.43	35.41	20	11°19′		
20	36.40	37.39	25	14°02′		
21	38.39	39.39	30	16°42′		
22	40.40	41.42	35	19°45′		
23	42.45	43.38	40	21°48′		
24	44.52	45.57	45	24°14′		
25	46.63	47.70	50	26°34′		

[1] The first figure refers to the height, the second to the horizontal distance; e.g. on a 1 : 2 smooth slope there is a climb (or descent) of 5 m for every 10 m traversed horizontally along a line parallel to the maximum gradient.

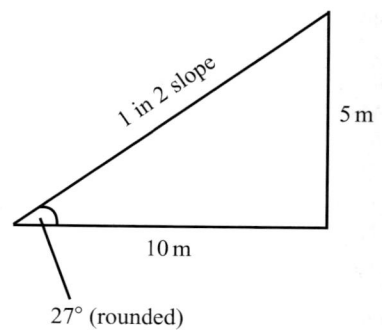

1 in 2 slope

5 m

10 m

27° (rounded)

Table A6.3 (*Continued*).

Degrees to percentages			Percentage to degrees	
Degrees	**Minutes**		**Per cent**	**Degrees and minutes**
	0'	**30'**		
26	48.77	49.86	55	28°49'
27	50.95	52.06	60	30°38'
28	53.17	54.30	65	33°01'
29	55.43	56.58	70	35°00'
30	57.74	58.91	75	36°52'
31	60.09	61.28	80	38°40'
32	62.49	63.71	85	40°22'
33	64.94	66.19	90	42°00'
34	67.45	68.73	95	43°32'
35	70.02	71.33	100	45°00'
36	72.65	74.00		
37	75.36	76.73		
38	78.13	79.54		
39	80.98	82.43		
40	83.91	85.41		
41	86.93	88.47		
42	90.04	91.63		
43	93.25	94.90		
44	96.57	98.27		

Percentage slope = natural tangent $\times$ 100 thus $1 : 2 = 50\%$.

Table A6.4 Guide to map scales and mapping resolution.

Map scale	Feet on ground per inch on map	Metres on ground per centimetre on map	Inches on map per mile on ground	Centimetres on map per kilometre on ground	True ground width of map lines of various thickness (mm). Ground widths are given in metres					
					0.10	0.13	0.25	0.35	0.50	1.00
1:500	41.67	5.00	126.72	200.00	0.05	0.07	0.13	0.18	0.25	0.50
1:600	50.00	6.00	105.60	166.67	0.06	0.08	0.15	0.20	0.30	0.60
1:1000	83.33	10.00	63.36	100.00	0.10	0.13	0.25	0.35	0.50	1.00
1:1250	104.17	12.50	50.68	80.00	0.13	0.16	0.31	0.44	0.63	1.25
1:1500	125.00	15.00	42.24	66.67	0.15	0.20	0.38	0.53	0.75	1.50
1:2000	166.67	20.00	31.68	50.00	0.20	0.26	0.50	0.70	1.00	2.00
1:2500	208.33	25.00	25.34	40.00	0.25	0.33	0.63	0.88	1.25	2.50
1:3000	250.00	30.00	21.12	33.33	0.30	0.39	0.75	1.05	1.50	3.00
1:5000	416.67	50.00	12.67	20.00	0.50	0.65	1.25	1.75	2.50	5.00
1:6000	500.00	60.00	10.56	16.67	0.60	0.78	1.50	2.10	3.00	6.00
1:10000	833.33	100.00	6.34	10.00	1.00	1.30	2.50	3.50	5.00	10.00
1:10560	880.00	105.60	6.00	9.47	1.06	1.37	2.64	3.70	5.28	10.56
1:12000	1000.00	120.00	5.28	8.33	1.20	1.56	3.00	4.20	6.00	12.00
1:20000	1666.67	200.00	3.17	5.00	2.00	2.60	5.00	7.00	10.00	20.00
1:24000	2000.00	240.00	2.64	4.17	2.40	3.12	6.00	8.40	12.00	24.00
1:25000	2083.33	250.00	2.53	4.00	2.50	3.25	6.25	8.75	12.50	25.00
1:40000	3333.33	400.00	1.58	2.50	4.00	5.20	10.00	14.00	20.00	40.00
1:48000	4000.00	480.00	1.32	2.08	4.80	6.24	12.00	16.80	24.00	48.00
1:50000	4166.67	500.00	1.27	2.00	5.00	6.50	12.50	17.50	25.00	50.00
1:63360	5280.00	633.60	1.00	1.58	6.34	8.24	15.84	22.18	31.68	63.36

Table A6.5 Map and AP scales: relation to actual lengths and areas (metric units).

Scale	Distance represented by 1 cm (m)	Distance representing 1 km (cm)	Area represented by 1 cm² (ha)[1]	Area representing 1000 ha (cm²)
1:1000	10	100	0.01	10^5
1:2500	25	40	0.0625	1.6×10^4
1:5000	50	20	0.25	4×10^3
1:7500	75	13	0.56	1786
1:10 000	100	10	1.00	10^3
1:15 000	150	6.7	2.25	444
1:20 000	200	5.0	4.00	250
1:25 000	250	4.0	6.25	160
1:30 000	300	3.3	9.00	111
1:40 000	400	2.5	16.0	62.5
1:50 000	500	2.0	25.0	40
1:100 000	1000 = 1 km	1.0	100 = 1 km²	10
1:250 000	2.5 km	0.4	6.25 km²	1.6
1:500 000	5 km	0.2	25 km²	0.4
1:750 000	7.5 km	0.1	56.25 km²	0.18
1:1 000 000	10 km	0.1	100 km²	0.1

[1] Effectively the minimum mappable area.

Table A6.6 Imperial units.

Scale	in/mi	mi/in	mi²/in²	ac/in²
1:1 000 000	0.063 4	15.782 8	249.097	159 422.09
1:500 000	0.126 7	7.891 4	62.274	39 855.36
1:253 400	0.250 0	4.000 0	16.000	10 240.00
1:250 000	0.253 4	3.945 7	15.569	9 964.16
1:126 720	0.500 0	2.000 0	4.000	2 560.00
1:125 000	0.506 9	1.972 8	3.892	2 490.88
1:100 000	0.633 6	1.578 3	2.491 0	1 594.22
1:63 360	1.000 0	1.000 0	1.000 0	640.00
1:62 500	1.013 8	0.986 4	0.973 0	622.72
1:50 000	1.267 2	0.789 1	0.622 7	398.56
1:31 680	2.000 0	0.500 0	0.250 0	160.00
1:30 000	2.112 0	0.473 5	0.224 2	143.49
1:25 000	2.534 4	0.394 6	0.155 6	99.64
1:10 560	6.000	0.166 6	0.027 7	17.728
1:10 000	6.336	0.157 8	0.024 9	15.942
1:5 000	12.672	0.078 9	0.006 23	3.986
1:2 534	25.000	0.040 0	0.001 60	1.024
1:2 500	25.344	0.039 5	0.001 56	0.996
1:1 250	50.688	0.019 7	0.000 389	0.249

References

Tennent, R. M. (ed.) (1971) *Science Data Book*. Oliver and Boyd, Edinburgh.

Index